TABLE 1 *(continued)*

Analyte	Fluid[a]	MGH Units	SI Units	Method	SI Units
Calcitonin	S			Immunoassay	1
Female		0–20 pg/mL	0–20 ng/L		
Male		0–28 pg/mL	0–28 ng/L		
Epinephrine	U	1.7–22.4 μg/day	9.3–122 nmol/day	Liquid chromatography	5.458
Supine	P	0–110 pg/mL	0–600 pmol/L	Liquid chromatography	5.458
Standing	P	0–140 pg/mL	0–764 pmol/L	Liquid chromatography	5.458
Norepinephrine	U	12.1–85.5 μg/day	72–505 nmol/day	Liquid chromatography	5.911
Supine	P	70–750 pg/mL	0.41–4.43 nmol/L	Liquid chromatography	0.005911
Standing	P	200–1700 pg/mL	1.18–10.0	Liquid chromatography	0.005911
Corticotropin (ACTH)	P	6.0–76.0 pg/mL	1.3–16.7 pmol/L	Immunoassay	0.2202
Cortisol	P			Immunoassay	27.59
Fasting, 8 a.m.–12 noon		5.0–25.0 μg/dL	138–690 nmol/L		
12 noon–8 p.m.		5.0–15.0 μg/dL	138–410 nmol/L		
8 p.m.–8 a.m.		0.0–10.0 μg/dL	0–276 nmol/L		
Cortisol, free	U	20–70 μg/day	55–193 nmol/day	Immunoassay	2.759
C peptide	S	0.30–3.70 μg/L	0.10–1.22 nmol/L	Immunoassay	0.33
1,25-Dihydroxyvitamin D	S	16–42 pg/mL	38–101 pmol/L	Immunoassay	2.4
Erythropoietin	S	<19 mU/mL	≤19 U/L	Immunoassay	1
Estradiol	S, P			Immunoassay	3.671
Female					
Premenopausal adult		23–361 pg/mL	84–1325 pmol/L		
Postmenopausal		<30 pg/mL	<110 pmol/L		
Prepubertal		<20 pg/mL	<73 pmol/L		
Male		<50 pg/ml	<184 pmol/L		
Gastrin	P	0–200 pg/mL	0–200 ng/L	Immunoassay	1
Growth hormone	P	2.0–6.0 ng/mL	2.0–6.0 μ/L	Immunoassay	1
17-Hydroxycorticosteroids	U			Colorimetry	2.759
Female		2.0–6.0 mg/day	5.5–17 μmol/day		
Male		3.0–10.0 mg/day	8–28 μmol/day		
17-Hydroxyprogesterone	S			Immunoassay	3.026
Female					
Prepubertal		0.20–0.54 μg/L	0.61–1.63 nmol/day		
Follicular		0.02–0.80 μg/L	0.61–2.42 nmol/day		
Luteal		0.90–3.04 μg/L	2.72–9.20 nmol/day		
Postmenopausal		<0.45 μg/L	<1.36 nmol/L		
Male					
Prepubertal		0.12–0.30 μg/L	0.36–0.91 nmol/day		
Adult		0.20–1.80 μg/L	0.61–5.45 nmol/day		
Insulin	S	0–29 μU/mL	0–208 pmol/L	Immunoassay	7.175
Parathyroid hormone	P	10–60 pg/mL	10–60 ng/L	Immunoassay	1
Prolactin	S			Immunoassay	1
Female		0–15 ng/mL	0–15 ng/L		
Male		0–10 ng/mL	0–10 μg/L		
Renin activity	P			Immunoassay	0.2778
Normal salt intake					
Recumbent 6 hr		0.5–1.6 ng/mL/hr	0.14–0.44 ng/(L·sec)		
Upright 4 hr		1.9–3.6 ng/mL/hr	0.53–1.00 ng/(L·sec)		
Testosterone, total, morning sample	P			Immunoassay	0.03467
Female		20–90 ng/dL	0.7–3.1 nmol/L		
Male		300–1100 ng/mL	10.4–38.1 nmol/L		
Testosterone, unbound, morning sample	P			Equilibrium dialysis	34.67
Female, adult		0.09–1.29 ng/dL	3–45 pmol/L		
Male, adult		3.06–24.0 ng/mL	106–832 pmol/L		
Thyroglobulin	S	0–60 ng/mL	0–60 μg/L	Immunoassay	1
Thyroid-stimulating hormone	S	0.5–5.0 μU/mL	0.5–5.0 μU/L	Immunoassay	1
Thyroxine, free	S	0.8–2.7 ng/dL	10–35 pmol/L	Direct equilibrium dialysis	12.87
Triiodothyronine, total (T₃)	S	75–195 ng/dL	1.2–3.0 nmol/L	Immunoassay	0.01536

Hematology and Coagulation

Analyte	Fluid[a]	MGH Units	SI Units	Method	SI Units
Hematocrit	WB			Automated cell counter	0.01
Female		37–48%	0.37–0.48		
Male		42–52%	0.42–0.52		
Hemoglobin	WB			Automated cell counter	0.6206
Female		12–16 g/dL	7.4–9.9 mmol/L		
Male		13–18 g/dL	8.1–11.2 mmol/L		
Iron	S	50–150 μg/dL	9.0–26.9 μmol/L	Colorimetry	0.1791
Platelet count	WB	150–350 × 10³/mm³	150–350 × 10⁹/L	Automated cell counter	1

[a]S, serum; WB, whole blood; P, plasma; U, urine.
Data from Case Records of the Massachusetts General Hospital. *N Engl J Med* 1992; 327(10): 718–724.

THE **George** F. Smith **Library**
OF THE HEALTH SCIENCES

UMDNJ
UNIVERSITY OF MEDICINE &
DENTISTRY OF NEW JERSEY

ESSENTIAL MEDICAL PHYSIOLOGY

THIRD EDITION

ESSENTIAL MEDICAL PHYSIOLOGY

THIRD EDITION

Edited by

LEONARD R. JOHNSON

Thomas A. Gerwin Professor and Chairman
Department of Physiology and Biophysics
University of Tennessee College of Medicine
Memphis, Tennessee

With Contributions By

JOHN H. BYRNE

JAMES M. DOWNEY

H. MAURICE GOODMAN

D. NEIL GRANGER

DIANNA A. JOHNSON

FRANK L. POWELL, JR.

JAMES A. SCHAFER

STANLEY G. SCHULTZ

NORMAN W. WEISBRODT

ELSEVIER
ACADEMIC
PRESS

Amsterdam Boston Heidelberg London New York Oxford
Paris San Diego San Francisco Singapore Sydney Tokyo

Academic Press
An imprint of Elsevier
525 B Street, Suite 1900, San Diego, California 92101-4495, USA
http://www.academicpress.com

Academic Press
84 Theobald's Road, London WC1X 8RR, UK
http://www.academicpress.com

Library of Congress Catalog Card Number: 2003109363

International Standard Book Number: 0-12-387584-6

PRINTED IN THE UNITED STATES OF AMERICA
03 04 05 06 07   9 8 7 6 5 4 3 2 1

*This edition of Essential Medical Physiology is dedicated to
Dr. Nancy Ann Dahl (1932–2002), Professor of Physiology and Cell Biology,
University of Kansas, and co-author of the Central Nervous System Physiology
section of the second edition this book—
For imparting her energy, enthusiasm, and passion for science
to her students and colleagues.*

Contents

PART III: RESPIRATORY PHYSIOLOGY

PART IV: RENAL PHYSIOLOGY

PART V: GASTROINTESTINAL PHYSIOLOGY

PART VI: ENDOCRINE PHYSIOLOGY

PART VII: CENTRAL NERVOUS SYSTEM PHYSIOLOGY

PART VIII: INTEGRATIVE PHYSIOLOGY

Contributors

Numbers in parentheses indicate the pages on which the authors' contribution begin.

John H. Byrne (71, 89, 97, 905) Department of Neurobiology and Anatomy, University of Texas Medical Center at Houston, Houston, Texas 77030

James M. Downey (157, 175, 187, 201, 215, 941) Department of Physiology, University of South Alabama, College of Medicine, Mobile, Alabama 36688

H. Maurice Goodman (11, 559, 573, 587, 607, 637, 659, 679, 701, 719, 737, 757, 953) Department of Physiology, University of Massachusetts, Worcester, Massachusetts 01655

D. Neil Granger (225, 235, 245) Department of Molecular and Cellular Physiology, Louisiana State University Health Sciences Center, School of Medicine in Shreveport, Shreveport, Louisiana 71130

Dianna A. Johnson (779, 789, 797, 807, 831, 839, 849, 861, 867, 877, 881, 889, 897) Department of Ophthamology, University of Tennessee Health Sciences Center, Memphis, Tennessee 38163

Leonard R. Johnson (465, 479, 497, 529) Department of Physiology and Biophysics, University of Tennessee College of Medicine, Memphis, Tennessee 38163

Frank L. Powell, Jr. (259, 277, 289, 299, 315, 933) White Mountain Research Station, Division of Physiology, Department of Medicine, University of California at San Diego, La Jolla, California 92093

James A. Schafer (333, 343, 361, 371, 391, 405, 423, 437, 447, 921) Department of Physiology and Biophysics, University of Alabama at Birmingham, Birmingham, Alabama 35294

Stanley G. Schultz (3, 37, 547, 949) Department of Integrative Biology and Pharmacology, University of Texas Medical School at Houston, Houston, Texas 77225

Norman W. Weisbrodt (123, 137, 145) Department of Integrative Biology and Pharmacology, University of Texas Medical School at Houston, Houston, Texas 77030

Preface

Physiology is a dynamic discipline, and as new information and concepts emerge and our understanding evolves, the material we teach must also change. Each chapter of the third edition of *Essential Medical Physiology* has been updated and revised to encompass this new information. The authors and I have tried to eliminate errors that appeared in the previous edition without introducing new ones. We are grateful to our colleagues and students for help in identifying mistakes and inconsistencies, and we ask for their continued aid in improving future editions. In addition, four new chapters have been added integrating material in the important areas of exercise, heart failure and circulatory shock, maternal adaptations to pregnancy, and diabetes.

Each of the authors is an acknowledged expert in the field he or she is writing about. This is the major strength of this text. In contrast to many texts, which are single-authored or written by members of a single department, the authors represent seven different medical schools. Each has more than 20 years of experience teaching medical students in his or her area of expertise. All are nationally and internationally known for their research and contributions in the field about which they are writing. This expertise is manifested in many different ways. Six authors have written monographs for medical or graduate students. Seven are, or have been, the chairs of their departments and responsible for the content of the course in medical physiology. Six have served as consultants for developing the Part I examination for the National Board of Medical Examiners, and three have chaired the Physiology Test Committee for that organization. This experience has made it possible for each author to select the material that is essential to the discipline of medical physiology. Each is an excellent writer and has provided a thorough, up-to-date treatment of this material. As editor, I am grateful to them for their willingness to contribute their time and abilities to this project.

Most modern textbooks contain features to make them easier for students to use and to identify important concepts. Each chapter begins with a page-indexed chapter outline and a bulleted summary of the major points covered. Basic items and concepts in the text appear in bold type the first time they are used and defined. Chapters contain clinical notes emphasizing the relationships between physiology and clinical medicine. This clinically-related material has been highlighted and serves to reinforce the physiology as well as identify clinical relevance.

With this edition, however, we have gone one step further and created a set of outlines integrated with the figures in the text. As our research has become more cellular and molecular the classical boundaries between basic science departments have disappeared. Most departments responsible for teaching physiology have faculty members who were not trained as medical physiologists. It is our hope that these outlines, presented as a compact disc, will aid in the organization of lecture notes or serve as the basis for student handouts.

Simply stated, physiology is the study of function. Function occurs at three levels: molecular, subcellular, and cellular. Events at these levels in turn determine the activities of tissues, organs, and systems. Understanding at each level is necessary to appreciate the overall function or dysfunction of an individual. The overriding principle of physiology is the integration of a variety of mechanisms. For example, how does the kidney produce urine? And how is this process regulated so that the volume and salt content of urine are matched to the salt and water intake of the individual and the loss of salt and water via other processes such as respiration and sweating? Each of the authors has presented and integrated in his or her section the material that is necessary to medical physiology as it is taught in most medical schools. Cell biology is integrated as part of the physiologic discipline as it arises, not in a separate section. Chapters often found in physiology textbooks that are better suited to disciplines such as histology, immunology, or biochemistry have been omitted. This textbook contains the information the authors and editor believe essential to providing a strong background in physiology for the practice of medicine. This is the basis of the title chosen for the text.

Physiology is the primary basic medical science. It is essential to pharmacology, the study of the effects of drugs on physiologic processes. Most of medicine itself is pathophysiology. We hope that each student will realize that the study of physiology should continue throughout his or her career. With that in mind, a few references are included at the end of each chapter. They cite well-written, recent articles that provide a wealth of information as well as extensive bibliographies for further, more detailed reading.

PART I

GENERAL PHYSIOLOGY

The Internal Environment

STANLEY G. SCHULTZ

KEY POINTS

- Physiologic processes have evolved to maintain the constancy or stability of the internal environment. This tendency toward maintenance of physiologic stability is *homeostasis*.
- Living things are open systems.
- The *total body water* is contained within two major compartments—the intracellular compartment and the extracellular compartment. The intracellular compartment is further subdivided into the *plasma* and the *interstitial* fluid compartments.
- The *intracellular compartment* is characterized by low intracellular Na^+ and Cl^- concentrations and a high K^+ concentration. The intracellular compartment comprises approximately 25 L,

which is approximately 35% of total body weight.
- The *extracellular compartment* is divided into the plasma and interstitial compartments; the former comprises approximately 3 L and the latter approximately 12 L. These compartments are characterized by high Na^+ and Cl^- concentrations and a low K^+ concentration.
- Four organs provide the interfaces between the extracellular fluid compartment and the outside world: the skin, alimentary canal, lung, and kidney. The proper interactions among these organs and the external environment is essential for *homeostasis*.

HOMEOSTASIS: THE SUBJECT OF PHYSIOLOGY

If one had to describe, with a single word, what physiology is all about, that word would be *homeostasis.* This word was coined by the great American physiologist, Walter B. Cannon, in his book entitled *The Wisdom of the Body* (1939) and refers to regulatory mechanisms by which biologic systems tend to maintain the internal stability necessary for survival while adjusting to internal or external threats to that stability. If homeostasis is successful, life continues; if it is unsuccessful, disease and perhaps death ensue.

Cannon's notion of homeostasis was an extension of the concept first introduced by one of the founders of modern physiology, Claude Bernard, a physician by training who made many seminal contributions to our early understanding of digestion, metabolism, vasomotor activity of nerves, neuromuscular transmission, and other areas of physiology. But perhaps his greatest lasting contribution was the notion that all physiologic processes are designed to maintain the internal environment, the *milieu interieur*, that bathes our cells, tissues, and organs. The following is from his opening lecture in a course in general physiology given at the College de France in 1887:

> The living body, though it has need of the surrounding environment, is nevertheless relatively independent of it. This independence which the organism has of its external environment derives from the fact that in the living being, the tissues are in fact withdrawn from direct external influences and are protected by a veritable internal environment which is constituted, in particular, by the fluids circulating in the body.

In short, he suggested that we exist within our own "hot houses" and that our existence depends on our abilities to maintain that "hot house."

Although this notion may seem trivial now, at the time when it was introduced it was revolutionary. It proved to be prophetic and has influenced all physiologic thinking to this day.

ORIGIN OF OUR *MILIEU INTERIEUR*

There is good evidence that our predecessors, the protovertebrates or prochordates, migrated from the seas into brackish or fresh water during the Cambrian period, almost 500 million years ago. During that migration, which lasted some 200 million years, they "locked" within themselves a fluid similar in composition to that of the seas from which they emerged. Some of these early vertebrates returned to the oceans, which had become saltier; others stayed in brackish or fresh water; and still others chose to live on land. But,

regardless of their final habitats, the "sea" within all vertebrates is remarkably similar in ionic composition to the salinity of the Cambrian seas from which they emerged. This sea that bathes all cells is rich in Na^+ and Cl^-, in contrast with the intracellular fluid, which, in all living forms including microorganisms, is rich in K^+ and poor in Na^+ and Cl^-.

LIVING THINGS ARE OPEN SYSTEMS

Maintaining a constant internal environment would pose no problem if living things were closed systems, like a solution in a sealed bottle. Instead, however, all freely living forms are open systems constantly exchanging matter and energy with the environment. The remarkable thing is that despite this constant exchange, essential for life, the compositions of our intracellular and extracellular fluids are maintained remarkably constant; this is what homeostasis is all about. How is this accomplished?

BODY FLUID COMPARTMENTS AND THEIR CONTACTS WITH THE OUTSIDE WORLD

The total body water (TBW) in higher animals is distributed among three major compartments: the blood plasma, the *interstitial* fluid (ISF), and the *intracellular* fluid (ICF). The plasma is separated from the ISF compartment by highly permeable capillaries; together, plasma and ISF constitute the *extracellular* fluid (ECF) compartment. This compartment is separated from the ICF compartment by cell membranes, which in most instances, as discussed in Chapter 3, are highly permeable to water but very selective with respect to the passage of solutes. A fourth, small compartment, called the transcellular fluid compartment, consists primarily of fluid in transit in the lumina of epithelial organs (*e.g.,* the gall bladder, stomach, intestines, and urinary bladder), as well as the cerebrospinal fluid and the intraocular fluid.

In an average human adult weighing approximately 70 kg, the TBW makes up approximately 60% of body weight or about 40 L. The distribution of this water among the plasma ISF and ICF compartments is shown in Fig. 1, and the way in which the sizes of these compartments are determined is discussed below. The *transcellular* fluid compartment in such an individual comprises approximately 2 to 4% of the TBW and contains approximately 1 to 2 L of water.

The ECF of all vertebrates is in intimate contact with four organs that interface with the external environment. One is a tube that runs from mouth to

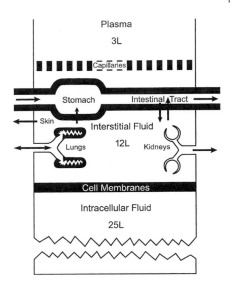

FIGURE 1   Sizes of the major body fluid compartments and the four organs that interface the extracellular fluid and the external environment. Total body water makes up approximately 60% of body weight.

anus—the alimentary canal. It is responsible for absorbing water, essential elements, and the metabolites that form our building blocks and fuel our activities; for the most part, however, it is indiscriminant with respect to what it will permit to enter the body. The second organ is the lungs, which are responsible for exchanging oxygen and carbon dioxide with the environment and, as discussed below, are a source of water loss. The third organ is the kidneys, which turn over the sea within us many times each day and correct for the indiscretions of our gastrointestinal tracts. In short, the composition of our body fluids is determined not by what the mouth takes in but by what our kidneys keep (Smith, 1961). The final organ in contact with the external environment is, of course, the skin, which plays a primary role in temperature regulation, but is also a source of water loss.

It is the precise interplay of these interfaces (particularly the alimentary canal, lungs, and kidneys) and the external environment that maintains the constancy of our *milieu interieur* and permits us to live a relatively free and independent life. A large portion of this text is devoted to considering how these organs perform those essential tasks. But, for the moment, the precision of these homeostatic processes can be illustrated by water balance in the normal human. As illustrated in Fig. 2, water is gained by the body via three sources: (1) water ingested in the form of liquids, (2) water contained in solid foods, and (3) water derived from oxidative metabolism of carbohydrates. A small amount of water (100–200 cc) is lost daily in the feces, and a rather substantial amount of water (approximately 900 cc) is lost by vaporization from the respiratory tract and skin; the latter is referred to as *insensible water loss* and must

be distinguished from perspiration. These two routes of water loss are relatively constant under conditions of normal activity in a moderate climate. The remaining water that must be lost to account precisely for the total intake is excreted by the kidneys. Were it not for this fine balance, an individual might gradually enter a state of negative water balance and become dehydrated or a state of positive water balance and become water loaded, both potentially serious conditions.

IONIC COMPOSITION OF THE MAJOR FLUID COMPARTMENTS

A comparison of the compositions of the three major body fluid compartments with respect to their major ionic components is shown in Fig. 3 and summarized in Table 1. The ionic compositions of the plasma and ISF are similar, as expected, inasmuch as they are separated by highly permeable capillaries. The major difference between these two compartments arises from the fact that the capillaries are only sparingly permeable to proteins, particularly albumin, so that the protein composition of the plasma is much greater than that of the ISF. Because of the Gibbs–Donnan effect discussed in Chapter 3, this asymmetric distribution of anionic proteins brings about small asymmetries in the distributions of the permeable ionic species, principally Na^+ and Cl^-. Thus, the concentration of Na^+ in the plasma is between 5 and 10% greater than that in the ISF, and the concentrations of Cl^- and bicarbonate in the ISF fluid are between 5 and 10% greater than those in the plasma.

The ICF, separated from the "Cambrian sea within us" by highly selective cell membranes, has a

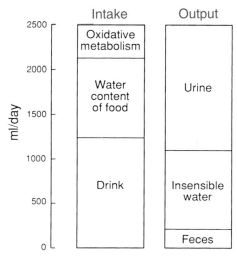

FIGURE 2   Balance between intake and output of water in a normal adult.

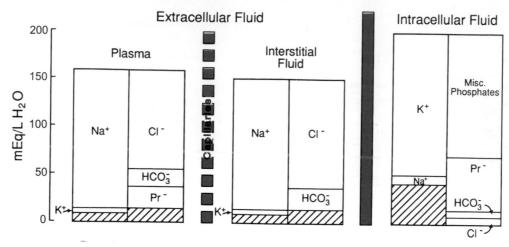

FIGURE 3   The major components of plasma, interstitial fluid, and intracellular fluid. Crosshatching refers to other or minor components. Pr⁻, anionic proteins.

composition resembling that found in all living cells down to the most primitive microorganisms. The predominant cation by far is K^+, whose positive charge is largely balanced by anionic proteins and by organic and inorganic phosphates; the next major cation is Mg^{2+}. The intracellular concentrations of Na^+, Cl^-, and bicarbonate are very low compared with those found in the ECF.

BODY FLUID OSMOLARITY AND pH

Inasmuch as capillaries and most plasma cell membranes are highly permeable to water, the osmolarity of the plasma, ISF, and ICF compartments are equal and have a value of approximately 295 mOsm/L. The pH of the ECF compartment is approximately 7.4, whereas, because of metabolic activities, that of the intracellular

compartment is somewhat more acidic and is approximately 7.1. The processes responsible for regulating the osmolarity and pH of the body fluids are discussed in Chapter 28.

MEASURING THE SIZES OF THE BODY FLUID COMPARTMENTS

There is one simple principle underlying all approaches to measuring the sizes of the fluid compartments—namely, the principle of conservation of matter. As illustrated in Fig. 4, if a certain quantity (Q) of some substance is injected into a fluid compartment having a volume (V), then, when an equilibrium (or steady-state) distribution is achieved, the concentration of that substance in the container will be Q/V.

$$\text{Concentration} = Q/V \qquad (1)$$

If we know Q and can determine the value of Q/V, it is a simple matter to determine V, the volume of distribution.

Many substances can be employed to determine the sizes of the body fluid compartments. The essential characteristics of all are that they (1) are nontoxic at the concentrations employed, (2) are neither synthesized nor metabolized, and (3) do not bring about shifts in fluid distribution among the compartments.

Plasma Volume

The simplest of all of the body fluid compartments to measure is the plasma volume (PV). A substance is used that does not readily permeate capillary walls so that,

---

TABLE 1   Important Components of Extracellular and Intracellular Fluid Compartments

	Extracellular[a]	Intracellular[b]
Ca^{2+} (mmol/L)	2.5	1×10^{-4}
Cl^- (mmol/L)	110	≈ 10
HCO_3^- (mmol/L)	20	≈ 10
K^+ (mmol/L)	4	120
Na^+ (mmol/L)	140	14
Osmolarity (mOsm/L)	295	295
pH	7.4	7.1–7.2

[a]These are reasonably approximate values for the plasma and interstitial fluid compartments. As noted in the text, because of differences in protein concentration, these two compartments display small differences in ion concentration.

[b]These are approximate values for many cell types.

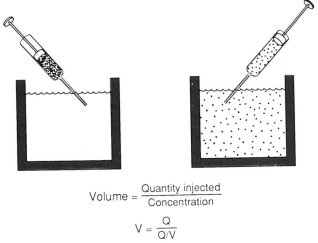

$$\text{Volume} = \frac{\text{Quantity injected}}{\text{Concentration}}$$

$$V = \frac{Q}{Q/V}$$

FIGURE 4　Principle underlying all methods for measuring the volumes of the body fluid compartments.

after intravenous injection into the plasma, it remains within the vascular bed at least for the time necessary to achieve a uniform distribution (Fig. 5A). A substance commonly employed for this purpose is serum albumin labeled with radioactive iodine (I-albumin).

Example: An amount of ^{131}I-albumin having 350,000 counts per minute (cpm) is injected intravenously. One hour later, 10 mL of whole blood are withdrawn and centrifuged. The whole blood consisted of 5.5 mL of plasma and 4.5 mL of packed blood cells. Then, 1 mL of the plasma is removed and found to contain 100 cpm. Thus,

$$PV = \frac{Q}{Q/V} \tag{2}$$

In this case, therefore,

$$PV = \frac{350,000}{100 \text{ cpm/L}} = 3.5\,\text{L} \tag{3}$$

Given these data, one can also determine the total blood volume employing a measure known as the hematocrit, which is defined as the fraction of the total blood volume comprised of cells, mainly red blood cells. In this case, the hematocrit (Hct) is

$$Hct = \frac{4.5}{10} = .45 \tag{4}$$

The total blood volume (TBV) is given by:

$$TBV = \frac{PV}{1 - Hct} \tag{5}$$

In this case, therefore,

$$TBV = \frac{3.5}{.55} = 6.4\,\text{L} \tag{6}$$

Another substance that is frequently employed to measure plasma volume is a dye, Evan's blue dye or T-1824, which binds avidly to serum albumin and can be determined colorimetrically.

Finally, one can employ the same principle to determine the volume of red cells in the plasma using red cells that have been labeled with chromium-51. In this case, an amount of chromium-labeled red cells is injected intravenously, but instead of counting the plasma collected after 1 hour, one discards the plasma and determines the concentration of chromium-51 in the packed red cells. This provides a measure of the

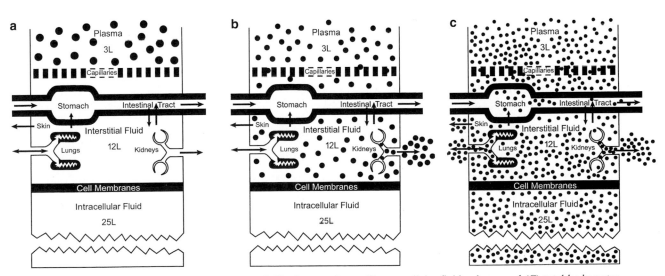

FIGURE 5　Principles underlying the measurement of (A) plasma volume, (B) extracellular fluid volume, and (C) total body water.

total volume of red blood cells in the circulation. Then, knowing the hematocrit, one can determine the total PV.

Extracellular Fluid Volume

To determine the volume of the ECF compartment, a substance is used that can readily permeate capillary walls but not cell membranes (Fig. 5B). The most commonly employed substance for this purpose is a polyfructoside, inulin, which has a molecular weight of approximately 5500 and a radius of approximately 15 Å, or 1.5 nm. This substance is inert and nontoxic. (Another important use of inulin is discussed in Chapter 26.)

The only minor complication that arises in determining the size of the ECF compartment (illustrated in Fig. 5B), stems from the fact that substances that can readily permeate the capillary wall are readily filtered by the renal glomeruli and may be excreted; this is particularly true for inulin, which, after filtration, is excreted completely. Thus, in applying the principle of conservation of mass, one must take into account the amount of the substance lost from the body in the course of the measurement.

Example: An individual's urinary bladder is catheterized and drained, then 2.4 g of inulin is injected intravenously; 6 hours later, a venous sample is withdrawn and found to have a plasma inulin concentration of 0.1 mg/mL. During those 6 hours, 200 mL of urine was produced having an inulin concentration of 5 mg/mL. Thus, 1.0 g of the original 2.4 g of inulin injected was lost. It follows that the ECF volume is given by:

$$ECF = \frac{Q(injected) - Q(lost)}{Q/V} \qquad (7)$$

In this case, therefore,

$$ECF = \frac{2400 - 1000}{0.1} = 14 \text{ L} \qquad (8)$$

Total Body Water

The volume of TBW can be determined using a substance that can rapidly permeate cell membranes and distribute itself uniformly throughout the ECF and ICF compartments. One commonly used substance for this purpose is antipyrine, which is highly lipid soluble and inert. Other substances that can be used for this purpose are isotopic forms of water such as deuterated water or tritiated water. As illustrated in Fig. 5C, this measurement is complicated because these substances are lost from the body via every route open for the loss

of water. Fortunately, however, only approximately 2 hours are needed for them to achieve a uniform distribution in the TBW and, during that period, the amount of water lost by the body is less than 1% of the TBW.

Interstitial and Intracellular Fluid Compartments

The sizes of the ISF and ICF compartments cannot be measured directly. The volume of the ISF compartment is the difference between those of the ECF compartment and the PV:

$$V_{ISF} = V_{ECF} - PV \qquad (9)$$

The volume of the ICF compartment is given by the difference between the volume of total body water and that of the extracellular fluid compartment:

$$V_{ICF} = TBW - V_{ECF} \qquad (10)$$

ALTERATIONS IN BODY FLUID COMPARTMENTS

The sizes of the body fluid compartments can be altered under a variety of normal and pathophysiologic circumstances. In attempting to analyze how this happens, several principles must be kept in mind. The first is that most cell membranes are highly permeable to water, so that at all times the body fluid compartments are isotonic with the plasma; the only exception is the renal medulla, which is discussed in Chapters 27 and 28. The second is that the driving force for water movement across biologic barriers is a difference in the *effective* osmolarity across those barriers. As will be discussed in Chapter 3, the effective osmolarity is determined only by those solutes that exert an osmotic force across the membrane.

Dehydration

Dehydration (water deprivation, intense perspiration, vomiting), which initially is a loss of water from the ECF compartment, causes an increase in the concentrations of NaCl (the predominant ions in that compartment) and plasma proteins and, hence, the *effective* osmolarity of that compartment. This increase results in water being drawn out of the intracellular compartment to maintain isosmolarity. The final result is a decrease in the volume of all of the fluid compartments.

Infusion of 2 L of Isotonic Saline (150 mmol/L NaCl)

Because of the pump-leak properties of the cell membranes discussed in Chapter 3, Na^+ and Cl^- are excluded from the cells. Because the infused fluid is isosmotic, almost all the infused fluid remains in the extracellular compartment and there is no change in the size of the intracellular compartment.

Infusion of Isotonic Urea (300 mmol/L)

Because urea permeates most cell membranes rapidly, the isotonic urea infusion distributes itself throughout the TBW, leading to an increase in the volumes of all fluid compartments.

Rapid Infusion of Pure Water

Because water permeates all cell membranes rapidly, all body fluid compartments expand.

Ingestion of 100 g of NaCl Tablets

The ECF becomes hyperosmotic and draws fluid from the intracellular compartment. There is no change in TBW, but water redistributes from the intracellular to the extracellular compartments.

Infusion of Isosmotic Saline (NaCl) Solution Containing 20% Albumin

The ECF compartment expands and the IFC compartment shrinks. This occurs because Na^+ and albumin distribute almost entirely in the extracellular fluid compartment, and the albumin results in increased osmotic pressure of that compartment. Because most albumin remains in the plasma, the PV expands much more than the ISF volume.

Infusion of Isosmotic (5%) Glucose Solution

Because the end products of glucose metabolism are carbon dioxide and water, the situation is not much different from infusion of water: all body fluid compartments expand.

Suggested Readings

Gamble JL. *Chemical anatomy, physiology and pathology of extracellular fluid.* Cambridge, MA: Harvard University Press, 1958

Edelman IS, Leibman J. Anatomy of body water and electrolytes. *Am J Med* 1959; 27:256–277

Smith HW. *From fish to philosopher.* Garden City, NY: Doubleday, 1961

2

Control of Cell Function

H. MAURICE GOODMAN

KEY POINTS

- Survival of a multicellular organism requires that different cells assume different functions.
- Genetic information is encoded in the nucleotide sequence of DNA. Complementary pairing of purine and pyrimidine bases permits accurate replication and transmission of stored information.
- Synthesis of a complementary strand of RNA is the first step in gene expression.
- The primary RNA transcript is converted to messenger RNA by removal of extraneous intervening sequences and splicing the remaining segments together. Alternate splicing may produce more than one messenger RNA template.

- Synthesis of proteins depends on base pairing of each transfer RNA with the mRNA template and the enzymatic activity of the ribosomes.
- Proteins can change their shape and consequently their activity when bound to other molecules or when covalently modified.
- Cells change their activity in response to signals received from the internal or external environment. The most common means of information exchange is through chemical signals.
- Target cells convert information they receive from their environment to intracellular biochemical reactions that produce changes in cellular behavior.

KEY POINTS (*continued*)

- Receptors bind signals with high specificity and affinity and undergo a conformational change sufficient to initiate a biochemical reaction. Receptor number and activity are subject to regulation.
- Receptors in the G-protein-coupled receptor superfamily are glycoproteins that thread through the plasma membrane seven times and transmit their signal by activating guanine nucleotide-binding proteins.
- The concept of the second messenger is that information from activated surface receptors is transmitted to intracellular effectors by molecules that initiate biochemical responses within the cell and its organelles. The ensuing responses are a function of the differentiated state of the cell rather than of the molecule

- that causes activation or inhibition of cellular responses.
- Receptors for steroid hormones, thyroid hormone, and vitamins A and D are transcription factors that bind directly to response elements in DNA and, when bound to their agonists, activate or repress expression of responsive genes.
- Secretory cells synthesize their products from simple precursor molecules taken up from the blood.
- For chemical signals to deliver information effectively, they must be eliminated from the vicinity of the receptor as soon as possible after the message has been received. This is accomplished by metabolic degradation, reuptake into secretory granules, or dilution.

SPECIALIZATION OF CELLULAR FUNCTION

All cells in a multicellular organism have the same basic needs for survival and, because they share a common genetic heritage, have the same potential for carrying out all of the various cellular functions. However, the "social contract" that governs cells in a complex organism requires that different cells selectively develop particular capabilities that allow them to perform or participate in the performance of certain specialized tasks on behalf of the organism as a whole. Red blood cells, for example, have the highly specialized ability to transport oxygen and carbon dioxide, while some cells in the gastric mucosa are specialized to secrete acid, and muscle cells are equipped with the structural and enzymatic apparatus that generates the contractile force that propels the organism toward its next meal or away from predators. Determination of which cellular capacities will be expressed and which will remain dormant resides in the province of the differentiation program to which a particular cell or cell lineage commits in the course of development. Such programs of differentiation require the timely expression of some genes and the silencing of others, with the result that different cells develop particular morphological, biochemical, and physical features in addition to those general housekeeping functions required for survival of all cells.

Specialized functions of cells and their arrangement into tissues and organs provides the whole organism with mechanisms to cope with a changing and often hostile environment in ways that ensure survival of the

individual and perpetuation of the species. The complex functions of organ systems that are discussed in subsequent chapters require precise coordination of the activities of the individual cells of which they are composed. Such coordination is critically dependent upon an exchange of information between cells, between cells and their environment, and between the organism and its environment. Control of cellular activity has two separate but closely related aspects; one aspect relates to determining the complement of the "machinery" that a cell has available to carry out a particular task (*i.e.*, what genes are expressed and to what degree), and the second aspect relates to governing the rate at which the machinery operates to perform that task.

Regulation of Gene Expression

Although selective expression of some combinations of genes during cellular differentiation accounts for the wide range of cellular phenotypes in complex organisms, regulation of gene expression does not end when cells become terminally differentiated. Except for red blood cells, which have no nuclei, all cells increase or decrease expression of some of their genes to bring functional capacity into alignment with changing physiological demands. We therefore consider briefly some pertinent, but general, aspects of gene regulation distilled from the wealth of information that has come to light in recent years. For a detailed understanding of this important topic, the student is urged to consult one of the many textbooks dedicated to molecular biology.

DNA Structure

We may consider the gene to be a stretch of *deoxyribonucleic acid* (DNA) that contains both the coded information that governs its regulation and the amino acid sequence of one, or sometimes more than one, protein. The sum total of all of the genes is called the *genome*, which in humans is estimated to contain between 30 and 35 thousand pairs of genes. DNA consists of a chain of many millions of deoxyribose (a five-carbon sugar) molecules linked together by phosphate groups that form ester bonds with the 3′ hydroxyl group of one sugar and the 5′ hydroxyl group of its neighbor (Fig. 1). Carbon 1 of each deoxyribose is attached by N-glycosidic linkage to an organic base, which may be adenine (A), guanine (G), thymine (T), or cytosine (C). A *nucleotide* is the fundamental unit of the nucleic acid polymer and consists of a base, a sugar, and a phosphate group. The information stored in DNA is encoded in the sequence of its constituent nucleotides. The ability of purine bases A and G to form *complementary pairs* with pyrimidine bases T and C on an adjacent strand of DNA is the fundamental property that permits accurate replication of DNA and transmission of stored information (Fig. 2). In the cell, DNA forms a double helix of two strands oriented in opposite directions, with each A on one strand pairing with a T on the complementary strand and each G pairing with a C. The long double strands of DNA are organized into *nucleosomes*, each of which consists of a stretch of about 180 nucleotides tightly wound around a complex of eight histone molecules. The nucleosomes are linked by stretches of about 30 nucleotides, and the whole strand of nucleoproteins is tightly coiled in a higher order of organization to form chromosomes. Each *chromosome* contains one double helix of DNA in which 50 to 250 million base pairs code for thousands of genes lined up like a string of beads. Humans have 23 pairs of chromosomes, with one member of each pair inherited from each parent.

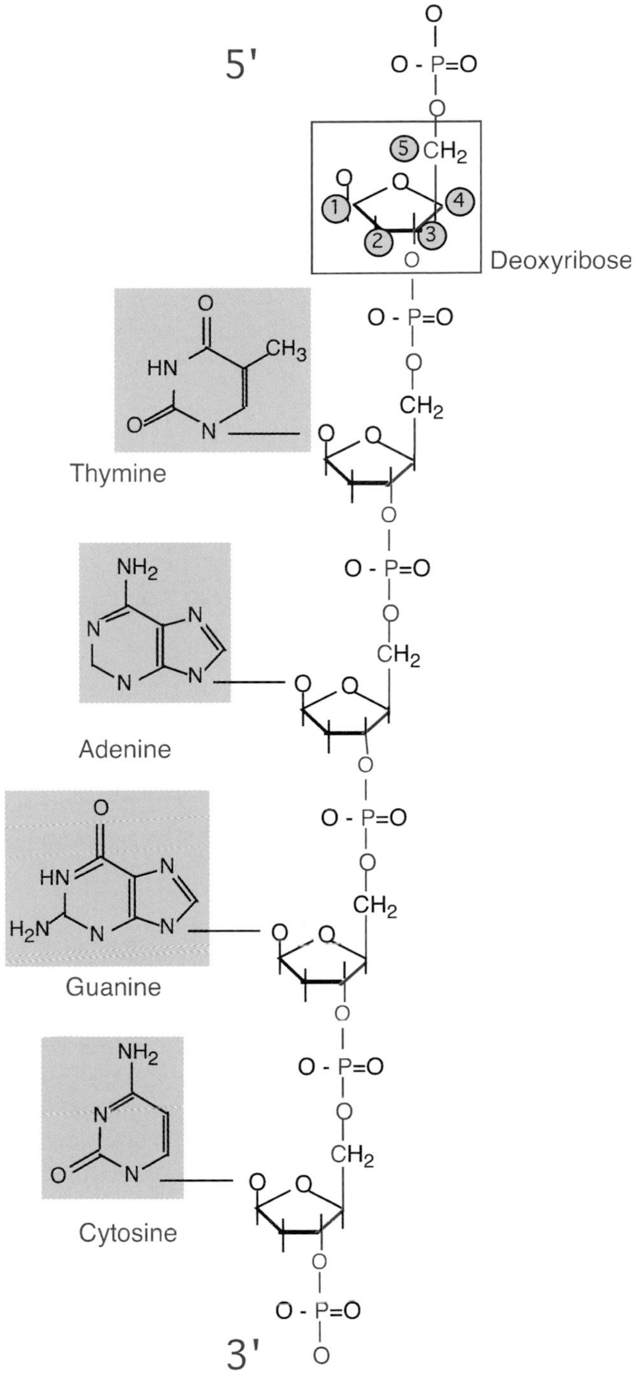

Gene Transcription

For the information encoded in the DNA to guide protein synthesis, a complementary strand of *ribonucleic acid* (RNA) must be synthesized to serve as the template that guides assembly of amino acids into proteins. The synthesis of RNA transfers encoded information from DNA to RNA and is therefore called *transcription*. RNA differs chemically from DNA in two ways: Its sugar is ribose instead of deoxyribose, and it contains *uracil* (U) instead of thymine. Only part of each gene is transcribed. Untranscribed portions include sequences of bases involved in regulation of transcription and

FIGURE 1 Composition of DNA. DNA is a polymer of the five-carbon sugar, *deoxyribose*, in diester linkage with phosphate (P) forming ester bonds with hydroxyl groups on carbons 3 and 5 on adjacent sugar molecules. The purine (*adenine* or *guanine*) and pyrimidine (*thymine* or *cytosine*) bases are linked to carbon 1 of each sugar. The numbering system for the five carbons of deoxyribose are shown in blue at the top of the figure. The chemical bonds forming the backbone of the DNA chain are also shown in blue. The 5′ or 3′ ends refer to the carbons in deoxyribose.

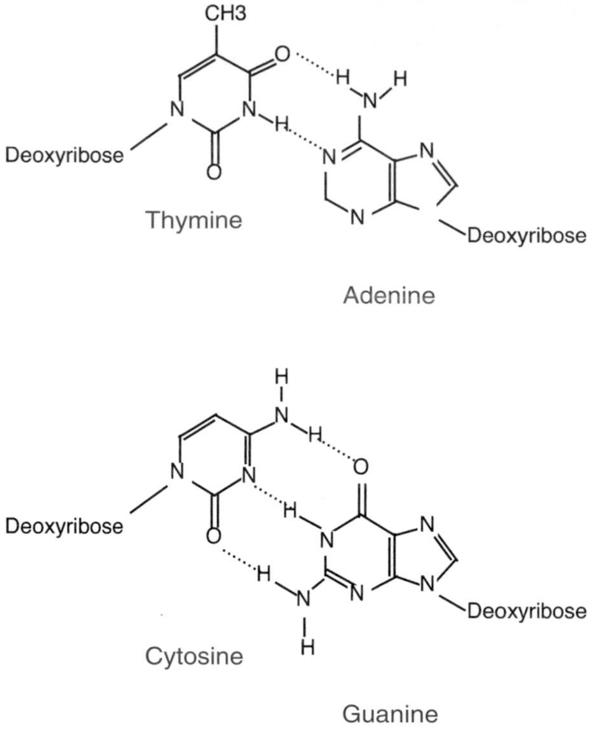

FIGURE 2 Complementary base pairing by the formation of hydrogen bonds between thymine and adenine and between cytosine and guanine. RNA contains uracil in place of the thymine found in DNA. Uracil and thymine differ in structure only by the presence of the methyl group (CH₃) found in thymine.

the sometimes very long stretches of DNA that lie between the regulatory sequences. The *promoter* is the particular nucleotide sequence located upstream of the 3' end of the transcribed portion that determines where the enzymatic machinery attaches and in which direction it will face. To initiate transcription, a multimolecular complex of nuclear proteins called *general transcription factors* and *RNA polymerase II* must be assembled on the promoter. Assembly begins with the binding of one of these transcription factors (called TFIID) to a particular nucleotide sequence, usually TATA, in the promoter. Sequential binding of other transcription factors aligns the RNA polymerase on the promoter, after which phosphorylation of the polymerase launches its movement along the DNA template as it assembles the RNA transcript. Access of the general transcription complex to a promoter generally requires loosening of the tightly coiled DNA strand and is accomplished in part by acetylation of nucleosomal histones. The rate of expression of a particular gene is controlled not only by proteins that bind to DNA sequences in the promoter, but also by proteins that bind to other sequences in more distant regulatory regions of DNA. Regulatory proteins are transcription factors that may act as *enhancers* or *repressors* of gene expression (Fig. 3).

There are hundreds of such proteins, which are expressed in different combinations in different cell types and at different times in their life cycles. In many cases, genes for regulatory proteins are themselves targeted for regulation.

RNA Processing

Genetic information is stored in the nucleus, but proteins are synthesized in the cytoplasm. The RNA transcript that emerges from the nucleus to direct synthesis of one or more proteins is called *messenger RNA* (mRNA), and the process of converting information carried by nucleic acids to proteins is called *translation*. Each amino acid encoded in mRNA is specified by a *codon* of three adjacent nucleotides. For example, alanine is encoded as CGA, while serine is specified as AGC. Not all of the nucleotides in the transcribed part of the gene have counterparts in mRNA. This is because the nucleotides that specify portions of the protein product or untranslated regulatory components of mRNA are interrupted by long stretches of DNA that have no known coding function. The portions of the gene that have complementary sequences in mRNA are called *exons* and the intervening stretches of DNA are called *introns* (see Fig. 3). The portions of RNA transcribed from introns are excised from the primary RNA transcript and destroyed. The remaining fragments representing the exons are spliced together by a complex apparatus of RNA and protein called a *spliceosome* to produce the mRNA. In processing the RNA transcribed from some genes, the spliceosomes may sometimes skip over stretches of RNA transcribed from one or more exons with the result that *alternately spliced* mRNAs are formed. This is one way that a single gene can give rise to several different proteins (Fig. 4).

Translation

Translation of mRNA requires two additional forms of RNA: *ribosomal RNA* and *transfer RNA* (tRNA). Each tRNA molecule reacts at one end with a specific amino acid, and at its other end it contains a triplet of nucleotides that are complementary to the codon in mRNA that specifies that amino acid. In this way, complementary base pairing between the tRNA and the mRNA lines up the amino acids in the appropriate sequence. The *ribosomes* are comprised of a large and a smaller subunit and are complexes of RNA and many different proteins. The ribosomes hold the reaction components together in the correct orientation and catalyze peptide bond formation. Initiation of protein synthesis requires a complex of regulatory and catalytic

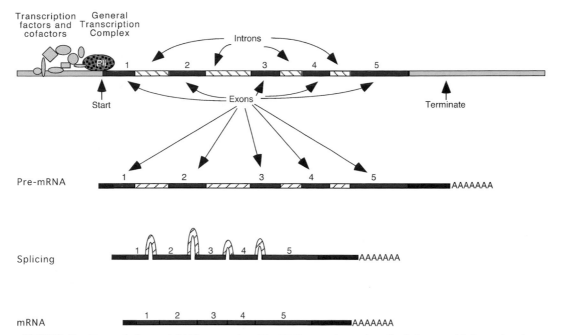

FIGURE 3   Transcription and RNA processing. The DNA strand contains all of the stored information for expression of the gene including the *promoter*, distant regulatory elements (not shown), binding sites (response elements) for regulatory proteins, and the coding for the sequence of the protein (**exons**) interrupted by intervening sequences of DNA (*introns*). Exons are numbered 1–5. The *primary RNA transcript* contains the complementary sequence of bases coupled to a poly A *tail* at the 3' end and a methyl guanosine *cap* at the 5' end. Removal of the introns and splicing the remaining exons together produce the *messenger RNA* (*mRNA*) that contains all of the information needed for translation including the codons for the amino acid sequence of the protein and untranslated regulatory sequences at both ends.

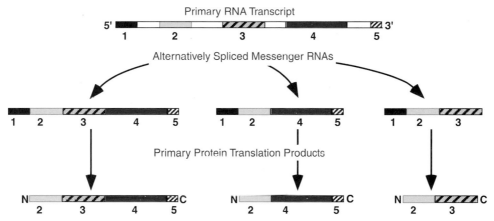

FIGURE 4   Alternative splicing of mRNA can give rise to different proteins. Numbers indicate exons. Exon 1 is untranslated. N, amino terminus; C, carboxyl terminus.

proteins called *initiation factors*, and its continuation requires *elongation factors*. For translation to begin, the two subunits of the ribosome must be assembled at a precise position on the mRNA because deviation by even a single nucleotide in either direction would specify a totally different protein. Protein synthesis begins when a special initiator tRNA carrying methionine binds to the AUG start codon in the mRNA and proceeds to

elongate the peptide chain by adding one amino acid at a time from the amino terminus toward the carboxyl terminus. The ribosome moves stepwise along the mRNA, catalyzing peptide bond formation and displacing the newly unloaded tRNA as the next amino acid is positioned by its tRNA (Fig. 5). When the ribosome reaches any one of three "stop" codons it releases the completed protein and dissociates from the mRNA.

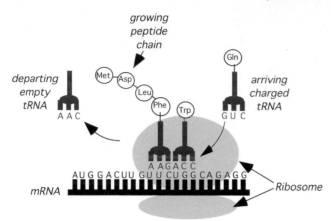

FIGURE 5   Translation. A molecule of transfer RNA (*tRNA*), charged with its specific amino acid, phenylalanine (*Phe*), and already linked to the growing peptide chain, is positioned on the *mRNA* by complementary pairing of its triplet of nucleotides with its codon of three nucleotides in the mRNA. A second molecule of tRNA charged with its specific amino acid, tryptophan (*Try*), has docked at the adjacent triplet of nucleotides and awaits the action of ribosomal enzymes to form the peptide bond with phenylalanine. Linking the amino acid to the peptide chain releases it from its tRNA and allows the empty tRNA to dissociate from the mRNA. A third molecule of tRNA, which brought the preceding molecule of leucine (*Leu*), is departing from the left, while a fourth molecule of tRNA, carrying its cargo of glycine (*Gly*), arrives from the right and waits to form the complementary bonds with the next codon in the mRNA that will bring the glycine in position to be joined to tryptophan at the carboxyl terminus of the peptide chain. The ribosome moves down the mRNA, adding one amino acid at a time until it reaches a stop codon. (Adapted from Alberts *et al.*, *Molecular biology of the cell*, New York: Garland Publishing, 1994.)

During rapid protein synthesis multiple ribosomes bind to a single mRNA separated by about 80 nucleotides to form *polysomes*. In this way, a single molecule of mRNA is used repetitively to synthesize multiple copies of a protein.

Posttranslational Processing

Most newly translated proteins must be modified and translocated to appropriate cellular or extracellular sites before they are ready to assume their biological functions. The peptide chain that emerges from the ribosome is subjected to various steps of *posttranslational processing* and directed toward an appropriate cellular compartment even before its synthesis is complete. For example, the N-terminal peptide of proteins destined to be secreted or integrated into membranes passes through a specialized pore in the membrane of the endoplasmic reticulum where the proteins are folded into their proper configurations and processed. Processing may include removal of the N-terminal "leader" sequence that provided the signal for entering the endoplasmic

reticulum, formation of disulfide bonds between cysteines on adjacent loops of the protein, proteolytic cleavage at one or more internal sites, attachment of carbohydrate or lipid moieties, and the formation of complexes with one or more products of other genes. Cleavage may release one or more biologically active proteins or peptides from a single precursor protein (see Fig. 6). Cleavage may also produce separate noncovalently attached subunits of a protein complex. Instructions for posttranslational processing and targeting to precise cellular loci are encoded in short sequences of amino acids within the protein itself. Mutations in these critical areas therefore can have profound effects on the stability or functionality of a protein even when amino acid sequences in catalytic or regulatory domains are unchanged. Similarly, how a protein folds or coils to assume its final shape is determined largely by its amino acid sequence, but proper folding is sometimes assisted by interaction with other proteins called *chaperones*.

From the foregoing, it is apparent that expression of the final product of any gene is controlled at multiple steps. The most thoroughly studied, and perhaps most

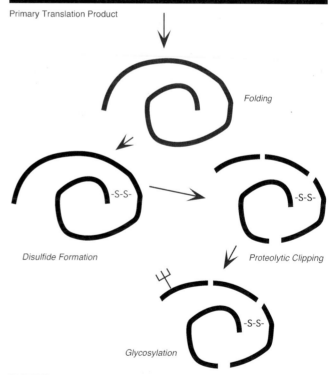

FIGURE 6   Examples of posttranslational processing. The straight chain primary translation product (A) folds into its proper conformation (B), either spontaneously or with the aid of cellular proteins called chaperons. Cysteines brought into alignment in the folded protein can now be oxidized to form disulfide bridges (C). Proteolytic clipping by processing proteases cleave peptide bonds (D), and one or more chains of carbohydrate is added (E).

important, of these regulated steps is the initiation of gene transcription, when the assembly of a complex array of regulatory factors provides a variety of opportunities for cell-specific and temporally dependent fine tuning. This is the principle site of regulation of gene expression by extracellular agents. After transcription, both the splicing process to form the mRNA and the stability of the mRNA are also subject to regulation, as is the process of mRNA transport out of the nucleus. Splicing may be cell specific, so that the same RNA transcript of a gene may give rise to different mRNAs in different cells. The complex reactions associated with the initiation of protein synthesis are also subject to control. Enzymatic processing of the translation product may determine whether or not an active protein or multiple proteins are formed and targeted to appropriate cellular loci. Proteolytic cleavage in different cell types can form different protein products from the same precursor depending on the availability and activity of processing enzymes of different specificities. In fact, as a result of alternate splicing of RNA and the complexities of posttranslational processing, it has been estimated that the 30 to 35,000 pairs of genes in the human genome may give rise to as many as 2,000,000 proteins.

Protein Degradation

Synthesis is not the only determinant of the complement of proteins present at any time within a cell. Proteins have limited life times and may be degraded within a few minutes to a few years after synthesis. Each cellular protein has a characteristic half-life, but changing physiological circumstances may shorten or lengthen life spans of some proteins; protein degradation is a regulated process. Most cellular proteins are destroyed by the *ubiquitin–proteosame* proteolytic pathway. Proteins destined for destruction are tagged by the covalent addition of multiple copies of a 76-amino-acid peptide called *ubiquitin*, which is named for its ubiquitous presence in all eukaryotic cells. Understanding of the factors that guide selection of any particular protein molecule for destruction is incomplete. Ubiquinated proteins attach to and are unfolded and threaded into a barrel-shaped complex of proteins called the *proteasome*. Proteolytic activity on the inner surface of the proteasome cleaves proteins into small fragments that are released into the cytosol along with intact ubiquitin, which can then be recycled. The fragments are taken up by the lysosomes and degraded to free amino acids. The overall process requires an input of energy derived from the breakdown of ATP and involves a series of enzymatic transformations. Lysosomes also degrade misfolded or otherwise impaired proteins immediately after synthesis. Extracellular proteins, including signal molecules and some membrane proteins, are taken up by endocytosis and transferred to lysosomes, where they are degraded.

Regulation of Protein Function

Much of cell function and its regulation depends upon two basic characteristics of proteins and nucleic acids. We have already seen that DNA and RNA bind with high affinity and great specificity to complementary sequences in other nucleic acid molecules. Similarly, proteins bind to nucleic acids and to other proteins with high specificity. Proteins can also bind specifically to small molecules, including ions and metabolites. This property depends upon the complementarity of shape, charge, and hydrophobicity of interacting surfaces, Van der Waals attractions, and the formation of hydrogen bonds. Such interactions may position a protein at a favorable or unfavorable location with respect to its substrate or regulatory proteins. The second important characteristic of proteins is their ability to assume different conformations when bound to other molecules or when modified covalently by the addition or removal of a small substituent. Such conformational changes may expose, reconfigure, or mask reactive surfaces and thereby profoundly influence the ability of the protein to bind other molecules, to catalyze biochemical reactions, to migrate to or become fixed at a particular cellular locus, and to provide cellular motility.

Protein Phosphorylation

The most important covalent modification of proteins in this regard is the reversible transfer of the bulky, negatively charged terminal phosphate group of ATP (adenosine triphosphate) to ester linkages with hydroxyl groups on serine, threonine, or tyrosine residues. Phosphorylation of proteins is catalyzed by enzymes called *protein kinases*, which typically contain regulatory as well as catalytic domains. There are hundreds and perhaps thousands of protein kinases that each recognize particular amino acid sequences or *motifs* in their substrates. Specificity of protein phosphorylation is conferred by the amino acid sequences of substrates and by the accessibility of substrates to kinases.

Protein phosphorylation was originally thought to result from formation of phosphate esters of serine and threonine residues. When it was later recognized that the hydroxyl group in tyrosine residues may also be phosphorylated, the term *tyrosine kinases* was coined to distinguish this class of enzymes from the protein kinases that phosphorylate serine and threonine residues. The enzymes that catalyze the dephosphorylation of proteins are known as *protein phosphatases* and *tyrosine phosphatases*. These enzymes are less diverse than their kinase counterparts, have a broader range of substrate specificities, and often are constitutively active, but there are important exceptions to these generalizations.

Phosphorylation/dephosphorylation reactions regulate a wide variety of cellular functions. Regulation by this means often involves a cascade of phosphorylation reactions, with each kinase serving as substrate for the next kinase in a series. Sequential phosphorylation, and thereby activation, of protein kinases rapidly amplifies a regulatory signal. For example, a single activated protein kinase molecule might catalyze the phosphorylation and hence activation of ten molecules of a second kinase, each of which might then phosphorylate ten molecules of a third kinase which might then phosphorylate and activate ten molecules of the final effector of the controlled function. In just three steps, the influence of the first activated kinase is amplified 1000-fold. It is important to note that this cascade of reactions can change cellular function within just a few seconds because it modifies the activity of molecules that are already present in the cell. In contrast, changing cell function by modifying gene expression requires many time-consuming steps and usually has a latency of 30 minutes or more.

The consequences of phosphorylation and dephosphorylation are not limited to conformational changes in the proteins that are the substrates for kinases and phosphatases (Fig. 7). Phosphorylation-related changes in one protein may increase or decrease its ability to bind to other proteins, which may thereby become activated or inhibited. In addition, the phosphate ester of tyrosine residues along with the adjacent amino acids in a loop of protein may provide a specific docking site to which a complementary peptide sequence in another protein may bind. These interactions are important in regulation of cell function by extracellular agents and in forming protein complexes required to carry out some particular cellular task. One important task that may be regulated in this way is control of gene expression through modification of the activity of nuclear regulatory proteins.

Partial Hydrolysis

Another means of regulating protein function is through proteolytic cleavage. Clipping off a segment of protein can expose a reactive surface on the remaining protein or may simply permit a conformational change to occur. Additionally, the activities of some proteins may be restrained by the presence of other proteins that

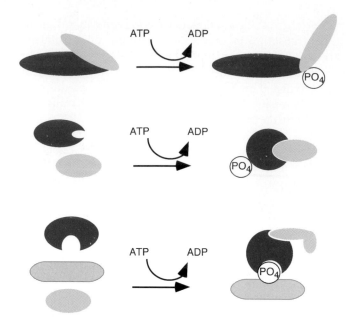

FIGURE 7 Phosphorylation of proteins produces changes in configuration that result in (A) increasing or umasking of catalytic activity or (B) directly enabling proteins to interact. These phenomena might also occur as a result of dephosphorylation. In (C), phosphorylation of one protein provided a docking site for a second protein which permitted interaction with a third protein.

act as inhibitors. Proteolytic destruction or inactivation of such an inhibitor allows expression of the activity. Alternatively, the activities of some proteins are limited by the binding of a portion of their peptide chains to a particular cellular locus which restricts their access to substrate. Cleaving a peptide bond in the portion of the protein that anchors it frees the rest of the protein to diffuse to its substrate.

COORDINATION OF CELLULAR ACTIVITY WITH CHANGING DEMANDS OF THE INTERNAL AND EXTERNAL ENVIRONMENT

Cells perceive a variety of chemical and physical factors as indicators of environmental change and respond to them in coordinated and characteristic ways. The final response, or output, of any specific cell and the signals, or inputs, to which it responds are determined by its particular differentiated state. Thus, a pancreatic beta cell perceives glucose and secretes insulin, a muscle cell perceives acetylcholine and contracts, and a retinal cell perceives light and stops releasing inhibitory neurotransmitter. Despite their different specialized functions, however, all cells draw upon a surprisingly limited set of molecular mechanisms to connect input and output.

Physiological Signals

Cells modify their activity in response to information received from their immediate environment, including such physical signals as light, heat, distortion of their surfaces by stretching or compression, and a vast array of chemical signals that may originate in the external environment or in other cells. Chemical signals are the most common form of information transfer that cells use to communicate with each other. Chemical signals may be simple substances such as ions, metabolites, or derivatives of amino and fatty acids, or they may be complex molecules such as peptides, proteins, glycoproteins, and steroids (Table 1). Signal molecules may be fixed on the surfaces of cells or the supporting extracellular matrix, or they may be dissolved in extracellular fluids. In general, signals confined to surfaces of cells or to extracellular matrix direct cellular behavior during development, wound healing, blood clotting, immunosurveillance, and other processes that involve cell migration. Soluble messages govern the various activities associated with maintenance of the internal environment, coping with the external environment, and reproduction. Mechanisms for information transfer, however, follow a similar pattern, regardless of whether the cell moves to contact a fixed signal or a mobile signal reaches a fixed cell. These activities are the substance of physiology; hence, the molecular bases for cellular communication are considered in detail below and in later chapters.

Secretory Products

Soluble signals have been categorized according to the cell of origin or the mode of delivery to the recipient, or *target*, cells. Generic terms for signal molecules are *agonists*, because they produce an effect, or *ligands*, because, in producing their effect, they bind to target cells. Secretions that are released from one cell and reach nearby target cells by diffusion through the extracellular space are called *paracrine*. Some secretory products may also act on the cells that produced them, in which case they are called *autocrine* secretions. Communication that depends purely on diffusion is highly efficient when the distances between communicating cells are short, but it becomes increasingly inefficient when distances exceed a few cell diameters. Nerve cells are specialized to conduct electrical impulses with great speed (as high as 120 m/sec) over the long distances they span and to release *neurotransmitters* from their terminals. Neurotransmitters diffuse across the synaptic spaces in a paracrine manner to reach adjacent target cells. Alternatively, some secretory products enter the circulation and are delivered to distant target cells by the blood stream. This mode is called *endocrine* secretion, and the secretory products are called *hormones*. Secretions of nerve cells that are carried by the blood to distant target cells are called *neurohormones*, and this category of communication is referred to as *neuroendocrine*. Secretions of cells of the immune system are called *cytokines* or *lymphokines* and may behave in a paracrine,

TABLE 1 Some Examples of Chemical Signals

Signal type	Signal molecule	Cell or tissue of origin	Target cell
Gas	Nitrous oxide	Vascular endothelium	Smooth muscle
Simple ions	Sodium		Kidney (macula densa)
	Calcium		Parathyroid chief cell
Metabolites or derivatives	Acetyl choline	Various nerves	Muscle, nerves
	Epinephrine	Adrenal medulla	Heart, liver, muscle, fat
	Histamine	Mast cells	Vascular smooth muscle
	γ-Aminobutyric acid	Nerve cells	Nerve
	Prostaglandins	Many different cells	Many different cells
	Thromboxane	Platelets	Vascular endothelium
Small peptides	Gastrin	Gastric mucosa	Gastric parietal cells
	Substance P	Nerve cells	Nerve cells
	Oxytocin	Hypothalamic neurons	Uterine smooth muscle, mammary myoepithelium
	Glucagon	Pancreatic alpha cells	Hepatocytes
	Platelet-derived growth factor	Platelets	Vascular endothelium
Proteins	Growth hormone	Pituitary somatotropes	Various
	Interleukin 1	Macrophages	Lymphocytes
Glycoproteins	Chorionic gonadotropin	Fetal trophoblast	Corpus luteal cells
Steroids	Aldosterone	Adrenal cortex	Renal tubular cells
	Progesterone	Corpus luteal cells Placenta	Uterine endometrium and myometrium

autocrine, or endocrine manner. Paracrine or autocrine secretions that stimulate cells to divide or differentiate are called *growth factors*. Pharmacologists refer to all of these secretions as *autocoids*. This array of terms arose from the independent discovery of chemical messages by workers in different disciplines. However, the terminology should not obscure the fact that, from the perspective of the target cell, all of the foregoing are merely chemical signals regardless of where they originate or how they reach their targets. In the ensuing sections we consider all of these secretions simply as chemical signals to emphasize the generality of the cellular processes involved.

Responses of Target Cells

The general sequence of events that occurs in cells when a signal is received is shown in Fig. 8. The *input* may be a physical signal or any of the classes of chemical signals listed in Table 1. To perceive a signal, the target cell must have a receptor for it. A *receptor* is a specialized molecule or complex of molecules that is capable of recognizing a specific signal and triggering the chain of events that produces a characteristic response. Interaction with a signal is thought to change the configuration of the receptor and thereby change the way the receptor molecule interacts with nearby molecules in the response pathway. *Transducer* is the term used to describe the molecular mechanism for

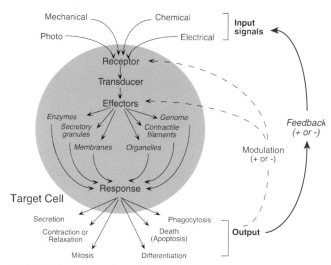

FIGURE 8 Events in cellular communication. Input signals are recognized by specific receptors and translated into a biochemical change by a transducer mechanism. The biochemical signal then acts on the cellular apparatus or effector to produce a physiological response or output. The output may feed back directly or indirectly to affect the source of the input signal and increase or decrease its intensity. The output may also act directly or indirectly to modulate the cellular response to a signal by augmenting or damping events at the level of the receptor, the transducer, or the effector apparatus.

converting the receptor–signal interaction into a biochemical change within the cell. The *effector* describes the cellular machinery that produces the cellular *response(s)*. Examples of cellular responses include secretion, contraction, relaxation, phagocytosis, cell division, or cell differentiation. Defective, unneeded, or unwanted cells may also receive signals that initiate reactions that lead to their death. The process of programmed cell death is called *apoptosis*.

Responses may be achieved by activation or inhibition of the enzymatic apparatus already present in the target cell or may require production of new enzymes, secretory products, or structures through changes in gene transcription or translation. A response or *output* from one cell may directly or indirectly become the input signal to another cell. Cellular input and output arranged in series may produce a feedback effect to shut down (*negative feedback*) or reinforce (*positive feedback*) production of the initial signal (see Chapter 37). Output from other cells may *modulate* the response of a target cell by modifying some aspect of the receptor, the transducer, or the effector.

Complex series of biochemical reactions referred to as *signaling pathways* or *transduction pathways* connect receptors with responses. Although thousands of signals produce highly specific responses in their target cells, relatively few families of signaling pathways may be utilized in different combinations in different cell types. In addition, the presence of multiple *isoforms* of key intermediate molecules contributes to the uniqueness of transduction pathways for different signals. By isoforms we mean closely related molecules that perform similar functions. Different isoforms of a protein may be products of different genes, or they may be products of the same gene and arise from alternative splicing of RNA or posttranslational protein processing. Even very small structural differences can confer significant differences in regulatory properties or in the specificity of interaction with other molecules.

Characteristics of Receptors

Because most chemical signals reach their target cells by way of extracellular fluids, they may be accessible to many different types of cells, but only certain cells respond to certain signals. This selectivity resides in the receptors; only those cells that have receptors for a signal can respond to it. All known receptors are proteins or glycoproteins. To function as a receptor, a molecule must have a domain that binds to the signal molecule with a high degree of selectivity and a domain that is sufficiently altered by agonist binding that its ability to interact with other molecules is changed. Similar considerations hold for photoreceptors and mechanoreceptors, for which photons of a specific wavelength or a mechanical perturbation produce a characteristic conformational change.

Photo-, mechano-, and most chemoreceptors reside on the cell surface and thus provide unimpeded access to signal molecules; however, receptors for some chemical signals are also found within the cytosol or the cell nucleus. To interact with receptors that are located in the cell interior, signal molecules must be able to cross the plasma membrane readily and resist degradation by intracellular enzymes. Receptors have very limited tolerance for variations in the structure or nature of signals. This property confers *specificity* and ensures that cells recognize and respond only to certain appropriate signals. Receptors for chemical signals can recognize and bind their agonists even though most agonists are present at very low concentrations, usually less than 0.1 μM ($10^{-7} M$). In addition to high *affinity*, receptors have a limited *capacity* and hence may become saturated when the concentration of agonist is high. Affinity and capacity of receptors for their agonists are two important determinants of the range of responses that a given cell can express.

The number of receptors in a cell is not fixed. Some receptors may only be expressed at certain stages of the life cycle of a cell or after a cell has been stimulated by other signals. Many cells adjust the number of receptors they express in accordance with the abundance of the signal that activates them. Frequent or intense stimulation may cause a cell to decrease the number of receptors expressed. This phenomenon is called *downregulation*. Conversely, cells may *upregulate* receptors in the face of rare or absent stimulation by their agonists or in response to another signal. Consequences of changes in receptor abundance are discussed in detail in Chapter 37. In addition, cells may temporarily activate or inactivate receptors by adding or removing phosphate or by sequestering them in vesicles to prevent access to agonists. Inactivation of receptors, called *adaptation* or *desensitization*, is called *homologous* desensitization when produced by its own ligand, and *heterologous* desensitization when caused by other agonists acting through their receptors.

Receptors that reside in the plasma membrane may be uniformly distributed over the entire surface of a cell or they may be confined to some discrete region such as the neuromuscular junction or the basal surface of renal tubular epithelial cells. Receptors in many cells are not fixed in place by attachments to the cytoskeleton, and can migrate laterally in the plane of the membrane. Some cell surface receptors are concentrated in specialized surface invaginations called *caveolae*. In some cells, receptors that are occupied by their ligands may cluster at one pole, a phenomenon known as *capping*. Clustering may be important for initiating some cellular responses. Also, membrane receptors are internalized either alone or bound to their ligands by a process called *receptor-mediated endocytosis*. Some cells recycle receptors between the plasma membrane and internal membranes and can vary the rate of transfer and hence the relative abundance of receptors on the cell surface. Receptors, like other cellular proteins, are broken down and replaced many times during the lifetime of a cell.

SIGNAL TRANSDUCTION THROUGH RECEPTORS THAT RESIDE IN THE PLASMA MEMBRANE

The G-Protein-Coupled Receptor Superfamily

The most frequently encountered cell surface receptors belong to a very large superfamily of proteins that couple with guanosine nucleotide binding proteins (G-proteins) to communicate with intracellular effector molecules. This ancient superfamily is widely expressed throughout eukaryotic phyla. G-protein-coupled receptors are crucial for sensing signals in the external environment, such as light, taste, and odor, as well as internal signals in the form of hormones, neurotransmitters, immune modulators, and paracrine factors (Table 2). So pervasive are the responses mediated by these receptors in this superfamily that about 30% of all effective pharmaceutical agents target actions mediated by them. Considerably more than 1000 different G-protein-coupled receptors are expressed in humans, with perhaps as many as 1000 expressed in the olfactory epithelium alone (see Chapter 54).

All G-protein-coupled receptors contain seven membrane-spanning α-helices comprised of about 25 amino acids each (Fig. 9). The single long peptide chain that constitutes the receptor threads back and forth through the membrane seven times, creating three extracellular and three intracellular loops. For this reason, these receptors are sometimes called *heptahelical* receptors, or *serpentine* receptors. The amino terminal tail is extracellular and along with the external loops may contain covalently bound carbohydrate. Outward facing components of the receptor, including parts of the α-helices, contribute to the agonist recognition and binding site. They tend to be more extensive in receptors for larger proteins and may also form all or part of the binding pocket for such agonists. The intracellular domain consists of the carboxyl-terminal peptide and the internal loops that connect the transmembrane helices. The intracellular domain also contains one or

TABLE 2   Some Examples of Molecules That Signal Through G Protein-Coupled Receptors

Agent	Target cell	Major effect on cell
Calcium	Parathyroid chief cell	Inhibits hormone secretion
Adenosine	Cardiac muscle	Decreases contractility
Epinephrine	Hepatocyte	Increases glucose production
Angiotensin II	Vascular smooth muscle	Increases contraction
Acetylcholine	Cardiac node cells	Slows heart rate
Interleukin-8	Lymphocytes	Increases cell migration
Thyroid-stimulating hormone	Thyroid follicle cell	Increases hormone synthesis and secretion
Glutamate	Hippocampal neurons	Induces long-term potentiation
Prostaglandin E_2	Uterine smooth muscle	Increases contractility
Somatostatin	Pituitary somatotrope	Inhibits secretion
Cholecystokinin	Pancreatic acinar cell	Stimulates enzyme secretion
Vasopressin	Renal tubular cell	Increases water permeability

more surfaces that interact with the G-proteins. In addition, the intracellular portion may contain amino acid sequences required for receptor internalization by endocytosis and phosphorylation sites that regulate receptor function.

G-proteins are heterotrimers comprised of alpha, beta, and gamma subunits, which are products of different genes. Lipid moieties covalently attached to the alpha and gamma subunits insert into the inner leaflet of the plasma membrane bilayer and tether the G-proteins to the membrane (Fig. 10). The alpha subunits are GTPases, enzymes that catalyze the conversion of guanosine triphosphate (GTP) to guanosine diphosphate (GDP). In the unactivated or resting state, the catalytic site in the alpha subunit is occupied by GDP. When the receptor binds to its ligand, a conformational change transmitted across the membrane allows its cytosolic domain to interact with the G-protein in a way that causes the alpha subunit to release GDP in exchange for a molecule of GTP and to dissociate from the beta/gamma subunits, which remain tightly bound

to each other. Though tethered to the membrane, the dissociated subunits can diffuse laterally along the inner surface of the membrane. In its GTP-bound state, the alpha subunit interacts with and modifies the activity of membrane-associated enzymes that initiate the physiological response. The liberated beta/gamma complex can also bind to cellular proteins and modify enzyme activities or membrane permeability to ions.

Hydrolysis of GTP to GDP restores the resting state of the alpha subunit, allowing it to reassociate with the beta/gamma subunits to reconstitute the heterotrimer. Because the hydrolysis of GTP is relatively slow, the alpha subunit may interact multiple times with effector enzymes before it returns to its resting state. In addition, because some G-proteins may be as much as 100 times as abundant as the receptors they associate with, a single liganded receptor may interact sequentially with multiple G-proteins before it dissociates from the agonist. These characteristics provide mechanisms for amplification of the signal. That is, interaction of a single signal molecule with a single receptor molecule may result in multiple signal-generating events within a cell.

Desensitization and Downregulation of G-Protein-Coupled Receptors

In addition to simple dissociation of an agonist from its receptor, signaling is often terminated by active cellular processes that may also desensitize receptors to further stimulation. G-protein-coupled receptors may be inactivated by phosphorylation of serine or threonine residues in one of their intracellular loops catalyzed by a special G-protein receptor kinase. Such phosphorylation uncouples the receptor from the α subunit and promotes binding to a cytoplasmic protein of the β-arrestin family. β-Arrestin was originally described in photoreceptors, where rapid arrest of the

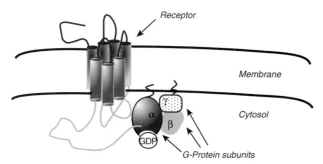

FIGURE 9   G-protein-coupled receptor. The seven transmembrane alpha helices are connected by three extracellular and three intracellular loops of varying length. The extracellular loops may be glycosylated, and the intracellular loops and C-terminal tail may be phosphorylated. The receptor is coupled to a G-protein consisting of a GDP-binding α subunit closely bound to a $\beta\gamma$ component. The α and γ subunits are tethered to the membrane by lipid groups.

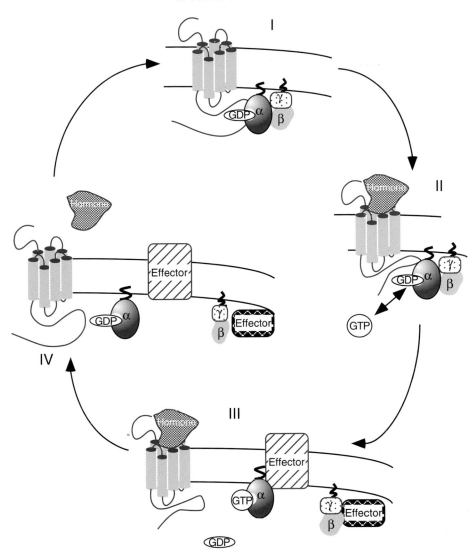

FIGURE 10 Activation of G-protein-coupled receptor. (I) Resting state. (II) Hormone binding produces a conformational change in the receptor that causes (III) the α subunit to exchange ADP for GTP, dissociate from the $\beta\gamma$ subunit, and interact with its effector molecule. The $\beta\gamma$ subunit also interacts with its effector molecule. (IV) The α subunit converts GTP to GDP which allows it to reassociate with the $\beta\gamma$ subunit, and the hormone dissociates from the receptor, restoring the resting state (I).

response to light is crucial for visual acuity. Similar proteins were subsequently found in other G-protein signaling systems where binding to β-arrestin may lead to receptor internalization and downregulation by sequestration in intracellular vesicles. Sequestered receptors may recycle to the cell surface, or, when cellular stimulation is prolonged, they may be degraded in lysosomes.

G-Proteins and Signal Transduction

At least 20 different isoforms of Gα have been identified and can be categorized into four different classes according to similarities in their amino acid sequences. Each class includes the products of several closely related genes, and, in general, each class signals through characteristic transduction pathways. Alpha subunits of the "s" (*stimulatory*) class (Gα_s) stimulate the transmembrane enzyme, adenylyl cyclase, to catalyze the synthesis of cyclic $3',5'$ adenosine monophosphate (cyclic AMP, or cAMP) from ATP (Fig. 11). Alpha subunits belonging to the "q" class stimulate the activity of the membrane-bound enzyme phospholipase C-beta (PLC-β) which catalyzes hydrolysis of the membrane phospholipid phosphatidylinositol 4,5-bisphosphate to liberate inositol 1,4,5 trisphosphate (IP$_3$) and diacylglycerol (DAG) (Fig. 12). Alpha subunits of the "i" (*inhibitory*) class (Gα_i) inhibit the

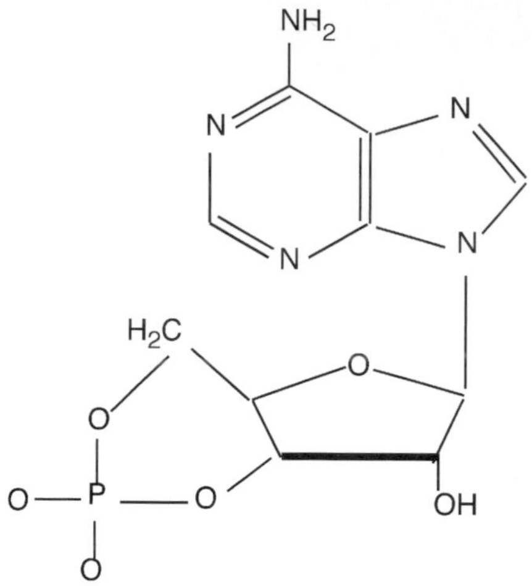

FIGURE 11   Cyclic adenosine monophosphate (cAMP).

activity of adenylyl cyclase and may also activate other enzymes. Beta/gamma subunits released from their association with this class of Gα subunits may interact with membrane proteins that allow passage of potassium (potassium channels), and activate PLC-β. Alpha subunits of the 12/13 class are understood less well, but appear to activate pathways that lead to gene transcription. The differences in transduction pathways activated by the different classes of G-proteins are not absolute, and many points of crossover of these pathways have been observed.

The Second Messenger Concept

For a chemical signal that is received at the cell surface to be effective its information must be transmitted to the intracellular organelles and enzymes that produce the cellular response. To reach intracellular effectors, the G-protein-coupled receptors rely on intermediate molecules called *second messengers*, which are formed and/or released into the cytosol in response to agonist stimulation (the first message). Second messengers activate intracellular enzymes and also amplify signals. A single

agonist molecule interacting with a single receptor may result in the formation of tens or hundreds of second messenger molecules, each of which might activate an enzyme that in turn catalyzes formation of hundreds of thousands of molecules of product. Most of the responses that are mediated by second messengers are achieved by regulating the activity of protein kinases that catalyze the phosphorylation of serine or threonine residues in effector proteins. Unlike responses that require synthesis of new cellular proteins, responses that result from these reactions occur very quickly; therefore, such second-messenger-mediated responses are turned on and off without appreciable latency. However, second messengers can also promote the phosphorylation of transcription factors and thus regulate expression of specific genes. These responses require considerably more time and are seen only after a delay.

Although a very large number of agonists act through surface receptors, to date only a few substances have been identified as second messengers. This is because receptors for many different extracellular signals utilize the same second messengers. When originally proposed, the hypothesis that the same second messenger might mediate the vastly different actions of many different agents was met with skepticism. The idea did not gain widespread acceptance until it was recognized that the special nature of a cellular response is determined by the particular enzymatic machinery with which a cell is endowed rather than by the signal that turns on that machinery. Thus, when activated, a hepatic cell makes glucose, and a smooth muscle cell contracts or relaxes.

The Cyclic AMP System

Cyclic AMP was the first of the second messengers to be recognized. The broad outlines of cAMP-mediated cellular responses to hormones are shown in Fig. 13. Cyclic AMP transmits a signal primarily by activating the enzyme protein kinase A (PKA). When cellular concentrations of cAMP are low, two catalytic subunits of PKA are firmly bound to a dimer of regulatory subunits which keeps them in an inactive state. An increase in cAMP leads to reversible binding of two molecules of cAMP to each regulatory subunit and liberates the catalytic subunits, which are now free to phosphorylate their substrates. Cyclic AMP that is not bound to regulatory subunits is degraded to 5'-AMP by the enzyme cyclic AMP phosphodiesterase, which, though subject to regulation, is usually constitutively active. As cAMP concentrations fall, bound cAMP separates from the regulatory subunits, which then reassociate with the catalytic subunits, thus restoring the basal activity of PKA. Constitutively active phosphatases rapidly remove phosphate groups from the

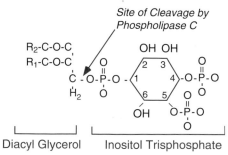

Diacyl Glycerol   Inositol Trisphosphate

FIGURE 12   Phosphatidylinositol-bisphosphate. When cleaved by phospholipase C, inositol 1,4,5 trisphosphate (IP$_3$) and diacylglycerol (DAG) are formed. R1 and R2 are long-chain fatty acids in ester linkage with glycerol.

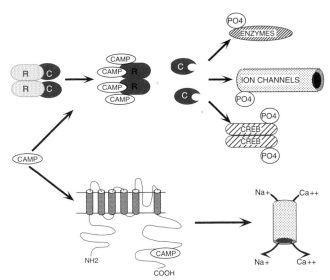

FIGURE 13 Effects of cAMP. Activation of protein kinase A accounts for most of the cellular actions of cAMP (upper portion of the figure). Inactive protein kinase consists of two catalytic units (C), each of which is bound to a dimer of regulatory units (R). When two molecules of cyclic AMP bind to each regulatory unit, active catalytic subunits are released. Phosphorylation of enzymes, ion channels, and transcription factors of the CREB family activates or inactivates these proteins. Cyclic AMP also binds to the α subunits of cyclic nucleotide-gated ion channels (lower portion of the figure), causing them to open and allow influx of sodium and calcium.

activated proteins and return the cell to its unstimulated state.

Protein-kinase-A-dependent phosphorylation of enzymes leads to rapid responses, including mobilization of metabolic fuels, secretion, muscle contraction or relaxation, and changes in membrane permeability. Cyclic AMP can also signal changes in gene expression. Some activated PKA migrates to the nucleus, where it catalyzes the phosphorylation of certain transcription factors that regulate the expression of certain genes by binding to nucleotide sequences called *response elements*, in the promoter region. Because regulation of these genes is sensitive to cAMP, the regulatory sequence is called the *cAMP response element* (CRE), and the phosphoprotein that binds to it is called a *cAMP response element binding* (CREB) protein. One or more forms of CREB are found in the nuclei of most cells and control expression of genes that are involved in such diverse processes as learning, glucose synthesis, and ion excretion.

Some effects of cAMP are independent of PKA. Cyclic AMP can also bind directly to certain plasma membrane proteins that function as *cation channels*. These proteins form tetrameric structures that span the entire thickness of membrane. When bound to cAMP, they change their configuration in such a manner that they form an open channel that allows sodium and

calcium ions to enter the cell. Activation of cAMP gated cation channels in olfactory receptor cells, for example, initiates the transmission of an electrical impulse in olfactory nerves. Cyclic AMP gated cation channels are widely distributed, but except for their role in sensory perception their function is not understood.

Calcium and Calmodulin

Although calcium is always abundantly available in extracellular fluid, it too can serve as a second messenger, largely because its concentration in cytoplasm can undergo abrupt dramatic changes. In the resting state, the intracellular free calcium concentration is about 10,000 times lower than that of the extracellular fluid. This enormous discrepancy is maintained by limited permeability of the plasma membrane to calcium, the presence of a large reservoir of intracellular proteins that can bind calcium and therefore buffer its concentration, and by membrane proteins that can transfer calcium out of the cell or sequester it in storage sites in the endoplasmic reticulum. The pumps and exchangers that carry out these tasks are discussed in Chapter 3. After stimulation by some agonists, the concentration of free calcium may increase tenfold or more. This dramatic change is brought about in part by release of calcium from within the endoplasmic reticulum, and in part by an influx of calcium from the extracellular fluid through channels that can be induced to open and allow calcium ions to diffuse across the membrane. Some calcium channels are *voltage sensitive* and open when the membrane depolarizes to some critical level (see Chapter 4). Opening of voltage-sensitive calcium channels requires the flow of current in the form of positively charged ions that cross the membrane through other ion-specific channels. Some calcium channels are controlled by phosphorylation–dephosphorylation reactions and some are even activated by calcium itself.

Increased calcium concentrations activate a variety of enzymes and trigger such events as muscular contraction, secretion, and polymerization of tubulin to form microtubules. Although calcium can directly activate some proteins, it generally does not act alone. Virtually all cells are endowed with a protein called *calmodulin*, which reversibly binds four calcium ions. When complexed with calcium, the configuration of calmodulin is modified in a way that enables it to bind to certain enzymes, usually protein kinases, and thereby activate them. Calcium/calmodulin-dependent protein kinase II (CAM-kinase II) is a widely distributed multifunctional protein kinase that may catalyze the phosphorylation of many of the same substrates as PKA, including CREB and other nuclear transcription factors. Several other proteins (e.g., troponin C) that are closely related to calmodulin also bind calcium and activate enzymes in some cells (see Chapter 7). Increased cytosolic calcium can also activate calcium-dependent protein phosphatases and proteases and thereby also modify the activities of some proteins. Free calcium is removed from the cytosol by proteins in the plasma membrane that transfer one molecule of calcium to the extracellular fluid in exchange for two molecules of sodium and by ATP-dependent membrane pumps that may resequester it in intracellular storage sites or transfer it to the extracellular fluid. As the intracellular concentration of free calcium is restored to its resting low level, calcium is released from calmodulin, which then dissociates from the various enzymes it has activated, and the resting state is reinstituted.

The DAG and IP₃ System

Both products of phospholipase-C-catalyzed hydrolysis of phosphatidylinositol 4,5 bisphosphate, DAG, and IP_3 behave as second messengers (Fig. 14). IP_3 diffuses through the cytosol to reach its receptors in the membranes of the endoplasmic reticulum and stimulate release of stored calcium into the cytoplasm. Because of its lipid solubility, DAG remains associated with the plasma membrane, promotes the translocation of another protein kinase (protein kinase C, or PKC) from the cytosol to the plasma membrane by increasing its affinity for phosphatidylserine in the membrane, and activates it. Protein kinase C has also been called the

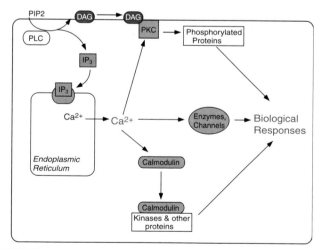

FIGURE 14   Signal transduction through the inositol trisphosphate (IP_3) diacylglycerol (DAG) second messenger system. Phosphatidylinositol 4,5 bisphosphate (PIP_2) is cleaved into IP_3 and DAG by the action of a phospholipase C (PLC). DAG activates protein kinase C (PKC), which then phosphorylates a variety of proteins to produce various cell-specific effects. IP_3 binds to its receptor in the membrane of the endoplasmic reticulum causing release of calcium (Ca^{2+}) which further activates PKC, directly activates or inhibits enzymes or ion channels, or binds to calmodulin, which then binds to and activates protein kinases and other proteins.

calcium, phospholipid-dependent protein kinase because the members of this enzyme family that were initially discovered require both phosphatidylserine and calcium to be fully activated. The simultaneous increase in cytosolic calcium concentration resulting from the action of IP_3 complements DAG in stimulating the catalytic activity of some members of the PKC family. Some members of the PKC family are stimulated by DAG even when cytosolic calcium remains at resting levels. Proteins phosphorylated by the various forms of PKC are involved in regulation of metabolism, membrane permeability, muscle contraction, secretion, and gene expression.

Inositol 1,4,5 trisphosphate is cleared from cells by stepwise dephosphorylation to inositol. DAG is cleared by addition of a phosphate group to form phosphatidic acid, which may then be converted to a triglyceride or resynthesized into a phospholipid. Phosphatidylinositides of the plasma membrane are regenerated by combining inositol with phosphatidic acid, which may then initiate stepwise phosphorylation of the inositol.

Arachidonic Acid Metabolites

The phosphatidylinositol precursor of IP_3 and DAG also contains a 20-carbon polyunsaturated fatty acid called *arachidonic acid* (Fig. 15). This fatty acid is typically found in ester linkage with carbon 2 of the glycerol backbone of phospholipids and may be liberated by the action of a diacylglyceride lipase from the DAG formed in the breakdown of phosphatidylinositol. Liberation of arachidonic acid is the rate-determining step in the formation of the thromboxanes, the prostaglandins, and the leukotrienes (see Chapter 40). These compounds, which are produced in virtually all cells, diffuse across the plasma membrane and behave as local regulators of nearby cells. Thus, the same ligand–receptor interaction that produces DAG and IP_3 as second messengers to communicate with cellular organelles frequently also results in the formation of arachidonate derivatives that inform neighboring cells that a response has been initiated. Phosphatidylinositol is only one of several membrane phospholipids that contain arachidonate. Arachidonic acid is also released from more abundant membrane phospholipids by the actions of the phospholipase A_2 class of enzymes that can be activated by calcium, by PKC-dependent phosphorylation, and by $\beta\gamma$ subunits of G-proteins.

G-Protein-Coupled Receptors and Ion Channels

Evidence is accumulating that G-proteins also link activated receptors to ion channels, which act as the effectors of agonist stimulation in much the same manner that adenylyl cyclase or phospholipase C are

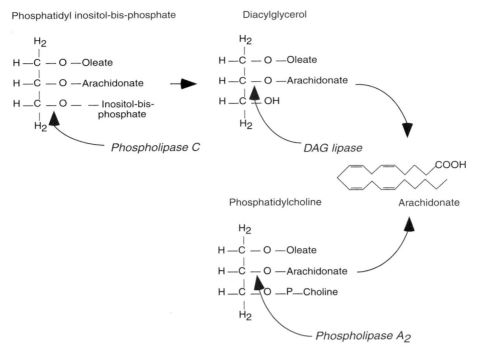

FIGURE 15   Diacylglycerol (DAG), formed from phosphatidylinositol 4,5 bisphosphate (PIP₂) by the action of phospholipase C, may be cleaved by DAG lipase to release arachidonate, the precursor of the prostaglandins and leukotrienes. Arachidonate is also released from other membrane phospholipids (*e.g.*, phosphatidylcholine) by the action of the enzyme phospholipase A2.

the effectors of responses to other G-protein receptors. In particular, it appears that the βγ subunit complex may activate potassium channels, and that α subunits may interact with calcium channels. The exact nature of the interaction of these proteins with ion channel proteins is not yet known. Because some membrane channel proteins are substrates for PKA and perhaps PKC, G-proteins also affect channels indirectly. Altering ion channel activity by the actions of G-protein-dependent agonists frequently results in modulating the responsiveness of target cells to the actions of other agonists.

Complexities of G-Protein-Related Transduction

We have described only a very limited set of generic mechanisms by which hundreds or perhaps thousands of diverse agonists arising in the external environment (e.g., light, odorants) and internal environment, including neurotransmitters and hormones, modify the behavior of their target cells through G-protein receptors. At first glance, there seem to be too few postreceptor mechanisms to accommodate the vast array of signals that impinge on cells. Actually, enormous diversity and complexity of responding systems can be achieved through selective, cell-specific expression of specific isoforms of almost every component of the response

systems described above. To date, 16 genes for the G-protein α subunits have been isolated. Including alternately spliced products, mammals express at least 20 different α subunits, each endowed with unique properties that shape its activity. Considering that 5 different β and 7 different γ subunits have also been described, the number of possible heterotrimeric G-proteins is potentially quite large. Eight different genes code for isoforms of adenylyl cyclase that are stimulated or inhibited by different isoforms of Gα and that have different sensitivities to phosphorylation (and hence modulation) by PKA, PKC, or CAM-kinase. Cyclic AMP can be degraded by phosphodiesterases belonging to four different families, one of which is activated by calcium/calmodulin and another by cyclic GMP. There are 2 major classes of PKA and at least 12 isoforms. Of the 9 isoforms of PKC, only 4 are considered to be conventional and require calcium for activation. Other forms are calcium independent, and all 9 are downregulated by prolonged stimulation.

Cyclic GMP

Cyclic guanosine 3′,5′-monophosphate (cGMP) is formed from guanosine triphosphate by the enzyme guanylyl cyclase which is not coupled to G-protein receptors. Though considerably less versatile than cAMP, cGMP plays an analogous second messenger role in many cells, particularly in smooth muscle cells, in platelets, and in the rods and cones of the visual system where it has been most thoroughly studied. Absorption of a photon of light causes a conformational change in rhodopsin, the visual pigment in rod cells, in much the same way that other G-protein linked receptors are activated by ligands (see Chapter 51). Activation of rhodopsin causes the α subunit of the heterotrimeric G-protein, *transducin*, to exchange its bound GDP for GTP and to dissociate from the $\beta\gamma$ subunits. Instead of protein kinase, the catalytic component activated by transducin is a cyclic nucleotide phosphodiesterase that converts cyclic GMP to 5'-GMP (Fig. 16). In the dark or resting state, cGMP binds to and activates cation channels in the membrane that allow sodium and calcium ions to enter. Activation of phosphodiesterase accelerates the hydrolysis of cGMP, lowering its concentration and allowing the channels to close. Closure of the channels changes the electrical properties of the cell membrane and the rate of neurotransmitter release. The decline in intracellular calcium concentration that results from closure of cGMP-gated cation channels de-inhibits guanylyl cyclase, and allows cGMP concentrations to return to the resting level.

In smooth muscle cells, guanylyl cyclase activity may either be an intrinsic component of some transmembrane receptors or reside in the cytosol as a soluble protein. Both the transmembrane and the soluble forms of guanylyl cyclase are directly stimulated by agonists without the intercession of G-proteins. Increased

formation of cGMP in these and other cells activates protein kinase G, which, like PKA, catalyzes the phosphorylation of serine and threonine residues in various regulatory proteins. In platelets, activation of protein kinase G prevents clot formation by interfering with a constellation of reactions that promote clotting. Cyclic GMP also activates phosphodiesterase in some cells, and at least some of its biological effects may be explained by accelerated degradation of cAMP.

Receptor Channels

Activation of G-proteins and formation of second messengers require several seconds and cannot account for the rapid transfer of information seen at neural synapses or the neuromuscular junction (see Chapter 6). Rapid transfer of information leading to electrical changes in membranes (depolarization or hyperpolarization) is mediated by a special group of receptors found in nerves and muscles. These receptors are bifunctional. They have both recognition and effector domains and do not require the intercession of transducer molecules. A good example is the *nicotinic receptor* for the neurotransmitter acetylcholine. This receptor is a large complex that spans the thickness of the membrane. It is also an ion channel comprised of five subunits, the products of four genes. When a molecule of acetylcholine binds to each of the two α subunits, an allosteric shift in configuration opens the channel for about a millisecond to allow passage of both sodium and potassium ions. Initially, influx of sodium exceeds the efflux of potassium, so that the membrane depolarizes rapidly, causing the muscle to contract. Other examples of *transmitter-gated ion channels* are the receptor channels in the central nervous system that open to admit sodium when stimulated by glutamate or serotonin and those that open to admit chloride when stimulated by glycine or γ-aminobutyric acid (GABA).

Receptors That Signal Through Tyrosine Kinase

A large group of agents, including hormones, cytokines, and growth factors, stimulate their target cells by exciting receptors that signal through activated tyrosine kinases (Table 3). The intracellular portions of the *receptor tyrosine kinases* have intrinsic tyrosine kinase activity. The *tyrosine-kinase-associated receptors* have no intrinsic tyrosine kinase activity, but the intracellular domain contains a sequence of amino acids that interacts with cytosolic tyrosine kinase molecules of the Janus kinase (JAK) family (Fig. 17). Although these receptors belong to several different structural families,

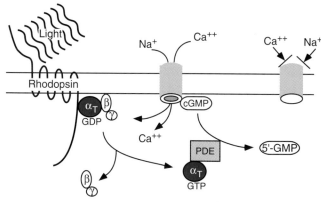

FIGURE 16 Role of cGMP in photoreception. In the dark, cyclic GMP (cGMP) binds to cyclic nucleotide-gated channels, keeping them open, so that both sodium (Na$^+$) and calcium (Ca^{2+}) enter the cell. In response to a photon of light, rhodopsin induces an allosteric change in its associated G-protein, transducin (T), causing the α subunit to exchange GDP for GTP and dissociate from the $\beta\gamma$ subunits. The free α subunit binds to cyclic nucleotide phosphodiesterase (PDE) and stimulates the degradation of cGMP. Decreased cGMP concentrations favor release of the bound nucleotide and closure of the cGMP-gated channels.

TABLE 3 Some Examples of Tyrosine-Kinase-Dependent Receptors

Receptor tyrosine kinases	Tyrosine kinase-associated receptors
Insulin	Erythropoietin (EPO)
Insulin-like growth factor I (IGF-1)	Growth hormone (GH)
Epidermal growth factor (EGF)	Prolactin
Fibroblast growth factor (FGF)	Interleukins 2,3,4,5,6,7,9,11,12,15
Platelet-derived growth factor (PDGF)	γ-Interferon
Nerve growth factor (NGF)	Leukemia inhibitory factor (LIF)

there are sufficient similarities in their modes of signal transduction to consider them as a group. These receptors rely on physical association between proteins (protein–protein interactions) and trigger formation of large aggregates of proteins to transmit the signal.

The tyrosine-kinase-dependent receptors have only a single membrane-spanning region and a seemingly less complex architecture than the G-protein receptors. The

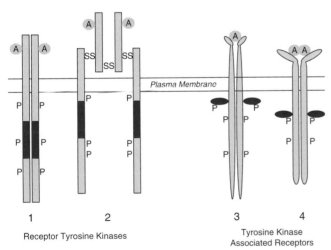

1   2   3   4

Receptor Tyrosine Kinases

Tyrosine Kinase
Associated Receptors

FIGURE 17 Prototypic tyrosine-kinase-dependent receptors. *Receptor tyrosine kinases* contain an intrinsic tyrosine kinase catalytic domain (indicated in blue) in the part of the receptor that extends into the cytosol. In the absence of ligand, they usually consist of a single peptide chain (1) which forms a dimer after binding to its agonist (A). Some receptor tyrosine kinases (2) are more complex, with subunits held together by disulfide bonds, whether or not the agonist is present. The *tyrosine-kinase-associated receptors* have no intrinsic catalytic activity but associate with cytosolic tyrosine kinases (indicated in blue). Some of these receptors (4) are composed of two or more dissimilar subunits in the absence of agonist, but all form dimers in the presence of agonist. Dimers of receptor tyrosine kinases contain two molecules of agonist, while dimers of the tyrosine-kinase-associated receptors contain only one molecule of agonist. Dimerization upon receptor binding leads to phosphorylation (P) on tyrosine residues in the receptor itself or perhaps on the associated kinase as well.

extracellular portions contain the ligand binding site. Agents that excite these receptors often promote cell division (*mitogens*) or differentiation. Their actions, which generally, but not exclusively, involve changes in gene expression, are slower in onset and more prolonged than most effects of ligands that act through G-protein-linked receptors. A notable exception is insulin (see Chapter 41), which may increase glucose transport across plasma membranes within just a few minutes.

For most of the tyrosine-kinase-dependent receptors, agonist binding causes two receptor molecules to come together to form a dimer as the first step in signal generation. Dimerization brings the cytosolic portions into close proximity to each other and permits the tyrosine kinase associated with each member of the dimer to phosphorylate tyrosine residues on the other. Dimerization generally produces a complex of two receptor tyrosine kinase molecules along with one or two agonist molecules. The insulin and insulin-like growth factor (IGF) receptors, which are dimeric even in the unliganded state (see Chapters 41 and 44), become activated upon binding two agonist molecules. The platelet-derived growth factor (PDGF) receptor complex is formed when a dimer PDGF binds to two receptor molecules. Tyrosine-kinase-associated receptors for growth hormone (see Chapter 44), prolactin (see Chapter 47), erythropoietin and many cytokines form dimers when two receptor molecules bind a single agonist molecule. Some of the tyrosine kinase-associated receptors have two or three subunits so that each half of the dimer contains an agonist binding subunit and a signaling subunit that binds to tyrosine kinase. Curiously, the same signaling subunit, called *gp130*, is found in receptors for several different cytokines.

The protein substrates for receptor-activated tyrosine kinases may have catalytic activity or may act only as scaffolds or adaptor proteins to which other proteins are recruited and positioned so that enzymatic modifications are facilitated. Phosphorylated tyrosines act as docking sites for proteins that contain so-called *Src homology 2* (SH2) domains. SH2 domains are named for the particular configuration of the tyrosine phosphate binding region originally discovered in v-Src, the cancer-inducing protein tyrosine kinase of the Rous sarcoma virus. SH2 domains represent one type of a growing list of modules within proteins that recognize and bind to specific complementary motifs in other proteins. Different SH2 groups recognize phosphorylated tyrosines in different contexts of adjacent amino acid residues. Typically, multiple tyrosines are phosphorylated so that several different SH2-containing proteins are recruited and initiate multiple signaling pathways.

Tyrosine-kinase-dependent receptors generally signal to the nucleus and regulate gene transcription through multiple pathways. One way that these receptors communicate with the genome is through activation of the mitogen-activated protein (MAP) kinase cascade (Fig. 18). MAP kinase is a cytosolic enzyme that is activated by phosphorylation of both serine and tyrosine residues and then enters the nucleus, where it phosphorylates and activates certain transcription factors. Activation of MAP kinase follows an indirect route that involves a small G-protein, called *Ras*, which was originally discovered as a constitutively activated protein present in many tumors. Ras proteins belong to a family of small G-proteins that function as biochemical switches to regulate such processes as entry of proteins into the nucleus, sorting and trafficking of intracellular vesicles, and cytoskeletal rearrangements. The small G-proteins are members of the same superfamily as the α subunit of the heterotrimeric G-proteins but do not form complexes with $\beta\gamma$ subunits or interact directly with receptors. The small G-proteins are GTPases that are in their active state when bound to GTP and in their inactive state when bound to GDP. Instead of liganded receptors, the small G-proteins are activated by proteins called *nucleotide exchange factors* that cause them to dissociate from GDP and bind GTP.

They remain activated as they slowly convert GTP to GDP, but inactivation can be accelerated by interaction with GTPase-activating proteins (GAPs).

One of the proteins that docks with phosphorylated tyrosine residues is the *growth factor binding protein 2* (Grb2). Grb2 is an adaptor protein that has an SH2 group at one end and other binding motifs at its opposite end which enable it to bind other proteins, including a nucleotide exchange factor called *Sos*. By means of these protein–protein interactions, the activated receptor can thus communicate with Sos, which activates Ras. Ras in turn activates the enzyme Raf kinase which phosphorylates and activates the first of a cascade of MAP kinases that ultimately result in phosphorylation of nuclear transcription factors.

The gamma isoform of phospholipase C is another effector protein that is recruited to tyrosine-phosphorylated receptors by way of its SH2 group. It is also a substrate for tyrosine kinases and is activated by tyrosine phosphorylation it catalyzes the formation of DAG and IP_3 in the same manner as already discussed for the beta isoforms associated with G-protein coupled receptors. In this manner, tyrosine-kinase-dependent receptors stimulate cellular changes that depend on protein kinase C and the calcium/calmodulin second messenger system, including phosphorylation of nuclear transcription factors by calmodulin kinase (see Fig. 18).

Another mechanism for modifying gene expression involves activation of a family of proteins called *Stat* (<u>s</u>ignal <u>t</u>ransducer and <u>a</u>ctivator of <u>t</u>ranscription) proteins. The Stat proteins are transcription factors that reside in the cytosol in their inactive state and have an SH2 group that enables them to bind to tyrosine-phosphorylated proteins. Upon tyrosine phosphorylation, Stat proteins dissociate from their docking sites, form homodimers, and migrate to the nucleus, where they activate transcription of specific genes (Fig. 19).

Yet another important mediator of tyrosine kinase signaling is phosphatidylinositol-3 (PI-3) kinase, which catalyzes the phosphorylation of carbon 3 of the inositol of phosphatidylinositol bisphosphate in cell and organelle membranes to form phosphatidylinositol 3,4,5 trisphosphate (PIP_3). PI-3 kinase consists of a regulatory subunit that contains an SH2 domain and a catalytic subunit. Binding of the regulatory subunit to phosphorylated tyrosines of the receptor-associated complex activates the catalytic subunit. Activation of PI-3 kinase plays a key role in transducing signals from tyrosine-kinase-dependent receptors to downstream events. The enzymes activated by PIP_3 are protein kinases that regulate cellular metabolism, vesicle trafficking, cytoskeletal changes, and nucleotide exchange factors that control the activity of the small G-proteins.

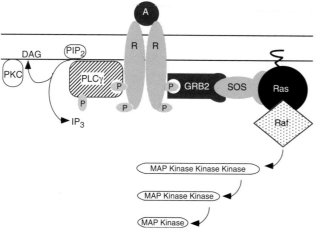

FIGURE 18 Phosphorylation of tyrosines on a receptor (R) following agonist (A) binding provides docking sites for the attachment of proteins that transduce the hormonal signal. The growth factor binding protein 2 (GRB2) binds to a phosphorylated tyrosine in the receptor and binds at its other end to the nucleotide exchange factor SOS, which stimulates the small G-protein Ras to exchange its GDP for GTP. Thus activated, Ras in turn activates the protein kinase Raf1 to phosphorylate mitogen-activated protein (MAP) kinase and initiate the MAP kinase cascade that ultimately phosphorylates nuclear transcription factors. The γ isoform of phospholipase C (PLCγ) docks on the phosphorylated receptor and is then tyrosine phosphorylated and activated to cleave phosphatidylinositol 4,5 bisphosphate (PIP$_2$), releasing diacylglycerol (DAG) and inositol trisphosphate (IP$_3$) and activating protein kinase C (PKC).

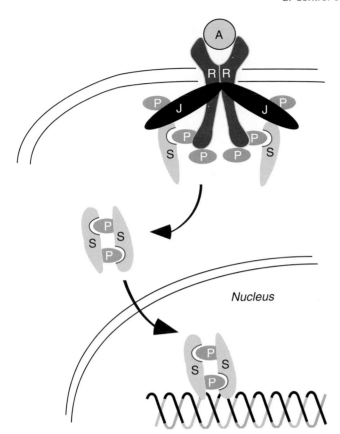

FIGURE 19 Hormone receptors (R) associate with the JAK family of cytosolic protein tyrosine kinases (J), and following binding of agonist (A) and dimerization become phosphorylated on tyrosines. Proteins of the STAT family (S) of transcription factors that reside in the cytosol in the unstimulated state bind to the phosphorylated receptor and, upon being tyrosine phosphorylated by JAK, dissociate from the receptor, form homodimers, and migrate to the nucleus where they activate gene transcription.

INTRACELLULAR RECEPTORS

To reach intracellular receptors, agonists must get across the plasma membrane. The steroid hormones and vitamin D (see Chapters 40, 43, and 45–47) are derivatives of cholesterol and readily diffuse through the lipid bilayer of the plasma membrane because they are highly lipid soluble. Similarly, the thyroid hormones (see Chapter 39), which are α-amino acids, have large nonpolar constituents and penetrate cell membranes both by diffusion and to some extent by carrier-mediated transport. Retinoic acid (vitamin A) is also quite lipid soluble, as is the gas nitrous oxide (NO). With the exception of the soluble form of guanylyl cyclase, which is an intracellular receptor for NO (see Chapter 17), all of the known intracellular receptors belong to the nuclear receptor superfamily of transcription factors that includes receptors for the steroid hormones, thyroid hormone, vitamins D and A, and derivatives of arachidonic acid. Because no ligands have yet been identified for some members of this superfamily, they are known as "orphan" receptors.

Nuclear receptors enhance or repress expression of some genes in a manner that is both agonist specific and cell specific. They are comprised of a single long peptide chain that ranges in length from about 350 amino acids for the thyroid hormone receptor (see Chapter 39) to more than 1000 amino acids for the mineralocorticoid receptor (see Chapter 40). Functionally, they can be divided into three domains. The N-terminal region, which accounts for most of the variability in length, contains an *activation function* (AF1) that regulates the transcription-promoting activity of the receptor as a consequence of its interaction with other transcription factors. It may be phosphorylated at multiple sites, usually on serine or threonine residues but sometimes also on tyrosine residues. Phosphorylation may increase its ability to interact with other transcription factors. The middle portion is the region with the greatest similarity of amino acid sequence among family members and is the part that binds to DNA. The carboxyl-terminal portion contains the hormone binding domain and a second activation function. It also contains a leucine-rich sequence that provides a surface for dimer formation with another nuclear receptor molecule.

In the unactivated state, the steroid hormone receptor subfamily of nuclear receptors may be located in the cytoplasm or in the nucleus. In the absence of ligand, the steroid hormone receptor is associated with a large complex of chaperon proteins (also called *heat shock proteins*) and is folded in such a way that the DNA binding surface is masked by the associated proteins. Ligand binding produces a change in shape that activates the receptor and causes it to dissociate from the chaperon proteins. Upon activation, the receptor forms a dimer with another activated receptor molecule and binds to specific nucleotide sequences in DNA called *hormone response elements*. These elements might be located in the immediate vicinity of the promoters of responsive genes or at appreciable distances away. The receptors activate or repress transcription of specific genes either by interacting directly with the basal transcription complex at the promoter or indirectly through interactions with other proteins that serve as accessory factors and coregulators (Fig. 20). It is likely that there are many coregulatory proteins, and that the complement of such factors available in any particular cell contributes to the determination of which genes will be affected. It is noteworthy that the receptors for several different steroid hormones may bind to the same hormone response elements, although the complement of genes activated is quite hormone specific. This

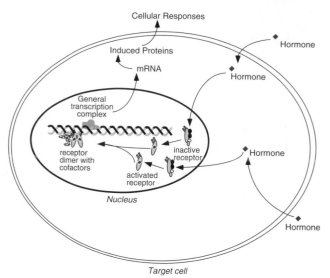

FIGURE 20 General scheme of steroid hormone action. Steroid hormones penetrate the plasma membrane and bind to intracellular receptors found largely in the nucleus (except adrenal steroid receptors). Hormone binding activates the receptor, which forms complexes with other proteins and binds to specific acceptor sites (hormone response elements, HRE) on DNA to initiate transcription and formation of the proteins that express the hormonal response. The steroid hormone is then cleared from the cell.

apparent paradox may be resolved if we assume that the role of the hormone response element is simply to anchor the receptor to the DNA in responsive genes, but whether or not a particular gene is transcribed is also determined by the combination of interacting transcription activators that associate with the receptor and the basal transcription complex. The thyroid hormones and retinoic acid act by a slight variation of this scheme; unoccupied receptors are already bound to their response elements in the DNA and do not form complexes with chaperons. Unoccupied receptors are either inactive until the ligand is bound or act as repressors in the absence of hormone. These receptors may form heterodimers with each other and in some cases with orphan receptors. Further discussion of signal transduction by this superfamily is found in chapters that discuss the actions of the various hormones.

INTEGRATION OF SIMULTANEOUS SIGNALS

As must already be quite obvious, binding of a signal molecule to its receptor sets in motion intracellular signaling pathways that are both intricate and complex. Cells express receptors for multiple signaling molecules and are simultaneously bombarded with excitatory, inhibitory, or a conflicting mixture of excitatory and inhibitory inputs from different agents whose signaling

pathways may run in parallel, intersect, coincide, diverge, and perhaps intersect again before influencing the final effector molecules. Some signaling pathways must compete for common substrates as well as for the final effector molecules that express the final alterations in cellular behavior. Target cells must integrate all inputs by summing them algebraically and sometimes geometrically and then respond accordingly. For example, in the hepatocyte, both glucagon and epinephrine stimulate adenylyl cyclase, each by way of its own G-protein-coupled receptor. The effects of these signals combine to produce a more intense activation of adenylyl cyclase than would result from either one alone. At the same time, these cells may also be receiving some input from insulin, an action of which is activation of cAMP phosphodiesterase, which breaks down cAMP. In the pancreatic beta cell, which secretes insulin, epinephrine binds to two classes of G-protein-coupled receptors: α_2 receptors, which couple to adenylyl cyclase through a $G\alpha_i$, and β receptors, which couple to adenylyl cyclase through a $G\alpha_s$. The two receptors thus transmit conflicting information; in this case, the inhibitory influence of the α_2 receptor is stronger and prevails. Figure 21 shows an example of how several signaling pathways might operate simultaneously.

Integration occurs at various levels along signal transduction pathways, with cross-talk among the various G-protein-receptor-mediated signaling pathways or among the various tyrosine-phosphorylation-dependent pathways and among G-protein- and tyrosine-kinase-mediated pathways and nuclear-receptor-mediated pathways. Integration thus is not limited to the rapidly expressed responses that result from phosphorylation/dephosphorylation reactions but can also occur at the level of gene transcription or may involve a mixture of the two. In fact, phosphorylation of nuclear receptors catalyzed by MAP kinase, CAM kinase II, or PKC alters their ability to bind to other nuclear factors and hence may increase or decrease their ability to influence gene transcription. In responding simultaneously to multiple inputs, cells are able to preserve the signaling fidelity of individual hormones even when their transduction pathways appear to share common effector molecules. Understanding of how cells accomplish this is still incomplete, but part of the explanation may derive from the findings that many of the protein molecules involved in complex intracellular signal transmission do not float around freely in a cytoplasmic "soup" but are anchored at specific cellular loci by interactions with the cytoskeleton or membranes of intracellular organelles. Protein kinase A, for example, may be localized to specific regions in the cell by specialized proteins called *AKAPs* (A kinase anchoring proteins), and various forms of PKC are localized by

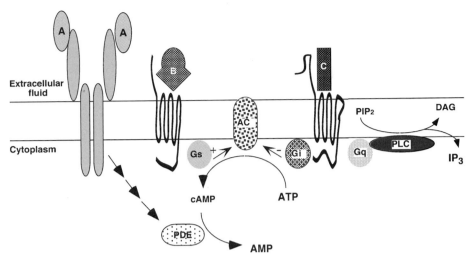

FIGURE 21 Cells may simultaneously receive inputs from agonists A, B, and C. Agonist B, acting through a G-protein-coupled receptor activates adenylyl cyclase (AC) through the a stimulatory subunit (α_s). Agonist C binds to its G-protein-coupled receptor which, through the inhibitory subunit (α_i), inhibits adenylyl cyclase and, through the α_q subunit, activates phospholipase C (PLC), resulting in the cleavage of phosphatidylinositol 4,5 bisphosphate (PIP$_2$) and the release of diacylglycerol (DAG) and inositol trisphosphate (IP$_3$). Agonist A, acting through a tyrosine kinase receptor, activates cAMP phosphodiesterase (PDE), which degrades cAMP. The cell must then sum all of these signals into an integrated response.

interacting with *RACKs* (receptors for activated C kinase). It is also possible that different agents use specific combinations of signal transduction pathways and that convergence of signals at critical reactions provides the reinforcement or dampening that enables cells to distinguish between agonists and appropriate cellular responses.

BIOSYNTHESIS, STORAGE, AND SECRETION OF CHEMICAL SIGNALS

Signal molecules generally are formed from simple precursors within the signaling cell, but occasionally common metabolites are used as signals without enzymatic modification. For example, glycine and glutamate are neurotransmitters in the brain. Sometimes only one or two enzymatic steps are needed to convert a common metabolite to a potent signal; histamine, for example, is formed by decarboxylation of the amino acid histidine, and adenosine is formed when adenosine monophosphate (5′ AMP) is dephosphorylated. A complex series of enzymatic reactions may be used to build steroid hormones from a simple metabolite such as acetate, or secretory cells may accomplish the same end simply by putting the finishing touches on cholesterol that they take up from the blood (see Chapters 40 and 45–47). Enzymatic reactions may take place in the cytosol or within cellular organelles. Often, the biosynthetic pathway meanders from one cellular compartment to another,

requiring energy-consuming, carrier-mediated transport to transfer precursors across intracellular membranes. As already described, all protein and peptide signal molecules, like other cellular proteins, are synthesized on ribosomes, processed in the endoplasmic reticulum and Golgi apparatus, and then packaged for later secretion by the process of exocytosis (Fig. 22). Specific reactions in biosynthesis and processing of signal molecules are discussed in subsequent chapters.

Secretory Granules and Vesicles

Cells usually maintain ample stores of the signal molecules they produce and therefore can respond rapidly and repeatedly to whatever changes in the internal or external environment might call forth secretion. Notable exceptions are the steroid hormones and the derivatives of arachidonic acid which are synthesized from stored precursors in the initial phase of the secretory process. Secretory products are usually segregated from the rest of the cell and stored in highly concentrated form in membrane-bound vesicles. In histological sections, stored secretory products appear as vesicular or granular inclusions for which the staining properties are often sufficiently distinctive to give the cell its name (e.g., *eosinophils, chromaffin cells*). At present, there is little reason to believe that synaptic vesicles that store neurotransmitters in nerve terminals are fundamentally different from the secretory granules of cells that secrete protein hormones or from the

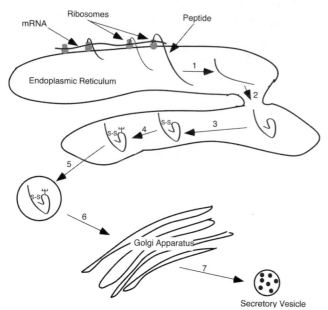

FIGURE 22 The leader sequence or signal peptide of proteins destined for secretion enters the cisternae of the endoplasmic reticulum even as peptide elongation continues. In the endoplasmic reticulum (1), the leader sequence is removed, (2) the protein is folded with the assistance of protein chaperons, sulfhydryl bridges may form (3), and carbohydrate may be added (glycosylation) (4). The partially processed protein is then entrapped in vesicles (5) that bud off the endoplasmic reticulum and fuse with the Golgi apparatus (6), where glycosylation is completed, and the protein is packaged for export in secretory vesicles (7) in which the final stages of processing take place.

granules in white blood cells that give rise to the name *granulocyte*.

Secretory vesicles are not inert receptacles; rather, they are active participants in the production, processing, and delivery of secretory products to the cell exterior. Their membranes are formed in the Golgi apparatus and are similar or identical in composition to the plasma membrane; they are selectively permeable and are endowed with the capacity for active transport of ions and complex molecules in either direction. Consequently, the composition of fluids in the interior of the vesicles is different from the composition of the cytosol. Carrier molecules in vesicular membranes account for the uptake of small signal molecules from the cytosol. Even the large amounts of signal molecules stored within vesicles have little osmotic impact because they are either complexed to macromolecules or precipitated out of solution.

The secretory granule provides a unique environment for enzymatic processing of secretory products and for protecting them from the degradative apparatus of the cell. Some enzymes required for synthesis or modification of secretory products may be intrinsic proteins in the vesicular membrane. In some cases, enzymes

necessary for final processing of protein and peptide signal molecules are packaged along with the signal precursor before the vesicle is released from the Golgi apparatus. More than 200 different proteins have been found within the secretory vesicles that store the hormone insulin, but only a few of these are precursors of insulin or byproducts of processing of insulin.

Exocytosis

The process by which the contents of secretory vesicles are delivered to the extracellular fluid is called *exocytosis*. Depending upon the type of cell, exocytosis may occur over the entire surface or be localized to a small discrete area. For example, nerve cells release their neurotransmitters only at synapses that comprise only a tiny fraction of their surface area. When an appropriate signal is received by the secretory cell, storage granules move to the cell surface generally triggered by an increased concentration of cytosolic calcium, and their surrounding membranes fuse with the plasma membrane. The area of fusion then breaks down and opens the storage vesicle to the extracellular fluid (Fig. 23). Membranes that surrounded secretory vesicles before fusion are retrieved from the plasma membrane by *endocytosis*. In some cells, particularly in nerve endings, these fragments of membrane are reformed into new secretory vesicles, refilled with signal molecules, and recycled many times over. In addition to providing some economy, reuptake of vesicular membranes or their equivalent prevents expansion of the cell surface area that would otherwise occur as secretory vesicles repeatedly fuse with the cell membrane.

During exocytosis, the entire contents of the secretory vesicle are released, including signal molecules, precursors, byproducts, and processing enzymes. In some instances (e.g., adrenal medullary cells, pituitary corticotropes), more than one of these products may have biological activity. Although one normally thinks of this process in terms of secretion of signal molecules, occasionally the processing enzyme is the critical secretory product. For example, the proteolytic enzyme *renin* is released by the kidney when blood volume decreases (see Chapter 29 and 40). Renin catalyzes the rate-determining step in the formation of the hormone angiotensin from its precursor, which circulates in the blood.

The cellular events that control exocytosis are still incompletely understood, but it has been recognized for many years that an increase in the concentration of free calcium in the cytosol is essential both for fusion of the vesicular membrane with the plasma membrane and for extrusion of vesicular contents. Movement of secretory granules to peripheral areas of

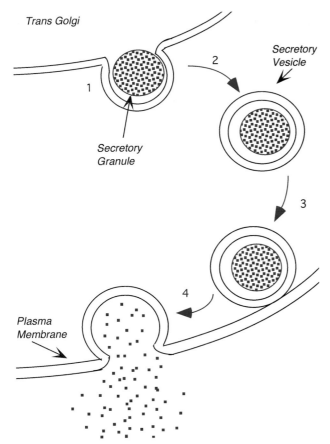

Trans Golgi

Secretory
Vesicle

Secretory
Granule

Plasma
Membrane

FIGURE 23 Exocytosis. Secretory vesicles bud off the trans-Golgi compartment (1) and move into the cytosol (2), where they await a signal for secretion. Secretion (3) is usually accompanied by increased cellular calcium which causes elements of the cytoskeleton to translocate secretory granules to the cell surface (4). The membrane surrounding the granule fuses with the plasma membrane, opening the secretory vesicle to the extracellular fluid and releasing the processed protein(s) along with enzymes and peptide fragments.

the cell is a necessary prelude to fusion and probably also requires calcium. Positioning of granules in the submembranous zone usually occurs before the actual secretory event. Some of the calcium-activated molecules and the proteins in the vesicular and plasma membranes that enable docking and fusion to occur have been identified, but a discussion of this complex topic is beyond the scope of this text; the interested student is encouraged to consult recent texts of cell biology.

Calcium that triggers secretion is released from intracellular organelles or gains access to the cytosol through calcium channels in the plasma membrane. In nerves and some glandular cells, the arrival of an electrical signal, the action potential, opens voltage-sensitive calcium channels. In some cells, calcium channels may open when they are phosphorylated by PKA or when they interact with an activated G protein.

These processes were considered above in the discussion of the molecular events set in motion after a signal molecule interacts with a receptor. Secretion, after all, is one of the common responses of cells to stimulation by a signal molecule.

It is obvious that synthesis must be coupled in some way with secretion, as cells rarely exhaust their supply of secretory product. It is likely that the same events that initiate secretion simultaneously initiate synthesis. It is also possible that cells have some mechanism for monitoring how much signal is stored and synthesize signal molecules when storage depots fall below some threshold level.

Clearance of Chemical Messages

For a molecule to function as a signal, it must be present in the environment of the target cell only during periods of information transfer. If the signal were not eliminated once the message was received, the target cell would remain in a state of constant activation or inhibition or would become exhausted and unresponsive to subsequent stimulation. Thus, elimination of signal molecules is as important as their secretion, and elaborate mechanisms have evolved to accomplish this task. In some cases effective signaling may be terminated simply by dilution followed by elimination at a distant site.

In general, signal molecules interact reversibly with their receptors and little is consumed or transformed in delivering their message. Target cells may take up signal molecules bound to their receptors (receptor-mediated endocytosis), separate signal from receptor in the *endosomes*, and then deliver the signal to the *lysosomes* for destruction and the receptor back to the cell membrane. Target cells may also produce hydrolytic enzymes on their surface in the vicinity of their receptors (e.g., acetylcholin esterase at the neuromuscular junction; see Chapter 6), thus efficiently destroying signal molecules that are not bound to receptors. One economical way to eliminate signal molecules is for the secreting cell to take them up again and repackage them in secretory vesicles for future use. This mechanism is utilized by terminals of sympathetic nerves and neurons in the brain. Most target cells rely on dilution of signal molecules in the large volume of extracellular fluid as they diffuse away from their receptors after the message has been delivered. Signal molecules are then destroyed by enzymes in extracellular fluid or blood or by specialized organs such as liver or kidney which alter the signals chemically and excrete the degraded products. Signals may be totally degraded or simply transformed so that they are no longer recognizable by receptors in target cells.

Suggested Readings

The subject matter of this chapter covers a broad range of topics that are among the most intensely investigated areas of modern biology and are therefore changing at a breath taking rate. Superb coverage of these and other relevant topics can be found in: Alberts B, Bray D, Lewis J, Raff M, Roberts K, and Watson JJ, *Molecular biology of the cell*, fourth ed., New York: Garland Publishing, 2002. Up-to-date reviews can also be found in *Cell, Annual Review of Biochemistry, Annual Review of Physiology, Endocrine Reviews*, and many other excellent journals.

Membrane Transport

STANLEY G. SCHULTZ

KEY POINTS

- Biologic membranes are composed of a phospholipid bilayer containing peripheral and integral proteins.
- Diffusion of nonelectrolytes takes place from a higher concentration to a lower concentration. The rate of diffusion across a membrane is a linear function of the concentration difference.
- The permeability coefficient for diffusion of a solute through the lipid matrix of the membrane increases with increasing solubility.
- *Water-soluble solutes* cross membranes by restricted diffusion though water-filled protein channels. The permeability coefficient for diffusion of uncharged solutes through channels decreases with increasing molecular radius.

- The permeability coefficients of ions depend on radius and charge.
- The rate of diffusion of *charged solutes* depends on both the concentration gradient and the electrical potential difference across the membrane.
- *Diffusion potentials* arise to preserve electroneutrality in the face of differences in ion mobility.
- The orientation of a diffusion potential is such that it accelerates the movement of less mobile ions and retards the movement of more mobile ions.
- The *Nernst potential* is the electrical potential difference at which there is no ionic flow in

KEY POINTS (*continued*)

spite of the presence of a concentration difference. An ion is said to be "actively transported" if its distribution cannot be described by the Nernst equation.

- The *osmotic pressure* is the hydrostatic pressure necessary to prevent volume displacement when there is a concentration difference of a solute across a semipermeable membrane. The osmotic pressure is a linear function of the concentration difference across the membrane.

- The *reflection coefficient* of a solute is a measure of the relative permeabilities of the membrane to water and solute or the ability of the membrane to distinguish between solute and solvent.

- The rate of volume displacement across a membrane is a linear function of the hydrostatic pressure difference across the membrane.

- *Facilitated diffusion* refers to carrier-mediated processes that are not coupled to a source of energy.

- *Primary active transport* refers to carrier-mediated processes directly coupled to a source of chemical energy. All primary active transporters in animal cells are ATPases.

- *Secondary active transport* refers to carrier-mediated processes indirectly coupled to a source of energy—for example, an ion gradient.

MEMBRANE COMPOSITION AND STRUCTURE

Historical Perspectives: The Dawning of Membrane Biology

The first suggestion of the existence of biological membranes is generally attributed to the botanist Carl Wilhelm Nageli (1817–1891), who became a pioneer in the application of microscopic techniques to the study of cell detail through his attempts to relate structure and function. In his classic work *Primordialschlauch*, published in 1855, he pointed out that the region immediately adjacent to the inner surfaces of the walls of plant cells appeared to differ from the underlying protoplasm and conjectured that this may be because the protoplasm becomes a firmer gel on contact with the extracellular fluid. Two decades later, another botanist, W. Pfeffer, noted that, when plant cells were placed in concentrated solutions, the protoplasm shrank away from the wall and appeared to be surrounded by a distinct structure, or "membrane." In his treatise dealing with these studies (*Osmotische Untersuchungen*), he argued that these membranes were discrete structures that could serve as selective barriers for the passage of substances into and out of cells. This was merely a guess, but one that had the virtue of being correct; that guess is sometimes referred to as *Pfeffer's postulate*.

It was not until 1890 that Ernst Overton confirmed Pfeffer's guess. Overton compared the solubility of a large number of solutes in olive oil with the ease with which they permeate (enter) cells and, as is discussed in greater detail in the following sections, concluded that Pfeffer's membrane behaved very much as if it were made up of lipids similar to olive oil and thus differed from the aqueous protoplasm.

This first clue to the composition of biological membranes takes us back many years, and, indeed, it would not be too much of an exaggeration to say that the earliest contribution in this area can be attributed to Pliny the Elder (A.D. 23–79) who noted, "Everything is soothed by oil, and this is the reason why divers send out small quantities of it from their mouths, because it smoothes every part which is rough" (*Natural History*, Book II, Section 234). Apparently, pearl divers made a habit of diving with a mouthful of oil when the surface of the water was rippled by winds; when they ejected the oil, it would "soothe the troubled waters" and increase underwater visibility.

In 1762, Benjamin Franklin was purportedly reminded of this calming effect of oil by an old sea captain, and, in 1765, when he was the American ambassador to the Court of St. James, he performed an experiment that is a landmark in the science of surface chemistry. Franklin poured olive oil onto a pond in Clapham Common (near London) and noted that the ripples caused by the wind were indeed calmed. He also astutely noted that a given quantity of oil would only spread over a definite area of the pond; the same quantity of oil would always cover the same area, and twice that volume of oil would cover twice that area. Knowing the volume of oil poured on the pond and the area covered, Franklin calculated that the thickness of the layer formed by the oil was about 25 Å (2.5 nm) and that it "could not be spread any thinner." Current measurements of the thickness of a monomolecular

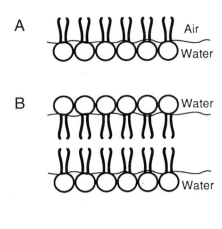

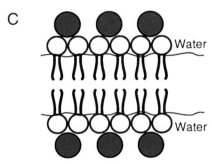

FIGURE 1 (A) Monolayer formed by phospholipids at an air–water interface. (B) A phospholipid bilayer separating two aqueous compartments. (C) A bimolecular lipoprotein membrane.

layer (Fig. 1A) of olive oil on water have not significantly improved on Franklin's estimate.

In 1925, Gorter and Grendel, stimulated by the findings of Franklin and Overton, performed a series of studies that had a major impact on all subsequent thinking dealing with membrane structure. These investigators extracted the lipids from erythrocyte membranes of a variety of species and calculated the area covered by these lipids when spread on water to form a monomolecular layer. They also approximated the total area of the membranes from which the lipid was extracted and concluded that the area of the monomolecular layer was twice the membrane area; that is, there was sufficient lipid to form a double or bimolecular layer of lipid around the cells, with each layer being about 25 Å thick (Fig. IB).

In 1935, Davson and Danielli modified the model proposed by Gorter and Grendel by including protein in the membrane structure. The bimolecular lipoprotein model they proposed is illustrated in Fig. 1C. The essential features of this model are (1) a bimolecular lipid core that is 50 Å (5 nm) thick and corresponds to the Gorter–Grendel model; and (2) inner and outer protein layers attached to the polar head groups of the lipids by ionic interactions. This model enjoyed 25 years of essentially universal acceptance.

Current Concepts: The Fluid-Mosaic Membrane

Since 1945, the application of increasingly sophisticated analytical and ultrastructural techniques to the study of the composition and structure of biologic membranes has met with remarkable success. First, the early notion that biologic membranes are made up of a mixture of lipids and proteins has been firmly established for all such barriers throughout the animal and plant kingdoms. The proportions of these two components differ among different cell membranes. In general, membranes that primarily serve as insulators between the intracellular and extracellular compartments and have few metabolic functions (e.g., myelin) are made up of a relatively high proportion of lipids compared to proteins. On the other hand, membranes that surround metabolic factories (e.g., hepatocytes, mitochondria) are relatively rich in protein content compared to lipid content.

Second, it is now generally accepted that the lipids that comprise biologic membranes are primarily from the group referred to as *phospholipids*. While cholesterol is present in the membranes of many eurokaryotic cells, it is not found in most prokaryotic cells. All phospholipids are derivatives of phosphatidic acid, which consists of a phosphorylated glycerol backbone to which two fatty acid tails are attached by ester bonds. The most prevalent phospholipids found in biologic membranes, such as phosphatidylcholine, phosphatidylserine, phosphatidylinositol, and phosphatidylethanolamine, result from esterification of the free phosphate group of phosphatidic acid with the hydroxyl groups of choline, serine, inositol, and ethanolamine, respectively. The important point is that all phospholipids contain a water-soluble (hydrophilic) head group (i.e., the phosphorylated glycerol backbone and its conjugates) and two water-insoluble (hydrophobic or lipophilic) tails. Such molecules are referred to as amphiphatic (or amphipathic). The prefix *amphi-* derives from both the Greek and Latin and means "having two sides"; e.g., amphibians live in air (on land) and in water. When these compounds are poured onto water, their hydrophilic head groups enter the aqueous phase and their hydrophobic tails simply wave in the air (Benjamin Franklin's observation; see Fig. 1A). But, when these molecules are confronted with two aqueous compartments, they will spontaneously form a bilayer (per the Gorter–Grendel model; see Fig. IB), with their water-soluble head groups immersed in the two aqueous phases and their lipid tails forming a hydrophobic core.

Finally, perhaps the most revolutionary advance in this area has been our understanding of the way in which these proteins and lipids are assembled in

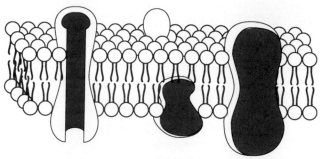

FIGURE 2 Schematic representation of the fluid-mosaic model proposed by Singer and Nicholson, featuring a phospholipid bilayer containing peripheral and integral proteins. Some proteins actually form channels that allow ions to move across the membrane.

biological membranes. It is now clear that the phospholipids and cholesterol form an "oily" fluid bilayer in which the adherent proteins are free to float around at will. Some of these proteins span the thickness of the bilayer; these so-called integral proteins have hydrophobic middles and hydrophilic ends, so that their ends protrude into the intracellular and extracellular watery compartments, while their middles are glued by hydrophobic bonds within the oil (Fig. 2). Other proteins are electrostatically attached to either the inner or outer surface of the bilayer; they are referred to as *peripheral proteins*. In stark contrast to the Davson–Danielli bimolecular lipoprotein model (Fig. 1C), which has a rather static appearance, this fluid-mosaic model is dynamic and possesses structural features that afford avenues for communication between the extracellular milieu and the intracellular compartments by virtue of signal transduction and the selective exchange of solutes. In short, the fluid-mosaic model provides an ideal basis for the correlation of membrane structure and function.

DIFFUSION OF NONELECTROLYTES

The term *diffusion* refers to the net displacement (transport) of matter from one region to another due to random thermal motion. It is classically illustrated by an experiment in which an iodine solution is placed at the bottom of a cylinder and pure water is carefully layered above this colored solution. Initially, there is a sharp demarcation between the two solutions; however, as time progresses, the upper solution becomes increasingly colored, and the lower solution becomes progressively paler. Ultimately, the column of fluid achieves a uniform color, and the diffusion of iodine ceases. This state of maximum homogeneity, or uniformity, will persist as long as the cylinder is undisturbed; a perceptible deviation from homogeneity

on the part of the entire column or any bulk (macroscopic) portion of the column will never be observed. The properties of the system, at this point, are said to be time independent, or, stated in another way, the system is said to have achieved a state of *equilibrium*.

The kinetic characteristics of diffusion can be readily developed by considering the simplified case illustrated in Fig. 3. Compartment o and compartment i both contain aqueous solutions of some uncharged solute (i) at concentrations C_i^o and C_i^i, respectively, where the superscripts designate the compartment and the subscript designates the solute. (An uncharged solute refers to a molecule that bears no net charge and thus includes solutes that have an equal number of oppositely charged groups such as zwitterions.) These compartments are separated by a sintered glass disk, which, because it possesses many large pores, may be treated as if it were an aqueous layer having a cross-sectional area A (the total area of the pores) and a thickness Δx; the disk simply serves as a barrier to prevent bulk mixing of the two solutions. We assume that each compartment is well stirred so that the concentrations C_i^o and C_i^i are uniform. Furthermore, for the sake of simplicity, we will also assume that both compartments have sufficiently large volumes so that their concentrations remain essentially constant during the period of observation, despite the fact that solute i may leave or enter these compartments across the disk.

Because the solute molecules are in continual random motion, due to the thermal energy of the system, there is a continual migration of these molecules across the disk in both directions. Thus, some of the molecules originally in compartment o will randomly wander across the barrier and enter compartment i. The rate of this process, because it is the result of random motions, is proportional to the likelihood that a molecule in compartment o will enter the opening of a pore in the

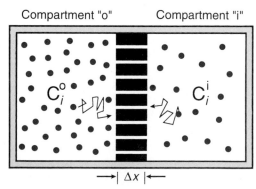

FIGURE 3 Random movement of solute i across a highly porous sintered disk having a thickness of Δx.

disk and is therefore proportional to C_i^o. Thus, we may write:

Rate of molecular migration (diffusion)
from o to $i = kC_i^o$

where k is a proportionality constant. Similarly, we may write:

Rate of molecular migration (diffusion)
from i to $o = kC_i^i$

The rate of net movement of molecules across the barrier is the difference between the rates of these two unidirectional movements so that:

Rate of net molecular migration (diffusion)
across barrier $k(C_i^o - C_i^i) = k\Delta C_i$

Thus, the net flow of an uncharged solute across a permeable barrier due to diffusion is directly proportional to the concentration difference across the barrier. In addition, the rate of transfer is directly proportional to the cross-sectional area A and inversely proportional to the thickness of the barrier Δx; that is, for a given concentration difference ΔC_i, doubling the area of the disk will result in a doubling of the rate of transfer from one compartment to the other, whereas doubling the thickness of the disk will halve the rate of transfer. Therefore, we may replace k with a more explicit expression that contains a new proportionality constant, D_i, where $k = AD_i/\Delta x$, and we obtain:

Rate of net migration (diffusion) across
the barrier $= AD_i\Delta C_i/\Delta x$

Dividing both sides of this expression by A, we obtain the rate of net flow per unit area, which is often referred to as the flux (or net flux). This is commonly symbolized by J_i, which is expressed in units of amount of substance per unit area per unit time (*e.g.*, mol/cm^2 hr). Thus,

$$J_i = D_i\Delta C_i/\Delta x \qquad (1)$$

where D_i is the diffusion coefficient of the solute i and is a measure of the rate at which i can move across a barrier having an area of 1 cm^2 and a thickness of 1 cm when the concentration difference across this barrier is 1 mol/L. The coefficient D_i is dependent on the nature of the diffusing solute and the nature of the barrier or the medium in which it is moving (interacting); we shall examine this dependence in further detail below. If J_i is

expressed in mol/cm^2 hr, Δx in cm, and ΔC_i in mol/1000 cm^3, then D_i emerges with the (somewhat uninformative) units of cm^2/hr.

Equation 1 describes diffusion across a flat or planar barrier having a finite thickness. Such systems are often called *discontinuous systems*, inasmuch as the barrier introduces a sharp and well-defined demarcation between the two surrounding solutions. Diffusion of iodine into water, in the experiment described above, represents a continuous system, as there is no discrete boundary between the two solutions. The general expression for diffusion in a continuous system can be derived from Eq. 1 by simply making the thickness of the disk vanish mathematically. Thus, as Δx approaches 0, $\Delta C_i/\Delta x$ approaches dC_i/dx so that:

$$J_i = D_i[dC_i/dx] \qquad (2)$$

Equation 2 was derived in 1855 by Fick, a physician, and is often referred to as *Fick's (first) law of diffusion*. It simply states that the rate of flow of an uncharged solute due to diffusion is directly proportional to the rate of change of concentration with distance in the direction of flow. The derivative dC_i/dx is referred to as the *concentration gradient* and is the driving force for the diffusion of uncharged particles. Thus, Eq. 2 states that there is a linear relation between the diffusional flow of i and its driving force, where D_i is the proportionality constant. Equation 2 is but one example of many linear relations between flows and driving forces observed in physical systems. For example, Ohm's law states that there is a linear relation between electrical current (flow) and its driving force, the voltage (electrical potential difference), where the proportionality constant is the electrical conductance g (recall that $g = 1/R$, where R is resistance). Thus, we can write:

$$I = gV \quad \text{or} \quad IR = V \qquad (3)$$

We will see below that the diffusion of charged particles (ions) can be described by a combination of Eqs. 2 and 3, where the driving force is a combination of the concentration difference (or gradient) (chemical force or potential) and the electrical potential difference (or gradient).

DIFFUSION THROUGH SELECTIVE BARRIERS

A sintered glass disk is highly porous, and the dimensions of the water-filled pores that penetrate the disk are extremely large compared with molecular dimensions. For this reason, the diffusing solute molecules behave as if they were passing through an

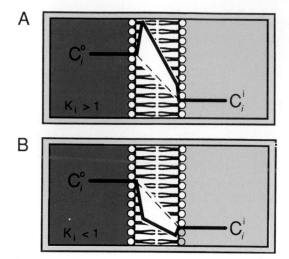

FIGURE 4 Partitioning of a lipophilic solute (A) and a hydrophilic solute (B) across the interfaces of a lipid membrane. The *solid lines* illustrate concentration profiles in the two solutions and within the lipid membrane.

aqueous layer and are unaffected by the presence of the barrier; in other words, they are not significantly influenced by either the openings or the walls of the pores through which they pass. The diffusion coefficient of a solute within the disk is the same as its diffusion coefficient in a continuous aqueous system (free solution), and the concentration of the solute just within the disk at each interface is the same as its concentration in the immediately adjacent aqueous solution. Thus, the sintered glass disk is a nonselective barrier because the properties of the solute within the disk are the same as those in the surrounding solutions.

If, as illustrated in Fig. 4, the sintered glass disk is replaced by a lipid membrane, the description of diffusion becomes slightly more complicated. First the concentration of the solute just within the membrane at each interface, in general, will not be equal to the concentration in the immediately adjacent aqueous solutions. If the solute is hydrophilic (or lipophobic; *e.g.*, an ion), it will prefer the aqueous phase to the lipid phase, and its concentration just within the membrane will be less than that in the adjacent aqueous solutions. On the other hand, if the solute is lipophilic (*e.g.*, fats, sterols, phospholipids), it will distribute itself so that the concentration at the interfaces just within the lipid barrier will exceed that in the adjacent aqueous solutions. The distribution of a given solute between the aqueous phase and the adjacent oil or lipid phase is described, quantitatively, by a unitless number termed a *partition coefficient* (also, a *distribution coefficient*). The partition coefficient of a given solute i between two solvent phases is determined in the following manner: The two immiscible solvents, for example, olive oil and

water, together with an arbitrary amount of solute are placed in a separately funnel. The funnel is then stoppered and shaken until an equilibrium distribution of solute i in the two phases is achieved. The olive oil–water partition coefficient is defined as:

$$K_i = \frac{\text{Equilibrium concentration of } i \text{ in olive oil}}{\text{Equilibrium concentration of } i \text{ in water}}$$

In this simple way, partition coefficients between lipid and aqueous phases have been determined for a large variety of solutes.

Returning to the example illustrated in Fig. 4A, it follows that if the solute in the two compartments is hydrophobic or lipophilic ($K_i > 1$), the concentration just within the membrane at the interface with compartment o will be $K_i C_i^o$, and the concentration just within the membrane at the interface with compartment i will be $K_i C_i^i$. Thus, the concentration difference within the membrane is K_i, ΔC_i, so that the partition coefficient has essentially amplified the driving force for the diffusion of the solute across the membrane. The flux of solute through the lipid membrane due to diffusion is now given by:

$$J_i = K_i D_i \Delta C_i / \Delta x \tag{4}$$

where the diffusion coefficient D_i now reflects the ease with which the solute can move through the lipid and will differ from the value of D_i in an aqueous solution. (We are assuming that diffusion through the membrane is slow, or rate-limiting, compared to partitioning into and out of the membrane at the two interfaces.) Conversely, if the solute is highly water soluble (*i.e.*, hydrophilic or lipophobic), then $K_i < 1$ and its concentration difference within the membrane will be attenuated.

Now, in general, K_i, D_i, and Δx cannot be readily determined. For this reason, these unknowns have been lumped together to give a new term, the *permeability coefficient*, P_i, which is defined as:

$$P_i = K_i D_i / \Delta x \tag{5}$$

Because K_i is unitless, when D_i is given in cm^2/hr and Δx in cm, then P_i has units of cm/hr. The permeability coefficient of a given membrane for a given solute is simply:

$$P_i = J_i / \Delta C_i \tag{6}$$

and can be readily determined experimentally. Clearly, the permeability coefficient of a membrane for an uncharged solute is the flow of solute (in moles per hour) that would take place across 1 cm^2 of membrane when the concentration difference across the membrane is 1 *M*.

We now recapitulate some of the laws that apply to diffusional movements across artificial as well as biologic membranes. First and foremost, the driving force or *sine qua non* for the net diffusion of an uncharged solute is a concentration difference. In the absence of a concentration difference, a net flux due to diffusion is impossible. In the presence of a concentration difference, diffusion will take place spontaneously and the direction of the net flux is such as to abolish the concentration difference. In other words, diffusion only brings about the transfer of net uncharged solutes from a region of higher concentration to a region of lower concentration; the reverse direction is thermodynamically impossible! For this reason, transport due to diffusion is often referred to as *downhill* (because the flow is from a region of higher concentration to one of lower concentration) or *passive* transport (because no additional energy need be supplied to a system to enable these flows to take place; the inherent thermal energy responsible for random molecular motion is sufficient). As we shall see, there are numerous biologic transport processes that bring about the flow of uncharged solutes from a region of lower concentration to one of higher concentration. These *uphill*, or *active*, transport processes cannot be due to diffusion and are dependent on an energy supply in addition to simple thermal energy.

Within the membrane, the link between the driving force, ΔC_i, and the flow, J_i, is the diffusion coefficient, D_i. This is the factor that determines the flow per unit driving force, and it is determined by the properties of the diffusing solute and those of the membrane through which diffusion takes place. At the molecular level, the diffusion coefficient is a measure of the resistance offered by the membrane to the movement of the solute molecule.

At the turn of the century, Einstein (1905) proposed that the resistance experienced by a diffusing particle results from the frictional interaction between the surface of the particle and the surrounding medium because the two, in essence, are moving relative to each other. Drawing on Stokes' law, which describes the friction experienced by a sphere falling through a medium having a given viscosity, Einstein demonstrated that the diffusion coefficient of a spherical molecule should be inversely proportional to both the radius of the molecule and the viscosity of the surrounding medium. Thus, when one is concerned with the diffusion of a variety of solutes through a single medium (fixed viscosity), the relative diffusion coefficients should be inversely proportional to their molecular radii. Einstein's contributions to our understanding of diffusion and the nature of the diffusion coefficient have

been repeatedly confirmed, and the Stokes–Einstein equation has proved to be a valuable approach to the calculation of molecular dimensions from measurements of diffusion coefficients.

We can now make several educated guesses about the permeability of biologic membranes to uncharged solutes. First, because biologic membranes are primarily composed of lipids, Eq. 5 suggests that the permeability coefficients of solutes having approximately the same molecular dimensions (same diffusion coefficients) should vary directly with their partition coefficients (*K*s); that is, the more lipid soluble the molecule, the greater its permeability coefficient. This deduction has been verified repeatedly and is a statement of *Overton's law*. In fact, as already noted, Overton's observation that lipid-soluble molecules penetrate biologic membranes more readily than water-soluble molecules of the same size predated the chemical analyses of membranes and was the first clue that biologic membranes are made up of lipids.

Second, Eq. 5 also predicts that the permeability coefficients of solutes having the same lipid solubilities (same *K*s) should vary inversely with their molecular sizes; that is, the larger the molecule, the lower its permeability coefficient. It should be noted that *K*s can vary over many orders of magnitude, whereas *P*s generally do not differ by more than a factor of five. Thus, the permeability coefficients for simple diffusion of uncharged solutes across biologic membranes are more strongly influenced by differences in lipid solubility than molecular size.

MEMBRANE PORES AND RESTRICTED DIFFUSION

By the early 1930s, the ability of Eq. 4 to describe diffusion across biologic membranes was experimentally well established, and the prevalent view was that cells are surrounded by an intact or continuous lipid envelope. This view was soon challenged by the exhaustive studies of Collander and his associates on the permeability of the plant cell *Chara* to a large number of solutes. The results of these studies are summarized in Fig. 5, where the permeability coefficients of a number of solutes (ordinate) are plotted against their oil–water partition coefficients (abscissa).

Clearly, there is, in general, a direct linear relationship between these parameters, as previously demonstrated by Overton, but Collander also noted that small water-soluble molecules tend to lie above the line describing the best fit of the data. In other words, small water-soluble molecules appear to permeate the membrane more rapidly than could be accounted for on

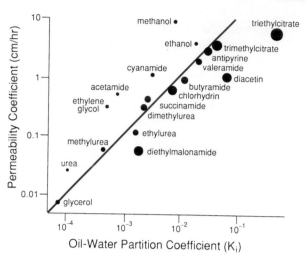

FIGURE 5 Relation between permeability coefficients and oil-water partition coefficients for a number of solutes. The sizes of the *circles* are proportional to the molecular diameters of the solutes.

the basis of their lipid solubilities alone. This astute observation led Collander to conclude:

> It seems therefore natural to conclude that plasma membranes of the *Chara* cells contain lipoids, the solvent power of which is on the whole similar to olive oil. But, while medium sized and large molecules penetrate the plasma membrane only when dissolved in the lipoids, the smallest molecules can also penetrate in some other way. Thus, the plasma membrane acts both as a *selective solvent* and as a *molecular sieve*.

In short, he proposed that small, highly water-soluble molecules can penetrate biologic membranes through two parallel pathways: (1) they can dissolve in the lipid matrix and diffuse through the membrane just as lipid-soluble molecules do but to a very much lesser extent, or (2) they can pass through channels or pores that perforate the lipid envelope.

During the years that have elapsed since Collander's suggestion of a mosaic lipid membrane (*i.e.*, a discontinuous lipid layer perforated by water-filled channels), a compelling body of evidence has accrued for the presence of proteinaceous channels that span the lipid bilayer and serve as the routes for the diffusional flows of highly water-soluble solutes across biologic membranes. These flows are often referred to as *restricted diffusion* inasmuch as their rates are dramatically influenced by molecular size.

DIFFUSION OF UNCHARGED SOLUTES

The prevalent view of diffusion of uncharged solutes through biologic membranes is that:

1. Lipid-soluble molecules readily penetrate the membrane by diffusion through the lipid bilayer.

The permeability coefficients of large lipid-soluble molecules may be quite high due to partition coefficients that strongly favor their entry into the lipid phase. These movements are often referred to as *simple diffusion*.

2. Small water-soluble molecules, which have little affinity for the lipid phase, may penetrate the membrane through pores. This restricted diffusion is, however, limited to small molecules; most cell membranes are essentially impermeable to water-soluble molecules having five or more carbon atoms. Most, if not all, essential metabolic substrates (*e.g.*, glucose, essential amino acids, water-soluble vitamins) cannot enter the cell to any appreciable extent by this mechanism. As we shall see, other mechanisms (carrier-mediated processes) are present that mediate the entry of these substances into cells at rates that are sufficiently rapid to sustain essential metabolic processes.

3. Inasmuch as water and small water-soluble uncharged molecules have finite lipid solubilities, it is often difficult to determine experimentally the extent to which they traverse biologic membranes by diffusing through the lipid phase and/or by restricted diffusion through pores. It is generally accepted that the permeation rates of water and other small, polar nonelectrolytes (*e.g.*, urea, glycerol) through biologic membranes (where determined) are too fast to be accounted for by diffusion through the lipid phase and that pores must be invoked. There is no doubt that inorganic ions traverse biologic membranes virtually exclusively via pores inasmuch as their solubilities in lipid are minute. Indeed, a large family of integral membrane proteins that serve as selective channels for water movements across biological membranes, named *aquaporins*, have been identified, isolated, and extensively studied at the molecular level. Some of these channels are almost exclusively permeable to water, while others are also permeable to small uncharged solutes such as urea. Further, highly selective channels have been identified, as have clones and sequences for virtually all major ions of physiological significance (*e.g.*, Na^+, K^+, Ca^{2+}), and they will be discussed later in this book.

DIFFUSION OF ELECTROLYTES

The previous section dealt with the diffusion of uncharged particles, but many of the fundamental properties of the diffusional process described therein also apply to the diffusion of charged particles. In both instances, net flow due to diffusion is the result of random thermal movements, and the diffusion coefficients of the particles are inversely proportional to their molecular or hydrated ionic radii. However, because ions bear a net electrical charge, the diffusion of a salt such as NaCl, which exists in aqueous solution in the form of oppositely charged dissociated ions, is somewhat more complicated for two reasons. First, because of the attractive force between particles bearing opposite electrical charges, the ions resulting from the dissociation of a salt do not diffuse independently. Second, under conditions of uniform temperature and pressure, only a difference in concentration can provide the driving force for the diffusion of uncharged particles; net flows do not occur in the absence of concentration differences. For the case of a charged particle, the driving force for diffusion is made up of two components: (1) a difference in concentration, and (2) the presence of an electrical field. The effect of an electrical field on the movement of charged particles is readily illustrated by the familiar phenomenon of electrophoresis. If an anode and a cathode are placed in a beaker that contains a homogeneous solution of NaCl, Cl^- will migrate toward the anode, and Na^+ will migrate toward the cathode. These net movements are due to attractive and repulsive forces that act on charged particles within an electrical field and take place despite the initial absence of a concentration difference. Each of these aspects of ionic diffusion is separately discussed below.

Origin of Diffusion Potentials

Consider the system illustrated in Fig. 3. If a solution containing equal concentrations of urea and sucrose is placed in compartment o and distilled water is placed in compartment i, both urea and sucrose will diffuse across the sintered glass disk independently. Because urea is smaller than sucrose, its diffusion coefficient will be greater than that of sucrose (i.e., $D_{urea} > D_{sucrose}$), and it will move more rapidly in response to the same driving force ($\Delta C_{urea} = \Delta C_{sucrose}$) so that, after sufficient time has elapsed, the solution in compartment i will contain more urea than sucrose. In essence, we have employed differential rates of diffusion as a separatory method. If the barrier is permeable to urea but impermeable to sucrose, only urea will appear in compartment i; this is the principle of separation by dialysis. Now, let us replace the solution in compartment o with an aqueous solution of potassium acetate (KAc). This salt dissociates into a relatively small cation (K^+) and a larger anion (Ac^-). If the diffusion of these ions were independent (as in the example with urea and sucrose), it would be possible, in time, to withdraw from compartment i a solution that contains more cation

than anion—that is, a solution that has excess K^+. This, in fact, has never been observed. Indeed, the failure to observe a separation of charge in solutions, as well as compelling theoretical arguments, are the bases of a universally accepted postulate known as the *law of electroneutrality*. In essence, this law states that any macroscopic or bulk portion of a solution must contain an equal number of opposite charges; that is, it must be electrically neutral. This is a very important statement that must be kept in mind whenever we consider the bulk movements of ions. Returning to our example of the diffusion of KAc, we may ask, "Why does K^+ not outdistance Ac^-? What is responsible for the maintenance of electroneutrality under these conditions?" The answer is that the tendency for K^+ to outdistance Ac^- because of the difference in ionic size (and, hence, diffusion coefficients) is counteracted by the electrostatic attractive forces that exist between these two oppositely charged particles. This mutual attraction, which tends to draw the two ions closer together, exactly balances the effect of the difference in ionic size, which would tend to draw them apart.

One may conceptualize the physical process as follows. In Fig. 6, we see the small ion (K^+) and the larger ion (Ac^-) lined up along a partition (the starting line). When the partition is removed, the race begins. Each ion initially moves from left to right at a rate determined by its ionic size; in other words, each ion initially diffuses from left to right at a rate determined by its inherent diffusion coefficient (denoted by D_K and D_{Ac}). Because $D_K > D_{Ac}$, there will be an initial separation of these oppositely charged ions, which, if unopposed, would eventually lead to a violation of the law of electroneutrality. However, the electrostatic attraction between these particles tends to hold them together. Thus, the attraction of Ac^- for K^+ tends to slow down the rate of diffusion of K^+, and, conversely, the attraction of K^+ for Ac^- tends to speed up the rate of diffusion of Ac^-. The net result is that the two ions move from left to right at the same rate but in an oriented fashion (K^+ in front of Ac^-) and thus form a small diffusing dipole (or ion pair). The diffusion coefficient of this ion pair, or dipole, is greater than that of Ac^- alone but less than that of K^+ ($D_K > D_{KAc} > D_{Ac}$).

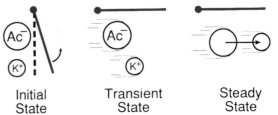

| Initial State | Transient State | Steady State |

FIGURE 6   Illustration of dipole (ion pair) formation as a result of the differential rates of diffusion of K^+ and Ac^-.

(The situation resembles what would happen if a fast swimmer and a slow swimmer were connected by an elastic cord; there would be an initial rapid separation of the two swimmers, but soon the pair would move together at a rate intermediate to the rate at which each could swim independently.)

Now let us apply this "dipole" concept to diffusion of KAc across a sintered glass disk. Consider the system illustrated in Fig. 7. We place a well-stirred solution of KAc in compartment o and a more dilute solution of KAc in compartment i; diffusion from o to i is initially prevented by the presence of a solid partition adjacent to the sintered glass disk (Fig. 7A). If we now pair up a K^+ ion with a neighboring Ac^- ion so as to form hypothetical dipoles (using the convention that the head of the arrow represents the positive ion), we find, as shown in Fig. 7A, that the dipoles are randomly oriented simply because the distribution of ions in a homogeneous solution is random. Thus, for every dipole

pointed in a given direction there will be another dipole of equal magnitude oriented in the exactly opposite direction. The total dipole moments in compartments o and i, as well as within the membrane, are therefore equal to zero, and if electrodes are inserted into the two compartments the electrical potential difference between these electrodes will be zero.

If we suddenly remove the solid partition, K^+ and Ac^- will diffuse from the higher concentration in compartment o to the lower concentration in compartment i (Fig. 7B). For the reasons discussed above, this diffusional process can be represented as a series of dipoles crossing the sintered glass disk in an oriented fashion. The random orientations of the dipoles, characteristic of the two well-stirred homogeneous solutions in compartments o and i, have been converted *within* the disk to an oriented distribution by the presence of a concentration difference *across* the disk. The sum of all these oriented dipoles can be represented by a single dipole whose positive end is pointed toward compartment i and whose negative end is pointed toward compartment o. If electrodes are inserted into compartments o and i, compartment i will be found to be electrically positive with respect to compartment o.

It is important to emphasize that there is *no bulk separation of charges* because the distance between the leading K^+ ion and the lagging Ac^- ion averages only a few (~10) angstrom units. The electrical potential difference across the membrane is not due to a bulk (chemically detectable) separation of charge but to the fact that the ions cross the disk in an oriented fashion rather than in a random fashion. In essence, the orientation of the dipoles within the disk can be viewed as having converted the disk into a battery with the positive pole facing compartment i. (Technically speaking, there is a small separation of charges sufficient to charge the capacitance of the membrane, which is completed very rapidly; thereafter, the movement of anions and cations across the membrane is one to one. Furthermore, the amount of charge separation is minute compared to the number of ions present in the two solutions so that bulk electroneutrality is preserved; see also Chapter 9.)

The electrical potential difference arising from the diffusion of ions derived from a dissociable salt from a region of higher concentration to one of lower concentration is referred to as a *diffusion potential*. It arises whenever the ions resulting from the dissociation of the salt differ with respect to their mobilities or diffusion coefficients. The orientation of the diffusion potential is such as to retard the diffusion of the ion having the greater mobility and to accelerate the diffusion of the ion with the lower mobility so as to maintain electroneutrality. Thus, the magnitude of a diffusion potential

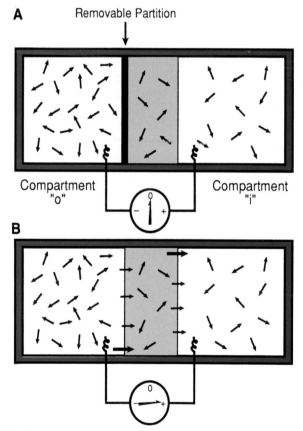

A

FIGURE 7 Generation of a diffusion potential as a result of oriented ionic diffusion through a uniform membrane separating two well-stirred compartments. The head of the *arrow* represents the positive ion. (A) With the partition in place, the dipoles are randomly oriented, and there is no electrical potential difference between compartments. (B) When the partition is suddenly removed, the concentration difference *across* the membrane causes the dipoles to orient *within* the membrane and diffuse into the compartment with the lower concentration.

will be directly dependent on the difference between the mobilities of the anion and cation.

It can be shown that the diffusion potential (V) arising from the diffusion of a salt that dissociates into a monovalent anion and a monovalent cation is given by:

$$V = \left\{\frac{D_+ - D_-}{D_+ + D_-}\right\} \times \left\{\frac{2.3RT}{F}\right\} \times \log\left\{\frac{C_i^o}{C_i^i}\right\} \qquad (7)$$

where D_+ is the diffusion coefficient of the monovalent cation; D_- is the diffusion coefficient of the monovalent anion; C_i^o and C_i^i are the concentrations of the salt in compartments o and i, respectively; R is the gas constant; T is the absolute temperature; and F is the Faraday constant. At 37°C ($T = 310K$), 2.3 $RT/F \simeq 60$ mV. Thus,

$$V = \left\{\frac{D_+ - D_-}{D_+ + D_-}\right\} \times 60 \log\left\{\frac{C_i^o}{C_i^i}\right\} (mV) \qquad (8)$$

Equation 8 discloses the following characteristics of diffusion potentials:

$$V = 0 \quad \text{when } C_i^o = C_i^i$$

Clearly, when $C_i^o = C_i^i$, there can be no net flow due to diffusion. Diffusion potentials can arise only in a system containing ion gradients:

$$V = 0 \quad \text{when } D_+ = D_-$$

Clearly, if both ions have equal mobilities (or sizes), there is no inherent leader and no inherent lagger, so dipoles are not formed. Another way of viewing this condition is that, when both ions have the same mobility, a one-to-one flow across the disk (i.e., bulk electroneutrality) is assured by the fact that both ions are driven by the same ΔC; a diffusion potential is not necessary for the preservation of bulk electroneutrality.

When C_i^o is not equal to C_i^i, the magnitude of V is directly proportional to the difference between the individual ionic diffusion coefficients. The orientation of V is such that it retards the more mobile ion and accelerates the less mobile ion. Thus, if the mobility of the anion exceeds that of the cation, the more dilute solution will be electrically negative with respect to the more concentrated solution. Conversely, if the cation has a greater mobility than the anion, the more dilute solution will be electrically positive with respect to the concentrated solution.

A particularly interesting situation arises when the barrier is impermeant to one of the ionic species—say,

for example, the cation. When $D_+ = 0$, Eq. 8 reduces to:

$$V = -60 \log\left\{\frac{C_i^o}{C_i^i}\right\} (mV) \qquad (9)$$

The same expression, with the opposite sign, is obtained if we set $D_- = 0$.

Equation 9 states that, when one of the ionic species of a dissociated salt cannot penetrate the barrier, V is dependent only on the concentration ratio across the membrane and is independent of the permeability (or diffusion coefficient) of the ion that can penetrate the barrier. The reason for this independence becomes evident when we recall the law of electroneutrality. If the membrane is impermeant to one of the ions, there can be no net flow of the other, permeant, ion across the membrane, or bulk electroneutrality would be violated. Thus, despite the presence of a concentration difference, there can be no net diffusion of salt across the barrier, and the system is in a state of equilibrium. Because the net flow of the permeant ion is prohibited, the mobility of this ion is of no importance.

The following example may serve to clarify this point and the roles played by concentration differences and electrical potential differences in the overall driving force for the diffusion of ions. Let us place a 0.1-M solution of K^+ proteinate in compartment o and a 0.01-M solution of the same salt in compartment i. If the barrier is impermeant to the large proteinate anion, the electrical potential difference across the barrier at 37°C will be:

$$V = 60 \log\{0.1/0.01\} = 60 \text{ mV}$$

with the dilute solution electrically positive compared to the concentrated solution. Now we may ask, "Why is there no diffusion of K^+ from compartment o to compartment i down a tenfold concentration difference despite the fact that the membrane is highly permeable to K^+?" The answer derives from the fact that there are two forces acting on the K^+ ion. There is a *chemical* force arising from the fact that the concentration of K^+ in compartment o is ten times that in compartment i; this force tends to drive K^+ from compartment o to compartment i. In addition, Eq. 9 tells us that there is a 60-mV electrical potential difference between compartments o and i, with compartment i electrically positive. This electrical potential difference tends to drive the positively charged K^+ ion from compartment i to compartment o. These two oppositely directed driving forces (the *chemical* force and the *electrical* force) exactly balance each other so that there is no net driving force for the diffusion of K^+ and, hence, no net movement.

Equation 9 was derived by the great German physical chemist Walther Hermann Nernst (1864–1941) from thermodynamic considerations and is referred to as the *Nernst equation* (also under these conditions it is often referred to as the *Nernst potential* or the *Nernst equilibrium potential*). In essence, it embodies the fact that (under conditions of uniform temperature and pressure) there are only two driving forces that influence the diffusion of charged particles: a force arising from concentration differences and a force arising from electrical potential differences. When solutions having different concentrations of the same dissociable salt are placed on opposite sides of a barrier that is impermeant to one of the dissociation products, there will be no net movement of salt across the barrier despite the concentration difference. Net movement of the permeant ion is prevented by the development of an electrical potential difference across the barrier whose magnitude and orientation are such that they exactly cancel the driving force arising from the chemical concentration difference across the barrier. (A more detailed discussion of this point is beyond the scope of this presentation. In essence, the two forces that influence the movements of charged particles—concentration differences and electrical potential differences—can be converted into the same units of force per ion or force per mole [*e.g.*, dynes/mol], and a tenfold concentration ratio at 37°C exerts the same force on a monovalent ion as does a 60-mV electrical potential difference. When, as in the above example, the two forces are oriented in opposite directions, the net force is zero, and there can be no net flow.)

Criteria for Active Transport of Ions

In a previous section we distinguished between passive (or downhill) and active (or uphill) transport of uncharged molecules. The sole criterion for this distinction is the relation between the direction of the net movement of the molecule and the direction of the concentration gradient. If an uncharged molecule moves or is transported from a region of higher concentration to one of lower concentration, the transport process is said to be passive (or downhill) because the flow is in the direction of the driving force and thus can be attributed entirely to the thermal energy prevalent at any ambient temperature. Conversely, if the molecule moves or is transported from a region of lower concentration to one of higher concentration, the flow is termed active (or uphill) because thermal energy alone cannot account for this movement, and additional forces must be involved.

The distinction between active and passive transport of ions is slightly more complicated because both concentration differences and electrical potential differences can provide driving forces for the diffusional movements of charged particles. This important point may be clarified by the following example. Consider the movement of K^+ across a membrane separating compartment o from compartment i (see Fig. 3) under the following conditions: (1) the concentration of K^+ in compartment o is 0.1 M; (2) the concentration of K^+ in compartment i is 0.2 M; and (3) compartment i is electrically negative with respect to compartment o by 60 mV. Thus, the chemical force acting on K^+ is a twofold concentration ratio tending to drive K^+ from compartment i to compartment o. The electrical force acting on K^+ is a 60-mV electrical potential difference tending to drive it from compartment o to compartment i. As we noted previously, a 60-mV electrical potential difference is equivalent to a tenfold concentration ratio; thus, in this example, the electrical driving force exceeds the chemical driving force, and K^+ will diffuse from compartment o to compartment i spontaneously, even though the direction of net movement is against the concentration gradient.

Now, in general, we can determine the equivalent electrical force, E_i, corresponding to a given concentration ratio by using the Nernst equation; that is,

$$E_i = (RT/zF) \ln(C_i^o/C_i^i) \tag{10}$$

where z is the valence of the ion i, and E_i is the equivalent electrical potential of ion i across the membrane (compartment i minus that of compartment o). Thus, if compartment o is arbitrarily chosen as the ground state and is defined as having a zero electrical potential, then E_i is the equivalent electrical potential difference corresponding to a given concentration ratio across the membrane with the magnitude and sign of compartment i. (A voltmeter does not measure an electrical potential but rather the difference between the electrical potentials of the two points to which its electrodes are connected. Thus, when we refer to the electrical potential of one compartment, it must be with reference to that of the other compartment, which is chosen as the ground or zero potential compartment. The universally accepted convention today is that the extracellular, or outer [*o*], compartment is designated to be the ground or reference in electrophysiologic studies.)

At 37°C, Eq. 10 reduces to:

$$E_i = (60/z) \log(C_i^o/C_i^i)(\text{mV}) \tag{11}$$

Now, using Eq. 11, we can determine the direction in which an ion will diffuse in the artificial system illustrated in Fig. 3, knowing the concentrations of the ion in compartments o and i and the electrical potential difference across membrane V_m, where V_m is the electrical potential of compartment i with respect to that of compartment o. Let us illustrate this point by considering some examples.

Example 1: $C_i^o = 10$ mmol/L; $C_i^i = 100$ mmol/L; $z = +1$; and $V_m = -60$ mV. Substituting these values into Eq. 11, we obtain $E_i = 60 \log(1/10) = -60 \log(10/1) = -60$ mV. Thus, the force of the tenfold concentration ratio which is driving the cation to flow from compartment i to compartment o is equivalent to 60 mV. But, at the same time, compartment i is 60 mV *negative* with respect to compartment o and this force tends to "hold" the cation in compartment i. Thus, the net force on the cation is zero, and there will be no net movement across the membrane in either direction. The system is at equilibrium. In short, when $E_i = V_m$, the chemical driving force arising from the concentration difference (ratio) across the membrane is exactly counterbalanced by the electrical driving force across the membrane, and there will be no flow.

Example 2: $C_i^o = 5$ mmol/L; $C_i^i = 100$ mmol/L; $z = +1$; and $V_m = -60$ mV. Now, from Eq. 11, $E_i = -78$ mV. Thus, the chemical driving force for the flow of i from compartment i to compartment o is 78 mV, but the electrical holding force is only 60 mV. Thus, i will diffuse across the membrane from compartment i to compartment o.

Example 3: $C_i^o = 20$ mmol/L; $C_i^i = 100$ mmol/L; $z = +1$, and $V_m = -60$ mmol/L. In this case, $E_i = -42$ mV. Thus, the chemical driving force tending to push i from compartment i to compartment o is less than the electrical driving force tending to pull the cation from compartment o to compartment i. The net result will be the diffusion of i from compartment o into compartment i despite the fact that this flow is against a concentration difference.

In summary, an uncharged substance is said to be actively transported if net movement is directed against a concentration difference. An ion is said to be actively transported only if its net movement is directed against a *combined* concentration and electrical potential difference; flow of an ion from a region of lower concentration to one of higher concentration is not by itself inconsistent with simple diffusion. The Nernst equation permits us to determine whether the movement of a charged solute is passive or active.

For an uncharged solute, the rate and direction of diffusion across a membrane can be described by the equation:

$$J_i = P_i(C_i^o - C_i^i) \tag{12}$$

where we define J_i as positive when the flow takes place from compartment o to compartment i; thus, when $C_i^i > C_i^o$, the flow is from compartment i to compartment o and is negative.

For an ion, J_i is determined by the permeability of the membrane and the total electrochemical driving force. Now, it is often convenient to express J_i as a current given by $I_i = zFJ_i$, where z_i is the valence of the ion, and F is the Faraday constant, which has the value of 96,500 coulombs (C)/mol. Thus, if J_i is in units of moles per unit area per unit time (*e.g.*, mol/cm^2/hr), then $I_i = z_iFJ_i$ is in units of C/cm^2/hr. Because electrical current flow is given in amperes, which is in units of C/sec (*i.e.*, the amount of charge flowing per second), I_i is in units of amperes per unit membrane area, or A/cm^2.

Now, the equation describing the rate of diffusion of an ion is:

$$z_iFJ_i = I_i = G_i(V - E_i)$$
$$= G_i\left[V - \frac{2.3RT}{z_iF}\log\frac{C_i^o}{C_i^i}\right] \tag{13}$$

where the total *electrochemical* driving force is $V - E_i$ (analogous to ΔC_i for an uncharged particle), which takes into account the two forces acting on the charged particle, and G_i is the conductance of the membrane to i (the inverse of the resistance of the membrane to the flow of i) and is analogous to P_i.

Equation 13 has the form of Ohm's law (i.e., $I = GV$), which, as discussed above, has the same form as Fick's (first) law of diffusion, which relates the rate of diffusion of an uncharged solute to its driving force.

ION DIFFUSION THROUGH BIOLOGIC CHANNELS

To this point, we have considered the diffusion of ions across barriers without specifying the nature of the pathway(s) traversed. Because all of the reasoning was based on thermodynamic principles, the conclusions arrived at in that chapter are valid regardless of the nature of the barrier; that is, it could be a sintered glass disk, a sheet of cellophane, a sheet of filter paper, a lipid bilayer—whatever! Turning now to biologic membranes, it has long been recognized that because ions are scarcely soluble in lipids, they can only cross those barriers by reversibly combining with some mobile component of the membrane and/or by diffusion through aqueous channels. There is now undeniable evidence for both types of mechanisms, and in this section we focus on the latter.

Let us start by inquiring how rapidly ions might be expected to traverse a lipid bilayer through an aqueous pathway assuming that the diffusion coefficient is the same as that in free aqueous solution. Let us assume that the ion has a "naked" or crystal radius, $r_i = 0.15$ nm, and that its diffusion coefficient (D_i) = 2×10^{-5} cm^2/sec; these are close to the values for K$^+$. Let us further assume that the channel has a radius (r_p) of only 0.2 nm, that the bilayer has thickness $\Delta x = 5$ nm, and that the concentration difference across the channel (ΔC_i) = 100 mmol/L. From Eq. 1, it follows that:

$$J = (\pi r_p^2 D_i \Delta C_i / \Delta x) = 4 \times 10^{-18} \text{mol/sec}$$

and, multiplying by Avogadro's number, $J_i = 2.5 \times 10^6$ ions/sec. If the ion is monovalent, then, because the electronic charge is $e = 1.6 \times 10^{-19}$ C, the current attributable to this flux of ions is 0.4×10^{-12} C/sec, or 0.4 pA. The important point of this exercise is that ballpark estimates are consistent with the notion that more than a million ions can diffuse per second across a lipid bilayer through a very snug pore; this is sufficient to generate a current that can be readily measured using available amplifiers.

Since the mid-1970s, measurements of bursts of ionic currents across biologic membranes have proven to be entirely consistent with the above estimate and, together with other evidence, leave little doubt that these bursts are the results of ion movements through pores or channels that are integral membrane proteins. During the past two decades, many such channels have been identified, cloned, sequenced, expressed, and subjected to extensive study of structure–function relations.

The two methods that are employed to measure single-channel activities of ion channels are illustrated in Fig. 8. The first (Fig. 8A) was introduced in 1976 and is referred to as the *patch-clamp* technique. Briefly, a polished glass micropipette tip having a diameter of approximately 1 μm is pressed against a patch of cell membrane, forming a seal that is essentially leakproof to ions (seal resistance is greater than 1 billion ohms). Thus, if there happens to be an ion channel in the membrane patch, ionic currents into or out of the cell will be entirely constrained within the pipette and the connected circuitry. One may also mechanically rip (excise) the patch of membrane together with the channel off the cell and examine single-channel properties under artificial but well-defined and easily controlled conditions.

The other technique involves reconstituting or incorporating membrane vesicles containing channels, or purified channel proteins, into planar artificial lipid bilayers separating two easily accessible and controlled solutions (Fig. 8B).

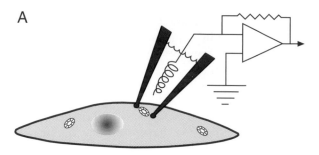

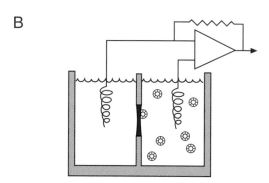

FIGURE 8   (A) Measurement of ion currents through a single channel by "clamping" a patch of membrane with a micropipette. (B) Apparatus for reconstituting vescicles containing channels into a planar lipid bilayer. The current-voltage (*i-V*) converter measures the small membrane currents by determining the voltage and developed across a very large resistor.

A recording of the activity of a single ion channel is illustrated in Fig. 9. (Note that the ordinate is in pA [*i.e.*, 10^{-12} A] and the abscissa is in milliseconds.) The openings are characterized by abrupt upward deflections from the baseline that reach a plateau and then abruptly close. An important characteristic of single-channel activity is that the openings and closings are stochastic or random processes. That is, when a channel will open and how long it will remain open under fixed conditions are independent of its past history (*e.g.*, how long it was closed before this opening or how long it was open the last time it was open); the channel does not have a memory. This characteristic enables one to analyze single-channel activity using statistical or probability

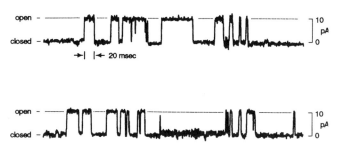

FIGURE 9   Bursts of current accompanying the random openings of an ion channel.

theory. For example, by analyzing a large number of opening and closing events, such as those shown in Fig. 9, one can determine the open-time probability of the channel, P_o; that is, the likelihood that the channel will be found in the open or conducting state at any given moment or the fraction of time the channel is open. Furthermore, if the channel can only randomly switch between two states due to thermal motion, then the transition between these states is given by:

$$(\text{closed state}) \underset{k_{-1}}{\overset{k_1}{\rightleftharpoons}} (\text{open state})$$

where the ks are rate constants analogous to those employed in chemical kinetics. In this instance, k_1 is the probability of transitions from the closed to the open state in unit time so that $(1/k_1)$ is the mean duration of the closed states or the mean closed time (τ_c). Likewise, k_{-1} is the probability of transitions from the open state to the closed state in unit time, so that $(1/k_{-1})$ is the mean open time (τ_o). It follows that the open-time probability will equal the mean open time divided by the mean total time; hence, $P_o = \tau_o/(\tau_o + \tau_c)$.

Now, because of the random nature of the openings and closings of a channel, the mean duration of either the open or closed states is determined only by the rate constants acting *away* from that state. Thus, by analogy with a simple first-order chemical reaction, we can write $d(\text{open})/dt = k_{-1}(\text{open})$, where the term in parentheses, roughly speaking, refers to the duration of a single channel in the open state. Integrating this equation, we obtain:

$$(\text{open})_t = (\text{open})_o e^{-tk_{-1}} = (\text{open})_o e^{-t/\tau_o} \qquad (14)$$

which states that once a channel opens the likelihood or probability that it will remain open decays exponentially with a rate determined by k_{-1} (or τ_o); the probability that a channel, once opened, will remain open for t msec or longer is given by e^{-t/τ_o}.

We can determine the mean open time (τ_o) by examining a large number of open events of a single channel and sorting them into "bins" based on their durations. For example, the first bin could contain the number of events when the channel was open for at least 0.5 msec; the second bin could contain all of the open events lasting at least 1 msec; the third, all of the open events lasting at least 1.5 msec; and so on in 0.5-msec increments. We can then construct a histogram as shown in Fig. 10, where the ordinate is the number or frequency of events per 0.5 msec and the abscissa is open time (t, in msec). If the channel conforms to the two-state model (*i.e.*, either fully closed or fully open),

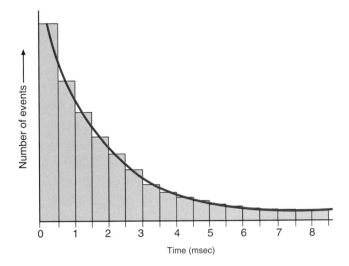

FIGURE 10   Histogram of number of open events versus cumulative duration of opening for a single channel observed over a long period of time. The curve corresponds to an exponential decay where $\tau_o \approx$ 2 msec. Thus, ~63% of the openings had durations less than 2 msec and ~37% had longer open times.

then the midpoints of the histogram bars can be joined by a curve described by a single exponential decay with increasing time and k_{-1} or τ_o can be derived by simple curve-fitting procedures. A similar approach can be employed to determine the mean closed time, τ_c. If, however, the channel's behavior is more complicated, the open-time and closed-time histograms will not conform to single exponentials. For example, a channel could reside in one of three possible states—very closed, not quite as closed but still not open, and open. In this case, one exponential might describe the open-time probability histogram but two might be needed to describe the closed-time histogram.

Finally, by determining single-channel activity when the membrane is clamped at a number of different electrical potential differences one can obtain valuable information regarding the conductance and ionic selectivity of the channel. For example, Fig. 11 illustrates the relation between the size of ionic currents flowing through a single channel (i_c) and the membrane potential, V_m, when the solution facing the outer surface of the channel contains 15 mmol/L KCl and that facing the inner surface contains 150 mmol/L KCl. Note that in this example, the relation between i_c and V_m is linear (or ohmic) and can be described by a relation analogous to Eq. 13:

$$i_c = g_c(V_m - V_r) \qquad (15)$$

where g_c, the slope, is the conductance of the single channel. (Recall that the accepted convention is that the

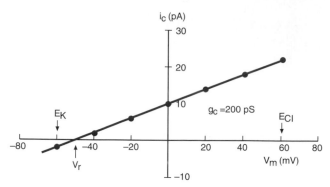

FIGURE 11   Relationship between single channel current, i_c, and the electrical potential difference across the channel, V_m, in the presence of asymmetric KCl solutions. E_K and E_{Cl} are the menst equilibrium potentials for K^+ and Cl^- and V_r is the electrical potential difference across the channel when current flow is zero (i.e., the "reversal" or "equilibrium" potential for the channel under the given conditions). g_c is the conductance of the channel.

outer or extracellular solution is considered ground [zero potential] so that V_m is the electrical potential of the inner compartment with respect to that of the outer compartment and that the flow of cations [i.e., a positive current] from the inner or intracellular compartment to the outer compartment results in an upward or positive deflection.)

Now, using the Nernst equation (11), we can calculate the equilibrium potentials for K^+ (i.e., E_K) and Cl^- (i.e., E_{Cl}); recall that E_K is the value of V_m at which there is no ionic flow if the membrane is permeable to K^+ but impermeable to Cl^-, and E_{Cl} is the value of V_m at which there is no ionic flow if the membrane is permeable to Cl^- and impermeable to K^+. As indicated on Fig. 11, for this tenfold concentration ratio, $E_K = -60$ mV and $E_{Cl} = 60$ mV. Note, however, that the observed value of V_m at which the current is zero (referred to as the *zero-current* or *reversal* potential, V_r) is -50 mV. The fact that V_r is not equal to either E_K or E_{Cl} indicates that the channel is not exclusively permeable to either K^+ or Cl^-. Moreover, the observation that V_r is much closer to E_K than to E_{Cl} indicates that the channel is much more permeable to K^+ than to Cl^-. The actual ratio of P_K/P_{Cl} can be obtained using the expression for the *diffusion potential* (Eq. 8), which can be written as follows:

$$V_r = \left\{ [(P_K/P_{Cl}) - 1]/[(P_K/P_{Cl}) + 1] \right\} \times 60 \log\left\{ (KCl)_o/(KCl)_i \right\} \quad (16)$$

Solving this equation for the condition, $(KCl)_o = 15$ mmol/L, $(KCl)_i = 150$ mmol/L, and $V_r = -50$ mV yields $(P_K/P_{Ci}) = 11$.

Clearly, if the channel were ideally permselective for K^+, the current–voltage relationship shown in Fig. 11

would have been described by the equation $i_K = g_K(V_m - E_K)$ and the intercept on the abscissa would be where $V_m = -60$ mV.

Determinants of Channel Selectivity

One amazing property of many biological ion channels is their ability to sharply distinguish among ions of the same charge whose dimensions differ by less than 0.1 nm. For example, as we will consider in greater detail below, some channels in nerve and muscle membranes may be more than 1000 times more permeable to K^+ than to Na^+ in spite of the fact that the former has a crystal radius of 0.133 nm, and the latter is *actually smaller*, having a radius of 0.095 nm!

The best explanation for the exquisite ability of many channels to discriminate among ions with a high degree of selectivity is the *closest fit* theory advanced by Hille. It has long been recognized that because water is a polar molecule, ions float around in aqueous solution in association with a cloud or shell of water molecules; that is, ions in aqueous solution are hydrated. Furthermore, the electrostatic attractions between ions and water molecules are quite strong (recall that the heat of solution of salts can be quite large), so a considerable amount of energy is needed to dehydrate an already hydrated ion. Now, Hille argues, suppose the steric arrangement of fixed charged groups in the region of the channel that determines ionic selectivity is such that an ion traversing that region can be stripped of its water of hydration without its knowing it. In other words, if, as illustrated in Fig. 12 for the case of a cation, the water of hydration can be replaced by the negative poles, or dipoles, of amino acids that line the channel

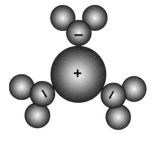 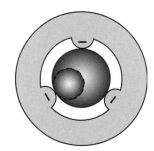

"Free" solution   Channel

FIGURE 12   The left side shows a monovalent cation in free solution with the hydration shell of three water molecules. On the right, the same cation is in a cylindrical channel where the interaction with electronegative fixed charges in the wall of the pore exactly mimics the interactions with the water dipoles. The dashed circle represents a smaller cation that cannot fit in the channel while retaining its hydration shell but is energetically uncomfortable without it.

(*e.g.*, carbonyl groups) in such a way that the cation is as energetically "comfortable" in the channel as it is in water, then the ion would not recognize the fact that it left its aqueous environment for that of the channel. (In the language of thermodynamics, the energy needed to partition into or out of the channel would be negligible so that, according to the Boltzman distribution, the probability of being in the channel is equal to the probability of being in the aqueous solution; this is analogous to an amphipathic molecule whose oil–water partition coefficient is unity so that it is just as comfortable in an aqueous solution as in a lipid solvent.) Clearly, if the radius of the dehydrated ion is too large to be accommodated by the channel, it will be excluded. But, if the radius of the dehydrated ion is too small to fit snugly (dashed circle in Fig. 12), it will be energetically disadvantaged compared to the ion with the closer fit; it will not be as willing to shed its comfortable coat of water molecules for the more foreign environment of the channel, and its partitioning into the channel will be energetically less favorable than that of the ion with the closer fit. An elegant discussion of the structure–function relations of a prototypical K^+ channel may be found in the paper by Doyle et al. (1998).

Regulation of Channel and Membrane Currents

Up to this point, we have considered single-ion channels whose currents are given by an equation having the form of Eq. 11. We also observed that a single channel is not open all of the time, but, instead, undergoes spontaneous transitions between open and closed states. It follows that the total current of an ion, *i*, across a unit area of membrane containing an ensemble of channels selective for *i* that are opening and closing at random, is given by:

$$I_i = N_i P_o i_i = N_i P_o g_i(V_m - E_i) \qquad (17)$$

where N_i is the total number of single channels per unit membrane area and P_o is the open-time probability. Comparing Eq. 17 with Eq. 13, we see that the total conductance of a membrane to *i* (G_i) is simply $N_i P_o g_i$. Furthermore, because (as we will soon see) P_o and g_i may be functions of V_m, a more general expression of Eq. 17 is:

$$I_i = N_i P_{o(V_m)} i_i = N_i P_o g_{i(V_m)}[V_m - E_i] \qquad (18)$$

Now, under steady-state conditions when C_i^o and C_i^i are constant so that E_i is constant (see Eq. 13), changes in I_i can result only from changes in N_i, P_o, and/or g_i. Under conditions where N_i is constant, physiologic

regulation of channel activity or (J_i) is the result of changes in P_o; that is, the channel is either nonconducting or fully conducting at fixed values of V_m and E_i and the parameter that is regulated is the fraction of time that the channel is in either of these two states.

The two major physiologic determinants of P_o are V_m and/or chemical regulators. Thus, channels are roughly categorized as voltage-gated or ligand-gated. It should be emphasized, however, that this classification is not ironclad; many channels that are predominantly considered voltage-gated are also influenced by chemical regulators, and some ligand-gated channels are influenced by V_m.

OSMOSIS

Osmosis refers to the flow or displacement of volume across a barrier due to the movements of matter in response to concentration differences. Although, in principle, any substance (solutes as well as solvents) may contribute to the volume of matter displaced during osmotic flow, the term *osmosis* has come to have a much more restricted meaning when applied to biologic systems. Because biologic fluids are relatively dilute aqueous solutions in which water comprises more than 95% of the volume, osmotic flow across biologic membranes has come to imply the displacement of volume resulting from the flow of water from a region of higher water concentration (a dilute solution) to a region of lower water concentration (a more concentrated solution).

Osmosis and the diffusion of uncharged solutes are closely related phenomena. Both are spontaneous processes that involve the flow of matter from a region of higher concentration to one of lower concentration; both are the results of random molecular movements and, hence, are dependent only on the thermal forces inherent in any system; and the end result of both processes is the abolition of concentration differences.

How do diffusion and osmosis differ? As we shall see, the answer to this question carries with it the key to understanding osmosis. Let us start by reexamining the definitions of these two processes. *Osmosis* refers to the flow of matter that results in a displacement of volume, and *diffusion* refers to a flow of matter in which displacements of volume are not involved. As it turns out, the key to understanding osmosis is the answer to the question, "Why is there no displacement of volume in diffusion?"

The answer to this question emerges when we carefully reconsider the molecular events involved in the diffusion of uncharged solutes. Referring once more to Fig. 3, when we say that the concentration of an uncharged solute in compartment *o* is greater than that in compartment *i*, we are, at the same time, implying that the concentration of water (or solvent) in compartment *i* is greater than that in compartment *o*. When there is a concentration difference across a membrane for one component of a binary solution (one solute and one solvent), there must also be a concentration difference for the other component. Thus, there are two concentration differences, two driving forces, and two diffusional flows: solute diffuses from compartment *o* to compartment *i* and water diffuses from compartment *i* to compartment *o*. In short, when we say that there is diffusion of a solute down a concentration difference we are describing only one half of the mixing process and are overlooking the fact that diffusion (or mixing) in the closed system illustrated in Fig. 3 is, in fact, interdiffusion. Clearly, in the closed system illustrated in Fig. 3, if the membrane is rigid, mixing or interdiffusion of solute and solvent must take place without any change in the volumes of compartments *o* or *i*.

To appreciate how diffusion can result in the displacement of volume, let us consider the system illustrated in Fig. 13. Compartment *o* is open to the atmosphere and contains pure water. Compartment *i* is closed by a movable piston and contains an aqueous solution of some uncharged solute. The membrane separating the two compartments is assumed to be freely permeable to water but impermeable to the solute (i.e., a semipermeable membrane). Although there are two concentration differences, there can be only one flow. Water will flow from compartment *o* to compartment *i*, driven by its concentration difference. The volume of water associated with this flow, however, cannot be counterbalanced by a flow of solute, so there will be a net displacement of volume; the volume of compartment *o* will decrease, whereas that of compartment *i* will increase. This volume flow, referred to as *osmosis* or *osmotic flow*, arises because the properties of the barrier or membrane are such as to prevent interdiffusion; mixing, which is the end to which all spontaneous processes are directed, can now only come about as the result of one flow rather than two flows.

Now, we can prevent the flow of volume from compartment *o* to compartment *i* by applying a sufficient pressure on the piston. The pressure that must be applied to prevent the flow of volume is defined as the *osmotic pressure*. When the solutions on both sides of the membrane are relatively dilute (as in the case of biologic fluids) and when the membrane is absolutely impermeable to the solute (i.e., interdiffusion is completely prevented), the osmotic pressure is given by the following expression, which was derived by the Dutch Nobel Laureate Jacobus Henricus van't Hoff (1852–1911):

$$\Delta\pi = RT\Delta C_i \qquad (19)$$

where $\Delta\pi$ denotes the osmotic pressure; R is the gas constant; T is the absolute temperature; and ΔC_i, is the concentration difference of the impermeant, uncharged solute across the membrane in moles per liter. At 37°C ($T = 310K$), this expression reduces to:

$$\Delta\pi = 25.4\Delta C_i \text{ (atm)}$$

Thus, if the concentration difference across the membrane is 1 mol/L, the pressure that must be applied on the piston to prevent osmotic flow of water is 25.4 atm.

At this point, we digress briefly to consider the full meaning of the term ΔC_i in Eq. 19. The ability of a solution of any solute to exert an osmotic pressure across a semipermeable membrane (i.e., a membrane permeable to the solvent [water] but ideally impermeable to the solute) is a *colligative property* of the solution that is dependent on the concentration of individual solute particles (other such colligative properties are the vapor-pressure depression, boiling-point elevation, and freezing-point depression of solutions). We must therefore introduce a new unit of concentration that is a measure of the number of free particles in the solution and thus reflects the osmotic effectiveness of a dissolved solute. For a nondissociable solute such as urea (molecular weight, 60), when we dissolve 60 g (i.e., 1 g weight) of this solute in a sufficient amount of water to yield a total volume of 1 L, we have a 1-*M* solution of urea that contains about 6×10^{23} individual particles (Avogadro's number). However, when we dissolve 58 g of NaCl (molecular weight, 58) in a sufficient amount of water to yield a total volume of 1 L, we obtain a 1-*M* solution of NaCl, but the number of individual particles in solution will be twice that of a 1-*M* solution of urea. Thus, one must distinguish between the molarity (the number of gram weights of a solute in a liter of aqueous solution) and the osmolarity (the number of individual particles per liter that results from dissolving 1 g weight of a solute in that volume). A 0.15-*M* solution of NaCl

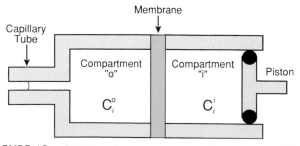

FIGURE 13   Apparatus for determining osmotic pressure and flow.

that consists of 0.15 mol/L Na^+ particles and 0.15 mol/L Cl^- particles has approximately the same osmolarity as a 0.3-M sucrose solution (*i.e.*, 300 mOsm/L). Indeed, historically, this observation provided the crucial evidence for the Arrhenius theory of the dissociation of salts in solution into their constituent ions.

In short, the osmolarity or the osmotically effective concentration of a solution of a dissociable salt will be n times its molarity, where n represents the number of individual ions (particles) resulting from the dissociation of the salt.

Finally, before considering osmotic flow across biologic membranes, it is important to gain deeper insight into the nature of osmotic pressure. When, as in the example given above, sufficient pressure is applied to the piston to prevent volume flow, the system is in equilibrium. There will be no flow of water despite the presence of a concentration difference for water across the membrane. This is because the pressure applied to compartment i exerts the same driving force for the flow of water as does the difference in water concentration, but in the opposite direction. Therefore, when the osmotic pressure is applied, there is no longer a net driving force for the movement of water, and volume flow ceases.

One may find this easier to understand by arbitrarily dividing the water flows into two hypothetical streams. One stream is directed from compartment o into compartment i and derives its driving force from the difference in the concentration of water across the membrane. The second stream is directed from compartment i to compartment o and is driven by the pressure applied by the piston. Flow ceases when these two oppositely directed streams are of equal magnitude; the pressure needed to achieve this equilibrium state is given by van't Hoff's law (Eq. 19). This balancing of driving forces is analogous to the condition discussed with reference to the Nernst equilibrium potential, where an electrical potential difference counterbalances the driving force arising from a concentration difference of a permeant ion when the counterion is impermeant.

Osmotic Flow and Osmotic Pressure Across Nonideal Membranes

Up to this point, we have considered the case of osmotic flow across a semipermeable membrane—that is, one that is permeable to solvent but impermeable to the solute so that interdiffusion is completely prevented. What would happen if the membrane in Fig. 13 could not distinguish at all between a solvent molecule and a solute molecule; that is, both could cross with equal ease? Suppose, for example, that compartment o

contains pure water so that the concentration of water in that compartment $(C_{H_2O}) = 55.6$ M. And, suppose that compartment i contains a 1-M solution of deuterium oxide (D_2O, or heavy water) in water. Under these circumstances, $C_{H_2O} = 54.6$ M, so $\Delta C_{H_2O} = \Delta C_{D_2O} = 1$ M. Thus, H_2O would diffuse from compartment o to compartment i driven by a concentration difference of 1 M, and D_2O would diffuse from compartment i to compartment o driven by an equal but oppositely directed concentration difference. Because both species can cross the membrane with equal ease, interdiffusion or mixing would not be restricted and would take place with no displacement of volume. In short, when the membrane cannot distinguish between solute and solvent, osmotic flow, $J_v = 0$ and, obviously, $\Delta \pi = 0$.

Clearly, the situation in which the membrane is ideally impermeable to the solute and the one in which the membrane cannot distinguish between solute and solvent are extreme examples. In most instances, we are faced with a situation somewhere between these two extremes—that is, a situation in which interdiffusion is to some extent, but not completely, restricted. Under this condition, for a given concentration difference the volume displacement, or osmotic flow, will be somewhere between zero and the maximum that would be observed with an ideal semipermeable membrane. Also, the osmotic pressure (the pressure necessary to abolish osmotic flow) will be between zero and that predicted by van't Hoff's law.

In 1951, the Dutch physical chemist Staverman provided a quantitative expression for the osmotic pressure across nonideal membranes, which was based on the following reasoning: Let us perform the ultrafiltration experiment illustrated in Fig. 14. In the upper cylinder, we place an aqueous solution of a solute at a concentration of C_i. We then apply pressure to the piston, force fluid through the membrane, and collect the filtrate. If the membrane does not restrict the movement of solute relative to that of water, then the concentration of solute in the filtrate (C_f) will be equal to C_i; that is, both components passed through the membrane in the same proportion as they existed in the solution of origin. At the other extreme, if the membrane is impermeable to the solute, the filtrate will be pure water (i.e., $C_f = 0$). Between these two extremes is the condition where the membrane partially restricts the movement of the solute relative to that of water, and under these conditions C_f will be lower than C_i but greater than zero. We can now define a parameter that tells us something about the relative ease with which water and solute i can traverse the membrane:

$$\sigma_i = 1 - \{C_f/C_i\} \qquad (20)$$

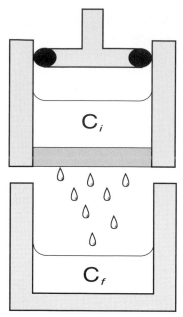

FIGURE 14   Determination of the reflection coefficient of a solute i, employing ultrafiltration through a membrane.

where σ_i is the *reflection coefficient* of the membrane to i, because it is a measure of the ability of the membrane to "reflect" the solute molecule i; that is, it tells us how perfect the membrane is as a molecular sieve. Clearly, the reflection coefficient must have a value between zero (for the case where the membrane does not distinguish between the solute and water) and unity (for a membrane that is absolutely impermeant to the solute). Staverman then showed that if the same membrane used in the ultrafiltration experiment is mounted in the apparatus shown in Fig. 13, and if two solutions of the same solute at two different concentrations are placed in compartments o and i, the *effective osmotic pressure* necessary to prevent volume flow is given by:

$$\Delta\pi_{\text{eff}} = \sigma_i RT\Delta C_i \qquad (21)$$

Because σ_i, must have a value between zero and unity, the effective osmotic pressure across a real membrane must fall between zero and that predicted by van't Hoff's law for ideal semipermeable membranes; the exact value depends on the concentration difference and the relative permeability of the membrane to water and the solute i given by σ_i.

In short, Eq. 21 permits quantitation of the effect of interdiffusion between solutes and solvent (water) across membranes on the effective osmotic pressure that is exerted across those membranes. If a membrane is equally permeable to the solvent (water) and the solute (i), then $\sigma_i = 0$, and the presence of a concentration difference for i across the membrane will not generate an osmotic pressure. Conversely, if the membrane is impermeant to i, then interdiffusion is prohibited, and the effective osmotic pressure across the membrane will be given by van't Hoff's equation.

Volume Flow in Response to a Difference in Pressure

Volume flow (J_v) across a membrane in response to a difference in hydrostatic pressure, ΔP, is given by the linear relation:

$$J_v = K_f\Delta P \qquad (22)$$

where K_f is a proportionality constant referred to as the hydraulic conductivity of the membrane (or, the filtration coefficient); Eq. 22 is analogous to Fick's law of diffusion (Eq. 2) inasmuch as it describes a linear relation between a flow and its driving force, which in the case of volume flow is the pressure difference across the barrier. Now, ΔP can be the difference in hydrostatic pressure, the difference in osmotic pressure, or a combination of both. For example, referring to Fig. 13, if the concentration in compartment o is equal to that in compartment i so that $\Delta\pi = 0$, application of pressure to the piston will bring about the flow of volume from i to o given by Eq. 21. Alternatively, if $C_i^o < C_i^i$ and no pressure is applied to the piston, there will be a flow of volume from o to i given by:

$$J_v = K_f\sigma_i\left(C_i^o - C_i^i\right) \qquad (23)$$

An important empirical observation made many years ago is the equivalence of osmotic and hydrostatic pressure as the driving forces for volume flow. That is, J_v across a given membrane will be the same when it is driven by a hydrostatic pressure difference as when driven by the equivalent osmotic pressure difference. In other words, for a given membrane, the same value of K_f applies to both forces. Thus, we can combine Eqs. 22 and 23 and derive a general equation that describes the situation when there are both osmotic and hydrostatic pressure differences across the membrane:

$$J_v = K_f(\Delta\pi_{\text{eff}} - \Delta P) \qquad (24)$$

where $\Delta\pi_{\text{eff}} = \sigma_i RT\left(C_i^o - C_i^i\right)$.

Thus, when $\Delta P = \Delta\pi_{\text{eff}}$, $J_v = 0$; this is the definition of osmotic pressure. When $\Delta P \neq \Delta\pi_{\text{eff}}$, there will be a flow from one compartment to the other, driven by the difference.

Equation 24 provides a general description of the effects of hydrostatic and osmotic forces on volume flow across all membranes. In the physiologic sciences, it is often referred to as the Starling equation, after the great British physiologist Ernest Starling, who applied it to the study of fluid movements across the walls of capillaries.

CARRIER-MEDIATED TRANSPORT

During the first quarter of this century, compelling evidence began to surface that life would not be possible if diffusion were the only mechanism available for the exchange of solutes across cell boundaries. There are two sets of observations that most strongly implied the necessity for additional transport mechanisms:

1. Most biologic membranes are virtually impermeable to hydrophilic molecules having molecular radii significantly greater than 4 Å or that have five or more carbon atoms. Thus, virtually no essential nutrients and building blocks (*e.g.*, glucose, amino acids) can penetrate biologic membranes to any significant extent by diffusion, so other mechanisms are necessary to provide for their entry into cells. Similarly, biologic membranes are generally impermeant to essential multivalent ions such as phosphate, so their movements across cell membranes must also be mediated by mechanisms other than diffusion.
2. The intracellular concentrations of many water soluble solutes differ markedly from their concentrations in the extracellular medium bathing the cells. For example, as discussed in Chapter 1, a characteristic of virtually every cell in the animal and plant kingdoms is that the intracellular K^+ concentration greatly exceeds that in the extracellular fluid (in some cases by a factor of more than 1000 to 1), and in cells from higher animals the intracellular Na^+ concentration is much less than that in the bathing media (often by a factor greater than 10). This ionic asymmetry is essential for a number of vital processes and, as we shall see, is the basis of many bioelectric phenomena that play an essential role in nerve conduction and muscle contraction. Diffusional processes alone cannot be responsible for the production and maintenance of these asymmetries.

To accommodate these two sets of observations, a concept referred to as *carrier-mediated transport* or the *carrier hypothesis* evolved. This hypothesis is generally attributed to Osterhout, who, in 1933, suggested that biologic membranes contain components ("carriers") that are capable of binding a solute molecule at one side of the membrane to form a carrier–solute complex,

which then crosses the membrane, dissociates, and discharges the transported solute on the other side.

Since then an overwhelming body of evidence for the role of membrane components, or carriers, in biologic transport processes has accumulated. Carriers have been implicated in the transport of a wide variety of solutes, and the specific properties of numerous carrier systems have been described in considerable detail. Needless to say, a comprehensive discussion of biologic carriers is beyond the scope of this presentation, and we will limit ourselves to a brief consideration of some of the general characteristics of carrier-mediated transport processes. Specific systems will be considered in later sections of this volume dealing with specific tissues or organs.

SOME GENERAL CHARACTERISTICS OF CARRIER-MEDIATED TRANSPORT

The following features are so widely characteristic of carrier-mediated transport processes that they are generally considered sufficient and often necessary criteria for the implication of carriers in the transport of a given solute:

1. Virtually all carriers appear to display a high degree of structural specificity with regard to the substances they will bind and transport. For example, the carriers responsible for the transport of glucose into animal cells are highly stereospecific; they will rapidly bind and transport the dextrorotary form (D-glucose) but have little affinity for the levorotary form (L-glucose). Conversely, the carriers responsible for the transport of amino acids into animal cells possess a high degree of selectivity in favor of the L-stereoisomer and little affinity for the D-stereoisomer.
2. All carrier-mediated transport processes exhibit *saturation kinetics*; that is, the rate of transport gradually approaches a maximum as the concentration of the solute transported by the carrier increases. Once this maximum rate is achieved, a further increase in the solute concentration has no effect on the transport rate. Plots of the rate of transport against concentration often closely resemble the hyperbolic plots characteristic of Michaelis–Menten enzyme kinetics, and, under these conditions, the kinetics of the transport process can be described by defining the maximum transport rate (J_{max}) and the substrate concentration at which the transport rate is half-maximum (K_t). Thus, $J_i = [J_{max}(C_i)/(K_t + C_i)]$, where C_i is the concentration of the solute.

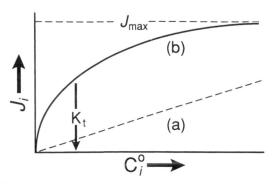

FIGURE 15 Linear relation between flux and concentration characteristic of diffusional processes (a). Relation between flux J_i, and concentration C_i for a saturable process illustrating J_{max} and K_t (the concentration at which J_i is one half J_{max} (b).

A graph illustrating the saturation kinetics characteristic of carrier-mediated transport is shown in Fig. 15b. In contrast, as illustrated by Fig. 15a, transport due to simple diffusion is usually (but not always) characterized by a linear relation between transport rate and solute concentration as predicted by Eq. 4. It should be emphasized that ionic diffusion through channels may exhibit saturation but usually only when concentrations are well beyond the physiologic range.

The saturation phenomena observed in carrier-mediated transport processes reflect the presence of a fixed and limited number of carrier molecules or binding sites in the membrane. When the solute concentration is sufficiently high so that all of the carrier sites are occupied (or complexed), a further increase in concentration cannot elicit a further increase in transport rate.

One consequence of the presence of a limited number of carrier molecules for a given class of transported solutes is the phenomenon of competitive inhibition. This is observed when two or more solutes that are capable of being transported by the same carrier are present simultaneously, competing with one another for the limited number of available binding sites. This phenomenon is closely analogous to competitive inhibition in enzyme–substrate interactions and often may also be described by classic Michaelis–Menten kinetics.

In addition to providing evidence for carrier-mediated transport, the phenomenon of competitive inhibition has proved to be extremely useful for the purpose of defining the transport specificity of a given carrier mechanism. Thus, if two solutes A and B are each transported by carriers and exhibit mutual, classic *competitive inhibition,* one may conclude that the same carrier mechanism is involved; if they in no way compete with each other, at least two distinct carrier mechanisms must be involved. For example, all of the D-hexoses that are absorbed by the small intestine mutually compete with one another for the same limited transport system.

When glucose and galactose are separately present in the intestinal lumen at high concentrations, they are each absorbed at approximately the same maximum rate. On the other hand, if the same concentrations of glucose and galactose are instilled into the intestinal lumen in the form of a mixture, each will be absorbed at a rate significantly lower than that observed when they were present separately. The total rate of sugar transport will be equal to the maximum rates observed when each sugar was present separately, indicating that the two sugars are competing for the same carrier system and are sharing in the saturation of the total number of available sites. On the other hand, the transport of glucose is not inhibited by the presence of hexoses that are not subject to carrier-mediated transport.

Nature of Membrane Carriers

The characteristics of carrier-mediated transport processes that we have just described—the high degree of structural specificity and saturation kinetics and competitive inhibition—strongly resemble the characteristics of enzyme–substrate interactions. After the introduction of the carrier hypothesis, it was suspected that carriers were enzyme-like molecules that comprise part of the protein portion of the lipoprotein membrane. However, for many years, these carriers defied isolation and characterization, and until recently there was a relatively large group of investigators who doubted, and even denied, their existence, but the results of numerous studies during the past few decades have dispelled these doubts. The development of techniques for isolating cell membranes and gently detaching their protein components has led to the isolation of integral proteins that are capable of specifically binding transported solutes. In many instances, these purified proteins have been reinserted into artificial lipid membranes, and these artificial (reconstituted) systems are capable of mediating the transport of specific solutes.

In short, considerable progress has been made toward defining the biochemical and/or molecular basis of carrier-mediated transport. The precise mechanism(s) by which the transported solutes are translocated across the membrane after binding, however, remains a mystery. But, in light of our current understanding of the assembly of proteins in biologic membranes, it is certain that the notions that carriers are ferry boats or that integral proteins flip-flop across the lipid bilayer are incorrect. It is more likely that carriers are integral proteins that in many respects resemble channels and that binding and translocation of solutes from one side of a gate to the other takes place within these channels.

Facilitated Diffusion and Active Transport

As discussed above, the two functions that membrane carriers must fulfill are (1) to provide a mechanism by which otherwise impermeant solutes can enter or leave cells across membranes, and (2) to provide a mechanism by which substances can be actively transported into or out of cells. The two classes of carrier-mediated transport processes that fulfill these functions are referred to as *facilitated diffusion* and *active transport,* respectively.

Facilitated Diffusion

Facilitated diffusion is the term reserved for carrier-mediated processes that are only capable of transferring a substance from a region of higher concentration to one of lower concentration. These processes are sometimes also referred to as *equilibrating carrier systems,* inasmuch as net transport ceases when the concentrations of the transported solute are the same on the two sides of the membrane—that is, when the system is equilibrated with respect to the solute in question. Thus, facilitated diffusion resembles noncarrier-mediated diffusional processes in that the direction of net flow is always downhill. It differs from them in that it exhibits all of the characteristics of carrier-mediated processes and often results in the transmembrane transfer of a solute that could not otherwise permeate the membrane; indeed, the latter is its sole function.

The classic example of facilitated diffusion is glucose transport across the membranes surrounding many animal cells, such as erythrocytes, striated muscle, and adipocytes. The glucose concentrations in these cells are much lower than the glucose concentrations in the extracellular fluid because glucose is rapidly metabolized by these cells after gaining entry. The only well-documented exceptions to this statement are renal proximal tubular cells, small intestinal epithelial cells, and the cells of the choroid plexus. These cells are responsible for transepithelial glucose absorption, and their intracellular glucose concentrations may exceed those in the extracellular fluid; as is discussed below, mechanisms other than facilitated diffusion are responsible for the uptake of hexoses by these cells. Thus, for most cells, the problem is not that of transporting glucose against a concentration difference but of transporting glucose rapidly across an essentially impermeant barrier. This is accomplished by a carrier mechanism, illustrated in Fig. 16. Glucose (represented by the small circle labeled S) combines with the carrier from one side of the membrane to form a glucose–carrier complex. This is followed by a change in conformation of the complex that permits glucose to

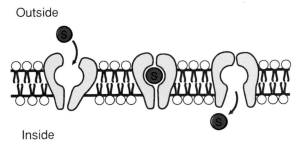

FIGURE 16   Model for carrier-mediated facilitated diffusion of a solute *S.*

dissociate from the carrier and enter the solution on the other side of the membrane. The free carrier site is then available for another passenger. Because there are a limited number of carriers, the process is saturable and subject to competitive inhibition.

One of the important features of the carrier model for facilitated diffusion, as illustrated in Fig. 16, is that the carrier itself is unaltered during the translocation process, and only thermal energy is required for the conformational change that exposes the binding site to one or the other side of the membrane. Thus, the transport process is symmetrical, and it is just as easy for the solute (S) to move from the extracellular fluid into the cell as in the opposite direction. Consequently, when the concentrations of solute on the two sides of the membrane are equal, the system is entirely symmetrical, and the carrier-mediated flows in both directions will be equal. This is the reason why this transport process is not capable of bringing about net transport from a region of lower concentration to one of higher concentration and why net transport ceases when the solute distribution is equilibrated.

To date, five carrier proteins capable of mediating the facilitated diffusion of glucose have been identified; in the jargon of molecular biology, they are referred to as GLUT1-5. All five consist of a polypeptide chain composed of approximately 500 amino acids and possess a high degree of homology, including 12 putative transmembrane-spanning segments. Their specific properties and differences are pointed out in other chapters.

Active Transport

Active transport is the term reserved for carrier-mediated transport processes that are capable of bringing about the net transfer of an uncharged solute from a region of lower concentration to one of higher concentration or the transfer of a charged solute against combined chemical and electrical driving forces. Thus, active transport processes are capable of counteracting or reversing the direction of diffusion, a spontaneous process, and therefore are capable of performing work. The concept that active transport processes perform work may be difficult to grasp for those who have not had some acquaintance with thermodynamics. Because this is an extremely important concept, we digress for a moment and attempt to provide it an intuitive basis.

The direction of all natural change in the universe is for systems to move from a state of higher energy to one of lower energy. Thus, an unsupported weight will fall from a position of higher gravitational (or potential) energy to one of lower gravitational energy; electrons will flow through a conductor from a region of electronegativity (the cathode of a battery) to one that is electropositive (the anode); uncharged solutes will diffuse from a region of higher concentration to one of lower concentration. All of these processes are spontaneous inasmuch as they are accompanied by a decrease in the free energy of the system and do not require any external assistance or intervention; they will occur in a completely isolated system. It is a universal experience (and one of the basic tenets of thermodynamics) that once a spontaneous change has taken place, the initial conditions cannot be restored without an investment of energy; that is, the only way one can reverse a spontaneous process is by performing work. In the examples cited above, mechanical work is required to restore the weight to its original height, and electrical work is needed to recharge the battery. A thoroughly mixed solution can be unmixed by ultracentrifugation, ultrafiltration through an appropriate molecular sieve, or distillation, but, whatever means are chosen, it is clear that unmixing will never occur spontaneously and that the result of diffusion can only be reversed through the investment of energy. The means by which a biologic cell reverses diffusional flows is referred to as active transport; here, too, work is performed, and energy derived from metabolism must be invested.

It follows that the ability of a cell to carry out active transport processes is dependent on an intact supply of metabolic energy, and all active transport processes can be inhibited by deprivation of essential substrates or through the use of metabolic poisons. Indeed, the *sine qua non* of active transport is the presence of a direct or indirect linkup or *coupling* between the carrier mechanism and cell metabolism; when an active transport process is initiated or accelerated, there is a concomitant increase in the metabolic rate (as measured by glucose utilization, oxygen consumption, etc.), and inhibition of active transport results in a decrease in the metabolic rate. The two classes of active transport processes are *primary active transport* and *secondary active transport,* which differ in the ways they derive (or are coupled to) a supply of energy.

Primary Active Transport.

Primary active transport implies that the carrier mechanism responsible for the movement of a solute against a concentration difference or a combined concentration and electrical potential difference (for the case of ions) is directly coupled to metabolic energy. The best-studied primary active transport processes in animal cells include:

1. The carrier mechanism found in virtually all cells from higher animals that is responsible for maintaining their low cell Na^+ and high cell K^+ concentrations
2. Carrier mechanisms found in sarcoplasmic reticulum and many plasma membranes responsible for active transport of Ca^{2+}
3. Carrier mechanisms capable of actively extruding protons from the cells present in the gastric mucosa and renal tubule and actively pumping protons into intracellular organelles (*e.g.*, lysozomes)

These carrier proteins have been purified, and all possess adenosine triphosphatase (ATPase) activity; that is, the same protein that is involved in the binding and translocation of Na^+ and K^+ is also capable of hydrolyzing adenosine triphosphate (ATP) and using the chemical energy released to perform the work of transport. The same holds for the Ca^{2+} pumps and the proton pumps. The precise mechanisms whereby the chemical energy of the terminal phosphate bond of ATP is converted into transport work is not clear.

Secondary Active Transport.

Secondary active transport refers to processes that mediate the uphill movements of solutes but are not directly coupled to metabolic energy; instead, the energy required is derived from coupling to the downhill movement of another solute. Let us illustrate such systems by considering Fig. 17. Figure 17A portrays a rotating carrier molecule (C) within a membrane that has two binding sites, one for Na^+ and the other for a solute (S), which, at this instant, are shown facing compartment *o*. Let us assume that the carrier can rotate only when both binding sites are empty or filled and is immobile when only one site is filled. Thus, it can transport Na^+ and S from compartment *o* to compartment *i* in a one-to-one fashion. Now let us assume that there is no (or very little) Na^+ in compartment *i* and that every Na^+ that enters from compartment *o* is removed.

Clearly, under these conditions, S can only move (to any appreciable extent) from compartment *o* to compartment *i*; because there is no Na^+ in compartment *i*, S that enters this compartment cannot move back out and is trapped. In time, the concentration of S in

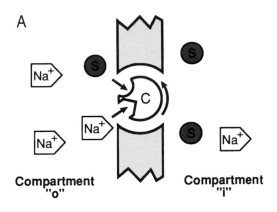

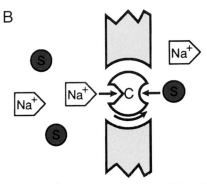

FIGURE 17   Na$^+$-coupled co-transport (A) and countertransport (B). S, solute; C, carrier molecule.

us further assume that the carrier can rotate only when both sites are either empty or filled but not when only one site is filled. Clearly, this mechanism can bring about a one-to-one exchange of Na$^+$ for S across the membrane. If there is little or no Na$^+$ in compartment i, the system will only be able to exchange Na$^+$ in compartment o for S in compartment i and not vice versa. Thus, the downhill movement of Na$^+$ from o to i can bring about uphill flow of S from i to o; this mechanism is referred to as *secondary active counter-transport*. And, once more, the trick is the removal of Na$^+$ from compartment i, which is accomplished by the primary active Na$^+$ transport mechanism that is energized by ATP hydrolysis. Two examples of such a countertransport system are (1)Na$^+$–H$^+$ exchange and (2) Na$^+$–Ca^{2+} exchange; both of these mechanisms have been found in a wide variety of cell types. In both instances, H$^+$ and Ca^{2+} are extruded from the cell coupled to the downhill influx of Na$^+$ into the cell.

Cellular models of these co- and countertransport processes are illustrated in Fig. 18. The essential common feature of these transport processes is that metabolic energy is directly invested into the operation of the (Na$^+$–K$^+$) pump, a primary active transport mechanism. The operation of this pump results in a cell Na$^+$ concentration that is much lower than that in the extracellular fluid. This transmembrane Na$^+$

compartment i will exceed that in compartment o so that the system will have actively transported S without a direct linkup to metabolic energy. This system is referred to as *secondary active cotransport*.

How is this possible? We have stipulated that every Na$^+$ that enters compartment i is removed. In animal cells, which is accomplished by the (Na$^+$–K$^+$) pump that is directly linked to (energized by) ATP hydrolysis. Thus, in essence, energy is directly invested into a primary active transport mechanism that is responsible for extruding Na$^+$ from the cell (in exchange for K$^+$) and thereby maintaining a low intracellular Na$^+$ concentration. The Na–S cotransport mechanism can then bring about the uphill movement of S energized by the downhill flow of Na$^+$. There are many Na$^+$-coupled secondary active cotransport processes in animal cell membranes. They include sugar and amino acid uptake across the apical membranes of small intestinal and renal proximal tubule cells; the uptake of many L-amino acids by virtually all nonepithelial cells; and Cl$^-$ uptake by a variety of epithelial and nonepithelial cells.

Figure 17B illustrates another mechanism for secondary active transport that operates along similar principles. Let us assume that the carrier C has two sites, one facing compartment o and the other compartment i. Let

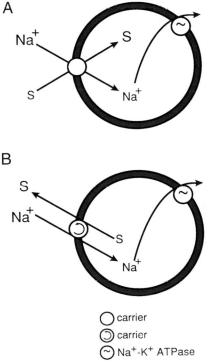

○ carrier
◎ carrier
◉ Na$^+$-K$^+$ ATPase

FIGURE 18   Cellular models of Na$^+$-coupled co-transport (A) and countertransport (B).

gradient in turn provides the energy for many Na^+-coupled secondary active co- and countertransport processes. (The terms *symport* and *antiport* are sometimes employed to describe co- and countertransport processes, respectively.)

THE PUMP–LEAK MODEL, THE ORIGIN OF TRANSMEMBRANE ELECTRICAL POTENTIAL DIFFERENCES, AND THE MAINTENANCE OF CELL VOLUME

All biologic membranes possess an assortment of pathways that permit the diffusion of water-soluble solutes, mainly ions (*i.e.*, "leak" pathways) and carriers. These components, acting in concert, are responsible for the uptake of essential nutrients and building blocks by cells, the extrusion of the end products of some metabolic processes from cells, the maintenance of a near-constant (time-independent, or steady-state) intracellular composition and volume, and the establishment of transmembrane electrical potential differences.

In this section, we illustrate the interactions among carrier-mediated *pumps* and channel-mediated *leaks* by considering the processes responsible for the maintenance of the high intracellular concentrations of K^+ and the low intracellular concentrations of Na^+ characteristic of virtually all cells in higher animals. We have chosen this system as a prototype for other pump–leak systems, not only because of its ubiquity but also because of the essential role it plays in energizing a wide variety of other secondary active pumps, in essential bioelectric processes, and in the maintenance of cell volume.

The (Na^+–K^+) Pump

The fact that cells from higher animals contain a high K^+ concentration and a low Na^+ concentration compared to the extracellular fluid was established in the 20th century shortly after analytic techniques for measuring these elements were developed. In the period from 1940 to 1952, Steinbach demonstrated that, when frog striated muscle is incubated in a K^+-free solution, the cells simultaneously lose K^+ and gain Na^+, a process that could be reversed by the addition of K^+ to the extracellular fluid. In the ensuing decade, abundant evidence accrued for the presence of carrier-mediated processes in biologic membranes that bring about the extrusion of Na^+ from cells obligatorily coupled to the uptake of K^+ by cells that are directly dependent on ATP; this mechanism is referred to as the (Na^+–K^+) pump. Furthermore, the stoichiometry of this process is that three Na^+ ions are

extruded in exchange for two K^+ ions for each ATP consumed.

In 1958, Skou identified an ATPase in a homogenate of crab nerve tissue whose hydrolytic activity was dependent on the simultaneous presence of Na^+ and K^+ in the assay medium. In addition, ATPase activity in the presence of Na^+ and K^+ could be inhibited by glycosides derived from the wild flower *Digitalis purpurea* (foxglove) (*e.g.*, ouabain), known since 1953 to be potent inhibitors of the carrier-mediated transport mechanisms responsible for the active extrusion of Na^+ from cells coupled to the active uptake of K^+ (*i.e.*, the [Na^+–K^+] pump).

During the past five decades, innumerable studies have incontrovertibly established that the (Na^+–K^+) pump and the (Na^+, K^+)-ATPase are one and the same. Furthermore, this (pump) ATPase has been isolated, purified, and reconstituted in active form in artificial lipid vesicles. It is now clear that it consists of two subunits: α and β. The α subunit has a molecular weight of approximately 100,000 Da, is minimally, if at all, glycosylated, and is the subunit that possesses the ATPase (catalytic) activity as well as the ability to bind Na^+, K^+, and digitalis glycosides such as ouabain. The β subunit has a molecular weight of about 55,000 Da, of which approximately two-thirds can be attributed to polypeptides and one-third to glycosylation; the β subunit has no ATPase activity, and its function may be to direct and insert (or anchor) the α subunit to the plasma membrane. Coassociation of the α and β subunits is necessary for pump activity.

The results of studies on the biochemical behavior of this ATPase are consistent with the very simplified sequence of events illustrated in Fig. 19. In the presence of intracellular Na^+ and Mg^{2+}, the (Na^+, K^+)-ATPase (E) is capable of hydrolyzing ATP to form the high-energy intermediate (E–P), which is capable of binding Na^+. The interaction between (E–P) and extracellular K^+ results in its hydrolysis, thereby reforming E and completing the cycle. Thus, the recycling of this enzyme-pump from the E stage through (several) high-energy

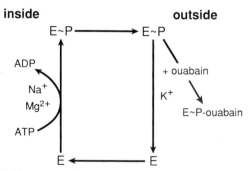

FIGURE 19   Simplified schematic of the partial reactions involved in coupled Na^+ – K^+ transport by (Na^+, K^+)-ATPase.

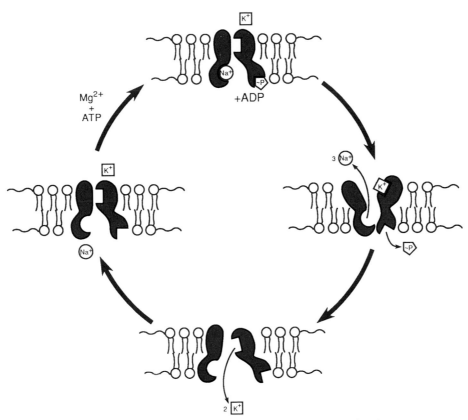

FIGURE 20 The possible operation of the $Na^+ - K^+$ pump, (Na^+, K^+)-ATPase.

(E–P) stages back to the original E stage requires the simultaneous presence of Na^+ in the intracellular compartment and K^+ in the extracellular compartment. Digitalis glycosides (such as ouabain) appear to bind tightly with (E–P) at or near the site where K^+ interacts with the enzyme and thereby prevent the conversion of (E–P) to E, which aborts the cycle. This sequence of events is shown in Fig. 20.

The Pump–Leak Model

The plasma membranes surrounding cells of higher animals not only contain (Na^+-K^+) pumps but are also traversed by channels that permit the diffusional flows (leaks) of Na^+ and K^+ across those barriers. The interaction between these pumps and leaks is illustrated in Fig. 21. Briefly, the (Na^+-K^+) pumps extrude Na^+ from the cells and simultaneously propel K^+ into the cells at the expense of metabolic energy (ATP). This results in a low intracellular Na^+ concentration and a high intracellular K^+ concentration, which in turn sets the stage for Na^+ diffusion into the cell and K^+ diffusion out of the cell through their respective leak pathways. The final result of the interactions between the pump and leaks is a time-independent or steady-state condition, where the movements of Na^+ and K^+

mediated by the pumps are precisely balanced by the oppositely directed flows of these ions through their leak pathways.

This pump–leak system is present in all cells in higher animals and is responsible for maintaining the low cell Na^+ and high cell K^+ concentrations characteristic of those cells. As discussed previously, the resulting Na^+ gradient (or concentration difference)

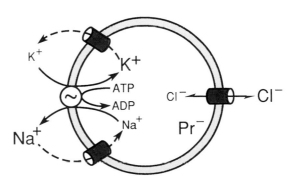

FIGURE 21 Model of a cell containing an ATP-dependent $(Na^+ - K^+)$ exchange pump, leak pathways for Na^+ and K^+, and a pathway for Cl^- diffusion across the membrane; Pr^- denotes negatively charged intracellular macromolecules (mainly proteins). Reasonable values for intracellular and extracellular K^+ are 140 and 5 mmol/L, respectively; for intracellular and extracellular Na^+, 15 and 140 mmol/L, respectively.

across the plasma membrane can serve to energize a number of secondary active transport processes, such as the Na^+-coupled accumulation of amino acids (cotransport), the extrusion of H^+ produced by metabolic processes via the Na^+–proton countertransport mechanism, and the regulation of cell Ca^{2+} by the Na^+–Ca^{2+} countertransport mechanism. In the final analysis, all of these secondary active transport processes derive their energy from the ATP hydrolyzed by the (Na^+–K^+) pump—a truly remarkable design.

Transmembrane Electrical Potential Differences

All biologic membranes are characterized by transmembrane electrical potential differences, which are, in almost all instances, oriented so that the cell interior is electrically negative with respect to the extracellular compartment. The size of the electrical (membrane) potential difference, V_m, ranges from -10 mV to as high as about -100 mV. Furthermore, in a number of cell types, V_m is variable, and this variation is responsible for the propagation of signals by nerve tissue, contraction of muscle, and stimulus-secretion coupling in exocrine and endocrine secretory cells.

What Is the Origin of V_m?

To appreciate how transmembrane electrical potential differences arise as a consequence of the interaction between pumps and leaks, let us consider a hypothetical cell, such as that illustrated in Fig. 22, which contains a coupled potassium acetate (KAc) pump, energized by the hydrolysis of ATP, and leak pathways for K^+ and Ac^-. Let us assume that this cell is initially filled with distilled water and is then dropped into a solution with a concentration $[KAc]_o = 10$ mmol/L. Initially, KAc will enter the cell, some via the pump and some via diffusion through the leaks. When the intracellular concentration reaches 10 mmol/L, diffusion of K^+ and Ac^- into the

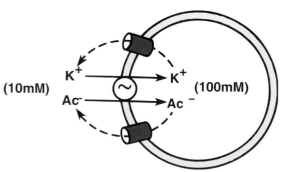

FIGURE 22 Hypothetical cell containing an energy-dependent pump for KAc and leaks for K^+ and Ac^-.

cell will cease. But, because the pump continues to operate, in time the concentration of KAc concentration in the cell will exceed that in the extracellular solution, and then K^+ and Ac^- will diffuse out of the cell through their leak pathways. At a subsequent time, a point will be reached when the rate at which KAc is pumped into this cell is precisely balanced by the rates at which K^+ and Ac^- diffuse out of the cell; let us say, for the sake of discussion, that when this steady state is reached the concentration of KAc in the cell ($[KAc]_i$) = 100 mmol/L. Furthermore, because the stoichiometry of the pump is one to one, the rates at which K^+ and Ac^- diffuse out of the cell must also be equal or the law of electroneutrality will be violated.

Because the concentration differences for K^+ and Ac^- across the membrane are equal—$\Delta C_K = \Delta C_{Ac} = 90$ mM—if the membrane is equally permeable to K^+ and Ac^- (i.e., $P_K = P_{Ac}$), equal rates of diffusion out of the cell (i.e., $J_K = J_{Ac}$) would not pose a problem. But, what if the membrane is more permeable to the small ion, K^+, than to Ac^-? Then, K^+ will tend to leave the cell faster than Ac^- and possibly cause a violation of the law of electroneutrality. This is prevented by the establishment of an electrical potential difference across the membrane oriented such that the cell interior is electrically negative with respect to the extracellular solution. This retards the diffusion of K^+ out of the cell and accelerates the outward diffusion of Ac^- so that $J_K = J_{Ac}$.

This situation is precisely analogous to that which we considered above dealing with the diffusion of KAc across an artificial membrane separating two solutions having different concentrations. The only differences are as follows:

1. In the artificial system, we provided the "muscle" (energy) to set up the concentration difference, whereas in our hypothetical cell, the energy is provided by the pump at the expense of ATP.
2. The artificial system will run down in time as KAc diffuses out of compartment o into compartment i, whereas in our hypothetical cell, the KAc leaving the cell is constantly being replenished by the pump; this is the essence of a steady state displaced from equilibrium by the investment of metabolic energy.

Finally, we can estimate V_m by employing Eq. 8, which describes the diffusion potential across a membrane arising from a concentration difference of a dissociable salt and differences in the permeabilities of the membrane to the resulting ions; that is,

$$V_m = [P_K - P_{Ac}]/[P_K + P_{Ac}] \times 60 \log([KAc]_o/[KAc]_i) \quad (25)$$

(Note: In Eq. 8, we employed diffusion constants, but the use of permeability coefficients as defined in Eq. 5 is also correct.) Thus, if $P_K = P_{Ac}$, then $V_m = 0$. But, if $P_K > P_{Ac}$, then, because $[KAc]_i > [KAc]_o$, V_m will be oriented such that the cell interior is electrically negative with respect to the extracellular solution. Furthermore, because $([KAc]_o/[KAc]_i) = 0.1$ (*i.e.*, 10 mmol/L/100 mmol/L), if P_K is very much greater than P_{Ac}, then V_m will approach -60 mV, that is, the Nernst potential for K^+. If $P_{Ac} > P_K$, then the cell interior will be electrically positive with respect to the extracellular solution; in this case, V_m can approach $+60$ mV if P_{Ac} is very much larger than P_K. Thus, V_m can assume any value between approximately $+60$ mV and approximately -60 mV, depending on the relation between P_K and P_{Ac}. Recall that in no instance is bulk electroneutrality violated; indeed, it is V_m that assures that $J_K = J_{Ac}$, thereby preventing a bulk separation of charge.

Now let us examine the behavior of the somewhat more realistic model illustrated in Fig. 21. Suppose that the only two ions subject to active transport are Na^+ and K^+ and that a one-to-one coupled carrier mechanism is involved; Cl^- is assumed to cross the cell membrane only by diffusion through water-filled pores. Under steady-state conditions, (1) Na^+ must diffuse into the cell at a rate equal to the rate of its carrier-mediated extrusion, and (2) K^+ must diffuse out of the cell at a rate equal to the rate of its carrier-mediated uptake. If the Na^+ and K^+ carrier mechanism is coupled such that for every K^+ pumped into the cell one Na^+ is extruded, the diffusional flows of these ions in opposite directions must also be equal; otherwise, electroneutrality would be violated. Now, in the example given, the concentration differences for Na^+ and K^+ across the membrane are approximately equal and opposite (this is approximately the case for many cells in higher animals) so that, if the permeabilities of the membrane to Na^+ and K^+ were also equal, the rates of diffusion in opposite directions would be equal and electroneutrality would be preserved. However, if the membrane is much more permeable to K^+ than to Na^+ (as is the case for most cells), the rates of net Na^+ and K^+ diffusion would not be equal in the absence of a transmembrane electrical potential difference. Thus, a V_m is generated that it is oriented such as to retard the diffusion of the more permeant ion (K^+) and accelerate the diffusion of the less permeant ion (Na^+) so that the inward diffusion of Na^+ is equal to the outward diffusion of K^+. Under these conditions, the cell interior will be electrically negative with respect to the exterior. To repeat, a diffusion potential arises across the membrane that maintains equal diffusion rates and preserves electroneutrality despite the fact that the permeabilities of the membrane to the two ions are not equal.

Several important points should be noted:

1. The example dealing with Na^+ and K^+ diffusion in opposite directions is formally analogous to our earlier example (Fig. 22) in which K^+ and Ac^- diffused in the same direction, because, as far as electroneutrality is concerned, the flow of a cation in one direction is equivalent to the flow of an anion in the opposite direction. In both examples, when K^+ is the more permeable ion, the orientation of V_m will be the same (*i.e.*, so as to retard the flow of K^+ and accelerate the flow of the less permeable ion).

2. The electrical potential differences (V_m) described in both examples are diffusion potentials resulting from the diffusional flows of ions down concentration differences. Although V_m is dependent on ion pumps, the relation is indirect; the pumps merely serve to establish the ionic concentration differences that provide the driving forces for the diffusional flows.

The orientation of V_m is always such as to retard the diffusional flow of the more permeant ion and accelerate the flow of the less permeant ion, and the magnitude of V_m is dependent on the individual permeabilities and concentrations of these ions (because these are the direct determinants of the diffusional flows that must be equalized by this electrical potential difference). We can generate an expression that defines the magnitude and orientation of V_m using an argument similar to that employed when we considered the diffusion potential generated by KAc diffusion out of the hypothetical cell shown in Fig. 22. Thus, if P_K is much greater than P_{Na}, then V_m will approach the Nernst potential for K^+—that is, $60 \log ([K^+]_o/[K^+]_i)$—and, because $[K^+]_o < [K^+]_i$, the cell interior will be electrically negative with respect to the extracellular compartment (*i.e.*, $V_m < 0$). Conversely, if P_K is much less than P_{Na}, then V_m will approach the Nernst potential for Na^+—that is, $60 \log ([Na^+]_o/[Na^+]_i)$—and, because $[Na^+]_o > [Na^+]_i$, under these conditions $V_m > 0$. These two extremes are satisfied by Eq. 26, which is sometimes referred to as a *double Nernst equation*:

$$V_m = 60 \log \left\{ \frac{P_K[K]_o + P_{Na}[Na]_o}{P_K[K]_i + P_{Na}[Na]_i} \right\} \tag{26}$$

This equation can be rewritten in the form:

$$V_m = 60 \log \left\{ \frac{[K]_o + \alpha[Na]_o}{[K]_i + \alpha[Na]_i} \right\} \tag{27}$$

where $\alpha = (P_{Na}/P_K)$. When P_K is much greater than P_{Na}, α is very small and V_m approaches the Nernst potential for K^+; conversely, when P_K is much less than P_{Na}, α is very large and V_m approaches the Nernst potential for Na^+. Thus, V_m can have any value between these two extremes depending on the value of α.

An expression having the form of Eq. 26 was first formally derived by Goldman in 1943. It was rederived by Hodgkin and Katz in 1949 and, discussed in Chapter 4, was first applied successfully to the analysis of the ionic basis of the resting and action potential of the squid axon. During the past three decades, it has provided the basis for understanding the origin of V_m across a wide variety of biologic membranes.

Finally, it should be noted that the stoichiometry of the (Na–K) pump is three Na^+ for two K^+ (not one to one). All this means is that, when a steady state is achieved, three Na^+ ions must diffuse into the cell for every two K^+ ions that diffuse out of the cell. But, if $P_{Na} < P_K$ the final result will still be that V_m is oriented such that the cell interior is electrically negative with respect to the extracellular fluid.

Before concluding this section, let us consider the distribution of Cl^- resulting from this pump–leak system. If, as stipulated above, the Cl^- distribution is determined solely by diffusion, then, when there is no net movement of Cl^- across the membrane (i.e., the composition of the cell is constant), the chemical potential forces acting on Cl^- must be balanced by the electrical forces acting on this anion. Thus, if V_m is negative (with respect to the outer solution), then the intracellular concentration of Cl^- (i.e., $[Cl]_i$) will be less than that in the extracellular solution (i.e., $[Cl]_o$). The precise relation among V_m, $[Cl]_i$, and $[Cl]_o$ is given by the Nernst equation, which, as discussed above, describes the condition for the balance or equality of chemical and electrical forces:

$$V_m = -60 \, \log[Cl]_o/[Cl]_i \qquad (28)$$

or, transposing,

$$[Cl]_i/[Cl]_o = 10^{[V_m/60]} \qquad (29)$$

When V_m is negative, the right-hand side of Eq. 29 has a value less than 1 so that $[Cl]_i < [Cl]_o$. Thus, if V_m is approximately -60 mV, then $[Cl]_i$ will be approximately one-tenth $[Cl]_o$.

Electrogenic (or Rheogenic) Ion Pumps

It is important to stress once more that the transmembrane potential differences we have discussed to

this point do not arise directly from the (Na^+–K^+) pump but indirectly from the ionic asymmetries that are generated by this pump. It is now clearly established, however, that under most circumstances, the (Na^+–K^+) pump is not neutral (i.e., one to one) but actually extrudes three Na^+ from the cell in exchange for two K^+. Thus, the pump itself brings about the movement of charge across the membrane (i.e., one positive charge from inside to outside for every cycle) and can be viewed as a current generator. Clearly, inasmuch as the pump generates a current across a membrane with a resistance, its action must directly result in an electrical potential difference given by the product of the pump current and the membrane resistance.

Thus, in general, transmembrane electrical potential differences have two origins. By far the largest fraction of these potential differences is attributable to diffusion potentials due to ionic asymmetries established by (Na^+–K^+) pumps. But, definite contributions to these potential differences arise directly from the current generated by the (Na^+–K^+) pump.

Transmembrane Distribution of Solutes Under Steady-State Conditions

The above considerations provide us with the principles that permit us to deduce important information with regard to the nature of the distributions of metabolically inert solutes across cell membranes when the cell composition is constant (i.e., in a steady state). Thus, let us assume that we can measure the electrical potential difference across a cell membrane, V_m, and at the same time determine the intracellular and extracellular concentrations (or, more properly, activities) of any solute i (i.e., C_i^i and C_i^o, respectively).

Now, the Nernst equation provides us with the criterion for determining whether the ratio of the concentrations (activities) of i across the membrane can be attributed entirely to thermal (passive) forces or whether additional (active) forces are necessary. Thus, if (at 37°C),

$$V_m = E_i = (60/z_i) \log\left[C_i^o/C_i^i\right](mV) \qquad (30)$$

or, multiplying both sides of Eq. 30 by z_i,

$$z_i V_m = 60 \log\left[C_i^o/C_i^i\right](mV) \qquad (31)$$

then, the steady-state distribution of i across the membrane can be considered passive. If this equality does not hold, then forces in addition to thermal energy must be involved in establishing the observed *distribution ratio* of i.

Now, we can rearrange Eq. 31 to provide the somewhat more useful expressions:

$$[C_i^o/C_i^i] = 10^{[z_i V_m/60]}$$

or

$$C_i^i = C_i^o 10^{[-z_i V_m/60]} \qquad (32)$$

Thus, if we know V_m and the extracellular concentration of solute i, C_i^o, we can predict the intracellular concentration C_i^i that would be consistent with a passive distribution of i across the membrane and then compare that predicted value with the actual measured value.

For a neutral solute (*i.e.*, $z_i = 0$), Eq. 32 states that the distribution of i across the cell membrane is independent of V_m and that if $C_i^i = C_i^o$, this distribution can be accounted for by passive transport processes that do not require direct or indirect coupling to a source of metabolic energy. If $C_i^i > C_i^o$, then metabolic energy must be invested into the transport process to pump i into the cell. Conversely, if $C_i^i < C_i^o$, then energy must be invested to extrude i from the cell. Thus, for a neutral, inert solute, $C_i^i \neq C_i^o$, transport across the cell membrane cannot be attributed to diffusion or carrier-mediated, facilitated diffusion.

Now let us turn to charged solutes and illustrate this approach by considering a cell bathed by a plasma-like solution, where $[Na^+]_o = 140$ mmol/L, $[K^+]_o = 4$ mmol/L, $[Cl^-]_o = 120$ mmol/L, and $[Ca^{2+}]_o = 2$ mmol/L. The electrical potential difference across this membrane is determined to be -60 mV, and the intracellular concentrations of Na^+, K^+, Cl^-, and Ca^{2+} are determined to be 10, 120, 12, and 10^{-3} mmol/L, respectively.

With respect to Na^+, Eq. 32 predicts that the intracellular concentration consistent with a passive distribution should be $[Na^+]_i = 10[Na^+]_o = 1400$ mmol/L, but the observed value of $[Na^+]_i$ was only 10 mmol/L. Thus, the distribution of Na^+ across the cell membrane cannot be attributed to passive transport processes. Instead, energy must be invested by the cell to extrude Na^+ and thereby lower its intracellular concentration to a level well below that predicted for a simple passive distribution.

Turning to K^+, Eq. 32 predicts that if K^+ is passively distributed across the membrane, its intracellular concentration should be $[K^+]_i = 10[K^+]_o = 40$ mmol/L. The observed intracellular concentration (120 mmol/L) is much greater than this predicted value so that energy must be invested to actively pump K^+ into the cell.

With respect to Cl^-, the predicted value for $[Cl^-]_i$ is $0.1[Cl^-]_o$, or 12 mmol/L. This agrees with the actual measured value so that one can conclude that the distribution of Cl^- across the cell membrane is the result of passive transport processes that do not require an investment of energy on the part of the cell.

Finally, applying Eq. 32 to the case of Ca^{2+}, we see that its predicted intracellular concentration is $[Ca^{2+}]_i = 2 \times 10^2 [Ca^{2+}]_o = 200$ mmol/L. This predicted value is much greater than the observed value of only 10^{-3} mmol/L so that the cell must invest energy into active transport processes to extrude Ca^{2+} from its interior. In short, the application of the Nernst equation, which defines the thermal balance of chemical and electrical forces across a membrane, permits us to determine whether the distribution of any inert solute across a cell membrane is passive or active.

Two caveats with regard to the application of the Nernst equation to the steady-state distributions of solutes across cell membranes must be noted. First, nonconformity with the predictions of the Nernst equation simply means that the observed distribution of a solute cannot be attributed to passive forces alone, but such nonconformity provides no insight into the detailed mechanism(s) responsible for this active distribution. Second, this line of reasoning does not apply to solutes that are either produced or utilized by cells. For example, the steady-state glucose concentration in most cells is much lower than that in the extracellular fluid because this nutrient is rapidly metabolized after gaining entry into the cells. As another example, the steady-state concentration of urea in most cells is greater than that in the extracellular fluid inasmuch as it is produced by protein catabolism and exits the cell by passive transport processes; thus, under steady-state conditions, there must be a concentration difference across the membrane that provides the driving force for urea exit at a rate equal to that at which it is produced. Clearly, an uncritical application of the criteria discussed above to these two neutral solutes would suggest that glucose is actively extruded from the cell and that urea is actively accumulated by these cells.

GIBBS–DONNAN EQUILIBRIUM, ION PUMPS, AND MAINTENANCE OF CELL VOLUME

So far we have considered three of the functions that are fulfilled by the operation of $(Na^+–K^+)$ pumps. First, they are responsible for the high K^+ and low Na^+ concentrations characteristic of the intracellular fluid of higher animals. A number of enzymes involved in intermediary metabolism and protein synthesis appear to require relatively high concentrations of K^+ for optimal activity and are inhibited by high concentrations of Na^+; the activities of these enzymes would be markedly impaired if the intracellular Na^+ and K^+

concentrations were the same as those in the extracellular fluid. Second, these ionic asymmetries are largely
responsible for establishing the electrical potential
differences across membranes and the bioelectrical
phenomena essential for the functions of excitable cells
such as nerve and muscles as well as many cells that are
not in this category. Finally, the Na^+ gradients established by these pumps energize the secondary active
transport of a number of solutes whose movements are
coupled to those of Na^+ (cotransport or countertransport).

(Na^+–K^+) pumps fulfill yet another, extremely vital
function in cells that do not possess a rigid cell wall;
namely, they are in part responsible for maintaining the
intracellular osmolarity equal to that of the extracellular
fluid, thereby preventing osmotic water flow into the
cells and, in turn, cell swelling. To appreciate why ion
pumps are necessary for the maintenance of cell volume,
we should first consider the artificial system illustrated
in Fig. 23. The two compartments illustrated are
assumed to be closed to the atmosphere and separated
by a membrane that is freely permeable to Na^+, Cl^-,
and water but is impermeable to proteins. A solution of
NaCl is added to compartment o, and a solution of the
sodium salt of a protein (Na^+–proteinate [NaP]) is
added to compartment i; assume that at the outset
$[Na^+]_o = [Na^+]_i$. The system is then left undisturbed for
a sufficiently long time until equilibrium is achieved.

Let us now consider the characteristics of this final,
time-independent, equilibrium condition, which was
derived by Josiah Willard Gibbs (1839–1903) and

Frederick George Donnan (1870–1956) and is often
referred to as the *Gibbs–Donnan equilibrium, Donnan
equilibrium,* or *Donnan distribution.* Because the
membrane is permeable to Na^+ and Cl^-, but there is
no Cl^- initially present in compartment i, some Cl^-
must diffuse from compartment o into compartment i,
but this must be also accompanied by an equal amount
of Na^+; otherwise, electroneutrality would be violated.

At all times, the system must obey a balance of
electrical charges such that:

$$[Na^+]_o = [Cl^-]_o \quad \text{and} \quad [Na^+]_i = [Cl^-]_i + z_p[P^-]_i \quad (33)$$

where z_p is the net anionic valence of the protein
molecule.

When equilibrium is achieved, the system will be
characterized by three properties:

1. Because we started out with equal Na^+ concentrations in both compartments and Na^+ and Cl^-
subsequently diffused into compartment i (at equal
rates), the equilibrium condition must be characterized by asymmetrical distributions of both of these
permeant ions across the membrane. As we have
already learned, if there is an asymmetric distribution
of a passively transported ion across a membrane,
then, at equilibrium, there must be an electrical
potential difference, V_m, across that membrane that
balances the concentration difference and is given by
the Nernst equation. Thus, one value of V_m must
simultaneously satisfy the equilibrium distributions
of both Na^+ and Cl^-, namely,

$$V_m = 60 \log([Na^+]_o/[Na^+]_i) = 60 \log([Cl^-]_i/[Cl^-]_o) \quad (34)$$

2. It follows from Eq. 34 that:

$$([Na^+]_o/[Na^+]_i) = ([Cl^-]_i/[Cl^-]_o) \quad (35)$$

If we consider the initial and final (equilibrium)
conditions illustrated in Fig. 23 together with Eq. 33,
it should be clear that when equilibrium is achieved
$[Na^+]_o < [Na^+]_i$ and $[Cl^-]_i < [Cl^-]_o$) with the precise
relation given by Eq. 35. Furthermore, it follows
from Eq. 34 that compartment i will be electrically
negative with respect to compartment o.

3. Finally, if we consider the initial and final conditions
together with Eq. 33, it should be clear that when
equilibrium is achieved, the concentration of osmotically active solutes in compartment i will be greater
than that in compartment o, so that a pressure will
have developed across the membrane given by van't
Hoff's law, Eq. 19. Furthermore, if instead of being

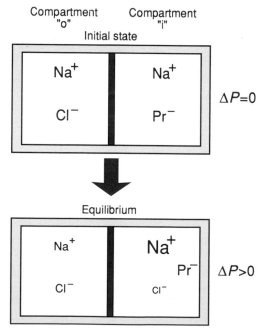

FIGURE 23 The development of the Gibbs-Donnan equilibrium
condition. ΔP, difference in hydrostatic pressure; Pr, protein anion.

rigid the membrane is distensible, it would bulge into compartment o.

We chose the initial condition $[Na^+]_o = [Na^+]_i$ simply to make it easier for students to comprehend the evolution of the asymmetries that characterize this equilibrium condition, but these equilibrium characteristics can be formally generalized to any set of initial conditions. Thus, if compartment i has a greater concentration of impermeant charged species than compartment o, the three additional asymmetries that will characterize the Gibbs–Donnan equilibrium when it is reached are as follows:

1. There will be an asymmetric distribution of all permeant monovalent cations (C_+) and anions (C_-) that conforms to the relation:

$$C_+^o/C_+^i = C_-^i/C_-^o = r$$

   where r is often referred to as the *Donnan ratio* and is, in part, a function of the difference in total charge between compartments o and i borne by impermeant ions. If compartment i contains a preponderance of impermeant anions, then $C_+^o < C_+^i$ and $C_-^i < C_-^o$; if the impermeant species are predominantly cationic, then these relations will be reversed.

2. There will be an electrical potential difference across the membrane given by the relation:

$$V_m = 60 \log(C_+^o/C_+^i) = 60 \log(C_-^i/C_-^o) = 60 \log r$$

   If compartment i contains a preponderance of impermeant anions then V_m will be oriented such that compartment i is electrically negative with respect to compartment o; if the impermeant species are predominantly cationic, then this orientation will be reversed.

3. Regardless of the sign of the total charge carried by the preponderance of impermeant species in compartment i, at equilibrium that compartment will contain a greater number of osmotically active particles than compartment o. Thus, there will be an osmotic driving force for the movement of water into compartment i. If the membrane is rigid then an osmotic pressure would balance that driving force; if the membrane is distensible, then it would bulge into compartment o.

We can appreciate the function of ion pumps in the maintenance of cell volume by considering what would happen if there were no ion pumps. Because the intracellular concentration of charged, largely anionic, impermeant macromolecules is much greater than that in the extracellular fluid, cells lacking ion pumps would

resemble the passive system illustrated in Fig. 23 and would move toward the direction of achieving a Gibbs–Donnan equilibrium. The total osmotic activity of intracellular solutes would exceed that in the surrounding fluid, and there would be a driving force for osmotic water flow into the cell. If the cells possess rigid cell walls that prevent any increase in cell volume, an osmotic pressure difference would develop across the cell walls and the cell interiors would be subjected to a pressure greater than the extracellular fluid ("turgor"). If, as in the case of animal cells, the membrane is distensible, water would flow into the cell, leading to cell swelling and, perhaps, rupture.

The (Na^+–K^+) pumps in animal cell membranes serve to reduce the intracellular content of osmotically active solutes, thereby counteracting the osmotic effect of intracellular macromolecules. The pumps extrude three sodium ions in exchange for two potassium ions and also establish an electrical potential difference across the membrane (negative cell interior) that reduces the steady-state intracellular concentrations of permeant, passively distributed anions (mainly Cl^-).

If the (Na^+–K^+) pumps are inhibited by digitalis glycosides or metabolic poisons, the cells will lose K^+ and gain Na^+ and Cl^-; in many cells, three Na^+ plus one Cl^- will be gained for every two K^+ lost, so that the total amount of Na^+ and Cl^- gained by the cell exceeds the amount of K^+ lost, and the total intracellular solute concentration will increase. This will result in an osmotic uptake of water and cell swelling and may lead to the destruction of the integrity of the cell membrane. (Recall that inhibition of the pump would not only lead to the dissipation of the asymmetric distributions of Na^+ and K^+, but also the V_m (cell interior negative) arising from these asymmetries. According to Eq. 32, if V_m becomes less negative, then intracellular Cl^- will increase. Also note that the redistribution of ions following inhibition of the pump does not, indeed cannot, violate the law of bulk electroneutrality.)

In summary, the membranes that surround most animal cells are distensible and highly permeable to water. If these cells are immersed in a hypertonic fluid, water rapidly leaves the intracellular compartment and they shrink. If these cells are immersed in a hypotonic solution, water flows rapidly into them and they swell, possibly rupturing. In higher animals, the osmolarity of the extracellular fluid is carefully regulated by the kidneys in response to neurohormonal stimuli so that it normally remains within very narrow limits. However, the maintenance of isotonicity between the intracellular and extracellular fluids depends, in part, on the presence of ion, particularly (Na^+–K^+), pumps in the cell membranes. These pumps serve to lower the intracellular concentration of permeant solutes and thereby balance and offset the osmotic effects of impermeant intracellular macromolecules. (Other transport mechanisms come into play when the preservation of cell volume is threatened by conditions that lead to swelling and, in some instances, shrinking; a discussion of these mechanisms is beyond the scope of this introductory text but can be found in the suggested readings.)

Suggested Readings

Bretscher MS. The molecules of the cell membrane. *Sci Am* 1985; 253(4):100–108

Darnell J, Lodish H, Baltimore D. *Molecular cell biology*, 2nd ed. New York: Scientific American Books, Chapter 13, 1990

Doyle DA, Cabral JM, Pfuetzner RA, Kuo A, Gulbis JM, Cohen SL, Chait BT, MacKinnon R. The structure of the potassium channel: molecular basis of K^+ conductance and selectivity. *Science* 1998; 280:69–77

Finkelstein A. *Water movement through lipid bilayers, pores, and plasma membranes: theory and reality*. New York: Wiley-Interscience, 1987

Hille B. *Ionic channels of excitable membranes*, 2nd ed. Sunderland, MA: Sinauer Associates, 1992

Reuss L, Russell JM, Jennings ML. *Molecular biology and function of carrier proteins*. New York: The Rockefeller University Press, 1992

Schultz SG. *Basic principles of membrane transport*. Cambridge: Cambridge University Press, 1980

Schultz SG, Andreoli TE, Brown AM *et al. Molecular biology of membrane transport disorders*. New York: Plenum Press, 1996

Semenza G, Kinne R. Membrane transport driven by ion gradients. *Ann N Y Acad Sci* 1985; Vol. 456

Stein WD. *Transport and diffusion across cell membranes*. Orlando: Academic Press, 1986

4

Resting Potentials and Action Potentials in Excitable Cells

JOHN H. BYRNE

KEY POINTS

- Action potentials are very brief electrical events and are elicited and propagated in an all-or-nothing fashion.
- Information about intensity is encoded as the frequency of firing of action potentials.
- The resting potential is due to a high resting permeability to K^+ and a low resting permeability to Na^+.
- The Goldman–Hodgkin–Katz equation can be used to predict the resting potential.
- Na^+ is essential for the nerve action potential.
- The initiation of the action potential is due to the opening of voltage-dependent Na^+ channels.

- Na^+ inactivation and a delayed opening of K^+ channels underlie the repolarization phase of the action potential.
- The slow recovery of the delayed increase in K^+ conductance underlies the hyperpolarizing afterpotential.
- No significant changes in the intracellular concentrations of Na^+ or K^+ occur during an action potential.
- Action potentials have absolute and relative refractory periods, and are affected by neurotoxins.

Resting potentials are characteristic features of all cells in the body, but nerve cells and other excitable cells, such as muscle cells, not only have resting potentials but are capable of altering these potentials for the purpose of communication, in the case of nerve cells, and for the purpose of initiating contraction, in the case of muscle cells. The material that follows introduces the ionic mechanisms that endow excitable membranes with this ability.

EXTRACELLULAR RECORDING OF THE NERVE ACTION POTENTIAL

The existence of "animal electricity" had been known for more than 200 years, but the first direct experimental evidence for it was not provided until the development of electronic amplifiers and oscilloscopes. Figure 1 illustrates one of the earliest recordings that demonstrated the ability of nerve cells to alter their electrical activity for the purpose of coding and transmitting information. In this experiment, performed in 1934 by Hartline, extracellular recordings were made from the optic nerve of an invertebrate eye. Details of the techniques and interpretation of these extracellular recordings are described in Chapter 5, but for now it is sufficient to know that it is possible to place an electrode on the surface of a nerve axon and record electrical events that are associated with potential changes taking place across the axonal membrane. In the experiment illustrated in Fig. 1, light flashes of different intensities were delivered to the eye. With a very weak intensity light flash, there was no change in the baseline electrical activity. When the intensity of the

light flash was increased, however, small spike-like transient events associated with the onset of the light were observed. Increasing the intensity of the light flash produced an increase in the rate of these spike-like events. These spike-like events are known as nerve *action potentials*, impulses, or, simply, spikes.

Even though this experiment was performed more than 60 years ago, it nonetheless illustrates three basic properties of nerve action potentials and how they are used by the nervous system to encode information. First, nerve action potentials are very brief, having a duration of only about 1 msec (1 msec = 10^{-3} sec). Second, action potentials are initiated in an *all-or-nothing* manner. Note that the amplitude of the action potentials does not vary during a sustained light flash. Third, and related to the above, with increasing stimulus intensity it is not the size of action potentials that varies but rather their number or frequency. This is the general means by which intensity information is coded in the nervous system, and it is true for a variety of peripheral receptors. Specifically, the greater the intensity of a physical stimulus (whether it be a stimulus to a photoreceptor, a stimulus to a mechanoreceptor in the skin, or a stimulus to a muscle stretch receptor), the greater is the frequency of nerve action potentials. This finding has given rise to the notion of the *frequency code* for stimulus intensity in the nervous system.

Most of the information transmitted to the central nervous system from the periphery is mediated by nerve action potentials. Moreover, all the motor commands initiated in the central nervous system are propagated to the periphery by nerve action potentials, and action potentials produced in muscle cells are the first step in the initiation of muscular contraction (see Chapter 6). Action potentials are therefore quite important, not only for the functioning of the nervous system, but also for the functioning of muscle cells, and for this reason it is important to understand the ionic mechanisms that underlie the action potential and its propagation.

INTRACELLULAR RECORDING OF THE RESTING POTENTIAL

The action potentials illustrated in Fig. 1 were recorded with extracellular electrodes. To examine the properties of action potentials in greater detail, it was necessary to move from these rather crude extracellular techniques to *intracellular recording techniques*. Figure 2 illustrates schematically how it is possible to record the membrane potential of a living cell. The upper left of Fig. 2A is an idealized nerve cell, composed of a cell body with a portion of its attached axonal process. Outside the nerve cell in the extracellular medium is

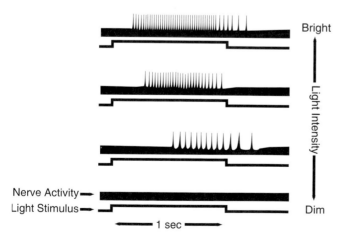

FIGURE 1   Action potentials recorded from an invertebrate optic nerve in response to light flashes of different intensities. With dim illumination no action potentials are recorded, but with more intense illumination the number and frequency of action potentials increase. (Modified from Hartline HK. *J Cell Comp Physiol* 1934; 5:229.)

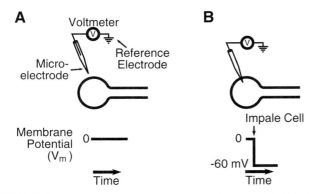

FIGURE 2 Intracellular recording of the resting potential. (A) One input to a voltmeter is connected to a microelectrode, and the second input is connected to a reference electrode in the extracellular medium. No potential difference is recorded when the tip of the microelectrode is outside the cell. (B) When the tip of the microelectrode penetrates the cell, a resting potential of −60 mV is recorded. (Modified from Kandel, ER. *The cellular basis of behavior*. San Francisco: Freeman, 1976.)

a glass microelectrode that is connected to a suitable voltage recording device, such as a voltmeter, a pen recorder, or an oscilloscope. A *glass microelectrode* is nothing more than a piece of thin capillary tubing that is stretched under heat to produce a very fine tip having a diameter less than 1 μm. The electrode is then filled with an electrolyte solution such as KCl to conduct current. Initially, with the microelectrode in the extracellular medium, no potential difference is recorded, simply because the extracellular medium (the extracellular fluid) is isopotential. If, however, the microelectrode penetrates the cell membrane so that the tip of the microelectrode is inside the cell, a sharp deflection is obtained on the recording device (Fig. 2B). The potential suddenly shifts from its initial value of 0 mV to a new value of −60 mV. The inside of the cell is negative with respect to the outside. The potential difference that is recorded when a living cell is impaled with a microelectrode is known as the resting potential. The resting potential remains constant for indefinite periods of time as long as the cell is not stimulated or no damage occurs to the cell with impalement. The resting potential varies somewhat from nerve cell to nerve cell (−40 to −90 mV), but a typical value is about −60 mV.

INTRACELLULAR RECORDING OF THE NERVE ACTION POTENTIAL

The techniques for examining resting potentials can be extended to study the action potential. Although nerve action potentials are normally initiated by mechanical, chemical, or photic stimuli to classes of

specialized receptors or by a process known as *synaptic transmission* (see Chapter 6), it is possible to elicit action potentials artificially in nerve cells and study their underlying ionic mechanisms in considerable detail and in a controlled fashion.

Figure 3 shows another idealized nerve cell with its cell body and attached axon. One microelectrode has penetrated the cell membrane so that the tip of the electrode is inside the cell. This electrode will be used to monitor the potential difference between the outside and inside of the cell. When this electrode penetrates the cell, a resting potential of about −60 mV is recorded. The cell is also impaled with a second microelectrode that will be used to alter the membrane potential artificially. This second electrode, called the *stimulating electrode,* is connected to a suitable current generator (in the simplest case, this current generator can be considered a battery). Obviously, there are two ways that a battery can be connected to any circuit. The battery can be inserted so that either its positive pole or negative pole is connected to the electrode. A switch is placed in the circuit so that the battery can be connected to and disconnected from the circuit at will.

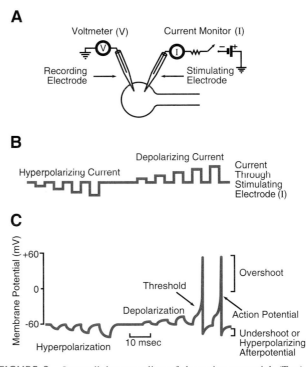

FIGURE 3 Intracellular recording of the action potential. (Top) A second intracellular microelectrode is used to hyperpolarize or depolarize the cell artificially. (Middle) Hyperpolarizing and then depolarizing current pulses of increasing amplitudes were passed into the cell. (Bottom) If the magnitude of the depolarization current is sufficient to depolarize the membrane potential to threshold, an action potential is initiated. (Modified from Kandel ER. *The cellular basis of behavior*. San Francisco, Freeman, 1976.)

Assume that a small battery is inserted and its negative pole is connected to the stimulating electrode. With the switch open, a resting potential of −60 mV is recorded. As a result of closing the switch, however, the negative pole of the battery is connected to the stimulating electrode, which tends to artificially make the inside of the cell more negative relative to the external solution. There is a slight downward deflection of the recording trace.

If this experiment is repeated using a slightly larger battery, more current flows into the cell, and a larger increase in the negativity of the cell is recorded. Larger batteries produce even greater increases in the potential. Any time the negativity of the cell interior is increased, the potential change is known as a hyperpolarization. The membrane is more polarized than normal.

Now consider the consequences of repeating this experiment with the positive pole of the battery connected to the stimulating electrode. Turning on the switch now makes the inside of the cell artificially more positive than the resting potential. The polarized state of the membrane is decreased. Increasing the size of the battery produces a greater decrease in the negativity of the cell, and over a limited range the resultant potential is a graded function of the size of the stimulus that is used to produce it. Any time the interior of the cell becomes more positive, the potential change is known as a depolarization. These hyperpolarizations and depolarizations that are artificially produced are known as electrotonic, graded, or passive potentials (some additional features of electrotonic potentials are discussed later in Chapter 5); however, the point to note here is that, within a limited range of stimulus intensities, hyperpolarizing and depolarizing electrotonic potentials are graded functions of the size of the stimulus used to produce them.

An interesting phenomenon occurs when the magnitude of the battery used to produce the depolarizing potentials is increased further. As the size of the battery and thus the amount of depolarization is increased, a critical level is reached, known as the *threshold,* wherein a new type of potential is produced that is different in its amplitude, duration, and form from the depolarizing pulse used to produce it. This new type of potential change elicited when threshold is reached is known as the action potential, which is elicited in an all-or-nothing fashion. Stimuli below threshold fail to elicit an action potential; stimuli at threshold or above threshold successfully elicit an action potential. Increasing the stimulus intensity beyond threshold produces an action potential identical to the action potential produced at the threshold level. In this experiment, the duration of the depolarization is so short that only a single action potential could be initiated. If the duration is longer, multiple action potentials are initiated, and their frequency depends on the stimulus intensity. This is simply a restatement of the all-or-nothing law of action potentials presented earlier. Below threshold, no action potential is elicited; at or above threshold, an all-or-nothing action potential is initiated. Increasing the stimulus intensity still further produces the same amplitude action potential; only the frequency is increased.

Not only are action potentials elicited in an all-or-nothing fashion, but, as described in Chapter 5, they also propagate in an all-or-nothing fashion. If an action potential is initiated in the cell body, it will propagate along the nerve axon and eventually invade the synaptic terminals and initiate a process known as *synaptic transmission* (see Chapter 6). Unlike action potentials, electrotonic potentials do not propagate in an all-or-nothing fashion. Electrotonic potentials do spread but only for short distances (see Chapter 5).

There are several interesting features of the action potential. One is that the polarity of the cell completely reverses during the peak of the action potential. Initially, the inside of the cell is −60 mV with respect to the outside but, during the peak of the action potential, the potential reverses and approaches a value +55 mV inside with respect to the outside. The region of the action potential that varies between the 0-mV level and its peak value is known as the *overshoot.* Another interesting characteristic of action potentials is their repolarization phase (the return to the resting level). The action potential does not immediately return to the resting potential of −60 mV; there is a period of time when the cell is actually more negative than the resting level. This phase of the action potential is known as the *undershoot* or the *hyperpolarizing afterpotential.*

As indicated earlier, nerve potentials are the vehicles by which peripheral information is coded and propagated to the central nervous system; motor commands initiated in the central nervous system are propagated to the periphery by nerve action potentials, and the action potential is the first step in the initiation of muscular contraction.

IONIC MECHANISMS OF THE RESTING POTENTIAL

Although the major focus of this chapter is to explain the ionic mechanisms that underlie the action potential, it is first necessary to review the ionic mechanisms that underlie the resting potential, because the two are intimately related. The basic principles have been introduced in Chapter 3.

Bernstein's Hypothesis for the Resting Potential

In 1902, Julius Bernstein proposed the first satisfactory hypothesis for generation of the resting potential. Bernstein knew that the inside of cells have high K^+ and low Na^+ concentrations and that the extracellular fluid has low K^+ and high Na^+ concentrations. In addition, there appeared to be large negatively charged molecules, presumably proteins, to which the cell was impermeable. Bernstein also knew (a critical piece of information) that cells were highly permeable to K^+ but not very permeable to other ions. Furthermore, Bernstein knew of the work of the physical chemist, Nernst. Bernstein therefore suggested that the resting potential could be predicted simply by applying the *Nernst equilibrium equation* for potassium:

$$V_m \overset{?}{=} E_{Na} = 60 \log \frac{[K^+]_o}{[K^+]_i} \text{ (mV)} \qquad (1)$$

where V_m is the membrane potential and E_K the potassium equilibrium potential (see also Chapter 3).

Although Bernstein's hypothesis was interesting, it could not be directly tested at the time (hence, the question mark in the equation) because microelectrode recording techniques had not been developed. It was not until the 1930s and 1940s and the advent of microelectrode recording techniques that it became possible to test the hypothesis directly. The testing of Bernstein's hypothesis was done primarily by Hodgkin and Huxley and their colleagues in England. As a result of this work, a general theory was developed for the generation of the resting potential that appears to be applicable to most cells in the body.

Testing Bernstein's Hypothesis

How would one go about testing Bernstein's hypothesis? If the membrane potential (V_m) is equal to the K^+ equilibrium potential (E_K), one should be able to substitute the known outside and inside concentrations of K^+ into the Nernst equation and determine the equilibrium potential (E_K), which should equal the measured membrane potential (V_m). Furthermore, because of the logarithmic relationship in the Nernst equation, if the outside K^+ concentration is artificially manipulated by a factor of 10, then the equilibrium potential will change by a factor of 60 mV. If the *membrane* potential is governed by the K^+ *equilibrium* potential, then the membrane potential should also change by 60 mV.

Figure 4 illustrates one direct experimental test of Bernstein's hypothesis performed by Hodgkin and

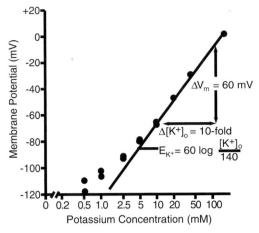

FIGURE 4   Effects of altered extracellular concentrations of K^+ on the membrane potential: (•), measured membrane potential at each of a variety of different concentrations of K^+; the straight line is the potential predicted by the Nernst equation. The value of 140 in the Nernst equation is the estimated intracellular concentration of K^+ for the cell used in the experiment. (Modified from Hodgkin AL, Horowicz P. *J Physiol* 1959; 148:127.)

Horowicz. A cell was impaled with a microelectrode and the resting potential was measured. The extracellular K^+ concentration was systematically varied, and the change in the resting potential was monitored. When the K^+ concentration was changed by a factor of 10, the resting potential changed by a factor of 60 mV. The straight line on the plot is the relationship predicted by the Nernst equation (note that it is a straight line because these data are plotted on a semilog scale).

The fit is not perfect, however, and the experimental data deviate from the predicted values when the extracellular K^+ concentration is reduced to low levels. If there is a deviation from the Nernst equation, the membrane must be permeable, not only to K^+, but to another ion as well. That other ion appears to be Na^+. As indicated earlier, Na^+ has a high concentration outside the cell and a low concentration inside the cell. If the cell has a slight permeability to Na^+, Na^+ will tend to diffuse into the cell and produce a charge distribution across the membrane so that the inside of the membrane will be positive with respect to the outside. This slight increase in the positivity on the inside surface of the membrane will tend to reduce the negative charge distribution produced by the diffusion of K^+ out of the cell. The slight permeability of the membrane to Na^+ will tend, therefore, to make the cell slightly less negative than would be expected were the membrane only permeable to K^+. If a membrane is permeable to more than one cation, the Nernst equation cannot be used to predict the resultant membrane potential. However, in such a case the Goldman equation can be used.

GOLDMAN–HODGKIN–KATZ EQUATION

The *Goldman equation* is also known as the *Goldman–Hodgkin–Katz (GHK) equation* because Hodgkin and Katz applied it to biologic membranes. As has already been seen in Chapter 3, the GHK equation can be used to determine the potential developed across a membrane permeable to Na^+ and K^+. Thus,

$$V_m = 60 \log \frac{[K^+]_o + \alpha [Na^+]_o}{[K^+]_i + \alpha [Na^+]_i} \text{ (mV)} \qquad (2)$$

where V_m is the membrane potential in millivolts and α is equal to the ratio of the Na^+ and K^+ permeabilities (P_{Na}/P_K). This equation looks rather complex at first, but it can be simplified by examining two extreme cases. Consider the case when the Na^+ permeability is equal to zero. Then, α is equal to zero, and the Goldman–Hodgkin–Katz equation reduces to the Nernst equation for K^+. If the membrane is highly permeable to Na^+ and has a very low K^+ permeability, α will be a very large number, which causes the Na^+ terms to be very large so that the K^+ terms can be neglected and the Goldman–Hodgkin–Katz equation reduces to the Nernst equation for Na^+. Thus, the Goldman–Hodgkin–Katz equation has two extremes. In one case, when Na^+ permeability is zero, it reduces to the Nernst equation for K^+; in the other case, when Na^+ permeability is very high, it reduces to the Nernst equation for Na^+. The GHK equation allows one to predict membrane potentials between these two extreme levels, and these membrane potentials are determined by the ratio of K^+ and Na^+ permeabilities. If the permeabilities are equal, the membrane potential will be intermediate between the K^+ and the Na^+ equilibrium potentials.

Figure 5 illustrates a test of the ability of the Goldman–Hodgkin–Katz equation to fit the same experimental data shown in Fig. 4. The straight line is generated by the Nernst equation, whereas the curved trace is generated by the Goldman–Hodgkin–Katz equation. The value of α that gives the best fit is 0.01. Thus, although there is some Na^+ permeability at rest, it is only one-hundredth that of the K^+ permeability. To a first approximation, the membrane potential is due to the fact that there is unequal distribution of K^+, and the membrane is selectively permeable to K^+ and to a large extent no other ion. Therefore, the membrane potential can be roughly predicted by the Nernst equilibrium potentials for K^+. However, there is a slight Na^+ permeability that tends to make the inside of the cell more positive than would be predicted, based on the assumption that the cell is permeable only to K^+. The GHK equation can be used to calculate or predict

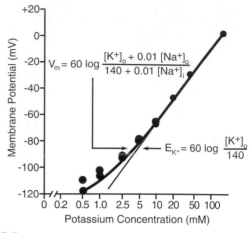

FIGURE 5   Same experiment as Fig. 4, but the graph also contains the prediction of the change in membrane potential obtained with the Goldman–Hodgkin–Katz equation with a value of α equal to 0.01. (Modified from Hodgkin AL, Horowicz P. *J Physiol* 1959; 148:127.)

the membrane potential knowing the ratio of Na^+ and K^+ permeabilities and the individual extracellular and intracellular concentrations of Na^+ and K^+.

THE SODIUM HYPOTHESIS FOR THE NERVE ACTION POTENTIAL

Is it possible to specify ionic mechanisms that account for the action potential just as it was possible to do so for the resting potential? It is interesting to note that Julius Bernstein in 1902, when proposing his theory for the resting potential, also proposed a theory for the nerve action potential. Bernstein proposed that during a nerve action potential, the membrane suddenly became permeable to all ions. Bernstein predicted, based on this theory, that the membrane potential would shift from its resting level to a new value of about 0 mV. (Can you explain why?) However, from Fig. 3 it is clear that the potential changes during the action potential do not range from a value of −60 to 0 mV, but actually go well beyond 0 mV and approach a value of +55 mV. So, whereas Bernstein's hypothesis for the resting potential was nearly correct, his hypothesis for the action potential clearly missed the mark.

At the same time that Bernstein proposed theories for resting potentials and action potentials, Overton, another early physiologist, made some interesting observations about the critical role of Na^+. Overton observed that Na^+ in the extracellular medium was absolutely essential for cellular excitability. In general, in the absence of extracellular Na^+, nerve axons cannot propagate information, and skeletal muscle cells are unable to contract.

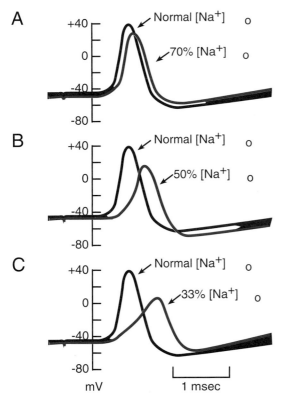

FIGURE 6 Changes in the amplitude of the action potential in the squid giant axon as a function of the extracellular concentration of Na^+ reduced to 70% (A), 50% (B), and 33% (C) of its normal value. (Modified from Hodgkin AL, Katz B. *J Physiol* 1949; 108:37.)

membrane was switching from its state of being highly permeable to K^+ at rest to being highly permeable to Na^+ at the peak of the action potential. If a membrane is highly permeable to Na^+ at the peak of the action potential (for the sake of simplicity we assume that the membrane is solely permeable to Na^+ and no other ions), what potential difference would one predict across the cell membrane? If the membrane is only permeable to Na^+, the membrane potential should equal the Na^+ equilibrium potential (E_{Na}), and

$$V \overset{?}{=} E_{Na} = 60 \log \frac{[Na^+]_o}{[Na^+]_i} \ (mV) \qquad (3)$$

Indeed, when the known values of extracellular and intracellular Na^+ concentrations for the squid giant axon are substituted, a value of $+55$ mV is calculated. This is approximately the peak amplitude of the action potential. Is this simply a coincidence? It is possible that the membrane is permeable to other ions as well. Perhaps the action potential is due to an increase in Ca^{2+} permeability; Ca^{2+} is in high concentration outside and low concentration inside the cell, so part of the action potential might be due to a selective increase in Ca^{2+} permeability. How can this issue be resolved? If the peak amplitude of the action potential is determined by E_{Na}, one would expect that as the extracellular levels of Na^+ are altered, the peak amplitude of the action potential would change according to the Nernst equation. Furthermore, because of the logarithmic relationship in the Nernst equation, if the extracellular Na^+ concentration is changed by a factor of 10, the Na^+ equilibrium potential and the peak amplitude of the action potential should change by a factor of 60 mV.

Figure 7 illustrates a test of this hypothesis. The peak amplitude of the action potential, shown on the vertical axis, is measured as a function of the extracellular Na^+ concentration. The dots on the graph represent the peak amplitude of the action potential recorded at various extracellular concentrations of Na^+. The straight line is the relationship that describes the Na^+ equilibrium potential as a function of extracellular Na^+.

Although there are some deviations between the predicted Na^+ equilibrium potential and the peak amplitude of the action potential (the action potential never quite reaches the value of the Na^+ equilibrium potential), the critical observation is that the slopes of these two lines are nearly identical. For a tenfold change in the extracellular Na^+ concentration, there is approximately a 60-mV change in the peak amplitude of the action potential. These experiments, therefore, provide strong experimental support for the hypothesis that

Overton, like Bernstein, could not test his hypothesis experimentally because microelectrodes were not available. Just as Hodgkin and his colleagues critically tested Bernstein's hypothesis for the resting potential, they also examined and extended Overton's observations. One of the earlier experiments performed by Hodgkin and Katz is illustrated in Fig. 6.

Hodgkin and Katz repeatedly initiated action potentials in the squid giant axon while they artificially altered the extracellular Na^+ concentration. When the extracellular Na^+ concentration was reduced to 70% of its normal value (Fig. 6A), there was a slight reduction in the amplitude of the action potential. Reducing the Na^+ concentration to 50% and 33% of its normal value produced further reductions in the amplitude of the action potential. These experiments, therefore, directly confirmed Overton's initial observations that Na^+ is essential for the initiation of action potentials. (Exceptions to this are action potentials in cardiac and smooth muscle cells; see Chapters 8 and 11.)

Hodgkin and his colleagues took these observations one step further. They suggested that, during an action potential, the membrane behaved as though it was becoming selectively permeable to Na^+. In a sense, the

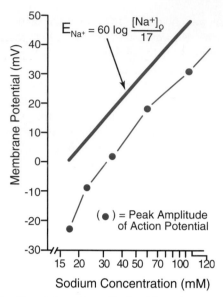

FIGURE 7   Plot of the peak amplitude of the action potential vs. the extracellular concentration of Na$^+$. The solid line is a prediction of the Nernst equation with an estimated intracellular Na$^+$ concentration for this cell of 17 mmol/L. (Modified from Nastuk WL, Hodgkin AL. *J Cell Comp Physiol* 1950; 35:39.)

during the peak of the action potential the membrane suddenly switches from a high permeability to K$^+$ to a high permeability to Na$^+$.

APPLYING THE GOLDMAN–HODGKIN–KATZ EQUATION TO THE ACTION POTENTIAL

There are two important positively charged ions (K$^+$ and Na$^+$), and the membrane potential appears to be governed by the relative permeabilities of these two ions.

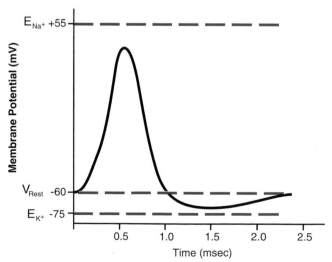

FIGURE 8   Sketch of a nerve action potential.

As a result, the Goldman–Hodgkin–Katz equation can be used:

$$V_m = 60 \log \frac{\left[K^+\right]_o + \alpha\left[Na^+\right]_o}{\left[K^+\right]_i + \alpha\left[Na^+\right]_i} \; (mV)$$

where $\alpha = P_{Na}/P_K$.

Figure 8 is a sketch of an action potential. One important observation is that the action potential traverses a region that is bounded by E_{Na} on one extreme and E_K on the other. Because the action potential traverses this bounded region, it is possible to use the Goldman–Hodgkin–Katz equation to roughly predict any value of the action potential simply by adjusting the ratio of the Na$^+$ and K$^+$ permeabilities. For the resting level, we have already seen that the ratio of Na$^+$ and K$^+$ permeabilities is 0.01. Thus, we can substitute these values into the Goldman–Hodgkin–Katz equation and calculate a value of approximately −60 mV.

Assume that the Na$^+$ permeability is very high. Then α is a very large number, and the Na$^+$ terms dominate the Goldman–Hodgkin–Katz equation. In the limit, the Goldman–Hodgkin–Katz equation reduces to the Nernst equation for Na$^+$. So, when there is a high Na$^+$ permeability and a low K$^+$ permeability, we can calculate a potential that approximates the peak amplitude of the action potential. During the repolarization phase of the action potential, we can simply assume that the ratio of Na$^+$ and K$^+$ permeabilities returns back to normal, substitute this value into the Goldman–Hodgkin–Katz equation, and calculate a membrane potential of −60 mV. The hyperpolarizing afterpotential could be accounted for by a slight decrease in Na$^+$ permeability to less than its resting level or by a K$^+$ permeability greater than its resting level. The important point is that by adjusting the ratio of Na$^+$ and K$^+$ permeabilities, it is possible to predict the entire trajectory of the action potential.

THE CONCEPT OF VOLTAGE-DEPENDENT NA$^+$ PERMEABILITY

Even though the Goldman–Hodgkin–Katz equation gives a good qualitative fit to the trajectory of the action potential, it fails to provide any insight into the fundamental question of how the presumed switch in permeability takes place. How can a membrane at one instant in time be highly permeable to K$^+$ and a short time thereafter be highly permeable to Na$^+$? Hodgkin and Huxley proposed that there is a voltage-dependent change in Na$^+$ permeability; Na$^+$ permeability is low at rest, but as the cell is depolarized, Na$^+$ permeability increases (Fig. 9A).

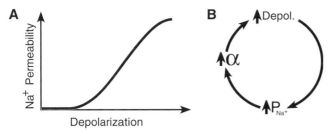

FIGURE 9   Relationship between depolarization and Na^+ permeability critical for the initiation of an action potential. (A) Na^+ permeability increases as the membrane potential becomes more depolarized. (B) The increase in Na^+ permeability produced by some initial depolarization can lead to greater depolarization.

Assume that the cell is depolarized by some stimulus. As a result of the depolarization, there will be an increase in Na^+ permeability. If Na^+ permeability is increased (assume for the moment that K^+ permeability remains unchanged), α in the Goldman–Hodgkin–Katz equation will be increased. If α is increased, the Na^+ terms are multiplied by a larger value, and they will tend to dominate the Goldman–Hodgkin–Katz equation. The membrane will become less negative (*i.e.*, more depolarized), but, as the cell depolarizes, Na^+ permeability increases further and α increases further. A positive feedback cycle is entered (Fig. 9B) such that once the cell is depolarized to a critical level, the cell will rapidly depolarize further in a regenerative fashion. Eventually, the membrane potential will approach the Na^+ equilibrium potential. Thus, a voltage-dependent relationship between membrane potential and Na^+ permeability can in principle completely account for the initiation of the action potential.

TESTING THE CONCEPT OF VOLTAGE-DEPENDENT Na^+ PERMEABILITY: THE VOLTAGE-CLAMP TECHNIQUE

So far we have just a theory. The critical hypothesis is that Na^+ permeability is regulated by the membrane potential. The simple way of testing this hypothesis is to depolarize the cell to various levels and measure the corresponding Na^+ permeability. The problem, however, is that as soon as the cell is depolarized, Na^+ permeability changes, an action potential is initiated, and because of practical reasons, there is insufficient time to measure the permeability change. This was a major obstacle in the further analysis of the ionic mechanisms that govern the action potential.

The major breakthrough came when Hodgkin and Huxley used an electronic feedback device known as a *voltage-clamp* amplifier to hold the membrane potential at various levels for indefinite periods of time (Fig. 10).

The voltage-clamp amplifier takes the difference between the actual recorded membrane potential and the desired level and generates sufficient hyperpolarizing or depolarizing current to minimize the difference. The amount of current necessary to hold the membrane potential fixed at the desired level is proportional to the membrane permeability, or conductance, at that particular voltage-clamp level. For example, by measuring the ionic current as a function of time, $I(t)$, and knowing the potential difference (which is constant), the conductance as a function of time, $G(t)$, can be determined simply by using Ohm's law (conductance for our purpose can simply be considered an electrical measurement of permeability, so we will use permeability and conductance interchangeably):

$$G(t) = I(t)/\Delta V \tag{5}$$

By changing the potential difference with the voltage-clamp amplifier, the corresponding conductances at a variety of different potentials can be determined.

Figure 11 illustrates some typical results. The procedure is as follows. Initially, the membrane potential is at its resting level of -60 mV. It is then artificially changed from the resting level to a new depolarized level (*e.g.*, -35 mV) and held there for 5 msec or longer. The membrane potential is then returned back to its resting level, and the membrane is stepped or clamped to a new depolarized level of -20 mV. By performing a sequence of these voltage-clamp measurements, changes in Na^+ permeability as a function of both voltage and time can be determined.

In the upper part of Fig. 11, the horizontal axis shows the time and the vertical axis shows the measured Na^+ conductance. As the membrane potential is forced to various depolarized levels from the resting level, there is a graded increase in Na^+ permeability. The greater the

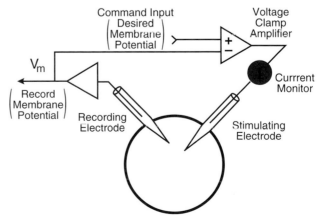

FIGURE 10   Schematic diagram of the voltage-clamp apparatus.

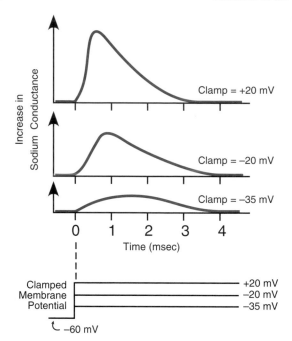

FIGURE 11   Changes in Na⁺ conductance produced by voltage steps to three depolarized levels. The greater the depolarization, the greater the amplitude of Na⁺ conductance. These data provide experimental support for the relationship depicted in Fig. 9A. (Modified from Hodgkin AL, Huxley AF. *J Physiol* 1952; 117:500.)

level of depolarization, the greater the Na⁺ permeability. This experiment therefore provides strong experimental support for the proposal that Na⁺ permeability is voltage dependent and demonstrates the existence of a mechanism that could explain the rising phase (initiation) of the action potential.

MOLECULAR BASIS FOR THE REGULATION OF NA⁺ PERMEABILITY

At the molecular level, the relationship between the membrane potential and Na⁺ permeability (Fig. 9A) is due to the existence of membrane channels that are selectively permeable to Na⁺ and that are opened or gated by the membrane potential. This discovery was made possible by the *patch-clamp technique*, which allows the conductance of individual channels to be measured. With the patch-clamp technique, a micropipette with a tip several microns in diameter is positioned so that the tip just touches the outer surface of the membrane. A high-resistance seal develops that allows the electrode and associated electronic circuitry to measure the current and thus the conductance of a small number of Na⁺ channels or, indeed, a single Na⁺ channel (Fig. 12). One of the major conclusions that has been derived from these studies is that, in response to membrane depolarization, single Na⁺ channels open in

an all-or-nothing fashion. A single channel has at least two states—open and closed—and once opened it cannot open further in response to depolarization. The gating process is subserved by a membrane-bound protein that is charged, such that when the membrane is depolarized, a conformational change in the protein takes place that results in the channel becoming more permeable to Na⁺ (see also below).

In response to membrane depolarization (Fig. 13A), individual channels open briefly and then close (Fig. 13B). The opening of single channels is a probabilistic function of time and voltage; however, when the opening of many channels is averaged, the averaged conductance predicts the conductance change of the entire population of channels (Fig. 13C). Thus, the time course of the changes of Na⁺ permeability (Fig. 11) is a reflection of the average opening and closing times of many individual Na⁺ channels. At the molecular level, the voltage dependence of the total membrane Na⁺ permeability (Fig. 9A) can be viewed as the probability that a depolarization will open single Na⁺ channels; the more the cell is depolarized, the greater the number of individual Na⁺ channels that will be opened, each in its characteristic all-or-nothing fashion.

STRUCTURE OF THE VOLTAGE-GATED NA⁺ CHANNEL

The principal structural and functional unit of the voltage-gated Na⁺ channel consists of a single polypeptide chain exhibiting four homologous domains (I–IV, Fig. 14), with each domain having six hydrophobic

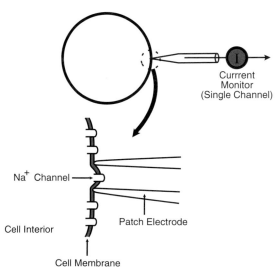

FIGURE 12   Schematic diagram of the apparatus used for single-channel recording. A micropipette is positioned to touch the surface of the membrane. A tight seal develops, and the current flowing through individual channels can be monitored.

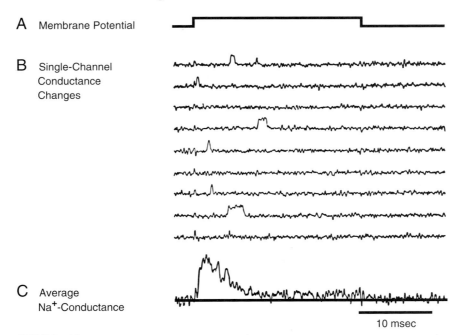

FIGURE 13 Single-channel changes in Na$^+$ conductance. In response to a pulse depolarization, the probability of single Na$^+$ channels opening is increased. (A) Voltage step used to depolarize the membrane. (B) Successive traces obtained in response to multiple presentations of the voltage step in A; with any single trace the relationship is not particularly clear. (C) When the traces are averaged, a conductance change resembling that of the macroscopic conductance (see Fig. 11) change is observed. (Modified from Sigworth FJ, Neher E. *Nature* 1980; 287:447–449.)

membrane-spanning regions designated S1 to S6. The functional significance of specific regions has been elucidated. For example, region S4 contains a high density of positively charged residues and is believed to represent the voltage sensor of the channel. The channel pore is believed to be formed by the four homologous regions between S5 and S6 (also designated the SS1 and SS2 regions, the H5 region, or the P region). Channel inactivation (see below) seems to be associated with the region that links domains III and IV. In particular, a hinged-lid structure formed by the amino terminus of domain IV has properties consistent with an ability to

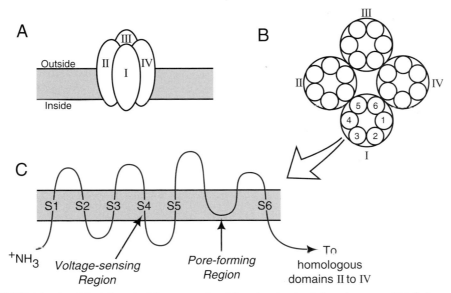

FIGURE 14 Model of the principle structural and functional unit of the voltage-gated Na$^+$ channel.

move into the channel pore and cause its blockade. Finally, the Na$^+$ channel can be regulated by *protein phosphorylation*. Specifically, the region that links domains I and II contains phosphorylation sites for cyclic adenosine monophosphate (cAMP)-dependent protein kinase, whereas the region that links domains III and IV contains a phosphorylation site for protein kinase C.

NA$^+$ INACTIVATION

There is one other important aspect of the data illustrated in Fig. 11. Even though the membrane potential is depolarized throughout the duration of each trace, Na$^+$ permeability spontaneously falls back to its resting level. Thus, although Na$^+$ permeability depends on the level of depolarization, it does not remain elevated and is only transient. Once it reaches its maximum value, it spontaneously decays back to its resting level. The process by which Na$^+$ permeability spontaneously decays back to its resting level (even though the membrane is depolarized) is known as *inactivation*. At the molecular level, the process of inactivation can be considered to be a separate voltage- and time-dependent process regulating the Na$^+$ channel. In Fig. 15, the Na$^+$ channel is represented as having two regulatory components: an activation gate and an inactivation gate. For the channel to be open, both the activation and inactivation gates must be open. At the resting potential, the activation gate is closed, and even though the inactivation gate is open, channel permeability is zero (Fig. 15A). With depolarization, the activation gate opens rapidly (Fig. 15B), and the channel becomes permeable to Na$^+$. Depolarization also tends to close the inactivation gate, but the inactivation process is slower. With depolarization occurring over a longer time, the inactivation gate closes, and even though the activation gate is still open, channel permeability is zero (Fig. 15C).

What is the physiologic significance of Na$^+$ inactivation? Let us return to the positive feedback cycle once again. Depolarization increases Na$^+$ permeability, and the increase in Na$^+$ permeability depolarizes the cell. Eventually, as a result of this regenerative cycle, the cell is depolarized rapidly up to a value near E_{Na}. The problem with this mechanism is how to account for the repolarization phase of the action potential. Based on the relationship between Na$^+$ permeability and membrane potential, one would predict that once the membrane potential moves to E_{Na} it would stay there for an indefinite period of time. The steep relationship between voltage and Na$^+$ permeability is only transient, however. After approximately 1 msec, Na$^+$ permeability

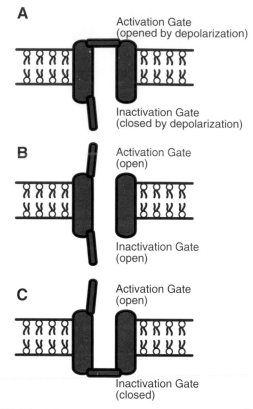

FIGURE 15 Schematic diagram of three states of the Na$^+$ channel: (A) rest; (B) peak g_{Na}; and (C) inactivated.

spontaneously decays. If Na$^+$ permeability decays, because of inactivation, the potential would move closer to E_K or, stated in a slightly different way, it will become less depolarized. Depolarization would be reduced, and the reduction in depolarization would produce a further reduction in Na$^+$ permeability (*i.e.*, deactivation) because of the basic relationship between voltage and Na$^+$ permeability (Fig. 9A). As a result, a new feedback cycle is initiated that would tend to repolarize the cell.

It is therefore intriguing to think that simply by accounting for (1) the voltage-dependent increase in Na$^+$ permeability and (2) the process of Na$^+$ inactivation, both the initiation and the repolarization phases of the action potential could be explained fully. However, there are at least two problems with this hypothesis. First, the duration of the action potential is only about 1 msec, yet the Na$^+$ permeability change takes 4 msec or so to return to the resting level (fully inactivated). So, by extrapolating these voltage-clamp measurements, one might expect that the action potential would be somewhat longer in duration than 1 msec. Second, it is difficult to explain the hyperpolarizing afterpotential. From the voltage-clamp data (Fig. 11), it is clear that Na$^+$ permeability increases dramatically and then inactivates. To explain the hyperpolarizing afterpotential, Na^+ permeability would have to be less than its initial value (α would have to be less than 0.01). So, based on the observed changes in Na^+ permeability, it would be impossible to account for the hyperpolarizing afterpotential.

ROLE OF VOLTAGE-DEPENDENT K^+ CONDUCTANCE IN REPOLARIZATION OF THE ACTION POTENTIAL

Not only does Na^+ permeability change, but there is also a change in K^+ permeability during the course of an action potential. Figure 16 illustrates these results. The upper portion of the illustration simply reviews the experimental data of Fig. 11. When the membrane is depolarized and held fixed at various levels, there is an increase in Na^+ permeability that is proportional to the depolarization. Keep in mind that during this entire sequence of events, the membrane potential is held depolarized by the voltage clamp at the levels indicated. The traces in Fig. 16 illustrate that in addition to

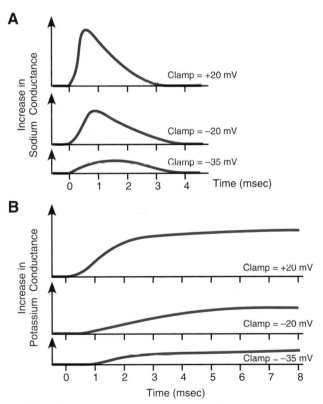

FIGURE 16 Simultaneous changes in Na^+ (A) and K^+ (B) conductance produced by voltage steps to three depolarized levels. Note the marked differences between the changes in Na^+ and K^+ conductance. (Modified from Hodgkin AL, Huxley AF. *J Physiol* 1952; 117:500.)

changes of Na^+ permeability, there are also voltage-dependent changes in K^+ permeability. The greater the level of depolarization, the greater the increase in K^+ permeability. There are two important differences between these two permeability systems. First, the changes in K^+ permeability are rather slow. It takes some time for K^+ permeability to begin to increase, whereas the changes in Na^+ permeability begin to occur immediately after the depolarization is delivered. Second, whereas Na^+ permeability exhibits inactivation, K^+ permeability remains elevated as long as the membrane potential is held depolarized.

Now that it is clear that there are changes in both Na^+ and K^+ permeabilities, how can this information be used to better account for the entire sequence of events that underlies the action potential? The initial explanation for the rising phase of the action potential is unaltered, simply because for a period of time less than about 0.5 msec, there is no major change in K^+ permeability (*i.e.*, the change in K^+ permeability is slow). In later phases of the action potential (at times greater than roughly 0.5–1 msec), we not only have to consider Na^+ permeability changes but also changes in K^+ permeability. What would be the consequences of not only having a fall in Na^+ permeability (due to inactivation), but also a simultaneous increase in K^+ permeability? Let us return to the Goldman–Hodgkin–Katz equation. At the peak of the action potential (about 0.5–1 msec from its initiation), there is very high Na^+ permeability. At this time, K^+ permeability begins to increase significantly. Thus, at any time after about 0.5 to 1 msec, not only will there be a certain increase in Na^+ permeability, but there will also be a K^+ permeability that is greater than its resting level. As a result, the value of α will be smaller than if only changes in Na^+ permeability were occurring. If α is smaller, then the Na^+ terms make less of a contribution to the Goldman–Hodgkin–Katz equation. Stated in a slightly different way, the K^+ terms make more of a contribution, and the membrane potential will be more negative. Thus, by incorporating the finding that there is a delayed increase in K^+ permeability, the membrane potential will be more negative for any given time (greater than about 0.5–1 msec) than it would have been without the changes in K^+ permeability. The delayed changes in K^+ permeability will tend to make the membrane potential repolarize more quickly because now there are two driving forces for repolarization. The first is Na^+ inactivation, and the second is the delayed increase in K^+ permeability. By incorporating the simultaneous changes in K^+ permeability, we can in principle account for a shorter duration action potential.

Can the changes in K^+ permeability help explain the hyperpolarizing afterpotential? The key is to understand the time course of the changes in K$^+$ permeability. Note that the changes in K$^+$ permeability are very slow to turn on. They are also slow to turn off. As the action potential repolarizes to the resting level, Na$^+$ permeability returns back to its resting level. Because the K$^+$ permeability system is slow, however, K$^+$ permeability is still elevated. Therefore, α in the Goldman–Hodgkin–Katz equation will actually be less than its initial level of 0.01. If α is less than 0.01, the contributions of the Na$^+$ terms become even more negligible than at rest, and the membrane potential approaches E_K. Thus, because Na$^+$ permeability decays rapidly and K$^+$ permeability decays slowly, during the later phases of the action potential K$^+$ permeability is elevated, and the hyperpolarizing afterpotential is produced.

Figure 17 summarizes the time course of the changes in Na$^+$ and K$^+$ permeability underlying the nerve action potential. Assume that by some mechanism the cell is depolarized to threshold. The depolarization initiates the voltage-dependent increase in Na$^+$ permeability. That voltage-dependent increase in Na$^+$ permeability produces a further depolarization, resulting in further increases in Na$^+$ permeability. This positive-feedback cycle leads to rapid depolarization of the cell toward E_{Na}. At the peak of the spike, which occurs about 3/4 msec from initiation of the action potential, two important processes contribute to the repolarization. First, there is the process of Na$^+$ inactivation. As a result of the decay of Na$^+$ permeability, the membrane potential begins to return to the resting level. As the membrane potential returns to the resting level, the Na$^+$ permeability decreases further (*i.e.*, deactivates), which

further speeds the repolarization process. A new *feedback cycle* is entered that moves the membrane potential in the reverse direction. Second, there is the delayed increase in K$^+$ permeability. At the point in time when the action potential reaches its peak value, there is a rather dramatic change in K$^+$ permeability. This change in K$^+$ permeability tends to move the membrane potential toward E_K. Therefore, there are two independent processes that contribute to repolarization of the action potential. One is Na$^+$ inactivation, and the other is the delayed increase in K$^+$ permeability. Note that when the action potential returns to its resting level of about −60 mV or so, the Na$^+$ permeability has reached its resting level; while Na$^+$ permeability has returned to its resting level, K$^+$ permeability remains elevated for a period of time. Thus, the ratio of the two permeabilities will be less than it was initially, and the membrane potential will move closer to E_K. Over a period of time, K$^+$ permeability gradually decays back to its resting level, and the action potential terminates.

In summary, the initiation of the action potential can be explained by the voltage-dependent increase in Na$^+$ permeability and the repolarization phase of the action potential by (1) the process of Na$^+$ inactivation and (2) by the delayed increase in K$^+$ permeability. Finally, the hyperpolarizing afterpotential can be explained by the fact that K$^+$ permeability remains elevated for a period of time after the Na$^+$ permeability has returned to its resting level.

Students frequently question the necessity for such an elaborate series of steps to generate short-duration action potentials. This question brings us back to a point raised at the beginning of the chapter. Recall that the nervous system codes information in terms of the number of action potentials elicited; the greater the stimulus intensity, the greater the frequency of action potentials. To encode and transmit more information per unit time, it is desirable to generate action potentials at a high frequency. With short-duration action potentials, a new action potential can be initiated soon after the first, and this requirement can be met.

The analysis of Hodgkin and Huxley, originally performed on the squid giant axon, has proved generally applicable to action potentials that are initiated in nerve axons and in skeletal muscle cells. The concept of voltage-dependent ion channels is now universal. What varies from cell to cell is the particular ion to which the channel is permeable. For example, a significant component of action potentials in smooth muscle and cardiac muscle cells is due to voltage-dependent Ca^{2+} channels. Many different types of voltage-dependent K$^+$ channels have also been described.

The structure of voltage-gated Ca^{2+} channels is similar to that of the Na$^+$ channel (see Fig. 14).

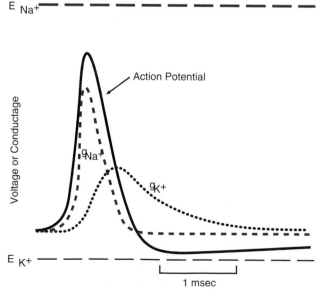

FIGURE 17   Time course of changes in Na$^+$ and K$^+$ conductance that underlie the nerve action potential.

Voltage-activated K$^+$ channels have similar six-membrane-spanning regions, but they differ in that the polypeptide chain does not contain multiply repeated domains. Rather, K$^+$ channels are formed by the functional association of four separate peptides, termed α subunits, to form an ionophore. A specific region of the N-terminal domain of the α subunit appears to be essential for proper aggregation of the subunits to form the tetrameric structures of the functional channel. Variations in the structure of individual subunits as well as different combinations of the subunits contribute to the great diversity of K$^+$ channel properties that has been observed in excitable membranes. For example, some K$^+$ channels exhibit inactivation similar to Na$^+$ channels. For these inactivating K$^+$ channels, the amino terminal sequence of the polypeptide appears to act as a plug to close the channel. Another important class of K$^+$ channels is activated by intracellular levels of Ca^{2+}.

SPECIFICITY OF ION CHANNELS UNDERLYING THE ACTION POTENTIAL

Other types of experiments further confirm the Hodgkin–Huxley hypothesis for the initiation and repolarization phases of the action potential. Compounds have been discovered that can be used selectively to block or inhibit these voltage-dependent permeability changes. One of these substances is known as *tetrodotoxin* (TTX), and the other is known as *tetraethylammonium* (TEA). TTX, a toxin that is isolated from the ovaries of the Japanese puffer fish, blocks the voltage-dependent changes in Na$^+$ permeability but has no effect on voltage-dependent changes in K$^+$ permeability. In contrast, TEA has absolutely no effect on the voltage-dependent changes in Na$^+$ permeability but completely abolishes the voltage-dependent changes in K$^+$ permeability. Thus, one substance (TTX) is capable of blocking the voltage-dependent Na$^+$ permeability, and another (TEA) is capable of blocking the voltage-dependent K$^+$ permeability.

Given the effects of TEA and TTX on permeabilities, how would one expect these substances to affect the action potential? If the voltage-dependent change in Na$^+$ permeability is blocked, one would expect that no action potential could be initiated or propagated. If the voltage-dependent change in K$^+$ permeability is blocked, one would expect the action potential to be somewhat longer in duration and, in addition, it should not have a hyperpolarizing afterpotential. Figure 18 illustrates these results; Fig. 18A is a normal action potential, and Fig. 18B is an action potential recorded in TEA. In TEA, the initiation and rising phase of the

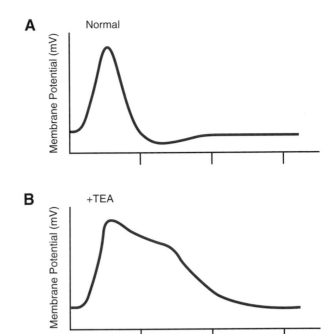

FIGURE 18 Effects of TEA on action potentials: (A) normal, and (B) in the presence of TEA.

action potential as well as its peak value are unaffected, but there is a dramatic increase in the spike duration and an absence of the hyperpolarizing afterpotential. Thus, the use of TEA confirms the Hodgkin–Huxley theory that the delayed increase in K$^+$ permeability contributes to the repolarization phase of the action potential and to the undershoot. In the presence of TEA, the process of Na$^+$ inactivation accounts entirely for the repolarization. When one perfuses the axon with TTX, one finds that no action potential can be elicited and no action potential can be propagated. Thus, the use of TTX confirms the Hodgkin–Huxley theory for the critical role that the increase in Na$^+$ permeability plays in initiating the action potential in nerve axons.

These experiments with TTX and TEA are also interesting in another respect because they demonstrate that the voltage-dependent changes in Na$^+$ permeability are mediated by completely different membrane channels from the voltage-dependent change in K$^+$ permeability because it is possible to selectively block one but not the other.

DO CHANGES IN NA$^+$ AND K$^+$ CONCENTRATIONS OCCUR DURING ACTION POTENTIALS?

It is clear that, as a result of the voltage-dependent increase in Na$^+$ permeability, some Na$^+$ ions will flow from the outside of the cell to the inside of the cell.

A frequently asked question is whether the flux of Na^+ that occurs during the action potential produces a concentration change on the inside of the cell. Alternatively, because of the increase in K^+ permeability during an action potential, there is a tendency for some K^+ to flow out of the cell. Does that flux of K^+ cause a change in the intracellular K^+ concentration? Although some Na^+ does indeed enter the cell with each action potential and some K^+ leaves the cell with each action potential, these fluxes are generally minute compared with the normal intracellular concentrations. For example, as a result of an action potential, the change in Na^+ concentration for a 1-cm^2 surface area of the membrane is equal to only approximately 1 pM (1×10^{-12} M) and that concentration change is restricted to the inner surface of the membrane. Therefore, even though some Na^+ does enter the cell with each action potential, the concentration change is minute compared with the normal millimolar concentration of Na^+ within the cell. Some K^+ also leaves the cell during an action potential, but the concentration change again is minute compared with the normal K^+ concentration. Indeed, if the (Na^+–K^+) exchange pump is blocked in the squid giant axon, it is possible to initiate more than 500,000 action potentials without any noticeable change in either the resting potential or the amplitude of the action potential.

The role of the membrane (Na^+–K^+) pump is to provide long-term maintenance of the Na^+ and K^+ concentration gradients. Eventually, if one were to generate more than 500,000 action potentials, there would be a change in ionic distribution, but this is a long-term phenomenon, and in the short term the (Na^+–K^+) pump is not essential. Even with the (Na^+–K^+) exchange pump blocked, a cell is capable of initiating a large number of action potentials without any major change in either the resting potential or the peak amplitude of the action potential. If the resting potential is unchanged, it can be inferred that the K^+ equilibrium potential is also unchanged. Similarly, if the peak amplitude of the action potential is unchanged, the Na^+ equilibrium potential is also unchanged.

THRESHOLD, ACCOMMODATION, AND ABSOLUTE AND RELATIVE REFRACTORY PERIODS

The Hodgkin–Huxley analysis not only described quantitatively the mechanisms that account for the initiation and repolarization of the action potential but also provided the explanation for some phenomena that had been known for some time but were poorly understood. Four of these phenomena are threshold, accommodation, and the absolute and relative refractory periods.

Threshold

The voltage dependence of Na^+ permeability explains the initiation of the action potential but by itself does not explain completely the threshold phenomenon, because the relationship between depolarization and Na^+ permeability, although steep, is a continuous function of membrane depolarization (e.g., see Fig. 9A). Threshold can be explained by taking into account the fact that K^+ permeability (both resting and voltage-dependent) tends to oppose the effects of increasing Na^+ conductance in depolarizing the cell and initiating an action potential. Threshold is the point at which the depolarizing effects of the increased Na^+ permeability just exceed the counter (hyperpolarizing) effects of K^+ permeability. Once the inward flow of Na^+ exceeds the outward flow of K^+, threshold is reached, and an action potential occurs through the positive feedback cycle.

Absolute and Relative Refractory Periods

The *absolute refractory period* refers to that period of time after the initiation of one action potential when it is impossible to initiate another action potential no matter what the stimulus intensity used. The *relative refractory period* refers to that period of time after the initiation of one action potential when it is possible to initiate another action potential but only with a stimulus intensity greater than that used to produce the first action potential. At least part of the relative refractory period can be explained by the hyperpolarizing afterpotential. Assume that a cell has a resting potential of −60 mV and a threshold of −45 mV. If the cell is depolarized by 15 mV to reach threshold, an all-or-nothing action potential will be initiated, followed by the associated repolarization phase and the hyperpolarizing afterpotential. What happens if one attempts to initiate a second action potential during the undershoot? Initially, the cell was depolarized by 15 mV (from −60 to −45 mV) to reach threshold. However, if the same depolarization (15 mV) is delivered during some phase of the hyperpolarizing afterpotential, the 15 mV depolarization would fail to reach threshold (−45 mV) and would be insufficient to initiate an action potential. If, however, the cell is depolarized by more than 15 mV, threshold can again be reached and another action potential initiated. Eventually, the hyperpolarizing afterpotential would terminate, and the original 15-mV stimulus would again be sufficient to reach threshold. The process of Na^+ inactivation also contributes to the relative refractory period (see below).

The absolute refractory period refers to that period of time after an action potential when it is impossible to initiate a new action potential no matter how large the stimulus. This is a relatively short period of time that varies from cell to cell but roughly occurs approximately 1/2 to 1 msec after the peak of the action potential. To understand the absolute refractory period, it is necessary to understand Na⁺ inactivation in greater detail. In Fig. 19, a membrane initially at a potential of −60 mV is voltage clamped to a new value of 0 mV (pulse 1, Fig. 19A). With depolarization, there is a rapid increase in Na⁺ permeability, followed by its spontaneous decay. When this first pulse is followed by an identical pulse (pulse 2) to the same level of membrane potential soon thereafter (Fig. 19B), there is still an increase in Na⁺ permeability, but the increase is much smaller than it was for the first stimulus. Indeed, when the separation between these pulses is reduced further, a point is reached where there is absolutely no change in Na⁺ permeability produced by the second depolarization (Fig. 19C). The two pulses must be separated by several milliseconds before the change in Na⁺ permeability is equal to that obtained initially (Fig. 19A). How do we explain these results, and what do they have to do with the absolute refractory period? Just as it takes a certain amount of time for the Na⁺ channels to inactivate, it also takes some time for these channels to recover from the inactivation and be able to respond again to a second depolarization. Therefore, as a result of initiating

one action potential, if one attempts to initiate a second action potential soon after the first, the Na⁺ permeability will not have recovered from inactivation, making a second depolarization ineffective in initiating a voltage-dependent change in Na⁺ permeability. If there is no voltage-dependent change in Na⁺ permeability, no action potential will be produced. Thus, the absolute refractory period is understood most simply in terms of this process of recovery from Na⁺ inactivation. The recovery from Na⁺ inactivation may contribute to the relative refractory period as well. During this recovery time, the threshold for an action potential will be higher because greater depolarization will be required to activate sufficient Na⁺ influx to exceed the K⁺ efflux.

Accommodation

Accommodation is defined as a change in the threshold of an excitable membrane when slow depolarization is applied. In the previous examples, rapid depolarization is applied, and the threshold occurs at a relatively fixed membrane potential (*e.g.*, Fig. 3). When a slowly developing depolarization is applied, however, the threshold is frequently at a more depolarized level and, indeed, if the depolarization is slow enough, no action potential will be initiated despite the level of depolarization. The process of Na⁺ inactivation contributes to the phenomenon of accommodation. Essentially, slow depolarization provides sufficient time for the Na⁺ channels to inactivate before they can be sufficiently activated. In terms of the molecular model of Fig. 15, an insufficient number of Na⁺ channels are in state B because they are already in state C.

Bibliography

Aidley DJ. *The physiology of excitable cells*, 3rd ed. Cambridge: Cambridge University Press, 1989

Hille B. *Ionic channels of excitable membranes*, 3rd ed. Sunderland, MA: Sinauer Associates, 2001

Kandel ER. *The cellular basis of behavior*. San Francisco: Freeman, 1976

Kandel ER, Schwartz JH, Jessell TM. *Principles of neural science*, 4th ed. New York: McGraw-Hill, 2000

Katz B. *Nerve muscle and synapse*. New York: McGraw-Hill, 1966

Nicholls JG, Martin AR, Wallace BG. *From neuron to brain*, 4th ed. Sunderland, MA: Sinauer Associates, 2001.

Schmidt RF, Ed. *Fundamentals of neurophysiology*, 2nd ed. New York: Springer-Verlag, 1978

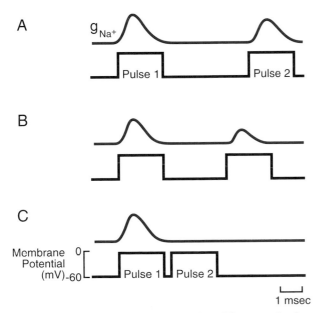

FIGURE 19 Recovery from inactivation. The second of two depolarizing pulses activates less Na⁺ conductance as the interval between the pulses decreases. Once the Na⁺ channels become inactivated by the first pulse, several milliseconds are required before they recover completely.

Suggested Readings

Anderson, PA and Greenberg, RM. Phylogeny of ion channels: clues to structure and function. *Comp Biochem Physiol B Biochem Mol Biol* 2001; 129:17–28

Bernstein J. Untersuchungen zur Thermodynamik der bioelektrischen Strome. Erster Theil. *Pflug Arch Ges Physiol* 1902; 92:521

Catterall, WA. From ionic currents to molecular mechanisms: the structure and function of voltage-gated sodium channels. *Neuron* 2000; 26:13–25

Choe, S. Potassium channel structures. *Nat Rev Neurosci* 2002; 3:115–121

Geddes LA. A short history of the electrical stimulation of excitable tissue. *Physiologist* 1984; 27(Suppl):S1–S47

Hartline HK. Intensity and duration in the excitation of single photoreceptor units. *J Cell Comp Physiol* 1935; 5:229

Hodgkin AL, Horowicz P. The influence of potassium and chloride ions on the membrane potential of single muscle fibres. *J Physiol* 1959; 148:127

Hodgkin AL, Huxley AF. Currents carried by sodium and potassium ions through the membrane of the giant axon of *Loligo*. *J Physiol* 1952a; 116:449

Hodgkin AL, Huxley AF. The components of membrane conductance in the giant axon of *Loligo*. *J Physiol* 1952b; 116:473

Hodgkin AL, Huxley AF. The dual effect of membrane potential on sodium conductance in the giant axon of *Loligo*. *J Physiol* 1952c; 116:497

Hodgkin AL, Huxley AF. A quantitative description of membrane current and its application to conduction and excitation in nerve. *J Physiol* 1952d; 117:500

Hodgkin AL, Katz B. The effect of sodium ions on the electrical activity of the giant axon of the squid. *J Physiol* 1949; 108:37

Keynes RD. The ionic movements during nervous activity. *J Physiol* 1951; 114:119

Li M, West JW, Numann R *et al*. Convergent regulation of sodium channels by protein kinase C and cAMP-dependent protein kinase. *Science* 1993; 261:1439–1442

Moreno Davila, H. Molecular and functional diversity of voltage-gated calcium channels. *Ann NY Acad Sci* 1999; 868:102–117

Nastuk WL, Hodgkin AL. The electrical activity of single muscle fibres. *J Cell Comp Physiol* 1950; 35:39

Overton E. Beitrage zur allgemeinen Muskel-und Nerve-physiologie. *Pflug Arch Ges Physiol* 1902; 92:346

Sigworth FJ, Neher E. Single Na^+ channel currents observed in cultured rat muscle cells. *Nature* 1980; 287:447–449

West JW, Patton DE, Scheuer T *et al*. A cluster of hydrophobic amino acid residues required for fast Na^+-channel inactivation. *Proc Natl Acad Sci USA* 1992; 89:10910–10914

Wu CH. Electric fish and the discovery of animal electricity. *Am Sci* 1984; 72:598–607

Yellen, G. The voltage-gated potassium channels and their relatives. *Nature* 2002; 419:6902:35–42

Propagation of the Action Potential

JOHN H. BYRNE

KEY POINTS

- Propagation of action potentials occurs by the movement of charge along the inner and outer surface of axons.
- Propagation velocity depends on the passive properties of the axonal membrane.
- The velocity of propagation is directly proportional to the space constant and inversely proportional to the time constant.
- Myelination of axons increases the propagation velocity of action potentials.

- Action potentials can be recorded by placing electrodes near the surface of the nerve.
- The magnitude of the signal extracellularly recorded depends on the number of axons that are active.
- Peaks that occur at different times after a stimulus reflect differences in the propagation velocity of action potentials in the different axons.

Up to this point, we have been considering the ionic mechanisms that underlie the action potential in a dimensionless nerve cell. However, one of the important features of action potentials is that, not only are they *elicited* in an all-or-nothing fashion, but they also are *propagated* in an all-or-nothing fashion. If an action potential is initiated in the cell body, it will propagate without decrement along the axon and eventually invade the synaptic terminal. An action potential recorded in the nerve axon has an amplitude and time course identical to the action potential that was initiated in the

cell body. It is the ability of action potentials to propagate in this all-or-nothing fashion that endows the nervous system with the capability to transmit information over long distances.

BASIC PRINCIPLES

To begin to consider the mechanisms that account for propagation of the action potential, it is useful to examine the charge distribution that is found in an

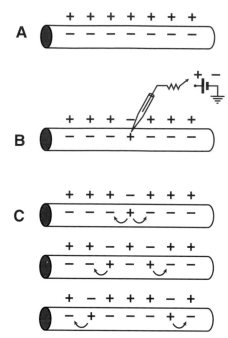

FIGURE 1 Schematic diagram of sequence of steps underlying propagation of the action potential. (A) Rest, (B) initiation, (C) propagation.

isolated section of a nerve axon. When an axon is at rest, the potential of the inside is negative with respect to the outside (Fig. 1A). The distribution of charge is simply due to the tendency of K^+ to diffuse from its region of high concentration inside the axon to its region of low concentration outside the axon. Consider what will happen if at some point along the axon an action potential is initiated. At the peak of the action potential the inside of the cell will be positive with respect to the outside (Fig. 1B). At this point, there is a new charge distribution at a localized portion of the membrane. Adjacent regions of the axon, however, still have their initial charge distribution (inside negative). Opposite charges attract each other, so the positive charge produced by the action potential will tend to move toward the adjacent region of membrane (still at rest), which has a negative charge (Fig. 1C). As a result of this positive charge movement, the adjacent region of the axon will become depolarized. If sufficient charge moves to depolarize the adjacent portion of the membrane to threshold and elicit voltage-dependent changes in Na^+ permeability, a new action potential at this adjacent region will be initiated. So, as a result of an action potential in one portion of an axon and the subsequent charge transfer along the surface of the membrane, a new action potential will be generated. This new action potential then will cause charge transfer to its adjacent region, causing, in a sense, another new action potential to be initiated. It should be clear that this process, once

initiated, will propagate all the way to the end of the axon.

DETERMINANTS OF PROPAGATION VELOCITY

What are the factors that determine the rate of propagation of the action potential? To address this question, some details of the passive properties of axonal membranes must be examined. Two important passive properties that are directly related to the rate at which an axon can propagate action potentials are the *space* (or *length*) *constant* and the *time constant*. The space and time constants are known as passive properties because they do not depend *directly* on metabolism or any voltage-dependent permeability changes such as those that underlie the action potential. These are intrinsic properties that are reflections of the physical properties of the neuronal membrane. Indeed, they are properties of all membranes.

Time Constant

To consider the time constant, first consider a very simple thermal analog. Take the case where a block of metal that is initially at 25°C is placed on a hot plate that is at 50°C. Assume that the hot-plate temperature is constant. What will be the consequences of placing the block on the hot plate? It is obvious that over a period of time the temperature of the block will change from its initial value of 25°C to a final value of 50°C, but it will not do so instantaneously. It will take a certain period of time for the heat transfer to occur. If the temperature of the block is measured as a function of time, the temperature will change as an exponential function of time, approaching a final value of 50°C.

A similar phenomenon is observed in membranes when one applies an artificial depolarizing or hyperpolarizing stimulus. This phenomenon is illustrated in Fig. 2. A nerve cell is impaled with one electrode to record the membrane potential and another electrode to depolarize or hyperpolarize the cell artificially (Fig. 2A). The cell is initially at its resting potential of −60 mV. At time zero, the stimulating electrode is connected to a battery. The size of the battery is such that the stimulus *eventually* will depolarize the cell by 10 mV (Fig. 2B). As a result, the membrane potential will change from its initial value of −60 mV to a final value of −50 mV. Note that even though the current flow (*i.e.*, stimulating current) is instantaneous (and constant), the membrane potential does not change instantaneously. There is a period of time during which the membrane potential charges to its new final level of −50 mV. This charging process is an exponential

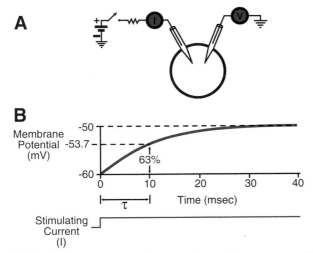

FIGURE 2 Measurement of the membrane time constant. (A) Experimental setup; (B) change in membrane potential as a function of time after the delivery of a constant step of depolarizing current. (Modified from Kandel ER. *The cellular basis of behavior.* San Francisco: Freeman, 1976.)

function of time. Just as the temperature in the block changed exponentially as a function of time, the change in potential in the neuron produced by an applied depolarization (or hyperpolarization) also follows an exponential time course. For such exponential functions of time it is possible to define what is known as a time constant (τ). The time constant refers to the time it takes for the potential change to reach 63% of its final value.

The general equation for a response that changes as an exponential function of time is:

$$\% \text{ repsonse} = 100(1 - e^{-t/\tau}) \qquad (1)$$

where t is time and τ is the time constant. At time 0, e^{-0} is equal to 1, and the % response is zero. At infinite time, $e^{-\infty}$ is equal to zero, so the % response is 100. A special case occurs when $t = \tau$. Then, $e^{-1} = 0.37$, and the % response = 63%. A detailed understanding of the mathematics is not important. What is important is that the time constant is a simple index of how rapidly a membrane will respond to a step stimulus.

The time constant for the cell illustrated in Fig. 2B is 10 msec. Thus, in 10 msec the potential has changed from −60 mV to −53.7 mV (63% of its final displacement). The smaller the time constant of the cell, the more rapidly the cell will respond to an applied stimulus. If the time constant is 1 msec, the potential change would occur very rapidly, and the potential would reach −53.7 mV in just 1 msec. If the time constant is 40 msec, the potential change would occur very slowly, and the potential would reach −53.7 mV in 40 msec.

A simple formula that describes the time constant in terms of the physical properties of the membrane is:

$$\tau = R_m C_m \qquad (2)$$

Here, R_m simply reflects the resistive properties of the membrane and is equivalent to the inverse of the permeability, because the less permeable the membrane the higher the resistance. C_m represents the *membrane capacitance*. This is a physical parameter that describes the ability of a membrane to store charge. It is equivalent to the ability of the metal block to store heat. The larger the size of a block, the better able it is to store heat. Similarly, the larger the membrane capacitance, the better able it is to store charge.

Space Constant

Before discussing how the time constant is related to propagation velocity, the other passive membrane property, the space (or length) constant, will be discussed. To introduce this phenomenon, it is useful to turn again to a thermal analog. Instead of considering a small block on a hot plate, consider what might happen when one end of a long metal rod touches the hot plate. The hot plate is at constant 50°C, and the rod is initially at 25°C. If the one end of the rod is placed in contact with the hot plate and a sufficient period of time elapses for the temperature changes to stabilize, what will be the temperature gradient along the rod? It is obvious that the temperature at the end of the rod in contact with the hot plate will be 50°C (*i.e.*, the same temperature as the hot plate). The temperature of the rod, however, will not be 50°C along its length. The temperature near the hot plate will be 50°C, but along the rod the temperature will gradually fall, and if the rod is long enough the temperature may still be 25°C at its other end. If the temperature of the rod at various distances from the hot plate is measured, the temperature will be found to decay as an exponential function of distance.

Just as there is a spatial degradation of temperature in a long rod, there is also a spatial degradation of potential along a nerve axon, which is referred to as electronic conduction. Figure 3A illustrates how it is possible to demonstrate this. One electrode is in the cell body and will be used to depolarize the cell artificially. A number of other electrodes are placed at various distances along the axon to record the potential gradient as a function of distance from the cell body. Initially, the cell body and all regions of its axon are at the resting potential of −60 mV. A sufficient *subthreshold* depolarization is then applied to the cell body to depolarize the cell body to −50 mV. Just as the temperature of the end

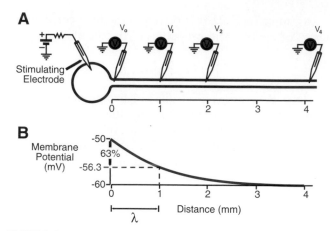

A

Stimulating
Electrode

0   1   2   3   4

B

Membrane
Potential
(mV)

-50
63%
-56.3
-60

0   1   2   3   4

λ   Distance (mm)

FIGURE 3   Measurement of the space constant. (A) Experimental
setup; (B) changes in membrane potential as a function of distance
along the axon. A constant depolarizing current is applied to the cell
body to depolarize the cell body from rest (−60 mV) to −50 mV.
(Modified from Kandel ER. *The cellular basis of behavior.*
San Francisco: Freeman, 1976.)

of the rod placed on the 50°C heat source became the
same as the hot plate, the cell body is forced from its
resting potential to a potential of −50 mV. After waiting
a sufficient period of time for the potential changes to
stabilize, the measurements are made. Very near the cell
body the potential is −50 mV (Fig. 3B). However, the
membrane potential sampled at various points away
from the cell body has changed from the value of
−50 mV at the cell body to more negative values distal
from the cell body. Measurements made a great enough
distance from the cell body reveal that the potential
recorded is not changed from the resting potential
(−60 mV). The potential profile is an exponential
function of distance and can be used to define a space
constant (denoted by the symbol λ). The space constant
is the distance it takes for the depolarizing displacement
(*i.e.*, 10 mV) to decay by 63% of its initial value. In this
particular cell, the space constant is 1 mm. This means
that 1 mm away from the cell body the potential would
have changed from its value of −50 mV in the cell body
to a value of −56.3 mV in the axon. The greater the
space constant, the greater will be the extent of the
propagation of this electrotonic potential. If the space
constant is 2 mm, this potential profile would decay less
so that at 2 mm the potential would be at −56.3 mV.

Just as it is possible to provide a formula for the time
constant in terms of the physical properties of the
membrane, it is also possible to derive a formula for the
space constant. The space constant is equal to:

$$\lambda = \sqrt{\frac{dR_m}{4R_i}} \qquad (3)$$

where R_m once again refers to the resistive properties
of the membrane (the inverse of the membrane permeability), R_i is a term that refers to the resistive properties
of the intracellular medium (the resistance of the
axoplasm to the flow of ions), and d is the diameter of
the axon.

Relationship Between Propagation Velocity and the Time and Space Constants

It is possible to make some qualitative predictions
about the way in which the space and time constants
affect propagation velocity. If the space constant is
large, a potential produced at one portion of an axon
will spread greater distances along the axon. Because the
potential will spread a greater distance along the axon, it
will bring distant regions to threshold sooner. Thus, the
greater the space constant, the greater will be the
propagation velocity. The time constant is a reflection of
the rate that a membrane can respond to an applied
stimulus current. The smaller the time constant, the
greater will be the ability of a membrane to respond
rapidly to stimulus currents. Action potentials will be
initiated sooner, and the propagation velocity will be
greater. Therefore, the smaller the time constant, the
greater will be the propagation velocity.

Thus, the *propagation velocity* is directly proportional
to the space constant but inversely proportional to the
time constant. Because relationships for both the space
and time constants are known, it is possible to derive a
new formula that describes the propagation velocity:

$$\text{Velocity} \propto \frac{\sqrt{dR_m/4R_i}}{R_m C_m} \qquad (4)$$

Thus,

$$\text{Velocity} \propto \frac{1}{C_m}\sqrt{\frac{d}{4R_m R_i}} \qquad (5)$$

It is desirable to have axons that have high propagation velocities because there is great survival value to
rapid information transmission. For example, to initiate
a motor response to some noxious stimulus, such as
touching a hot stove, action potentials must propagate
rapidly along sensory and motor axons.

Given that the propagation velocity can be described
in terms of the physical properties of nerve axons, we can
begin to examine strategies used by the nervous system to
endow axons with high propagation velocities. One of
the simplest and most obvious ways of doing this is to
increase the diameter of the axon. By increasing the
diameter, the propagation velocity is increased. This is

exactly the strategy that has been used extensively by many invertebrate axons, of which the squid giant axon is the prime example. The giant squid axon has a diameter of about 1 mm, which endows it with perhaps the highest propagation velocity of any invertebrate axon. A severe price is paid, however, when the propagation velocity is increased in this way. The key to understanding this problem is the square root relationship in the formula for propagation velocity. The square root relationship requires that to double the propagation velocity the fiber diameter must be quadrupled.

To get moderate increases in propagation velocity, therefore, one has to increase axons to very large diameters. Although this is frequently observed in invertebrates, it is not generally used in the vertebrate central nervous system. For example, it is known that the propagation velocity of axons in the optic tract is about the same as that of the squid giant axon. If all the axons in the optic tract were the size of the squid giant axon, however, the optic tract by itself would take up the space of the entire brain.

Conduction in Myelinated Axons

Clearly, there must be another means available by which axons can increase their propagation velocity without drastic changes in fiber diameter. You will note from the relationship for propagation velocity that by changing the membrane capacitance velocity can be affected directly without involving the square root relationship. It is possible to decrease the membrane capacitance simply by coating the axonal membrane with a thick insulating sheath. This is exactly the strategy used by the vertebrates. Many vertebrate axons are coated with a thick lipid layer known as *myelin*. As a result of myelin, the capacitance is greatly reduced and propagation velocity is greatly increased. In principle, there is one severe problem with increasing the propagation velocity with this technique. Coating the axon with a lipid layer would tend to cover the channels or pores in the membrane that endow the axon with the ability to initiate and propagate action potentials. The nervous system has solved this problem by coating only portions of the axon with myelin. Certain regions called nodes are not covered. At these bare regions, voltage-dependent changes in membrane permeability take place that generate action potentials.

The process of conduction in myelinated fibers is illustrated in Fig. 4. The dashed lines show a nerve axon that is covered with a layer of myelin. Note that the myelin does not cover the entire axon; there are bare regions or nodes where voltage-dependent changes in permeability can take place and action potentials can be elicited. Assume an action potential is elicited at the

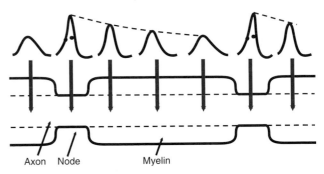

FIGURE 4 Saltatory conduction in myelinated axon. The electrical activity jumps from node to node. Between the nodes the action potential is propagated electronically with little decrement. At the nodes, where ionic current can flow, the electrotonic potential emerges from the internodal region, reaches threshold in the nodal region, and triggers an action potential. The process is then repeated. Arrows indicate regions of the axon where the potentials are recorded.

node to the left. As a result of the action potential, there is a large depolarization. The inside of the cell becomes positive with respect to the outside. The action potential cannot propagate along the myelinated region via the active process that was described earlier, simply because the voltage-dependent changes in permeability cannot take place; however, the action potential can conduct passively. That conduction will occur rapidly because the membrane capacitance is reduced. Because of the small amount of decrement, the potential that "emerges" at the next node will be of a sufficient level to depolarize the next node to threshold. A new action potential will be initiated, and the process will be repeated.

The type of propagation that occurs in myelinated fibers is known as *saltatory conduction* because the action potential appears to jump from node to node. At the nodes, there are voltage-dependent changes in membrane permeability, whereas in the internodal regions the potential is conducted in a passive fashion. No voltage-dependent changes in permeability take place in the internodal region.

EXTRACELLULAR RECORDINGS FROM EXCITABLE MEMBRANES

In many cases, it is not possible to monitor the electrical activity of nerve cells using intracellular microelectrode recording techniques; however, it is possible to measure a reflection of that intracellular activity from the immediate extracellular environment of excitable cells. Typical examples of this type of recording are the electromyogram (EMG), electrocardiogram (ECG), electrooculogram (EOG), and electroencephalogram (EEC). Each of these recordings reflects the intracellular activity from a host of simultaneously excited cells.

Extracellular Recording From a Single Axon

To introduce the principles of *extracellular recording,* it is useful to begin with the simplest case—the extracellular recording of activity from a single nerve axon. Figure 5 illustrates both the general strategy and typical results. Two metal electrodes, A and B, are placed in close proximity to an isolated nerve axon. They do not impale the axon but are very near its outside surface. The electrodes are connected to a suitable electronic amplifier that is designed so that the voltage displayed is the difference between the voltage sensed at electrode B (V_B) and the voltage sensed at electrode A (V_A):

$$V = V_B - V_A \qquad (6)$$

The recording device measures voltages B and A and subtracts the two to generate a visual display. In this example, the display is such that an upward deflection on the recording device reflects a positive difference (*e.g.*, $V_B - V_A = +$), and a downward deflection reflects a negative difference (*e.g.*, $V_B - V_A = -$).

Let us first examine the case where no action potential is propagating along the nerve axon (Fig. 5, trace 1). At rest, the membrane potential of the axon is approximately -60 mV inside with respect to the outside. Although this is not the usual convention, another way of stating it is that the axon is $+60$ mV outside with respect to the inside. Thus, the electrodes placed in the immediate environment of the cell membrane will both sense equal positive charges on the outside of the membrane. By taking the difference between the potentials at the recording electrodes, the electronic amplifier will record a potential difference of 0. This will correspond to no deflection on the recording device. To summarize, at rest,

$$V = V_B - V_A$$

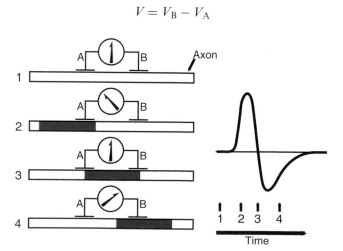

FIGURE 5   Experimental arrangement for the extracellular recording of electrical activity from a single nerve axon.

at point 1 (Fig. 5) and

$$V = (+) - (+) = 0 = \text{no deflection}$$

Let us now initiate an action potential at some distant point to the left of the axon illustrated in Fig. 5 and allow that action potential to propagate along the nerve axon toward the recording electrodes. Consider the consequences of the changes in charge distribution when the wave of depolarization representing the propagating action potential first comes near the region of electrode A. Assume for the moment that we can freeze the action potential at this point in time. We have learned that during the peak of the action potential, the potential inside the cell becomes approximately $+55$ mV with respect to the outside. Stated in a different way, the outside of the axon will be negative with respect to the inside. So, electrode A will now sense a negative outside surface charge, but electrode B will sense the same positive outside charge (because the action potential has not yet propagated to electrode B). Thus, at point 2 (Fig. 5), the electronic amplifier will take the difference between the two potentials and determine a positive difference. The positive difference will correspond to an upward deflection of the recording device. To summarize, at point 2 (Fig. 5),

$$V = V_B - V_A$$
$$V = (+) - (-) = (+) = \text{upward deflection}$$

If we now unfreeze the action potential and allow the wave of depolarization to continue to propagate, it eventually will reach a point where the action potential is in the region of both electrodes A and B (trace 3, Fig. 5). We now freeze the action potential at the point at which each electrode senses equal parts of the action potential. The regions under both electrodes will be positive inside with respect to the outside or negative outside with respect to the inside. So, electrode B will sense a negative charge, and electrode A will sense the same negative charge. The electronic amplifier will subtract the two and determine that there is no difference. The recording device will return to its initial state and display no deflection. To summarize, at point 3 (Fig. 5),

$$V = V_B - V_A$$
$$V = (-) - (-) = 0 = \text{no deflection}$$

The potential recorded at point 3 is an interesting case. The recording device is back to its initial state, but that does not mean that the recorded potential is equal to the resting potential. Rather, it means that electrodes A and B are recording the same potential.

What happens as the action potential continues to propagate along the axon? The area of excitation under the two electrodes will move away from electrode A and eventually will reach a point at which the excited region (the action potential) is predominantly in the vicinity of electrode B. As a result, the region under electrode A will repolarize so that the inside of the axon will become negative with respect to the outside, or the outside will be positive with respect to the inside (as it was initially). The region under electrode B will still be excited so that it will be negatively charged. Thus, at point 4 (Fig. 5), the difference between the potentials at the two electrodes is negative, and a negative difference corresponds to a downward deflection of the recording device: At point 4,

$$V = V_B - V_A$$
$$V = (-) - (+) = (-) = \text{downward deflection}$$

Eventually, the excited region will move away from the electrodes, and both electrodes will sense positive charge. The difference between the two will be zero, and the recording device will return to its initial state.

Thus, by using extracellular recording techniques, it is possible to obtain signals that correspond to the passage of an action potential along a nerve axon. Note that the form of the measured potential is very different from the action potential recorded with an intracellular microelectrode. In addition, it is only a rough reflection of the magnitude and duration of the underlying membrane permeability changes. This type of recording is used primarily as a simple index of whether or not an action potential has occurred.

Extracellular Recording From a Nerve Bundle

Figure 6 illustrates stimulation and recording of a nerve bundle rather than a single axon. Recall that a nerve bundle, or nerve, is made up of a group of individual axons. Stimulating electrodes are placed near one end of the nerve bundle to depolarize the axons to threshold and initiate action potentials, and recording electrodes are placed at the other end to record the resulting extracellular potential changes produced by the action potentials propagating along the nerve. Trace 1 of Fig. 6 shows that when a very weak stimulus is delivered no action potential is recorded. The small transient diphasic deflection is due to interference picked up by the recording electrodes from the stimulating electrodes; this small initial deflection is known as a *stimulus artifact*.

Increasing the stimulus intensity further produces a larger artifact, but still no action potential is produced

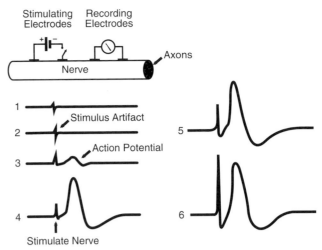

FIGURE 6   Extracellular recording of the electrical activity from a nerve. One pair of electrodes is placed so that the nerve can be electrically stimulated while a second pair of electrodes records the neural activity. With weak intensity stimulation (traces 1 and 2), only stimulus artifacts are produced, and no neural activity is observed. As the stimulus intensity to the nerve is increased, the amplitude of the action potential increases (traces 3–5) until a point is reached where further increases in intensity (although producing a larger artifact) do not produce a larger action potential (traces 5 and 6). (Modified from Katz B. *Nerve muscle and synapse*. New York: McGraw-Hill, 1966.)

(trace 2); however, as the intensity is increased further, a point is reached at which a diphasic action potential is recorded (trace 3, Fig. 6). Note that there is a delay between the stimulus artifact and the diphasic action potential, which is a reflection of the time it takes for the action potential to propagate from its site of initiation (at the stimulating electrodes) to the recording electrodes. (Knowing the time delay and the distance between the stimulating and recording electrodes, the propagation velocity of the action potentials can be calculated.)

An interesting observation occurs when the stimulus intensity is increased still further. Now the amplitude of the extracellular action potential is increased (trace 4, Fig. 6). Increasing the stimulus intensity further produces yet a greater increase in the size of the action potential (trace 5, Fig. 6). Eventually, a point is reached at which increasing the stimulus intensity produces no further increase in the size of the action potential.

At first, these results may seem somewhat paradoxical because they appear to contradict the all-or-nothing law of the action potential. Recall that an action potential not only is propagated in an all-or-nothing fashion but is also initiated in an all-or-nothing fashion. We learned earlier (Fig. 3 in Chapter 4) that an increase in stimulus intensity beyond threshold produces an action potential identical in its amplitude and time course to the action potential produced with a threshold stimulus. In Fig. 6 the action potential amplitude increases as a function of the stimulus intensity. The key to understanding this apparent paradox is that we are now dealing with a nerve bundle that contains many different axons. The individual axons in the nerve have different diameters and therefore different thresholds to extracellularly applied stimulating currents. As a result, when the stimulus intensity is increased, the first axon to initiate an action potential is the nerve axon that has the lowest threshold. That action potential is then sensed by the recording electrodes and displayed on the recording device. As the stimulus intensity is increased further, more axons are brought to threshold so that there are multiple action potentials propagating along the nerve bundle, and these individual action potentials will each make a contribution to the signal sensed by the recording electrodes. Eventually, a stimulus intensity is reached that brings all the axons above threshold so that further increases in the stimulus intensity initiate no additional action potentials, and the size of the extracellularly recorded action potential remains at its peak value. The action potential that is recorded with extracellular electrodes from a nerve bundle is known as a *compound action potential* (compound because it is a summation of contributions from the individual action potentials in the axons that compose the nerve bundle).

One additional variant on this theme is illustrated in Fig. 7. This is a case in which the stimulating and recording electrodes are separated by a considerable distance. An interesting observation is that the potential change recorded can be quite different from the shape of the potential changes seen in Figs. 5 and 6. Previously, there was a relatively smooth rise, fall, and then recovery—a diphasic action potential. In this new case, the action potential has multiple peaks (in some cases, there can be three or more different peaks). What is the origin of these different peaks? The important point is that a nerve bundle may contain many different axons, many of which have different diameters. In addition, some of these axons may be myelinated and others unmyelinated. As a result, action potentials that are initiated in these axons will have different propagation velocities. The action potentials will reach the recording electrodes at different times so the different peaks reflect differences in the time of arrival of action potentials. The situation is somewhat analogous to a fast runner

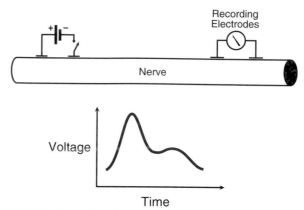

FIGURE 7   Recording the compound action potential at a point distant from the stimulus site. Multiple peaks can be observed due to the contributions of fibers with different conduction velocities.

and a slow runner in a race. Initially, they are both at the starting line, but over a period of time the faster runner outdistances the slower runner and reaches the finish line sooner. The stimulating electrodes are analogous to the starting line, whereas the recording electrodes are analogous to the finish line.

Bibliography

Adley DJ. *The physiology of excitable cells*, 3rd ed. Cambridge: Cambridge University Press, 1989

Hodgkin AL. *The conduction of the nervous impulse.* Springfield, IL: Charles C Thomas, 1984

Jack JJB, Nobel D, Tsien RW. *Electric current flow in excitable cells.* Oxford: Clarendon Press, 1975

Katz B. *Nerve muscle and synapse.* New York: McGraw-Hill, 1966

Nicholls JG, Martin AR, Wallace BG. *From neuron to brain*, 4th ed. Sunderland, MA: Sinauer Associates, 2001

Ruch T, Patton HD, Eds. *Physiology and biophysics.* Philadelphia: W.B. Saunders, 1982

Shepherd GM, Ed. *The synaptic organization of the brain*, 4th ed. New York: Oxford University Press, 1998

Suggested Readings

Huxley AF, Stampfli R. Evidence for saltatory conduction in peripheral myelinated nerve-fibres. *J Physiol* 1949a; 108:315

Huxley AF, Stampfli R. Saltatory transmission of the nervous impulse. *Arch Sci Physiol* 1949b; 3:435

Rushton WAH. A theory of the effects of fibre size in medullated nerve. *J Physiol* 1951; 115:101

Tasaki I. Conduction of the nerve impulse. *Handbook of physiology. Section 1: Neurophysiology*, Vol. 1. Baltimore: Williams & Wilkins, 1959

Neuromuscular and Synaptic Transmission

JOHN H. BYRNE

KEY POINTS

- Synaptic transmission occurs by electrical and chemical synapses; chemical synapses are the predominant type in the nervous system.
- The end-plate potential is due to the release of ACh from the presynaptic terminals of motor axons.
- ACh is removed from the synaptic cleft by diffusion and hydrolysis by AChE.
- ACh opens channels in the postsynaptic membrane that are equally permeable to Na^+ and K^+.
- Ca^{2+} is essential for the release of neurotransmitter; the presynaptic terminal contains voltage-dependent Ca^{2+} channels, and Ca^{2+} influx during an action potential in the presynaptic terminal promotes the exocytosis of vesicles of transmitter.

- Transmitter release is quantized.
- Synaptic potentials can be either excitatory or inhibitory and are mediated by ionotropic and metabotropic receptors.
- Synaptic responses mediated by ionotropic receptors are generally fast because they involve the direct opening of a channel by the transmitter.
- Synaptic responses mediated by metabotropic receptors are generally slow because they can involve the activation of second messenger pathways.
- Integration of synaptic potentials occurs through temporal and spatial summation.

Nerve cells are capable not only of initiating action potentials but also of communicating directly with other cells. Transferring information from one nerve cell to another or from one nerve cell to a muscle cell is a process known as *synaptic transmission*. Synaptic transmission is mediated by specialized junctions called *synapses*.

TYPES OF SYNAPTIC TRANSMISSION

There are two categories of synaptic transmission: electrical and chemical. Each category is associated with a number of characteristic morphologic and physiologic properties. Features of these two types of synaptic transmission are illustrated in Fig. 1. *Electrical synaptic transmission* is mediated by specialized structures known as gap junctions. The gap junctions associated with electrical synapses provide a pathway for cytoplasmic continuity between the presynaptic and the postsynaptic cells. As a result, a depolarization (or a hyperpolarization) produced in the presynaptic terminal produces a change in potential of the postsynaptic terminal. The potential change in the postsynaptic cell is due to the direct ionic pathway between the presynaptic and postsynaptic cell. For electrical synapses, there is a minimal synaptic delay. As soon as a potential change is produced in the presynaptic terminal, there is a reflection of that potential change in the postsynaptic cell.

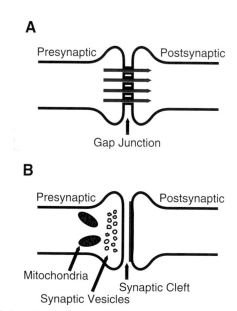

A

Presynaptic   Postsynaptic

Gap Junction

B

Presynaptic   Postsynaptic

Mitochondria
          Synaptic Cleft
Synaptic Vesicles

FIGURE 1   Schematic diagram of two types of synaptic junctions. (A) Electrical synapse; transmission takes place through gap junctions that provide an ionic pathway between the pre- and postsynaptic junctions. (B) Chemical synapse; neurotransmitter is released from the presynaptic terminal, which diffuses across the synaptic cleft and interacts with receptor sites on the postsynaptic membrane.

Electrical junctions are found not only in the nervous system but also between other excitable membranes, such as smooth muscle and cardiac muscle cells. In these muscle cells, they provide an important pathway for the propagation of action potentials from one muscle cell to another.

For *chemical synapses,* there is a distinct cytoplasmic discontinuity that separates the presynaptic and post-synaptic membranes. This discontinuity is known as the *synaptic cleft*. The presynaptic terminal of chemical synapses contains a high concentration of mitochondria and synaptic vesicles, and there is a characteristic thickening of the postsynaptic membrane. As a result of a depolarization or an action potential in the presynaptic terminal, chemical transmitters are released from the presynaptic terminal, which diffuse across the synaptic cleft and bind to receptor sites on the postsynaptic membrane. This leads to a permeability change that produces the postsynaptic potential. For chemical synapses, there is a delay (usually, approximately 0.5–1 msec in duration) between the initiation of an action potential in the presynaptic terminal and a potential change in the postsynaptic cell. The *synaptic delay* is due to the time necessary for transmitter to be released, diffuse across the cleft, and bind with receptors on the postsynaptic membrane. Chemical synaptic transmission is generally unidirectional. A potential change in the presynaptic cell releases a transmitter that produces a postsynaptic potential, but a depolarization in the postsynaptic cell does not produce any effects in the presynaptic cell because no transmitter is released from the postsynaptic cell at the synaptic region. The most predominant type of synapse is the chemical synapse, and for that reason this chapter will focus on chemical synaptic transmission.

SYNAPTIC TRANSMISSION AT THE NEUROMUSCULAR JUNCTION

The chemical synapse from which most of our information about synaptic transmission has been derived is the synaptic contact made by a spinal motor neuron with a skeletal muscle cell. Figure 2 is a schematic diagram of some of the general features of this synapse. The myelinated axon of a motor neuron whose cell body is located in the ventral horn of the spinal cord innervates a number of individual muscle fibers (Fig. 2A). At each muscle fiber, the axon branches further and forms a series of contacts with the muscle cell. An expanded view of one such synaptic contact, which illustrates some of the characteristic morphologic features of the chemical synapse, is shown in Fig. 2B. There are (1) a large concentration of small vesicles and

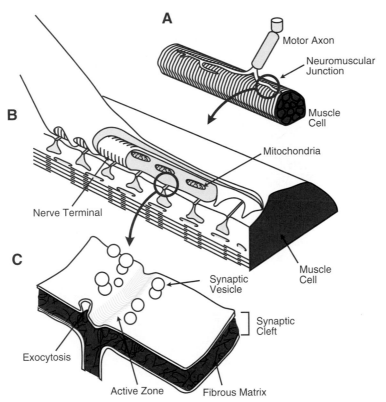

FIGURE 2 Sketch of the skeletal neuromuscular junction. (Modified from Lester HA. *Sci Am* 1977; 236:106; Hille B. *Ionic channels of excitable membranes.* Sunderland, MA: Sinauer Associates, 1984.)

mitochondria in the presynaptic terminal, and (2) a distinct synaptic cleft that separates the presynaptic terminal from the postsynaptic cell (in this case, the skeletal muscle cell). This particular synapse, because of its characteristic shape, is known as the *motor end plate.* It is also known as the *neuromuscular junction* because it is the junction made by a motor axon with a muscle cell. Additional morphologic features of the synapse at the neuromuscular junction are illustrated by the expanded view in Fig. 2C. Here, it is seen that: (1) the synaptic vesicles are clustered in distinct regions of the terminal known as the *active zone;* (2) the active zone is opposite an invagination of the muscle membrane known as the *junctional fold;* and (3) there is a characteristic thickening of the postsynaptic membrane, perhaps due to the high density of postsynaptic receptors for acetylcholine (ACh). In addition, the synaptic cleft is filled with a fibrous matrix and the enzyme acetylcholinesterase (AChE). The functions of ACh and AChE are described below. Figure 3 is an electron micrograph of the synaptic junction illustrating the high concentration of vesicles and mitochondria in the presynaptic terminal, the fibrous matrix in the synaptic cleft, and the characteristic thickening of the postjunctional membrane.

END-PLATE POTENTIAL

It is possible to study various aspects of chemical synaptic transmission in a reduced preparation. One can dissect a skeletal muscle cell (with its intact neural innervation) from an animal and place it in an experimental solution, where it can remain viable for long periods of time (Fig. 4). The postsynaptic muscle cell is impaled with a microelectrode to record potentials in the muscle cell, and the motor axon is stimulated to initiate action potentials in the axon.

Figure 5A illustrates a typical result. At the arrow, an electrical stimulus is delivered to the motor axon. This stimulus elicits an action potential in the axon, which then propagates down the axon, invades the synaptic terminal, and leads to the release of chemical transmitter. The transmitter diffuses across the cleft and binds with receptor sites on the postsynaptic membrane to trigger the illustrated sequence of potential changes. Note that there is a distinct delay between the application of the electrical stimulus and the production of any potential change in the muscle cell. The delay is due to two factors: (1) it takes time for the action potential to propagate from its site of initiation down the motor axon, and (2) there is a delay due to the time necessary

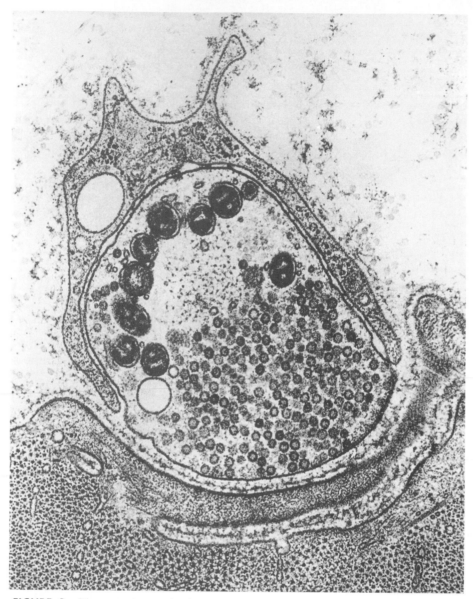

FIGURE 3   Electron micrograph of a synaptic contact at the neuromuscular junction. The presynaptic terminal (upper portion) contains many small vesicles and larger mitochondria. There is a distinct separation (the synaptic cleft) between the membranes of the pre- and postsynaptic cells. (Micrograph produced by Dr. John E. Heuser of Washington University, St. Louis, MO.)

for the chemical transmitter substance to be released from the presynaptic terminal, diffuse across the synaptic cleft, and produce the permeability changes that trigger the potential changes recorded in the postsynaptic muscle cell.

There are two components to the potential changes in the muscle that are produced as a result of stimulating the motor axon. At first, there is a relatively slow rising potential. At a potential of about −50 mV, there is a sharp inflection at which a second potential is triggered. This second potential, the action potential, quickly reaches a peak value and then rapidly decays back to the resting potential. In the discussion here, the major focus

is not the action potential but the somewhat slower initial underlying event that triggers the action potential.

An important chemical substance that has facilitated the analysis of synaptic transmission at the neuromuscular junction is *curare*. (Curare is derived from plants and is used by some South American Indians for arrow poison.) If a low dose of curare is added (Fig. 5B) and the motor axon is again stimulated, the slower underlying event is reduced in amplitude but is still capable of depolarizing the muscle cell to threshold and initiating an action potential (assume that threshold in this cell is about −50 mV). If a somewhat higher

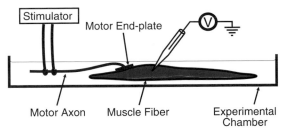

FIGURE 4 Schematic diagram of the preparation used to study features of synaptic transmission at the skeletal neuromuscular junction.

dose of curare is added, the slower underlying event is reduced further (Fig. 5C) and now fails to reach threshold. This underlying postsynaptic potential that is produced as a result of stimulation of the presynaptic motor axon and release of chemical transmitter substance is known as the *end-plate potential* (EPP). It is called the end-plate potential because it is the potential that is recorded at the motor end plate. One of the clear results of this experiment is that curare reduces the amplitude of the EPP. If sufficient curare is added, the EPP is completely abolished. A person poisoned with curare will asphyxiate because curare blocks neuromuscular transmission at the respiratory muscles. Note that if the muscle cell is artificially depolarized to threshold in the presence of a high dose of curare, an action potential can still be initiated that is indistinguishable from an action potential produced artificially in the absence of curare. Thus, although curare is effective in blocking the EPP, it has no *direct* effect on the action potential. Because curare does not affect the action potential, it does not affect the voltage-dependent Na^+ and K^+ channels that underlie the action potential.

The EPP is the critical event underlying the initiation of an action potential in a muscle cell. For this reason,

we will explore some of the properties of EPPs in considerable detail.

PROPAGATION OF THE END-PLATE POTENTIAL

One obvious question is whether the EPP is propagated in an all-or-nothing fashion like the action potential. Figure 6 shows a simple experiment designed to answer this question. In this experiment, the motor axon is electrically stimulated, and multiple intracellular recordings are made from the muscle cell at 1-mm intervals from the end plate. There is a large EPP (V_o) at the region of the motor end plate; however, at more distant regions from the end plate, the amplitude of the EPP becomes smaller. In fact, the decay is an exponential function of distance. This indicates that the EPP is not propagated in an all-or-nothing fashion; rather, it spreads with decrement, just as weak, artificially produced hyperpolarizations or depolarizations (subthreshold) spread with decrement along a nerve axon. As shown in Fig. 6, an agent such as curare is used so that the EPP is reduced in size and fails to trigger an action potential. When the EPP triggers an action potential (as occurs in the absence of curare), the action potential propagates without decrement along the muscle cell.

ACETYLCHOLINE HYPOTHESIS FOR SYNAPTIC TRANSMISSION AT THE NEUROMUSCULAR JUNCTION

One of the major breakthroughs in the characterization of the sequence of events that underlie synaptic transmission at the neuromuscular junction was the identification of the chemical transmitter substance. In

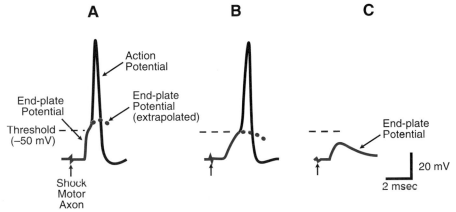

FIGURE 5 The recording of the end-plate potential (EPP) and the effects of curare. (A) Normal; (B) low dose of curare; (C) high dose of curare.

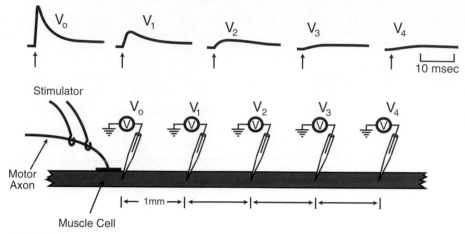

FIGURE 6   Propagation of the EPP. A muscle cell is impaled at multiple points from the end plate, and the motor axon is stimulated. The amplitude of the EPP is maximal at the motor end plate. As recordings are made more distant from the end plate, the amplitude of the EPP becomes smaller. (Modified from Fatt P, Katz B. *J Physiol* 1951; 115:320–370.)

1936, Dale and his colleagues found that electrical stimulation of motor axons led to an increase in the concentration of a substance in the perfusing solution, which they identified as ACh. When the isolated ACh was injected into the arterial supply it was capable of producing large muscular contractions. Subsequent studies on the transmitter substance used by the neuromuscular junction have confirmed and greatly extended the original observations. As a result, it is now possible to describe the total sequence of events under-lying synaptic transmission at the neuromuscular junction (Fig. 7).

ACh is synthesized and stored in the presynaptic terminals of motor axons. As a result of a nerve action potential that invades the presynaptic terminal, ACh is released into the synaptic cleft. Acetylcholine diffuses across the synaptic cleft and combines with receptors on the postsynaptic or the postjunctional membrane. When ACh binds with these receptors, there is an increase in Na^+ and K^+ permeabilities. The increase in Na^+ and K^+ permeabilities depolarizes the postjunctional, or the postsynaptic, membrane. This depolarization is known as the EPP. Normally, the EPP is approximately 50 mV in amplitude, so the threshold level of the muscle is easily reached and an action potential is triggered. The action potential in the muscle cell leads to muscular contraction. The EPP is a transient event that persists for about 10 msec. There are two reasons for its transient nature. First, ACh diffuses away from the synaptic cleft and produces no further permeability changes. Second, there is a substance in the synaptic cleft known as AChE. AChE hydrolyzes ACh into inactive substances. In the following sections, this sequence of events is examined in greater detail.

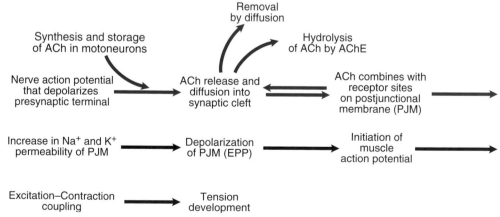

FIGURE 7   Summary of the sequence of events underlying synaptic transmission at the skeletal neuromuscular junction. PJM, postjunctional membrane; EPP, end-plate potential; ACh, acetylcholine; AChE, acetylcholinesterase.

Role of Acetylcholinesterase

One aspect of the ACh hypothesis for chemical synaptic transmission at the neuromuscular junction involves the action of AChE. If AChE acts to break down ACh into inactive forms, one would predict that preventing this hydrolysis would allow ACh to act in the synaptic cleft for a longer time and produce a larger and longer lasting EPP. A group of substances have been identified, one of which is known as *neostigmine,* that block the action of AChE. Figure 8 illustrates the effects of neostigmine on the EPP. The lower trace shows a normal EPP. Once again the EPP is reduced (*e.g.,* by a low dose of curare) such that it fails to trigger an action potential. The upper trace shows the effect of adding neostigmine to the extracellular medium. In the presence of neostigmine, the amplitude and duration of the EPP are increased.

Iontophoresis of Acetylcholine

Another prediction of the ACh hypothesis is that it should be possible to mimic the release of ACh from the presynaptic terminal by artificially applying some ACh to the vicinity of the neuromuscular junction. A simple technique called *iontophoresis* is available to precisely deliver very small amounts of ACh to restricted regions of a cell. The technique is illustrated in Fig. 9. One intracellular microelectrode is used to record the membrane potential. Another extracellular microelectrode is filled with ACh and is placed close to the neuromuscular junction (but it does not impale either the motor axon or the muscle). ACh is a positively charged molecule. If the positive pole of a battery is connected to the ACh electrode, the positive charge of the battery will force ACh out of the electrode. The small amount of ACh ejected from the tip of the micropipette could then bind with any ACh receptors on the muscle cell. If ACh is the transmitter at the neuromuscular junction, iontophoretic application of

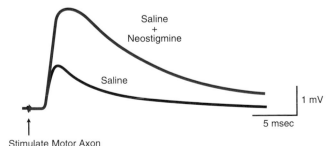

FIGURE 8   Effects of neostigmine on the EPP. In a saline solution containing an agent such as a small dose of curare to keep the EPP subthreshold, a small EPP is produced. When neostigmine (an agent that blocks the actions of AChE) is also added to the bath, the amplitude and duration of the EPP increase.

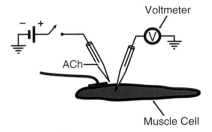

FIGURE 9   Procedure for iontophoretic application of ACh. Standard intracellular recordings are made from a skeletal muscle cell while a second micropipette filled with ACh is positioned near the neuromuscular junction. By applying a positive voltage to the ACh-filled micropipette, small quantities of ACh can be ejected in the vicinity of the motor end plate.

ACh should produce a potential change in the muscle cell similar to that produced by stimulation of the motor axon.

Figure 10 illustrates the results. The upper trace is a normal EPP produced by stimulating the motor axon. The small initial downward deflection is the stimulus artifact that indicates the point in time when the motor axon is electrically stimulated. The lower trace shows the result of the iontophoretic application of ACh. The potential changes produced are nearly identical. Experiments similar to Fig. 10 have demonstrated a number of other aspects of synaptic transmission at the neuromuscular junction. When ACh is applied in the presence of neostigmine, the size of the potential is enhanced. These results are consistent with the cholinergic nature of synaptic transmission. If the preparation is perfused with neostigmine, ACh is not broken down by AChE, more receptors are bound, the permeability changes are greater, and the resultant potential change is greater. The iontophoretic potential is also affected by curare. By adding curare to the experimental preparation, the size of the iontophoretic response is reduced. In addition, the iontophoretically produced response is not affected by tetrodotoxin (TTX). We have already learned that TTX blocks the voltage-dependent change in Na^+ permeability underlying the action potential. Because TTX has no effect on the iontophoretic application of ACh, channels in the membrane that are sensitive to ACh must be different from the channels in the membrane that underlie the action potential. It has also been observed that injecting ACh into the muscle cell produces no potential change. Thus, the receptors for ACh must be on the outer surface of the muscle cell. Iontophoretic applications of ACh to points along the muscle distant from the end plate yield no potential changes. A potential change due to local application of ACh is obtained only in the immediate vicinity of the motor end plate. Presumably, there is no potential change produced by ACh at sites more distant

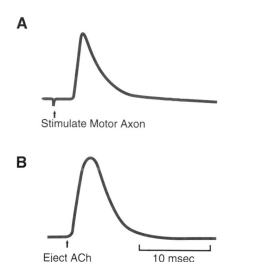

A

Stimulate Motor Axon

B

Eject ACh   10 msec

FIGURE 10   Comparison of the EPP (A) and the response to iontophoretic application of ACh (B).

from the motor end plate because there are no receptors for ACh at these more distant sites. The receptors for ACh are located at the neuromuscular junction. One additional insight has been obtained from the iontophoretic application of ACh. As indicated above, curare blocks the EPP. The action of curare could be at two basic sites, however. Curare could be acting to reduce the release of chemical transmitter from the presynaptic terminal, or, alternatively, it could have a postsynaptic effect. When ACh is artificially applied, the possibility of presynaptic changes is eliminated. Therefore, the reduction of the iontophoretic potential with curare is due to a postsynaptic action. As a result of these and other studies it is now known that curare is a competitive inhibitor of ACh. Curare binds with the same receptor site on the postjunctional membrane as does ACh. Although it binds with the receptors, it does not produce the resultant permeability changes.

IONIC MECHANISMS UNDERLYING THE END-PLATE POTENTIAL

How does ACh produce the permeability change responsible for the EPP? In the early 1950s, Bernard Katz and his colleagues proposed that the binding of ACh with receptors on the postjunctional membrane led to a simultaneous increase in Na^+ and K^+ permeabilities. If a membrane is permeable only to Na^+ and K^+, then the Goldman–Hodgkin–Katz (GHK) equation is applicable:

$$V_m = 60 \log \frac{[K^+]_o + \alpha[Na^+]_o}{[K^+]_i + \alpha[Na^+]_i} \, (mV)$$

where $\alpha = P_{Na}/P_K$. If $\alpha = 1$ and the K^+ and Na^+ concentrations on the inside and outside of the cell are substituted into the equation, a membrane potential of about 0 mV is predicted.

Even though the EPP is due to the opening channels that have equal permeability to Na^+ and K^+, an EPP with a peak value of 0 mV does not occur for two reasons. First, as the muscle membrane depolarizes toward 0 mV, threshold is reached, and an action potential is initiated or triggered and the EPP is obscured. Second, the membrane channels opened by ACh are only a small fraction of the ion channels in the muscle cell. These other channels (not affected by ACh) tend to hold the membrane potential at the resting potential and prevent the membrane potential from reaching the 0-mV level. (When the EPP triggers an action potential, a potential more positive than 0 mV is reached, but this is a reflection of the ionic mechanisms underlying the action potential and not the ionic mechanisms underlying the EPP.)

During the past 15 years, considerable evidence has accumulated about the molecular events associated with the actions of ACh. Using patch recording techniques (Fig. 11), it has been possible to measure the ionic current flowing through single channels opened by ACh. When the micropipette contains normal saline without ACh, no electrical changes are observed. However, when ACh is added to the electrode, small, step-like fluctuations are observed which are the result of ions flowing through specific membrane channels that are opened by ACh. The electrical events associated with the opening of a single channel are extremely small and, as a result, any single channel makes a small contribution to the normal EPP.

As a result of these patch recording techniques, three general conclusions can be drawn. First, ACh causes the opening of individual ionic channels (for a channel to open, two molecules of ACh must bind to the receptor). Second, when an ACh-sensitive channel opens, it does so in an all-or-nothing fashion. Increasing the concentration of ACh in the patch microelectrode does not increase the permeability of the channel; rather, it increases its probability of opening. And, finally, when a larger region of the muscle, and thus more than one channel, is exposed to ACh the net permeability is larger because more individual channels are opened, each in their characteristic all-or-nothing fashion. It is this summation of the permeability of many individual open channels that gives rise to the normal EPP. The properties of ACh-sensitive channels and voltage-sensitive Na^+ channels are similar in that both channels open in an all-or-nothing fashion and, as a result, the macroscopic effects that are recorded are due to the summation of many individual open ion channels.

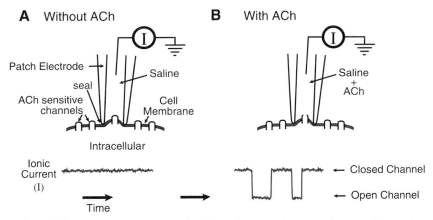

FIGURE 11 Techniques for measuring the conductance change produced by the opening of single ACh-sensitive membrane channels. (A) When the solution in the patch electrode contains no ACh, the membrane channel is closed. (B) With ACh in the pipette the channel opens and closes in a probabilistic fashion. (Modified from Neher E, Sakmann B. *Nature* 1976; 260:779–802.)

The two types of channels differ, however, in that one is opened by a chemical agent, whereas the other is opened by depolarization.

Structure of the Ligand-Gated Acetylcholine Receptor

Unlike the Na^+ channel, which consists of a single polypeptide chain with four homologous domains (each of which has six membrane-spanning regions; see Chapter 4, Fig. 14), the cation channel of the ACh receptor consists of five subunits, four of which are from separate gene products. The subunits have been designated α, β, γ, ε, and δ, and in fetal muscle are arranged in the stoichiometry of $\alpha^2\beta\gamma\delta$ (Fig. 12A,B), whereas in adult muscle the stoichiometry is $\alpha^2\beta\varepsilon\delta$. The subunits are approximately 50% homologous at the amino acid level, indicating that they evolved from a common ancestral gene. A common feature to all subunits is their four membrane-spanning regions (designated M1 to M4; Fig. 12B,C). The pore of the channel is believed to be formed by the proper alignment of the M2 regions of each of the five subunits (Fig. 12B). A key difference between the α subunit and the other subunits is its

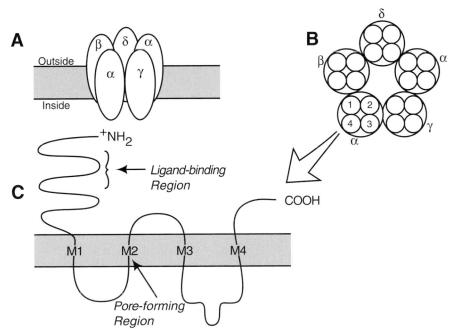

FIGURE 12 Model of the fetal ligand-gated ACh receptor/channel.

extended amino terminal (NH$_2$) domain that contains the extracellular binding site for ACh (Fig. 12C). The long cytoplasmic loop found between the M3 and M4 regions is present in all the ACh subunits. It may function as a binding site to the cytoskeleton.

PRESYNAPTIC EVENTS UNDERLYING THE RELEASE OF NEUROTRANSMITTER

Voltage-Dependent Release of Neurotransmitter

One of the most interesting aspects of synaptic transmission is the mechanism by which an action potential in the presynaptic terminal triggers the release of the chemical transmitter substance. One possibility is that transmitter release is due to some aspect of the sequence of permeability changes underlying the action potential in the presynaptic terminal. The action potential is associated with a slight influx of Na$^+$ and a slight efflux of K$^+$. Perhaps in some way either the small influx of Na$^+$ or efflux of K$^+$ disturbs the intracellular environment and causes the release of transmitter.

Figure 13 illustrates an experiment to examine this hypothesis. Rather than using the neuromuscular junction, a more advantageous experimental preparation, the squid giant synapse, is used. The squid giant synapse is so large that the presynaptic terminal can be impaled with two electrodes: one to record the presynaptic potential and the other to depolarize the terminal artificially (Fig. 14) (this is not possible at the skeletal neuromuscular junction). A third electrode is used to record the potential changes in the postsynaptic cell. To examine the possible roles of Na$^+$ influx and K$^+$ efflux in triggering release, the preparation is exposed to TTX (to block Na$^+$ influx) and tetraethylammonium (TEA) (to block K$^+$ efflux). By passing brief current pulses of various amplitudes into the stimulating electrode, the membrane potential of the presynaptic terminal can be depolarized to a variety of different membrane potentials without initiating action potentials (Fig. 13). When the presynaptic cell is depolarized by a small amount, there is no postsynaptic potential. A striking observation is made, however, when greater levels of depolarization are applied. In this case, postsynaptic potentials are produced, and their amplitudes depend on the level of depolarization. Thus, this experiment indicates that transmitter release is voltage dependent and that artificial depolarization of the presynaptic terminal is able to produce a postsynaptic potential. Because the preparation has been treated with TTX and TEA, this experiment clearly demonstrates that the voltage-dependent changes in Na$^+$ and K$^+$ permeability that

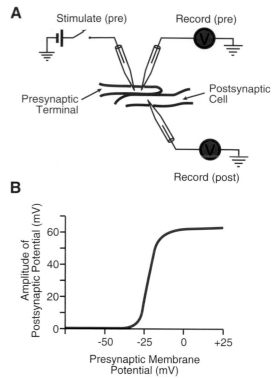

FIGURE 13 The relationship between depolarization of the presynaptic terminal and the release of neurotransmitter. (A) Experimental setup; (B) as the presynaptic terminal is depolarized to greater levels, transmitter release (as measured by the amplitude of the postsynaptic potential) increases. Because the results were obtained in the presence of TEA and TTX, these results indicate that the Na$^+$ influx or K$^+$ efflux that occurs during the action potential is not necessary for the release of neurotransmitter. (Modified from Katz B, Miledi R. *J Physiol* 1967; 192:407–443.)

underlie the action potential have no *direct* effect on the release of chemical transmitter substance from the presynaptic terminal. One merely has to depolarize the cell to get the release of transmitter. Normally, the action potential is essential because it is the vehicle by which the terminal is depolarized. Although changes in Na$^+$ and K$^+$ permeability are important for producing an action potential, the direct influx of Na$^+$ and efflux of K$^+$ are, by themselves, not causally related to the release of the chemical transmitter substance.

ROLE OF CA^{2+} IN TRANSMITTER RELEASE

If Na$^+$ and K$^+$ are not directly involved in the release of chemical transmitter, might other ions be involved? It has been known for many years that Ca^{2+} in the extracellular medium is important for chemical transmission. If the concentration of Ca^{2+} in the extracellular medium is decreased, synaptic transmission is reduced

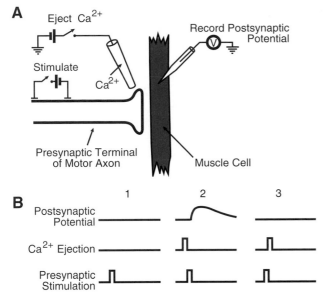

FIGURE 14   Role of Ca^{2+} in synaptic transmission. (A) Experimental setup. (B1) In a Ca^{2+}-free medium a depolarization of the presynaptic terminal does not lead to any transmitter release; (B2) when a small amount of Ca^{2+} is ejected just before a second depolarization, a postsynaptic potential is produced; (B3) if Ca^{2+} is ejected after the depolarization, no potential is produced. (Modified from Katz B, Miledi R. *J Physiol* 1967; 192:407–443.)

(increasing the concentration of Mg^{2+} in the extracellular medium also reduces transmitter release). An experiment that examines the role of Ca^{2+} in the release of chemical transmitter substances is illustrated in Fig. 14. Here the neuromuscular junction is used; one electrode is placed in the muscle cell to record the postsynaptic potential, and stimulating electrodes are placed close to the presynaptic terminal but not inside the terminal. The stimulating electrodes are sufficiently close to the presynaptic terminal to depolarize the terminal by extracellular means. A third electrode filled with $CaCl_2$ is positioned near the presynaptic terminal. The preparation is perfused with a Ca^{2+}-free medium. Figure 14B illustrates the results. In panel 1, stimulation

of the presynaptic terminal produces no postsynaptic potential. Thus, in a Ca^{2+}-free medium, transmitter release is abolished. In panel 2 (Fig. 14B), the presynaptic stimulation is preceded by a brief ejection of Ca^{2+} from the electrode containing $CaCl_2$. Calcium ions are positively charged, so it is possible to eject a small amount of Ca^{2+} from the Ca^{2+} electrode simply by connecting the electrode to the positive pole of a battery. As a result of the brief ejection of Ca^{2+} prior to the stimulus, the depolarizing stimulus now produces a postsynaptic potential in the muscle cell. If Ca^{2+} is ejected after depolarization of the presynaptic terminal, no postsynaptic potential is produced in the muscle cell (panel 3, Fig. 14B). This experiment clearly indicates that Ca^{2+} is absolutely essential for the release of chemical transmitters. Furthermore, it illustrates that Ca^{2+} must be present just before (or during) depolarization of the presynaptic terminal. Based on these and other experiments, Katz and his colleagues proposed the *Ca^{2+} hypothesis for chemical transmitter release.*

Calcium ions are in high concentration outside the cell and in low concentration inside the cell. Furthermore, the inside of the cell is negatively charged with respect to the outside. As a result, there is a large chemical and electrical driving force for the influx of Ca^{2+}. Normally, the cell is relatively impermeable to Ca^{2+}, so even though there is a large driving force, Ca^{2+} does not enter the cell. It is proposed that there is a voltage-dependent change in Ca^{2+} permeability (Fig. 15A). Thus, depolarization of the presynaptic membrane results in an increase in Ca^{2+} permeability, and Ca^{2+} moves down its electrical and chemical gradient and flows into the synaptic terminal. The resultant elevation of intracellular Ca^{2+} concentration leads to the release of chemical transmitter (Fig. 15B).

Thus, according to this hypothesis, Ca^{2+} is the critical trigger for the release of chemical transmitter. One can artificially depolarize the presynaptic terminal to produce a voltage-dependent change in Ca^{2+} influx and release of transmitter. Normally, however, the

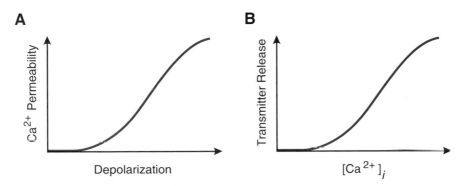

FIGURE 15   The Ca^{2+} hypothesis for the release of neurotransmitters.

action potential invades the terminal, depolarizes it, and leads to the subsequent voltage-dependent increase in Ca^{2+} permeability, influx of Ca^{2+}, and chemical transmitter release.

Quantal Nature of Acetylcholine Release

What is the mechanism by which Ca^{2+} causes the release of chemical transmitter? Figure 16 illustrates an experiment that provided some initial insights. As discussed above, the EPP is a relatively large event, 50 mV or so in amplitude under normal situations. Most scientists who had been recording the EPP at the neuromuscular junction were doing so with the amplification of their oscilloscope set at a relatively low level. Katz and his colleagues increased the amplification and examined the background noise in the absence of any stimulation. (This would be equivalent to tuning your stereo receiver to a portion of the band that has no station and maximally increasing the volume to listen to the noise.) Figure 16 illustrates the type of electrical measurements made at the neuromuscular junction under these conditions. The traces in Fig. 16A are a continuous recording from the same experiment going

from left to right and down the illustration. (It is important to keep in mind that in this experiment the motor axon is not stimulated and that the recordings are made from the unstimulated muscle cell.) It is evident that without stimulation there is no absence of activity. Indeed, numerous small voltage deflections (on the average only about 0.5 mV in amplitude; Fig. 16B) are observed. They are small compared with the relatively large potential changes observed in response to a presynaptic action potential. The potentials occur in a random fashion at an average rate of about once every 50 msec.

The small potential changes have some interesting features. When the recording electrode is placed in regions distant from the motor end plate, the potentials disappear. For this reason, these potential changes are called *miniature EPPs*, or MEPPs, because they are EPPs that have occurred in the vicinity of the motor end plate and are much smaller than the EPP normally produced by stimulating the motor axon. If curare is added to the extracellular medium, the size of the MEPPs decreases. If neostigmine is added to the extracellular medium, the size of the MEPPs increases. Based on these considerations, it appears that MEPPs are due to the spontaneous and random release of ACh from the presynaptic terminal. Acetylcholine presumably is in relatively high concentration in the presynaptic terminal. So, by chance, some of that ACh might diffuse out of the presynaptic terminal, bind with postsynaptic receptors, and produce small permeability changes.

How much ACh is necessary to produce a MEPP? The simplest answer is that a MEPP is produced by one ACh molecule binding to a receptor and opening a single ACh-sensitive channel. When small amounts of ACh are ejected into the vicinity of the neuromuscular junction, however, potentials much smaller than MEPPs are produced. Indeed, the single-channel recording techniques (*e.g.*, Fig. 11) indicate that the opening of a single ACh-sensitive channel produces a potential change of approximately 0.4 μV. This indicates that the MEPPs are not due to the release of a single molecule of ACh. Because the opening of a single ACh-sensitive channel produces a potential change of approximately 0.4 μV, it would take at least 1000 molecules of ACh to produce a potential change the size of one MEPP (0.4 mV). Indeed, because of loss due to diffusion in the synaptic cleft and the fact that two ACh molecules are necessary to open a channel, approximately 10^4 ACh molecules are required to produce a MEPP. Therefore, it appears that it is not a single molecule of ACh that is spontaneously and randomly released from the presynaptic terminal, but a package of 10^4 molecules of ACh. It is now well established that the morphologic locus of this package of 10^4 ACh molecules

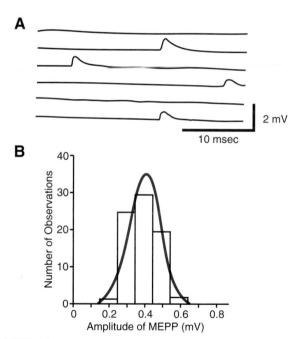

FIGURE 16   The miniature end-plate potential (MEPP). (A) In the absence of nerve stimulation, small spontaneously occurring potentials are recorded. The successive traces are continuous recordings from a neuromuscular junction. The MEPPs are approximately 0.4 mV in amplitude and occur about once every 50 msec. (Modified from Liley AW. *J Physiol* 1956; 133:571–587.) (B) Histogram representing the number of times a MEPP of a given amplitude is recorded. There is a variation in amplitude, but MEPPs appear to be derived from a single population and have an average amplitude of 0.4 mV. (Modified from Boyd IA, Martin AR. *J Physiol* 1956; 132:74–91.)

is the synaptic vesicle found in high concentration in a presynaptic terminal. Chemical analyses have revealed that the synaptic vesicles in the motor axon contain ACh.

Is the normal EPP due to the release of these packages of ACh as well? One observation consistent with this hypothesis is that when Ca^{2+} is injected into the presynaptic terminal, the frequency of occurrence of the MEPPs is increased. This indicates that the release of the vesicles is Ca^{2+} dependent. As a result, it is attractive to think that, when a larger amount of Ca^{2+} enters the presynaptic terminal during an action potential, a large number of vesicles is released synchronously.

Figure 17 illustrates a test of this hypothesis. To make the interpretation of the experiment easier, the

A

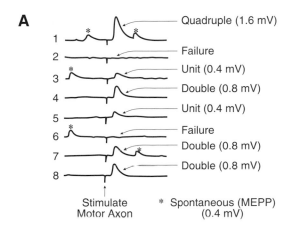

B

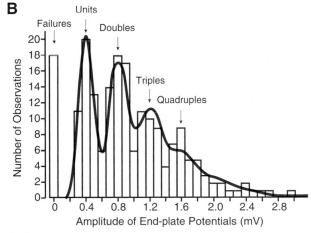

FIGURE 17 Quantal nature of transmitter release. (A) In a low-Ca^{2+} solution, the motor axon is repeatedly stimulated, and the amplitude of the EPP is monitored. In such a solution, the amplitude of the EPP is small and variable. Asterisks indicate spontaneous MEPPs. (Modified from Liley AW. *J Physiol* 1956; 133:571–587.) (B) A histogram showing the number of times EPPs of various amplitudes are recorded in the low-Ca^{2+} solution. There are many EPPs that have an amplitude identical to the amplitude of the MEPP (0.4 mV). Other peaks are multiples of the first. (Modified from Boyd IA, Martin AR. *J Physiol* 1956; 132:74–91.)

Ca^{2+} concentration in the extracellular medium is reduced to a point at which the evoked EPPs are quite small (approximately the same size as the MEPPs). The left side of the traces in Fig. 17A illustrate some of the spontaneous MEPPs (approximately 0.4 mV in amplitude [asterisks]). The small vertical lines on the traces illustrate the point in time at which the stimulus is applied to the motor axon. When the motor axon is stimulated in the low-Ca^{2+} solution, very small EPPs are produced, but there is considerable variation in their size. For example, sometimes the EPP is about the same size as the MEPP (spontaneously occurring) (trace 3, Fig. 17A). Such a response is called a *unit EPP*. Sometimes no EPP is produced by the stimulus (traces 2 and 6, Fig. 17A) and at other times an EPP is produced that is about twice the size of the unit (*double*) (traces 4, 7, and 8, Fig. 17A). In other cases, EPPs are produced that are about three times the size of the unit or the spontaneous EPP. Sometimes an EPP is produced that is about four times the size of the unit EPP and is referred to as a *quadruple* (trace 1, Fig. 17A).

By measuring the amplitude of all the evoked EPPs, a plot can be drawn of the number of times the EPPs of a given amplitude are observed. Similarly, an amplitude distribution of the spontaneous MEPPs can be made (*e.g.*, Fig. 16B). As we have already learned, the spontaneous MEPPs have an average value of approximately 0.4 mV. The plot of the evoked EPPs produced in low Ca^{2+} reveals a remarkable finding; namely, there is a multimodal distribution of sizes (Fig. 17B). There is a large proportion of EPPs that have an average value of approximately 0.4 mV, another large percentage that have an average value of approximately 0.8 mV, and some that have an average value of approximately 1.2 mV. The critical finding is that the size of the first peak of the evoked EPP (0.4 mV) is nearly identical to the size of the spontaneous MEPPs and that subsequent peaks are multiples of the first.

Based on these observations, Katz and his colleagues proposed the *quantal nature of transmitter release*. The hypothesis was developed at the neuromuscular junction, but the quantal nature of release appears to be widely applicable to chemical synapses. Transmitter release at some synapses may be nonvesicular, however. According to the quantal hypothesis, the size of the EPP fluctuates because different numbers of packages (quanta) of ACh are released with each stimulus. The smallest EPP is due to the release of a single package of ACh containing 10^4 molecules. An EPP twice that size is due to the release of two packages, and so forth. The EPP of about 50 mV produced in normal concentrations of Ca^{2+} is a compound of the individual contributions of many packages of ACh. Because each package of ACh produces an amplitude of approximately 0.4 mV, the normal EPP is due to the simultaneous release of more than 100 packages of ACh, each containing 10^4 molecules and each quanta producing a potential change by itself of 0.4 mV. Calcium ions appear to be the critical step in causing the release of the vesicles. In low extracellular Ca^{2+}, an action potential in the presynaptic terminal results in a small amount of Ca^{2+} influx and a small number of vesicles released. When the extracellular Ca^{2+} concentration is high, a large amount of Ca^{2+} flows into the terminal and causes a larger number of vesicles to be released (and therefore a larger EPP). The Ca^{2+} that enters the cell promotes the binding of vesicles with the inside membrane of the presynaptic terminal to promote exocytosis (Fig. 18). Considerable progress has

been made in identifying the specific proteins involved in the docking, fusion, and regulation of vesicular release.

SUMMARY OF THE SEQUENCE OF EVENTS UNDERLYING SYNAPTIC TRANSMISSION AT THE SKELETAL NEUROMUSCULAR JUNCTION

As a result of work during the past several decades, a fairly complete understanding of the sequence of events underlying synaptic transmission at the skeletal neuromuscular junction is available (Fig. 19; see also Table 1). The resting potential at both the presynaptic terminal of the motor axon and the postsynaptic muscle cell is due

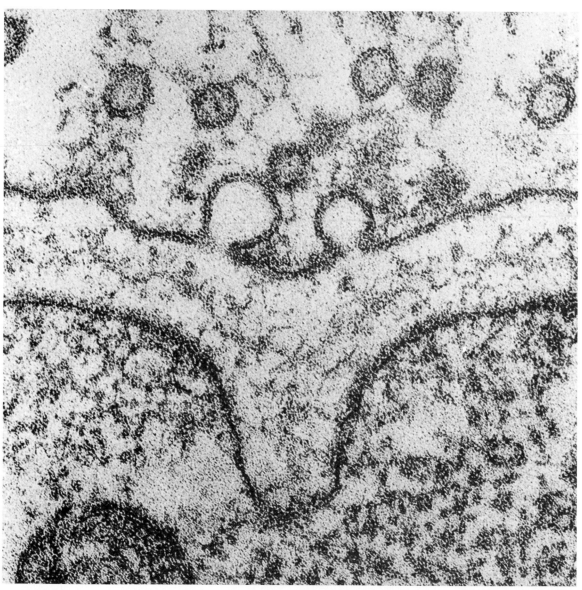

FIGURE 18   Exocytosis. Transmitter vesicles caught in the process of excocytosis. The top portion of the photograph is the presynaptic ending. (Micrograph produced by Dr. John E. Heuser of Washington University School of Medicine, St. Louis, MO.)

Presynaptic   **Postsynaptic**

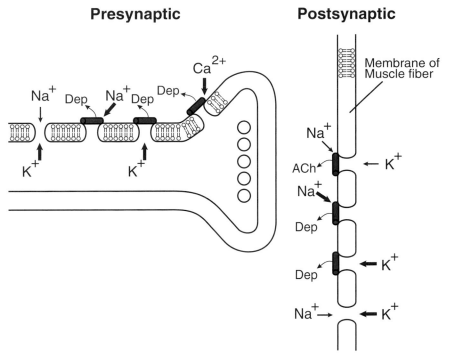

FIGURE 19   Summary diagram of the membrane channels important for the process of synaptic transmission at the skeletal neuromuscular junction. Dep, depolarization.

to the high resting K^+ permeability. At rest, a small amount of transmitter release occurs because of the spontaneous release of vesicles (quanta) from the presynaptic terminal. The spontaneous release is presumably due to the low basal levels of Ca^{2+} in the presynaptic terminal. The resultant MEPPs are only approximately 0.4 mV and are insufficient to trigger an action potential in the muscle cell. When an action potential propagates down the motor axon, it eventually invades the presynaptic terminal. As the action potential invades the terminal, the terminal is depolarized, and

that depolarization leads to the opening of voltage-dependent Ca^{2+} channels. Calcium ions move down their electrochemical gradient and enter the terminal. The resultant increase in the intracellular Ca^{2+} concentration leads to the release of 100 or so synaptic vesicles, each containing 10^4 molecules of ACh. The released ACh diffuses across the synaptic cleft and binds to receptors on the postsynaptic membrane. The binding of ACh with ACh-sensitive channels causes individual channels that are normally closed to open in an all-or-nothing fashion. The chemically gated channels opened

TABLE 1   Properties of Action Potentials and Synaptic Potentials at the Skeletal Neuromuscular Junction

	Synaptic potential (End-plate potential)	Action potential
Changes in membrane conductance	Initiated by ACh	Initiated by depolarization (the end-plate potential)
Duration, rising phase (initiation)	Simultaneous chemically gated increase in g_{Na} and g_k	Specific voltage-dependent increase in g_{Na}
Duration, falling phase	Passive decay; diffusion and acetylcholinesterase (AChE)	Specific increase in g_k and decrease in g_{Na}
Duration	10–20 msec	1–3 msec
Equilibrium potential	Reversal potential close to 0 mV	ENa (+55 mV)
Pharmacology	Blocked by curare; enhanced by neostigmine; not blocked by tetrodotoxin (TTX)	Blocked by TTX but not by curare; not affected by neostigmine
Propagation	Propagates with decrement	All-or-nothing
Other features	No evidence for regenerative action or refractory period	Regenerative rise, followed by absolute and relative refractory periods

by ACh are distinct from the voltage-gated channels (Na^+ and K^+) that underlie the action potential in the axon and skeletal muscle cell. The channel opened by ACh is equally permeable to Na^+ and K^+, and as a result the postsynaptic membrane is depolarized. This postsynaptic potential (the EPP), if sufficiently large (as it normally is), brings the muscle membrane potential to threshold. A new potential, the action potential, is initiated in the muscle cell by the voltage-dependent changes in Na^+ and K^+ permeabilities. The muscle action potential propagates along the muscle cell membrane and leads, by a process known as *excitation–concentration coupling*, to the development of muscle tension.

SYNAPTIC TRANSMISSION IN THE CENTRAL NERVOUS SYSTEM

The study of synaptic transmission in the central nervous system (CNS) is both an opportunity to learn more about the diversity and richness of mechanisms underlying this process and an opportunity to learn how some of the fundamental signaling properties of the nervous system, such as action potentials and synaptic potentials, work together to process information and generate behavior. One of the simplest behaviors controlled by the central nervous system is the knee-jerk

or stretch reflex. The tap of a neurologist's hammer to a ligament elicits a reflex extension of the leg, illustrated in Fig. 20. The brief stretch of the ligament is transmitted to the extensor muscle and is detected by specific receptors in the muscle and ligament. Action potentials initiated in the stretch receptors are propagated to the spinal cord by afferent fibers. The receptors are specialized regions of sensory neurons with somata located in the dorsal root ganglia just outside the spinal column. The axons of the afferents enter the spinal cord and make at least two types of excitatory synaptic connections. First, a synaptic connection is made to the extensor motor neuron. As the result of its synaptic activation, the motor neuron fires action potentials that propagate out of the spinal cord and ultimately invade the terminal regions of the motor axon at neuromuscular junctions. There, ACh is released, an EPP is produced, an action potential is initiated in the muscle cell, and the muscle cell is contracted, producing the reflex extension of the leg. Second, a synaptic connection is made to another group of neurons, *interneurons* (nerve cells interposed between one type of neuron and another). The particular interneurons activated by the afferents are inhibitory interneurons, because activation of these interneurons leads to the release of a chemical transmitter substance that inhibits the flexor motor neuron. This inhibition tends to prevent an uncoordinated (improper) movement (*i.e.*, flexion) from occurring.

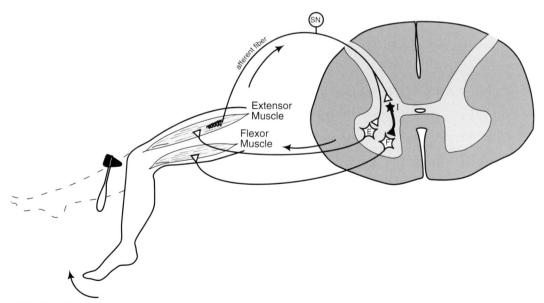

FIGURE 20   Features of the vertebrate stretch reflex. Stretch of an extensor muscle leads to the initiation of action potentials in the afferent terminals of specialized stretch receptors. The action potentials propagate to the spinal cord via afferent fibers (sensory neurons , SN). The afferents make excitatory connections with extensor motor neurons (E). Action potentials initiated in the extensor motor neuron propagate to the periphery and lead to the activation and subsequent contraction of the extensor muscle. The afferent fibers also activate interneurons (I) that inhibit the flexor motor neurons (F).

Mechanisms for Excitation, Inhibition, and Integration of Synaptic Potentials

The stretch reflex provides a suitable model system to study the properties of excitatory and inhibitory synaptic transmission in the CNS and to illustrate how these properties are used to process information and generate simple behavior. Figure 21A illustrates procedures that can be used to examine experimentally some of the components of synaptic transmission in the reflex pathway for the stretch reflex. Intracellular recordings are made from one of the sensory neurons, the extensor and flexor motor neurons, and an inhibitory interneuron. Normally, the sensory neuron is activated by stretch to the muscle, but this step can be bypassed by simply injecting a pulse of depolarizing current of sufficient magnitude into the sensory neuron to elicit an action potential. The action potential in the sensory neuron leads to a potential change in the motor neuron known as an *excitatory postsynaptic potential* (EPSP) (Fig. 21B). The potential is excitatory because it increases the probability of firing an action potential in the motor neuron; it is postsynaptic because it is a potential that is recorded on the receptive (postsynaptic) side of the synapse.

Postsynaptic potentials (PSPs) in the CNS can be divided into two broad classes based on mechanisms and duration of these potentials. One class arises from the *direct* binding of a transmitter molecule with a receptor–channel complex; these receptors are *ionotropic*. The resulting PSPs are generally short lasting and hence are sometimes called fast PSPs; they have also been referred to as *classical* because these were the first synaptic potentials that were recorded in the CNS. Fast EPSPs also resemble the synaptic potentials at the neuromuscular junction (*i.e.*, the EPP). One typical feature of the classical synaptic potentials is their time course. Recall that the duration of an EPP is about 20 msec. An EPSP, or an inhibitory postsynaptic potential (IPSP), recorded in a spinal motor neuron is of approximately the same duration.

The second class of PSPs arises from the *indirect* effect of a transmitter molecule binding with a receptor. For these PSPs, one common coupling mechanism is an alteration in the level of a second messenger. The receptors that produce these PSPs are *metabotropic*. The responses can be long lasting and are therefore referred to as slow PSPs. The mechanisms for fast PSPs mediated by ionotropic receptors will be considered first.

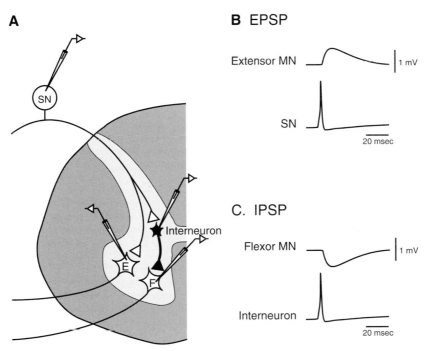

FIGURE 21 Excitatory and inhibitory postsynaptic potentials in spinal motor neurons. (A) Intracellular recordings are made from a sensory neuron (SN), interneuron, and extensor (E) and flexor (F) motor neurons. (B) An action potential in the sensory neuron produces a depolarizing response in the motor neuron (MN). This response is called an excitatory postsynaptic potential (EPSP). (C) An action potential in the interneuron produces a hyperpolarizing response in the motor neuron. This response is called an inhibitory postsynaptic potential (IPSP).

Postsynaptic Mechanisms Produced by Ionotropic Receptors

Ionic Mechanisms of EPSPs

Mechanisms responsible for fast EPSPs mediated by ionotropic receptors in the CNS are fairly well known. Specifically, for the synapse between the sensory neuron and the spinal motor neuron, the ionic mechanisms for the EPSP are essentially identical to the ionic mechanisms at the skeletal neuromuscular junction. Transmitter substance released from the presynaptic terminal of the sensory neuron diffuses across the synaptic cleft, binds to specific receptor sites on the postsynaptic membrane, and leads to a simultaneous increase in permeability to Na^+ and K^+, which makes the membrane potential move toward a value of 0 mV.

Although this mechanism is superficially the same as that for the neuromuscular junction, two fundamental differences exist between the process of synaptic transmission at the sensory neuron–motor neuron synapse and the motor neuron–skeletal muscle synapse. First, these two different synapses use different transmitters. The transmitter substance at the neuromuscular junction is ACh, but the transmitter substance released by the sensory neurons is an amino acid, probably *glutamate*. (Although beyond the scope of this chapter, many different transmitters are used by the nervous system— up to 50 or more, and the list grows yearly. Thus, to understand fully the process of synaptic transmission and the function of synapses in the nervous system, it is necessary to know the mechanisms for the synthesis, storage, release, and degradation or uptake, as well as the different types of receptors for each of the transmitter substances. The clinical implications of deficiencies in each of these features for each of 50 transmitters cannot be ignored. Fortunately, most of the transmitter substances can be grouped into four basic categories: ACh, monoamines, amino acids, and the peptides.)

A second major difference between the sensory neuron–motor neuron synapse and the motor neuron–muscle synapse is the amplitude of the postsynaptic potential. Recall that the amplitude of the postsynaptic potential at the neuromuscular junction was approximately 50 mV and that a one-to-one relationship existed between an action potential in the spinal motor neuron and an action potential in the muscle cell. Indeed, because the EPP must only depolarize the muscle cell by approximately 30 mV to initiate an action potential, there is a safety factor of 20 mV. In contrast, the EPSP in a spinal motor neuron produced by an action potential in an afferent fiber has an amplitude of only 1 mV.

The small amplitude of the EPSP in spinal motor neurons (and other cells in the CNS) poses an interesting question. Specifically, how can an ESP with an amplitude of only 1 mV drive the membrane potential of the motor neuron (*i.e.*, the postsynaptic neuron) to threshold and fire the spike in the motor neuron that is necessary to produce the contraction of the muscle? The answer to this question lies in the principles of temporal and spatial summation.

Temporal and Spatial Summation

When the ligament is stretched (see Fig. 20), many stretch receptors are activated. Indeed, the greater the stretch, the greater the probability of activating a larger number of the stretch receptors present; this process is referred to as *recruitment*. Activation of multiple stretch receptors is not the complete story, however. Recall the principle of frequency coding in the nervous system (see Chapter 4). Specifically, the greater the intensity of a stimulus, the greater the number per unit time or frequency of action potentials elicited in a sensory receptor. This principle applies to stretch receptors as well. Thus, the greater the stretch, the greater the number of action potentials elicited in the stretch receptor in a given interval and, therefore, the greater the number of EPSPs produced in the motor neuron from that train of action potentials in the sensory cell. Consequently, the effects of activating multiple stretch receptors add together (spatial summation), as do the effects of multiple EPSPs elicited by activation of a single stretch receptor (temporal summation). Both these processes act in concert to depolarize the motor neuron sufficiently to elicit one or more action potentials, which then propagate to the periphery and produce the reflex.

Temporal Summation. Temporal summation can be illustrated by considering the case of firing action potentials in a presynaptic neuron and monitoring the resultant EPSPs. For example, in Fig. 22A and B, a single action potential in SN1 produces a 1-mV EPSP in the motor neuron. Two action potentials in quick succession produce two EPSPs, but note that the second EPSP occurs during the falling phase of the first, and the depolarization associated with the second EPSP adds to the depolarization produced by the first. Thus, two action potentials produce a summated potential that is 2 mV in amplitude. Three action potentials in quick succession would produce a summated potential of 3 mV. In principle, 30 action potentials in quick succession would produce a potential of 30 mV and easily drive the cell to threshold. This summation is strictly a passive property of the cell. No special ionic conductance mechanisms are necessary.

A thermal analog is helpful to understand temporal summation. Consider a metal rod that has thermal properties similar to the passive electrical properties of

A

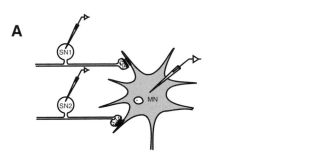

B Temporal Summation

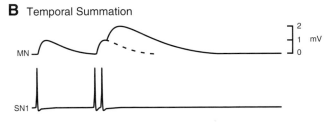

C Spatial Summation

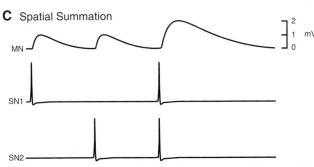

FIGURE 22   Temporal and spatial summation. (A) Intracellular recordings are made from the two sensory neurons (SN1 and SN2) and a motor neuron (MN). (B) Temporal summation; a single action potential in SN1 produces a 1-mV EPSP in the MN. Two action potentials in quick succession produce a dual-component EPSP, the peak amplitude of which is approximately 2 mV. (C) Spatial summation; alternative firing of single action potentials in SN1 and SN2 produce 1-mV EPSPs in the MN. Simultaneous action potentials in SN1 and SN2 produce a summated EPSP, the amplitude of which is approximately 2 mV.

a dendrite and the flame of a cigarette lighter that will generate heat equivalent to a depolarization generated by a synaptic current at the synapse. Now consider the consequences of presenting the flame to one end of the rod on the temperature measured in the middle of the rod. Obviously, a temperature change would be produced; assume, for the sake of the example, that the temperature change is 1°C. If the flame is presented twice in quick succession, a greater temperature change would be recorded, because the heat generated would produce temperature changes that would summate because of the passive properties (heat capacity) of the rod. If the two flames were presented in quick succession, a temperature change of 2°C would be produced. If a long time elapses between the presentation of the two

flames, there would be little or no summation, because the initial 1°C increase in temperature would have dissipated. Similarly, potentials in dendrites summate because of the passive properties of the nerve cell membrane. Specifically, the membrane has a capacitance and can store charge. Thus, the membrane can temporarily store the charge of the first EPSP, and the charge from the second EPSP can be added to that of the first.

Spatial Summation. Spatial summation (Fig. 22C) involves a consideration of more than one input to a postsynaptic neuron. An action potential in SN1 produces a 1-mV EPSP, just as it did in Fig. 22B. Similarly, an action potential in a second sensory neuron (SN2) by itself also produces a 1-mV EPSP. Now, consider the consequences of action potentials elicited simultaneously in SN1 and in SN2. The net EPSP is equal to the summation of the amplitudes of the individual EPSPs. Here the EPSP from SN1 is 1 mV, the EPSP from SN2 is 1 mV, and the summated EPSP is 2 mV (Fig. 22C). Thus, spatial summation is a mechanism by which synaptic potentials generated at different sites can summate. Spatial summation in nerve cells is determined by the space constant, which is the ability of a potential change produced in one region of a cell to spread passively to other regions of a cell (see Chapter 5).

Again, a thermal analog is useful to help understand this phenomenon. Consider the metal rod as an analogy for a dendrite. As before, temperature is recorded in the middle, but now heat sources can be delivered to either end separately or both ends at the same time. A flame presented to only one end of the metal rod produces a 1°C increase in temperature at the middle of the metal rod; if a flame is presented simultaneously to each end of the metal rod, a 2°C increase in temperature is produced. The temperature changes induced by the two flames presented simultaneously will summate because of the ability of heat to spread passively along the rod.

In summary, whether a neuron fires in response to synaptic input is dependent, at least in part, on how many action potentials are produced in any one presynaptic excitatory pathway, as well as how many individual convergent excitatory input pathways are activated. The final behavior of the cell is also due to the summation of other kinds of synaptic inputs—specifically, inhibitory synaptic inputs.

Inhibitory Postsynaptic Potentials

Some synaptic events *decrease* the probability of generating spikes in the postsynaptic cell. Potentials produced as a result of these events are called *inhibitory postsynaptic potentials* (IPSPs). Consider the inhibitory interneuron illustrated in Fig. 21C. Normally, this interneuron would be activated by summating EPSPs from converging afferent fibers. These EPSPs would summate in space and time such that the membrane potential of the interneuron would reach threshold and fire an action potential. This step can be bypassed by artificially depolarizing the interneuron to initiate an action potential. The consequences of that action potential from the point of view of flexor motor neurons are illustrated in Fig. 21C. The action potential in the interneuron produces a transient increase in the membrane potential of the motor neuron. This transient hyperpolarization (the IPSP) looks very much like the EPSP, but it is reversed in sign.

What are the ionic mechanisms for these fast IPSPs and what is the transmitter substance? Because the membrane potential of the flexor motor neuron is approximately -65 mV, one might expect an increase in the conductance to some ion (or ions) with an equilibrium potential (reversal potential) more negative than -65 mV. One possibility is K^+; indeed, the K^+ equilibrium potential in spinal motor neurons is approximately -80 mV. A transmitter substance that produced a selective increase in K^+ conductance would lead to an IPSP. The K^+-conductance increase would move the membrane potential from -65 mV toward the K^+ equilibrium potential of -80 mV. Although an increase in K^+ conductance mediates IPSPs at some inhibitory synapses, it does not do so at the synapse between the inhibitory interneuron and the spinal motor neuron. At this particular synapse, the IPSP seems to be due to a selective increase in Cl^- conductance. The equilibrium potential for Cl^- in spinal motor neurons is approximately -70 mV. Thus, the transmitter substance released by the inhibitory neuron diffuses across the cleft and interacts with receptor sites on the postsynaptic membrane. These receptors are coupled to a special class of receptor channels that are normally closed, but when opened they become selectively permeable to Cl^-. As a result of the increase in Cl^- conductance, the membrane potential moves from a resting value of -65 mV toward the Cl^- equilibrium potential of -70 mV.

Like the case of the sensory neuron–spinal motor synapse, the transmitter substance released by the inhibitory interneuron in the spinal cord is an amino acid, but in this case the transmitter is *glycine*. Indeed, glycine is used frequently in the CNS as an inhibitory transmitter, most often in the spinal cord. The most common transmitter associated with inhibitory actions in many areas of the brain is *gamma aminobutyric acid* (GABA). The agents *bicuculline* and *picrotoxin* are specific blockers of the actions of GABA. *Strychnine* is a specific blocker of the actions of glycine.

General Features of Ionotropic-Mediated PSPs and the Molecular Biology of the Receptor–Channel Complexes

The general features of *ionotropic receptors* in the CNS are similar to those of the ligand-gated ACh receptor-channel found in the skeletal muscle. In particular, ionotropic receptors in the CNS for ACh, glycine, and GABA are made up of multiple subunits (usually a pentameric structure), with each of the subunits having four membrane-spanning regions (see Fig. 12).

Neuronal nicotinic ACh receptors and GABA receptors have been particularly well characterized. Unlike the skeletal muscle ACh receptor, which is made up of four types of subunits (α, β, δ, and ε or γ), the *neuronal* ACh receptor is made up of only two types of subunits (α and β). A considerable diversity of channel properties is possible, however, because there are at least eight different α subunits and three different β subunits. Indeed, it has been estimated that the combinatorial possibilities could result in more than 4000 different types of receptors. The ionotropic GABA receptor (called GABA$_A$) are generally comprised of a combination of three different subunits, α, β, and γ, in a stoichiometry of $2\alpha2\beta1\gamma$. Each subunit type has multiple isoforms; therefore, the considerable diversity of GABA$_A$ receptors observed *in vivo* appears to be due to the various stoichiometric combinations of different subunit isoforms.

Glutamate is the predominant transmitter with excitatory actions in the CNS, and in recent years insights into the structural characteristics of its receptors have been obtained. Like many other ion channels, ionotropic glutamate receptors are homo- or heteromultimeric proteins consisting of four subunits. Each subunit consists of an extracellular amino-terminal domain, three transmembrane domains, a re-entrant loop that forms the channel pore, and a cytoplasmic carboxy terminal. Although the structure of the full-length channel has not yet been resolved, the crystal structure of the amino-terminal ligand-binding region has been resolved and conformational changes associated with ligand binding have been identified. However, the mechanisms by which ligand binding drives channel opening are not fully understood. Ionotropic glutamate receptors can be divided into two broad classes based on their sensitivity to the agonist *N*-methyl-D-aspartate (NMDA) and are referred to as NMDA and non-NMDA receptors. Both types are located throughout the nervous system, but their relative abundance differs. Non-NMDA glutamate receptors have the functional properties described previously for fast ionotropic-mediated synaptic actions at the sensory neuron motor neuron synapses (e.g., Fig. 21B). Specifically, as a result

A Non-NMDA

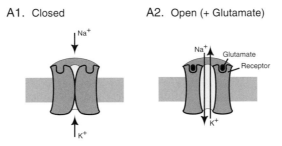

B NMDA

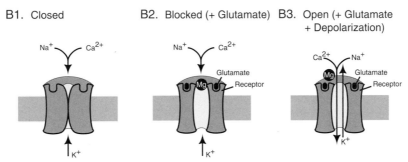

FIGURE 23 Features of non-NMDA and NMDA glutamate receptors. (A) Non-NMDA receptors: (A1) In the absence of agonist, the channel is closed; (A2) glutamate binding leads to channel opening and an increase in Na^+ and K^+ permeability, which depolarizes the cell to produce an EPSP. (B) NMDA: (B2) The presence of agonist leads to a conformational change and channel opening, but no ionic flux occurs because the pore of the channel is blocked by Mg^{2+}; (B3) in the presence of depolarization, the Mg^{2+} block is removed and the agonist-induced opening of the channel leads to changes in ion flux (including Ca^{2+} influx into the cell).

of glutamate binding, a channel opens that is highly permeable to both Na^+ and K^+ (Fig. 23A). The agent α-amino-3-hydroxy-5-methylisoxazol-4-propinoic acid (AMPA) is a specific agonist of non-NMDA receptors. Inhibitors of non-NMDA receptors include the quinoxaline derivatives 6-cyano-7-nitroquinoxaline-2,3-dione (CNQX) and 1,2,3,4-tetrahydro-6-nitro-2,3,-dioxo-benzo[f]quinoxaline-7-sulfonamide (NBQX).

The NMDA glutamate receptors differ from non-NMDA receptors in four ways. First, they are selectively blocked by the agent 2-amino-5-phosphonovalerate (APV). Second, they have a high permeability to Ca^{2+} as well as Na^+ and K^+. Third, at normal values of the resting potential, the pore of the channel is blocked by Mg^{2+}. Thus, even if glutamate binds to the receptor, there will be no ionic flow (and no EPSP) because the channel is blocked. The block can be relieved by depolarization, which presumably displaces the Mg^{2+} from the pore (Fig. 23B). A fourth and final difference between non-NMDA and NMDA channels is a glycine-binding site on the NMDA channel. Glycine must be present for NMDA receptors to function. The

physiologic role of this binding in regulating the channel is unclear, however. Basal levels of glycine in the vicinity of cells with NMDA receptors seem to be sufficiently high to maintain function of the NMDA receptor. Several of these unique features of the NMDA channel have important physiologic consequences. First, activation of the NMDA receptor results in more than a simple EPSP. The influx of calcium associated with the channel opening can induce a cascade of biochemical reactions, including activation of Ca^{2+}-dependent phosphatases, kinases, and proteases. Second, the dual regulation of the channel by glutamate and voltage (depolarization) allows other synaptic inputs through electrotonic propagation and spatial summation to profoundly regulate the ability of an NMDA-mediated synaptic input to affect a postsynaptic cell.

Slow Synaptic Potentials Produced by Metabotropic Receptors

A common feature of the types of synaptic actions described above is the direct binding of the transmitter with the receptor–channel complex. An entirely separate class of synaptic actions is due to the indirect coupling of the receptor with the channel. So far, two types of coupling mechanisms have been identified. These include a coupling of the receptor and channel through an intermediate regulatory protein or coupling through a diffusible second messenger system. Receptors that activate the latter mechanism are called *metabotropic,* because they involve changes in the metabolism of second messengers or other compounds such as ATP and phospolipids. Because they are the most common, attention will be focused on the responses mediated by metabotropic receptors.

A comparison of the features of direct, fast iono-tropic- and indirect, slow metabotropic-mediated synaptic potentials is shown in Fig. 24. Slow synaptic

potentials are not observed at every postsynaptic neuron, but Fig. 24A illustrates an idealized case in which a postsynaptic neuron receives two inputs, one that produces a conventional fast EPSP and the other that produces a slow EPSP. An action potential in neuron 1 leads to an EPSP in the postsynaptic cell whose duration is approximately 30 msec (Fig. 24B1). This is the type of potential that might be produced in a spinal motor neuron by an action potential in an afferent fiber. Neuron 2 also produces a postsynaptic potential (Fig. 24B2), but its duration (note the calibration bar) is more than three orders of magnitude longer than that of the EPSP produced by neuron 1.

How can a change in the postsynaptic potential of a neuron persist for many minutes as a result of a single action potential in the presynaptic neuron? Possible answers to this question include a prolonged presence of the transmitter due to continuous release, slow degradation, or slow re-uptake of the transmitter; but the mechanism here involves a transmitter-induced change in metabolism of the postsynaptic cell. Figure 25 compares the general mechanisms for fast and slow synaptic potentials. Fast synaptic potentials are produced when a transmitter substance binds to a channel and produces a conformational change in the channel, causing it to become permeable to one or more ions (both Na^+ and K^+ in Fig. 25A). The increase in permeability leads to a depolarization associated with the EPSP (Fig. 25A3). The duration of the synaptic event is critically dependent on the amount of time the transmitter substance remains bound to the receptors. The transmitters that have already been mentioned (ACh, glutamate, and glycine) remain bound for only a very short period of time. These transmitters are either removed by diffusion, enzymatic breakdown, or re-uptake into the presynaptic cell; therefore, the channel closes rapidly.

One mechanism for a slow synaptic potential is shown in Fig. 25B. In contrast to the fast PSP for which the receptors are actually part of the ion–channel complex, the channels that produce the slow synaptic potentials are not directly coupled to the transmitter receptors. Rather, the receptors are physically separated and exert their actions indirectly through changes in the metabolism of specific *second messenger systems*. Figure 25B illustrates one type of response that involves the cyclic adenosine monophosphate (cAMP)–protein kinase A system, but other slow PSPs use other second messenger–kinase systems (*e.g.*, the protein kinase C system). In the case of cAMP-dependent slow synaptic responses, transmitter binding to membrane receptors activates G proteins and stimulates an increase in the synthesis of cAMP. Cyclic AMP then leads to the activation of

A

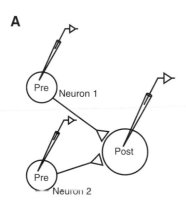

B1 Fast EPSP

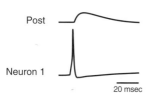

B2 Slow EPSP

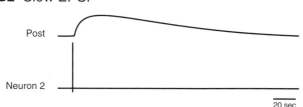

FIGURE 24   Fast and slow synaptic potentials. (A) Two neurons (1 and 2) make synaptic connections with a common postsynaptic follower cell (Post). (B1) An action potential in neuron 1 leads to a conventional fast EPSP with a duration of about 30 msec. (B2) An action potential in neuron 2 also produces an EPSP in the postsynaptic cell, but the duration of this slow EPSP is more than three orders of magnitude greater than that of the EPSP produced by neuron 1. Note the change in calibration bar.

Ionotropic

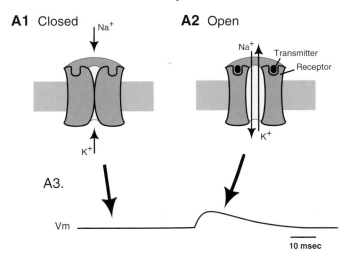

A1 Closed

A2 Open

A3.

Metabotropic

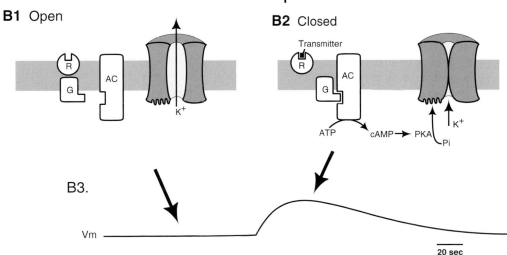

B1 Open

B2 Closed

B3.

FIGURE 25 (A) Ionotropic receptors and mechanisms of fast EPSPs: (A1) Fast EPSPs are produced by the binding of transmitter to specialized receptors that are directly associated with an ion channel (*i.e.*, a ligand-gated channel); when the receptors are unbound, the channel is closed. (A2) Binding of transmitter to the receptor produces a conformational change in the channel protein such that the channel opens. In this example, the channel opening is associated with a selective increase in the permeability to Na^+ and K^+. The increase in permeability results in the EPSP shown in A3. The duration of the EPSP is directly related to the time that the transmitter remains bound to the receptor. (B) Metabotropic receptors and mechanisms of slow EPSPs; unlike fast EPSPs that are due to the binding of a transmitter with a receptor–channel complex, slow EPSPs involve the activation of receptors (metabotropic) that are *not* directly coupled to the channel. Rather, the coupling takes place through the activation of one of several second-messenger cascades (in this example, the cAMP cascade). Binding (B) of the transmitter to the receptor leads to activation of a G protein (G) and adenylyl cyclase (AC). The synthesis of cAMP is increased, cAMP-dependent protein kinase (PKA) is activated, and a channel protein is phosphorylated. The phosphorylation leads to closure of the channel; the subsequent depolarization associated with the slow EPSP is shown in panel B3. The decay of the response is due to both the breakdown of cAMP by cAMP-dependent phosphodiesterase and the removal of phosphate from channel proteins by protein phosphatases (*not shown*).

cAMP-dependent protein kinase, protein kinase A (PKA), which phosphorylates a channel protein or protein associated with the channel. A conformational change in the channel is produced which then leads to a change in ionic conductance. Thus, in contrast to a direct conformational change produced by the binding of a receptor to the receptor–channel complex, in this case a conformational change is produced by protein phosphorylation. Indeed, phosphorylation-dependent

channel regulation is one of the general features of slow PSPs.

Another interesting feature of slow synaptic responses is that they are sometimes associated with decreases rather than increases in membrane conductances. For example, the particular channel illustrated in Fig. 25 is selectively permeable to K^+ and is normally open. As a result of the activation of the second messenger, the channel closes and becomes less

Clinical Note

Myasthenia Gravis

Myasthenia gravis is a debilitating neuromuscular disease associated with weakness and fatigability of skeletal muscle. The condition is aggravated by exercise. The muscular weakness and fatigability are in turn associated with EPPs that are smaller in amplitude than normal. Recall that there is generally a one-to-one relationship between an action potential in a motor axon and an action potential in the skeletal muscle cell. This is so because the EPP is normally quite large (50 mV in amplitude). As a result, an EPP in a muscle cell with a resting potential of approximately −80 mV is always capable of depolarizing the muscle cell past threshold (approximately 50 mV) and initiating a skeletal muscle action potential. Note that to reach threshold, an EPP need only have an amplitude of 30 mV (e.g., −80 mV + 30 mV = −50 mV). Thus, there is normally a safety factor of approximately 20 mV. In myasthenia gravis, the EPP is smaller and there is less of a safety factor. In severe cases, the EPP is so small that it fails to reach threshold, and no muscular contraction is produced.

The fatigability associated with myasthenia gravis is explained by the tendency for transmitter release to depress with repeated activation of a motor neuron. Such depression may be due to depletion of the pool of synaptic vesicles in the presynaptic terminal. In normal healthy adults, depression of transmitter release is of little consequence because the safety factor is so large. For example, the one-to-one relationship between motor neuron activation and skeletal muscle contraction can be maintained even if the EPP is reduced from an amplitude of 50 mV to an amplitude of 30 mV. In patients with myasthenia gravis, however, the EPP is already reduced. As a result, when the motor neuron is repeatedly

activated and transmitter release is depressed, the EPP becomes subthreshold and fails to elicit an action potential in the muscle cell. Thus, whereas initial motor responses in myasthenic patients may be relatively normal, they fatigue rapidly as the EPP falls below threshold.

At the molecular level it is now known that the reduction of the EPP in patients with myasthenia gravis is due to reduction in the number of ACh receptors in the postjunctional membrane. Because of a reduced number of receptors, the postsynaptic permeability changes and the EPP are smaller. Current evidence indicates that the reduction of ACh receptors is due to an auto-immune response to the ACh receptor.

There is no known cure for myasthenia gravis. A common treatment, however, is the use of neostigmine. Neostigmine, by blocking the actions of AChE, makes more ACh available to bind with postjunctional ACh receptors and thus partially compensates for the reduced number of receptors in the myasthenic patient.

Tetany

Tetany is a pathologic condition accompanying hypocalcemia (low extracellular Ca^{2+}) that is associated with hyperexcitability of nerve and muscle cells. We have just learned that low Ca^{2+} tends to reduce chemical transmitter release, but low Ca^{2+} also has an effect on the postsynaptic cells as well (in this case, the muscle cell). Lowering the Ca^{2+} concentration tends to reduce the threshold for initiating action potentials. The combined effect of low Ca^{2+} is to make the membrane of the muscle cell easier to depolarize and thus better able to initiate action potentials, despite the decreased chemical transmitter release.

permeable to K^+. The resultant depolarization may seem paradoxical, but recall that the membrane potential is due to a balance of the resting K^+ and Na^+ permeabilities. The K^+ permeability tends to move the membrane potential toward the K^+ equilibrium potential (-80 mV), whereas the Na^+ permeability tends to move the membrane potential toward the Na^+ equilibrium potential ($+55$ mV). Normally, the K^+ permeability predominates and the resting membrane potential is close to, but not equal to, the K^+ equilibrium potential. If K^+ permeability is decreased because some of the channels close, the membrane potential will be biased toward the Na^+ equilibrium potential and the cell will depolarize.

At least one reason for the long duration of slow PSPs is that second messenger systems are slow (second to minutes). Take the cAMP cascade as an example. Cyclic AMP takes some time to be synthesized but, more importantly, cAMP levels can remain elevated for a relatively long period of time (minutes) after synthesis. The duration of the elevation of cAMP depends on the actions of cAMP-phosphodiesterase, which breaks down cAMP. The duration of an effect could even outlast the duration of the change in the second messenger because of persistent phosphorylation of the substrate protein(s). Phosphate groups are removed from the substrate proteins by protein phosphatases. Thus, the net duration of a response initiated by a metabotropic receptor is dependent upon the actions of not only the synthetic and phosphorylation processes, but also the degradative and dephosphorylation processes.

The activation of a second messenger by a transmitter can have a localized effect on membrane potential through phosphorylation of membrane channels near the site of synthesis. The effects can be more widespread and even longer lasting than depicted in Fig. 25, however. For example, second messengers and protein kinases can diffuse and affect more distant membrane channels. Moreover, a long-term effect can be induced in the cell by altering gene expression. For example, PKA can diffuse to the nucleus, where it can activate proteins (*e.g.*, transcription factors) that regulate gene expression.

Bibliography

Aidley DJ. *The physiology of excitable cells*, 3rd ed. Cambridge: Cambridge University Press, 1989

Cooper JR, Bloom FE, Roth RH. *The biochemical basis of neuropharmacology*, 7th ed. New York: Oxford University Press, 1996

Eccles JC. *The understanding of the brain*, 2nd ed. New York: McGraw-Hill, 1977

Hille B. *Ionic channels of excitable membranes*, 3rd ed. Sunderland, MA: Sinauer Associates; 2001

Kandel ER, Schwartz JH, Jessell TM. *Principles of neural science*, 4th ed. New York: McGraw-Hill, 2000

Karlin A. Emerging structure of the nicotinic acetylcholine receptors. *Nat Rev Neurosci* 2002; 3:102–114

Katz B. *The release of neural transmitter substances*. Springfield, IL: Charles C Thomas, 1969

Krogsgaard-Larsen P, Hansen J, Eds. *Excitatory amino acid receptors*. New York: Ellis Horwood Press, 1992

Nicholls JG, Martin AR, Wallace BG. *From neuron to brain*, 4th ed. Sunderland, MA: Sinauer Associates, 2001

Schmidt RF, Ed. *Fundamentals of neurophysiology*, 2nd ed. New York: Springer-Verlag, 1978

Shepherd GM, Ed. *The synaptic organization of the brain*, 4th ed. New York: Oxford University Press, 1998

Siegel GJ, Agranoff BW, Albers RW, Molinoff PB. *Basic neurochemistry*, 6th ed. Philadelphia: Lippincott-Raven, 1999

Suggested Readings

Attwell D, Barbour B, Szatkowski M. Nonvesicular release of neurotransmitter. *Neuron* 1993; 11:401–407

Augustine GJ, Burns ME, DeBello WM, Hilfiker S, Morgan JR, Schweizer FE, Tokumaru H, Umayahara K. Proteins involved in synaptic vesicle trafficking. *J Physiol* 1999; 520(Pt. 1):33–41

Barnard EA. Receptor classes and the transmitter-gated channels. *Trends Biochem Sci* 1992; 17:368

Bennett MVL. Electrical transmission: a functional analysis and comparison to chemical transmission, in: Kandel ER, Ed., *Handbook of physiology, Section 1: The nervous system*, Vol. 1, Pt. 1. Bethesda: American Physiological Society, 1977, pp. 357–416

Boyd IA, Martin AR. The end-plate potential in mammalian muscle. *J Physiol* 1956; 132:74–91

Burke RE, Rudomin P. Spatial neurons and synapses, in: Kandel ER, Ed., *Handbook of physiology, Section 1: The nervous system*, Vol. 1, Pt. 2. Bethesda: American Physiological Society, 1977, pp. 877–944

Chapman ER. Synaptotagmin: a Ca(2+) sensor that triggers exocytosis? *Nat Rev Mol Cell Biol* 2002; 3(7):498–508

del Castillo J, Katz B. Quantal components of the end-plate potential. *J Physiol* 1954a; 124:560–573

del Castillo J, Katz B. Statistical factors involved in neuro-muscular facilitation and depression. *J Physiol* 1954b; 124:574–585

Dermietzel R, Spray DC. Gap junctions in the brain: where, what type, how many and why? *TINS* 1993; 16:186–192

Fatt P, Katz B. An analysis of the end-plate potential recorded with an intracellular electrode. *J Physiol* 1951; 15:320–370

Galzi JL, Revah F, Beisis A, Changeux J-P. Functional architecture of the nicotinic acetylcholine receptor: from electric organ to brain. *Annu Rev Pharmacol* 1991; 31:37–72

Katz B, Miledi R. The release of acetylcholine from nerve endings by graded electrical pulses. *Proc Roy Soc London (Biol)* 1967a; 167:23–38

Katz B, Miledi R. The timing of calcium action during neuromuscular transmission. *J Physiol* 1967b; 189:535–544

Katz B, Miledi R. A study of synaptic transmission in the absence of nerve impulses. *J Physiol* 1967c; 192:407–443

Lester HA. The response to acetylcholine. *Sci Am* 1977; 236:10

Liley AW. The quantal components of the mammalian end-plate potential. *J Physiol* 1956; 133:571–587

Lin RC, Scheller RH. Mechanisms of synaptic vesicle exocytosis. *Annu Rev Cell Dev Biol* 2000; 16:19–49

Madden DR. The structure and function of glutamate receptor ion channels. *Nat Rev Neurosci* 2002; 3:91–101

Martin AR. Junctional transmission II. Presynaptic mechanisms, in: Kandel ER, Ed., *Handbook of physiology, Section 1: The nervous system*, Vol. 1, Pt. 1. Bethesda: American Physiology Society, 1977, pp. 329–355

Neher E, Sakmann B. Single-channel currents recorded from membrane of denervated frog muscle fibres. *Nature* 1976; 260:779–802

Role LW, Diversity in primary structure and function of neuronal nicotinic acetylcholine receptor channels. *Curr Opin Neurobiol* 1992; 2:254

Schuetze S. Understanding channel gating: patch-clamp recordings from cloned acetylcholine receptors. *Trends Neurosci* 1986; 9:140–141

Spencer WA. The physiology of supraspinal neurons in mammals, in: Kandel ER, Ed., *Handbook of physiology, Section 1: The nervous system*, Vol. 1, Pt. 2. Bethesda: American Physiological Society, 1977, pp. 969–1022

Stroud RM, Finer-Moore J. Acetylcholine receptor structure, function, and evolution. *Annu Rev Cell Biol* 1985; 1:317–351

Takeuchi A. Junctional transmission I. Postsynaptic mechanisms, in: Kandel ER, Ed., *Handbook of physiology, Section 1: The nervous system*, Vol. 1, Pt. 1. Bethesda: American Physiological Society, 1977, pp. 295–327

Turner KM, Burgoyne RD, Morgan A. Protein phosphorylation and the regulation of synaptic membrane traffic. *Trends Neurosci* 1999; 22(10):459–464

Zucker RS. The role of calcium in regulating neurotransmitter release in the squid giant synapse, in: Feigenbaum J, Hanani M, Eds., *Presynaptic regulation of neurotransmitter release: a handbook*. London: Freund Publishing House 1991, pp. 1:153–195

7

Striated Muscle

NORMAN W. WEISBRODT

KEY POINTS

- *Actin* and *myosin,* arranged in thin and thick filaments respectively, are the main contractile proteins in muscle.
- *Myosin crossbridges* interact with actin during contraction to effect sliding of thick and thin filaments over one another, thus effecting force development and muscle cell shortening.
- Contractions are initiated by increases in calcium that occur in response to action potentials.
- Skeletal muscle action potentials are initiated by sudden increases in sodium permeability of the sarcolemma.
- Cardiac muscle action potentials are initiated by sudden increases in sodium permeability followed by slower and longer lasting increases in calcium permeability of the sarcolemma.
- The force of muscle contraction *in vitro* is a function of muscle length before stimulation. The velocity of shortening is a function of the afterload against which the muscle is contracting.

- The force and velocity of contraction of skeletal muscle *in vivo* are regulated by recruitment of muscle fibers and by alterations in the frequency of stimulation of those fibers.
- The force and velocity of cardiac muscle contraction *in vivo* depend on length; in addition, force and velocity at any given length and afterload can vary due to changes in contractility.
- ATP is used to fuel crossbridge cycling, calcium re-uptake, and the maintenance of ionic gradients across the sarcolemma. The rate of ATP use is highest in fast-twitch skeletal muscle, lower in slow-twitch skeletal and cardiac muscle, and lowest in smooth muscle.
- ATP is generated by both glycolysis (in most fast-twitch muscle cells) and oxidative phosphorylation (in most slow-twitch and cardiac muscle cells).

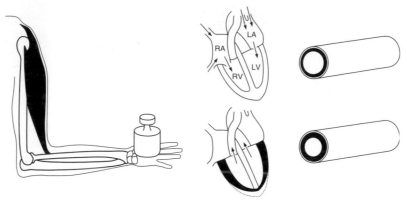

FIGURE 1   (Left) The attachment of a skeletal muscle. Contractions of the muscle result in movement of the arm, thus lifting the weight. (Center) Relaxation of cardiac muscle (top) allows the heart to fill with blood. Contraction of cardiac muscle (bottom) reduces the size of the ventricals, thus expelling blood into the pulmonary artery and the aorta. (Right) A small artery whose wall is comprised mostly of smooth muscle. Relaxation (top) and contraction (bottom) of the muscle increases and decreases, respectively, the diameter of the artery, thus altering the resistance to blood flow.

Many cells can change shape and/or move about. For example, intestinal and renal epithelial cells have microvilli that contract, perhaps to mix lumenal contents, and white blood cells can move between capillary endothelial cells and into tissues to combat invading organisms. However, for some cells, the ability to change shape is developed to a high degree. These cells are arranged such that their contractions result in the movement of attached structures (*e.g.*, the skeleton) or in a change in shape of the structure they in part comprise (*e.g.*, the heart, and arteries) (Fig. 1). These cells are muscle cells, and in humans such cells compose 50 to 70% of the lean body mass.

Muscle cells can be classified on the basis of structure, location, and function. All muscle cells can be divided into two groups—striated and smooth—based on their microscopic structure (see below). On the basis of location, striated muscle can be divided into three subgroups—skeletal, cardiac, and visceral—whereas smooth muscle can be divided into many subgroups (*e.g.*, arterial, intestinal, uterine). Functional classifications of muscle cells are based on contractile behavior (*e.g.*, fast twitch or slow twitch, tonic or phasic, unitary or multiunit) and biochemical activities (*e.g.*, oxidative or glycolytic).

BASIC CONTRACTILE UNIT OF MUSCLE

In perhaps no other cell type is function explained so well by structure and biochemical interactions as it is in muscle cells. Contraction of muscle cells is associated both qualitatively and quantitatively with specific structural arrangements of specific proteins. *Actin* and *myosin* are proteins that form the basic structural

characteristic of muscle and are arranged in filaments—actin in thin filaments and myosin in thick filaments (Fig. 2). Each actin molecule (G-actin or globular actin) is a globular protein with a molecular weight of approximately 45,000 Da. In the thin filaments, the monomers are polymerized together like two strands of

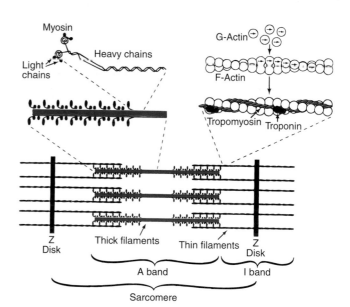

FIGURE 2   Longitudinal view of the organization of contractile proteins into myofilaments and of myofilaments into a sarcomere. The proteins actin and myosin comprise the major portions of the thin and thick filaments, respectively. Thin filaments project outward from both sides of the Z disks. Thick filaments lie between and overlap with thin filaments from two adjacent Z disks. Regions of overlap appear darker when viewed microscopically; thus, this region is referred to as the *anisotropic* or *A band*. Regions that contain only thin filaments appear lighter and are referred to as *isotropic* or *I bands*. The structure from one Z disk to another is a *sarcomere* and is the basic contractile unit of striated muscle. G, globular; Γ, filamentous.

pearls that are twisted in an alpha helix to form F-actin (filamentous). Each myosin molecule comprises six monomers: two proteins, each with a globular head and a filamentous tail (the myosin heavy chain) and each with a molecular weight of approximately 200,000 Da, and four smaller, globular proteins (the myosin light chains), each with molecular weights of 17,000 to 24,000 Da. The light chains associate with the globular heads of the heavy chains by ionic bonds and Van der Waals forces. Large portions of the tail segments from many heavy chains interact with one another to form the backbone of the thick filament. Part of their tails and the globular heads protrude from the backbone to form structures called *crossbridges*.

Actin and myosin are found in all muscle cells. However, each protein monomer exists in multiple isoforms that are present in different proportions, depending on the particular type of muscle cell (*e.g.*, striated or smooth, skeletal or cardiac, fast twitch or slow twitch) and on its stage of development (*e.g.*, embryonic or adult). The functional consequences of some of the isoforms are discussed below.

The way in which the thick and thin filaments are arranged within the cells defines in large part the two main types of muscle: striated and smooth. In striated muscle, the thick and thin filaments are very ordered in their anatomic arrangement within the cell. Thin filaments extend in opposite directions from protein structures called Z disks. In relaxed muscle, the thin filaments from two opposing Z disks extend toward each other but do not touch or overlap. Bridging the gap between the thin filaments, and overlapping with them, are the thick filaments. This arrangement—Z disk, thin filament, thick filament, thin filament, Z disk—defines the functional unit called a *sarcomere* (see Fig. 2). In striated muscle, sarcomeres are arranged in transverse registry, thus accounting for the characteristic banding pattern or striations. As shown in Fig. 2, the sarcomere can be characterized in minute detail. This fine structure was discovered long before the mechanisms responsible for the shortening were elucidated, thus accounting for the descriptive terminology used to describe the sarcomere.

Sarcomeres not only have ordered structure when viewed in the longitudinal axis, but they also exhibit symmetry when viewed in cross section (Fig. 3). Each thick filament is surrounded by six thin filaments in a pattern that is repeated such that there are two thin filaments for every thick filament (each thin filament interacts with three thick filaments). Also, the crossbridges are arranged around the thick filament, separated by angles of 60°, in such a way that they can associate with the actin monomers in the thin filaments. In smooth muscle, thick and thin filaments are present,

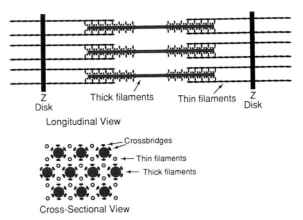

FIGURE 3 Longitudinal and cross-sectional views of the thick and thin filaments. The cross section was made in a region of overlap of the filaments. The crossbridges extend from the thick filaments toward the thin filament. Each thick filament is surrounded by six thin filaments, and each thin filament interacts with crossbridges from three thick filaments.

but they have no such structured arrangement and, thus, give a homogeneous appearance under the microscope (see Chapter 8).

The specific way in which the proteins actin and myosin are arranged gives muscle cells the ability to shorten. When striated muscles contract, crossbridges from the thick filaments attach to specific regions on the actin molecules (Fig. 4). The crossbridge heads then change angles, causing the thick and the thin filaments to slide over one another. The crossbridges then release, and their angles assume the resting positions. They now are ready to attach to a different actin molecule, thus repeating the cycle until the stimulus to contract ceases. Because two opposing sets of thin filaments are associated with a single set of thick filaments, filament sliding results in movement of the Z disks toward one another without either the thick or the thin filaments changing length. Also, because the Z disks and the thin filaments are linked with other cytoskeletal elements, movement of the Z disks toward one another results in shortening of the muscle cell.

BIOCHEMICAL INTERACTIONS OF ACTIN AND MYOSIN

The myosin molecule is an adenosine triphosphatase (ATPase), and adenosine triphosphate (ATP) is required for crossbridge cycling in intact muscle. Thus, muscle contraction requires energy expenditure. Actin and myosin can be extracted and purified. If purified striated muscle myosin is dissolved under the ionic conditions found in muscle cells, it exhibits a low level of ATPase activity. Upon addition of striated muscle actin,

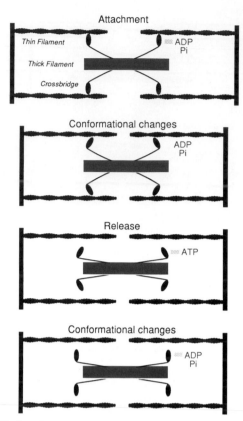

FIGURE 4 The events during a crossbridge cycle. First, activated crossbridges attach to binding sites on the actin (*attachment*). Then, the crossbridge heads tilt due to *conformational changes*. This tilt causes movement of the thin filaments and the Z disks. The crossbridges then *release*. Finally, the crossbridges are reactivated to assume the position conducive for attaching to actin (conformational changes). If cytosolic calcium levels are still high, the cycle will repeat. *ATP hydrolysis during a crossbridge cycle:* The activated crossbridges that attach to actin (attachment) contain bound adenosine diphosphate–inorganic phosphate (*ADP-Pi*). Binding to actin results in release of the ADP-Pi products and conformational changes in the crossbridge heads. ATP binding to the attached crossbridge is necessary to disassociate the crossbridge from the actin (release). Once disassociated, ATP is hydrolyzed, but the products are not released. This induces conformational changes to produce activated crossbridges.

however, the ATPase activity increases dramatically. In the cell, the cycling of the myosin crossbridge with actin also results in the splitting of ATP. As depicted in Fig. 4, this splitting takes place in several steps. Upon activation of the muscle, (1) the charged myosin crossbridges present in resting muscle bind to the actin molecules; (2) the crossbridges undergo a conformational change to slide the thick and thin filaments over one another, which also results in the loss of adenosine diphosphate (ADP) from the myosin heads; (3) ATP binds with the actin–myosin complexes to bring about a dissociation; and (4) ATP is hydrolyzed to bring the crossbridges to their original charged conformation. Because ATP is required for the dissociation of actin and myosin,

a depletion of ATP results in muscle stiffness, not relaxation. This is the cause of the rigor that is seen shortly after death.

Because purified striated muscle actin and myosin are always ready to interact to hydrolyze the ever-present ATP and to cause contraction, regulatory mechanisms must exist to ensure that muscle will contract only when excited. *Tropomyosin* and *troponin* are two protein–protein complexes located on the thin filament that mediate the regulation (see Fig. 2). Tropomyosin is a filamentous protein that lies close to the groove between the two strands of actin. Troponin is a complex of three globular proteins—troponin T, troponin I, and troponin C. This complex is bound to tropomyosin periodically along the thin filament. The tropomyosin–troponin complex interferes with the interaction of the myosin crossbridges with actin. In the resting muscle, the crossbridge binding sites on the actin molecules are protected. When the muscle is stimulated to contract, a conformational change in the thin filament uncovers the binding sites, and crossbridge cycling takes place. The conformational change comes about because of the interaction of calcium with troponin C of the tropomyosin–troponin complex.

PIVOTAL ROLE OF CALCIUM

Whether a muscle is contracting or relaxed depends on the level of cytosolic calcium available to interact with the regulatory proteins. In relaxed muscle, the level of free cytosolic calcium (calcium that is not bound to other structures such as sarcoplasmic reticulum, mitochondria, or nuclei) is low ($< 10^{-7}$ M). Upon stimulation of the muscle, the calcium level increases into micromolar or higher ranges to initiate contraction. In striated muscle, calcium binds directly to troponin C of the tropomyosin–troponin complex to bring about a conformational change in the complex. Once the stimulus for muscle contraction ceases, free calcium levels decrease and calcium dissociates from the regulatory proteins. The muscle then relaxes. The relation between free calcium levels and contraction force is complex and indicates cooperation among the contractile proteins once calcium binds to troponin to initiate contraction.

Free ionic calcium is the mediator between the events of the cell membrane that indicate excitation and the protein interactions that result in contraction. Thus, the events that describe calcium movements in muscle cells often are referred to as *excitation–contraction (E–C) coupling*. E–C coupling differs among the various types of muscles; similar to the contractile process, it is closely linked to structure.

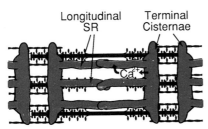

FIGURE 5   Sarcoplasmic reticulum (SR) in skeletal muscle. The SR consists of two components. The terminal cisternae serve as reservoirs of calcium during relaxation. Upon stimulation, Ca^{2+} is released to interact with the troponin complex. Once released, Ca^{2+} is avidly taken up by the second component, the longitudinal SR. Uptake is mediated by a calcium ATPase.

E–C Coupling in Skeletal Muscle

In relaxed skeletal muscle, all the calcium that normally takes part in E–C coupling is stored inside the cell in the sarcoplasmic reticulum (SR) (Fig. 5). As might be expected, the SR of these cells is highly developed and extensive. Structurally, it is of two types: the longitudinal SR, which runs parallel to the thick and thin filaments, and the terminal cisternae, which form large pouch-like structures at each end of the longitudinal SR. Functionally, the terminal cisternae serve as storage places for calcium during muscle relaxation. Upon muscle excitation, calcium moves into the cytoplasm down a large concentration gradient. Once in the cytoplasm, calcium interacts with the tropomyosin–troponin complex to allow full activation of the contractile proteins. Calcium then is taken up by the longitudinal SR by an active process that involves a calcium ATPase. This pump has a high affinity for calcium and, if no calcium is being released from the terminal cisternae, can quickly lower cytosolic calcium to levels that do not support contraction. Calcium taken up by the longitudinal SR returns to the terminal cisternae, from which it can be released again. The SR is so well developed in most skeletal muscles that intracellular calcium is conserved to the point that these muscles can contract normally for some time, *in vitro*, even in a medium devoid of calcium.

E–C Coupling in Cardiac Muscle

E–C coupling in cardiac muscle differs only slightly from that in skeletal muscle, but the differences are functionally extremely important. In cardiac muscle, the SR is not as highly developed in that the terminal cisternae are not as large (see Chapter 11). One consequence of this is that variable amounts of calcium are released from the SR to allow for variable activation of the contractile proteins. However, the calcium that is released into the cytoplasm acts just as it does in skeletal muscle. That is, it binds to tropomyosin–troponin complexes to initiate contraction. Another functionally important difference between skeletal and cardiac muscle is that some calcium must enter cardiac muscle cells from the extracellular space in order to initiate contraction (see Chapter 11). Although this amount is small, cardiac muscle cells *in vitro* soon stop contracting if placed in a medium that does not contain calcium. For relaxation to occur, most of the calcium is taken up into the longitudinal SR by an active process involving a calcium ATPase just as in skeletal muscle. However, between contractions, some calcium must leave the cell. This is accomplished mainly by a sodium–calcium exchange mechanism present on the sarcolemma. There can be temporary imbalances between the amount of calcium entering and the amount leaving, such that the stores of releasable calcium in the SR increase and decrease. These imbalances are important in regulating cardiac contractile activity (see below).

EXCITATION OF THE MUSCLE CELL

All muscle cells have resting membrane potentials in the range of -70 to -90 mV. As in nerve, this potential is due to the presence of ionic concentration gradients (with K^+ being greater intracellularly and Na^+ greater extracellularly) and to the resting membrane being much more permeable to K^+ than to Na^+. Muscle cells are excitable due to the presence of voltage-dependent ion channels in their cell membranes. However, the type of channels present, the manner in which channels are activated, and the way in which their activation leads to E–C coupling vary among muscle types.

Skeletal Muscle

Skeletal muscle closely resembles nerve in that the voltage-dependent channel is a sodium channel. Once the sarcolemma is depolarized to the threshold for opening of sodium channels, a positive feedback opening of channels occurs and the membrane potential reverses toward the sodium equilibrium potential. Opening of the sodium channel is time dependent; the channel quickly inactivates and the membrane repolarizes. The entire action potential lasts about 1 msec. Skeletal muscles do not exhibit spontaneous action potentials because there are no inherent mechanisms to depolarize the cells to threshold. Also, there are no connections between individual skeletal muscle cells to allow for conduction of activity from one cell to another. Each cell depends on innervation by a motor nerve. These nerves release acetylcholine (ACh) at their junctions with the muscle cell (neuromuscular junction). The ACh binds to specific receptors on the muscle cell membrane to bring about an increase in permeability of that portion of the cell membrane to Na$^+$ and K$^+$. This results in depolarization of adjacent areas of the membrane to threshold, at which point an action potential ensues (see Chapter 6).

Skeletal muscle cells are large, multinucleated cells that result from the embryonic fusion of many myoblasts. Thus, sarcomeres and SR in the center of the muscle are many microns away from the cell surface. This would make it impossible for sarcolemmal action potentials to have any effect on these structures, if it were not for the fact that the sarcolemma invaginates to make contact with each and every terminal cisterna. If a muscle cell is sectioned properly, a sarcolemmal invagination, a *transverse (T) tubule*, can be seen running between two terminal cisternae at right angles to the SR. This complex of tubules is called a *triad* (Fig. 6). Although there is intimate contact at the triad, the T tubule and the SR retain their individual membranes so that lumenal separation is maintained. The T tubules allow action potentials to penetrate into the interior of the cell and to influence the terminal cisternae of the SR. The action potential does not propagate into the SR, but it does bring about the release of calcium. Because the action potential spreads rapidly over the entire sarcolemma, including the T tubules, the entire muscle can be activated almost simultaneously.

Cardiac Muscle

The events in cardiac muscle differ from those in skeletal muscle; in addition, there are differences among the various types of cardiac muscle cells. As discussed in

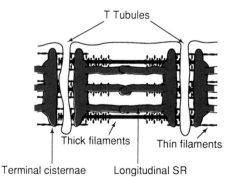

FIGURE 6   Transverse (T) tubules in skeletal muscle. The sarcolemma invaginates to reach the interior of the muscle cell such that it contacts the terminal cisternae of the SR. Thus, events initiated by the membrane action potential are brought into close proximity to the SR to release CA^{2+} by mechanisms that are not understood. The rapid, almost simultaneous propagation of the action potential down the T tubules allows for almost simultaneous initiation of the contractile event.

Chapter 11, cardiac muscle cells in the sinoatrial and atrioventricular nodes exhibit action potentials spontaneously and act as pacemakers, whereas other muscle cells are specialized to conduct action potentials throughout the ventricles. These cells are not discussed in this chapter because their importance is not related to their contractile activity. The cells described here are the ventricular myocardial cells, which contract to pump blood through the pulmonary and systemic circuits. In these cells, the resting membrane potential has the same basis as that in nerve and skeletal muscle. However, when the membrane is depolarized to threshold, two types of voltage-dependent channels open—a sodium channel that opens and closes within a few milliseconds (similar to what is found in skeletal muscle) and a calcium channel that opens more slowly and can stay open for several hundred milliseconds. It is this second channel that accounts for calcium entry from the extracellular space. Like the sodium channel, opening of the calcium channel is also time dependent. It eventually closes to allow for repolarization of the cell.

Myocardial cells normally do not exhibit spontaneous action potentials; however, they are connected to one another via low-resistance membrane contact points. If the myocardial cells are connected to pacemaker cells and conducting cells, the action potentials initiated in the pacemaker cells will rapidly propagate throughout the myocardial cells. Conduction from cell to cell is by way of electrotonic spread and does not require a chemical transmitter such as ACh.

Cardiac muscle cells are smaller than skeletal muscle cells; however, they do have an extensive T tubule system. In fact, in cardiac muscle, the T tubules are larger. They not only invaginate to run perpendicular to the SR, but they also turn and run parallel alongside the SR. Perhaps it is these structural characteristics that allow the T tubules not only to participate in the release of calcium from the terminal cisternae but also to facilitate the entry of significant amounts of calcium during the action potential. As in skeletal muscle, T tubules allow for spread of excitation into the interior of the cells so that all portions are excited almost simultaneously (see Chapter 11 for further discussion of these points).

MECHANICAL RESPONSE OF MUSCLE *IN VITRO*

When muscle contracts, it normally develops force and shortens; however, much has been learned by studying muscle under conditions that permit the study of each function separately. A contraction that generates only force, with no muscle shortening, is called an *isometric contraction*. One that results in shortening against a constant force is called an *isotonic contraction*. Isometric and isotonic contractions can be studied *in vivo*—for example, by straining against an immovable object (isometric contraction) or by moving a load (isotonic contraction); however, contractions are more easily studied *in vitro* using isolated strips of muscle.

Skeletal Muscle

Isometric Contraction

A small skeletal muscle, composed of hundreds of individual muscle cells, can be placed in physiologic salt solution with one end connected to a fixed point and the other end connected to a transducer so that the time course and force of isometric contractions can be recorded (Fig. 7). If single, short-duration electric pulses of increasing intensity are delivered to the muscle, a threshold intensity will be reached at which a mechanical response is elicited. This contraction is due to the activation of one or a few of the most excitable muscle cells. As stimulus pulse intensity is increased, contractile force increases up to a point at which further increases in intensity result in no further increase in force. Such pulses are referred to as *maximal* or *supramaximal*. This behavior is due to the fact that individual muscle cells

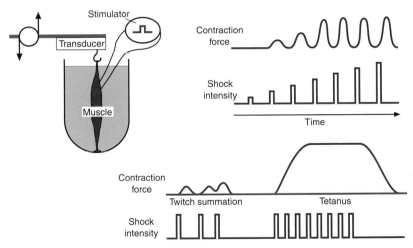

FIGURE 7   Recording isometric contraction *in vitro*. A small bundle of skeletal muscle fibers is placed in a warmed, oxygenated physiologic salt solution. One end is anchored to the bottom of the bath and the other end to a movable force transducer. Wires from a stimulator are placed so that the muscle can be excited with electrical pulses. (Top) The muscle is stretched to its *in vivo* resting length and stimulated with single shocks of increasing intensity. Once a threshold is reached, a single, longer lasting contractile response, called a *twitch*, results. As simultaneous intensity is progressively increased, contraction forces increase up to a point at which further increases in intensity result in no further increase in force. (Bottom) If shocks that cause a single twitch of maximal force are delivered at a higher frequency, the muscle does not have a chance to relax fully before the second stimulus. This results in the second response generating a greater force (summation). If a series of closely spaced shocks are delivered, muscle force rises smoothly to a maximum and stays elevated (tetanus) until stimulation ceases.

within a muscle bundle differ from one another in their excitabilities, and that excitation of one muscle cell does not spread to other muscle cells. Threshold stimuli induce only the most excitable cells to contract. As stimulus intensity is increased, more cells are recruited until all of the cells comprising the muscle are recruited by maximal and supramaximal stimuli. Although *recruitment* can be demonstrated in this manner *in vitro*, as discussed below, recruitment *in vivo* occurs at the level of the nervous system.

If, instead of using only a single supramaximal pulse, a second supramaximal pulse is delivered before the muscle relaxes from the first stimulation, the second contraction will be larger, a phenomenon known as *summation*. The magnitude of summation is increased by decreasing the time between successive pulses (*i.e.*, increasing pulse frequency). Eventually, a pulse frequency will be reached that is so high that muscle responses to individual pulses cannot be detected; at this point, a smoothly shaped contraction to a maximum force is obtained that is referred to as a *tetanic contraction*. Once reached, further increases in pulse frequency will not increase peak contractile force.

The phenomena of summation and tetanus occur because: (1) the time courses of excitation, excitation–contraction coupling, and activation of the contractile proteins are different; and (2) structures with elastic properties exist within and between the cycling crossbridges and the ends of the muscle cells. Each short, supramaximal electrical pulse elicits a single action potential in each muscle cell (Fig. 8). Even though each single action potential lasts only about 1 msec, it results in the release of enough calcium to fully activate the contractile proteins. However, the duration of each calcium pulse is only a few milliseconds. Such a short,

but complete, activation of the contractile proteins does allow some crossbridge cycling and sliding of the thick and thin filaments over one another, but such short activation does not allow the full force of this interaction to be transmitted to the ends of the muscle. Also, during an isometric contraction, there is no shortening of the ends of the muscle even though some filament sliding has taken place. How can this be? Muscle proteins have elastic properties much like that of a spring. These structures transmit the forces generated by crossbridge cycling to the ends of the muscle and are stretched as force is developed at the ends of the muscle. This stretching takes energy and time. In response to a single action potential, crossbridge cycling lasts too short a time to stretch these structures fully, so full force is not seen at the ends of the muscle. Although full force is not seen, not all the energy expended in stretching the elastic structures is lost. Much of the energy is stored and released during and after crossbridge cycling, so that the contractile response at the muscle ends has a much slower rise time and a much longer duration than either the action potential or the rise in calcium.

As pulse frequency is increased to produce summation, each pulse still produces a single action potential and an associated rise and fall in calcium. However, the contractile proteins are activated a second time before the elastic structures of the muscle return to their resting state. Thus, the energy of the second contractile protein interaction can stretch the elastic elements more fully and can develop more force at the ends of the muscle. During tetanus, each pulse still is eliciting an action potential, but now calcium levels do not fall between action potentials. The maximum calcium level is no higher than with a single, isolated action potential, but it is maintained for a longer time. The continual level of

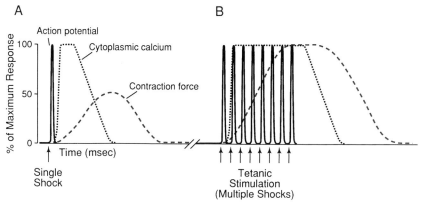

FIGURE 8   Action potentials, cytoplasmic calcium levels, and contraction forces of skeletal muscle in response to a single shock (A) and tetanic stimulation (B). Note that each shock is followed by an action potential of identical amplitude and duration and by an increase in calcium to its maximal level. However, the time that calcium levels are increased is greater during the tetanus. This is in part the cause of the increased force during tetanus. (Modified from Berne RM, Levy MN, *Physiology*, 2nd ed. St. Louis: Mosby, 1988.)

calcium allows for continual activation of the contractile proteins, so that elastic elements can be stretched completely and the full force of crossbridge cycling can be realized at the ends of the muscle.

The force of contraction also depends on the resting length of the muscle. If the relaxed muscle is hung in the bath so that it is at the shortest length at which it hangs straight, no resting force is recorded by the transducer. If the relaxed muscle is slowly stretched by moving the force transducer, the resting force remains at almost imperceptible levels over the initial changes in length, then force increases exponentially with further stretch. The curve generated by such stretching describes the *length–passive force* (sometimes called *passive length–tension*) relationship (Fig. 9). Note that this curve is similar to that seen for any elastic body. Another curve will be obtained if the experiment is repeated, only this time stretching the muscle in steps and, at each level of stretch, stimulating with pulses of supramaximal intensity at a frequency that will produce tetanus. If the peak force during tetanus at each length is plotted, the *length–total force* relationship is obtained. Total force at each length is the sum of the passive force (defined

previously) and the active force developed by crossbridge cycling. Thus, a *length–active force* relationship can be obtained by subtracting the passive force from the total force at each length. By inspecting the length–active force curve, it is obvious that there is one length at which active force is maximal. This length is the optimal length (L_o). Going to either shorter or longer lengths results in less active force.

The length–force relationships can be explained on structural grounds. The length–passive force relationship is due to the elastic behavior of the cell membranes and of the connective tissue between the muscle cells. These structures resist the forces applied to a resting muscle. The length–active force relationship is due to the arrangements of the actin and myosin molecules within the sarcomere. As described previously, the core of the thick filament is made up of the tail regions of the myosin molecules, and the crossbridges are due to protrusion of a portion of the tail and of the globular heads of the myosin (see Fig. 2). Additionally, each thick filament is bilaterally symmetrical in the direction of protrusion of the crossbridges. There is a bare zone in the middle with no crossbridges. On each side of this zone, crossbridges extend outward and upward toward the thin filaments. At L_o, the thin and thick filaments overlap such that almost every myosin crossbridge is capable of interacting with an active site on the actin (Fig. 9B). This allows for maximum active force to be generated. Stretching the muscle lessens the overlap. Recall that the thin filaments are attached to the Z disks and that the Z disks are in turn attached to the cell membrane. Stretching the membrane results in the Z disks moving away from each other, dragging the thin filaments with them. This results in not every crossbridge being able to reach an actin molecule (Fig. 9C). Thus, active force is reduced. If stretched to the point of no thick and thin filament overlap, no active force can be generated. At lengths shorter than L_o, the mechanisms responsible for the decreased force are less clear. A likely explanation is that, because cell volume must remain constant, the lateral distance between thick and thin filaments increases as the cell, and thus sarcomere, length shortens. This reduces the possibility of strong myosin crossbridge interactions with sites on actin.

In summary, the force of skeletal muscle contraction under isometric conditions is influenced by: (1) the number of cells in the muscle that are stimulated, (2) the frequency at which they are stimulated, and (3) the length of the muscle cells before stimulation.

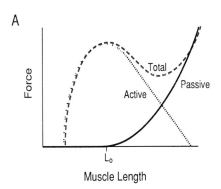

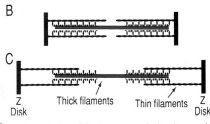

FIGURE 9 The relationship between skeletal muscle length and isometric forces developed by the muscle. (A) The length–*passive* force curve represents the force exerted by the unstimulated muscle at each length. The length–*total* force curve represents the force exerted by the muscle during a tetanus at each length. The length–*active* force curve is obtained by subtracting the passive from the total force at each length. This represents the force developed by the crossbridges. The magnitude of the active force at each length depends mostly on the overlap of the thick and thin filaments at each muscle length. (B) The length at which maximum force occurs is the length at which there is optimal overlap (L_o). (C) As the muscle is stretched, some crossbridges cannot interact with the thin filaments, and active force falls.

Isotonic Contraction

Isotonic contractions can be recorded *in vitro* using a movable lever (Fig. 10A) instead of the force transducer

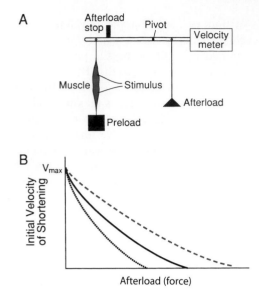

FIGURE 10   The relationship between the load against which a skeletal muscle must shorten (afterload) and the initial velocity of shortening. (A) A muscle is connected to a lever to which different weights (afterload) can be attached in such a way that the muscle does not bear the weight until it begins to contract. The initial muscle length before stimulation can be set by attaching a preload and holding the muscle at that length until shortening begins. (B) At any given initial muscle length, the velocity of shortening decreases as the afterload is increased. If the dashed line indicates values obtained with an initial length of L_o, either increasing (solid line) or decreasing (dotted line) the initial length will change the velocity at any given afterload, except for the velocity at zero afterload (V_{max}). This velocity, V_{max}, depends primarily on the isoform of myosin present in the muscle.

used to monitor isometric contractions. This lever is arranged so that changes in muscle length over time can be monitored and so that a weight can be added such that the muscle must bear the weight only after the muscle is fully activated. Such a weight is called an *afterload*. The experiment begins by setting the muscle to a given length with a *preload*. Then an afterload is added to the lever. The muscle is stimulated at a supramaximal pulse intensity at a tetanic frequency. If the muscle can develop force greater than that exerted by the afterload, the muscle will shorten and the initial velocity of shortening can be recorded. By using different afterloads—which really are equal to the forces that the muscle must develop in order to shorten—a *force–velocity relationship* is determined. If the experiment is repeated using different preloads to set different resting lengths, a family of force–velocity curves is generated (Fig. 10B). Note that all these curves have one point in common. If the curves are extrapolated back to zero afterload, a single velocity, the *maximal velocity* (V_{max}), is obtained.

The force–velocity relationship can be explained partly by structure and partly by biochemical features of the muscle. For any particular muscle, the difference

in velocity at any given afterload depends on the number of crossbridges taking part in the contraction; the more taking part, the higher the velocity, because each crossbridge will have to bear less of the load. As discussed above, the number of crossbridges taking part depends on the initial length of the muscle. Thus, the shortening velocity of an afterloaded muscle depends on muscle length. On the other hand, V_{max} depends on the rapidity with which a single crossbridge can cycle unimpeded by any load. Theoretically, as long as one crossbridge is cycling under no load, the muscle will shorten at its maximal velocity. That is why V_{max} is not affected by changes in initial muscle length. V_{max} is determined by molecular properties of the contractile proteins (*e.g.*, the ATPase activity of myosin).

Cardiac Muscle

As might be expected from their many structural and biochemical similarities, cardiac muscle exhibits many of the same mechanical characteristics as skeletal muscle; however, there are some important differences. Cardiac muscle cells, unlike skeletal muscle cells, are electrically coupled (see Chapter 11). Thus, once a stimulus pulse intensity is reached that effectively excites one cardiac muscle cell, the action potential spreads to all cells. Further increases in stimulus pulse intensity have little effect on contractile force. Also, unlike the situation seen in skeletal muscle, summation and tetanus do not occur in cardiac muscle. Action potential durations in cardiac muscle are almost as long as the contractile responses. Thus, the muscle relaxes before it is possible to stimulate it again because of the refractory period of the membrane (see Chapter 11). Also, the long duration of each action potential, through its associated rise in cytosolic calcium, allows for prolonged activation of the contractile proteins. This provides a longer time for crossbridge cycling to stretch the elastic structures of the muscle.

Cardiac muscle exhibits length–force and force–velocity relationships similar to those of skeletal muscle (Fig. 11): (1) isometric force is a function of muscle length, (2) the velocity of contraction at any initial length is a function of the afterload, and (3) the velocity of contraction of an afterloaded muscle is a function of the initial length of the muscle. The explanations for these relationships in cardiac muscle are similar to those given for skeletal muscle. However, in cardiac muscle that normally functions *in vivo* at lengths less than L_o (see Chapter 13), the sensitivity of myofilament interaction to calcium may be modulated by more than the lateral distance between thick and thin filaments. Length effects on the affinity of troponin C for calcium also have been proposed.

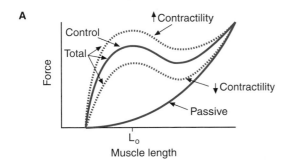

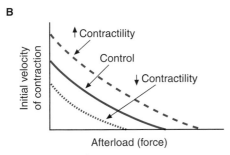

FIGURE 11 The mechanical responses of cardiac muscle. Note the qualitatively similar shapes of the length–passive force, length–total force, and afterload–velocity curves to those of skeletal muscle (see Figs. 9 and 10). (A) The length–passive force curve indicates that there is significant passive force at L_o. Also, note that there is more than one total force curve, due to the presence of more than one state of contractility. Thus, at any given length, active force can vary. (B) The afterload–force curves indicate that the entire relationship, including V_{max}, can be shifted by changes in contractility. Thus, for any given afterload, at any given initial length velocity can vary.

Although there are many similarities in mechanical function between cardiac and skeletal muscle, there are some important quantitative and qualitative differences. Quantitatively, cardiac muscle exhibits greater passive force at all lengths. At L_o, skeletal muscle exhibits almost no passive force, whereas cardiac muscle exhibits passive force equal to approximately 20% of the total force. This difference probably is related to differences in connective tissue composition. Qualitatively, cardiac muscle can exhibit moment-to-moment differences in: (1) the isometric force developed at any given length, (2) the velocity of shortening against an afterload at any given muscle length, and (3) the maximum velocity (V_{max}) of an isotonic contraction.

Such changes in V_{max}, in isometric force, and in velocity of afterloaded contractions that occur at constant thick and thin filament overlap are called changes in *contractility*. Some changes in contractility are due to differences in the amounts of calcium that are released with each action potential. As stated above, approximately the same level of cytosolic calcium is reached in skeletal muscle in response to each action potential, and this level of calcium is more than enough to activate fully the contractile proteins. This is not the case in cardiac muscle. Both the influx and efflux of calcium during the cardiac action potential and the amount of calcium released from the cardiac SR can change from contraction to contraction. Thus, cytosolic calcium levels can fluctuate from contraction to contraction. This means that during some contractions not all the crossbridges that proximate actin binding sites will be cycling because the binding sites will still be protected by tropomyosin–troponin complexes that are not binding calcium. Alterations in the number of cycling crossbridges will lead to differences in isometric force and in velocities of shortening against afterloads. The strength and velocity of crossbridge interaction also are influenced by phosphorylation of contractile proteins brought about through the actions of norepinephrine and epinephrine.

In summary, the force of cardiac muscle contraction is influenced by the length of the muscle cells before stimulation and by changes in contractility. Force is not influenced by variations in the number of excited cells nor by the occurrence of summation and tetanus.

MECHANICAL RESPONSE OF MUSCLE *IN VIVO*

Skeletal Muscle

Contractions of skeletal muscles are graded in force and in duration. For example, many of the same muscles are used to pick up a pencil as to pick up this book, but much more force will have to be developed to lift the book. The ways in which muscle contractions can be graded have been presented. However, not all the mechanisms are used by all muscles. Most skeletal muscles in the body are innervated by somatic nerves and are under voluntary control. In each nerve are many axons of *α-motor neurons*. Each of these axons branches to innervate a number of skeletal muscle cells. An α-motor neuron and the muscle cells it innervates comprise a *motor unit*. Motor units are small in those muscles over which fine motor control exists, and large in muscles without such fine control. Because the force generated by a whole muscle will depend on the number of its motor units that are active at any one time, the central nervous system can regulate muscle contractions by regulating the number of motor units activated and by the frequency of stimulation of each motor unit.

Length–tension relationships are not important as determinants of skeletal muscle force *in vivo*. Muscles are attached near fulcrum points of the skeleton at lengths near L_o (Fig. 12). Such attachment limits length changes such that muscle cells seldom are at more than ±15% of L_o. Attachment of muscle near fulcrum points

FIGURE 12   The effects of skeletal attachment on contractions of skeletal muscle. By having insertions near the joints, changes in muscle length are amplified at the end of the bone being removed. This enables the muscle to cause large movements of the skeleton while always functioning near its own L_o. On the other hand, the muscle must develop forces larger than those that must be overcome at the end of the bone that the muscle must move. (Modified from Guyton AC. *Textbook of medical physiology*, 7th ed. Philadelphia: W.B. Saunders, 1986.)

also means that these muscles must generate forces greater than those that would be needed to directly affect the load at the end of the bone being moved; however, such an attachment has the advantage that small changes in muscle length are translated into large movements of the load.

Skeletal muscles differ from one another in their force–velocity relationships. Some (*e.g.*, extensor digitorum longus) contract more quickly than others (*e.g.*, soleus). This difference is due to variations in the number and types of muscle cells that make up the various muscles in the body. Although there is a spectrum of velocities among various muscle cells, they have been divided into two main groups—fast twitch and slow twitch. Fast-twitch muscle cells generally are found in those muscles associated with rapid movement; slow-twitch cells are found in muscles associated more with endurance and posture. Many muscles are composed of a mixture of fast- and slow-twitch cells. Fast- and slow-twitch muscle cells differ in the

contractile protein isoforms that are present and in the ATPase activities of the myosin isoforms; other differences are given in Table 1.

Although force–velocity relationships in any given muscle do not change acutely, they can change over long periods of time (days to years). This is seen during development from embryo to adult and in muscle that undergoes hypertrophy. In these conditions, the contractile protein isoforms that are expressed by the cells change. This in turn affects velocity of shortening.

Cardiac Muscle

Cardiac muscle contractions are graded in both frequency and force such that the cardiac output can vary from approximately 5 L/min at rest to 20 L/min during exercise. Factors that affect frequency influence the pacemaker and conducting cells of the heart (discussed in Chapter 11). The potential mechanisms that alter force have been presented above; however, as with skeletal muscle, not all mechanisms are operative. Force of cardiac contraction is not influenced by variations in the number of stimulated myocardial cells in the heart because the cells are electrically coupled and are excited as a unit (or syncytium), nor is force altered by the occurrence of summation and tetanus because these cannot occur. On the other hand, force is influenced by the length of the muscle before contraction and by changes in contractility.

Myocardial cells are not attached to rigid structures such as the skeleton, but form chambers (the atria and ventricles) that can be distended. Thus, myocardial cell length can change significantly. The length of cardiac muscle cells before contraction is determined by the volume of blood contained in the cardiac chambers at the end of diastole. In the normal case, the more blood, the greater the length; the greater the length, the more forceful the contraction. For this to be true, the volume of blood must never be enough to stretch the cells to lengths greater than L_o. Such extensive stretch is prevented by the appreciable passive force at lengths at and below L_o. The importance of this length–force relationship in regulating cardiac output is discussed in Chapter 14.

The force and velocity of myocardial cell contractions also are affected by the level of contractility that exists during the contraction. The level of contractility normally is determined by heart rate, by the levels of circulating hormones such as epinephrine, and by the level of activity of the sympathetic nervous system. Contractility is decreased in diseases such as congestive heart failure. Regulation of myocardial contractility also is discussed in more detail in Chapters 13 and 14.

TABLE 1   Skeletal Muscle Cell Types

	Fast twitch	Slow twitch
V_{max}	High	Low
Myosin ATPase	High	Low
Glycolytic metabolism	High	Low
Oxidative metabolism	Low	High
Mitochondrial content	Low	High
Myoglobin content	Low	High

ENERGY BALANCE IN MUSCLE CELLS

Crossbridge Cycling

Energy in the form of ATP is used to conduct many functions of striated muscle cells. In addition to the energy required for synthesis of structural components of the cell, large amounts are needed to support the events associated with contraction—crossbridge cycling, calcium translocations, and membrane electrical activity.

A large amount of ATP is used during contraction because each cycle of each crossbridge involves the hydrolysis of one ATP molecule; the higher the rate of crossbridge cycling, the higher the rate of ATP use. As discussed earlier, crossbridge cycling rate determines the velocity of shortening, which is determined by the afterload placed on the muscle. Thus, more ATP is consumed by a muscle cell that is shortening rapidly against a light afterload than is consumed during an isometric contraction. Also, muscle cells that have myosin isoforms with high ATPase activity (fast-twitch skeletal) will use more ATP during contraction than will muscle cells that have myosin isoforms with lower ATPase activity (slow-twitch skeletal and cardiac).

Cycling of the Cytoplasmic Calcium

Adenosine triphosphate is also used during cycling of the cytoplasmic calcium that couples cell excitation to crossbridge cycling. In this case, ATP hydrolysis is not needed directly to effect the rise in intracellular calcium. This rise is due to the rapid release of calcium from the SR and, in the case of cardiac muscle, to the influx of calcium from the extracellular fluid. Both release and influx occur down large concentration gradients for calcium. ATP hydrolysis is required to lower cytosolic calcium levels to allow the cell to relax. This requires moving calcium against a large electrochemical gradient to maintain the calcium gradients between the extracellular fluid and sarcoplasmic reticulum, on the one hand, and the cytoplasm, on the other. In skeletal muscle, cytosolic calcium is reduced by the uptake of calcium into the lateral SR. This involves a calcium pump, with each cycle requiring hydrolysis of one ATP molecule. In cardiac muscle, most of the calcium is handled just as in skeletal muscle. However, some calcium must be removed from the cell to balance, over the long term, that which enters during the action potential. This calcium is removed through the activity of a sodium–calcium countertransport. Sodium–calcium countertransport does not use ATP directly. It is driven by the gradient in sodium that exists across the membrane as a result of the sodium pump. This pump uses ATP to remove sodium from the cell while bringing in potassium.

Muscle Resting and Action Potentials

As with the events responsible for the rise in intracellular calcium, the events associated with muscle cell membrane resting and action potentials also do not use ATP directly; rather, the ionic fluxes responsible for these potentials are due to the opening and closing of channels to allow ions to move down their electrochemical gradients. Resting potential is due mostly to gradients in potassium, whereas action potentials are due mostly to gradients in sodium and potassium in skeletal muscle and to gradients in sodium, potassium, and calcium in cardiac muscle cells. As in other cells, gradients of sodium and potassium are maintained by the sodium pump; as discussed previously, this pump also helps maintain the calcium gradient in cardiac muscle cells.

Resting ATP levels in muscle cells are enough to sustain contraction only for brief periods. ATP must be replenished as rapidly as it is hydrolyzed so that ATP levels fall minimally during periods of contraction. Like most other cells, muscle can derive ATP from both glycolysis and oxidative phosphorylation; however, muscle cells also have a more immediate precursor for the generation of ATP. Creatine phosphate is a high-energy molecule that can rephosphorylate ADP to ATP. Creatine phosphate concentrations do fall early during a period of contraction and, if not replenished, can supply enough energy for only a few contractions; however, it does supply energy early on until the required increases in glycolysis and/or oxidative phosphorylation take place.

Glycolysis and Oxidative Phosphorylation

The proportions of ATP generated by glycolysis and oxidative phosphorylation vary from cell to cell. Human skeletal muscle cells are divided mainly into two types, based on both the myosin isoform present and on metabolic pathways (see Table 1). One type, fast-twitch glycolytic, possesses a myosin isoform with high ATPase activity. As expected, this muscle exhibits a large V_{max} and derives most of its ATP from glycolysis. The other type, slow-twitch oxidative, possesses a myosin isoform with a lower ATPase activity and a concomitant lower V_{max}. It generates most of its ATP via oxidative phosphorylation. Cardiac muscle cells have their own myosin isoform, which also has low ATPase activity. Thus, it more closely resembles that found in slow oxidative skeletal muscle cells. These cells also derive most of their ATP from oxidative phosphorylation.

The slow-twitch oxidative skeletal muscles and cardiac muscle contain myoglobin. This heme-containing protein aids in the transfer of oxygen from the hemoglobin of blood to the mitochondria of the muscle cells, thus facilitating ATP generation through oxidative metabolism. Myoglobin also imparts a red color to the muscle. On the other hand, fast-twitch glycolytic skeletal muscles do not contain myoglobin and, thus, lack the red color. This difference accounts for the latter being referred to as *white muscles*.

Individual skeletal muscles vary in the proportion of the slow oxidative and fast glycolytic cells they contain; however, most muscles will contain both. Within each muscle, motor units are organized such that an individual α-motor neuron will innervate only one type of cell. In general, motor units composed of fast glycolytic cells will be smaller than those composed of slow oxidative

cells. When the muscle is called on to do work, the type of motor unit activated will vary depending on the type of work. Short, rapid movements use mainly fast glycolytic units, whereas slower, more sustained, movements use slow oxidative units.

Suggested Readings

Berne RM, Levy MN. *Physiology*, 4th ed. St. Louis: Mosby, 1998

Berne RM, Sperelakis N, Eds. *Handbook of physiology, Section 2: The cardiovascular system, Vol. 1: The heart*. Bethesda, MD: American Physiological Society, 1979

Engel AG, Franzini-Armstrong C, Eds. *Myology, basic and clinical*, 2nd ed. New York: McGraw-Hill, 1994

Gordon AM, Homsher E, Regnier M. Regulation of contraction in striated muscle. *Physiol Rev* 2000; 80:853–924

Swynghedauw B. Developmental and functional adaptation of contractile proteins in cardiac and skeletal muscles. *Physiol Rev* 1986; 66:710–771

8

Smooth Muscle

NORMAN W. WEISBRODT

KEY POINTS

- The *isoforms* of actin and myosin in smooth muscle differ from those in striated muscle.
- Filaments are not arranged in *transverse registry* and the proportion of thin filaments is higher than that found in striated muscle.
- *Contraction* is initiated by the phosphorylation of the 20,000-Da light chains of myosin by the enzyme myosin light-chain kinase, which is activated by increases in intracellular-free calcium. Contraction is due to crossbridge cycling and requires ATP.
- *Relaxation* occurs when kinase activity decreases and the light chains are dephosphorylated by phosphatase; for relaxation to occur, calcium is actively transported back into the sarcoplasmic reticulum and/or out of the cell.
- In many smooth muscles, excitation is due to increases in calcium permeability in response to the opening of voltage-gated channels and/or ligand-gated channels.

- In *unitary smooth muscles*, individual muscle cells are electrically coupled to one another and respond as a unit; in *multiunit smooth muscles*, cells are not coupled.
- Smooth muscles exhibit length–force and force–velocity relationships qualitatively similar to those seen for striated muscle.
- Maximal force per unit of smooth muscle is as great as, if not greater than, that of striated muscle; V_{max} of smooth muscle is much lower.
- Many smooth muscles are supplied by both excitatory and inhibitory nerves.
- Consumption of ATP is low compared with striated muscle because of the relatively lower myosin ATPase activity in smooth muscle; many smooth muscles can sustain contraction, once developed, with little crossbridge cycling or use of ATP.

BASIC CONTRACTILE UNIT

Actin and *myosin* are the major contractile proteins in all smooth muscles. As in striated muscle, these proteins are arranged in two sets of filaments: actin in thin filaments and myosin in thick filaments. Although there are many similarities between the smooth muscle contractile proteins and their counterparts in striated muscle, there are also qualitative as well as quantitative differences. Smooth muscle actin is a globular protein with a molecular weight of approximately 50,000 Da. However, there are two, perhaps three, isoforms in smooth muscle, and they all differ from the predominant ones found in striated muscle. Most of the actin is polymerized into thin filaments, but some appears to be present in globular form. Each molecule of smooth muscle myosin consists of two myosin heavy chains and two pairs of light chains; however, each of these monomers differs in isoform from those found in skeletal muscle. Myosin molecules are thought to be arranged in the thick filaments in the same way as those in skeletal muscle, with crossbridges extending to make contact with the actin filaments.

Although thick and thin filaments are present, their arrangement in smooth muscle cells is different from that seen in striated muscle (Fig. 1). The most striking difference is that, in smooth muscle, filaments are not organized into sarcomeres, thus giving a homogeneous appearance under the light microscope. There are many more thin filaments than thick, with the ratio being closer to 10:1 than to the 2:1 seen in skeletal muscle. Also, not every thin filament is in close proximity to a thick filament. In fact, there may be two or more populations of thin filaments: those associated with thick filaments and those associated with other actin-binding proteins and the cytoskeleton. Thin filaments are attached to elements of the cytoskeleton, but these attachments bear little anatomic resemblance to the Z disks found in striated muscle. Most common are thin filaments anchored to protein structures known as *dense bodies*.

Even though thick and thin filaments are not arranged in sarcomeres, the basic contractile model given for striated muscle—the sliding of one filament over the other due to crossbridge cycling—is thought to hold for smooth muscle. The lack of a rigid structure may account for some of the quantitative differences seen in contractions of smooth muscle compared to those of striated muscle (see below).

BIOCHEMICAL INTERACTIONS OF ACTIN AND MYOSIN

Smooth muscle myosin is also an adenosine triphosphatase (ATPase), and energy is required for muscle contraction. Like skeletal muscle myosin, pure smooth muscle myosin exhibits little ATPase activity. Unlike skeletal muscle myosin, smooth muscle myosin ATPase activity is not increased upon addition of actin alone, unless one set of the myosin light chains is phosphorylated. In smooth muscle cells, adenosine triphosphate (ATP) hydrolysis due to the cycling of phosphorylated myosin crossbridges with actin is thought to be similar to what occurs in striated muscle cells (see Chapter 7).

Regulation of smooth muscle contraction is mediated via the thick filaments rather than via the thin filaments. Tropomyosin is present in smooth muscle and is thought to reside on the thin filaments; however, there is little or no troponin complex. Thus, regulatory mechanisms must differ from those seen in striated muscle (Fig. 2). The main mechanism appears to involve phosphorylation of the 20,000-Da light chains of myosin. In resting smooth muscle, phosphorylation is low. *Myosin light-chain kinase* is the enzyme that, on stimulation, is activated and quickly catalyzes the phosphorylation of the myosin light chains. This results in actin-activated ATPase activity and contraction. When the stimulus to contract ceases, kinase activity decreases, myosin light chains are dephosphorylated by phosphatases, and the muscle relaxes. Although the regulatory proteins are different in smooth muscle, the sequence of events leading to contraction is due to the actions of calcium.

Pivotal Role of Calcium

As in striated muscle, contraction and relaxation of smooth muscle are regulated by changes in the amount of cytosolic calcium available to interact with the regulatory protein. In relaxed muscle, the level of *free*

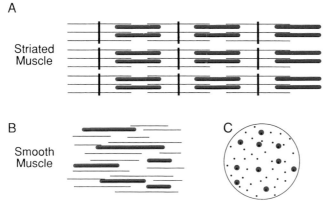

A

Striated
Muscle

B

Smooth
Muscle

C

FIGURE 1   (A) Schematic longitudinal view of the sarcomeric structure of striated muscle. (B) Schematic longitudinal view of the filaments of smooth muscle. (C) Schematic cross-sectional view of the filaments of smooth muscle.

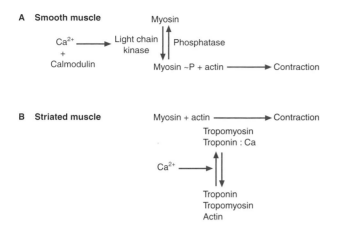

FIGURE 2 (A) Pathways for calcium regulation of contraction in smooth muscle. (B) Pathways for calcium regulation in striated muscle.

cytosolic calcium (calcium that is not bound to other structures such as sarcoplasmic reticulum [SR], mitochondria, and nuclei) is low ($< 10^{-7} M$). Upon stimulation of the muscle, the level increases into the micromolar, or higher, range to initiate contraction. Calcium binds first with *calmodulin* (one of the calcium-binding proteins found in many tissues), and then the calcium–calmodulin complex binds to and activates the myosin light-chain kinase. Once the stimulus for muscle contraction ceases, free calcium levels decrease, and calcium dissociates from the regulatory proteins. The muscle then relaxes.

The sources and sinks for calcium (and, therefore, excitation–contraction coupling) vary markedly from one smooth muscle to another (Fig. 3). Some have an abundant SR. When these cells are excited, events initiated at the cell membrane cause the release of calcium from the SR. The manner in which this release comes about is not clear; however, at least two

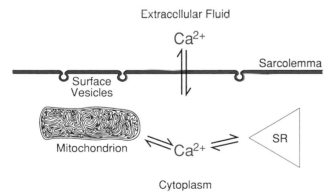

FIGURE 3 Intracellular free calcium levels are the result of (1) the influx and efflux of Ca^{2+} from the extracellular fluid, (2) the release and reuptake of Ca^{2+} from the sarcoplasmic reticulum (SR), and, perhaps, (3) the release and reuptake of Ca^{2+} from the mitochondria.

mechanisms have been demonstrated. During excitation of the cell membrane, an influx of small amounts of extracellular calcium through calcium channels may trigger the release of the internal stores. Additionally, receptor activation by ligands may stimulate the intracellular production of second messengers, such as *inositol trisphosphate*, which in turn cause the release of SR calcium. Other smooth muscle cells have almost no SR; these cells must rely on the entry of enough calcium through membrane calcium channels to activate their contractile proteins.

As in striated muscle, cytoplasmic free calcium must be decreased to allow for relaxation. In those cells with abundant SR, most of this calcium is pumped back into the SR via a calcium ATPase. However, in these cells, and especially in those cells with little SR, calcium must also be expelled from the cell across the cell membrane. Presumably, this is accomplished by a sodium–calcium exchange mechanism and perhaps by a membrane-bound calcium ATPase.

No smooth muscle has as well developed an SR as that seen in striated muscle. Thus, as in cardiac muscle, not enough calcium is normally present within smooth muscle cells to activate fully the contractile proteins. As in cardiac muscle, smooth muscles rely to varying degrees on an influx of extracellular calcium for contraction. During steady-state conditions, influx will be matched by efflux. However, as in cardiac muscle, varying rates of influx and efflux make it possible for there to be moment-to-moment net gains and losses of calcium available to initiate contraction.

Excitation of the Muscle Cell

Smooth muscles vary in the electrical events exhibited by their cell membranes. During relaxation, all are polarized, exhibiting resting membrane potentials of -40 to -80 mV. The basis for this potential is primarily the same as in striated muscle (Fig. 4). In many smooth muscles the membrane potential in relaxed cells is not stable. For example, in intestinal smooth muscle, cyclic depolarizations and repolarizations of 10 to 15 mV occur regularly (Fig. 5). The importance of these fluctuations is discussed in Chapter 33. Although different types of smooth muscle differ only slightly in resting potential, they differ markedly from one another and from striated muscle in the types of potentials exhibited when they are excited (Fig. 6). Most smooth muscles appear to have sarcolemmal calcium channels that open upon stimulation of the cell. Many of these are voltage dependent, like those in cardiac muscle. However, some cells have calcium channels that are activated not by voltage changes but by the combination of a ligand with its receptor on the cell surface or by

Extracellular Fluid

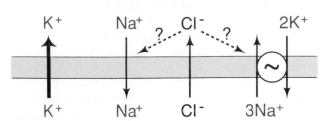

FIGURE 4   Schematic of the ionic events that may be responsible for the resting membrane potential in smooth muscle. The shaded horizontal bar indicates the cell membrane. The direction of the arrows indicates the direction of the movement, and the thickness of the arrows indicates the relative permeabilities. The coupling between Na^+ and K^+ indicates an active transport. The role of Cl is not understood, as indicated by the question marks.

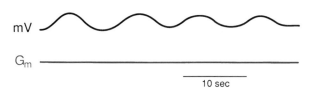

FIGURE 5   Electrical (mV) and contractile (G_m, where G_m indicates that the force is measured in grams) activities of intestinal smooth muscle. The muscle exhibits spontaneous slow depolarizations and repolarizations of membrane potential that cause little change in force.

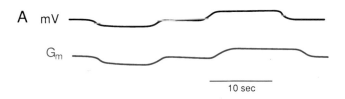

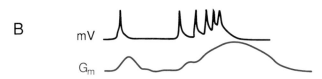

FIGURE 6   Electrical (mV) and contractile (G_m, where G_m indicates that the force is measured in grams) activities of two different smooth muscles. (A) One muscle exhibits stepwise depolarizations and repolarizations that cause increases and decreases in force, respectively. (B) The other muscle exhibits rapid transients that arise from a more stable baseline potential. These transients can occur alone or in bursts such that they induce a muscle twitch, summation, or tetany.

stretch of the membrane. Action potentials do not occur in all cells that are activated by ligand–receptor interaction or stretch. In these cells, calcium influx may be matched by an efflux of potassium, resulting in small, or no, changes in membrane potential.

Smooth muscle cells also vary in the manner in which they are excited. As stated above, many have membrane

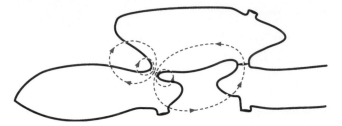

FIGURE 7   Schematic of three smooth muscle cells that are connected by gap junctions. The dashed lines indicate the pathways and spread of excitation.

potentials that fluctuate rhythmically to periodically reach threshold levels. Others have stable resting potentials. Most can respond to neurotransmitters, hormones, or other ligands that combine with membrane receptors, and many can even respond to mechanical stimuli such as stretch. Each particular type of smooth muscle usually will exhibit one of these behaviors to a higher degree. Examples of each type are found in those chapters that deal with specific organs such as the intestine, uterus, and blood vessels.

Smooth muscle cells differ in their degrees of coupling to one another. In some organs, such as the intestine, membranes of adjacent smooth muscle cells make intimate contact with one another so that low-resistance electrical pathways exist (Fig. 7). Thus, excitation of one cell quickly will spread and a group of cells will contract in unison. Such muscle is referred to as *unitary smooth muscle*. In other organs, such as the vas deferens, no such intercellular pathways exist, and each cell must be excited individually. Such muscle is referred to as *multiunit smooth muscle* (Fig. 8).

Smooth muscle cells lack well-developed transverse (T) tubules. Some have pockmarked indentations of their membrane and collections of SR vesicles just under the membrane, but these lack the structure of the T

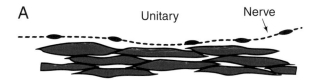

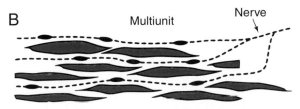

FIGURE 8   Schematic of unitary (A) and multiunit (B) smooth muscle. Note the difference between the two types of muscle cells in innervation and contact.

tubule system seen in striated muscle. Thus, excitation of the cell membrane is not quickly transmitted to the interior of the cell. Although sarcolemmal excitation may not spread rapidly into the cell interior, the small size of smooth muscle cells may enable the events that lead to the increase in cytosolic free calcium necessary for adequate activation of the contractile mechanism. On the other hand, the lack of a well-developed T tubule system also may be part of the reason why smooth muscle contractions are rather slow in comparison with those seen in striated muscle.

MECHANICAL RESPONSE OF SMOOTH MUSCLE *IN VITRO*

Pure isometric and isotonic contractions of smooth muscle are difficult to study under standard conditions. Many smooth muscles are tonically contracted or contract phasically, even when no discernible stimulus is being applied (Fig. 9). Others are relaxed until stimulated. Many smooth muscle tissues contain neurons or other cells that contain active chemicals that are released when the tissue is stimulated. These chemicals may either enhance or inhibit contractions of the muscle. Also, some smooth muscles are difficult to stimulate electrically because of their membrane properties. Despite these difficulties, many smooth muscles have been characterized.

Contractions of most smooth muscle vary in force with variations in stimulus intensity; however, the mechanisms may differ among muscles. Multiunit muscle behaves more like skeletal muscle in that increases in intensity are accompanied by increases in force as more and more cells become active. Unitary

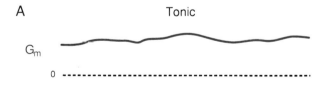

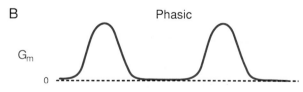

FIGURE 9   Recordings of isometric forces from a tonically active (A) and a phasically active (B) smooth muscle. The dashed lines indicate 0 force when the muscles are relaxed completely. G_m indicates that the force is measured in grams. Note that, during the time of these recordings, the tonic muscle never relaxed, while the phasic muscle went through two cycles of contraction and relaxation.

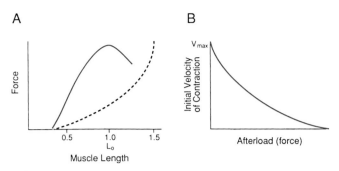

FIGURE 10   (A) Length–force relationship for a smooth muscle. The solid line indicates active force, and the dashed line indicates passive force. Length is expressed as a fraction of optimal length (L_o). (B) Afterload–velocity relationship for a smooth muscle. Note that the general shapes of these relationships are the same as for skeletal muscle.

smooth muscle behaves more like cardiac muscle in that once an adequate intensity is reached to stimulate a small group of cells, the excitation spreads through low-resistance pathways to excite the entire tissue. Increasing stimulus intensity affects force in these muscles by enhancing excitation–contraction coupling. The response to variations in stimulus frequency in most smooth muscles is similar to that of skeletal muscle: Summation and tetany do occur (see Fig. 6). All smooth muscles exhibit length–force (tension) relationships similar to those seen in striated muscle, even though sarcomeres are absent in smooth muscle (Fig. 10). This similarity of length–force relationships is evidence for the contention that a sliding filament mechanism for contraction is present in smooth muscle; however, there are some quantitative differences when compared with striated muscle. Smooth muscle cells can develop active force over greater variations in muscle length, and many can generate greater force than skeletal muscle. The bases for these differences are not known.

All smooth muscles also exhibit force–velocity relationships that are similar to those seen in striated muscle. A major quantitative difference is that V_{max} is much lower. This is reflected in, and probably due mostly to, the low ATPase activity of the myosin isoforms present in smooth muscles. Finally, many smooth muscles resemble cardiac muscle in that changes in contractility occur. This probably is due to the varying amounts of calcium that enter and/or are released with a single action potential or other excitation event.

MECHANICAL RESPONSE OF SMOOTH MUSCLE *IN VIVO*

Most smooth muscles form, along with other tissues, the walls of hollow organs such as the gastrointestinal tract, the uterus, and the blood vessels. In these organs, smooth muscle contractions serve many purposes such as the movement of lumenal contents, the regulation of lumenal volume, and the alteration of the resistance to flow through the lumen. For some of these functions, contractions must be phasic to allow the lumens to refill between contractions. For others, contractions must be tonic to resist continual distending forces. Thus, it is not surprising to learn that many factors regulate smooth muscle contractions *in vivo* and that the exact factors responsible for variations in the contractile activities of most individual smooth muscles are unknown.

Regulation of smooth muscle contraction is difficult to study *in vivo* for many reasons. Most smooth muscles are multiply innervated. Many have membrane receptors for circulating hormones and locally released paracrines and autocoids. In addition many smooth muscles respond directly to stretch of their membranes. Also, in contrast to what occurs in striated muscle, certain ligand–receptor interactions in smooth muscle lead to inhibition of contraction rather than excitation. Thus, at any one time, a given smooth muscle cell will be receiving multiple inputs, some excitatory and some inhibitory.

Response to Stretch

Quick or sustained stretch of many smooth muscles results in contraction of that muscle. In many instances, contraction occurs even in the absence of nerves; thus, it is not due to activation of a neural reflex as in skeletal muscle. Contraction most likely results from membrane depolarization or from the opening of stretch-activated calcium channels. Such responsiveness may explain why organs such as the stomach, bladder, and small arterioles contract to oppose distension. In some smooth muscles, especially those of some larger blood vessels, a quick stretch is followed by a temporary increase in wall tension; however, this increase is quickly followed by a relaxation toward the original wall tension referred to as *stress relaxation* (Fig. 11). The opposite, *reverse stress relaxation,* occurs when the muscle is allowed to shorten; that is, if an external stress is removed, wall tension temporarily decreases. However, the decrease is followed quickly by muscle contraction to return to the original wall tension. Such responses allow the larger blood vessels to accommodate to different blood volumes while maintaining transmural pressure nearly constant (see Chapter 14).

Response to Nerve Stimulation

There is wide diversity in the type and degree of innervation of smooth muscle tissue in the body. For

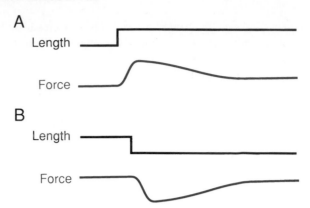

FIGURE 11   Length and isometric force changes of a strip of smooth muscle *in vitro*. (A) Stress relaxation; when the muscle is stretched (seen as a quick increase in length), force increases at first but then returns to almost the previous level. (B) Reverse stress relaxation; when the muscle is allowed to shorten (seen as a quick decrease in length), force decreases at first but then returns to almost the previous level.

example, smooth muscle of the gastrointestinal tract is innervated by both the parasympathetic and sympathetic divisions of the autonomic nervous system, as well as by postganglionic nerves from the enteric nervous system. On the other hand, many blood vessels receive only sympathetic innervation. In addition to differences in the degree and type of innervation, the response of smooth muscle to a given neurotransmitter varies from one smooth muscle to another. For example, norepinephrine causes a marked contraction of vascular smooth muscle; however, this same neurotransmitter causes inhibition of most gastrointestinal smooth muscles. Also, many smooth muscles respond to neurotransmitters, even if not innervated by nerves possessing those transmitters. Thus, it is not surprising to find that dysfunctions of the autonomic nervous system and pharmacologic agents that alter the function of the autonomic nervous system markedly affect many smooth muscle tissues.

Smooth muscles differ in several ways from skeletal muscle in their response to nerve stimulation. In innervated skeletal muscle, a single nerve impulse leads to release of enough neurotransmitter (acetylcholine, ACh) to cause an end-plate potential that will invariably lead to an action potential and contraction of the muscle fiber. For innervated smooth muscle, there are more types of neurotransmitters (such as ACh, norepinephrine, and serotonin); not every nerve impulse leads to a mechanical response of the muscle, and the response of the smooth muscle membrane may be either a depolarization or a hyperpolarization (Fig. 12). Thus, integration takes place at the level of the smooth muscle cell and not just within the central nervous system, as with skeletal muscle.

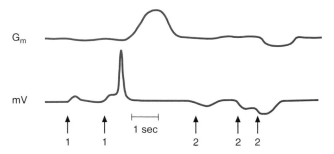

FIGURE 12 Recordings of contractile (G_m, where G_m indicates that the force is measured in grams) and intracellular electrical activities (mV) from an innervated smooth muscle. Arrows indicate stimulation of excitatory (1) and inhibitory (2) nerves. Mild stimulation of an excitatory nerve leads to a small depolarization but no change in tension. Stronger stimulation leads to an action potential and contraction. Weak stimulation of an inhibitory nerve leads to a membrane hyperpolarization but no change in resting tension. Stronger stimulation leads to a larger hyperpolarization and relaxation.

Response to Circulating and Locally Released Chemicals

In addition to responding to many neurotransmitters, most smooth muscles respond to many hormones, paracrines, autocoids, and tissue metabolic products. Just a few examples can illustrate that the function of most organ systems is influenced in this manner.

1. In the cardiovascular system, circulating epinephrine and angiotensin cause marked contraction of vascular smooth muscle, whereas local tissue metabolites such as adenosine cause arteriolar dilatation.
2. Both the growth and contractile activity of uterine smooth muscle are influenced by the hormones estrogen and progesterone. Also, it is believed that the initiation of labor is brought about by hormonal changes that induce uterine contractions.
3. Airway smooth muscle contracts dramatically in response to the local release of histamine and relaxes when circulating levels of epinephrine are increased (note the diametrically opposite effects that epinephrine has on airways as compared to vascular smooth muscle).
4. Both glomerular filtration and tubular reabsorption of fluid are influenced dramatically by the state of

contraction of the smooth muscle in the afferent and efferent arterioles of the kidney. Although little understood, there are local and circulating chemicals that markedly affect the smooth muscle of these arterioles.

An important result of this responsiveness is that practically every disease state has a component that is characterized by altered smooth muscle function. Furthermore, just about every pharmacologic intervention that is attempted will, in one system or another, affect smooth muscle.

ENERGY BALANCE IN SMOOTH MUSCLE

For smooth muscle as for striated muscle (Chapter 7), ATP is required for many processes—for example, activation of the contractile proteins, crossbridge cycling, calcium sequestration and removal, membrane electrical activity, and synthesis of structural components. Although in many instances the processes are the same in both smooth and striated muscles, both qualitative and quantitative differences exist.

One ATP-consuming process is essential in smooth muscle but relatively unimportant in striated muscle. As described previously, activation of the contractile proteins in smooth muscle, but not in striated muscle, involves phosphorylation of one of the myosin light chains. This requires ATP. Once phosphorylated, each crossbridge can cycle many times, with each cycle also consuming ATP. At first glance, it would seem that more ATP would be required during smooth muscle contraction, but the opposite is true: Less ATP is used to generate contractions with forces equal to those developed by striated muscle. This economy is due partly to the myosin isoforms found in smooth muscle. These isoforms have a low ATPase activity that imparts low velocities of contraction but somehow enables smooth muscle to develop force with low ATP consumption. Also, many smooth muscles have the ability to maintain a tonic contraction with even lower ATP use than that needed during the initial stages of the contraction. During the early stages of force

---

Clinical Note

Immune-mediated responses in many organs lead to altered smooth muscle function. The breathing in of an antigen by an individual who is sensitized will result in the release of various cytokines that induce contraction of airway smooth muscle, thus increasing airway resistance. On the other hand, antigen-induced responses taking place within or in proximity to blood vessels can lead to vascular smooth muscle relaxation and a decrease in arterial blood pressure.

development, myosin phosphorylation and crossbridge cycling are relatively high; however, during the tonic phases of the contraction, both myosin phosphorylation and crossbridge cycling fall, whereas force development remains high. Thus, force can be maintained with minimal expenditure of ATP. The exact basis for this economy is not known.

As in striated muscle, smooth muscles use both glycolysis and oxidative phosphorylation for ATP generation. The importance of each process varies from smooth muscle to smooth muscle, with some being more dependent on oxygen than others; however, there do not appear to be the clear-cut distinctions among smooth muscles that exist among striated muscles.

Suggested Readings

Bohr DF, Somlyo AP, Sparks HV, Eds. *Handbook of physiology, Section 2: The cardiovascular system.* Bethesda, MD: American Physiological Society, 1980

Bolton TB, Prestwich SA, Zholos AV, Gordienko DV. Excitation–contraction coupling in gastrointestinal and other smooth muscles. *Annu Rev Physiol* 1999; 61:85–115

Kamm KE, Stull JT. The function of myosin and myosin light chain kinase phosphorylation in smooth muscle. *Annu Rev Pharmacol Toxicol* 1985; 25:593–620

Autonomic Nervous System

NORMAN W. WEISBRODT

KEY POINTS

- The autonomic nervous system can be divided anatomically into the *sympathetic* and *parasympathetic* divisions, with each division having central (brain and spinal cord) and peripheral components consisting of preganglionic and postganglionic nerves, as well as an *enteric nervous system.*
- Most visceral organs are innervated by both divisions; the effects that stimulation of each division has on a particular organ usually oppose each other.
- Transmission across synapses occurs via chemicals called *neurotransmitters*, the two most important being *acetylcholine* and *norepinephrine.*
- Neurotransmitters exert their effects by binding to postsynaptic receptors. Acetylcholine binds to nicotinic and muscarinic receptors; norepinephrine binds to α and β adrenergic receptors.
- The intensity of response of an effector organ depends in large part on the concentration of neurotransmitters in the synapse.

- Each innervated organ usually receives continuous but varying input from the autonomic nervous system. This input generally modulates rather than initiates organ activity.
- Alterations of autonomic nervous system activity can be *en masse* (fight or flight) or discrete.
- The *enteric system* is comprised of afferent nerves, interneurones, and efferent nerves, all contained within effector organs, mainly of the gastrointestinal tract.
- The enteric system regulates many functions of these organs without the need of extrinsic innervation; however, the enteric system is both innervated and influenced by the sympathetic and parasympathetic divisions.
- *Autonomic ganglia* appear to be not just relay sites but also sites at which integration of inputs can take place. Also, chemicals other than acetylcholine and norepinephrine appear to be present within the autonomic nervous system and to modulate autonomic activity.

The peripheral nervous system can be divided functionally and anatomically into two subsystems: the *somatic nervous system* and the *autonomic nervous system* (ANS). The efferent motor neurons of the somatic system innervate the striated muscle fibers of skeletal muscles. Because the contractions of many of these muscles are consciously controlled, the system is often referred to as the voluntary nervous system. The ANS innervates organs and tissues, many of which are located within the viscera, that normally do not require control at the conscious level; thus, it often is referred to as the visceral or involuntary nervous system. The ANS, in turn, is composed of three major components: the sympathetic and parasympathetic divisions (Fig. 1), and the enteric nervous system. Acting in an integrated manner, the ANS is one of the primary effectors of homeostasis, in part regulating the function of those organs involved in maintaining the constancy of the internal environment of the body.

THE SYMPATHETIC AND PARASYMPATHETIC DIVISIONS

The *sympathetic* and *parasympathetic* divisions have both central and peripheral components. Clusters of nerve cells that contribute to the control of many organs via the ANS are located in the hypothalamus and brain stem (Fig. 2). Many of these clusters seem to be involved in regulating specific functions. For example, direct stimulation of certain areas can induce an increase in arterial blood pressure, whereas stimulation of other areas alters body temperature, gastrointestinal activity, and bladder function. To a great extent, these autonomic regions in the lower brain stem are influenced by neurons originating from cortical regions of the brain; thus, many human behavioral responses include autonomic responses that are mediated through the hypothalamus and reticular formation. Examples of such autonomic responses include increased gastric acid secretion upon the sight and smell of food and increased heart rate and dilatation of the pupils of the eye upon being frightened.

In contrast to the peripheral efferent component of the somatic nervous system, which consists only of the axons of spinal motor neurons, the peripheral efferent components (Figs. 1 and 3) of both the sympathetic and parasympathetic divisions of the ANS are rather complex, consisting of *preganglionic axons, autonomic ganglia,* and *postganglionic neurons*. The somata of efferent sympathetic preganglionic neurons are located in the intermediolateral horn of the spinal cord in regions from the first thoracic to the third or fourth lumbar segment (see Fig. 1). Axons of these preganglionic

neurons leave the spinal cord and are traditionally thought to make synaptic contact with postsynaptic neurons located in either the paravertebral or prevertebral ganglia of the sympathetic nervous division. Axons of these postganglionic neurons then innervate via synapses the various organs to be controlled. Preganglionic axons of the parasympathetic nervous division arise from neurons whose somata are located within the motor nuclei of cranial nerves and the sacral portion of the spinal cord (see Fig. 1). These preganglionic axons make synaptic contact with postganglionic neurons located in parasympathetic ganglia. These ganglia are located near or in each innervated effector organ.

Most organs that are influenced by the ANS are innervated by both its sympathetic and parasympathetic divisions. The actions of these two divisions on a given organ are usually antagonistic. For example, an increase in sympathetic neural input to the sinoatrial node of the heart causes an increase in heart rate, whereas an increase in parasympathetic input to the node causes a reduction in heart rate. At first glance, such an arrangement may seem to work at cross purposes; however, the divisions usually function in a way such that an increase in the input of one is accompanied by a decrease in input from the other. On the other hand, some organs are innervated by only one division of the ANS. Two notable examples of this arrangement are innervation by the sympathetic division of the sweat glands and the peripheral blood vessels.

TRANSMISSION ACROSS SYNAPSES

Transmission across synapses in the ANS is mediated by chemicals known as *neurotransmitters*. These chemicals are synthesized within the nerves and in most instances are stored in synaptic vesicles in the nerve endings. Upon activation of the nerve, nerve action potentials invade the nerve ending, causing the release of a portion of the stored neurotransmitter. Once released, the neurotransmitter binds to specific receptors located on postsynaptic structures. Only those cells that possess specific receptors for that neurotransmitter will be affected. Shortly after its release, the neurotransmitter is then inactivated.

As is the case in the brain, a multitude of neurotransmitters act within the ANS. The two about which most is known are acetylcholine and norepinephrine; thus, this chapter will deal mainly with them. *Acetylcholine* (ACh) acts as the neurotransmitter at synapses between somatic motor nerves and skeletal muscle. ACh also acts at the synapses between preganglionic and postganglionic neurons in both the sympathetic

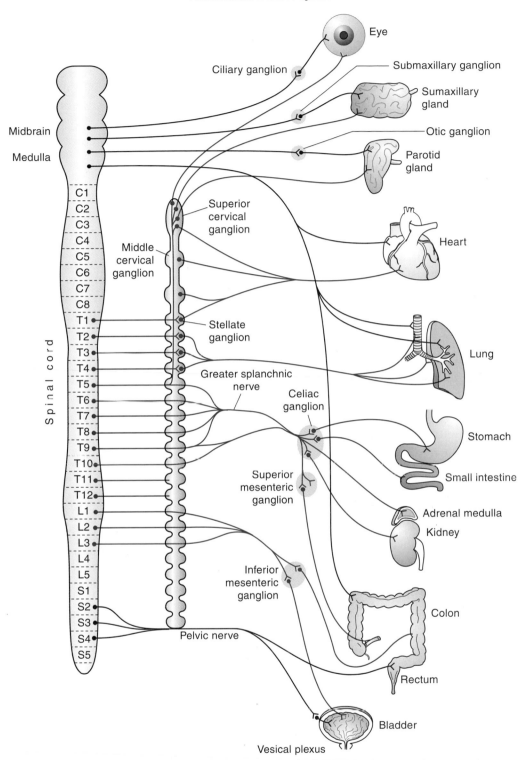

FIGURE 1 Organization of the sympathetic and parasympathetic divisions of the autonomic nervous system. Parasympathetic pathways are denoted by dark blue lines, whereas sympathetic pathways are illustrated by light blue lines.

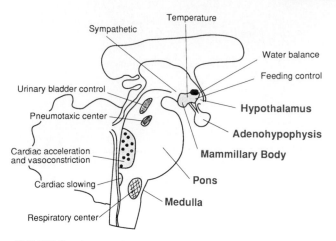

FIGURE 2 Autonomic control centers located in the brain stem.

and parasympathetic divisions. Finally, ACh is the transmitter acting between many postganglionic parasympathetic nerves and effector organs. *Norepinephrine* acts as the neurotransmitter between many postganglionic sympathetic nerves and the effector organs they innervate.

Receptors are identified according to the specific chemicals with which they interact (Table 1). *Cholinergic receptors* are those that bind acetylcholine; *adrenergic receptors* are those that bind norepinephrine and epinephrine. In addition to binding and reacting to the naturally occurring neurotransmitters, these receptors also bind a number of other structurally related chemicals. Many of these chemicals mimic the naturally occurring neurotransmitters and initiate cellular reactions; these chemicals are referred to as *agonists*. On the other hand, some chemicals bind with receptors but do not induce a cellular reaction. Furthermore, their binding to a receptor inhibits the ability of an agonist, when present, to bind with and/or activate the receptor; such chemicals are referred to as *antagonists*.

The cholinergic receptors present on skeletal muscles and in autonomic ganglia are activated by nicotine and are blocked by the antagonists tubocurarine and hexamethonium. On the other hand, cholinergic receptors on many end organs are activated by muscarine and are blocked by the antagonist atropine. Thus, these receptors are referred to as *nicotinic-cholinergic* and *muscarinic-cholinergic* receptors, respectively.

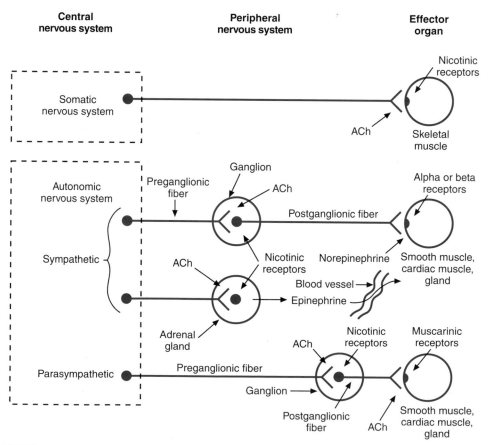

FIGURE 3 Overview of synaptic pathways and mediators of the somatic, sympathetic, and parasympathetic neural subsystems.

TABLE 1 Agonists and Antagonists at Specific Receptors

Receptor	Agonist	Antagonist
Nicotinic-cholinergic	ACh, nicotine	Tubocurarine, hexamethonium
Muscarinic-cholinergic	ACh, muscarine	Atropine
Alpha-adrenergic	Epinephrine, norepinephrine, isoproterenol[a]	Phentolamine, phenoxybenzamine
Beta-adrenergic	Isoproterenol, epinephrine, norepinephrine[a]	Propranolol

[a]In order of potency, the most potent first.

On the adrenergic side, two major classes of receptors have been defined based on the potency of epinephrine, norepinephrine, and the synthetic compound isoproterenol to act as agonists. Receptors for which the order of potency is epinephrine > norepinephrine ≫ isoproterenol are known as *α-adrenergic receptors,* and two antagonists selective for these receptors are phentolamine and phenoxybenzamine. Receptors for which the order of potency is isoproterenol > epinephrine ≫ norepinephrine are known as *β-adrenergic receptors.* Propranolol is an example of a β-adrenergic antagonist. One should note that the pharmacology of cholinergic and adrenergic receptor–agonist/antagonist interactions is considerably more varied than just described. For example, β-adrenergic receptors have been subdivided into β_1 and β_2 receptors. However, the general concepts presented here form much of the basic foundation upon which traditional autonomic physiology and pharmacology are based.

EFFECTS OF NEUROTRANSMITTERS

The binding of ACh to nicotinic receptors located on postganglionic neurons induces changes in membrane potential called *fast excitatory postsynaptic potentials.* Fast excitatory postsynaptic potentials, if strong enough and frequent enough, in turn induce action potentials that are transmitted to the synaptic terminals of the postganglionic nerves. These action potentials in turn result in the release of transmitter onto the effector organs.

Terminals of postganglionic fibers of the parasympathetic system release ACh, which interacts with muscarinic receptors located on the effector. It is impossible to assign any general effector response such as excitation or inhibition to this type of cholinergic receptor. Stimulation of muscarinic receptors in the atria of the heart decreases both heart rate and the force of contraction, whereas activation of muscarinic receptors on intestinal smooth muscle increases the number and force of muscle contractions.

Terminals of postganglionic fibers of the sympathetic system release norepinephrine, which interacts with α- and β-adrenergic receptors located on the effectors. Once again, it is impossible to assign a general response to agonist–receptor interaction. Stimulation of β-adrenergic receptors on the heart increases both heart rate and the force of contraction, whereas activation of β-adrenergic receptors on intestinal smooth muscle decreases the number and force of contractions.

A specific listing of responses that can be induced in each organ of the body by the sympathetic and parasympathetic nervous systems or by drugs that act on autonomic receptors is not presented in this chapter; such information is presented in the chapters on individual organ systems. Also, additional information is available in pharmacology texts. Many pharmacologic agents have been developed to act preferentially on specific receptors located in the tissue or tissues of one or more organs. The β-adrenergic receptor antagonist propranolol, for example, is often prescribed for the treatment of certain forms of cardiac arrhythmias because it competes with the binding of endogenously released catecholamines to β-adrenergic receptors located in the heart. However, because propranolol will also block β-adrenergic receptors located on airway smooth muscle, resulting in airway constriction, it is not desirable to prescribe propranolol in patients with airway disease. This lack of selectivity has led to the development of drugs such as metoprolol that preferentially block cardiac β_1-adrenergic receptors over the β_2-adrenergic receptors of the respiratory system.

Thus, the diversity and specificity of action of the ANS on the various organs of the body are due to several factors (see Fig. 3): (1) the discrete anatomic organization and projection of its sympathetic and parasympathetic divisions, (2) the presence of multiple neurotransmitters within each division, and (3) the presence of multiple receptors for those transmitters within effector organs.

REGULATION OF NEUROTRANSMISSION

The level of control exerted by either division of the ANS at any instant depends largely on the concentration of transmitter available for binding to the receptors located on the effector. Axons of sympathetic postganglionic neurons release norepinephrine from varicosities located at their terminals. These terminal regions resemble strings of pearls in that the varicosities are pearl-like expansions distributed along the axon. A single varicosity is represented in Fig. 4. Several factors

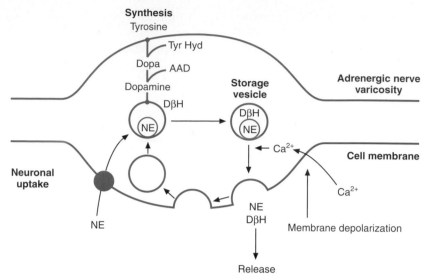

FIGURE 4   Synthesis, release, and reuptake of norepinephrine at sympathetic (adrenergic) nerve varicosities. AAD, aromatic ʟ-amino acid decarboxylase; DβH, dopamine β-hydroxylase; NE, norepinephrine; Tyr Hyd, tyrosine hydroxylase. (Modified from Vanhouette PM, *Fed. Proc.* 1978, 37:181–186.)

affect the temporal release of norepinephrine from varicosities. The frequency of action potentials arriving at a varicosity is a primary determinant of the flux of calcium ions into that varicosity. This influx of calcium in turn controls the vesicular release of norepinephrine from the varicosity. The amount of norepinephrine available for release depends on its synthesis and on its reuptake by the varicosity. The steps in the synthesis pathway are illustrated in Fig. 4. It should be noted that dopamine is converted into norepinephrine by the enzyme dopamine β-hydroxylase. This enzyme is also packaged into the norepinephrine-containing vesicles of the varicosities and is released along with norepinephrine into the extracellular space. Because dopamine β-hydroxylase is not degraded as rapidly as norepinephrine, it is often used as an indicator of the level of overall sympathetic activity.

The amount of norepinephrine released from a varicosity in response to a specific pattern of axonal action potentials is modulated by receptors located on that varicosity. A varicosity having presynaptic receptors located on its surface is illustrated in Fig. 5. A variety of compounds, such as histamine, 5-hydroxytryptamine, and acetylcholine, may act to decrease the amount of norepinephrine released by a train of action potentials, whereas other substances such as angiotensin II may augment its release. The concentration of norepinephrine within the junctional cleft may itself modulate release of norepinephrine. Varicosities with autoreceptors (α_2-adrenergic receptors) inhibit the release of norepinephrine when these receptors bind with norepinephrine. This negative feedback mechanism

tends to conserve norepinephrine and stabilize its concentration within the cleft.

In addition to the rate of norepinephrine release, the concentration of norepinephrine available for binding to receptors is modulated by several other factors, which are illustrated in Fig. 6. These include neuronal uptake, extraneuronal uptake and subsequent degradation by some effector cells, and diffusion into capillaries.

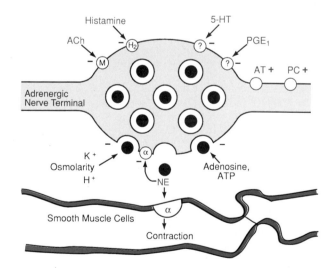

FIGURE 5   Various excitatory (+) and inhibitory (−) receptor mechanisms located on sympathetic varicosities. ACh, acetylcholine; α, α-adrenergic receptor; AT, angiotensin II; H_2, histamine$_2$ receptor; 5-HT, 5-hydroxytryptamine; M, muscarinic receptor; NE, norepinephrine; PC, prostacyclin; PGE$_1$, prostaglandin E$_1$; ?, unknown mechanism. (Modified from McGrath MA, Vanhouette PM, in Vanhouette PM, Leusen S, Eds., *Mechanisms of vasodilation.* Basel: S. Karger, 1978, pp. 248–257.)

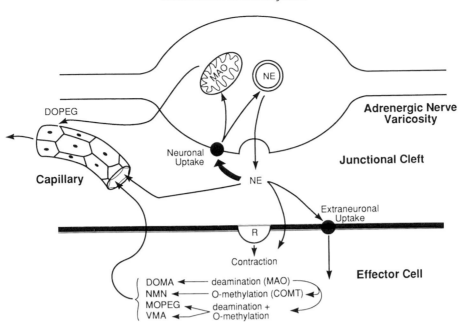

FIGURE 6 Mechanisms affecting the concentration of norepinephrine (NE) within the junctional cleft available for binding to a receptor (R) on an effector cell. Norepinephrine is removed by (A) neuronal uptake, in which some is degraded by monoamine oxidase (MAO) to 3,4-dihydroxyphenylglycol (DOPEG), but most is restored in vesicles; (B) diffusion into capillaries; and (C) uptake into effector cells and subsequent degradation by the enzymes MAO and catechol-*O*-methyltransferase (COMT) to 3,4-dihydroxymandelix acid (DOMA), normetanephrine (NMN), 3-methoxy-4-dihydroxyphenylglycol (MOPEG), and 3-methoxy-4-hydroxymandelic acid (VMA). These metabolites are inactive and diffuse into the extracellular fluid and the capillaries. Neuronal uptake of norepinephrine is mediated by an active transport.

Diffusion of transmitter away from the target cells is enhanced by the fact that the junctional cleft between the varicosities and the effector cells is large compared with the junctional cleft between somatic neurons and end plates of skeletal muscles.

The mechanisms available for modulating the concentration of ACh in the junctional clefts between terminals of parasympathetic postganglionic neurons and effector cells are not as varied as those associated with sympathetic adrenergic neurotransmission. Some parasympathetic postganglionic axons have terminal regions lined with varicosities, and others have single boutons at their ends. A bouton-type ending and its mechanisms for regulating ACh in the junctional cleft are illustrated in Fig. 7. Action potentials arriving at the bouton induce an influx of Ca^{2+} ions, triggering the release of ACh from the synaptic vesicles into the cleft. ACh in the cleft is degraded by acetylcholinesterase into choline and acetate. Some of the choline is taken up by the bouton and resynthesized into ACh.

PATTERNS OF NEUROTRANSMISSION

Neural input to various organs via the postganglionic neurons of the sympathetic and parasympathetic divisions is varied in two general ways. The behavior of many of the individual peripheral autonomic pathways can be altered in concert such that several organs and organ systems are regulated simultaneously to meet specific physiologic demands placed on the body. The reaction of the overall autonomic nervous system to imposed demands is often referred to as *mass action* or the *fight-or-flight* mechanism. Alternatively, the neural firing pattern of individual postganglionic pathways to specific organs can be varied independently such that the behavior of a single effector or group of effectors can be altered without producing changes in the behavior of other organs.

The fight-or-flight mechanism occurs in individuals acutely concerned for their physical safety, as in the case of someone about to be physically attacked. The autonomic system immediately and simultaneously increases heart rate, decreases blood flow to organs not needed for rapid body movement, increases blood flow to many skeletal muscles, and increases blood glucose. A similar set of responses occurs when someone is in a state of rage. In the fight-or-flight mode of operation, there is a generalized increase in the firing rate of sympathetic neurons and a generalized decrease in the firing rate of parasympathetic neurons.

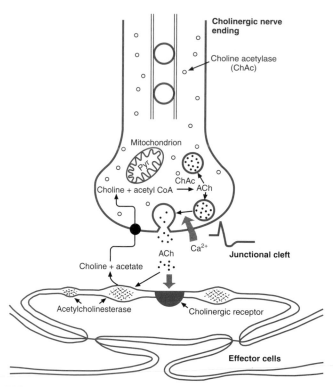

FIGURE 7 Synthesis, release, and degradation of ACh at a bouton of a parasympathetic postganglionic axon.

Individuals do not, however, spend most of their time in the fight-or-flight state. During the more mundane physical and emotional states associated with typical daily events, the firing rates of various sympathetic pathways may increase, whereas the firing rates of other sympathetic pathways decrease. A similar lack of homogeneity is usually associated with the activity of parasympathetic neurons. Furthermore, most of the neural circuitry of the autonomic nervous system is organized to produce specific reflexes. For example, the smell of food initiates the secretion of digestive juices even before food enters the mouth. Low blood pressure in the carotid arteries causes a decrease in the firing rate of baroreceptors located in these arteries, which in turn causes appropriate changes to occur in the cardiovascular system to increase carotid arterial pressure and,

hence, maintain blood supply to the brain. These are only two examples of the many *autonomic reflexes* that function to regulate the behavior of organ systems.

Autonomic regulation of organ function is usually modulatory and not initiatory. The heart, for example, pumps blood in the absence of sympathetic or parasympathetic inputs; thus, autonomic inputs do not initiate cardiac contractile activity but only modulate the existing autogenic capability of the heart. This modulation is accomplished by the parasympathetic and sympathetic postganglionic fibers maintaining a variable, but omnipresent, discharge of action potentials referred to as *tonic discharge* or, simply, *tone*. An increase in sympathetic tone to the heart will, for example, increase heart rate. Caution, however, is warranted when one encounters or uses the phrase *sympathetic tone;* unless the context in which the phrase is used is clear, it is often impossible to determine if one means tone along all efferent pathways of the sympathetic nervous system or only along specific pathways to one or more effectors.

ENTERIC SYSTEM

The enteric division of the ANS consists of the neural plexuses that are contained within the walls of the esophagus, stomach, small intestine, and colon (Fig. 8). These plexuses within the gastrointestinal system are indispensable in the control of gastrointestinal motility, secretion, and transport. They form a complex nervous system that is capable of functioning independently of any extrinsic neural input from the sympathetic or parasympathetic divisions. Most, if not all, of the transmitters found in the brain are also found within neurons of the enteric system. The neuronal mechanisms and neural circuits that form the plexuses are capable of high-level information processing and receive sensory information directly from sensory receptors located within the gastrointestinal tract. Such processing is possible because of the many neurons present. In fact, there are more neurons in the enteric nervous system

---

Clinical Note

Diabetic Autonomic Neuropathy
A significant percentage of patients with diabetes mellitus exhibit a syndrome described as diabetic autonomic neuropathy. The neuropathy involves the loss of preganglionic and postganglionic autonomic nerve fibers and is characterized by

hypotension upon standing, constipation and/or diarrhea, decreased gastric emptying, impotence, and abnormal sweating. The etiology of the disorder is not known; however, both metabolic defects and defects in blood flow to the nerves have been implicated.

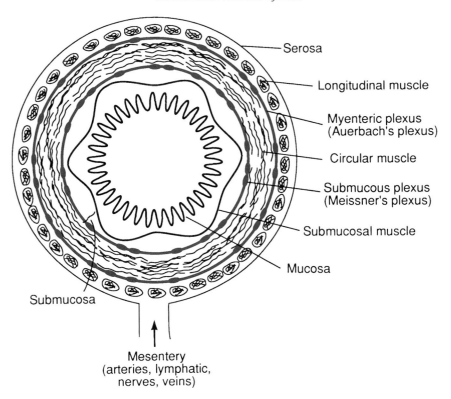

Serosa

Longitudinal muscle

Myenteric plexus
(Auerbach's plexus)

Circular muscle

Submucous plexus
(Meissner's plexus)

Submucosal muscle

Mucosa

Submucosa

Mesentery
(arteries, lymphatic,
nerves, veins)

FIGURE 8 Simplistic representation of the neural plexuses of the enteric division of the autonomous nervous system, located within the small intestine (see Chapter 32, Fig. 3, for more detail).

than in the spinal cord. The enteric system is in turn innervated by the sympathetic and parasympathetic divisions. This extrinsic innervation appears to serve primarily a modulatory function to alter the intrinsic activity of the enteric neural network.

The intrinsic activity of the enteric nervous system (ENS) is important in regulating most of the functions of the gastrointestinal system. For example, during fasting, contents of the stomach and small intestine are swept toward the colon by a complex of contractions that periodically starts in the stomach and systematically progresses along the length of the bowel. This migrating motor complex cycles such that a new one begins in the stomach approximately every 90 minutes. This entire pattern of spatiotemporal contractions of the intestinal musculature is produced and controlled by the ENS. Immediately after ingestion of a meal, a fed pattern of intestinal contractile activity is generated. It appears that this pattern also is preprogrammed within the ENS and that it is activated by hormones such as cholecystokinin, interacting with command neurons within the intrinsic neural plexuses. Cholecystokinin is released by the intestine in response to the meal. Both the migrating motor complex and the fed pattern of contractile activity are modulated by the sympathetic and parasympathetic components of the autonomic nervous system.

ADDITIONAL COMPLEXITIES

The conceptual approach to the sympathetic and parasympathetic divisions presented here is consistent with the presentation of this topic in most general textbooks and forms the basic foundation underlying much of autonomic pharmacology; however, this foundation omits many basic mechanisms that function in autonomic ganglia. For example, traditionally, sympathetic ganglia have been thought to function primarily as relays in which neural information coming from the spinal cord via preganglionic fibers is simply relayed to postganglionic neurons. The synaptic mechanism implementing this relay was thought to be the ACh-to-nicotinic cholinergic receptor interaction that produces fast excitatory postsynaptic potentials; however, it is now known that neural terminals in these ganglia contain numerous other substances that appear to be neurotransmitters. Many of these substances are polypeptides that produce slow synaptic potentials that contribute to the synaptic integration of neural information within sympathetic ganglia. The cell bodies of neurons located within prevertebral sympathetic ganglia also receive input directly from sensory receptors located in most organs of the abdominal viscera. Thus, these ganglia use a variety of synaptic mechanisms to combine peripheral

Clinical Note

Hirschsprung's Disease

Hirschsprung's disease is a congenital disorder characterized morphologically by a narrowed segment of distal colon and a hypertrophic segment of intestine orad to the narrowing. Histologically, the narrowed segment is characterized by an absence of enteric nerves. Thus, it appears as if enteric nerves are primarily inhibitory, and their absence results in a failure of the gut smooth muscle to relax. The etiology of the disease appears to be a failure of neuroblasts from the neural crest to migrate properly during fetal development, perhaps because of a defect in the microenvironment of the developing bowel. Treatment usually involves surgical removal of the aganglionic segment.

sensory information with neural signals from the spinal cord to form the neural commands sent to effector organs via postganglionic fibers. Although it is now clear that these ganglia are not simply relays, little is understood about what kind of information is sent to these ganglia, what rules are used to integrate this information, and what roles postganglionic neural commands play in the regulation of organ function.

Suggested Readings

Furness JB, Bornstein JC, Kunze WAA, Clerc N. The enteric nervous system and its extrinsic connections, in Yamada T, Ed., *Textbook of gastroenterology*. Philadelphia: Lippincott/Williams & Wilkins, 1999

Hoffman BB, Taylor P. Neurotransmission: the autonomic and somatic motor nervous system, in Hardman JG, Limbird LE, Gilman AF, Eds., *Goodman and Gilman's the pharmacological basis of therapeutics*. New York: McGraw-Hill, 2001

Korczyn AD, Ed. *Handbook of autonomic nervous system dysfunction*. New York: Marcel Dekker, 1995

McGrath MA, Vanhouette PM. Vasodilation caused by peripheral inhibition of adrenergic neurotransmission, in Vanhouette PM, Leusen S, Eds. *Mechanisms of vasodilation*. Basel: Karger, 1978, pp. 248–257

Vanhouette PM. Adrenergic neuroeffector interaction in the blood vessel wall. *Fed. Proc.* 1978, 37:181–186

PART II

CARDIOVASCULAR PHYSIOLOGY

10

Hemodynamics

JAMES M. DOWNEY

KEY POINTS

- The *left heart* pumps oxygenated blood into the systemic circulation from which the venous blood returns to the *right heart;* the right heart pumps deoxygenated blood into the pulmonary circulation from which the venous blood returns to the left heart.
- *Hydrostatics* refers to the physics of fluid at rest; *hydrodynamics* refers to the physics of fluid in motion.
- *Pressure* is a scalar quantity with units of force per unit area; pressure in a fluid system increases in proportion to depth because of the force of gravity.
- *Velocity* refers to the linear rate at which the molecules of a fluid are moving. In a tube the mean velocity is determined by the flow rate divided by the cross-sectional area. The

pressure within the fluid is reduced as the fluid flows faster because of a kinetic energy component.
- Fluid in a tube resists motion due to viscous drag. *Poiseuille's equation* describes the relationship between tube dimension and this resistance to flow.
- Although blood flow in the body is approximated by *Ohm's law*, it is not exactly described by it because (1) the viscosity of blood which is not a newtonian fluid changes with its velocity, and (2) blood vessels are not rigid tubes and change their dimensions as a function of internal pressure.
- Pressure differs in each vessel type; arteries have the highest pressure and veins the lowest.

KEY POINTS (continued)

- Arterial pressure is pulsatile, being highest during ejection of the ventricle (systolic pressure) and lowest between beats (diastolic pressure). Mean arterial pressure can be estimated by adding 1/3 of the pulse pressure to the diastolic pressure.

- Tension in the blood vessel wall is proportional to the pressure inside the lumen times the radius of the vessel (law of LaPlace).
- Most of the blood volume resides in the venous system.

HISTORICAL PERSPECTIVE

Although the vital nature of the cardiovascular system was clearly apparent to the ancients, its actual function was a mystery. An early attempt to explain the function of the heart and the blood was made by the Greek physician and writer Galen (130–201 A.D.). He taught that there was an ebb and flow of fluid between the heart and the abdominal viscera, where the "natural spirits" were formed; between the heart and the brain, where the "animal spirit" was created; and between the heart and the lungs, where "vital spirits" entered the body through the trachea. This erroneous concept remained a major cornerstone of medical thought for the next 14 centuries.

In 1628, William Harvey, an English physician and anatomist, published an article in which he proposed from "demonstrations and logical arguments" that blood flows in a completely closed circulation. Even though Harvey could not actually see the microcirculation, he proposed its existence to complete the circuit from arteries to veins. Harvey's application of the scientific method finally gave rise to a modern era of understanding and investigation of cardiovascular physiology. Although Harvey appreciated the fact that the heart functioned as two pulsatile pumps connected in series to move blood around the system, it would still be another 200 years before the French physiologist Claude Bernard proposed that complex, multicellular organisms live in both an external and an internal environment. In the 1850s, he introduced the concept of the *milieu intérieur* to describe the fact that there is an internal fluid environment that surrounds and bathes all tissues of the body. The circulating blood, via a complex network of microscopic capillaries, provides for the exchange of ions, nutrients, wastes, heat, and chemical messengers with every cell in the body. He further recognized that the temperature and ionic and chemical composition of the internal environment are maintained remarkably constant even though the external environment that we live in is subjected to major fluctuations.

The constant state of this internal environment is maintained by intricate control systems with many sensors and effectors. In 1939, the American physiologist W.B. Cannon introduced the term *homeostasis* to refer to this maintenance of a state of uniformity in the fluid matrix of the body by the integrated activity of various physiological controllers. Many of these control systems work directly through the cardiovascular system and are explained in detail in subsequent chapters.

OVERVIEW OF THE CARDIOVASCULAR SYSTEM

Conceptually, it is useful to subdivide the cardiovascular system into the following four functional components: blood, vessels, heart, and its associated control systems. As shown in Fig. 1, blood circulates within a closed series of vessels consisting of a systemic and a pulmonary circuit that are in series with each other. The blood functions as the carrier for transporting substances to and from the various tissues. Although simple diffusion can be an effective transporter of material, it is practical only over relatively short distances. The circulating blood keeps the internal fluid matrix well stirred such that most of the transport is by convection rather than by diffusion. Once a substance arrives at the tissues, it diffuses across the capillary walls as dictated by the concentration gradients. For example, the blood picks up oxygen in the lungs, nutrients in the intestine, and hormones in the various endocrine organs, where each of these substances is in excess. These substances will then be delivered to tissues where their concentrations are low on a subsequent pass through the circulation. Blood also acts to remove noxious products from the cells. Carbon dioxide produced by metabolizing cells is picked up by the circulating blood and transported to the lungs, where it is removed. Metabolic wastes and excess electrolytes are removed from the blood at the kidneys. The blood even removes excess heat from metabolizing

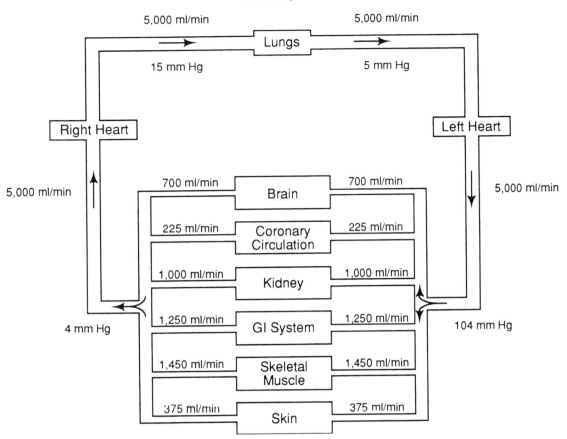

FIGURE 1   Diagram of the human cardiovascular system illustrating how blood flow is distributed to the various organ systems.

tissues and transports it to the skin where it can be radiated into the environment.

The cardiovascular circuitry depicted in Fig. 1 illustrates that the right heart, lungs, left heart, and systemic vasculature are in series with one another; thus, the volume of blood flowing through any of these four regions per unit of time must be equal. Flow leaving either the right or left heart is *cardiac output,* whereas flow returning to the heart is *venous return*. At a steady state, cardiac output and venous return must be equal. In Fig. 1, cardiac output and venous return are illustrated as being 5 L/min. The right heart, pulmonary artery, lungs, and pulmonary veins are collectively referred to as the *pulmonary circulation*. The left heart, systemic arteries, peripheral capillary beds, and systemic veins make up the *systemic circulation*.

The study of the physical principles governing the movement of blood within the cardiovascular system is known as *hemodynamics*. In its simplest form, the cardiovascular system can be considered to be a pump (the heart) in series with a system of tubes through which it pumps blood. In this chapter, we examine the physical principles that dictate the motion of fluids and see how they apply to the cardiovascular system.

HYDROSTATICS: THE PHYSICS OF FLUID AT REST

A liquid within a container exerts a force on the walls of the container that we call *pressure* and has the units of force per unit area. Gravitational forces have a profound effect on the pressure within a fluid system because fluid at the bottom of a container is compressed by the weight of the fluid above it. The pressure (P) at any point in a container of fluid that is open to the atmosphere depends on the vertical distance (h) between that point and the surface of the fluid, the density of the fluid (p), and the acceleration due to gravity (g). This relationship is given as:

$$P = p \times g \times h \qquad (1)$$

For the example given in Fig. 2, a tank is filled with water having a density of 1 g/cm^3. At a depth of 136 cm, the pressure on a side tube would be 133,280 dynes/cm^2 because the acceleration of gravity is 980 cm/sec^2. Because the fluid is not moving, the pressure is equal in all directions and a second gauge oriented vertically at the same point records the same pressure. The gauges in

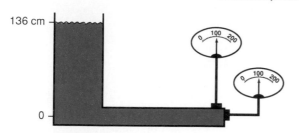

FIGURE 2   In an open system in which there is no flow, the pressure exerted by a column of water 136 cm high is 100 mm Hg. Also, side pressure and end pressure are equal.

the figure show 100 rather than 133,280, because the units dynes/cm^2 are seldom used in medicine. Rather, pressure is usually expressed relative to that required to push a column of mercury up a manometer tube. A *mercury manometer* makes for a simple yet accurate measuring device in the physician's office. A direct conversion is possible because mercury is 13.6 times as dense as water (100 mm × 13.6 = 1360 mm, or 136 cm). Thus, a 100-mm-high column of mercury (10 cm) will exert a pressure of 13.6 × 980 × 10, or 133,280 dynes per cm^2. That means that the 136-cm-high column of water produces a pressure equal to a 100-mm-high column of mercury. In this new system, we would express this pressure as simply 100 mm Hg. Any other force that acts on the fluid will either add to or subtract from the pressure caused by the gravitational acceleration. For example, in Fig. 3, a compressional force of 100 mm Hg is applied to the surface of the tank. That pressure will be added to the preexisting pressures throughout the tank so that a pressure of 200 mm Hg will now be seen at the bottom of the tank.

HYDRODYNAMICS: THE PHYSICS OF FLUID IN MOTION

The movement of fluid is referred to as *flow*. Flow can be viewed in two ways. One way is to consider the displacement of a volume of fluid per unit of time. This is what is conventionally meant by the term *flow* and the

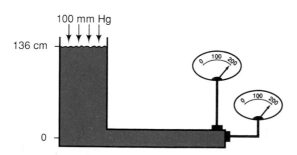

FIGURE 3   If an additional compressional force equal to 100 mm Hg is applied to the system described in Fig. 1, both end and side pressure will equal 200 mm Hg.

units of liters/minute or gallons/hour are familiar. Another way to appreciate flow is to consider the linear displacement of the individual particles of the fluid. As the fluid flows, each particle is moving at a finite *velocity*. Although one does not normally think of fluid movement in terms of velocity, fluid velocity is actually the major determinant of the distribution of forces within a moving fluid.

As a fluid flows through a tube, the mean velocity of flow (v) must be directly proportional to the flow rate (Q) and inversely proportional to the cross-sectional area (A) at that point. Thus:

$$\bar{v} = Q/A \qquad (2)$$

In Fig. 4, a tube has three different diameters along its length. The flow rate will be the same in all three regions (assuming that fluid is neither created nor destroyed within the tube). If flow is set to 200 mL/sec, the velocity of flow will be different in each region. The individual fluid particles will be moving through region 1 at an average of 100 cm/sec, through region 2 at 20 cm/sec, and through region 3 at 200 cm/sec. The narrower the tube the faster the individual fluid particles must travel to accommodate the flow. It should be emphasized that these velocities represent an average. As seen below, the velocities of the individual fluid particles are actually heterogeneous, with some traveling faster than this average and others traveling more slowly.

RELATIONSHIP BETWEEN PRESSURE AND VELOCITY

The distribution of forces (pressure) must be equal in all directions in a fluid at rest. When fluid movement occurs, however, that distribution of forces changes. As expected, the magnitude of that change depends on the velocity of flow. One way to appreciate this change in force is to consider the energy within the fluid. First, consider the pressures in an ideal system in which there is no frictional resistance to fluid flow. In such a system the total energy (TE) is divided between potential energy (PE) and kinetic energy (KE):

$$TE = PE + KE \qquad (3)$$

In a fluid at rest all that energy exists as potential energy and is responsible for the hydrostatic pressure, which is equal in all directions. Once the fluid starts to move, however, part of the potential energy is converted to kinetic energy. This conversion of energy actually causes the pressure in the flowing fluid to be reduced.

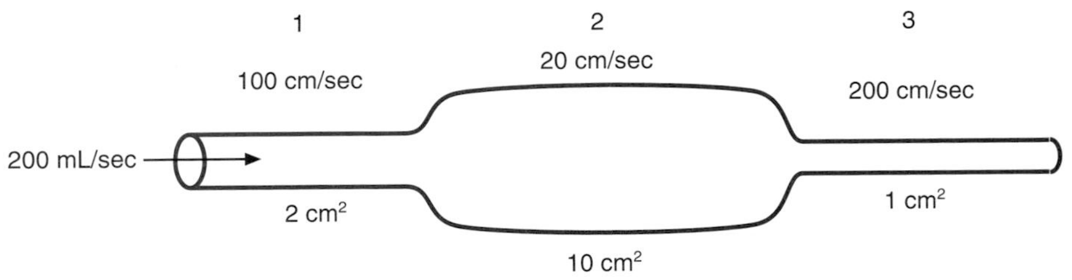

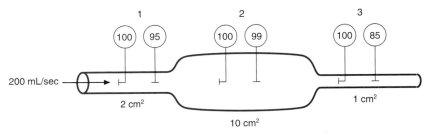

FIGURE 4   When a fluid is forced to flow through a system of tubes of differing cross-sectional areas, the velocity of flow in each tube depends on the cross-sectional area. The larger the diameter of the tube, the slower will be the velocity.

How much it is reduced depends on the velocity of the fluid. In Fig. 5, we can consider the pressures that would be measured in the system described in Fig. 4. One sensor is arranged to face upstream so that any fluid particles striking it will be stopped. The deceleration results in the conversion of kinetic energy back to potential energy. This *end pressure,* therefore, measures total energy. The other sensor is arranged to open perpendicular to the axis of the flow. This sensor is said to sense the *side pressure,* which would be the same as that experienced by the walls of the vessel. Because we are ignoring frictional losses in our example, the end pressure is the same at all three points. Note that the side pressure is always lower than the end pressure at any point and that this difference increases with the velocity.

The difference between the end and side pressures in a flowing fluid is due to the kinetic component. The kinetic component can be calculated if one considers that the kinetic energy is equal to:

$$KE = 1/2pv^2 \qquad (4)$$

In Fig. 4, the flow through region 3 was 200 cm/sec. Thus:

$$1/2 \times 1\,\text{g/cm}^3 \times 40,000\,\text{cm}^2/\text{sec}^2 \\ = 20,000\,\text{dynes/cm}^2 \qquad (5)$$

To convert from dynes/cm^2 to mm Hg, we divide by 1333. Dividing 20,000 by 1333 yields approximately 15 mm Hg, which is the difference between the end pressure and the side pressure in region 3. The concept that pressure decreases as the velocity of flow increases always seems illogical to the beginning student, but this is predicted by the *Bernoulli principle,* which, among other things, keeps airplanes aloft. It should be pointed out that for the most part the kinetic component is small in the cardiovascular system, reaching a maximum value of only 5 to 10 mm Hg at the aortic root where blood velocity reaches its highest value in the body. Thus, any measurement error due to an inappropriately positioned catheter would normally be negligible. In the case of high cardiac output syndromes, however, such as arteriovenous fistulas, regurgitant valves, or heavy exercise, the kinetic components can be much greater and then must be taken into account when directly measuring pressures in such individuals. Also, be aware that it is the side pressure that stretches the vessel wall and forces blood down an artery.

RESISTANCE TO FLOW

In the ideal system described above, friction during flow was ignored. In a real system the fluid particles will interact with each other to resist each other's movement.

FIGURE 5   In an idealized system in which there is no frictional resistance to flow, end pressures are equal all along the system of tubes. However, the difference between end pressure and side pressure in each segment depends on the velocity of flow in that segment. The greater the velocity, the larger will be the difference in side and end pressures. The velocity of flow depends, in turn, on the cross-sectional area of the tube.

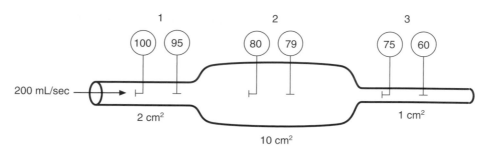

FIGURE 6   When friction is considered, end pressure falls along the length of the tube due to the resistance to flow; however, at each point, side pressure will still be less than end pressure as determined by the velocity of flow.

Thus, the total energy in a system must include the frictional component as well as that which is lost as heat:

$$TE = PE + KE + \text{heat} \qquad (6)$$

Because heat energy cannot be converted back to potential or kinetic energy, the total energy as reflected in the end pressure will drop along the path of flow. Thus, a real system with friction included would be more like that shown in Fig. 6 where the end pressure drops from left to right. Also note that in each location, side pressure is less than the end pressure, as dictated by the velocity, and that those differences remain the same as those calculated under the previous conditions where friction was ignored.

The frictional losses a tube imposes on a flowing stream can be expressed as *resistance* to flow. Several factors determine resistance. One is the character of the flow, which can be either *laminar* (in other words, streamlined) or *turbulent*. In laminar flow, the particles of the fluid move in straight lines (*laminae*) parallel to each other. Although fluid particles in different laminae all move in a parallel direction, they do not move

at the same velocity, as shown in Fig. 7. In a round tube, an infinite number of concentric circular lamina are formed where all particles in a given lamina move at the same velocity. The fluid immediately adjacent to the walls wets the walls and does not move at all. Moving toward the center of the vessel each lamina goes slightly faster than the lamina that surrounds it. This arrangement causes the fluid in the center of the stream to move the fastest. An equilibrium will be achieved with a *parabolic velocity profile* across the vessel having a peak flow in the center equal to exactly twice the average velocity. It is often remarked that resistance to flow results from the friction between the fluid and the walls of the vessel. This is wrong, of course, as it was pointed out above that the lamina of fluid adjacent to the wall does not move. The friction results from the relative motion between each lamina. *Viscosity* refers to the force generated by the particles of the fluid that resists any relative motion between them. The greater the viscosity of a fluid, the more it resists such movement. The viscosity of blood is the source of the frictional losses as it resists movement between adjacent laminae.

Let us now consider laminar flow through a round tube such as a blood vessel. Blood moves through the blood vessels along pressure gradients in the cardiovascular system. Flow through a tube is proportional to the pressure difference across it, ΔP. That flow is impeded by the resistance (R) of the tube such that:

$$Q = \Delta P / R \qquad (7)$$

Resistance is determined in part by the viscosity (η) of the fluid. The more viscous a fluid, the more resistance the fluid will encounter passing through a tube, making resistance proportional to viscosity:

$$R \propto \eta \qquad (8)$$

Resistance is also affected by the length (L) of the tube through which the fluid flows. The longer the length, the more resistance the fluid will encounter so

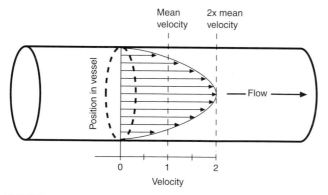

FIGURE 7   As fluid flows through a round tube, an infinite number of concentric lamina will be formed, with each lamina flowing slightly faster than that surrounding it. Velocity of the fluid is zero at the vessel wall and increases parabolically until a maximum velocity is achieved at the center. If the flow rate is increased, the laminar flow profile may break down, resulting in turbulent flow.

that resistance is proportional to the length:

$$R \propto L \qquad (9)$$

Finally, resistance is affected by the radius of the tube. The larger the radius, the more easily fluid can pass through the vessel. Although one might think that resistance should be related to the cross-sectional area and thus follow r^2, it actually is inversely proportional to the fourth power of the radius (r) of the tube.

$$R \propto 1/r^4 \qquad (10)$$

In the late 18th century, a French physician named Poiseuille found that all these terms could be grouped together into one equation by including the appropriate constants. The following is known as *Poiseuille's equation*:

$$R = \frac{\eta \times L \times 8}{\pi \times r^4} \qquad (11)$$

Poiseuille's equation also can be rearranged to solve for flow rather than resistance:

$$Q = \Delta P/R = \frac{\Delta P \times \pi \times r^4}{\eta \times L \times 8} \qquad (12)$$

Fluid Flowing in Tubes Is Described by Ohm's Law

The equations that describe current flow in electrical circuits can also be adapted to describe fluid flow. By substituting flow for current and pressure for voltage, the analogy is almost exact. *Ohm's law* for electrical circuits states that the current through a device can be determined by dividing the voltage across it by the resistance of the device:

$$\text{Current} = \Delta \text{ voltage/resistance} \qquad (13)$$

Ohm's law can be applied to the circulation as well. Note that the relationship among flow through a segment (Q), the difference in pressures (ΔP) across the segment, and the resistance to flow (R) across the segment given above in Eq. 7 is identical to Ohm's law, where Q is substituted for current and ΔP for Δ voltage.

A powerful feature of Ohm's law is that if one knows values for any two of the variables, the third can be calculated. Also, remember that R is taken directly from Poiseuille's equation above and includes terms for both the viscosity of the fluid and geometry of the vessel. For example, if flow is to be kept constant when the length of a tube is doubled, then the driving pressure (AP) must be twice as great (Fig. 8). It is important to remember that R (during laminar flow) is determined only by tube radius, length, and fluid viscosity and not by Q or ΔP. In the body, the length of the blood vessels or the viscosity of the blood cannot be easily changed from moment to moment. On the other hand, the radius of smooth muscle in the walls of the vessels is constantly changing. Because resistance varies according to the fourth power of the radius, small changes in radius have a profound effect on flow. Thus, if the radius of the tube is

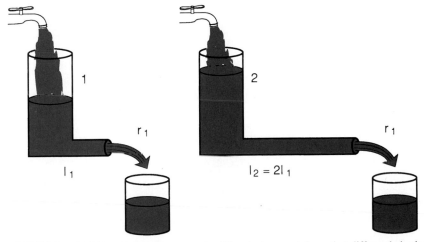

FIGURE 8   Fluid is added at the same rate of flow to two containers that differ only in the lengths (*l*) of their outflow tubes such that the length in example 2 is twice that in example 1. The radii (*r*) of tubes 1 and 2 are equal. The column of water in each container will rise until the pressure exerted by that column is large enough to cause outflow to equal inflow. Because the resistance to flow in example 2 is twice that in example 1, column height will be twice as great in example 2 at steady state.

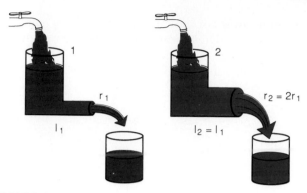

FIGURE 9 Fluid is added to two containers that differ only in the radii of their outflow tubes. Because the radius of tube 2 is twice that of tube 1, the resistance to flow through tube 2 is 1/16 of that through tube 1. Thus, flow in tube 2 must be 16 times as great to maintain the same column height.

doubled, Q will increase 16-fold for the same ΔP (Fig. 9). In the cardiovascular system, the dimensions of the tubes seldom are known well enough to calculate resistance directly. More often, both Q and ΔP are measured and R is calculated.

Flow is proportional to driving pressure only under laminar flow conditions. If the velocity of the fluid particles becomes too great, turbulence may develop. During turbulent flow the orderly pattern of flow is lost and the particles frequently change direction abruptly and move between lamina rather than along them. This results in additional energy loss and an increased resistance to flow. Generally, flow changes from laminar to turbulent as velocity increases. The physicist, Osborne Reynolds, empirically determined the factors that lead to turbulence and expressed them quantitatively in *Reynolds equation*:

$$\mathrm{Re} = \frac{D \times \bar{v} \times p}{\eta} \tag{14}$$

where D is the diameter of the tube, v is the mean velocity, p is the density, and η is the viscosity of the fluid. Turbulence can be expected whenever the Reynolds

number, Re, exceeds a value of about 2000. It should be noted that some texts use radius in the derivation of Re rather than diameter. When that is done, the critical value of Re becomes 1000. While laminar flow is silent, turbulent flow is audible. In practical terms Reynolds numbers are highest where there is a deformity in the vessel wall, such as at a branch point or in the root of the aorta where velocity is very high, but normally turbulence does not occur in a blood vessel unless its shape has been modified by disease.

Resistance in Series and Parallel Circuits

Collectively, the arteries, arterioles, capillaries, venules, and veins are arranged in series. Also, each vessel type, with the exception of the aorta and the pulmonary artery, finds itself in parallel with other vessels of the same type because of the extensive branching of the vasculature. Thus, resistances to blood flow are arranged in both series and parallel circuits. As will be discussed in other chapters, this arrangement allows for each region of the body to be perfused with blood in a regulated manner.

One can appreciate the consequences of whether the resistance is in series or in parallel with the circuit by again using the electrical analogy. Consider the following example of three tubes with values of 5, 25, and 100 resistance units (RU). If these tubes are arranged in series (Fig. 10), total resistance (R_t) is given as their simple sum because the flowing fluid must overcome all resistances in the circuit:

$$R_t = R_1 + R_2 + R_3$$
$$R_t = 5 + 25 + 100 = 130 \text{ RU} \tag{15}$$

Because the same fluid must pass through all resistances in a series circuit, changing the resistance in one segment changes the flow in all segments. For example, constricting the arterioles in an organ equally reduces flow through the arteries, capillaries, and veins associated with that organ. When calculating the

---

Clinical Note

Bruits

Turbulence can occur in the cardiovascular system at points where the flow velocity is high, such as where a vessel is abruptly narrowed by an atherosclerotic plaque. Turbulent flow at such points can often be heard with a stethoscope; when such a sound is heard in the peripheral circulation it is referred to as a *bruit*. Bruits are not

infrequently heard in the carotid arteries of the elderly and are usually accompanied by signs of impaired cerebration because of inadequate cerebral blood flow. The Reynolds number is also influenced by the viscosity of the fluid. When the viscosity of the blood is reduced by severe anemia, bruits may be heard at branch points in the large vessels even without any vascular disease.

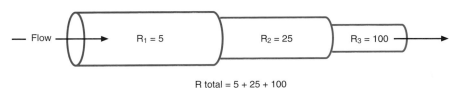

FIGURE 10   Three tubes of different radii offer different resistances (R) to flow. If these tubes are arranged in series, the resistance offered by the combination of tubes is equal to the sum of their individual resistances. Note that the total resistance will always be greater than the largest single resistance.

equivalent of several resistances in series, the total must be greater than the greatest resistance in the circuit. If you do not find that to be the case, then a calculation error has been made.

If the same tubes are arranged in parallel (Fig. 11), the fluid is offered alternative paths that make it easier to flow through the circuit. Total resistance must be calculated by adding reciprocals:

$$\frac{1}{R_t} = \frac{1}{R_1} + \frac{1}{R_2} + \frac{1}{R_3}$$

$$\frac{1}{R_t} = \frac{1}{5} + \frac{1}{25} + \frac{1}{100} = \frac{25}{100} = \frac{1}{4} \qquad (16)$$

$$R_t = 4 \text{ RU}$$

Thus, it would take a ΔP 32.5 times as great to maintain the same Q through the three tubes in series compared with the tubes in parallel. Note that in parallel circuits the overall resistance must be less than the smallest individual resistance. In parallel beds, a change in an individual resistance affects only the flow through that leg. For example, flow to the intestine is in parallel with that to the kidneys. An increased resistance to flow in the intestine will not affect flow to the kidneys as long as aortic pressure does not change. The cardiac output would be reduced, however. The arterial pressure in the systemic circulation perfuses many parallel vascular beds.

The principle of hydrodynamics can be applied to understand blood flow in the cardiovascular system; however, additional complications are added because: (1) blood is a complex fluid, and (2) blood vessels are distensible and many can change their radii by contraction of the muscle in their walls. Each of these complications affects one or more of the parameters of resistance, pressure, and flow.

BLOOD IS NOT A NEWTONIAN FLUID

Blood is comprised of water, electrolytes, proteins, and cells. Many of these components can change in various physiologic and pathologic states and some changes affect the overall viscosity of the blood. The two components with the most influence on viscosity are the plasma proteins and the red blood cells. Figure 12 depicts the viscosity of blood (relative to water) as a

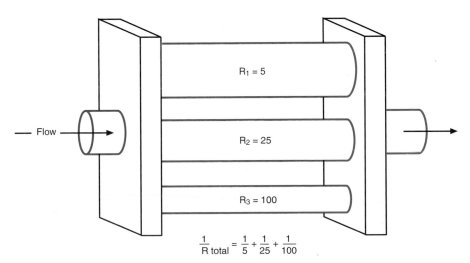

FIGURE 11   If the same three tubes described in Fig. 10 are arranged in parallel, the reciprocal of the resistance offered by the combination of tubes is equal to the sum of the reciprocals of their individual resistances. Note that this total resistance will always be less than the smallest single resistance.

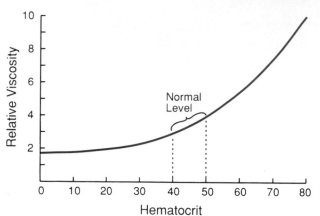

FIGURE 12   The viscosity of blood (relative to water) is a function of hematocrit. Note that at 0 hematocrit (*plasma*), viscosity due to plasma proteins is about twice that of water. For a normal hematocrit of 45–50%, blood viscosity is about three times that of water.

function of *hematocrit* (the percent of total blood volume comprised of red blood cells). Even with no cells present, blood plasma is about twice as viscous as water because of the presence of proteins, mainly albumin; however, at a normal hematocrit of 45%, blood is about three times as viscous as water. Viscosity will increase in conditions in which the number of red blood cells is increased above normal (*polycythemia*) and will decrease in conditions in which the hematocrit is decreased below normal (*anemia*).

Unlike water, which is a newtonian fluid, the viscosity of blood changes with flow. This can be seen by observing the difference in driving pressure that it takes to maintain flows at various rates through a rigid glass tube (Fig. 13). For a newtonian fluid, flow would

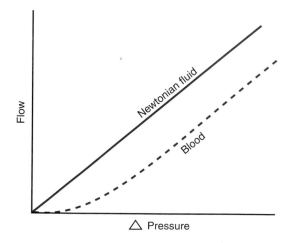

FIGURE 13   When a newtonian fluid flows through a rigid tube, the flow is proportional to the pressure gradient across the tube. With blood, however, the viscosity appears to increase at low flow rates, causing a nonlinear pressure flow curve. The formed elements in the blood are responsible for this anomalous viscosity.

always be exactly proportional to ΔP. A doubling of ΔP would double Q no matter what the rate of flow. For blood, however, the pressure flow curve becomes quite nonlinear at low flow rates. Pure plasma is a solution, and therefore, fact newtonian. The variable viscosity of blood occurs because at low flow rates the formed elements, including the red cells and platelets, begin to stick to one another. This non-newtonian property of blood, referred to as *anomalous viscosity,* in practice causes only a small deviation from Ohm's law at normal blood flow rates and usually can be ignored. Nevertheless, anomalous viscosity can significantly increase resistance to flow when perfusion pressure is low, as occurs in shock and other hypotensive states, and can exacerbate low blood flow state. This is sometimes referred to as *blood sludging*.

In the microcirculation, the formed elements cause the viscosity of blood to be dependent on the diameter of the vessel through which it is flowing. This phenomenon is known as the *Fahraeus–Lindquist effect*. As shown in Fig. 14, the viscosity of blood appears to actually decrease as the vessel diameter decreases below 50 μm, which is just opposite of what might be imagined. The explanation for the Fahraeus–Lindquist effect appears to be due to the proclivity of red cells to concentrate in the center of the blood vessel in a process termed *axial streaming*. Because the velocity is highest in the center of the vessel, the pressure is lowest in that region and the cells are literally sucked into the center. On the other hand, it can be shown that the geometry of a round tube causes most of the viscous drag to be concentrated nearest to the vessel wall. In the small vessels where axial streaming is pronounced, the blood in the critical region near the vascular wall is left with a low hematocrit and thus a low viscosity. The net result is that overall viscosity appears to drop. Another consequence of the axial streaming of blood cells is that they literally pass through an organ faster than the plasma. This causes the intra-organ hematocrit always to be somewhat lower than that of the central circulation.

BLOOD VESSELS DO NOT BEHAVE AS RIGID TUBES

In the discussion of hydrodynamics, only tubes with rigid walls were considered. Blood vessels, on the other hand, are distensible. If the diameter of a vessel increases with increasing perfusing pressure, then its resistance will fall. Such an effect will cause resistance to be dependent on perfusion pressure, thus causing additional nonlinearity in the pressure–flow relationship. Such distensibility is very beneficial in the

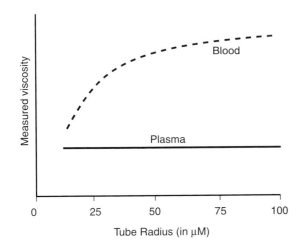

FIGURE 14 The viscosity of blood appears to decrease as the diameter of the tube it flows through decreases. Because most of the peripheral resistance in the body resides in vessels with radii of 50 μm or less, the Fahraeus–Lindquist effect allows blood to flow more easily through the tissues. (Modified from Fahraeus R, Lindquist T, *Am J Physiol* 1931; 96:562.)

pulmonary system, where a large increase in cardiac output causes only a small increase in pulmonary arterial pressure. On the other hand, arterioles are not very distensible. Because most of the peripheral resistance resides in arterioles, flow through most organs remains a remarkably linear function of the perfusion pressure as depicted by Ohm's law. The basis for the distensibility of a blood vessel can be seen in Table 1. The walls of different blood vessel types consist of varying amounts of connective tissue and muscle. Also, the vessels differ in radius and wall thickness. Thus, distensibility differs among vessel types. Distensibility of blood vessels is measured as *compliance* (*C*), a change in volume (*V*) in response to a change in distending pressure (*P*):

$$C = \Delta V / \Delta P \qquad (17)$$

Because of the compliance relationship, the pressure in a vessel cannot be raised without moving blood into it. As a result, pressure changes within the cardiovascular system are usually accompanied by shifts in the blood distribution. For example, if arterial pressure is increased, blood must be translocated into the arterial system at the expense of some other compartment such as the veins, whose pressure and volume must fall. Figure 15 reveals that there is a large difference between the compliance of arteries and · veins. At any given pressure, much more volume is contained in the veins. This large capacitance contributes to the reservoir function of the veins. Thus, in the above example, a large rise in arterial pressure would usually be accompanied by a smaller drop in venous pressure. Notice that at zero pressure the vessels still contain appreciable blood. This volume is referred to as the *unstressed volume*. Thus, the total volume of blood in a vessel is given by:

$$(P \times C) + \text{unstressed volume} \qquad (18)$$

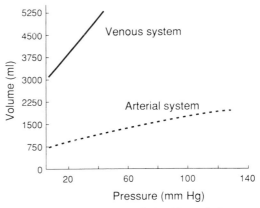

FIGURE 15 Volume-pressure relationships of the venous and arterial systems indicate that the venous compliance (indicated by the slopes of the lines) is much greater than that of the arteries. Also, the unstressed volume (the volume at zero pressure) is much larger for the veins.

TABLE 1   Characteristics of Blood Vessels

Characteristic	Aorta	Arterioles	Capillaries	Venules	Vena cava
Number in the body	1	10	10^{10}	3×10^8	2
Total cross-sectional area (cm²)	9.6	300	5000	4000	18
Individual radius (mm)	17.5	0.010	0.004	0.02	17
Wall thickness (mm)	2	0.020	0.001	0.002	1.5
Elastin	+ + +	+	0	+	+
Smooth muscle	+ +	+ + +	0	+	+ +
Collagen	+ + +	+	0	+	+
Transmural pressure (mm Hg)	110	70	20	10	5
Peak velocity of flow (cm/sec)	50	0.3	0.017	0.02	4.6

+, low abundance; + +, moderate abundance; + + +, high abundance.

PRESSURES AND RESISTANCES IN THE CARDIOVASCULAR SYSTEM

The resistance to blood flow differs at each level in the vasculature. Resistance in the aorta and the network of large conducting arteries is minimal because of their large radii. On the other hand, resistance is quite high in the arteriolar network because of the small radius of each vessel, even though there are many arterioles arranged in parallel. This illustrates the major importance of vessel radius. The capillaries offer much less resistance than the arterioles. Although the radius of each capillary is less than that of an arteriole, the capillaries as a group have a low resistance because there are so many of them in parallel. Each arteriole supplies many capillaries. This hierarchical arrangement lowers capillary resistance and at the same time increases the surface area for exchange of metabolites. Because of their large radii and their parallel arrangement, the veins offer minimal resistance to blood flow.

Although both length and radius of the vessel contribute to its resistance, it is the radius that is altered to modulate flow. Smooth muscle cells in the vessels, especially the arterioles, are arranged such that their contraction or relaxation results in reductions and increases in vessel radius, resp., causing major changes in the resistance to blood flow. Figure 16 illustrates the change in resistance in a single vascular bed that occurs during stimulation of its sympathetic nerves. The perfusion pressure is held constant. Note that at increasing stimulation rates the flow falls, indicating an increase in resistance. Now let's extrapolate this experiment to the intact body under two conditions. First, if there were generalized sympathetic stimulation to *all* the organs, and if cardiac output (*CO*) remained constant, then *mean arterial blood pressure* would increase. That is because the total peripheral resistance

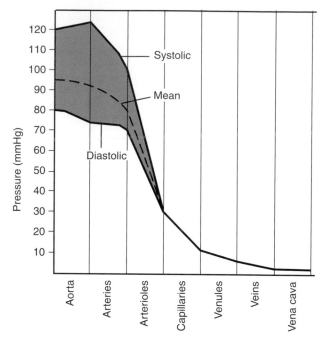

FIGURE 17   As blood flows from the root of the aorta to the vena cava, the pressure within the vascular system decreases. The decrease is due to the resistance to blood flow. As this figure illustrates, the greatest decrease in pressure (due to large resistance) occurs in the arterioles. There is little resistance to flow in the large arteries or in the veins.

(*TPR*) would increase. By Ohm's law:

$$\text{Mean arterial pressure} = CO \times TPR \qquad (19)$$

If, on the other hand, stimulation were confined to a single organ, the effect on TPR and thus mean arterial blood pressure might be small. Blood flow to the stimulated area would be reduced, whereas that to all other tissues would remain the same. Modulation of vascular resistance is a primary means of controlling pressures and flow in the cardiovascular system.

Pressure Differs in Each Vessel Type

Because total energy for blood flow through the blood vessels comes from the heart, total energy (expressed as end pressure) is highest in the aorta. As blood flows through the vessels, energy is lost because of friction, and total energy will be lowest in the vena cava. However, as one might suspect from the preceding discussion of resistances in the system, the drop in energy is not the same across each vessel type. Figure 17 illustrates the mean pressure recorded at all levels throughout the system. Because the vessel types are essentially in series with each other in each organ, the resistance of each vessel type will be proportional to the drop in pressure across it. As mentioned previously, little pressure is lost as blood moves through the larger

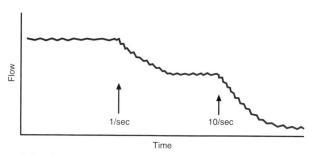

FIGURE 16   Arterial blood under constant pressure is introduced into an artery supplying an organ. The sympathetic nerves to that organ are stimulated at various rates. Note that flow falls because the resistance to flow increases with stimulation rate. These nerves cause constriction of vascular smooth muscle in the arterioles which decreases their caliber.

Clinical Note

Measuring Blood Pressure

Blood pressure is measured in the human noninvasively with a device called a *sphygmomanometer*. Basically, an inflatable cuff is wrapped around the upper arm and inflated to about 170 mm Hg. A stethoscope is placed over the brachial artery just below the cuff and the air is slowly let out of the cuff. To begin with, the brachial artery is collapsed by the pressure. As the pressure in the cuff falls below the systolic blood pressure, the vessel pops open briefly during each systole. As the flow starts and stops it is turbulent and audible. As cuff pressure falls below diastolic pressure, the vessel remains patent throughout the cardiac cycle and the sounds disappear. The cuff pressure where the sounds are first heard is taken to be the patient's systolic pressure and the pressure at which they disappear as the diastolic pressure.

---

arteries and veins. Those vessels serve as efficient conduits, bringing blood to and away from all parts of the body. The greatest pressure drop occurs across the arterioles. These vessels also have abundant smooth muscle in their walls that can change their resistance to flow moment to moment. Because they constitute a major portion of the resistance to flow of an organ, arterioles are the ideal site for local regulation of blood flow.

Arterial Pressure is Pulsatile

The major energy source for blood flow in the cardiovascular system is the heart. During systole the heart ejects blood into the aorta, raising the pressure. During the rest of the cycle (diastole), the aortic pressure is falling as the heart refills with blood from the veins. The shaded areas depicted in Fig. 17 are meant to illustrate the pulsations in pressure. Note that the pulsations are damped out in the microcirculation. These pulsations are better appreciated in Fig. 18 where pressure in an artery is plotted against time. The difference between the peak pressure (*systolic*) and the minimum pressure (*diastolic*) over the cycle is referred to as the *pulse pressure*. The *mean pressure* refers to the average pressure with respect to time as indicated by the dotted line labeled 100 in Fig. 18. If the aortic pressure varied as a perfect sign wave then the mean aortic pressure would be simply halfway between the systolic and diastolic pressure. Unfortunately, the aortic pressure has a very peculiar shape. In practice, the mean aortic pressure can be closely approximated by adding 1/3 of the pulse pressure to the diastolic pressure.

The basis for the pulse pressure is as follows. The equation for compliance states that the pressure in the distensible arterial tree at any instant depends on the volume of fluid in those vessels divided by their capacitance. Because arterial compliance changes little

from moment to moment, the factor that determines arterial pressure is the volume of blood in the arterial tree. This volume depends on the balance between inflow from the heart and outflow through the resistance vessels. Inflow occurs primarily during the rapid ejection period of the heart (see Chapter 13) while outflow depends on the peripheral resistance and the aortic pressure and occurs throughout the cardiac cycle. Because the rapid ejection period of systole is so brief, little outflow occurs during that period, and it is as if the entire stroke volume suddenly appears in the aorta. Thus, pulse pressure (*PP*) will be closely approximated by:

$$PP = SV/C \qquad (20)$$

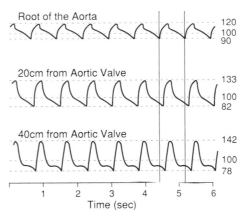

FIGURE 18 Pressures recorded from three sites in the aorta. At all three sites pressure fluctuates between a maximum systolic pressure and a minimum diastolic pressure. The difference between systolic and diastolic pressure is the pulse pressure. The amplitude and shape of the pressure signal is altered as distance from the heart increases. Note that pulse pressure in the abdominal aorta (bottom panel) is higher than that in the aortic root. Mean pressure (100 mm Hg) are the same at all levels. The vertical time lines reveal that the systolic rise in pressure occurs much later in the abdominal aorta than in the root.

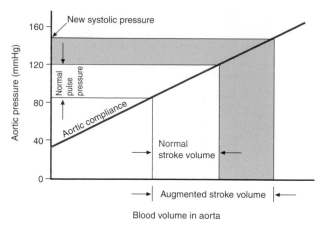

FIGURE 19   The influence of a change in stroke volume on pressures recorded in the aorta. The slope of the line relating pressure and volume defines compliance. With a normal stroke volume, aortic pressure fluctuates between 120 mm Hg and 80 mm Hg. When stroke volume increases, pulse pressure increases as well, as indicated by the shaded area.

where SV = stroke volume and C = arterial compliance. In Fig. 19, arterial volume is plotted as a function of arterial pressure, and the diagonal line indicates vascular compliance. At the end of ejection the arterial system is distended with the stroke volume and pressure rises to 120 mm Hg. Over the ensuing diastole, volume decreases and pressure falls to 85 mm Hg, giving a pulse pressure of 35 mm Hg. If stroke volume is increased as shown by the shaded area, a greater pulse pressure will be achieved. Note that this increase

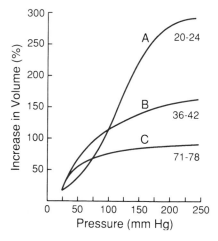

FIGURE 20   Aortas from individuals of three age groups (20–24, 36–42, and 71–78 years) were studied by filling them to various pressures with fluid. At each pressure, the volume was noted. The slope of each line relating volume to pressure describes the compliance for each group. Note that compliance is much higher in the youngest group. In older individuals, compliance is less, especially at higher aortic pressures. (Modified from Hallock P, Benson JC, *J Clin Invest* 1937; 16:596.)

in stroke volume has resulted in an increase in both the mean pressure and the pulse pressure, even though compliance remained constant.

Changes in capacitance also affect pulse pressure. Figure 20 illustrates the volume–pressure relationships for aortas that were removed from individuals who had died at various ages. The slope of the line relating volume and pressure indicates compliance and is relatively high for vessels taken from younger individuals. With age, compliance decreases because of a change in the composition of the aortic wall. The effect of a loss of arterial compliance, such as that associated with aging, is shown in Fig. 21. In this figure, a person with normal arterial compliance is compared to a person with low compliance. If heart rate, ejection time, and resistance are the same for both situations, it follows that for any given stroke volume the pulse pressure will be greater in the individual with the lower compliance. Note that in this example a change in compliance does not affect mean arterial pressure, just the pulse pressure. The normal pulse pressure for a 60-year-old person averages 60 mm Hg, whereas that for a 20-year-old person averages only 40 mm Hg. Because the physical work done by the heart is the volume pumped times the systolic pressure, this puts an added burden on the elderly heart, even though the cardiac output and the mean arterial pressure may be the same as that in the young individual.

Distensibility of the arterial tree also allows for a more efficient use of the energy generated by the heart in the maintenance of blood flow. If blood vessels were rigid, blood flow and pressures would be very high during systole, but both would fall to zero during diastole. Such a flow pattern would require excessive work by the heart because it would have to pump against an extremely high pressure to force the entire cardiac output through the circulation during only a small fraction of the cardiac cycle. Much of the energy expended by the heart during ejection is stored in the distended walls of the arterial system. This stored energy serves to force blood into the periphery during the remainder of the cardiac cycle. A good analogy is the bagpipe player who blows into the bag in short spurts. The energy from those spurts of air is stored in a compliant bag so that air exits through the pipes in a steady stream, giving an uninterrupted tone even while the player is taking a breath.

The pressure pulse, like the blood, is transmitted down the aorta to the peripheral vessels. The vertical lines in Fig. 18 reveal that the pressure pulse takes about 0.25 sec to travel 40 cm down the aorta. The pressure profile becomes distorted as it traverses the aorta. Interestingly, reflected waves in the aorta cause the pulse pressure to actually increase in the distal

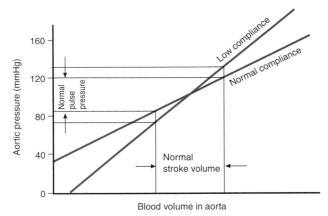

FIGURE 21 The influence of a change in compliance on pulse pressure in the aorta. The slopes of the lines relating pressure and volume define compliance. When compliance is reduced, pulse pressure increases for the same stroke volume.

aorta. Note, however, that the mean pressure and thus the total energy are unchanged as the distance from the heart increases. As the pressure pulse progresses through the arterioles, the combination of arteriolar resistance and compliance attenuates the pulse such that it is almost completely damped out at the level of the capillaries and beyond (see Fig. 17).

Forces in the Blood Vessel Wall

The side pressure exerted on the wall of a blood vessel acts to distend the vessel. The hydrostatic pressure in the interstitium that would oppose that distension is normally very low. This distending force must, therefore, be balanced by forces within the vessel wall or else the vessel would rupture. The magnitude of the tension (T) in the wall opposing the transmural pressure (P_t) is influenced by vessel radius (r). The relationship among the three factors is given by the *law of LaPlace*:

$$T = P_t r \qquad (21)$$

Note that the wall tension increases not only with the internal pressure, which is obvious, but also with the radius of the vessel, which is not so obvious. The LaPlace relationship can be used to explain many of the structural features found in the cardiovascular system. For large arteries such as the aorta, both pressure and radius are large so that a very large force is generated in the aortic wall. However, the aortic wall is thick and composed of strong connective tissue and muscle. In the thin-walled veins, the radii are still large but the wall tension is low because the pressures are low.

The capillaries pose an interesting case. The walls of these vessels are a single layer of endothelial cells devoid of connective tissue. Although the pressures are much higher than those found in veins, this higher pressure is offset by the very small radii of the capillaries. Because of the small radius, the wall tension is very low. As a result, the capillaries seldom burst, whereas rupture of the thick-walled aorta is a relatively common medical event.

Distribution of Flow and Volume in the Vascular System

The volume of flow must be the same in all divisions of the cardiovascular system. That is, cardiac output must equal the volume of flow through the aorta, the arteries, arterioles, capillaries, and veins. The velocity of flow, on the other hand, differs markedly. This is because the total cross-sectional area of each of the divisions is different. The flow velocity is highest in the root of the aorta. As Table 1 indicates, the cross-sectional area of the aorta in the body is 9.6 cm^2. According to Eq. 2, if the cardiac output is 5000 cm^3/min (83 cm^3/sec), then the mean velocity will be approximately 8.65 cm/sec. The flow velocity is the slowest in the capillaries. The total cross-sectional area of all the capillaries (remember that the capillaries are in parallel with each other) is about 5000 cm^2. Because flow through the capillary beds must also be 83 cm^3/sec,

---

Clinical Note

Narrow Pulse Pressure
Patients with a very narrow pulse pressure are a cause for concern. Equation 20 indicates that a narrow pulse pressure could result from either a reduced stroke volume or increased compliance of the aorta. Reduced stroke volume can result from inadequate blood volume as might occur after hemorrhage or from a failing heart.

A common cause of increased aortic compliance is the presence of a distended region of the aorta, called an *aneurysm*. Aneurisms tend to be unstable because, once distended, the wall tension further increases because of the law of LaPlace, putting the patient at risk of a catastrophic rupture of the aorta.

velocity is only approximately 0.017 cm/sec in the capillaries. This low velocity allows time for exchange of oxygen, carbon dioxide, nutrients, and waste to take place between the tissues and the blood in the capillaries. If the kinetic energy component for the flow in the root of the aorta is calculated from the above figures, we arrive at only 0.028 mm Hg. The calculation assumed that the cardiac output was delivered over the entire cardiac cycle. Actually peak velocity during the rapid ejection period exceeds 50 cm/sec, which would yield a peak kinetic component of 0.54 mm Hg.

The heart and blood vessels contain approximately 5000 mL of blood, approximately 4000 mL in the systemic vascular system and 1000 mL in the pulmonary system at any instant. The distribution of blood among the various vessel types in both systems depends on the side pressures existing in the vessels and the vessel capacitances at any instant. As it turns out, the large capacitance and unstressed volume of the veins causes most of the blood volume to be found in the veins (Fig. 22).

Because of their large blood volume, the veins serve an important *reservoir* function in the body. Smooth muscle in the walls of the veins allows them to change their compliance as needed. As this compliance changes, so does venous pressure, which is the filling pressure for the heart. Modulation of venous compliance is an important mechanism for controlling cardiac output. Stimulating the sympathetic nerves to the veins causes smooth muscle in their walls to contract, which raises the venous pressure. As we will see in Chapter 13, filling of the heart is determined by the venous pressure. The more the heart fills, the more it will eject with every beat. Thus, the body can easily increase the cardiac output by simply constricting the veins.

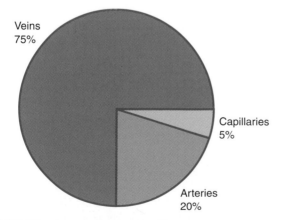

FIGURE 22   At any instant, the volume of blood in the systemic circulation is distributed as indicated. Owing to their large unstressed volume and capacitance, the veins contain most of the blood, even though venous pressures are low.

GRAVITY CREATES HYDROSTATIC COLUMNS WITHIN THE CARDIOVASCULAR SYSTEM

Flow and pressure in the vasculature, especially in the veins, are affected by gravity (Fig. 23). First consider an individual who is lying down. In this position, the only force affecting pressure in the blood vessels is that generated by flow through them because all levels of the cardiovascular system are at about the same vertical height. Upon standing, however, the pressures generated by the hydrostatic columns of blood in the vessels become quite large. At the beginning of this chapter, we stated that 100 mm Hg are generated by every 136 cm (54 inches) of water (or in this case blood, which is only slightly more dense than water). Consider the arterial system of a 6-foot-tall individual. When standing, the hydrostatic column will cause the pressure in an artery at the base of his foot to be 133 mm Hg higher than that in an artery in the top of his head. That is remarkable considering that mean arterial pressure at the heart is only about 100 mm Hg. Because of the position of the heart high in the chest, arterial blood pressure is high enough to force blood to the head because there is only about a 30-mm Hg hydrostatic column between the two structures. If the gravitational acceleration were increased, such as occurs in a high-performance aircraft during a tight turn, the arterial pressure loss to the head due to hydrostatic columns can be a problem. At three times the acceleration of gravity, the arterial pressure to the brain approaches zero and the pilot loses consciousness. Hydrostatic columns augment the pressure in the arteries below the heart. Figure 23 reveals that, upon standing, an additional 100 mm Hg are added to the arterial pressure at the feet.

Gravity markedly affects function of the veins as well. Pressure in the central veins near the heart is normally less than 5 mm Hg. Thus, upon standing, pressure in the veins in the head and neck tends to become negative, which causes these vessels to collapse. That is why the jugular pulse is not normally seen in a person in the erect posture but can be seen when supine. Inside the cranium, rigid venous sinuses actually allow venous pressure to become slightly negative at the top of the head, as shown in Fig. 23. If sinuses are ruptured during head trauma, air can actually be drawn into them. Figure 23 also shows that the hydrostatic column in the veins below the heart would cause venous pressure in the feet to increase to more than 100 mm Hg if compensatory mechanisms did not come into play. Such a high venous pressure would cause blood to pool in the dependent veins because of their high compliance. Also, the increased venous pressure would be reflected back into the

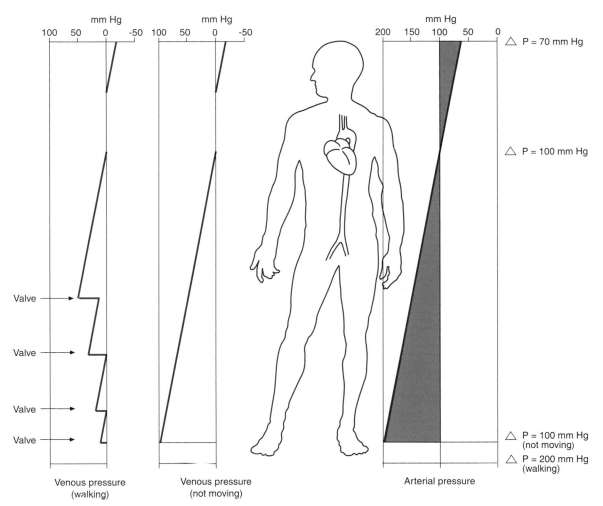

FIGURE 23   Schematic diagram of the effects of gravity on pressure in the arterial and venous systems. Note that, in both, the pressure increases below the heart and decreases above the heart. The left panel shows that valves in the veins in conjunction with the muscle pump help to break these columns in the dependent limbs, which reduces venous pressure in that region. The numbers to the right show the ΔP between the arteries and the veins.

capillaries and cause massive filtration across the capillaries (see Chapter 16). Fortunately, structural and functional mechanisms exist to counteract the effects of gravity. One-way valves in the veins prevent the back flow of blood. The veins course through the skeletal muscles of the legs and, when these muscles contract, they squeeze the veins, actively forcing blood across the valves toward the heart. When the muscles relax the valves periodically break the column of blood between the dependent limbs and the heart, as shown in the left-most panel of Fig. 23. The valves then allow the empty venous segments to

Clinical Note

Venous Pooling

Standing without periodically contracting the leg muscles can lead to venous pooling of blood to the point at which cardiac return is reduced enough to make an individual faint. This often occurs in military parades where soldiers are asked to stand at attention for long periods of time. If the erect posture is prolonged, the effects of venous pooling can even prove fatal. This was the mechanism for execution by crucifixion in ancient times.

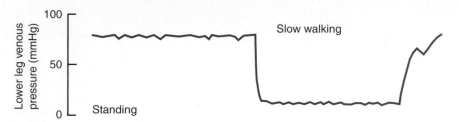

FIGURE 24   Venous pressure is measured in a patient's foot. While the patient stands quietly, pressure approaches 80 mm Hg because of the hydrostatic column. When the muscle pump is evoked by slow walking, however, pressure quickly falls to less than 20 mm Hg. (Modified from Noddeland H, Ingemansen R, Reed RK, *Clin Physiol* 1983; 3:573.)

refill from the periphery. This "muscle pump" is very important. Figure 24 shows venous pressure in the foot of an individual standing quietly and while walking. Note how walking causes the muscle pump to greatly reduce venous pressure in the foot and that the muscle pump also increases the ΔP for flow through the leg muscles by lowering venous pressure as they exercise.

Suggested Readings

Fahraeus R, Lindquist T. The viscosity of the blood in narrow capillary tubes. *Am J Physiol* 1931; 96:562

Hallock P, Benson IC. Studies on the elastic properties of human isolated aorta. *Clin Invest* 1937; 16:596

Noddeland H, Ingemansen R, Reed RK. A telemetric technique for studies of venous pressure in venous blood from the human foot in orthostasis. *Clin Physiol* 1983; 3:573

Electrical Activity of the Heart

JAMES M. DOWNEY

KEY POINTS

- Heart muscle cells contract in response to an *action potential* on their membrane.
- Cells in the *SA node region* of the atrium depolarize spontaneously and act as the heart's pacemaker.
- Action potentials spread electronically from cell to cell over the entire heart.
- Slow conduction through the *AV node,* the only excitation pathway from the atria to the ventricles, allows atrial contraction to precede ventricular contraction.
- Rapidly conducting *Purkinje fibers* cause rapid activation of the ventricle.

- Cardiac muscle has a long *plateau* in the action potential to maintain a prolonged period of activation.
- The plateau (phase 2) is maintained by opening of slow voltage-gated calcium channels and closure of potassium channels.
- The plateau is preceded by *rapid depolarization* (phase 0) because of opening of voltage-gated sodium channels, and followed by *repolarization* (phase 3) because of the spontaneous closure of calcium channels and a reopening of potassium channels.

KEY POINTS (*continued*)

- Between beats (phase 4) the sodium and calcium channels are closed and potassium channels are open.
- Action potentials cause contraction by liberating calcium from the SR into the cytosol.

- Autonomic nerves modulate the heart rate. *Sympathetic stimulation increases* the rate of SA node firing and the rate of conduction through the AV node. *Parasympathetic stimulation decreases* the rate of SA node firing and the rate of conduction through the AV node.

CARDIAC MUSCLE IS SIMILAR TO OTHER MUSCLE TYPES

Cardiac muscle, like skeletal muscle, is a striated muscle and much of the mechanism of contraction is similar between the two muscle types. The electrophysiology of the two muscles differs dramatically, however. In skeletal muscle an action potential in the synaptic terminal of a motor neuron coming from the central nervous system releases a transmitter substance, acetylcholine, which triggers a very brief action potential on the muscle cell, causing it to contract. Skeletal muscle contracts in an all-or-none fashion and the force generated, for any given length, will be the same for every twitch (for a review, see Chapter 7). The action potential is so short in skeletal muscle that a single action potential generates an insignificant amount of force. Usable force can only be achieved by stimulating the fiber repeatedly with a train of neural discharges (temporal summation). Individual motor units can be stimulated within a muscle as a further means of control. Cardiac muscle differs in several important respects. First, the cardiac action potential is not initiated by neural activity. Instead, specialized cardiac muscle tissue in the heart itself initiates the action potential, which then spreads directly from muscle cell to muscle cell. Neural influences have only a modulatory effect on the heart rate. Second, the duration of the cardiac action potential is quite long. As a result, the full force of cardiac contraction results from a single action potential. The force of contraction is not the same for every beat of the heart and can be modulated by the cardiac nerves. Finally, all cells in the heart contract together as a unit in a coordinated fashion with every beat.

This chapter examines the electrical properties of the cardiac muscle cells. We begin by looking at the different types of cells in the heart. The properties and ionic mechanism of cardiac action potentials, the processes of excitation–contraction coupling, and the regulation of the heart rate will be explained.

EXCITATION ORIGINATES WITHIN THE HEART MUSCLE CELLS

Two broad types of cells are found within the heart: (1) *contractile* cells and (2) *conductile* cells. Contractile cells are the cells of the working myocardium and constitute the bulk of the muscle cells that make up the atria and the ventricles. An action potential in any one of these cells leads to a mechanical contraction of that cell. Furthermore, an action potential in one cardiac muscle cell will stimulate neighboring cells to undergo an action potential, such that activation of any single cell will be propagated over the whole heart. Conductile cells are specialized muscle cells that are involved with the initiation or propagation of action potentials but have little mechanical capability. The principal conductile cells are indicated in Fig. 1. Of critical importance is the *sinoatrial* (SA) node. The SA node (sometimes called the *sinus node*) lies in the right atrium near the vena cava. SA nodal cells generate spontaneous action

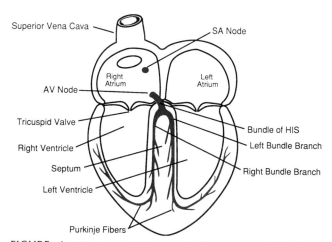

FIGURE 1   Structure of the conduction system of the heart. Structures colored blue are those responsible for generating and propagating the wave of excitation that leads to contraction of the heart. (Modified from Katz AM. *Physiology of the heart.* New York: Raven Press, 1992.)

potentials and act as the normal *pacemaker* for the heart. Because the SA node is located in the atria, action potentials will first be propagated over the atria, making them the first structures in the heart to contract.

Action potentials spreading across the atria eventually reach another conducting structure known as the *atrioventricular* (AV) node. The AV node is located in the interatrial septum between the ostium of the coronary sinus and the septal leaflet of the tricuspid valve. The AV node serves two important functions. One is to relay the wave of depolarization from the atria to the ventricles. A sheet of connective tissue associated with the atrioventricular valves separates the atria from the ventricles, and the AV node is the only conductive link between the atria and the ventricles. The second function is to delay the spread of excitation from the atria to the ventricles. The AV node cells are specialized to conduct very slowly. This delay permits the atria to eject their blood and therefore fill the ventricles before the latter begin to contract.

The fibers of the AV node give rise to fibers of the *AV bundle* (*common bundle* or *bundle of His*), which in turn divides into two major branches, the *left bundle branch* and the *right bundle branch*. These branches then divide into an extensive network of *Purkinje fibers*. The Purkinje fibers are conductile cells that conduct action potentials very rapidly. They are interwoven among the contractile cells of the ventricles and serve to quickly spread the wave of excitation throughout the ventricles. Because they conduct so rapidly, all muscle cells of the ventricles appear to contract in unison. It is important to emphasize that all of these conductile structures (e.g., SA node, AV node, Purkinje network) are not nervous tissue but, rather, specialized cardiac muscle cells.

EXCITATION IS CONDUCTED FROM CELL TO CELL THROUGH GAP JUNCTIONS

Figure 2 is a micrograph of adjacent contractile cells from the ventricle. Like all striated muscle, ventricular muscle has the typical pattern of cross striations because of its highly organized contractile

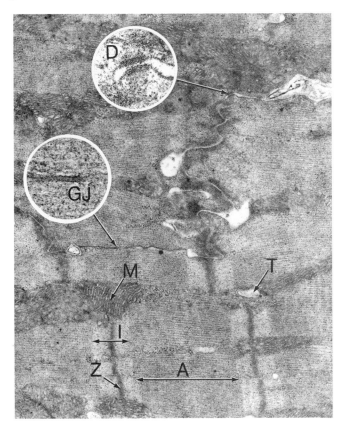

FIGURE 2 An electron micrograph of ventricular muscle. The sliding filaments display the I, A, and Z bands. A T tubule is indicated by *T* and the mitochondria are indicated by *M*. Detail *D* shows a *desmosome* on the transverse aspect of the intercalated disc between two myocytes. *Inset GJ* shows a gap junction on the longitudinal aspect of the intercalated disc. Electrical activation spreads from one cardiac cell to the next via gap junctions. (Modified from Goodman SR. *Medical cell biology*. Philadelphia: JP Lippincott, 1994.)

proteins (see Chapter 7 and Fig. 3). The region at which the cell membranes of adjacent myocytes adjoin is termed the *intercalated* disc. The myocytes are arranged in the heart in a staggered pattern, much like bricks in a wall. The intercalated discs occur at the junction between the myocytes and would be analogous to the mortar between the bricks. The discs run transversely where the ends of two adjacent myocytes abut. The disc then turns 90° and runs longitudinally between the myocytes until another end-abutment begins. The

---

Clinical Note

If the heart were to beat in response to motor nerve stimulation like a skeletal muscle, the transplant procedure would be virtually impossible because surgical reconnection of nerves is not feasible. The surgeon would not be able to attach the nerves and have a beating heart at the end of the operation. Fortunately, the transplanted heart brings its own pacemaker with it.

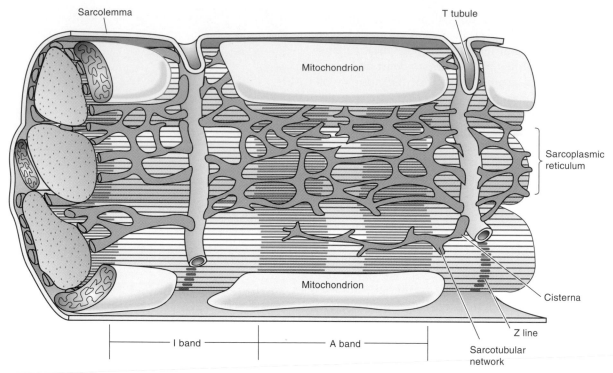

FIGURE 3 Ultrastructure of a contractile cell. A contractile cell in the heart is very similar to skeletal muscle in its basic cellular organization. (Modified from Katz AM. *Physiology of the heart.* New York: Raven Press, 1992.)

transverse aspect of the intercalated disc is filled with structures called *desmosomes*. The desmosomes make strong mechanical attachments between the cells and transmit the force of contraction. The transverse aspect as well as the longitudinally oriented region of the intercalated disc are rich in low-resistance connections between the cells called *gap junctions*. Small pores in the center of each gap junction allow ions and even small peptides to flow from one cell to another along their electrochemical gradients. Cardiac muscle gap junctions serve an identical function as the gap junctions at electrical synapses between nerve cells (Chapter 6) and nexuses in smooth muscle cells (Chapter 8). An action

potential in one muscle cell is propagated to adjacent muscle cells via direct *electrotonic propagation across the gap junctions*. The gap junctions cause every cell in the heart to be electrically coupled to its neighboring cells and that is what causes the heart to behave like a single motor unit. Theoretically, an ion inside an SA nodal cell could travel through every cell in the heart without ever having to enter the extracellular space.

The Purkinje cells are still muscle cells but contain fewer contractile proteins than contractile cells. They are specialized not for contraction but, rather, for fast electrical propagation. The large diameter of the Purkinje cells gives them a high conduction velocity.

Clinical Note

Any cell in the heart is in theory capable of a spontaneous action potential. The SA node happens to have the fastest intrinsic rate and, hence, is the pacemaker. If the AV node is injured by disease, conduction through it can be lost and ventricular cells will be isolated from the pacemaker in the atrium. In that case the cells with the next highest rhythm, usually in the bundle of His, will keep the ventricles

beating but at a much reduced rate. In this condition, called *complete heart block,* the atria and ventricles beat independently of one another. While the appearance of a ventricular pacemaker keeps the heart beating, the reduced ventricular rate is usually too slow to maintain a suitable cardiac output and the patients have severe symptoms such as fatigue and even loss of consciousness.

(Recall from Chapter 5 that the greater the diameter of a nerve axon, the greater the propagation velocity of nerve action potentials.) A Purkinje fiber has a particularly high density of gap junctions between its cells, and they will conduct action potentials at about 4 m/sec over the ventricle. Atrial and ventricular muscle cells on the other hand conduct at only 1 m/sec.

Cells of the SA and AV nodes, like the Purkinje fibers, also have a reduced quantity of contractile proteins. As discussed later, nodal cells lack fast sodium channels and that reduces the rate at which they depolarize. They also are much smaller than either the contractile or Purkinje cells. Both of these properties cause them to have a low conduction velocity. AV nodal cells also have a reduced density of gap junctions that even further depresses their conduction velocity to only about 0.05 m/sec. It is the low propagation velocity of the AV nodal cells that provides the delay between atrial and ventricular contraction.

The ultrastructural features of a typical cardiac muscle cell are illustrated in Fig. 3. At first glance, the ultrastructure appears much like a skeletal muscle cell (see Chapter 7). Common features include characteristic *A, I,* and *Z bands,* a *T-tubule system,* and *sarcoplasmic reticulum* (SR). Some subtle differences exist, however, between cardiac muscle and the skeletal muscle ultrastructure. One difference lies in the T tubules (which stands for "transverse" tubules because they are transversely oriented to the long axis of the cell). They are centered on the Z band with only one tubule per sarcomere. Mammalian skeletal muscle is modified to have two T tubules per sarcomere, which reduces the distance over which calcium must diffuse to reach the sliding filaments. This results in an extremely fast twitch that can provide an important survival benefit. Cardiac muscle does not require such a fast activation time and, hence, a single T tubule per sarcomere is sufficient.

The SR in cardiac contractile cells consists of two types of structures: (1) the *sarcotubular network,* making up the bulk of the SR, is in proximity to the contractile machinery, and (2) the *subsarcolemma cisternae,* the region at which the SR abuts the T tubules. The subsarcolemma cisternae are equivalent to the terminal cisternae of skeletal muscle cells.

Finally, cardiac myocytes contain a large quantity of *mitochondria* reflecting the aerobic nature of cardiac muscle metabolism. Most of the metabolic energy of the heart comes from oxidative metabolism of fatty acids and lactate, with glucose accounting for only a small fraction of the energy source. Although the heart can derive some energy by anaerobic glycolysis of glucose, it is not enough to keep up with the high energy demands of a beating heart. As a result, interruption of the heart's oxygen supply will cause a cessation of mechanical activity within less than 1 min and irreversible injury of the cells will begin within 20 min if the oxygen supply is not restored.

CARDIAC MUSCLE EXPERIENCES AN ACTION POTENTIAL

Figure 4 shows an action potential typical of what would be recorded from a contractile cell in the ventricle. Superficially, this action potential appears similar to the action potentials seen in nerve, skeletal, and smooth muscle cells (Chapters 6, 7, and 8). One major difference, however, is its duration. Note that the action potential remains in a depolarized state for about 300 msec (almost a third of a second) giving it a shape like a *plateau.* In contrast, an action potential in a nerve or skeletal muscle cell lasts only about 1 msec. It is the plateaued action potential that keeps the heart activated long enough to develop a forceful contraction from a single action potential. The rapid phase of depolarization is termed *phase 0.* Phase 1 is a trasient repolarization following the overshooting of Phase 0. Phase 1 is followed by a long period during which the membrane potential remains depolarized, the *plateau phase* or *phase 2.* After the plateau phase, a phase of repolarization occurs during which the membrane potential returns to its resting level. This phase is termed *phase 3.* The resting potential between beats is referred to as *phase 4.*

CARDIAC MUSCLE EXHIBITS BOTH DIVERSITY AND SPECIALIZATION

Just as there is morphologic diversity and specialization of individual cells in the heart, there is also electrical

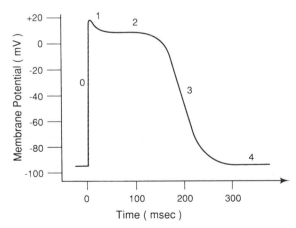

FIGURE 4   Phases of the action potential in a contractile cell. The action potential has 5 distinct phases.

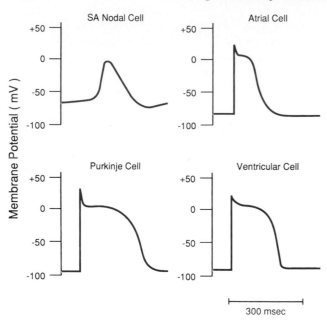

FIGURE 5 Diversity of action potentials in four difference cardiac cells. The action potentials in difference regions of the heart have characteristic features. Note the significant differences between an action potential in an SA nodal cell and an action potential in a ventricular cell. (Modified from Katz AM. *Physiology of the heart.* New York: Raven Press, 1992.)

diversity and specialization. Figure 5 illustrates action potentials recorded in several different regions of the heart. Action potentials in contractile and Purkinje cells are associated with a very rapid depolarizing phase and a broad plateau region. Both the rapidity of depolarization as well as the plateau phase are less prominent in action potentials recorded from nodal cells. None of these cells experiences a hyperpolarizing afterpotential, as is the case for action potentials in nerve and skeletal muscle cells. Finally, note that the phase 4 baseline is normally stable in contractile and Purkinje cells, but is not stable in the SA nodal cells.

PROLONGED OPENING OF CALCIUM CHANNEL CAUSES THE PLATEAU PHASE OF THE ACTION POTENTIAL

In a nerve cell or a muscle cell, a rapid increase in Na^+ permeability (actually conductance, because we are talking about charge movement) is seen that depolarizes the cell, followed by an increase in K^+ conductance that repolarizes it. The peculiar shape of the cardiac action potential suggests that the ionic mechanisms are different in heart cells, as indeed they are. Figure 6A shows an action potential from a normal contractile cell, whereas Fig. 6B shows the action potential after treating the cell with tetrodotoxin (TTX). TTX selectively blocks

the *voltage-gated fast sodium channels.* TTX does not completely block the action potential, as it does in nerve or muscle; rather, it only attenuates the rate of phase 0 depolarization. Clearly, a sustained increased Na^+ conductance is not what prolongs the duration of the cardiac action potential.

The long plateau phase is actually due to a prolonged increase in Ca^{2+} conductance through *voltage-dependent calcium channels,* often referred to as *L-type calcium channels.* These channels are similar to those seen in the presynaptic terminal (Chapter 6) or in smooth muscle (Chapter 8). An increase in Ca^{2+} permeability will depolarize the cell just as an increase in Na^+ permeability does. The concentration of Ca^{2+} is much higher outside the cell than inside, with an equilibrium potential of about +120 mV. Therefore, if the membrane were freely permeable to Ca^{2+}, the voltage inside the cell would then approach +120 mV.

Compare the action potential in the TTX-treated contractile cell of Fig. 6 to the nodal cells in Fig. 5. The slow phase 0 and lack of overshoot in the nodal cells is due to the fact that fast sodium channels are not expressed in these cells. The action potential in nodal tissue is maintained entirely by the L-type calcium channels. As expected, TTX has no effect on the action potential in nodal cells.

PROLONGED REFRACTORY PERIODS PREVENT THE HEART FROM BEING TETANIZED

The fast sodium channels trigger action potentials in contractile and Purkinje cells. They have actually two gates that control passage of Na^+ through the channel. The *activation gate* normally blocks the channel. Because it is voltage-gated it opens only when the membrane potential reaches threshold creating phase 0. Shortly after the activation gate opens a second part of the molecule, the *inactivation gate* closes the channel, causing phase 0 to end. The inactivation gate is voltage sensitive and does not reopen until the membrane repolarizes. Only when the inactivation gate finally

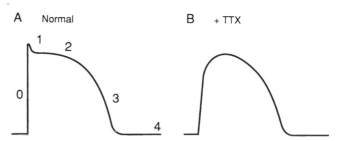

FIGURE 6 Effects of sodium channel blocker, tetrodotoxin (TTX), on action potential in a contractile cell. TTX affects the early phases (phases 0 and 1), but has little effect on phases 2 and 3.

opens will the cell be receptive to restimulation. This causes an *absolute refractory period* all through phase 2 and into phase 3, making it impossible to tetanize cardiac muscle. The refractory period ensures that the heart will relax between beats, allowing it to refill with the blood that will be pumped on the next beat. As the inactivation gates begin to open, the heart goes through a *relative refractory period* late in phase 3 where it can be restimulated but the threshold required for stimulation is elevated. Also an attenuated contraction will result because not all cells have their inactivation gates open yet. In nodal tissue lacking fast sodium channels the L-type calcium channels are also refractory to restimulation well into electrical diastole.

As the wave of excitation finishes traversing the ventricles, the entire heart is in a refractory period, causing the wave to die out and the heart to relax. Sometimes conduction can be slowed in a diseased region to the point where the rest of the heart is out of the refractory period by the time the wave emerges from the depressed segment. When that occurs the heart can be restimulated by this delayed impulse causing a *reentrant rhythm,* in which the cycle repeats itself over and over. Under those conditions the heart may beat very rapidly or even fibrillate.

IONIC FLUXES IN THE CARDIAC MUSCLE CELLS

Figure 7 summarizes some of the key changes in ion permeability that contribute to the cardiac action potential. The upper trace shows a typical cardiac action potential and the lower traces illustrate the changes in permeability to Na^+, K^+, and Ca^{2+}. As a consequence of a depolarizing stimulus, a voltage-dependent increase in Na^+ conductance occurs. A regenerative cycle is initiated, which tends to depolarize the cell toward the equilibrium potential of Na^+. This regenerative increase in Na^+ conductance rapidly moves the membrane potential to its peak level of +20 mV. Thus, the voltage-dependent increase in

Na^+ conductance underlies phase 0 and 1 of the action potential. The inactivation gates in the sodium channels spontaneously close within a millisecond or two after activation gate opening. Although the cell remains depolarized, Na^+ conductance quickly returns to its resting level.

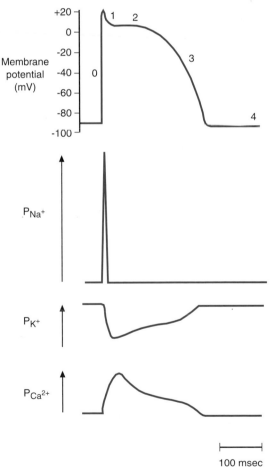

FIGURE 7 Simplified diagram of the sequence of some of the permeability changes contributing to the action potential of contractile cells. (Modified from Katz AM. *Physiology of the heart.* New York: Raven Press, 1992.)

Changes in Ca^{2+} permeability are shown in the bottom panel of Fig. 7. The initial depolarization of the membrane causes Ca^{2+} conductance to increase in a voltage-dependent manner. Unlike the sodium channels, the calcium channels are slow to close and there is a sustained increase in Ca^{2+} conductance, which maintains the plateau of the cardiac action potential. Another factor contributing to the plateau of the cardiac action potential is the change in K^+ conductance (Fig. 7). At rest, the K^+ conductance of cardiac cells is high, just as it is in nerve and skeletal muscle cells. The similarity, however, ends here. When nerve and muscle cells repolarize, a transient increase is observed in K^+ conductance over the resting value, which quickly terminates the action potential and momentarily hyperpolarizes the cell. In contrast, in cardiac muscle K^+ conductance decreases with depolarization (Fig. 7), and during repolarization K^+ conductance simply returns to its phase 4 value. That is why cardiac muscle has no hyperpolarizing afterpotential. We should mention that there are many different types of potassium channels in the cardiomyocyte and their individual functions in controlling K^+ conductance through the action potential are complex. A detailed description of these is beyond the scope of this chapter.

EXCITATION–CONTRACTION COUPLING IS ACCOMPLISHED BY CALCIUM IONS

Cardiac muscle cells, like skeletal muscle cells, have an extensive SR (see Fig. 3). Excitation–contraction coupling in cardiac muscle cells is similar to that in skeletal muscle cells. Specifically, an action potential travels down the T tubules and causes the release of Ca^{2+} from the SR, which in turn activates the contractile machinery. Three pools of Ca^{2+} are important to the cardiac muscle cell, as shown in Fig. 8: (1) in the extracellular fluid, (2) in the SR, and (3) in the cytoplasm. Only the cytoplasmic pool is available to bind with the troponin-binding sites and initiate contraction. Consider the consequences of an action potential. Ca^{2+} entry through the sarcolemma increases the concentration of Ca^{2+} in the cytoplasm. Because the amount of Ca^{2+} entering is relatively small, it accounts for only a fraction of the activation of the contractile proteins and, hence, is indicated by a dashed line in Fig. 8. In skeletal muscle, action potentials on the T tubules electrically stimulate the SR to release calcium. That is not the case in cardiac muscle however. Rather the small amount of Ca^{2+} entering the cell through the L-type calcium channels actually triggers the release of the sequestered Ca^{2+}

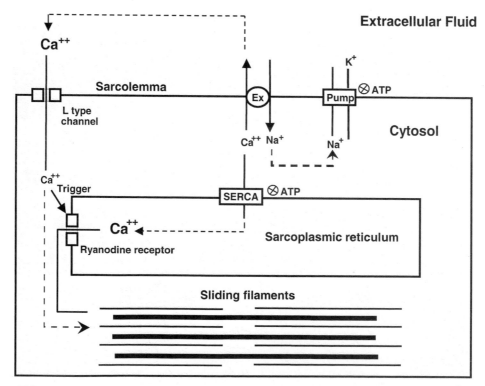

FIGURE 8   Simplified model of calcium handling in cardiac muscle cells. EX, Na^+-Ca^{2+} exchanger. See text for details.

within the SR. The trigger calcium acts at *ryanodine receptors* located on the SR. These are actually calcium-gated calcium channels that open in the presence of cytosolic calcium. The name derives from the original observation that they bind the toxin ryanodine, which blocks these release channels in the SR and hence contraction. The importance of this trigger calcium is evidenced by the fact that heart muscle will not contract when the influx of Ca^{2+} across the sarcolemma is blocked, even though adequate stores of Ca^{2+} are still present in the SR.

RELAXATION IS ACCOMPLISHED BY REMOVING Ca^{2+} FROM THE CYTOSOL

In cardiac muscle, as in skeletal muscle, there is a *Ca^{2+} pump* in the SR membranes termed the *sarco-endoplasmic reticulum Ca^{2+}-ATPase (SERCA)*. SERCA (Fig. 8) removes Ca^{2+} from the contractile pool and pumps it into the SR. When enough Ca^{2+} is removed from the cytosol, the muscle relaxes. With each action potential some additional Ca^{2+} moves into the cell during phase 2. After a period of time, the intracellular Ca^{2+} concentration, if left unchecked, would be the same as that outside. Clearly, a mechanism is needed to remove Ca^{2+} from the cell. The *Na^+-Ca^{2+} exchange system* in the sarcolemma is primarily responsible for removing calcium from the cytosol. The exchanger, which is not a pump, derives its energy from the Na^+-K^+ pump. The exchanger will pass three Na^+ ions in one direction for one Ca^{2+} ion in the other direction, but, being an exchanger rather than a pump, will do so only along a favorable energy gradient. Because the Na^+-K^+ pump maintains a strong transmembrane gradient for sodium, any free Ca^{2+} in the cytosol will be favorably exchanged for three Na^+ ions in the extracellular fluid. There are also true calcium pumps in the sarcolemma, but they account for only a small percentage of the calcium flux. At a steady state, the exact same amount of Ca^{2+} that entered the cell is removed with each beat. In this system the sarcolemmal exchanger and the SERCA compete for cytosolic Ca^{2+}

during diastole. The amount of Ca^{2+} available for release in the SR is therefore dependent on the net outcome of this competition.

STRENGTH OF CONTRACTION CAN BE MODULATED IN CARDIAC MUSCLE

The force generated by cardiac muscle cells depends on the cells' *contractility,* sometimes called the *inotropic state. An increased contractility means that for any given length of the muscle it contracts with a greater force.* It is important to note that the amount of Ca^{2+} released from the SR with each action potential is not sufficient to fully cover all of the troponin binding sites and thus activate all of the contractile proteins. Therefore, any manipulation that leads to enhanced release of Ca^{2+} from the SR will result in more troponin binding sites being occupied by Ca^{2+} and hence more force being generated by the muscle. The Ca^{2+} fluxes can, in turn, be altered by a variety of physiological control systems. Three commonly encountered modulators of contractility will be considered next: (1) stimulation frequency, (2) catecholamines, and (3) cardiac glycosides.

Stimulation Frequency

When the frequency of contraction increases, so does the tension generated by each contraction. This phenomenon, known as the *positive staircase effect,* was first described by Bowditch, an early cardiovascular physiologist, and is illustrated in Fig. 9. When Bowditch stimulated the heart, he found that a different tension was produced for each rate of stimulation. Stimulation at low rates was associated with a low tension and when the frequency of the stimulation was increased, the tension increased as well. The tension associated with a new rate of stimulation is not achieved instantaneously; it instead takes a period of time and appears to develop in a stepwise fashion. Hence the name "staircase." The positive staircase effect is an example of a *positive inotropic effect,* because it is associated with increased

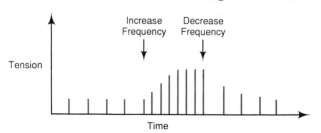

FIGURE 9 Positive staircase effect. As the frequency of stimulation increases (*first arrow*), the tension produced by each contraction increases. Several contractions at the new rate occur before the new steady state is reached. When the frequency is returned back to its starting level (*second arrow*), the tension produced by each contraction gradually returns to the initial level.

contractility. The model of excitation–contraction coupling illustrated in Fig. 8 can explain the positive staircase phenomenon. Increasing the heart rate increases the number of action potentials per unit of time and, thus, the rate of Ca^{2+} influx from the extracellular compartment. At the same time, it shortens the duration of diastole when the exchanger removes Ca^{2+}. As a result, more cytosolic Ca^{2+} is pumped into the SR than is removed by the Na^+-Ca^{2+} exchange mechanism, causing more Ca^{2+} to be available for release from the SR with each beat. Because the new equilibrium will take several beats to be reached, the increased contractility is seen to increase stepwise.

Catecholamines

When the catecholamines *norepinephrine* and *epinephrine* secreted by the sympathetic nerve endings

and by the adrenal medulla bind to the heart's β_1-*adrenergic receptors*, they exert a profound effect on the cardiac its contractile state. Receptor binding, as shown in Fig. 10, causes *adenylyl cyclase* to make *cyclic adenosine monophosphate* (cAMP), which in turn activates *cAMP-dependent kinase* (PKA). Catecholamines exert their effect by influencing both of the sarcolemmal and SR Ca^{2+} fluxes. PKA phosphorylates the L-type calcium channels, which causes more Ca^{2+} to enter the cell with each action potential. If more Ca^{2+} enters the cell, more will be available to be pumped into the SR, causing more to be available for release with subsequent action potentials (the same basic argument was used to explain the positive staircase effect discussed earlier). The second action of catecholamines is to increase the activity of SERCA. This is accomplished when PKA phosphorylates a protein called *phospholamban* on the SR. If the activity of SERCA is increased, a greater portion of the cytosolic Ca^{2+} will be pumped into the SR, making a greater amount available for release by a subsequent action potential. Catecholamines also tend to shorten phase 2 and thus shorten the duration of systole.

Inhibition of the Sodium Pump by Cardiac Glycosides

Cardiac glycosides such as *digitalis* were traditionally used to treat *congestive heart failure*. The rationale for their use as a therapeutic agent at first may seem questionable, because at the molecular level cardiac glycosides block the Na^+-K^+ pump in all of the body's

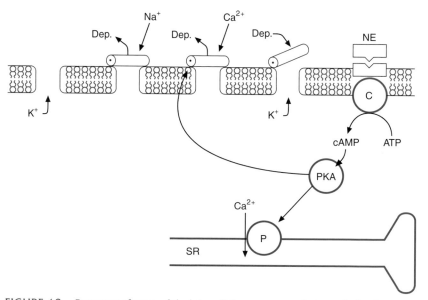

FIGURE 10 Summary of some of the intracellular messengers that couple the β-receptor to L-type calcium channels. *P*, phospholamban; *C*, adenylyl cyclase; *Dep.*, indicates voltage-dependent channel; *PKA*, cyclic AMP-dependent protein kinase; NE, norepinephrine.

cells, including those in the heart, and are therefore potent poisons. The key strategy in their use is to carefully adjust the dose so that only a partial block of the Na^+-K^+ pump occurs. When this is done, cardiac glycosides lead to enhanced contractility of the heart by the following mechanism. Due to decreased activity of the Na^+-K^+ pump, the intracellular concentration of Na^+ in cardiac cells increases, decreasing the driving force for Na^+ to enter the cell. Because the Na^+ gradient is the energy source for the Na^+-Ca^{2+} exchanger, the latter will remove less Ca^{2+} from the cell. Intracellular levels of Ca^{2+} will increase, with more Ca^{2+} available to be pumped into the SR.

PACEMAKER CELLS CONTROL THE HEART RATE

The SA node is the normal pacemaker of the heart. The action potentials in the SA node are somewhat different from the action potentials in contractile cells. The most obvious difference is that the phase 4 resting potentials are unstable, as shown in Fig. 11. Note that the resting potential starts from its maximum value of about −60 mV, then slowly depolarizes until it reaches threshold and undergoes a regenerative action potential. On completion of the action potential, the cell again begins to depolarize and another action potential is initiated. This process occurs about 60–100 times each minute, resulting in the cardiac rhythm. The progressive depolarization of the phase 4 potential is known as the pacemaker potential. The ionic mechanism underlying the pacemaker potential is not fully understood, but it is thought to be due to an inward sodium leakage current.

Cells in the AV node also have pacemaker potentials but they are slower (only about 40 beat/min) and thus the SA node dominates with its faster rhythm. An isolated strip of Purkinje fibers will spontaneously generate action potentials with a frequency of about 25 action potentials per minute, a rate slower than that of either SA or AV nodal cells. The AV node and Purkinje fibers are called latent pacemakers because they will assume the pacemaker role should the signal from the SA node be interrupted. Atrial and ventricular cells have virtually no pacemaker activity.

AUTONOMIC NERVES MODULATE PACEMAKER ACTIVITY

The sympathetic and parasympathetic divisions of the autonomic nervous system have profound influences on the heart rate. The transmitter substance of the parasympathetic division, acetylcholine (ACh), reduces the heart rate, a negative chronotropic effect, whereas the transmitter substances of the sympathetic division, epinephrine and norepinephrine, increase the heart rate, a positive chronotropic effect. The dashed line in Fig. 11 illustrates the effect of ACh on the SA node. ACh binds to G_i-coupled muscarinic cholinergic receptors, which results in an increase in K^+ conductance.

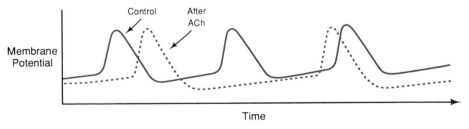

FIGURE 11   Effects of ACh on SA nodal cells. ACh hyperpolarizes nodal cells; thus, it takes a longer time for the pacemaker potential to reach threshold. The rate of rise of the pacemaker potential is also slowed. Consequently, the interval between action potentials is increased. (Modified from Katz AM. *Physiology of the heart.* New York: Raven Press, 1992.)

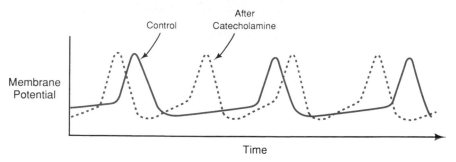

FIGURE 12   Effects of catecholamines on SA nodal cells. The catecholamines epinephrine and norepinephrine increase the rate of rise of the pacemaker potential so that it takes a shorter time for the pacemaker potential to reach threshold. Consequently, the interval between action potentials is decreased, increasing the heart rate. (Modified from Katz AM. *Physiology of the heart.* New York: Raven Press, 1992.)

The increase in K$^+$ conductance hyperpolarizes the cell so that it takes a longer time to reach threshold. ACh also slows the rate of rise of the pacemaker potential itself. Both of the preceding effects tend to decrease the firing rate. If the vagus nerve is intensely stimulated, the cells in the SA node will be so hyperpolarized that the pacemaker potential will not reach threshold to fire the cells and the heart will stop beating. After a short time, however, the heart will resume beating with *escape beats.* The escape rate is much slower because the latent pacemakers in the AV node or the Purkinje network have now taken control.

Catecholamines from the sympathetic nerves increase the heart rate. The dashed line in Fig. 12 shows a recording made after the addition of a catecholamine. Note that the catecholamine increases the rate at which the pacemaker potential approaches threshold. The ionic mechanism causing this increased slope in the pacemaker potential is thought to be an increased Na$^+$ conductance. For many years the channels responsible for this current were not fully characterized and so the current became known as the "funny" current, I_f. It is now thought that I_f occcurs through the hyperpolarizing-activated cyclic nucleotide-gated cation channel (HCN).

Transmitter substances of the autonomic nervous system also affect the cell-to-cell conduction velocity in both the atria and the ventricles. ACh slows the rate of propagation, and catecholamines increase the rate of propagation. Intense vagal stimulation can easily depress conduction through the AV node to the point at which conduction to the ventricle fails, a condition called *heart block.* The ionic mechanisms that underlie these changes in propagation velocity are still not well understood.

Suggested Readings

Katz AM. *Physiology of the heart.* Philadelphia: Lippincott Williams & Williams, 2001.

Sperelakis N, ed. *Heart physiology and pathophysiology.* London: Academic Press, 2001

The Electrocardiogram

JAMES M. DOWNEY

KEY POINTS

- Deflections in the ECG correspond to electrical events on the heart. Action potentials on cardiac muscle cells create extracellular voltages. The thorax acts as a volume conductor so that the voltages generated by the cells are conducted to the body surface.
- The *P wave* represents atrial depolarization; the *QRS complex* represents ventricular depolarization; the *T wave* represents ventricular repolarization.
- The *PR interval* is influenced by the conduction time in the AV node.
- The *QT interval* is influenced by the duration of ventricular action potentials and conduction time over the ventricle.

- There are six limb leads, which can be used to calculate the electrical vectors of the waves in the frontal plane.
- *Unipolar leads* are used to explore the chest, which gives front-to-back information about the electrical vectors.
- *Bundle branch block* or *hypertrophy* of a chamber can alter the angle of the QRS wave's electrical vector, which is normally in the left lower quadrant.
- *Myocardial ischemia* produces a current of injury that moves the ST segment off of the baseline.

KEY POINTS (*continued*)

- The heart rate (in beats per minute) can be determined by multiplying the reciprocal of the R-R interval (in seconds) times 60.
- Failure of conduction through the AV node causes heart blocks.

- Ectopic foci in the atria or ventricles cause premature contractions of the heart.
- Electrical impulses can reenter to cause fibrillations.

ACTION POTENTIALS CREATE EXTRACELLULAR VOLTAGES

Action potentials from the heart's myocytes cause minute voltages to occur on the surface of the thorax, which can be measured with a machine called an *electrocardiograph*. Normally, transmembrane voltages are measured with a *microelectrode* placed inside the cell. However, *extracellular electrodes* can also detect action potentials within a tissue as illustrated in Fig. 1. Imagine the rectangle to represent a block of cardiac muscle containing many cells. One electrode is placed at one end of the muscle, and another electrode at the opposite end. These electrodes are connected to an amplifier that measures the difference in potential between the two electrodes. One electrode is labeled "positive" and another is labeled "negative." The output of the amplifier is connected to a pen recorder, such that an upward deflection occurs whenever the voltage on the electrode marked "positive" is more positive than that on the electrode marked "negative." Similarly, a downward deflection indicates that the voltage on the "positive" electrode is negative with respect to that at the "negative" electrode.

In the resting state (Fig. 1, panel A1), all of the cells in the muscle strip have a transmembrane voltage difference of about 90 mV with the extracellular region positive with respect to the intracellular region. Thus, at rest, the electrode on the left senses the same positive potential as the electrode on the right so that the difference between the two is zero and no deflection is recorded by the pen recorder (panel B, point A1).

Now consider the consequence of stimulating the strip near the left electrode, causing the cells in that region to depolarize, as shown in Fig. 1, panel A2. With depolarization, positive charge enters the cells, leaving the extracellular space negatively charged. At this time, the right electrode becomes more positive than the left electrode, and the pen is deflected in the positive direction, as illustrated in Fig. 1, panel B, point A2. As the depolarization is conducted from cell to cell,

the entire strip is eventually depolarized, as shown in panel A3. At this time, there is again no potential difference between the electrodes, and the pen returns to the center, as shown in panel B, point A3. Soon, the cells at the left side will begin to repolarize, returning the extracellular fluid in that region to a positive state, as shown in panel A4. At that time, the electrode on the left finds itself more positive than the electrode on the right so that the pen deflects downward, as shown in panel B, point A4. Finally, when the entire strip repolarizes, all potential differences disappear and the pen again returns to the center position.

The following two points are important to remember:

1. Extracellular electrodes experience a potential difference only when part of the heart is in a depolarized state and part of the heart is polarized.
2. The tissue ahead of a wave of depolarization is positively charged and that behind the wave of depolarization is negatively charged.

THE THORAX ACTS AS A VOLUME CONDUCTOR

The tissues of the body are bathed in an ionic fluid that is conductive to electricity. Therefore, the thorax becomes a *volume conductor* in which currents are free to flow in three dimensions. This can be contrasted with a *linear conductor* such as a wire in which current flow is limited to movement along its length. Current flows within a volume conductor whenever a *dipole* exists within the volume conductor. A dipole can be defined as a separation of charge in space. It is a vector quantity whose magnitude is the charge difference times the distance between the two points of charge (poles). Its direction is determined by the orientation of a line between the two poles. An ordinary flashlight battery is a good example of a dipole. One end is positively charged and the other end is negatively charged. If the battery were to be suspended in a tank of saltwater, current would flow from the positive end to the negative

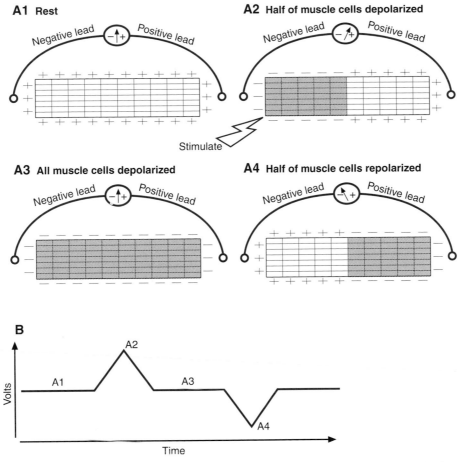

FIGURE 1   Potentials generated from an isolated strip of ventricular muscle at sequential time periods. (A1) The muscle at rest. The small blocks represent individual muscle cells within the strip of muscle tissue. (A2) Half the fibers in the strip activated. (A3) All fibers activated. (A4) Half the fibers repolarized. (B) ECG record from the muscle strip depicted in panel A. The labels (points A1–A4) on the trace indicate the potential corresponding to the time periods depicted in panels A1, A2, A3, and A4, respectively.

end, setting up a voltage field in the water around the battery. Imagine a plane through the center of the battery perpendicular to its long axis. All the water on the positive pole-side of that plane would have a positive voltage, whereas everything on the negative pole-side would have a negative voltage. The voltage within the plane would be zero; hence, it is called the *isoelectric plane*. Finally, the voltage fields become weaker as the distance from the battery increases because the current paths tend to be most concentrated near the battery.

When part of the heart is depolarized (a region at which the extracellular fluid has a negative charge) and part is polarized (a region in which the extracellular fluid has a positive charge), then a dipole is created within the heart and current will flow between the two regions. The resulting electrical field can be detected on the surface of the chest with electrodes, and the intensity of the voltage will depend on the orientation of the electrodes with respect to that of the dipole. The amplitude of the

dipole, as might be expected, will be proportional to the mass of the tissue involved.

DEFLECTIONS IN THE ECG CORRESPOND TO ELECTRICAL EVENTS IN THE HEART

The record of the heart's electrical activity from the surface of the body is called an *electrocardiogram* (ECG). A typical ECG trace is shown in Fig. 2. For this recording, the positive electrode was placed on the left shoulder and the negative electrode was placed on the right shoulder, which is the standard *lead I* electrode placement. The activation sequence of the heart begins with action potentials within the *sinoatrial (SA) node*. No contribution from the firing of action potentials in the SA node can be seen in the ECG tracing because the mass of the SA node is too small. After activation of the SA node, action potentials propagate across the atria.

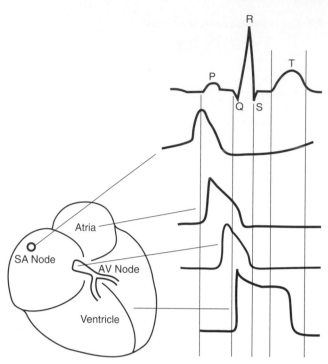

FIGURE 2 The ECG. Major waves (P, QRS, and T) of the ECG are indicated as well as the timing of action potentials from the heart's key conductile structures.

The atria are large enough to produce a low-amplitude signal called the *P wave*. The P wave is the first deflection of the ECG and represents depolarization of the atria. The termination of the P wave does not represent relaxation of the atria (a common misconception), but rather the time when all the muscle cells in the atria are depolarized and contracting (analogous to panel A3 in Fig. 1). Thus, the duration of the P wave indicates the speed of propagation of action potentials through the atria.

Soon after action potentials begin to spread across the atria, they reach the *atrioventricular (AV) node*. Action potentials propagate through the AV node and the bundle of His very slowly; this continues even after the entire atrium is depolarized. There are no potential changes in the ECG during this period because of the small size of the AV node/common bundle apparatus; therefore, the deflection briefly returns to baseline after the P wave.

The *QRS complex* represents the propagation of the wave of depolarization across the ventricles. If the trace begins with a negative deflection it is called a *Q wave*. The Q wave is absent if the first deflection is positive rather than negative. The first positive deflection, if one is present, is named the *R wave*. If a negative deflection follows an R wave, then it is called an *S wave*, even if it is the first negative deflection. As explained later, the shape of the QRS complex depends on where the

electrodes are attached. As a result, a trace may have only an R wave or only a Q wave comprising the entire QRS complex. The QRS complex normally has a greater voltage than the P wave because the mass of the ventricles is much larger than that of the atria. *The QRS complex is also shorter than the P wave (< 0.1 sec) because of the presence of the Purkinje network, which rapidly propagates the wave of excitation to all the cells in the ventricles.* The QRS complex terminates when all cells in the ventricles are depolarized.

After the QRS complex the deflection again returns to baseline, indicating no potential differences. This, of course, is because the entire ventricle is depolarized and all action potentials are in their phase 2 plateau phase. As the ventricle begins to repolarize, the *T wave* is created. Figure 1 would lead us to predict that the T wave should be a mirror image of the QRS complex, but instead it usually has the same polarity as the QRS complex. That is because the repolarization, unlike depolarization, is not conducted from cell to cell. Rather, each cell repolarizes independently whenever it is ready to do so. Depolarization begins at the endocardium, where the Purkinje fibers are located, and spreads to the epicardium. For an unknown reason, action potentials in the subepicardium are shorter in duration than those in the subendocardium. Thus, repolarization appears to progress in a reverse fashion from the epicardium back toward the endocardium; hence, the upright T wave. The termination of the T wave represents the point in time when all the cells in the ventricles have returned to their resting potential and the heart is fully relaxed. There is no component in the ECG that corresponds to the repolarization phase of the atria. The repolarization of the atrial cells is disorganized and sets up negligible net voltage. The voltage between beats is referred to as the *baseline* or sometimes the *TP segment* and should be a flat line.

SEGMENTS ARE VOLTAGES AND INTERVALS ARE TIMES

The student should distinguish between *segments* and *intervals* in the ECG tracing. A segment refers to a part of the ECG where the tracing is expected to be relatively flat and is measured in terms of voltage (its vertical position). Intervals indicate time periods and are measured in seconds on the horizontal scale. For example, the *ST segment* is the flat region after the end of the QRS complex and the start of the T wave. It is normally at the same voltage as the baseline. The *QT interval* is defined as the time interval between the beginning of the QRS complex and the end of the T wave. It is typically about 0.3–0.4 sec in duration. The *R-R interval* is the time interval between the R waves in two consecutive beats. Heart rate (in beats per minute) can be calculated from the ECG by dividing 60 by the R-R interval in seconds.

ATTACHING THE ELECTRODES TO THE PATIENT

The thorax can be considered to be an equilateral triangle, with the two shoulders and pubis representing the apices and the heart in its center. When electrodes are attached to the wrists and ankles, the limbs act as extensions of the wires such that the voltage at the wrist (or ankle) is identical to that at the shoulder (or pubis). The right leg is always attached to an electrical "ground" to reduce electrical interference. The electrocardiograph is normally calibrated to deflect the pen 1 cm vertically for each millivolt difference between the electrodes. The paper moves at 25 mm/sec and a 1-mm grid is printed on the paper. Thus, each small box is 0.1 mV high and 0.04 sec wide. The electrocardiograph can record the voltage differences between any two electrodes with the aid of an electrical switch. When *lead I* is selected, voltages are being measured between the left and right arms, as shown in Fig. 3. *Lead II* is between the left foot and the right arm and *lead III* is between the left foot and the left arm. Notice that each of these leads is oriented at 60° to the other two leads. The other three leads in the figure, aV_L, aV_R, and aV_F are explained later in the chapter as are the chest leads, $V_1–V_6$.

VECTORS CAN BE DETERMINED FROM THE LIMB LEADS

As action potentials are conducted through the heart, the resulting dipoles are constantly changing in both magnitude and orientation. Because of the physical orientation of the limb leads, they are sensitive to the orientation of the dipoles generated by the heart. Consider a vertically oriented dipole whose positive pole points down toward the patient's foot. This would cause the pubis to be positively charged with respect to either shoulder. Thus, the pen would be deflected upward in leads II and III because the positive lead of the ECG is attached to the left foot in both of those leads. On the other hand, lead I would show no deflection because this lead is oriented at right angles to the dipole and hence parallel to the dipole's isoelectric plane.

Figure 4 shows the wave of conduction spreading across the heart at eight time points. As the wave of conduction spreads, it sets up a dipole in the heart that is the net sum of all potential differences in the heart at that moment. Note how the orientation and magnitude of the dipole are continuously changing throughout the cardiac cycle. Each lead will be influenced by the vector component of the dipole that is parallel to that lead at each time point. Conversely, we could reverse the process and reconstruct the entire history of the heart's electrical dipole by simply having a computer analyze the ECG traces from several leads and determine the dipole's magnitude and orientation at all times throughout the cardiac cycle. This is the basis of *vector cardiography*. However, such an intricate analysis has been found to be of little diagnostic value. It is easier and more useful to just determine the *mean electrical axis* of the ventricles.

The mean electrical axis is simply the average direction of the dipole throughout the QRS complex. It can easily be determined by estimating the area under the QRS complex in any two leads and plotting those quantities on the triaxial system shown in Fig. 5. A number of approaches exist for estimating the area under a QRS complex, but the simplest is to assume that the areas are proportional to the height of the deflections. One then simply sums all positive deflections and subtracts all negative deflections from a given trace to determine the net deflection. The units for measuring the deflections are not important because only the angle of the vector—not its magnitude—is of interest. When plotting vectors one must be careful to observe the polarity, however. Deflections above the baseline are plotted toward the positive end of the lead and those below the baseline toward the negative end of the lead. Refer to Figs. 3 and 5 to determine the polarity of each lead configuration.

Figure 5 shows how to plot a vector on the triaxial system. For each of the three sides of the triangle, draw a line perpendicular to that side bisecting it at its midpoint (see dashed lines in Fig. 5). Do this for all three sides. All three lines should intersect at the center of the triangle as shown in Fig. 5. That intersection is the tail of the vector. Next plot the magnitude of the deflection for each of the three leads along the leg of the triangle corresponding to that lead. Start at the center of the leg and measure a distance along it equal to the net deflection of the QRS complex for that lead and make a mark at that point. If the deflection is positive, make the measurement toward the positive end of the leg (+). If the deflection is negative, the measurement should be toward the negative end (−). Then on each of the three legs draw a second perpendicular line going through the mark on that leg. Where those three new lines intersect is the head of the vector. Because all three lines should converge at a single point, all that is really needed to plot a vector is a calculation from any two leads.

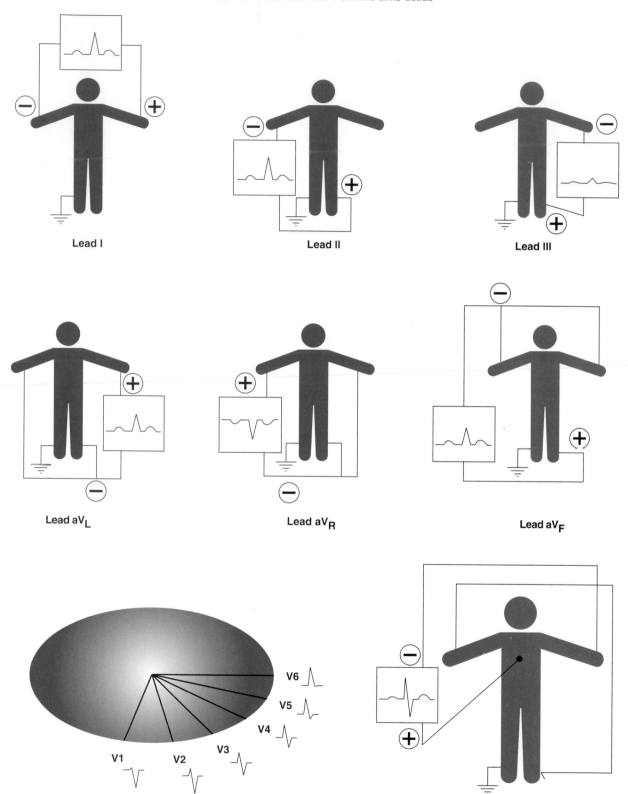

FIGURE 3   A diagram of the 12-lead placements commonly used for a diagnostic ECG. The lower left diagram represents a cross-sectional view of the chest as seen from above with the anterior surface of the chest is at the bottom.

Progression of depolarization

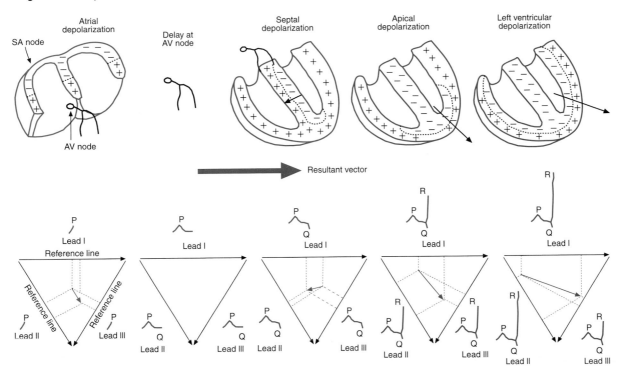

End of depolarization and repolarization

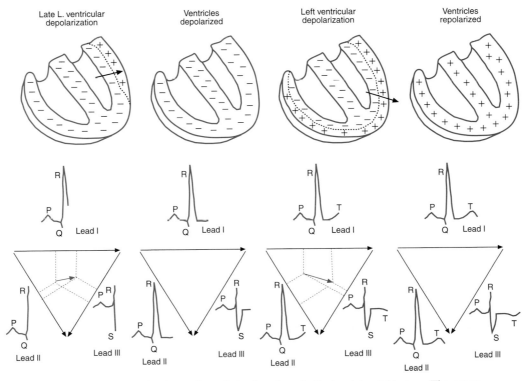

FIGURE 4   Progression of depolarization and repolarization through the heart in eight steps. The arrow indicates the resultant dipole at each step with its length indicating its magnitude. The influence of that dipole on leads I, II, and III is illustrated as well.

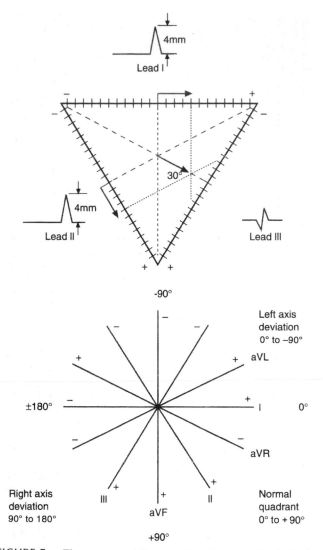

FIGURE 5 The upper panel illustrates how the mean electrical axis can be determined by plotting the deflection of any two leads vectorially. Only leads I, II, and III are shown for simplicity. Note that lead III shows a QRS complex that is of net-zero voltage and thus has a vector of zero magnitude. A resultant mean electrical axis of +30° was calculated. The lower panel shows the orientation and polarities of all six limb leads. Note that a lead is oriented every 30°.

THE NORMAL AXIS IS IN THE LOWER LEFT QUADRANT

Electrocardiographers use a special axis convention. A horizontal line from the heart to the patient's left side is 0°, as shown in Fig. 5. The angle increases in a clockwise manner as viewed from the front of the patient. Thus, an orientation pointing directly down from the heart would be +90° and horizontal to the right of the heart +180°. Moving counterclockwise from 0°, the numbers become negative. An orientation pointing vertically above the heart would be −90°. This continues to −180°, which is the same as +180°.

It is interesting to note the peculiar way in which the polarities of the attachments for leads I, II, and III are arranged. This convention was adopted early in the development of electrocardiography simply because it caused a positive QRS deflection to occur in all three leads in most people. It is easy to appreciate that a mean electrical axis down and to the left, which is the normal direction for most individuals, will point toward the positive electrode in all three leads. The top panel of Fig. 5 shows an example of a mean electrical axis of 30° and what the QRS complexes in leads I, II, and III would look like. Note that lead III is parallel to the isoelectric plane.

UNIPOLAR LEADS ARE USED TO EXPLORE THE CHEST

A *neutral reference lead* can be made by connecting all three of the limb leads and then connecting this to the "negative" lead of the ECG amplifier, as shown in the bottom right panel of Fig. 3. This reference lead approximates the voltage at the center of the chest because it is the average of the two shoulders and the pubis. The "positive" lead is then free to explore the body surface. The positive electrode is moved from the anterior chest to the axilla in six equal steps to make leads V1 through V6. The *V* derives from the original designation as a voltage lead. Although the limb leads detect vectors only in the frontal plane, the chest leads can determine the front-to-back orientation of the mean electrical axis. This is called *sense* and normally the electrical sense is toward the patient's back.

The bottom left panel of Fig. 3 shows how the chest leads V1 through V6 are orientated around a cross section of the chest. A normal trace for each lead is also shown. Notice how lead V4 is isoelectric with as much positive deflection as negative. The mean axis must therefore be at right angles to lead V4. Because lead V1 is at right angles to lead V4 and shows a net-negative QRS complex (a deep S wave), the vector must point to the left rear of the chest. If the axis had been to the front, then an R wave would have been seen in lead V1 rather than the normal S wave. Note that the amplitudes are not comparable between the chest leads and the limb leads, and thus an accurate three-dimensional vector calculation is not possible.

AUGMENTED LIMB LEADS PROVIDE THREE MORE LIMB LEADS

Although any two leads should be sufficient to calculate the mean electrical axis in the frontal plane, in fact, six leads are routinely examined in a standard ECG. We have already considered leads I, II, and III. Using the unipolar lead system described earlier, it is possible to put the exploring electrode on any of the three limbs. Imagine that the exploring electrode is placed on the left foot. This lead would be oriented from the pubis to the center of the chest (the neutral lead) and thus have a vertical orientation, halfway between leads II and III. For technical reasons, eliminating the lead from the left foot to the reference connection augments the signal to a level comparable to that in leads I through III. Such a lead is termed an *augmented unipolar lead* from the foot, or aV_F. Augmented unipolar leads can also be similarly made for the left and right arms, thus giving aV_L and aV_R, respectively. The advantage of the augmented leads is that they can be used along with leads I through III for calculating the electrical vectors. Figure 3 shows the connections for these leads and a normal trace for each of them.

The lower panel of Fig. 5 shows the orientation of the augmented limb leads with respect to leads I, II, and III. The augmented limb leads aV_R, aV_L, and aV_F bisect the angle between standard limb leads I and II, I and III, and II and III, resp. Thus, there is a limb lead oriented every 30° in the frontal plane, which makes it much easier to calculate vectors. For example, lead I can now be plotted against aV_F in the conventional Cartesian coordinate system because they are at right angles to each other. *Note also that a patient with a mean electrical axis in the normal quadrant should have a QRS complex that is zero or net-positive in both lead I and aV_F. If the QRS complex is net-negative in either of those, then the mean axis is outside the normal quadrant of 0° to 90°.* Note that aV_F is indeed net-negative in both Figs. 6 and 7. Also, with leads oriented at six equally spaced angles around the heart, one lead will invariably be nearly perpendicular to the mean electrical axis and hence be isoelectric. This usually allows the physician to estimate the mean electrical axis at a glance without even plotting the vectors. The direction of the vector is simply determined by noting the polarity of the axis, which is at right angles to the lead having the minimal deflection. Use the lower panel of Fig. 5 to determine the orientation and polarity of the six limb leads. Note that aV_R is nearly isoelectric in Fig. 6. That would make the vector parallel to lead III. Because III is strongly negative the vector must be −60° rather than +120°.

DISEASES CAN AFFECT THE MEAN ELECTRICAL AXIS

Because the normal sequence of activation is down and to the left (from the SA node to the apex of the

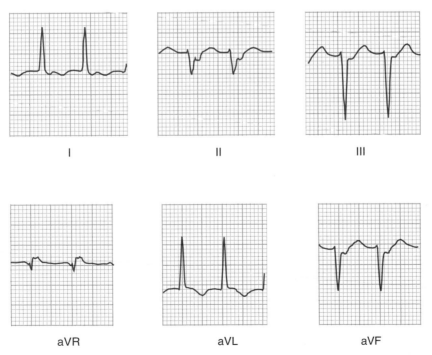

FIGURE 6   An ECG from a patient with left ventricular hypertrophy. The mean electrical axis is approximately −60°. Note that P waves are absent in this trace because the heart was responding to a pacemaker near the bundle of His.

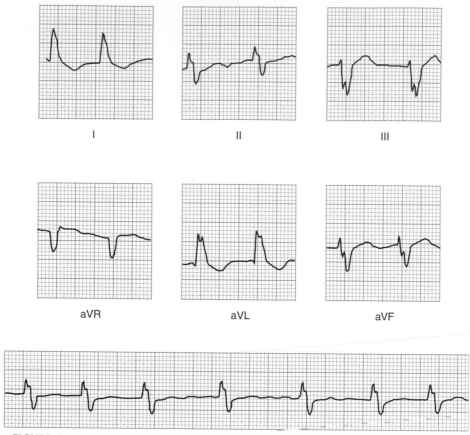

I   II   III

aVR   aVL   aVF

FIGURE 7 An ECG from a patient with left bundle branch block. The mean electrical axis is approximately −30°, a left axis deviation. Note the long duration of the QRS complex, reflecting the delayed conduction to the left ventricle. This patient also was in atrial fibrillation, as indicated by the rhythm strip below.

heart), the electrical axis tends to have the same orientation. As shown in Fig. 5, the normal range for the mean electrical axis is from 0° to +90°. Certain pathologic conditions can cause a change in this axis. For example, if the left ventricle were to become *hypertrophied* (enlarged), then the mass of tissue would be increased on the left side of the heart. This could cause the dipoles generated from activation of the left ventricle to be much greater in magnitude than those from the right ventricle. Left ventricular hypertrophy might therefore shift the mean electrical axis to the left. When the shift is sufficient to move the mean electrical axis out of the normal range, it is called a *left axis deviation*. Thus, a mean electrical axis of −45° should alert the physician that left ventricular hypertrophy may be present. That diagnosis, of course, would have to be confirmed by additional tests such as an echocardiogram. Figure 6 shows an ECG from a patient with left ventricular hypertrophy. Hypertrophy of the right ventricle can cause a *right axis deviation* and thus an electrical axis between +90° and +180°.

An axis deviation is commonly caused by failure of one of the *bundle branches* to conduct. If the left bundle is blocked, then conduction over the left ventricle will be slower than normal because the fast Purkinje fibers will not be used for conduction. The slowed activation will abnormally prolong the duration of the QRS. The electrical axis will also be shifted to the left because the wave of depolarization will continue on the left ventricle long after it has been completed on the right ventricle. Figure 7 shows an ECG from a patient with a left bundle branch block. The distinguishing feature between Figs. 6 and 7 is that the QRS interval in Fig. 6 is about 0.08 sec, whereas that in Fig. 7 is about 0.16 sec. Similarly, an abnormally prolonged QRS duration and a right axis deviation would characterize a right bundle branch block. Such blocks may occur as a congenital defect and be of little consequence. If a bundle branch block were instead to be acquired, as was the case in the patient in Fig. 7, it would indicate the electrical conduction tissue was being interrupted by either progressive fibrosis as is seen in advanced age or by necrosis as seen in myocardial infarction.

MYOCARDIAL ISCHEMIA PRODUCES A CURRENT OF INJURY

Myocardial *ischemia* occurs when coronary blood flow is inadequate to meet the requirements of the heart. This commonly results from atherosclerotic obstruction of one of the coronary arteries. When a region of the heart becomes ischemic, it quickly loses its ability to maintain transmembrane ionic pumps and the cells become depolarized. This movement of positive charge into the cells causes the ischemic region of the heart to become negative with respect to the rest of the heart. Thus, a dipole is created during diastole and it is termed the *current of injury*. The amplifier in the electrocardiograph is not DC coupled, which means that there is always an ambiguity as to where zero voltage actually is on the recording. As a result, it is impossible to tell whether a current of injury is present by simply examining whether the baseline is at zero volts. A little logic, however, can overcome this problem. Where the QRS complex ends is called the *J point*. At that time, all of the ventricular cells are depolarized and no voltage differences exist on the heart. For that reason, the voltage in the ECG returns to zero for a moment before the onset of the T wave. The J point is more commonly called the *ST segment* (the flat region between the end of the QRS and the start of the T wave). In a normal ECG, the ST segment should always be at the same level as the baseline, zero volts.

Now consider the case in which there is an ischemic region on the heart. Because the injured region is always depolarized, it generates a current of injury between beats when the rest of the heart is polarized. At the J point, however, all of the heart is depolarized, including the injured region. Therefore, the voltage from the heart at the J point is truly zero. *If the ST segment is above or below the baseline in any lead, then the above described injury is present.* The actual location of the injured region on the heart and thus the coronary branch involved can be determined by plotting the ST segment shifts vectorially.

Consider the example in Fig. 8. The greatest ST segment shift is seen in aV_F and in that lead the baseline

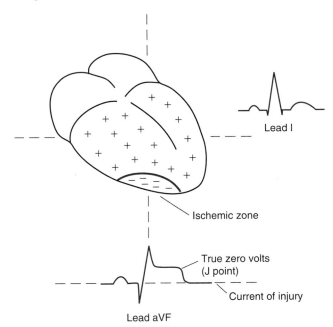

FIGURE 8 A heart with an ischemic region on its inferior surface. The injured region causes a current of injury that shifts the ST segment above the baseline. The location of the injured region can be determined by vector analysis of the ECG traces.

is below the ST segment. This means that the current of injury caused the lower half of the body to be negative between beats. Because the injured region is negatively charged during diastole, that region must be on the inferior surface of the heart. Note that no ST segment shift is seen in lead I, which is at right angles to the current of injury's dipole.

THE ELECTROCARDIOGRAM CAN DETECT CHANGES IN THE HEART'S RHYTHM

In the first part of this chapter we discussed factors that affect the electrical vectors generated by the heart and how they can be detected by an ECG. In the remainder of the chapter, we will discuss factors that affect the timing of events in the heart, the *arrhythmias*.

Clinical Note

The common heart attack occurs when a blood clot forms in a coronary artery, thus depriving a region of the heart muscle of its nutrition. This is accompanied by intense chest pain and a compromised ability of the heart to pump blood. To confirm that the pain is cardiac in origin, the physician will look for shifted ST segments in the ECG. ST segment changes become apparent after only a minute or two of interrupted blood flow.

DEPRESSED CONDUCTION THROUGH THE ATRIOVENTRICULAR NODE CAUSES HEART BLOCKS

Conduction through the atrioventricular (AV) node is vulnerable to disease. When this is depressed we refer to it as a *heart block*. In heart block's mildest form, conduction can be simply slowed. The *PR interval* represents the time from the start of the P wave to the first appearance of the QRS complex. The PR interval is the sum of conduction time from the SA node to the AV node plus the time required to traverse the AV node. Conduction through the AV node comprises most of this time. We are forced to use the PR interval as an index of AV nodal conduction because there is no way from the ECG tracing to determine precisely when the depolarization wave enters the AV node. Normal values for the PR interval range from 0.12–0.21 sec. If the PR interval is longer than 0.21 sec and if every P wave is followed by a QRS complex, we would suspect abnormally slow conduction through the AV node, which is termed *first-degree heart block*. Figure 9 shows a normal ECG in panel A and a first-degree heart block in panel B. If the process affecting the AV node intensifies, a *second-degree heart block* can occur. In second-degree block, conduction through the AV node intermittently fails so that some but not all of the P waves are followed by a QRS complex. Panel C shows a second-degree heart block. Finally, injury can be so extensive to the AV node that it does not conduct at all. In that case, we call it a *third-degree* or *complete heart block*. Panel D shows a complete heart block. Notice that the atria (P waves) and ventricles (QRS complexes) are beating independently of each other. The ventricles will beat more slowly because they are responding to a latent pacemaker in the ventricle. If complete heart block is acquired as an adult, the heart rate will be too low to maintain an adequate cardiac output. Interestingly patients born with complete heart block increase their stroke volume so that they have normal cardiac outputs despite the low heart rate.

Although the preceding discussion might imply that the AV node's vulnerability makes it a liability, that is hardly the case. Not only does the AV node provide a delay between atrial and ventricular contraction to aid ventricular filling, but it also protects the ventricle from

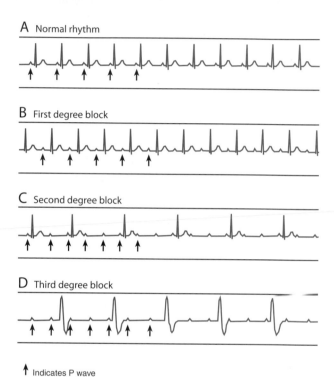

A Normal rhythm

B First degree block

C Second degree block

D Third degree block

↑ Indicates P wave

FIGURE 9   ECG tracings from (A) a normal individual, (B) a patient with first-degree heart block, (C) a patient with second-degree heart block, and (D) a patient with third-degree heart block. All four recordings are from lead I.

atrial tachyarrhythmias. The long refractory periods in the AV node prevent rapid reactivation of the ventricle. At a normal heart rate the AV node is barely out of its refractory period before the next beat. When the discharge rate of the SA node is increased by sympathetic stimulation, the refractory period of the AV node is also reduced, which allows the ventricle to follow the increased rate.

AN ACCESSORY PATHWAY CAN SHORTEN THE PR INTERVAL

Occasionally, a muscle bridge exists between the atria and ventricles as a congenital defect. This is termed an *accessory pathway* which can activate the ventricles before the signal can traverse the AV node. In this case the PR interval will be abnormally short (less than 0.12 sec). This condition, termed *Wolff-Parkinson-White syndrome,* puts the patient at risk because he or she now lacks the long refractory period of the AV node. Should an atrial arrhythmia such as atrial fibrillation or flutter occur, the ventricle would try to follow it, resulting in too fast a ventricular rate for efficient pumping, often with disastrous consequences. Another problem with these patients is that the accessory pathway may be a source for reentry. Impulses can pass down the AV node and then back up the accessory pathway to restimulate the atria. Because the reentry path is short, the heart will beat very fast, causing a condition termed *atrial tachycardia.* Atrial tachycardia is common in these patients. The condition is treated by surgically ablating the accessory pathway.

ECTOPIC FOCI CAUSE PREMATURE CONTRACTIONS OF THE HEART

The SA node is the normal pacemaker for the heart. Occasionally cells outside the SA node can become irritated and elicit an action potential before the SA node fires. When this occurs, it is termed an *ectopic focus* and that action potential will conduct over the entire heart, causing a premature contraction often referred to as an *extrasystole.* Extrasystoles can originate in the atria, the ventricles, or the AV node region. *Premature atrial contractions* have an early P wave, usually with a modified shape because the pacemaker is outside the SA node. The QRS complex may be normal or may be altered if part of the ventricle is still partially a refractory. Because the SA node will be reset, the following beat will usually occur with a normal R-R interval.

Figure 10 shows an extrasystole that originated from an *ectopic focus* located in the ventricles, often referred to as a *premature ventricular contraction* (PVC). Because the SA node did not initiate this beat, there is no P wave preceding it. Furthermore, because the ectopic focus did not activate the ventricles by traveling down the Purkinje fibers, conduction is slow and the QRS complex is consequently very broad and unusually shaped. The AV node usually does not conduct well in the retrograde direction; hence, the SA node often is not reset by a PVC. This causes the ectopic beat to be followed by a *compensatory pause* while the ventricle waits for the next conducted impulse from the SA node. If the ectopic focus is in the AV node region, a near normal QRS complex will be seen since the normal conduction system is used to activate the ventricle. It will be early, however, and not preceded by a P wave. PVCs occur occasionally in all healthy hearts and are not a cause for alarm. When they suddenly become

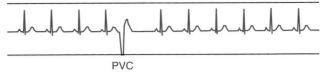

FIGURE 10   ECG tracings showing a premature ventricular contraction.

numerous in patients with an impaired coronary circulation, however, they can signal increased irritability of the heart and possibly impending ventricular fibrillation (see next section).

ELECTRICAL IMPULSES CAN REENTER TO CAUSE FIBRILLATION

Normally, the action potentials spread over the entire heart and die out with each beat. In disease states, however, conduction through a region may be depressed to the point that when the conducted wave exits that region, it finds the surrounding myocardium out of its refractory period and ready to respond with another action potential. If this occurs, the wave is said to have *reentered* and the wave of excitation can go around and around the heart in a condition called *fibrillation.* In this state the muscle only quivers rather than rhythmically contracting and relaxing.

If the atria fibrillate, the AV node will be stimulated in rapid succession. The AV node's long refractory period, however, prevents the ventricle from being stimulated too rapidly to efficiently pump blood. Whenever the AV node comes out of its refractory period, the next wave of excitation will then be conducted to the ventricle, resulting in a normal QRS complex but at irregular intervals. Figure 7 and panel A in Figure 11 both show *atrial fibrillation. Note that P waves are absent from the tracings and that the R-R interval is irregularly irregular.* Atrial fibrillation is not a serious condition and is common in the elderly and in patients with mitral valve disease. Unfortunately, blood

A Atrial fibrillation

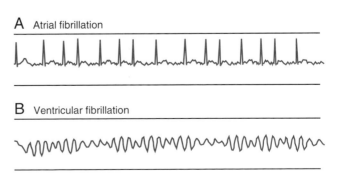

B Ventricular fibrillation

FIGURE 11   ECG tracing from a patient with atrial fibrillation (A) and a patient with ventricular fibrillation (B).

Clinical Note

Forty years ago, ventricular fibrillation was a major cause of in-hospital mortality in heart attack patients because the infarcted heart is very prone to fibrillation until it heals. That problem has been largely solved now by the creation of the intensive care unit, where the patient's ECG is continuously monitored. Should ventricular fibrillation occur, it is detected by the monitoring equipment which alerts the staff to quickly use a defibrillator to resuscitate the patient.

---

can stagnate in the nonbeating atria and become a source for blood clots, putting these patients at risk of stroke. Patients with chronic atrial fibrillation are often put on anticoagulation medicines.

The heart continues to pump blood in atrial fibrillation because a coordinated atrial contraction is not vital to pumping by the ventricle. That is not the case if the ventricles fibrillate, however. *Ventricular fibrillation* causes an immediate cessation of the rhythmic cycle of contraction and relaxation of the ventricles. Rather they simply quiver. Cardiac output falls to zero with rapid loss of consciousness and ensuing death of the victim. Panel B in Figure 11 shows that all coordinated electrical activity is lost in ventricular fibrillation and only an irregular baseline is seen. The most common causes of ventricular fibrillation are occlusion of a coronary artery (myocardial ischemia), advanced heart failure, drug overdose, or electrocution. If detected early enough, ventricular fibrillation can often be stopped by a high-voltage shock to the heart from a device called a *defibrillator*. That shock puts all of the heart cells into a refractory period simultaneously so that a coordinated beat can resume. Interestingly, although the electrical energy from a defibrillator can shock a fibrillating heart into a normal rhythm, a higher-voltage shock, it can also cause a normally beating heart to fibrillate.

Suggested Reading

Marriot HJL. *Practical electrocardiography*. Baltimore: Williams and Wilkins, 1988

C H A P T E R

13

The Mechanical Activity of the Heart

JAMES M. DOWNEY

KEY POINTS

- The heart is an example of a reciprocating pump.
- During *diastole*, the semilunar valves (aortic on the left and pulmonic on the right) are closed and systemic and pulmonary venous blood flows through the respective atria and fills the relaxed ventricles. The ventricular pressure is essentially venous pressure.
- During *systole*, the atrioventricular (mitral on the left and tricuspid on the right) valves are closed and the contracting heart ejects blood through the semilunar valves. Left ventricular pressure is essentially aortic pressure.
- The beating heart normally makes four sounds. Two are audible and two are usually inaudible, requiring instrumentation to appreciate. The heart may make other sounds, including *murmurs* or *clicks*.
- The mechanics of the ventricle are best analyzed by the pressure–volume plot on which an ejection loop can be drawn.
- The force with which the ventricle contracts is proportional to its filling (Frank-Starling relationship).

KEY POINTS (*continued*)

- The stroke volume is determined by the contractility, the filling pressure, the arterial pressure, and the ventricle's diastolic compliance. Hence, small changes in venous pressure profoundly affect the stroke volume and thus the cardiac output.
- Because the right heart is in series with the left heart, the output of one generates the filling pressure for the other, which keeps the two outputs synchronized.

- The systolic pressure-volume area is proportional to the oxygen consumption of the heart.
- Wall stress in the heart is proportional to the pressure in the ventricle times its radius (law of LaPlace).
- The *ejection fraction* is the most commonly used clinical index to assess contractility. The *end-systolic volume-pressure relationship* is the most accurate index available.

THE HEART IS AN EXAMPLE OF A RECIPROCATING PUMP

The heart can be classified as a simple reciprocating pump. The mechanical principles of a reciprocating pump are illustrated in Fig. 1. The pumping chamber has a variable volume and input and output ports. A one-way valve in the input port is oriented such that it opens only when the pressure in the input chamber exceeds pressure within the pumping chamber. Another one-way valve in the output port opens only when pressure in the pumping chamber exceeds the pressure in the output chamber. In the example, the rod and crankshaft cause the diaphragm to move back and

forth. The chamber's volume changes as the piston moves, causing the pressure within to rise and fall. In the heart, the change in volume is the result of contraction and relaxation of the cardiac muscle that makes up the ventricular walls. One complete rotation of the crankshaft in Fig. 1 will result in one pump cycle. Each cycle, in turn, consists of a filling phase and an ejection phase. The filling phase occurs as the pumping chamber's volume is increasing and drawing fluid through the input port. During the ejection phase, the pumping chamber's volume is decreasing and fluid is ejected through the output port. The volume of fluid ejected during one pump cycle is referred to as the *stroke volume*. The volume of fluid pumped each minute can be determined by simply multiplying the stroke volume times the number of pump cycles per minute.

FIGURE 1   A schematic representation of a reciprocating type pump having a pumping chamber and input and output ports with oppositely oriented valves.

CARDIAC CYCLE

The heart is actually composed of two separate pumps, one on the right side that supplies the pulmonary circulation and one on the left that supplies the systemic circulation. The principles that regulate flow into and out of the heart's ventricles are somewhat different from the pump in Fig. 1, which has a fixed stroke volume. The temporal relationships between ventricular contraction and blood flow in the heart are illustrated in Fig. 2. When the ventricular muscle is relaxed, a period referred to as *diastole,* pressures within the veins and atria exceed the pressures in the ventricles, causing blood to flow into the ventricles through the *atrioventricular* (*mitral* on the left and *tricuspid* on the right) valves. Unlike the mechanical pump in Fig. 1, however, the relaxed ventricle cannot create a negative pressure to pull blood into it. Instead, the ventricular lumen can only be distended passively with blood under a positive pressure. That pressure must be generated in

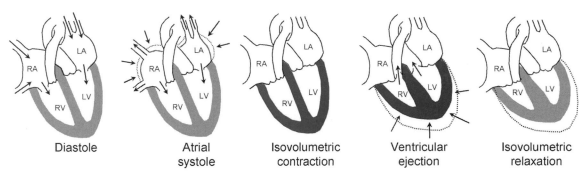

FIGURE 2 Mechanical events in the heart during a single cardiac cycle. Blood flow and wall motion are indicated by arrows. Atrial and ventricular activation are symbolized by color in the respective wall. RA, LA, RV, and LV refer to right atrium, left atrium, right ventricle, and left ventricle, respectively.

the veins that feed the heart. Because ventricular filling is in proportion to venous pressure, the heart's stroke volume is quite variable. At the end of diastole, the atria contract. Because there are no valves between the atria and the veins, much of the atrial blood is actually forced back into the veins. Nevertheless, some additional blood is also pushed forward into the ventricles, causing further increases in ventricular pressure and volume. Although the benefit of atrial contraction at normal heart rates may be negligible, it can substantially increase ventricular filling at high heart rates when diastolic filling time is curtailed.

As the ventricular musculature contracts, a period termed *systole,* the force in the walls is transmitted to the blood within the ventricular lumen. Ventricular pressure increases and, as it rises above venous pressure, the atrioventricular valves close. The heart now begins a period of *isovolumetric* contraction as pressure builds in the lumen. No blood can enter or leave the ventricle because both the inflow and the outflow valves are closed, hence the term isovolumetric. When pressure in the ventricular lumen finally exceeds that in the outflow vessel (the aorta for the left heart and the pulmonary artery for the right heart), the semilunar valves (aortic on the left and pulmonic on the right) open and blood is ejected.

As systole ends, the ventricular musculature relaxes and the force exerted on the blood in the ventricular lumen subsides. Ventricular pressure falls below outflow pressure in the outflow vessel and the semilunar valves close. At this point both the semilunar and the atrioventricular valves are closed so that a second isovolumetric period occurs. Atrial blood will not flow into the ventricles until relaxation has proceeded to the point when ventricular pressure falls below atrial pressure. When that occurs, the atrioventricular (AV) valves open and the filling phase of the cardiac cycle once again repeats itself.

PRESSURES WITHIN THE CARDIOVASCULAR SYSTEM

The physician can best appreciate the events of the cardiac cycle by measuring the pressures at various locations in the cardiovascular system with a catheter. Cardiac catheterization has become a powerful tool for the diagnosis of cardiac disease and the student must, therefore, become thoroughly familiar with the pressure profiles in the atria, ventricles, and great vessels. Formally, seven distinct phases during a single cardiac cycle are recognized. Figure 3 illustrates how aortic pressure, left ventricular pressure, left atrial pressure, left ventricular volume, and the ECG are temporally correlated throughout these seven phases.

Period A in Fig. 3 represents *atrial systole*. Note that contraction of the left atrium causes both the left ventricular and left atrial pressure to rise by a few mm Hg. This rise in the atrial pressure is called the *A wave*. As the atrium begins to relax atrial pressure falls causing the X wave. The volume of blood present in the ventricle at the end of atrial systole is termed the *end-diastolic volume*. In period B, the *isovolumetric period of contraction,* ventricular pressure is seen to separate from atrial pressure because of closure of the mitral valve. The upward movement of the mitral valve into the atrium causes the *C wave*. This is followed by a second fall in atrial pressure, the *X' wave*. The isovolumetric period ends as left ventricular pressure reaches arterial pressure and the aortic valve opens. During period C, most of the stroke volume is ejected into the aorta, as shown by the volume trace; hence the term *rapid ejection phase*. The next phase, D, is termed the *reduced ejection period*. During both ejection periods, the aortic valve is open, making the aorta and left ventricle a common chamber and the pressure within them nearly equal. Therefore, the atrial pressure appears to closely follow the ventricular pressure trace throughout the two ejection periods.

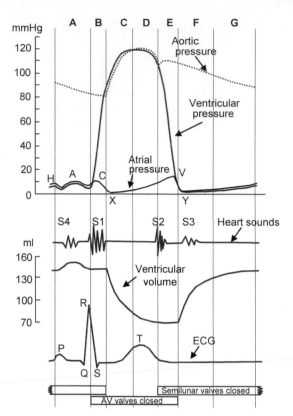

FIGURE 3   Phases of the cardiac cycle are correlated with pressure, auditory, volume, and electrical events in and around the heart. The seven phases of the cardiac cycle are as follows: (A) atrial systole, (B) isovolumetric ventricular contraction, (C) rapid ventricular ejection, (D) reduced ventricular ejection, (E) isovolumetric ventricular relaxation, (F) rapid ventricular filling, and (G) reduced ventricular filling.

During rapid ejection the velocity at which blood is being ejected is increasing, causing ventricular pressure to slightly lead that in the aorta by a few mm Hg. As the rate of ejection slows during the reduced ejection period, the inertia of the decelerating column of blood traveling down the aorta reverses the gradient causing aortic pressure to slightly lead ventricular pressure.

As the ventricle begins to relax, pressure in the ventricle falls. As blood begins to flow backward across the aortic valve, it closes its leaflets. That momentary retrograde flow of blood at the aortic valve and its abrupt deceleration as the valve snaps closed cause a small rebound in the aortic pressure trace called the *dicrotic notch*. The volume of blood left in the ventricle at aortic valve closure is termed the *end-systolic volume*. During the isovolumetric period of relaxation, E, left ventricular and aortic pressure separate and ventricular pressure continues to fall. The isovolumetric relaxation period ends when ventricular pressure reaches left atrial pressure and the mitral valve opens. Although the mitral valve is closed during systole, blood continues to return to the left atrium, causing its pressure to steadily rise, generating the *V wave* in the atrial pressure tracing. This

elevated pressure causes blood to surge into the ventricle as soon as the mitral valve opens. For that reason, period F is called the *rapid filling phase*. The abrupt fall in atrial pressure during the rapid filling phase gives rise to the *Y wave*. During the remainder of diastole, the *reduced ventricular filling period,* the pressure within the ventricle has equilibrated with atrial pressure, and little additional blood enters the ventricle. As atrial blood fills the ventricle, atrial pressure rises once more as the *H wave*.

The pressure in the aorta is the arterial blood pressure. The peak pressure during ejection is referred to as the *systolic pressure,* whereas the lowest pressure just prior to aortic valve opening is called the *diastolic pressure*. Blood pressure is usually reported as systolic/diastolic, so for the example in Fig. 3 the patient's blood pressure would be 120/80 mm Hg.

The diagram in Fig. 3 is for the left heart and the aorta. The pressure relationships within the right heart and pulmonary artery are virtually identical to those shown in Fig. 3, with the exception that the pressures are only about one-fifth as great. The student should meticulously study this diagram until the mechanisms underlying these correlations are conceptually obvious.

HEART SOUNDS

Closure of the valves and rapid movement of blood within the heart give rise to sounds that can be heard at the chest wall. *The first heart sound,* S1 in Fig. 3, accompanies the start of the isovolumetric contraction phase and is caused primarily by tensing of the atrioventricular valves as they close and the redistribution of blood within the ventricles. Note that antegrade flow through the semilunar valves does not contribute to S1. As atrial systole forces blood through the AV valves, the Bernoulli forces tend to keep the valve cusps close to each other. Thus, as ventricular contraction occurs, very little blood regurgitates back into the atrium before the AV valves close. If ventricular contraction is delayed, however, as in first-degree heart block, then the cusps may have moved away from each other before ventricular systole and an exaggerated first heart sound will result because the valves must stop a considerable regurgitant flow. S1's pitch is low because the large atrioventricular valves and associated ventricular structures have a low resonant frequency. S1 sounds like the word *lub*. Because the vibrations involve ventricular structures, the first heart sound is heard best low in the chest near the apex of the heart. The *second heart sound,* S2 in Fig. 3, is associated with the closing of the aortic and pulmonic valves and is caused by the resulting vibration of the valves and the large arteries near the valves as blood is abruptly decelerated in the outflow tract. The second heart sound occurs at the beginning of the isovolumetric relaxation phase. The semilunar valves, the aortic and the pulmonic, are small and have a higher resonant frequency. The second heart sound sounds like the word *dub*. Because the second heart sound involves vibrations of the aorta and pulmonary artery, it is best heard high in the chest.

The *third heart sound* (S3) results from turbulence associated with the rapid filling of the ventricles; the *fourth heart sound* (S4) results from the movement of blood during atrial systole. The third and fourth heart sounds are not usually audible with a stethoscope, but can be seen when the sound waves are recorded electronically and displayed as in Fig. 3. They are best appreciated when the stethoscope is placed near the apex of the heart. When S3 or S4 is heard, the sound is referred to as a *gallop* because its cadence in conjunction with S1 and S2 sounds like a galloping horse. S3 is normally audible in children and some narrow-chested adults. In certain disease states, the third or fourth heart sound may be significantly augmented. The third heart sound is most commonly audible when the ventricle is dilated and atrial pressure is high as in heart failure. An audible fourth heart sound occurs most commonly in hearts in which the ventricle is very stiff, as is the case in a hypertrophied heart.

Although the mitral valve normally closes slightly before the tricuspid, the separate components are not easily discerned. Because the right ventricle is normally weaker than the left ventricle, the pulmonic valve closes significantly later than the aortic, causing *splitting* of the second heart sound. If the venous return is selectively increased to the right ventricle by a deep inspiration, pulmonic valve closure will be further delayed and the splitting will be accentuated. We call this response *physiologic splitting*. This is of diagnostic significance because a failing left ventricle, which has abnormally low contractility, will eject blood more slowly and valve closure will be delayed. If the left ventricle were so depressed that the aortic valve closed after the pulmonic valve, then a deep inspiration would reduce the splitting rather than accentuate it. This is referred to as *paradoxical splitting* and it may occasionally be heard in patients with left ventricular failure. Left bundle branch block can also delay aortic valve closure and produce paradoxical splitting of the second heart sound.

FRANK-STARLING RELATIONSHIP

An efficient way for the body to alter its cardiac output is to change the heart's stroke volume. Stroke volume depends both on how much blood is in the heart at the end of diastole and how much remains at the termination of ejection. The German physiologist Otto Frank was the first investigator to systematically examine how stroke volume is controlled. His research demonstrated that the amount of ventricular pressure actively generated by a frog's contracting heart depends on the volume of fluid within the ventricle at the end of diastole. A fluid-filled balloon was placed within the ventricle of a beating frog's heart and the pressure inside the balloon was measured with a catheter. The balloon was then filled to different volumes and ventricular pressure was recorded at each volume both during diastole and systole. The left panel of Fig. 4 shows the time course of the ventricular pressure for four different volumes. The right panel shows a plot of the pressures against the volume in the balloon. The solid curve is for the peak pressure achieved during systole and the dashed curve is for the pressure during diastole. The plot reveals that over a wide range the greater the distention of the heart during diastole, the more forcibly

Clinical Note

Certain valve lesions and other anatomic defects can give rise to sounds other than the four just mentioned. For example, if the aortic valve fails to close properly (aortic regurgitation), blood will be forced across it in a retrograde manner during diastole. This flow will be turbulent and result in additional cardiac sounds collectively referred to as a *murmur*. A murmur is by definition a sustained sound that it not normally present. Although a regurgitant aortic valve will cause a diastolic murmur, a regurgitant mitral valve will cause a murmur during systole, because the inappropriate flow would be during the systolic period. The timing and quality of the murmurs help to identify the nature of the defect causing them. Sometimes the turbulent flow caused by a murmur is so intense that the low-frequency vibrations can be felt with the fingers on the chest wall. This is called a *thrill*. Other conditions, such as mitral value prolapse, can cause clicking sounds termed *clicks*.

it contracts during systole. This length dependency is a consequence of the overlap between the actin and myosin filaments inside the cardiac muscle cells. The British physiologist Ernest Starling, studying the canine heart, extended Frank's concept to demonstrate that this length dependency acts to ensure that the heart will pump a volume of blood equal to that which it receives. Because of the pioneering work of these two physiologists, we now refer to the concept that the force of ejection is proportional to the length of the muscle fibers as the *Frank-Starling law*.

PRESSURE-VOLUME RELATIONSHIPS IN THE VENTRICLE

Let us examine the basis for Frank's pressure-volume plot. Ventricular muscle cells are organized into a hollow spheroid with the muscle fibers oriented parallel to the surface of that sphere. When the myocytes develop tension, the ventricular walls are caused to move inward, elevating the pressure of the blood in the lumen of the ventricle. Because of this geometry, tension in the individual ventricular myocytes is proportional to

the pressure in the lumen. Similarly, the length of each myocyte will be proportional to the volume of blood contained within the lumen. These relationships are actually somewhat nonlinear because of the LaPlace relationship (as discussed later) but, for now, a simple linear relationship can be assumed. As a result of the preceding relationships, the length-tension curve that was presented for a cardiac muscle fiber in Chapter 7 can be relabeled, as shown in Fig. 4B. The vertical axis is changed from tension to pressure and the horizontal axis is changed from length to volume, but the shapes of the curves are unchanged. This approach yields two pressure-volume plots: one for the contracted state (the upper curve) and one for the relaxed state (the lower curve). As discussed in Chapter 10, the relationship between pressure and volume for a distensible container is described by its *compliance*. The more compliant a structure is, the more its volume will change for a given change in filling pressure. Figure 4 reveals that during diastole the heart is very compliant because a small increment in filling pressure will cause a large increment in ventricular volume. Note that during systole, the slope of the pressure-volume curve increases dramatically, making it much less compliant.

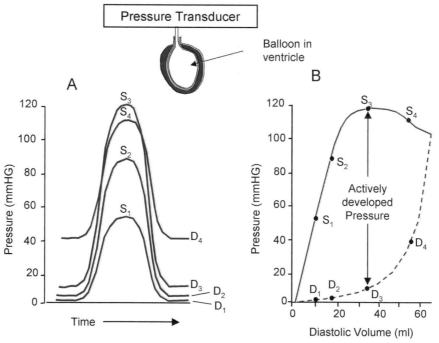

FIGURE 4   A schematic representation of the frog-heart preparation used by Otto Frank to demonstrate the relationship between the force of contraction and ventricular end-diastolic volume. Four different end-diastolic volumes were selected and the corresponding diastolic and systolic pressures generated by the heart are denoted as D_1–D_4 and S_1–S_4, respectively. (A) The ventricular pressures for a complete cardiac cycle are shown for the four volumes. (B) The diastolic (dashed line) and systolic (solid line) pressures from that heart are plotted against the balloon volume, which gives a pressure-volume plot for the ventricle in both systole (upper line) and diastole (lower line). The difference between the two lines at any volume is the actively developed pressure.

AN EJECTION LOOP CAN BE PLOTTED ON THE PRESSURE VERSUS VOLUME GRAPH

When Frank tried to calculate how much blood the frog heart would pump, he found that the relationship was quite complex because the frog heart begins to relax before the end of ejection. It was assumed that the human heart behaved in a similar fashion until the studies of Kichi Sagawa in the 1970s. Sagawa found that the mammalian heart, unlike the frog heart, does completely eject before it begins to relax and, as a result, a relatively simple analysis, the *ejection loop,* can be employed to calculate the heart's output. Figure 5 depicts a pressure-volume curve for the human heart in both systole (upper curve) and diastole (lower curve). The ventricle fills with blood passively during diastole. The filling pressure for the left ventricle is the *pulmonary venous pressure* and for the right ventricle the *systemic venous pressure*. These pressures typically will be from 3–7 mm Hg. Note in Fig. 5 that a filling pressure of 5 mm Hg causes the ventricle to fill to an *end-diastolic volume* of 150 mL (point A). The filling pressure is often termed the *preload* because this is the load on the muscle fibers before contraction.

When the heart is activated, it moves from the diastolic compliance curve to the systolic curve. Because the pressure rises isovolumetrically in the ventricle, the trajectory is a vertical line drawn between point A and B. At a volume of 150 mL, the ventricle shown would be capable of developing 200 mm Hg of pressure (point E), but as soon as the ventricular pressure exceeds the pressure in the aorta, the aortic valve opens and the ventricle will begin to eject blood into the aorta. At point B in Fig. 5, the mode of muscle contraction changes abruptly from isometric (A to B) to isotonic (B to C). The *afterload* for the contraction is the aortic pressure, which for simplicity is depicted in these examples as being constant through ejection. Actually, aortic pressure rises and falls during ejection (see Fig. 3) so that the contraction is not strictly isotonic from B to C but is actually auxotonic, as discussed later in this chapter.

Because of the Frank-Starling relationship, as the ventricular volume gets smaller during ejection the potential for the ventricle to develop pressure also falls. At point C, a stable equilibrium has been reached. Note that any further ejection into the aorta (moving to the left) would put the fibers at a length that could not support the aortic pressure. The heart then stays at point C, the *end-diastolic volume,* until the action potential subsides and the heart begins to relax.

At the end of ejection, the aortic valve closes and the heart relaxes isovolumetrically so the plot moves vertically from C to D, where it rejoins the diastolic compliance curve. As pressure in the ventricle falls below atrial pressure, the mitral valve opens, and the ventricle refills with venous blood and returns to the starting point A.

STROKE VOLUME CAN BE CALCULATED FROM THE EJECTION LOOP

We have just described an ejection loop for the heart that makes it easy to calculate the output of the heart. For example, the stroke volume is given by:

$$\text{Stroke volume} = \text{end-diastolic volume} - \text{end-systolic volume}. \quad (1)$$

In the example shown in Fig. 5, stroke volume is calculated as the difference between points B and C, or 75 mL. The fraction of the contents that the heart ejects is called the *ejection fraction* and is calculated as the stroke volume divided by the end-diastolic volume, and in this case would be 50% (75/150). Of course, the shape of the systolic and diastolic compliance curves will vary from heart to heart, but if the curves are known for any heart, then its output can accurately be calculated. The next several sections reveal that only three variables affect stroke volume in the normal heart: contractility, venous pressure, and aortic pressure.

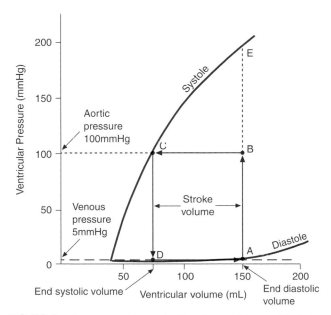

FIGURE 5 A pressure-volume plot for a human left ventricle during both systole and diastole. Note that the ventricle is very compliant during diastole (shallow slope) and becomes very stiff during systole (steep slope). The four points A, B, C, and D describe an ejection loop. See text for details.

FACTORS THAT AFFECT STROKE VOLUME

Contractility

When heart muscle is activated, insufficient calcium is released into the cytosol to cover all the troponin C calcium binding sites. As a result, the force generation is less than maximal. Furthermore, the calcium concentration can be modulated by signal transduction systems such as occurs when the β-adrenergic receptors are stimulated by norepinephrine. If the amount of calcium released into the cytosol is increased, the force generation for any given muscle length is increased, causing an upward shift of the systolic pressure volume curve, as shown in Fig. 6. This is termed an increase in *contractility*. The effect of an increase in contractility on the ejection loop is to move point C to the left (toward a smaller end-systolic volume). If filling pressure, which is determined by the venous pressure, is unchanged, then end-diastolic volume will be unchanged. The reduced end-systolic volume will result in an increased stroke volume. Conversely, decreasing contractility tends to decrease stroke volume.

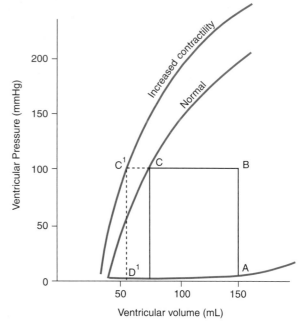

FIGURE 6   The effect of an increase in contractility on the ejection loop is shown by the dashed lines. Note that stroke volume increases because of a decrease in end-systolic volume.

Venous Filling Pressure

The end-diastolic volume is determined by the venous filling pressure. Because the ventricle is very compliant, small changes in venous filling pressure have a major effect on end-diastolic volume and thus stroke volume. This relationship makes venous filling pressure a primary determinant of cardiac output, which will be discussed in further detail in Chapter 14, where the control of cardiac output is considered. Because of its pivotal role in controlling cardiac output, the central venous pressure is of great importance clinically and is often monitored closely in critically ill patients, particularly those who have lost blood volume. Figure 7 shows that an increase in venous pressure increases stroke volume entirely by changing end-diastolic volume and does not affect the end-systolic volume. Factors that increase venous filling pressure tend to increase cardiac output and those that decrease venous pressure decrease cardiac output.

Aortic Pressure

It is easy to see in Fig. 8 that the end-systolic volume is determined by the aortic pressure. As aortic pressure rises, equilibrium point C will occur at a larger end-systolic volume. Increasing end-systolic volume reduces stroke volume. Thus, a rise in aortic pressure tends to decrease stroke volume, whereas a falling aortic pressure tends to increase it. As was seen with contractility, changes in aortic pressure affect stroke volume by changing only the end-systolic volume.

Clinical Note

Heart failure is a relatively common condition in which left aterial pressure is abnormally high. This is often caused by a depressed contractility. A number of diseases can lead to heart failure including hypertension, valve lesions, such as aortic or mitral regurgitation coronary atherosclerosis, or myopathic processes within the heart muscle. In patients with failure related to depressed contractility, the primary consequence is reduced stroke volume and inadequate cardiac output. The condition is often difficult to treat and the prognosis is poor for advanced cases. Cardiac transplantation has emerged as an effective treatment.

Diastolic Compliance

Although the body can change the heart's contractility from moment to moment, no such control system exists for changing the diastolic compliance curve, because in the normal heart the myocytes are fully relaxed and their stiffness is determined wholly by the structures that make up their parallel elastic element. In some patients, e.g., those with inadequate coronary blood flow, the heart may be slow to relax during systole and further impede filling as shown in Fig. 9.

AUXOTONIC BEAT

In the preceding examples, we simplified the ejection phase of the loop by drawing it as a horizontal line. Examination of the aortic pressure profile shown in Fig. 3 reveals that pressure in the aorta actually rises by about 40 mm Hg from the time when the aortic valve opens to the end of the rapid ejection phase. This means that the afterload for the contracting ventricle is not

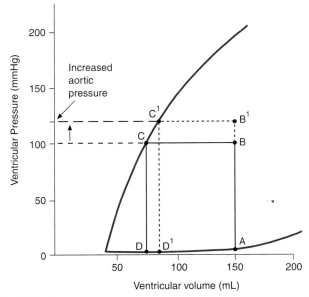

FIGURE 8   The effect of an increase in aortic pressure on the ejection loop is shown by the dashed lines. Note that stroke volume decreases because of an increase in end-systolic volume.

constant and thus the contraction cannot be called *isotonic* in the strict sense of the word. When the load on a muscle changes as it shortens, it is called an *auxotonic* contraction. Figure 10 shows how the ejection loop can be redrawn to depict accurately the auxotonic conditions. This simply means that end-systolic volume is determined by the aortic pressure at the time of closure of the aortic valve.

COORDINATION BETWEEN THE LEFT AND RIGHT VENTRICLES

The heart consists of two separate sides, the left and the right. These sides are in series with one another so that blood pumped through the lungs by the right side returns to the left side to be pumped through the systemic circulation. Obviously, in the steady state the stroke volume of one side must exactly equal that of the other. This coordination is maintained by changes in systemic and pulmonary venous pressure. If, for example, the left ventricle were to reduce its output, then blood pumped by the right ventricle would accumulate

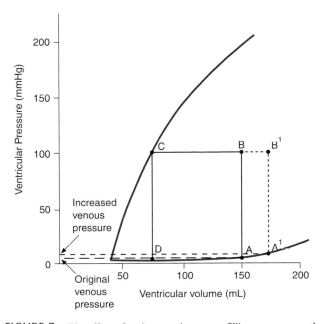

FIGURE 7   The effect of an increase in venous filling pressure on the ejection loop is shown by the dashed lines. Note that stroke volume increases because of an increase in end-diastolic volume.

---

in the pulmonary veins, raising the pulmonary venous pressure. This rise in venous pressure raises the end-diastolic volume of the left ventricle and, thus, the left ventricular stroke volume. At the same time, the reduced output of the left ventricle will reduce systemic venous return and thus the filling pressure for the right ventricle will fall. An equilibrium will soon be met at which the stroke volume will again be equal on both sides.

ASSESSMENT OF CONTRACTILITY

The contractile state of the ventricle is of much interest to the physician because many disease states involve abnormal contractility. In the *hyperthyroid* patient, contractility may be elevated to the point at which the heart may be caused to hypertrophy from the hyperkinetic state. Conversely, in the patient with

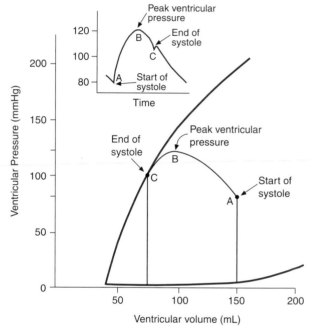

FIGURE 10 An ejection loop drawn to reflect the auxotonic conditions of an actual ejection. The inset shows the aortic pressure tracing for that cardiac cycle.

cardiac failure, contractility may be reduced to the point at which an adequate cardiac output cannot be achieved. Because of the heart's inaccessible position in the chest, it has been surprisingly difficult to measure the heart's contractility in the clinical setting. Over the years, several indices of cardiac contractility have been introduced; the student should have some understanding of the major ones, including the ventricular function curve, the velocity index dP/dt, the ejection fraction, and the end-systolic pressure-volume plot. Each of these is discussed in the following subsections.

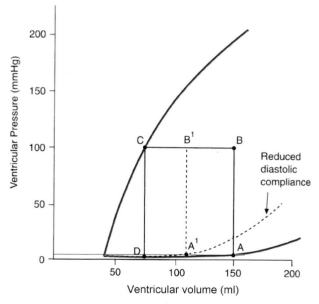

FIGURE 9 The effect of a decreased compliance on the ejection loop is shown by the dotted lines. Note that stroke volume decreases because the increased diastolic compliance causes end-diastolic volume to be reduced for any given filling pressure.

Ventricular Function Curve

The most obvious approach to assessing contractility is to find a way to relate the heart's pump performance to its contractility. The problem with that approach is that two other factors, venous pressure and aortic pressure, also affect stroke volume. That problem was cleverly solved by Stanley Sarnoff in the 1960s. To isolate the effect of venous pressure, he proposed simply to plot stroke volume as a function of venous pressure. If one examines the ejection loop shown in Fig. 5, it can be seen easily that if contractility and aortic pressure are held constant, then a simple relationship exists between stroke volume and venous pressure. That relationship is depicted in Fig. 11A. Similarly, it can easily be appreciated that increasing the contractility of the heart will shift that line upward as indicated in the figure and decreasing contractility will shift it downward.

The problem with the preceding approach is that the stroke volume–venous pressure curve also depends on the aortic pressure. As shown in Fig. 11B, increasing

aortic pressure shifts the curve downward and decreasing aortic pressure shifts it upward. Thus, it is not possible to tell if a shift in the stroke volume–venous pressure curve resulted from a change in contractility or aortic pressure. This ambiguity can be eliminated by plotting *stroke work* rather than stroke volume on the vertical axis. In a fluid system, physical work is given by the volume pumped times the pressure it is pumped against, in this case the product of stroke volume and aortic pressure. Stroke work for the ventricle is fairly independent of the aortic pressure because aortic pressure and stroke volume are reciprocally related. As a result, their product is surprisingly constant over a wide range of pressures. Thus, the stroke work–venous pressure plot shown in Fig. 11C, usually called the *ventricular function curve,* is affected only by changes in contractility. In practice, the venous pressure is changed over several beats by rapid infusion of intravenous fluids or suddenly lifting the patient's legs. Stroke volume and aortic pressure are monitored and from that data the plot is made. Any factor that shifts that plot is said to have changed contractility. The shortcomings are that very invasive procedures are required to obtain such data, which limits the clinical application.

Velocity-Related Indices of Contractility

As contractility increases, two factors change in cardiac muscle: the isometric force and the velocity of isotonic shortening. The velocity of shortening can be assessed by measuring the rate at which ventricular pressure rises (dP/dt) during the isovolumetric period of contraction. In practice, the ventricular pressure recording from a catheter is differentiated electronically so that a continuous recording of ventricular dP/dt can be monitored. Actually, ventricular filling pressure also affects the dP/dt signal. As ventricular filling rises so does the dP/dt. As will be seen in a subsequent chapter, however, a rise in contractility usually evokes a fall in venous pressure. Thus, an increase in dP/dt can be interpreted unambiguously as an increase in contractility if the end-diastolic pressure is either unchanged or falls. The major shortcoming of dP/dt is that it is imprecise and requires insertion of a ventricular catheter. It is often used in animal experiments in which a simple index of contractility is to be continuously monitored.

Ejection Fraction

The ejection fraction, as mentioned earlier, is calculated by dividing the stroke volume by the end-diastolic volume. It is literally the fraction of the ventricular volume that is ejected with each beat. Ejection fraction

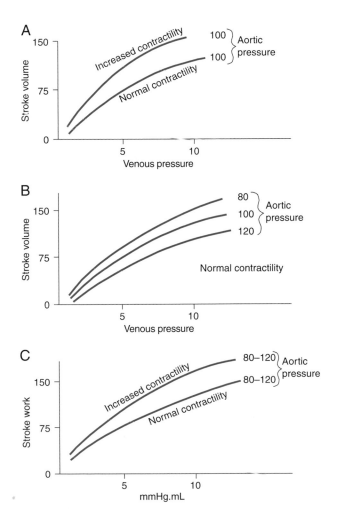

FIGURE 11 Evolution of the ventricular function curve. When venous filling pressure is plotted against stroke volume it is sensitive to both changes in (A) contractility and (B) aortic pressure. (C) If stroke work is chosen for the vertical axis, however, only changes in contractility can shift the curve.

can be measured by a variety of methods in the human heart. These now include *ultrasound, nuclear medicine methods,* and *x-ray angiography.* Although the ejection fraction is strongly influenced by contractility, the ejection loop analysis reveals that it is also influenced by filling pressure and aortic pressure. Nevertheless, the ejection fraction is the most important index of contractility in the clinic today, primarily because of its ease of measurement. Normally, the ejection fraction should be about 0.6 for a healthy heart. Ejection fractions below 0.5 suggest disease and those below 0.3 are associated with high mortality.

End-Systolic Pressure-Volume Relationship

The best measurement of contractility would be to analyze the systolic pressure-volume curves for that ventricle. It is now possible to obtain a continuous ventricular volume recording with what is called a *conductance catheter.* By having a computer plot the ventricular volume signal against ventricular pressure, an ejection loop for each beat can be displayed. If aortic pressure is varied over several beats, the pressure and volume at the end of systole can be determined for each of those beats, as shown in Fig. 12. Connecting those points plots a segment of the systolic pressure-volume curve. The end-systolic pressure-volume relationship is clearly the most accurate measurement of contractility available today. Unfortunately, the method still requires relatively invasive instrumentation, which limits its clinical utility.

LAW OF LAPLACE

In the beginning of the chapter, we stated that the tension on the individual muscle fibers is proportional to the pressure in the ventricular lumen. The relationship between wall tension and luminal pressure for a spheroid is given by the *law of LaPlace.* When applied

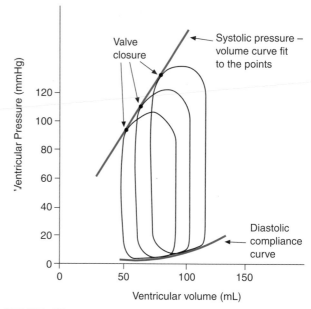

FIGURE 12 Determination of the end-systolic pressure-volume relationship using a conductance catheter. Ventricular volume from the conductance catheter is displayed on the horizontal axis and ventricular pressure on the vertical axis. Each loop depicts data from a single beat. Filling pressure is elevated for several beats by injecting saline into the left atrium so that several different ejection loops are traced. A line is then fitted to the end-systolic points indicated by the solid dots. The steeper the slope, the greater the contractility of the ventricle.

to thin-walled structures like soap bubbles, the tension is given by:

$$\text{Tension} = PR/2. \qquad (2)$$

where P is the pressure in the lumen and R is the radius of the sphere. In a thick-walled structure like the ventricle, one must divide by the wall thickness as well. The LaPlace relationship not only predicts that tension is proportional to pressure but also to the radius as well. Muscle fibers in the ventricle generate a pressure in the lumen that is in the proportion to their force of

Clinical Note

Angina pectoris is a condition in which one or more coronary arteries have become narrowed because of atherosclerosis. This causes the blood flow to the heart muscle to be inadequate (ischemia), resulting in chest pain and impaired contractility of the affected regions. Such attacks are usually brought on by physical exertion that transiently increases the oxygen requirements of the heart. Many drugs have been identified that can decrease the incidence and severity of these attacks. Virtually all of these act either to reduce the contractility of the heart (beta blockers and calcium antagonists) or to reduce the blood pressure by venous dilation (nitrates). The systolic pressure-volume area analysis predicts that either of these will reduce the oxygen demand of the heart.

contraction divided by the radius of the lumen. Therefore, increasing end-diastolic volume acutally diminishes the heart's ability to convert its fiber tension into pressure. In the physiologic range, this disadvantage is greatly overshadowed by the increase in force generation that results from the longer fiber length. Thus, the normal heart responds to an increased chamber volume with a more forceful contraction. If the radius of the lumen becomes excessive, however, as in the dilated heart, the effect of increasing fiber length can be overwhelmed (see clinical note on page 212).

SYSTOLIC PRESSURE-VOLUME AREA PREDICTS THE OXYGEN CONSUMPTION OF THE HEART

Virtually all of the heart's energy is derived from oxidative metabolism. Substrate is primarily fatty acids and to a much lesser extent carbohydrates. As a result, a linear relationship exists between the heart's energy production and its oxygen consumption. Because most of the heart's energy expenditure goes into mechanical work, it is not surprising that there is a close correlation between the oxygen consumption of the heart and its mechanical activity. Figure 13 reveals this relationship. The external physical work done by the heart (pressure times volume) is exactly given by the area within the ejection loop. That area, however, correlates poorly with the heart's oxygen requirements. If, however, an internal work component as shown in Fig. 13 is added to the external work, that sum correlates almost perfectly with the heart's energy demands. This *systolic pressure-volume area* index leads to some interesting and surprising predictions. For example, notice that raising the aortic pressure with no change in venous pressure decreases stroke volume and may even decrease external work but always increases oxygen demand. The preceding analysis explains why high blood pressure is so stressful to the

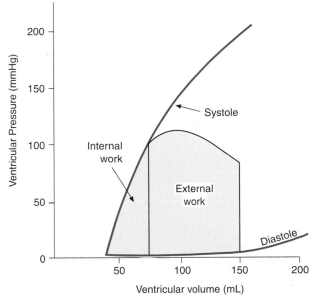

FIGURE 13   The oxygen consumption per beat is very accurately predicted by the area encompassed by the ejection loop (external work) plus the triangle between the systolic and diastolic compliance curves (internal work). (Adapted from Suga and Sagawa, *Circ Res* 1974;35:117–126.)

heart. Conversely, reducing aortic pressure increases cardiac output and decreases oxygen demand. The analysis also explains why increased contractility increases oxygen demand.

Suggested Readings

Frank O. On the dynamics of cardiac muscle (translated by Chapman CB and Wasserman E). *Am Heart J* 1959;58:282–378.

Suga H, Hisano R, Goto Y, *et al.* Effect of positive inotropic agents on the relation between oxygen consumption and systolic pressure volume area. *Circ Res* 1983;53:306–318.

Suga H, Sagawa K. Instantaneous pressure-volume relationships and their ratio in the excised and supported canine left ventricle. *Circ Res* 1974;35:117–126.

Regulation of Venous Return and Cardiac Output

JAMES M. DOWNEY

KEY POINTS

- Venous pressure is a major determinant of cardiac output because it is the filling pressure for the heart. Venous pressure at any cardiac output can be predicted by the vascular function curve.
- Venous pressure tends to be reciprocally related to arterial pressure. Because *venous capacitance* is much larger than arterial capacitance, large changes in arterial pressure cause much smaller changes in venous pressure as blood shifts from the arteries to the veins.
- The periphery interacts with the heart through changes in venous pressure because venous pressure falls as cardiac output increases.
- The crossing point of the cardiac and the venous function curves determines cardiac output.

- Increasing contractility increases cardiac output and reduces venous pressure.
- Increasing blood volume increases cardiac output and venous pressure.
- Increasing venous tone increases cardiac output and venous pressure.
- Increasing arteriolar tone decreases cardiac output but has an unpredictable effect on venous pressure.
- Heart rate affects cardiac output in a triphasic manner. Very low heart rates (bradycardia) limit cardiac output. In the normal range of heart rates (50–150 beats/min), stroke volume changes reciprocally with heart rate so that little effect is seen on cardiac output. High heart rates (tachycardia) limit cardiac output because of inadequate time for ventricular filling.

VENOUS PRESSURE IS A MAJOR DETERMINANT OF CARDIAC OUTPUT

The heart and the vascular system provide all body tissues with a continuous flow of blood that closely matches their metabolic demand. To overperfuse the tissues would require the heart to perform unnecessary work, and to underperfuse would cause ischemic injury to the tissues. As the metabolic demands change from moment to moment, so must the cardiac output. The question arises: How can the body's control systems alter the cardiovascular system to change the cardiac output? In Chapter 13, we learned that stroke volume of the heart is sensitive to changes in venous pressure, the heart's filling pressure. Because venous pressure is determined by the peripheral vascular system, changes in the periphery can have profound effects on cardiac output. This chapter analyzes the complex interaction between the heart and the periphery so that a basic understanding of cardiac output control can be appreciated.

VENOUS PRESSURE IS RECIPROCALLY RELATED TO ARTERIAL PRESSURE

Venous return refers to the amount of blood flowing back to the heart per unit of time. At steady state, venous return and cardiac output must be equal. Because the atria lack valves, their contraction contributes little to diastolic filling and blood flows into the heart passively as a result of the pressure in the great veins. Thus, systemic venous pressure can be considered to be the filling pressure for the right ventricle. To develop a conceptual understanding of what affects

venous pressure, consider a situation in which a patient's heart is replaced by a heart-lung machine for which cardiac output can be set to any desired level and is not affected by changes in venous pressure. A simplified view of this arrangement is presented in Fig. 1. The cardiovascular system now consists of a heart-lung machine that is pumping blood out of the veins and into the aorta. Pressure in the arterial compartment forces blood from the arterial compartment across the microcirculation where most of the *total peripheral resistance* (TPR) is located and into the veins. Blood in the veins then returns to the heart-lung machine. A steady state will be soon be achieved wherein output of the heart-lung machine (cardiac output) equals the flow across the TPR, which, in turn, equals venous return. The system is self-regulating. If flow through the microcirculation were less than cardiac output, then blood would start to accumulate in the arterial vessels. As the additional blood stretches the arterial walls, the arterial pressure rises, which in turn increases the flow into the microcirculation until the equilibrium is again reached.

Let's assume we start with a mean arterial pressure of 102 mm Hg and a venous pressure of 2 mm Hg. This pressure difference of 100 mm Hg results in a flow of 5 L/min across the TPR (TPR in this case equals 20 mm Hg $XL^{-1} \times min^{-1}$). If the heart-lung machine were suddenly stopped, blood would continue to flow across the microcirculation, propelled by the recoiling arterial capacitance. Arterial pressure would fall as blood leaves the arterial compartment and *venous pressure would rise as blood is shifted into the venous compartment.* Flow through the microcirculation will stop only when arterial and venous pressures are equal. The equilibrium pressure that is reached throughout the cardiovascular

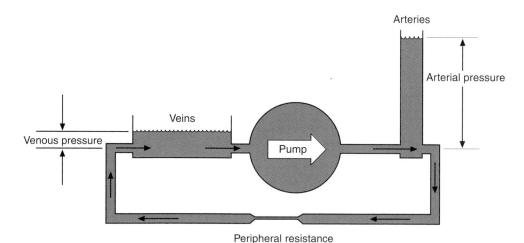

FIGURE 1   The circulation can be modeled by a heart-lung machine (pump) connected to a circulatory system consisting of an arterial capacitance that is small, a peripheral resistance, and a venous capacitance that is large. All components are connected in series. See text for details.

system when cardiac output is stopped is called the *mean circulatory filling pressure.* Mean circulatory filling pressure is thought to be approximately 7 mm Hg and is determined by the amount of blood within the cardiovascular system and the pressure-volume relationship (capacitance) of the blood vessels.

Withdrawing blood from a patient will decrease the mean circulatory filling pressure. If blood were removed until the mean circulatory filling pressure fell to zero (no stretch on the blood vessel walls), the system would still contain an appreciable amount of blood. That volume is called the *unstressed volume.* The unstressed volume can be varied by the body's control systems by changing the contractile state of the smooth muscle in the walls of the veins. Venoconstriction decreases unstressed volume and thus increases mean circulatory filling pressure, whereas venodilation increases unstressed volume and decreases the mean circulatory filling pressure.

Venous Capacitance Is Much Larger Than Arterial Capacitance

In Chapter 10, we learned that the ratio of volume to pressure for a compartment defines its capacitance. The capacitance of the venous compartment is approximately 19 times greater than that of the arterial compartment. We have tried to illustrate the differences in capacitance in the shape of the arterial and venous reservoirs in Fig. 1. The arterial reservoir is tall and narrow so that a small increment in volume will greatly increase the height of fluid in the reservoir and thus its pressure. Conversely, the venous side is broad so that the same increment in volume would have much less effect on pressure. Movement of a volume of blood from the arterial side to the venous side will cause arterial pressure to decrease and venous pressure to increase. Because of the differences in capacitances, however, the arterial pressure change will be 19 times greater than that in the veins. Thus, if the output of the heart-lung machine were reduced from 5 to 2.5 L/min, arterial pressure would fall to 54.5 mm Hg and venous pressure would rise to 4.5. Although arterial pressure fell by 47.5 mm Hg, venous pressure rose by only 2.5 mm Hg, 19 times less. Note that any change in arterial pressure will be accompanied by a reciprocal but smaller change in venous pressure. Figure 2 shows the change in venous and arterial pressures that occurs when cardiac output goes from 0 to 1 L/min. At zero cardiac output, arterial and venous pressures are equal and at the mean circulatory filling pressure. As flow begins arterial pressure rises markedly, whereas venous pressure falls only slightly. Although these changes in venous pressure may seem small, they are of paramount importance because venous pressure is the filling pressure for the

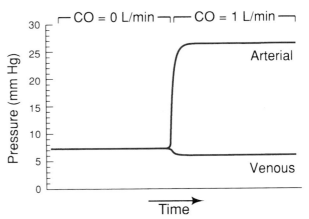

FIGURE 2   Arterial pressure and venous pressure are plotted as a function of time when the output of the heart-lung machine depicted in Fig. 1 is increased from 0 to 1 L/min.

heart and in the previous chapter we learned that filling pressure is a primary determinant of stroke volume.

The Relationship between Cardiac Output and Venous Pressure Is Termed the Vascular Function Curve

When the steady-state venous pressure is plotted as a function of cardiac output, one observes an inverse linear relationship (Fig. 3). Venous pressure is equal to the mean circulatory filling pressure when cardiac output is zero (point A on Fig. 3). As cardiac output is increased, venous pressure decreases linearly until zero venous pressure is attained. At that point, so much blood is sequestered in the arterial compartment that the veins are at their unstressed volume. Zero venous pressure

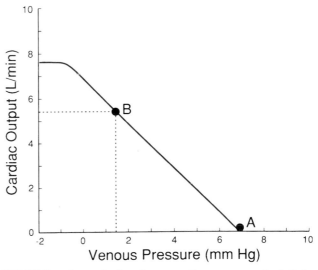

FIGURE 3   A vascular function curve. Venous pressure is plotted as a function of cardiac output.

represents the maximum possible cardiac output because any further fall in pressure would result in a negative pressure and the central veins would collapse. Hence, the curves become flat in the range of negative venous pressures. The plot in Fig. 3 is known as the *vascular function curve*. By convention, plots of the vascular function curve have cardiac output (the independent variable) on the *y* axis and venous pressure (the dependent variable) on the *x* axis. Thus, any point (e.g., B) on the vascular function curve is associated with a specific cardiac output and a venous pressure. This convention facilitates a graphic analysis that is useful in determining cardiac output, as explained later.

Periphery Interacts with the Heart through Changes in Venous Pressure

In Chapter 13, we saw that venous filling pressure was an important determinant of stroke volume. If arterial pressure, heart rate, and contractility are held constant, cardiac output will be a positive function of venous pressure. The relationship between venous pressure and cardiac output for the heart is shown by the *cardiac function curve* in Fig. 4. The cardiac function curve presented here is similar to the *ventricular function curve* described in Chapter 13 except that cardiac output rather than stroke work is plotted on the vertical axis. We chose cardiac output so that the curve can be co-plotted with the vascular function curve. Keep in mind that unlike the ventricular function curve, the cardiac output/venous pressure plot is affected by changes in both contractility and aortic pressure.

Although venous pressure determines the output of the heart, venous pressure is, in turn, inversely related to cardiac output, as shown by the vascular function curve in the same figure. Thus, a constant interplay occurs between the heart and blood vessels at the level of venous pressure. As cardiac output tries to increase, venous pressure decreases, thus bringing cardiac output back to its original value in a self-regulating fashion. To calculate the equilibrium cardiac output, we must treat the two relationships as simultaneous equations and solve for the one cardiac output and venous pressure pair that would satisfy both curves. In Fig. 4, we have solved the simultaneous equations graphically by co-plotting them. Where the two curves cross is their equilibrium point. If the heart described by the cardiac function curve in Fig. 4 is connected to the vascular system depicted by the venous function curve in Fig. 4, then the cardiac output will be 5.5 L/min and the venous pressure will be 1 mm Hg. It is important for the student to understand that the system cannot operate off the equilibrium point. *The only way that cardiac output can be altered is to change one or both of the curves.*

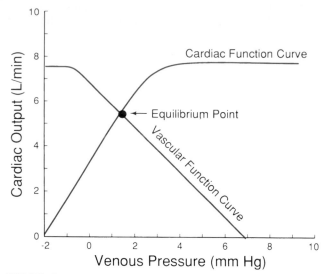

FIGURE 4   Simultaneous plots on the same axis of the vascular function and cardiac function curves. The intersection of the two curves reveals the steady-state cardiac output and venous pressure that must exist in the system.

CARDIAC OUTPUT CAN BE CHANGED ONLY BY ALTERING EITHER THE CARDIAC OR THE VENOUS FUNCTION CURVE

Effect of Changing Contractility

In Fig. 5, stimulation of the cardiac sympathetic nerves has shifted the cardiac function curve upward and to the left as a result of increased ventricular contractility and heart rate. As predicted (see Chapter 13) this results in a greater cardiac output for any given filling pressure.

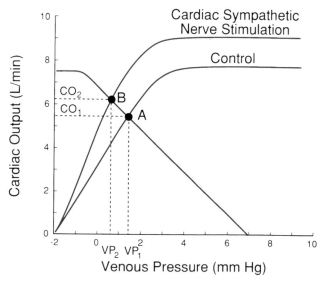

FIGURE 5   Effect of cardiac sympathetic nerve stimulation on the cardiovascular state as depicted on a simultaneous plot of the cardiac and vascular function curves.

Note that the equilibrium point has also been shifted from A to B. Cardiac output will be increased as expected, but the increased cardiac output caused venous pressure to fall, thus blunting the increase in output.

Effect of Changing Blood Volume

It was noted previously that mean circulatory filling pressure is solely determined by the blood volume and the capacitance of the blood vessels. Mean circulatory filling pressure (cf. the venous function curve's *x* intercept in Fig. 6) would be increased by increasing

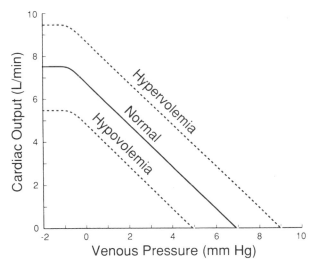

FIGURE 6 Effect of altering blood volume on the vascular function curve.

the blood volume, resulting in hypervolemia. Similarly, mean circulatory filling pressure would be decreased by reducing the blood volume, resulting in hypovolemia. Assuming that TPR remains constant, the difference in arterial and venous pressure required to force any given cardiac output across the TPR will be unchanged regardless of the blood volume and, therefore, the slope of the curves remains the same. *Changes in blood volume shift the vascular function curve in a parallel fashion,* as indicated in Fig. 6.

An increase in blood volume resulting from a blood transfusion (or perhaps renal fluid retention) would shift the vascular function curve up and to the right, as shown in Fig. 7. Note the change in the equilibrium point from A to B in Fig. 7. *Hypervolemia causes cardiac output to increase with a concomitant increase in venous pressure.* Contrast the effect of hypervolemia with increasing contractility in Fig. 5, in which cardiac output increases but venous pressure falls.

Effect of Changing Venous Tone

Venoconstriction decreases the unstressed volume of the venous compartment and shifts the function curve in the same manner as transfusion. Whether one makes the container smaller (venoconstriction) or the contents larger (transfusion), the effect will be the same: a greater venous pressure for any given cardiac output. *Venodilation increases* the unstressed volume and shifts the vascular function curve in the same manner as

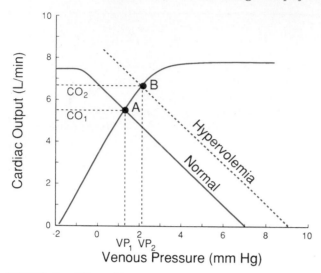

FIGURE 7   Effect of hypervolemia on the cardiovascular state as depicted on a simultaneous plot of the cardiac and vascular function curves.

hemorrhage. Cardiac output is carefully controlled in the body primarily in association with blood pressure control. The body adjusts both blood volume and venous tone as an integral part of this control system. The former is associated with long-term control, primarily by the kidney over hours to days, whereas the latter can be adjusted by the central nervous system in a moment-to-moment fashion. These control systems are explained in detail in Chapter 15.

Effect of Changing Arteriolar Tone

Arteriolar constriction or relaxation has a different effect on the vascular function curve from that seen with changing blood volume. At zero cardiac output, altering the TPR will not affect the mean circulatory filling pressure because the volume of blood contained within the arterioles is negligible. However, except when cardiac output is zero, an increase in arteriolar constriction increases the pressure difference required to drive the cardiac output across the TPR. To illustrate this, let us return to our original example in which the heart has been replaced by a mechanical pump. If the output of the heart-lung machine is held constant, say, at 4 L/min, a sudden increase in TPR will temporarily decrease blood flow across the microcirculation. The mismatch between cardiac output and flow through the microcirculation will cause blood to accumulate in the arterial compartment until arterial pressure becomes high enough to force the entire cardiac output across the increased TPR. The additional blood in the arterial compartment came from the venous compartment. Therefore, increasing the TPR will cause venous

pressure to be lower for any cardiac output, causing a counterclockwise rotation of the vascular function curve about the mean circulatory filling pressure, as is shown in Fig. 8. Arteriolar dilation decreases the TPR and has the opposite effect, a clockwise rotation.

Increasing the TPR will cause the equilibrium point to move toward reduced cardiac output and venous pressure (A to B in Fig. 8). Decreasing the TPR will likewise increase the cardiac output and the venous pressure (A to C in Fig. 8). But there is more to the story. Because changes in TPR have a marked effect on arterial blood pressure, they also shift the cardiac function curve. In Chapter 13, we learned that end-systolic volume is directly related to the arterial pressure. As a result, stroke volume decreases when arterial pressure increases. We therefore have to draw a new cardiac function curve down and to the right (equivalent to that seen with reduced contractility) when TPR is increased, as shown in Fig. 8. This shift of the ventricular function curve further exaggerates the effect on cardiac output beyond what would be expected by the change in venous pressure alone. Therefore, the equilibrium point actually moves from A to D (in Fig. 8) after an increase in TPR. Decreasing the TPR has the opposite effect and increases stroke volume, moving the cardiac function curve up and to the left, so the equilibrium point actually moves from A to E. Unlike the effect on cardiac output, the change in venous pressure resulting from a change in TPR will be ambiguous because both curves are altered. Depending on the exact characteristics of the individual, venous pressure may either rise or fall with a change in TPR.

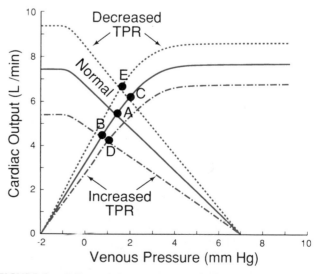

FIGURE 8   Effects of changes in total peripheral resistance on the vascular and cardiac function curves.

Intrathoracic Pressure Affects the Cardiac Function Curve

Intrathoracic pressure is the pressure presented to the outside of the heart. The pressure surrounding the heart acts in parallel with the compliance of the cardiac musculature to oppose ventricular distension by the venous pressure. End-diastolic volume will, therefore, be determined by the difference between venous pressure and intrathoracic pressure. As a result, changing intrathoracic pressure shifts the cardiac function curve horizontally along the x axis. Normally, intrathoracic pressure is 2–3 mm Hg negative because of the breathing mechanics. Increasing the intrathoracic pressure to atmospheric by opening the chest will reduce the cardiac output and increase the venous pressure because it will shift the ventricular output curve about 2–3 mm to the right.

RIGHT HEART AND LEFT HEART PUMP AS ONE UNIT

In our analysis we have considered systemic venous pressure to be the filling pressure of the heart. Systemic venous pressure is, of course, the filling pressure only for the right heart. In actuality, the output of the left heart will have to equal the output of the right heart since they are in series. If the left heart's output were to lag behind that of the right, blood would accumulate in the pulmonary veins and raise filling pressure to the left heart. The increase in stroke volume would quickly correct any mismatch. Thus, the two ventricles are in tandem and really do work as one unit. This relationship is maintained even if one of the chambers is diseased and begins to fail. Our analysis suggests that venous pressure could never exceed mean circulatory pressure. However, if only one side of the heart fails, the healthy ventricle can actively pump blood into the veins feeding the failing side, raising venous pressure on that side to 20–30 mm Hg. That is enough pressure to force fluid out of the capillaries and cause edema upstream of the failing ventricle.

HEART RATE AFFECTS CARDIAC OUTPUT IN A TRIPHASIC MANNER

Cardiac output is calculated as heart rate times stroke volume. It would be reasonable, then, to assume that heart rate is an important determinant of cardiac output. Changes in heart rate within the physiologic range, however, have surprisingly little effect on cardiac output, as is shown in Fig. 9. That is because venous return strongly governs cardiac output in the normal heart. If the cardiac output starts to rise with

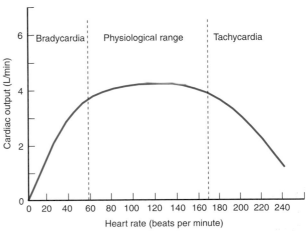

FIGURE 9　The heart rate is varied by electrical pacing and the effect on cardiac output is plotted. Note that cardiac output is relatively insensitive to changes in rate within the physiologic range, but cardiac output falls dramatically at both the high and low extremes.

an accelerated heart rate, venous pressure simply falls and stroke volume is reduced proportionately. If the heart is electrically paced to a very high rate (tachycardia) or decreases to a very low rate (bradycardia), cardiac output is severely impaired. At heart rates below 50 beats/mm, as might occur in a patient with complete heart block, stroke volume is maximal and can no longer compensate for the reduced rate. At heart rates above 150 beats/min, as might occur in a patient in ventricular tachycardia, the filling time is so short that the heart fails to fill adequately between beats. It is interesting to note that when the heart rate is increased by sympathetic stimulation, the duration of the action potential is reduced in the ventricle so that the systolic period is shortened, thus allowing more time for filling between beats. That phenomenon allows the heart to pump efficiently at rates in excess of 200 beats/min during exercise.

REGULATION OF CARDIAC OUTPUT

Any change in cardiac output must be the result of a change in either the vascular or the cardiac function curves. Neural, endocrine, and other regulatory mechanisms constantly change cardiac output to meet the body's metabolic needs by altering those curves. Often, simultaneous changes in several parameters work in concert or in opposition. For example, a decrease in blood volume may have little or no effect on mean circulatory filling pressure if the decreased blood volume is accompanied by venoconstriction and a resulting decrease in unstressed volume.

It is useful for the student at this point to consider the cardiovascular system in three levels. The first level is to understand the three determinants of the heart's stroke volume: filling pressure, contractility, and afterload. The second level is to appreciate the mechanical interaction between the vascular system and the heart as described by the co-plotting of the cardiac and venous function curves. The final level is to appreciate how the body can capitalize on that interaction to actually modulate the system through the various control systems present in the body. These control systems have only four variables that they can alter—venous tone, TPR, contractility, and blood volume—to effect the desired change in cardiac output. As will be learned in the following chapters, there are very few blood flow sensors in the system, so that the body's control systems actually monitor arterial pressure and then change cardiac output appropriately to control that pressure.

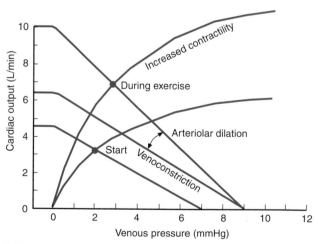

FIGURE 10   Alterations in the cardiac and vascular function curves produced by exercise. Before exercise, this patient's cardiac output is 3.3 L/min, and during exercise, it increases to 7 L/min. The venous function curve is shifted up and to the right due to venoconstriction and the further rotation indicated by the *arrow* is due to arteriolar dilation in the exercising muscles, which decreases TPR. The cardiac function curve is shifted up and to the left due to an increased contractility.

Exercise Moves Both the Cardiac and Ventricular Function Curves

Consider exercise as an example of this control. The onset of exercise causes a sudden drop in TPR as the exercising muscles dilate their arterioles through a phenomenon termed *active hyperemia* in order to increase their perfusion (see Chapter 17). Hence, the venous function curve is rotated clockwise, as shown in Fig. 10. The decreased TPR causes a drop in blood pressure, which is sensed by pressure sensors called *baroreceptors* (see Chapter 15) that reflexively cause venoconstriction, causing a rightward shift of the venous function curve that further increases cardiac output. Finally, the heart's contractility is increased as part of the baroreflex, shifting the ventricular function curve in Fig. 10 upward and to the left, further increasing cardiac output. To promote even more perfusion of the muscles, the central nervous system will increase the set point for arterial pressure so that the baroreflex will actually overcompensate for the transient fall in arterial pressure and elevated blood pressure will occur during the exercise period.

CONCLUSION

In summary, the vascular function curve is directly altered by changing one or more of the following: (1) TPR, (2) venous tone, and/or (3) blood volume. The cardiac function curve is directly altered by changing (1) heart rate, (2) cardiac contractility, (3) TPR, and/or (4) external cardiac pressure. The only parameter directly affecting *both* relationships is TPR. All changes in cardiac output must involve alteration of at least one of the two function curves. In the following chapters we learn how the body alters these parameters to cause meaningful and appropriate changes in cardiac output.

Suggested Readings

Green JF. Determinants of systemic blood flow. *Int Rev Physiol Cardiovasc Physiol III* 1979;8:33–65.

Guyton AC, Jones CE, Coleman TG. *Circulatory physiology: cardiac output and its regulation.* Philadelphia: WB Saunders, 1973.

Levy MN. The cardiac and vascular factors that determine systemic blood flow. *Circ Res* 1979;44:739–746.

Regulation of Arterial Pressure

D. NEIL GRANGER

KEY POINTS

- The *vasomotor center* integrates information from sensory receptors that monitor various cardiovascular parameters and sends efferent commands to regulate cardiac function and arteriolar tone.
- *Vasomotor tone* is a partial state of vascular smooth muscle contraction sustained by continuous repetitive firing of sympathetic vasoconstrictor fibers.
- *Baroreceptors* are stretch receptors that provide the central nervous system with information regarding the blood pressure. The firing frequency of the baroreceptors is proportional to the pressure applied to them.
- The baroreceptor reflex allows the vasomotor center to respond to a reduced arterial pressure by increasing the firing rate of sympathetic vasoconstrictor fibers to the arterioles so that peripheral resistance increases.
- The *carotid baroreceptors* monitor arterial pressure within the range of 50–200 mm Hg, whereas the *aortic baroreceptors* monitor pressure between 100 and 200 mm Hg.
- The baroreceptors adapt to increases in arterial pressure after 1–2 days; hence, the baroreceptor system does not contribute to the long-term regulation of arterial pressure.
- Low-pressure receptors in the atrium help control blood volume, rather than blood pressure, through the reflex release of antidiuretic hormone.
- The reduced fluid loss from the kidneys initiated by antidiuretic hormone increases blood volume and consequently raises blood pressure.
- Changes in the chemical environment within the carotid and aortic chemoreceptors activate the vasomotor center to elicit vasoconstriction when arterial pressure falls below 80 mm Hg.

Essential Medical Physiology, Third Edition

KEY POINTS (*continued*)

- The most important long-term controller of arterial pressure is the kidney, which makes this contribution through its ability to regulate blood volume. The kidney uses adjustments in urine output as a means for regulation of blood volume and arterial pressure.

OVERVIEW OF ARTERIAL PRESSURE REGULATION

In the previous chapter, the contributions of various cardiovascular parameters, including peripheral vascular resistance, venous tone, and contractile state, to the regulation of cardiac output were examined. In this chapter, we explore the mechanisms that are directed toward regulating arterial blood pressure. The overall strategy of the cardiovascular system is to *provide all organs with a constant perfusion pressure and to allow each individual organ to regulate its blood flow in accordance with the local needs of the tissue.* In Chapter 11, it was noted that virtually all capillary beds perfusing the various tissues of the body are arranged in parallel and, thus, the driving force for blood flow, the mean arterial pressure (MAP), is the same for each tissue. Increased demand for blood flow within a particular organ activates local mechanisms that result in a reduction in arteriolar resistance. As long as MAP remains constant, blood flow into that tissue will increase as mandated by Ohm's law (see Chapter 11). MAP, however, remains constant only if cardiac output exhibits an equal increment. If cardiac output does not increase accordingly, a reduction in the vascular resistance of one organ would decrease total peripheral resistance, which in turn would decrease MAP and cause blood flow to fall in all of the other organs. The overall effect would be to divert a greater fraction of cardiac output through one organ at the expense of others. *Hence, to prevent changes in blood flow to one organ from interfering with flow to the other organs, MAP must be invariant.*

Cardiovascular control is designed so that MAP is closely monitored and normally held at approximately 100 mm Hg. If, for example, a decrease in MAP is sensed, the control systems will alter cardiac function and vascular resistance so that MAP is quickly returned to its normal value. To meet the moment-to-moment changes in local tissue metabolism, regional vascular resistances are continuously being adjusted, usually by local mechanisms that do not involve the central nervous system. Central control mechanisms, on the other hand, are directed toward sensing any change in MAP and reacting to such changes by adjusting the activities of the heart and vascular smooth muscle, thereby promptly restoring MAP to its normal value. The regulation of MAP involves both neural reflexes, which can restore MAP within seconds, and renal-hormonal pathways that have time constants measured in hours or even days. In the following sections, each component of blood pressure control is examined and its relative importance explained.

NEURAL EFFECTORS FOR BLOOD PRESSURE CONTROL

As was mentioned in the previous chapter, the heart receives efferent innervation from both the sympathetic and parasympathetic nervous systems. *Sympathetic cardiac efferents* release norepinephrine, which acts via various β-adrenergic receptors to increase heart rate and contractility. *Parasympathetic cardiac efferents* release acetylcholine, which stimulates muscarinic receptors and reduces heart rate and, to a lesser extent, myocardial contractility.

Both arteriolar and venous smooth muscles are innervated by *sympathetic efferent fibers* that release norepinephrine, which acts via α-adrenergic receptors to produce smooth muscle contraction. Sympathetic stimulation of arteriolar smooth muscle increases resistance to blood flow, whereas activation of sympathetic constrictor fibers to venous smooth muscle decreases unstressed volume and promotes venous return to the heart. Although some vascular beds, like sweat glands and some skeletal muscles, are innervated via *sympathetic dilator fibers,* they do not participate in blood pressure control.

All autonomic efferent (postganglionic) neurons arise from peripheral ganglia, where they receive synaptic input from preganglionic neurons of central origin. In the sympathetic nervous system, the peripheral ganglia are in the *sympathetic chain,* whereas parasympathetic ganglia are near or in the organ they innervate (Chapter 9). The soma of parasympathetic preganglionic neurons are located in the *vagal nuclei,* whereas the soma of sympathetic preganglionic neurons are located in the *intermediolateral column* of the spinal cord. Both sympathetic and parasympathetic preganglionic neurons innervating cardiovascular efferent neurons receive input from nerve tracts originating in the vasomotor center.

NEURAL INTEGRATION IN THE VASOMOTOR CENTER

The *vasomotor center* is located in the reticular substance of the medulla and a portion of the pons. Its anatomy and functional organization is complex and still not completely understood; however, it is considered to have at least four important functional areas: (1) the *vasoconstrictor region* (often referred to as C-l), which is located in the upper medulla and lower pons, sends fibers into the spinal cord where they activate sympathetic vasoconstrictor neurons; (2) the *vasodilator region* (often referred to as A-l), which is located in the lower half of the medulla, sends axons into the vasoconstrictor region that appear to inhibit neurons within C-l; (3) a *sensory region,* which is located in the tractus solitarius of the medulla and the pons, has neurons that receive sensory input from glossopharyngeal and vagal fibers; and (4) the *cardiac center* regulates heart rate and contractility. In terms of the cardiac center, direct electrical stimulation of one part of it, the *cardiac accelerator area,* increases heart rate and contractility via the sympathetic nerves; stimulation of another part of the cardiac center, the *cardiac inhibitory area,* decreases the heart rate via vagal fibers.

Within the vasomotor center, spontaneous activity is integrated with information from sensory receptors that monitor various cardiovascular parameters. Efferent commands are then generated as a result of this integration to regulate cardiac and vascular function in a purposeful manner. Neurons emanating from the vasoconstrictor region are always active. This ongoing activity produces a continuous repetitive firing of sympathetic vasoconstrictor fibers that sustains a partial state of vascular smooth muscle contraction, referred to as *vasomotor tone.* Interrupting that basal sympathetic activity by spinal anesthesia will reduce MAP from 100 mm Hg to about 50 mm Hg, as shown in Fig. 1. Vasomotor tone is important because the sympathetic constrictor fibers are usually not paired with corresponding dilator fibers. The vasomotor tone, however, provides the vasoconstrictor center with the ability to dilate blood vessels by simply withdrawing sympathetic tone.

Arterial Pressure Sensors

Baroreceptors are stretch receptors located within the walls of the heart and some blood vessels. Baroreceptors

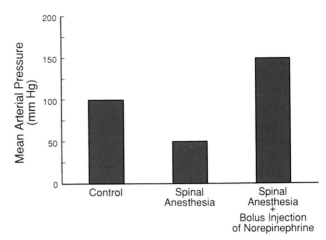

FIGURE 1   Sympathetic maintenance of mean arterial pressure. Effects of spinal anesthesia and spinal anesthesia plus intravenous injections of norepinephrine.

provide the central nervous system with information regarding the blood pressure. When they are stretched by pressure within the vessel wall, they elicit action potentials in the sensory neurons that innervate them. The firing frequency of all baroreceptors is proportional to the pressure applied to them.

One set of baroreceptors is within the walls of the *carotid sinuses* near the bifurcation of the common carotid arteries. Sensory fibers from the carotid baroreceptors enter the central nervous system via the *sinus nerve (Hering's nerve),* which is a branch of the glossopharyngeal nerve. These afferent nerves go to the sensory region of the vasomotor center, in particular the *tractus solitarius.*

Baroreceptor Reflex

Cutting the sinus nerves causes an immediate rise in MAP. Similarly, electrical stimulation of the central end of the bisected sinus nerve produces a reduction in MAP. These experiments, in conjunction with others that have examined associated alterations in autonomic control of the heart and the vasculature, have characterized the *carotid baroreceptor reflex.* As illustrated in Fig. 2, an increase in MAP stretches the walls of the carotid sinus. This stretch increases the firing rate of sensory fibers in the carotid sinus nerve, which informs the vasomotor center that arterial pressure is above 100 mm Hg, which seems to be the body's *set point* for MAP. The vasomotor center responds by increasing the firing rate of cardiac vagal fibers and decreasing the firing rate of cardiac sympathetic fibers so that heart rate and contractility are decreased. Simultaneously, the vasomotor center will decrease the firing rate of sympathetic vasoconstrictor fibers

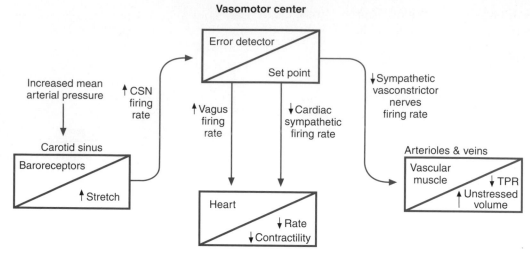

FIGURE 2   Functional organization of the carotid sinus reflex. CSN, carotid sinus nerve.

to the arterioles and veins so that peripheral resistance falls and the unstressed volume increases. All of these changes will act to reduce arterial pressure. Cutting the sinus nerves would be interpreted as a precipitous drop in MAP by the vasomotor center, hence, the pressor response. Similarly, repetitive stimulation of the nerve would be interpreted as an increased MAP.

A *negative feedback* relationship exists between carotid sinus pressure and MAP as a result of the baroreceptor reflex. This negative feedback can be appreciated by dissociating carotid sinus pressure from MAP. This can be accomplished easily by independently perfusing the carotid sinus via a cannula, while MAP is monitored. When the carotid sinus is independently perfused, MAP becomes inversely related to pressure in the carotid sinus, as shown in Fig. 3.

Additional baroreceptors, the *aortic baroreceptors,* are located in the wall of the aortic arch. Pressure

information from these receptors is carried to the sensory region of the vasomotor center via vagal afferent fibers. Both the carotid and aortic baroreceptors monitor arterial pressure, and the information from each is combined in the vasomotor center. The dynamic ranges of these two receptor sets are different because of the structure of the receptors themselves As shown in Fig. 4, the carotid baroreceptors provide the vasomotor center with information regarding arterial pressure within the range of 50–200 mm Hg, whereas the aortic baroreceptors can only provide pressure information within the range of 100–200 mm Hg.

Both the carotid and aortic baroreceptors are *rate sensitive,* which means that a rising pressure causes a more rapid rate of firing than a steady pressure. As a result, *the carotid and aortic baroreceptors provide the vasomotor center with information regarding both MAP and arterial pulse pressure.* Recordings from single baroreceptor afferents demonstrate that information pertaining to both of these pressures is encoded within the pattern of action potentials generated by the baroreceptors. It has been further demonstrated that the vasomotor center uses both mean and pulse pressure information to modulate cardiovascular effectors. Figure 5 illustrates this principle. As in Fig. 3, the carotid sinus is perfused by a pump and MAP is measured. The family of curves represents different pulse pressures in the carotid sinus. Note that *the more pulsatile the pressure in the carotid sinus, the greater the inhibitory effect seen on arterial pressure.* Thus, the functional relationship between carotid sinus pressure and MAP is shifted when carotid sinus pressure is pulsatile, and that effect is most pronounced at lower pressures.

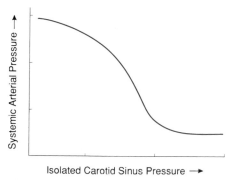

FIGURE 3   Functional relationship between carotid sinus pressure and mean arterial pressure in an isolated carotid sinus preparation.

Baroreceptor Dynamic Range

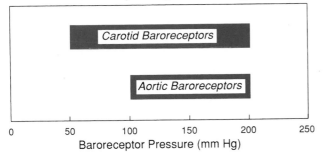

FIGURE 4   Baroreceptor dynamic range. Pressure ranges over which the carotid and aortic baroreceptors can monitor arterial blood pressure.

Importance of the Baroreceptor Reflex

The carotid sinus nerves and the vagal nerves from the aortic arch are often called *buffer nerves* because, by monitoring arterial pressure, they help the vasomotor center to buffer any change in MAP associated with daily activity. This buffer function can be readily demonstrated with the frequency distribution of arterial pressure over a 24-hr period for a normal dog and the pressure distribution recorded in the same dog several weeks after baroreceptor denervation (Fig. 6). Figure 6 reveals that when the buffer nerves are intact, MAP rarely deviates more than 10 or 15 mm Hg away from the 100-mm Hg set point. However, when the buffer nerves are sectioned, pressure varies widely throughout the day, often 50 mm Hg or more above or below the 100-mm Hg set point. Note from the figure that denervation of the baroreceptors does not raise the average blood pressure on a long-term basis. The reason is that the baroreceptor reflex adapts to the pressure

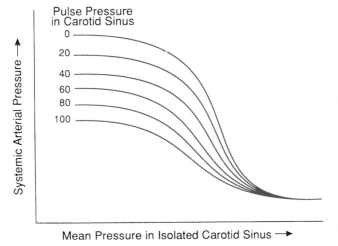

FIGURE 5   Pulse pressure modulation of the functional relationship between mean arterial pressure and systemic vascular resistance resulting from the baroreceptor reflexes.

input after a few days. This experiment vividly demonstrates that long-term control of blood pressure is regulated by other mechanisms, as explained later.

It is not a coincidence that the baroreceptors are located in the upper thorax and neck. It is vital that the brain and heart have a continuous blood flow because of their high metabolic rates. Interruption of that flow for even a brief period can spell disaster for an individual. To protect those two circulations, the baroreceptors are located close to the brain (carotid sinuses) and the heart (aortic baroreceptors). In an erect adult, there may be a 100-mm Hg difference between the pressure in the cerebral arteries and those in the foot because of the hydrostatic column between them. The strategic location of the baroreceptors prevents pressure differences within the vascular system from depriving the brain and heart of an adequate perfusion pressure.

As we will see in Chapter 17, the blood flow of many organs is closely coupled to their metabolic needs via local control systems. Organs exhibiting such a tight local control include the brain, the heart, the digesting intestines, and exercising skeletal muscles. Other organs such as the skin, resting skeletal muscle, or nondigesting intestines tend to be well perfused with respect to their nutritional needs, and their blood flows are greatly influenced by the autonomic nerves. The latter vascular beds participate in baroreceptor reflexes from the vasomotor center by sacrificing their flow when MAP falls. When MAP begins to fall, vascular resistances in the gut and skeletal muscle increase to buffer the changes in MAP, while local control mechanisms in the heart and brain override any central influences on their blood flows. Skeletal muscle and gut are more readily recruited into the baroreceptor reflex when they are resting, with most neural commands being overridden when they are active and a high blood flow is required.

The cardiovascular control system is also able to handle perturbations associated with stressful situations such as an extreme drop in cardiac output resulting from hemorrhage. Under such a stress, peripheral sympathetic stimulation will be so intense that even the vascular beds under strong local control (except the brain and heart) will constrict. The resulting increase in total peripheral resistance will be at the expense of tissues that can tolerate a period of reduced blood flow and divert vital flow to the brain and heart. If that condition persists, however, the organs that sacrifice their flows will experience ischemic tissue injury.

Low-Pressure Receptors

The atria and pulmonary arteries also contain stretch receptors that are often referred to as *low-pressure receptors* because they are responsive to pressures several times lower than those in the systemic arteries. These atrial receptors are also often referred to as *volume receptors* because venous pressure is primarily dependent on total blood volume and because their primary role is to help control blood volume rather than arterial pressure. But, as you will recall, blood volume is a primary determinant of cardiac output and, thus, arterial pressure. A fall in atrial pressure initiates a reflex release of *antidiuretic hormone* (also known as *vasopressin*) from the hypothalamus. Antidiuretic hormone causes the kidney to reabsorb filtered fluid at the level of the collecting ducts. Stimulation of the low-pressure receptors also causes constriction of the afferent renal arterioles, and consequently reduces the glomerular filtration rate, by a neural reflex. Both effects *reduce fluid loss at the kidneys, which in turn increases blood volume*. Vasopressin also constricts arteriolar and venous smooth muscles, particularly in the splanchnic circulation, producing an increase in peripheral vascular resistance and a reduction in unstressed volume, respectively. All of these effects tend to raise blood pressure. The importance of the low-pressure receptors can best be appreciated by denervation studies. When both the arterial baroreceptors and the low-pressure baroreceptors are denervated, a transfusion of blood will cause MAP to increase more than twice as much as would occur if only the arterial baroreceptors were denervated.

Heart rate can also be affected by stimulation of atrial baroreceptors. This reflex, called the *Bainbridge reflex*, can increase heart rate by 75%. Receptors eliciting the reflex are located in both the right and left atria. Their afferent fibers are located in the vagus nerves, and the efferent limbs of the reflex are present in both sympathetic and parasympathetic cardiac nerves. The magnitude and even the direction of the reflex are influenced by the prevailing heart rate. When heart rate is low, an increase in atrial pressure tends to increase heart rate, whereas at high basal heart rates increasing atrial volume may actually slow the heart. The Bainbridge reflex is purposeful in that failure of the heart to keep up with venous return will result in elevated atrial pressure and stimulation of the heart. On the other hand, it opposes the baroreceptor reflex under most conditions in which filling pressure to the heart is reciprocally related to cardiac stimulation. Finally, the Bainbridge reflex is extremely selective in that sympathetic activity to the heart is increased with virtually no change in arteriolar and venous smooth muscle tone in the periphery. The one exception is the dilatation of renal arterioles, which promotes fluid loss through enhanced glomerular filtration.

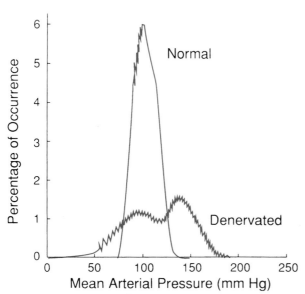

FIGURE 6   Variations in mean arterial pressure over a 24-hr period in a normal dog before and several weeks after baroreceptor denervation. (Modified from Guyton AC, Hall JE. *Textbook of medical physiology,* 9th ed. Philadelphia: WB Saunders, 1996.)

OTHER NEURAL REFLEXES INVOLVED WITH BLOOD PRESSURE REGULATION

Although the baroreceptor reflex is the primary moment-to-moment controller of aortic pressure, other mechanisms exist that come into play *under periods of extreme cardiovascular stress*. These are discussed in the following subsections.

Chemoreceptors

More severe reductions in mean arterial pressure (i.e., when MAP falls below 80 mm Hg) can also be detected by *chemoreceptors* located in the *carotid* and *aortic bodies*. These structures are located near the bifurcation of the common carotid arteries and along the aortic arch. They receive blood via a small nutrient artery and have an extremely high rate of oxygen consumption for their size. Each body contains cells that are sensitive to levels of oxygen, carbon dioxide, and hydrogen ion. Although much of their sensory input goes to the respiratory control centers, some of it goes to the vasomotor center as well. A decrease in MAP causes blood flow to these bodies to fall with a corresponding reduction in oxygen delivery (hypoxia) and an accumulation of CO_2 and H^+. Information regarding these changes in the chemical environment within the carotid and aortic bodies is transmitted via the carotid sinus and vagus nerves to the vasomotor center. The excited vasomotor center then generates the same responses that it would issue for a fall in carotid sinus pressure, with a resultant rise in arterial pressure. The chemoreceptors are thought to be more intimately involved in detecting changes in blood oxygenation than blood pressure. Hence, only when MAP is drastically reduced do the chemoreceptors reinforce the carotid baroreceptors' efforts to stimulate the vasomotor center to restore the blood pressure.

Cerebral Ischemia

If MAP is reduced to 60 mm Hg or less, blood flow to the brain becomes significantly impaired. When flow to the brain is reduced below an adequate level, ischemia results. In cerebral ischemia, tissue levels of CO_2 and hydrogen ion increase. Chemosensitive cells within the vasomotor center respond directly by sending efferent commands that strongly stimulate both the vagal and sympathetic nerves. Blood flow to virtually all other tissues including the kidneys is reduced in a last ditch attempt to maintain the required flow to the brain and heart. What is peculiar about the cerebral ischemic response is that simultaneous stimulation of both the vagus and the sympathetic nerves results in an increase in myocardial contractility but a net slowing of the heart rate, because the vagus has the stronger influence at the sinoatrial node. Thus, the cerebral ischemic response is characterized by a profound bradycardia.

It is extremely important to recognize that reduced blood flow to the brain may occur for reasons other than a reduced MAP. If cerebrospinal fluid pressure begins to increase, the vessels in the brain will be compressed and cerebral blood flow will fall. A common cause of increased cerebrospinal pressure is bleeding within the skull. In an attempt to restore blood flow to the brain after a cerebral hemorrhage, blood pressure may exceed 250 mm Hg. The pressor response induced by an increase in intracranial pressure is referred to as the *Cushing reaction*. Although the Cushing reaction may temporarily restore cerebral flow, it can also cause further hemorrhage. Thus, a positive feedback situation can occur in which death rapidly ensues unless the intracranial pressure is relieved.

Hormonal Control Mechanisms

Four hormonal mechanisms are clearly identified as important in blood pressure control: (1) epinephrine and norepinephrine from the *adrenal medulla*, (2) vasopressin released from the hypothalamus, (3) the *renin-angiotensin system*, and (4) atrial natriuretic peptide released in the atria.

The adrenal medulla is a functional extension of the sympathetic nervous system. Activation of the preganglionic fibers to the adrenal medulla triggers the release of both epinephrine and norepinephrine directly into the circulation. Norepinephrine is primarily an α-agonist, whereas epinephrine is almost equally divided between α and β effects. Activation of the adrenal medulla usually occurs in concert with activation of the sympathetic nerves to the cardiovascular effectors, and the circulating catecholamines extend and augment their action. The vasoconstrictor effects of circulating catecholamines persist for only 2–3 min, which is the time required to degrade the adrenergic transmitters. Certain vessels not innervated by adrenergic neurons, such as the metarterioles, can be constricted via this hormonal pathway.

In the section on atrial baroreceptors, it was noted that atrial stretch and increased atrial pressure reduce the hypothalamically controlled secretion of antidiuretic hormone from the posterior pituitary gland. Antidiuretic hormone functions in both the acute and chronic regulation of blood pressure. As the name implies, the antidiuretic action of the hormone plays an important role in regulating the production of urine by the kidney. Although the hormone can contribute to the pressor responses observed with hemorrhage, it is not thought to be involved in moment-to-moment blood pressure control. Antidiuretic hormone is more often released in response to *osmoreceptors* in the hypothalamus, which signal that extracellular fluid is becoming hypertonic.

Another important consequence of increased filling of the right atrium and the resultant increase in right atrial pressure is the release of atrial natriuretic peptide (ANP) from the myocytes that bear this increased mechanical load. ANP acts on the peripheral vasculature and kidneys in a manner that favors a reduction in arterial blood pressure. The smooth muscle surrounding arterioles relax when exposed to ANP, resulting in a reduction in vascular resistance. ANP also diminishes the barrier function of endothelial cells lining capillaries and postcapillary venules, which favors a redistribution of plasma volume to the extravascular space. Natriuresis, diuresis, and a reduction in renin release are important responses of the kidney to ANP. The combined actions of ANP to cause vasodilation and reduce plasma volume account for the tendency of elevated plasma ANP levels to exert a hypotensive effect. ANP is metabolized by neutral endopeptidases (NEP) that are found in blood, the kidneys, lungs, the central nervous system, and other tissues. Because drugs that inhibit NEP increase plasma ANP levels several times normal and for several hours, NEP inhibitors are gaining recognition as potential therapeutic agents for the management of patients with chronic arterial hypertension.

The kidney functions as an endocrine organ as well as a blood purifier. When renal arterial pressure decreases, the *juxtaglomerular cells* within the wall of the afferent arterioles secrete *renin*. Renin secretion can also be stimulated from these cells by activation of their sympathetic innervation. Renin is an enzyme that converts plasma *angiotensinogen* into *angiotensin I*. As blood passes through the lungs, angiotensin I is converted into *angiotensin II* by a *converting enzyme* produced by the pulmonary endothelial cells. Angiotensin II is one of the most potent vasoconstrictors known. On a weight basis, it is four to eight times more active than norepinephrine. Angiotensin II increases total peripheral resistance and elicits an intense venoconstriction. The renin-angiotensin system represents a mechanism for regulating renal blood flow at the expense of mean arterial pressure. For example, if blood flow to the left kidney is compromised as a consequence of atherosclerotic narrowing of the left renal artery, the kidney will interpret that flow reduction as a low blood pressure and therefore secrete renin in an effort to elevate blood pressure. In restoring the pressure downstream from the atherosclerotic stenosis back to a normal level, the previously underperfused kidney will elevate MAP to an abnormally high level. This condition, which exemplifies the potential impact of the renin-angiotensin system in regulating blood pressure, is called *renovascular hypertension* and can often be cured by simply bypassing the narrowed artery with a grafted vessel.

Endothelial Mechanisms

Endothelial cells contribute to the regulation of arterial blood pressure by regulating vascular tone through activation or inactivation of various circulating vasoactive substances, and by producing numerous agents that can act locally and/or systemically to affect vascular tone. Nitric oxide (NO) is a potent vasodilator that is produced in endothelial cells via the oxidation of L-arginine by the enzyme NO synthase. Inhibitors of NO synthase elicit a reduction in blood flow to some tissues and a concomitant rise in arterial blood pressure, suggesting that NO produced by endothelial cells exerts a tonic inhibitory influence on arteriolar tone, thereby contributing to basal blood pressure. Nitric oxide also mediates the endothelium-dependent vasodilation that is elicited by a variety of agents, including acetylcholine and bradykinin. Patients with some forms of chronic arterial hypertension exhibit an impaired endothelium-dependent vascular relaxation, further supporting the possibility that an inability of endothelial cells to produce NO can lead to dysregulation of blood pressure.

Endothelin-1 (ET-1) is a potent vasoconstrictor peptide that is produced by vascular endothelial cells. Upon release by endothelial cells, ET-1 interacts with specific receptors, designated as ET_A and ET_B, on adjacent smooth muscle cells to mediate contraction. Elevated plasma levels of ET-1 are associated with hypertension, while treatment with antagonists directed against ET_A lowers blood pressure, suggesting that ET-1 contributes to the regulation of blood pressure. Furthermore, the vasoconstrictor responses to ET-1 are exaggerated, and there is an increased expression of ET-1 in the arteries of hypertensive patients.

LONG-TERM REGULATION OF ARTERIAL PRESSURE

Although neural reflexes may buffer moment-to-moment changes in MAP, they can quickly adapt to a continued stimulus, thereby negating their effectiveness in long-term control of blood pressure. *The most important long-term controller of arterial pressure is the kidney, which makes this contribution through its ability to regulate blood volume.* Fluid loss by the kidney is determined directly by the pressure perfusing it. The higher the kidney's perfusion pressure, the more fluid will be filtered, which ultimately contributes to the urine. When perfusion pressure is low, less urine is formed and less salt and water will be lost. Also, the renin-angiotensin system amplifies the relationship between renal perfusion pressure and fluid loss. Angiotensin II causes the kidneys to selectively retain

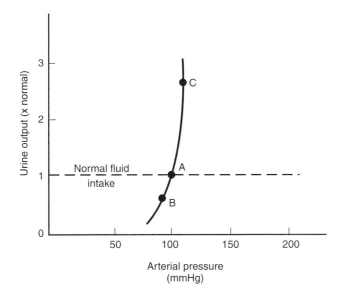

FIGURE 7 Relationship between arterial pressure and urine output in a normal person. The intersection of the broken line with the curve (point A) represents the point at which fluid intake equals urine output—normal arterial pressure. Point B illustrates the influence of a reduced arterial pressure on urine output, and point C shows the urine output response to an increased arterial pressure.

salt and water, both through a direct effect and through its effect on the release of *aldosterone,* a hormone that controls salt excretion (see Chapter 32 for a detailed account of aldosterone's action). By retaining salt and water when MAP is low, blood volume and the mean circulatory filling pressure are increased, which, in turn, increases cardiac output and finally restores the MAP. This renal-blood volume mechanism for regulating blood pressure requires a few hours to manifest a significant response; however, unlike the baroreceptor reflex, it never fatigues or adapts, and it will eventually restore blood pressure all the way back to normal.

Figure 7 shows the effect of arterial pressure on urine output in a normal person and it demonstrates the sensitivity of the renal responses to changes in arterial blood pressure. In a normal individual with a MAP of 100 mm Hg, urine output over a 24-hr period will equal the amount of fluid intake (point A in Fig. 7). However, if this individual experiences significant blood loss, the arterial pressure will initially fall (with an effort by neural mechanisms to raise the pressure toward normal) and the kidneys will produce less urine (point B in Fig. 7). If fluid intake remains normal, then the fluid retention that accompanies the reduced urine output will eventually (requiring days to weeks) restore blood volume back to normal, leading to reestablishment of arterial pressure and urine output. On the other hand, if arterial pressure is elevated above normal, the renal

output of water (and salt) increases profoundly above fluid intake (point C in Fig. 7); the resulting fluid loss reduces blood volume, and arterial pressure eventually returns to normal. The steepness of the relationship between urine output and arterial pressure illustrates how effectively the kidneys can respond to even small changes in arterial pressure and in doing so use adjustments in urine output as a means for precise regulation of arterial pressure.

OTHER BLOOD VOLUME-RELATED MECHANISMS INVOLVED IN BLOOD PRESSURE REGULATION

Although the kidneys play a dominant role in restoring blood pressure to normal values over the long term, there are other blood volume-related mechanisms that can contribute to arterial pressure regulation. Two notable mechanisms that become activated within minutes to hours after a pressure change and whose effects can last for days are (1) the *capillary fluid shift mechanism* and (2) the *stress-relaxation mechanism*. The former mechanism involves a role for capillary hydrostatic pressure in promoting the filtration or absorption of fluid in the microvasculature during periods of high or low arterial pressure, respectively. Vascular beds that most likely contribute to this process include skeletal muscle and the splanchnic circulation. For example, if capillary pressure rises in response to increased blood volume or arterial pressure, then fluid will be filtered out of skeletal muscle capillaries at an accelerated rate. The resultant reduction in blood volume that occurs over hours will tend to restore arterial pressure and capillary pressure back toward their control values.

The stress-relaxation mechanism represents an effort by the blood vessels to fit around the existing blood volume. This phenomenon is related to the intrinsic ability of smooth muscle to return to its original force of contraction after it has been elongated or shortened. Stress-relaxation is more pronounced in the visceral smooth muscle surrounding hollow organs like the urinary bladder, but it does occur to some extent in vascular smooth muscle. Hence, if blood volume and arterial pressure become too high, the blood vessels are stretched and then slowly relaxed as though to accommodate a larger intravascular volume. As a result, the pressure within the arterial tree will fall toward normal. With hemorrhage, on the other hand, a reverse stress-relaxation will cause blood vessels to contract around the reduced blood volume, thereby tending to buffer the fall in arterial pressure.

Clinical Note

A common disease in Western cultures is *hypertension* (high blood pressure). With hypertension the MAP is elevated 10 or more mm Hg above 100 mm Hg. Unfortunately, the body does not tolerate hypertension well. The high MAP puts an increased workload on the heart, causing it to enlarge dangerously; it also promotes accelerated atherosclerosis of blood vessels. Hypertension is an established risk factor for coronary artery disease, stroke, and renal failure. Although some patients have renal hypertension as described earlier, most (>90%) suffer from *essential hypertension*. The word *essential* simply means that the cause is unknown; however, there is a strong hereditary tendency for this disease. The treatment strategies for hypertension illustrate how the basic physiology of blood pressure control can be manipulated. One of the approaches is to limit salt intake and to give the patient a *natriuretic* drug like furosemide, which causes the excretion of both salt and water, thus reducing blood volume. Another approach has been to give a *cardiac depressant,* like propranolol, which blocks norepinephrine's effect on the heart and reduces cardiac output. Still another approach has been to give a *vasodilator* like hydralazine to reduce peripheral resistance and unstressed volume. One of the more popular therapies today is to target angiotensin II bioactivity using either *angiotensin-converting enzyme inhibitors,* which block the formation of angiotensin II, or angiotensin II receptor blockers. All of these approaches have undesirable side effects but have been effective in controlling a deadly disease. It is hoped the defect in blood pressure control in these patients will someday be identified so the disease can be treated at its source.

Suggested Readings

Cowley AW. Long-term control of arterial pressure. *Physiol Rev* 1992,72:231.

Guyton AC, Hall JE. *Textbook of medical physiology,* 10th ed. Philadelphia: WB Saunders, 2000.

Loscalzo J, Creager MA, and Dzau VJ. *Vascular medicine: A textbook of vascular biology and diseases,* 2nd ed. Boston: Little, Brown & Co., 1996.

Oparil S, Chen YF, Naftilan AJ, Wyss JM. Pathogenesis of hypertension. In Fozzard HA, ed., *The heart and cardiovascular system,* 2nd ed. New York: Raven Press, 1991, pp 295–331.

Sagawa K. Baroreflex control of systemic arterial pressure and vascular bed. In *Handbook of physiology: the cardiovascular system-peripheral circulation and organ blood flow,* Sec 2, Vol 3. Bethesda, MD: American Physiological Society, 1983.

Zucker IH, Gilmore JP. *Reflex control of the circulation.* Boca Raton, FL: CRC Press, 1991.

Capillary Exchange

D. NEIL GRANGER

KEY POINTS

- The microstructure of the capillary wall varies between tissues, with the highest rate of exchange of water and proteins noted in tissues perfused by discontinuous capillaries, followed by continuous and fenestrated capillaries, respectively.
- Capillaries in most tissues behave like negatively charged filters; that is, they retard the transport of negatively charged plasma proteins while enhancing the exchange of positively charged proteins.
- The magnitude of solute exchange across a vascular bed is determined by the number of perfused capillaries and the patency of the endothelial cell junctions.
- The rate and direction of fluid movement across capillaries is governed by the balance of hydrostatic and oncotic pressures exerted across the capillary wall.
- *Conditions that elevate capillary hydrostatic pressure,* such as arteriolar dilation or an increased venous pressure, favor the filtration of fluid out of the capillaries. *Conditions that*

elevate plasma oncotic pressure favor the absorption of fluid from the interstitium to the capillary lumen.
- The *lymphatic system* provides the only means for removal of plasma proteins that escape the blood and enter the interstitium.
- The principal driving force for the entry of interstitial fluid into the terminal lymphatics is the *interstitial fluid pressure.*
- When capillary pressure is increased, the increased interstitial hydrostatic pressure and lymph flow and reduced interstitial oncotic pressure that result help prevent excessive fluid accumulation in the interstitium.
- *Edema* generally occurs when the rate of fluid filtration out of a capillary bed exceeds the ability of its lymphatic drainage to return the filtered fluid to the vascular system.
- The three most common vascular abnormalities that result in interstitial edema are an increased capillary hydrostatic pressure, a reduction in plasma oncotic pressure, and an increased capillary permeability.

Essential Medical Physiology, Third Edition

The cardiovascular system exists for one fundamental purpose: the delivery of nutrients to, and removal of metabolic end products from, tissues. The exchange of solutes and water between blood and the interstitial compartment largely occurs in capillaries, which are ideally designed for this function. The numerous thin-walled and porous capillaries that are found in all tissues provide a huge surface area for exchange of gases (oxygen, carbon dioxide), water, nutrients (glucose, amino acids, fatty acids), and hormones, whereas other elements of blood (such as platelets and plasma proteins) do not leave capillaries readily. Although the magnitude and direction of capillary exchange is generally described in terms of simple physical principles such as diffusion and convection (bulk flow), active changes in the shape and function of vascular smooth muscle and endothelial cells in arterioles and capillaries can exert a profound influence on these exchange processes. Hence, capillary exchange is a dynamic process that often changes in relation to organ function. Furthermore, many diseases are associated with abnormalities in capillary exchange that may become so severe as to impair organ function (e.g., pulmonary edema) and cause death.

FUNCTIONAL ANATOMY OF THE MICROCIRCULATION

Figure 1 illustrates an extensive capillary network and its associated structures (arterioles and venules) that constitute the microcirculation. Blood enters the microcirculation through *arterioles,* which are surrounded by a thick, continuous layer of smooth muscle. Contraction of smooth muscle reduces the internal diameter of this microvessel and consequently increases the resistance to blood flow in the entire vascular bed. This feature makes the arteriole the major resistance element in the circulation and the principal determinant of the total peripheral resistance. Arteriolar smooth muscle tone also governs the amount of pressure transmitted from arteries to veins; hence, capillary pressure falls when arterioles constrict and rises when arterioles dilate. Blood flows from arterioles into a narrower vessel, the *metarteriole,* which is surrounded by a discontinuous smooth muscle layer. Capillaries branch off from the metarteriole. The density of capillaries, which is an important determinant of the total area available for exchange between blood and tissue, varies significantly between organs, with the lung exhibiting the largest capillary area (3500 cm^2/g) compared with muscle (100 cm^2/g). The junction between the metarteriole and some capillaries is encircled by a single band of smooth muscle called the *precapillary sphincter*. These sphincters determine the percentage of capillaries open to blood perfusion. Although all capillaries are normally open to perfusion (and exchange) in tissues like the heart, only 20–30% of capillaries are normally open in skeletal muscle and intestine. In the latter tissues, relaxation of the precapillary sphincter allows for the recruitment of more open capillaries and hence greater transcapillary exchange. Capillaries coalesce into a *venule,* which

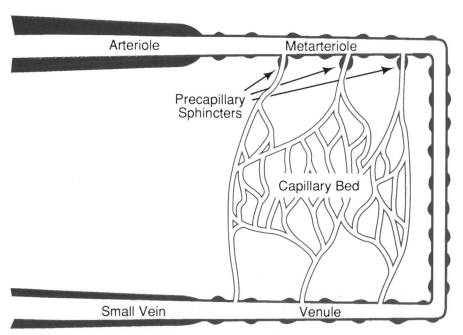

FIGURE 1   Schematic diagram of the microcirculation.

possesses a discontinuous, thin coat of smooth muscle that drains into small veins. Changes in venular smooth muscle tone can exert a significant influence on capillary exchange inasmuch as constriction of venules leads to an increased capillary pressure, whereas dilation of venules exerts the opposite effect.

PATHWAYS AND MECHANISMS OF TRANSCAPILLARY SOLUTE EXCHANGE

The term *exchange vessel* is often used instead of capillary by microvascular physiologists to denote the fact that exchange processes occur on both sides of the anatomic capillary bed, with oxygen readily diffusing across metarterioles and proteins leaking across the smallest venules. Nonetheless, a substantial component of fluid and solute exchange does indeed take place across capillaries; our discussion of this process will be largely confined to this component of the microcirculation. The microstructure of the capillary wall varies between tissues, with capillaries classified into three ultrastructural types based on unique characteristics of the endothelial cell and its underlying basement membrane: continuous, fenestrated, and discontinuous. Some of the characteristics shared by all three types of capillaries include a single layer of highly attenuated endothelial cells that are less than 0.2 μm thick (except in the region of their nuclei); thin slits (*clefts*) that are created by the close apposition of adjacent endothelial cells; a *basement membrane* lying beneath the endothelial cells, which is composed of fine fibrillar material capable of retarding the passage of macromolecules; and *micropinocytotic vesicles* that are formed on the luminal and abluminal surfaces of the endothelial cells and found within their cytoplasm.

Continuous capillaries (Fig. 2a), the most abundant and widespread type, are found in muscle, skin, lung, connective tissue, and the nervous system. Lipid-soluble (lipophilic) solutes, including dissolved oxygen and carbon dioxide, can take the *transcellular route,* which involves simple diffusion through the two cell membranes that make up the capillary wall. Hydrophilic solutes and water can cross the capillary wall through intercellular junctions (clefts), which in most capillary beds have a width of approximately 60 Å. These clefts can be easily traversed by water, ions, and small organic solutes dissolved in the plasma; however, albumin (with a 70-Å diameter) and other plasma proteins have a difficult time penetrating the clefts. Continuous capillaries in the nervous system exhibit a functionally closed cleft between adjacent endothelial cells, giving rise to vessels that are impermeable to passive diffusion of solutes as small as glucose and some electrolytes.

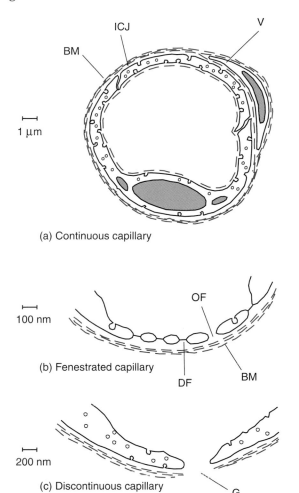

(a) Continuous capillary

(b) Fenestrated capillary

(c) Discontinuous capillary

FIGURE 2   Schematic diagram of a cross section through the wall of a continuous-, fenestrated-, and discontinuous-type capillary. The following routes of transcapillary exchange are noted: *BM,* basement membrane; *ICJ,* intercellular junction; *V,* vesicles; *OF,* open fenestrae; *DF,* diaphragmmed fenestrae; *G,* gap.

A potentially important route for transfer of plasma proteins between blood and the interstitial compartment is the micropinocytotic vesicle. Plasma proteins are thought to cross the capillary wall by endocytosis at the luminal surface, vesicular transport through the cytoplasm, and exocytosis at the interstitial side of the endothelial cell.

Fenestrated capillaries (Fig. 2b) are about 10 times more permeable to water and small (hydrophilic) solutes than most continuous capillaries. The unique structural feature that accounts for the increased permeability of these microvessels is the perforations (fenestrae) in the endothelium with diameters between 500 and 600 Å. The fenestrae are mostly open holes, but in some vascular beds (intestine) the endothelial perforations are bridged by a thin membrane or diaphragm, which can offer significant restriction to the passage of plasma proteins. The existence of these diaphragms accounts for the fact that fenestrated capillaries appear to be as equally impermeable to plasma proteins as are many continuous-type capillaries, but they are far more permeable to water. Hence, it is not surprising that fenestrated capillaries are found in tissues specializing in fluid exchange (renal glomeruli, exocrine glands, intestinal mucosa, choroid plexus).

In discontinuous (sinusoidal) capillaries (Fig. 2c), adjacent endothelial cells do not meet to form clefts at all. As a consequence, intercellular gaps as large as 1000–10,000 Å are formed. These capillaries are highly permeable to water and plasma proteins. They are found in organs in which red blood cells are often seen to migrate between blood and tissue, that is, spleen, bone marrow, and liver.

Anatomic and physiologic studies indicate that there is an increasing capacity for water and solute exchange across endothelial cells from the arterioles to the postcapillary venules. This axial gradient in vascular permeability in continuous and fenestrated capillaries can be explained by larger cleft widths on the venous end of capillaries and in small venules, compared with arterial segments of the capillary. In fenestrated capillaries, the density of fenestrae that pierce the endothelium also increases in frequency from arterial to venous ends of the capillary. A consequence of the axial differences in exchange pathway dimensions and frequency is the tendency to lose more plasma proteins from the venous side of the microcirculation, both under normal conditions and in pathologic states accompanied by increased protein extravasation (e.g., inflammation).

The principal mechanism responsible for the movement of small solutes between the blood and interstitium is simple diffusion down chemical activity gradients. Dissolved gases and small lipophilic molecules can rapidly diffuse across the entire endothelial cell membrane, with the rates of exchange within a tissue governed primarily by their respective concentration gradients and the number of capillaries that are perfused and thus open for exchange. Although small hydrophilic solutes like glucose and amino acids largely cross the capillary wall through a more restricted area of the capillary wall, that is, the interendothelial clefts, their rates of diffusion are extremely rapid, reaching an equilibrium across the capillary wall within a fraction of a second. Another mechanism that contributes to transcapillary solute exchange is convection or bulk flow. This process represents the transfer of dissolved solutes across a membrane via the movement of its solvent, for example, filtered plasma. As the fluid filtration rate across the capillaries increases, so does the convective transfer of solutes dissolved in plasma. The diffusional exchange of small solutes across the capillaries is much

faster than that which can be achieved by convection; however, the latter process contributes significantly to the movement of plasma proteins from blood to interstitium. In some tissues (e.g., intestine) convection accounts for almost half of the plasma protein extravasation observed under normal conditions and essentially all of the protein exchange that occurs in conditions associated with a very high capillary fluid filtration rate.

The rate of efflux of plasma proteins from the capillaries can also be profoundly influenced by the electrostatic charges of the protein and the interendothelial cleft. Albumin is a negatively charged plasma protein that appears to permeate capillaries at a rate that is 1/100 the rate observed for hemoglobin, a protein of nearly identical molecular size to albumin but which is almost electrostatically neutral. Physiologic studies in organs such as the kidney and lung have revealed that the capillaries in these tissues behave like a negatively charged filter, that is, they retard the transport of negatively charged plasma proteins while enhancing the exchange of positively charged proteins. Hence, it is generally assumed that the fixed negative charges associated with mucopolysaccharides found on the endothelial cell surface and in interendothelial clefts normally serve to minimize the loss of proteins from the blood. Pathologic conditions that result in the loss of these negative surface charges on capillaries, such as glomerular capillaries in patients with nephrotic syndrome, are characterized by excessive protein loss and interstitial edema.

The exchange of solutes across capillaries is a very dynamic process, with large fluctuations often occurring in association with accelerated organ function (e.g., in skeletal muscle during exercise and in intestine after ingestion of a meal) and with organ dysfunction (e.g., liver cirrhosis). Two structural elements of the microcirculation represent major centers of control over the magnitude of solute exchange in a vascular bed: the precapillary sphincter and the endothelial cell junctions. By virtue of its ability to determine the number of perfused capillaries (and the surface area available for exchange) at any given moment, the precapillary sphincter can govern the overall flux of solutes from blood into the interstitium, and vice versa, within an organ. Contraction of the sphincters will reduce the surface area for exchange and thereby limit overall solute (and water) transfer, whereas dilation of the sphincter will increase the number of perfused capillaries and enhance overall solute transfer. The contractile state of the endothelial cell can also exert a profound influence on the rate of solute (and water) exchange across a vascular bed. Endothelial cells possess the capacity to contract through a process that involves rearrangement of cytoskeletal elements such as *F-actin* and *myosin*. When endothelial cells do contract in response to a physiologic stimulus, this results in an opening of the cleft between adjacent cells, with a consequent increase in capillary permeability. Bradykinin and histamine are examples of endogenous inflammatory agents that can interact with their receptors on endothelial cells and subsequently cause cell contraction and an increased transfer of solutes across the microvasculature. These agents and many others appear to exert their permeability-increasing effects predominantly on the venous side of the capillary and on small venules, suggesting that the density of their receptors is greater on endothelial cells in this segment of the microcirculation.

Fluid Movement across the Capillary

In most organs, fluid is normally filtered across capillaries, passes through the interstitial spaces, and returns to the bloodstream via the lymphatic system. This process of fluid filtration is driven by a small imbalance in the hydrostatic and osmotic pressures that are normally exerted across the walls of capillaries. Inasmuch as 2–3 L of fluid are filtered across all capillaries in the body each day, regulation of this process is critical for the maintenance of normal plasma and interstitial volumes, with abnormalities of capillary filtration forces often resulting in abnormal fluid distribution and organ dysfunction.

Fluid movement across the capillary wall is a passive process that can be described by the relation (Starling equation):

$$J_v = K_f[(P_c - P_{if}) - \sigma(\pi_c - \pi_{if})],$$

where J_v is the rate of fluid flow, K_f is the *filtration coefficient* of the capillaries, P_c is the *capillary hydrostatic pressure*, P_{if} is the hydrostatic pressure of the interstitial fluid surrounding the capillaries, σ is the *reflection coefficient* of the capillaries to plasma proteins, π_c is the oncotic (colloid osmotic) pressure generated by plasma proteins within the capillaries, and π_{if} is the oncotic pressure generated by proteins in interstitial fluid. The term $[(P_c - P_{if}) - \sigma(\pi_c - \pi_{if})]$ is referred to as the *net filtration pressure* (NFP) when it has a positive value, and the *net absorptive pressure* (NAP) when it is a negative value. When an NFP exists, J_v is also positive and fluid flow across the capillary wall is directed from the plasma to the interstitial fluid (termed *filtration*). An NAP indicates that J_v is negative and fluid flows from the interstitial compartment to the plasma (termed *absorption*).

Permeability and Exchange Surface Area Determinants of Capillary Filtration

Two components of the Starling equation relate transcapillary fluid flow to the number and size of capillary pores that are available for fluid exchange: K_f and σ. The filtration coefficient, expressed as milliliters per minute per mm Hg, is an index of the capacity of a membrane to transport water for a 1-mm Hg pressure gradient across that membrane. In tissues, K_f (LpS) is determined by the number and size of pores in endothelial cells of each capillary (Lp) as well as by the number of perfused capillaries (S). Hence, any condition that results in either an increased capillary permeability (increased pore size), an increased number of perfused capillaries, or both will lead to an elevated rate of transcapillary fluid flow. The capillary reflection coefficient (σ) is a membrane parameter that describes the fraction of plasma and interstitial oncotic pressures transmitted across the capillary wall. When plasma proteins are too large to permeate (100% reflection) the capillary pores (e.g., blood–brain barrier), σ is 1.0, whereas σ is equal to 0 when proteins penetrate and cross the pores as easily as water. If capillary permeability is increased, σ falls and the effectiveness of π_c and π_{if} in influencing the rate of capillary fluid flow is diminished.

The values for K_f and σ differ considerably between tissues. Organs (liver, spleen) perfused by capillaries of the discontinuous (sinusoidal) type generally exhibit the highest values for K_f and the lowest values for σ, indicating that the very large interendothelial clefts in these capillaries can sustain high fluxes of both water and proteins. On the opposite end of the capillary permeability spectrum are the organs (muscle, lung) perfused by continuous capillaries, which tend to exhibit much lower values for K_f and higher values for σ, with the brain representing the extreme case in this group. The substantial differences in K_f that are noted between some organs in this group, such as skeletal muscle versus lung, are largely due to corresponding differences in capillary density, rather than permeability (pore size). In all organs, there appears to be a gradient in K_f that increases progressively from arterial to venous ends of the capillary.

A number of physiologic and pathologic conditions are associated with changes in K_f and/or σ. For example, the capillary recruitment (increased number of perfused capillaries) that occurs in exercising skeletal muscle leads to an increased K_f and a corresponding increase in capillary filtration. Ingestion of a lipid meal appears to affect both K_f and σ in capillaries of the intestinal mucosa. The meal-initiated rise in K_f and fall in σ suggest that there is an increased number of perfused capillaries (K_f) as well as an increased capillary permeability, which may serve to facilitate the entry of newly absorbed lipids into the blood and lymph circulations. Inflammatory conditions are known to be associated with the release of vasoactive agents that can profoundly increase both capillary permeability (decrease σ) and perfused capillary density (increase K_f), with a resultant enhancement of capillary fluid filtration that often leads to interstitial edema.

Hydrostatic Forces Acting on the Capillary Wall

The gradient of hydrostatic pressures between the capillary lumen (P_c) and its surrounding interstitial compartment (P_{if}) is a major determinant of the rate and direction of transcapillary fluid flow. P_c tends to push fluid out of the capillaries, whereas *interstitial fluid pressure* opposes this outward movement of plasma water. A number of factors govern the pressure generated within capillaries, most notably arterial pressure, arteriolar resistance, venous pressure, and venous resistance. Consequently, there are significant differences in average capillary pressure between tissues, with organs like the kidney and skeletal muscle exhibiting higher P_c than lung or liver. Typically, there is a gradient of pressure along the length of the capillary, with pressures ranging between 25 and 45 mm Hg on the arterial end and between 10 and 15 mm Hg on the venous end of the capillary. This axial gradient in hydrostatic pressure largely accounts for the view that capillaries tend to filter fluid on the arterial end and then reabsorb some of the filtered fluid on the venous end (with the remaining fluid entering the lymphatics). Conditions that elevate P_c, such as arteriolar dilation or an increased venous pressure, would favor the filtration of fluid along the entire length of the capillary. In most tissues, changes in venous pressure and/or venous resistance have a more profound effect on P_c than comparable changes on the arterial side of the circulation. For example, with a 10-mm Hg increment in venous pressure, 8 mm Hg is transmitted to the capillaries, whereas a similar increase in arterial pressure will result in less than a 1-mm Hg rise in P_c. However, a significantly greater fraction of arterial pressure is transmitted to capillaries when arterioles dilate, as seen in inflammatory conditions in which P_c is elevated secondary to an intense vasodilatory response.

The P_{if} in an organ is determined by the interstitial fluid volume and the compliance of its interstitial compartment. Interstitial volume varies widely between tissues, from as little as 10% of total tissue volume in skeletal muscle to more than 30% in skin. Similarly, the absolute values of P_{if} differ between tissues, with

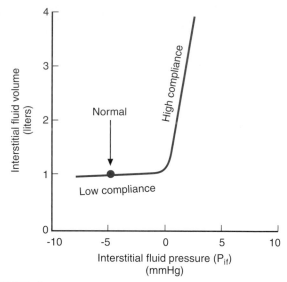

FIGURE 3 Relation between the interstitial fluid pressure and interstitial fluid volume. Note that small increments in interstitial volume elicit large increases in interstitial pressure in the low-compliance region, whereas much larger increases in volume are required to elevate pressure further when interstitial compliance is high.

subatmospheric values measured in lung, skeletal muscle, and subcutaneous tissue, whereas positive values (1–3 mm Hg) are obtained in liver, kidney, and intestine. These resting values of P_{if} can increase or decrease dramatically if fluid is added to or removed from the interstitial compartment. The fact that adding small volumes of fluid into the normal interstitium causes large increases in P_{if} suggests that interstitial compliance is normally low (Fig. 3). This allows for some control of the capillary filtration rate because the sudden increases in J_v that accompany a rapid rise in P_c can be offset by a simultaneous rise in P_{if}. However, if there is excessive accumulation of fluid in the interstitium, the interstitial matrix components are disrupted and interstitial compliance rises. With an elevated compliance in the interstitium, large volumes of fluid can accumulate without major changes in P_{if}. Thus, the capacity for increases in P_{if} to oppose further capillary filtration is virtually lost when the compliance of the interstitial compartment increases greatly above normal.

Osmotic Forces Acting across the Capillary Wall

As discussed in Chapter 2, molecules that fail to penetrate the pores of a semipermeable membrane will exert an osmotic pressure across that membrane proportional to the concentration of the impermeant molecule. Proteins are the only dissolved constituents of plasma and interstitial fluid that do not readily penetrate the pores of the capillary membrane. Hence, these proteins are largely responsible for the osmotic pressures (commonly referred to as *colloid osmotic pressures* or *oncotic pressures*) that are exerted across the capillary wall. The concentration of proteins in plasma (7.0 g/dL) can generate an osmotic pressure of about 28 mm Hg. A similar protein concentration and oncotic pressure is found in interstitial fluid of tissues perfused by discontinuous-type capillaries (liver, spleen). Consequently, the transcapillary oncotic pressure gradient, which is essentially zero, does not influence the rate of fluid filtration across these capillaries. However, in tissue perfused by continuous or fenestrated capillaries, the interstitial protein concentration (and oncotic pressure) is only one-third to one-half that of plasma. In these organs, a significant oncotic pressure exists across the capillary wall, with plasma oncotic pressure favoring the withdrawal of fluid from the interstitial compartment.

A number of factors can influence the oncotic pressures generated in plasma and interstitial fluid. Because plasma oncotic pressure is generated by proteins dissolved in plasma, any factor that affects plasma protein concentration can alter π_c. Sudden increases in plasma volume, as might occur in patients receiving large volumes of physiologic salt solutions, will cause π_c to fall, with a resultant increase in capillary fluid filtration. A sudden increase in P_c can also alter the transcapillary oncotic pressure gradient because the accompanying accumulation of filtered fluid in the interstitium will dilute the interstitial proteins and reduce π_{if}. A consequence of the dilution-related fall in π_{if} is a braking effect on capillary filtration because π_{if} normally acts to facilitate fluid movement from blood to interstitium. If capillary permeability is increased, then plasma proteins leak out of the blood into the interstitial compartment of the affected tissue, thereby leading to a rise in π_{if} and a dissipation of the normal transcapillary

oncotic pressure gradient. Here again, the ultimate outcome is an accelerated rate of capillary fluid filtration. Another perturbation that will cause π_{if} to rise is obstruction of lymph drainage from a tissue. The lymphatic system provides the only means for the removal of plasma proteins that escape the blood and enter the interstitium. If lymph drainage is obstructed, then the interstitial protein concentration will eventually equilibrate with the plasma protein concentration and the normal transcapillary oncotic pressure gradient will be dissipated.

LYMPHATIC DRAINAGE SYSTEM

All of the fluid and protein that normally accumulate in the interstitial compartment as a consequence of capillary filtration are efficiently removed and carried back to the bloodstream via the lymphatic system. It is estimated that for a 70-kg man, the lymphatics return nearly 3 L of fluid and approximately 120 g of protein to the bloodstream every 24 hr. Because the recycling of plasma proteins is essential for maintenance of the normal oncotic pressure gradient between plasma and interstitial fluid, any injury or blockage of the main lymphatic channels may be life threatening.

Figure 4 illustrates some of the major functional elements of the lymphatic system. The sac-like *terminal lymphatics* are blind-ended endothelial tubes suspended in the interstitium by connective tissue anchoring filaments. The gaps between the endothelial cells of the terminal lymphatics are very large, which, in the absence of a basement membrane, allows ready access of interstitial fluid, along with its suspended particles (lipoproteins) and cells (lymphocytes), to the lymphatic vessel lumen. Hence, the composition of lymph is considered to be identical to that of interstitial fluid.

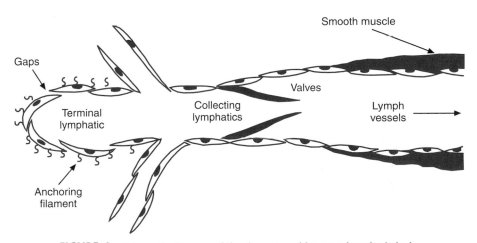

FIGURE 4   Schematic diagram of the elements making up a lymphatic bed.

The terminal lymphatics coalesce into larger vessels, *collecting lymphatics,* that possess one-way valves similar to those in the veins. These lymph channels ultimately drain into the thoracic duct, which drains the lower part of the body, and the right lymph duct, which drains the upper body, and in turn empties into the venous system at the internal jugular and subclavian veins.

The principal driving force for the entry of interstitial fluid into the terminal lymphatics is the P_{if}. As fluid filtered from capillaries accumulates in the interstitium, P_{if} rises, thereby creating a hydrostatic pressure gradient between the interstitial compartment and the lymphatic vessel lumen. Expansion of the interstitium also causes the anchoring filaments to pull open the terminal lymphatic. This allows the elevated P_{if} to readily propel interstitial fluid into the lymph vessel (Fig. 5). Stretch-mediated, active contractions of lymphatics also favor the propulsion of lymph. When a collecting lymphatic becomes distended with interstitial fluid, a myogenic response is elicited that causes the lymphatic smooth muscle to automatically contract. The one-way pumping action of collecting lymphatics is the most likely explanation for the subatmospheric P_{IF} measured in some tissues, because it literally sucks fluid out of the interstitial compartment. Additional lymph propulsive forces are generated when external compression causes local, transient increases in P_{if}. Because lymph vessels have valves that prevent backflow, external intermittent compression phenomena such as muscular contraction, respiratory movements, and even arterial pulsations tend to push the lymph toward the veins into which the lymphatics terminate. Despite these mechanisms to facilitate lymph flow, in most tissues lymph flow rate

cannot exceed about 20 times normal (Fig. 5). Thus, the capacity for lymph drainage to increase is limited and can be overloaded with high capillary fluid filtration rates. When this occurs, interstitial edema is likely to follow.

INTERACTIONS OF LYMPH FLOW AND INTERSTITIAL FORCES DURING PERIODS OF ENHANCED CAPILLARY FILTRATION

From the preceding discussion it is evident that the direction and magnitude of fluid movement across the capillary wall is determined by the balance of hydrostatic and oncotic forces that are exerted on this membrane. As illustrated in Fig. 6A, the balance of these forces across a typical resting capillary is such that a small (0.2 mm Hg) NFP exists. However, the small NFP is sufficient to generate a capillary fluid flow (J_v) of 0.02 mL/min (derived from the product of K_f and NFP). This low rate of capillary filtration is matched by an equal rate of lymph flow from the tissue and, consequently, the interstitial fluid volume remains constant at a normal level. If, however, the capillary hydrostatic pressure is suddenly increased by 10 mm Hg because of intense arteriolar dilation, the balance of forces across the capillary wall and lymph flow is dramatically altered (Fig. 6B). The rise in P_c results in an instantaneous and comparable increase in NFP. However, as fluid rapidly enters the interstitial compartment, interstitial volume increases, which in turn causes P_{if} to rise and interstitial oncotic pressure to fall (because of dilution of interstitial proteins by filtered water). The elevated P_{if} serves to drive the capillary filtrate into the lymphatics and produce an 18-fold increase in lymph flow. The ultimate impact of these adjustments in P_{if} and π_{if} is an attenuated increase in the net filtration pressure and the creation of a new steady state, wherein the interstitium is slightly more hydrated and lymph flow continues to drain the interstitial fluid at a rate equal to the elevated capillary filtration rate. This condition will continue as long as capillary pressure remains elevated. Once capillary pressure returns to its normal value, the elevated lymph flow will remove the excess capillary filtrate and interstitial volume will be eventually restored to its original value.

INTERSTITIAL EDEMA

In the example presented in Fig. 6, the sudden and substantial elevation in P_c did not lead to an excessive accumulation of fluid in the interstitial compartment because of the adjustments in interstitial hydrostatic and

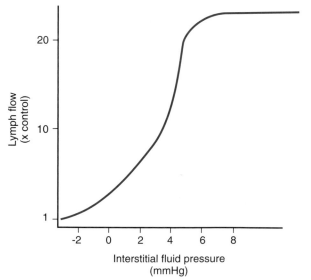

FIGURE 5   Relation between lymph flow and interstitial fluid pressure.

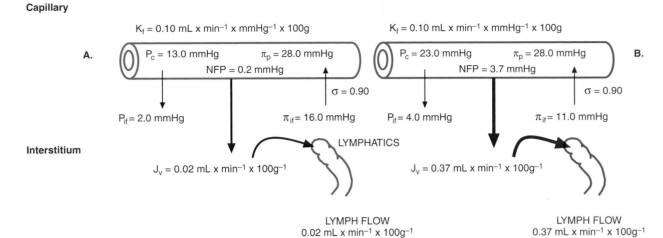

FIGURE 6 Balance of the hydrostatic (P_c and P_{if}) and oncotic (π_p and π_{if}) pressures acting (A) across a normal capillary and (B) after an acute (10 mm Hg) elevation in capillary hydrostatic pressure. Note that capillary fluid filtration rate (J_v) and lymph flow increase almost 20-fold in response to the elevated capillary pressure.

oncotic pressures, as well as the elevated lymph flow. These compensatory responses that prevent excessive fluid accumulation in the interstitium are called the *safety factors* against edema. When these safety factors are exhausted, the fluid filtered from capillaries can accumulate in the interstitium in an unimpeded fashion. This results in *edema* (derived from the Greek word meaning "swelling"), a condition in which there is an abnormal and substantial increase in the interstitial fluid volume. Edema can be localized (to a finger, a limb, or the lungs) or it can be widespread. Edema occurs in a number of diseases and it is a valuable diagnostic sign for the physician. Furthermore, edema can be a life-threatening condition when it occurs in tissues such as the lung, where excessive interstitial fluid accumulation can seriously compromise gas exchange in alveoli. Edema in the brain is also dangerous because the elevated P_{if} can compress and collapse cerebral microvessels, thereby causing ischemia.

Edema generally occurs when the rate of fluid filtration out of a capillary bed exceeds the ability of its lymphatic drainage to return the filtered fluid to the vascular system. Thus, edema results either from an abnormally increased rate of capillary fluid filtration and/or a decreased rate of lymphatic drainage. The three most common vascular abnormalities that result in interstitial edema are an increased P_c, a reduction in plasma oncotic pressure, and an increased capillary permeability.

Hydrostatic edema is the term frequently used to describe the edema resulting from an increased capillary hydrostatic pressure. A common cause of hydrostatic edema is venous occlusion or congestion. Left-sided heart failure, resulting from a myocardial infarction, can lead to increases in left atrial and pulmonary venous pressures. The subsequent elevation in pulmonary capillary hydrostatic pressure can lead to pulmonary edema. In right-sided heart failure, vena caval pressure is elevated, resulting in an increased P_c, particularly in the lower extremities and digestive system. If the edema fluid generated by the intestine and liver accumulates in the peritoneal cavity, it is called *ascites*. An example of a more localized form of hydrostatic edema is the swelling of the lower leg noted in patients with thrombophlebitis of the deep vein in the calf muscle.

Low plasma oncotic pressure is another common cause of interstitial edema. Plasma oncotic pressure is frequently reduced in severe liver disease (the plasma proteins are synthesized by the liver), malnutrition (particularly an inadequate dietary intake of proteins), and renal diseases, resulting in excessive filtration of proteins into the urine. In these conditions, the edema will be widespread. The bulging abdomen of the starving individual generally reflects the accumulation of ascites in the peritoneal cavity secondary to a nutrition-dependent loss of plasma proteins.

Permeability edema results when capillary permeability to water and plasma proteins is profoundly increased. An increased capillary permeability leads to excessive fluid filtration as a consequence of the resultant increase in the capillary K_f and a reduction in the capillary reflection coefficient. The most common causes of an increase in capillary permeability are (1) *burns* and (2) *allergic* and *inflammatory reactions*. Burns can physically damage capillary walls, but when the burn is not severe enough to destroy the epidermis (second-degree burn), an increased capillary permeability in distant organs (lungs) can result from circulating factors released from burned tissue. Allergic and inflammatory reactions are generally associated with mast cell degranulation and the release of histamine and other substances (cytokines) that increase both capillary permeability and P_c (resulting in arteriolar dilatation). A common manifestation of allergic reactions is the formation of subcutaneous *wheals* or *hives,* which are localized regions of edema. Severe generalized allergic reactions such as those seen with *anaphylaxis* may be accompanied by life-threatening edema of the airways. In addition, fluid shifts into the interstitial compartment may be so large as to reduce blood volume to the point at which cardiac output is compromised and *shock* ensues.

Edema resulting from impaired lymphatic drainage is almost always the result of obstructed lymphatic vessels. This can result from malignancies of the lymph nodes or from surgical removal or irradiation of the lymph nodes. In the tropics, a common cause of obstruction of lymph nodes is *filariasis,* which results from infestation of the lymphatics by the parasitic nematode *Wuchereria (Filaria) bancrofti* or *W. malayi.* The adult worm resides in the lymph nodes, where it causes fibrosis and eventual obstruction of lymph flow. This may result in enormous, grotesque swelling of the legs and/or scrotum, aptly described by the word *elephantiasis* (like an elephant).

Suggested Readings

Guyton AC, Hall JE. *Textbook of medical physiology,* 9th ed. Philadelphia: WB Saunders, 1996.

Michel CC. Fluid movements through capillary walls. In *Handbook of physiology, Section 2, The cardiovascular system–microcirculation,* Vol 4. Bethesda, MD: American Physiological Society, 1984.

Shepro D, D'Amore PA. Physiology and biochemistry of the vascular wall endothelium. In *Handbook of physiology, Section 2, The cardiovascular system–microcirculation,* Vol 4. Bethesda, MD: American Physiological Society, 1984.

Taylor AE. Capillary fluid filtration, Starling forces and lymph flow. *Circ Res* 1981;49:557–575.

Taylor AE, Granger DN. Exchange of macromolecules across the microcirculation. In *Handbook of physiology, Section 2, The cardiovascular system–microcirculation,* Vol 4. Bethesda, MD: American Physiological Society, 1984.

Regulation of Regional Blood Flow

D. NEIL GRANGER

KEY POINTS

- The *myogenic theory* of blood flow regulation invokes a role for arteriolar wall tension, rather than blood flow, as the controlled variable in the vasculature.
- The *metabolic theory* of blood flow regulation links blood flow to the metabolic activity of the tissue, with vasodilator metabolites (e.g., adenosine) and low tissue P_{O_2} serving as mediators of metabolic hyperemias.
- Activation of sympathetic nerves, via release of norepinephrine, increases resistance to blood flow in vascular beds that are richly innervated with α-adrenergic nerve fibers.
- Blood flow to the heart is governed almost entirely by intrinsic metabolic factors.
- The influence of autonomic nerve stimulation on coronary blood flow is dictated by the action of the released neurotransmitter on cardiac work, with sympathetic stimulation resulting in an increased flow and parasympathetic stimulation in a reduced flow.
- Cerebral arterioles respond with dilation to increases in metabolic activity in the brain.

- Cerebral arterioles are exquisitely sensitive to the vasodilatory action of an increased arterial blood P_{CO_2}.
- Exercise can elicit a profound increase (20-fold) in skeletal muscle blood flow that is mediated by vasodilator metabolites.
- The increased O_2 demand that is associated with exercise is met by both an increased blood flow and an increased O_2 extraction.
- Gastrointestinal blood flow increases after ingestion of a meal, with digested nutrients, metabolic factors, gastrointestinal hormones, and neuropeptides mediating the active hyperemia.
- Splanchnic blood flow is profoundly reduced during sympathetic activation and with increased blood levels of angiotensin II and vasopressin.
- The amount of heat lost from the body is regulated by the amount of blood flowing through the skin.
- Blood flow through the skin is primarily controlled by sympathetic fibers, whose activity is linked to body temperature.

Essential Medical Physiology, Third Edition

Chapter 15 explained how the cardiovascular system has multiple control systems, all directed at maintaining a constant mean arterial pressure. An advantage of this effort to maintain a constant blood pressure is that each organ system is ensured a high and relatively stable pressure to drive blood flow through its vascular bed. Nonetheless, the burden of responsibility for regulating the rate of blood perfusion in an organ lies within that tissue. Placing the responsibility for blood flow control at the local level appears appropriate in view of the widely varying needs of different tissues for perfusion. In some tissues, such as skin and kidney, the blood flow at any given moment is related largely to the special functions of those organs. However, in most organs, the metabolic needs of the tissue dictate the level of blood perfusion, and several intrinsic control mechanisms exist to ensure that these needs are met by the vasculature. Figure 1 illustrates how cardiac output (total body blood flow) is distributed between tissues under resting conditions. It is readily apparent that blood flow, whether expressed as the total flow per organ or as a fraction of cardiac output, varies substantially between organs.

This chapter explores the factors that enable the different organs in the body to regulate blood flow in accordance with their specific functional and metabolic needs. Special attention is given to those organ systems that exhibit unique blood flow requirements or that regulate blood flow through special mechanisms. (Note that local regulation of blood flow in the lung and kidney are discussed in Chapters 20 and 24, respectively.) The unique anatomic arrangement and blood flow distribution of the fetal circulation are also discussed.

REGULATION OF BLOOD FLOW

As discussed in Chapter 10, the flow of blood through a vascular bed depends on the pressure gradient across it and its resistance to flow. Because arterial and venous pressures are normally maintained within narrow limits, regulation in flow through an organ is achieved by changing the internal diameter of the major resistance vessels, that is, the arterioles. Vascular resistance within many organs is regulated by systems that are *intrinsic* to the organ, as well as by *extrinsic* influences, such as the autonomic nerves and hormones. Intrinsic and extrinsic influences at times may act to induce similar changes in blood flow. At other times, they must oppose one another and balance the needs of the organ with the needs of the entire body.

For most organs, blood flow is normally regulated to ensure the adequate delivery of oxygen (O_2), which is usually the rate-limiting metabolite delivered by the blood. Two vascular elements contribute to the regulation of O_2 delivery to metabolically active tissues: arterioles and precapillary sphincters. Although the arterioles govern the convection (or bulk flow) of O_2 to the organ by regulating blood flow, the precapillary sphincters modulate the diffusive exchange of O_2 by regulating the number of capillaries open to perfusion and hence the surface area available for O_2 exchange. Thus, in some tissues, both arterioles and precapillary sphincters will dilate when O_2 demand increases. As a result of these vascular responses, O_2 supply and demand are matched. The mechanisms responsible for this matching are intrinsic to the tissue and no neural or endocrine influence is involved. The intrinsic ability of a tissue to regulate blood flow is usually defined in terms of the intensity of vasoregulatory phenomena such as pressure flow autoregulation, active (functional) hyperemia, and reactive hyperemia.

The most extensively studied and characterized intrinsic vasoregulatory phenomenon in most organs is *pressure flow autoregulation*. This term refers to the ability of a tissue to maintain a relatively constant blood

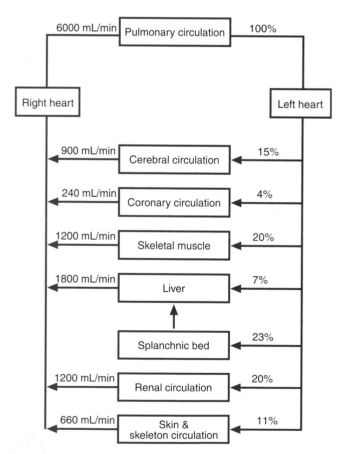

FIGURE 1   The distribution of cardiac output to different organs. Organ blood flows are shown as percentage of cardiac output and as absolute values for the organ or organ system.

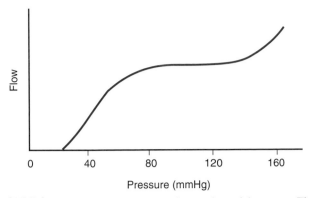

FIGURE 2   Relation between blood flow and arterial pressure. The phenomenon of autoregulation illustrates the ability of an organ to maintain a relatively constant blood flow over a wide range of arterial pressures. In this example, the autoregulatory range lies between 60 and 140 mm Hg.

flow over a wide range of arterial pressures. The phenomenon of autoregulation is illustrated in Fig. 2, which demonstrates that arterial pressure must be reduced below a value of 60 mm Hg before blood flow is significantly compromised. Similarly, increases in arterial pressure above a normal value of 100 mm Hg do not profoundly affect blood flow until arterial pressure exceeds 140 mm Hg. This ability of the tissue to autoregulate blood flow requires that arterioles dilate at low arterial pressures, whereas the vessels must constrict in order to maintain flow when arterial pressure is increased. Although many organs (heart, brain, kidney, intestine, and skeletal muscle) exhibit pressure flow autoregulation, the intensity of this phenomenon varies between tissues, with the more metabolically active organs often exhibiting more precise control of blood flow when arterial pressure is changed.

A second vasoregulatory phenomenon that is generally attributed to intrinsic control of blood flow is *active*

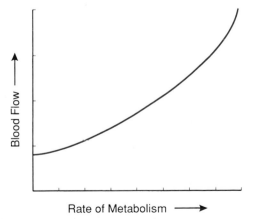

FIGURE 3   The phenomenon of active hyperemia. An organ is being perfused at a constant arterial pressure. As the rate of metabolism of the organ increases, blood flow increases to deliver the additional O_2 needed to meet the metabolic needs.

(or *functional*) *hyperemia*. Figure 3 shows that blood flow to an organ tends to be proportional to its metabolic activity. Hence, when an organ becomes active, this leads to an increased metabolic demand and blood flow rises in an effort to deliver more of the O_2 and nutrients needed to sustain the increased level of activity. In some tissues (e.g., myocardium), increasing blood flow is the only means for enhancing O_2 delivery to more active tissue because opening additional capillaries to facilitate diffusive O_2 exchange is not possible. (In myocardium, nearly all capillaries are open under resting conditions and there are no closed capillaries in reserve.) However, other organs, like skeletal muscle and intestine, are able to meet the increased O_2 demand associated with enhanced organ function (exercise and nutrient absorption, respectively) by increasing both blood flow (arteriolar dilation) and the number of perfused capillaries (precapillary sphincter dilation).

The third phenomenon that supports the existence of intrinsic blood flow control mechanisms is called *reactive hyperemia*. As shown in Fig. 4, when an artery supplying an organ is occluded temporarily, a transient increase occurs in blood flow above the preocclusion level after release of the arterial occlusion. The longer the occlusion, the larger and longer the subsequent increase in blood flow when the occlusion is released. It would appear that the previously ischemic tissue is attempting to repay the blood flow debt that was incurred during the occlusion periods. Clearly, the repayment of this O_2 or blood flow debt is made possible by a progressively more intense dilation of the arterioles as the ischemic duration increases.

The responses associated with the three intrinsic vasoregulatory phenomena described above are consistent with mechanisms geared toward regulating either blood flow and/or the delivery of O_2 (or some other critical nutrient). However, the possibility has been raised that these phenomena are linked more closely to stretch-related events that are intrinsic to smooth muscle. The *myogenic theory* invokes a role for arteriolar wall tension, rather than blood flow, as the controlled variable in the vasculature. This concept evolved from observations that vascular smooth muscle contracts in response to stretch and relaxes when smooth muscle tension is reduced. According to the law of LaPlace, $T = P \times r$, where T is the vessel wall tension, P is the pressure, and r is the radius of the vessel. If T is indeed the controlled variable in arterioles, then one would expect the vessel to dilate (r increases) when pressure is reduced, whereas the opposite would occur when pressure is increased, that is, r must fall in order to maintain a constant T. These properties of smooth muscle may well explain why vascular resistance falls in response to a reduction in arterial pressure

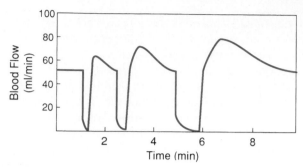

FIGURE 4 The phenomenon of reactive hyperemia. An organ is being perfused at a constant arterial pressure. Blood flow is stopped for 20, 40, and 60 sec. Note that blood flow increased transiently on release of each occlusion. Also note that the magnitude and duration of the hyperemic response increased with the duration of arterial occlusion.

(pressure flow autoregulation), as well as the sudden vasodilation observed in tissues after release of a temporary arterial occlusion (reactive hyperemia). Indeed, there is strong evidence that myogenic responses occur in the vasculature of organs like the intestine. However, the importance of myogenic factors in the regulation of blood flow appears to diminish when the metabolic demands of the organ are increased (e.g., in the intestine after ingestion of a meal). Thus, it would appear that metabolic factors can override the myogenic mechanism when the nutritive needs of the tissue are increased.

The *metabolic theory* has been proposed to explain the tight coupling of blood flow to tissue metabolism. This mechanism is particularly attractive as an explanation for the active hyperemia observed in a number of organs. Many metabolites are known to be vasoactive. Among those thought to be important for regulating vascular resistance are carbon dioxide (CO_2), H^+, O_2, K^+, and adenosine. Of these, all but O_2 relax smooth muscle and act as vasodilators. An increased O_2 tension is generally associated with vascular smooth muscle contraction, whereas reducing O_2 tension will relax smooth muscle. The vasoactive properties of molecular O_2 and metabolites such as adenosine may provide the link that exists between metabolism and blood flow for conditions like pressure flow autoregulation, reactive hyperemia, and active hyperemia. If blood flow is suddenly increased in response to an elevation in mean arterial pressure, the increased flow will promote the washout of vasodilatory products of metabolism, such as adenosine, and it will increase the O_2 tension around arterioles. The increased tissue P_{O_2}, coupled with the decline in the concentration of vasodilator metabolites, will induce smooth muscle contraction, increase vascular resistance, and reduce blood flow until the normal balance between blood flow, metabolite concentration, and tissue P_{O_2} is once again achieved.

The metabolic theory can also explain reactive and active hyperemias. If the metabolic activity and O_2 consumption of an organ are increased, tissue P_{O_2} will decrease and the levels of metabolites (adenosine) will increase. This will result in dilation of both arterioles and precapillary sphincters. The resulting increases in blood flow and the number of perfused capillaries will then enhance both the convective and diffusive delivery of O_2 to the more active tissue. Finally, if an artery perfusing an organ is occluded, ongoing metabolic activity without blood flow will result in a decrease in tissue P_{O_2} and an accumulation of vasodilator metabolites. As a consequence of these changes, the arterioles will already be dilated when the occlusion is released and blood flow will rapidly increase above the preocclusion level (reactive hyperemia). The increased flow or hyperperfusion will be temporary, gradually subsiding as tissue P_{O_2} and vasodilator metabolite concentrations return to their normal values.

Adenosine has received considerable attention as a metabolic vasodilator. When flow is too low to meet the O_2 requirements of a tissue, adenosine triphosphate (ATP) breakdown exceeds its synthesis. Adenosine, an end product of ATP degradation, can readily exit cells and enter the interstitial fluid, which bathes vascular smooth muscle. Adenosine causes smooth muscle to dilate by stimulating A_2-adenosine receptors on the smooth muscle cell. On restoration of blood flow, adenosine production is diminished and that which was released is quickly degraded or washed away. Although adenosine is thought to contribute to metabolic control of blood flow, it cannot explain phenomena such as pressure flow autoregulation because this occurs in some tissues even when A_2 receptors are completely blocked.

Extrinsic Control

Although intrinsic mechanisms normally ensure that an organ will receive a blood supply adequate to meet its metabolic needs, in many instances blood flow through particular organs serves a purpose other than simply supplying nutrients and removing metabolites. Under those circumstances, blood flow regulation is often mediated via extrinsic pathways such as the autonomic nervous system or circulating hormones. Norepinephrine is the principal neurotransmitter released at sympathetic nerve terminals. Inasmuch as norepinephrine is a potent vasoconstrictor in some vascular beds (skin and intestine), its release from sympathetic nerve terminals can exert a profound influence on blood flow. Most vascular beds are innervated by adrenergic nerves; however, the density of innervation varies markedly between organs. For example, the arterioles of the skin are heavily innervated by sympathetic fibers, whereas innervation to the coronary arterioles is quite sparse. As might be expected, mild to moderate activation of the skin's sympathetic nerves will markedly increase vascular resistance and decrease skin blood flow, as illustrated in Fig. 5. Various vascular beds also differ in their density of receptors for vasoactive hormones such as epinephrine and angiotensin II. Thus, release of these hormones can greatly alter the distribution of blood flow. Under such circumstances, flow will be diverted from those beds with high receptor density, such as skin and intestines, to those beds whose vessels have fewer receptors, such as the coronary artery.

Parasympathetic neural influences on blood flow have received less attention than sympathetic influences because only a small proportion of the arterioles in the body receive parasympathetic fibers. However, in tissues such as the salivary glands and intestines, stimulation of cholinergic fibers can elicit intense vasodilation. Acetylcholine, a neurotransmitter released at cholinergic nerve terminals, is known to produce a vasodilatory response that is endothelium dependent; that is, when the endothelial lining in small arteries is rubbed away, the vessels will no longer relax in response to acetylcholine. The apparent mediator of this endothelium-dependent vasodilator response to acetylcholine is *nitric oxide*

(NO). NO is synthesized in endothelial cells via the oxidation of the amino acid, L-arginine, by an enzyme called NO synthase. Inhibition of this enzyme by stable analogs of L-arginine causes a rise in mean arterial pressure and a reduction in blood flow in different vascular beds, suggesting that the NO produced by endothelial cells exerts a tonic inhibitory influence on arteriolar resistance. Endothelial cell-derived NO appears to diffuse to the underlying smooth muscle in arterioles, where it activates the enzyme guanylyl cyclase, which in turn generates the vascular smooth muscle relaxant, cyclic guanosine monophosphate. In addition to acetylcholine, agents such as bradykinin, substance P, and adenosine diphosphate, as well as an increased shear stress on the wall of blood vessels, lead to an increase in production of NO by endothelial cells.

SPECIAL CIRCULATIONS

In this section, we address how blood flow regulation in several representative organs illustrates the principles discussed above and we also describe the additional features of blood flow control that are unique to certain tissues and conditions (fetus).

Coronary Circulation

The coronary circulation must deliver O_2 at a rate that keeps pace with cardiac demand. Under resting conditions, blood flow to the heart is about 70 mL/min per 100 g, and it increases five- to sixfold during maximal cardiac work. Blood flow in the left ventricle is about 80 mL/min per 100 g, which is twice the flow in the right ventricle and four times the flow in the atria. Myocardial O_2 consumption is also very high (8 mL/min per 100 g) at rest (which is 20 times the resting value in skeletal muscle), and it increases about fivefold during maximal cardiac work.

Blood flow to the heart is governed almost entirely by intrinsic factors. The rate of O_2 use by the myocardium is proportional to cardiac work; hence, factors such as systolic pressure (afterload), heart rate, and stroke volume all influence myocardial O_2 consumption. As cardiac work increases, the elevated metabolic activity results in dilation of the coronary arterioles, thus

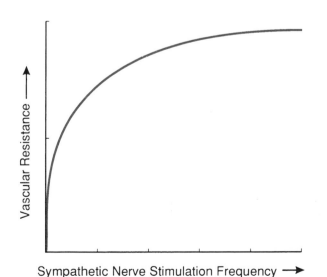

FIGURE 5 The response of vascular resistance to sympathetic nerve stimulation. As the frequency of stimulation is increased, vascular resistance increases.

increasing blood flow to heart muscle (active hyper-emia). The coronary vasculature normally exhibits excellent pressure flow autoregulation and brisk reactive hyperemias. Hormones and neurotransmitters exert only a small direct influence on the coronary blood flow because of the low density of receptors on coronary vessels. The low density of adrenergic receptors enables the heart to respond to sympathetic stimulation with vasodilation. Although sympathetic nerves would ordi-narily tend to constrict coronary arteries, the increased cardiac work that accompanies sympathetic stimulation also results in an elevated O_2 demand. Thus, the feeble vasoconstriction elicited by α-adrenergic stimulation is easily overridden by the corresponding metabolic vaso-dilation. The opposite scenario is seen with vagal stimulation. Although acetylcholine tends to directly dilate the coronary arterioles, cholinergic stimulation reduces cardiac O_2 demand by slowing the heart rate. Here again, metabolic factors dominate coronary vascular tone and the result of vagal stimulation is a reduction in blood flow.

Although the average flow through the coronary circulation is regulated by intrinsic factors, the moment-to-moment flow of blood through the coronary vessels is strongly influenced by the mechanical activity of the heart. Figure 6 demonstrates the mechanical effects on coronary blood flow. When the left ventricle starts to

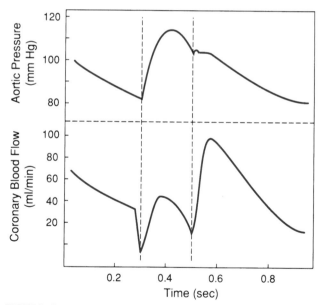

FIGURE 6   Blood flow through the left coronary artery and aortic pressure during the cardiac cycle. With the onset of ventricular contraction (from *just before the first dashed line*), blood flow decreases sharply as the coronary vessels are compressed by the ventricular muscle. During the ejection phase (*between the dashed lines*), only minimal flow occurs. Most coronary flow occurs during ventricular relaxation (from the *second dashed line on*). Flow is highest in early diastole when perfusion pressure is high and compressive forces are low.

contract, flow through the left coronary arteries falls abruptly and, under conditions of increased contracti-lity, may even reverse for a brief period. This reduction of flow is due to compression of the coronary vessels that pass through the left ventricular muscle. During the ejection period of the cardiac cycle, the aortic pressure is rising and some forward flow occurs. The compressional forces are highest near the endocardium and diminish near the epicardium. As a result, coronary flow during systole is nonuniformly distributed across the heart wall, with almost no blood flow reaching the inner layers. Figure 6 also reveals that most of the coronary flow takes place during early diastole; when the ventricular muscle is relaxed, the mechanical compression forces are low and aortic pressure is still high. At this time, the blood flow gradient across the ventricular wall reverses, with the inner muscle layers receiving most of the flow. Although a similar pattern of blood flow changes is observed in the right coronary artery during the cardiac cycle, the changes in flow to the right ventricle are relatively small because the compression forces gener-ated in the right ventricle during systole are comparably small.

It is interesting that the average blood flow over the entire cardiac cycle is uniformly distributed across the wall of the left ventricle. Although the intrinsic mecha-nisms are too slow to keep blood flow constant through the cardiac cycle, they can adjust flow so *that the average flow meets the metabolic needs of the myocardium.* To compensate for the blood flow deficit during each systole, the arterioles in the inner layers are simply kept in a more dilated state so these muscles will be compensated with extra perfusion during each diastole. Unfortunately, this compensatory response diminishes the *dilatory reserve* of the inner muscle layers, and as a result *the subendocardium is always the most vulnerable to ischemia whenever the coronary arteries are obstructed, as occurs in coronary artery disease.*

Another unique feature of the coronary circulation is the high O_2 *extraction* normally seen across this vascular bed. The myocardium extracts $>80\%$ of the O_2 from coronary blood, which compares with the whole body average of 25% at rest. The high O_2 extraction can be attributed to the great demand of this tissue for O_2 under resting conditions. It is not surprising, therefore, that all capillaries are open to perfusion in the resting myocardium. In other organs (e.g., skeletal muscle), recruitment of additional perfused capillaries is an important mechanism for enhancement of O_2 delivery by both increasing the surface area available for O_2 exchange and by reducing the diffusion distance for O_2 between the blood and myocytes. The high O_2 extrac-tion in the heart would indicate that this tissue is on the verge of underperfusion. However, because of the tight coupling of coronary blood flow with oxidative metabolism, the O_2 requirements of the heart always appear to be precisely met by changes in blood flow. Of the chemical factors that have been invoked to explain metabolic vasodilation, adenosine has received the most attention in the heart.

If coronary blood flow is interrupted, *myocardial ischemia* will quickly ensue. The heart normally oxidizes fatty acids and lactate as its energy source. However, in the absence of O_2, cardiac myocytes switch to anaerobic glycolysis (using its glycogen store), which is inefficient and cannot maintain energy balance. Within 30 sec, lactic acid buildup results in cellular acidosis, while the accumulation of inorganic phosphate inhibits the function of actin-myosin filaments. A consequence of these cellular events is an inhibition of muscle contraction that occurs almost immediately after cessation of blood flow. Although ATP falls at a relatively slow rate, such that the myocytes will not begin to die until after at least 20 min of severe ischemia, the loss of mechanical function of the heart can be disastrous. If only a small branch of the coronary system is involved, the remaining well-perfused myocardium can maintain cardiac output. If a major branch is involved, however, the ventricle can be so compromised that blood pressure will fall, rendering the rest of the heart ischemic, and circulatory collapse will soon follow. In this positive feedback situation, death of the individual quickly ensues.

Collateral vessels connect adjacent branches of coronary arteries so that occlusion of one arterial branch, as occurs in a heart attack, does not completely stop blood flow to the affected tissue. Some animals, such as the guinea pig, have well-developed collaterals and occlusion of a coronary branch is tolerated with no tissue necrosis. Other animals such as the pig have few collaterals and the same occlusion will result in *infarction* (death of tissue) of the entire downstream region. Humans are somewhere between these two extremes. If the occlusion of a coronary branch is gradual in onset, that is, over weeks and months, the collateral vessels will actually grow in response to the decreased blood flow. It is not known what factors mediate collateral growth, but it is probably related to the synthesis and release of *angiogenic* factors (e.g., adenosine) from underperfused tissue. Some coronary patients have such well-developed collateral vessels that an entire arterial branch may be completely occluded without compromising blood flow to the muscle tissue that would normally be perfused from that branch.

Cerebral Circulation

Under resting conditions, blood flow to the brain is about 50 mL/min per 100 g, accounting for about 15% of cardiac output. O_2 consumption by the human brain averages about 3.5 mL/min per 100 g. Gray matter has a very high rate of oxidative metabolism and its flow rate is up to six times higher than that of white matter. The brain, particularly gray matter, is exquisitely sensitive to hypoxia and consciousness is lost in humans after as little as 10 min of ischemia, with irreversible cell damage occurring within minutes. Thus, the primary function of the cerebral circulation is to ensure an uninterrupted supply of O_2 to the brain.

As in the coronary circulation, cerebral resistance vessels are predominantly under the control of intrinsic factors. Cerebral arterioles respond with dilation to increases in metabolic activity in the brain. Cerebral blood flow falls when brain function and metabolism are reduced during sleep or anesthesia. Any increase in metabolism in a specific area of the brain will result in an increased blood flow to that region. Unlike any other organ, the brain is able to safeguard its own blood supply by controlling the vascular resistance of other organs through its autonomic outflow. Even though sympathetic nerves can be demonstrated on the cerebral blood vessels, the constrictor response to stimulation of these nerves is weak, presumably reflecting a low density of adrenergic receptors on cerebral arterioles. Parasympathetic stimulation exerts a similarly weak vasodilatory response in the brain. The cerebral vessels also respond poorly to blood-borne vasoactive agents because of the highly restrictive nature of cerebral capillaries. This functional adaptation of the cerebral microvasculature is called the *blood–brain barrier*. Although O_2 and CO_2 readily cross the blood–brain barrier, glucose (the primary energy source for the brain) and amino acids use bidirectional transport systems in the endothelium of cerebral capillaries.

Pressure flow autoregulation is well developed in the brain, which maintains a normal cerebral blood flow at arterial pressures between 60 and 140 mm Hg. This autoregulatory range is shifted to the right (blood flow maintained constant over a range of higher pressures) by chronic arterial hypertension and during acute sympathetic stimulation. Both metabolic and myogenic mechanisms have been implicated in cerebral autoregulation. Unlike the heart, however, cerebral arterioles appear to be more sensitive to changes in arterial blood P_{CO_2} than to changes in P_{O_2}. Figure 7 shows that even a slight elevation in arterial P_{CO_2} (hypercapnia) is accompanied by a large increase in cerebral blood flow. This observation has led to the proposal that CO_2 is an important metabolite that mediates intrinsic blood flow regulation in the brain. The vasoactive effects of CO_2 have been attributed to its effect on tissue pH (increased P_{CO_2} leads to a reduced pH). Indeed, the hypercapnic vasodilation seen in the brain is considered to be a

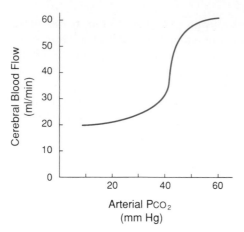

FIGURE 7 Relation between cerebral blood flow and the partial pressure of CO_2 (normal $P_{CO_2} = 40$ mm Hg). Note that elevating P_{CO_2} elicits a profound increase in blood flow.

mechanism for maintaining a constant pH in neuronal tissue, which is an important objective in view of the profound inhibitory effects of acidosis on neuronal activity.

Skeletal Muscle Circulation

The circulations of skeletal muscle and cardiac muscle have much in common; however, the two vascular beds differ in several respects. The more notable differences include a fiber type (tonic versus phasic)-based heterogeneity of O_2 demand and blood perfusion in skeletal muscle, a greater role for extrinsic factors in regulating skeletal muscle blood flow, and a larger potential for increasing O_2 extraction and the number of perfused capillaries in skeletal muscle. Blood flow in resting skeletal muscle consisting of mixed fiber types (tonic and phasic) is about 3 mL/min per 100 g, with a corresponding low basal O_2 consumption (0.3 mL/min per 100 g). Flow in resting phasic (glycolytic, rapid twitch) muscle is about 20% the value of resting tonic (oxidative, postural) muscle, but the former can increase more than 20-fold during exercise. The differences in resting flow to the two muscle types can be attributed to the greater need for O_2 by tonic muscles, a premise supported by the correspondingly larger capillary density surrounding tonic fibers.

Skeletal muscle arterioles are richly innervated by sympathetic vasoconstrictor fibers, whose tonic discharge contributes to resting arteriolar tone, as evidenced by an increased muscle blood flow after sympathetic denervation. Further stimulation of sympathetic nerves to skeletal muscle results in an increased arteriolar resistance and a reduction in blood flow. In this manner, resting skeletal muscle can contribute to the regulation of total peripheral resistance and arterial pressure because of the shear mass of skeletal muscle in the body. During periods of exercise, however, vascular resistance falls in skeletal muscle because of a dominance of metabolic vasodilation over vasoconstrictor signals derived from the autonomic nervous system. The active hyperemia observed in exercising skeletal muscle is due almost entirely to local metabolic vasodilation, rather than the relatively modest increase in arterial pressure.

Unlike the heart, resting skeletal muscle extracts only 25–30% of the O_2 from blood. The low O_2 extraction and low blood flow in resting skeletal muscle appears more than sufficient to meet the low O_2 demand of this tissue. However, during severe exercise, blood flow increases up to 20-fold, the number of perfused capillaries increases 3- to 4-fold, and O_2 extraction can reach 90%. All of these responses are geared toward meeting the 50- to 75-fold increase in O_2 demand of skeletal muscle known to occur in healthy, untrained humans during strenuous exercise.

When skeletal muscle contracts, it compresses the intramuscular arteries and arterioles sufficiently to reduce blood flow, as shown in Fig. 8. However, between contractions, blood flow is high because of the dilated arterioles, and the transient hyperemic condition is enough to meet the elevated O_2 demand. If a strong contraction is sustained, then the fibers become hypoxic and lactate accumulates, causing ischemic pain, which will eventually remind the individual to relax the muscle momentarily so that its O_2 debt can be repaid. The massaging effect of rhythmic muscle contractions on the deep veins of the calf muscle appears to assist limb perfusion because this muscle pumping action lowers venous pressure and consequently provides a larger pressure gradient for blood perfusion.

Clinical Note

Physicians must be aware of the unique permeability properties of the blood–brain barrier in order to treat diseases of the nervous system effectively. For example, the antibiotics chlortetracycline and penicillin enter the brain to a very limited degree. Erythromycin, on the other hand, enters quite readily.

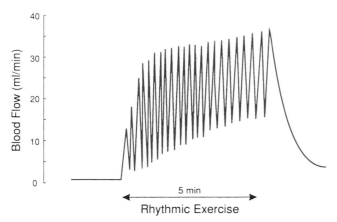

FIGURE 8   Blood flow through the vascular bed of a contracting skeletal muscle. The muscle begins to contract rhythmically approximately once every 13 sec. Note that this causes an increase in mean blood flow because of active hyperemia. Also note that each contraction of the skeletal muscle transiently inhibits its blood flow by mechanically compressing the blood vessels inside the muscle.

Splanchnic Circulation

The splanchnic circulation includes the vasculature of the gastrointestinal tract, liver, spleen, and pancreas. The principal functions of this vascular bed are to transport absorbed nutrients to the liver and systemic circulation and to mobilize blood to the systemic circulation during periods of whole body stress. Some of the features that distinguish the splanchnic circulation from other regional circulations include an active hyperemia elicited by food ingestion, an exquisite sensitivity of splanchnic arterioles to extrinsic control mechanisms, and reciprocal blood flow control between the liver and other splanchnic organs.

After ingestion of a meal, all organs of the digestive system become metabolically active. The increased acid secretion in the stomach, nutrient absorption and increased motility in the small bowel, and formation of digestive juices in the pancreas all impose a need for more blood flow to the respective tissues. The postprandial hyperemia in the gastrointestinal tract appears to be mediated by multiple factors, many of which are unique to specific regions of this organ system. For example, in the proximal small intestine, ingested long-chain fatty acids are the most potent stimulants for the active hyperemia, whereas bile acids dominate as mediators of the hyperemia in the distal ileum. Metabolic factors, gastrointestinal hormones, and neuropeptides released from cholinergic nerve terminals have all been implicated in the increased blood flow that is elicited by a meal. Figure 9 illustrates the time course of changes in resistance of the splanchnic vessels during and after ingestion of a meal. There is a slowly developing decrease in vascular resistance over the first hour after eating. As the meal is digested and absorbed over a period of several hours, vascular resistance and blood flow return to their resting levels.

Splanchnic blood vessels are frequently called on to sacrifice their blood flow for the remainder of the circulation. The ability of the splanchnic circulation to contribute in this way toward whole-body homeostasis is made possible by the dense innervation of splanchnic arterioles by sympathetic nerves and by the highly responsive nature of these vessels to circulating vasoconstrictors, such as vasopressin and angiotensin II. Sympathetic stimulation elicits an intense reduction in intestinal blood flow that is blunted only somewhat by intrinsic metabolic factors. The combined vasoconstrictor effects of norepinephrine, vasopressin, and angiotensin II likely account for the intense reduction in splanchnic blood flow observed after severe hemorrhage.

The liver receives blood from two sources, the portal vein and the hepatic artery. The portal vein, with its low O_2 content (much of the O_2 is extracted in the gastrointestinal tract), normally provides 75% of liver blood flow, whereas the hepatic artery, with its fully O_2 saturated blood, delivers the remainder. Blood flows in the hepatic artery and portal vein vary reciprocally. When liver blood flow increases through the portal vein, arterioles derived from the hepatic artery constrict, thereby preventing sudden increases in flow and pressure in downstream liver sinusoids. Because the sinusoids are extremely permeable to water and plasma

---

Clinical Note

The pale, cold skin characteristic of hypovolemic shock reflects a rise in cutaneous vascular resistance that appears to help support arterial blood pressure. During World War I, it was noticed that men rescued quickly and warmed in blankets (producing cutaneous vasodilation) were less likely to survive than men who could not be reached for some time and who retained their natural cutaneous vasoconstriction. Hence, patients in shock should not be warmed to the point at which their body temperature rises.

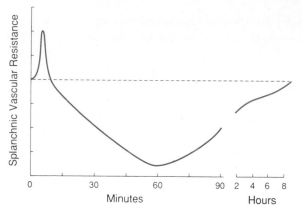

FIGURE 9   Active hyperemia in the gastrointestinal tract. Vascular resistance in the splanchnic circulation before, during, and after ingesting a meal is plotted. During eating, resistance increases transiently but is followed quickly by a slowly developing and long-lasting decrease in resistance that increases blood flow to the gut. As the meal is digested and absorbed over the next several hours, vascular resistance (and blood flow) return to the levels seen during ingestion of a meal.

proteins, the hepatic arterial constrictor response to increases in portal flow serves to prevent excessive filtration across this microvascular bed.

Cutaneous Circulation

The metabolic requirements of the skin are modest and blood flow through this tissue is controlled primarily by sympathetic fibers, whose activity is linked to body temperature. The amount of heat lost from the body is regulated by the amount of blood flowing through the skin. The skin's ability to serve as the body's radiator is due largely to the existence of arteriovenous anastomoses (AVAs) beneath the skin surface. These AVAs are low-resistance shunt pathways that are richly innervated by sympathetic vasoconstrictor fibers. When body core temperature is low, thermosensitive regions of the hypothalamus induce an increased sympathetic activity to the fibers innervating the AVAs. The resulting reduction in skin blood flow leads to a diminished dissipation of heat from blood to the environment and body temperature rises. Conversely, AVAs are dilated when core temperature is high because of a reduced sympathetic drive. In this instance, the cutaneous vasodilation favors the delivery of more blood to the skin, where heat readily crosses the vessel walls to reduce body temperature to a normal level. These temperature-induced changes in cutaneous vascular resistance account for the observation that blood flow can vary between 1 and 200 mL/min per 100 g in human skin.

Cutaneous vessels are exposed to an additional vasodilatory stimulus when increased body temperature

is accompanied by sweating. The heat-induced parasympathetic stimulation of sweat glands results in the liberation of bradykinin, a potent vasodilator that acts on AVAs situated in proximity to the sweat glands. Figure 10 illustrates the effect of changes in ambient temperature on skin blood flow in the hand of an individual at rest and during exercise. In the resting state, an increased ambient temperature results in increased skin blood flow. During exercise, however, comparable increases in ambient temperature elicit more profound increments in hand blood flow. The difference between the two conditions is that the rise in body core temperature is greater in the exercising state and thus the stimulus for vasodilation is greater.

Fetal Circulation

Several of the anatomic and functional characteristics of the fetal circulation distinguish it from the adult circulation. Because the lungs of the fetus are collapsed and nonfunctional, some of the unique features of the fetal circulation are directed toward ensuring the absence of lung perfusion. In the fetus, the placenta serves as the functional equivalent of the lung, providing the oxygen and nutrients needed for survival, growth, and development of the fetus. Oxygenated fetal blood

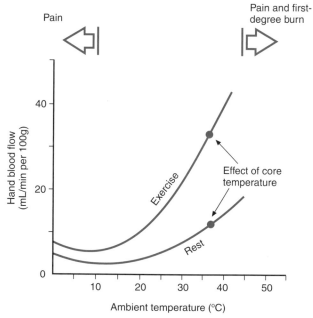

FIGURE 10   Relation between hand blood flow and ambient temperature in a healthy volunteer at rest and during leg exercise. Ambient temperature was varied by hand immersion in water at different temperatures. The difference in the two curves can be attributed to a higher internal heat load (body core temperature) during exercise. (Modified from *Handbook of physiology: Cardiovascular system, Vol 3, Peripheral circulation, Part 2.* Bethesda, MD: American Physiological Society, 1963:1325–1352.)

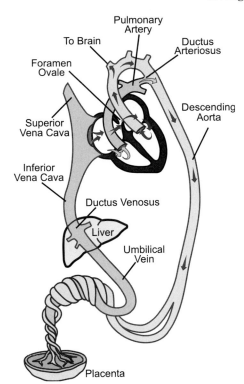

FIGURE 11   Schematic of the normal fetal circulation.

from the placenta is delivered directly to the liver via the umbilical vein (Fig. 11). A shunt called the *ductus venosus* allows some of the umbilical vein blood to bypass the liver and empty into the inferior vena cava. There is little mixing of the poorly oxygenated blood from the lower inferior vena cava with the well-oxygenated blood from the ductus venosus. Hence, the two streams remain distinct from one another in the inferior vena cava, and as they enter the right ventricle the larger, well-oxygenated bloodstream from the ductus venosus is shunted to the left atrium via an opening, the *foramen ovale*, in the septum. (The foramen is covered by a flap that allows for right-to-left atrial flow, but it opposes flow in the opposite direction.) This diversion of well-oxygenated blood directly into the left heart, and into the ascending aorta, ensures that the heart and brain receive an adequate supply of oxygen.

Another shunt that diverts blood away from the fetal lungs is the *ductus arteriosus*. This large opening between the pulmonary artery and aortic arch diverts most of the right ventricular output into the descending aorta, rather than into the lungs. The mixture of poorly oxygenated inferior vena caval blood and superior vena caval blood that is pumped from the right ventricle traverses the ductus arteriosus and mixes with blood pumped out of the left ventricle. This arrangement, therefore, ensures that blood with the lowest O_2 content perfuses the viscera and lower extremities.

There are several important physiological consequences of the cardiovascular arrangement in the fetus. Pulmonary blood flow is low and pulmonary vascular resistance is high. Pulmonary artery pressure is higher than aortic pressure because of the high pulmonary vascular resistance as well as the low peripheral vascular resistance resulting from the substantial distribution of cardiac output through the low-resistance placental circulation. Another consequence of the low pulmonary blood flow is a left atrial pressure that is lower than right atrial pressure.

At birth, the fetal circulatory and respiratory systems undergo profound changes to ensure survival in the extrauterine environment. Clamping the umbilical cord and the taking of the first breath produce dramatic effects on the circulation. Interruption of the umbilical circulation leads to closure of the ductus venosus, and systemic vascular resistance almost doubles by removing the low-resistance placental circulation to produce an increase in arterial pressure. Inflation of the lungs with the first breath leads to a profound fivefold reduction in pulmonary vascular resistance and increased lung blood flow. The resultant increase in left atrial filling elevates left atrial pressure by a few mmHg, which is sufficient to reverse the pressure gradient between left and right atria and abruptly close the valve over the foramen ovale. The reduction in pulmonary artery pressure, coupled to the elevated aortic pressure, reverses flow in the ductus arteriosus, which then closes gradually over a period of 1–2 days. The net result of these changes after birth is the establishment of an adult pattern in the circulation.

Suggested Readings

Abramson DI, PB Dobrin. *Blood vessels and lymphatics in organ systems*. Orlando, FL: Academic Press, 1984.

Downey JM. Extravascular coronary resistance. In Sperelakis N, ed., *Physiology and pathophysiology of the heart*. Boston: Kluwer, 1989, pp 939–954.

Handbook of physiology: The gastrointestinal system, Vol 1, Motility and circulation, Part 2. Bethesda, MD: American Physiological Society, 1989.

Mohrman DE, LJ Heller. *Cardiovascular physiology*, 3rd ed. New York: McGraw-Hill, 1991.

PART III

RESPIRATORY PHYSIOLOGY

Structure and Function of the Respiratory System

FRANK L. POWELL, JR.

KEY POINTS

- The primary function of the lung is *gas exchange*—transporting oxygen from the environment into the blood and eliminating carbon dioxide from the blood to meet the metabolic demands of the tissues.
- The lung is a series of branching tubes consisting of conducting airways (trachea to terminal bronchioles) and respiratory airways (respiratory bronchioles to alveoli).
- Transport of O_2 from small air sacs in the lung, called *alveoli*, into pulmonary capillary blood occurs by diffusion. The total surface area for diffusion in the lung is as large as a tennis court because there are 300 million alveoli.

- Quantitative descriptions of gas exchange depend on relatively simple applications of the principle of mass balance and the ideal gas law.
- *Ventilation* brings O_2 into the lungs by bulk flow. Total lung volume is more than 10 times larger than a normal breath, providing a large reserve capacity for increased ventilation.
- The pulmonary circulation is in series with the systemic circulation, so the lungs receive the entire cardiac output and pulmonary blood flow equals total systemic blood flow.
- In the pulmonary circulation, compared with the systemic circulation, pressures are lower,

OVERVIEW OF THE RESPIRATORY SYSTEM

The primary function of the respiratory system is *gas exchange*—delivering oxygen (O_2) from the environment to the tissues and removing carbon dioxide (CO_2) from the tissues. Generally, the respiratory system acts as a servant to the rest of the body by delivering enough O_2 and removing sufficient CO_2 for metabolic demands. As O_2 demand increases, the body responds with a variety of mechanisms to ensure an adequate supply of O_2. These physiologic mechanisms include the unique functions of several cell types in the lung, pulmonary circulation, mechanics of the respiratory system, transport of O_2 and CO_2 in blood, respiratory gas exchange, and coordination of all of these mechanisms by the respiratory control system. Each of these topics is covered in subsequent chapters.

It is important that the respiratory system respond to changes in O_2 supply. Many of the responses to changes in O_2 supply are the same as those used to respond to changes in O_2 demand. If the supply-and-demand relationship for O_2 is disrupted by disease, then physical activity is limited and organ failure may occur. Diagnosing and treating respiratory disease requires understanding how the structure, function, and control of the respiratory system interact to match supply and demand at each step in the chain of O_2 transport.

Oxygen Cascade

Gas exchange between cells and the environment occurs by a series of physiologic transport steps across different structures, as shown in Fig. 1. (This figure uses a standard set of symbols developed for quantitative descriptions of respiratory physiology, which are defined in Table 1 and in the section titled "Physical and Chemical Principles in Respiratory Physiology.") Breathing movements bring fresh air into the lungs, and

the heart pumps O_2-poor blood to the lungs. O_2 diffuses from the lung gas into the blood, and this O_2-rich blood is returned to the heart via the pulmonary circulation. Arterialized blood is pumped to the various organs and tissues of the body via the systemic circulation. Finally, O_2 diffuses out of the systemic capillaries to metabolizing tissues and ultimately to the mitochondria inside cells. CO_2 moves out of the cells to the environment through these same steps in the opposite direction from O_2. Figure 1 also shows normal values for the levels of O_2 and CO_2, expressed as partial pressures, P_{O_2}, and P_{CO_2} (see later discussion). Note that the O_2 level decreases at each step in this series of transport processes, so this transport chain is frequently called the *O_2 cascade*.

The physiologic principles governing O_2 supply-and-demand relationships can be studied by measuring maximal O_2 consumption while varying the supply. Figure 2 shows how decreasing oxygen levels in the

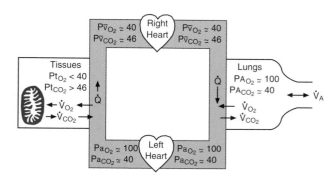

FIGURE 1   Oxygen uptake into the body ($\dot{V}_{O_2}$ in mL O_2/min) and carbon dioxide elimination ($\dot{V}_{CO_2}$ in mL CO_2/min) occur in a series of physiological transport steps from the atmosphere to mitochondria in the tissues. Typical O_2 and CO_2 partial pressures (P in mm Hg) are shown for gas in the alveoli of the lungs (A), arterial blood (a) and mixed-venous blood ($\bar{v}$) and tissues (t). $\dot{V}_A$, alveolar ventilation; $\dot{Q}$, blood flow rate. Symbols and units used in this section are defined in Table I.

TABLE 1 Symbols in Respiratory Physiology

Primary Variables and Their Units	
C	Concentration or content (mL/dL or mmol/L)
D	Diffusing capacity ($mL_{O_2}/(min.\cdot mmHg$)
F	Fractional concentration in dry gas (dimensionless)
P	Gas pressure or partial pressure (mmHg or cm H_2O)
$\dot{Q}$	Blood flow or perfusion (L/min)
R	Respiratory exchange ratio (dimensionless)
RQ	Respiratory quotient (dimensionless)
T	Temperature (°C)
V	Gas volume (L or mL)
$\dot{V}$	Ventilation (L/min)

Modifying Symbols

A	Alveolar gas
B	Barometric
D	Dead space gas
E	Expired gas
Ē	Mixed-expired gas
I	Inspired gas
L	Lung or transpulmonary
T	Tidal gas
aw	Airway
w	Chest wall
es	Esophageal
pl	Intrapleural
rs	Transrespiratory system (total system)
a	Arterial blood
b	Blood (general)
c	Capillary blood
c'	End-capillary blood
t	Tissue
v	Venous blood
v̄	Mixed-venous blood

Examples

P_{AO_2}	Partial pressure of O_2 in alveolar gas
P_{aO_2}	Partial pressure of O_2 in arterial blood
F_{E-CO_2}	Fraction of CO_2 in dry mixed, expired gas
$\dot{V}_{O_2}$	O_2 consumption per unit time
$\dot{V}_A$	Ventilation of the alveoli per unit time

Interface between the Respiratory System and the Environment

High O_2 levels are desirable in the respiratory system because O_2 transport ultimately occurs by passive diffusion down an O_2 gradient. Ventilatory and cardiovascular pumps are used to deliver O_2 by bulk flow in air or blood to large surface areas, where diffusion of O_2 can occur. The actual interface between the environment and the body's *internal milieu* occurs deep in the lungs at the pulmonary blood–gas barrier. Figure 3A illustrates the branching pattern of airways used to deliver gas to the terminal lung units called *alveoli*. The trachea divides into two smaller primary bronchi to each lung, and the primary bronchi divide into still smaller bronchi to the different lobes of the lung. The pulmonary blood vessels undergo a similar branching pattern so the alveoli are virtually covered with pulmonary capillaries (Fig. 3B).

The effect of this branching is to increase greatly the surface area available for gas exchange by diffusion. If the lungs were two simple spheres of about 3 L each, they would have a total internal surface area of only 0.1 m^2. However, the lungs branch more than 20 times, so the lungs contain more than 300 million alveoli of only about 300 μm in diameter. The surface-to-volume ratio increases as the diameter of a sphere decreases, so dividing the large volume of the lung into millions of smaller units results in an extremely large area available for gas exchange. The total surface area of the lung's blood–gas barrier is 50–100 m^2, or about the area of a tennis court!

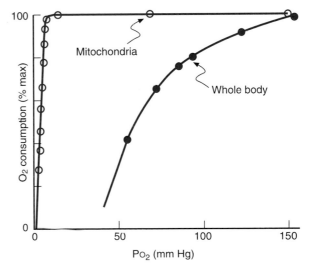

FIGURE 2 Maximal O_2 consumption is less sensitive to decreased O_2 supply in isolated mitochondria than in the whole body, because O_2 levels decrease between the environment and the mitochondria. P_{O_2} is a measure of O_2 level in mitochondrial suspension medium or inspired air. (After Winzler, *J Cell Comp Physiol* 1941;17:263, and Pugh *et al.*, *J Appl Physiol* 1964;19:431.)

environment affect maximal O_2 consumption, equivalent to the maximum capacity for aerobic exercise in a healthy individual. Maximal O_2 consumption for the whole body starts to decrease at relatively high levels of O_2, in comparison with mitochondria. Isolated mitochondria can continue to function until O_2 levels decrease to almost zero, whereas whole-body O_2 consumption is limited when environmental O_2 decreases only 25%. This is not an artifact of the *in vitro* methods used to measure mitochondrial O_2 consumption, but a consequence of the decrease in O_2 levels occurring at each step in the O_2 transport chain. A healthy respiratory system functions to preserve high O_2 levels at each step in the transport chain.

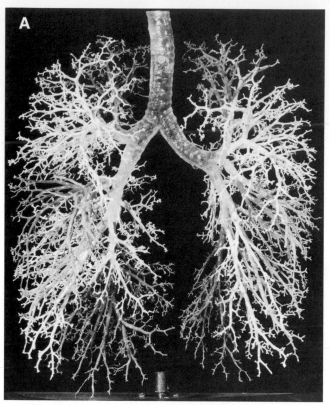

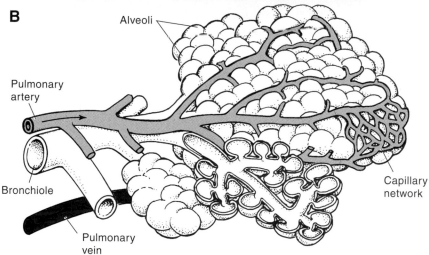

FIGURE 3   (A) Plastic cast of the lung airways showing extensive branching. (B) Diagram of terminal airway branches in an acinus, which is the functional unit of gas exchange in the lung. Pulmonary arterioles travel next to the bronchi to the level of respiratory bronchioles and branch extensively to cover the alveoli with pulmonary capillaries. Pulmonary veins are further from the bronchioles.

Physical and Chemical Principles in Respiratory Physiology

Quantitative descriptions of gas exchange depend on relatively simple applications of the principle of conservation of mass (or mass balance) and the ideal gas law using the symbols defined in Table 1. The symbols may appear complicated at first, but they are based on a few simple conventions. *Primary variables* are symbolized with a *capital letter,* and a *dot* over the variable indicates the *first derivative with respect to time* (e.g., $\dot{Q}$ = blood flow, or quantity of blood per unit time in liters per minute). *Modifiers are small capitals for the gas phase* (e.g., $\dot{V}_A$ = alveolar ventilation) and *lowercase letters for liquid or tissues* (e.g., Pa = partial pressure in arterial blood). Finally, a specific *gas species* is indicated with a *subscript* (eg, C_{AO_2} = O_2 concentration in arterial blood in mL O_2/dL blood).

The principle of *conservation of mass* is simply that matter is neither created nor destroyed. This principle was applied to physiologic transport by the German physiologist Fick in the last century. The *Fick principle* states that the amount of a substance consumed or produced by an organ is the difference between the amount of the substance entering the organ and the amount leaving the organ. To calculate whole-body O_2 consumption ($\dot{V}_{O_2}$ in milliliters per minute), one measures the difference between the amount of O_2 inspired and the amount of O_2 expired from the lungs per unit time. The amount of O_2 inspired = $\dot{V} \cdot F_{IO_2}$, where $\dot{V}$ = ventilation (in liters per minute) and F_{IO_2} = fractional concentration of O_2 in inspired gas (0.21 for room air). The amount of O_2 expired = $\dot{V} \cdot F_{EO_2}$, where F_{EO_2} is the fractional concentration of O_2 in mixed-expired gas. If inspired ventilation equals expired ventilation, then:

$$\dot{V}_{O_2} = \dot{V}_E(F_{IO_2} - F_{EO_2}).$$

In a steady state, $\dot{V}_{O_2}$ is equal at each of the steps in the O_2 cascade, so a similar equation can be written for the cardiovascular system. The amount of O_2 consumed by the body equals the difference between arterial O_2 delivery and venous O_2 return:

$$\dot{V}_{O_2} = \dot{Q}(C_{aO_2} - C\bar{v}_{O_2}),$$

where $\dot{Q}$ = cardiac output (in liters per minute), C_{aO_2} = O_2 concentration in arterial blood (mL O_2/dL blood), and $C\bar{v}_{O_2}$ = O_2 concentration in mixed venous blood. The Fick principle can be used to calculate cardiac output ($\dot{Q}$) from measurements of whole-body O_2 consumption and arterial and venous O_2 concentrations by rearranging the preceding equation.

Chapter 21 considers more applications of the Fick principle, which describes gas transport by *convection,* or *bulk flow* of air or blood. In contrast, *diffusion* is the mechanism of O_2 transport across the blood–gas interface in the lungs and across systemic capillaries in metabolizing tissues. Fick also quantified diffusive gas transport with *Fick's first law of diffusion:*

$$\dot{V}_{O_2} = \Delta P_{O_2}D,$$

where ΔP_{O_2} is the average O_2 partial pressure gradient between two compartments, and D is a diffusing capacity, as defined in Chapter 21.

Note that the diffusive transport of respiratory gases occurs down a partial pressure gradient. *Partial pressure* of a gas is defined by *Dalton's law,* which states that the partial pressure of gas x in a mixture of gases is equal to the pressure that gas x would exert if the other gases were not present. Therefore:

$$P_x = F_x(P_{tot}),$$

where F_x is the fractional concentration of gas x in a *dry* gas sample. For example, P_{O_2} in dry air at sea level is 160 mm Hg (= 0.21 · 760 mm Hg, where the O_2 concentration is 21% in air and barometric pressure is 760 mm Hg at sea level). Partial pressure is also expressed in units of Torr (1 Torr = 1 mm Hg) or SI units of kilopascals (1 kPa = 7.5 mm Hg) in physiology.

For calculating partial pressure in the gas phase, it is important to specify the total *dry* gas pressure because of the effects of water vapor pressure in humidified gases. *Water vapor pressure* is determined only by the temperature and relative humidity of a gas, and it is independent of total pressure. Inside the lungs, temperature is generally 37°C and relative humidity is 100%. Saturated water vapor pressure at 37°C = 47 mm Hg, so the total gas pressure available for O_2 and CO_2 inside the body is reduced by this amount. Assuming barometric pressure equals 760 mm Hg:

$$P_{dry} = (760 - 47) = 713 \text{ mmHg.}$$

Therefore, P_{O_2} in inspired gas, which is saturated with water vapor at body temperature, is only 150 mm Hg (= 0.21 · 713 mm Hg) at sea level.

Gases dissolved in fluids also exert a partial pressure, and diffusion of gases also occurs down partial pressure gradients between fluids. For example, O_2 diffuses from O_2-rich blood in capillaries toward mitochondria where it is near zero. The partial pressure of gas x in solution equals P_x in a gas mixture that would be in equilibrium with that solution. *Henry's law* describes the linear relationship between the concentration (C in mL/dL or mmol/L) and partial pressure (P in mm Hg) of gas x dissolved in solution:

$$C_x = \alpha_x P_x,$$

where α_x is the physical *solubility* of gas x in the solution. The relationship between O_2 and CO_2 concentration and partial pressure in blood is more complex because of chemical reactions between these physiologic gases and blood (Chapter 20).

The volume of a gas sample depends on temperature and pressure according to the *ideal gas law:*

$$PV = nRT,$$

where n is the number of moles, R is the universal gas constant, and T is temperature in degrees kelvins. *Avogadro's law* specifies that 1 mol of an ideal gas occupies 22.4 L at standard temperature (0°C) and standard pressure (760 mm Hg) when dry. Such volumes are called *standard temperature and pressure dry* (STPD), and can be used instead of moles to quantify the amount of a gas. For example, $\dot{V}O_2$ and $\dot{V}CO_2$ are generally expressed as mL_{STPD}/min.

Ventilation and lung volumes are not usually dry gas volumes measured at 0°C and 760 mm Hg, however. Lung volumes occur at body temperature, actual barometric pressure, and saturated with water vapor. Such physiologic volumes are called *body temperature and pressure saturated* (BTPS) and they can be converted to STPD volumes as follows:

$$V_{STPD} = V_{BTPS} (273 \text{ K}/T_{BODY} \text{ in K})$$
$$(P_B/760) ([P_B - P_{H_2O}]/P_B).$$

At 37°C and $P_B = 760$ mm Hg, this simplifies to:

$$V_{STPD} = V_{BTPS} (273°/310°) (713/760)$$
$$V_{STPD} = V_{BTPS} (0.826).$$

This equation derives from two special applications of the ideal gas law. *Boyle's law* states that volume is inversely proportional to pressure at constant temperature:

$$V_1/V_2 = P_2/P_1.$$

Charles' law states that volume is directly proportional to temperature at constant pressure:

$$V_1/V_2 = T_1/T_2.$$

Volumes are measured frequently at ambient temperature and pressure, or *ambient temperature and pressure saturated* (ATPS) conditions. For normal values of $P_B = 760$ mm Hg and $T = 37°C$, ATPS can be converted to STPD or BTPS by:

$$V_{STPD} = V_{ATPS} (0.885)$$
$$V_{BTPS} = V_{ATPS} (1.086).$$

LUNG AIRWAYS AND VENTILATION

Airways

The lung is a series of branching tubes leading from the *trachea* to small terminal air sacs at the ends of the airways called *alveoli*. At each branch point in this tree, the daughter branches are smaller in diameter but there are at least two more of them. For example, the single

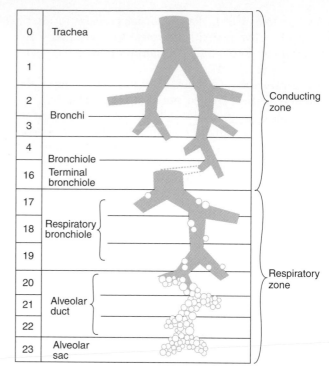

FIGURE 4   The airway tree consists of two functional zones: The conducting zone includes the first 16 orders of branching and does not participate in gas exchange; the respiratory zone includes the last 7 orders of bronchial branching, which have alveoli that are responsible for gas exchange. (After Weibel, *Morphometry of the human lung.* Berlin: Springer, 1963.)

trachea divides into two slightly smaller main stem (or primary) bronchi to the right and left lungs. Figure 4 shows the types of airways in the 23 orders of bronchial branching that occur in a human lung. Note that all bronchi beyond the 16th generation contain some alveoli, so the number of alveoli (300 million) is even greater than the number predicted for one at the end of each branch in a tree with 23 bifurcations (2^{23}; 8.4 million). Although each airway generation is smaller, their number is increasing exponentially so the total cross-sectional area of the airways increases dramatically with each generation. The total cross-sectional area of the first 10 generations of airways is relatively constant at a few square centimeters. However, by the 17th generation, the total cross-sectional area of the airways is more than 200 cm^2.

This dramatic increase in cross-sectional area has an important functional consequence for gas exchange. Once the total area is large enough, and the distances to the gas exchanging airways are short enough, the forward velocity of bulk flow decreases and diffusion becomes an effective mechanism for gas transport. Hence, diffusion is the primary mechanism of gas transport in airways after the terminal bronchioles. The airways and alveoli served by a terminal bronchiole are called an *acinus* (Fig. 3B). There are about 150,000 acini, with a path length for gas transport of about 5 mm in human lungs. The acinus is the functional unit of gas exchange because diffusion is so effective at mixing and equilibrating gas in it.

Figure 4 shows how the airways can be divided into a *conducting zone* and a *respiratory zone*. Airways in the respiratory zone are structurally stronger and function to distribute air by convection to peripheral airways. There is no significant uptake of O_2 or elimination of CO_2 across the walls of conducting airways. The respiratory zone consists of peripheral airways that function to equilibrate blood with lung gases, and this zone contains most of the lung volume. Airways and alveoli in the respiratory zone can be extremely delicate to facilitate diffusion from lung gas to blood because they are not subject to the larger stresses associated with ventilation and bulk flow.

Figure 5 shows how the structure and cellular biology of the airways change between the conducting and respiratory zones. These changes involve three principles of organization: (1) The airways form a barrier between gas and the body consisting of *layers* of epithelium, interstitium, and endothelium; (2) airway cells are *differentiated* according to their hierarchy in the tree of bronchial branching; and (3) a *mosaic* of airway cell types changes between the conducting and airway zones.

In the trachea and large bronchi, the airway walls are very thick and include *cartilage* and smooth muscle to provide structural support. The airway *epithelial cells* form a confluent (or continuous) sheet that is anchored by *basement membrane* and covers the entire internal surface of the airways, from the trachea to the alveoli. *Tight junctions* between epithelial cells limit and control molecular transport across this barrier. The mosaic of *columnar* epithelial cells in the bronchi includes *ciliated cells* and superficial *Goblet cells* that secrete mucous glands. Other mucous cells form invaginations in the airway that function as submucosal secretory glands. The surface cilia move secreted mucus toward the mouth to remove foreign objects from the airways. Neuroendocrine cells that secrete mediators into the bloodstream are relatively rare. Endothelial cells in the bronchi are part of the systemic circulation, which supplies the nutrient demands of large airways by the bronchial circulation.

In the transition zone, there is no cartilage, but helical bands of *bronchial smooth muscle* surround the airways. This smooth muscle controls bronchial caliber, airway resistance, and the local distribution of ventilatory gas flow. Ciliated epithelial cells are smaller and *cuboidal* in the transition zone. Goblet cells on the airway surface secrete mucus to be moved up and out of the airways by the ciliated epithelial cells. *Clara cells* are another type

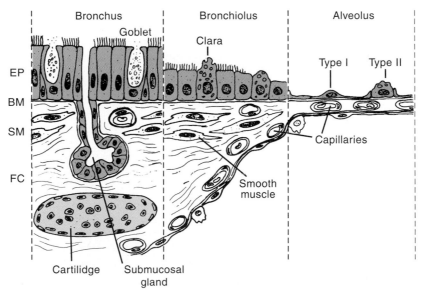

FIGURE 5   Cellular structure of airways showing how the layering and cellular forms at different levels from the trachea (left) to the alveoli (right). Both nonciliated epithelial cells (goblet cells) and submucosal glands secrete mucus in the bronchi. Secretory Clara cells occur in the bronchioles. The alveoli do not contain cilia but do contain type II epithelial cells that secrete surfactant. The epithelium (EP), basement membrane (BM), and interstitium (IN) are very thin in the alveoli to allow effective diffusion of O_2 from alveolar gas through pulmonary capillary endothelium and into blood. (After Burri and Weibel, *Rontgendiagnostik der lunge,* Huber, 1973.)

of secretory epithelial cell in the small bronchi but their function is not known.

In the respiratory zone, there is no smooth muscle and little connective tissue. *Type I alveolar epithelial cells* are *squamous* (or flattened), with long and thin cytoplasmic extensions and no cilia so the epithelial barrier to gas exchange is as thin as possible. The interstitial layer and pulmonary capillary endothelial cells are also very thin, reducing the barrier to gas exchange. *Type II alveolar epithelial cells* are specialized cells that synthesize and secrete surfactant, a substance that influences the mechanical properties of the lung, as described in Chapter 19. Type II cells are also precursors for type I cells and important for repair in lung injury.

Lung Volumes

The volume of gas in the lungs can be divided into different components, and these individual volumes can be useful in diagnosing certain pulmonary diseases. Figure 6 shows the different volumes that can be measured with a spirometer. A *spirometer* measures the volume inspired or expired by a subject through a mouthpiece connected to a container with a water seal (as in Fig. 6) or a collapsible bellows (which is more common today). The patient wears a nose clip to ensure that an entire inhalation or exhalation is collected.

Total lung capacity (TLC) is the maximum volume that can be contained by the lungs *in vivo,* and it includes several different volumes. The convention is that *capacities are composed of volumes* that can be measured independently. *Residual volume* (RV) is the one volume that cannot be measured with a spirometer because it is

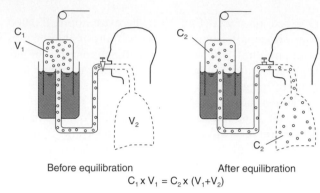

Before equilibration   After equilibration

$$C_1 \times V_1 = C_2 \times (V_1 + V_2)$$

FIGURE 7   Measurement of FRC by helium dilution as described in the text. V_1, spirometer volume; V_2, FRC; C_1, initial helium concentration; C_2, final helium concentration.

the volume of gas remaining in the lungs after a maximal expiratory effort. Therefore, absolute values of TLC and the *functional reserve capacity* (FRC) cannot be measured with a spirometer. FRC is the volume of gas left in the lungs at the end of a normal passive expiration.

Tidal volume (V_T) is the normal volume inspired and expired with each breath. *Inspiratory reserve volume* (IRV) is the maximum volume that can be inspired above the end of a normal inspiration, and the *expiratory reserve volume* (ERV) is the maximum volume that can be forcibly exhaled after a normal expiration (i.e., below FRC). *Inspiratory capacity* (IC) is the maximum volume that can be inspired from FRC and the *vital capacity* (VC) is the maximum volume that can be inhaled or exhaled *in vivo.*

One method used to measure RV or FRC is that of gas dilution, which is another application of the principle of conservation of mass as illustrated in Fig. 7. The initial

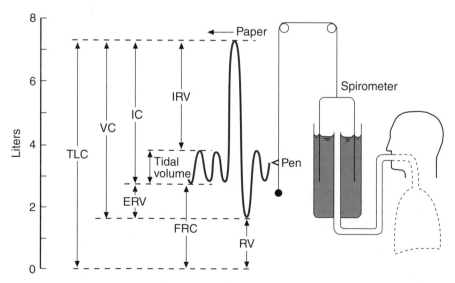

FIGURE 6   Only lung volumes that do not include residual volume (RV) can be measured with a spirometer. TLC, total lung capacity; VC, vital capacity; IC, inspiratory capacity; ERV, expiratory reserve volume; IRV, inspiratory reserve volume; FRC, functional reserve capacity.

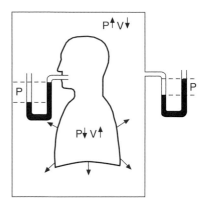

FIGURE 8 Measurement of FRC with a body plethysmograph. Pressure (P) and volume (V) changes measured when the subject attempts to inspire against a closed airway are used to calculate lung volume as described in the text ($P \cdot V = k$).

lung volume (V_2), which can be FRC or RV, contains none of a tracer gas such as helium. After the mouthpiece is opened to the spirometer, the individual rebreathes in and out of the spirometer until the tracer gas concentration equilibrates in the lung volume and the spirometer. O_2 can be added to the spirometer to replace that consumed and CO_2 can be removed by CO_2 absorbents. If the final tracer gas concentration (C_2) is measured, and the initial spirometer volume (V_1) and tracer gas concentration (C_1) are known, then one can solve for the unknown lung volume (V_2).

Another method used to measure FRC is that of a body *plethysmograph* as illustrated in Fig. 8. A plethysmograph is a sealed box in which an individual can sit and breathe through a mouthpiece connected outside the box. The mouthpiece is sealed at a specified lung volume, such as FRC, and the person makes an inspiratory effort. First, Boyle's law is used to calculate the change in box volume (ΔV) from the known initial box volume (V_1) and pressure (P_1), and the final box pressure (P_2) measured during inspiratory effort:

$$P_1 V_1 = P_2(V_1 - \Delta V).$$

Next, Boyle's law is applied to the lung where FRC is the unknown initial lung volume and P_3 and P_4 are the initial and final lung pressures measured at the mouth:

$$P_3 \text{FRC} = P_4(\text{FRC} + \Delta V).$$

The decrease in box volume equals the increase in lung volume (ΔV), and the equation can be solved for FRC.

The measurement of lung volumes can be a useful diagnostic tool for pulmonary disease. For example, FRC increases with emphysema or chronic obstructive lung disease. However, the methods for measuring lung volumes can also be affected by disease. Gas dilution

depends on the tracer gas reaching all parts of the lung volume, which may not occur with lung disease and gas trapping in obstructed distal airways. In contrast, the plethysmograph method will also measure trapped gas volumes within the thorax.

Normal lung volumes also provide important lessons about respiratory physiology in healthy individuals. TLC is more than 10 times larger than the normal V_T, so there is a tremendous reserve capacity for increased ventilation with increased O_2 demand or reduced supply. This is part of the reason why pulmonary gas exchange is usually not a limiting factor in O_2 uptake at sea level in anyone except highly trained elite athletes. It is also important that the FRC is over three times larger than normal V_T. Ventilation is tidal (i.e., in and out), so oscillations in alveolar gas composition can occur during the breathing cycle. However, the large FRC relative to V_T dilutes these oscillations in alveolar gas composition so P_{O_2} and P_{CO_2} in blood leaving the pulmonary capillaries is almost constant (± 2–4 mm Hg).

Ventilation

Ventilation is defined as the volume of air moved into or out of the lungs in a given time, and it can be changed by changing either the volume of a breath (V_T) or the *respiratory frequency* (f_R). In an average healthy human, V_T is about 500 mL and f_R is about 12 breaths/min, so *expired ventilation* ($\dot{V}_E$) is about 6 L_{BTPS}/min (= 0.5 L/breath · 12 breaths/min). If either V_T or f_R are doubled, then $\dot{V}_E$ will double. Similar reductions in f_R or V_T also have equivalent effects on $\dot{V}_E$.

Because ventilation is a tidal process and the conducting airways do not participate in gas exchange, not all of the inspired volume is effective at gas exchange. Figure 9 shows that only a part of V_T is

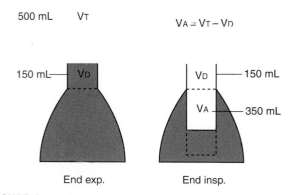

FIGURE 9 Tidal volume (V_T) includes a volume that reaches the alveoli and is effective at gas exchange ($\dot{V}_A$), and dead space volume that remains in the conducting airways and is not effective at gas exchange (V_D).

effective at bringing fresh gas to the alveoli because (1) the first gas inspired is gas left in conducting airways from the last expiration, which has already had O_2 removed and CO_2 added to it in the lungs; and (2) the last part of an inhalation does not get past the conducting airways and into the gas exchanging alveoli. The volume in the conducting airways that is not effective at gas exchange is called *anatomic dead space* (VD). The portion of ventilation effective for gas exchange is that portion actually reaching the alveoli, or alveolar ventilation ($\dot{V}A$). In a normal individual, anatomic VD is about 150 mL (or about 1 mL/lb body mass) so $\dot{V}A$ is only about 4.2 L/min:

$$\dot{V}A = f_R(V_T - V_D).$$

The effects of changing f_R on $\dot{V}A$ are different from those of changing V_T, in contrast to the case for $\dot{V}E$. Doubling or halving f_R will double or halve $\dot{V}A$, respectively. However, doubling V_T will more than double $\dot{V}A$ if V_D is constant. Similarly, decreasing V_T can decrease $\dot{V}A$ disproportionately. Differential effects of V_T and f_R on $\dot{V}A$ have important implications for artificial ventilation. It also means that the optimal breathing pattern depends on gas exchange, in addition to respiratory mechanics.

Anatomic dead space can be measured with a *single breath method* as shown in Fig. 10. Nitrogen concentration is continuously measured at the mouth of a person who inspires a breath of pure O_2, and then slowly exhales to RV. Expired volume is measured at the same time. The first gas expired from the dead space contains the pure O_2 that was just inhaled. After the dead space

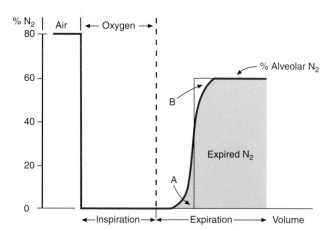

FIGURE 10 Fowler's method for measuring dead space as described in the text. After a single inspiration of 100% O_2, expired N_2 concentration is plotted against expired volume. A vertical line drawn so area A = area B, and this line intersects the volume axis at the anatomic deadspace volume (VD). In real life, the alveolar plateau may slope and have inflection points from uneven ventilation between lung regions (see Chapter 21).

gas, alveolar gas that still contains nitrogen is expired. There is not a perfectly sharp transition between the anatomic dead space gas and alveolar gas because of diffusive mixing at the interface between conducting and respiratory airway zones. However, the average volume at this interface can be determined as shown in Fig. 10, and this equals the anatomic dead space volume.

In reality, anatomic VD can increase with VT as conducting airways lengthen and dilate during inspiration, and vice versa during expiration. Also, as discussed in Chapter 21, $\dot{V}A$ may be reduced even more by *physiologic dead space,* which can exceed anatomic dead space.

PULMONARY CIRCULATION

There are important differences in structure and function between the pulmonary and systemic circulations. The first difference involves the correct use of the words *arteries* and *veins* as delivery and return vessels, respectively. Arteries and veins cannot be defined by the O_2 content of the blood they contain. Systemic venous blood returns to the right atrium of the heart, and the right ventricle pumps this deoxygenated blood to the lungs through the *pulmonary arteries*. Oxygenated blood from the pulmonary capillaries flows into the *pulmonary veins*. The pulmonary veins return blood to the left atrium and the left ventricle pumps blood through the systemic circulation again.

Another important difference is that the lung is the only organ to receive the entire *cardiac output*. Because the amount of blood pumped by the right and left ventricles is equal, and because the pulmonary and systemic circulation are in series, the lungs receive the same amount of blood flow as the rest of the body. This places the lung in a unique position to process blood. Also, many of the structure–function relationships in the pulmonary circulation are explained by the fact that the lungs must handle high rates of blood flow.

Bronchial Circulation

The *bronchial circulation* is part of the systemic circulation, and it serves the metabolic needs of the large airways and blood vessels. The bronchial circulation does not extend to the respiratory zone, which is served by the pulmonary circulation. Bronchial arteries arise from the aorta and intercostal arteries, and the bronchial circulation returns blood to the heart by two pathways. Bronchial veins from large airways return about half the bronchial blood flow to the right heart via the azygos vein. The other half of the bronchial circulation drains directly into the *pulmonary* circulation. This adds deoxygenated blood to the oxygenated blood returning to the left heart, and constitutes an *anatomic shunt*. Blood flow through the bronchial circulation is only 1–2% of cardiac output in normal individuals, so this anatomic shunt has a small effect on arterial O_2 levels. However, the anatomic shunt can increase with inflammatory airway disease and cause significant reductions in arterial O_2 levels.

Pulmonary Vascular Pressures

Table II shows that the pressures in the pulmonary circulation are generally lower than in the systemic circulation, for at least two reasons. First, the pulmonary circulation supplies only a single organ, so a large pressure head is not necessary to distribute blood flow to multiple organs at different distances from the heart. The right ventricle only needs to generate sufficient pulmonary artery pressure to lift blood to the top of the lung. Low pressures mean that the pulmonary artery and its branches can have relatively thin walls, and they have much less connective tissue and smooth muscle compared with systemic arteries and arterioles. Pulmonary vascular pressures are so low that they are often measured with units of cm H_2O, instead of mm Hg (1.3 cm H_2O = 1 mm Hg). Second, *pulmonary capillaries* are not supported on the outside by tissue. Pulmonary capillaries are exposed to open gas spaces in the alveoli, so they are more susceptible than systemic capillaries to *stress failure,* or bursting open if their internal hydrostatic pressure is too high. All capillaries must be extremely thin to allow effective diffusion of gases.

Table II also shows that the pressure drop from artery to vein is more uniform in the pulmonary circulation than in the systemic circulation. Direct and indirect measurements indicate that pulmonary capillary pressure is near the mean of the average pulmonary arterial and venous pressures. Pulmonary capillaries contribute to more of the total pressure drop from artery to vein than do systemic capillaries. This means that capillaries are more important determinants of total resistance in the pulmonary circulation, compared with the systemic circulation.

Pulmonary vascular pressures can be altered by a variety of physiologic and pathologic conditions. For example, mean pulmonary artery pressure can increase to more than 35 mm Hg during exercise, and pulmonary venous pressure can exceed 25 mm Hg in patients with congestive heart failure. Pressures in the pulmonary circulation also vary a small amount with the ventilatory cycle. This is because the heart is surrounded by intrapleural pressure, which decreases on inspiration and increases on expiration (Chapter 19). Consequently, vascular pressures tend to fall on inspiration.

Pulmonary Vascular Resistance

The hydraulic analogy of *Ohm's law* can be used to define the relationship between pulmonary vascular pressure, flow, and resistance:

$$\Delta P = \dot{Q} \cdot PVR,$$

where ΔP is the pressure gradient between the inlet and outlet of a vessel (in mm Hg or cm H_2O), $\dot{Q}$ is blood flow (in liters per minutes), and PVR is *pulmonary vascular resistance*. PVR is by definition the resistance for both lungs and is about 1.7 (mm Hg · min)/L for a normal cardiac output of 6 L/min with an average pressure drop of 10 mm Hg from the pulmonary artery to left atrium.

The resistance to flow through a vessel obviously depends on its dimensions. The dimensions of pulmonary vessels are strongly influenced by several external forces, which is different from the situation for rigid pipes in a plumbing system, or even systemic arteries. The fundamental geometry of the pulmonary capillary network is also different from pipes or systemic capillaries, as illustrated in Fig. 11. The numerous capillaries in the alveolar wall constitute an almost continuous sheet for blood flow between two flat membranes held together by numerous posts. This is called *sheet flow,* and the resistance to sheet flow can be less than the resistance to flow through a network of tubes. Therefore, Poiseuille's law (Chapter 19) cannot be used to calculate pulmonary capillary resistance from capillary dimensions. Still, PVR increases with the length and decreases by a power function with the internal size of pulmonary capillaries.

The primary determinant of vessel size is the *transmural pressure*, which depends on the pressure difference between the inside and outside of the vessel:

$$P_{\text{transmural}} = P_{\text{inside}} - P_{\text{outside}}.$$

TABLE II   Mean (or Systolic/Diastolic) Pressures in the Pulmonary and Systemic Circulations (mm Hg)

	Pulmonary	Systemic
Ventricle	Right (25/0)	Left (120/80)
Artery	15 (25/8)	100 (120/80)
Arteriole	12	30
Capillary	10	20
Venule	8	10
Atrium	Left 5	Right 2

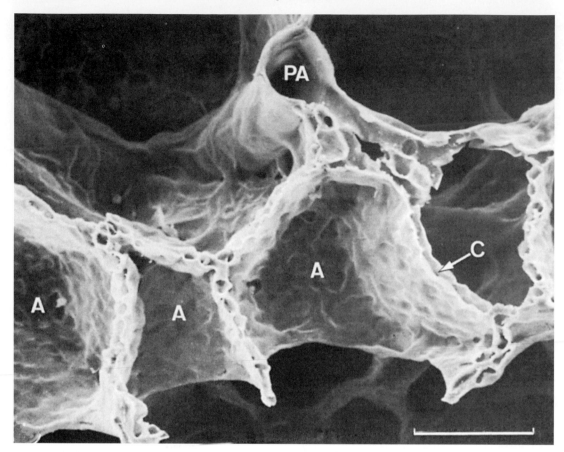

FIGURE 11   Blood flow in pulmonary capillaries (C) surrounds the alveoli (A) like a "sheet" of blood flow.
PA, pulmonary arteriole; marker = 50 μm. (From Weibel, Chap. 82 in Crystal *et al.*, eds., *The lung: Scientific
foundations.* Philadelphia: Lippincott-Raven, 1997.)

Therefore, increasing pulmonary arterial pressure will increase flow by two mechanisms: (1) the pressure gradient for Ohm's law is increased and (2) the transmural pressure is increased, which increases vessel size and decreases PVR. Figure 12 shows how PVR becomes even smaller when pulmonary arterial pressure is increased. Increasing pulmonary venous pressure also decreases PVR, because some of this pressure increase is transmitted to the capillaries. As discussed earlier, capillary dimensions significantly affect PVR. Hence, pressure affects resistance and vice versa in the pulmonary circulation, in contrast to the systemic circulation in which resistance primarily affects pressure.

Alveolar pressure is the outside pressure for calculating transmural pressure in most pulmonary vessels. Alveolar pressure varies with the ventilatory cycle, but it is generally near zero (i.e., atmospheric pressure; see Chapter 19). Therefore, vascular pressure is the primary determinant of transmural pressure in pulmonary vessels. However, large positive alveolar pressures can occur with some forms of artificial ventilation, and this will tend to collapse pulmonary capillaries.

Increasing transmural pressure can affect capillary dimensions by two mechanisms: *recruitment* and *distention*. At very low pressures, some capillaries may be closed, and increasing pressure will open them by recruitment. At higher pressures, capillaries are already open, but they may be distended or stretched by increased transmural pressure. Together, recruitment and distention increase the effective size of the pulmonary capillaries and reduce PVR.

Another important determinant of pulmonary vessel size is lung volume, but this effect differs for different types of vessels. *Extra-alveolar vessels* are surrounded by lung parenchyma, which acts as a tether or support structure to hold the vessels open. Therefore, lung volume is more important than alveolar pressure for determining the dimensions of extra-alveolar vessels. At high lung volumes, the extra-alveolar vessels are pulled open by tissues outside the vessels. At low lung volumes, this tethering effect is reduced and the extra-alveolar vessels narrow. Also, extra-alveolar vessels have smooth muscle and elastic tissue that tend to collapse the vessels at low lung volumes. The effects of lung volume on

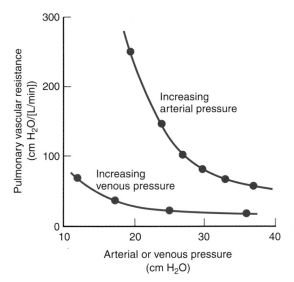

FIGURE 12   Pulmonary vascular resistance decreases with increasing pulmonary arterial or pulmonary venous pressure, while the other pressure is held constant. This is because increasing either pressure increases capillary pressure and causes recruitment and distention of pulmonary capillaries. (After West, Chap. 58 in Fenn and Rahn, eds., *Handbook of physiology, Section 3, Respiration*. Bethesda, MD: American Physiological Society, 1965.)

alveolar vessels are generally opposite those on extra-alveolar vessels. At high lung volumes, the alveolar wall is stretched and becomes thinner, reducing the size of pulmonary capillaries and alveolar vessels. At low lung volumes, the alveolar wall is not stretched and the capillaries relax open to a wider dimension.

Distribution of Blood Flow

The distribution of pulmonary blood flow throughout the pulmonary vascular tree and to different parts of the lung is not uniform. This was first shown in humans with a technique measuring pulmonary blood flow at different heights in the erect human lung using an insoluble radioactive gas. A saline solution containing radioactive xenon was infused in a vein, so the gas would enter the lung in proportion to blood flow (similar to CO_2 elimination from the blood). Radioactive counters were placed at different heights outside the chest to determine relative blood flow rates at different heights in the lung. Relative blood flow increased progressively from top to bottom of the lung.

The effect of gravity on pulmonary vascular pressures is a major factor determining the regional differences in blood flow in the upright human lung. The pulmonary vasculature can be considered a continuous hydrostatic column that is about 30 cm tall in the upright human lung. This means there is a hydrostatic pressure difference of 30 cm H_2O (or 23 mm Hg) between vessels at the top and bottom of the lung. This pressure

difference is nearly as large as the pulmonary artery pressure, so it has profound effects on regional distribution of blood flow. Evidence for a gravitational mechanism includes a reduction in the gradient of blood flow in erect persons during exercise when pulmonary arterial pressure increases, and a reduction in the gradient of blood flow in the supine posture. A dorsal-ventral gradient can be measured in people lying supine and the vertical gradient is reversed in persons suspended upside down.

Figure 13 illustrates these effects using the zone model for pulmonary blood flow. This model conceptually divides the lung into three zones to explain how gravity affects blood flow through alveolar vessels at different heights up the lung. *Zone 1* would occur at the top of the lung, where the pulmonary arterial pressure may not be sufficient to pump blood to the top of the lung. In this case, pulmonary arterial pressure is less than the hydrostatic pressure column between the heart and the top of the lung. Alveolar pressure, even if 0, is greater than arterial pressure so the capillaries collapse. Normally, zone 1 does *not* occur because the normal pulmonary arterial pressure (30 cm H_2O) is greater than the height of a water column between the heart and top of the lung (about 15 cm).

Zone 2 occurs near the middle of the lung, where pulmonary arterial pressure is increased by the hydrostatic column, and blood flow occurs. However, venous pressure is less than alveolar pressure because these veins may be below the level of the heart. Intravascular pressure decreases from the arterial to venous level along the capillary, and at some point the alveolar

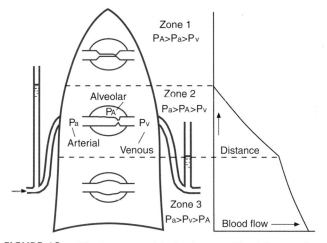

FIGURE 13   West's zone model of pulmonary blood flow predicts increasing blood flow down the lung because of the effects of gravity on pressures, as explained in the text. Pa, arterial pressure; PA, alveolar pressure; Pv, venous pressure. (After West, Chap. 58 in Fenn and Rahn, eds., *Handbook of physiology, Section 3, Respiration*. Bethesda, MD: American Physiological Society, 1965.)

pressure exceeds capillary pressure. This tends to collapse the capillary and reduce flow. If flow actually stops, then pressure in the capillary rises toward the arterial level until the capillary is reopened and flow resumes. In zone 2, the relevant pressure gradient driving blood flow is the arterial-alveolar difference and venous pressure is not important in determining zone 2 flow. Systems with flow determined by upstream and outside (instead of downstream) pressures are called *Starling resistors*. Figure 13 shows flow progressively increasing down zone 2 because the hydrostatic column increases arterial pressure while alveolar pressure is constant. Both capillary recruitment and distention can contribute to increased flow in zone 2.

Zone 3 occurs near the bottom of the lung, where venous pressure is increased sufficiently by the hydrostatic column to exceed alveolar pressure. Therefore, the arterial-venous pressure difference determines blood flow in zone 3. Figure 13 shows flow increasing down zone 3 because the hydrostatic column distends the capillaries. Some data suggest a *zone 4,* with decreased flows at the very bottom of the lung. It was hypothesized that high intravascular pressure leads to edema and vascular compression by the interstitium. However, zone 4 can be measured even after animals are inverted, suggesting that factors other than gravity may be involved.

Other methods of measuring the distribution of pulmonary blood flow suggest that factors other than gravity are important also. Radioactive microaggregates of albumin or plastic microspheres can be injected to the pulmonary circulation and they will lodge in capillaries in proportion to local blood flow. Gradients in blood flow have been measured between the center and the periphery of the lung at a given height up the lung. Local stresses and the anatomic details of vascular branching may contribute to such *intraregional hetero-geneity* of blood flow. Intraregional heterogeneity may explain up to half the total heterogeneity of blood flow in the lungs.

Control of Pulmonary Blood Flow

The most important physiologic mechanism that *actively* controls blood flow in the lungs is *hypoxic pulmonary vasoconstriction*. Hypoxic pulmonary vaso-constriction is a direct response of vascular smooth muscle in pulmonary arterioles to decreased alveolar P_{O_2}. The cellular mechanism involves potassium channels in the pulmonary artery endothelium that are sensitive to O_2 level. Hypoxic pulmonary vasoconstriction can be reduced by low concentrations of inhaled nitric oxide (20 ppm NO) in humans. NO also relaxes systemic vessels through a cyclic guanosine monophosphate (cGMP) pathway. Hypoxic pulmonary vasoconstriction is not

a reflex response, and it can even be induced in rings of pulmonary arterioles *in vitro*. This is opposite the vasodilatory effect of hypoxia in systemic arterioles.

A direct vasoconstrictor response to local alveolar P_{O_2} allows blood flow to be selectively diverted away from poorly ventilated regions of the lung. Hence, hypoxic pulmonary vasoconstriction is important for matching unequal distributions of ventilation and blood flow throughout the lungs. However, this response might be more important during the transition in the circulatory pattern at birth. Pulmonary blood flow is only 15% of the cardiac output in the fetus. P_{O_2} is relatively low in amniotic fluid, so hypoxic pulmonary vasoconstriction helps keep the pulmonary vascular resistance high in the fetus. At birth, the alveolar P_{O_2} increases with the onset of air breathing, and pulmonary vascular resistance decreases dramatically so the lungs can handle 100% of the cardiac output. Figure 14 shows the effect of P_{O_2} on pulmonary vascular resistance in a newborn animal; note that the main effect of P_{O_2} occurs at very low levels. However, blood pH has an effect even at high P_{O_2}, so the increase in pH with the onset of air breathing will also reduce pulmonary vascular resistance in the newborn.

Other physiologic factors capable of influencing the pulmonary circulation include a weak vasoconstrictor effect from the sympathetic nervous system and potent vasoconstriction by endothelins, which are peptides released by pulmonary epithelial and endothelial cells (e.g., ET-1).

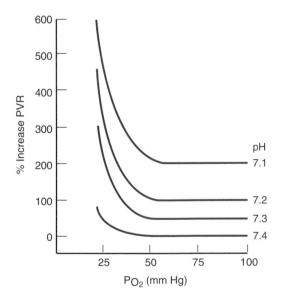

FIGURE 14 Deceasing O_2 in inspired gas (P_{O_2}) causes hypoxic vasoconstriction and increases pulmonary vascular resistance (PVR). Arterial acidosis exaggerates this effect in newborns and may be important in helping toe establish the adult pattern of circulation. (After Rudolph and Yuan, *J Clin Invest* 1966;45:399.)

Lung Fluid Balance

The pulmonary capillaries are extremely thin and contain pores that allow fluid to move across their walls. *Starling's law* describes the forces that govern fluid flux across capillary walls (see Chapter 16), and an understanding of these forces is necessary to understand both normal and pathologic lung fluid balance. Starling's law states that the net fluid flux across the capillary depends on a balance of *hydrostatic forces* (P) and *colloid osmotic* (or *oncotic*) *forces* (π):

$$\text{Net fluid flux} = K_{fc}[(P_c - P_i) - \sigma(\pi_c - \pi_i)],$$

where K_{fc} is *a filtration coefficient* that depends on the total surface area of the capillary and on the number and size of pores in the capillary. Hydrostatic pressure in the capillary (P_c) tends to move fluid out, and interstitial pressure (P_i) tends to move fluid into the capillary. Conversely, capillary osmotic pressure (π_c) tends to hold fluid in the capillary, and interstitial osmotic pressure (π_i) tends to draw fluid out of the capillary. The osmotic reflection coefficient (σ) describes the effectiveness of osmotic pressure at moving fluids, and it can range from 0 to 1. Conceptually, σ compares the size of the pore to the osmotically active solute: $\sigma = 0$ if the solute can move freely through the pore, and $\sigma = 1$ if the solute cannot move through the pore.

Normally, the balance of forces results in net *filtration,* or the movement of a few milliliters per hour of fluid *out* of the capillaries. Normal $P_c \approx 10$ mm Hg and normal P_i in the lungs is subatmospheric so there is a positive hydrostatic force moving fluid out of the capillaries. The *interstitial space* around alveolar capillaries is not compliant, so filtration in this region tends to increase local interstitial pressure. This local pressure increase is thought to provide a gradient moving filtrate toward the interstitium around the extra-alveolar vessels. Filtrate in this extra-alveolar region can be reabsorbed by the bronchial circulation or collected by lymphatics, which also return the fluid to the vascular system.

Normal *plasma protein* concentration is about 7.5 g/dL (mainly albumin); this exerts an osmotic pressure of about 28 mm Hg. The interstitium contains only about 5 g/dL of protein with an osmotic pressure of 15–20 mm Hg. Therefore, the osmotic forces promote absorption. Also, osmotic forces provide a natural feedback system, in which increased filtration dilutes the interstitial space. This reduces the osmotic gradient pulling fluid out of the capillaries. Recall that the osmotic pressure depends on the number of molecules in solution.

When this normal balance of forces is disturbed, filtration can exceed the capacity of reabsorption, and lymphatic drainage and fluid accumulates in the interstitium. *Edema* is the accumulation of excess filtrate outside the capillaries. Pulmonary edema fluid accumulates first in the peribronchiolar and perivascular spaces; this is called *interstitial edema*. Interstitial edema can alter local ventilation and perfusion and make gas exchange inefficient (Chapter 21). *Alveolar edema,* or flooding of excess filtrate into the alveolar spaces, is more serious because it can totally block ventilation and cause blood flow shunts in affected lung regions (Chapter 21). The exact mechanisms resulting in alveolar edema are not known, but they involve exceeding the lung's capacity for lymphatic drainage and changes in solute and fluid transport across airway epithelial cells. Alveolar type II epithelial cells normally transport NaCl to the basolateral surface, and water follows, keeping the alveoli dry.

Edema fluid can have a low or high protein concentration. Hydrostatic edema, which may occur with elevated pulmonary capillary pressures in congestive heart failure, results in filtrate with low protein concentrations. Other lung injuries, such as adult respiratory distress syndrome, may alter the permeability of the capillary endothelium (i.e., changes) and produce a protein-rich edema fluid.

NONRESPIRATORY FUNCTIONS OF THE LUNG

Airway Defense Mechanisms

The exchange surface area of the lung is the largest interface between the body and the environment. Therefore, the lungs have an important set of mechanisms to defend the body from foreign matter. The first line of defense is the *upper airways,* including the mouth and nose. A major function of the upper airways is to warm and humidify air entering the respiratory system, which prevents drying and cooling of the delicate epithelial barrier in the lungs. Complex air passages in the nose, called *turbinates,* also help trap large inhaled particles. Inhaled air enters the *pharynx* from the oronasal cavities, then passes to the *larynx* and through the *vocal cords,* and finally enters the trachea. Food and drink are kept out of the lungs by the *epiglottis,* which moves over the entrance to the larynx during swallowing. The lung is protected from very small particles suspended in the air, called *aerosols,* by three mechanisms. Large aerosols, with diameters of 1 μm or more, are removed from inhaled gas by *impaction* in the nose and pharynx as just described . Impaction traps aerosols when they fail to turn a corner with gas flow, and inertia carries the particle onto a wet mucosal surface. Medium-sized aerosols are trapped in the airways by *sedimentation,* as the particles fall out of the airflow under their own weight. Sedimentation occurs in the terminal and respiratory bronchioles because the total cross-sectional area of the airways greatly increases, and the forward velocity of inhaled air decreases (see Fig. 4). This is the area at which most soot and coal dust is deposited. The smallest aerosols, with diameters of 0.1 μm or less, can actually reach the alveoli by *diffusion.*

Particles that deposit in the lungs and airways are removed by two mechanisms. First, *mucociliary transport* removes foreign particles from the conducting airways (Fig. 15). Particles deposited in mucus are moved toward the mouth by the continuous beating of *cilia* on the airway epithelial cells. The cilia beat about 20 times/sec in a coordinated manner to move mucus upward out of the large airways at a speed of about 1–3 cm/min. When mucus reaches the pharynx, it can be swallowed, so deposited particles are removed from the respiratory system. Ciliary function can be impaired by smoking and pollutants such as sulfur and nitrogen oxides. Mucus is actually a complex secretion from the airway epithelium consisting of a gel layer and a sol layer. The top layer, or gel layer, is viscous and sticky to trap particles deposited on the airways. It contains macromolecules, such as *mucin.* The bottom layer is a less viscous secretion that bathes the 5- to 7-μm-long *cilia* on the airway epithelium. Therefore, the cilia can move easily in the sol layer, and the gel layer floating on top is moved up and out of the airways. Diseases such as cystic fibrosis and chronic bronchitis can affect mucous secretion. Second, alveolar *macrophages* provide an additional mechanism for removing particles deposited deeper in the lungs, where the blood–gas barrier must be very thin for gas exchange. Macrophages originate in the bone marrow and circulate in the blood as monocytes before settling in the respiratory zone of the lungs, where the epithelium is not ciliated. They roam the airway surfaces by ameboid action and engulf foreign particles by *phagocytosis.* Most foreign substances are destroyed by *lysozymes* inside the macrophage. However, carbon and mineral particles may be stored in residual bodies in the macrophage, which then settles in the interstitium. The effects of mineral dusts are especially insidious, leading to a progressive destruction of lung tissue, and even lung cancer in the case of asbestosis. Normal macrophages that do not settle in the interstitium leave the lung by the mucociliary transport or the lymphatics.

Neutrophils can leave the pulmonary circulation and provide a secondary line of phagocytic defense in the alveoli. Phagocytes, as well as immune cells (see later discussion), may release reactive oxygen species that can cause tissue damage. However, such damage is limited by the antioxidant glutathione, which occurs in surfactant at levels 100 times higher than in other tissues. Pulmonary cells have evolved efficient mechanisms to

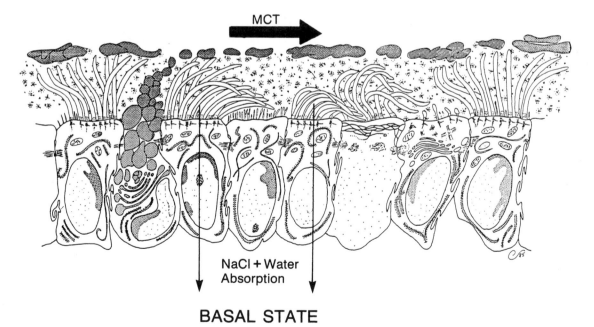

BASAL STATE

FIGURE 15   Cross-sectional schematic of conducting airway showing ciliated epithelial cells interspersed with mucous secretory cells. Mucociliary transport (MCT) moves the sticky surface layer of mucus (gel layer = shaded globules) up the airways by the beating motion of cilia in the less viscous, lower layer of mucus (sol layer = stars). (From Gabriel and Boucher, Chap. 20 in Crystal *et al.,* eds., *The lung: Scientific foundations.* Philadelphia: Lippincott-Raven, 1997.)

Clinical Note

Cystic fibrosis (CF) is the most common lethal genetic disease in Caucasians (1 in 2500 live births). It involves the lungs, sweat glands, pancreas, intestine, liver, and reproductive system, but pulmonary disease causes 90% of the deaths in CF patients. CF is characterized by abnormally thick mucous secretion, which impairs mucociliary transport in the airways and leaves the lungs vulnerable to bacterial infections. Chronic airway infection leads to abnormalities in gas exchange and the pulmonary circulation and ultimately to death by respiratory failure. Treating gastrointestinal complications and enhancing nutrition is important for preserving pulmonary function in CF. Improved treatment of infections has increased the life expectancy for CF patients from infancy to young adulthood, but recent results offer the promise for curing CF with gene therapy.

Abnormal chloride transport occurs in all tissues affected by CF, and this is explained by mutations in the CF *transmembrane conductance regulator* (CFTR). CFTR is involved in several aspects of mucous secretion in the airways: (1) Normal CFTR acts as a channel for chloride secretion at the apical surface of airway epithelial cells and (2) normal CFTR regulates sodium conductance through different channels by an unknown cell-signaling mechanism. Figure 15 shows that NaCl and water are *normally* absorbed out of the airways, but secretagogues can stimulate NaCl and mucous secretion into the airways. Normally the NaCl pulls water into the sol layer of the mucus so it is not too sticky and ciliary motion is efficient; this process is impaired in CF.

An autosomal recessive mutation results in deletion of a single amino acid from CFTR in 70% of CF patients. This mutant CFTR does not fold properly in the endoplasmic reticulum and it is catabolized before it can be inserted in the cell membrane. New therapeutic trials have tried (1) correcting ion transport by blocking sodium absorption or stimulating sodium secretion with non-CFTR mechanisms, (2) reducing the stickiness of mucus with enzymes, and (3) rescuing mutant CFTR from defective cell trafficking and stimulating normal membrane insertion. However, the most exciting possibility is that of introducing a normal copy of the CFTR gene into the airway in early life. Lung epithelial cells are easily accessible to gene transfer agents and adenoviruses have been used to deliver normal CFTR and restore chloride secretion in the nasal mucosa of patients for up to 3 weeks. Airway submucosal glands, which are normally rich in CFTR, are not as accessible to such gene transfer. Hence, the future of this therapy depends on delivering the gene to enough cells in the respiratory tract to normalize function throughout the airways and on lengthening the duration of expression of the normal gene product.

---

defend against oxidant injury, presumably because P_{O_2} is so high in the airways compared to other body compartments.

Immune System Defense Mechanisms

The lung is similar to all other organs by containing *lymphocytes* (T and B cells) in the interstitium. These defense cells originate from the bone marrow and lymph nodes and respond to foreign invaders with cellular (acquired antibody) mechanisms. Dendritic cells present antigens to the lymphocytes. Considering that up to 10^{10} antigenic particles may reach the alveoli every day, the challenge for the pulmonary immune system is to process this foreign material and not overamplify an inflammatory response. Basic immune mechanisms in the lungs are similar to the rest of the body and are not covered here.

Lymphatics are the main pathway for removal of immune cells that have already responded to foreign substances or cells in the lungs. The lymphatic drainage also removes excess fluid filtration from the pulmonary capillaries as described earlier. Pulmonary lymphatics start as blind end vessels in the acini, where they collect fluid and lymphocytes through a leaky endothelium. Lymphatics do not occur at the level of the alveoli. Lymph flow is always directed out of the lungs by numerous valves in a network of vessels that generally follows the large airways and blood vessels. *Lymph nodes* occur along this network and function as biological filters. Macrophages and immune defense cells collect in lymph nodes and present antigens to immunocompetent cells in the nodes. This programs new immune system defenses for future invasions. Ultimately, pulmonary lymph returns to the venous system near the junction of the subclavian and jugular veins.

Biosynthetic Functions

The lung is ideally situated to carry out many biosynthetic functions on substances in the blood because it receives the entire cardiac output. Three major mechanisms are used by the lung to process substances in blood: (1) synthesis or addition, (2) degradation or removal, and (3) activation or conversion, i.e., biotransformation. Table III summarizes the effect of the lungs on substances in blood that pass through the pulmonary circulation.

Biogenic amines are removed from the pulmonary circulation in varying degrees. *Serotonin* (or 5-hydroxytryptamine) occurs in the lungs as a product of alveolar macrophages and pulmonary mast cells, in addition to arriving by the circulation. A carrier-mediated uptake process in the pulmonary endothelium almost completely removes serotonin from the blood. *Norepinephrine* occurs in the lungs from local activation of sympathetic nerve endings. Norepinephrine is removed from the blood by a carrier-mediated process into pulmonary endothelial cells, which contain enzymes to degrade the neurotransmitter. However, this pulmonary uptake mechanism is not effective at controlling systemic norepinephrine levels. *Histamine* is stored in large amounts in pulmonary mast cells in the airway walls and epithelium. Histamine can be released from these cells by allergic reactions and causes bronchial smooth muscle contraction and pulmonary vasoconstriction. Enzymes for degrading histamine occur in the lung, but the pulmonary circulation is not effective at removing histamine, perhaps because a cellular uptake mechanism does not occur.

Peptides are degraded or activated by the pulmonary circulation. *Angiotensin II* is a vasoconstrictor produced in the lungs by *angiotensin-converting enzyme* (ACE). ACE occurs in pulmonary capillary endothelial cells and creates angiotensin II by cleaving two amino acids from a precursor decapeptide called angiotensin I. Angiotensin I is not a strong vasoconstrictor and it is not changed by the pulmonary circulation. ACE is the main example of activation by the pulmonary circulation and it is extremely effective, with almost 100% conversion in a single pass of the blood through the lungs. *Bradykinin* is a potent vasodilator that is not synthesized in the lung, but is largely inactivated by the pulmonary circulation. Enzymatic inactivation of bradykinin occurs at the same endothelial site and uses the same enzyme (ACE) used to activate angiotensin I.

Arachidonic acid metabolites, also called *eicosanoids,* are either almost completely removed from blood by the pulmonary circulation or they are not affected. Arachidonic acid is made in the lungs and other tissues by breakdown of phospholipid membranes. *Prostaglandins* (PG) and *thromboxanes* (Tx) are synthesized from arachidonic acid by cyclo-oxygenase and peroxidase, and can be released in pathologic states such as embolism and anaphylaxis. $PGF_2\alpha$ is a bronchoconstrictor and vasoconstrictor, PGE_2 is a vasoconstrictor, and PGE_1 is a vasodilator, but they affect only the local airways and vessels where they are produced because they are metabolized by the pulmonary circulation. TxA_2, a bronchoconstrictor, vasodilator, and platelet-aggregating substance, is also metabolized by the pulmonary circulation. On the other hand, PGI_2 (or prostacyclin), which is a vasodilator and inhibitor of platelet aggregation, is not affected by the pulmonary circulation so it can exert systemic effects also.

Leukotrienes (LTs) are synthesized from arachidonic acid by a different pathway, and they are part of the airway inflammatory response. LTC_4, LTD_4, and LTE_4 may play a role in asthma. These LTs are 1000 times more potent than histamine as bronchoconstrictors, they stimulate mucous production, and they increase vascular permeability, which can lead to edema. LTB_4 induces leukocyte chemotaxis, increased vascular permeability, and vasodilation. LTs are removed almost completely by the pulmonary circulation, so their effects are local.

TABLE III   Effect of the Pulmonary Circulation on Substances in Blood

Biogenic Amines:

- Serotonin: 95% removed
- Norepinephrine: Up to 40% removed
- Histamine: No change

Peptides:

- Angiotensin I: Converted to angiotensin II by ACE
- Angiotensin II: No change
- Bradykinin: Up to 80% inactivated

Arachidonic Acid Metabolites:

- Prostaglandin E_1, E_2, F_{2a}: 90% removed
- Prostaglandin A_1, A_2: No change
- Prostacyclin (PGI): No change
- Leukotrienes: 90% or more removed
- Thromboxane: Removed

Suggested Readings

Crystal RG, West JB, Weibel ER, Barnes PJ, eds. *The lung: Scientific foundations,* 2nd ed. Philadelphia: Lippincott-Raven, 1997.

Davis PB, Drumm M, Konstan MW. Cystic fibrosis. *Am J Respir Crit Care Med* 1996;154:1229–1256.

Fishman AP, ed. Circulation and nonrespiratory functions, Vol. I. In *Handbook of physiology, Section 3, The respiratory system.* Bethesda, MD: American Physiological Society, 1985.

Grippi MA, Metzger LF, Krupinski AV, Fishman AP. Pulmonary function testing. In Fishman AP, ed. *Pulmonary diseases and disorders,* 2nd ed. New York: McGraw-Hill, 1988, pp 2469–2522.

Wanner A, Salathe M, O'Riordan TG. Mucociliary clearance in the airways. *Am J Respir Crit Care Med* 1996;154:1868–1902.

Mechanics of Breathing

FRANK L. POWELL, JR.

KEY POINTS

- Skeletal muscle contractions generate pressure gradients to pull fresh air into the lung; the diaphragm is the most important muscle for *inspiration.*
- *Expiration* at rest is passive, but expiratory muscles are recruited at higher levels of ventilation (e.g., exercise).
- *Lung volume* is determined by the pressure difference between the airways inside the lungs and intrapleural space outside the lungs. Intrapleural pressure is subatmospheric at most lung volumes.
- *Compliance* quantifies the pressure–volume relationship of the lungs, which changes with both physiologic and pathologic conditions.
- *Pulmonary surfactant,* a lipoprotein that lines the alveoli and reduces surface tension, stabilizes the alveoli, reduces the pressures required for ventilation, and helps prevent pulmonary edema.
- The *functional residual capacity* of the lungs is determined by a balance of lung and chest wall forces, and is a sensitive indicator of lung compliance changes with disease.
- The main site of *airway resistance* is in the medium-sized bronchi. Pulmonary disease usually starts in small airways, so it is difficult to detect increased airway resistance in early stages of disease.
- *Maximum expiratory flow* is independent of muscular effort at most lung volumes less than total lung capacity (TLC) and is very dependent on lung compliance. In contrast, maximum inspiratory flow rate is proportional to pressures generated by respiratory muscles at most lung volumes.
- Local differences in lung compliance and airway resistance are the main causes of *regional differences in ventilation.*
- The *oxygen cost of breathing* is small in healthy individuals, ranging from 1–3% of total oxygen consumption at rest and less than 15% at maximal exercise.

Respiratory mechanics explain how the forces generated by respiratory muscles cause effective ventilation of the alveoli. *Resistance* and *compliance* are two physical factors important in determining the relationship between pressure and flow in the lungs, and in the distribution of ventilation to different regions of the lungs. The effectiveness of ventilation depends not only on the amount of fresh air reaching the alveoli, but also on the matching of ventilation blood flow in different regions of the lung. Understanding how the mechanical properties of individual components of the respiratory system affect the behavior of the whole system is useful in diagnosing and treating many pulmonary diseases, such as asthma, emphysema, and infant respiratory distress syndrome.

RESPIRATORY MUSCLES

Ventilatory flow is driven by pressure differences between the alveoli and the atmosphere. In normal individuals at rest, active contraction of skeletal muscle generates this pressure difference during inspiration. In contrast, expiration at rest results from passive elastic recoil of the lungs.

The *diaphragm* is the main muscle of inspiration and, therefore, the most important muscle for resting breathing. The diaphragm is innervated by the *phrenic nerve,* which originates from the third, fourth, and fifth cervical spinal nerves. Figure 1 shows the normal domed configuration of the diaphragm separating the

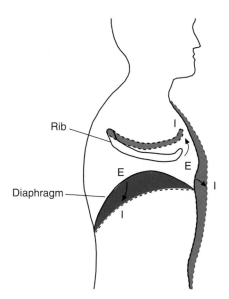

FIGURE 1   During inspiration (I) the diaphragm contracts and flattens, and external intercostal muscles raise the ribs to expand the thoracic volume. During expiration (E) the diaphragm relaxes to its domed shape and external intercostal muscles pull the ribs down to decrease thoracic volume.

thoracic and abdominal cavities. Contraction of the diaphragm flattens the floor of the thoracic cavity so that the lung increases in height. The *external intercostal muscles* also contribute to inspiration. Intercostal muscles are innervated by intercostal nerves from thoracic segments of the spinal cord. Figure 1 also shows how the external intercostals move the ribs upward and outward during inspiration to increase the diameter of the thoracic cavity.

The *abdominal muscles* are the most important muscles of expiration. These include the rectus abdominis, internal and external oblique muscles, and transversus abdominis, innervated by lower thoracic and lumbar spinal nerves. Abdominal muscles pull the ribs downward and squeeze the abdomen to increase abdominal pressure below the diaphragm. This moves the diaphragm upward and decreases thoracic volume. The abdominal muscles are also important for the respiratory act of coughing and nonrespiratory acts of vomiting and defecating. *Internal intercostal muscles* contribute to expiration by pulling the ribs downward (Fig. 1).

Accessory respiratory muscles include both inspiratory and expiratory muscles that are recruited at high levels of ventilation (e.g., during exercise). These muscles can also be important for ventilation in patients with abnormal diaphragm function. Accessory respiratory muscles include the scalene muscles, which lift the first two ribs, and the sternomastoids, which lift the sternum.

Upper airway muscles in the *pharynx* and *larynx* also contract in phase with breathing and are important in determining airway resistance. Inspiratory activation of pharyngeal muscles innervated by the *hypoglossal nerve* stiffens the soft palate and holds the tongue out of the way for breathing. Laryngeal muscles innervated by the *recurrent and superior laryngeal nerves* dilate the airway during inspiration, mainly at the level of the vocal cords. Dilation of the nares by the *alae nasi* can also decrease resistance to airflow, but larger decreases can be achieved by switching to mouth breathing.

STATIC MECHANICS OF THE RESPIRATORY SYSTEM

The first step in understanding respiratory mechanics is understanding the determinants of static lung volumes. *Static lung volumes* refer to conditions of no flow (no ventilation). The lungs are elastic structures and their volume depends on the pressure difference between the inside and outside, similar to a balloon. Figure 2 illustrates the important pressure differences determining lung volume. Alveolar pressure (P_A) is the pressure inside the lungs. With an open airway and no

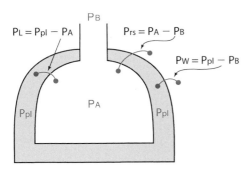

FIGURE 2   Three transmural pressures in the respiratory system should be considered: (1) transrespiratory system, $Prs = PA - PB$; (2) transpulmonary pressure, $PL = PA - Ppl$; (3) trans-chest wall, $Pcw = PA - Ppl$. PA, alveolar pressure; PB, barometric pressure; Ppl, intrapleural pressure. The importance of these pressure differences for determining lung volume are described in the text.

flow, PA equals atmospheric or barometric pressure (PB). Intrapleural pressure (Ppl) is the pressure outside the lungs, so static transpulmonary pressure (PL) and lung volume are determined only by Ppl with open airways and no air flow.

The intrapleural space is a sealed space filled with a very thin layer (10–20 μm) of pleural fluid. This provides lubrication between the visceral pleura surrounding the lungs and the parietal pleura lining the chest wall, and allows the lungs to move easily against the chest wall surface during volume changes. Ppl is determined by the elastic properties of the lungs and chest wall, which vary with volume, as described later. Ppl is negative relative to barometric pressure (i.e., subatmospheric) at normal lung volumes because the lungs tend to collapse inward from the chest wall. However, the pleural surfaces cannot separate because the intrapleural space is sealed. The situation is similar to two pieces of glass with a layer of water between them: The panes cannot be pulled apart easily, but they can slide over one another. Therefore, the positive PL distending the lungs results from a *negative* pressure outside the lungs and zero pressure inside the lungs. The lungs expand whenever inside pressure exceeds outside pressure ($PA > Ppl$). If the lungs were exposed to atmospheric pressure outside the thorax, then the comparable volumes could be achieved by blowing the lungs up with equivalent positive pressure.

Compliance

Figure 3 illustrates the pressure–volume relationship of a lung and how this changes under different conditions. The curves demonstrate three important features: (1) The pressure–volume relationship is nonlinear and changes with volume, (2) the air curves show *hysteresis*, i.e., a difference between inflation and deflation, and (3) the curves are different for inflation

with air and saline. These observations can all be described in terms of *compliance,* which quantifies the volume change for a given pressure change in distensible elastic structures. Compliance is the slope of the pressure–volume curve:

$$C = \Delta V/\Delta P,$$

where ΔV is the change in lung volume and ΔP is the change in pressure distending the lungs.

Absolute values of compliance depend on body size because lung volume changes with size, but pressures do not. (In physical terms, volume is an extensive, or size-dependent, property whereas pressure is an intensive, or size-independent, property.) *Specific compliance,* or compliance per unit lung volume, can be used to compare lungs of different size. *Elastance* is a measure of the stiffness of a structure and is defined as the inverse of compliance.

Compliance of the air-filled lungs shown in Fig. 3 could be measured by blowing up a lung outside the body. However, compliance can be measured physiologically by estimating intrapleural pressure from *esophageal pressure* (Pes). The entire thoracic contents outside of the lung are essentially in equilibrium with Ppl. Therefore, Pes measurements at different static lung volumes can be used to estimate Ppl and measure compliance *in vivo.*

The air inflation compliance curve in Fig. 3 is nonlinear. Volume does not increase on inflation until a critical pressure level is achieved. A *critical opening*

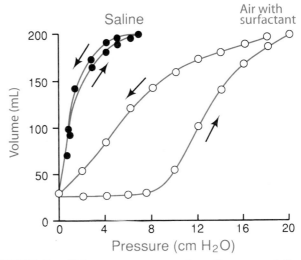

FIGURE 3   Volume versus pressure (compliance curves) for an animal lung filled and emptied with air or saline. Surface tension decreases compliance ($\Delta V/\Delta P$) in the air-filled lung compared to the saline-filled lung. Compliance would be even lower in an air-filled lung without surfactant (not shown). (After Clements, *Physiologist* 1968;5:11.)

pressure is not observed in lungs inflated with saline, indicating that surface tension contributes to this behavior (see later discussion). After the critical opening pressure is reached, large increases in lung volume occur with relatively small increases in pressure. Eventually, compliance decreases as the lung becomes fully distended and cannot stretch anymore. The lung has reached its elastic limits, set by the biomechanical properties of cartilage in conducting airways and a network of connective tissue in the alveolar walls. This network has rubber-like *elastin* bands that can stretch to a point at which the twine-like *collagen* fibers are fully extended and can stretch no further.

Deflation of air-filled lungs also shows changes in compliance with volume (Fig. 3). Lung volume decreases relatively little for a given drop in pressure near maximum volumes, but then compliance increases at lower lung volumes. At the end of deflation, there is no distending pressure but some volume remains in the lungs. This is *airway closure, or* trapping of gas in terminal spaces distal to collapsed airways. Airway closure is at its worst with lung compliance changes caused by aging or disease (see later discussion).

Inflating and deflating the lungs with saline eliminates hysteresis between the inflation and deflation curves (Fig. 3). This indicates that the interface between a gas phase and the wet alveolar surface contributes to hysteresis. The consequences of this for lung mechanics are considered next.

Surface Tension

Surface tension is a force generated in the surface of a liquid at a gas–liquid interface. This occurs in the lungs between the alveolar gas and the fluid lining the alveoli, which is necessary to keep these delicate structures moist. Surface tension has the units of force per unit length in the surface of the interface (e.g., dynes per centimeter) because it is the force holding the liquid molecules together at the interface. It is the force that forms a skin on water when a glass is filled slightly above the rim. Surface tension tends to shrink the surface area of the interface, so the tendency for the alveolar surface area to shrink tends to decrease alveolar volume. Therefore, surface tension acts like a pressure outside the alveoli and tends to collapse them. The *law of LaPlace* relates the surface tension (T) and pressure (P) in a gas-filled, liquid-lined sphere with radius r:

$$P = 4T/r.$$

Figure 4 shows why surface tension and the law of LaPlace are important in the lung. Pressures inside the alveoli are inversely proportional to their radius.

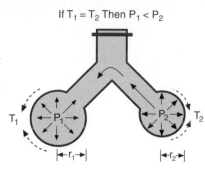

If $T_1 = T_2$ Then $P_1 < P_2$

FIGURE 4   The law of Laplace predicts instability between alveoli of different radii (r) because pressure (P) = $2T/r$, where T = tension. Therefore, small alveoli will empty into larger alveoli unless surface tension is lower in small alveoli.

Therefore, smaller alveoli would collapse and empty into larger alveoli if the surface tension were equal in both alveoli. This is an oversimplification because alveoli are not simple bubbles, but are an interdependent network of air cells resembling a sponge. The tendency of one alveolus to collapse is opposed by elastic forces in adjacent alveoli resisting further expansion. Still, the law of LaPlace predicts a mechanical advantage if surface tension could be decreased in smaller alveoli, so they would not tend to empty into larger alveoli.

Pulmonary surfactant is a biologic secretion lining the alveoli that reduces surface tension in the lung. Figure 5 shows two effects of surfactant on surface tension. First, surfactant reduces the surface tension below that of water (which would otherwise line the alveoli), and second, surface tension of a surfactant layer depends on surface area, in contrast with either water or detergent. The low surface tension of surfactant reduces the pressure necessary to inflate the lungs and reduces the work of breathing. Lung compliance is reduced and the work of breathing is increased if disease or injury depletes surfactant from the lungs (see Clinical Note later in this chapter on respiratory distress in newborns).

Detergent also has a lower surface tension than water; this explains why bubbles in soap solutions last longer than bubbles in a pure water. However, the surface tension of detergent does not change with surface area as it does for surfactant (Fig. 5), and this dependence on area is physiologically important. Surfactant promotes alveolar stability by decreasing surface tension in small alveoli, and the law of LaPlace predicts that this will reduce the difference in pressures between alveoli of different radii (see Fig. 4).

Surfactant is a lipoprotein synthesized in *lamellar bodies* in alveolar type II cells and turns over rapidly in the lungs. It is 90–95% phospholipids with a glycerol backbone on a polar head and two nonpolar fatty acid chains. *Dipalmitoylphosphatidylcholine* (DPPC) is the main phospholipid, and it spontaneously forms a

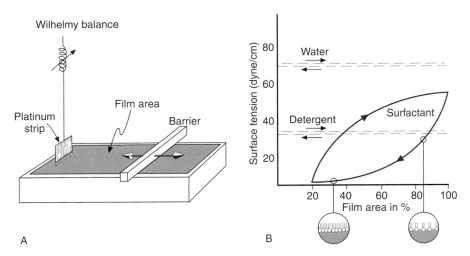

FIGURE 5   (A) A Wilhelmy balance can be used to measure surface tension on a film of water as the surface area changes. Decreasing area, by moving the barrier to the left, increases surface tension on the film. This pulls down the platinum strip, so surface tension can be measured as the force on a string supporting the strip. (B) The surface tension of surfactant depends on surface area and shows hysteresis, unlike water or detergent. The insets show how molecular organization of surfactant changes with different surface areas. (After Clements and Tierney, Chap. 69 in Fenn and Rahn, eds., *Handbook of Physiology, Section 3, Respiration.* Bethesda, MD: American Physiological Society, 1964.)

monomolecular firm on fluid surfaces. This behavior is called *adsorption*. The hydrophobic fatty acid ends of DPPC molecules face the surface and the polar phospholipid heads contact the fluid (see Fig. 5). When the surface film is compressed, this regular alignment of surfactant molecules squeezes out water molecules from the surface and effectively lowers surface tension.

The *apolipoproteins* in surfactant include large collagen-like glycoproteins. Their function is not completely understood, but they appear to be involved in *tubular myelin* formation and surfactant removal and recycling. Surfactant-associated proteins with antioxidant and anti-inflammatory properties have also been isolated (e.g., SP-A and SP-D). Surfactant is stored in 100-Å-thick stacks inside lamellar bodies, and it can be transformed to tubular myelin during secretion from type II cells onto the alveolar surface. Tubular myelin is probably the direct precursor for the surface active film lining the alveoli. Surfactant may be recycled into type II cells or removed from the alveoli by macrophages.

The hysteresis for surfactant explains the hysteresis in lung compliance curves. The plot of surface tension versus area in Fig. 5 is analogous to the plot of the pressure–volume curve in Fig. 3 except the axes are reversed; surface area is proportional to volume, and surface tension is proportional to pressure. As alveolar surface area decreases during expiration, alveolar surface tension is lower than at an equivalent area during inspiration. Hence, less pressure is necessary to maintain a given lung volume during expiration than during inspiration and this helps prevent the lung from

collapsing to very low volumes during normal breathing. The work of breathing is minimized because inspiratory muscles do not have to generate large critical opening pressures on every breath, which would be necessary to expand the lung from minimal volume.

Surfactant is also important for fluid balance in the lung. Surface forces tending to collapse the alveoli will lower interstitial pressure around capillaries in the alveolar walls, and tend to draw fluid out of capillaries (see Starling's law, Chapter 18). Therefore, the effect of surfactant to reduce alveolar surface tension helps to prevent edema.

Determinants of Functional Residual Capacity

Lung volume is determined by the interaction of elastic and surface tension forces in the lung and by the elastic properties of the chest wall. This interaction is illustrated in a relaxation *pressure–volume curve,* obtained from a person breathing from a spirometer (Fig. 6). At specified volumes, the mouthpiece is closed, the person relaxes the respiratory muscles, and pressures are measured. Figure 6 shows pressure–volume curves plotted against the appropriate transmural pressure for the lung alone (PL), the chest wall alone (Pcw), and intact respiratory system (Prs). The lung curve is a function of Ppl, estimated from esophageal pressure as described earlier, and equals the curve that would be measured using positive pressure inflation on an excised

Clinical Note

Infant respiratory distress syndrome (RDS) occurs in about 1% of deliveries worldwide, but it is most common in preterm infants, and occurs in 75% of babies delivered at 26–28 weeks of gestational age. Infants with RDS have rapid shallow breathing, hypoxemia, and acidosis, and can die without treatment within 72 hr. In 1959, Avery and Mead showed that extracts from infants with RDS had abnormal surface tension. This led to our current understanding of RDS as a disease resulting primarily from abnormal surfactant and to our current treatments with surfactant replacement.

Surfactant production normally begins *in utero* and term infants are born with lung surfactant levels greater than adults. Such high surfactant levels probably help protect the airway epithelium from damage by high shear forces from amniotic fluid during breathing movements *in utero*, and improve lung mechanics during the transition from fluid to air breathing upon birth. Normal surfactant metabolism is assured only after 36 weeks' gestation, and preterm infants appear to suffer from surfactant dysfunction as much as surfactant deficiency. The normal cycle of lamellar body secretion, formation of tubular myelin, and adsorption of surfactant to the alveolar surfaces does not occur until after full-term birth. *Hyaline membranes,* or a coagulation of cellular debris and plasma proteins, line the alveoli in infant RDS, which is also known as hyaline membrane disease.

Surface tension is high with RDS, so normal inflation pressures of 5–10 cm H_2O are ineffective at expanding the lung. The stiff lung of RDS and the compliant chest wall of the newborn decrease FRC. RDS infants breathe with an expiratory grunt, by contracting the vocal cords during expiration to maintain lung volume. These factors increase the work of breathing and can lead to a metabolic acidosis. Rapid shallow breathing increases dead space ventilation. Abnormal surfactant also leads to hypoxemia from edema and *atelectasis* (alveolar collapse) by increasing alveolar surface tension, and probably by changing epithelial permeability.

Surfactant replacement therapy has increased infant RDS survival from 30% in the 1970s to 90% today. Exogenous surfactant can be instilled in the trachea and it spreads into the lungs and alveoli during artificial ventilation. Surfactant can be isolated from late gestational human amniotic fluid or animal lungs, synthesized in the laboratory, or produced from bacteria using genetic engineering. Synthesizing surfactant that contains the normal apoproteins may be important for correcting defects other than increased surface tension, and the genetic engineering approach holds the most promise for therapies in other forms of RDS, such as acute RDS in adults.

---

lung. Total respiratory system is a function of pressure in the closed airway during relaxation at any volume. (At FRC, the airway can be open because FRC is by definition the relaxed lung volume.) Chest wall behavior is calculated as the difference between the respiratory system and lung curves, both of which can be measured physiologically. In theory, the chest wall curve could be measured by positive pressure inflation of an empty thoracic cage.

FRC occurs at the end of a normal expiration when respiratory system transmural pressure (Prs) is zero. Figure 6 shows how this results from an interaction between the lung and chest wall so transmural pressure across the lung (PL) and chest wall (Pw) are equal and opposite at FRC. The tendency for the lung to collapse is perfectly balanced by the tendency for the chest wall to expand at FRC. At volumes below FRC, the chest wall tends to expand even more. Residual volume (RV) occurs when the expiratory muscles cannot generate

sufficient force to decrease volume further. Figure 6 shows that RV is greater than the minimal volume of the lung (MV). MV includes gas in large stiff airways and gas trapped in alveoli distal to small collapsed airways. The lungs can never reach MV in an intact respiratory system but a *pneumothorax,* or leak between the intrapleural space and atmosphere, can cause total deflation of the lungs and seriously compromise gas exchange. A pneumothorax eliminates the subatmospheric Ppl that holds the lungs open.

At high lung volumes, the chest wall has a tendency to collapse. Therefore, a positive Prs is necessary across both the lung and chest wall to achieve high lung volumes. Total pressure to inflate the respiratory system increases as the sum of lung and chest wall pressures (Prs = PL + Pw). TLC occurs at the elastic limits of the chest wall, but this is also very near the limits of the lungs. Lung compliance decreases sharply at high volumes (see Figs. 3 and 6), so volume could not

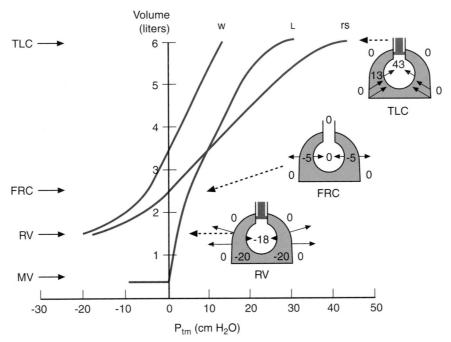

FIGURE 6   Relaxation pressure–volume curves of lungs (L), chest wall (w), and total respiratory system (rs). Arrows on diagrams by RV, FRC, and TLC indicate magnitude and direction of lung and chest wall elastic forces, and numbers indicate pressures. Prs equals the sum of PL and Pcw, which are equal and opposite at FRC. Lung minimal volume (MV) cannot be measured in the intact respiratory system. (After Rahn *et al., Am J Physiol* 1946;146:161.)

increase much more even without the constraints of the chest wall.

Pulmonary disease frequently affects lung compliance without affecting the chest wall; this explains the particular changes in lung volume associated with different diseases. *Emphysema* increases lung compliance by destroying elastic tissue in the lungs. The disease disrupts the normal balance between elastases in the lung (e.g., from neutrophils) and endogenous elastase inhibitors. For example, emphysema occurs in individuals with a genetic deficiency for α_1-antitrypsin. The normal elastase activity of trypsin destroys lung tissue when it is not inhibited by endogenous α_1-antitrypsin, as is the case in normal subjects. Smoking also contributes to emphysema by inhibiting α_1-antitrypsin. FRC increases in emphysema because the lung has less elastic recoil to balance chest wall expansion, so FRC moves toward the equilibrium position of the chest wall at a higher volume.

Interstitial pulmonary fibrosis, a so-called restrictive disease, can have the opposite effect. Fibroblasts lay down thick bundles of collagen in the alveolar walls, and this decreases lung compliance and volumes. FRC is a useful diagnostic measurement because it is a static volume that merely requires a person to relax after a normal exhalation. However, changes in airway resistance and the distribution of ventilation are the most important mechanical factors in lung disease.

AIRWAY RESISTANCE AND THE DISTRIBUTION OF VENTILATION

Ventilatory air flow occurs down pressure gradients. Upon inspiration, the respiratory muscles generate subatmospheric Ppl, which are transmitted to the alveoli. Air flows into the lungs down a pressure gradient from the atmosphere to the alveoli. During expiration, the respiratory muscles compress the intrapleural space, creating positive pressure in the alveoli that force air out to the atmosphere. *Airway resistance* is the primary determinant of the pressure gradient necessary to produce a given flow rate.

Physical Determinants of Resistance

Under conditions of streamlined or *laminar flow*, the relationship between the pressure gradient (ΔP) and flow ($\dot{V}$) is directly proportional to airway resistance (R):

$$\Delta P = \dot{V}R.$$

This is the hydraulic equivalent of *Ohm's law,* which is used to describe pulmonary vascular resistance also. Resistance to laminar flow through a straight and rigid tube depends on the physical properties of the fluid and the tube as described by Poiseuille's law:

$$R = 8\eta l/\pi r^4,$$

where η is viscosity, l is tube length, and r is tube radius. Notice that the radius is much more important than length in determining the resistance to laminar flow through a tube.

Under conditions of *turbulent flow,* a larger pressure gradient is necessary, and it is approximately proportional to the square of the flow rate:

$$\Delta P = \dot{V}^2 R.$$

The physical factors determining whether flow is laminar or turbulent include tube radius (r), gas velocity (v), density (d), and viscosity (η), which are used to define the *Reynolds number* (Re):

$$Re = 2rvd/\eta.$$

The Reynolds number quantifies the ratio of inertial to viscous forces, and turbulence occurs when this number exceeds 2000. In the airways, turbulence occurs when velocities are high relative to airway radius, or mainly in the upper airways before extensive branching increases total cross-sectional area and decreases linear gas velocity. Air flow is laminar in the small airways, where velocity is low; Re approaches 1 in the terminal bronchioles.

Flow in most of the airways is *transitional,* or somewhere between being laminar and turbulent. In general, the pressure gradient for ventilation is described by the flow rate *and* the flow rate squared:

$$\Delta P = K_1\dot{V} + K_2\dot{V}^2,$$

where K_1 and K_2 are constants. This is Rohrer's equation, and under most conditions the second term is relatively small, so the equation simplifies to Ohm's law ($K_1 = R$). Factors increasing the importance of turbulence and increasing the pressures necessary for ventilation include high gas velocities, such as may occur during exercise, or breathing high-density gases, for example, during deep-sea diving.

Measures of Airway Resistance

The physical principles just described can be used to predict the main site of airway resistance in the

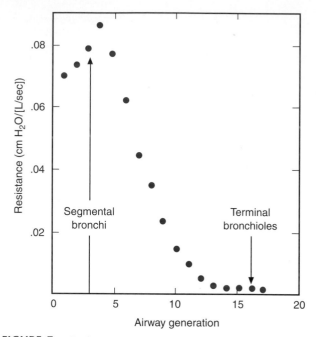

FIGURE 7   Resistance to air flow changes with airway generation, where 0 is the trachea. Resistance is greatest in the segmental bronchi and decreases rapidly in smaller but more numerous peripheral airways. (After Pedley *et al., Respir. Physiol.* 1970;9:387.)

bronchial tree, and this can be verified on experiments with isolated lungs or models of lungs. Figure 7 shows that the main site of airway resistance is the large airways, even though radius has a dominant influence on resistance (cf. Poiseuille's law). This is because the small cross-sectional area in the upper airway branches results in high velocities and turbulent flow. Small airways contribute less than 20% of airway resistance because their large number reduces their total resistance. Unfortunately, many airway diseases start in the small airways, where resistance changes have to be very large before they can be detected by physiologic tests. This level of the bronchial tree is called the *silent zone.*

Airway resistance can also be approximated by solving Ohm's law with measurements of flow and the pressure gradient between alveoli and the mouth. Flow is relatively easy to measure with a spirometer (see Chapter 18, Fig. 6) or a pneumotachometer (a physiologic gas flow meter). Mouth pressure is also easy to measure, but P_A cannot be measured directly. P_A can be measured indirectly with a body plethysmograph by solving Boyle's law for lung pressure instead of volume (see Chapter 18, Fig. 8).

P_A can also be estimated from P_{pl} measurements. Figure 8 shows two plots for P_{pl}: The solid line is the pressure during resting ventilation, and the dashed line is the pressure that could maintain the given lung volume without flow. The dashed line is the same data used to plot static compliance curves (see Fig. 3). The

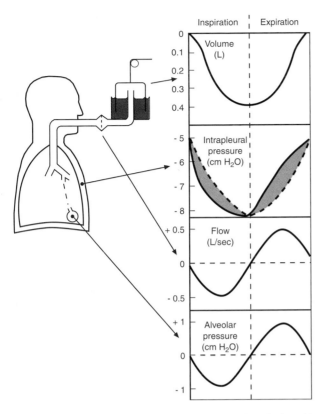

FIGURE 8   Lung volume, pressures, and flow during the breathing cycle. The solid line is the intrapleural pressure during ventilation; the broken line is the intrapleural pressure that would occur with no flow and the same lung volume. The additional intrapleural pressure change during flow (shaded area) equals alveolar pressure and is necessary to overcome airway and tissue resistance.

shaded area between the solid and dashed lines shows the additional Ppl change necessary to overcome resistance to flow. This pressure difference equals P_A at any time during flow if there is no tissue resistance (which is a good approximation; see later discussion).

The plots of P_A in Fig. 8 illustrate two important points. First, these pressures are small for resting breathing in healthy individuals. P_A changes most during maximal flow rates, near the midpoints of inspiration or expiration, but the change is only about 1 cm H_2O. This results in normal airway resistance values of only 1.5–2.0 (cm H_2O · sec)/L. However, even such low resistances require substantially larger P_A for high ventilatory flow rates (e.g., during heavy exercise). Second, the P_A and flow curves have essentially the same shape in healthy resting individuals. This is expected if airway resistance is constant throughout the breathing cycle. However, as described later, several physiologic and pathologic factors may affect airway resistance, even within a single breath.

Measuring airway resistance from P_A is too difficult for routine diagnosis, so *forced expiratory volumes* (FEV) are more commonly measured as indexes of airway resistance. For example, $FEV_{1.0}$ is the volume that can be exhaled with maximum effort in the first 1.0 sec of an expiration.

Physiologic Determinants of Airway Resistance

Several physiologic factors are also important in determining airway resistance, in addition to the physical factors discussed earlier. *Tissue resistance* requires additional changes in Ppl to generate a given flow rate. The elastic behavior of the chest wall and the network of elastin and collagen in the lungs create a resistance to volume changes during ventilation. This resistance can be measured by getting a true measure of P_A with a plethysmograph by independently solving Boyle's law for pressure instead of volume (as described in Chapter 18, Fig. 8). Tissue resistance is typically only 20% of pulmonary resistance, where *pulmonary resistance* is the term used to distinguish total resistance from airway resistance.

Lung volume is another important determinant of airway resistance. As lung volume increases from RV to TLC, airway resistance decreases nonlinearly from 4.0 to 0.5 (cm H_2O sec)/L. *Airway conductance* is the inverse of airway resistance, and it increases linearly over the same volume range from near 0 to almost 2 L/(sec cm H_2O). This decrease in resistance with increasing volume occurs because radial traction pulls the airways open as lung volume increases. The opposite occurs at low lung volumes, and small airways may even collapse, at the bottom of the lung.

Bronchial smooth muscle is also important in determining airway caliber and resistance. Smooth muscle in the airways is tonically active and is under the control of the autonomic nervous system, as described in Chapter 22. Histamine released from mast cells in the airways, or even reaching the lungs through the pulmonary circulation, causes powerful bronchoconstriction. Alveolar P_{CO_2} also affects airway resistance. Decreased P_{CO_2} causes bronchoconstriction by a direct effect on smooth muscle, and this tends to divert ventilation to other parts of the lung that are not receiving as much ventilation.

Dynamic Compression of Airways

Figure 9 is a *flow–volume curve* that shows how lung volume affects inspiratory and expiratory flow rates with different ventilatory efforts. (Ventilatory effort can be quantified as the Ppl change generated by respiratory muscles.) During normal breathing at rest, the small circle near the center of Fig. 9 shows that peak

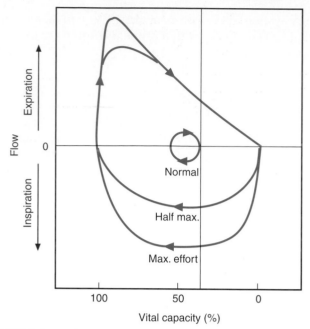

FIGURE 9 Flow–volume loops. Maximal and half-maximal efforts produce similar expiratory flows at low lung volumes, indicating flow limitation, but inspiratory flow is proportional to effort at most lung volumes. (After Mead and Agostoni, Chap. 14 in Fenn and Rahn, eds., *Handbook of Physiology, Section 3, Respiration.* Bethesda, MD: American Physiological Society, 1964.)

inspiratory and expiratory flows are similar, and both occur near the midpoint of the volume change above FRC. This contrasts sharply with the outer curves in Fig. 9 for maximal ventilatory effort. Maximum inspiratory flow occurs over a wide range of lung volumes between RV and TLC, but maximum expiratory flow occurs only near TLC and it decreases progressively with lung volume. Examining the curves for a half-maximal ventilatory effort reveals further differences between inspiration and expiration. Inspiratory flow appears to increase in proportion to inspiratory effort, but expiratory flow is independent of effort at lung volumes less than about half of TLC.

Extensive measurements show that this envelope of the expiratory flow–volume curve cannot be exceeded, and expiratory flow at these lung volumes is referred to as *effort independent. Dynamic compression* of airways is the physiologic mechanism determining maximum expiratory flow rate at low lung volume, and it depends on airway transmural pressure (Ptm). Factors affecting airway Ptm during expiration include (1) expiratory effort, which raises Ppl outside the airway; (2) lung volume and compliance, which affect pressures inside and outside the airways by elastic recoil, as described in Fig. 6; and (3) airway resistance, which causes a pressure gradient inside the airways from the alveoli and the mouth. Figure 10 illustrates how these factors interact

to collapse the airways at certain lung volumes, independent of further increases in expiratory efforts.

Figure 10A shows the conditions before inspiration with no air flow. A positive 5 cm H_2O Ptm gradient distends the alveoli *and* the entire length of the airways. Inside pressure equals atmospheric pressure (0 cm H_2O) from the alveoli to the mouth with no flow. Outside pressure equals the subatmospheric Ppl, caused by the lung's tendency to collapse, and the tendency of the chest wall to expand, at FRC (see Fig. 6). In Fig. 10B, the respiratory muscles expand the thoracic cavity and decrease the Ppl. This Ppl decrease is transmitted to the alveoli, so flow occurs down a pressure gradient within the airways, from 0 at the mouth to −2 cm H_2O in the alveoli. Notice that this gradient of pressure inside the airways that is driving flow during inspiration can actually increase the positive Ptm and distend the airways at some point.

The effects during expiration are very different. At the end of inspiration (Fig. 10C), the alveoli and airways are distended to a larger volume by a larger Ptm difference. The Ptm is the same at all points along the airways in this condition with no flow. A maximum expiratory effort is initiated (Fig. 10D) by contraction of expiratory muscle, which increases Ppl. P_A increases and expiratory flow occurs down the pressure gradient inside the airways toward the mouth. However, the decrement in airway pressure along the airways means that outside (intrapleural) pressure can exceed inside (airway) pressure at some point and cause a negative Ptm, which collapses the airways. This is called *dynamic compression,* and the point of airway collapse is called the *equal pressure point.*

Dynamic compression occurs at low lung volumes because the elastic recoil of the lung contributes less to

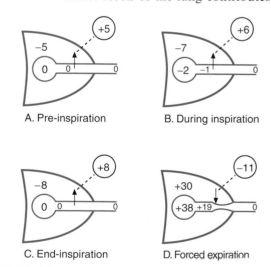

FIGURE 10 Dynamic compression of airways during forced expiration, caused by a negative transmural airway pressure (D). See text for details.

PA at these volumes, so more positive Ppl are needed to drive expiratory flow. Any increase in Ppl tends to collapse the airways more, and makes expiratory flow independent of effort at low lung volumes. Lung compliance will influence dynamic compression by a similar mechanism. If compliance is increased by lung disease, the lungs have less elastic recoil to generate the positive PA needed to drive expiratory flow.

Finally, increased airway resistance can increase dynamic compression by causing larger gradients in pressure inside the airways. Large peripheral airway resistance will tend to move the equal pressure point closer to the alveoli. This exacerbates dynamic compression because peripheral airways have weaker walls and are more likely to collapse. Therefore, two mechanisms contribute to increased dynamic compression and limited maximal expiratory flow in patients with obstructive diseases like emphysema. Lung tissue damage increases lung compliance and airway obstruction increases airway resistance. In contrast, dynamic compression is not such a problem in the *restrictive* diseases like fibrosis. Maximum expiratory flow is maintained, if it is normalized for the reduction in lung volumes, in patients with interstitial pulmonary fibrosis. Recall that this disease reduces lung compliance, so the airways are less likely to collapse.

Distribution of Ventilation

The distribution of ventilation among the hundreds of airways in the bronchial tree is not perfectly uniform. This can be either harmful or helpful to gas exchange. Any tendency to mismatch air and blood flow is deleterious to gas exchange, as described in Chapter 21. On the other hand, an uneven distribution of ventilation would be helpful if it tended to match the distribution of pulmonary blood flow, as described in Chapter 18. Fortunately, gravity is an important factor in determining the distribution of both ventilation and pulmonary blood flow, so ventilation and blood flow are fairly well matched in the lungs.

Gravity affects the distribution of ventilation by creating a gradient in Ppl between the top and bottom of the lung (Fig. 11). The increase in Ppl at the bottom relative to the top of the lung is caused by two mechanisms: (1) The weight of the column of intrapleural fluid creates a hydrostatic column, similar to the blood in the pulmonary capillaries; and, more importantly, (2) the weight of the lung pulls the lung away from the top of the thorax and compresses the lung on the diaphragm at the bottom. This local distortion in volume results in different regions of the lung operating on different parts of the compliance curve, as shown in Fig. 11.

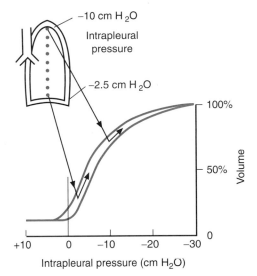

FIGURE 11   Regional differences in ventilation at FRC are caused by a gradient of intrapleural pressure down the lung as explained in the text. The base of the lung operates on a more compliant (steeper) part of the pressure–volume curve, so volume changes more at the bottom than at the top of the lung for the same intrapleural pressure change. (After West, *Ventilation/blood flow and gas exchange*, New York: Blackwell Scientific, 1990.)

Figure 11 shows how this difference in Ppl between the top and bottom of the lung and the nonlinear compliance curve increases regional ventilation at the bottom of the lung at FRC. If the whole-lung static compliance curve is assumed to apply to any region of the lung (a reasonable assumption for this discussion), then the top of the lung is relatively more expanded and less compliant than the bottom of the lung. The bottom regions of the lung operate on a steeper portion of the compliance curve, so an inspiratory effort is more effective at increasing volume in the bottom of the lung. The decrease in Ppl with inspiratory muscle contraction is the same at the top and bottom of the lung.

At lower lung volumes, the situation may change. At RV, for example, Ppl may be positive at the bottom, but still negative at the top of the lung. The effects on ventilation at RV can be estimated from the compliance curve in Fig. 11 assuming a Ppl of +3.5 cm H_2O at the bottom and −4.0 cm H_2O at the top of the lung at RV. At RV, the compliance curve is flat for the bottom of the lung, so no volume change would occur, and all the volume change would occur in higher parts of the lung with greater compliance. Such *airway closure* traps gas in alveoli behind small collapsed airways at the base of the lungs and only occurs at very low lung volumes in healthy young individuals. However, as lung compliance decreases with age or pulmonary disease, reduced elastic recoil tends to increase Ppl. This makes the lungs more susceptible to airway closure, and ventilation to the base of the lungs may be intermittent.

Similar to the case for pulmonary blood flow, height and gravity cannot explain all aspects of a nonuniform distribution of ventilation. Ventilation is more uniform in the supine lung than in the upright lung because the vertical height is reduced. However, considerable differences in ventilation persist between the apex and base of the supine lung, and these are called *intraregional differences* in ventilation. Such intraregional differences in ventilation are caused by local differences in resistance (R) and compliance (C) of terminal airways, which determines the airways *time constant* (τ):

$$\tau = RC.$$

Time constants are a measure of the time it takes to fill a given volume, so terminal airways with greater resistances or compliance are ventilated more slowly than airways with lower resistance or compliance. Lung disease can alter resistance and/or compliance and increase time constants fivefold. Because lung disease is frequently patchy in the lungs, this leads to large intraregional differences in ventilation. The harmful effects of such ventilatory inequality on gas exchange are explained in Chapter 21.

The *single breath nitrogen test* can be used to assess the degree of ventilatory inequality in humans. A single breath of 100% oxygen (O_2) is inspired, and the N_2 concentration is plotted as a function of expired volume on the following expiration; this is the same procedure used to measure dead space by Fowler's method (see Chapter 18, Fig. 10). During the inspiration, O_2 will dilute the resident N_2 more in better ventilated lung regions. During a slow expiration, different lung regions will empty at different rates and the alveolar plateau of N_2 concentration will slope upward (unlike the flat plateau shown for an ideal lung in Chapter 18, Fig. 10). Frequently, a large increase in nitrogen concentration will occur at the end of the slow expiration, as *airway closure* collapses airways in the bottom of the lung, and the more poorly ventilated apical regions of the lung contribute more to the expired gas.

WORK OF BREATHING

Mechanical work can be defined as the area inside a pressure–volume curve. Therefore, the mechanical work of breathing can be calculated from a plot of lung volume versus alveolar pressure, measured under the dynamic conditions of flow, as shown in Fig. 8. This calculation would include work done by muscular contraction and work done on the system by elastic recoil of the lungs and chest wall. Work done on the respiratory system is ultimately fueled by muscular effort to inhale. Hence, O_2 consumption by respiratory muscles, called the O_2 cost of breathing, is the physiologically relevant measure of the work of breathing.

The O_2 cost *of breathing* is low in healthy individuals at rest, comprising only 1–2% of resting $\dot{V}O_2$. The O_2 cost of breathing depends on factors influencing the pressure requirements for ventilation, such as lung and chest wall mechanics, and airway and tissue resistance, as discussed earlier. It also depends on the efficiency of skeletal muscles converting chemical energy into mechanical work. *Efficiency* is defined as the ratio of effective mechanical work to the energy input, or O_2 cost, and it is about 20–25%, for aerobic muscle activities such as cycling and breathing. Muscle efficiency is also influenced by the degree of overlap between myosin and actin filaments (which influences the length–tension relationship of skeletal muscle), so posture and the resting length of sarcomeres in the diaphragm can also influence the O_2 cost of breathing.

During high levels of ventilation, the work of breathing increases considerably and the O_2 cost can reach 15% of the total at maximal exercise. At high levels of ventilation, it becomes more important to minimize mechanical work by adjusting the breathing pattern to optimize the pressure–volume and pressure–flow relationships. As shown in Fig. 9, resting breathing usually occurs around FRC, and maximum ventilation tends to occur around midlung volumes too. Ppl can change up to a maximum of 100 cm H_2O during inspiration and 150 cm H_2O during expiration. Extreme pressures and volumes do not usually occur during physiologic breathing movements, and this reduces the work of breathing. Pulmonary disease increases the mechanical work and O_2 cost of breathing not only by increasing airway resistance, but also by altering lung volume and compliance.

Suggested Readings

Hawgood, S. Surfactant: Composition, structure and metabolism. In Crystal RG, West JB, Weibel WR, Barnes PJ, eds. *The lung: Scientific foundations,* Vol 1, 2nd ed. Philadelphia: Lippincott-Raven, 1997;557–572.

Mead J, Macklem PT, eds. Mechanics of breathing, Vol III. In *Handbook of physiology: Section 3, The respiratory system.* Bethesda, MD: American Physiological Society, 1986.

Milic-Emili J, D'Angelo E. Work of breathing. In Crystal RG, West JB, Weibel WR, Barnes PJ, eds. *The lung: Scientific foundations,* Vol 1, 2nd ed. Philadelphia: Lippincott-Raven, 1997;1437–1446.

Rahn H, Otis AB, Chadwick LE, Fern WO. The pressure–volume diagram of the thorax and lung. *Am J Physiol* 1946;148:161–178.

Oxygen and Carbon Dioxide Transport in the Blood

FRANK L. POWELL, JR.

KEY POINTS

- A small amount of O_2 is physically dissolved in blood, but most of the O_2 in blood is chemically bound to hemoglobin.
- O_2 concentration in blood is a nonlinear function of O_2 partial pressure. Hemoglobin is responsible for the S-shape of the blood-O_2 equilibrium curve, which is advantageous for O_2 loading in lungs and O_2 unloading in tissues.
- Concentration of O_2 in blood can vary with hematocrit for a given P_{O_2} and O_2 saturation of hemoglobin.
- CO_2, pH, temperature, and organic phosphates can influence the position and shape of the blood-O_2 equilibrium curve and affect O_2 exchange.
- CO_2 is more soluble than O_2 in blood, and there are smaller differences in CO_2 partial

  pressure in the body, compared with the range of O_2 partial pressures.
- The blood-CO_2 equilibrium curve is influenced by hemoglobin-O_2 saturation, and this effect facilitates gas exchange.
- Carbonic anhydrase facilitates ion and gas exchange between red blood cells, plasma, and tissues.
- Blood is a bicarbonate and protein buffer system, and the relationship between blood pH and CO_2 partial pressure can be described by the Henderson-Hasselbalch equation.
- Both the lungs and kidneys are involved in blood pH homeostasis. The lungs can compensate for chronic disturbances in acid–base balance caused by the kidneys, and vice versa, but the lungs work faster than the kidneys

Essential Medical Physiology, Third Edition

This chapter describes *blood oxygen* (O_2) and *carbon dioxide* (CO_2) *equilibrium curves* (also called *dissociation curves*), and the physiologic factors that determine the shape and position of these curves. Equilibrium curves quantify O_2 and CO_2 carriage in blood as graphs of concentration versus partial pressure. It is necessary to consider *both* partial pressure *and* concentration because partial pressure gradients drive diffusive gas transport in lungs and tissues, but concentration differences determine convective gas transport rates in lungs and the circulation (see Chapters 18 and 21).

This subject would be much simpler if O_2 and CO_2 were physiologically *inert* and occurred in blood only as physically dissolved gases. The concentration of a dissolved gas in a liquid is directly, and linearly, proportional to its partial pressure according to Henry's law ($C = \alpha P$, where α = solubility; see Chapter 18). However, O_2 and CO_2 also enter into chemical reactions with blood. These reactions (1) increase O_2 and CO_2 concentrations in blood, (2) allow physiologic modulation of O_2 and CO_2 transport in blood, and (3) make respiratory CO_2 exchange an important mechanism of acid–base balance in the body. The physiologic and pathologic consequences of the these three factors for O_2 and CO_2 exchange are considered in Chapter 21.

OXYGEN IN BLOOD

Normal O_2 concentration in arterial blood (Ca_{O_2}) is about 20 mL/dL. (The usual units for O_2 and CO_2 concentration in blood are mL/dL, also called volume %; 1 mL/dL 0.45 mmol/L.) However, only 0.3 mL/dL is physically dissolved gas; normal arterial P_{O_2} (Pa_{O_2}) is 100 mm Hg and the physical solubility of O_2 in blood is 0.003 mL/(dL · mm Hg) at 37°C. If arterial blood contained only dissolved O_2, then cardiac output would have to be 100 L/min to deliver enough O_2 to the tissues for a normal metabolic rate of 300 mL O_2/min! Hemoglobin increases O_2 concentration in blood so a cardiac output of only 6 L/min is sufficient for resting metabolic demands.

Hemoglobin

Hemoglobin (Hb) is a large molecule (molecular weight = 64,485) consisting of four individual polypeptide chains, each with a *heme* (iron-containing) protein that can bind O_2 with iron in the *ferrous* (Fe^{2+}) form. *Methemoglobin* occurs when the iron is in the *ferric* form (Fe^{3+}), and it cannot bind O_2. Small amounts of methemoglobin normally occur in blood and slightly reduce the amount of O_2 that can be bound to hemoglobin. One gram of pure human hemoglobin

can bind 1.39 mL of O_2 when fully saturated, but methemoglobin reduces this value to 1.34–1.36. Hemoglobin is concentrated inside red blood cells, or erythrocytes. This cellular packaging is important for the biophysics of the microcirculation, and it provides physiologic control of O_2 binding through cellular changes in the hemoglobin microenvironment.

The four subunits of hemoglobin include two alpha and two beta chains, and variations in the amino acid sequence of these polypeptides explain the differences in Hb-O_2 affinity between species, at different stages of development, and with some genetic diseases. *Fetal hemoglobin* (HbF) has a high affinity for O_2, which is important for O_2 transport across the placenta *in utero*. In the first year after birth, HbF is gradually replaced by *adult hemoglobin,* which has a lower affinity for O_2. *Hemoglobin S* (HbS) results from a single amino acid substitution on the beta chain in *sickle-cell anemia*. Deoxygenated HbS tends to crystallize within red blood cells, distorting them into a crescent or sickle shape. Sickle cells are more fragile and less flexible than normal erythrocytes, and they tend to plug capillaries.

The three-dimensional shape of a hemoglobin molecule, which is determined by the *allosteric* interactions of its four subunits, causes the O_2-equilibrium curve to be S shaped, or *sigmoidal* (Fig. 1). O_2-equilibrium curves for individual alpha and beta chains are not sigmoidal, but simple convex curves like the O_2-equilibrium curve for myoglobin. *Myoglobin* occurs in muscle and has only a single polypeptide chain with one heme group. The

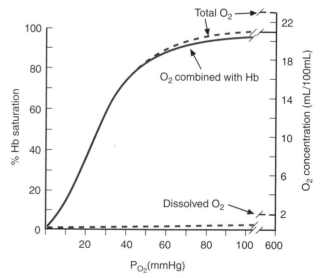

FIGURE 1   Standard human blood-O_2 equilibrium (or dissociation) curve at pH = 7.4, P_{CO_2} = 40 mmHg, and 37°C. Left ordinate shows O_2 saturation of hemoglobin *available* for O_2 binding; right ordinate shows absolute O_2 concentration in blood. Most O_2 is bound to hemoglobin, and dissolved O_2 contributes very little to total O_2 concentration.

sigmoidal shape of the O_2-Hb equilibrium curve facilitates O_2 loading on blood in the lungs, and O_2 unloading from blood in the tissues, as explained in Chapter 21 on tissue gas exchange.

Blood-Oxygen Equilibrium Curves

Figure 1 shows the two forms of the O_2-equilibrium curve: (1) saturation of hemoglobin with O_2 (So_2) versus O_2 partial pressures (Po_2), and (2) O_2 concentration in blood (Co_2) versus Po_2. Saturation quantifies the amount of O_2 in blood as the percentage of the total hemoglobin sites *available* for *binding* O_2 that actually bind O_2 at a given Po_2. Therefore, saturation equilibrium curves are independent of hemoglobin concentration in blood. In contrast, concentration curves quantify the absolute amount of O_2 in a volume of blood with a given Po_2 and they depend on the amount of hemoglobin available in blood for O_2.

O_2 *capacity* (O_2cap) is defined as the O_2 concentration in blood when hemoglobin is 100% saturated with O_2. Pure hemoglobin binds 1.39 mL O_2/g Hb, and Fig. 1 shows that the O_2cap for normal blood with a hemoglobin concentration of 15 g/dL is 20.85 mL/dL ($= 1.39 \cdot 15$). Physically dissolved O_2 also contributes a small amount to O_2 concentration. Therefore, total O_2 concentration in blood (in mL O_2/dL blood) is calculated as:

$$Co_2 = (O_2cap[So_2/100]) + (0.003Po_2),$$

where O_2cap = O_2 capacity, So_2 = saturation, and 0.003 = physical solubility for O_2 in blood. This equation and Fig. 1 show that dissolved O_2 is not a large component of O_2 concentration at Po_2 levels in arterial or venous blood (i.e., 100–40 mm Hg).

The shape of the O_2-Hb equilibrium curves is complex, and it can be generated only experimentally or by sophisticated mathematical algorithms. However, remembering only four points on the curve allows one to solve many common problems of O_2 transport:

1. Po_2 = 0 mm Hg, So_2 = 0% (the origin of the curve)
2. Po_2 = 100 mm Hg, So_2 = 98% (normal arterial blood, which is almost fully saturated)
3. Po_2 = 40 mm Hg, So_2 = 75% (normal mixed venous blood)
4. Po_2 = 26 mm Hg, So_2 = 50% (P_{50}, defined as the Po_2 at 50% saturation)

The P_{50} quantifies the affinity of Hb for O_2. For example, a decrease in P_{50} indicates an increase in O_2 affinity because O_2 saturation or concentration is greater for a given Po_2. In adult human blood, P_{50} = 26 mm Hg under standard conditions of Pco_2 = 40 mm Hg, pH = 7.4, and 37°C.

Modulation of Blood-Oxygen Equilibrium Curves

The O_2-equilibrium curve in blood can be physiologically modulated in three ways: (1) The vertical *height* of the concentration curve (but not the saturation curve) can change, indicating a change in O_2cap; (2) the horizontal *position* of saturation *and* concentration curves can change, indicating a change in Hb-O_2 affinity; and (3) the *shape* of saturation *and* concentration curves can change, indicating a change in the chemical reaction between O_2 and hemoglobin. The maximum height of the saturation curve cannot change by definition; the maximum is always 100% when O_2 is bound to all available hemoglobin sites. However, changes in hemoglobin concentration [Hb] will change the maximum height of the *concentration* curve, according to the relationship between O_2cap and [Hb] described earlier. *Mean corpuscular Hb concentration* (MCHC) quantifies [Hb] in red blood cells, and *hematocrit* (Hcrit) quantifies the percentage of blood volume that is red blood cells. Therefore [Hb], in g/dL of blood, depends on both of these factors:

$$[Hb] = MCHC \cdot Hcrit.$$

Typical human values of MCHC = 0.33 and Hcrit = 45%, and [Hb] = 15 g/dL are used for Fig. 1, which shows normal O_2 concentration in arterial blood (Cao_2) = 20 mL/dL and 15 mL/dL in mixed venous blood ($C\bar{v}o_2$). If [Hb] decreases, for example with decreased hematocrit in *anemia*, then O_2cap and concentration decreases at any given Po_2. The O_2cap increases when [Hb] increases, for example, by the stimulation of red blood cell production in bone marrow by the hormone *erythropoietin*. Erythropoietin transcription is a regulated hypoxia inducible factor (HIF-1α) and is released from cells in the kidneys in response to decreases in arterial O_2 levels. Polycythemia, or increased hematocrit, occurs with chronic hypoxemia in healthy people (e.g., during acclimatization to altitude) and with disease.

The horizontal position of Hb-O_2 equilibrium curves reflects the affinity of Hb for oxygen, and changes in horizontal position are quantified as changes in P_{50}. A decrease in P_{50} is referred to as a *left shift* of the equilibrium curve, and indicates increased Hb-O_2 affinity; O_2 saturation or concentration is increased for a given Po_2. Similarly, increased P_{50} or a *right shift* reflects decreased Hb-O_2 affinity. Figure 2 shows the three most important physiologic variables that can modulate P_{50}: pH, Pco_2, and temperature.

The *Bohr effect* describes changes in P_{50} with changes in blood Pco_2 and pH. Decreased Pco_2 causes Hb-O_2

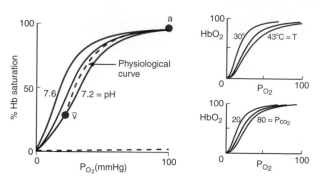

FIGURE 2 Effects of pH and P_{CO_2} (i.e., Bohr effect) and temperature on the position of the O_2-hemoglobin equilibrium curve. The "physiologic" curve connects the arterial (a) and mixed-venous points ($\bar{v}$) so the *in vivo* curve is steeper than the standard curve at pH = 7.4. (After Roughton, Chap. 31 in Fenn and Rahn, eds., *Handbook of physiology, Section 3, Respiration*. Bethesda, MD: American Physiological Society, 1964.)

affinity to increase (decreased P_{50}), and increased P_{CO_2} causes Hb-O_2 affinity to decrease (increased P_{50}). As described later, pH decreases when P_{CO_2} increases and vice versa, and pH changes explain most of the Bohr effect with P_{CO_2} changes in blood. H^+ binds to histidine residues in hemoglobin molecules, and this changes the conformation of hemoglobin and the ability of heme sites to bind O_2. However, CO_2 also has a small independent effect on Hb-O_2 affinity if pH is held constant. The physiologic advantage of the Bohr effect is that it facilitates O_2 loading in the lungs, where CO_2 is low and pH is high (see later discussion). In muscles, the opposite occurs and increased CO_2 causes pH to decrease and facilitates O_2 unloading from hemoglobin to the tissues.

The affect of temperature on Hb-O_2 affinity also has physiologic advantages. Warm temperatures in intensely exercising muscles will increase P_{50}, and decrease Hb-O_2 affinity to facilitate O_2 unloading to tissues.

A *physiologic O_2-blood equilibrium curve* can be defined as the curve showing the change in blood O_2 concentration when P_{O_2} decreases from arterial to venous levels in the tissues or increases in the opposite direction in the lungs. Figure 2 shows how the increase in P_{CO_2}, decrease in pH, and potential increase in temperature between arterial and venous points makes the physiologic curve steeper than individual curves. This is an advantage for gas exchange because it increases the change in O_2 concentration for a given change in P_{O_2}, thereby increasing O_2 uptake or delivery. (The consequences of the shape of the O_2-blood equilibrium curve for gas exchange are explained in more detail in Chapter 21.)

Hb-O_2 affinity is also affected by organic phosphates, with 2,3-diphosphoglycerate (2,3-DPG) being most important in humans. 2,3-DPG is produced during glycolysis in red blood cells, and increases P_{50} by

interacting with hemoglobin beta chains to decrease their O_2-binding affinity. Physiologic stimuli that lead to enhanced O_2 delivery, for example, chronic decreases in blood P_{O_2} levels, typically increase the concentration of 2,3-DPG and promote O_2 delivery to tissues. In blood stored in blood banks, 2,3-DPG is generally decreased and the increased Hb-O_2 affinity can lead to problems in O_2 delivery after blood transfusion.

Carbon Monoxide

Carbon monoxide (CO) is a deadly gas that also modulates the Hb-O_2 dissociation curve by changing the shape and position of concentration or saturation curves. The affinity of Hb for CO is 240 times greater than it is for O_2, so even very small amounts of CO greatly reduce the capacity for hemoglobin to bind O_2. Therefore, CO decreases the O_2cap, or maximum height of the concentration curve (Fig. 3). However, CO also decreases the P_{50} and makes the Hb-O_2 curve less sigmoidal. Figure 3 illustrates these changes and compares the effects of CO poisoning with anemia, when O_2cap is reduced the same amount in both cases. CO causes a left shift of the curve by altering the ability of the hemoglobin molecule to bind O_2. This means blood O_2 concentration remains high until P_{O_2} decreases to very low levels, which impairs O_2 unloading from blood to tissues. CO poisoning also has direct effects on cellular cytochromes, which contribute to its deadliness. CO is particularly dangerous because it is colorless, odorless, and the decrease it causes in arterial O_2 concentration is not sensed by respiratory control systems (which respond only to O_2 partial pressure as explained in Chapter 22). *Hyperbaric O_2 exposure* is used to treat CO poisoning, because only very high P_{O_2} levels are effective at competing with CO for Hb-binding sites and driving CO out of the blood.

CARBON DIOXIDE

Blood-CO_2 equilibrium (or dissociation) curves are nonlinear, but they have a different shape and position than O_2-blood equilibrium curves. Figure 4 shows how blood holds more CO_2 than O_2 and this is, in part, because CO_2 is carried by blood in three forms (see later discussion). Also, the blood CO_2 equilibrium curve is steeper than the O_2 curve. This results in a smaller range of P_{CO_2} values in the body, compared with the range of P_{O_2} values, although the differences between arterial and venous concentrations are similar for CO_2 and O_2 (about 5 mL/dL of blood). The resulting *physiologic CO_2 dissociation curve* between the arterial and venous points is much more linear than the physiologic O_2 dissociation curve (Fig. 4).

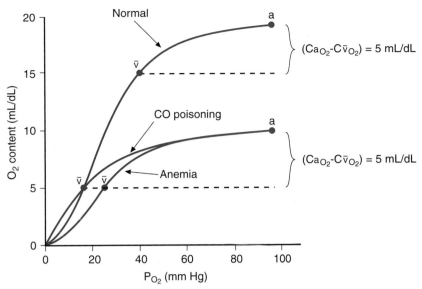

FIGURE 3   Effects of carbon monoxide (CO) and anemia on blood-O_2 equilibrium curves. Anemia (7.5 g/dL total hemoglobin) decreases O_2 concentration but does not change the shape of the curve. CO poisoning (7.5 g/dL hemoglobin *available* for O_2 binding) decreases O_2 concentration *and* shifts the curve to the left. $P\bar{v}_{O_2}$ must decrease to maintain the arterial-venous O_2 concentration difference for a given Pa_{O_2} with CO. (After Roughton, Chap. 31 in Fenn and Rahn, eds., *Handbook of physiology, Section 3, Respiration.* Bethesda, MD: American Physiological Society, 1964.)

Forms of Carbon Dioxide in Blood

Physically dissolved CO_2 is a function of CO_2 solubility in water or plasma, which is 0.067 mL/(dL mm Hg) and 20 times more soluble than O_2. Still,

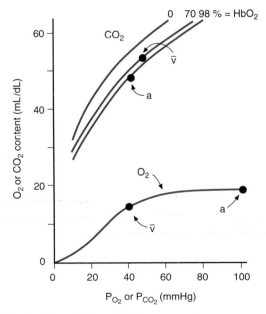

FIGURE 4   Blood-CO_2 equilibrium curve shown on same graph with O_2 equilibrium curve. Differences between the curves result in higher CO_2 concentrations in the blood, and smaller Pco_2 differences between arterial and venous blood. Hemoglobin-O_2 saturation affects the position of the CO_2 equilibrium curve (i.e., Haldane effect).

dissolved CO_2 contributes only about 5% of total CO_2 concentration in arterial blood.

Carbamino compounds comprise the second form of CO_2 in blood. These compounds occur when CO_2 combines with terminal amine groups in blood proteins, especially with the globin of hemoglobin. However, this chemical combination between CO_2 and hemoglobin is much less important than Hb-O_2 binding, so carbamino compounds comprise only 5% of the total CO_2 in arterial blood.

Bicarbonate ion (HCO_3^-) is the most important form of CO_2 carriage in blood. CO_2 combines with water to form carbonic acid, and this dissociates to HCO_3^- and H^+:

$$CO_2 + H_2O \rightleftharpoons H_2CO_3 \rightleftharpoons HCO_3^- + H^+.$$

Carbonic anhydrase is the enzyme that catalyzes this reaction, making it almost instantaneous. Carbonic anhydrase occurs mainly in red blood cells, but it also occurs on pulmonary capillary endothelial cells and it accelerates the reaction in plasma in the lungs. The uncatalyzed reaction will occur in any aqueous medium, but at a much slower rate, requiring more than 4 min for equilibrium. The rapid conversion of CO_2 to bicarbonate results in about 90% of the CO_2 in arterial blood being carried in that form.

Figure 5 shows the carbonic acid reactions in plasma and red blood cells, and it illustrates important ion

Clinical Note

Currently, there is tremendous commercial interest in the development of *blood substitutes* as an alternative to real blood for transfusions during scheduled surgery or emergency resuscitation. A 1994 World Health Organization survey found that 90 million units of blood are collected globally each year, yet blood shortages still occur and the cost of banking blood has increased with the need to screen for new blood-borne diseases such as human immunodeficiency virus. Besides increasing supply, blood substitutes could eliminate the need to cross-match blood types during transfusions, because blood type is determined by antigens on the surface of red blood cells.

Various hemoglobin solutions have been tested as blood substitutes since the early 1970s, when scientists made advances in purification and chemical modification of hemoglobin. However, the effectiveness of these substitutes is still not established and, surprisingly, some problems with hemoglobin blood substitutes relate to O_2 delivery. When a person loses blood, peripheral blood vessels constrict (e.g., in muscle), and this helps sustain O_2 delivery to vital organs. Because blood flow in the microcirculation is also controlled by local P_{O_2} (Chapter 17), hemoglobin blood substitutes may not be effective at reopening microvessels *if too much O_2 is supplied*; high P_{O_2} can increase microvascular resistance. Also, the iron in hemoglobin is an extremely effective scavenger of *nitric oxide* (NO), which is a powerful vasodilator also known as endothelial derived relaxing factor (EDRF). Finally, when blood flow and O_2 delivery are reestablished, hemoglobin solutions can increase the production of oxygen radicals and exacerbate so-called reperfusion injury of tissues.

Recent experiments have discovered new chemical reactions between NO and hemoglobin, and have suggested possible solutions for problems with O_2 delivery and the hypertension that occur with hemoglobin blood substitutes. NO can form S-nitrosothiols (RSNOs) with cysteine residues in hemoglobin, and these RSNO compounds retain vasoactive properties like NO. Furthermore, hemoglobin oxygenation modifies this reaction, so O_2 exerts allosteric control over NO transport. This promotes a cycle of NO delivery to the microvasculature in tissues (from deoxygenated Hb) and NO loading in the lungs (onto oxygenated hemoglobin). The cysteine residue on the (3-Hb chain responsible for this reaction is highly conserved in all mammals and birds, indicating that it has an important physiologic function that has changed little during evolution. Incorporating this chemistry into hemoglobin used for blood substitutes may increase their clinical usefulness.

---

fluxes occurring with CO_2 transport in blood. CO_2 rapidly enters red blood cells from the plasma because it is soluble in cell membranes. Carbonic anhydrase catalyzes the rapid formation of HCO_3^- and H^+ in the cells, and some of this HCO_3^- is transported out by an electrically neutral bicarbonate-chloride exchanger. The *Hamburger* (or *chloride) shift* is an increased intracellular chloride with increased CO_2, or vice versa. The H^+ produced from CO_2 reacts with hemoglobin and affects both the O_2 equilibrium curve (Bohr effect) and CO_2 equilibrium curve, as described later.

Modulation of Blood-Carbon Dioxide Equilibrium Curves

Hb-O_2 saturation is the major factor affecting the position of the CO_2 equilibrium curve. The *Haldane effect* increases CO_2 concentration when blood is deoxygenated, or decreases CO_2 concentration when blood is oxygenated, at any given P_{CO_2} (see Fig. 4). The Haldane effect is actually another view of the same molecular mechanism causing the Bohr effect on the O_2 equilibrium curve (see earlier discussion). H^+ ions from CO_2 can be thought of as competing with O_2 for hemoglobin binding. Hence, increasing O_2 decreases the affinity of hemoglobin for H^+ and blood CO_2 concentration (Haldane effect), and increased $[H^+]$ decreases the affinity of hemoglobin for O_2 (Bohr effect). Figure 5 summarizes these interactions.

The physiologic advantages of the Haldane effect are that it promotes unloading of CO_2 in the lungs when blood is oxygenated and CO_2 loading in the blood when O_2 is released to tissues. The Haldane effect also results in a steeper physiologic CO_2-blood equilibrium curve (see Fig. 4), which has the physiologic advantage of increasing CO_2 concentration differences for a given P_{CO_2} difference.

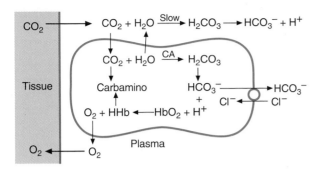

FIGURE 5 CO_2 and O_2 reactions in blood and tissues; the opposite reactions occur in the lungs. CA, carbonic anhydrase. See text for details.

ACID–BASE BALANCE

The conversion of CO_2 to H^+ and HCO_3^- ions has tremendous implications for acid–base physiology. Every day, resting metabolism produces more than 15,000 mmol of CO_2, or 15,000 mmol/L of carbonic acid, and this acid leaves the body through the lungs. By comparison, the kidneys typically excrete only 100 mmol/L of acid per day. The ability to change blood P_{CO_2} levels rapidly by changing ventilation has a powerful effect on blood pH, so acid–base balance depends on the integrated function of respiratory and renal systems.

Henderson-Hasselbalch Equation

The *Henderson-Hasselbalch equation* describes the relationship between P_{CO_2} pH, and $[HCO_3^-]$ in human blood as follows:

$$pH = 6.1 + \log([HCO_3^-]/0.03 \cdot [P_{CO_2}]),$$

where $[HCO_3^-]$ is the bicarbonate concentration in millimoles per liter, and 0.03 P_{CO_2} is the total dissolved CO_2 concentration, also in millimoles per liter (0.03 mmol/L per mm Hg is the solubility of CO_2 in water).

As described in Chapter 31, the Henderson-Hasselbalch equation is based on the chemical equilibrium between carbonic acid and its dissociation products:

$$H_2CO_3 \rightleftharpoons H^+ + HCO_3^-$$

The dissociation constant for a weak acid (K_a), like carbonic acid, is calculated from the law of mass action as:

$$K_a = [H^+] \cdot ([HCO_3^-]/[H_2CO_3]).$$

By definition, pH $= -\log_{10}[H^+]$, and $pK_a = -\log_{10}(K_a)$. Therefore, taking the logarithm$_{10}$ of both

sides and rearranging the preceding equation results in a general form of the Henderson-Hasselbalch equation:

$$pH = pK_a + \log[HCO_3^-]/[H_2CO_3].$$

For human blood, $pK_a = 6.1$, which is essentially constant in human blood under physiologic conditions, and $[H_2CO_3] = [0.03 \cdot P_{CO_2}]$, because total carbonic acid is proportional to dissolved CO_2 in blood.

The normal human value for *arterial pH* (pHa) is 7.4, and this can be calculated from the Henderson-Hasselbalch equation using other normal values. In human blood, $pK_a = 6.1$, and it is essentially constant under physiologic conditions. $P_{CO_2} = 40$ mm Hg and $[HCO_3^-] = 24$ mmol/L in normal arterial blood. Therefore:

$$pH = 6.1 + \log(24/[0.03 \cdot 40])$$
$$= 6.1 + \log(20)$$
$$= 7.4.$$

At a normal *arterial pH* of 7.4, the $[H^+]$ is only 40 mmol/L, or significantly less than many other important ions in the body, such as Na^+, Cl^-, or HCO_3^-, which occur in the millimole per liter range. Small changes in pH, corresponding to very small changes in $[H^+]$ (see Table 1 in Chapter 31), can lead to large changes in physiologic function. The effects of pH on organ function are discussed in individual chapters throughout this book.

The Henderson-Hasselbalch equation shows how the physiologic control of pH depends on the ratio of $[HCO_3^-]$ to $[0.03 P_{CO_2}]$. Notice that a normal pH of 7.4 could occur with a variety of $[HCO_3^-]$ and P_{CO_2} values, so pH = 7.4 does not necessarily indicate normal acid–base status (see later discussion).

Arterial P_{CO_2} (Pa_{CO_2}) is determined by alveolar ventilation at any given metabolic rate (as described in Chapter 21). Increasing ventilation will decrease Pa_{CO_2} and increase pHa; decreasing ventilation will have the opposite effects. Therefore, ventilation is an extremely effective mechanism for changing pHa quickly, and ventilatory reflex responses to pH (described in Chapter 22) are the most important physiologic mechanisms for rapid control of pH. The kidneys can also control pH by changing $[HCO_3^-]$ independent of CO_2 changes, as described in Chapter 31. The kidneys process large amounts of $[HCO_3^-]$ filtered from the plasma, so blood $[HCO_3^-]$ depends on bicarbonate reabsorption and generation of new bicarbonate in renal tubules. Also, the kidneys are responsible for processing H^+ from *fixed acids* (other than carbonic acid, which is considered a volatile acid because CO_2 is a gas). Finally, renal processing of other ions, such as K^+, can affect blood pH. Acid–base balance is described here in terms of the Henderson-Hasselbalch equation, but another approach emphasizes the control of pH by concentration differences between strong ions, such as sodium and chloride. Details on the strong ion approach to acid–base diagnosis can be found in the suggested readings listed at the end of this chapter.

Bicarbonate-pH Diagram

The *bicarbonate-pH diagram*, also called a *Hastings-Davenport diagram*, illustrates the three primary variables in the Henderson-Hasselbalch equation (Fig. 6A). This diagram plots $[HCO_3^-]$ as a function of pH. The blue lines are called P_{CO_2} *isobars,* and they show the various combinations of $[HCO_3^-]$ and pH values that can occur with any given P_{CO_2}. Moving *along* a P_{CO_2} isobar shows how increasing $[HCO_3^-]$ buffers H^+ ions and increases pH for a given amount of carbonic acid. Moving *between* P_{CO_2} isobars shows how $[HCO_3^-]$ and pH change with changes in total carbonic acid. If carbonic acid is increased (moving from point A to B in Fig. 6A), then pH not only decreases, but the amount of base, i.e., $[HCO_3^-]$, increases also. Conversely, decreasing carbonic acid (point A to C in Fig. 6A) increases pH but decreases $[HCO_3^-]$. This is a graphical alternative to the system of equations describing acid–base control in Chapter 31, and readers can use whichever method they find most helpful.

Respiratory pH changes result from changes in P_{CO_2}. Pure respiratory changes in pH occur along a *blood-buffer line,* shown as the dark line connecting points A–B–C in Fig. 6A. If the blood-buffer line is steeper, then pH will change less for a given change in acid (i.e., P_{CO_2}). Because hemoglobin is a major buffer of H^+ in blood, increasing [Hb] causes the slope of the blood-buffer line to increase. Therefore, the slope of the blood-buffer line is steeper than the slope of the buffer line for plasma. However, the slope of the blood-buffer line measured *in vivo,* by sampling pHa in a person at different Pa_{CO_2} levels, is less than the *in vitro* slope, measured by changing P_{CO_2} in a blood sample in a test tube. This is because the whole body includes interstitial fluid and other fluid compartments that do not contain hemoglobin and cannot buffer P_{CO_2} changes as well as blood.

Metabolic pH changes refer to changes in $[H^+]$ and $[HCO_3^-]$ that are not caused by P_{CO_2} changes. As described in Chapter 31, this generally involves renal processing of HCO_3^- and fixed acids. Pure metabolic changes in pH will shift the blood-buffer line up or down without changing its slope on the bicarbonate-pH diagram (moving between points A and D on Fig. 6A). *Base excess* is used to quantify the magnitude of a metabolic change in pH, and it equals the vertical displacement of the blood-buffer line. Conceptually,

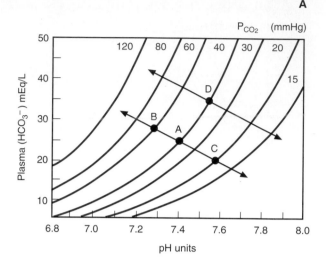

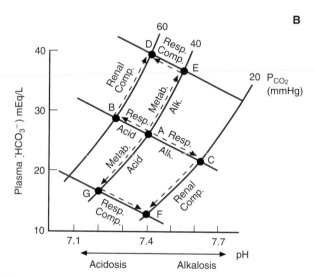

FIGURE 6 Bicarbonate-pH diagrams. (A) Blood-buffer line (B–A–C) shows pH and $[HCO_3^-]$ changes with changes in P_{CO_2}; blue lines are P_{CO_2} isobars. Base excess equals the vertical distance between blood-buffer lines (e.g., A and D). (B) pH and $[HCO_3^-]$ changes with respiratory and metabolic acidosis and alkalosis. See text for details. (After Davenport, *The ABC of acid base chemistry.* Chicago: University of Chicago Press, 1974.)

base excess is the amount of base necessary to titrate a blood sample with a metabolic disturbance back to pH = 7.4 at P_{CO_2} = 40 mm Hg. A negative base excess (which occurs with a metabolic alkalosis as described later) may be referred to as a *base deficit.*

In reality, it is rare to find a pure respiratory, or a pure metabolic, pH disturbance. Ventilatory and renal control mechanisms tend to return arterial pH to the normal value of 7.4. There are four primary forms of pH disturbance, and compensatory responses accompany each of them. These primary disturbances and secondary compensations follow characteristic pathways on the pH-bicarbonate diagram (Fig. 6B), as described later. The reader is referred to Chapter 31 for more details about renal mechanisms.

Respiratory Acidosis

A *respiratory acidosis* occurs when P_{CO_2} increases. Increased P_{CO_2} decreases the $[HCO_3^-]/P_{CO_2}$ ratio and decreases pH along the blood buffer line, between points A and B in Fig. 6B. *Hypercapnia,* or an increase in arterial P_{CO_2}, occurs when ventilation is not sufficient for the existing level of CO_2 production. Ventilation may be depressed at rest, for example, with a drug overdose that depresses respiratory centers (see Chapter 22), or ventilation may not increase sufficiently for increased metabolism during exercise (see Chapter 21).

If an *acute* respiratory acidosis is not corrected, then it is said to be a *chronic* respiratory acidosis, and renal mechanisms will increase HCO_3^-, as described in Chapter 31, to return pH toward normal. The movement from point B to D on Fig. 6B is termed a *metabolic compensation* (or *renal compensation)* for a chronic respiratory acidosis, and it represents an increase in base excess. The final acid–base status at point D (Fig. 6B) is called a *compensated respiratory acidosis.* Metabolic compensation takes 1–2 days, and pH is not restored to a completely normal value of 7.4 in reality. A hallmark of an acute or compensated respiratory acidosis is an increase in $[HCO_3^-]$ and $[H^+]$.

Respiratory Alkalosis

A *respiratory alkalosis* occurs when P_{CO_2} decreases. This increases the $[HCO_3^-]/P_{CO_2}$ ratio and increases pH along the blood-buffer line, between points A and C in Fig. 6B. A respiratory alkalosis increases pH, but $[HCO_3^-]$ is decreased because less total acid is present. *Hypocapnia,* or a decrease in arterial P_{CO_2}, occurs when ventilation is in excess of that necessary for CO_2 production, for example, when ventilation is stimulated by low O_2 levels at high altitude (Chapter 22).

If the respiratory alkalosis is *chronic,* then renal mechanisms will decrease $[HCO_3^-]$, as described in Chapter 31, to return pH toward normal. The movement from point C to F on Fig. 6B is termed a *metabolic* or *renal compensation* for a chronic respiratory alkalosis, and base excess is decreased. The final acid–base status at point F (Fig. 6B) is called a *compensated respiratory alkalosis.* Renal compensation for a respiratory alkalosis can occur more quickly than compensation for a respiratory acidosis, but it still takes a day or more and does not return pH completely to normal. Both $[HCO_3^-]$ *and* $[H^+]$ are decreased in a compensated respiratory alkalosis.

Metabolic Acidosis

A *metabolic acidosis* occurs when the kidneys do not excrete sufficient fixed acid, or retain sufficient bicarbonate, to maintain the normal $[HCO_3^-]/P_{CO_2}$ ratio when $P_{CO_2} = 40$ mm Hg. Metabolic acidosis increases $[H^+]$ but decreases $[HCO_3^-]$, and causes a parallel shift of the blood-buffer line from points A to G in Fig. 6B. Base excess is decreased (or base deficit is increased) in a metabolic acidosis. Physiologic mechanisms causing a metabolic acidosis are detailed in Chapter 31.

With a chronic metabolic acidosis, the respiratory control system senses decreased pHa and causes a reflex increase in ventilation (see Chapter 22). Increased ventilation causes P_{CO_2} to decrease, increases the base excess, and returns pHa toward normal (point G to F in Fig. 6B). In such a *compensated metabolic acidosis,* pH may not return completely to normal because decreased Pa_{CO_2} may limit the increase in ventilation necessary for complete compensation (see Chapter 22). With an acute, or compensated, metabolic acidosis, $[H^+]$ is increased but $[HCO_3^-]$ is decreased.

Metabolic Alkalosis

A *metabolic alkalosis* occurs when the kidneys excrete excess fixed acid, or retain too much bicarbonate, for the normal $[HCO_3^-]/P_{CO_2}$ ratio when $P_{CO_2} = 40$ mm Hg. Metabolic alkalosis increases pH and $[HCO_3^-]$, and causes a parallel shift of the blood-buffer line from points A to E in Fig. 6B as base excess increases. Physiologic mechanisms causing a metabolic alkalosis are detailed in Chapter 31.

Respiratory compensation for a chronic metabolic alkalosis involves decreased ventilation, which increases P_{CO_2} (points E to D in Fig. 6B). However, increased Pa_{CO_2} and possible decreases in arterial O_2 levels (see Chapter 22), prevent a complete return to normal pH with a *compensated metabolic alkalosis.* With an acute, or compensated, metabolic alkalosis, $[H^+]$ is decreased, *but* $[HCO_3^-]$ is increased.

Diagnosing Acid–Base Disturbances

The primary cause of a chronic acid–base disturbance cannot be determined from a bicarbonate-pH diagram, or P_{CO_2}, pH, and $[HCO_3^-]$ data alone. Notice that the data are similar for a compensated respiratory acidosis and a compensated metabolic alkalosis (point D, Fig. 6B), or a compensated respiratory alkalosis and compensated metabolic acidosis (point F, Fig. 6B). Therefore, other details of the patient's history, pulmonary function, or blood chemistry (see Chapter 31) must be obtained for a proper diagnoses.

Suggested Readings

Bauer C. Structural biology of hemoglobin. In Crystal RG, West JB, Weibel ER, Barnes PJ, eds. *The lung: Scientific foundations,* Vol 1, 2nd ed. Philadelphia: Lippincott-Raven, 1997, pp 1615–1624.

Davenport HW. *The ABC of acid base chemistry,* 6th ed. Chicago: University of Chicago Press, 1974.

Jian L, *et al.* S-nitrosohaemoglobin: A dynamic activity of blood involved in vascular control. *Nature* 1996;380:221.

Roughton FJW. Transport of oxygen and carbon dioxide. In Perm WO, Rahn H, eds. *Handbook of physiology, Section 3, Respiration.* Bethesda, MD: American Physiological Society, 1964, pp 767–826.

Stewart PA. Independent and dependent variables of acid–base control. *Respir Physiol* 1978;33:9.

Pulmonary Gas Exchange

FRANK L. POWELL, JR.

KEY POINTS

- O_2 moves from the atmosphere to tissues by a series of convective (bulk flow) and diffusive transport steps, which decrease P_{O_2} at each step.
- Alveolar P_{CO_2} is inversely related to alveolar ventilation at any given metabolic rate. Physiologic dead space is the functional difference between total ventilation and effective ventilation of the alveoli.
- O_2 diffusion from alveolar gas to pulmonary capillary blood is effective at equilibrating alveolar and arterial P_{O_2} in normal individuals under most conditions.
- O_2 delivery by the cardiovascular system depends on cardiac output and the arterial-venous O_2 difference. The shape of the

blood-O_2 equilibrium curve is advantageous for O_2 transport to tissues.
- The pathway for O_2 diffusion is longer in tissues than in the lung, so P_{O_2} differences between blood and tissue are much larger than the blood–gas P_{O_2} difference in the lung.
- There are four main causes of arterial hypoxemia. *Hypoventilation* differs from other causes of hypoxemia because it does not increase the alveolar-arterial P_{O_2} difference, and it can occur in patients with normal lungs.
- *Diffusion limitation* can result from thickening of the blood–gas barrier in lung disease, but oxygen breathing rapidly alleviates this form of arterial hypoxemia.

PHYSIOLOGIC MECHANISM OF OXYGEN TRANSPORT

Transport of respiratory gases is the primary function of the respiratory system. *Pulmonary gas exchange* describes the process of oxygen (O_2) uptake and carbon dioxide (CO_2) elimination by the lungs. This chapter focuses on pulmonary gas exchange and on how the lungs load adequate O_2 in the blood to meet tissue O_2 demand. It also covers the physiology of O_2 and CO_2 transport by the cardiovascular system and *tissue gas exchange*. The relation between all of these elements is illustrated by the O_2 cascade, which shows how P_{O_2} decreases along the O_2 transport chain between the environment and the mitochondria (Fig. 1).

Figure 1 shows normal P_{O_2} values that could be measured in gas, blood, and tissue samples from a resting individual at sea level. The P_{O_2} drop between dry room air and tracheal gas is due to the humidification of inspired gas. Ventilation is the primary determinant of the much larger drop in P_{O_2} between inspired and alveolar gas. Diffusion across the blood–gas barrier, and other factors such as shunts and the matching of ventilation to pulmonary blood flow, explains the relatively small decrease in P_{O_2} between alveolar gas and arterial blood. The circulation and O_2 diffusion from capillaries to tissues cause the large decreases in P_{O_2} between arterial and venous blood, and between blood and mitochondria in the tissues.

Changes in O_2 supply or demand can have different effects on the O_2 cascade, depending on the nature of the change. Because O_2 transport occurs in a *series* of steps, changes at one level will affect P_{O_2} at other levels. Therefore, understanding the physiologic factors that determine the P_{O_2} at any step in the O_2 cascade is useful for diagnosing abnormalities in gas exchange.

Some General Principles of Gas Exchange

Before covering each step of O_2 transport in detail, it is helpful to introduce some general principles that apply to all steps. Chapter 18 introduced the principles of mass balance that will be used here to predict P_{O_2} values for each step in the O_2 cascade. The reader should refer to Chapter 18 for derivations of the simple models used here to describe O_2 transport by (1) *convection,* that is, "bulk flow" transport by ventilation and blood flow; and (2) *diffusion* across blood–gas or blood–tissue barriers. It is important to note that O_2 is *not* actively transported or secreted across membranes in the human body.

Quantitative descriptions of gas exchange are also referred to as "models" of gas exchange. Gas exchange models in this chapter assume a *steady state,* defined as an equal and constant rate of gas transport at each step in the O_2 cascade. Steady-state conditions do *not* necessarily imply resting conditions. O_2 transport can be elevated but still equal at every step in the O_2 cascade, for example, during steady-state exercise. However, non-steady-state conditions occur frequently, for example, at the onset of exercise.

Gas exchange models are useful for quantifying respiratory function. For example, in an "*ideal*" model of alveolar gas exchange, arterial blood equilibrates with

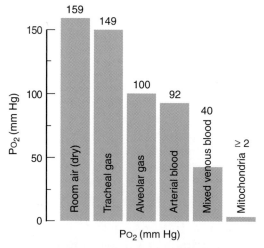

FIGURE 1   The oxygen cascade in a healthy subject breathing room air at sea level shows the pattern of P_{O_2} decrease between the different steps of O_2 transport as described in the text. The difference between alveolar and arterial P_{O_2} occurs because of pulmonary gas exchange limitations.

alveolar gas, so the *ideal alveolar-arterial* P_{O_2} *difference* equals zero. However, Fig. 1 shows that, in reality, the alveolar-arterial P_{O_2} difference exceeds zero, even in healthy young adults. Several factors can increase the alveolar-arterial P_{O_2} difference, relative to the ideal value of zero, and these are called gas exchange *limitations*. The alveolar-arterial P_{O_2} difference quantifies the net effect of these limitations on pulmonary gas exchange. Gas exchange limitations do not affect O_2 consumption at rest, but they will lower maximal O_2 consumption and P_{O_2} values along the O_2 cascade in a steady state.

The alveolar-arterial P_{O_2} difference is useful for diagnosing O_2 exchange limitations. The physiologic mechanisms responsible for an alveolar-arterial P_{O_2} difference increase gradually with age or dramatically with lung disease. However, different mechanisms respond differently to simple tests such as O_2 breathing. This means that changes in the alveolar-arterial P_{O_2} difference with such simple tests can be used to diagnose pulmonary disease.

ALVEOLAR VENTILATION

Ventilation is the first step in the O_2 cascade, and the level of alveolar ventilation ($\dot{V}_A$) is the most important physiologic factor determining arterial P_{O_2} for any given inspired P_{O_2} and level of O_2 demand ($\dot{V}_{O_2}$) in healthy lungs. As described in Chapter 18, *anatomic dead space* reduces the fraction of the tidal volume that reaches the alveoli:

$$\dot{V}_A = f_R(V_T - V_D),$$

where $\dot{V}_A$ is alveolar ventilation, f_R is respiratory frequency, V_T is tidal volume, and V_D is anatomic dead space. Anatomic dead space can be measured with the single breath method (see Chapter 18, Fig. 10) but gas exchange principles can be used to obtain a more direct measure of the effective, or functional, alveolar ventilation.

Alveolar Ventilation Equation Predicts $P_{A_{CO_2}}$

The Fick equation (see Chapter 18) defines CO_2 elimination from the lungs ($\dot{V}_{CO_2}$) as:

$$(\dot{V}_{CO_2}) = (\dot{V}_A F_{A_{CO_2}}) - (\dot{V}_I F_{I_{CO_2}}),$$

where $\dot{V}_{CO_2}$ is the difference between the CO_2 expired from the alveoli and the amount of CO_2 inspired to the alveoli. $F_{I_{CO_2}}$ is nearly zero, so the inspired terms can be dropped.

The *alveolar ventilation equation* is obtained by substituting $P_{A_{CO_2}}$ for $F_{A_{CO_2}}$ and rearranging the Fick equation:

$$\dot{V}_A = (\dot{V}_{CO_2}/P_{A_{CO_2}})K,$$

where K is a constant ($= 0.863$) to convert F_{CO_2} to P_{CO_2} in mm Hg, and $\dot{V}_{CO_2}$ in mL_{STPD}/min to $\dot{V}_A$ in L_{BTPS}/min. In practice, arterial P_{CO_2} is substituted for alveolar P_{CO_2} because the two values are equal in normal lungs and an arterial blood sample is usually taken to measure arterial P_{O_2} for evaluating gas exchange.

The most important thing to remember about the alveolar ventilation equation is that $\dot{V}_A$ and $P_{A_{CO_2}}$ (or $P_{a_{CO_2}}$) are inversely related for any given metabolic rate. For example, if $\dot{V}_A$ is doubled, $P_{A_{CO_2}}$ is halved, regardless of the exact values for either variable. Therefore, the effectiveness of ventilation can be judged by the $P_{a_{CO_2}}$ at any given metabolic rate. *Hyperventilation* is defined by a decrease in $P_{a_{CO_2}}$ from the normal value, implying excess $\dot{V}_A$ for the given $\dot{V}_{CO_2}$. Increased ventilation does not always mean hyperventilation. For example, $\dot{V}_A$ must increase to maintain normal $P_{a_{CO_2}}$ when $\dot{V}_{CO_2}$ increases during exercise. *Hypoventilation* is defined by an increase in $P_{a_{CO_2}}$ and this occurs when $\dot{V}_A$ is lower than normal for a given $\dot{V}_{CO_2}$.

Physiologic Dead Space

Physiologic dead space is a functional measure of "wasted ventilation," and it is always greater than anatomic dead space. Physiologic dead space is defined from another rearrangement of the Fick principle applied to CO_2 elimination by the lungs. In a steady state, $\dot{V}_{CO_2}$ measured in mixed-expired gas must equal $\dot{V}_{CO_2}$ measured from alveolar gas:

$$(\dot{V}_E \bar{F}_{E_{O_2}}) - (\dot{V}_I F_{I_{CO_2}}) = (\dot{V}_A F_{A_{CO_2}}) - (\dot{V}_I F_{I_{CO_2}}).$$

The inspired terms can be subtracted from both sides, and ventilation is converted to volume by dividing both sides by respiratory frequency. This yields:

$$V_T \bar{F}_{E_{CO_2}}(V_T - V_D)F_{A_{CO_2}},$$

where $V_T - V_D = V_A$. As illustrated in Fig. 2, *mixed-expired* F_{CO_2}, which can be measured by collecting all inspired gas in a bag or a spirometer, includes gas exhaled from the alveoli *and* dead space. Therefore, physiologic dead space can be defined as a ratio of mixed expired and alveolar gas CO_2 levels:

$$V_D/V_T = (F_{A_{CO_2}} - \bar{F}_{E_{CO_2}})/F_{A_{CO_2}}.$$

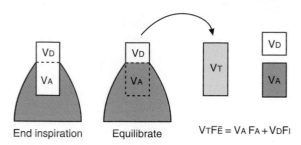

$$V_T F\bar{E} = V_A F_A + V_D F_I$$

FIGURE 2   Not all of the tidal volume (V_T) is effective at bringing fresh gas into the alveoli during inspiration because of dead space (V_D). CO_2 in the mixed expired gas (F_E) is a mixture of dead space (inspired gas that has not undergone exchange, F_I) and alveolar gas (F_A). The equation can be rearranged to calculate V_D/V_T as described in the text.

This measure of physiologic dead space is also called *Bohr's dead space* after the Danish physiologist who developed the method. In practice, P_{CO_2} is substituted for F_{CO_2} so measured Pa_{CO_2} can be substituted for alveolar P_{CO_2} as follows:

$$V_D/V_T = (Pa_{CO_2} - P\bar{E}_{CO_2})/Pa_{CO_2}.$$

Factors that may increase physiologic dead space relative to anatomic dead space are considered in detail later (see $\dot{V}_A/\dot{Q}$ Mismatching between Different Alveoli section).

Alveolar Gas Equation Predicts P_{AO_2}

Alveolar P_{O_2} (P_{AO_2}) can be predicted from inspired P_{O_2} (P_{IO_2}) and alveolar P_{CO_2} (PA_{CO_2}) by the *alveolar gas equation:*

$$P_{AO_2} = P_{IO_2} - (PA_{O_2}/R) + F,$$

where R is the respiratory exchange ratio (see below), and $F = [PA_{CO_2} \cdot F_{IO_2}(1 - R)/R]$. Note that F increases P_{AO_2} about 2 mm Hg under normal conditions ($F_{IO_2} = 0.21$ and $PA_{CO_2} = 40$ mm Hg), so F *can be neglected under normal conditions.* The alveolar gas equation is *only* valid if inspired $P_{CO_2} = 0$, which is a reasonable assumption for room air breathing ($F_{ICO_2} = 0.0003$). Alveolar ventilation is a major determinant of P_{AO_2} because $\dot{V}_A$ determines PA_{CO_2}, according to the alveolar *ventilation* equation described earlier.

P_{IO_2} is less than ambient P_{O_2} because air is warmed and humidified in the respiratory system. The vapor pressure of water = 47 mm Hg at 37°C, and the total pressure available for O_2 is decreased by this amount, whereas the fractional concentration of O_2 in dry gas remains constant (Chapter 18). Ambient $P_{O_2} = 160$ mm Hg ($0.21 \cdot 760$ mm Hg) but inspired $P_{O_2} = 150$ mm Hg [$0.21 \cdot (760 - 47) = 0.21 \cdot 713$ mm Hg].

The *respiratory exchange ratio* (R) is the ratio of O_2 uptake by the lungs to CO_2 elimination by the lungs:

$$R = \dot{V}_{CO_2}/\dot{V}_{O_2}.$$

Under steady-state conditions, R equals the *respiratory quotient* (RQ), which is the ratio of CO_2 production to O_2 consumption in metabolizing tissues. RQ averages 0.8 but can range from 0.67 to 1, depending on the relative amounts of fat, protein, and carbohydrate being metabolized. However, R can exceed this range in nonsteady states, for example when R exceeds 1 during hyperventilation or at the onset of exercise. CO_2 stores in the body are much greater than O_2 stores because of bicarbonate in blood and tissues. R can increase because it takes longer to wash out the CO_2 stores than it does to charge up the much smaller O_2 stores in the body.

Substituting normal values predicts that $P_{AO_2} = 100$ mm Hg in normal individuals breathing room air ($P_{AO_2} = 150 - 40/0.8$). Increases in $\dot{V}_A$ (hyperventilation) increase P_{AO_2} by decreasing PA_{CO_2}, whereas decreases in $\dot{V}_A$ (hypoventilation) decrease P_{AO_2}. The alveolar gas equation calculates ideal alveolar P_{O_2}, which is greater than measured arterial P_{O_2} (Fig. 1). Reasons for this difference are explained later.

DIFFUSION

Diffusion of O_2 from alveoli to pulmonary capillary blood is the next step in the O_2 cascade after alveolar ventilation. However, it is important to note that blood leaving the pulmonary capillaries is in equilibrium with alveolar gas in healthy lungs under normal resting conditions. Hence, the small decrease between P_{AO_2} and Pa_{O_2} shown in Fig. 1 is *not* caused by diffusion, but by ventilation-perfusion mismatching in healthy lungs under normal conditions (see Ventilation-Perfusion Mismatching section).

O_2 moves from the alveoli into pulmonary capillary blood barrier according to Fick's first law of diffusion (Chapter 18):

$$\dot{V}_{O_2} = \Delta P_{O_2} \cdot D_{O_2},$$

where ΔP_{O_2} is the average P_{O_2} gradient across the blood–gas barrier and D_{O_2} is a "diffusing capacity" for O_2 across the barrier. Some readers may find it helpful to note the analogy between Fick's law for O_2 flux and Ohm's law for the flow of electrons (current = voltage/resistance). $\dot{V}_{O_2}$ is analogous to current and ΔP_{O_2} is analogous to the potential energy difference of voltage. However, D_{O_2} is analogous to a *conductance,* which is the inverse of resistance (current = voltage · conductance). O_2 flux can be increased either by increasing the P_{O_2} gradient or increasing the O_2 conductance (D_{O_2}).

D_{O_2} depends on both the molecular properties of O_2 and the geometric properties of that membrane:

$$D_{O_2} = (\text{solubility}/\text{MW})_{O_2} \cdot (\text{area}/\text{thickness})_{\text{membrane}}.$$

Solubility is important because gas molecules have to "dissolve" in a membrane before they can diffuse across it and, once dissolved, low-molecular-weight (MW) molecules move more quickly by the random motions of diffusion. Large surface areas increase the probability that an O_2 molecule will come into contact with the membrane through random motion, but membrane thickness increases the distance over which O_2 molecules must travel.

Pathway for Oxygen

Figure 3 shows the pathway for an O_2 molecule diffusing from alveolar gas to hemoglobin inside an erythrocyte (red blood cell). This is the anatomic basis for D_{O_2}. The total area of the blood–gas barrier is nearly 100 m^2 in the human lung (Chapter 18). The barrier is also extremely thin but variable in thickness, and it consists of several different layers. The "thin" side of a pulmonary capillary (0.3 μm) separates gas from plasma with (1) thin cytoplasmic extensions from type I alveolar epithelial cells, (2) a thin basement membrane, and (3) thin cytoplasmic extensions from capillary endothelial cells. The thicker side of a capillary has collagen in the interstitial space to provide mechanical strength in the alveoli. Epithelial and endothelial cell bodies are in alveolar corners between capillaries, to minimize further the thickness of the gas exchange barrier. Finally, O_2 has to diffuse through plasma and across the red blood cell membrane before it can combine with hemoglobin. The diffusing capacity for O_2 between the alveolar gas and hemoglobin is called the *membrane diffusing capacity for O_2, or Dm_{O_2}.*

After O_2 diffuses into red blood cells, the finite *rate of reaction between O_2 and hemoglobin* (abbreviated with the symbol θ) offers an additional "resistance" to O_2 uptake. The magnitude of this chemical resistance depends on θ and the total amount of hemoglobin, which is a physiologic function of pulmonary capillary volume (V_C). This chemical resistance is in series with the membrane resistance, so the total resistance to O_2

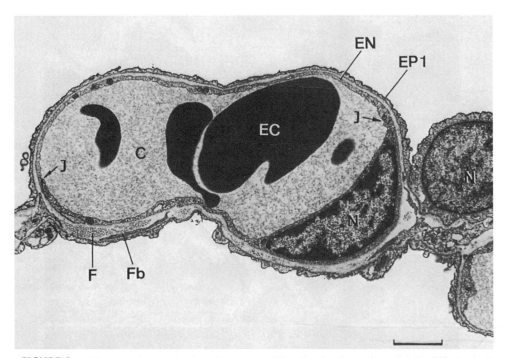

FIGURE 3   Electron micrograph of a pulmonary capillary showing the pathway for O_2 diffusion in the lung. O_2 diffuses from the alveolar space (open areas above and below capillary, C) through epithelial cell (EP1), interstitial space and endothelial cell (EN), and plasma before combining with hemoglobin in the erythrocyte (EC). Collagen fibers (F) and fibroblasts (Fb) thicken the interstitial space on one side of the capillary. N, nucleus; J, endothelial cell junctions. Scale marker = 2 μm. (From Weibel, Chap. 82 in Crystal *et al.*, eds., *The lung: Scientific foundations.* Philadelphia: Lippincott-Raven, 1997.)

diffusion in the lung can be defined as follows:

$$1/\text{D}_{LO_2} = 1/\text{D}_{MO_2} + 1/(\theta\,\text{V}_C),$$

where D_{LO_2} *is the lung diffusing capacity for* O_2. (D_{LO_2} is a conductance; recall that conductance is the inverse of resistance and resistors in series are additive.)

It is estimated that membrane and chemical reaction resistances to O_2 diffusion are about equal in normal lungs. Both D_{MO_2} and V_C are under physiologic control through pulmonary capillary recruitment and distension. Therefore, D_{LO_2} increases with exercise by recruitment of D_{MO_2} and $\theta\,\text{V}_C$. Methods for measuring D_{LO_2} are described in a later section.

P_{O_2} Changes along the Pulmonary Capillary

The ΔP_{O_2} value used in Fick's law is an average value, corresponding to the *mean partial pressure gradient operating over the entire length of the pulmonary capillary.* Alveolar P_{O_2} is constant everywhere outside the capillary because diffusion is rapid in the gas phase and effectively mixes O_2 in the small alveolar spaces. However, P_{O_2} *in* the capillary blood must increase from mixed-venous levels at the beginning to arterial levels at the end of the capillary. Figure 4 shows the normal time course of P_{O_2} changes along the capillary in a healthy lung at normal P_{O_2} levels.

An average *capillary transit time* of 0.75 sec (Fig. 4) is calculated from a cardiac output of 6 L/min and capillary

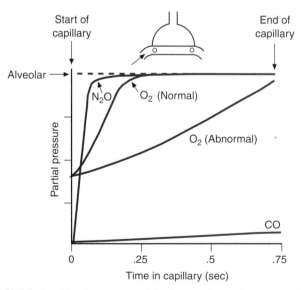

FIGURE 4 The time course of the increase in partial pressure for different gases diffusing from alveolar gas into pulmonary capillary blood. N_2O and O_2 under normal conditions equilibrate very quickly, but CO or O_2 under abnormal conditions does not equilibrate in the time it takes blood to flow through the capillary (0.75 sec).

volume of 75 mL (time = volume/flow rate). Note that diffusion equilibrium normally occurs between blood and gas in only 0.25 sec, providing a threefold safety factor. However, if D_{LO_2} is decreased sufficiently with lung disease, then capillary P_{O_2} may not equilibrate with P_{AO_2} during the transit time (Fig. 4, abnormal O_2 curve). In the abnormal case, $P_{C'O_2} < P_{AO_2}$, which is defined as a *diffusion limitation for* O_2, where $P_{C'}$ is used to designate end-capillary partial pressure.

Only two conditions lead to diffusion limitation for O_2 in healthy individuals: (1) *Elite athletes at maximal exercise,* with very high O_2 consumption and cardiac outputs, can have transit times that are too short for O_2 diffusion equilibrium. Capillary volume increases by recruitment and distension with elevated cardiac output (Chapter 18), but this is not sufficient to balance the huge increase in flow rate that occurs in elite athletes. (2) *Normal individuals exercising at altitude* may not achieve diffusion equilibrium because transit time *and* P_{AO_2} decrease. Transit time will decrease with elevated cardiac output during exercise, but capillary volume recruitment and distension can preserve enough time for O_2 diffusion equilibrium to occur *if* P_{AO_2} is normal. However, P_{AO_2} is decreased at altitude, and this slows O_2 diffusion in two ways.

First, decreased P_{AO_2} slows O_2 diffusion by decreasing the P_{O_2} gradient driving diffusion. For example, at an altitude of 3050 m (10,000 feet), the barometric pressure is only 523 mm Hg and P_{IO_2} is 100 mm Hg [= 0.21 (523 − 47) mm Hg]. In a normal individual doing mild exercise at this altitude, P_{AO_2} is measured to be about 55 mm Hg. $P\bar{v}_{O_2}$ decreases much less than this because of the shape of the O_2-blood equilibrium curve (see below also, the Cardiovascular and Tissue Oxygen Transport section). Measurements show $P\bar{v}_{O_2}$ is 24 mm Hg at altitude, compared with 30 mm Hg at sea level, with this amount of exercise. Therefore, the P_{O_2} gradient at the beginning of the capillary decreases from 70 mm Hg with mild exercise at sea level (= 100 − 30 mm Hg) to 31 mm Hg (= 55 − 24 mm Hg) at altitude. The exact values in this example are not as important as the general concepts that *the shape of the O_2-blood equilibrium curve maintains $P\bar{v}_{O_2}$ in exercise at altitude* (see also Cardiovascular and Tissue Oxygen Transport section), and *this decreases the P_{O_2} gradient for diffusion in the lung.*

Decreasing P_{AO_2} also slows the rate of rise in P_{O_2} because gas exchange is occurring on the steep portion of the O_2-blood equilibrium curve. This means that a given increase in O_2 concentration is not effective at increasing P_{O_2} toward the alveolar equilibrium value. In contrast, the flat shape of the O_2-blood equilibrium curve around normal P_{AO_2} levels promotes diffusion equilibrium. Small amounts of O_2 diffusing from alveolar gas into capillary blood cause large increases in P_{O_2}, so capillary P_{O_2} rapidly approaches equilibrium with P_{AO_2} in normoxia.

Diffusion- and Perfusion-Limited Gases

Figure 4 shows dramatic differences in the time course of diffusion equilibrium for different gases in the lung. The anesthetic gas nitrous oxide (N_2O) achieves equilibrium rapidly, whereas carbon monoxide (CO) never comes close to diffusion equilibrium. Understanding the differences between these gases is not only important for anesthesiology and emergency medicine, but it also helps one understand the physiologic mechanisms for O_2 diffusion limitations.

The uptake of a gas that achieves diffusion equilibrium depends on the magnitude of pulmonary blood flow. For example, N_2O diffuses rapidly from the alveoli to capillary blood (Fig. 4), so the only way to increase its uptake is to increase the amount of blood flowing through the alveolar capillaries. N_2O is an example of a *perfusion-limited gas*. Changes in the diffusing capacity have no effect on the uptake of a perfusion-limited gas or its partial pressure in the blood and body. All anesthetic and *"inert" gases that do not react chemically with blood* are perfusion limited. Notice that under normal resting conditions, O_2 is a perfusion-limited gas also.

The uptake of a gas that does *not* achieve diffusion equilibrium could obviously increase if the diffusing capacity increased. CO is an example of such a *diffusion-limited gas* (Fig. 4). Hemoglobin has a very high affinity for CO, so the effective solubility of CO in blood is large. Therefore, increases in the CO concentration in blood are not effective at increasing P_{CO}. This keeps blood P_{CO} lower than alveolar P_{CO}, and results in a large disequilibrium and diffusion limitation. Under abnormal conditions, O_2 may become a diffusion-limited gas (Fig. 4).

The reason that some gases are perfusion limited and some are diffusion limited is not related solely to their solubility in blood. All anesthetic gases are perfusion limited, yet they have a wide range of solubility in blood. A gas is a diffusion-limited gas when its effective solubility in blood is significantly greater than its solubility in the blood–gas barrier (which is approximated by its solubility in water or plasma). This is because a gas has to dissolve in the barrier before it can diffuse across the barrier and dissolve in the blood. An analogy is putting a liquid into different-sized containers with different-sized openings. The supply of liquid is assumed to be unlimited, analogous to a constant alveolar partial pressure. The size of the container represents solubility in blood and the size of the opening represents solubility in the membrane. It takes a long time to fill a large container through a small opening.

However, a larger container will not take longer to fill if it also has a larger opening.

In practice, *the only diffusion-limited gases are CO and O_2 under hypoxic conditions*. All other gases are perfusion limited, including O_2 under normoxic conditions in healthy lungs. CO is diffusion limited because it is always much more soluble in blood than in the blood–gas barrier. This is also the case for O_2 in hypoxia, when the slope of the O_2-blood equilibrium curve is much steeper than the slope for physically dissolved O_2 in plasma (see Chapter 20, Fig. 1). However, at high P_{O_2} levels, the slope of the blood-O_2 equilibrium curve equals the slope of the solubility curve in plasma, and O_2 behaves like inert gases.

Measures of Diffusing Capacity

The diffusing capacity of the lung can be measured from the uptake of a diffusion-limited gas like CO. If very low levels of CO are inspired (about 0.1%), then hemoglobin saturation with CO is very low, arterial oxygenation is not disturbed, and there are no toxic effects. Also, the amount of CO entering the capillary blood does not increase blood P_{CO} significantly because CO is so soluble in blood. Therefore, the lung-diffusing capacity for CO (D_{LCO}) can be defined by Fick's first law of diffusion as follows:

$$D_{LCO} = \dot{V}_{CO}/P_{ACO},$$

where $\dot{V}_{CO}$ = CO uptake by the lung and the gradient driving CO diffusion equals P_{ACO} because average capillary P_{CO} = 0. In theory, D_{LCO} could be used to calculate the D_L for O_2 by correcting for physical factors that determine diffusing capacity (MW and solubility). However, only D_{LCO} is reported clinically.

In the *steady-state D_{LCO} method,* the individual breathes a low level of CO for a couple of minutes. Then CO uptake is calculated from the Fick principle using measurements of ventilation and inspired and expired P_{CO}. Alveolar P_{CO} can be estimated from expired P_{CO}. In the *single breath D_{LCO} method,* an individual takes a breath with a low concentration of CO and holds the breath for 10 sec. This method also requires a simultaneous measurement of lung volume (e.g., by helium dilution; see Chapter 18, Fig. 7) to calculate $\dot{V}_{CO}$ from P_{CO} changes in the lung. Alveolar P_{CO} is estimated from expired P_{CO} and corrected for the change that occurs during the breath hold.

A normal D_{LCO} is about 25 mL/(min · mm Hg). D_{LCO} can increase two- to threefold with exercise, as expected for capillary recruitment and distension. D_{LCO} also changes with O_2 level because the rate of chemical reaction between CO and hemoglobin is decreased by hemoglobin oxygenation, and D_{LCO} has a chemical reaction rate component ($\theta\, V_C$) similar to D_{LO_2}. In lung disease, D_{LCO} may be affected by other factors such as unequal distributions of alveolar volume, pulmonary blood flow, and diffusing properties. Such factors explain why *morphometric estimates of diffusing capacity,* based on anatomic measurements of the blood–gas barrier surface area, thickness, etc., are about twice as large as functional measurements of diffusing capacity. Because other factors can affect D_{LCO} measurement, it is sometimes referred to as a *transfer factor* instead of the diffusing capacity.

CARDIOVASCULAR AND TISSUE OXYGEN TRANSPORT

The cardiovascular system is responsible for transporting O_2 to metabolizing tissues after it has diffused into pulmonary capillary blood. The heart pumps O_2-rich arterial blood to the tissues, where O_2 leaves the systemic capillaries and moves to the mitochondria through tissue gas exchange. The heart also pumps O_2-poor venous blood back to the lungs, where it is reoxygenated. The magnitude of the P_{O_2} decrease between arterial and venous blood (see Fig. 1) depends on both the cardiovascular O_2 supply and O_2 demand in the tissues.

Cardiovascular Oxygen Transport

O_2 *supply,* or *delivery,* to tissues can be defined as the product of cardiac output and arterial O_2 concentration ($\dot{Q}\, Ca_{O_2}$). The tissues will extract enough O_2 to meet their metabolic demands, as long as O_2 supply is sufficient. Hence, O_2 supply and demand determine venous O_2 levels. These factors are related by the *Fick principle* (Chapter 18), which describes O_2 transport by the cardiovascular system as follows:

$$\dot{V}_{O_2} = \dot{Q}(Ca_{O_2} - C\bar{v}_{O_2}),$$

where $\dot{V}_{O_2}$ is O_2 consumption, $\dot{Q}$ is cardiac output, and the last term is the arterial-venous O_2 concentration difference. This equation can be rearranged and used to calculate cardiac output from measurements of $\dot{V}_{O_2}$ and blood O_2 concentrations.

The Fick principle can be used to predict the *arterial-venous O_2* concentration difference from normal resting values for O_2 consumption ($\dot{V}_{O_2}$ = 300 mL/min) and cardiac output (Q = 6 L/min): (300 mL of O_2/min)/(6000 mL of blood/min) = 5 mL O_2/dL blood. If the normal value for Ca_{O_2} = 20 mL of O_2/dL

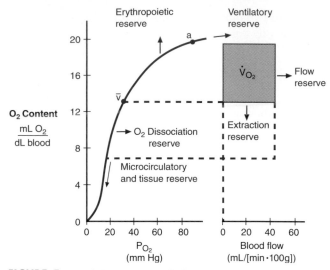

FIGURE 5   Left-blood-O_2 equilibrium curve showing arterial (a) and mixed-venous ($\bar{v}$) points. Right-graphical representation of the Fick principle for cardiovascular O_2 transport. Horizontal axis is blood flow normalized to body mass; vertical axis is O_2 concentration from the left panel; shaded area is O_2 consumption. "Reserves," which can increase $\dot{V}_{O_2}$, are described in the text. (After Woodson, *Basics of Respiratory Disease,* 5:1, 1977.)

of blood, then $C\bar{v}_{O_2}$ = 15 mL/dL. Notice that venous O_2 level is determined by (1) the ratio of metabolism to blood flow and (2) arterial O_2 concentration, which is determined by alveolar P_{O_2} and the blood-O_2 equilibrium curve. The O_2-blood equilibrium curve (see Chapter 20, Fig. 1) is used to convert $C\bar{v}_{O_2}$ (15 mL/dL) to mixed-venous O_2 saturation (75%) and $P\bar{v}_{O_2}$ (40 mm Hg).

Figure 5 shows the cardiovascular Fick principle graphically, to illustrate the importance of the shape of the O_2-blood equilibrium curve. In the right panel, the height of the shaded rectangle represents the arterial-venous O_2 concentration difference, and its width represents cardiac output (normalized to 100 g of body mass). The area of the rectangle is the product of these two factors, and represents $\dot{V}_{O_2}$.

Changes in $\dot{V}_{O_2}$ can be achieved by increasing cardiac output ("flow reserve") and/or increasing the arterial-venous O_2 difference. The dashed lines on Fig. 5 show the consequences of increasing $\dot{V}_{O_2}$ by increasing venous O_2 extraction (*extraction reserve*). Changes in $P\bar{v}_{O_2}$, are minimized with large decreases in venous O_2 concentration by the shape of the O_2-blood equilibrium curve. A right shift of the O_2-blood equilibrium curve can increase $P\bar{v}_{O_2}$ for a given $C\bar{v}_{O_2}$ (*O_2-dissociation reserve*). Maintaining a high $P\bar{v}_{O_2}$, is important for tissue gas exchange and the *microcirculatory and tissue reserve,* as discussed later. All of these reserves are important mechanisms for meeting increased O_2 demands during exercise.

Increases in O_2 *delivery* are achieved primarily through increases in cardiac output in normoxic conditions. Increasing alveolar and arterial Po_2 is not effective at increasing Cao_2 in normoxic conditions because the slope of the O_2-blood equilibrium curve is flat at normal Pao_2 values (*ventilatory reserve*; Fig. 5). However, changes in Pao_2 are much more effective at changing O_2 delivery when Po_2 is low, for example, at altitude, when exchange occurs on a steeper part of the O_2-blood equilibrium curve. Changes in hematocrit and hemoglobin concentration, which occur with chronic hypoxia (see Chapter 20), can increase O_2 delivery by increasing total O_2 concentration for any given Po_2 (*erythropoietic reserve*). The erythropoietic reserve is the physiologic basis for the questionable practice of "blood doping," which uses blood transfusions or artificial erythropoietin in attempts to increase maximal O_2 consumption and athletic performance.

Tissue Gas Exchange

O_2 moves out of systemic capillaries to the mitochondria in cells by diffusion. Therefore, O_2 transport in tissues is described by *Fick's first law of diffusion*, similar to diffusion across the blood–gas barrier in the lung:

$$\dot{V}^-o_2 = \Delta Po_2 \cdot Dto_2,$$

where ΔPo_2 is the average Po_2 gradient between capillary blood and the mitochondria, and Dto_2 is a tissue-diffusing capacity for O_2. Interestingly, anatomic estimates of Dto_2 for the whole body are similar to anatomic estimates of Dmo_2 in the lung.

The main difference between O_2 diffusion in tissue and in the lung is that diffusion pathways are much longer in tissue. Tissue capillaries may be 50 μm apart, so the distance from a capillary surface to mitochondria can be 50 times longer than the thickness of the blood–gas barrier (< 0.5 μm). Long diffusion distances can lead to significant Po_2 gradients in tissues. Also, the Po_2 gradient varies along the length of a capillary as O_2 leaves the blood, and capillary Po_2 decreases from arterial to venous levels. A mathematical model called the *Krogh cylinder* can be used to predict Po_2 profiles in metabolizing tissue. This model predicts that Po_2 in cells farthest away from a capillary, or at the venous end of the capillary, may be zero when O_2 demand is increased. However, mitochondria function normally until Po_2 decreases below a few mm Hg (see Chapter 18, Fig. 1), so metabolism will continue under all but these most extreme conditions.

During increased O_2 demand (e.g., exercise in skeletal muscle, absorption in the gut, nervous activity in the brain), additional capillaries may be recruited (Chapter 17). This helps maintain adequate O_2 supply by decreasing diffusion distances. Factors increasing, or maintaining, $P\bar{v}o_2$ also help tissue O_2 diffusion by enhancing the Po_2 gradient. These factors include the steep shape of the O_2-blood equilibrium curve in the venous range and right shifts of the curve by temperature, Pco_2, and pH changes, for example, in exercising muscle. $P\bar{v}o_2$ is sometimes used as an index of tissue O_2 exchange because it represents the minimum pressure head driving O_2 diffusion in the body.

Myoglobin may facilitate O_2 diffusion in muscle by shuttling O_2 to sites far away from a capillary. Recent measurements show that Po_2 in skeletal muscle is much more uniform than predicted by the Krogh cylinder model and that myoglobin may shuttle O_2 to the venous end. The implications of this finding for differences in O_2 transport between muscle (e.g., during myocardial ischemia) and brain (e.g., during a stroke) remain to be determined.

LIMITATIONS OF PULMONARY GAS EXCHANGE

Gas exchange limitations in the lungs can reduce Po_2 throughout the O_2 cascade. Recall that limitations do not decrease resting $\dot{V}o_2$, although they may limit maximal O_2 consumption during exercise. *Hypoxemia* is defined as a decrease in blood Po_2, and arterial hypoxemia, or decreased Pao_2, indicates a limitation of pulmonary gas exchange. Gas exchange limitation does not imply a decrease in resting O_2 consumption, because Po_2 will adjust throughout the O_2 cascade to maintain O_2 consumption in a steady state. For example, Pvo_2 (and $\dot{Q}$) will change as necessary to satisfy the cardiovascular Fick equation when O_2 consumption increases.

Gas exchange limitations not only decrease Pao_2, but they also can increase the *alveolar-arterial* Po_2 *difference*. The concept of an "ideal" lung without limitations was introduced earlier, and under such ideal conditions, $Pao_2 - Pao_2 = 0$ mm Hg. However, in reality, Pao_2 calculated from the alveolar gas equation is greater than Pao_2, measured from an arterial blood sample, and this alveolar-arterial Po_2 difference increases with *some, but not all*, mechanisms that limit gas exchange.

There are four kinds of pulmonary gas exchange limitations: (1) hypoventilation, (2) diffusion limitations, (3) pulmonary blood flow shunts, and (4) mismatching of ventilation and blood flow in different parts of the lung. The following subsections explain how each of these limitations decreases Pao_2 and how the alveolar-arterial Po_2 difference is useful for diagnosing the causes of hypoxemia in a patient.

Hypoventilation

Hypoventilation is the *only* pulmonary gas exchange limitation that does not increase the alveolar-arterial P_{O_2} difference. Therefore, hypoxemia with an alveolar-arterial P_{O_2} difference in the normal range is diagnostic of hypoventilation.

The magnitude of hypoxemia caused by hypoventilation is predicted by the alveolar gas equation:

$$P_{AO_2} = P_{IO_2} - (P_{ACO_2}/R).$$

Hypoventilation increases P_{ACO_2}, according to the inverse relationship between $\dot{V}_A$ and P_{ACO_2} described by the alveolar ventilation equation. This is easiest to understand when $R = 1$, and P_{AO_2} decreases 1 mm Hg for every 1 mm Hg increase in P_{ACO_2}. Conceptually, P_{IO_2} represents the total amount of gas inspired, and gas exchange replaces each molecule of O_2 consumed with one molecule of CO_2. Hence, P_{AO_2} is simply the difference between P_{IO_2} and P_{ACO_2} when $R = 1$. However, a normal value for R is 0.8, and this magnifies the effects of hypoventilation and increased P_{ACO_2} on hypoxemia.

The two primary classes of problems that cause hypoventilation are (1) mechanical limitations and (2) ventilatory control abnormalities. Abnormal respiratory mechanics, such as increased airway resistance or decreased compliance with lung disease, may limit the effectiveness of the respiratory muscles in generating volume changes and $\dot{V}_A$. Also, the respiratory muscles themselves may be damaged and ineffective at generating the pressures necessary for normal ventilation. In all of these cases, the ventilatory control system may be normal, in terms of sensing P_{AO_2} and P_{ACO_2} changes and sending neural signals to the respiratory muscles, to increase ventilation. However, abnormal control of ventilation can also occur, as described in Chapter 22.

Diffusion Limitations

Pulmonary *diffusion limitation* is defined as disequilibrium between the partial pressure of a gas in the alveoli and in the blood leaving the pulmonary capillaries. Therefore, diffusion limitations decrease P_{AO_2} by increasing the alveolar-arterial P_{O_2} difference. This occurs when (1) the pressure head driving O_2 diffusion across the blood–gas barrier (P_{AO_2}) is too low, or (2) the lung's diffusing capacity for O_2 (D_L) is not sufficient for the O_2 demands of the body.

As described earlier, diffusion limitations can occur in normal individuals when P_{AO_2} is decreased at high altitude *and* O_2 demand is increased during hard exercise. Increased O_2 demand alone can increase the alveolar-arterial P_{O_2} difference and cause arterial hypoxemia in some elite athletes during maximal exercise at sea level. In lung disease, the measured D_{LCO} can decrease with destruction of surface area and capillary volume (e.g., emphysema) or thickening of the blood–gas barrier (e.g., interstitial lung disease) but D_{LCO} has to decrease to less than 50% of normal before arterial hypoxemia is observed in resting patients. Hypoventilation and ventilation-perfusion mismatch can also lower P_{AO_2} and decrease the pressure gradient driving diffusion.

Arterial hypoxemia caused by a diffusion limitation can be relieved rapidly by increasing inspired O_2 (within several breaths). This increases the driving pressure for O_2 from the alveoli into the blood.

Shunts

The ideal models used to analyze alveolar ventilation and diffusion have considered gas exchange occurring in a single compartment, so arterial P_{O_2} equals P_{O_2} in the blood leaving the pulmonary capillaries ($P_{AO_2} = P_{C'O_2}$). In reality, arterial blood is *not* pure pulmonary capillary blood; it also includes shunt flow. *Shunt* is defined as deoxygenated venous blood flow that enters the arterial circulation without going through ventilated alveoli in the pulmonary circulation. This kind of shunt is also called *right-to-left shunt,* to distinguish it from left-to-right shunt, which shunts systemic arterial blood into pulmonary artery flow with some congenital heart defects, and in the three-chambered hearts of some lower vertebrates. Right-to-left shunt decreases P_{AO_2} by diluting end-capillary blood with deoxygenated venous blood.

Shunt is calculated by applying the principle of mass balance to a two-compartment model, which splits total cardiac output ($\dot{Q}t$) between a shunt flow to an unventilated compartment ($\dot{Q}s$) and flow to a normally ventilated alveolar compartment (Fig. 6). O_2 delivery in arterial blood must equal the sum of O_2 delivery from the two compartments:

$$\dot{Q}t\,Ca_{O_2} = \dot{Q}s\,C\bar{v}_{O_2} + (\dot{Q}t - \dot{Q}s)Cc'_{O_2},$$

where $Cc'_{O_2} = O_2$ concentration in blood at the end of the pulmonary capillaries. This can be rearranged to the Berggren *shunt equation* defining shunt flow as a fraction of total cardiac output:

$$\dot{Q}s/\dot{Q}t = (Cc'_{O_2} - Ca_{O_2})/(Cc'_{O_2} - C\bar{v}_{O_2}).$$

The value of $\dot{Q}s/\dot{Q}t$ can be calculated in practice by measuring arterial and mixed-venous blood samples in an individual during 100% O_2 breathing. This removes

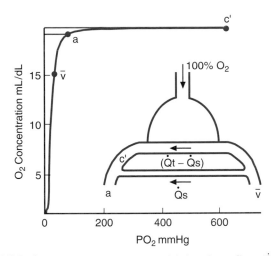

FIGURE 6   Two compartment model for shunt flow ($\dot{Q}s$) and effective pulmonary blood flow ($\dot{Q}_T - \dot{Q}s$). Blood-O_2 equilibrium curve illustrates how small shunt flows of mixed-venous blood ($\bar{v}$) significantly decrease Po_2 in arterial blood (a) relative to Po_2 in end-capillary blood leaving the alveoli (c').

any diffusion limitation in ventilated alveoli, so $Pc'o_2$ = PAo_2. Cao_2 and $C\bar{v}o_2$ are measured directly. $Cc'o_2$ is estimated from PAo_2 using an O_2-blood equilibrium curve, where PAo_2, is calculated from the alveolar gas equation. (Note that Fio_2 appears in the constant F in the alveolar gas equation, and this constant should not be neglected when $Fio_2 = 1.0$.)

Figure 6 shows the effect of shunt on Pao_2. Alveolar and end-capillary Pao_2 are predicted to be more than 600 mm Hg during pure O_2 breathing. However, shunt significantly decreases Pao_2 because of the shape of the O_2-blood equilibrium curve. The large increase in Po_2 with O_2 breathing does not increase $Cc'o_2$ enough to offset the low level of $C\bar{v}o_2$. Therefore, persistent hypoxemia during 100% O_2 breathing indicates a shunt if all the alveoli are effectively ventilated with 100% O_2. (Exceptions to this condition may occur with ventilation-perfusion mismatching in lung disease, as described later.) If Pao_2 can be increased above 150 mm Hg during O_2 breathing, and cardiac output is normal, then 1% shunt increases the alveolar-arterial Po_2 difference about 20 mm Hg.

In healthy individuals, shunt during O_2 breathing averages less than 5% of cardiac output, including (1) venous blood from the *bronchial circulation* that drains directly into the pulmonary veins and (2) venous blood from the coronary circulation that enters the left ventricle through the *Thebesian veins*. If shunt is calculated during room air breathing, it is called *venous admixture*. Venous admixture is larger than the shunt during O_2 breathing because it is an "as if" shunt, which includes the effects of low Po_2 from poorly ventilated alveoli. This occurs even in healthy lungs with

ventilation-perfusion mismatching as described in the next section.

Ventilation-Perfusion Mismatching

Mismatching of ventilation and blood flow in different parts of the lung is the most common cause of alveolar-arterial Po_2 differences in health and disease. It is also the most complicated mechanism of hypoxemia, and will be approached in two steps. First, the effect of the *alveolar ventilation-perfusion ratio* ($\dot{V}A/\dot{Q}$) ratio on PAo_2 is described for an ideal lung. Second, the mechanisms resulting in *different $\dot{V}A/\dot{Q}$ ratios in different parts of real lungs* and the effect of this on arterial Po_2 are considered. It will be important to understand that only this second factor increases the alveolar-arterial Po_2 difference.

Ventilation-Perfusion Ratio

The effect of changing $\dot{V}A/\dot{Q}$ on Pao_2, has already been introduced in the section on Alveolar Ventilation. PAo_2 increases with $\dot{V}A$ according to the alveolar ventilation equation and alveolar gas equations. $\dot{V}A/\dot{Q}$ adds the concept of blood flow. The effect of $\dot{V}A/\dot{Q}$ on PAo_2 can be understood by thinking of $\dot{V}A$ as bringing O_2 into the alveoli, and $\dot{Q}$ as taking it away. If $\dot{Q}$ suddenly increases and removes more O_2 from the alveoli (recall that O_2 is normally a perfusion-limited gas), then PAo_2, will decrease. However, if $\dot{V}A$ increases O_2 delivery to match increased O_2 removal (returning the $\dot{V}A/\dot{Q}$ ratio to normal), then PAo_2 will return to normal. Decreasing $\dot{V}A/\dot{Q}$ has the opposite effect and decreases PAo_2.

The O_2-CO_2 *diagram* of Fig. 7 shows the effects of changing $\dot{V}A/\dot{Q}$ in an ideal lung, modeled as a single alveolus in a steady state, with no shunts or diffusion limitations. The ventilation-perfusion ratio ($\dot{V}A/\dot{Q}$) is defined with alveolar ventilation to eliminate the effects of dead space. The $\dot{V}A/\dot{Q}$ line on the CO_2-O_2 diagram shows *all* possible Pco_2-Po_2 combinations that could occur in this lung with $\dot{V}A/\dot{Q}$ ratios ranging from 0 to infinity. When $\dot{V}A/\dot{Q} = 0$, this indicates a shunt. The shunt alveolus is not ventilated, so it will equilibrate with mixed-venous blood and the $\dot{V}A/\dot{Q} = 0$ point corresponds to $P\bar{v}o_2$ and $P\bar{v}co_2$. When $\dot{V}A/\dot{Q}$ is infinite, this indicates dead space. There is no blood flow to dead space, so this alveolus equilibrates with inspired gas, and the infinite $\dot{V}A/\dot{Q}$ point corresponds to Pio_2 and $Pico_2$. Normal $Paco_2$ and Pao_2 values are shown for a normal $\dot{V}A/\dot{Q}$ of 0.8.

The important point to notice about the $\dot{V}A/\dot{Q}$ line is *that changes in $\dot{V}A/\dot{Q}$ around the normal value affect PAo_2 more than $Paco_2$* (notice the different CO_2 and O_2 scales

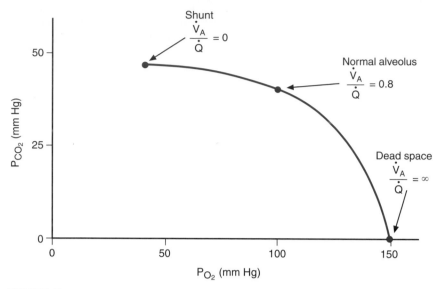

FIGURE 7 O_2-CO_2 diagram. The ventilation-perfusion curve describes all possible P_{O_2}-P_{CO_2} combinations in the alveoli (in an ideal lung, P_{AO_2} = P_{aO_2}). Mixed-venous values occur in alveoli with no ventilation (shunt) and inspired values occur in alveoli with no perfusion (dead space). (After Rahn and Fenn, Chap. 31 in Renn and Rahn, eds., *Handbook of physiology, Respiration.* Bethesda, MD: American Physiological Society, 1964.)

in Fig. 7). This generalization holds even if the $\dot{V}_A/\dot{Q}$ line is altered by changing mixed-venous or inspired gas (which changes the end points), or by physiologic changes in the blood-O_2 or CO_2 equilibrium curves (which determine the exact shape of the $\dot{V}_A/\dot{Q}$ line). The $\dot{V}_A/\dot{Q}$ line is calculated from the ventilation-perfusion equation: $\dot{V}_A/\dot{Q}$ = $8.63R(C_{aO_2} - C_{\bar{v}O_2})/P_{aCO_2}$ with $\dot{V}_A$ in L_{BTPS}/min, $\dot{Q}$ in L/min, O_2 concentrations in mL/dL, and P_{CO_2} in mm Hg. This equation is *not* generally used in clinical medicine, but is included here for reference.

$\dot{V}_A/\dot{Q}$ Mismatching between Different Alveoli

In real lungs, total alveolar ventilation and cardiac output must be distributed between some 300 million alveoli, and this distribution is *not* perfectly uniform. This results in $\dot{V}_A/\dot{Q}$ *heterogeneity,* or different $\dot{V}_A/\dot{Q}$ ratios in different parts of the lung; $\dot{V}_A/\dot{Q}$ mismatching refers to spatial $\dot{V}_A/\dot{Q}$ heterogeneity between functional units of gas exchange in real lungs—*not* a mismatch between total $\dot{V}_A$ and $\dot{Q}$ or a deviation of the overall $\dot{V}_A/\dot{Q}$ from 1. Changes in the $\dot{V}_A/\dot{Q}$ ratio in an ideal lung, or in a single alveolus, change P_{aO_2} as described earlier (see Fig. 7), but this does *not* change the alveolar-arterial P_{O_2} difference from the ideal value of zero. In contrast, $\dot{V}_A/\dot{Q}$ heterogeneity between lung units does decrease P_{aO_2} *and* it increases the alveolar-arterial P_{O_2} difference.

Regional differences in alveolar ventilation occur because of the mechanical properties of the lung, as described in Chapter 19. Briefly, gravity tends to distort

the upright lung so alveoli in the apex are more expanded than those in the base of the lung. This results in basal alveoli operating on a steeper part of the lung's compliance curve, so $\dot{V}_A$ is greater at the bottom than at the top of the lung (see Chapter 19, Fig. 11). $\dot{V}_A$ per unit lung volume differs by a factor of 2.5 between the top and bottom of the upright human lung (Fig. 8).

Regional differences in blood flow occur because of the effects of gravity on the pulmonary circulation, as described in Chapter 18. Briefly, capillary pressure is greater at the bottom than at the top of the upright lung, which reduces local vascular resistance at the bottom of

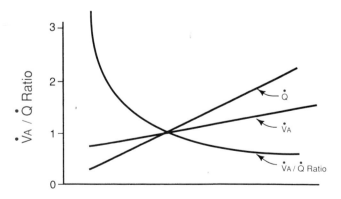

FIGURE 8   Gravity results in regional differences in alveolar ventilation ($\dot{V}_A$) and blood flow between the apex and base of the lung, as described in the text. This causes the $\dot{V}_A/\dot{Q}$ ratio to decrease about 2.5-fold from the top to the bottom of the lung. (After West, *Ventilation/blood flow and gas exchange.* New York: Blackwell Scientific, 1990.)

the lungs and increases regional blood flow (see Chapter 18, Fig. 13). Figure 8 shows that $\dot{Q}$ per unit lung volume changes by a factor of 6 between the top and bottom of the upright human lung, or relatively more than $\dot{V}_A$. The net result is a large decrease in $\dot{V}_A/\dot{Q}$ between the top and bottom of the upright lung (Fig. 8).

This $\dot{V}_A/\dot{Q}$ heterogeneity between different regions of the lung leads to regional differences in P_{AO_2} and P_{ACO_2} corresponding to differences predicted by the $\dot{V}_A/\dot{Q}$ line on the CO_2-O_2 diagram (see Fig. 7). For example, $\dot{V}_A/\dot{Q}$ in the upright lung at rest ranges from 3.3 at the top of the lung to 0.6 at the bottom. This decreases P_{AO_2} from 132 mm Hg at the top of the lung to 89 mm Hg at the bottom. P_{ACO_2} increases from 28 mm Hg at the top of the lung to 42 mm Hg at the bottom. These regional differences in alveolar gas cause O_2 uptake and CO_2 elimination to decrease from the top to the bottom of the lung. However, O_2 uptake decreases more than CO_2 elimination, corresponding to the larger decrease in P_{AO_2} and this decreases R (the respiratory exchange ratio) between the top and the bottom of the lung. Exercise reduces regional heterogeneity of $\dot{V}_A/\dot{Q}$, alveolar gases, and gas exchange by increasing blood flow at the top of the lung.

$\dot{V}_A/\dot{Q}$ heterogeneity also causes heterogeneity in P_{O_2} of the end-capillary blood (Pc'_{O_2}), because alveolar-arterial P_{O_2} equilibrium occurs in any region with a normal diffusing capacity. Figure 9 illustrates how this leads to hypoxemia. Diffusing capacity is assumed normal, so Pc'_{O_2} equals P_{AO_2} in each functional unit of gas exchange with a different $\dot{V}_A/\dot{Q}$. Arterial blood is a mixture of blood draining each unit, so O_2 concentration in the "mixed" arterial blood is a flow-weighted average of blood from individual units. The "high" $\dot{V}_A/\dot{Q}$ unit contributes relatively little to total blood flow, and Cc'_{O_2} is not increased significantly by the high P_{AO_2}, because the blood-O_2 equilibrium curve is flat at high P_{O_2}. The "low" $\dot{V}_A/\dot{Q}$ unit contributes relatively more to total blood flow, and Cc'_{O_2} is decreased significantly because the blood-O_2 equilibrium curve is steep at low P_{O_2}. Consequently, Ca_{O_2} is weighted toward the level in the low $\dot{V}_A/\dot{Q}$ units, and P_{AO_2} is lower than the numerical average of P_{AO_2} from all three alveoli.

Figure 9 also shows how $\dot{V}_A/\dot{Q}$ heterogeneity increases the measured alveolar-arterial P_{O_2} difference, without increasing alveolar-arterial P_{O_2} difference in any single gas exchange unit. P_{O_2} in the mixed alveolar gas can be calculated as a flow-weighted mixture of the gas expired from all the units. Increases in P_{O_2} can effectively balance decreases in P_{O_2} in the gas phase, because partial pressure and concentration (or fraction, P_{O_2}) are linearly related for O_2 in the gas phase, unlike O_2 in blood. However, mixed P_{AO_2}, exceeds the numerical average of P_{O_2} from the three units because

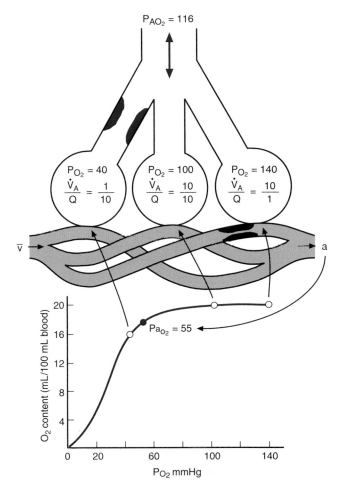

FIGURE 9 A three-compartment model showing how $\dot{V}_A/\dot{Q}$ differences between lung units can increase the difference between *mixed*-alveolar P_{O_2} (P_{AO_2} = 116 mm Hg) and *mixed*-arterial P_{O_2} (Pa_{O_2} = 55 mm Hg). Inspired P_{O_2} is assumed normal (150 mmHg). Alveolar P_{O_2} in any individual unit is assumed equal to the P_{O_2} in end-capillary blood from that unit (open circles on O_2 dissociation curve in lower panel). However, P_{O_2} in mixed arterial blood is weighted toward P_{O_2} in the low $\dot{V}_A/\dot{Q}$ units, and P_{O_2} in mixed alveolar gas is weighted toward P_{O_2} in the high $\dot{V}_A/\dot{Q}$ units. The shape of the O_2 dissociation curve also contributes to the large alveolar-arterial P_{O_2} difference as described in the text. (After West, *Ventilation/blood flow and gas exchange.* New York: Blackwell Scientific, 1990.)

the high $\dot{V}_A/\dot{Q}$ unit contributes more volume to mixed alveolar gas.

Therefore, $\dot{V}_A/\dot{Q}$ heterogeneity decreases Pa_{O_2}, *and* increases P_{AO_2} from ideal values, so the alveolar-arterial P_{O_2} is increased. This mechanism increases the alveolar-arterial partial pressure difference for any gas, including CO_2 and anesthetic gases. However, $\dot{V}_A/\dot{Q}$ affects O_2 more than other gases because the shape of the blood-O_2 equilibrium curve depresses Pa_{O_2}. High and low $\dot{V}_A/\dot{Q}$ regions can offset the effects of each other more effectively for CO_2 and anesthetic gases because the blood-equilibrium curves are relatively linear for these gases. Consequently, increasing overall ventilation is effective at overcoming Pa_{CO_2} increases from $\dot{V}_A/\dot{Q}$ heterogeneity, and the ventilatory control reflexes are quite effective at maintaining normal Pa_{CO_2} by this mechanism (see Carbon Dioxide Exchange and Chapter 22).

$\dot{V}_A/\dot{Q}$ heterogeneity occurs in normal lungs and explains the normal alveolar-arterial P_{O_2} in healthy young individuals. However, only half the $\dot{V}_A/\dot{Q}$ heterogeneity necessary to explain the normal alveolar-arterial P_{O_2} difference is caused by gravitational-dependent differences in $\dot{V}_A$ and $\dot{Q}$ at different heights in the lung. Significant *intraregional $\dot{V}_A/\dot{Q}$ heterogeneity* occurs between functional units of gas exchange (*acini;* see Chapter 18) at any given height in the lung. $\dot{V}_A/\dot{Q}$ heterogeneity is also the most common cause of hypoxemia in lung disease (see Clinical Note later in this chapter).

Several methods are available for measuring the exact nature of $\dot{V}_A/\dot{Q}$ heterogeneity, but they are generally restricted to the research laboratory and not useful clinically. $\dot{V}_A/\dot{Q}$ heterogeneity can be diagnosed clinically by eliminating other causes of hypoxemia. Hypoventilation can be ruled out if the measured alveolar-arterial P_{O_2} difference is greater than the normal predicted value. Diffusion limitation can be ruled out if the measured D_{LCO} is at least 50% of normal, or if breathing high inspired O_2 relieves hypoxemia and decreases the alveolar-arterial P_{O_2} difference. However, 100% O_2 breathing will also eliminate hypoxemia from $\dot{V}_A/\dot{Q}$ heterogeneity if *all of the alveoli equilibrate with inspired P_{O_2}.* With pure O_2 breathing, only O_2 and CO_2 (plus water vapor) are in the alveolar gas, so Pa_{O_2} is at least 600 mm Hg in all alveoli. In practice, $\dot{V}_A/\dot{Q}$ heterogeneity includes poorly ventilated lung units, so it may take up to 30 min to wash nitrogen out of all the alveoli during O_2 breathing. Consequently, O_2 breathing improves hypoxemia from $\dot{V}_A/\dot{Q}$ heterogeneity, but not nearly as quickly as it does with a pure diffusion limitation (which requires < 1 min). If shunt is present, 100% O_2 breathing will never resolve the hypoxemia or decrease the alveolar-arterial P_{O_2} difference.

Other clinical measures of $\dot{V}_A/\dot{Q}$ heterogeneity include physiologic shunt and dead space. Low $\dot{V}_A/\dot{Q}$ gas exchange units and shunt have similar effects on Pa_{O_2}. This means that *physiologic shunt* (or *venous admixture*) can be used to quantify $\dot{V}_A/\dot{Q}$ heterogeneity. Physiologic shunt is measured with the Berggren shunt equation in an individual breathing less than 100% O_2, and usually room air (see Shunts section). $\dot{V}_A/\dot{Q}$ heterogeneity causes hypoxemia "as if" there were an increase in shunt, so it increases physiologic shunt. Similarly, the effects of high $\dot{V}_A/\dot{Q}$ units on Pa_{O_2} resemble the effects of dead space. Therefore, *physiologic dead space* (see Physiologic Dead Space section) can be used to quantify the effects of $\dot{V}_A/\dot{Q}$ heterogeneity on Pa_{O_2} and Pa_{CO_2} "as if" there were an increase in anatomic dead space. Note that both physiologic shunt and dead space will be less than the actual amounts of blood flow or ventilation going to abnormal $\dot{V}_A/\dot{Q}$ units. This is because shunt and dead space represent the extremes of the $\dot{V}_A/\dot{Q}$ ratio, and actual $\dot{V}_A/\dot{Q}$ ratios between 0 and infinity will have smaller effects on alveolar and arterial P_{O_2} and P_{CO_2} (see Fig. 7).

Note that the alveolar ventilation equation introduced earlier does *not* accurately quantify $\dot{V}_A$ when arterial P_{CO_2} is substituted for alveolar P_{CO_2} if $\dot{V}_A/\dot{Q}$ heterogeneity is present. $\dot{V}_A/\dot{Q}$ heterogeneity causes alveolar P_{CO_2} to be less than arterial P_{CO_2}. Hence, the total ventilation to all alveoli is underestimated if arterial P_{CO_2} is used in the alveolar ventilation equation. Because it is difficult to measure *mixed* alveolar P_{CO_2}, the alveolar ventilation equation is not used on people with lung disease.

CARBON DIOXIDE EXCHANGE

Physiologic CO_2 transport follows the same general principles described for O_2 in the earlier sections. For example, the Fick principle was used earlier to calculate Pa_{CO_2}. The Fick principle can also be used to calculate the normal arterial-venous CO_2 concentration

Clinical Note

Most lung disease is patchy in nature, and leads to considerable differences in ventilation ($\dot{V}_A$) and perfusion ($\dot{Q}$), and *$\dot{V}_A/\dot{Q}$ heterogeneity*. This increases the alveolar-arterial P_{O_2} and, consequently, it can be difficult to distinguish between diffusion limitations and $\dot{V}_A/\dot{Q}$ heterogeneity.

Unfortunately, it is not possible to routinely measure $\dot{V}_A/\dot{Q}$ heterogeneity for diagnostic purposes, although rapid progress is being made in lung imaging techniques. However, sophisticated experimental techniques have been applied in special clinical studies and these

Clinical Note (continued)

have revealed some important insights about the exact causes of hypoxemia in certain lung diseases, as well as how they respond to treatment.

Diffuse interstitial pulmonary fibrosis is a thickening of the alveolar walls with collagen and scarring in the interstitium. As expected, fibrosis decreases the diffusing capacity (D_{LCO}) experiments hypoxemia in patients with fibrosis at rest can be explained by $\dot{V}_A/\dot{Q}$ heterogeneity. Uneven scarring of the lungs results in local changes in compliance and resistance (airway obstructions), leading to regional differences in time constants and ventilation (see Chapter 19). Local scarring may also affect resistance to pulmonary blood flow but not always in the same way it affects ventilation, leading to $\dot{V}_A/\dot{Q}$ heterogeneity. *Diffusion limitation* for oxygen only becomes significant in patients with fibrosis during exercise.

Another pulmonary disease that might be expected to cause a diffusion limitation is *adult respiratory distress syndrome* (ARDS), which can result from blunt trauma to the chest and lungs (e.g., in a car crash). ARDS begins with pulmonary edema and leads to interstitial fibrosis (which eventually reverses if the patient recovers). Oxygen therapy should relieve hypoxemia caused by a diffusion limitation from a thickening of the interstitium, but oxygen can actually decrease Pa_{O_2} in ARDS. This is because *shunt* can increase during oxygen breathing by a mechanism called *absorption atelectasis*. Alveoli with low $\dot{V}_A/\dot{Q}$ ratios are especially susceptible to atelectasis, or collapse, during oxygen breathing because a large diffusion gradient is driving all of the oxygen into pulmonary capillary blood, and there is no nitrogen left in the alveoli to hold them open. Experiments confirm that blood flow to low $\dot{V}_A/\dot{Q}$ regions in the lungs of ARDS patients is converted to shunt during O_2 breathing. Hence, other treatments such as O_2 administered with *positive end-expiratory pressure* (PEEP) artificial ventilation may be necessary.

Pulmonary thromboembolism occurs when a blood clot obstructs part of the pulmonary circulation, leading to increased pulmonary artery pressure and high $\dot{V}_A/\dot{Q}$ ratios in parts of the lungs that are distal to the clot and poorly perfused. This $\dot{V}_A/\dot{Q}$ heterogeneity leads to

hypoxemia but *shunt* may occur also. *Surfactant* production requires substrates delivered by the pulmonary circulation; this is impaired in poorly perfused parts of the lungs, leading to increased *surface tension* and alveolar collapse. Also, pulmonary capillary pressures increase in perfused regions of the lungs, with the high pulmonary artery pressure, and this can lead to *edema* and alveolar flooding that causes shunt.

Chronic obstructive pulmonary disease (COPD) is a general term for patients with *emphysema* (see Chapter 19) and *chronic bronchitis*. $\dot{V}_A/\dot{Q}$ heterogeneity is the main cause of hypoxemia in this disease too, but at least two different patterns of $\dot{V}_A/\dot{Q}$ distributions are seen that correlate roughly with the arterial P_{CO_2}. Some COPD patients maintain Pa_{CO_2} in the normal range by increasing total ventilation in the face of $\dot{V}_A/\dot{Q}$ mismatching as described in the text. Other COPD patients *hypoventilate*, so their Pa_{O_2} is low and Pa_{CO_2} is elevated (suggesting a problem with ventilatory control; see Chapter 22). The COPD patients with normal Pa_{CO_2} tend to have less chronic bronchitis and high $\dot{V}_A/\dot{Q}$ regions in their lungs, consistent with emphysema destroying part of the alveolar capillary bed. In contrast, COPD patients with elevated Pa_{CO_2} typically have advanced chronic bronchitis and low $\dot{V}_A/\dot{Q}$ regions in the lungs, consistent with increased airway resistance in the inflamed airways.

Asthma is the other main obstructive pulmonary disease. Asthma causes severe *bronchoconstriction* leading to low $\dot{V}_A/\dot{Q}$ regions in the lung but no shunt. A big advance in the treatment of asthma was the development of *selective β_2-adrenergic agents* (e.g., albuterol) to stimulate bronchodilation. Older, non-selective β-adrenergic agents (e.g., isoproterenol) stimulated β_1 receptors on the heart also, which increased blood flow to low $\dot{V}_A/\dot{Q}$ regions of the lung and could actually decrease Pa_{O_2}. The benefits of relieving bronchoconstriction and increasing ventilation generally outweighed the negative side effects of β-adrenergic agents on gas exchange, but the selective β_2-adrenergic agents completely obviate these side effects.

difference:

$$\dot{V}_{CO_2} = \dot{Q}(Ca_{CO_2} - C\bar{v}_{CO_2}).$$

For normal values of $\dot{V}_{CO_2}$ = 240 mL of CO_2/min and $\dot{Q}$ = 6 L/min, the arterial-venous CO_2 concentration difference is 4 mL/dL. This differs from the normal value for O_2 only by the difference in $\dot{V}_{CO_2}$ and $\dot{V}_{O_2}$ (or by their ratio, which equals R).

Differences between CO_2 and O_2 exchange result mainly from (1) differences in the O_2 and CO_2 dissociation curves and (2) differences in the effects of CO_2 and O_2 on ventilatory control reflexes. As explained in Chapter 20, P_{CO_2} differences are much smaller than P_{O_2} differences between arterial and venous blood, although the concentrations are similar, because CO_2 is more soluble in blood. As explained in Chapter 22, Pa_{CO_2} is the most important value in determining the resting level of ventilation, and ventilatory reflexes tend to increase $\dot{V}_A$ as much as necessary to restore normal Pa_{CO_2} when gas exchange is altered.

Hypoventilation has almost the same effect on both O_2 and CO_2. Differences between decreases in Pa_{O_2} and increases in Pa_{CO_2} with hypoventilation are explained by the effect of the normal respiratory exchange ratio (R) in the alveolar gas equation. A normal R = 0.8 magnifies Pa_{O_2} changes for a given Pa_{CO_2} change.

Diffusion limitation affects CO_2 and O_2 similarly, but normal ventilatory control reflexes will increase $\dot{V}_A$ and return Pa_{CO_2} to normal. Calculating a diffusing capacity for CO_2 is less certain than for O_2 because resistances from the chemical reactions of CO_2 in the blood are more uncertain. Membrane-diffusing capacity for CO_2 is greater than Dm_{O_2} because CO_2 solubility is much greater than O_2 solubility, but other factors are similar or identical. [Recall that D = (solubility/MW)$_{gas}$ · (area/thickness)$_{membrane}$]. However, under normal circumstances, the rate of equilibration between alveolar gas and capillary blood is estimated to be similar for CO_2 and O_2, requiring approximately 0.25 sec for equilibrium. CO_2 equilibrium is slower than expected with its large membrane-diffusing capacity, in part because chemical reactions for CO_2 in blood are slow. Also, both membrane and blood solubility are high for CO_2, and the membrane-to-blood solubility ratio determines rate of diffusion equilibrium (see the Diffusion- and Perfusion-Limited Gases section).

The effects of *shunts* and *ventilation-perfusion mismatching* on Pa_{CO_2} are similar, and relatively small for two reasons: (1) Pa_{CO_2} changes little when shunt or low $\dot{V}_A/\dot{Q}$ units increase CO_2 concentration, because the CO_2-blood equilibrium curve is so steep (see Chapter 20, Fig. 4); and (2) the linearity of the physiologic CO_2 dissociation curve (see Chapter 20, Fig. 4) means that increases in CO_2 concentration can be offset by increasing $\dot{V}_A$, which decreases Pa_{CO_2}. As explained earlier, $\dot{V}_A/\dot{Q}$ mismatching will increase the alveolar-arterial partial pressure difference for any gas, including CO_2. However, the normal ventilatory control system will increase the overall $\dot{V}$ as necessary to restore Pa_{CO_2} to normal. In fact, some patients with shunts and low $\dot{V}_A/\dot{Q}$ units may actually have decreased Pa_{CO_2} if hypoxemia is severe enough to override the normal control of Pa_{CO_2} arid induce hyperventilation (see Chapter 22).

The extra ventilation necessary to compensate for shunt and low $\dot{V}_A/\dot{Q}$ units contributes to *physiologic dead space* calculated by the Bohr method. The difference between physiologic and anatomic dead space is sometimes called *alveolar dead space*. This represents an "as if" amount of wasted ventilation that could explain measured CO_2 exchange if there was no shunt or $\dot{V}_A/\dot{Q}$ mismatching in the lung.

Suggested Readings

Farhi LE, SM Tenney, eds. Gas exchange, Vol IV. In *Handbook of physiology: Section 3, The respiratory system.* Bethesda, MD: American Physiological Society, 1987.

Taylor CR, Karas RH, Weibel ER, Hoppler H. Adaptive variation in the mammalian respiratory system in relation to energetic demand. *Respir Physiol* 1987;69:1–127.

Wagner PD. Diffusion and chemical reaction in pulmonary gas exchange. *Physiol Rev* 1977;57:257.

West JB, Wagner PD. Ventilation-perfusion relationships. In Crystal RG, West JB, Weibel WR, Barnes PJ, eds. *The lung: Scientific foundations,* Vol 1, 2nd ed. Philadelphia: Lippincott-Raven, 1997;1693–1710.

Control of Breathing

FRANK L. POWELL, JR.

KEY POINTS

- *Negative feedback systems* control ventilation to maintain normal arterial blood gases and minimize the work of breathing in response to changes in the environment, activity, and lung function.
- The basic respiratory rhythm is generated by neurons in the brain stem, and this rhythm is modulated by ventilatory reflexes.
- Arterial P_{CO_2} is the most important factor determining ventilatory drive in resting humans.
- The ventilatory response to hypoxia depends strongly on arterial P_{CO_2} and it is not large in normal individuals until arterial P_{O_2} drops below 60 mm Hg.
- *Arterial chemoreceptors* sense changes in arterial P_{O_2}, P_{CO_2}, and pH, but *central chemoreceptors* sense changes only in arterial P_{CO_2}.

- Vagal nerve endings sensitive to stretch in the lungs mediate reflexes that fine-tune the rate and depth of breathing.
- Vagal nerve endings sensitive to mechanical and chemical irritation of the airways and lungs stimulate coughing, mucous production, bronchoconstriction, and rapid shallow breathing.
- The autonomic nervous system and vagal sensory nerves in the airways are involved in local control of airway function.
- The most common cause of increased ventilation in healthy humans is exercise, but this cannot be explained by negative feedback ventilatory reflexes.
- The ventilatory response to arterial blood gases changes with time during chronic hypoxemia at high altitude or with lung disease.

CONTROL OF VENTILATION

The body contains several physiologic control systems to maintain arterial pH (pHa) within normal limits, to meet the oxygen demands of the tissues, to minimize the mechanical work of breathing, and to prevent lung injury by environmental agents. This means that limitations in lung function or gas exchange can be masked by the physiologic control systems acting to maintain *homeostasis*. Therefore, knowledge of the normal physiology of respiratory control is critical for understanding respiratory responses to activity, the environment, and pulmonary disease. This knowledge is also necessary to distinguish abnormalities in the respiratory control system from primary disturbances in lung function.

Reflexes and Negative Feedback

Breathing originates from a respiratory rhythm generated in the brain, as described next in the Respiratory Rhythm Generation section. However, homeostasis requires reflexes to modulate the timing and amplitude of the basic breathing rhythm in response to changes in respiratory system function.

The *reflex* control system has three components (Fig. 1). First, a sensory system detects and transmits information about the level of some physiologic variable in the body, called the *physiologic stimulus*. This sensory information is called *afferent input*. Afferent nerves code sensory information with frequency of action potentials (see Chapter 49). Second, *central integration* processes afferent input from *multiple* sensory systems and determines an appropriate response. Such complex

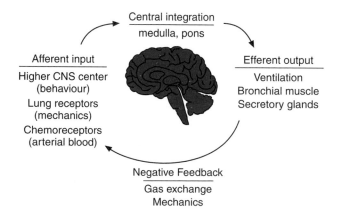

FIGURE 1   Three components of negative feedback ventilatory control by reflexes as described in the text. Lung receptors and chemoreceptors sense changes in mechanics and arterial blood, and cause reflex changes in ventilation and airways. These efferent responses affect gas exchange and lung mechanics and also provide negative feedback.

information processing occurs in the central nervous system (CNS) and involves the neurophysiologic and neurochemical mechanisms of synaptic transmission (Chapter 6). Central integration of respiratory reflexes occurs in the *medulla* and *pons,* in proximity to the areas that generate the respiratory rhythm. Third, *efferent output* occurs when an effector system executes the appropriate response. *Negative feedback* occurs when the efferent output has an effect on the original stimulus and afferent input, such as ventilation changing arterial P_{CO_2}.

Figure 1 shows the important components of respiratory reflexes. Reflexes are involuntary by definition, but notice that *breathing can be changed voluntarily* by neural commands for behaviors from higher centers in the CNS. For example, breathing may be changed during speech, which requires a voluntary command from the cortex. However, such voluntary commands are integrated with all other afferent information to generate an appropriate efferent response. This explains why it is so difficult to speak in a normal voice immediately after hard exercise, when other factors are strongly stimulating ventilation.

Two main classes of sensory systems convey afferent information about the function of the respiratory system: the chemoreceptors and the mechanoreceptors (see Fig. 1). Arterial P_{O_2}, P_{CO_2}, and pH are collectively referred to as *arterial blood gases,* and they provide a good index of gas exchange and lung mechanics, as described in previous chapters. *Chemoreceptors* monitor changes in arterial blood gases and cause reflex changes in ventilation that return arterial blood gases toward normal values. For example, increases in Pa_{CO_2} or decreases in Pa_{O_2} stimulate reflex increases in ventilation. *Mechanoreceptors* monitor pressure and volume in the lungs and airways to provide afferent information about pulmonary mechanics. Generally, mechanoreceptors induce reflex changes in the rate and depth of breathing to minimize the work of breathing under different mechanical conditions and at different levels of ventilation. Mechanoreceptors (and other sensory nerves from the lungs and airways) are also involved in airway smooth muscle and secretory responses that defend the lungs from environmental insult. Note that, in this context, the term *receptor* refers to a specialized sensory nerve ending and not a neurotransmitter or drug receptor.

All of the reflexes described in this chapter are examples of *negative feedback control*. Negative feedback describes the effect of the response (efferent output) on the stimulus. For example, increased Pa_{CO_2} will stimulate chemoreceptors to cause a reflex increase in ventilation and this, in turn, decreases Pa_{CO_2}. Negative feedback *reduces* the original deviation in the stimulus from its normal level. In control systems terminology, the measured input from a *regulated variable* (Paco₂) is held constant by changes in the output of a *controlled variable* (ventilation). In most physiologic control systems, increases *or* decreases in a regulated variable stimulate the appropriate response from a controlled variable to minimize the initial disturbance. For example, decreases in $Paco_2$ will decrease chemoreceptor stimulation and cause a reflex decrease in ventilation. The resulting increase in $Paco_2$ is also an example of negative feedback.

Note that the most common stimulus for increased ventilation is exercise. However, despite years of study, we still cannot explain exercise hyperpnea based on the reflexes described in this chapter. The topic is discussed more in the chapter on exercise.

Respiratory Rhythm Generation

Although breathing is similar to the heartbeat in terms of being automatic and being continuous during sleep or general anesthesia, the skeletal muscles driving ventilation do not contract spontaneously like cardiac muscle does. Rhythmic breathing results from periodic activation of the ventilatory muscles by motor nerves from the CNS. A so-called *central pattern generator,*

composed of networks of neurons, generates this basic respiratory rhythm. The central pattern generator is located in the medulla near other respiratory centers that integrate afferent information for ventilatory reflexes to fine-tune the motor output to ventilatory muscles.

Figure 2 shows the areas of the brain that include the (1) central pattern generator of respiration, (2) the motor nuclei for respiratory muscles, and (3) the sites of central integration for respiratory reflexes. The neural structures responsible for these functions have been identified by experiments on animals and studies of patients with various brain injuries. Modern imaging techniques hold promise for a better understanding of human respiratory centers, but the current technology does not have the resolution necessary to study these small complexes of neurons in the brain stem.

The central pattern generator for ventilation has been isolated in experimental animals to a very small region of the medulla called the *pre-Bötzinger complex*. It has been known for centuries that ventilation begins in the CNS. Galen observed that Roman gladiators with injuries in the neck stopped breathing, but breathing continued if the same injury was below the neck. In the 1800s, physiologists identified most of the important

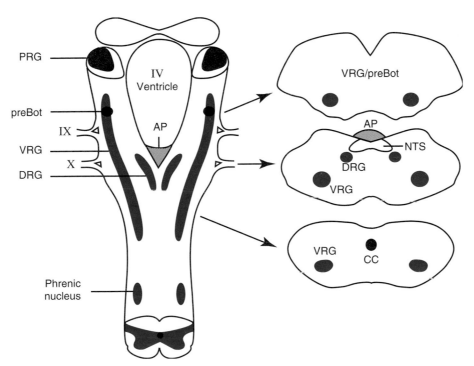

FIGURE 2   Respiratory centers in the brain stem with functions described in text. Left, dorsal view of the pons and medulla with the cerebellum removed; right, transverse sections from three levels of the medulla; PRG, pontine respiratory group; preBot, pre-Botzinger complex, responsible for generating the normal respiratory rhythm; VRG, ventral respiratory group; DRG, dorsal respiratory group; IX cranial nerve (glossopharyngeal); X cranial nerve (vagus); AP, area postrema; NTS, nucleus tractus solitarius; CC, central canal.

respiratory centers by making discrete lesions in the *pons* and *medulla* in experimental animals. Recently, neuroscientists targeted a specific class of neurons in the pre-Bötzinger complex, which have *neurokinin 1 (NK-1) receptors* that are responsive to the neurotransmitter *substance P*. When these neurons are selectively killed by a toxin targeted to the NK-1 receptor in rats, the animals develop an abnormal and unstable ventilatory pattern over several days, resulting in severe arterial hypoxia and hypercapnia. However, the animals survive, suggesting that other parts of the brain can compensate for the loss of the central pattern generator in the pre-Bötzinger, although the compensation is less than perfect.

Experiments can be done on the pre-Bötzinger *in vitro*, using slices of the brain from experimental animals that continue to generate a basic respiratory rhythm. Such studies have shown that the major inhibitory neurotransmitters, *GABA* and *glycine*, are not necessary for rhythm generation. This rules out a reciprocal inhibition network model for rhythm generation. Current evidence supports a *hybrid network model*, which incorporates pacemaker neurons and synaptic interactions that fine-tune the duration of, and transition between, inspiration and expiration. *Glutamate* is the major excitatory neurotransmitter within the central pattern generator, as well as between the rhythm generating neurons and *bulbospinal (premotor) neurons*, and at *spinal (e.g., phrenic) motor neurons*.

Other respiratory centers include the *ventral and dorsal respiratory groups* in the medulla. The *ventral respiratory group* (VRG) is a column of neurons that fire action potentials in phase with respiration. It includes neurons depolarizing during inspiration *(inspiratory or I neurons)* and expiration *(expiratory or E neurons)*. The pre-Bötzinger is in the rostral VRG. The dorsal respiratory group (DRG) contains mainly I neurons and is part of the nucleus of the solitary tract (or *nucleus tractus solitarius,* NTS). The afferent input from arterial chemoreceptors and lung mechanoreceptor synapses on neurons in the NTS near the DRG. The *pons* includes a group of neurons in the pontine respiratory group, which are involved in the phase transition between inspiration and expiration, and the reflex effects of lung mechanoreceptors on ventilation. *Apneusis* (abnormally long inspiration) can occur if the pons is lesioned in humans. All of these respiratory centers are bilaterally symmetrical on the right and left sides of the brain stem.

Efferent Pathways

Normal resting ventilation consists of the rhythmic contraction of the diaphragm during inspiration and

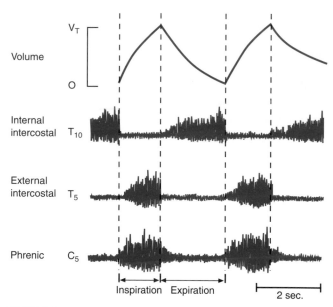

FIGURE 3 Ventilatory volume changes and the pattern of motor nerve activity to an expiratory muscle (internal intercostal, recorded from 10th thoracic spinal nerve), and two inspiratory muscles (external intercostal, recorded from 5th thoracic spinal nerve; and diaphragm, recorded from phrenic or 5th cervical spinal nerve). (Nerve activity after Hlastala and Berger, *Physiology of respiration,* Oxford, UK: Oxford University Press, 1996.)

expiration by passive elastic recoil of the lung and chest wall (see Chapter 19). Figure 3 shows this pattern of activity for the phrenic nerve (lower trace), which is the motor nerve innervating the diaphragm. The rhythm from the central pattern generator is synaptically transmitted to phrenic motor neurons at the third to fifth cervical levels of the spinal cord (C3–C5). Injuries that disrupt this normal flow of information between the medulla and C3–C5 spinal cord result in respiratory paralysis, so high quadriplegics require artificial ventilation. Phrenic nerve activity reflects basic features of the central pattern generator, including (1) a sudden onset of inspiratory activity, (2) a ramp-like increase in activity during inspiration, and (3) a relatively sudden termination of activity at the onset of expiration. Low levels of phrenic activity at the onset of expiration (Fig. 3) allow the diaphragm to smooth the transition to passive expiration.

Inspiratory and expiratory intercostal muscles are innervated by spinal nerves from all levels of the thoracic spinal cord (T1–T11). The pattern of electrical activity in the external intercostal nerves is similar to that in the phrenic nerve, whereas the internal intercostals are activated during expiration (see Fig. 3). Remember that expiration is passive at rest, so ventilation is elevated in Fig. 3 to illustrate intercostal activity. Electrical activity in lower thoracic and upper lumbar spinal nerves, which innervate the *abdominal expiratory muscles,* is similar to the internal intercostal activity shown in Fig. 3. Inspiratory and expiratory activity in the vagus and *hypoglossal (XII cranial) nerves,* which innervate the upper airway muscles in the *larynx* and *pharynx,* is also similar to the patterns shown in Fig. 3.

The patterns of motor neuron discharge, which determine ventilatory flow profiles and tidal volume, are shaped by central integration of respiratory reflexes in the brain stem. However, some central integration does occur in the spinal cord through *γ-efferent control of muscle spindles,* as described in Chapter 55. Afferent feedback from muscle spindles, and the modulation of this feedback by the *γ*-efferent system, is more important for the intercostal muscles than the diaphragm. This reflex may stabilize ventilation when mechanical loads are applied to the chest wall, and act in concert with reflexes from the lungs that fine-tune the pattern of ventilation to minimize the work of breathing under different conditions.

In addition to the ventilatory muscle pathways described, respiratory reflexes can stimulate responses from the *autonomic nervous system,* which are described separately in the Reflexes from the Lungs and Airways section.

VENTILATORY RESPONSE TO ARTERIAL P_{O_2}, P_{CO_2}, AND pH

The chemical control of ventilation is a negative feedback system that monitors arterial blood gases as the output from gas exchange in the lungs and causes reflex changes in ventilation that tend to keep arterial blood gases at a normal level. Arterial P_{O_2}, P_{CO_2}, and pH are sensed directly by arterial chemoreceptors, and Pa_{CO_2} is also sensed indirectly by chemoreceptors in the CNS. There is no strong evidence for mixed-venous or alveolar chemoreceptors that can affect breathing. Normally, the overall level of ventilation is determined by arterial blood gases, with Pa_{CO_2} being the most important stimulus. Differences in the ventilatory response to Pa_{O_2}, Pa_{CO_2}, and pH can be explained by differences in the sensory mechanisms for each stimulus, which are described next.

Central Chemoreceptors

A ventilatory response to increases in Pa_{CO_2} can be observed in experimental animals that have no afferent input to the CNS from any peripheral sensory nerves. This response to Pa_{CO_2} is mediated by *central chemosensitive areas* of the medulla, including the retrotrapezoid nucleus, raphé, nucleus of the solitary tract, and the ventral surface of the medulla at the fourth ventricle.

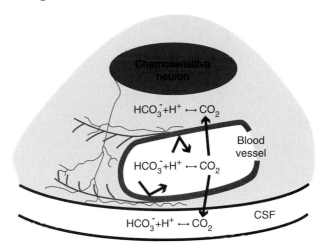

FIGURE 4 Central chemoreceptor cells sense changes in Pa_{CO_2} by a H^+ mechanism, but central chemoreceptors are not affected by changes in arterial pH because the blood–brain barrier prevents H^+ ion movement across capillaries in the brain. In contrast, P_{CO_2} easily crosses brain capillaries and changes pH in the chemoreceptor cells, in the interstitial fluid surrounding the cells, and in the CSF between the surface of the brain and the pia mater.

Many of the sites are near the respiratory centers described earlier (see the Respiratory Rhythm Generation section). Clearly delineated chemosensory organs, analogous to the arterial chemoreceptors described later, have *not* been identified in the CNS but experiments show certain neurons in these central chemosensitive areas are sensitive to changes in P_{CO_2} and pH.

A common feature of central chemosensitive neurons is that they have dendrites with endings near cerebral blood vessels. As shown in Fig. 4, these vessels and nerve endings are also frequently near the surface of the brain, which is bathed in *cerebral spinal fluid* (CSF). Specialized chemosensitive nerve endings depolarize in response to decreased intracellular pH, which occurs when arterial P_{CO_2} increases. CO_2 is very soluble in lipids so it moves easily across the capillaries and membranes in the brain and generates H^+ inside central chemosensitive neurons, as well as in the extracellular space and CSF around these neurons. In contrast, the so-called *blood–brain barrier* prevents polar molecules like H^+ (and certain drugs) from moving across the capillaries, or through tight junctions between capillary endothelial cells, in the brain. Hence, changes in arterial pH are not sensed by central chemosensitive neurons because the H^+ stimulus cannot reach the chemosensitive dendrites. This means that the ventilatory response to changes in arterial pH requires other chemoreceptors, i.e., the arterial chemoreceptors described later.

Although the mechanism is unknown, central chemosensitive neurons may decrease CO_2 sensitivity in response to long-term increases in arterial P_{CO_2} (e.g., in patients with chronic lung disease) and increase CO_2

sensitivity in response to chronic decreases in arterial P_{CO_2} (e.g., in normal subjects experiencing chronic hypoxia at altitude). Previously, it was thought that changes in central CO_2 sensitivity could be explained by metabolic compensations for respiratory disturbances in CSF pH. However, the pH of CSF does not show complete compensation to its normal value (7.3) during chronic respiratory acidosis or alkalosis, but remains acidotic or alkalotic. Renal compensation of respiratory changes in arterial pH could not correct CSF pH because metabolic acids do not cross the blood–brain barrier (see Fig. 4). Nevertheless, ventilatory reflexes from central chemoreceptors can adapt to chronic changes in P_{CO_2} "as if" the H^+ stimulus at central chemoreceptors changed in parallel with metabolic compensations of arterial pH, at least under some conditions (see Ventilatory Responses to Pa_{CO_2} section later).

Arterial Chemoreceptors

Arterial chemoreceptors is a generic term for both the *carotid body chemoreceptors* and *aortic body chemoreceptors*. The carotid bodies are small (diameter ≈ 2 mm) sensory organs located near the carotid sinus at the bifurcation of the common carotid artery at the base of the skull. The aortic bodies are on the aortic arch near the aortic arch baroreceptors. Afferent nerves travel to the CNS from the carotid bodies in the *glossopharyngeal (IX cranial) nerve,* and from the aortic bodies in the *vagus (X cranial) nerve.* Most of our knowledge about arterial chemoreceptors is based on studies of the carotid body, and the aortic bodies will be considered similar throughout this chapter unless noted otherwise.

Carotid body chemoreceptors respond to changes in arterial P_{O_2}, P_{CO_2}, and pH. These organs have the highest blood flow per unit mass, and this rich blood supply makes them efficient at sensing changes in arterial blood gases. Carotid bodies are the most important chemoreceptors for ventilatory sensitivity to Pa_{O_2} and they respond to changes in O_2 partial pressure, but *not* changes in O_2 content or hemoglobin saturation. This is critical to understanding clinical problems such as anemia or carbon monoxide poisoning. In these cases, ventilation will not be stimulated because Pa_{O_2} is normal, or even elevated, and there is no sensory system for decreases in O_2 content of hemoglobin saturation.

The mechanism of P_{O_2} sensing in the carotid bodies is not completely understood but it depends on specialized neurosecretory cells, called *glomus* (or *chief*) *cells.* The carotid body is a complex organ, consisting of glomus cells, glial-like *sustentacular* (or *sheath*) *cells,* capillary endothelial cells, afferent nerve endings from the glossopharyngeal nerve, and even sympathetic efferent nerve endings. Candidates for molecular O_2 sensors are specialized NADP(H) oxidases or cytochromes, or O_2-sensitive potassium channels, and O_2-sensing appears to involve signaling by reactive oxygen species. The glomus cells contain several types of neurotransmitters and neuropeptides. Hypoxia depolarizes these cells, causing release of an (unknown) excitatory neurotransmitter and excitation of the afferent nerve ending that sends action potentials to respiratory centers in the brain.

Changes in Pa_{O_2} are coded as changes in the frequency of action potentials in the afferent nerves from the arterial chemoreceptors. Figure 5 shows that carotid body afferent nerve activity has a low level of tonic activity with normal Pa_{O_2} (100 mm Hg) and Pa_{CO_2} (40 mm Hg). Arterial chemoreceptor activity does not increase until Pa_{O_2} falls below normal levels *if* Pa_{CO_2} is *normal.* However, carotid body chemoreceptors are also sensitive to Pa_{CO_2}. If Pa_{CO_2} is below normal levels, then Pa_{O_2} must decrease even further below normal to excite

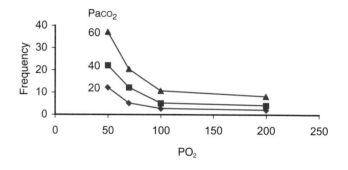

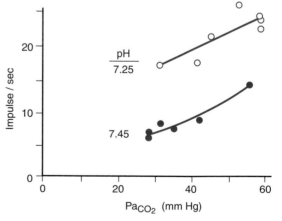

FIGURE 5   Carotid body chemoreceptor response (action potential frequency) to Pa_{O_2} and Pa_{CO_2} (upper) and Pa_{CO_2} and pHa (lower) for two *different* fibers. Increasing P_{CO_2} increases the response to P_{O_2}, and the effect of the two stimuli together is more than the sum of the individual effects, as explained in the text. Decreasing pH increases the response at any P_{CO_2}, but P_{CO_2} still has an effect when pH is held constant. (After Biscoe *et al., J Physiol* 1970;208:121; and Hornbein in Torrance, ed., *Arterial chemoreceptors.* New York: Blackwell Scientific, 1968.)

carotid body chemoreceptors. Conversely, carotid body chemoreceptors can be stimulated at higher P_{O_2} levels if Pa_{CO_2} is increased. The interaction between P_{O_2} and P_{CO_2} at carotid body chemoreceptors is synergistic, so hypoxia and hypercapnia increase the effect of each other as chemoreceptor stimuli. This synergistic, or multiplicative, effect at arterial chemoreceptors is important because it explains the multiplicative effect of Pa_{O_2} and Pa_{CO_2} on ventilation, as described later (see Ventilatory Response to Pa_{O_2} section).

The lower panel of Fig. 5 shows the effect of pH and P_{CO_2} on carotid body chemoreceptors. Pa_{CO_2} changes can affect chemoreceptor activity even if pHa is held constant, and vice versa. Aortic bodies in humans do not respond to pHa changes, and this is one exception to remember about the aortic bodies being qualitatively different from the carotid bodies. Therefore, the carotid bodies are the *only* chemoreceptors that respond to metabolic changes in pHa when Pa_{CO_2} is constant.

The mechanism of CO_2 and pH sensitivity may be a common response to intracellular pH. P_{CO_2} can diffuse into chemoreceptor cells and cause large changes in intracellular pH to stimulate the cells. Extracellular pH changes in blood cause smaller changes in intracellular pH, consistent with a smaller arterial chemoreceptor response to a metabolic pH change, compared with an equivalent respiratory pH change.

Arterial chemoreceptors respond rapidly within seconds to changes in Pa_{O_2}, Pa_{CO_2}, and pH. Changes in arterial blood gases that occur in phase with breathing, especially during conditions such as exercise, can be sensed by arterial chemoreceptors and may stimulate ventilation. This rapid response explains how ventilation can be altered within a single breath when arterial blood gases change. Carotid body chemoreceptors contain *carbonic anhydrase*, which will increase the speed of response to Pa_{CO_2} according to the intracellular pH sensing mechanisms described earlier.

Ventilatory Response to Pa_{CO_2}

In healthy humans under normal resting conditions, Pa_{CO_2} is the most important stimulus for ventilatory reflexes. Figure 6 shows the ventilatory response to changes in inspired CO_2 for normal individuals and how the response varies between individuals. The *hypercapnic ventilatory response* is plotted as expired ventilation versus Pa_{CO_2}, which is a noninvasive measure of the actual physiologic stimulus, i.e., Pa_{CO_2}. Ventilation approximately doubles when Pa_{CO_2} increases 5 mm Hg above the control value (0% inspired CO_2). The average increase in ventilation is about 2 L/(min · mm Hg) over the physiologic range of 40–50 mm Hg Pa_{CO_2}. The two dashed lines in Fig. 6 show the ventilatory response to

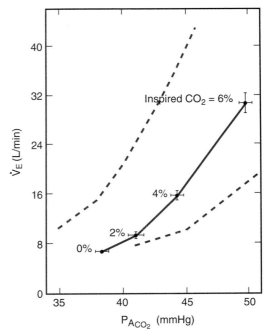

FIGURE 6   Ventilatory response to increases in inspired CO_2 plotted as a function of Pa_{CO_2}. Solid line is average for 33 normal subjects with standard deviation error bars. Dashed lines show the responses for subgroups of the whole population to demonstrate the large range of individual variability in normal subjects. (After Kellogg, Chap. 20 in Fenn and Rahn, eds., *Handbook of physiology, Respiration*. Bethesda, MD: American Physiological Society, 1964.)

CO_2 for two subgroups of the population used to establish the average curve. The range of differences between these groups illustrates the large variability that can occur in physiologic control systems between individuals, which may be under genetic control.

The hypercapnic ventilatory response can be measured in patients more easily by the *rebreathing method*. A person breathes in and out of a bag filled with 7% CO_2 and 93% O_2, while continuously measuring ventilation and end-expired P_{CO_2} ($\approx Pa_{CO_2}$ or Pa_{CO_2}). Metabolic CO_2, which is expired into the bag, progressively increases P_{CO_2} inspired from the bag, causing a progressive increase in Pa_{CO_2} and ventilation. High O_2 levels in the bag ensure high Pa_{O_2} to eliminate any potential effects of hypoxia on ventilation. The slope of the ventilation versus P_{CO_2} plot quantifies the hypercapnic ventilatory response.

Both tidal volume and frequency increase with Pa_{CO_2}, but tidal volume reaches a maximum before frequency. Further increases in ventilation at high Pa_{CO_2} levels depend on increases in respiratory frequency and this is less effective than increasing tidal volume at increasing alveolar ventilation (because of dead space; see Chapter 21). Hence, the hypercapnic ventilatory response is most efficient at low levels of Pa_{CO_2}, and larger increases in total expired ventilation are necessary at high levels of

Pa_{CO_2} when only frequency increases. This helps explain the increase in slope of the ventilatory response to CO_2 at high of levels CO_2 (see Fig. 5).

It is difficult to decrease Pa_{CO_2} below normal in an individual without changing other ventilatory stimuli, but ventilation can decrease if this does occur. This indicates that chemoreceptors are tonically active at normal Pa_{CO_2} levels (see Fig. 5), so hypocapnia decreases chemoreceptor stimulation and ventilatory drive. However, ventilatory sensitivity to low levels of Pa_{CO_2} is relatively low, as shown by the low slope of the hypercapnic ventilatory response between 0% and 2% inspired CO_2, compared with higher levels (see Fig. 6). Ventilation may actually cease if Pa_{CO_2} is lowered enough to go below the so-called *apneic CO_2 threshold*. It is estimated that about two-thirds of the drive to breathe in normal resting humans is from chemoreceptor stimulation and the balance is from mental state (i.e., *wakefulness*).

The ventilatory response to CO_2 is mainly a function of central chemoreceptor stimulation, which explains 75% of the total response. The rest of the response is explained by arterial chemoreceptors, which depend on stimulation by arterial P_{CO_2}, as well as H^+ from respiratory changes in arterial pH. The direct effect of P_{CO_2} on arterial chemoreceptors probably explains less than 10% of the total ventilatory response. The independent effect of arterial pH on ventilation and important interactions between Pa_{O_2} and Pa_{CO_2} as ventilatory stimuli are described later in separate sections on pH and P_{O_2}.

The time course of the ventilatory response to Pa_{CO_2} begins rapidly (within a few breaths of inhaling CO_2) from stimulation of arterial chemoreceptors. It requires a few minutes for ventilation to reach a steady value after Pa_{CO_2} stabilizes at a new level. This probably reflects the time it takes central chemoreceptors to respond to Pa_{CO_2} by an indirect H^+-sensing mechanism (Fig. 4). When Pa_{CO_2} is experimentally increased in normal subjects over hours to weeks, ventilation remains near the level measured during the first few minutes of exposure to CO_2. This occurs despite a reduction in arterial chemoreceptor stimulation, as H^+ decreases with renal compensation for the respiratory acidosis. Also, it is different than the case in patients with lung disease and chronically high Pa_{CO_2} (see Integrated Ventilatory Responses section later).

Ventilatory Response to Arterial pH

The ventilatory response to changes in arterial pH is the result of arterial chemoreceptor stimulation, and only the carotid bodies contribute to this response in humans. Patients without normal carotid body innervation do not have a response to metabolic changes in pHa. H^+ cannot cross the blood–brain barrier, so a metabolic acidosis or alkalosis cannot stimulate central chemoreceptors. The ventilatory response to physiologic changes in pHa of ±0.1 pH units is very small (less than 10% of resting ventilation). Larger increases in pHa (to 7.6) also only cause small decreases in ventilation. However, larger decreases in pH have a stronger effect, so resting ventilation may double at pHa = 7.2. In severe acidosis, ventilation increases even without the involvement of arterial chemoreceptors. Very large increases in arterial $[H^+]$ might allow some H^+ to cross the blood–brain barrier, or this could represent a generalized response to stress.

Increases in ventilation with *chronic* metabolic changes in pH are essentially the same as the increase with acute changes. For example, in a diabetic patient with long-term ketoacidosis, ventilation remains elevated despite (1) pH being compensated toward normal, which reduces H^+ stimulation of arterial chemoreceptors; (2) decreased P_{CO_2} stimulation of arterial chemoreceptors, after respiratory compensation of the metabolic acidosis; and (3) decreased central chemoreceptor stimulation from the low Pa_{CO_2}. Recall that metabolic H^+ changes in the blood are not sensed by the central chemoreceptors so the small arterial acidosis (from incomplete compensation) stimulates arterial chemoreceptors and provides the only known stimulus for ventilation. However, chronic decreases in Pa_{CO_2} may increase the sensitivity of central chemoreceptors and restore some of that ventilatory drive, at least in the case of normal subjects acclimatizing to chronic hypoxia (see Integrated Ventilatory Responses section later).

Ventilatory Response to Pa_{O_2}

The ventilatory response to Pa_{O_2} is called the *hypoxic ventilatory response,* and it is notable for its nonlinearity. Increases in Pa_{O_2} have relatively little effect on ventilation, whereas decreases in Pa_{O_2} below about 60 mm Hg cause large increases in ventilation (Fig. 7). This nonlinearity is the reason why Pa_{CO_2} is more important than Pa_{O_2} at controlling ventilation in normal individuals. Increases *or* decreases in Pa_{CO_2} from a normal value of 40 mm Hg are effective at changing ventilation (see Fig. 6). Normally, the hypoxic ventilatory response is not large until Pa_{O_2} falls to a level at which O_2-hemoglobin saturation starts decreasing significantly (Chapter 20) and ventilation is a linear function of arterial O_2 saturation. However, this linear relationship is a coincidence and not a mechanistic explanation because arterial chemoreceptors respond only to O_2 partial pressure, and *not* O_2 content or saturation (see Arterial Chemoreceptors section). The hypoxic

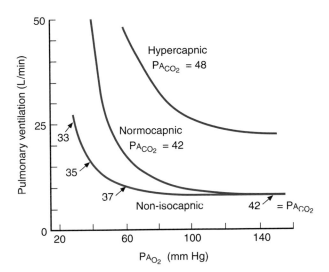

FIGURE 7 Ventilatory response to hypoxia plotted as a function of Pa_{O_2} ($\approx Pa_{O_2}$ in these normal subjects). Nonisocapnic curve: Pa_{CO_2} decreases when hypoxia stimulates increases in ventilation. Normocapnic curve: Hypoxic ventilatory response is increased if Pa_{CO_2} is held constant, by adding CO_2 to inspired gas. Hypercapnic curve: The ventilatory response to increased Pa_{CO_2} and decreased Pa_{O_2} is greater than the sum of the individual stimuli, as explained in the text. (After Loeschke *et al.*, *Pflug Arch Ges Physiol* 1958;267: 460.)

ventilatory response is explained completely by arterial chemoreceptor stimulation. In fact, patients with both of their carotid bodies removed have no hypoxic ventilatory response.

Figure 7 illustrates the important interactions that occur between Pa_{O_2} and Pa_{CO_2} as ventilatory stimuli. The lower response curve shows a normal ventilatory response to hypoxia where Pa_{CO_2} is decreasing as a result of hypoxic ventilatory stimulation. This nonisocapnic hypoxic ventilatory response would occur, for example, during exposure to progressively higher altitudes. As Pa_{O_2} decreases below 60 mm Hg, the arterial chemoreceptors are stimulated and cause a small increase in ventilation. This decreases Pa_{CO_2} and, thereby, ventilatory drive from central and arterial chemoreceptor stimulation. The net effect on ventilation is the result of stimulation by hypoxia and inhibition from decreased Pa_{CO_2} (i.e., 37 mm Hg). As Pa_{O_2} decreases more, hypoxic ventilatory stimulation increases, and this causes further reductions in Pa_{CO_2}. Ventilation remains a compromise between hypoxic stimulation and hypocapnic inhibition on the nonisocapnic hypoxic ventilatory response.

In contrast, if Pa_{CO_2} is held constant during hypoxia, for example, by increasing inspired P_{CO_2}, then the hypoxic ventilatory response is increased. This is shown on the middle normocapnic curve in Fig. 7. Although this curve is normocapnic, in the sense of maintaining Pa_{CO_2} at a normal value (42 mm Hg in this case), note that it is not the normal hypoxic ventilatory response;

hypoxic stimulation of ventilation normally decreases Pa_{CO_2}, as shown on the nonisocapnic curve. The increase in ventilation between the nonisocapnic and normocapnic curves is explained by the removal of hypocapnic inhibition and reflects the pure excitatory response to hypoxia.

Increasing Pa_{CO_2} above normal reveals a synergistic interaction between Pa_{O_2} and Pa_{CO_2}. On Fig. 7, the pure effect of increasing Pa_{CO_2} from 42 to 48 mm Hg equals the change in ventilation between the normocapnic and hypercapnic curves, at least at high Pa_{O_2} levels when there is no hypoxic stimulation. This 6 mm Hg increase in Pa_{CO_2} causes ventilation to increase, equaling 16 L/min at $Pa_{O_2} = 140$ mm Hg. The independent effect of decreasing Pa_{O_2} from 140 to 60 mm Hg equals the increase in ventilation along the normocapnic curve, which is about 10 L/min in Fig. 7. The combined effect of hypercapnia and hypoxia equals the increase in ventilation between the normocapnic and hypercapnic curves as Pa_{O_2} decreases from 140 to 60 mm Hg. The combined effect equals 42 L/min, which is greater than the sum of the individual effects (26 L/min). Therefore, the Pa_{O_2}–Pa_{CO_2} interaction is said to be *synergistic*, or *multiplicative*. The physiologic advantage of this multiplicative effect is that combined hypoxia and hypercapnia, which usually occur together with physiologic or pathologic limitations of gas exchange, increases ventilation more than either stimulus could alone.

The time course of the ventilatory response to changes in Pa_{O_2} begins within a single breath of breathing a low O_2 mixture, as expected, given the rapid response of arterial chemoreceptors. If hypoxemia persists for several minutes, ventilation may show a decrease relative to the acute increase but this varies between individuals. This secondary decrease in ventilation occurs when hyperventilation decreases Pa_{CO_2}, which induces slower changes in ventilation (see Ventilatory Response to Pa_{CO_2} section). However, even when Pa_{CO_2} is held constant, this secondary decrease in ventilation is observed, and it involves a process called *hypoxic ventilatory decline*. The physiologic mechanism of hypoxic ventilatory decline is not known yet. The time course of the ventilatory response to hours to years of hypoxia (at altitude) are considered in the Integrated Ventilatory Responses section.

REFLEXES FROM THE LUNGS AND AIRWAYS

Reflexes from the upper airways and lungs are primarily defense reflexes, which protect the lung from injury and environmental insults to the body. Pulmonary reflexes also adjust frequency and tidal volume to stabilize ventilation. The vagus nerve forms the afferent

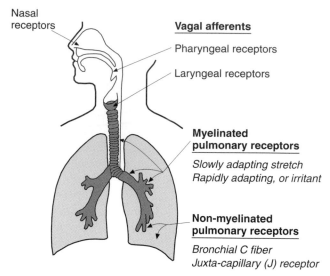

FIGURE 8   Mechanoreceptors and chemoreceptors in the upper airways and lungs that are important for respiratory reflexes. All of these sensory nerves travel in the vagus nerve, except the nasal receptors.

limb for most of these reflexes (Fig. 8). The following sections describe reflexes associated with different types of pulmonary receptors, where the term *receptor* refers to a specialized sensory nerve ending in the lungs—not a neurotransmitter or drug receptor. The efferent pathways for ventilation were described earlier, and autonomic efferent pathways in the lung are described at the end of this section.

Nose and Upper Airways

The *nose* has sensory nerves that transmit afferent information to the respiratory centers via the *trigeminal (V) cranial nerve* (Fig. 8). The ends of these sensory nerves respond to mechanical and chemical irritants in the nasal mucosa, so they are called *mechanoreceptors* and *chemoreceptors,* respectively. The *sneeze reflex* occurs when nasal mechanoreceptors are stimulated, for example, by inhaled dust, or nasal chemoreceptors are stimulated by noxious gases. Sneezing is a strong inspiration that is followed immediately by strong expiration, which directs air mainly through the nose to remove the offending stimuli. Stimulation of nasal chemoreceptors with water elicits the *diving reflex*. The diving reflex causes *apnea* (breath holding), laryngeal closure, and bronchoconstriction to protect the airways from water inhalation. A secondary cardiovascular reflex response to arterial chemoreceptor stimulation during apnea also slows the heart rate and diverts blood flow to vital organs such as the brain. This is considered an important part of the diving reflex because it conserves O_2 supplies in the body until breathing can resume safely.

The *pharynx* and *epipharynx* (the nasal passages just above the pharynx) contain vagal mechanoreceptors (Fig. 8). Mechanical irritants in the epipharynx stimulate the *aspiration,* or *sniff reflex,* consisting of several short and strong nasal inspirations in rapid succession. This sniffs material down into the pharynx where it can be coughed out or swallowed. Mechanoreceptor stimulation in the pharynx causes the *swallowing reflex.* Swallowing inhibits inspiration and closes the larynx, which protects the lungs, while the tongue and other muscles move food or liquid into the esophagus.

The *larynx* contains mechanoreceptors and chemoreceptors from the *recurrent laryngeal* and *superior laryngeal nerves,* which are branches of the vagus (Fig. 8). The laryngeal chemoreceptors are sensitive to inhaled noxious gases (e.g., ammonia and sulfur dioxide) and smoke, which stimulate coughing and the expiratory reflex. The *expiratory reflex* is a short and strong expiratory effort, but *coughing* also involves inspiratory activity. Liquid can also stimulate laryngeal chemoreceptors to cause apnea, which protects the lungs from inhaling fluids.

Laryngeal mechanoreceptors respond to changes in airway pressure, upper airway muscle contraction, and temperature. Airway temperature can change with inspired gas temperature, ventilation rate and the velocity of air flow, and mouth versus nose breathing. Stimulation of laryngeal mechanoreceptors causes reflex changes in upper airway muscle tone, which decrease *airway resistance* and prevent *upper airway collapse* with negative pressures during inspiration. A short and strong inspiratory effort results in either a *sigh* or a *hiccup,* depending on whether or not the upper airway muscles are simultaneously activated. Changes in upper airway reflexes during sleep can cause *snoring* and *sleep apnea* (see Clinical Note).

Lungs and Lower Airways

Figure 8 show how the vagal sensory nerves from the lungs and lower airways fall into two functional groups: *myelinated* and *nonmyelinated pulmonary receptors.* Myelinated nerves conduct action potentials rapidly and are generally involved in fine motor control and rapid defense responses. Nonmyelinated nerves conduct action potentials more slowly and are involved in slower defense reflexes.

Pulmonary Stretch Receptors

The vagus contains myelinated afferents called *slowly adapting pulmonary stretch receptors* (sometimes abbreviated PSR or SAR), which are stimulated by changes in lung volume (see Fig. 8). These mechanoreceptors are located in the smooth muscle of the trachea and intrapulmonary airways. Stretch depolarizes these receptors, sending action potentials to respiratory centers in the brain via the vagus nerve. If volume is increased rapidly and maintained at a new level, the frequency of action potentials increases rapidly and then settles to a slightly lower frequency. However, the steady-state frequency is proportional to the steady-state volume, and the receptors are described as slowly adapting because frequency does not completely adapt back to the basal rate. Slowly adapting pulmonary stretch receptors are tonically active at functional residual capacity (FRC), so they can send afferent information about increases *or* decreases in lung volume to the CNS.

Slowly adapting pulmonary stretch receptors are involved in the control of tidal volume and respiratory frequency through the *Hering-Breuer,* or *inflation inhibitory, reflex.* Increasing lung volume causes increased action potential frequency from pulmonary stretch receptors. This afferent signal inhibits further inspiratory nerve activity and terminates an inspiration through synaptic mechanisms in the pons and medulla. Hence, the inflation inhibitory reflex limits a breath from being larger or longer than necessary to achieve a given level of ventilation. This reflex is of historic interest because it was the first description of negative feedback in a physiologic control system in 1868. It is of physiologic interest because it explains the effect of the vagus on the pattern of breathing.

Cutting the vagus nerves in anesthetized animals results in slow, deep breathing, as predicted by the inflation inhibitory reflex. In contrast, removing all pulmonary afferent input in adult humans with a total lung transplant does not cause slow deep breathing. There is evidence that the inflation inhibitory reflex is not important in awake adults unless tidal volumes exceed FRC by more than 1 L. However, Fig. 9 shows that pulmonary stretch receptor activity can be important in adults during sleep. The inflation inhibitory reflex is greatly reduced in lung transplant patients during sleep, compared with normal individuals. This could reflect changes in central integration of respiratory reflexes during the sleep state or changes in chest wall mechanics. The inflation inhibitory reflex is more powerful in neonates and experimental animals, which have more compliant chest walls than adult humans.

Figure 9 also shows that the pattern of ventilation is more irregular in lung transplant patients. This also occurs in *awake* lung transplant patients and experimental animals with pulmonary denervation. Hence, *pulmonary stretch receptors decrease breath-to-breath variations in tidal volume and frequency* for a given level of ventilation. This fine-tuning of the ventilatory pattern is hypothesized to reduce the mechanical work of breathing.

Normal subject

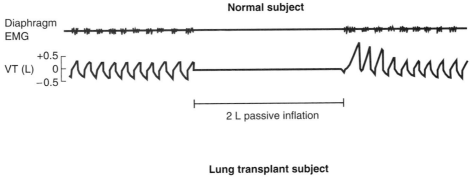

Lung transplant subject

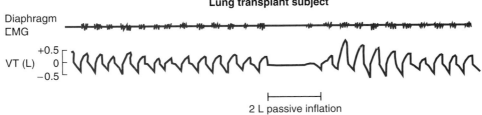

FIGURE 9 Ventilatory efforts (diaphragm EMG) and ventilation (tidal volume, Vт changes) in a normal subject (upper) and a bilateral lung transplant patient (lower panel) during sleep. Passive lung inflation inhibits ventilatory efforts for 40 sec (length of inflation bar) in normal patient by inflation inhibitory reflex; lung inflation is removed at the first sign of ventilatory effort on the EMG so breathing resumes. The vagus nerves are cut during lung transplantation so this reflex is virtually absent in the patient. Vagal denervation also increases variability in the breathing pattern of the lung transplant patient compared to the normal subject. (After Iber *et al., Am J Respir Crit Care Med* 1995;152:217.)

Pulmonary stretch receptors are also involved in the *deflation reflex,* which increases respiratory rate at low lung volumes. Recall that pulmonary stretch receptors are tonically active at FRC, so at low lung volumes the afferent input to respiratory centers is decreased.

Pulmonary Irritant Receptors

Pulmonary *irritant receptors* are the second type of myelinated vagal afferent in the lung (see Fig. 8). These sensory nerves are in the airway epithelium and respond to inhaled chemical irritants such as smoke, dust, and ammonia vapor. Irritant receptors also respond to endogenous chemical stimuli such as *histamine,* which is released from *mast cells* in the airways. It is hypothesized that irritant receptors may play a role in asthma because histamine is released during asthma. The reflex response to stimulating irritant receptors in the trachea and large bronchi includes (1) coughing, (2) mucous production, and (3) bronchoconstriction. This three-part response serves to remove noxious inhaled substances and protect the lungs from further exposure.

Irritant receptors also respond to mechanical stimuli, such as changes in lung volume. However, in contrast to slowly adapting pulmonary stretch receptors, irritant receptors are more sensitive to dynamic changes in volume than absolute volumes. As lung volume is increased, action potential frequency from irritant receptors initially increases but then *rapidly adapts* back toward the basal rate, despite the volume change being maintained. Hence, another name for irritant receptors is *rapidly adapting pulmonary stretch receptors* (sometimes abbreviated RAR). It is possible that the response of irritant receptors to some chemical stimuli results from mechanical changes in the receptor's microenvironment.

Mechanical stimulation of irritant receptors includes decreased lung compliance and deflation of the lung. Hence, irritant receptors may be involved in the *deflation reflex* described earlier. Stimulating irritant receptors in small bronchi can elicit the *gasp reflex.* This is an unusual case of positive feedback, in which irritant receptor stimulation by large rapid increases in lung volume causes a further increase in tidal volume, or a gasp. The physiologic significance of this reflex is not clear.

Bronchial C-Fibers

C-fiber is another term for a nonmyelinated fiber, where the *C* designates conduction velocity in an alphabetical system ($C \leq 2.5$ m/sec). Pulmonary C-fibers can be defined further by their blood supply. *Bronchial C-fibers* are supplied by the systemic circulation, in contrast to juxtacapillary C-fibers, which are supplied by the pulmonary capillaries (see Fig. 8).

Bronchial C-fibers can be stimulated by chemicals injected in the bronchial circulation, such as capsaicin (the hot ingredient in red chilies) and phenyldiguanide (a serotonin receptor agonist). Physiologically, bronchial C-fibers are probably stimulated by local release of cytokines, such as histamine, prostaglandin, and bradykinin in the airways.

The reflex response to bronchial C-fiber stimulation is an airway defense reflex, including rapid shallow breathing, bronchoconstriction, and mucous secretion. Bronchial C-fibers may contribute to the cough reflex also. Finally, bronchial C-fibers are involved in bronchoconstriction and changes in vascular permeability in the airways with airway inflammation. Bronchoconstriction involves both autonomic effectors (see section titled Autonomic Nervous System in the Lungs) and a local axon reflex. The *axon reflex* is a local response to a neuropeptide (substance P) released from a sensory nerve ending.

Juxtacapillary Receptors

Pulmonary vagal C-fibers that can be stimulated by chemicals in the pulmonary circulation are called *juxtacapillary receptors,* or *J-receptors,* because of their presumed location next to the pulmonary capillaries. J-receptors can be stimulated by capsaicin injected in the pulmonary artery, and the reflex response is *tachypnea,* or rapid shallow breathing. *Apnea* (no breathing) may precede tachypnea depending on the dose and timing of chemical stimulation. The cardiovascular system also responds with *bradycardia* (a decrease in heart rate) and hypotension.

Physiologically, J-receptor stimulation occurs with pulmonary embolism, congestion, and edema, and causes the rapid shallow breathing observed in these conditions. J-receptor stimulation probably explains the tachypnea and sensation of breathlessness *(dyspnea)* with interstitial lung disease also. Nonmyelinated vagal afferents are responsible for all sensation, including pain, from the lower airways.

Autonomic Nervous System in the Lungs

The airways in the lungs receive both *parasympathetic* and *sympathetic* innervation, which controls bronchial smooth muscle constriction, mucous secretion, vascular smooth muscle, fluid transport across the airway epithelium, and vascular permeability in the pulmonary and bronchial circulations. In normal conditions, parasympathetic control of bronchial smooth muscle tone is the most important of these functions. Acetylcholine released from the vagus nerve causes bronchoconstriction, and tonic vagal activity determines bronchial smooth muscle tone. Sympathetic innervation of the lung is less important. Circulating epinephrine from the adrenal medulla causes bronchodilation through *β-adrenergic receptors* on airway smooth muscle. Norepinephrine released from sympathetic nerves in the airways can cause bronchodilation indirectly via *α-adrenergic receptors* on parasympathetic ganglia in the lung. Activating these α-adrenergic receptors inhibits parasympathetic activity and cholinergic bronchoconstriction.

These different autonomic mechanisms provide a physiologic basis for treating bronchoconstriction in lung disease. Chronic bronchoconstriction from chronic obstructive pulmonary disease is treated with *acetylcholine receptor* antagonists to reduce the effects of vagal tone. Acute and severe bronchoconstriction during an asthma attack is treated with β₂-adrenergic agonists. β₂-Adrenergic agonists are selective for bronchial smooth muscle and have fewer cardiac effects.

Neuropeptides are also important in controlling airway function. The bronchoconstriction from substance P released directly from sensory nerves by the axon reflex was described earlier. This is also called the *excitatory nonadrenergic, noncholinergic system (e-NANC)* to distinguish it from autonomic control. In contrast, the neuropeptide *VIP (vasoactive intestinal peptide)* causes bronchodilation. VIP is released from

nerves arising from parasympathetic ganglia, which are called the *inhibitory nonadrenergic, noncholinergic system (i-NANC)*. Nitric oxide (NO) may be involved in i-NANC bronchodilation also.

INTEGRATED VENTILATORY RESPONSES

Integrated ventilatory responses to changes in activity and the environment illustrate many important interactions between elements of the respiratory control system. Such interactions are typical in patients, and the physician needs to understand them to make an intelligent diagnosis and provide appropriate treatments. The most common stimulus to increase ventilation in healthy subjects is exercise. Relatively common problems in control of ventilation during sleep are described in the Clinical Note on sleep apnea. The next sections compare and contrast the integrative response to chronic hypoxemia in normal subjects during acclimatization to high altitude and patients with lung disease.

Acclimatization to High Altitude

Questions about how the body responds to decreased O_2 supply at high altitude have fascinated respiratory

---

Clinical Note

Sleep apnea is a group of conditions characterized by pauses or reductions in breathing during sleep that last 10 sec or more. In most patients this leads to arterial hypoxemia and CO_2 retention. When hypoxic or hypercapnic stimuli reach a sufficiently high level, breathing efforts increase and this arouses or awakens the individual. Apneas and arousals can occur repeatedly and cause sleep fragmentation and excessive daytime sleepiness. Cognitive and neurological effects may result from sleepiness or direct effects of intermittent hypoxia during recurrent apneas.

The two general classes of sleep apnea are central and obstructive. *Central apneas* occur when the efferent signal to the diaphragm is insufficient to trigger inspiration. Central apneas occur in all sleep stages but are more common in the early stages of *NREM sleep* (NREM stands for non-rapid eye movement). During NREM sleep (also called *quiet sleep*), a generalized decrease in neural and metabolic activity occurs. In NREM

sleep the threshold for the ventilatory response to CO_2 increases. Coupled with a drop in metabolism, this pushes $Paco_2$ below the apneic threshold (see Ventilatory Response to CO_2 section) and causes central apneas. $Paco_2$ increases during apnea and will eventually exceed the apneic threshold, so breathing resumes. Patients with central apnea also tend to have abnormally low ventilatory responses to CO_2 during wakefulness. The most extreme example of a central sleep apnea is Ondine's curse, which is the complete cessation of breathing in a patient whenever he goes to sleep. Pure central sleep apnea is relatively rare, occurring in only 5% of patients suffering from sleep apnea.

Obstructive sleep apnea is more common and it occurs when the upper airways collapse during inspiratory efforts. Patients with obstructive sleep apnea are frequently sleepy during the day and almost always have a history of loud snoring. Obstructive sleep apnea occurs in almost 5% of

Clinical Note (continued)

the middle-aged working population, and can occur in 35–50% of some groups, such as diabetics, the morbidly obese, or the elderly (>65 years of age). Snoring is a mild form of upper airway obstruction, where the soft palate vibrates during inspiration at 40–60 Hz and impedes airflow. Obstructive apneas can occur during all sleep stages, however, often they are longer and result in more severe hypoxemia during REM (rapid eye movement) sleep. REM sleep is also called active sleep and is the sleep stage during which dreaming occurs. During REM, there is atonia (lack of muscle tone) and a general inhibition of sensory input, which inhibits afferent information from the lungs and ventilatory muscles, causing breathing to be more irregular.

Obstructive apnea can occur when reflexes do not respond to the normal negative pressure in the upper airways during inspiratory flow and induce the normal contractions of upper airway muscles to support the airways in an open position. Increased inspiratory effort, for example, in response to chemoreceptor stimulation, makes the tendency for airway collapse worse if the upper airways do not respond also. Therefore, the fundamental problem in obstructive apnea is a lack of coordination between the inspiratory and upper airway muscles and this is worst in REM sleep. The most effective treatment for obstructive apneas is nasal continuous positive airway pressure

(nasal CPAP). This treatment applies a positive pressure to the upper airways by a mask fitted over the nose of the sleeping patient, to counteract the decrease in airway pressure during inspiration and support the airways in an open position.

Sleep apnea also occurs in other conditions also. High-altitude sleep apnea may occur in normal subjects when sleeping on the mountains because of instabilities in the ventilatory control system. The ventilatory drive from P_{aO_2} is increased at high altitude, but the ventilatory drive from P_{aCO_2} is decreased (because hypoxia stimulates hyperventilation). These changes can lead to periodic central apneas when other ventilatory drives are decreased during sleep. Abnormal interactions in ventilatory control may also be involved in sudden infant death syndrome (SIDS). SIDS, or crib death, refers to the unexplained death of an infant during sleep. SIDS probably results from multiple causes, but an immature ventilatory control system that fails to arouse an infant during sleep apnea is certainly one cause. The risk for SIDS decreases with age, as breathing becomes more regular. Periodic breathing (recurrent apneas) occurs in 40–50% of premature infants, and in 90% of babies delivered at 28–29 weeks of gestational age. There is currently no way to predict SIDS, and treatment consists of carefully monitoring infants who have shown signs of sleep apnea.

---

physiologists for more than a century. The seminal studies of Bert, a French physiologist and contemporary of Claude Bernard in the late 1800s, established that the respiratory response to altitude is a response to hypoxia, or decreased inspired P_{O_2}. Barometric pressure decreases with altitude, but the body does not respond to hypobaric conditions per se. Rather, physiologic control systems monitor the decrease in partial pressure of O_2, which decreases with barometric pressure: $P_{IO_2} = F_{IO_2} (P_B - P_{H_2O})$, where P_B = barometric pressure and $F_{IO_2} = 0.21$ in air at any altitude.

Acclimatization occurs during prolonged exposures to altitude and is a physiologic response in an individual, as opposed to evolutionary or genetic adaptation occurring over generations. Acclimatization to hypoxia involves multiple systems, including changes in hemoglobin (Hb-O_2 affinity, Hb concentration, and

erythropoiesis; see Chapter 20), the microcirculation, and cellular metabolism. This section focuses on changes in the ventilatory control system with chronic hypoxia.

Ventilatory acclimatization is a time-dependent increase in ventilation occurring over hours to weeks of continuous hypoxic exposure. Figure 10 shows how ventilation and P_{aCO_2} change during 8 hr of continuous exposure to $P_{IO_2} = 88$ mm Hg, corresponding to an altitude of 13,500 ft (4115 m) above sea level. Initially, P_{aO_2} decreases and stimulates arterial chemoreceptors, which cause a rapid reflex increase in ventilation. Metabolic rate is unchanged in humans resting at such altitudes, so increased ventilation causes a decrease in P_{aCO_2}. This decrease in P_{aCO_2} inhibits ventilatory drive, which limits the hypoxic ventilatory response (see Fig. 7). Hence, the acute response to altitude is a compromise

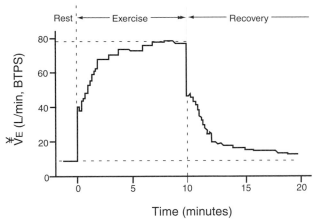

Time (minutes)

FIGURE 10 The time course of changes in ventilation and P_{aCO_2} in a normal male subject during 8 hr of breathing 12% O_2, to simulate an exposure to a 13,500-ft (4115-m) altitude. Values measured during normoxia (before and after hypoxia) are circled. Note the immediate increase in ventilation and decrease in P_{aCO_2} at onset of hypoxia, and slower changes during sustained hypoxia, indicating ventilatory acclimatization. When normoxia is restored, ventilation remains elevated and P_{aCO_2} remains low, but these values return toward the original control levels over a similar time course to the onset of ventilatory acclimatization (not shown).

between ventilatory stimulation by hypoxia and ventilatory inhibition by decreased P_{aCO_2}. *Hypoxic ventilatory decline* (see Ventilatory Response to P_{aO_2} section) will also limit the increase in ventilation during early exposure to altitude.

If hypoxia is sustained for several hours, ventilation increases further so P_{aCO_2} decreases more. An additional 5 mm Hg decrease in P_{aCO_2} (e.g., Fig. 10) will increase P_{aO_2} about 6 mm Hg according to the alveolar gas equation (Chapter 21). This is an obvious benefit in hypoxia and it is remarkable that ventilation increases when hypoxic stimulation of arterial chemoreceptors actually decreases and hypocapnic (low P_{aCO_2}) inhibition of arterial and central chemoreceptors increases. If the hypoxic exposure is extended beyond 8 hr, then ventilation continues to increase toward a plateau value over 2 days, but 2 weeks are required for full ventilatory acclimatization. Hence, the stimulus–response relationships of the arterial chemoreflexes are enhanced by chronic hypoxia so ventilation is significantly greater than during acute hypoxia.

The physiologic mechanisms of ventilatory acclimatization to hypoxia are not completely understood, but it is clear that they involve both arterial chemoreceptors and the CNS. Animal experiments show that carotid body chemoreceptors become more sensitive to P_{aO_2} during prolonged exposure to hypoxia. This will increase afferent input (action potential frequency) for any given level of P_{aO_2} and contribute to an increased hypoxic ventilatory response, and resting ventilation, after a few days at altitude. A second factor increasing

arterial chemoreceptor stimulation during prolonged hypoxia is renal compensation for primary respiratory alkalosis that occurs with hyperventilation (see Chapter 20). Arterial pH will decrease back toward the normal value over hours to days, increasing arterial chemoreceptor stimulation for any given level of P_{aO_2} and P_{aCO_2} (see Fig. 7). Finally, changes appear to happen in the CNS responsiveness to sensory input from arterial chemoreceptors during sustained hypoxia, which results in a greater ventilatory response for any given afferent input. The mechanism for this increased responsiveness of respiratory centers in the brain during chronic hypoxia is not known.

Changes in P_{aCO_2} during acclimatization to hypoxia also support increased CNS responsiveness. When normoxia is restored after prolonged hypoxia, P_{aCO_2} remains lower than normal (see Fig. 10) despite removal of the hypoxic stimulus. Arterial chemoreceptors should not be stimulating ventilation at such high P_{aO_2} and low P_{aCO_2} levels so a change in central chemoreceptor function is likely. As discussed earlier in the Central Chemoreceptors section, changes in CSF pH cannot explain this increase in CO_2 sensitivity. However, the ventilatory response to P_{aCO_2} during acclimatization to altitude changes "as if" the H^+ stimulus at central chemoreceptors changed in parallel with metabolic compensations of the respiratory alkalosis.

In contrast to the increased hypoxic ventilatory response described for acclimatization to hypoxia, individuals who live at high altitude for years show a *decreased* ventilatory response to hypoxia, at least until P_{aO_2} reaches extremely low levels (Fig. 11). This change in the ventilatory response in high-altitude residents and natives is called *hypoxic desensitization,* or *blunting of the hypoxic ventilatory response.* In high-altitude natives there is a genetic component to the low hypoxic ventilatory response, and it is hypothesized that evolution reduced ventilation and the work of breathing because other steps in the oxygen transport chain adapted to hypoxia. However, hypoxic desensitization in lowlanders who live at high altitude for years indicates that these changes can occur by physiologic, as well as genetic, mechanisms.

Chronic Lung Disease

Chronic hypoxemia can result from several forms of heart and lung disease. One example is *chronic obstructive pulmonary disease* (COPD), which is a general term for chronic bronchitis and emphysema. Both of these pathologies can lead to airway obstruction, $\dot{V}/\dot{Q}$ mismatching, and hypoxemia (Chapter 21). This severely limits the ability to exercise and patients with $P_{aO_2} < 60$ mm Hg may qualify for supplemental

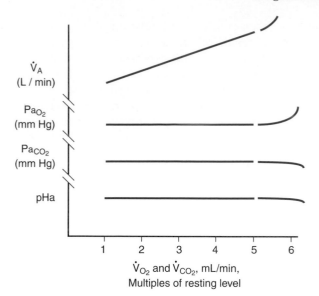

FIGURE 11   Hypoxic ventilatory response measured in normal
adults living at low altitude (controls), adults who were born at low
altitude but who have lived at 3100-m altitude for 3–39 years (high-
altitude residents), and adults who were born at 3100-m altitude and
lived there all of their lives (high-altitude natives). The hypoxic
ventilatory response is "blunted" between 100 and 60 mm Hg Pa_{O_2} in
both high-altitude groups. This must be a physiologic mechanism in
high-altitude residents, but a genetic component may be involved in
high-altitude natives. (After Weil *et al., J Clin Invest* 1971;50:186.)

oxygen, called *long-term oxygen therapy* (LTOT). To put
this in some perspective, a healthy person has $Pa_{O_2} < 60$
mm Hg when breathing room air at rest at 12,500 ft
(3800 m) above sea level.

Ventilatory control in COPD patients is dominated
by hypoxic stimulation of arterial chemoreceptors, in
contrast to Pa_{CO_2} being the most important determinant
of resting ventilation in healthy subjects. For example, if
a patient with COPD has an acute health problem that
further impairs gas exchange (e.g., pneumonia), then it is
often counterproductive to administer supplemental
oxygen. In some patients, supplemental oxygen causes
ventilation to *decrease*, presumably by relieving hypoxic
stimulation of arterial chemoreceptors. This will
increase Pa_{CO_2} but it does not appear to stimulate
central chemoreceptors as it would in normal
subjects. The central chemoreceptors seem to be desensitized to increases in Pa_{CO_2}. Extremely high P_{CO_2} levels

(80 mm Hg) may lead to *respiratory failure* and further
depression through a narcotic effect of CO_2.

Patients with COPD can be classified into two
groups: normal and elevated Pa_{CO_2}. *CO_2 retention*
(chronic increases in Pa_{CO_2}) is often associated with a
decreased ventilatory response to CO_2. It is difficult to
determine if this represents a time-dependent decrease in
the responsiveness of central chemoreceptors, a deficiency in the motor output from the respiratory centers,
or reduced function of ventilatory muscles. Increasing
airway resistance decreases the ventilatory response to
Pa_{CO_2} in normal individuals, suggesting that patients
with obstructive lung disease might retain CO_2 to
minimize the work of breathing. Ventilation in patients
with COPD and CO_2 retention behaves "as if" the
increased Pa_{CO_2} is no longer increasing H^+ ion in the
central chemoreceptors. However, pH in the CSF, at
least, remains acidotic in people with chronic increases
in Pa_{CO_2}, so the mechanism of decreased central
chemoreceptor responsiveness in some COPD patients
remains unknown.

Ventilatory acclimatization would be expected to
reduce the physiologic consequences of chronic hypoxemia, but it is not known if the changes described for
ventilatory control in the previous section occur in
patients with chronic hypoxemia or not. It is possible
that abnormal ventilatory acclimatization in only some
patients contributes to differences in the degree of
hypoxemia and CO_2 retention between patients.
Certainly genetic differences exist in the acute ventilatory responses to arterial blood gases in healthy
individuals.

Suggested Readings

Cherniack NS, Widdicombe JG, eds. Control of breathing, Vol II. In
    Handbook of physiology: Section 3, The respiratory system.
    Bethesda, MD: American Physiological Society, 1986.
Dempsey JA, Forster HV. Mediation of ventilatory adaptations.
    Physiol Rev 1982;62:262–346.
Dempsey JA, Pack AI, eds. Regulation of breathing, Vol 79, 2nd ed. In
    Lung biology in health and disease. New York: Marcel Dekker,
    1995.
Phillipson EA. Disorders of ventilation. In Isselbacher KJ, Braunwald
    E, Wilson JD, *et al.,* eds., *Harrison's principles of internal medicine,*
    13th ed. New York: McGraw Hill, 1994, pp 1234–1240.

PART IV

RENAL PHYSIOLOGY

23

Functional Anatomy of the Kidney and Micturition

JAMES A. SCHAFER

KEY POINTS

- The *cortex* of the kidney is the outer, richly vascular layer; the *medulla* is the inner, less vascular layer.
- The *ureter* exits the kidney in the same region as the *renal artery* and *renal vein* and conveys urine to the bladder.
- The *glomerular capillaries* and the *peritubular capillaries* form the microcirculation of the kidney. Fluid is filtered from the high-pressure glomerular capillaries, whereas solutes and fluid are absorbed from the interstitial space by the low-pressure peritubular capillaries.
- The *afferent arterioles* and *efferent arterioles* control pressure in the capillaries.
- Each of the nephrons (there are about 1 million in each kidney) is divided into functional units, beginning with *Bowman's space* in the glomerulus, where the initial filtrate collects and enters the *proximal tubule*.
- The *ascending loop of Henle* in the nephron returns from the medulla and extends to

   contact the arterioles in the glomerulus of its origin. This specialized region of contact is called the *juxtaglomerular apparatus* and is involved in regulating secretion of the hormone renin.
- *Superficial nephrons* are so named because their glomeruli lie in the outer region of the cortex and they have short loops of Henle, whereas *juxtamedullary nephrons* arise from glomeruli in the minor cortex and have long loops of Henle.
- The kidney is essential for the conversion of vitamin D_3 to its most active form, *1,25-dihydroxycholecalciferol*, and for the production of *erythropoietin*, an endocrine hormone that stimulates the production of red blood cells. The kidney also releases *angiotensin II* and the prostaglandins PGE_2 and PGI_2, which play important rolls in the maintenance of extracellular fluid volume and blood pressure.

The kidney is of paramount importance in the maintenance of homeostasis. Despite its relatively small mass, the kidney processes enormous amounts of solute and water daily and finely regulates the excretion of most substances to match their rates of ingestion and production by metabolism.

Data compiled for the 2002 Annual Report of U.S. Renal Data System (http://www.usrds.org/adr.htm) show that in 2000 (the latest year for which complete data were available) more than 96,000 people in the United States developed *end-stage renal disease* (ESRD), a condition in which the patient has little or no kidney function and would die without hemodialysis or a kidney transplant. In that same year, more than 275,000 patients received hemodialysis and more than 103,000 were sustained by a successful renal transplant. Treatment of ESRD in 2000 cost more than $12.3 billion in Medicare and even more in private, charitable, and other public health care funds! The increasing prevalence and incidence of ESRD are especially disappointing in view of the fact that diabetes mellitus (especially type 2 diabetes) and hypertension accounted for, respectively, ~44% and ~25% of the cases. Early treatment in both types of cases would greatly reduce the enormous human and monetary costs of ESRD. Furthermore, ESRD represents only the tip of the iceberg among chronic and acute renal diseases. Many nonfatal conditions such as urinary incontinence and other disorders of urination severely compromise the quality of life for millions.

One usually thinks of the kidney in terms of its role in ridding the body of metabolic by-products, and this function is certainly important. When the operation of the kidneys is compromised to a sufficient extent, numerous metabolites such as urea, creatinine, uric acid, sulfate, and phosphate begin to accumulate in the body fluids. The accumulation of these and other substances in blood leads to the syndrome called uremia, which signals the progression of renal failure. The kidneys are also a primary component of the body's defense against toxins and other foreign substances in the environment. The kidney excretes many of these substances and their metabolites, whether they are taken into the body directly or produced by the metabolism of other substances.

Despite the importance of the excretion of metabolic waste products and potential toxins, the threat to life in renal failure typically comes not from the accumulation of metabolic wastes or environmental toxins, but from the loss of the body's ability to balance the daily intake of salts and water by an appropriate rate of excretion. In the patient with renal failure, a frequent presenting symptom is the edema that results from the resulting retention of salt and water. The expansion of the blood plasma volume increases the workload of the heart and eventually leads to heart failure and pulmonary edema. These events are often complicated by *acidosis* and *hyperkalemia* (a high blood K^+ concentration), which result from the inability of the kidneys to excrete acids and potassium at the proper rates.

To understand the progression of these disorders and their management, it is necessary to understand the normal functions of the kidney and the location and mechanisms that underlie those functions. But even more important is an understanding of how these processes are regulated by a variety of hormonal, neural, and local control mechanisms to respond appropriately to various factors that tend to disturb homeostasis.

FUNCTIONAL ANATOMY OF THE KIDNEY

A person's kidney is about the size of a clenched fist. When examined in cross section as shown in Fig. 1, the kidney is easily divided into two regions: the *cortex* and the *medulla*. The blood, lymphatic, and neural supply of the kidney enter through the *hilus* together with the *ureter,* which carries the urine from the kidney to the bladder, where it is stored until emptied by *micturition* (urination).

In the human kidney, the medulla terminates in several conical structures called *papillae*. The *renal pelvis* is essentially an enlarged extension of the ureter. It is divided into individual cup-shaped structures called *calyxes* that surround each papilla and convey the effluent urine into the ureter.

A layer of tough connective tissue called the *capsule,* which protects the more delicate parenchyma, covers the kidney. About 1 million nephrons constitute the

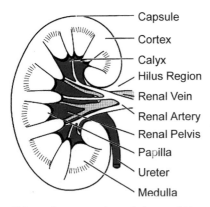

FIGURE 1   Schematic cross section of a human kidney. The ureter, renal artery, and renal vein exit the kidney in the hilus region. The calyces collect the urine as it flows out of the ducts of Bellini (see Fig. 4) at the tips of the papillae and convey it into the ureter, which carries it to the urinary bladder.

majority of the parenchyma in each human kidney. As discussed in detail in subsequent chapters, the nephron is the functional unit that produces an initial ultrafiltrate of the plasma at its point of origin in the glomerulus and modifies that ultrafiltrate by the processes of reabsorption and secretion to control the rate of excretion of solutes and water.

Vasculature

Although the two kidneys together constitute less than 0.5% of the body mass, they receive almost 25% of the total cardiac output when the body is at rest. This blood flows into each of the two kidneys via the right and left renal arteries, which branch directly from the abdominal aorta. On entering the hilus, the renal artery divides into several interlobar arteries that radiate from the hilus toward the cortex (Fig. 2). Near the boundary of the cortex and the medulla, the interlobar arteries divide into arcuate arteries that run parallel to the arc of the corticomedullary junction and give off the radial arteries. These major vessels of the arterial side are paralleled in their course by the interlobar, arcuate, and radial veins. The interlobar veins come together to form the renal vein, which exits from each kidney at the hilus and carries the venous blood to the vena cava.

The network of arterioles, capillaries, and venules connecting the arterial and venous sides of the renal circulation is quite different from that found in other

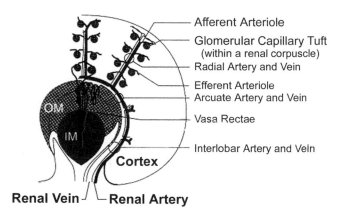

FIGURE 2 Schematic diagram of the major blood vessels of the kidney. All of the glomeruli are located in the cortex and each is supplied by an afferent arteriole. The efferent arteriole conveys the blood from the glomerular capillary tuft to the peritubular capillary circulation, which is not shown in the figure. The peritubular capillary network arising from glomeruli in the outer regions of the cortex is confined to the cortex in close association with the nephron of origin. The peritubular capillary network arising from glomeruli near the corticomedullary junction penetrates deeply into the medulla in a series of long hairpin loops referred to as vasa recta. OM, outer medulla; IM, inner medulla. (Modified from Kriz W. A standard nomenclature for structures of the kidney. *Am J Physiol Renal Fluid Electrolyte Physiol* 1988;254:F1–F8.)

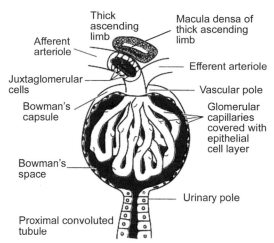

FIGURE 3 Schematic illustration of the glomerulus. The afferent arteriole distributes into the glomerular capillary network. The pressure in these capillary loops is higher than in other systemic capillaries, which results in filtration of fluid from the capillaries into Bowman's space, from which it passes to the proximal convoluted tubule. The blood that remains after filtration flows from the glomerular capillaries into the efferent arteriole, from which it proceeds to the peritubular capillary network. The thick ascending limb from the same nephron is closely associated with the afferent arteriole to form the juxtaglomerular apparatus. This structure is comprised of the macula densa cells of the thick ascending limb and the specialized juxtaglomerular smooth muscle cells of the afferent arteriole.

organs and is responsible for many of the functional characteristics of this organ. The unique feature of the microcirculation is the presence of two capillary beds in series. *Afferent arterioles* branch off the radial arteries and feed the first of the two capillary networks, the glomerular capillaries. As discussed later, the blood pressure in these capillaries is relatively high, so it serves to filter fluid. The glomerular capillary bed is effectively embedded in the epithelial layer referred to as *Bowman's capsule* (which forms the initial portion of the nephron, as described later), and fluid filtered from the glomerular capillaries enters Bowman's space (Fig. 3). The complex of the glomerular capillaries and Bowman's capsule is termed the renal corpuscle, or, more frequently, the *glomerulus*.

In contrast to other capillary beds, the glomerular capillary network empties not into a venule but into another resistance vessel, the *efferent arteriole*. The efferent arteriole gives rise to a second capillary bed, the *peritubular capillaries*. As the name implies, this capillary network lies adjacent to the tubular components of the nephron. In the cortex, the peritubular capillaries form a dense plexus surrounding all the tubular components. Efferent arterioles from glomeruli that lie near the corticomedullary junction give rise to a different type of capillary network that supplies the renal medulla as long, hairpin-shaped capillary loops.

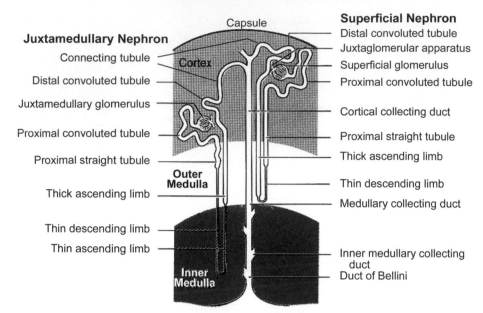

FIGURE 4 Primary structural elements of the nephron. A juxtamedullary nephron is shown on the left, and a superficial nephron on the right. Note the longer loop of Henle, which occurs in many but not all juxtamedullary nephrons. Also, in the juxtamedullary nephron the connecting tubule (CNT) is elongated in structures called *arcades* to reach from the deeper cortex to the superficial cortex where it merges with other CNTs, including shorter CNTs of superficial nephrons, to form the cortical collecting duct. (Modified from Kriz WA. Standard nomenclature for structures of the kidney. *Am J Physiol Renal Fluid Electrolyte Physiol.* 1988;254:F1–F8.)

Blood in these capillaries descends first downward into the medulla via the *descending vasa recta* and then returns to the cortex via the *ascending vasa recta*. Blood pressure in both the cortical and medullary regions of the peritubular capillaries is relatively low and, thus, these capillaries can take up the fluid and solutes that have been reabsorbed by the renal tubules from the ultrafiltrate formed by the glomerulus.

Glomeruli and Nephrons

The nephrons begin as an extension of Bowman's space, which surrounds the glomerular capillaries. All glomeruli are located in the cortex, but those that lie near the cortical surface give rise to what are referred to as *superficial nephrons,* whereas those near the cortico-medullary junction are referred to as *juxtamedullary nephrons.* Although the pattern is not invariant, superficial nephrons give rise to short loops of Henle that end in the outer medulla and have at most a short, thin ascending limb. A larger proportion of the juxtamedullary nephrons have long loops of Henle that extend to varying depths in the inner medulla and even to the tip of the papilla. The various regions of the nephron have quite different structural characteristics that reflect differences in their metabolism and function.

Throughout its length, the nephron is comprised of a single epithelial cell layer with an underlying basal lamina (basement membrane). However, the cells that constitute the different tubular segments of the nephron shown in Fig. 4 differ both anatomically and functionally. To simplify these details, the nephron is discussed in subsequent chapters as if it were divided into six functional regions: (1) the proximal tubule, comprised of the convoluted and straight tubules; (2) the thin descending limb of the loop of Henle; (3) the thin and thick segments of the ascending limb of the loop of Henle; (4) the distal convoluted tubule; (5) the connecting tubule; and (6) the collecting duct, including the initial cortical collecting tubule, and the cortical and medullary collecting ducts.

As shown in Fig. 4, the *proximal convoluted tubule* arises from Bowman's space and folds in a complex series of turns in the region of its own glomerulus. The proximal straight tubule is a straight continuation of the convoluted segment, and the convoluted and straight segments are called collectively the *proximal tubule.* The straight segment runs radially through the cortex into the outer medulla, and in this region, its morphology changes to become the *thin descending limb of the loop of Henle.*

The thin descending limb of the loop of Henle runs radially into the medulla, where it makes a hairpin turn at a level determined, in part, by the location in the cortex of its glomerulus of origin (see later discussion). The *thin ascending limb of the loop of Henle* then returns toward the outer medulla. Both these regions of the nephron are comprised of a very thin epithelial cell layer, thus giving rise to their names. The thin ascending limb changes in the outer medulla to an epithelium with taller cells and numerous mitochondria, which is called the *thick ascending limb,* but is sometimes referred to as the distal straight tubule. The thick ascending limb proceeds through the outer medulla into the cortex, where it comes into close contact with the afferent and efferent arterioles associated with the glomerulus from which the nephron originated. This region of contact is referred to as the *juxtaglomerular apparatus* (see Figure 3).

The juxtaglomerular apparatus consists of specialized regions of both the cortical thick ascending limb and the afferent arteriole. The thick ascending limb cells in this region of contact form a plaque of taller and larger cells referred to as the *macula densa* (see Fig. 3), which monitors the flow and composition of the tubular fluid. Specialized cells of the afferent arteriole in the juxtaglomerular apparatus, referred to as granular cells or juxtaglomerular cells, store a hormone, renin, that can be released into the circulation. The functions of this hormone and the juxtaglomerular apparatus are discussed in Chapters 24 and 29.

The distal convolution lies between the macula densa and the collecting ducts, and it consists of two segments, the *distal convoluted tubule* and the *connecting tubule* (see Fig. 4), which are structurally and functionally distinct. Beyond the macula densa, the thick ascending limb cells continue for a short distance until they are abruptly replaced by the distal convoluted tubule cells (*DCT cells*). DCT cells have the greatest density of mitochondria and Na^+-K^+-ATPase of any nephron segment and the basolateral membrane is highly folded and interdigitated with adjacent cells. The connecting tubule, which is populated primarily by connecting tubule cells (*CNT cells*), follows the distal convoluted tubule but the transition is gradual and CNT cells begin to appear together with DCT cells near the end of the distal convoluted tubule. Finally, the connecting tubule at its distal end has the same structural and functional characteristics as the cortical collecting duct, and therefore, it is referred to as the *initial collecting tubule.*

In the human kidney, individual nephrons do not begin to merge until the *cortical collecting duct,* which is characterized by yet another cell type, the *principal cell*. Principal cells begin to appear among CNT cells at the transition from the connecting tubule to the initial collecting tubule. On the average, the confluence of 10–12 initial collecting tubules near the cortical surface forms a cortical collecting duct, which then runs unbranched through the cortex and the outer medulla until it reaches the inner medulla. The inner medullary collecting ducts fuse successively to form the ducts of Bellini, each of which carries the urine originating from approximately 2800 glomeruli.

Yet one additional cell type has not been discussed. The *intercalated cells,* which are responsible for urinary acidification, begin to appear in the early part of the connecting tubule where they are intermingled in varying proportions with CNT cells. In the cortical collecting duct, the proportion of intercalated to principal cells is about 1:2, but this proportion decreases in the medullary collecting duct until intercalated cells disappear in the inner medullary collecting duct.

The localization of different cells along the distal nephron segments is of more than histological interest because the cell type determines both the transport processes that occur and the hormones that regulate them. For example, as will be discussed in detail in Chapter 27 in the section titled Transport in the Distal Tubule and Collecting Duct, CNT and principal cells respond to the hormones aldosterone and vasopressin that regulate, respectively, Na^+ reabsorption via an electrogenic Na^+ channel and water reabsorption. In contrast, DCT cells reabsorb Na^+ and Cl^- by an electroneutral cotransporter that is inhibited by thiazide diuretics, and they do not reabsorb water in the presence or absence of vasopressin.

ENDOCRINE FUNCTIONS OF THE KIDNEY

In addition to its role in the regulation of solute and water excretion, the kidney has several endocrine functions, including the synthesis of erythropoietin, the release of angiotensin I and angiotensin II, the conversion of vitamin D_3 to its final form, and the production of several autocrine and paracrine agents. Vitamin D_3 from the diet and that produced by the skin must be hydroxylated to be fully active. Enzymes in the kidney are responsible for the final conversion of vitamin D_3 to its most active form *1,25-dihydroxycholecalciferol* [$1,25(OH)_2D_3$; see Chapter 43].

Erythropoietin is interesting because it is a growth factor (or cytokine) that behaves like a hormone. Erythropoietin is a 34-kDa glycoprotein that is secreted by interstitial cells in the kidney, and its synthesis and release is increased by a low hematocrit or a fall in blood oxygen carriage. Erythropoietin acts on erythroid progenitor cells in bone marrow as a colony-stimulating factor and increases the production of red blood cells. As might be expected, renal failure is accompanied by anemia as one of its many side effects. However, the fall in erythropoietin production is not the only reason for the anemia. Disseminated tissue hemorrhaging and a decrease in hemoglobin and red blood cell production accompany uremia. Nevertheless, the anemia of renal failure can be reduced effectively by the administration of erythropoietin, which was one of the first human hormones to be produced industrially by recombinant DNA technology. Recombinant erythropoietin is also used extensively to boost red blood cell production in patients who wish to donate their own blood before surgery.

As discussed in more detail in the next chapter, juxtaglomerular cells in the afferent arterioles release renin, which results in the production of angiotensin I and angiotensin II. Both within the kidney and in the general circulation, angiotensin has broad effects as a pressor agent by increasing the total peripheral resistance of the circulation and by maintaining or increasing extracellular fluid volume. Several organs including the adrenal cortex, vascular smooth muscle, the brain, and the heart have receptors for circulating angiotensin II (see also Chapter 40).

The kidney produces several substances whose actions are limited to neighboring cells (*paracrine agents*) or on the cells that secrete them (*autocrine agents*). For example, prostaglandin E_2 (PGE_2) and prostacyclin (PGI_2) are produced throughout the kidney, especially by medullary interstitial cells. They have various autocrine and paracrine functions within the kidney; however, these actions are usually manifest only when other stimuli such as circulating catecholamines and sympathetic nerve input to the kidney are high. In this setting, prostaglandins help to maintain renal blood flow, glomerular filtration rate, and salt excretion. When the production of prostaglandins, especially PGE_2, is high, they can also act as systemic vasodilators and tend to decrease blood pressure; however, they are rapidly converted to inactive metabolites.

MICTURITION

The urine that enters the calyces from the papillae is the final urine that will be excreted from the body, its composition having been determined by the functional characteristics of the nephrons. Urine is carried from the kidneys via the ureter to the bladder, where it is stored until it exits the body via the urethra (Fig. 5). The composition of the urine is altered little if at all by any of these structures. Nevertheless, the renal pelvis, ureters, bladder, and urethra are frequently the origin of urinary tract pathology. Obstruction of the urinary tract by calculi (stones) or tumors can lead to impaired renal function. The lower urinary tract is also frequently the site of infection, which may ascend the ureters and infect the kidneys. Finally, many neurologic and anatomic abnormalities can affect voiding of the

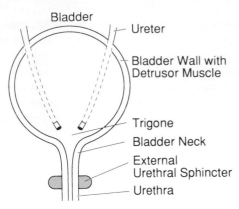

FIGURE 5   Schematic illustration of the bladder. The bladder wall is heavily invested with smooth muscle layers that provide the contractile force in voiding. The ureters enter the bladder on its posterior side, above the bladder neck in the region of the trigone called the ureterovesicular junction. The tension of the detrusor muscle in the bladder wall keeps the ureteral openings closed so that urine does not reflux toward the kidneys. When the pressure of a ureteral peristaltic wave exceeds the pressure in the bladder, this functional sphincter opens transiently, allowing urine to flow into the bladder. The muscle in the trigone region also forms part of the internal sphincter of the bladder neck. Beyond this, the urethra passes through the urogenital diaphragm, where voluntary muscle fibers form the external sphincter. Both the external and internal sphincters prevent urine movement out of the bladder until micturition is initiated.

bladder (micturition) and lead either to incontinence or to urinary retention.

Urine that collects in the hollow structure of the renal pelvis is propelled through the ureter toward the bladder by peristaltic contractions of the renal pelvis and ureter, both of which are invested with smooth muscle in their walls. As in the case of other visceral smooth muscle, this action is enhanced by parasympathetic and inhibited by sympathetic innervation. The walls of the pelvis and ureters are also invested with pressure receptors that convey pain sensations to the central nervous system if pressure rises in these structures. This is the origin of the extreme pain associated with obstruction of the urinary tract by kidney stones. These receptors are also the sensory input to a reflex arc, the ureterorenal reflex, which produces a sympathetic efferent discharge to the kidney that reduces the rate of glomerular filtration.

As shown in Fig. 5, the ureters enter the bladder through the *detrusor muscle* in the trigone region. The normal tone of this muscle tends to occlude the ureter as it passes through the bladder wall, thus forming a functional valve referred to as the *ureterovesicular valve*, which prevents the backflow of urine from the bladder to the ureter. Each peristaltic wave along the ureter increases the pressure within the ureter so that the region within the bladder wall opens, allowing urine to flow into the bladder.

As urine enters the bladder, it distends. The volume of the bladder can increase from essentially zero after micturition to a maximum of about 500 mL in the adult. This is the largest volume change of any hollow structure in the body, and it is accompanied by thinning and an increase in the tension in the bladder wall that is proportional to the radius of the bladder according to the law of LaPlace. The increase in bladder wall tension as the bladder expands stimulates the firing of stretch receptors on afferent nerves. Expulsion of urine from the bladder is driven by contraction of the detrusor muscle. Normally this muscle is flaccid and allows the bladder to expand as it fills. At the tip of the trigone, the urethra exits from the bladder in a region called the *bladder neck*. Leakage of urine from the bladder during filling is prevented by contraction of the muscle fibers in what is referred to as the *internal urethral sphincter*. The internal sphincter is not a discrete anatomic structure but represents the combined effects of circular smooth and striated muscle fibers in the urethra to keep its lumen closed. The relative importance of different regions of the urethra in performing this sphincter function also differs between males and females. At the point where the urethra passes through the urogenital diaphragm, the striated muscles of the pelvic floor act as an *external urethral sphincter*. Voluntary or reflex contractions of these muscles increase the outflow resistance of the urethra and can prevent voiding even during strong constrictions of the detrusor muscle in the bladder wall.

Clinical Note

Urinary Tract Calculi (Kidney Stones)

The formation of kidney stones is a relatively common and potentially serious renal disorder. Kidney stones are crystalline precipitates of solutes that are present in the urine, and they usually form in the calyces because the urine is the most concentrated there. However, the presence of high concentrations of ions with low solubility products, such as calcium and phosphate, is not the sole reason for the formation of calculi. A normal individual may have even higher average concentrations of such solutes in his or her urine than an individual who is susceptible to kidney stones, and yet never have kidney stones. Although the biochemistry of the processes remains poorly understood, it is thought that endogenous substances are produced by renal epithelia that inhibit crystallization from supersaturated urine. Stone formers may be deficient in these agents. In other cases, such as the formation of oxalate, uric acid, or cystine stones, the underlying cause may be a metabolic disturbance that leads to abnormally high concentrations of these solutes in the urine.

Infection is also a frequent cause of stone formation, especially in women with frequent urinary tract infections (UTIs). Some bacteria, especially *Proteus,* metabolize urea and increase the urine concentration of ammonium ion, which can result in the formation of a triple phosphate stone, calcium magnesium ammonium phosphate.

Kidney stones can remain in the renal calyxes, or they may be small enough that they produce no symptoms when excreted. However, when larger stones are passed through the ureters to the bladder, extreme acute pain can result that brings most patients to the emergency room. The danger of stone formation is that persistent occlusion of the ureter may lead to deterioration of renal function and eventual destruction of the renal parenchyma. Kidney stones are diagnosed by severe pain, usually extending from the loin to the groin, and accompanied by microscopic hematuria and often crystals in the urine sediment. These crystals can be recognized by appearance or chemically analyzed, and include (in order of most to least frequent) calcium oxalate, calcium oxalate and phosphate, triple phosphate, uric acid, and cystine.

In many cases, kidney stones can be passed without intervention other than pain medication and fluids to increase urine flow. When the stones are too large, extracorporeal shockwave lithotripsy has proved to be very effective. Shockwaves from, for example, a shock generator or electromagnetic plate generator, are concentrated under x-ray visualization through the skin and tissues on the stone, and are used to break the stone into pieces small enough to be passed in the urine. In the extreme, kidney stones can be too large for even this treatment. For example, "staghorn" calculi, so-called because of a shape that conforms to the shape of the calyxes, usually have to be removed surgically.

Neural Regulation of Micturition

Bladder function is normally regulated by a coordinated switching between two states: a storage phase and an emptying phase. As shown in Fig. 6, during the storage phase the volume of urine in the bladder can increase to 200–300 mL with little rise in pressure due to the remarkably high compliance of the bladder wall. The bladder wall is made up of an extracellular fibrous matrix of collagen and elastin, and the detrusor muscle fibers, which can lengthen more than threefold during bladder filling. Nevertheless, as the radius of the bladder increases, the wall tension increases and activates stretch receptors. The resulting afferent firing is conveyed to the lumbar spinal cord via the hypogastric nerve. When the pressure in the bladder increases above about 10 cm H_2O, this afferent input causes reflex firing of parasympathetic cholinergic efferent fibers that run from the sacral cord via the pelvic nerves to the detrusor muscle. This efferent input causes the muscle to contract, resulting in transient pressure waves and heightened afferent firing that, when conveyed to higher brain centers, gives the sensation of bladder fullness and the urge to urinate.

When the bladder is only partially filled, these pressure waves relax spontaneously, the detrusor muscle ceases to contract, and the pressure falls back to the baseline, as shown by the spikes of pressure (dashed lines) in Fig. 6. In the storage phase, continence is maintained by the influence of descending fibers from the pons that suppress the firing of parasympathetic efferents to the detrusor muscle while augmenting the output of parasympathetic efferents to the internal urethral sphincter. The pressure of the filling bladder on the pelvic floor also activates reflex arcs that increase the tension of the external sphincter via cholinergic efferents in the pudendal nerves. Contraction of the external sphincter is an important part of the "guarding reflex" that prevents urinary leakage during the bladder pressure waves. The voluntary augmentation of this reflex and constriction of other muscle groups in the pelvic floor are important for maintaining continence when the bladder is filled and afferent output is high or when other regulatory pathways are defective.

In the absence of any input from higher nervous centers as occurs, for example, after spinal cord transection, the bladder can fill and empty spontaneously, a condition referred to as an *automatic bladder*. When the bladder reaches a certain volume, the reflex contraction of the detrusor muscle becomes sufficiently strong to force open the internal sphincter and urine distends the bladder neck, initiating a reflex inhibition of the external sphincter. However, voiding that occurs in the absence of input from the pons is sporadic and incomplete because the afferent output from the bladder drives parasympathetic output that constricts both the detrusor muscle and the internal sphincter muscle.

Even in infants and unconscious adults, the emptying phase of the bladder is usually coordinated by input from the micturition center in the pons. When afferent output from the bladder due to distension reaches some threshold, the pontine center reverses its influence on the spinal reflexes by decreasing efferent output to the internal and external sphincters and augmenting the output to the detrusor muscle. However, the decrease in urethral outflow resistance precedes contraction of the detrusor muscle by a few seconds, thereby decreasing the pressure necessary to expel the urine from the bladder. When urine enters the urethra, afferent output acts to maintain detrusor contraction and sphincter relaxation until the bladder has completely emptied and flow ceases.

During development of the nervous system in the toddler, voluntary control of voiding is established by input to the pontine center primarily from the right frontal cerebral cortex. Older children and adults can voluntarily stop voiding with a partially emptied bladder. It has been found that such voluntary stoppage is accomplished by constriction of the internal and external sphincters followed by detrusor muscle relaxation. When voiding is reinitiated voluntarily, detrusor muscle contraction is preceded by relaxation of the sphincters. Thus, the voluntary regulation and timing of urination is exerted by facilitory or inhibitory input to

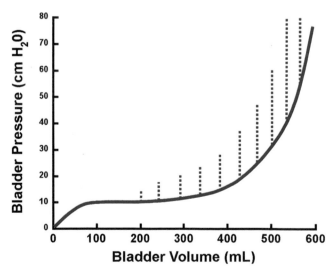

FIGURE 6   Relationship between pressure in the bladder and bladder volume. As filling of the bladder progresses beyond 200 mL, pressure waves begin to appear. These transient contractions of the bladder detrusor muscle are shown by the vertical dotted lines superimposed on the smooth pressure curve. These waves of contraction contribute to the sensation of bladder fullness and increase in frequency and intensity as the bladder continues to fill.

Clinical Note

Urinary Tract Infection

An otherwise healthy 21-year-old woman is seen at an outpatient clinic. She complains of a burning sensation upon urination and an increase in her frequency and urgency of urination. A review of her medical history reveals no significant medical problems, but she has had a recent increase in sexual activity. Physical examination reveals no abnormalities. Her blood pressure and pulse are normal and she has no fever; however, laboratory results show that her urine sediment has abundant bacteria and white blood cells. The diagnosis is cystitis, that is, an infection of the bladder.

Cystitis is a frequent medical complaint of sexually active young women, with an incidence of up to 5% yearly in women of reproductive age. Women are particularly susceptible to UTI because of the typical bacterial flora of the vagina and the shorter length of the female urethra. The incidence of UTI in males is much less also because the prostate and seminal vesicles secrete antibacterial agents. However, when prostatic or other diseases interfere with complete voiding, cystitis is a frequent presenting symptom in the male as well. UTI is usually confined to the bladder and urethra because of the barriers provided by the ureterovesicular valve and the normal descending flow of urine from the kidneys to the bladder. However, congenital abnormalities and incomplete development can lead to failure of this valve, such that contraction of the bladder during micturition produces reflux of the urine from the bladder to the kidneys. Reflux can lead to pyelonephritis, an infection that ascends from the bladder and reaches the kidneys. Pyelonephritis is a more serious disease and is usually accompanied by fever, chills, and tenderness and pain in the flank region over either or both kidneys. In addition, the white blood cells in the urinary sediment frequently form long cylindrical clumps called *casts* because they take the cylindrical form of the nephrons in which they form. These characteristics usually allow the physician to discriminate between cystitis and pyelonephritis based on the history and physical, and an examination of the urinary sediment.

the pontine center rather than by direct stimulation of efferents to the internal or external sphincter. However, even when the bladder is painfully full and the pontine center cannot be suppressed, voluntary constriction of the pelvic floor muscles can delay or greatly reduce voiding.

Suggested Readings

Blok BFM. Central pathways controlling micturition and urinary continence. *Urology* 2002;59(Suppl 5A):13–17.

Chancellor MB, Yoshimura N. Physiology and pharmacology of the bladder and urethra. In Walsh, PC, ed., *Campbell's urology*, 8th ed. Philadelphia: Saunders, 2002, pp 831–886.

Kriz W, Kaissling B. Structural organization of the mammalian kidney. In Seldin DW, Giebisch G, eds., *The kidney: Physiology and pathophysiology*, Vol 1, 3rd ed. Philadelphia: Lippincott Williams & Wilkins, 2000, pp 587–654.

Reilly RF, Ellison DH. Mammalian distal tubule: Physiology, pathophysiology, and molecular anatomy. *Physiol Rev* 2000;80:277–313.

Tisher CC, Madsen KM. Anatomy of the kidney. In Brenner, BM, ed., *Brenner & Rector's the kidney*, 6th ed. Philadelphia: Saunders, 2000, pp 3–67.

Renal Blood Flow and Glomerular Filtration

JAMES A. SCHAFER

KEY POINTS

- *Renal blood flow* (RBF) is 1.0–1.2 L/min or about 20% of the cardiac output. At a normal hematocrit, the *renal plasma flow* (RPF) is ~670 mL/min.
- Glomerular filtration refers to the *ultrafiltration* of a protein-free fluid from the blood plasma into Bowman's space, the beginning of the nephron.
- Reabsorption by the nephron restores most of the filtered water and solutes to the blood, but in a carefully regulated manner.
- The afferent and efferent arterioles are the primary resistance vessels in the renal circulation. Their independent regulation determines RBF and the *glomerular filtration rate* (GFR).
- The GFR is normally about 130 mL/min, or more than 180 L/day. The *filtration fraction* (FF)

is the ratio of GFR to RPF and is approximately 0.2 (20%).
- The forces favoring glomerular filtration are the high hydrostatic pressure in the glomerular capillaries and the small or absent *colloid osmotic pressure* (COP) in Bowman's space. The forces opposing filtration are the higher COP of the glomerular capillary plasma and the hydrostatic pressure in Bowman's space.
- Increased afferent arteriolar resistance decreases both RBF and GFR. A small increase of efferent arteriole resistance may augment GFR, but larger increases diminish RBF and GFR.
- The filtration barrier comprises the endothelial cells, the basement membrane, and the epithelial cells of Bowman's capsule. The *ultrafiltration coefficient* (K_f) relates GFR to the net

---

KEY POINTS (*continued*)

filtration driving force. The hydraulic conductivity of the filtration barrier and the surface area of the filtering capillaries determine K_f.

- *Renin* is released from the juxtaglomerular cells by a fall in blood pressure or by neural and hormonal signals that indicate a decrease in extracellular fluid volume. Renin catalyzes the cleavage of angiotensin I from circulating angiotensinogen. *Converting enzyme* catalyzes the conversion of angiotensin I to its active circulating form, angiotensin II.

- *Angiotensin II* has three major effects: (1) It acts systemically as a pressor to increase blood pressure; (2) in the proximal tubule, it increases salt and water reabsorption; and (3) it stimulates the adrenal cortex to secrete aldosterone, which increases Na^+ reabsorption and K^+ secretion in the distal nephron.

- Because of intrinsic renal *autoregulation*, RBF and GFR remain constant over a wide range of fluctuations in the systemic blood pressure. Autoregulation involves a *myogenic mechanism* and *tubuloglomerular feedback*.

The two kidneys are supplied with massive blood flow in comparison with other organs. In the young healthy adult, the *renal blood flow* (RBF) ranges from 1000–1200 mL/min, or ~20% of the resting cardiac output. At a hematocrit of 0.40 to 0.45, the *renal plasma flow* (RPF) is about 670 mL/min. Because the metabolic cost of the many transport processes that are constantly occurring in the nephrons is considerable, the kidneys, despite their small mass, account for ~8% of the total body oxygen consumption. Nevertheless, the very high blood flow rate is far in excess of that needed to meet these metabolic demands, resulting in the small difference of 1.7 mL/dL between the oxygen levels in arterial and venous blood. Instead of merely supplying oxygen and metabolic substrates, this large rate of blood flow is required to produce the high rate of glomerular filtration needed for excretion of metabolic by-products such as urea, uric acid, and creatinine.

Glomerular filtration, the initial event in the formation of the urine, results in the movement of a large volume of fluid from the glomerular capillaries into Bowman's space. Most solutes that are smaller than proteins pass freely across the glomerular membranes, and thus glomerular filtration results in the production of an *ultrafiltrate* of the plasma. This ultrafiltrate contains very little protein, but has approximately the same concentrations of the smaller solutes as plasma. *Reabsorption* is the regulated transport of water and solutes from the urine to the blood that occurs along the nephrons. Given the high rate of filtration, solutes that are to be removed from the body can be excreted in large amounts merely by a failure to reabsorb them along the nephron. On the other hand, the high rate of filtration imposes a high energy cost in order to reabsorb those ions and metabolic substrates that must be conserved by the body.

Filtration and reabsorption are functions not only of the nephron; they are coordinated with the unique properties of the vasculature. Ultrafiltrate is formed from the glomerular capillaries, but solutes and water that are reabsorbed from the nephron must be returned to the circulation by uptake into the peritubular capillaries. The same Starling forces that apply to all capillary beds govern the movement of fluid into and out of the renal capillaries. However, the series arrangement of two capillary beds in the kidney is unique. The high hydrostatic pressure in the capillary network of the glomerulus produces rapid net filtration of fluid, whereas the higher colloid osmotic pressure (COP) and lower hydrostatic pressure of the peritubular capillaries allow for net fluid uptake. The importance of these functions is reflected by a phenomenon referred to as *autoregulation,* which permits glomerular filtration to be maintained relatively constant over a wide range of systemic blood pressures and despite other extrarenal changes. Autoregulation depends primarily on alterations in the resistance of the afferent arteriole in response to such external factors. However, changes in glomerular filtration rate away from the autoregulated set point can also be produced in response to certain physiologic stimuli by neurally or hormonally induced changes in the resistance of the afferent and efferent arterioles.

This chapter describes the basic mechanisms involved in the regulation of blood flow, glomerular filtration, and peritubular capillary uptake in the kidney and how these processes may be altered under normal as well as pathologic conditions.

BLOOD FLOW VELOCITIES AND PRESSURES IN THE RENAL VASCULATURE

The glomerulus is a unique structure designed to produce rapid ultrafiltration of fluid from the plasma across the glomerular capillary endothelium and into

arteries to −60 mm Hg in the glomerular capillaries. Because of the lower transmural pressure, the smooth muscle of the efferent arteriolar is thinner and distributed more irregularly, but it drops the pressure by a further 33–35 mm Hg.

Glomerular Capillaries

The relative resistance of these two arterioles determines the average pressure within the glomerular capillaries, and it is estimated to be about 60 mm Hg in humans. This pressure is considerably higher than in the capillary beds of other organs. Because of the large cross-sectional surface area provided by the numerous capillary loops (Fig. 1), the blood pressure falls by at most 2 mm Hg along the glomerular capillaries and the flow velocity is relatively slow.

The tone of the smooth muscle in the arterioles is the primary determinant of their resistance. However, it is conjectured that their resistance might also be altered by *mesangial cells,* which are interstitial cells located between the afferent and efferent arterioles within the glomerulus and just external to it. Mesangial cells contract or relax in response to various hormones. Contraction of the mesangial cells could augment the resistance of the arterioles and possibly change the number of open capillary loops in the glomerular tuft, but there is presently no definitive information about the role of these cells in regulating arteriolar resistances.

On leaving the glomerulus, the efferent arteriole branches to form the peritubular capillary network. The densest region of the peritubular capillary network is in the cortex. Each efferent arteriole supplies capillaries near the nephron of its glomerular origin, and about 90% of the postglomerular blood flows only through the cortex. The remaining 10% of the blood flow supplies the medulla via the elongated capillary loops formed by the ascending and descending *vasa rectae*. Vasa rectae

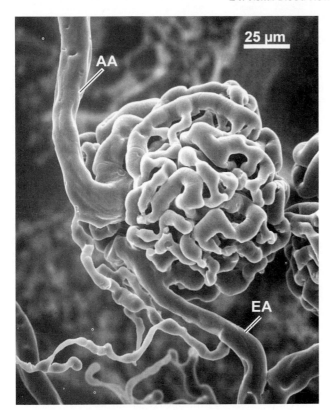

FIGURE 1 Scanning electron micrograph showing a vascular cast of a glomerular capillary tuft and its associated afferent arteriole (AA) and efferent arteriole (EA). Within the glomerular tuft smaller capillaries are seen interconnecting larger loops. (Electron micrograph courtesy of Dr. Andrew P. Evan, Indiana University Medical Center, Indianapolis, IN.)

Bowman's space. As shown by the scanning electron micrograph in Fig. 1, the glomerular capillaries are supplied by the *afferent arteriole.* However, the blood that leaves the glomerulus does not flow out through a venule as in most capillary systems, but through a second resistance vessel, the *efferent arteriole.* Because of this arrangement, most of the total drop in pressure from the arterial to the venous side of the circulation is divided between the two arterioles, resulting in a higher blood pressure in the glomerular capillary network and a lower blood pressure in the peritubular capillary network. The distribution of the fall in hydrostatic pressure is given in Table 1.

As can be seen in Table 1, the largest drop in pressure from the mean systemic arterial pressure occurs in the afferent and efferent arterioles, which have roughly equal resistances. Because of the higher transmural pressure in the afferent arteriole, thick rings of vascular smooth muscle that provide the contractile force to maintain and alter the resistance surround it. The afferent arteriole provides about 35% of the total resistance to flow within the kidney, causing a fall in blood pressure from 90 mm Hg in the medium-sized

TABLE 1 Pressures and Resistances in the Renal Vasculature

Vessel	Pressure in Vessel (mm Hg)		Relative Resistance
	Beginning	End	%
Renal artery	92	~92	~0
Interlobular, arcuate, and arcuate arteries	92	90	2
Afferent arteriole	90	61	34
Glomerular capillaries	61	59	2
Efferent arteriole	59	25	40
Peritubular capillaries	25	6	22
Major renal veins	6	~6	~0

arise primarily from efferent arterioles of juxtamedullary nephrons (see Chapter 23, Fig. 2).

The average blood pressure in the peritubular capillaries is only 15–20 mm Hg. This is considerably lower than in the capillaries of other organs because the two series resistances of the afferent and efferent arterioles have dropped it. This lower pressure means that the sum of the Starling forces favors the uptake of water and solutes that have been reabsorbed by the renal tubules, which is the primary function of this capillary bed.

GLOMERULAR FILTRATION

The high hydrostatic pressure of the blood in the glomerular capillaries is responsible for a net driving force favoring ultrafiltration from the glomerular capillaries. In the young male adult, the *glomerular filtration rate* (GFR) averages about 130 mL/min, which amounts to a whopping 187 L/day. The high rate of glomerular filtration means that as plasma flows through the kidneys at a rate of 670 mL/min, 130 mL/min, or ~20%, is filtered. This 20% represents what is referred to as the *filtration fraction* (FF). The FF is calculated as the ratio of the GFR to the RPF:

$$FF = GFR/RPF. \qquad (1)$$

As discussed later, the filtration fraction, which is usually given as a percent rather than a fraction, determines the effect of the renal blood flow on the GFR.

The GFR, as with most extensive parameters of body function, is proportional to body size and is correspondingly lower in females. Thus, reported values of the normal GFR vary over a wide range, but for the purposes of the following discussion, we will consider the average GFR to be 130 mL/min, or an average of ~180 L/day. This should be regarded as the optimal value that would be found in a young adult with healthy kidneys. GFR declines with age and in renal disease because of a decrease in the number of functioning glomeruli. Therefore, measurement of the GFR is an important index of renal function and the progression or amelioration of renal disease. The following section considers the individual driving forces that determine the net force for glomerular filtration and how those forces are modified by physiologic or pathologic mechanisms.

Forces Driving and Opposing Glomerular Ultrafiltration

As in all capillaries, the movement of fluid into or out of the glomerular and peritubular capillaries is

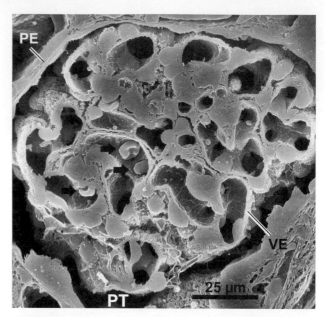

FIGURE 2   Scanning electron micrograph of a fractured glomerulus revealing the beginning of the proximal tubule (PT) at the urinary pole (lower edge). The cross-sectional plane of the fracture has exposed the lumens of several glomerular capillaries. These capillaries have attenuated endothelial cells. In addition, one can see the visceral epithelium (VE) that covers the capillary loops, and the parietal epithelium (PE) that forms the outer wall of Bowman's space. Red blood cells (arrow) can be seen in some of the capillary loops. (Electron micrograph courtesy of Dr. Andrew P. Evan, Indiana University Medical Center, Indianapolis, IN.)

governed by the balance of the so-called Starling forces (see Chapter 16), that is, the hydrostatic and osmotic pressure differences between the capillary lumen and the region surrounding the capillary. In the case of the peritubular capillaries, this region is the interstitial fluid, whereas the glomerular capillaries are surrounded by Bowman's space, as shown in Fig. 2. According to the Starling Law, the rate of fluid movement (J_v) will be determined by the following equation:

$$J_v = L_p \cdot A \cdot (\Delta P - \Delta \Pi), \qquad (2)$$

where L_p is the hydraulic conductivity of the capillary wall, A is its area, ΔP is the difference in hydrostatic pressure, and $\Delta \Pi$ is the difference in osmotic pressure across the capillary wall. The osmotic pressure is produced only by macromolecules because both the peritubular and glomerular capillaries are virtually impermeable to plasma proteins, but they are quite permeable to smaller solute molecules. In other words, the reflection coefficients of solutes smaller than proteins are zero (see Chapter 3). Therefore, the relevant osmotic pressure difference is the difference in COP.

In the case of the glomerular capillary, because plasma proteins normally appear in the glomerular ultrafiltrate only in extremely low concentrations, the COP of Bowman's space can be regarded to be zero. However, in diseases affecting the glomeruli, the permeability of the filtration barrier to proteins may rise and their concentration in Bowman's space can become appreciable, thus reducing the difference in COP between the plasma and the space.

It is difficult to estimate accurately the area of most capillary beds including the glomerular capillary network; consequently a parameter referred to as the *ultrafiltration coefficient* (K_f) is used to represent the product of the hydraulic conductivity and the area. Taking all of the preceding factors into consideration, Eq. [2] can be modified to give the rate of fluid movement (J_v), which in this case represents the GFR, as follows:

$$J_v = \text{GFR} = K_f \cdot (P_c - P_b - \pi_c), \qquad (3)$$

where P_c is the average glomerular capillary pressure, P_b is the pressure in Bowman's space, and π_c is the *average* COP of the blood along the glomerular capillaries. As discussed later, COP rises in the glomerular capillaries because of the fluid lost during filtration.

The directions and usual magnitudes of these three forces—the glomerular capillary pressure, the hydrostatic pressure of Bowman's space, and the capillary COP—are shown diagrammatically in Fig. 3. The values

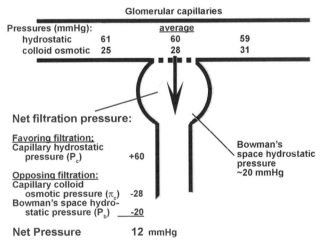

FIGURE 3   Balance of hydrostatic and osmotic pressures governing glomerular filtration. The hydrostatic pressure in the glomerular capillaries is considerably higher than in other systemic capillaries and falls only modestly along the length of the glomerular capillary. On the other hand, the plasma COP rises along the glomerular capillary because of the filtration of fluid into Bowman's space. The pressure in Bowman's space is about 20 mm Hg; and this pressure is required to propel the urine along the nephron. The net filtration pressure is given as the difference between the capillary hydrostatic pressure and the sum of the capillary COP and the hydrostatic pressure in Bowman's space.

presented in this figure are not known directly in the human, but have been extrapolated from measured values in experimental animals. These estimates are used to illustrate the relative magnitudes of the driving forces producing glomerular filtration. Under normal conditions, the glomerular capillary hydrostatic pressure averages 60 mm Hg and falls little along the length of a capillary loop because of the large cross-sectional area available for flow. The hydrostatic pressure in Bowman's space, which is estimated to be 20 mm Hg, opposes this hydrostatic pressure. The colloid osmotic pressure of the plasma rises along the glomerular capillary because as the protein-free fluid is filtered, the protein concentration in the plasma remaining in the capillaries rises. Thus, the average COP in the glomerular capillaries is higher than that of systemic plasma, and is estimated to be about 28 mm Hg. As shown in Fig. 3, the net filtration pressure is the algebraic sum of these three separate forces and is approximately 12 mm Hg. The factors that influence each of these individual forces, and thus govern the net filtration pressure, are considered in the following sections.

Glomerular Capillary Pressure

As discussed earlier, the relative resistances of the afferent and efferent arterioles determine glomerular capillary pressure. These resistances are in turn determined by hormonal and neural input to the smooth muscles of these arterioles. As illustrated by Fig. 4, when the resistance of the afferent arteriole is greater than that of the efferent arteriole, a greater fraction of the total arteriovenous pressure drop occurs across the afferent arteriole, and thus the glomerular capillary pressure is lower. Conversely, when the efferent arteriolar resistance is greater than that of the afferent, the glomerular capillary pressure is higher than normal. In some circumstances, the changes in the two resistances are offsetting so that glomerular pressure remains constant while blood flow decreases, as they do whenever the afferent and efferent arteriolar resistances increase about equally. It might be expected that when glomerular pressure rises due to an increase in the resistance of the efferent arteriole (as shown in the third panel of Fig. 4) the GFR would increase, but the increased resistance also decreases RPF. The decreased flow rate causes the FF and thus the glomerular capillary COP to rise. In other words, ultrafiltration from the smaller blood flow produces a larger increase in the concentration of the remaining proteins. As discussed later, the rise in COP may result in a lower net effective filtration pressure, despite the increased capillary hydrostatic pressure.

Equal Resistances

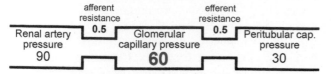

Higher Afferent Resistance

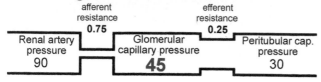

Higher Efferent Resistance

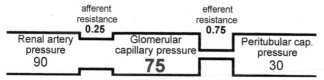

FIGURE 4   Effects of changes in the relative resistances of the afferent and efferent arteriolar resistances. For simplicity, it is assumed that the total resistance of the two arterioles together is constant in the three situations. Thus, in each panel, the total pressure drop from the renal artery (90 mm Hg) to the peritubular capillaries (30 mm Hg) and the RBF remains constant. When afferent and efferent arteriolar resistances are nearly equal (upper panel), the pressure drop across each arteriole is 30 mm Hg. If the afferent arteriolar resistance is greater (middle panel), the glomerular capillary pressure is lower. If the efferent arteriolar resistance is greater (lower panel), the glomerular capillary pressure is increased.

Hydrostatic Pressure in Bowman's Space

The hydrostatic pressure of 20 mm Hg in Bowman's space is required to drive the flow of urine along the nephron against the resistance presented by the long, thin, tubular segments leading eventually to the ducts of Bellini. These ducts empty into the renal calyces, which are approximately at atmospheric pressure. Based on Poiseuille's law (see Chapter 10), the pressure in Bowman's space will be proportional to the resistance of the nephron and the rate of urine flow in the nephron. When urine flow is increased, for example, by diuretics, or when there is an obstruction of the lower urinary tract, the pressure in Bowman's space rises. This increase in pressure in Bowman's space decreases the net filtration pressure, and thus decreases the GFR.

The length and inside diameter of each nephron segment is relatively invariant and, therefore, its resistance to flow is constant. However, pathologic processes may affect the outflow resistance. This is frequently seen in elderly men with benign prostatic hypertrophy. The resulting urethral outflow resistance often leads to incomplete emptying of the bladder and to increased pressure in the bladder, which is transmitted up the urinary tract. This results in a higher pressure in the calyces and the necessity for a higher pressure in Bowman's space to drive urine flow. Partial or complete obstruction of the lower urinary tract by kidney stones or tumors also leads to increases in the pressure in Bowman's space. These conditions produce a decrease in the net filtration pressure and, thus, a decrease in GFR.

Colloid Osmotic Pressure in the Glomerular Capillaries

As protein-free fluid is filtered from the glomerular capillary plasma, the protein concentration of the remaining plasma rises, producing a corresponding rise in the COP, as illustrated in Fig. 5. When RPF falls to a greater degree than the GFR, the COP rises more rapidly along the glomerular capillaries. In the extreme, the COP rises sufficiently to reduce the net filtration pressure to zero in the distal end of the glomerular capillary, as shown in Fig. 5 (lower panel). The resulting balance of forces is referred to as *filtration equilibrium,* and it results in the loss of filtration in the distal portion of the capillary. The rise of COP in the glomerular capillary is inversely related to the FF as shown in Fig. 6. Because in humans FF is normally about 20%, filtration equilibrium probably never occurs under physiologic conditions. However, if renal blood flow is reduced by proportional changes in the resistances of the afferent and efferent arterioles so that the net hydrostatic pressure driving force remains the same, FF rises. Figure 6 indicates that filtration equilibrium would be attained at any filtration fraction in excess of about 37%. It is for this reason that the RPF has an important influence on the net filtration pressure and hence on GFR.

Absorption of Fluid in Peritubular Capillaries

In the peritubular capillaries, the situation is much different. Because of the resistance of the efferent arteriole, the hydrostatic pressure in the peritubular capillaries is much lower than in other capillary networks in the body. Conversely, because of the filtration of protein-free fluid from the glomerular capillaries, the plasma entering the peritubular capillaries has a much higher COP. Both the low hydrostatic pressure and the high COP of the peritubular capillary plasma favor the reabsorption of fluid from the interstitial spaces surrounding the nephrons, which is the primary function of the peritubular capillary network.

Reabsorption by the peritubular capillaries can be altered by changes in either the hydrostatic or the COP

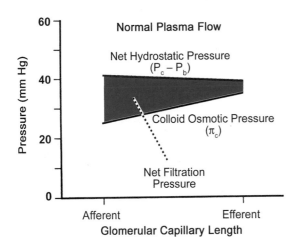

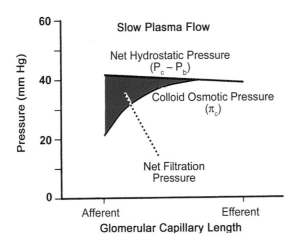

FIGURE 5 Effect of glomerular capillary plasma flow on net filtration pressure. At normal flow rates (upper graph), the COP of the plasma in the glomerular capillaries rises as fluid is filtered from the capillaries. If the plasma flow rate is slowed (lower graph) by increased vascular resistance, the plasma COP in the capillaries rises more steeply and may become equal to the net hydrostatic pressure, which is the difference between the glomerular capillary pressure ($P_c = 60$ mm Hg) and the pressure in Bowman's space ($P_b = 20$ mm Hg). In this situation, called *filtration equilibrium,* the net filtration pressure becomes zero before the end of the glomerular capillary network, and no filtration occurs beyond that point.

THE FILTRATION BARRIER

As discussed earlier, the process of glomerular filtration produces an ultrafiltrate of plasma, that is, a fluid that has approximately the same composition as the blood plasma but virtually no protein. It is estimated that in humans much less than 1% of the plasma albumin that enters the kidney via the renal artery is filtered at the glomerulus, and this small amount of filtered protein is reabsorbed by the proximal nephron. Only trace amounts of the larger globulins are filtered and reabsorbed. On the other hand, substantial amounts of smaller proteins such as myoglobin, hemoglobin, and light chains from immunoglobulins can be filtered. However, these small proteins appear in the plasma only under pathologic conditions, for example, when a transfusion reaction causes the release of hemoglobin from damaged erythrocytes.

Because of their small molecular dimensions, solutes with molecular weights of less than 5000 Da are freely filtered; this includes all components of the plasma except proteins and those solutes that bind to plasma proteins. Because of the Gibbs-Donnan effects of the plasma proteins (see Chapter 3), the concentrations of cations in the glomerular ultrafiltrate will be 4–5% lower than in the plasma and the concentrations of anions will

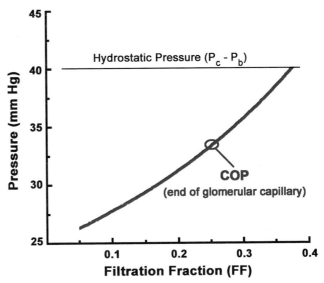

FIGURE 6 Rise in the COP at the end of the glomerular capillaries with increasing filtration fraction. Because the glomerular filtrate is protein free, the COP of the plasma remaining in the glomerular capillary is increased by ultrafiltration. If the COP rises to the same level as the net hydrostatic pressure ($P_c - P_b$; see Eq. [3]) at some point along the length of the capillary, there is no filtration beyond that point. The graph assumes that the net hydrostatic pressure is 40 mm Hg and the COP of the plasma entering the glomerular capillary is 25 mm Hg (as in Fig. 3). For a filtration fraction greater than 0.37, the net filtration pressure would fall to zero before the end of the glomerular capillary.

across them. For example, if the COP of the peritubular capillary plasma is reduced because of a lower FF, fluid will not be reabsorbed from the interstitium as rapidly, and hydrostatic pressure will rise in the interstitium until it is sufficient to drive uptake into the capillaries at the same rate at which it is reabsorbed by the nephrons. However, this interstitial pressure increases only when the interstitial volume has increased. The absorption of fluid by peritubular capillaries is considered further in Chapters 26 and 29 in connection with the regulation of salt and water reabsorption by the proximal nephron.

be 4–5% higher, but for practical purposes their concentrations in the ultrafiltrate are considered here to be the same as in plasma. On the other hand, some anions (such as phosphate) and some cations (especially the divalent cations Ca^{2+} and Mg^{2+}) bind in substantial amounts to plasma proteins, and their rate of filtration will be proportional to their free ionized concentration. For example, in the case of Ca^{2+}, approximately 50% of the total amount in the plasma is bound to plasma proteins and, consequently, the rate of Ca^{2+} filtration is only ~50% of the rate calculated from the product of the GFR and the total Ca^{2+} concentration in the plasma.

Bowman's space normally contains negligible concentrations of plasma proteins, so that the COP difference across the filtration barrier is simply the average glomerular capillary COP. However, in pathologic states that lead to the presence of smaller proteins in the plasma, or that increase the permeability of the filtration barrier to albumin and globulins, proteins in Bowman's space can produce a significant COP. This would reduce the net COP that normally opposes glomerular filtration and, thus, the GFR may actually increase in the early stages of renal diseases in which the filtration barrier becomes permeable to proteins. The GFR diminishes as these diseases progress because of a decrease in the number of functioning glomeruli.

The anatomic barrier between the plasma in the glomerular capillary and Bowman's space is a composite of the three structures shown in Fig. 7: the capillary endothelium, the basement membrane or basilar lamina covering the capillary endothelium, and the epithelium surrounding Bowman's space. The endothelium of renal

---

Clinical Note

Glomerular Diseases

Glomerular diseases are the primary cause of chronic renal failure that often progresses, either rapidly or more typically over 10–20 years or more, to end-stage renal disease (ESRD). ESRD is defined by a sufficient reduction of GFR (usually to ≤10% of normal) such that dialysis or kidney transplantation is required to maintain life. The glomerular filtration barrier is the site of primary pathology, which is usually often produced or exacerbated by the immune system. Immune damage can be due either to an abnormal reaction against the normal components of the glomerular filtration barrier, so-called autoimmune disease, or to the deposition of immune complexes against exogenous antigens. Because of the rapid rate of blood flow through the glomerular capillaries, the relatively permeable architecture of the endothelial cells, and the high rate of ultrafiltration, glomerular cells and the basement membrane are particularly susceptible to leukocyte infiltration, and the accumulation of immunoglobulins and immune complexes. The hallmark of glomerular damage is the presence of significant amounts of protein in the urine (proteinuria), which indicates failure of the filtration barrier.

Often the pattern of damage is not one of acute inflammation, and there may be histological evidence only of immunoglobulins trapped in the endothelial and basement membrane layers, with no cell proliferation. Early in such disease, there is often little evidence of glomerular capillary damage. This pattern is typical of the nephrotic syndrome. In some cases, the patient is asymptomatic, and the disease is first revealed only by proteinuria in a routine urine sample. More often, the disease is characterized by marked edema due to the loss of plasma proteins and renal salt retention. However, early in the course of the nephrotic syndrome, RBF and GFR can be normal and there is no uremia.

In contrast, the more prevalent glomerular disease, glomerulonephritis, is characterized by mild to severe proteinuria, with the presence of red and white blood cells, and epithelial cell casts in the urine. These are signs of an active inflammatory process in the glomeruli that is accompanied by proliferation of inflammatory cells among the endothelial, mesangial, and epithelial cells, and often within Bowman's space. Glomerulonephritis may occur acutely, for example, as the result of immune complex deposition after a streptococcal infection, or in a slow and patchy fashion. However, the GFR is usually reduced from the outset because of the closure of glomerular capillary loops by immune cells.

Both the nephrotic and nephritic forms of glomerular disease usually progress with destruction of the glomeruli by fibrotic infiltration, called glomerulosclerosis, resulting in a progressive decline in the number of functioning nephrons and hence in the RBF and GFR.

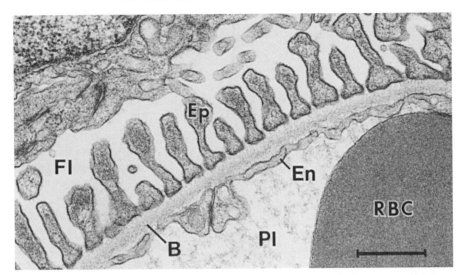

FIGURE 7   Electron micrograph of the filtration barrier. The blood space of the capillary lumen is separated from the ultrafiltrate in Bowman's space by three layers: the endothelial cell layer of the capillary (En), the glomerular basement membrane (B), and the visceral epithelial layer of Bowman's capsule (Ep). The cells in this epithelial layer (Ep) are characterized by the foot processes (podocytes) that lie against the basement membrane (B). A portion of a red blood cell (RBC) is seen in the lower right-hand corner of the micrograph surrounded by blood plasma (Pl). The plasma contains significant amounts of protein, as indicated by the diffuse, electron-dense, granular regions. In contrast, the ultrafiltrate (Fl) seen in Bowman's space contains no electron-dense material. The scale bar is 0.5 μm in length; magnification is 54,000×. (Electron micrograph courtesy of Dr. Dale R. Abrahamson, University of Kansas Medical Center, Kansas City, KS.)

glomerular capillaries, much like the liver, has fenestrations occupying nearly 10% of its surface area that allow plasma proteins to reach the basement membrane. On the other side of the basement membrane, the plasma membrane of the epithelial cells is folded intricately into so-called foot processes or podocytes (shown in cross section in Fig. 7) where they contact the basement membrane.

Although good evidence suggests that damage to the basement membrane can lead to proteinuria, recent studies of the *congenital nephrotic syndrome* have provided evidence that the podocyte layer is at least of equal importance. In contrast to the usual forms of glomerular pathology discussed in the Clinical Note on glomerular diseases, the congenital nephrotic syndrome is an autosomal recessive disorder that is characterized by severe proteinuria, which begins at birth with no underlying immune damage. Although this disorder is relatively rare worldwide, in Finland it has an incidence of 1 in 8000, and it is associated the usual morbidity and mortality associated with rapidly progressive glomerulonephritis. Recently, the gene defect responsible for the congenital nephrotic syndrome has been identified. The gene encodes a protein, named *nephrin,* which is truncated by a mutation in affected individuals. Nephrin is a member of an immunoglobulin family that is involved in cell adhesion and it is found exclusively in the junctional

complexes between glomerular epithelial cells in the region of the podocyte layer. Thus, it appears that an abnormality in a single protein results in sufficient disruption of the cell-to-cell contacts in the epithelial layer to allow substantial protein filtration.

The thickness of the glomerular basement membrane (about 50 nm) is considerably more than that of other capillaries, because this basement membrane is formed by a merging of the capillary and epithelial basement membranes during development. Like all basement membranes, the glomerular basement membrane is a gel-like structure formed from collagenous and noncollagenous glycoproteins called *proteoglycans*. Laminin is one of the more important of these proteoglycans in the glomerular basement membrane. The proteoglycans have a net negative charge because of their sialic and dicarboxylic amino acids. Podocyte membranes also have a high density of negative charge due to the presence of other glycoproteins and, together with the basement membrane, they electrostatically repel negatively charged plasma proteins. Therefore, the ability of macromolecules to permeate the filtration barrier is determined by their size and their valence. These two factors have been well demonstrated by experiments in humans and animals in which the excretion of test macromolecules has been measured as shown in Fig. 8. These test substances have included synthetic polymers

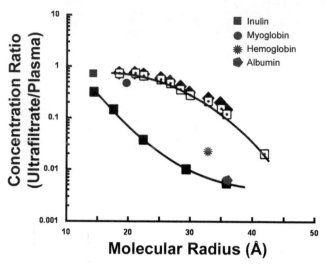

FIGURE 8   Effect of molecular size and charge on filtration of macromolecules in the glomerulus. The ratio of the concentrations of the macromolecule in the ultrafiltrate compared to the plasma is plotted as a function of the molecular radius of the test molecule. These experiments were performed with different sizes of uncharged polymers of dextran (▲) or polyvinylpyrrolidone (PVP; □) and with the negatively charged polymer dextran sulfate (■). Note that concentration ratios for dextran sulfate were significantly lower throughout the size range tested, and that the ratio for the largest molecule of dextran sulfate tested closely approximated that of albumin, which is also negatively charged. For comparison, the concentration ratios for inulin, myoglobin, and hemoglobin are also included. (Replotted data from Chang *et al.*, *Kidney Int* 1975;8:212–218; Lambert *et al.*, *Pflügers Arch* 1975;359:1–22; Vanrenterghem *et al.*, *Clin Sci* 1980;58:65–75; and Whiteside, Silverman, *Am J Physiol Renal Fluid Electrolyte Physiol* 1983;245:F485–F495.)

that are available in a wide range of molecular sizes such as polyvinylpyrrolidone or dextran (a sucrose polymer to which anionic sulfate residues can be bonded), as well as naturally occurring protein molecules.

The ultrafiltration coefficient, K_f, can be altered not only by changes in the thickness or composition of the basement membrane and the podocyte slits, but also by their effective areas. It has been proposed that this filtration area may be regulated by humorally regulated contraction or relaxation of mesangial cells within the glomerulus. Disease processes can also change the number and size of the endothelial fenestrae, the shape and size of the epithelial podocytes, or the thickness and properties of the basement membrane. Any of these processes may change the permeability characteristics of the filtration barrier, leading to changes in K_f as well as the loss of proteins into the filtrate. When the filtration of proteins exceeds the ability of the proximal tubule to reabsorb them, they appear in the final urine, a condition known as *proteinuria*. Excessive albumin losses in the urine may exceed the capacity of the liver to synthesize albumin, leading to a reduction in plasma albumin concentration and a decrease in plasma COP. This can

contribute to the severe edema (see Chapter 16), which is a characteristic of the nephrotic syndrome (see earlier Clinical Note).

RENIN-ANGIOTENSIN SYSTEM

Many factors regulate RBF and GFR, but perhaps the most important is the renin-angiotensin system (RAS). Not only does the RAS regulate renal hemodynamics, but it is also more globally involved in regulating the ionic composition and volume of the extracellular fluid, and the systemic blood pressure. Blood pressure within the kidney is tightly regulated and is sensed by mechanisms that feed back and regulate systemic blood pressure through the RAS. Modified vascular smooth muscle cells, referred to as *granular cells* or *juxtaglomerular cells,* are present within the afferent arteriole. These cells store the enzyme *renin* in numerous cytoplasmic secretory granules, and they are extensively coupled to mesangial and smooth muscle cells by gap junctions. Juxtaglomerular cells release renin in response to an increase in intracellular cAMP or a fall in cytoplasmic Ca^{2+}. The effect of cAMP is understandable because it is the intracellular second messenger for receptors that respond to sympathetic afferent nerves and circulating catecholamines. On the other hand, the response to intracellular Ca^{2+} is opposite from most systems in which an increase in Ca^{2+} stimulates secretion. However, decreased stretch in the afferent arteriole and juxtaglomerular cells, as would occur when the blood pressure falls, closes Ca^{2+} channels and lowers intracellular Ca^{2+} in these cells. As discussed later, the resulting release of renin counteracts the fall in blood pressure, thus completing a negative feedback system.

Renin itself has no direct hemodynamic effects, however, as shown in Fig. 9, it acts as a protease that cleaves the decapeptide angiotensin I from a specific α_2-globulin (angiotensinogen) circulating in the blood. A second protease, *angiotensin-converting enzyme* (ACE), in turn cleaves two amino acids from angiotensin I to produce the active octapeptide angiotensin II. ACE is found primarily in the lungs, but it is also present in the kidneys and other tissues where it rapidly converts angiotensin I to angiotensin II.

As illustrated in Fig. 9, angiotensin II has two major systemic effects: (1) It is a direct vasoconstrictor that acts on resistance vessels throughout the body, resulting in an elevation of total peripheral resistance and blood pressure; and (2) it acts via receptors in the adrenal cortex to increase secretion of the hormone aldosterone. As discussed later in relation to the function of the distal nephron (see Chapter 27), aldosterone increases the

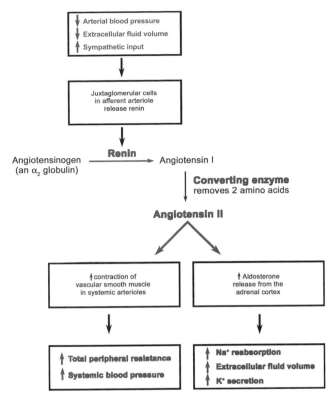

FIGURE 9   Production of angiotensin II by the release of renin from juxtaglomerular cells. A variety of signals associated with falling blood pressure and/or extracellular fluid volume loss cause release of renin from juxtaglomerular cells. Renin acts as an enzyme to cleave angiotensin I from an α_2-globulin, angiotensinogen. Angiotensin I is changed to angiotensin II by the cleavage of two amino acids, a process catalyzed by ACE. Angiotensin II acts on vascular smooth muscle to increase systemic blood pressure. Angiotensin II also acts on the adrenal cortex to release aldosterone, which increases renal Na^+ reabsorption and K^+ secretion.

reabsorption of Na^+ as well as the secretion of K^+ and protons. Thus, the release of renin has two actions that counteract any reduction of blood pressure and blood volume. Its pressor effects elevate the blood pressure, and its effects via aldosterone conserve Na^+ and thus expand the extracellular fluid volume, which also indirectly elevates blood pressure and cardiac output (see also Chapter 29).

As might be expected if the RAS were involved in feedback regulation of blood pressure and volume, renin is released in response to signals that indicate a decrease in extracellular fluid volume, such as increased afferent sympathetic nerve input to the kidney. However, renin is also released in response to a fall in blood pressure that is sensed directly by the afferent arteriole. A fall in systemic blood pressure would cause renin release; however, a fall in the *renal* arterial pressure, independent of any change in systemic blood pressure, also causes renin release (see the Clinical Note on renovascular hypertension).

Renin-mediated hypertension may also be treated by drugs referred to as *ACE inhibitors*, which inhibit the formation of angiotensin II from its inactive precursor angiotensin I. The actions of angiotensin II can also be blocked by drugs that block its receptors. There are at least two angiotensin II receptors and the drugs now in common use block both. Investigators are currently working to identify differences in the actions of the two types of receptors, and type-specific receptor blockers have been developed, but it is presently unclear if there are advantages to blocking one or the other receptor type. Both ACE inhibitors and receptor antagonists have been extremely important antihypertensive drugs, especially in treating salt-sensitive hypertension with elevated plasma renin and hypertension in diabetic patients, who frequently develop kidney failure.

Because ACE is present in the kidney, renin secretion results in the production of angiotensin II in the renal cortex where it can act as a paracrine factor. Angiotensin II increases proximal tubule volume absorption (see Chapter 26), and it acts directly on the afferent and efferent arterioles to increase their resistance, thus decreasing RBF and GFR. The RAS also responds to changes in salt delivery to the ascending limb of the loop of Henle and, thus, to the macula densa cells. In situations where the GFR is elevated or proximal tubule reabsorption of NaCl and water is diminished, for example, with isotonic volume expansion, there is an increased delivery of both to the loop of Henle. This results in an increased NaCl concentration at the end of the loop of Henle. The increased NaCl delivery is sensed by the macula densa cells, which signal adjacent juxtaglomerular cells to decrease renin secretion. A resulting decrease is seen in circulating angiotensin II levels and a corresponding decrease in aldosterone production, both of which would contribute to natriuresis and diuresis.

AUTOREGULATION

As shown in Fig. 10, RBF and GFR remain relatively constant over a wide range of systemic blood pressures. To prevent changes in blood flow and GFR with changes in systemic blood pressure that occur normally during the day with changes in activity, the resistance to flow must change appropriately. As systemic blood pressure increases, total renal vascular resistance increases so that blood flow and GFR remain constant. Autoregulation, which is exhibited in the circulation of many organs, refers to an intrinsic adjustment of vascular resistance that counterbalances any extrinsic factors that would change flow other than by a direct effect on the vascular resistive elements themselves.

Clinical Note

Renovascular Hypertension

Because the juxtaglomerular cells release renin in response to lower renal vascular pressures and not systemic blood pressure, stenosis of a renal artery or one of its branches stimulates renin secretion. Even if renin secretion is augmented in only one portion of one kidney, it can be sufficient to raise systemic plasma renin levels. The resulting formation of angiotensin II has a direct pressor effect by increasing the resistance of arterioles throughout the body. Furthermore, as discussed in more detail in Chapters 27 and 29, angiotensin II stimulates aldosterone secretion, which drives salt retention by its effect on Na^+ reabsorption in the connecting tubule and collecting duct.

Patients with atherosclerotic plaques or tumors that compress the renal artery and produce renal artery stenosis usually present clinically because of their hypertension. Laboratory work often reveals hypokalemia secondary to high plasma aldosterone levels because aldosterone drives K^+ secretion. More specialized laboratory work may reveal an elevated plasma renin level, but in some cases the increased renin can be observed only by measuring unilateral renal vein renin levels, in which case the renin activity on the stenotic side is elevated but depressed in the contralateral renal vein. Renal arteriography is used to confirm the stenosis, its origin, and to determine the method of treatment.

Because renovascular hypertension (sometimes called *renoprival hypertension*) is treatable by relief of the occlusion (e.g., by angioplasty or surgery), it is imperative that this cause of hypertension be excluded early in the normal workup of the hypertensive patient.

In the kidney, autoregulation is more precise, and it appears to act primarily to regulate GFR rather than blood flow, because GFR remains constant not only with increased mean arterial pressure, but also with changes in venous pressure, increased ureteral pressure, and increased systemic plasma COP. For example, an increase in COP of the systemic plasma might be expected to decrease GFR but not to affect RBF, yet it has been found that GFR is maintained by a compensating increase in RBF. Similarly, obstruction of the lower urinary tract would decrease GFR by increasing the hydrostatic pressure in Bowman's space but, again, unless the obstruction is severe, GFR is usually maintained nearly constant by an increase in RBF.

The maintenance of a constant GFR in the face of changes in mean arterial pressure, venous pressure, or obstruction (i.e., renal autoregulation) appears to be accomplished largely by changes in the resistance of the afferent arteriole. Decreasing afferent arteriolar resistance increases the effective filtration pressure and counteracts events tending to reduce GFR. Alternatively, increasing afferent arteriolar resistance counteracts increases in GFR. Two mechanisms, the myogenic response and tubuloglomerular feedback, contribute to this regulation of afferent arteriolar resistance.

When the pressure in the afferent arteriole increases, stretches the vessel wall, and this stretch directly stimulates the arteriolar smooth muscle to contract. The resulting constriction of the arteriole increases its resistance, thereby counteracting the tendency of increased blood pressure to increase flow. This intrinsic regulation is referred to as a *myogenic response*. Although this mechanism operates in many organs (see Chapter 17), it cannot be the only autoregulatory mechanism in the kidney because, as described earlier, GFR is the primary parameter that appears to be regulated, even at the sacrifice of a constant RBF. Thus, it seems reasonable that the parameter sensed and regulated is GFR or some factor dependent on GFR, rather than RBF.

The tubuloglomerular feedback mechanism is responsible for this sensitive autoregulation of GFR and it

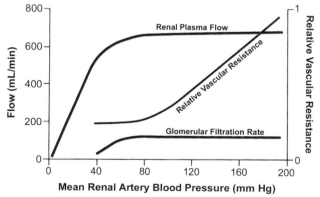

FIGURE 10   Autoregulation of RPF and GFR with changes in arterial blood pressure. The relative resistance of the renal vasculature increases as the mean blood pressure in the renal artery increases, thereby maintaining relatively constant RPF and GFR over mean arterial blood pressures ranging from 80–180 mm Hg.

depends of the effect of changes in the GFR on the delivery to NaCl to the macula densa. When, for example, the filtration rate of a single nephron (the so-called single nephron GFR, or snGFR) increases, the rates of fluid and NaCl delivery to the loop of Henle and the macula densa increase in direct proportion. This increase in NaCl delivery is found to cause a decrease in the snGFR of that nephron, thus forming a negative feedback loop. In animal experiments, it has been shown that perfusion of the distal nephron with a maximal flow rate reduces snGFR by 40–80%. Alternatively, when distal nephron flow is reduced, snGFR is increased. This behavior is called *tubuloglomerular feedback*, and it provides a mechanism by which each nephron maintains a constant snGFR. Recent studies, such as the one illustrated in Fig. 11, have shown that increased NaCl delivery to the macula densa causes the cells to swell. This swelling causes the release of adenosine triphosphate (ATP) and/or adenosine, which acts via receptors on the juxtaglomerular cells in the afferent arteriole to increase its resistance by contraction.

Taking all nephrons together, the result of tubuloglomerular feedback would be the maintenance of a constant total kidney GFR. For example, if arterial pressure were to increase, the initial increase in glomerular capillary pressure would cause an increase in snGFR of all nephrons. The resulting increase in NaCl and water delivery to the macula densa would then produce an increase in afferent arteriolar resistance that would decrease glomerular capillary pressure and snGFR would fall toward normal. The increase in afferent arteriolar resistance would also maintain a constant RBF despite the increase in blood pressure.

Tubuloglomerular feedback depends on signaling between the macula densa cells that sense a change in NaCl delivery and the afferent arteriole, which is mediated by the release of either ATP or adenosine from the macula densa cells in response to increased NaCl delivery. However, students (and professors) are often confused by what seem to be contradictory effects of changes in salt delivery to the macula densa. According to the tubuloglomerular feedback mechanism, decreased NaCl delivery to the macula densa should produce relaxation of the afferent arteriole and an increase in GFR. However, as noted in the previous section, decreased NaCl delivery to the macula densa is a stimulus for renin release by the juxtaglomerular cells. The resulting increase in angiotensin II should constrict afferent and efferent arterioles and decrease the renal blood flow and GFR.

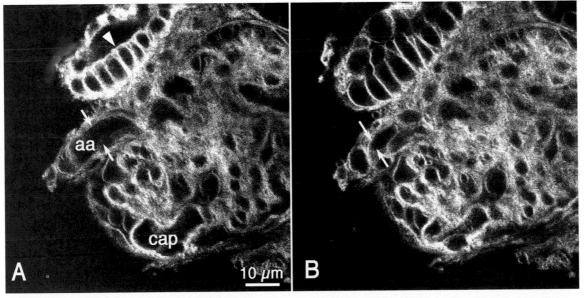

FIGURE 11   Effect of increased NaCl concentration on macula densa cells and afferent arteriole. Both panels show the same isolated glomerulus and adhering components of the juxtaglomerular apparatus. The cell membranes were visualized using a fluorescent dye, and optical sections were obtained by two-photon laser scanning fluorescence microscopy. (A) A micropipette was used to perfuse the thick ascending limb (upper left, arrowhead in lumen) with a solution containing 25 mmol/L NaCl. The macula densa cells form a plaque of cells on the side of the tubule toward which the arrowhead points. The diameter of the lumen of the afferent arteriole (aa) is indicated by the space between the two arrows. A portion of a capillary loop (cap) is shown at the lower edge of the glomerulus. (B) When the solution perfusing the thick ascending limb in the same preparation was changed to one containing 135 mmol/L NaCl, the macula densa cells swelled dramatically, while the lumen of the afferent arteriole was almost obliterated by constriction. The increased afferent arteriolar resistance and the reduction in blood flow are reflected by the reduction in the diameter of the capillary. (From Petri-Peteroli, J. *et al.* (2002). *Am. J. Physiol.* 283:F197–F201. With permission of the journal and the author.)

There are two important points to note in addressing this apparent contradiction. First, the tubuloglomerular feedback mechanism is mediated by direct signaling from macula densa cells to the afferent arteriole. This signaling does not involve renin release or angiotensin II production. Second, tubuloglomerular feedback is directed at the rapid regulation of the filtration rate of individual nephrons, whereas the release of renin when NaCl delivery is reduced has systemic effects but only when it involves many nephrons. The subsequent production of angiotensin II and aldosterone, and their effects to elevate systemic blood pressure and extracellular fluid volume, can be viewed as an emergency response to a general reduction in the filtration rate of all nephrons, which might occur in dehydration or when systemic blood pressure falls.

In summary, the tubuloglomerular feedback mechanism is a rapidly responding local mechanism for the regulation of snGFR, whereas the effect of flow and NaCl delivery to the macula densa on renin release is a more slowly responding system that has significant effects only when multiple nephrons are involved. Furthermore, flow and NaCl delivery to the macular densa is one of only several factors that regulate renin release see Chapter 29.

PHYSIOLOGIC AND PATHOPHYSIOLOGIC CHANGES IN RENAL BLOOD FLOW AND GLOMERULAR FILTRATION RATE

Although GFR and RPF are normally relatively constant, they can change from their normal set point when influenced by other signaling systems. Under normal conditions, the GFR is nearly maximal so that most of the normal physiologic changes occur in the direction of decreasing GFR. Presuming there are no changes in plasma protein concentration, this could be accomplished by changes in afferent or efferent arteriolar resistance, or in the ultrafiltration coefficient K_f. The effects of arteriolar resistance changes on RBF are predictable as in any capillary bed. However, because the relative resistances of the afferent and efferent arterioles affect both glomerular capillary pressure and the average glomerular capillary plasma COP, the effects of resistance changes on GFR are more complex. It is also likely that K_f may be altered by changes in the level of arteriolar contraction, and by the configuration of mesangial cells surrounding the glomerular capillaries, which may alter the capillary area available for filtration.

Both the arterioles and mesangial cells constrict in response to circulating catecholamines, and they receive sympathetic adrenergic innervation and constrict in response to increased firing. Other hormones, including vasopressin (also called antidiuretic hormone or ADH), angiotensin II, and glucocorticoids (the latter probably by the increase in circulating angiotensin II it produces), also cause constriction of the arterioles and mesangial cells. In addition to these circulating hormones, the arterioles and mesangial cells respond to agents that are produced locally in the kidney rather than in extrarenal endocrine glands. These substances are referred to as *autacoids* when they act on the same cells that produce them, and as *paracrine factors* when they act on neighboring cells.

Effect of Changes in Afferent and Efferent Arteriolar Resistances

As illustrated previously in Fig. 4, if the efferent arteriolar resistance is constant, the glomerular capillary pressure is inversely proportional to the afferent arteriolar resistance. Alternatively, if the afferent arteriolar resistance is constant, the glomerular capillary pressure is directly proportional to the efferent arteriolar resistance. However, if there is no offsetting decrease in the resistance of the other arteriole, as in the examples in Fig. 4, increases in either afferent or efferent resistance cause a decrease in RPF. Furthermore, as shown in Figs. 5 and 6, when RPF decreases without a proportional decrease in GFR, the resulting increase in FF leads to an increase in the average COP in the glomerular capillaries, which diminishes the net filtration pressure.

When the afferent arteriolar resistance decreases with no change in efferent resistance, both GFR and RPF rise, and both fall when the resistance increases, as shown in Fig. 12. This can be understood easily because changes in afferent arteriolar resistance affect net filtration pressure in a direct way that parallels the effect on RPF. For example, when afferent arteriolar resistance rises, the glomerular capillary pressure falls, as does the RPF. Therefore, changes in only afferent arteriolar resistance give rise to large and parallel changes in both GFR and RPF.

The effects of changes in efferent arteriolar resistance on GFR, although sometimes difficult to understand, are predictable. As shown in Fig. 13, a decrease in the efferent arteriolar resistance with no change in afferent resistance leads to a rise in RPF and a fall in GFR, because the fall in efferent arteriolar resistance causes a fall in glomerular capillary pressure and thus a fall in net filtration pressure. GFR increases with a small increase in the efferent arteriolar resistance, but only slightly. At higher efferent arteriolar resistances, GFR falls even though glomerular capillary pressure rises. This apparent paradox occurs because the rise in COP

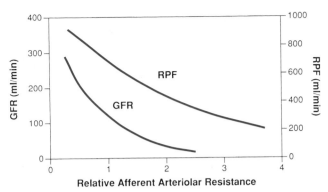

FIGURE 12   Effect of changes in the relative resistance of the afferent arteriole on GFR and RPF. The normal resistance is taken to be 1.0 so that, for example, a relative resistance of 2.0 is two times the normal resistance. The efferent arteriolar resistance in this example is presumed to remain constant. (Modified from Navar LG, Bell PD, Evan AP. The regulation of glomerular filtration rate in mammalian kidneys. In Andreoli TE *et al.*, eds., *The physiology of membrane disorders,* 2nd ed. New York: Plenum, 1986, pp 637–667.)

caused by the decreased RPF is greater than the rise in glomerular capillary pressure, and the net filtration pressure falls. Therefore, there is only a limited capacity to increase GFR by increasing efferent arteriolar resistance because of the opposing effects on FF and COP.

In the two examples of Figs. 12 and 13, it is assumed that only one or the other arteriolar resistance changes alone with no other changes except for COP. In most circumstances, both resistances change in parallel. Generally, when both resistances increase by 20% or less, as occurs with a moderate increase in sympathetic

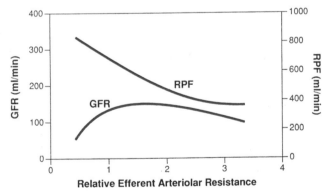

FIGURE 13   Effect of changes in the relative resistance of the efferent arteriole on GFR and RPF. As in Fig. 10, the normal resistance is taken to be 1.0. At this resistance, GFR is nearly maximal so that an increase in resistance gives only a small increase in GFR and then a decrease for the reasons discussed in the text. The afferent arteriolar resistance in this example is presumed to remain constant. (Modified from Navar LG, Bell PD, Evan AP. The regulation of glomerular filtration rate in mammalian kidneys. In Andreoli TE *et al.*, eds., *The physiology of membrane disorders,* 2nd ed. New York: Plenum, 1986, pp 637–667.)

firing via the renal nerves, GFR changes little and RPF falls. When both afferent and efferent resistances increase more than about 20%, both RPF and GFR fall because of the effects of the increased FF on the average COP.

Renal Nerves and Circulating Vasoactive Hormones

Nerve fibers from the sympathetic celiac plexus supply the kidney via renal nerves with both nonmyelinated and myelinated fibers, although the latter are not found in glomerular structures. The nerve endings are primarily α-adrenergic and secrete norepinephrine. These nerve endings are found in association with the afferent and efferent arterioles, with mesangial cells, and with the juxtaglomerular apparatus. In addition, there are nerve endings in the epithelial cell layer of both proximal and distal regions of the nephron. Because these nephron segments have receptors that respond to adrenergic and dopaminergic agonists and antagonists, the sympathetic nervous system is also involved in the regulation of reabsorptive and secretory functions, as well as with the regulation of vascular resistance.

Intravenous infusions of epinephrine and norepinephrine decrease RBF and GFR. However, under normal conditions renal nerve activity is so low that it exerts only minimal effects on the GFR. However, in this setting, it has been shown that a small increase in sympathetic stimulation that has little effect on RBF augments renin release. Modest increases in renal nerve activity or the infusion of low concentrations of adrenergic agonists also have little effect on GFR, although they reduce RBF because they produce equivalent constriction of both afferent and efferent arteriolar resistances. Increased sympathetic stimulation, such as that produced by occlusion of a carotid artery, increases both afferent and efferent arteriolar resistances, and RBF and GFR decrease significantly. It has also been proposed that increased renal nerve activity may decrease K_f, thus contributing to the fall in GFR seen with moderate renal nerve stimulation or epinephrine infusion.

Strong stimulation via the renal nerves can markedly increase both resistances. In the extreme, such stimulation leads to marked diminution in RPF and a cessation of GFR. This is obviously an extreme response; however, renal "shutdown" such as this is often observed in severe hemorrhage, as sequelae of surgical procedures (e.g., during coronary bypass), shock, or trauma. These events are common precipitants of ischemic damage to the kidney, resulting in acute renal failure.

Autocrine and Paracrine Regulators of Renal Blood Flow and Glomerular Filtration Rate

Table 2 lists several endogenous substances that have been shown to produce changes in afferent and efferent arteriolar tone. Many of these substances are autacoids and paracrine factors that are released by the glomerular capillary endothelium and that act locally on their associated arterioles and mesangial cells. Nitric oxide (NO), previously referred to as endothelial-derived relaxing factor, is also released by endothelial cells in response to increased shear stress or pressure. NO is a potent vasodilator in most arterioles including the afferent and efferent renal arterioles. The binding of acetylcholine, bradykinin, or histamine to endothelial cells results in the production of NO and decreased arteriolar resistance. Prostaglandins, particularly PGE_2 and PGI_2 (prostacyclin), which are produced from arachidonic acid in vascular smooth muscle, mesangial, and epithelial cells, are also important vasodilators, but they act primarily by moderating the effects of strong vasoconstrictor stimulation to the kidney. For example, prostaglandins have little vasodilator effect when administered to a normal individual, but when sympathetic tone to the kidney is high or when circulating levels of angiotensin II are reducing RBF and GFR, they have a marked antagonistic effect, as illustrated by the clinical note in this section on the effect of NSAIDs. Prostaglandin production by the kidney is stimulated by

TABLE 2 Circulating Hormones, Paracrine Factors, and Autacoids That Alter Resistances of Afferent and Efferent Arterioles

Vasoconstrictors	Vasodilators
Circulating Hormones	Circulating Hormones
Catecholamines	Dopamine
Angiotensin II	
Vasopressin (ADH)	
Glucocorticoids	
Paracrine Factors & Autacoids	Paracrine Factors & Autacoids
Endothelin	Nitric oxide (NO)
Thromboxane A_2	Acetylcholine
Leukotrienes	PGE_2
	PGI_2 (prostacyclin)
	Bradykinin

angiotensin II and, conversely, renin release is inhibited by PGE_2.

The most potent vasoconstrictor of endothelial origin is endothelin. Endothelin actually represents a family of 21-amino-acid peptides that are synthesized by various endothelial cells. Endothelin acts on neighboring vascular smooth muscle cells in arterioles to increase intracellular Ca^{2+} by releasing it from intracellular stores and thus increasing arteriolar resistance. The production of endothelin has been implicated in the nephrotoxicity of some drugs such as cyclosporine,

---

Clinical Note

Effect of NSAIDs in the Presence of Stimuli for Renal Vasoconstriction

Nonsteroidal anti-inflammatory drugs (NSAIDs; e.g., ibuprofen, naproxen, aspirin, and acetaminophen) exert analgesic, antipyretic, and anti-inflammatory effects because they inhibit the enzyme cyclooxygenase, which is necessary for the synthesis of prostaglandins from arachidonic acid. In normal individuals, the use of these drugs has little or no effect on RBF or GFR. However, in patients who have hemorrhaged or who are dehydrated (volume contracted), or who are otherwise physically stressed, these drugs may exacerbate damage to the kidney and increase the likelihood of acute renal failure. In these situations, the pathologic event increases the activity of the renal nerves, which is accompanied by the release of renin and the production of angiotensin II. Both increased renal nerve activity and angiotensin II increase afferent and efferent arteriolar resistance and decrease both RBF and GFR. When severe, these responses can diminish RBF to the point of renal infarction. This extreme situation is normally countered by the release of prostaglandins whose vasodilatory actions oppose the vasoconstrictors. When the patient is being treated with NSAIDs, prostaglandin synthesis is inhibited and, consequently, the likelihood of acute renal failure is increased with hemorrhage, dehydration, or surgical stress. For this reason, NSAIDs are counterindicated before many surgical procedures, or when circulating renin or catecholamine levels, or renal sympathetic nerve activity, are expected to be high. For example, NSAIDs must be used with caution or avoided in patients with chronic renal failure, congestive heart failure, cirrhosis of the liver with ascites, or before surgery.

which is used to inhibit the immune response in transplant recipients.

Redistribution of Renal Blood Flow

The factors regulating RBF and GFR may differ in various regions of the kidney so that the distribution of glomerular filtrate can be shifted between superficial and juxtamedullary nephrons. Superficial nephrons have lower rates of glomerular filtration than do midcortical nephrons, and juxtamedullary nephrons have higher filtration rates than either. Juxtamedullary nephrons appear to have a greater intrinsic capacity to reabsorb salt and water, and glomerular filtration may be shifted to these nephrons when salt and water conservation is dictated. There may be an increase in flow to the superficial nephrons during periods of increased salt intake and extracellular fluid volume expansion. Although such changes in the distribution of RBF and glomerular filtration could play an important role in the renal response to volume expansion or contraction, there is no conclusive evidence about the extent to which they contribute to regulation of salt and water excretion.

Suggested Readings

Arendshorst WJ, Navar LG. Renal circulation and glomerular hemodynamics. In Schrier RW, Gottschalk CW, eds., *Diseases of the kidney*; Boston: Little-Brown, 1997, pp 59–106.

Dworkin LD, Sun AM, Brenner BM. Biophysical basis of glomerular filtration. In Seldin DW, Giebisch G, eds., *The kidney: Physiology and pathophysiology*, Vol 1, 3rd ed. Philadelphia: Lippincott Williams & Wilkins, 2000, pp 749–770.

Moss NG, Colindres RE, Gottschalk CW. Neural control of renal function. In Windhager EE, ed., *Handbook of physiology, Section 8, Renal physiology*, Vol. I. New York: Oxford, 1992, pp 1061–1128.

Navar LG. Regulation of renal hemodynamics. *Adv Physiol Educ* 1998;275:S221–S235.

Navar LG, Inscho EW, Majid DSA, Imig JD, Harrison-Bernard LM, Mitchell, KD. Paracrine regulation of the renal circulation. *Physiol Rev* 1996;76:425–536.

Schnerrman J, Briggs JP. Function of the juxtaglomerular apparatus: Control of glomerular hemodynamics and renin secretion. In Seldin DW, Giebisch G, eds., *The kidney: Physiology and pathophysiology*, Vol 1, 3rd ed. Philadelphia: Lippincott Williams & Wilkins, 2000, pp 945–980.

Mass Balance in Body Homeostasis and Renal Function

JAMES A. SCHAFER

KEY POINTS

- *Mass flow* is the amount of solute or water that moves per unit of time. Mass flow is calculated as the product of volume flow and the concentration of the solute in that volume.
- *Mass balance* in the body refers to the relationship between the rates at which substances are taken into the body and/or produced by metabolism, and the rate at which they leave the body. A *steady state* exists when the rate of intake plus production and the rate of output are equal.
- *Homeostasis* refers the maintenance of a constant and optimal body content for each substance. The constancy requires the maintenance of a steady state despite relatively high rates of intake, production, and loss for some substances, and changes in those rates.

- The fundamental mass balance relationship for the nephrons is that the rate of excretion is equal to the sum of the rate at which a substance is filtered and the rate at which it is secreted, minus the rate at which it is reabsorbed.
- The *filtration rate* of a solute is the mass flow defined as the product of the glomerular filtration rate (GFR) and the solute concentration in the glomerular ultrafiltrate. The *excretion rate* of a solute is the product of the urine flow rate and the solute concentration in the urine.
- *Inulin* is a good indicator of the GFR because it is neither reabsorbed nor secreted; thus its rate of excretion is equal to its rate of filtration.
- The *clearance* of a substance is an index of the rate (in milliliters per minute) at which the solute is being removed from the renal plasma flow.

Understanding the role of the kidney in maintaining the constancy of the internal environment requires an appreciation of the quantities of substances that are taken into the body, produced by metabolism, and excreted from the body, as well as the quantities exchanged among body fluid compartments. Because substances are exchanged continuously and rapidly among these compartments, static parameters such as the concentration of a solute do not provide sufficient information about the *rates* of material movement or the regulatory adjustments that are occurring constantly. To appreciate the rate of movement into or out of a compartment, it is necessary to think in terms of *mass flow,* that is, in terms of an amount per minute. This flow can be expressed, as examples, in terms such as milliliters of water per minute, millimoles (or milliequivalents) of Na^+ per minute, or milligrams of creatinine per minute.

The total amounts of all solutes and water entering the kidney via the renal artery leave in the combined flows of the urine, the renal vein, and the lymphatics. Some of these solutes and water are filtered from the plasma in the glomerulus. Solutes and water are returned from the tubular fluid to the peritubular capillary circulation by the process of *reabsorption.* For some solutes, the amount filtered may be augmented by transport from the peritubular circulation into the tubular fluid, a process referred to as *secretion.* Thus, the rate at which a particular solute is excreted is given by the sum of the rate at which it is filtered and the rate at which it is secreted, minus the rate at which it is reabsorbed.

EQUATION OF MASS FLOW

It is easy enough to calculate a mass flow when talking about the amount of a solid material ingested. For example, if someone ingests 7.2 g of NaCl over a 24-hr period, the average rate of ingestion is 300 mg/hr or 5 mg/min (equivalent to about 0.09 mmol of Na^+ per min). On the other hand, when the substance is present in solution, as is usually the case, the mass flow is calculated as the product of the concentration of the solute and the volume flow of the solution in which the solute is present, or

$$\text{Mass flow} = \text{concentration} \cdot \text{volume flow}. \quad (1)$$
$$\text{amount/min} \qquad \text{amount/vol} \qquad \text{(vol/min)}$$

Equation [1] provides an easy and direct way of computing the rate of mass flow for solutes in a number of different settings by using any combination of convenient units as shown by the following.

With careful attention to maintaining consistent units, the same principles can be applied to any solute mass flow, for example, to the rate of ingestion of Na^+ in liquids. *This principle is applied to quantifying and comparing mass flows throughout this examination of renal function, so Eq. [1] is fundamental to a complete understanding of the material in the chapters that follow.*

MASS BALANCE IN THE TOTAL BODY

The body normally maintains a constant total content of all substances including water, solutes, and solid material stores. This constancy of the internal environment, with optimal amounts of all relevant substances, is referred to as *homeostasis* (see Chapter 1). To maintain a constant composition, every substance introduced into the body by either ingestion or parenteral infusion must leave the body at the same rate. However, the form of the eliminated substance is not necessarily the same as that taken in. For example, glucose in the diet is normally eliminated as equimolar amounts of carbon dioxide and water, but the atoms of carbon, hydrogen, and oxygen of the glucose molecule exit from the body at the same rate as they enter the body in the form of glucose.

Examples

1. What is the mass flow of Na^+ excreted in the urine if the rate of urine flow (UF) is 1.5 mL/min and the Na^+ concentration in that urine is 78 mmol/L? To apply Eq. [1], it is first necessary to convert the volume flow and concentration into *consistent volume units.* The Na^+ concentration can be expressed as 0.078 mmol/mL, or the UF can be expressed as 0.0015 L/min. The rate of Na^+ mass flow in the urine, referred to as the excretion rate, can then be calculated: 1.5 mL/min · 0.078 mmol/mL or 0.117 mmol/min.

2. If 2 L of urine are produced over a 24-hr period, and the creatinine concentration in that urine volume is 1.0 mg/mL, then the rate of creatinine excretion is (2 L/day) · (1000 mg/L) = 2000 mg/day or 83.3 mg/hr or 1.4 mg/min.

Other substances may not normally be ingested but are produced by metabolism. In this case, the rate at which a particular substance is eliminated must equal the rate at which it is produced in order for the total amount of the substance in the body to remain constant. In the most general terms, balance for any given substance can be maintained only when the rate of output from the body is equal to the sum of the rates at which the substance is ingested (or administered parenterally) and produced by metabolism:

$$\text{Rate of output} = \text{rate of intake} + \text{rate of production.} \quad (2)$$

This equality is referred to as a condition of *steady state* and is shown diagrammatically in Fig. 1. Although there may be a continuous flow of a substance into the body, a steady state (and, thus, a constant total amount of the substance in the body) will occur as long as the rate of output equals this intake plus any additional metabolic production.

MASS BALANCE IN THE KIDNEY

The same principles of mass balance apply to the kidney as apply to the whole body. If any given substance leaves the kidney at a rate different from that at which it enters, the renal content of that substance would constantly increase or decrease. The route of input into the kidney is the renal artery, but there are three routes of output: the renal vein, the renal lymphatics, and the urine. (Note that, for simplicity, any losses or gains due to metabolism within the kidney are not considered here, but in the case of some substances these changes may be important.) The renal vein and the lymphatics return substances to the systemic circulation,

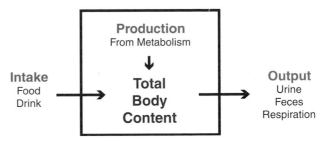

FIGURE 1   Mass balance relations in the whole body. The total body volume is indicated by the box. For the total body content of any substance to remain constant, the rate of output must equal the rate of intake plus the rate of production.

but the amounts in the urine represent a net loss from the systemic circulation and from the whole body. Of course, this process of excretion and its regulation constitutes the essential function of the kidney. Considered in terms of mass balance, the relevant equation is:

$$\text{Input rate to kidney} = \text{output rate from kidney.} \quad (3)$$
$$\text{(via renal artery)} \qquad \text{(via renal vein, lymphatics,}$$
$$\text{and urine)}$$

This relationship and the routes of mass flow into and out of the kidney are shown diagrammatically in Fig. 2.

The rate of mass flow for a substance by each of these four routes can be calculated from Eq. [1] as the product of the rate of plasma, lymphatic, or urine flow times the concentration of the substance in that flow.

MASS BALANCE CONSIDERATIONS ALONG THE NEPHRON

The same considerations of maintaining mass balance apply to the nephron as apply to the whole body and the kidney. To maintain a steady state, the water and solutes

Examples

1. A typical adult human ingests of 8–10 g about NaCl each day. There is no metabolic production of this inorganic compound, so for a steady state to be maintained (as is further discussed in Chapter 29) NaCl must be excreted (by the sum of urinary, sweat, and fecal routes) at the same rate of 8–10 g per day.

2. Water is typically ingested at a rate of 1–2 L/day with about 150–250 mL/day added from metabolism of various substrates such as

glucose. Water loss, which occurs via the same routes as for NaCl, must equal this daily intake.

3. Little urea is ingested in the diet, but it is synthesized in the liver at a rate of about 30 g/day or 20 mg/min from the HCO_3^- and NH_4^+ produced by the degradation of amino acids. Thus, for a steady state to exist, urea must be eliminated at the same rate, and this occurs primarily via urinary excretion.

that flow into the nephron must flow out. There are two routes of inflow into the nephron: the filtered mass flow and contributions due to secretion along the nephron. Water enters the nephron only by filtration. On the other hand, solutes may not only be filtered at the glomerulus but also secreted into various nephron segments by epithelial transport processes that are described in the two following chapters. Solutes and water that enter the nephron by these routes must leave the nephron in equal amounts by two routes: the mass flow excreted in the final urine and absorption across the epithelium of the various nephron segments.

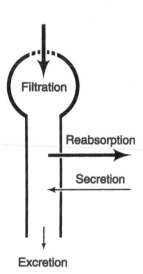

FIGURE 3　Mass balance relations along the nephron. The rate of excretion must equal the rate of filtration plus the rate of secretion minus the rate of reabsorption.

Figure 3 shows diagrammatically the operation of all 2 million nephrons together as if they were one giant nephron. For any given substance, inflow to the nephron must equal outflow, or, in terms of the rates of these processes:

$$\text{Filtration} + \text{secretion} = \text{excretion} + \text{reabsorption.} \quad (4)$$

If this equation is rearranged, the rate of excretion of a substance can be calculated directly as:

$$\text{Excretion} = \text{filtration} - \text{reabsorption} + \text{secretion.} \quad (5)$$

Equation [5] is the fundamental mass balance equation that must be appreciated in understanding

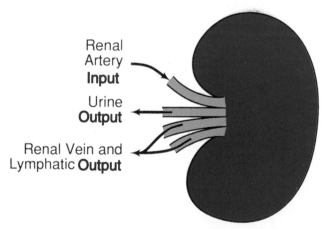

FIGURE 2　Mass balance relations for the kidney. In the steady state, the output of any substance from the kidney via the urine, renal vein, and lymphatic flows must be equal to the rate of input via the renal artery. For simplicity, this neglects any production or metabolism of the substance within the kidney, which is negligible for most substances.

the components that regulate the final excretion of water and solutes by the kidney, as shown in Fig. 3. A solute is filtered at the glomerulus at a certain rate. This may be augmented by secretion and diminished by reabsorption, resulting in the final rate of excretion. This view does not specify where the reabsorptive and secretory processes occur. Some solutes may be reabsorbed in one region of the nephron and secreted in another; some solutes are only reabsorbed and not secreted, and some are neither reabsorbed nor secreted.

When a solute is reabsorbed across the epithelium of the nephron, it is taken up into the peritubular capillaries; this subtracts from its rate of excretion in the urine and adds to its outflow from the kidney via the renal vein. When a substance is secreted, it is transported by the epithelium from the renal interstitium into the tubular fluid, whereas the interstitial concentration of the substance is maintained by passive diffusion of the solute out of the peritubular capillaries. Thus, secretion of a solute enhances its rate of excretion via the urine and reduces its rate of outflow from the kidney via the renal vein. These processes are shown diagrammatically in Fig. 4.

As for all mass flow processes, the rates of filtration and excretion are equal to the product of the volume flow rate and the solute concentration in that volume (see Eq. [1]). In the case of glomerular filtration, the volume flow is the glomerular filtration rate (GFR). For most substances that are freely filtered and not bound to plasma proteins, their concentration in the glomerular ultrafiltrate will be equal to their plasma concentration P_x, where x stands for any solute meeting these criteria. The filtration rate of the substance x can be calculated as

$$\text{Filtration rate} = \text{GFR} \cdot P_x. \tag{6}$$

Similarly, the excretion rate for a given substance x can be calculated from the rate of UF and the concentration of the substance in the urine (U_x):

$$\text{Excretion rate} = \text{UF} \cdot U_x. \tag{7}$$

These calculations become important when one attempts to determine quantitatively the effectiveness of the kidney in filtering and excreting substances, as illustrated by the determination of the glomerular rate.

Determination of the Glomerular Filtration Rate

The GFR is an important clinical index of the extent and rate of progression of renal disease. In most renal diseases, there is a progressive loss of filtering glomeruli due to sclerosis and destruction. Consequently, the GFR falls as the disease progresses. The clinician tries to obtain an indication of the GFR first to determine the extent of renal damage that has already occurred and then, at intervals, to determine the rate at which the disease is abating or progressing.

Ideally, one would like to be able to measure the GFR directly. This could be accomplished if one could measure the rate of excretion of a substance that is neither reabsorbed nor secreted. If such a substance were freely filtered, the rate of its excretion would be precisely equal to the rate of its filtration (see Eq. [5] and Fig. 5). One substance that meets the criteria is inulin. *Inulin* is a polyfructose molecule with a molecular weight of 5500 Da. The molecular size of inulin is sufficiently small that it is freely filtered at the glomerulus. However, inulin is too large to be reabsorbed passively by the nephron, and there are no active transport processes that reabsorb or secrete inulin. This relationship can be expressed quantitatively as follows, using Eqs. [6] and [7]:

Rate of inulin filtration = rate of inulin excretion

$$\text{GFR} \cdot P_{in} = \text{UF} \cdot U_{in}, \tag{8}$$

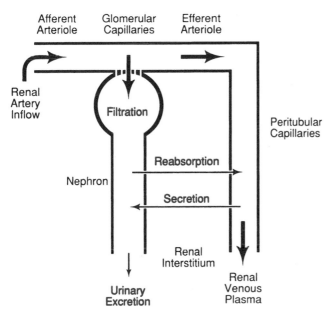

FIGURE 4 Schematic diagram of solute movements between the nephron and renal capillary networks. An ultrafiltrate of plasma is formed by filtration out of the glomerular capillaries. The amount of a given substance that is filtered can be reduced by reabsorption along the nephron, which returns solute through the renal interstitium to the peritubular capillaries and, thus, to the systemic circulation. Secretion adds to the amount of the substance excreted by moving the substance through the renal interstitium from the peritubular capillaries into the lumen of the nephron and, thus, reduces the amount of the substance that returns to the systemic circulation via the renal venous plasma.

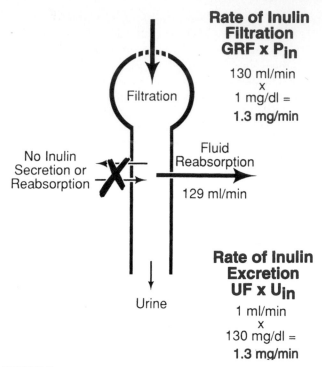

**Rate of Inulin
Filtration
GRF x P$_{in}$**

130 ml/min
x
1 mg/dl =
1.3 mg/min

Filtration

No Inulin
Secretion or
Reabsorption

Fluid
Reabsorption

129 ml/min

**Rate of Inulin
Excretion
UF x U$_{in}$**

1 ml/min
x
130 mg/dl =
1.3 mg/min

Urine

FIGURE 5   The rate of inulin excretion is equal to the rate of inulin filtration at the glomerulus because inulin is neither secreted nor reabsorbed.

where the subscript "in" refers to inulin. Rearranging gives:

$$GFR = (UF \cdot U_{in})/P_{in}. \qquad (9)$$

The right-hand side of Eq. [9] is referred to as the *clearance of inulin.* Conceptually, it represents the milliliters per minute of plasma that flow through the kidney from which inulin is completely removed. In the case of this substance, which is neither reabsorbed nor secreted, this clearance equals the GFR.

Clearance Concept

Plasma clearance, although a complete abstraction, is widely used to describe the effectiveness with which the kidney excretes various substances, and the student should study carefully the following discussion and quantitative relationships to understand the concept. Consider the relationship between the flows of fluid and their inulin concentrations in Fig. 6. In this example, it is assumed that the inulin concentration in the plasma is 1 mg/dL. Of the approximately 130 mL/min of fluid that are filtered at the glomerulus, more than 99%, or about 129 mL/min, is normally reabsorbed, resulting in a UF of 1 mL/min. Because inulin is neither reabsorbed nor secreted, its concentration in the tubular fluid rises along

the nephron until it reaches a maximum concentration 130-fold higher than in the plasma: 130 mg/dL. This figure is determined by rearranging Eq. [9] to give U$_{in}$ = (GFR $\cdot$ P$_{in}$)/UF.

On the other hand, the reabsorbed fluid that is returned to the peritubular circulation contains no inulin and, thus, the inulin concentration of the renal venous plasma is diluted. The concentration of inulin in the renal venous plasma can be calculated for the example in Fig. 6 by dividing the amount of inulin flowing out of the kidney via the renal vein by the volume flow rate. As seen in Fig. 6, if we neglect the small amount of inulin that leaves the kidney by the lymphatics, the amount of inulin flowing out of the kidney by the renal vein will be equal to the amount flowing in by the renal artery minus the amount that is excreted. The urea inflow via the renal arteries is 6.9 dL/min $\cdot$ 1 mg/dL = 6.9 mg/min, and the amount that is filtered (and therefore, excreted) is 0.01 dL/min $\cdot$ 130 mg/dL = 1.3 mg/min. Thus, the amount of inulin leaving the kidney via the renal vein will be (6.9 − 1.3) = 5.6 mg/min. The renal venous plasma flow equals the arterial inflow minus the UF, or 689 mL/min (6.89 dL/min). Thus, the inulin concentration in the renal venous blood will be (5.6 mg/min)/(6.89 dL/min) = 0.81 mg/dL. This flow would be equivalent to a combined flow of 560 mL/min containing 1 mg/dL of inulin (equivalent to the flow through the efferent arterioles after filtration)

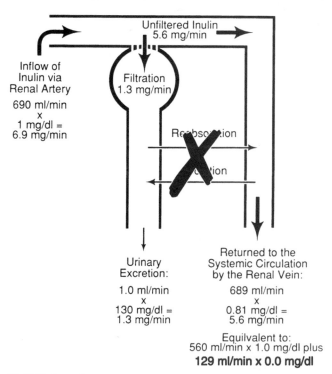

Unfiltered Inulin
5.6 mg/min

Inflow of
Inulin via
Renal Artery

690 ml/min
x
1 mg/dl =
6.9 mg/min

Filtration
1.3 mg/min

Reabsorption

Urinary
Excretion:

1.0 ml/min
x
130 mg/dl =
1.3 mg/min

Returned to the
Systemic Circulation
by the Renal Vein:

689 ml/min
x
0.81 mg/dl =
5.6 mg/min

Equivalent to:
560 ml/min x 1.0 mg/dl plus
129 ml/min x 0.0 mg/dl

FIGURE 6   The clearance of inulin is equivalent to the renal plasma flow from which inulin has been completely removed plus the urine flow rate (UF). See text for explanation.

plus a flow of 129 mL/min containing no inulin (equivalent to the reabsorbed fluid).

These considerations lead directly to the clearance concept. The clearance of inulin is equal to the flow of renal venous blood that can be considered to be free (cleared) of inulin plus the UF that is lost: 129 mL/min + 1 mL/min = 130 mL/min. This is equal to the clearance of inulin calculated using Eq. [9]. Thus, in general, the clearance of any substance (C_x) from the plasma is:

$$C_x = (UF \cdot U_x)/P_x. \qquad (10)$$

If the urine and plasma concentrations in Eq. [10] are in the same units (e.g., mg/dL or mmol/L), the clearance has the same units as the urine volume flow rate. Thus, clearances are traditionally reported in units of milliliters per minute.

A substance that is reabsorbed along the nephron has a clearance that is less than the GFR because some of the filtered amount returns to the peritubular capillaries and adds to the amount flowing out of the renal vein. Alternatively, any substance that is secreted along the nephron has a clearance that is higher than the GFR because the peritubular blood flow loses some of the solute due to secretion, resulting in a lesser amount of solute flowing out of the kidney via the renal vein.

Practical Indicators of the Glomerular Filtration Rate

The importance of inulin, a substance that is neither reabsorbed nor secreted, is that it allows the GFR to be determined from three parameters that are measured relatively easily: UF, the plasma concentration of inulin, and the urine concentration of inulin. The only problem is that inulin does not occur naturally in the mammalian body and, therefore, to measure its clearance, inulin must be infused intravenously to produce readily measurable plasma concentrations.

Because of its large size, inulin distributes only in the extracellular space. To obtain a plasma inulin concentration of 1 mg/dL, as in the preceding example, if the extracellular fluid volume is 15 L, one has to infuse a priming dose of 150 mg. However, because of the renal clearance of the inulin, its plasma concentration immediately begins to fall. To maintain a constant plasma inulin concentration while the clearance is being measured, inulin has to be constantly infused at the same rate at which it is being lost in the urine, for example, at 1.3 mg/min in the preceding example. Obviously, the measurement of an inulin clearance can be conducted only in the hospital, where intravenous infusions are possible. In practical terms, this is seldom done in a patient except as part of a research protocol when the best estimates of GFR are required. Other substances that are neither reabsorbed nor secreted can be used clinically, but, as with inulin, they must be infused because they are not normally present in the body.

To avoid the complications of intravenous infusion, it is preferable to measure the clearance of an endogenous substance that approximates the GFR. Historically, physicians sometimes measured the clearance of urea as an approximation. However, urea is reabsorbed to a substantial degree along the nephron, and its rate of reabsorption is quite dependent on the UF. Consequently, the urea clearance not only underestimates the true GFR substantially, but it is dependent on the UF.

A better approximation of the GFR is obtained by measuring the clearance of creatinine. Creatinine is produced in the body by the metabolism of creatine and phosphocreatine derived from the breakdown of muscle. Muscle tissue is constantly being broken down and rebuilt in the body, and meats (i.e., animal muscle) provide an additional source in the usual Western diet. Almost no creatinine is directly ingested; therefore, in order for the body to maintain a constant creatinine content, the rate of creatinine excretion must equal its rate of metabolic production, which in a 70-kg man is somewhat less than 2 g/day. Creatinine is actively secreted into the urine in the late proximal nephron, but at a relatively slow rate in comparison with its rate of glomerular filtration. Consequently, the rate of creatinine excretion is normally only about 10% higher than its rate of filtration, and the creatinine clearance provides a useful measure, albeit a slight overestimate, of the GFR.

In practical terms, the rate of creatinine production and, thus, its rate of excretion, varies in the course of a day. Production increases about 1 hr after a meal and wanes during the night. Therefore, in order to estimate the GFR of a kidney patient, the nephrologist will usually have the patient collect his or her entire urine volume over a 24-hr period. The creatinine concentration in that urine is an average of the urine concentration over a 24-hr period, the so-called *clearance period*. A blood sample is usually drawn at some point during or just at the end of the clearance period, when the patient comes to the physician's office with his or her sample of urine. Knowing both the urine and plasma creatinine concentrations, one needs only the average UF rate during the clearance period in order to calculate the creatinine clearance according to Eq. [10]. The average UF rate is simply the quotient of the total urine volume and the total minutes over which it was collected, with the final answer given in units of milliliters per minute.

However, in following a patient over a long period of chronic renal failure, the physician may only occasionally measure the creatinine clearance. Instead, the physician may more conveniently follow the *plasma creatinine concentration* (P_{cr}) as an index of the GFR. The reason that this concentration reflects the GFR can be understood from the following analysis of the mass balance concepts developed in this chapter.

As noted earlier, creatinine is produced by metabolism at a rate of $\sim$2 g/day. To maintain mass balance of creatinine, the body must match this rate of production by the excretion rate. However, the creatinine excretion rate is approximately equal to the rate of creatinine filtration, if one neglects the small contribution of secretion. Thus, from Eq. [8]:

$$GFR \cdot P_{cr} = UF \cdot U_{cr} \qquad (11)$$
$$= 2g/day = 1.39 \ mg/min$$

where the subscript "cr" indicates creatinine. Actually, because of the small amount of creatinine secreted, the rate of creatinine filtration is closer to 1.3 mg/min. Using this approximation and rearranging the equation, one can see that the plasma creatinine concentration is inversely proportional to the GFR:

$$P_{cr} = \frac{1.3 \ mg/min}{GFR(mL/min)} \qquad (12)$$

Therefore, for a normal GFR of 130 mL/min, one would predict that the plasma creatinine concentration would be (1.3 mg/min)/(1.3 dL/min) = 1.0 mg/dL. (As in the case of many plasma nonelectrolytes, the creatinine concentration is given in clinical units of milligrams per deciliter.) Thus, the equation predicts what is regarded as the normal plasma creatinine concentration rate of 0.8–1.2 mg/dL. Using Eq. [12], we can examine what happens to the plasma creatinine concentration when the GFR is progressively decreased by renal disease.

If the GFR were halved to 65 mL/min by a loss of functioning nephrons, Eq. [12] predicts that the plasma concentration of creatinine would double to 2 mg/dL. The increase in plasma creatinine concentration would occur merely because of the necessity of the rate of filtration to increase to match the rate of creatinine production. With GFR reduced by half, Eq. [6] indicates that the same rate of creatinine filtration can be maintained only by a doubling of the creatinine concentration in the plasma. If the GFR fell progressively over a period of years, as in chronic renal failure, the plasma creatinine concentration would slowly rise, so that at any given time the rate of creatinine excretion

would equal the rate of production. There would be only imperceptible transient periods in which the rate of excretion lagged behind the rate of production resulting in a rise in the plasma creatinine concentration.

It is again most important to realize that an elevated P_{cr} does not indicate that the kidneys are not excreting a normal amount of creatinine. When a physician sees a normal patient or one with renal damage, either patient is usually approximately in balance with respect to creatinine, that is, he or she is excreting it at the rate at which it is being produced. The problem is that the patient with the damaged kidney requires a higher plasma creatinine concentration to maintain this normal rate of excretion, which is determined primarily by filtration.

If the loss of functioning nephrons progresses in the patient, the creatinine concentration will rise in inverse proportion to the fall in GFR. According to Eq. [12], when the GFR is 25% of normal, the creatinine concentration will be 4 mg/dL; at a GFR of 12.5%, 8 mg/dL, etc. Thus, as illustrated in Fig. 7, the plasma creatinine concentration rises geometrically as the GFR falls.

Because the rate of urea excretion is also primarily dependent on the GFR, measurements of plasma urea concentration are also followed. The urea concentration is traditionally reported in the rather obscure clinical units of blood urea nitrogen (BUN, mg/dL). The BUN actually is the concentration of the *nitrogen* atoms associated with urea. Because there are two nitrogen

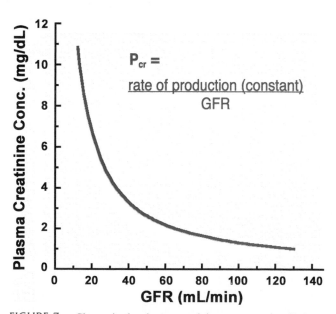

FIGURE 7 Change in the plasma creatinine concentration (P_{cr}) as a function of the glomerular filtration rate (GFR). In this example, the rate of creatinine production, which is assumed to remain constant, was $\sim$1.8 g/day, or 1.3 mg/min.

atoms per urea molecule, a 1.0 mmol/L urea concentration is equivalent to a BUN of 2.8 mg/dL. The normal BUN is in the range of 10–20 mg/dL, which is, therefore, equivalent to a urea concentration of 3.6–7.1 mmol/L.

Note from Fig. 7 that even a doubling of the normal plasma creatinine concentration indicates a significant fall in the GFR to one-half normal, but the diagnosis of renal damage should never be based on a single such

Clinical Note

Creatinine Balance and the Plasma Creatinine Concentration

Consider the following hypothetical and idealized example to understand why a patient with reduced renal function remains in creatinine balance, but with a higher plasma creatinine concentration.

An 80-kg patient has a normal metabolism and diet and produces 1.8 g of creatinine per day. He then undergoes a uninephrectomy to remove a renal carcinoma. Presume that after the uninephrectomy the function of his remaining kidney is unchanged with one half of his normal total (two-kidney) GFR, and that his metabolism and diet also remain unchanged.

- How is body creatinine balance normally maintained so that the total body creatinine and plasma creatinine concentration remain unchanged from day to day?
- If, before the uninephrectomy, the patient had a plasma creatinine concentration of 1.0 mg/dL, what was his GFR?
- What is the patient's rate of creatinine excretion immediately after removal of the other kidney and why?
- What happens to the plasma creatinine concentration in the hours immediately after surgery?
- When is a new steady state of creatinine balance achieved, and what are the plasma creatinine concentration and the daily rate of creatinine excretion at this new steady state?

These questions relate to the normal renal handling of creatinine, the necessary equality of production and excretion in the steady state, and the effect of the GFR on the plasma creatinine concentration. Because creatinine is assumed to be neither reabsorbed nor secreted, creatinine excretion is equal to the rate of filtration. The rate of creatinine production depends on the rate of turnover of endogenous muscle and that in the diet, and is relatively constant on a fixed diet unless the individual is in a markedly anabolic or catabolic state.

If the excretion of creatinine is decreased by a fall in GFR, the plasma creatinine concentration will rise rapidly. This can be appreciated semi-quantitatively by considering that retention of one day's production of creatinine (1.8 g), which would distribute primarily in the extracellular volume compartment ($\sim$15 L in this person), would raise the plasma creatinine concentration by about 12 mg/dL. In other words, even if the GFR decreases abruptly, there is only a short transient period during which the rate of production exceeds the rate of excretion. The resulting rise in plasma creatinine concentration causes a proportional increase in filtration and thus in excretion until a new steady state is achieved with an elevated plasma creatinine concentration.

It is important to understand that partial renal failure does not mean that the rate of excretion of solutes is chronically less than the rate of ingestion or production. Rather, the patient reaches a new steady state but one that is not normal or optimal. Only when the renal disease reaches end stage is it no longer possible to achieve a steady state, leading to death in the absence of intervention.

Here are the answers to the questions:
- The rate of filtration of creatinine (= GFR · P_{cr}) is equal to the rate of excretion of creatinine (= UF · U_{cr}), which is equal to the rate of production = $\sim$1.8 g/day; thus, the GFR can be calculated to be $\sim$180 L/day.
- *Immediately* after uninephrectomy, the rate of excretion is one-half normal, or $\sim$0.9 g/day.
- Because the rate of filtration and thus the rate of excretion is less than the rate of production initially, the P_{cr} rises.
- A new steady state is achieved when the rate of filtration of creatinine again becomes equal to the rate of production of creatinine, that is, when:

$$P_{cr} = \frac{1.8 \text{ g/day}}{90 \text{ L/day}} = 20 \text{ mg/L} = 2 \text{ mg/dL}$$

determination. A higher-than-normal plasma creatinine concentration can be due to abnormal rates of production (e.g., catabolic processes associated with an underlying disease or excessive meat intake) or even to lab error. Similarly, a low plasma creatinine concentration can sometimes be observed in the severely ill or fasting patient (e.g., a patient who has been unable to eat or retain food because of gastrointestinal problems). For these reasons, the plasma creatinine concentration or BUN must be regarded with thoughtful interpretation in trying to estimate renal function and the patient's GFR. Although an elevated plasma creatinine concentration and BUN in a routine blood sample are indicators of deficient renal function that should be followed up immediately, these parameters are best used to follow the progress of renal disease in the relatively stable patient, rather than as quantitative estimates of GFR in an acute illness.

Suggested Reading

Levinsky NG, Lieberthal W. Clearance techniques. In Windhager EE, ed. *Handbook of Physiology, Section 8: Renal Physiology,* Vol I. New York: Oxford, 1992, pp 226–247.

Rose BD Rennke HG. *Renal pathophysiology—the essentials.* Baltimore: Williams & Wilkins, 1994.

Schuster VL, Seldin DW. Renal clearance. In Seldin DW, Giebisch G, eds. *The kidney: Physiology and pathophysiology,* Vol 1, 2nd ed. New York: Raven Press, 1992, pp 943–997.

Reabsorption and Secretion in the Proximal Tubule

JAMES A. SCHAFER

KEY POINTS

- The proximal tubule reabsorbs two-thirds of the volume of fluid filtered by the glomeruli. The fluid reabsorbed by the proximal tubule is essentially an isotonic salt solution; however, the proximal tubule preferentially reabsorbs solutes that are useful for metabolism and secretes those that are by-products of metabolism.
- All nephron segments consist of an *epithelial cell layer* with an underlying *basement membrane*. Solutes may move across this epithelium *transcellularly* by crossing both the *apical membrane* and the *basolateral membrane* of the epithelial cells, and/or through the *junctional complexes* and *lateral intercellular spaces* between epithelial cells.
- The most important primary active transporter in the proximal tubule, as in all nephron segments, is the Na^+,K^+-ATPase. Because this transporter is present only in the basolateral

KEY POINTS (*continued*)

membrane, it is constantly transporting Na^+ actively from the cell toward the interstitial fluid and the plasma.

- The *luminal membrane* contains specialized transporters that allow Na^+ entry into the epithelial cell down its electrochemical potential gradient. This polarization of Na^+ transporters—passive influx at the luminal membrane and active efflux at the basolateral membrane—imparts the directionality, i.e., reabsorption, to net Na^+ transport all along the nephron.

- Solute reabsorption slightly reduces the osmolality of the tubular fluid and increases the osmolality of the interstitial fluid, which provides a small osmotic pressure difference to drive nearly *isosmotic volume reabsorption*.

- The active reabsorption of solutes and water by the proximal tubule causes a positive hydrostatic pressure of ~ 7 mm Hg in the renal interstitium.

- Protons (H^+) are actively secreted in the proximal tubule and in the most distal nephron segments. Intracellular HCO_3^- rises due to the loss of H^+ from the cytoplasm, and produces an electrochemical gradient for HCO_3^-

transport across the basolateral membrane, thus returning this base to the plasma.

- In the proximal tubule, an Na^+/H^+ antiporter couples the energy from downhill Na^+ entry across the luminal membrane to drive *secondary active H^+ secretion*.

- As water is reabsorbed from the proximal tubule fluid, the concentrations of passively reabsorbed solutes rise, which favors their *passive reabsorption* by diffusion. The rate of passive reabsorption of a given solute is proportional to its permeability across the proximal cell membranes or through the junctional complexes.

- The preferential reabsorption of many solutes such as glucose and amino acids is driven by secondary active transport driven by Na^+ cotransport down its electrochemical potential gradient. Each of these transporters is characterized by a *transport maximum* (T_m), which limits the amount of solute that can be reabsorbed.

- *Active secretion* of metabolic by-products, drugs, and potential toxins in the proximal straight tubule accelerates their excretion.

Glomerular filtration delivers approximately 130 mL of plasma ultrafiltrate to the nephrons every minute. Approximately two-thirds of this volume is reabsorbed by the proximal tubules and returned to the blood via the peritubular capillaries. Although it is important that the proximal tubule reabsorb most of the filtered load of water and solutes, this is a selective process: Useful substrates for metabolism, such as glucose and amino acids, are rapidly reabsorbed, whereas some by-products of metabolism, such as creatinine and urea, are reabsorbed slowly or not at all. The proximal tubule also actively secretes many organic solutes that must be excreted rapidly from the body, including metabolic by-products and "foreign substances" such as toxins and drugs.

The proximal tubule is a "mass transporter" in that it transports the largest fraction of most substances handled by the nephron. Segments that are more distal are better adapted for the fine regulation of reabsorption and secretion that matches the final excretion rate to the rate of intake plus production for each substance.

The proximal tubule is the major reabsorber of the massive quantities of salts and water that are filtered and must be returned to the plasma. The rate at which the proximal tubule reabsorbs NaCl and water is a major determinant of the rate of their excretion. Thus, the regulation of this reabsorptive process is of utmost importance in maintaining a normal extracellular fluid volume (as discussed in Chapter 29).

In this chapter, the general function of the proximal tubule is presented as if this segment of the nephron were a homogeneous structure. Thus, the convoluted and straight segments of superficial and juxtamedullary nephrons are referred to collectively as the *proximal tubule* when considering both their morphology and function. In fact, the proximal tubule segments differ in length and size between superficial and juxtamedullary nephrons. There are also three cell types in the proximal tubule. However, these details are ignored here, except in those instances in which the localization of certain reabsorptive or secretory processes may be important. In addition, this chapter examines the collective operation of all proximal tubules, in other words, the total effect of all two million proximal tubules in reabsorbing substances from the total glomerular ultrafiltrate or adding substances to it.

EPITHELIAL CHARACTERISTICS OF THE NEPHRON

All nephron segments, from the glomerulus to the ducts of Bellini, consist of *epithelial cells* that are joined in a continuous layer by specialized structures called *junctional complexes,* as illustrated in Fig. 1. Although some epithelia consist of multiple cell layers, a single cell layer forms the epithelium of the nephron. On the basal side of all epithelia a continuous basement membrane composed of collagen, laminin, and other extracellular matrix proteins provides structural support that is essential for the organized growth, development, and organization of the epithelial cells. Epithelia separate the extracellular fluid compartment of the body on their basal side from a compartment with a different composition on the apical side (usually called the luminal side in the nephron). For example,

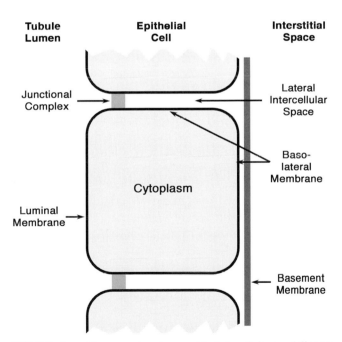

FIGURE 1   Components of an epithelial cell layer. Adjacent epithelial cells are joined by junctional complexes ("tight junctions"), which form the closed ends of the lateral intercellular spaces between cells. The luminal membrane borders the urinary (luminal) side of the nephron, whereas the basolateral membrane consists of the lateral membrane bordering intercellular space and the membrane on the basal (interstitial) side of the epithelial cell. The basement membrane separates the basal side of the epithelial cell layer from the extracellular fluid in the interstitial space.

epithelial layers of the skin and respiratory tract separate the extracellular fluid from, respectively, the external environment and inspired air. Most epithelia, e.g., the nephron and the gastrointestinal tract, serve not only as a barrier that protects the extracellular fluid, but they also modify the composition of the fluid on the apical or luminal side by absorptive and secretory processes that transport solutes and water, respectively, out of or into the apical compartment.

The epithelium of the nephron is somewhat different from other epithelia because it consists of the single-cell epithelial layer and a basement membrane, with no other cell layers on the basal (or serosal) side. Epithelia in most other tissues have a thick serosal matrix consisting of stromal fibers and other tissues such as muscle layers. The absence of these serosal elements in a nephron allows the length of the path from the lumen to the peritubular capillaries running through the interstitium to be quite short, on the order of 20–30 µM, and permits an efficient exchange of large amounts of solutes and water between the lumen and the capillaries.

Solutes and water can move between the lumen of the nephron and the interstitium by one or both of two routes: through the cells or between the cells. Most substances that utilize the *transcellular route* require specialized transporters that facilitate their movement across the apical and the basolateral membranes, and these transporters are generally regulated to produce physiologic changes in the rates of reabsorption or secretion. The *paracellular route* involves movement through the junctional complex and the lateral intercellular space. The lateral intercellular space is a free solution layer that provides little resistance to solute or water movement in the nephron. The junctional complexes are often referred to as "tight junctions" because they typically serve as relatively impermeant barriers to the movement of most solutes and even water. However, in many epithelia such as the proximal tubule, the junctional complexes are quite permeable to many ions and small solutes such as urea. The junctional complexes in such epithelia are often referred to as "leaky," and they have a low electrical resistance.

Junctional complexes have long been assumed to be relatively nonselective, such that the permeability of a solute is determined solely by its size and valence. However, more recent evidence has shown that the junctional complexes in some epithelia may provide selective channels through which specific ions can move. This is the case with paracellin, which, as will be discussed in more detail in Chapter 27, allows Mg^{2+} to be reabsorbed through the paracellular pathway in the thick ascending limb of the loop of Henle.

General Characteristics of Active Na$^+$ Reabsorption in the Nephron

The active reabsorption of Na$^+$ is the most important active transport process in the nephron. It occurs by the same basic mechanism in all nephron segments and it accounts for the majority of oxygen consumed by the kidney. As in the case of many epithelia, Na$^+$ reabsorption is driven by Na$^+$,K$^+$-ATPase in the basolateral membranes. As discussed in Chapter 3 ("Membrane Transport" section on "The Na$^+$-K$^+$ Pump"), Na$^+$,K$^+$-ATPase actively transports three Na$^+$ ions out of the cell in exchange for the transport of two K$^+$ ions into the cell. The operation of this ion pump maintains the normal low intracellular Na$^+$ and high intracellular K$^+$ concentrations as well as the transmembrane electrical potential difference, which is normally oriented with the cell negative by 60–90 mV with respect to the extracellular fluid. The energy required for the operation of the ion pump is provided by ATP hydrolysis.

The net absorptive transport of Na$^+$ and other solutes across the nephron and other epithelia is dependent on the asymmetrical distribution of membrane transporters between the apical and basolateral membranes. This *polarization* of the membrane transporters, which is illustrated in Fig. 2, is a fundamental characteristic of epithelia that underlies their ability to produce transepithelial active transport. The

Na$^+$,K$^+$-ATPase transporters are found exclusively in the basolateral membranes of all nephron segments; therefore, Na$^+$ can be actively extruded from the cell only into the lateral intercellular space and the interstitial space that borders the basolateral membrane, and not into the lumen. The low intracellular Na$^+$ concentration maintained by the Na$^+$,K$^+$-ATPase and the negative intracellular electrical potential difference provide a large electrochemical potential difference that favors the passive entry of Na$^+$ down its electrochemical gradient into the cell. Throughout the nephron, Na$^+$ enters the cells much more easily across the luminal than across the basolateral membrane, and then is actively transported into the interstitium by the Na$^+$,K$^+$-ATPase.

Differences in the details of Na$^+$ reabsorption among nephron segments relate only to the mechanism by which Na$^+$ enters the cells across the luminal membrane, i.e., the X transporter in Fig. 2. The movement of Na$^+$ into the cell is "downhill" and embodies a considerable amount of potential energy—energy that can perform work. In the proximal tubule, this energy is used to drive the reabsorption and secretion of other solutes by various specialized transporters in the luminal membrane as discussed later in this chapter. In fact, Na$^+$ is the only quantitatively important substance whose transport is directly coupled to metabolic energy production in the proximal tubule. The energy required for the transport of most other solutes into or out of the proximal tubule is derived from the Na$^+$ electrochemical potential difference that is generated by the Na$^+$,K$^+$-ATPase.

The transporters responsible for reabsorptive and secretory processes in the nephron were first identified only by their functional characteristics, but during the past 20 years the genetic messages for most channels and carriers in the kidney and other epithelia, including Na$^+$,K$^+$-ATPase and the various luminal membrane Na$^+$ transporters, have been determined by DNA sequencing. Knowledge of the amino acid sequence (from the DNA sequence) in turn permits models of the structure of the transport proteins to be developed. These analyses have revealed marked similarities among transport proteins, allowing them to be grouped into families whose members have a high degree of homology, that is, similarity of structure and function. Furthermore, the homology extends across species, ranging from bacteria to humans. This conservation of structure during evolution is a clear indication of the importance of such transporters to survival. Most transport proteins have 2–12 regions (domains) consisting of stretches of relatively lipophilic amino acids. The lipophilic regions allow the protein to reside as a complex structure within the membrane, where it provides the route for transmembrane transport. In

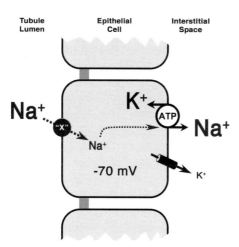

FIGURE 2   Fundamental mechanism of Na$^+$ reabsorption in all nephron segments. Na$^+$ is extruded from the cell by the Na$^+$,K$^+$-ATPase located in the basolateral membrane. This transporter pumps three Na$^+$ ions out of the cell in exchange for two K$^+$ ions, which enter the cell to maintain the high intracellular K$^+$ concentration. Na$^+$ enters the cell passively by moving down its electrochemical potential gradient from the higher concentration in the lumen to the lower concentration in the cell. The greater negative voltage of the cell interior also adds to the electrochemical potential gradient for Na$^+$ entry. The mechanism of Na$^+$ entry across the luminal membrane (indicated by "X") is left unspecified in this diagram, and occurs by a variety of transporters depending on the nephron segment.

most cases the molecular mechanisms that allow a carrier to transport a specific substance have not been established, but it is clear that these specialized proteins possess specific binding sites for the molecules they transport.

Transepithelial Voltages

Another characteristic of many epithelia is the presence of a transepithelial voltage, or an *electrical potential difference* (PD), between their luminal and interstitial sides. A transepithelial voltage implies that there must be a difference between the voltage across the apical membrane and that across the basolateral membrane. The voltage across the basolateral membrane in most segments of the nephron, as in nonepithelial cells, is determined by the dominance of K^+-selective channels (see Fig. 2), as discussed in Chapter 3 ("Membrane Transport,: in the section "What Is the Origin of V_m?") and Chapter 4 (section on "Ionic Mechanisms of the Resting Potential"). Thus, voltage between the cytoplasm and the interstitial (extracellular) fluid is near the K^+ diffusion potential, or -60 to -90 mV. The voltage across the apical membrane of individual nephron segments, however, varies considerably and depends on ion conductances that are present in the apical membrane. For example, if the luminal Na^+ transporter in Fig. 2 were a Na^+ channel, it would cause the voltage across that membrane to be depolarized in much the same way that a nerve membrane is depolarized by the opening of Na^+ channels. (However, epithelial Na^+ channels, in contrast to those in nerves, are not voltage inactivated and remain open much longer.) Thus, for example, if the voltage from the lumen to the cytoplasm is only 30 mV, whereas the voltage from the cytoplasm to the interstitium is -70 mV (as in Fig. 2), the transepithelial voltage could be as high as -40 mV. Why not exactly -40 mV? The presence of a transepithelial voltage also depends on the electrical resistance of the paracellular pathway. If the resistance of the junctional complexes is very high, i.e., if the junctions are truly tight, the transepithelial voltage approaches the maximum of -40 mV in this example. However, if the junctions are very leaky to ions as is the case in the proximal tubule, ion movement through the paracellular pathway effectively short-circuits the epithelium and diminishes the transepithelial voltage.

SALT AND WATER REABSORPTION IN THE PROXIMAL TUBULE

As shown in the following sections, it is important to think of the transport of each substance along the nephron in terms of its rate of movement rather than by its concentration in the tubular fluid. The same considerations of mass balance that apply to the kidney as a whole also apply to the proximal tubule. In the case of the proximal tubule, the rate of input of substances is determined by the product of the glomerular filtration rate (GFR) and the concentration of the substance in the glomerular ultrafiltrate. The rate at which the substance is delivered to the next segment of the nephron, the loop of Henle, is determined by the rate of filtration minus the rate at which the substance is reabsorbed. In the case of substances that are secreted, the rate of delivery to the loop of Henle will be greater than the rate of filtration by an amount equal to the rate of secretion. Thus, comparison of the rate at which a substance is delivered to the loop of Henle with its rate of filtration indicates whether there has been net addition (secretion) or loss (reabsorption) of the substance along the proximal tubule. However, as in the case of the whole nephron, it does not indicate whether only reabsorption or only secretion is occurring, or the mechanism by which these processes occur.

Isosmotic Volume Reabsorption

Quantitatively, the most impressive event in the proximal tubule is the reabsorption of more than two-thirds of the filtered load of salt and water. Thus, with a GFR of $\sim$130 mL/min, the proximal tubule reabsorbs more than 85 mL/min and passes less than 45 mL/min on to the loop of Henle. This rate of reabsorption is substantially larger than occurs across any other epithelium in the body. The reabsorption is driven by a massive transport of solutes from the lumen of the proximal tubule to the interstitial fluid and, thus, to the peritubular capillary circulation. The cells of the proximal tubule exhibit morphology consistent with these high transport rates, as shown in Fig. 3. On the apical surface of the cells, the luminal membrane forms a dense carpet of microvilli, which, as in the small intestine, greatly expands the surface area available for transport. The surface area of the basolateral membrane is also dramatically amplified by being folded into long undulating ridges like a very full drapery, and the basal membrane has finger-like microvilli that abut the basement membrane. As in the case of other leaky epithelia such as the distal small intestine and the gallbladder (see Chapter 36), the reabsorption occurs with virtually no change in the osmolality of the luminal fluid. Thus, the process has come to be referred to as *isosmotic volume reabsorption*.

If there is no change in the osmolality of the tubular fluid, how can one determine that any solutes or water have been reabsorbed? This question is answered most

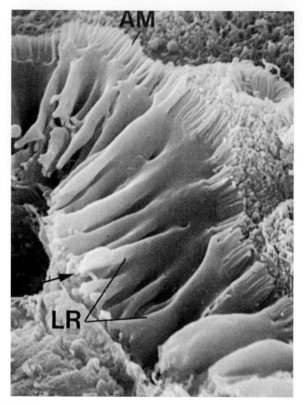

FIGURE 3   Scanning electron micrograph showing several cells in the midportion of a proximal tubule. A frozen segment of a proximal tubule was fractured to expose the lateral aspects of the cells with their luminal surfaces toward the upper right and the basement membrane toward the lower left. These cells possess an elaborate apical (luminal) surface of microvilli (AM). The lateral cell membrane is folded into long lateral ridges (LR), and the basal membrane also forms microvilli that lie on the basement membrane. (Courtesy of Dr. Andrew P. Evan, Indiana University Medical Center, Indianapolis, IN.)

rise in the inulin concentration. If the fluid flow to the loop of Henle is only one-third of the GFR, the inulin concentration must become three times higher than that in the plasma.

Historically, this is how the rate of fluid reabsorption was determined experimentally by micropuncture experiments in individual proximal tubule segments. In these experiments, inulin was infused into an experimental animal to maintain a constant plasma concentration. Very fine glass micropipettes were then used to collect samples of proximal tubule fluid, and the inulin concentration (TF_{in}) in these samples (1–5 nl in volume) was measured and compared with the inulin concentration in the plasma (P_{in}). From the ratio of these concentrations, the rate of fluid delivery to the point of sampling (V_x) could be calculated using Eq. [1] as:

$$V_x = GFR \cdot P_{in}/TF_{in} \qquad (2)$$

Thus, because inulin is neither reabsorbed nor secreted, volume reabsorption in the proximal tubule is matched by a proportional increase in inulin concentration. It was found that the inulin concentration in the proximal tubule fluid samples was always greater than that in the plasma, indicating that the volume flow at the point of sampling was less than the GFR. In other words, the higher concentration of inulin in the proximal tubule fluid sample indicated that some of the volume flow had been reabsorbed between the glomerulus and the point of sampling. Furthermore, as shown in Fig. 4, it was observed that the ratio of the inulin concentration in the

easily if we consider what would happen to a solute that is not reabsorbed or secreted. Inulin is ideal in this regard, as discussed in connection with the determination of the GFR (see Chapter 25). If inulin were present in the plasma at a concentration of 1 mg/dL (0.01 mg/mL), it would enter all the proximal tubules collectively at a rate equivalent to the product of the GFR times the plasma concentration, or (130 · 0.01) = 1.3 mg/min. Because inulin is neither reabsorbed nor secreted, it must leave the proximal tubule to enter the loop of Henle at the same rate as it is filtered. If the rate of volume flow to the loop of Henle is given as V_L, and the concentration of inulin in the tubular fluid at the end of the proximal tubule is TF_{in}, the mass flow equality can be expressed as:

$$GFR \cdot P_{in} = V_L \cdot TF_{in}, \qquad (1)$$

where P_{in} is the plasma inulin concentration. From this relation, it can be seen that as V_L is reduced due to fluid reabsorption, there must be an inversely proportional

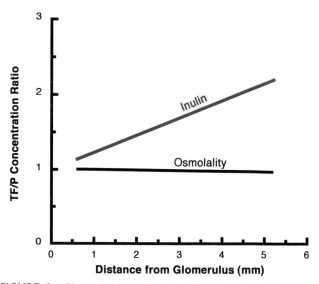

FIGURE 4   Change in the TF/P ratio of inulin and osmolality with distance along the proximal tubule. The ratio of the inulin concentration or osmolality in tubular fluid (TF) samples to that in the plasma (P) is plotted as a function of the distance along the proximal tubule from the glomerulus.

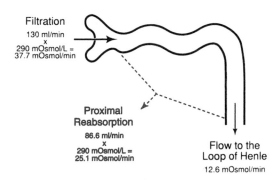

FIGURE 5   Reabsorption of solutes and water in the proximal tubule. Approximately two-thirds of the filtered solute and water is reabsorbed along the proximal tubule. The remainder is passed to the descending limb of the loop of Henle.

fluid sample to that in the plasma rose as the distance of the sampling point from the glomerulus increased, whereas the osmolality remained nearly the same as the plasma; i.e., the reabsorption was isosmotic. However, micropuncture can be used to sample only from those regions of the proximal convoluted tubule that are accessible on the surface of the cortex just beneath the capsule of the kidney. Thus, it measures the operation of only superficial proximal tubules, and only the convoluted segments of these superficial nephrons. However, if one extrapolates the rise in inulin concentration in the superficial proximal convoluted tubules to the proximal straight tubule, and assumes similar rates of reabsorption in the juxtamedullary proximal tubules, it can be calculated that the (TF/P) concentration ratio for inulin reaches 3 by the end of the proximal tubule. Thus, from Eq. [2], one can calculate that the rate of fluid delivery to the loop of Henle is one-third of the rate of glomerular filtration and, thus, that two-thirds of the GFR must be reabsorbed in the proximal tubule.

Because the tubular fluid osmolality remains nearly the same as that of the plasma, the total amount of solute reabsorbed must be in proportion to the water reabsorbed. Considered from this perspective as shown in Fig. 5, the total rate of solute filtration at the glomerulus would be approximately 37.7 mOsm/min (290 mOsm/L · 130 mL/min), and the rate of reabsorption would be two-thirds of this, or 25.1 mOsm/min, leaving 12.6 mOsm/min to flow on into the loop of Henle, as shown schematically in Fig. 3. The next sections consider what solutes comprise the 25.1 mOsm/min reabsorbed and how these reabsorptive processes drive water (volume) reabsorption.

Transepithelial Voltage in the Proximal Tubule

As Na$^+$ is actively reabsorbed along the proximal tubule, an equivalent amount of anions must

accompany it to maintain electrical neutrality. In the early part of the proximal tubule, the active reabsorption of Na$^+$ leads to the development of a lumen-negative electrical potential difference—the lumen of the early proximal convoluted tubule is negative with respect to the interstitium by about −5 mV. This potential difference is relatively low because the electrical resistance of the junctional complexes in the proximal tubule epithelium is very low, which essentially "short-circuits" the transepithelial voltage. Nevertheless, even this small lumen-negative voltage across the epithelium serves as an effective driving force for passive Cl$^-$ reabsorption across the junctional complexes in the early proximal convoluted tubule.

In the more distal portions of the proximal tubule, the lumen-negative electrical potential decreases and becomes positive in the proximal straight tubule. This change is due to slower rates of Na$^+$ reabsorption in the later regions of the proximal tubule and to the development of diffusion potentials as concentration differences of the anions develop. As discussed in subsequent sections, the concentration of Cl$^-$ rises along the proximal tubule because HCO$_3^-$ is preferentially reabsorbed. Therefore, there is a concentration gradient of Cl$^-$ from the lumen to the interstitium that imposes a diffusion potential that is lumen positive by as much as 5 mV, as shown in Fig. 6. Because of the change in the electrical potential difference from lumen negative to lumen positive, and because this potential difference is small, it is most convenient to regard the average electrical potential difference in the proximal

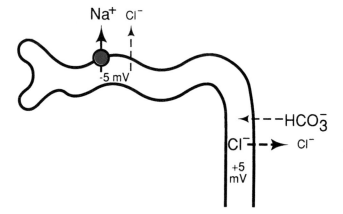

FIGURE 6   Change in the transepithelial voltage along the proximal tubule. In the early regions of the proximal tubule, the active, electrogenic transport of Na$^+$ out of the lumen causes development of a lumen-negative voltage that favors passive Cl$^-$ diffusion to accompany Na$^+$. Owing to the preferential reabsorption of HCO$_3^-$ along the proximal tubule, the HCO$_3^-$ concentration in the tubular fluid falls, whereas the Cl$^-$ concentration rises. Because Cl$^-$ is more permeant than HCO$_3^-$, a lumen-positive diffusion potential develops in the late proximal tubule, but the concentration gradient for Cl$^-$ favors its continued reabsorption.

tubule to be negligible. Using this assumption, the net passive movement of ions across the proximal tubule is determined simply by their concentration differences.

WATER REABSORPTION DRIVEN BY SOLUTE REABSORPTION

As described earlier, reabsorption of Na^+ with its accompanying anions is driven by the Na^+,K^+-ATPase. This movement is also accompanied by the reabsorption of other solutes linked to Na^+ reabsorption including glucose, amino acids, and other organic solutes. The net effect of the reabsorption of all of these solutes is to deplete the tubular lumen of solutes while adding them to the lateral intercellular spaces and interstitium. Therefore, the fluid in the proximal tubule should become relatively dilute, whereas that in the lateral intercellular space and interstitium becomes concentrated. This conclusion seems to be contradicted by the observation that the osmolality of the fluid in the proximal tubule remains the same as that of the plasma, giving rise to the term *isosmotic volume absorption* to describe the process. The explanation for the contradiction is that the proximal tubule is extremely permeable to water; in fact, the water permeability of the proximal tubule is higher than any other epithelium. Because of its high water permeability (or hydraulic conductivity), water flow across the epithelium can be driven quite rapidly by a very small difference in the osmolality of the tubular fluid compared with the interstitium (Fig. 7). A difference of 5–20 mOsm/kg H_2O has been shown to develop and to provide the driving force for water reabsorption.

The high water permeability of the proximal tubule is due to the presence of water channels that are formed by the protein aquaporin-1. Aquaporin-1 belongs to a large family of aquaporins, some members of which transport ions and small nonelectrolytes in addition to water. Aquaporin-1 transports only water, and it is found to be abundant in both the luminal and basolateral membranes of the proximal tubule. Aquaporin-1 is the most widely distributed member of the aquaporin family. Not only is it expressed in many other epithelia, but it is one of the most abundant proteins in the red blood cell in which it also serves to provide a high water permeability of the plasma membrane.

Capillary Uptake of Reabsorbed Fluid: Peritubular Factors

Due to the solute concentration gradient across the epithelium and the high hydraulic conductivity in the

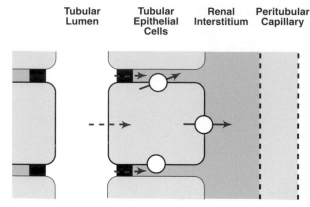

FIGURE 7   Development of a transepithelial osmolality difference in the proximal tubule. Active transport of Na^+ (indicated by the circles with arrows) and the passive movement of anions and other solutes into and through the cells and junctional complexes (dashed arrows) cause the luminal fluid to become slightly dilute and the interstitial fluid to become slightly concentrated relative to the cells or the plasma (as indicated by the differential shading). This small difference in osmolality between the luminal fluid and the interstitium serves as the osmotic driving force for water reabsorption.

proximal tubule, water follows the reabsorption of solute in nearly exact proportion. The reabsorbed fluid must be returned to the general circulation via the peritubular capillaries. As fluid enters the interstitial space, it distends it; thus, the interstitial fluid pressure rises to about +7 mm Hg, in contrast with the negative interstitial fluid pressure found in most tissues. As discussed in Chapter 24, the hydrostatic pressure of the blood in the peritubular capillaries is lower than in most capillary beds, and the colloid osmotic pressure is higher. Thus, the Starling forces across the peritubular capillary endothelium are poised for rapid uptake of the fluid reabsorbed by the proximal tubule. The colloid osmotic pressure is high, and the net opposing hydrostatic force (capillary blood pressure minus interstitial pressure) is low. Normally, this allows fluid to be taken up by the capillaries as rapidly as it is reabsorbed by the proximal tubule, with little expansion of the interstitial fluid. However, when the Starling forces favoring uptake are reduced, the interstitial fluid may expand, with a corresponding rise in its hydrostatic pressure. The rise in interstitial fluid pressure with volume expansion may also cause morphologic changes in the epithelium of the proximal tubule, opening junctional complexes between proximal tubule cells and allowing a backflux of solutes and water, with a resulting decrease in their net reabsorption (Fig. 8). These changes in Starling forces and interstitial pressure are often referred to as *peritubular factors,* and they may alter the rate of volume reabsorption.

Clinical Note

Infusion of Isotonic Saline Solution

Perhaps the most frequently administered intravenous fluid is isotonic saline (0.9% = 0.9 g of NaCl/dL). Isotonic saline is used as a replacement for extracellular salt and water that may be depleted by dehydration, or when a patient is not eating and drinking normally. Thus, unless there is reason to suspect that a patient is actually volume overloaded, which may occur with congestive heart failure or liver or kidney failure, ample isotonic saline is given as a means of maintaining normal extracellular fluid volume. If the kidneys are operating normally, it is difficult to expand the extracellular fluid volume significantly because the kidney increases its excretion of salt and water to match the increased input. Most of this adaptation occurs quickly and is due to decreased salt and water reabsorption in the proximal tubule.

One explanation for the decreased proximal tubule reabsorption is the effect of the saline infusion on the uptake of salt and water into the peritubular capillaries. Because the isotonic saline is a protein-free fluid, the plasma colloid osmotic pressure drops when it is added to the blood. This favors an increased GFR, but as discussed in Chapter 24, autoregulatory adjustments in afferent and efferent arteriolar resistances generally act to maintain a constant GFR, and no change is observable. However, the colloid osmotic pressure of the blood in the peritubular capillaries is also reduced by the saline expansion, giving a lesser net force for fluid uptake. For a short period after the saline infusion, the rate of fluid reabsorption by the proximal tubule will exceed uptake into the capillaries, leading to an increase in interstitial fluid volume and, because of distension, an increase in interstitial fluid hydrostatic pressure. The increase in interstitial fluid pressure counteracts the fall in colloid osmotic pressure, so that the net Starling force again rises until the uptake of fluid into the capillaries matches its reabsorption from the proximal tubule, and the steady state is restored but with a new balance of peritubular factors. However, this rise in interstitial fluid pressure may increase paracellular backleak and, thus, reduce volume reabsorption.

ACTIVE AND PASSIVE SOLUTE TRANSPORT

Up until this point, solute reabsorption in the proximal tubule has been discussed in general terms as

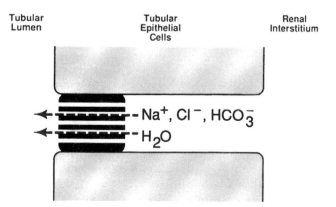

FIGURE 8   Effect of reduced plasma colloid osmotic pressure and increased interstitial fluid pressure on the permeability properties of the junctional complexes. It has been proposed that an increase in interstitial fluid pressure may be associated with an increased permeability of the junctional complexes between proximal tubule cells that would enable backflux diffusion of Na^+, Cl^-, HCO_3^-, and water from the interstitium to the lumen, thus reducing the net reabsorption of solutes and water.

if all filtered solutes were being reabsorbed equally. In fact, some solutes, such as amino acids and glucose, are reabsorbed much more rapidly due to their direct coupling to Na^+ reabsorption by specialized transport proteins in the luminal membrane. This *preferential reabsorption* causes the concentration of these substances in the proximal tubule fluid to fall below that in the plasma and the original ultrafiltrate. The movement of other substances down a concentration gradient from the lumen to the peritubular fluid results in *passive reabsorption*. This gradient develops as water is reabsorbed, producing a higher concentration of the passively reabsorbed solute in the lumen. The concentration of the passively reabsorbed solute increases in the tubular fluid until its rate of passive diffusion out of the tubular lumen is sufficient to match the rate of volume reabsorption. Therefore, the steady-state concentration of the passively reabsorbed solute will depend on its ability to pass through the tubular epithelium. The lower the permeability of the proximal tubule to the solute, the more its concentration must rise in the tubular fluid before it is reabsorbed at a sufficient rate to maintain a constant concentration in the lumen. Other substances are

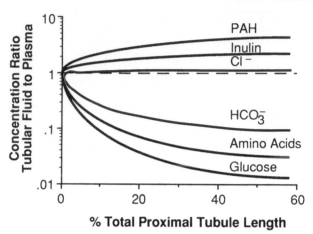

% Total Proximal Tubule Length

FIGURE 9   Changes in the ratio of tubular fluid concentration to plasma concentration for various solutes along the length of the proximal tubule. The concentration ratio (TF/P) is plotted as a function of the percent of the total proximal tubule length. The tubular concentrations of the preferentially absorbed solutes such as HCO$_3^-$, glucose, and amino acids fall rapidly as they are actively reabsorbed. The tubular concentrations of passively reabsorbed or nonreabsorbed substances such as Cl$^-$ and inulin rise as water is reabsorbed. The tubular concentration of an actively secreted substance such as PAH rises even more rapidly.

actively secreted into the proximal tubular fluid and, consequently, have higher concentrations in the tubular fluid.

Because of these specialized transport processes, the composition of the proximal tubule fluid is markedly altered along the length of the proximal tubule. As shown schematically in Fig. 9, at the beginning of the proximal convoluted tubule, the fluid is simply an ultrafiltrate of plasma, but by the end of the proximal tubule virtually all filtered amino acids, glucose, and other metabolic substrates have been reabsorbed. Consequently, the concentration of Na$^+$ salts is higher because they replace the solute deficit left by the preferentially reabsorbed solutes. Secreted solutes such as para-aminohippurate (PAH, discussed later; see also Chapter 24) develop TF/P ratios exceeding that of inulin (3.0) by the end of the proximal tubule; however, their actual concentrations are quite small, so they contribute little to the osmolality of the tubular fluid. The next sections describe the transport processes responsible for the preferential reabsorption of some solutes and secretion of others.

ACTIVE H$^+$ SECRETION AND HCO$_3^-$ REABSORPTION

As mentioned earlier, one of the primary changes in the composition of the proximal tubular fluid is a rise in its Cl$^-$ concentration, which is a consequence of the preferential reabsorption of HCO$_3^-$. This preferential reabsorption process causes the HCO$_3^-$ concentration of the proximal tubular fluid to fall, but, because electroneutrality must be maintained, there must be a corresponding rise in the concentration of the most abundant anion, Cl$^-$.

Active secretion of protons (H$^+$) into the lumen of the proximal tubule drives the preferential reabsorption of HCO$_3^-$. Protons entering the lumen are buffered by HCO$_3^-$, resulting in a loss of HCO$_3^-$ consumed in the formation of carbonic acid and its subsequent diffusion from the lumen as CO$_2$. This complex process needs to be considered in more detail because it is essential not only to reabsorptive processes in the proximal tubule but also to the regulation of body acid–base balance.

As shown in Fig. 10, both proximal tubule cells and intercalated cells, which are found in the connecting tubule and collecting duct, are capable of actively

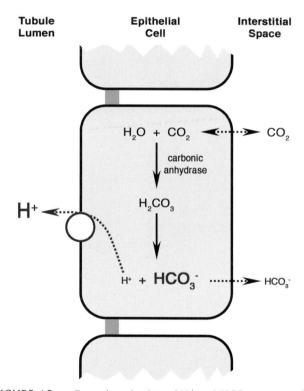

FIGURE 10   General mechanism of H$^+$ and HCO$_3^-$ transport by the nephron. In this model, which applies to both the proximal tubule cells and intercalated cells, which are found in the connecting tubule and collecting duct, the details of the active H$^+$ secretory mechanism and basolateral HCO$_3^-$ exit are left unspecified in order to focus on the intracellular reactions producing H$^+$ and HCO$_3^-$. Because of mass balance considerations, the active extrusion of H$^+$ from the cell into the lumen causes the flow of the reactions increasing the HCO$_3^-$ concentration in the cell. Because of this rise in the cell HCO$_3$ concentration, there is a favorable electrochemical potential gradient for the passive movement of HCO$_3^-$ from the cell, and this process occurs only across the basolateral membrane because of specialized HCO$_3^-$ transporters in that membrane.

secreting H^+ into the lumen. This H^+ is derived from CO_2 that is produced by metabolism or that enters the cells by diffusion from the extracellular fluid. As a gas, CO_2 permeates cell membranes and rapidly attains diffusion equilibrium. CO_2 in the cell is hydrated to form carbonic acid (H_2CO_3), a weak acid that partially dissociates to H^+ and HCO_3^- (see also Chapter 20). The reaction of CO_2 with water to form carbonic acid is normally at equilibrium because of the presence of the enzyme carbonic anhydrase in all cells. Proximal cells differ from intercalated cells in the mechanism used for active transport of H^+ from the cytosol into the lumen, but this secretory process lowers the H^+ concentration in the cell. As shown in Fig. 9, when intracellular pH rises (i.e., H^+ concentration falls) the CO_2/carbonic acid system in the cell favors the flow of equilibrium reactions to produce more product, that is, more H^+ and HCO_3^-. Thus, the active transport of H^+ from the cell decreases the intracellular H^+ concentration and increases the intracellular HCO_3^- concentration.

The HCO_3^- concentration in the cell would normally be low. If it were in electrochemical equilibrium with the extracellular fluid, its concentration would be only about 2 mmol/L. Because of the active H^+ secretion from the cell, the HCO_3^- concentration can be much higher, leading to an electrochemical potential difference favoring HCO_3^- movement out of the cell. In both proximal tubule cells and intercalated cells, HCO_3^- leaves the cells across the basolateral membrane because this membrane has facilitated exit transporters, whereas the luminal membrane does not. The net effect is that H^+ is actively transported into the lumen and HCO_3^- diffuses passively from the cell into the peritubular fluid.

The details of this process in the proximal tubule are shown in Fig. 11. The active H^+ transport across the luminal membrane in the proximal tubule is driven primarily by an exchange mechanism for Na^+, which is called an Na^+/H^+ antiporter. The active transport of H^+ out of the cell against its electrochemical potential difference is driven by the energy made available by Na^+ movement into the cell down its electrochemical potential difference. Just as the kinetic energy of moving water can be converted into useful work by a turbine, the kinetic energy of Na^+ moving into the cell can be converted into work to move H^+ into the lumen and acidify the tubular fluid. As H^+ enters the lumen, the pH of the tubular fluid drops. This fall in pH is buffered by the filtered HCO_3^- in the lumen. The HCO_3^- buffers H^+ as the pH falls, exactly the reverse of the process occurring in the cell. The resulting carbonic acid can dissociate to CO_2 and water. Carbonic anhydrase is bound to the luminal brush border membrane in the proximal tubule. It catalyzes the dissociation of carbonic acid to CO_2 and water, thereby allowing a rapid

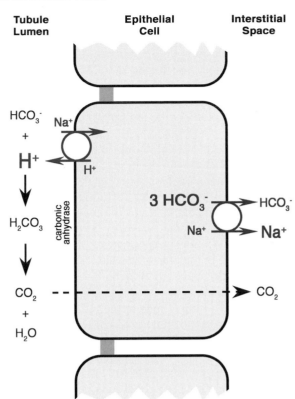

FIGURE 11 Mechanism of HCO_3^- reabsorption in the proximal tubule. Active H^+ secretion into the lumen is driven by exchange diffusion (antiport) for Na^+ entering the cell. Movement of HCO_3^- out of the cell across the basolateral membrane occurs by a specialized transporter involving cotransport of three HCO_3^- ions with one Na^+ ion. The high concentration of H^+ in the tubular fluid titrates filtered HCO_3^- to produce CO_2. This reaction is catalyzed by carbonic anhydrase enzyme bound to the luminal brush border membrane. HCO_3^- disappearing from the lumen is replaced by an equimolar amount of HCO_3^- leaving the cell across the basolateral membrane.

conversion of the secreted H^+ and the HCO_3^- buffer to CO_2. As this reaction proceeds, the CO_2 produced in the lumen rapidly diffuses down its concentration gradient back to the blood. The result is that HCO_3^- disappears from the lumen as the tubular fluid is acidified.

Given these reactions, why is the process referred to as bicarbonate reabsorption rather than HCO_3^- titration or some other term? Close inspection of the reaction scheme shown in Fig. 11 shows that the net result is the same as if the HCO_3^- were reabsorbed directly. For each H^+ that is secreted and titrated by one HCO_3^- in the lumen, one HCO_3^- is also formed in the cell and diffuses across the basolateral membrane to the blood. Thus, each bicarbonate that disappears from the lumen is matched by one that appears in the peritubular capillaries and is returned to the systemic circulation. In producing the H^+ and HCO_3^- within the cell, one CO_2 produced by metabolism within the proximal tubular cell or elsewhere in the body is consumed; however, one

CO_2 is produced in the lumen by the titration of HCO_3^- and returns to the blood. The net effect of this circuitous process is that the luminal pH falls, and there is a loss of bicarbonate from the lumen and the addition of bicarbonate to the blood. There is no net gain or loss of CO_2. Therefore, the process is equivalent to bicarbonate reabsorption with a corresponding luminal acidification.

As discussed in more detail in Chapter 31, this reabsorption of bicarbonate is essential to the maintenance of normal acid–base balance. Filtered HCO_3^- is reclaimed from the glomerular filtrate and returned to the blood, where it can serve as a buffer for the H^+ ions produced by metabolism.

The active H^+ secretory process in the proximal tubule is limited in the extent to which it can acidify the lumen and, thus, reabsorb bicarbonate by the energy available from the Na^+ electrochemical potential difference across the luminal membrane. It is also limited because the proximal tubule is a leaky epithelium and allows H^+ to diffuse back toward the blood. Both factors set a lower limit of 6.6–6.8 on the pH that can be achieved in the lumen of the proximal tubule under normal systemic acid–base conditions.

At a pH of 6.6, the equilibrium concentration for HCO_3^- in the lumen is about 3.8 mmol/L. Thus, the bicarbonate concentration falls from a normal filtered concentration of about 25 mmol/L to 3.8 mmol/L. However, about 95% of the filtered HCO_3^- is reabsorbed. How is this possible? Consider bicarbonate mass balance along the proximal tubule. If 130 mL/min of fluid is filtered at the glomerulus containing 25 mmol/L of bicarbonate, the rate of HCO_3^- filtration is $(0.13 \cdot 25)$ = 3.25 mmol/min. Because 66% of the filtered fluid is reabsorbed along the proximal tubule, the flow entering the loop of Henle is 44.2 mL/min. If the HCO_3^- concentration of this fluid is 3.8 mmol/L, 0.168 mmol/min of HCO_3^- $(0.0442 \cdot 3.8)$ leaves the proximal tubule. Thus, $(3.25 - 0.168)$ = 3.082 mmol/L, is reabsorbed along the proximal tubule, which is equivalent to a *fractional reabsorption* of 95% of the filtered load of HCO_3^-.

As discussed in Chapter 31, HCO_3^- reabsorption is regulated according to the requirements for acid–base balance. Merely by reducing the fractional reabsorption of bicarbonate in the proximal tubule, its excretion can be increased markedly, thus, causing the loss of base from the body.

PASSIVE REABSORPTIVE PROCESSES

The preferential reabsorption of bicarbonate represents a sizable fraction of the anion reabsorption that accompanies Na^+ reabsorption in the proximal tubule. As a consequence of the fall in the luminal bicarbonate concentration from 25 mmol/L to 3.8 mmol/L, the luminal Cl^- concentration rises a corresponding amount from about 105 mmol/L to 126 mmol/L so that the total concentration of these major anions in the luminal fluid remains nearly constant. Given our approximation that the average transepithelial electrical voltage is zero, the increase in the luminal Cl^- concentration gives rise to a driving force that favors the passive diffusion of Cl^- from the lumen to the blood side of the proximal tubule epithelium. The rate at which Cl^- is reabsorbed depends on the magnitude of its transepithelial concentration difference and its relatively high permeability across the epithelium. Therefore, the rate of Cl^- reabsorption increases as the Cl^- concentration in the lumen rises until, in the late proximal tubule, HCO_3^- reabsorption has ceased and the reabsorption of Cl^- equals that of Na^+. By means of this passive mechanism, about 60% of the filtered Cl^- is reabsorbed. This is less than the fractional volume reabsorption of 66%, accounting for the rise in the Cl^- concentration.

The same passive mechanism serves to drive K^+ reabsorption—as fluid is reabsorbed, the K^+ concentration of the luminal fluid rises, leading to the diffusion of K^+ from the lumen. Because K^+ is also relatively able to permeate the epithelium, this mechanism accounts for the reabsorption of about 60% of the filtered load of K^+, and the K^+ concentration in the lumen also rises slightly.

There is also no active reabsorptive mechanism for urea. Urea is a metabolic by-product of protein metabolism that must be eliminated from the body, therefore, active reabsorption would be counterproductive. Nevertheless, passive urea absorption does occur. Because the ability of urea to permeate the epithelium is less than that of either Cl^- or K^+, it is reabsorbed somewhat more slowly, and its tubular fluid (TF) concentration increases as water is reabsorbed. The normal concentration of urea in plasma and, thus, in the initial filtered fluid is 4 mmol/L [in clinical units this is equivalent to a blood urea nitrogen concentration (BUN) of 11.2 mg/dL]. By the end of the proximal tubule, the urea concentration rises to about 6 mmol/L, meaning that slightly less than 50% is reabsorbed. Because urea reabsorption is passive along the entire nephron, the extent to which it is reabsorbed depends on the rate of fluid flow along the nephron and, thus, on the GFR and final urine flow (UF). This flow dependence occurs because the urea is in contact with the tubular epithelium for a longer time at slower flow rates, giving more time for diffusional equilibration and, thus, increased reabsorption. At high rates of UF, the urea clearance approaches 60–70% of the GFR but it decreases markedly with slower UF rates. This effect gives rise to the fact that urea clearance by the kidney increases with more rapid UF.

Poorly Permeant Solutes and Osmotic Diuresis

Water reabsorption leads to an even greater increase in the luminal concentration of solutes that permeate the tubular epithelium poorly. Any such solute that is filtered in appreciable quantities becomes concentrated in the lumen and opposes fluid reabsorption because of its osmotic effect. Normally, no such solute is filtered in any appreciable concentration at the glomerulus. However, the effect can be used to advantage clinically by introducing a poorly reabsorbed substance into the circulation so that it will be filtered into the proximal tubule and oppose reabsorption. The result is the development of an osmotic diuresis as described in the following Clinical Note.

In mannitol diuresis, the reabsorption of Na^+ is decreased because its concentration in the lumen falls due to the water retained by the osmotic effect of the mannitol. This decrease in luminal Na^+ concentration has little effect on the active transport of Na^+ out of the lumen; however, there is a concentration gradient that favors the backflow of Na^+ from the interstitium to the lumen because of the reduced Na^+ concentration in the latter compartment. Historically, mannitol diuresis was used to demonstrate experimentally that Na^+ reabsorption must be active. It was difficult to establish that Na^+ was actively transported under normal conditions because the Na^+ concentration in the lumen remained about the same as that in the blood and interstitial fluid. However, in the presence of mannitol it was readily shown that net Na^+ reabsorption occurred against a concentration gradient and against an electrochemical potential gradient.

SECONDARY ACTIVE REABSORPTIVE PROCESSES

In addition to HCO_3^-, several other solutes are actively reabsorbed in the proximal tubules. Most of these are organic substrates such as glucose, amino acids, and other organic acids that must be reclaimed and returned to the blood for utilization by the mills of metabolism. As in the intestine, these active transport processes are driven not by the direct expenditure of energy, but by coupling to the Na^+ electrochemical potential gradient. A vast quantity of Na^+ is reabsorbed in the proximal tubule, and in the process it diffuses

---

Clinical Note

Osmotic Diuresis Produced by Mannitol Infusion

In certain clinical situations, such as congestive heart failure or liver failure, fluid reabsorption by the proximal tubule is excessive. In these settings, the proximal tubule may reabsorb up to 90% of the filtered load of salt and water so that diuretics that act on segments that are more distal may have relatively little effect. Therefore, one must interfere with the proximal reabsorptive process to effect a diuresis and natriuresis and rid the body of excess extracellular fluid volume.

Decreasing proximal reabsorption could be attempted by using a carbonic anhydrase inhibitor such as acetazolamide, which reduces proximal HCO_3^- reabsorption by slowing down the dehydration of carbonic acid to CO_2 and water, but carbonic anhydrase inhibitors are relatively weak diuretics and produce a metabolic acidosis as a side effect of their interference with HCO_3^- reabsorption. The solution is to use a freely filtered but poorly reabsorbed solute. A useful substance for this purpose is *mannitol*, a monosaccharide not normally found in mammals. It has a molecular weight of 182 Da so that it is freely filtered at the glomerulus, but it has a poor ability to permeate the proximal tubule epithelium due to its size and polarity, and the absence of specialized transporters for its reabsorption. Mannitol is also fairly water soluble [solutions of 25 g/dL (1400 mmol/L) are prepared for intravenous administration], so that infusion of a small volume can produce a high plasma mannitol concentration. Let us consider quantitatively what would happen along the nephron after the infusion of a typical diuretic dose of mannitol. The following numerical example illustrates not only the effects of an osmotic diuretic but gives some insight into the mechanism of isosmotic volume reabsorption and the rates of salt and water delivery to the distal nephron.

Numerical Example of Osmotic Diuresis Produced by Mannitol

Consider first the normal rates of salt and water reabsorption in the proximal tubule. Normal plasma osmolality is 290 mOsm/kg H_2O, which is primarily due to Na^+ present at a concentration of 140 mmol/L with its associated anions. With a normal GFR of 130 mL/min, the rate of Na^+ filtration would be $(0.13 \cdot 140) = 18.2$ mmol/min. By the end of the proximal tubule, two-thirds of the filtered fluid has been reabsorbed, leaving a flow rate of $(0.333 \cdot 130) = 43.3$ mL/min, containing approximately the same Na^+ concentration as the plasma so that the delivery of Na^+ to the loop of Henle would be $(0.0433 \cdot 140) = 6.1$ mmol/min. The rate of proximal reabsorption is the difference between the rate of filtration and the rate of delivery to the loop of Henle. For water, the reabsorption rate is $(130 - 43.3) = 86.7$ mL/min. For Na^+ the reabsorption rate is $(18.2 - 6.1) = 12.1$ mmol/min.

Consider now the consequences of giving an intravenous infusion of hypertonic mannitol sufficient to produce a plasma mannitol concentration of 30 mmol/L, which would give a plasma osmolality of 320 mOsm/kg H_2O, a typical increase in clinical practice. For convenience, we assume that the plasma Na^+ concentration remains the same. Note that there would be an equivalent increase in the osmolality of the intracellular fluid, even though mannitol would not enter the cells to any appreciable extent. Owing to the high water permeability of most cell membranes, water leaves the cells rapidly until intra- and extracellular osmolality are equal. Thus, the movement of water would expand the extracellular space at the expense of the intracellular fluid compartment.

Assuming the GFR also remains constant, the rate of Na^+ filtration would still be 18.2 mmol/min. The rate of mannitol filtration would be $(0.13 \cdot 30) = 3.9$ mmol/min. At the end of the proximal tubule, because of the osmotic effect of the mannitol, which is only poorly reabsorbed passively, assume that only 58% of the filtered water has been reabsorbed so that the flow rate is now $(0.42 \cdot 130) = 54.6$ mL/min. Assume that only 5% of the filtered mannitol has been reabsorbed so that its concentration has risen to $(3.9 \cdot 0.95/.0546) = 67.9$ mmol/L. Because the water permeability of the proximal tubule is very high, the tubular fluid always has nearly the same osmolality as the plasma, which in this case is 320 mOsm/kg H_2O. However, if the mannitol concentration is about 68 mmol/L, the remainder of the osmolality of the tubular fluid that is not due to mannitol would be $(320 - 68) = 252$ mOsm/kg H_2O. If Na^+ makes up approximately one-half of this osmolality, the Na^+ concentration at the end of the proximal tubule would be 126 mmol/L, and its delivery to the loop of Henle would be $(0.0546 \cdot 126) = 6.9$ mmol/min. [It is important to note that, even though the Na^+ concentration of the tubular fluid is much less than in normal circumstances, the Na^+ delivery to the loop of Henle is still increased by $(6.9 - 6.1) = 0.8$ mmol/min because of the increased flow rate!] The rate of volume reabsorption in the proximal tubule is $(130 - 54.6) = 75.4$ mL/min, and the rate of Na^+ reabsorption is $(18.2 - 6.9) = 11.3$ mmol/min. This illustrates that the reabsorption of both water and Na^+ are decreased because of the osmotic diuresis created by mannitol.

down its electrochemical potential gradient into the proximal tubule epithelial cells and is pumped out of the cells across their basolateral membrane into the interstitial space (see Fig. 2). The movement of Na^+ into the cells represents the loss of a considerable amount of potential energy. By coupling this downhill movement to the uphill transport of sugars, amino acids, and other solutes into the cells, the proximal tubule realizes a considerable energy savings. It requires no more energy investment than that already obligated for the Na^+,K^+-ATPase to transport Na^+ out of the cell.

The energy available from downhill luminal Na^+ entry is coupled to the transport of several organic solutes by specific membrane transporters that simultaneously transport the solute plus Na^+ across the membrane. This process is referred to as *cotransport* or *symport*. The transcellular movement of the cotransported solute is shown in Fig. 12. The solute enters across the luminal membrane against a concentration gradient, driven by the energy available from the cotransport of Na^+ down its electrochemical potential gradient. Owing to the higher concentration of the

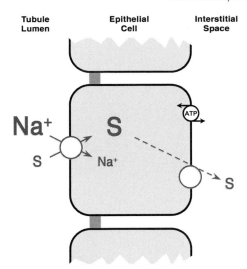

FIGURE 12 Active reabsorption of organic solutes by cotransport with Na$^+$. The active transport of many organic solutes (S) from the lumen into the proximal tubule cell is driven by cotransport with Na$^+$ on specific carrier molecules. The Na$^+$ moves into the cell down its electrochemical potential gradient, thus making energy available to concentrate S in the cell. The potential energy for Na$^+$ entry is maintained by the operation of the Na$^+$,K$^+$-ATPase in the basolateral membrane. The basolateral membrane contains a facilitated diffusion mechanism that allows S to diffuse out of the cell down its electrochemical potential gradient into the interstitial fluid and the plasma in the peritubular capillary network.

solute in the cell, there is a concentration gradient favoring its movement out of the cell. Other transport proteins facilitate passive solute exit across the basolateral membrane. These facilitated diffusion mechanisms allow the solute to move readily across the cell membrane to the interstitial space.

Glucose

All useful organic substrates are reabsorbed by Na$^+$ cotransport mechanisms. These include glucose, amino acids, lactate, acetate, citrate, succinate, oxalate, and various carboxylic acids. However, the carriers involved in a given cotransport process are limited in number, so there is a maximal rate of transport for each of these substrates based on the number of transporters available in the proximal tubule. The effect of this limitation on the overall renal handling of these solutes is best illustrated by glucose. As shown in Fig. 13, at plasma concentrations of glucose up to double the normal ~100 mg/dL, there is no glucose in the urine. (Actually, there are normally trace amounts that are not detected by the usual clinical laboratory methods.) In other words, for all practical purposes, 100% of the filtered glucose load is reabsorbed. However, when the plasma glucose concentration exceeds 200–220 mg/dL, readily detectable amounts of glucose appear in the urine.

The plasma glucose concentration at which glucose begins to appear in the urine is called the *renal plasma glucose threshold*. Above the threshold, the rate of excretion increases until it parallels the rate of filtration. The rate of glucose reabsorption can be calculated as the rate of filtration minus the rate of excretion, and as the plasma glucose concentration rises above about 350 mg/dL, the rate of reabsorption reaches a maximum of about 375 mg/min. This maximum, which is referred to as the *transport maximum* (T_m), is reached when all nephrons are reabsorbing glucose at their maximal rate and any additional amount filtered is excreted. The maximal rate of transport is a function of the total transport capacity of each transporter and their number. Because the number is proportional to the number of functioning nephrons, the T_m for glucose decreases in acute or chronic renal failure in proportion to the decrease in the number of functioning nephrons. However, the renal plasma glucose threshold remains relatively constant.

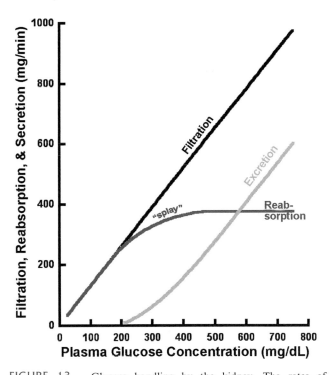

FIGURE 13 Glucose handling by the kidney. The rates of filtration, excretion, and reabsorption of glucose are plotted as a function of the plasma glucose concentration. It is assumed that the GFR remains constant at 130 mL/min so that the rate of glucose filtration rises linearly with the plasma glucose concentration (see Chapter 25, Eq. [6]). Up to a plasma glucose concentration of 200 mg/dL, the rate of reabsorption equals the rate of filtration, and thus there is no glucose excretion. Above this plasma glucose concentration (referred to as the plasma threshold), the excretion rate begins to rise, and the reabsorptive rate plateaus at a transport maximum (T_m) of approximately 375 mg/mm. Splay refers to the gradual increase in excretion and gradual plateauing of reabsorption as the plasma glucose concentration rises above the renal plasma threshold.

Clinical Note

Differential Diagnosis of Glucosuria

In normal individuals, no glucose is detectable in the urine by clinical tests. Mild glucosuria is usually not accompanied by any clinical signs, but if the excretion of glucose becomes significant, the presenting symptom is often complaints of thirst, frequent voiding, and nocturia (waking at night to urinate). Why? When glucose is not completely reabsorbed, it acts like an osmotic diuretic. Glucose has the same molecular size and polar hydroxyl groups as mannitol. Thus, the only reason that glucose is normally reabsorbed is that there is a special transporter for glucose but not for mannitol. The osmotic diuresis produced by hyperglycemia is the cause of the frequent voiding and nocturia, and the thirst is a consequence of persistent water loss in the urine, which can be confirmed by the demonstration of significant glucose in the urine.

The physician must next determine the cause of the glucosuria. Pregnancy may produce an *apparent* glucosuria (more accurately termed *glycosuria*) because of the presence of lactose and galactose in the urine. Also, in pregnancy a normal increase in GFR may raise the filtered load of glucose beyond the T_m. Other than pregnancy, Fig. 13 shows there are two potential reasons for the abnormal appearance of glucose in the urine: a plasma glucose concentration exceeding the plasma threshold of 200 mg/dL (which is often indicative of diabetes mellitus) or an abnormality in the glucose transporter such that significant amounts of glucose appear in the urine even at normal plasma concentrations. In the latter case, there may be a decreased renal plasma threshold for glucose, possibly in combination with a decreased T_m. This condition is referred to as *renal glucosuria*—that is, glucose appearance in the urine due to a renal defect. Most glucosuria is a consequence of elevated plasma glucose concentrations, LCA in although renal glucosuria is a rare hereditary disorder, it should be ruled out as a cause of glucosuria. A diagnosis of diabetes mellitus, on the other hand, can be confirmed by additional clinical laboratory tests.

A further note: Intestinal glucose intolerance and renal glucosuria provide an interesting side light on the genetics of glucose transport abnormalities. Glucose is also absorbed in the small intestine by a Na^+ cotransport mechanism, but two different Na^+/glucose cotransporters are encoded by separate genes. The first, SGLT1, is the more common and is found in the intestine as well as in the kidney. It has a relatively high affinity for glucose, and an even higher affinity for galactose. Mutations in SGLT1, which are inherited as an autosomal recessive trait, lead to a combination of mild glucosuria (due to increased splay) and an intestinal malabsorption of galactose and glucose, which produces severe and even fatal diarrhea in neonates. A rarer mutation in the second transporter, SGLT2, leads to severe glucosuria because of a reduction in T_m. SGLT2 is a low-affinity, high-capacity transporter that is present in the proximal convoluted tubule and the early proximal straight tubule, whereas SGLT1 is present only in the late proximal straight tubule. Because most of the filtered glucose is reabsorbed in the proximal convoluted tubule, it is not surprising that a defect in SGLT2 leads to severe glucosuria but no significant effects on intestinal sugar absorption.

The gradual rise in the rate of excretion with increasing glucose concentration is referred to as *splay*. In other words, the renal plasma glucose threshold is lower than the plasma glucose concentration at which maximal reabsorption is achieved. The reasons for this are twofold: (1) Various nephrons have different capacities to reabsorb glucose. Some reach their maximum at relatively low plasma glucose concentrations and others at high concentrations, but glucose will begin to appear in the urine as soon as the nephrons with the lowest capacity are saturated. However, it requires higher filtered loads to saturate all carriers. (2) The transporters are known to demonstrate enzyme-like kinetics; they require high luminal glucose concentrations to be fully saturated, but before they are saturated they let measurable amounts of glucose pass to the urine.

Amino Acids

Separate but similar transporters reabsorb all the organic solutes mentioned previously. Depending on the particular solute, the normal filtered load may or may not exceed the T_m. There are several different transporters for various groups of amino acids such as neutral, acidic, and basic, and subgroups of these. However, the reabsorption of most amino acids is not complete and, depending on the amino acid, 0.5–2% of the filtered load normally appears in the urine. This incomplete reabsorption does not indicate that the T_m for the amino acid transport systems are exceeded but rather that the plasma concentrations of the amino acids exceed their renal plasma thresholds. The filtered loads of the amino acids are normally much less than their T_m but the transport mechanisms exhibit considerable splay, meaning that the renal plasma amino acid thresholds occur at concentrations lower than the plasma concentrations at which the filtered loads exceed the T_m. There are several known congenital abnormalities in the reabsorption of various groups of amino acids that result from a lack of expression or mutation of the gene for specific transporters. This can result in the excretion of large amounts of those amino acids that are normally reabsorbed by the affected transporter.

Organic Acids

Most of the filtered weak organic acids that are useful metabolic substrates are reabsorbed by the proximal tubule. These include acetate, lactate, citrate, oxalate, and mono- and dicarboxylic acids such as the Krebs cycle intermediates. These organic acids are normally present at very low concentrations in the plasma, resulting in low filtered loads that are completely reabsorbed. However, under abnormal metabolic conditions such as diabetic ketoacidosis [in which the plasma contains high concentrations of acetoacetate and (3-hydroxybutyrate)] and lactic acidosis (in which the plasma has high concentrations of lactate), the filtered load of these particular solutes exceeds the renal plasma threshold and the T_m, resulting in excretion of a large amount of the solutes. This incomplete reabsorption can produce an osmotic diuresis, but has a salutary effect in delivering a large amount of buffer anions to the more distal nephron segments so that more buffered protons (titratable acidity; see Chapter 31) can be excreted in the urine, thus counteracting the acidosis.

Proteins and Peptides

T_m-limited transporters in the proximal tubule also reabsorb proteins and small peptides. Although the glomerular filtration barrier is impermeable to proteins with molecular weights in excess of 60,000 Da, a finite but small amount of albumin still remains in the glomerular filtrate. The concentration of albumin in the filtrate is normally less than 30 mg/dL, that is, less than 1% of the plasma concentration. Nevertheless, if the filtered load of albumin were not reabsorbed, up to (30 mg/dL · 180 L/day) = 54 g

---

Clinical Note

Proteinuria Due to Other Small Proteins

Other than the low concentrations of the hormones noted earlier, plasma normally contains no significant amounts of low molecular weight peptides or proteins. However, some disease states are associated with the production or release of small peptides into the plasma. These include (1) multiple myeloma, in which large concentrations of the light chains of immunoglobulins appear in the plasma; (2) hemoglobinemia, in which hemoglobin is released from red blood cells due to hemolysis, as in a transfusion reaction; and, (3) myoglobinemia, in which the myoglobin normally present in muscle cells is released, as occurs in trauma and in the presence of toxins including chemotherapeutic agents.

The presence of significant concentrations of any of these peptides in the plasma results in a filtered load that far exceeds the protein reabsorptive capacity of the proximal tubules. Consequently, these proteins appear in the urine, but because they become concentrated as fluid is reabsorbed along the tubules, they may precipitate and occlude the lumens of the nephrons. Protein precipitates can be seen as cylindrical casts when the urinary sediment is examined under the microscope. Also, as these proteins are reabsorbed in proximal tubule cells, they are degraded to smaller peptides and amino acids. For reasons that are not completely understood, the by-products of this degradation can be toxic to proximal tubule cells, especially in the case of certain light chains. Both the toxicity and the occlusion of nephrons by these small proteins can lead to acute renal failure as the number of functioning nephrons is reduced.

would be excreted per day. The kidney conserves this filtered protein by reabsorption so that the usual renal loss is less than 100 mg of albumin per day. This process is also T_m-limited. If the plasma albumin concentration increases above 6–7 g/dL, the rate of albumin excretion increases markedly and in proportion to the plasma concentration. Thus, the renal plasma albumin threshold is 6–7 g/dL, and the T_m is on the order of 30–40 mg/min. When the glomerular barrier is altered by disease so that more protein is filtered, massive amounts can appear in the urine, as seen in the nephrotic syndrome.

The kidney also reabsorbs the small amounts of low molecular weight peptides that are normally present in the plasma. These include hormones such as insulin, parathyroid hormone (PTH), glucagon, calcitonin, vasopressin, and angiotensin. The peptides are taken up by specific transporters located in the brush border membrane and are degraded to amino acids within the cell. Renal metabolism of the filtered hormones is one of the primary routes of their normal turnover.

Phosphate and Sulfate

Both phosphate and sulfate anions are reabsorbed in the proximal tubule by cotransport with Na^+, and their

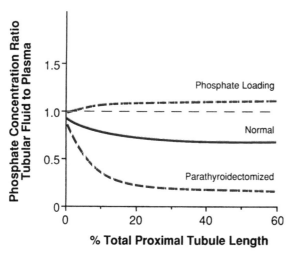

FIGURE 14   Regulation of phosphate reabsorption in the proximal tubule. The ratio of the phosphate concentration in the tubular fluid compared with plasma (TF/P) is plotted as a function of the percent of the proximal tubule length. Under normal conditions, somewhat more than 70% of the filtered phosphate is actively reabsorbed along the proximal tubule by a T_m-limited mechanism. Consequently, the tubular fluid phosphate concentration falls below that in the plasma along the length of the proximal tubule. After acute parathyroidectomy (lower line), an even greater fraction of the filtered phosphate is reabsorbed. Ingestion or infusion of large amounts of phosphate causes a decrease in phosphate reabsorption in the proximal tubule, with a corresponding rise in its luminal concentration.

reabsorption demonstrates the same T_m-limited transport as observed for the organic solutes. The reabsorption of phosphate is of particular interest because of its importance in maintaining a normal plasma phosphate concentration and, thus, a normal calcium-phosphate solubility product in the plasma. Furthermore, the plasma phosphate concentration (2–3 mmol/L) normally exceeds its renal plasma threshold, so that it is excreted continuously in the urine where it is normally the most important buffer anion. In fact, because the filtered load of phosphate is poised near the T_m, the excretion of phosphate rises in almost direct proportion to its plasma concentration. If the plasma phosphate concentration rises above normal levels, excretion increases, tending to reduce the excess. If the plasma phosphate concentration is too low, less is excreted. However, the T_m for phosphate can also be regulated physiologically. PTH (see Chapter 43) regulates phosphate reabsorption as shown in Fig. 14. Normally phosphate excretion is determined primarily by filtration, but high or low levels of PTH can alter this markedly. PTH has a phosphaturic action and reduces the T_m. In the patient with renal failure on hemodialysis, the management of plasma phosphate and calcium becomes one of the most difficult problems. Such a patient usually has a high plasma phosphate level that results in secondary hyperparathyroidism and osteodystrophy.

Sulfate is also reabsorbed by a T_m-limited mechanism. The usual plasma concentration of sulfate, 2–3 mmol/L, exceeds the renal plasma threshold so that sulfate appears in the urine. As in the case of phosphate, the filtered load is above or near the T_m so that the rate of sulfate excretion is proportional to its plasma concentration, and renal excretion serves as the main regulator of the plasma concentration. However, in contrast to phosphate, no known hormone regulates sulfate excretion.

SECRETORY PROCESSES IN THE PROXIMAL STRAIGHT TUBULE

The proximal tubule has an important function in secreting many substances that can be regarded as metabolic by-products or potential toxins. Given normal rates of production of some metabolic by-products, the body requires a renal secretory process to maintain acceptable plasma concentrations. Renal secretory processes serve a more important role in excreting exogenous toxic substances that are ingested in the diet. Secretion of these substances results in an excretion rate that exceeds their rate of filtration, and serves as a means of rapidly clearing them from the plasma. Secreted solutes include many organic acids and bases, creatinine, PAH, radiocontrast agents, and various drugs. In each case, the secretory process involves active accumulation of the solute into proximal tubule cells across the basolateral membrane from the plasma. This active accumulation is driven energetically in different ways, including Na^+ cotransport and other cotransport and exchange processes. The accumulation of the solute in the cell provides a favorable electrochemical potential gradient for the solute to move into the tubular lumen, and various transporters located in the brush border membrane facilitate this movement. Secretion of these organic solutes occurs almost exclusively in the straight portion of the proximal tubule, and can produce luminal concentrations that are much higher than their plasma concentrations.

All of these secretory processes, because they are mediated by a fixed number of transporters, are saturable and exhibit the same T_m-limited rates of transport as the active reabsorptive processes discussed earlier. However, it is the rate of secretion rather than reabsorption that reaches a maximum as the plasma concentration increases.

Secretion of Para-Aminohippurate: Measurement of Renal Blood Flow

Of particular interest are those substances that are the most avidly secreted, including para-aminohippurate (PAH) and certain radiocontrast agents. At arterial plasma PAH concentrations of less than 10 mg/dL, blood flowing out of the kidney via the renal vein contains almost no PAH because it is all excreted in the urine. If PAH were cleared only by filtration, the amount excreted could be at most 20% of the amount flowing into the kidney. The greater clearance of PAH is achieved by its secretion from the peritubular capillaries into the lumen of the proximal straight tubule. This secretory process has such a high affinity for PAH that it can reduce the PAH concentration on the blood side of the epithelium to almost zero. Actually, however, the amount of PAH flowing out of the kidney is reduced not to zero but to about 10% of the amount delivered to the kidney via the renal artery. This residual amount is thought to be due to a fraction of the renal blood flow that is not exposed to filtration or secretion because it perfuses nontransporting structures such as the renal capsule.

The significance of the avid secretion of PAH is that this agent can be used as an indicator of renal plasma flow. If we assume that 100% of the PAH entering the kidney is excreted in the urine, then the amount excreted must be equal to the amount entering the kidney, which is the product of the renal plasma flow (RPF, mL/min) and

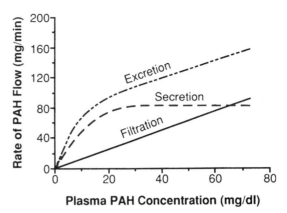

FIGURE 15   PAH secretion in the proximal tubule. The rates of PAH filtration, secretion, and excretion are plotted as a function of the plasma PAH concentration. In calculating the rate of filtration, a constant GFR of 130 mL/min is assumed. The excretion rate of PAH is always higher than the filtration rate because of the contribution of secretion in the proximal tubule; however, as the plasma PAH concentration rises above 10 mg/dL, the rate of secretion approaches a T_m of approximately 80 mg/mm. When the T_m is exceeded, the rate of excretion parallels the rate of filtration.

the arterial PAH concentration (P_{PAH}, converted to units of mg/mL). The amount excreted will be the product of the urine PAH concentration (U_{PAH}) and UF. Thus:

$$RPF \cdot P_{PAH} = UF \cdot U_{PAH}, \text{ or}$$

$$RPF = \frac{UF \cdot U_{PAH}}{P_{PAH}} = C_{PAH}. \qquad (3)$$

In other words, the clearance of PAH is equal to the renal plasma flow. Actually, it is an underestimate by 10–15% because not all of the PAH entering the kidney is excreted. Nevertheless, this is a good index of the rate of renal blood flow.

As noted earlier, PAH is excreted completely only at arterial concentrations of less than 10 mg/dL. This is because the secretory process is saturable, as illustrated in Fig. 15. At concentrations less than 10 mg/dL, the rate of PAH excretion rises linearly with plasma concentration, and virtually all of the PAH entering the kidney is excreted. The rate of secretion of PAH, which is determined as the difference between the rate of excretion and the rate of filtration, thus also rises linearly. But above this plasma concentration, the rate of secretion falls off and reaches a maximum of about 80 mg/min, referred to as the secretory T_m. The T_m is reached when all the basolateral membrane transporters involved in the secretion of PAH are saturated. This T_m, as is the case for reabsorptive processes, is a function of the total number of functioning nephrons and decreases with the loss of nephrons in renal failure.

The renal plasma flow is an important clinical index. It can be decreased by partial occlusion of the renal vasculature or by a reduction in renal mass as occurs in renal failure. However, because of the difficulty in determining plasma PAH concentrations in the clinical laboratory, it is used primarily for laboratory or clinical research protocols. In clinical practice, it is more convenient to use the excretion of radiocontrast agents as rough indicators of renal plasma flow. Because these agents are secreted, their rate of clearance is proportional to the renal plasma flow and the functional renal mass. For example, the time required to excrete a bolus of the radiocontrast agent given during renal angiography is measured and compared with a standard to obtain an index of renal plasma flow. Radioisotope scans are also used for this purpose.

Urate

The renal excretion of urate is also determined largely by secretion into the proximal tubule, although in the normal individual, the rate of urate excretion is lower than its rate of glomerular filtration, indicating that it is reabsorbed along the nephron. However, the excretion of urate actually depends on the balance of opposing reabsorptive and secretory transport processes in the proximal tubule. Both processes involve active urate transport, and the net excretion of urate is determined by which process dominates. The secretory process is the one that is physiologically regulated to maintain a normal plasma urate concentration in the face of changing dietary intake and metabolic production. Thus, when there is an excess of urate, secretion is stimulated, and it is decreased when there is a deficit.

Secretion of Organic Acids and Bases

Many other organic acids and bases, including many not normally found in the body, can be secreted into the proximal tubule. Coupled with hepatic metabolism, renal filtration and secretion processes are essential for clearing potentially harmful substances from the blood. However, many drugs are also cleared rapidly from the body by secretion in the proximal tubule. Sometimes this is helpful, as in the case of diuretics. Because of tubular secretion, the most widely used diuretics reach an effective concentration in the tubular fluid without requiring potentially harmful plasma concentrations. In the case of other drugs such as salicylates, the penicillin-like antibiotics, and sulfonamides, secretion tends to clear the drug rapidly and can be a problem in achieving an optimal plasma concentration. In such cases, probenecid, a competitive inhibitor of the secretory transporter, can be administered to reduce secretion.

Because the doses of most drugs are based on the assumption of normal renal function, care must be exercised in adjusting the doses for the patient with renal failure. Because of the reduction of functioning renal mass by disease, there is a decreased rate of filtration and secretion of drugs. For those drugs that are cleared from the plasma primarily by the kidney, dosages must be adjusted downward in consideration of the decreased renal clearance.

Suggested Reading

Berry CA, Ives HE, Rector FC Jr. Renal transport of glucose, amino acids, sodium, chloride, and water. In Brenner MB, ed. *The kidney*, 5th ed. Philadelphia: WB Saunders, 1996, pp 334–370.

Hamm LL, Alpern RJ. Cellular mechanisms of renal tubule acidification. In Seldin DW, Giebisch G, eds. *The kidney: Physiology and pathophysiology*, Vol 1, 3rd ed. Philadelphia: Lippincott Williams & Wilkins, 2000, pp 1935–1980.

Sackin H, Palmer LG. Epithelial cell structure and polarity. In Seldin DW, Giebisch G, eds. *The kidney: Physiology and pathophysiology*, Vol 1, 3rd ed. Philadelphia: Lippincott Williams & Wilkins, 2000, pp 533–568.

Schafer JA. Transepithelial osmolality differences, hydraulic conductivities and volume absorption in the proximal tubule. *Annu Rev Physiol* 1990;52:709–726.

Weinstein AM. Sodium and chloride transport: Proximal nephron. In Seldin DW, Giebisch G, eds. *The kidney: Physiology and pathophysiology*, Vol 2, 3rd ed. Philadelphia: Lippincott Williams & Wilkins, 2000, pp 1287–1332.

Reabsorption and Secretion in the Loop of Henle and Distal Nephron

JAMES A. SCHAFER

KEY POINTS

- The nephron beyond the proximal tubule is specialized to produce urine of variable composition that can be quite different from the plasma. *Urine osmolality* can vary from 50–1200 mOsm/kg H_2O, and the Na^+ excreted can vary from nearly zero to 2% of the amount filtered at the glomerulus.

- The *thin descending limb of the loop of Henle* is surrounded by medullary interstitial fluid, the osmolality of which can be as high as 1200 mOsm/kg H_2O in the papilla. Urea contributes approximately one-half of this osmolality, and NaCl contributes approximately one half.

- Due to the presence of *aquaporin-1*, the thin descending limb of the loop of Henle is very water permeable regardless of the final urine osmolality, but it is less permeable to solutes.

- In contrast to the descending limb, the water permeability of the *ascending limb of the loop of Henle* is nearly zero regardless of the final urine osmolality. In the thin segment of the ascending limb, NaCl reabsorption is due to passive diffusion from a higher concentration in the tubular fluid to the interstitium. Active reabsorption of NaCl in the thick segment is produced by a cotransporter carrying one Na^+, one K^+, and two Cl^- ions. The most potent diuretics, such as furosemide, directly block this transporter.

- NaCl reabsorption in the medullary thick ascending limb produces the hypertonicity of the interstitium, but in the cortex the continued active reabsorption of NaCl dilutes the tubular fluid to less than one-half the osmolality of the plasma.

KEY POINTS (*continued*)

- The *distal convoluted tubule* is also water impermeable so that continued NaCl reabsorption further dilutes the urine.
- However, beyond the distal convoluted tubule, the water permeability of the connecting tubule and the collecting duct is regulated by the hormone *vasopressin* (antidiuretic hormone or ADH), which regulates the number of aquaporin-2 water channels in the luminal membrane of the *CNT cells* in the connecting tubule and of the *principal cells* in the collecting duct.
- *DCT cells* in the distal convoluted tubule reabsorb NaCl via an electroneutral NaCl cotransporter that can be blocked by thiazide diuretics.
- The connecting tubule and the collecting duct comprise the *aldosterone-responsive distal*

nephron (ARDN). CNT cells and principal cells in these segments actively reabsorb Na^+ via a Na^+ channel (ENaC) in the luminal membrane.
- Principal cells and CNT cells in the ARDN also secrete K^+ by K^+-selective channels in the luminal membrane. Depolarization of the luminal membrane by ENaC channels and increased urine flow enhance the favorable driving force for K^+ secretion.
- *Aldosterone* increases Na^+ reabsorption and K^+ secretion in principal and CNT cells in the ARDN by three mechanisms: increased activity of Na^+ and K^+ channels in the luminal membrane, increased Na^+,K^+-ATPase transporter activity in the basolateral membrane, and increased synthesis of Krebs cycle enzymes.

Beyond the proximal tubule, the nephron is different structurally and functionally. The proximal tubule is a bulk reabsorber—most of the filtered salt and water is returned to the plasma in this region in a relatively nonselective fashion. Therefore, the concentrations of most of the major ions and the total osmolality of the proximal tubular fluid remain approximately the same as those of the plasma. Among the major filtered ions only Cl^- shows a significant change; its concentration increases as HCO_3^- is selectively reabsorbed. Thus, the fluid at the end of the proximal tubule resembles plasma, except that it contains no plasma proteins and that most of the organic substrates have been removed by specific transporters. Only trace amounts of glucose, amino acids, and organic acids are delivered to the loop of Henle. On the other hand, other substances that have been secreted into the proximal tubule, such as para-aminohippurate (PAH), are present in higher concentrations than in the plasma. However, these secreted substances are present in such low concentrations in the plasma that their concentrations are also negligible in the tubular fluid and their contribution to its osmolality can be ignored in discussing the function of the remaining regions of the nephron.

The loop of Henle and the distal portions of the nephron have specialized and tightly regulated transport characteristics that allow them to develop urine that can have a composition different from plasma. Urine osmolality can vary from one-fifth to four times that of plasma. Depending on the physiologic circumstances, the Na^+ excretion rate may vary between nearly zero to more than 2% of that filtered at the glomerulus, which is

called the *filtered load*. The thick ascending limb of the loop of Henle, the distal tubule, and the collecting duct regulate excretion of other important ions such as potassium, calcium, and magnesium. The urine also contains a high urea concentration in comparison with plasma. Furthermore, the phosphate that has been filtered but not reabsorbed by the proximal tubule is titrated by hydrogen ions that are secreted into the distal regions of the nephron, producing a final urine pH that may vary from 4.5–8.0.

The following sections describe the transport processes necessary to produce urine that can vary so markedly in composition from the plasma. Later chapters discuss how the final daily excretion of the most important ions is hormonally regulated to match their daily intake.

TRANSPORT IN THE THIN DESCENDING LIMB

As anatomically described, the descending limb of the loop of Henle begins with the proximal straight tubule. In the outer medulla, the epithelium of the proximal straight tubule transforms into the thin descending limb of the loop of Henle, which is named for its very thin epithelium. Functionally, the proximal straight tubule can be regarded as a continuation of the proximal tubule, whereas the thin descending limb is functionally quite different from the proximal tubule. The length of the thin descending limb varies among nephrons. Most superficial nephrons have relatively short loops of Henle, whereas many juxtamedullary nephrons have long ones that may extend to the tip of the papilla. In this discussion, both long and short thin descending limbs are considered collectively.

A unique feature of the loop of Henle is that it resides in an interstitial fluid having a composition completely different from any other tissue—one that, under typical conditions, is markedly hyperosmotic to the plasma. The osmolality of the medullary interstitial fluid rises from approximately isosmotic at the border between the cortex and the medulla and increases progressively to a maximum of about 1200 mOsm/kg H_2O at the papillary tip. The interstitial fluid at the papillary tip has a urea concentration of about 600 mmol/L and a NaCl concentration of about 300 mmol/L. Thus, approximately 50% of the osmolality is due to the osmotic effect of NaCl and 50% is due to urea.

The fluid flowing into the thin descending limb of the loop of Henle from the proximal tubule has an osmolality that is approximately the same as plasma—about 280 mOsm/kg H_2O. The concentrations of Na^+ and K^+ are approximately the same as in plasma, but Cl^- is the primary anion because of the reabsorption of most of the filtered HCO_3^- by the proximal tubule. The urea concentration is about 6 mmol/L. As this fluid flows through the thin descending limb, an osmotic pressure gradient occurs between this initially isosmotic tubular fluid and the hyperosmotic medullary interstitium.

Figure 1 illustrates schematically the transepithelial transport processes that occur as fluid flows from the outer medullary to the papillary regions of a thin descending limb. There is no active transepithelial transport in the thin descending limb of the loop of Henle. The cells that constitute this epithelium are thin and contain few mitochondria. The most important transport characteristic of the thin descending limb is that it is highly permeable to water because aquaporin-1 water channels are present in the luminal and basolateral membranes of thin descending limb cells as they are in the proximal tubule. The epithelium of the thin descending limb is also somewhat permeable to urea and to a lesser extent NaCl, which would allow the diffusion of these solutes from the interstitium to the tubular fluid, but these permeabilities are significantly less than that of water. Therefore, because of the hyperosmolality of the medulla, there is an osmotic water flow from the tubular fluid into the interstitium with a relatively small diffusional entry of urea and NaCl from the interstitial fluid. The net result is that the tubular fluid becomes concentrated as it flows down into the medulla, and this increase in osmolality develops primarily because of the water loss rather than solute gain.

Because of the loss of water, and to a much lesser extent the entry of urea and NaCl, the osmolality of the tubular fluid rises along the thin descending limb from

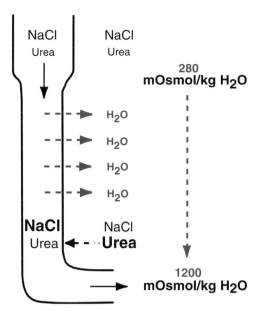

FIGURE 1   Reabsorption of water but not solute in the thin descending limb of the loop of Henle. The interstitial fluid in the medulla has an osmolality that rises from 280 mOsm/kg H_2O to 1200 mOsm/kg H_2O at the tip of the papilla, as shown by the shading in the figure. As fluid flows from the proximal tubule toward the papillary tip, it is concentrated by the movement of water out of the lumen into the hyperosmotic medullary interstitium. This causes the NaCl and urea concentrations to rise in the tubular fluid so that the total osmolality equals that of the adjacent interstitium. Although the osmotic equilibration occurs primarily by the loss of water to the interstitium, some urea also diffuses into the tubular fluid.

280 mOsm/kg H_2O at its proximal end to 1200 at the tip of the loop of Henle. In other words, osmotic equilibration of the tubular fluid with the surrounding medullary interstitium occurs continuously along the thin descending limb because of its high water permeability. Therefore, at the tip of the loop of Henle for the long-looped nephrons, only approximately 10% of the filtered volume remains, and its osmolality has risen to 1200 mOsm/kg H_2O. At this point, most of the total 1200 mOsm/kg H_2O osmolality of the tubular fluid is due to NaCl, and 50–100 mOsm/kg H_2O is due to urea. In the short-looped nephrons, about 25–30% of the filtered fluid volume remains, and the osmolality has increased to a lesser extent, depending on the level within the medulla at which the tip of the loop resides.

TRANSPORT IN THE THIN AND THICK ASCENDING LIMB

When the tubular fluid turns the bend at the tip of the loop of Henle, it enters the thin ascending limb of the loop of Henle. At this point, the epithelial transport properties change dramatically. Both the thin and the thick regions of the ascending limb of the loop of Henle are water impermeable because of the absence of aquaporin water channels, and there is no significant transepithelial water flow. On the other hand, these regions avidly reabsorb NaCl and can remove 20–25% of the filtered load of Na^+, which is more than two-thirds of the amount delivered to the loop of Henle from the proximal tubule.

Although differences exist in the transport characteristics of the thin and thick ascending limbs of the loop of Henle, and between the medullary and cortical regions of the thick ascending limb, this discussion concentrates on their common ability to reabsorb salt while leaving water behind in the tubular fluid, resulting in dilution of the urine. This is the first region of the nephron in which water and solute transport are dissociated so that the tubular fluid can have an osmolality that is significantly different from the surrounding interstitial fluid.

Because the urine becomes concentrated in the thin descending limb of the loop of Henle, the NaCl concentration of the tubular fluid is higher than that of the medullary interstitium, about 575 mmol/L compared with 300 mmol/L in the interstitium, as shown in Fig. 2. (Chapter 28 examines how the medullary interstitium is made hyperosmotic and how the osmolality gradient is maintained.) Thus, a large concentration gradient favors passive diffusion of NaCl from the lumen into the interstitial fluid. Because the thin ascending limb is permeable to Na^+ and Cl^-, this diffusion occurs rapidly, allowing an equilibration of the NaCl concentration between the tubular fluid and the adjacent medullary interstitium. There is also a gradient for urea that favors its entry into the ascending limb of the loop of Henle because of its higher concentration in the medullary interstitium; however, because this segment is relatively impermeable to urea, there is little entry.

As NaCl diffuses from the lumen of the thin ascending limb to the medullary interstitium, the tubular fluid becomes dilute compared with the interstitium. This osmolality difference would favor water movement out of the tubule, but this does not occur because of the low water permeability of this segment. The low water permeability, which is different from what is normally observed for most cell membranes and epithelia, results in a tubular fluid that is dilute with respect to the surrounding interstitium and becomes progressively more dilute as the urine flows into the less hyperosmotic regions of the medullary interstitium.

At the junction between the inner and outer medulla, the characteristics of the tubular epithelium change again. The thinner cells of the thin ascending limb (which contain relatively few mitochondria) disappear, the cells contain numerous mitochondria, and the epithelial cell layer thickens, giving the name to the

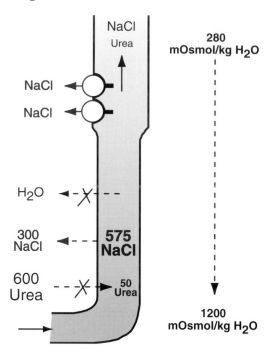

FIGURE 2   Reabsorption of NaCl in the thin and thick ascending limb of the loop of Henle. At the papillary tip, approximately one-half of the interstitial osmolality of 1200 mOsm/kg H_2O is due to 600 mmol/L urea, and half is due to 300 mmol/L NaCl (= 600 mOsm/kg H_2O). In the thin ascending limb of the loop of Henle, NaCl is reabsorbed passively by diffusion down its concentration gradient from the lumen, where its concentration is ~575 mmol/L to the inner medullary interstitium. In the thick ascending limb, NaCl is actively reabsorbed, which further dilutes the tubular fluid. Because both the thin and thick segments of the ascending limb are virtually water impermeable, both mechanisms of NaCl reabsorption cause the osmolality of the luminal fluid to become less than that of the adjacent interstitium (shown by the lighter shading in the lumen).

thick ascending limb. This segment also serves to transport NaCl from the tubular lumen into the interstitial fluid without accompanying water movement. However, in this region, the transport of NaCl out of the lumen is an active, energy-requiring process, as shown in Fig. 3. Furthermore, the transporters that carry out this active NaCl reabsorption are unique. When isolated perfused segments of thick ascending limb were first examined in the early 1970s, it was recognized that, in contrast with most epithelia, these segments developed a lumen-positive voltage proportional to the rate of transport. They also actively reabsorbed Na^+ and Cl^- and reduced the NaCl concentration in the tubular fluid. This meant that Cl^- was being transported against both a concentration gradient and an electrical potential difference (the positive luminal voltage). Initially, the process was referred to as active Cl^- transport. Later investigation showed, however, that this transport process ultimately depended on energy from ATP hydrolysis by the Na^+,K^+-ATPase. The mechanisms involved are shown in Fig. 3.

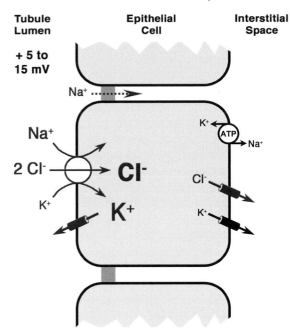

Tubule Lumen   **Epithelial Cell**   **Interstitial Space**

+ 5 to 15 mV

FIGURE 3   Active NaCl reabsorption in the thick ascending limb of the loop of Henle. The unique transporters in this nephron segment are highlighted in color. The movements of Na$^+$, K$^+$, and two Cl$^-$ ions across the luminal membrane are coupled by a furosemide-sensitive NaK2Cl cotransporter. Active Cl$^-$ accumulation in the cell is driven by the movement of Na$^+$ down its concentration gradient, which is maintained by the basolateral Na$^+$,K$^+$-ATPase. K$^+$ channels in the luminal membrane allow K$^+$ accumulated in the cell by the cotransporter to recirculate, and they give rise to the lumen-positive voltage in this region of the nephron, which serves as a driving force for passive Na$^+$ diffusion through the junctional complexes. Cl$^-$ leaves the cell primarily by Cl$^-$ channels in the basolateral membrane.

The unique characteristics of NaCl reabsorption by the thick ascending limb are imparted by the Na$^+$/K$^+$/2 Cl$^-$ cotransporter, which cotransports one Na$^+$, one K$^+$, and two Cl$^-$ ions. Thus, an electrically neutral combination of ions is transported into the cell from the tubular fluid. This is a "downhill" or passive transport process because of the favorable chemical potential gradients of Na$^+$ and Cl$^-$. (The electrical gradient is not of any consequence when one considers the movement of a neutral combination of ions coupled by a single transporter, as in this case.) Of course, energy is available for this process only as long as the cell maintains a low intracellular Na$^+$ concentration. This ultimately depends on extrusion of Na$^+$ from the cell across the basolateral membrane via the Na$^+$,K$^+$-ATPase. Because of the operation of the cotransporter, the Cl$^-$ concentration in the cell rises above its equilibrium value. Consequently, there is a favorable electrochemical potential difference across the basolateral membrane for Cl$^-$ diffusion out of the cell via Cl$^-$-selective channels, as shown in Fig. 3. K$^+$ transported into the cell by the luminal cotransporter and by the

Na$^+$,K$^+$-ATPase leaves the cell by K$^+$ channels in the luminal and basolateral membranes.

Another significant characteristic of the Na$^+$/K$^+$/2Cl$^-$ cotransporter in the thick ascending limb of the loop of Henle is its sensitivity to the most powerful diuretics. The class of diuretics that includes *furosemide* and *bumetanide* have a high affinity for the Cl$^-$ site on the cotransporter and block its operation. Because these diuretics are actively secreted into the proximal straight tubule, they have a much higher concentration in the lumen of the thick ascending limb than in the plasma. When the cotransporter is blocked, NaCl reabsorption by the thick ascending limb is greatly diminished, and increased loads of NaCl and isotonic fluid are delivered to the distal regions of the nephron. Because furosemide, bumetanide, and related drugs act on the medullary and cortical regions of the thick ascending limb, they are referred to as *loop diuretics,* and they interfere with the ability to concentrate or dilute the urine as will be discussed in Chapter 28.

The active NaCl reabsorptive mechanism described in Fig. 3 is present in both the medullary and cortical regions of the thick ascending limb and operates to dilute the tubular fluid in comparison with the medullary and cortical interstitium. However, the transporter is limited in its capacity to form a transepithelial NaCl gradient—as the NaCl concentration in the lumen falls, there is a gradient for diffusion of NaCl from the interstitium back into the tubular fluid. Thus, the medullary and cortical regions of the thick ascending limb can develop an NaCl gradient of approximately 75–100 mmol/L, resulting in a tubular fluid that is 150–200 mOsm/kg H$_2$O dilute in comparison with the adjacent medullary interstitium, as shown in Fig. 4.

The rate of NaCl reabsorption by the thick ascending limb increases when the rate of NaCl delivery to the loop of Henle is increased by an increased GFR or by a decrease in proximal tubule reabsorption; however, this increase in reabsorption does not fully compensate for the increased delivery. Consequently, when more NaCl is delivered to the loop of Henle, although the rate of NaCl reabsorption by the thick ascending limb increases, less dilution of the tubular fluid osmolality occurs.

The rate of NaCl reabsorption in the thick ascending limb is also hormonally regulated. *Vasopressin* (also known as *antidiuretic hormone* or *ADH*) stimulates NaCl reabsorption in the medullary regions of the thick ascending limb and thus increases the transport of NaCl into the medullary interstitium and increases the interstitial fluid osmolality. (The regulation of vasopressin release and its role in the concentration and dilution of the urine are considered in Chapter 28.) In addition, increased osmolality of the medullary interstitium decreases NaCl reabsorption, as do locally produced

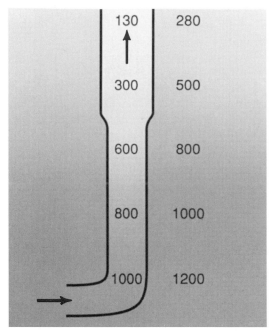

FIGURE 4 A reduction in tubular fluid osmolality is a consequence of NaCl absorption along the ascending limb of the loop of Henle. Because of passive NaCl reabsorption in the *thin segment* and active NaCl absorption in the *thick segment* (shown in Fig. 2), the osmolality of the tubular fluid falls to a steady-state value that is ~200 mOsm/kg H_2O less than the adjacent interstitial fluid.

prostaglandins. This constitutes a negative feedback arc such that rising medullary osmolality counteracts excessive NaCl reabsorption via the thick ascending limb both by a direct effect on the thick ascending limb cells and by the production of prostaglandins.

As fluid leaves the medulla in the thick ascending limb, it is always hypotonic to plasma and to the adjacent interstitium, which is approximately isotonic to plasma. Thus, the tubular fluid that flows from the thick ascending limb to the distal convoluted tubule has an osmolality of 100–150 mOsm/kg H_2O, with a urea concentration of 50 mmol/L or more. The collective rate of fluid delivery from the thick ascending limb to the distal convoluted tubule is ~15 mL/min, which is slightly more than 10% of the water volume filtered at the glomerulus (the GFR). The cortical region of the thick ascending limb is sometimes referred to as the *diluting segment* because it produces a tubular fluid that is dilute with respect to plasma, and this occurs regardless of the osmolality of the final urine.

TRANSPORT IN THE DISTAL CONVOLUTION AND COLLECTING DUCT

As discussed in Chapter 23 (see Fig. 5 in that chapter), the distal convolution, i.e., the portion of the nephron between the macula densa and the beginning of the cortical collecting duct, consists of the *distal convoluted tubule* and the *connecting tubule*. The initial portion of the distal convolution is a very short continuation of the thick ascending limb, which is followed by the abrupt appearance of distal tubule cells *(DCT cells)*, which have highly amplified basolateral membranes, as well as the highest density of mitochondria and Na^+,K^+-ATPase activity of any nephron segment. Beyond the middle of the distal convolution, connecting tubule cells *(CNT cells)* appear with increasing frequency as the nephron transitions to become the connecting tubule. Intercalated cells also appear as a minority cell type scattered among the CNT cells, and they appear in increasing frequency along the length of the connecting tubule. The density of intercalated cells rises to ~30% in the cortical collecting duct, but it then diminishes along the medullary collecting duct and falls to zero in the inner medullary collecting duct. The functional properties of the intercalated cells will be discussed in subsequent chapters concerning renal potassium and acid–base balance, whereas this section focuses on the functional properties of the primary cell types in these nephron segments.

Along a short segment at the end of the connecting tubule, *principal cells,* which are the major cell type in the collecting duct, replace the CNT cells and form the initial collecting tubule. These initial collecting tubules converge, with the confluence of 10–12 forming a *cortical collecting duct.* Cortical collecting ducts run through the cortex and enter the medulla, where they are referred to as *medullary collecting ducts.* Within the inner region of the medulla, these collecting ducts come together to eventually form the larger ducts of Bellini that empty into the renal calyces.

It is more convenient to discuss the functions of the distal convolution and collecting duct according to the cell types that populate it rather than by the traditional segmental divisions. In the early to mid-distal convoluted tubule, DCT cells are the majority cell type and they have distinct functional properties. Although the majority cell types in the connecting tubules (CNT cells) and collecting ducts (principal cells) have somewhat different morphology, they appear to be functionally quite similar. As discussed later, CNT and principal cells reabsorb Na^+ by an electrogenic channel mechanism in contrast to an electroneutral NaCl cotransporter found in DCT cells. The CNT and principal cells are also unique as the primary sites of action of the hormones aldosterone and vasopressin in the nephron. In this and subsequent chapters of this section, the connecting tubule and the collecting duct, which are populated primarily by CNT and principal cells, will be referred to collectively as the *aldosterone-responsive distal nephron* (ARDN).

Approximately 10% of the filtered load of water collectively enters the distal convolution and it contains less than 10% of the filtered load of NaCl and KCl and more than 50% of the filtered urea. Along the distal convolution, Na^+ is actively reabsorbed and K^+ is secreted. Because the net reabsorption of Na^+ is normally greater than the secretion of K^+, there is a net loss of solute from the lumen. Thus, the tubular fluid in the early distal convolution would continue to be diluted if the cells in all segments were water impermeable. This is true in the distal convoluted tubule because, just as with the thin and thick ascending limb, the epithelium formed by the DCT and intercalated cells is impermeable to water. However, the water permeability of the luminal membrane of CNT and principal cells in the ARDN is regulated by vasopressin, and it can range from negligible to very high. As will be discussed in detail in Chapter 28, when there is an excess of body water, the water permeability of the luminal membranes of CNT and principal cells is very low. In this condition, the final urine osmolality can be as low as 50 mOsm/kg H_2O or less with a urine flow rate as much as 10% of the GFR. In dehydration, the water permeability of the luminal membranes of the CNT and principal cells is high, allowing osmotic equilibration between tubular fluid and the adjacent interstitium. In the cortex, this results in a tubular fluid that becomes isosmotic to plasma. In the medullary regions of collecting duct, where the interstitium is hypertonic, the tubular fluid also becomes hypertonic as water is osmotically reabsorbed. In a maximum antidiuretic state, the urine osmolality can rise as high as 1200 mOsm/kg H_2O with a urine flow rate down to 0.5 mL/min.

NaCl Reabsorption by DCT Cells in the Distal Convoluted Tubule

An important function of the distal convoluted tubule is the reabsorption of NaCl. Active transport of Na^+ occurs by the same fundamental mechanism in other nephron segments. Na^+ enters the cell across the luminal membrane down an electrochemical potential gradient that is generated by the Na^+,K^+-ATPase in the basolateral membrane. This pump uses energy from the hydrolysis of ATP to drive Na^+ from the cell in a ratio of three Na^+ exiting for two K^+ entering the cells.

As shown in Fig. 5, Na^+ enters DCT cells across the luminal membrane by a mechanism that cotransports one Na^+ and one Cl^-, in contrast to the Na^+/K^+2Cl^- cotransporter in the thick ascending limb. The Cl^- that

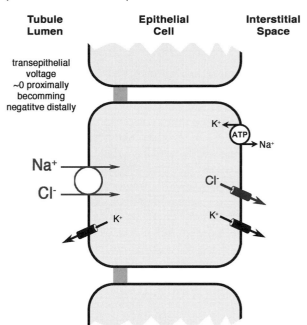

FIGURE 5 Active NaCl reabsorption by DCT cells in the distal convoluted tubule. The unique transporters in the DCT cells are highlighted in color. NaCl entry across the luminal membrane is mediated by a thiazide-sensitive NaCl cotransporter. As in the thick ascending limb, Cl^- leaves the cell primarily by selective channels in the basolateral membrane. There are K^+ conductive channels in the luminal and basolateral membrane.

enters the cell by this route exits the cell primarily by Cl^--selective channels in the basolateral membrane. The NaCl cotransporter has been cloned and sequenced, and it is different from the Na^+/K^+/2Cl^- cotransporter in the thick ascending limb. This difference is also reflected by the fact that different diuretics selectively inhibit the two transporters. Whereas the Na^+/K^+/2Cl^- cotransporter is inhibited by furosemide and other loop diuretics (see earlier discussion), the NaCl cotransporter is inhibited by thiazide diuretics such as hydrochlorothiazide. Because the thick ascending limb reabsorbs much more Na^+ than the early distal convoluted tubule, the loop diuretics are more effective than the thiazide diuretics.

Because the luminal entry of Na^+ in DCT cells is electroneutral, it does not depolarize the luminal membrane and, thus, the transepithelial voltage in the distal convoluted tubule is nearly zero. However, the voltage becomes progressively more lumen-negative as the transition to the connecting tubule is made. The luminal membrane of DCT cells also contains K^+ channels of the type found in the thick ascending limb and in CNT and principal cells. These channels can mediate K^+ secretion but probably only in the very late distal convoluted tubule where a rising lumen-negative transepithelial voltage provides the necessary driving force.

Ion Transport in the
Aldosterone-Responsive Distal Nephron

The mechanisms involved in NaCl reabsorption and K^+ secretion in CNT cells and principal cells, in the connecting tubule and collecting duct, are shown in Fig. 6. Na^+ enters the cells via a Na^+-selective channel (ENaC) in the luminal membrane and is pumped out of the cell by the Na^+,K^+-ATPase. The ENaC-mediated Na^+ conductance of the luminal membrane causes it to be depolarized relative to the basolateral membrane and, thus, it gives rise to a lumen-negative transepithelial voltage. As discussed in the previous section, this lumen-negative transepithelial voltage is small at the end of the distal convoluted tubule but it can be as high as –50 mV in the connecting tubule and collecting duct.

The epithelial Na^+ channel is highly selective to Na^+ compared with K^+, although it can also transport Li^+. These channels can also be selectively blocked by the diuretics amiloride and triamterene. Because the ARDN reabsorbs less than 2% of the filtered load of Na^+,

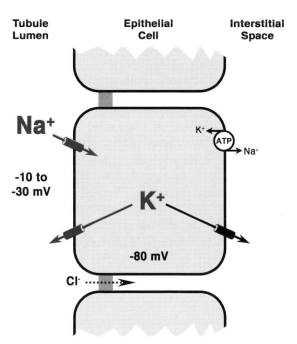

Tubule Lumen	Epithelial Cell	Interstitial Space

FIGURE 6 Na^+ reabsorption and K^+ secretion by CNT cells and principal cells. These cells are the majority cell type in, respectively, the connecting tubule and the collecting duct. Na^+ enters both types of cells by ENaC. This channel can be selectively blocked by the diuretics amiloride and triamterene. The luminal membrane also contains K^+ channels, but the predominance of Na^+ channels depolarizes the luminal membrane and gives rise to the lumen-negative transepithelial voltage. The depolarization of the luminal membrane is a significant driving force for the preferential efflux of intracellular K^+ across this membrane into the lumen. Thus K^+ secretion results from K^+ accumulation in the cell produced by the Na^+,K^+-ATPase and the favorable electrochemical potential gradient for movement from the cell to the lumen.

amiloride and triamterene are not nearly as potent in producing natriuresis (increased Na^+ excretion) as the loop or thiazide diuretics. However, they are often used in combination with the latter diuretics because, as discussed later, they counteract the tendency of those diuretics to produce hypokalemia.

The high K^+ concentration in CNT and principal cells favors the movement of K^+ from the cells into the lumen. This movement occurs through another type of channel that is specific for K^+ ions. The membrane potential opposes the movement of K^+ out of the cell, because the cell is negative with respect to the lumen. However, because of the depolarization produced by Na^+ channels in the luminal membrane, this voltage is lower than across most cell membranes. Thus, there is a net electrochemical driving force for the movement of K^+ into the lumen. This results in secretion of K^+ so that the tubular fluid K^+ concentration can become much higher than in the plasma.

Two important factors favor increased secretion of K^+. First, whenever fluid flow to the connecting tubule and collecting duct is increased, the secretion of K^+ causes less of a rise in the luminal concentration because that secretion occurs into a larger volume flow. Thus, the amount of K^+ secreted and finally excreted can be markedly increased by an increased flow out of the loop of Henle. Again, this is an illustration of the principle of mass flow: although the luminal K^+ concentration is lower, its mass flow (the product of the tubular fluid K^+ concentration and the volume flow rate) is greater. Second, the secretion of K^+ is also increased when the delivery of Na^+ to the connecting tubule and collecting duct is increased, because Na^+ entry into the cell depolarizes the luminal membrane and makes the lumen more negative. Thus, when Na^+ reabsorption is augmented, the more lumen-negative transepithelial voltage favors increased potassium secretion and excretion. These are important principles in considering the effects of diuretics on K^+ balance (see Clinical Note).

ALDOSTERONE REGULATES ION TRANSPORT IN CNT AND PRINCIPAL CELLS

The processes of Na^+ reabsorption and K^+ secretion in the connecting tubule and collecting duct are regulated by the mineralocorticoid hormone *aldosterone*. Aldosterone is produced by the adrenal cortex in the zona glomerulosa (see Chapter 40, the section on "Physiology of the Mineralocorticoids"). It increases Na^+ reabsorption and K^+ secretion by the CNT and principal cells, and H^+ secretion by the intercalated cells in the distal nephron. Aldosterone, like most steroids, is lipophilic and easily permeates cell membranes.

Clinical Note

Effect of Diuretics on Potassium Balance

Loop diuretics, that is, diuretics that act primarily on the $Na^+/K^+/2Cl^-$ cotransporter in the thick ascending limb of the loop of Henle, increase the delivery of salt and water to the more distal portions of the nephron because they interfere with Na^+ and water reabsorption in the loop of Henle (see also Chapter 28). For the reasons given in the text, the increase in Na^+ delivery and volume flow to the connecting tubule and collecting duct favor K^+ secretion and excretion, and can lead to a low plasma K^+ concentration (hypokalemia) and depletion of total body K^+. The thiazide diuretics have a similar effect because they act on the electroneutral NaCl cotransporter in DCT cells, but not on the Na^+ channel in the CNT and principal cells. Thus, hypokalemia is a frequent side effect of diuretic therapy. Mild to moderate hypokalemia is associated with muscle weakness, reduced reflexes, and fatigue. When severe, it can lead to coma and fatal cardiac arrhythmias. Patients who are given loop diuretics should be encouraged to eat foods rich in potassium such as fruit juices and bananas, but may still need to take potassium supplements.

In contrast, those diuretics that act on the Na^+ channel such as amiloride and triamterene, and spironolactone, which blocks the action of aldosterone (see next section), are referred to as "potassium-sparing" diuretics because they do not result in increased K^+ secretion. Amiloride and triamterene block the Na^+ channel in the luminal membrane of principal cells; therefore, they decrease the depolarization of the luminal membrane, resulting in decreased K^+ secretion.

Intracellularly, it binds to cytoplasmic and nuclear receptors that are present exclusively in the CNT and principal cells and produces the changes shown schematically in Fig. 7. First, the activity of Na^+ channels in the luminal membrane is increased. More Na^+ enters the cells, and the transepithelial transport of Na^+ rises. The increased luminal membrane Na^+ conductance also further depolarizes the luminal membrane and enhances K^+ secretion. Second, aldosterone increases the activity and synthesis of Na^+,K^+-ATPase and it can expand the basolateral membrane surface area by more than twofold. Increases in the activity and number of pumps and the basolateral membrane surface area enhance the ability of the CNT and principal cells to pump Na^+ out of the cell and K^+ in, again resulting in enhanced Na^+ reabsorption and K^+ secretion. Third, the hormone increases the synthesis of some of the Krebs cycle enzymes, which increases the ability of the cells to produce ATP and supply energy to the Na^+,K^+-ATPase. Finally, aldosterone may also increase the activity of K^+ channels in the luminal membrane, thus directly stimulating K^+ secretion.

The reabsorption of Na^+ and secretion of K^+ by the distal convolution and collecting duct are not entirely dependent on aldosterone. Even in the complete absence of aldosterone, as in *Addison's disease,* the kidney excretes a maximum of about 500 mmol of Na^+ per day, or about 2% of the filtered load, and 6–8% of the filtered load is still reabsorbed by the distal nephron. At the other extreme, in the presence of maximal aldosterone levels, the kidney excretes less than 0.1% of the filtered load of Na^+. Thus, the normal range of regulation of Na^+ excretion by aldosterone is 0.1–2% of the filtered Na^+ load.

Aldosterone secretion by the adrenal cortex is regulated in the manner of a negative feedback mechanism as described in more detail in Chapter 40. Table 1 lists the normal stimuli for increases and decreases in aldosterone secretion. As might be expected, extracellular fluid (ECF) volume contraction, which is usually the result of a reduction in total body Na^+ (see Chapter 29), is a stimulus to increased aldosterone secretion. However, the signal to the adrenal zona glomerulosa cells for the decreased ECF volume is an increase in plasma angiotensin II, which is the most potent stimulus of aldosterone production and secretion (see Chapter 24, the section on the "Renin-Angiotensin System"). A fall in extracellular fluid volume, a fall in renal arterial pressure, or increased sympathetic outflow to the kidneys increases the release of renin from juxtaglomerular cells in the afferent arteriole. Renin leads to an increase in plasma angiotensin II. On the other hand, increases in ECF volume inhibit the release of aldosterone. This effect appears to be mediated by the release of atrial natriuretic peptide (ANP) from distended atrial muscle in the heart, which directly inhibits zona glomerulosa cells (see also Chapters 29 and 40).

Small changes in the plasma K^+ concentration affect aldosterone secretion. Hyperkalemia stimulates aldosterone secretion, while hypokalemia inhibits it.

Tubule Lumen   **Epithelial Cell**   **Interstitial Space**

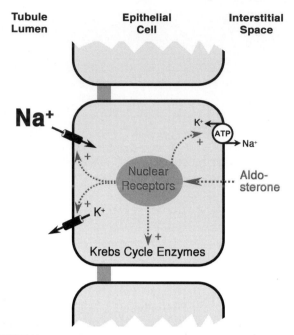

FIGURE 7   Actions of aldosterone on CNT and principal cells in the ARDN. Aldosterone, acting through nuclear receptors, increases the activity of Na^+ and K^+ channels in the luminal membrane, increases the activity of Na^+,K^+-ATPase, and increases the synthesis of Krebs cycle enzymes. Spironolactone is a competitive inhibitor of aldosterone binding to its receptor and thus acts as a diuretic by antagonizing the action of this hormone.

Again, this is a logical negative feedback mechanism because aldosterone release counteracts hyperkalemia by increasing K^+ secretion by the CNT and principal cells. Aldosterone release from the adrenal cortex is also increased by some less specific stimuli such as trauma, stress, fright, and general sympathetic discharge. In contrast with its role in regulating the glucocorticoids, adrenocorticotropic hormone (ACTH) has little effect on aldosterone secretion. Nevertheless, ACTH weakly

TABLE 1   Factors That Alter Aldosterone Secretion by the Zona Glomerulosa Cells of the Adrenal Cortex

Increased aldosterone secretion	Decreased aldosterone secretion
↑ Plasma angiotensin II (via angiotensin II receptors on zona glomerulosa cells)	↓ Plasma angiotensin II
↓ ECF volume, ↓ cardiac output, or *functional hypovolemia*; see Clinical Note (primarily via the renin-angiotensin system)	↑ ECF volume (via suppression of the renin-angiotensin system, and the release of ANP from the atria)
↑ Plasma K^+ (direct effect on zona glomerulosa cells)	↓ Plasma K^+
Trauma, stress (via generalized sympathetic discharge?)	↓ Plasma ACTH (loss of a small trophic effect on zona glomerulosa cells?)

stimulates aldosterone release, and aldosterone secretion is diminished when plasma ACTH levels are decreased. Interestingly, glucocorticoids also stimulate K^+ secretion but, in contrast to the mineralocorticoid aldosterone, this effect is indirect. Glucocorticoids enhance glomerular filtration and thus augment the delivery of Na^+ and fluid to principal cells in the distal nephron, which, as discussed earlier, increases K^+ secretion.

As discussed in Chapter 40, the aldosterone receptors in CNT and principal cells can be activated by the glucocorticoid cortisol almost as well as by aldosterone. Because the plasma levels of cortisol are usually several-fold higher than those of aldosterone and exhibit large diurnal variations, the aldosterone signal would be drowned out if it were not for the presence of the enzyme 11β-hydroxysteroid dehydrogenase, which converts cortisol to cortisone, a steroid that does not bind to the mineralocorticoid receptor. Only the CNT and principal cells in the nephron express this enzyme, but it is found in other epithelia that are regulated by aldosterone (e.g., the colonic epithelium and sweat ducts), and it is responsible for the selective response of these cells to aldosterone and not to cortisol.

PROTON SECRETION BY INTERCALATED CELLS IN THE CONNECTING TUBULE AND COLLECTING DUCT

Acidification of the urine also occurs in the connecting tubule and collecting duct. The residual amounts of bicarbonate are reabsorbed, and the pH of the final urine is lowered so that substantial amounts of titrated buffers can be excreted. Protons (H^+) are actively secreted by intercalated cells that appear among the principal cells from the midregion of the distal convoluted tubule into the medullary collecting duct, although they are absent in the inner medullary collecting duct. The regulation of proton secretion in the distal nephron is an important determinant of acid–base balance, as is discussed in Chapter 31. Figure 8 shows the luminal surfaces of intercalated cells surrounded by principal cells in the cortical collecting duct.

Intercalated cells acidify the urine by secreting protons in much the same way as the proximal tubule cell. As in most cells, the usual intracellular reaction between H^+, HCO_3^-, carbonic acid, CO_2, and water is catalyzed by carbonic anhydrase (see Chapter 26, Fig. 9). Thus, CO_2 from metabolism within the intercalated cell and diffusing from the plasma provides a ready supply of H^+ and HCO_3^- ions within the cell. As in the proximal tubule, protons are then secreted into the lumen by an active transport process, and HCO_3^- anions selectively diffuse across the basolateral membrane back

Clinical Note

Aldosteronism and Liddle's Syndrome

Bilateral hyperplasia or a discrete tumor of the adrenal cortex (adrenal adenoma) can result in the secretion of aldosterone at high and unregulated rates. In such patients, an elevated stimulus is observed in Na^+ reabsorption and K^+ secretion by CNT and principal cells, and in H^+ secretion by the intercalated cells (see later discussion). The resulting increases in K^+ and H^+ secretion lead to hypokalemia and alkalosis, whereas the decreased excretion of Na^+ leads to extracellular volume expansion. The resulting syndrome, referred to as *primary aldosteronism,* is characterized by mild to moderate hypertension from the expanded vascular volume and by muscle weakness, cramping, and cardiac arrhythmias from the hypokalemia and alkalosis. In diagnosing this condition, it is important to rule out diuretic use as a cause (see the preceding Clinical Note). Then, typically, the serum renin level is measured.

If the renin level is low, it indicates that the plasma angiotensin II level is also low, and the physician can rule out excessive aldosterone production being driven by excessive renin production, as might occur with atherosclerotic blockage of blood flow to one or both renal arteries. The physician can then order a measurement of the aldosterone level in a plasma sample taken after the patient is given intravenous saline to be sure that extracellular fluid volume contraction is not driving aldosterone production. If plasma aldosterone is high in this setting, it confirms a diagnosis of aldosteronism because all of the stimuli for aldosterone release are either absent or working in an opposing direction: Extracellular fluid volume is normal or expanded, the plasma Na^+ concentration is normal or slightly elevated, the plasma K^+ concentration is low, and angiotensin II levels are low. At this point, a computed tomography scan or adrenal venous catheterization can demonstrate the presence of a renal adenoma or may indicate bilateral adrenal hyperplasia. The syndrome can then be treated surgically or, if less severe, by

administering the aldosterone renal receptor blocker spironolactone.

In the early 1960s, the endocrinologist Grant Liddle described a patient with all of the symptoms of aldosteronism but with a *low* plasma aldosterone. Treatment of this patient with spironolactone, a competitive inhibitor for aldosterone receptors, was ineffective, but the hypertension and hypokalemia were diminished by the Na^+ channel blocker triamterene, leading Liddle to suggest that the cause was abnormally stimulated Na^+ reabsorption in the distal tubule and collecting duct. This turned out to be a prescient hypothesis.

This condition, which came to be known as *Liddle's syndrome* or *pseudoaldosteronism,* was found to be due to an autosomal dominant genetic disorder. More than 30 years after the first patient was described, the genetic messages coding the proteins of the Na^+ channel were cloned and sequenced. Three similar proteins called *subunits* (for epithelial Na^+ channel) *ENaC* were found to make up the channel and all affected members of the kindred had a mutation in one of these three proteins. This mutation prevents normal regulation of the channel so it remains in a permanently active state in the luminal membranes of CNT and principal cells, regardless of the aldosterone level. It is interesting to note that the resulting syndrome is an example of hypertension caused by a renal defect. In fact, the original patient later in her life underwent a kidney transplant because of end-stage renal disease and experienced a subsequent abatement of her hypertension.

Apparent mineralocorticoid excess is a syndrome similar to Liddle's syndrome and includes hypertension, hyperkalemia, and alkalosis in the presence of low plasma aldosterone. In this case, the defect is caused by a variety of mutations in the gene for 11β-hydroxysteroid dehydrogenase, which lead to a loss in its activity. The enzyme is also inhibited by some exogenous agents including licorice!

to the plasma. The primary difference between the proximal and the distal tubule is the large gradient of protons that can be developed across the epithelium by the intercalated cells in the distal nephron, which can

produce urine with a pH as low as 4.5. This is equal to a hydrogen ion concentration of 31.6 μM, compared with a concentration of 40 nM in the plasma—almost an 800-fold increase! Obviously, a high-energy active transport

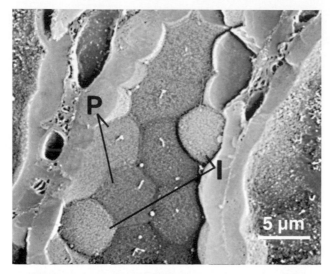

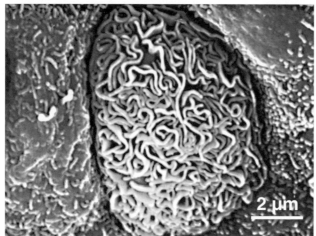

FIGURE 8   Scanning electron micrographs of epithelial cells in the cortical portion of the collecting duct. (Top) A collecting duct has been fractured along its axis to reveal the luminal surface of several cells. This segment of the nephron possesses two distinct cell types, principal (P) and intercalated (I) cells. The principal cell has only a few short microvilli on its luminal surface and a single cilium; the apical membrane of the intercalated cell is folded in numerous ridge-like microplicae. (Bottom) A higher magnification scanning electron micrograph showing the microplicae of an intercalated cell. This intercalated cell is surrounded by several principal cells. (Micrographs courtesy of Dr. Andrew P. Evan, Indiana University Medical Center, Indianapolis, IN.)

process is required to transport protons up such a steep electrochemical potential gradient. Furthermore, the epithelium of the distal nephron segments must be relatively impermeant to the back-leak of protons from the lumen to the plasma, which would dissipate the luminal acidity.

The mechanism used for active proton secretion across the luminal membrane of the proximal tubule is Na^+-H^+ exchange. However, the amount of energy available in the Na^+ electrochemical potential gradient is not sufficient to generate the steep proton

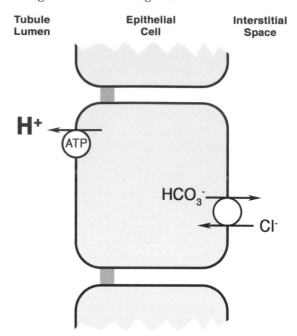

FIGURE 9 H^+ secretion by intercalated cells. H^+ is actively secreted from the cell into the lumen by an H^+-ATPase located in the luminal membrane. HCO_3^- exits across the basolateral membrane down its electrochemical potential gradient into the interstitium via an antiport mechanism in exchange for Cl^-.

concentration gradient observed in the distal nephron. As shown in Fig. 9, proton secretion by intercalated cells is directly coupled to ATP hydrolysis by the proton-activated H^+-ATPase in the luminal membrane.

As protons are secreted into the lumen, the proton concentration within the cells falls and drives the production of additional HCO_3^-. Therefore, just as in the proximal tubule cell, the secretion of H^+ results in a higher HCO_3^- concentration in the cells that favors the diffusion of HCO_3^- out of the cell down its electrochemical potential gradient. As in the proximal tubule, the efflux of HCO_3^- must be mediated across the cell membrane, and the only transporter available is located in the basolateral membrane. However, in contrast with the proximal tubule, this transporter is an anion exchanger. HCO_3^- movement out of the cell occurs in exchange for Cl^- movement into the cell, as shown in Fig. 9. The net result is that H^+ is secreted actively into the lumen and HCO_3^- diffuses passively into the plasma.

H^+ secretion by the intercalated cell is also sensitive to aldosterone. Aldosterone increases H^+ secretion, particularly in the outer medullary regions of the collecting duct, but the mechanisms involved in this stimulation are not yet fully understood.

The H^+-ATPase in the luminal membrane of the intercalated cell is of a type referred to as a vacuolar ATP that is also found in some intracellular organelles.

Tubule Lumen   **Epithelial Cell**   **Interstitial Space**

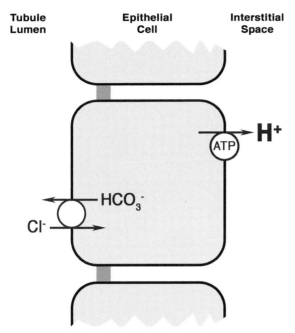

FIGURE 10   In severe chronic alkalosis, H$^+$ is actively transported across the basolateral membrane of the intercalated cell toward the interstitium, resulting in HCO$_3^-$ secretion across the luminal membrane.

In severe acidosis or hypokalemia, the intercalated cell also expresses another proton pump that is identical to that responsible for acid secretion in the stomach (see Chapter 34). This pump is the electroneutral H$^+$,K$^+$-ATPase that secretes H$^+$ into the lumen while actively transporting K$^+$ into the cell. Thus, this ATPase not only acidifies the lumen but also reduces K$^+$ secretion. In fact, during chronic potassium depletion the expression of the H$^+$,K$^+$-ATPase is enhanced sufficiently to produce net K$^+$ reabsorption in the distal nephron, thus conserving body K$^+$.

Up until this point, we have spoken primarily of H$^+$ secretion in the distal nephron, and this process

normally occurs in omnivores including the human. However, it is also possible for the distal tubule and collecting duct to reverse the net direction of H$^+$ transport. In severe chronic alkalosis, the urine is not acidified in the distal nephron; instead, HCO$_3^-$ is secreted. This conversion occurs primarily in the intercalated cells of the cortical collecting duct, in which the normal arrangement of the H$^+$-ATPase and the HCO$_3^-$/Cl$^-$ exchanger is reversed as shown in Fig. 10. This reversal of the transporters causes H$^+$ to be actively extruded toward the plasma while HCO$_3^-$ is secreted into the lumen. Both the H$^+$ and the HCO$_3^-$ secreting types of intercalated cells may differentiate from a common precursor cell, or individual intercalated cells may reorganize the location of their transporters.

Suggested Reading

Edwards A, Delong MJ, Pallone TL. Interstitial water and solute recovery by inner medullary vasa recta. *Am J Physiol* 2000;278:F257–F269.

Hamm LL, Alpern RJ. Cellular mechanisms of renal tubular acidification. In Seldin DW, Giebisch G, eds. *The kidney: Physiology and pathophysiology,* Vol 2, 3rd ed. Philadelphia: Lippincott Williams & Wilkins, 2000, pp 1935–1980.

Malnic G, Muto S, Giebisch G. Regulation of potassium secretion. In Seldin DW, Giebisch G, eds. *The kidney: Physiology and pathophysiology,* Vol 2, 3rd ed. Philadelphia: Lippincott Williams & Wilkins, 2000, pp 1575–1614.

Muto S. Potassium transport in the mammalian collecting duct. *Physiol Rev* 2001;81:85–116.

Reeves WB, Winters CJ, Andreoli TE. Chloride channels in the loop of Henle. *Annu Rev Physiol* 2001;63:631–641.

Reilly RF, Ellison DH. Mammalian distal tubule: Physiology, pathophysiology, and molecular anatomy. *Physiol Rev* 2000;80:277–313.

Schafer JA. Abnormal regulation of ENaC: Syndromes of salt retention and salt wasting by the collecting duct. *Am J Physiol* 2002;283:F221–F235.

C H A P T E R

28

Regulation of Body Fluid Osmolality

JAMES A. SCHAFER

KEY POINTS

- The kidneys maintain the normal constant osmolality of body fluids and normal constant body water content.
- Because the hydraulic conductivity of most cell membranes is high, all body fluid compartments have approximately the same osmolality.
- The nonapeptide hormone *vasopressin* (or *antidiuretic hormone,* ADH) regulates the water permeability of the collecting duct and thus controls the osmolality of the urine.
- *Osmoreceptors* in the supraoptic and para-ventricular nuclei of the hypothalamus respond rapidly to an increase in plasma osmolality (P_{osm}) and increase the secretion of vasopressin from their nerve endings in the posterior hypothalamus. *Thirst receptors* in the hypothalamus also respond to increased P_{osm} and lead to the sensation of thirst.
- In the absence of vasopressin, the luminal membrane of the collecting duct is water impermeable. Continued solute reabsorption without water can reduce the urine osmolality to 50 mOsm/kg H_2O. In the presence of high vasopressin, the collecting duct is very water permeable and the urine osmolality approaches 1200 mOsm/kg H_2O.
- *Osmolal clearance* (C_{osm}) is the clearance of all solutes calculated as $(UF \cdot U_{osm})/P_{osm}$. Typically, it is in the range of 1.5–2.5 mL/min. *Free water clearance* (C_{H_2O}) is a measure of the

Essential Medical Physiology, Third Edition

KEY POINTS (*continued*)

kidneys' ability to excrete water in excess of solute. C_{H_2O} is the difference between the urine flow (UF) and C_{osm}.

- The loop of Henle and the early distal convoluted tubule dilute urine to about one-third the osmolality of the plasma. Because the fluid entering the loop from the proximal tubule is nearly isosmotic, the loop reabsorbs a hyperosmotic fluid into the medulla, thus raising its interstitial osmolality.
- The slow medullary blood flow and the arrangement of the capillaries in the long hairpin loops (vasa recta) allow for *countercurrent exchange,* which minimizes dilution of the medullary interstitium by equilibration with the blood plasma.
- Although the ascending limb of the loop of Henle can generate at most a 200 mOsm/kg

H_2O transepithelial osmolality difference, *countercurrent multiplication* establishes the 900 mOsm/kg H_2O gradient of interstitial osmolality from the corticomedullary junction to the tip of the papilla.

- Urea provides half the osmolality of the medullary interstitium. The nephron is impermeable to urea from the tip of the loop of Henle to the inner medullary collecting duct. Thus, during antidiuresis the urea concentration in the urine in the inner medullary collecting duct exceeds 600 mmol/L, producing a favorable gradient for passive urea diffusion from the urine into the medulla. The urea permeability of the inner medullary collecting duct is increased by vasopressin, thus allowing urea reabsorption to occur.

One of the most important functions of the kidney is the regulation of the total body water content (volume) and the osmolality of the body fluid compartments. Because most cell membranes are highly water permeable, any water is rapidly distributed among the various body fluid compartments, and a water excess dilutes them equally. Similarly, in dehydration, the osmolality of all body fluid compartments rises equally. Changes in the total body water content are reflected by changes in the osmolality of the plasma. The osmolality of the body fluid compartments is determined by the total body solute content divided by the total body water volume. In response to changes in the plasma osmolality, the kidney regulates the osmolality of the urine to restore the normal plasma osmolality.

The volume of water excreted by the kidneys depends on the amount that is reabsorbed in the segments of the nephron from connecting tubule to the ducts of Bellini, which are the same nephron segments that constitute the aldosterone-responsive distal nephron (ARDN). Water reabsorption in these segments is controlled by *vasopressin* (more precisely, arginine vasopressin in man and most mammals; also called antidiuretic hormone or ADH), which regulates the *water permeability* of the luminal membranes of the CNT and principal cells. In addition, *thirst* regulates the intake of water. The same stimuli that produce changes in plasma vasopressin levels also regulate the thirst drive that increases water intake when the plasma

osmolality rises. This chapter considers those mechanisms involved in regulating thirst and vasopressin secretion, the action of vasopressin on the kidney, and the mechanism by which the kidney concentrates and dilutes the final urine.

BODY WATER BALANCE

As discussed in Chapter 25, the homeostatic regulation of any substance in the body requires a careful balance of input to the body, production by the body, and output from the body. The same is true with water balance. To maintain a constant amount of water, the body must eliminate as much water as is taken in and produced by metabolism. The intake of water can vary markedly from day to day, depending on our dietary habits and whether we are relatively sedentary or exercising.

The commonly accepted "classical" values for the daily intake and output of water by healthy adults under sedentary conditions in a temperate climate are given in Table 1. Data from studies before 1980 indicate that the average intake of water in this setting (as fluids that are drunk as well as the water in solid foods) ranges from 2200–2700 mL of water per day; however, as discussed in the following Clinical Note, the quantity of water drunk has increased in the last two decades. Metabolism of foodstuffs ultimately leads to the production of another ~300 mL, as well

TABLE 1 Daily Balance of Water Intake and Production in Sedentary Healthy Adults in a Temperate Climate

Intake and production	(mL/day)	Output	(mL/day)
Water intake:		Urine	1500–2000
in fluids	1200–1700		
drunk in solid foods	1000		
Metabolic production	300	Feces	100
		Insensible losses from the skin and lungs	900
TOTAL	2500–3000		2500–3000

Note: The values presented here are typical of those presented in other textbooks of physiology and medicine. The upper end of the range of fluids drunk, 1700 mL/day, is based on a survey conducted by the U.S. Department of Agriculture in 1977–1978. As discussed in the Clinical Note about how much water we should drink and as reported in a 1994–1996 survey, the average water consumption has increased to ~2200 mL/day, which may reflect a public perception that an increased water intake is desirable.

as CO_2 and other metabolic by-products such as urea and creatinine. The metabolic production of water varies with the diet, but it is generally a small fraction of the total intake. On the other hand, the intake of water in the diet can be changed considerably by our thirst drive. When an uncontrolled loss of water occurs, for example, when we are sweating due to exercise in a hot environment, or in diarrhea, the thirst mechanism comes into operation to increase our day's intake of water.

The primary route of output under sedentary conditions is via the urine. About 1.5–2 L of urine are produced per day, but this can also vary considerably depending on the dietary intake. At least 500 mL of urine per day is required just to excrete urea, creatinine, and other solutes. Thus, unless glomerular filtration is greatly diminished or ceases with volume contraction, the urine output does not go below 500 mL/day even in dehydration. On the other hand, with excessive water intake or in the absence of vasopressin, urine output can be as high as 10% of the glomerular filtration rate or approximately 18 L per day.

Considerable "insensible" losses of water from the skin and from the lungs also occur, even when resting at normal room temperature. Losses from the skin can increase up to as much as 3.5 L/hr during exercise in a hot environment. Just being exposed to a hot environment without exercise can result in losses of 1–2 L per day of water in sweat. Under these conditions, because urine volume can be decreased only to 500 mL/day, the only way body water balance can be achieved is if drinking increases intake.

The output of water from the lungs is also variable. Because inspired air has a lower water content, dependent on the relative humidity, and expired air is saturated with water at body temperature, a considerable amount of water is lost by respiration. This loss is increased markedly during exercise, with its accompanying hyperventilation, especially in hot, dry climates when the inspired air has a low water content.

The loss of about 100 mL/day of water in the feces is typical for a normal diet; however, gastrointestinal losses of water can be increased dramatically as a consequence of diarrhea or vomiting. Again, if these losses exceed the ability of the kidney to conserve water by concentrating the urine and reducing its volume, they must be replaced by increasing water intake.

Estimation of Plasma Osmolality

The regulation of total body water content is based primarily on a sensing system designed to respond to *plasma osmolality*. Plasma osmolality is determined by the total solute content in the plasma and the total plasma water volume. Because Na^+, Cl^-, and HCO_3^- are the major solutes in plasma, they are the primary determinants of plasma osmolality. However, in pathologic conditions, such as renal failure in which the blood urea nitrogen concentration (BUN) increases, or in diabetes with an elevated plasma glucose concentration (P_{glu}), urea and glucose may contribute significantly to the total plasma osmolality. Plasma osmolality is not a routine measurement by the clinical laboratory, but it is conveniently estimated from the plasma Na^+ concentration (P_{Na}), P_{glu}, and BUN, which can be provided by most clinical laboratories, by adding the osmotic contribution of each solute according to the following equation:

$$\text{Plasma osmolality} \cong 2 \cdot P_{Na} + \frac{P_{glu}}{18} + \frac{BUN}{2.8} \quad (1)$$

Equation [1] is a surprisingly accurate approximation of the plasma osmolality. The first term gives the contribution to the osmolality from Na^+ and accompanying anions such as Cl^- and HCO_3^-. Because most Na^+ salts in the plasma are monovalent, their contribution to the total osmolality is approximately equal to two times the plasma Na^+ concentration. The osmolality in mOsm/kg H_2O produced by the urea and glucose is approximately equal to their mmol/L concentrations because they are small nonelectrolytes; however, in contrast to Na^+, the plasma glucose concentration and the BUN are traditionally given in clinical units of mg/dL. In the case of glucose, P_{glu} in

Clinical Note

How Much Water "Should" We Drink?

In recent years, the popular media have promoted what appears to have become an urban myth: We hear repeatedly that one should drink at least eight 8-oz. glasses (i.e., one-half gallon or 1.9 L) of water every day for good health. Furthermore, we are told that caffeinated and alcoholic beverages should not be included in this total because they act as diuretics and cause water loss. The origins of this advice and its validity have been examined recently by Dr. H. Valtin, an authority on the physiology of body water balance (see reference in the "Suggested Readings" section), who found that the source is obscure and not supported by either physiologic or nutritional studies.

From Table 1 it appears that before 1980 the water intake of healthy adults in the United States was considerably below the minimum 8 × 8 oz. per day, especially if one takes into account that about half of the 1200–1700 mL of fluids drunk was in the form of caffeinated and alcoholic beverages. Apparently, however, the message has been heeded by the public. An analysis of daily water intake in the United States by the Department of Agriculture in 1998 indicates that the average daily water consumption (as fluids drunk) by adults in 1994–1996 was ~2200 mL, or the equivalent of more than two additional 8-oz. glasses in comparison with the "classical" data in Table 1. It is interesting to note that more than one-half of the increased intake was in the

form of caffeinated and alcoholic beverages. When one subtracts the latter, the average healthy adult is still lagging behind the eight glasses of water per day that is alleged to be needed. More recent studies, however, have shown that the diuretic effects of caffeinated beverages have been greatly overemphasized, and that these amounts, which constitute more than 50% of the water intake in a large proportion of the population, should be included as part of the daily water intake.

Certainly, some medical conditions may indicate a need for increased water consumption, and it is always important to increase water intake appropriately when exercising or when exposed to environmental conditions such as high heat and low humidity, high altitude (due to an increase in minute ventilation of dry air), or long airplane flights. On the other hand, in some medical conditions and with the use of some "recreational" drugs, excessive water intake can produce fatal hyponatremia. So, how much water should a normal healthy adult drink if he or she is just leading a relatively sedentary life in a temperate climate? Perhaps the best answer is just to let thirst be our guide. Some proponents of increased water consumption argue that when we become thirsty we are already significantly dehydrated. As will be discussed later in this chapter, this argument is false. The thirst drive is very sensitive to changes in hydration far smaller than those regarded to be of clinical significance.

---

units of mg/dL can be converted to mOsm/kg H_2O by dividing by the molecular weight of glucose (180 mg/mmol) times the unit conversion of 0.1 L/dL. Similarly, the BUN in mg/dL is divided by 2.8 in order to obtain the concentration of urea in mmol/L. The reason for the divisor of 2.8 may be difficult to understand at first. This factor takes into account that there are 28 g of *nitrogen* per mole of urea, or 28 mg/mmol.

As an example, let's consider the plasma osmolality that would be predicted by Eq. [1] for typical values of plasma solutes: a Na^+ concentration of 137 mmol/L, a glucose concentration of 100 mg/dL, and a BUN of 15 mg/dL. Using Eq. [1], the calculated plasma osmolality is approximately 285 mOsm/kg H_2O. Plasma osmolality is usually maintained in the range of 280–290 mOsm/kg H_2O by the regulation of thirst and the urine osmolality.

Osmoreceptors and the Regulation of Vasopressin Secretion

The system for the maintenance of a constant plasma osmolality is typical of most negative feedback systems. First, there are receptors that sense plasma osmolality, referred to as *osmoreceptors,* located in the supraoptic and paraventricular nuclei of the hypothalamus. Second, the plasma vasopressin concentration serves as a signaling mechanism that is regulated in response to changes in receptor activity. Finally, the water permeability of the region of the distal nephron that contains CNT and principal cells constitutes an effector mechanism, which is regulated by vasopressin and counteracts the stimulus to the osmoreceptor by changing the urine and thus the plasma osmolality.

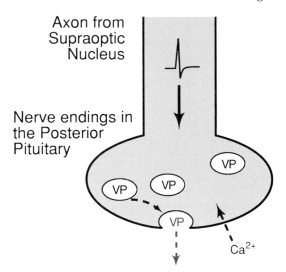

Axon from
Supraoptic
Nucleus

Nerve endings in
the Posterior
Pituitary

FIGURE 1   Release of vasopressin (VP) from nerve endings in the posterior pituitary. Action potentials conducted along axons from nerve bodies in the supraoptic nucleus trigger an influx in Ca^{2+}. Ca^{2+} entry is associated with increased fusion of secretory granules containing vasopressin with the plasma membrane, leading to the release of vasopressin into the circulation.

When the plasma osmolality rises, the osmoreceptor cells shrink. This serves as a signal for increased firing of nerve fibers running primarily from the supraoptic nucleus, with a limited contribution from the paraventricular nucleus, through the stalk of the pituitary gland to the posterior pituitary. Action potentials conducted along these axons change the permeability of the nerve endings in the posterior pituitary to allow Ca^{2+} entry. Vasopressin is stored in secretory vesicles that are located in the nerve endings. In response to Ca^{2+} entry, these vesicles fuse with the membrane and release vasopressin into the circulation, as shown schematically in Fig. 1. When the osmoreceptors sense a decreased plasma osmolality, firing of these nerve fibers is suppressed, and vasopressin release ceases.

A similar receptor mechanism serves to regulate thirst. Osmoreceptors are also located in the lateral preoptic hypothalamus near the supraoptic nucleus. This area is referred to as the *hypothalamic thirst center*. The sensation of thirst and the associated drinking behavior come about when these osmoreceptors are stimulated by a high plasma osmolality. The osmoreceptors that regulate both vasopressin release and thirst are stimulated most effectively by increases in the Na^+ concentration of the plasma. Small nonelectrolytes such as urea have a lesser effect on stimulating either thirst or vasopressin release because they can readily penetrate cell membranes, whereas Na^+ is confined to the extracellular space and has a greater osmotic effect on the receptor cells. Thus, NaCl or larger solutes such as glucose produce more shrinkage of the osmoreceptors.

Vasopressin (ADH)

The hormone released from the posterior pituitary by osmoreceptor firing is now most commonly referred to simply as *vasopressin*. This name reflects the history of our knowledge of the hormone. It was originally observed that pituitary extracts caused an increase in blood pressure because of an increase in total peripheral resistance. However, at normal physiologic levels, vasopressin is not a vasoconstrictor, and its action is confined to augment the water permeability of the connecting tubule and collecting duct. Thus, vasopressin is also referred to as *antidiuretic hormone* (ADH).

Vasopressin is a nonapeptide with a molecular weight of 1100 Da. Because of its small size, vasopressin is readily filtered at the glomerulus, but it is taken up by proximal tubule cells that degrade it to its constituent amino acids. The liver also metabolizes vasopressin, so its plasma concentration falls rapidly when its pituitary secretion ceases. Secretion of vasopressin in response to hyperosmotic plasma is also rapid, and plasma vasopressin levels rise accordingly. Because of this rapid secretion and clearance, plasma levels of vasopressin respond within a period of a few minutes to changes in plasma osmolality.

Cellular Action of Vasopressin

Vasopressin acts on CNT and principal cells located, respectively, in the connecting tubule and the collecting duct to increase the water permeability of their luminal membranes. Thus vasopressin operates on the same region of the nephron as does aldosterone, i.e., the ARDN. The basolateral membranes of the CNT and principal cells are always quite water permeable due to the presence of water channels of the aquaporin family (aquaporin-3 and -4), which are similar to aquaporin-1 in the proximal tubule and thin descending limb. However, the transepithelial movement of water is limited by the water permeability of the luminal membrane. In contrast to most cell membranes, the luminal membranes of CNT and principal cells are quite water impermeable in diuresis. However, the water permeability of these membranes is high in antidiuresis due to the action of vasopressin, which causes the insertion of yet another type of water channel, aquaporin-2, into the luminal membrane and thereby increases the transepithelial water permeability.

Aquaporin-2 is structurally quite similar to the other aquaporins already mentioned, but it is unique because it can be moved into and out of the luminal membrane of CNT and principal cells by the cytoskeleton. In the absence of vasopressin, aquaporin-2 channels are stored in vesicles that lie just under the luminal membrane. As shown in Fig. 2, when vasopressin is present in the

Tubule   Epithelial   Interstitial
Lumen   Cell   Space

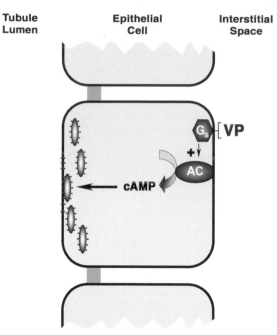

FIGURE 2   Vasopressin (VP) causes the insertion of aquaporin-2 water channels into the luminal membrane of CNT and principal cells. Vasopressin from the plasma binds to V_2-type receptors on the basolateral membrane. This receptor is coupled to a GTP-binding protein (G_s), which stimulates adenylate cyclase (AC), resulting in increased synthesis of cAMP from ATP. A cAMP-dependent kinase phosphorylates proteins, which are as yet unidentified, and causes the fusion of intracellular vesicles containing aquaporin-2 water channels with the luminal membrane thereby increasing the water permeability of the luminal membrane.

plasma, it binds to V_2 receptors located on the basolateral membranes of CNT and principal cells. (These receptors are different from the V_1 receptors found on vascular smooth muscle that mediate the pressor effect of vasopressin.) Upon binding of vasopressin to the V_2 receptor, the following sequence of events occurs. The receptors act through a GTP-binding protein (G_s) to stimulate adenylyl cyclase. The resulting increase in intracellular cAMP activates a protein kinase that phosphorylates yet unknown proteins. This phosphorylation causes the aquaporin-2-containing vesicles to fuse with the luminal membrane, thereby inserting the water channels into the membrane and raising its water permeability.

Although this mechanism is complicated, it is also rapid. When the vasopressin concentration of the plasma rises, vasopressin is released by the posterior pituitary and the water permeability of the luminal membrane of the CNT and principal cells increases within 10 min. On the other hand, when vasopressin levels in the plasma decrease, the aquaporin channels are rapidly retrieved from the luminal membrane by endocytosis. Therefore, changes in plasma osmolality result in rapid changes in urine osmolality.

Plasma Osmolality Normally Regulates Vasopressin Secretion

With a high plasma osmolality and increased vasopressin release, the water permeability of the connecting tubule and collecting duct is high, and urine osmolality rises as water is reabsorbed (Fig. 3). In maximal *antidiuresis,* the urine osmolality can be as high as 1200 mOsm/kg H_2O, and the urine volume flow can decrease to about 0.5 mL/min. Alternatively, under conditions of decreased plasma osmolality when vasopressin secretion is suppressed, the water permeability of the connecting tubule and collecting duct is low and little water is reabsorbed (Fig. 4). Because of continued NaCl reabsorption along the connecting tubule and collecting duct, the urine becomes even more dilute, resulting in a final urine osmolality that can be as low as

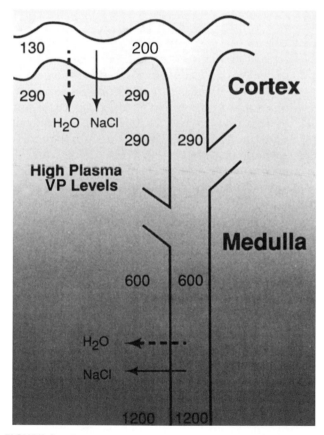

FIGURE 3   Concentration of the urine along the connecting tubule and collecting duct in the presence of high vasopressin levels. The urine enters the connecting tubule from the distal convoluted tubule dilute (upper left: shown as 130 mOsm/kg H_2O). In the presence of vasopressin (VP), the water permeability of the luminal membranes of CNT and principal cells is high, allowing water to be reabsorbed and the development of osmotic equilibrium between the tubular fluid and the interstitium. With maximal vasopressin levels, the osmolality of the tubular fluid rises to become isosmotic in the cortex, and the final urine is as concentrated as the interstitium at the tip of the papillae, 1200 mOsm/kg H_2O.

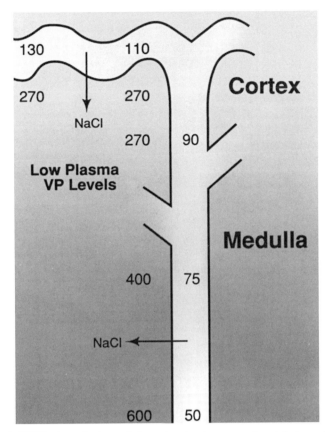

FIGURE 4 Dilution of the urine in the connecting tubule and collecting duct in the absence of vasopressin (VP). With low plasma vasopressin levels, the water permeability of the luminal membranes of the CNT and principal cells is low, allowing little water reabsorption. The continued reabsorption of salt further dilutes the urine to as low as 50 mOsm/kg H₂O. In this state of water diuresis, increased water delivery to the medulla dilutes the interstitium, as discussed in the text.

50 mOsm/kg H_2O but is usually in the range of 50–100 mOsm/kg H_2O. This is referred to as a *water diuresis*. During water diuresis, the medullary hypertonicity is reduced, as shown in Fig. 4, for reasons that are discussed later.

Plasma vasopressin levels are measured by radioimmunoassay. The normal range of plasma concentration is 0.5–20 pg/mL (0.5–18 pmol/L). The lowest plasma vasopressin level detectable by the assay is about 0.5 pg/mL. As expected, the plasma vasopressin concentration varies in proportion to plasma osmolality, as shown in Fig. 5. At a plasma osmolality of 270 mOsm/kg H_2O or less, plasma vasopressin levels are below detectable limits. When plasma osmolality rises above some threshold in the range of 270–280 mOsm/kg H_2O, detectable levels of vasopressin appear in the plasma and increase progressively with plasma osmolality, reaching the maximum of about 20 pg/mL when the plasma osmolality rises above 290–295 mOsm/kg H_2O. As plasma osmolality rises, thirst is

also experienced subjectively at a plasma osmolality in the range of 280–285 mOsm/kg H_2O and becomes maximal at about 290–295 mOsm/kg H_2O. The integration of these two effects of rising plasma osmolality into the negative feedback regulation of plasma osmolality is illustrated schematically in Fig. 6. Both the increased thirst and increased release of vasopressin act to restore plasma osmolality to normal. Similarly, as shown schematically in Fig. 7, a hypo-osmotic plasma suppresses thirst and vasopressin release and returns plasma osmolality toward normal levels.

Effect of ECF Volume Depletion on Vasopressin Secretion

Although plasma osmolality is the normal stimulus for vasopressin release, extracellular fluid volume is also an important factor. An increase in extracellular fluid volume, as evidenced by an increase in venous filling in the thorax, may produce a water diuresis. This can be seen, for example, with water immersion up to the neck, which results in compression of the veins in the extremities and increases central venous pressure. Alternatively, decreased vascular filling, as occurs during forceful exhalation and increased intrathoracic pressure, decreases the volume of the central veins and stimulates increased vasopressin release. Normally, these effects are relatively minor, and serve to enhance or decrease the effects of the vasopressin normally released in response to changes in plasma osmolality. However, under conditions of severe volume contraction, the volume stimulus may override the normal regulation of vasopressin release by plasma osmolality.

As shown in Fig. 8, plasma vasopressin levels may rise to much higher levels with severe decreases in extracellular fluid volume. When extracellular fluid volume falls by 5–10%, there is a marked stimulation

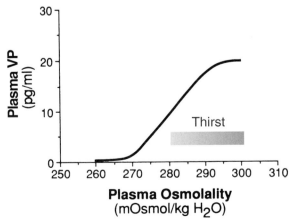

FIGURE 5 Effect of plasma osmolality on the plasma vasopressin (VP) concentration and thirst.

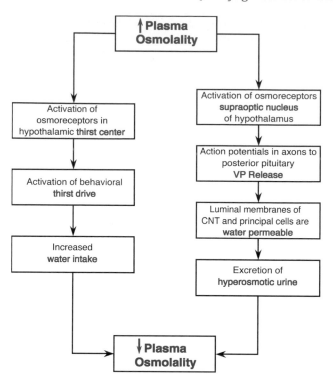

FIGURE 6   Normal physiologic response to a hyperosmotic plasma. Increased water ingestion and the excretion of a hyperosmotic urine cause a decrease in the plasma osmolality toward normal.

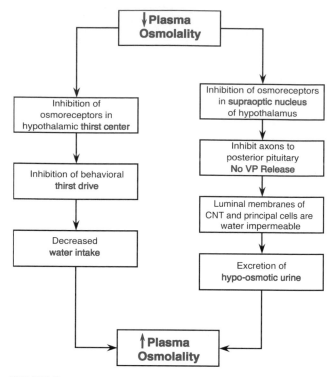

FIGURE 7   Normal physiologic response to a hypo-osmotic plasma. Decreased water ingestion and excretion of a hypo-osmotic urine cause an increase in the plasma osmolality toward normal.

of vasopressin release; larger fluid volume decreases can lead to extremely high plasma vasopressin levels, approaching 50 pg/mL. In clinical situations, the decrease in extracellular fluid volume can override the normal response to plasma osmolality and cause the urine to be concentrated even when the plasma osmolality becomes dangerously low. Consider the following Clinical Note.

QUANTIFYING RENAL CONCENTRATING ABILITY: FREE WATER CLEARANCE

The ability of the kidney to concentrate or dilute the urine decreases with age and may be considerably compromised in the elderly. Urinary concentrating and diluting ability is also disturbed in renal failure and, consequently, it can be a useful index of renal function.

Clinical Note

Hyponatremia Due to Gastrointestinal Fluid Losses
A patient has had a severe gastrointestinal disturbance during the past week or more, and has lost large volumes of salt and water through diarrhea and vomiting. Because of this illness, the patient has been unable to eat, but has been able to drink clear fluids. The large losses of salt and water deplete the patient's extracellular fluid volume. Under these circumstances, the dehydration is a strong volume signal to increase vasopressin release. The release of vasopressin causes urinary concentration with a resulting

dilution of the plasma. Thus, the patient conserves most of the water she or he ingests, which expands the extracellular fluid compartment and all other body fluid compartments. However, because the patient cannot eat solid food, plasma osmolality will decrease as water is retained because it is not matched by an equivalent salt intake. Normally, this decrease in plasma osmolality, which is reflected by a decrease in the plasma Na^+ concentration (called *hyponatremia* when the plasma Na^+ falls below 135 mmol/L), would decrease the

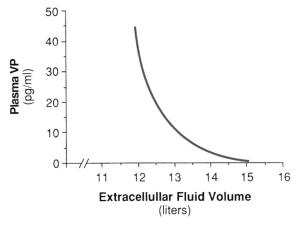

FIGURE 8   Increase in plasma vasopressin (VP) in response to reduction in extracellular fluid volume. As the extracellular fluid volume decreases below the normal level of 15 L (for a typical 70-kg person), there is a progressive increase in plasma vasopressin, *even if plasma osmolality is less than 280 mOsm/kg H_2O*.

Often, the presenting symptom of renal disease is a patient's complaint that he or she has to get up frequently during the night to urinate, a symptom referred to as *nocturia*. Nocturia often results when the kidneys are unable to concentrate the urine adequately and, thus, reduce its volume. However, nocturia and frequent voiding are symptoms of more common disorders such as urinary tract infection or inflammation, or prostate enlargement in the aging male.

The ability of the kidneys to dilute or concentrate the urine depends on the rate at which they excrete water in relation to solute. Therefore, to assess the efficacy of the kidney in producing dilute or concentrated urine, it is first necessary to quantify the rate of excretion of total solute. The parameter used for this purpose is referred to as the *osmolal clearance* (C_{osm}). The osmolal clearance is calculated in the same manner as the clearance of any substance (see Chapter 25), except that the concentrations of all solutes (which are reflected by the osmolality) are considered in the calculation. As shown in Eq. [2], the osmolal clearance is simply the product of the urine volume flow rate and the ratio of the urine osmolality to the plasma osmolality:

$$\text{Osmolal clearance} = C_{osm} = \frac{UF \cdot U_{osm}}{P_{osm}} \qquad (2)$$

The symbols in Eq. [2] have the same meaning as in Chapter 25. UF is the urine volume flow rate in mL/min; P_{osm} is the plasma osmolality and U_{osm} is the urine osmolality, both in mOsm/kg H_2O. The osmolal clearance is normally fairly constant in the range of 2 ± 0.5 mL/min because it represents the rate at which solutes are being removed from the plasma and excreted in the urine, and this depends in turn on the daily dietary intake, which is typically fairly constant. In effect, the osmolal clearance represents the rate of plasma flow from which solute is totally removed.

Example

If a person is excreting urine with an osmolality of 50 mOsm/kg H_2O at a rate of 10 mL/min, it is obvious that there is a greater than normal proportion of water to solute in the urine, and therefore this production of dilute urine is acting to concentrate the plasma. If the plasma osmolality is 300 mOsm/kg H_2O, the osmolal clearance can be calculated to be 1.67 mL/min. In other words, a urine flow of 1.67 mL/min is what would be required to excrete this quantity of solutes with a urine osmolality of 300 mOsm/kg H_2O. The actual urine flow of 10 mL/min is mathematically equivalent to this flow of 1.67 mL/min plus a flow of 8.33 mL/min of pure water. In this case, we consider the 8.33 mL/min to be the free water clearance, i.e., the volume of urine flow in excess of the amount required to excrete the solutes in the urine with the same osmolality as the plasma.

To compare the rate of solute excretion with the rate of water excretion, the concept of *free water clearance* has been developed. To understand the meaning of free water, consider the following example.

The free water clearance (C_{H_2O}) is calculated as a difference between the rate of urine volume flow and the osmolal clearance:

$$\text{Free water clearance} = C_{H_2O} = UF - C_{osm} \qquad (3)$$

Using Eq. [2], Eq. [3] can be rearranged to a form that is sometimes more convenient:

$$C_{H_2O} = UF \cdot \left[1 - \frac{U_{osm}}{P_{osm}}\right] \qquad (4)$$

Equation [4] indicates that the free water clearance will be greater than zero whenever the urine osmolality is less than the plasma osmolality. In other words, whenever the urine is dilute compared with the plasma, free water is being lost from the body and the plasma is being concentrated. On the other hand, whenever the urine osmolality is greater than the plasma osmolality, the free water clearance becomes negative. A negative free water clearance indicates that relatively more solute than water is being excreted and that water is being conserved and returned to the plasma. The negative free water clearance is sometimes given by a different symbol, T_{CH2O}; however, there is little justification for this additional term and it is simpler to refer to a negative C_{H2O} as indicating urinary concentration.

As described earlier, when the kidneys are excreting hypo-osmotic urine, the plasma is being concentrated. On the other hand, when hyperosmotic urine is being excreted, the kidneys are effectively diluting the plasma. To understand this more fully, the mass balance relations involved in urinary dilution and concentration are shown in the following examples. The few minutes spent working out and understanding the calculations in each example will be rewarding in appreciating the concentrating and diluting capacity of the kidneys.

From these two examples, it can be seen that the kidney is more effective in clearing free water from the plasma than in concentrating the urine and conserving water. The maximal rate of free water clearance can be as high as 10–12 mL/min in a maximal water diuresis. On the other hand, with maximal vasopressin levels, free water clearance can become only as negative as about 2.0 mL/min. The human kidney is less able to conserve water than are the kidneys of many other mammals. For example, desert animals can develop urine osmolalities up to 5000 mOsm/kg H_2O, and certain desert rats in Australia can develop urine osmolalities of 9000 mOsm/kg H_2O!

DEVELOPMENT AND MAINTENANCE OF MEDULLARY HYPEROSMOLALITY

Thus far, this chapter has discussed the effects of the hyperosmotic medullary interstitium on the abstraction of water from the thin descending limb and from the medullary collecting duct in the presence of vasopressin. How does this medullary hyperosmolality develop, and how is it maintained in the face of plasma flow through the medulla? In any other tissue, one would expect that the constant blood flow, as well as the constant water entry from the thin descending limb of the loop of Henle and the medullary collecting duct, would dilute the medullary hypertonicity, soon causing it to become isotonic. Water reabsorption from the thin descending limb actually does not dilute the medullary interstitium because salt is actively reabsorbed from the thick ascending limb, keeping the medullary salt concentration high. That is why the urine delivered to the distal convoluted tubule is always dilute; in essence, a hyperosmotic fluid is being reabsorbed from the loop

Numerical Examples

1. In the example shown in Fig. 9, we assume that the plasma is dilute at 270 mOsm/kg H_2O. In these circumstances, vasopressin release would be completely suppressed and plasma levels would be undetectable. If plasma flow into the kidney via the renal artery is 690 mL/min, then the rate of solute entry into the kidney via the renal artery would be (0.69 L/min · 270 mOsm/kg H_2O) = 186.3 mOsmol/min. (1 kg of water is equivalent to 1 L; however, the unit of kg H_2O has usually been used for expressing osmolalities.) For this example, assume that the resulting urine flow in the absence of vasopressin is 12 mL/min with an osmolality of 50 mOsm/kg H_2O. This would give a solute excretion rate of (0.012 L/min · 50 mOsm/kg H_2O) = 0.6 mOsmol/min. Because this volume of fluid is lost from the plasma flowing through the kidney, the plasma flow out of the renal vein would be (690 − 12) = 678 mL/min. The solute flow out of the renal vein would be the difference between the inflow via the renal artery and the loss in the urine; that is, (186.3 − 0.6) = 185.7 mOsm/min. If this solute flow rate is divided by the volume flow rate in the renal vein (185.7 mOsmol/min ÷ 0.678 L/min), the resulting osmolality of the plasma flowing out of the renal vein would be 274 mOsm/kg H_2O. In other words, by producing hypo-osmotic urine, the kidney returns a more concentrated plasma to the general circulation via the renal vein. If we calculate the free water clearance in this example, we obtain −9.78 mL/min. The plasma that flows through the kidney is being concentrated at a rate equivalent to that which would be produced if 9.78 mL/min of pure water were removed from the plasma each minute. The difference between this free water clearance and the urine volume flow rate is that fraction of the urine flow that can be regarded to be isosmotic to plasma.

2. An example for the case of urinary concentration is shown in Fig. 10. In this case, we assume that the plasma osmolality is 300 mOsm/kg H_2O, which would cause plasma vasopressin to be elevated. The rate of solute entry into the kidney via the renal artery would be (0.69 L/min · 300 mOsm/kg H_2O) = 207 mOsmol/min. Because of the high vasopressin levels, the urine flow rate is quite low at 0.5 mL/min, and the urine osmolality is maximal at 1200 mOsm/kg H_2O. Thus, the rate of solute excretion is the same as in the previous example, or 0.6 mOsmol/min. Because only 0.5 mL/min of fluid has been lost from the plasma flow, the rate of plasma flow out of the renal vein is 689.5 mL/min. The solute flow out of the renal vein would be (207 − 0.6) = 206.4 mOsmol/min. If we divide this by the renal vein plasma flow (206.4 mOsmol/min ÷ 0.6895 L/min), we obtain a renal vein plasma osmolality of 299.3 mOsm/kg H_2O. In other words, the production of hyperosmotic urine has resulted in a dilution of the plasma flowing through the kidney by 0.7 mOsm/kg H_2O. If we calculate the free water clearance from these data, we obtain a negative free water clearance of 1.5 mL/min. The kidney in this case is operating to dilute the plasma at a rate equivalent to the effect of adding 1.5 mL/min H_2O to the renal plasma flow.

of Henle and is deposited in the medullary interstitium. In other words, salt delivery to the medulla by the loop exceeds the delivery of water.

Medullary hyperosmolality is also conserved because most of the water reabsorbed by the connecting tubule and collecting duct during antidiuresis is reabsorbed in the cortex. This appears to be counterintuitive because the osmolality of the urine rises to only 300 mOsm/kg H_2O in the cortex, but to 1200 mOsm/kg H_2O in the medulla. However, consider the amount of water reabsorbed to achieve these increases in osmolality. Assuming a flow of 12 mL/min into the distal convolution, for the osmolality to rise from 100 mOsm/kg H_2O to 300 mOsm/kg H_2O, the flow must be reduced by at least two-thirds to 4 mL/min, which flows on into the medullary collecting ducts. This flow into the medulla is actually even less because NaCl is constantly reabsorbed in the cortical collecting tubule and must be accompanied by water in antidiuresis. This would reduce the flow into the medulla by at least another 1 mL/min. If 3 mL/min flows into the medulla and the osmolality rises from 300 mOsm/kg H_2O to 1200 mOsm/kg H_2O at a final urine flow rate of 0.5 mL/min, then 2.5 mL/min of water must have been reabsorbed. This is about one-third of the 8 mL/min reabsorbed in the cortex to raise the urine osmolality to 300 mOsm/kg H_2O. In other words, more water must be reabsorbed to raise the osmolality of a given volume of solution from

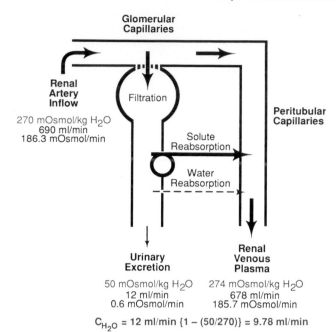

FIGURE 9　Example of positive free water clearance resulting in increased concentration of the plasma leaving the kidney. (See Numerical Example box for explanation.)

100 mOsm/kg H_2O to 300 mOsm/kg H_2O in the cortex than the additional water that must be reabsorbed in the medulla to further increase the osmolality to 1200 mOsm/kg H_2O.

It may also seem to be paradoxical that more water may be reabsorbed in the medulla in water diuresis than in antidiuresis. Although the permeability of the medullary collecting duct is very low in the absence of vasopressin, there is a large osmotic gradient between the dilute tubular fluid and the more concentrated medullary interstitium. Because of the larger osmolality difference, despite its very low water permeability more water may be reabsorbed in the medullary collecting duct in water diuresis than in antidiuresis. Furthermore, there is an increase in medullary blood flow during diuresis, and both increased blood flow and increased water entry in diuresis serve to dilute the osmolality of the medullary interstitium during diuresis, as illustrated in Fig. 4 compared to Fig. 3.

Medullary Blood Flow

Although the loop of Henle delivers more solute than water to the medullary interstitium as discussed earlier, it is still not readily apparent why the blood flow does not simply wash away solutes in the medulla and bring the osmolality back to isosmotic. One reason is that the blood flow is slow through the medulla. Only 5–10% of the renal plasma flow, or an average of about 50 mL/min, flows into the medulla.

Despite the blood flow, one would still expect that if a capillary bed in the medulla were organized as in other tissues, the interstitium would be diluted. Consider the *hypothetical* capillary arrangement shown in Fig. 11. In this example, blood is flowing into a capillary bed at a rate of 50 mL/min and has an osmolality of 300 mOsm/kg H_2O. If the capillary bed lies in a tissue region with a hyperosmotic interstitial fluid, one would expect a rapid diffusion of solutes into the isosmotic capillary from the hyperosmotic interstitial fluid, because of the high permeability of the capillary bed. (Little water flow would occur because small solutes exert no osmotic force across the capillary endothelium.) Thus, blood would flow out of the region with an osmolality that would be the same as the hyperosmotic interstitial fluid.

The reason why the renal medulla escapes this type of washout of solute is because of the organization of each capillary into a long loop, called the *vasa recta*. The blood first flows down the long descending limb of the vasa recta and then returns in a countercurrent flow up the ascending vasa recta. The result of this countercurrent flow is shown in Fig. 12. The total blood flow into the medulla is about 50 mL/min, and, for the sake of convenience, assume that the plasma osmolality is 300 mOsm/kg H_2O. As this isosmotic fluid flows down the descending limb, it is exposed to the hypertonicity of the medulla. Just as in the capillary shown in Fig. 11, the plasma in the descending vasa recta is concentrated by the entry of NaCl and urea and to a much lesser extent

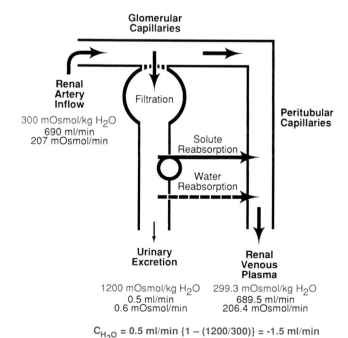

FIGURE 10　Example of negative free water clearance resulting in dilution of the plasma leaving the kidney. (See Numerical Example box for explanation.)

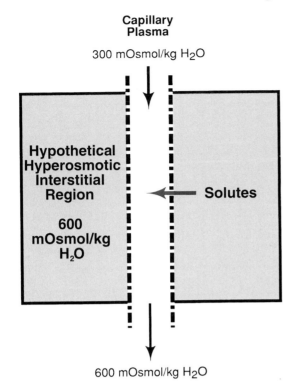

FIGURE 11   Hypothetical equilibration of isosmotic capillary plasma with a hyperosmotic interstitial region. Plasma flowing through a capillary bed in a hyperosmotic region would rapidly gain solutes by diffusion so that, on exiting, it would have the same osmolality as the hyperosmotic region.

by the loss of water to the hyperosmotic medulla. Thus, at the tip of a vasa recta, which extends to the papillary tip, the osmolality would be 1200 mOsm/kg H₂O and the plasma flow rate would be somewhat less than 50 mL/min.

The blood flow then reverses direction at the tip of the vasa recta and proceeds back up through the less concentrated regions of the medulla. As it flows upward through the medulla, NaCl and urea diffuse from the capillary into the interstitium and some water is regained from the less concentrated medullary interstitium. For this reason, solute recirculates in loops within the medullary interstitium, first entering the descending limb of the vasa recta and then exiting from the ascending limb as the blood flows back upward.

The process just described is referred to as a *countercurrent exchange*. However, it is not perfectly efficient. The osmolality of the blood flowing out of the ascending vasa recta into the cortex is still hyperosmotic to the normal plasma, but much less so than it was at the tip of the vasa recta. The rate of blood flow out of the ascending vasa recta is also more rapid than the flow in. This is a necessary consequence of mass balance. Because fluid is constantly being delivered to the medullary interstitium by water lost in

the descending limb of the loop of Henle and the medullary collecting duct, the medulla would continuously swell unless net fluid were removed by the blood flow. Consequently, the blood flow out of the medulla is 20–50% higher than the blood flow in, as shown in Fig. 12.

Countercurrent exchange explains how the medullary blood flow is organized so that it does not rapidly wash out the medullary hypertonicity. However, this still leaves the question of how this medullary hyperosmolality is generated in the first place. The high concentration of solutes in the medullary interstitium compared with the adjacent tissue of the cortex represents a nonequilibrium distribution of the solutes that requires energy to be maintained. The development and maintenance of the medullary hyperosmolality is a function of the unique permeability properties of the loop of Henle and the active NaCl reabsorptive process in the thick ascending limb of the loop of Henle.

Countercurrent Multiplication System

Clues to the mechanism involved in medullary concentration came early in this century with the observation that the ability of different mammals to concentrate the urine was correlated with the length of the papilla and, thus, of the loop of Henle. For example, the Australian desert rat mentioned previously has an

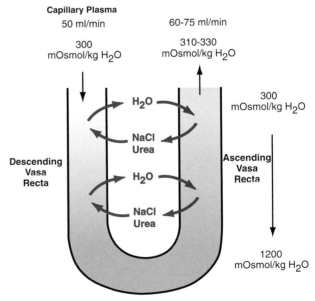

FIGURE 12   Countercurrent exchange in the vasa recta. The plasma flowing down the descending limb of the vasa recta becomes concentrated primarily because of solute entry. However, in the ascending limb of the loop of Henle, solute diffuses back outward. Note that the water movements are much less significant for osmotic equilibration than the solute movements.

extremely long loop of Henle. Early studies of the renal medulla of various mammals revealed the gradient of osmolality from most concentrated at the tip of the papilla to isosmotic at the corticomedullary junction. The final step in demonstrating where the medullary solute must come from and where the urine is concentrated occurred when micropuncture experiments showed that tubular fluid in the early distal convolution (the distal convoluted tubule) was always hypo-osmotic to plasma. Some samples were as low as 100 mOsm/kg H_2O, and most fell in the range of 100–150 mOsm/kg H_2O. This hypo-osmolality was observed regardless of whether the final urine osmolality was maximally dilute or maximally concentrated.

These micropuncture results demonstrated that in antidiuresis the final urine was concentrated by water reabsorption along the connecting tubule and the collecting duct. The results also showed that solute was added to the medullary interstitium in excess of water by the loop of Henle because the proximal tubular fluid samples were always isosmotic, whereas the samples from the early distal convolution were always hypo-osmotic. *In vitro* experiments with isolated perfused segments of the thick ascending limb also showed that active NaCl absorption could produce a tubular fluid that was dilute compared with the peritubular bathing solution. Nevertheless, even at slow flow rates, the maximal osmolality difference that could be generated was about 200 mOsm/kg H_2O. Therefore, how does the kidney manage to develop a total osmolality difference of 900 mOsm/kg H_2O from the corticomedullary junction to the tip of the papilla?

The development of this large osmolality gradient is a consequence of the hairpin arrangement of the descending and ascending limbs of the loop of Henle, much like the vasa recta. The effect of the countercurrent flow together with active NaCl absorption in the thick ascending limb of the loop of Henle can be best understood from the classic sequence of diagrams of Pitts, shown in Fig. 13. For the moment, consider the effects only in a short loop of Henle in which the hairpin turn occurs at the junction between the inner and the outer medulla. In this loop of Henle, there is no thin ascending limb, so the effects only of the actively absorbing thick ascending limb are significant. Figure 13 presents what occurs during the development of medullary hyperosmolality from a starting point with an isosmotic fluid in the loop of Henle (step 1) and in the interstitium. Due to active NaCl absorption, the osmolality of the fluid in the thick ascending limb will fall by approximately 100 mOsm/kg H_2O, and the interstitial fluid will rise by an equivalent amount to give a transepithelial osmolality gradient of 200 mOsm/kg H_2O. Because of the high water permeability of the

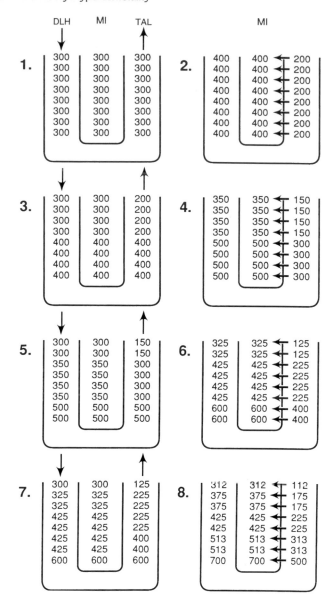

FIGURE 13 Countercurrent multiplication in the loop of Henle. Structures are the thin descending limb (DLH), the medullary interstitium (MI), and the thick ascending limb (TAL). (Based on Pitts RF. *Physiology of the kidney and body fluids,* 3rd ed. Chicago: Yearbook Medical Publications, 1974, p126.)

thin descending limb, it has the same osmolality as the medullary interstitium, as shown in step 2. In step 3, the effects of fluid flow into the thin descending limb are shown. This flow shifts new isosmotic fluid into the thin descending limb and shifts the hyperosmotic fluid around the hairpin turn. In step 4, active NaCl absorption again dilutes the osmolality of the fluid in the thick ascending limb and concentrates it in the medullary interstitium and the thin descending limb. However, in this case, there is a difference in osmolality along all of the structures due to the shift of isosmotic fluid into the descending limb.

The cycle is again completed in step 5 with an inflow of isosmotic fluid and in step 6 with the active transport process establishing a 200 mOsm/kg H_2O gradient.

In the steady state that is established by this *countercurrent multiplication,* one expects there to be a gradient of osmolality along the thin descending limb and the medullary interstitium from 300 mOsm/kg H_2O at the corticomedullary junction to 700 mOsm/kg H_2O at the tip of the loop for the example in Fig. 13. On the other hand, there is a gradient from 100–500 mOsm/kg H_2O along the thick ascending limb. This example is shown in a stepwise manner in Fig. 13, assuming separate steps of flow into the descending limb and active transport out of the thick ascending limb. Obviously, these processes normally go on concurrently and continuously in the development and normal maintenance of the osmolality gradient. It can be seen from the figure that the longer the loop of Henle, the greater will be the difference in osmolality from the corticomedullary junction to the tip of the loop.

Countercurrent multiplication also occurs in the inner medulla by NaCl absorption from the thin ascending limb. The only difference is that, in the case of the thin ascending limb, the NaCl absorption is not active but occurs down a concentration gradient, as discussed in Chapter 27. This NaCl gradient develops because water is reabsorbed from the thin descending limb of the loop of Henle and because half of the osmolality of the medullary interstitium in the inner medulla is made up of urea, whereas most of the hyperosmolality of the fluid in the thin descending and thin ascending limb is due to NaCl. This results in an effective *passive* countercurrent multiplication in the inner medulla (Fig. 14), which depends ultimately on the maintenance of a high urea concentration in this region of the kidney.

Urea Recycling in the Medulla

The final component of the medullary hyperosmolality is urea. This solute is delivered passively to the medulla and kept there in high concentration by countercurrent exchange. Figure 15 shows the primary permeability characteristics that allow urea to be concentrated in the inner medulla. As described in Chapter 26, approximately 50% of the filtered urea is reabsorbed passively in the proximal tubule. Because of the high urea concentration in the medulla, some urea diffuses passively into the thin descending limb of the loop of Henle. However, beyond the tip of the loop the thin and thick ascending limbs of the loop of Henle, the distal convoluted tubule, the connecting tubule, the cortical collecting tubule, and outer medullary collecting duct are all impermeable to urea. Consequently, as water is lost from the tubular fluid in the descending limb of the loop

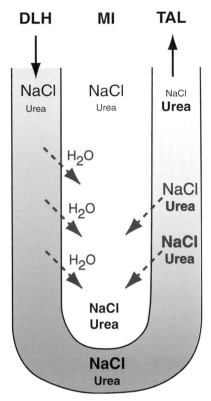

FIGURE 14 Passive countercurrent multiplication in long loops of Henle within the inner medulla. Water is abstracted from the thin descending limb of the loop of Henle (DLH) by the osmotic gradient and there is some urea diffusion from the medullary interstitium into the DLH. NaCl diffuses passively down its concentration gradient from the thin ascending limb of the loop of Henle into the medullary interstitium, leaving water and urea behind because they are impermeable. Thus, the tubular fluid flowing to the thick ascending limb (TAL) has the same urea concentration as at the tip of the medulla, but the NaCl concentration is progressively reduced.

of Henle and, in antidiuresis, in the connecting tubules and collecting ducts, urea reaches a high concentration by the time the tubular fluid reaches the inner medulla.

The collecting duct of the inner medulla is, however, permeable to urea, and this permeability is increased by vasopressin. Because of the higher urea permeability in antidiuresis, the urea that has been concentrated in the tubular fluid can diffuse from the concentrated tubular fluid into the inner medulla. In the presence of vasopressin, when there is considerable water reabsorption, the urea concentration equilibrates at about 600 mmol/L in both the urine and the medulla.

In diuresis, the tubular fluid flowing into the inner medulla has a lower urea concentration because less water reabsorption has occurred in the cortex and in the outer medulla. Thus, there is a gradient for urea to diffuse from the medullary interstitium into the urine. This is prevented to some extent by the lower urea permeability in the absence of vasopressin. But the loss

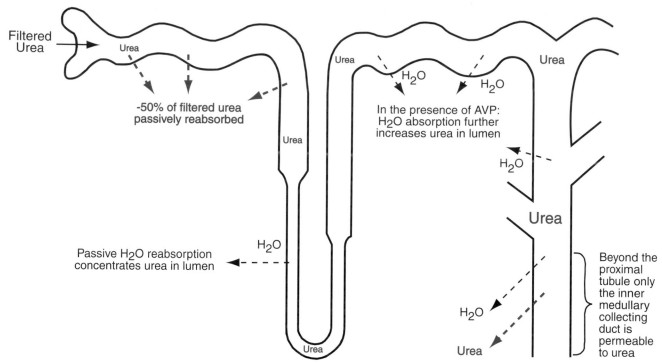

FIGURE 15   Sites of urea reabsorption along the nephron. Urea is reabsorbed passively in the proximal tubule; some urea diffuses into the thin descending limb of the loop of Henle (i.e., secreted; not shown). The nephron segments beyond the tip of the loop of Henle, with the exception of the inner medullary collecting duct, are impermeable to urea. Consequently, the urea concentration rises because of water reabsorption from the thin descending limb of the loop of Henle, and in the connecting tubule and collecting duct. This is shown by the increasing size of the letters "Urea." The inner medullary collecting duct is permeable to urea and that permeability is increased by vasopressin, therefore, in antidiuresis, urea diffuses passively into the medullary interstitium.

of urea into the tubular fluid, in combination with the increased water entry from the collecting duct and the increased medullary blood flow discussed earlier, is responsible for the fall in the inner medullary hyperosmolality during diuresis.

IMPAIRMENTS IN URINARY CONCENTRATING AND DILUTING ABILITY

Deficiencies in the ability of the kidney to regulate plasma osmolality by appropriately forming either dilute or concentrated urine may have several causes. There may be a defect in production or regulation of vasopressin release, an inability of the collecting duct to respond to vasopressin, or a failure to form a medullary osmolality gradient.

Diabetes Insipidus

Diabetes insipidus refers to high rates of production of dilute urine either because the posterior pituitary fails to release vasopressin or because the kidney does not respond to the hormone. This should be contrasted with diabetes mellitus in which increased urine flow is a consequence of the osmotic diuresis produced by the filtration and incomplete reabsorption of large amounts of glucose (see Chapter 26, the section on "Passive Reabsorptive Processes"). (The modifier *mellitus* refers to the fact that urine was sweet to the taste, whereas *insipidus* refers to a urine without taste using archaic diagnostic procedures!) The failure to produce or release vasopressin is referred to as *central diabetes insipidus*. This is a rare congenital disorder, but it may also be caused by surgical procedures (such as hypophysectomy to treat a pituitary tumor), infections, cerebrovascular accidents, or trauma. Removal of the pituitary does not always cause a total lack of vasopressin production because some of the nerves responsible for releasing the hormone end in the pituitary stalk and are not lost.

Inability to release vasopressin in response to plasma hyperosmolality leaves only the thirst mechanism to regulate plasma osmolality. Because the urine cannot be concentrated, dilute urine is continually produced, with volumes that can reach 18 L/day. To combat the

Clinical Note

Differential Diagnosis of Polyuria/Polydipsia

Often it is difficult to differentiate patients who have diabetes insipidus from patients who have a primary disorder of polydipsia, that is, who habitually ingest water in excessive quantities. Psychogenic polydipsia is a neurotic disorder in which the patient feels compelled to drink excessive amounts of water. In severe cases, the amount of water ingested (above 15 L/day!) can exceed the capacity of the kidneys to excrete it, resulting in severe and even lethal hyponatremia. Typically, patients with psychogenic polydipsia are secretive about their drinking behavior or believe they are drinking only when very thirsty.

The problem in the differential diagnosis is a classic example of which came first, the chicken or the egg: Does the patient have polydipsia because he or she has diabetes insipidus resulting in polyuria, or does the patient have polyuria because he or she has psychogenic polydipsia? In the first case, one might expect that the plasma osmolality would be elevated because of the urinary water loss and this would drive thirst. In the latter case, one might expect that the plasma osmolality would be low because of the water ingestion that is suppressing vasopressin secretion and leading to urine dilution. However, determination of plasma vasopressin levels alone usually does not provide a definitive diagnosis. Because of the large flux of water into and out of the body, the plasma vasopressin level may vary markedly depending on whether the patient has just drunk a lot of water, or whether she or he has been without water for a while.

To make a diagnosis, it is helpful to withhold water for at least a 12-hr test period. This water deprivation test requires a hospital stay with close monitoring of potential water sources (no flower vases in the room, etc.!). After water deprivation, it is found that patients with polydipsia have significant vasopressin levels in the plasma, and they excrete hyperosmotic urine. On the other hand, a patient with central diabetes insipidus still does not have detectable plasma vasopressin levels and does not concentrate urine despite the water deprivation, but will concentrate the urine if given vasopressin. The patient with nephrogenic diabetes insipidus develops high plasma vasopressin levels with water deprivation, but does not concentrate the urine even when given exogenous vasopressin. However, it should be recognized that even in the presence of vasopressin and normal kidney function, the urine osmolality might be only slightly above isosmotic in patients who have had polyuria and polydipsia before the water deprivation test. The reason is that diuresis reduces medullary hypertonicity (see next section), and it takes several hours or days for the kidneys to reestablish their medullary concentration gradient.

consequent concentration of the plasma, the thirst mechanism drives compensatory water ingestion. As long as the individual with diabetes insipidus has access to water, normal plasma osmolality can be maintained, the only inconvenience being thirst and polyuria day and night. However, individuals with diabetes insipidus can become critically dehydrated very rapidly when their water intake is impaired.

In some cases, the kidney fails to respond to vasopressin, although release of the hormone is normal. These instances are referred to as *nephrogenic diabetes insipidus*. The collecting duct may not respond appropriately to vasopressin for a variety of reasons. The disease may be congenital, or may be acquired due to the effects of toxins and some drugs such as derivatives of the antibiotic tetracycline, and lithium, which is used for treatment of bipolar disorder. Hypokalemia and hypercalcemia are also associated with a decreased response to vasopressin. Congenital nephrogenic diabetes has been shown to be due to genetic mutations of two types. In some pedigrees, the V_2 receptors have mutations that decrease vasopressin binding or prevent the receptor from increasing intracellular cAMP. In other pedigrees, mutations in the aquaporin-2 gene reduce the ability of the channel protein to transport water or interfere with its insertion into the luminal membrane of the collecting duct.

Loss of Medullary Concentrating or Diluting Ability

The concentration of the urine ultimately depends on the presence of a concentrated medullary interstitium, and dilution of the urine depends on active NaCl reabsorption without water in the thick ascending limb of the loop of Henle and in the distal convoluted tubule. These processes can be compromised in a variety of conditions. First, the loop diuretics such as furosemide directly inhibit active NaCl absorption by the medullary and cortical regions of the thick ascending limb of the loop of Henle. Consequently, they result in a reduction in medullary interstitial hypertonicity and impairment in the ability of the kidney to concentrate the urine. The thiazide diuretics inhibit NaCl cotransport in the distal convoluted tubule and result in a diminished ability to dilute the urine.

Medullary hypertonicity can also be compromised by an excessive delivery of fluid to the loop of Henle from the proximal tubule. For example, with saline diuresis (produced by excessive intake of salt and water and decreased proximal tubule reabsorption; see Chapter 29) or with osmotic diuresis, an increased amount of water is delivered to and reabsorbed by the descending limb of the loop of Henle. This dilutes the medullary hypertonicity and reduces the concentrating ability. Volume expansion and increased renal blood flow also increase medullary blood flow, which "washes out" the medullary hyperosmolality.

On the other hand, the development of a hyperosmotic medullary interstitium also depends on the delivery of adequate amounts of salt to the thick ascending limb in order to provide sufficient solute to produce the hyperosmolality. Therefore, when delivery from the proximal tubule is diminished by excessive volume reabsorption in that segment, medullary hypertonicity is compromised.

The passive countercurrent multiplication mechanism in the inner medulla also depends on urea recycling, as described in the previous section. This mechanism requires the delivery of adequate amounts of urea to the medullary collecting duct so that urea will diffuse into the medullary interstitium and contribute to the hyperosmolality that concentrates NaCl in the thin descending limb of the loop of Henle. Therefore, conditions of decreased urea production (as associated with malnutrition, especially with a deficit in dietary protein) result in decreased plasma levels of urea and a consequent decrease in the filtered load of urea. Thus, a smaller amount of urea is delivered to the medullary interstitium and the maximum urinary osmolality is reduced. It should, however, be noted that the decreased intake of food and production of urea reduces the necessity of maximal urinary concentration.

A decrease in concentrating and diluting ability is also associated with both age and renal failure. The ability to concentrate the urine maximally is reduced in infants because of their lower protein intake and relatively anabolic metabolism. Urinary concentration in the infant depends primarily on active NaCl absorption with little contribution of medullary urea. With old age and with renal failure, the number of functioning nephrons decreases and, consequently, there are fewer thick ascending limbs to concentrate NaCl in the medulla. The result is a decrease in both concentrating and diluting ability. Thus, a common presenting symptom associated with an elderly patient or a patient with compromised renal function is nocturia due to an impaired ability to concentrate the urine.

Suggested Readings

Agre P, Nielsen S, Knepper MA. Aquaporin water channels in mammalian kidney. In Seldin DW, Giebisch G, eds. *The kidney: Physiology and pathophysiology,* Vol 1, 3rd ed. Philadelphia: Lippincott Williams & Wilkins, 2000, pp 363–377.

Berliner RW. Formation of concentrated urine. In Gottschalk CW, Berliner RW, Giebisch GH, eds. *Renal physiology. People and ideas.* Bethesda, MD: American Physiological Society, 1987, pp 247–276.

Morello J-P, Bichet DG. Nephrogenic diabetes insipidus. *Annu Rev Physiol* 2001;63:607–630.

Robertson GL. Vasopressin. In Seldin DW, Giebisch G, eds. *The kidney: Physiology and pathophysiology,* Vol 2, 3rd ed. Philadelphia: Lippincott Williams & Wilkins, 2000, pp 1133–1152.

Rose BD, Post TW. *Clinical physiology of* acid–base and electrolyte disorders, 5th ed. New York: McGraw-Hill, 2001, pp 285–298.

Valtin H. "Drink at least eight glasses of water a day." Really? Is there scientific evidence for "8 × 8"? *Am J Physiol* 2002;283: R993–R1004.

Regulation of Sodium Balance and Extracellular Fluid Volume

JAMES A. SCHAFER

KEY POINTS

- *Extracellular fluid volume* is directly proportional to the total amount of Na$^+$ in the body.
- The plasma Na$^+$ concentration is not an indicator of total body Na$^+$ but of water balance. Changes in plasma Na$^+$ concentration occur only when water loss or gain from the body exceeds the capacity to regulate plasma osmolality by controlling water drinking through thirst and water excretion by the kidneys.
- A typical American diet contains 8–15 g NaCl, equivalent to 150–250 mmol of NaCl. An equal amount must be excreted to maintain a constant and normal extracellular fluid volume.
- About 25,000 mmol of Na$^+$ are filtered per day, but only 150–250 mmol are usually excreted; thus, more than 99% of the filtered Na$^+$ is reabsorbed by the nephron.

- The large veins and atria have *stretch receptors* that increase their rates of firing with changes in filling. *Arterial baroreceptors* are also stretch receptors that respond to changes in blood pressure and pulse pressure, and relay information about cardiac output.
- The kidney responds to changes in volume sensed by stretch receptors through three basic mechanisms: changes in the glomerular filtration rate (GFR), regulation of aldosterone secretion, and secretion of natriuretic hormones.
- Reabsorption of water and salt in the proximal tubule normally is a constant fraction of the filtered amount irrespective of the GFR. This is referred to as *glomerulotubular balance*. However, a small change in GFR results in a relatively large change in NaCl delivery to the

The volume of extracellular fluid (ECF) volume is kept constant by a variety of homeostatic mechanisms. This multiplicity of regulation is essential for survival because the ECF volume determines the circulating plasma volume and, thus, mean circulatory filling pressure and cardiac output. ECF volume is determined directly by Na^+ balance, which maintains a constant total body content of Na^+. Thus, the total *amount* of Na^+ in the body is the important regulated parameter, and not the Na^+ concentration.

The relative unimportance of the *concentration* of Na^+ in the plasma as an indicator of ECF volume is often confusing. However, as shown by Eq. [1] in Chapter 28, the plasma Na^+ concentration is kept constant by the regulation of free water excretion by the kidneys via alterations in vasopressin secretion. The vasopressin mechanism maintains a constant plasma osmolality, and thus a constant plasma Na^+ concentration. Therefore, changes in this concentration reflect a loss or gain of body water but not necessarily a loss or gain in the total amount of Na^+ in the body. Changes in the Na^+ concentration in the plasma occur only when fluid losses or gains have exceeded the capacity of the thirst mechanism and the renal concentrating and diluting mechanism to regulate water balance. This may occur with extreme loss or intake of water, or when the thirst or vasopressin mechanisms are impaired (see Chapter 28).

Why is the ECF volume so dependent on body Na^+ balance? Equation [1] illustrates the determinants of the plasma Na^+ concentration (P_{Na}). As with any solute concentration, it can be calculated as the total amount of solute divided by the volume of distribution. Because Na^+ is actively pumped out of most cells in the body, the volume of distribution of Na^+ is approximately the ECF volume.

$$P_{Na} = \frac{\text{amount of } Na^+ \text{ in ECF}}{\text{ECF volume}} \qquad (1)$$

As discussed earlier and in Chapter 28, vasopressin and thirst maintain a constant plasma osmolality, but because the primary determinant of the plasma osmolality is the concentration of Na^+ and its associated anions, a constant osmolality also requires a constant Na^+ concentration. However, the body could maintain a constant P_{Na} despite an increase of the total amount of Na^+ if extracellular fluid volume were increased. Conversely, Na^+ loss accompanied by an equivalent decrease in extracellular fluid volume would also lead to a constant P_{Na}.

The primary parameter of concern is the ECF volume, which can be expressed in terms of the other two parameters in Eq. [1] as follows:

$$\text{ECF volume} = \frac{\text{amount of } Na^+ \text{ in ECF}}{P_{Na}} \qquad (2)$$

Because P_{Na} is normally kept constant by the vasopressin and thirst mechanisms, the extracellular fluid volume is proportional to the total extracellular fluid Na^+ content. Thus, in order to maintain a constant ECF volume the total amount of Na^+ in the body must be regulated. This regulation revolves around the usual considerations of body mass balance in order to maintain homeostasis of the extracellular fluid. As a fixed cation, Na^+ is not produced by metabolism. Therefore, the maintenance of a constant amount of Na^+ in the body involves matching its daily output to daily intake primarily by adjustments in its urinary excretion.

DAILY Na$^+$ INTAKE AND OUTPUT

On a typical American diet, the average adult human ingests on the order of 8–15 g (0.3–0.5 oz., or 2 to more than 3 cooking teaspoons) of NaCl per day. This represents 150–250 mmol of Na$^+$ per day, which would result in a dramatic expansion of the ECF volume if it were not matched by an equivalent daily output of Na$^+$. For example, if one were to retain just 1 day's intake of Na$^+$ (let's say, 150 mmol), this would obligate the retention of 1 L of water in order to maintain an isotonic extracellular fluid. (A solution of 150 mmol Na$^+$ with accompanying monovalent anions in 1 L of water is approximately isosmotic to plasma.) This extra water comes from increased intake of water in response to thirst and the retention of water by the kidneys. The retention of just 1 day's Na$^+$ and an accompanying 1 L of water would increase the body weight by 1 kg (2.2 lb). The body weight of most individuals is relatively constant as long as their daily caloric intake is relatively constant. Therefore, weight gain or loss over a short period on a constant diet is an important indicator of the status of Na$^+$ balance. Patients with chronic renal failure and patients on dialysis are requested to follow their daily weight as an indicator of the extent to which they are retaining fluid and, thus, how much hemodialysis may be required to return them to a state of Na$^+$ balance.

The kidneys are typically the primary route of Na$^+$ output, but there are also losses from the gastrointestinal tract and from the skin. Normally the feces contain little Na$^+$, but gastroenteritis with diarrhea can produce dramatic losses of NaHCO$_3$. Vomiting also results in the loss of an isotonic NaCl solution plus a further Cl$^-$ loss as HCl. Sweat is approximately one-half isosmotic, and the usual daily insensible loss of perspiration by the skin is less than 50 mmol/day. However, with exercise or work, or in a hot environment, Na$^+$ losses in the sweat can be considerable.

The body normally has no difficulty matching Na$^+$ output to Na$^+$ input. However, with severe gastrointestinal losses, especially when these are not replaced by adequate salt and water intake, or with excessive sweating, net daily Na$^+$ losses may compromise the maintenance of ECF volume. The normal balance of Na$^+$ input and output is also disturbed when the ability of the kidney to retain Na$^+$ is compromised, for example, with the use of diuretics.

As noted earlier, the plasma Na$^+$ concentration is not a good indicator of a deficit or excess in ECF volume. Therefore, the clinician must rely on other signs of extracellular volume excess or deficit. *Extracellular volume contraction* is often indicated by a decrease in systemic blood pressure (hypotension). Although this may not be marked when the patient is lying in a hospital bed, it often occurs when the patient stands up. Because of cardiovascular regulatory mechanisms, which operate effectively in a well-hydrated individual, little change is typically seen in blood pressure upon standing. However, when plasma volume is significantly reduced, these reflex mechanisms cannot restore blood pressure upon standing, and both systolic and diastolic blood pressure fall and the pulse rate rises. This sign is referred to as *orthostatic hypotension* and is generally an indicator of an extracellular volume deficit.

Extracellular volume expansion, when moderate to severe, is usually indicated by the presence of edema. Edema is evidenced as a swelling of the lower extremities in the ambulatory patient, but it may be seen more frequently in the tissues of the back in the bedridden patient, because excess interstitial fluid pools in those parts of the body that are the lowest in relation to the heart. Although localized edema may be observed in several settings not associated with significant volume expansion, generalized edema is evident clinically only if the ECF volume is increased by 2.5–3 L. The presence of generalized edema is often the presenting symptom in patients with chronic renal failure or heart failure due to excessive salt and water retention.

As ECF volume expansion progresses, pulmonary edema may also develop. This is diagnosed by a chest X ray or by hearing fine rales upon auscultation of the lungs. These rales are due to the presence of fluid in the alveoli. Excess ECF volume is also indicated in the heart sounds by the presence of an S3 gallop, which occurs as venous return and venous congestion progressively increase.

The extent of distension of the large veins is also an indicator of ECF volume. With volume expansion, one expects to find an elevated central venous pressure. This rise in central venous pressure can also be inferred without direct measurement by the degree of distension of the neck veins. In the normal patient, as the upper body is raised from the recumbent to the sitting position, it is found that the neck veins (the internal jugular is the best indicator) collapse at a body angle of about 25° to 45° from the horizontal. However, in the severely volume-expanded patient, neck vein distension may occur even in the full upright position.

LOCATION AND AMOUNTS OF RENAL Na$^+$ REABSORPTION

Chapters 26 and 27 discussed the mechanisms of Na$^+$ reabsorption in the proximal tubule, the loop of Henle,

---

Clinical Note

Edema with Normal or Low ECF Volume

In certain settings such as hypoalbuminemia due to liver disease or in the nephrotic syndrome (see Chapter 24 Clinical Note), abundant evidence may point to generalized edema without venous distension. To explain this apparent contradiction, it is important to remember that the ECF encompasses both the interstitial fluid and the blood plasma, and that the Starling forces across the capillary walls determine the distribution of fluid between these two compartments. When the plasma albumin concentration and thus the plasma colloid osmotic pressure is decreased by liver disease or renal losses, a shift can be seen in ECF from the vascular to the interstitial compartment. Edema reflects the increased interstitial fluid volume, whereas the vascular volume is diminished. The same shift of fluid out of the vascular compartment into the interstitial compartment also occurs in burn patients. In these patients, the permeability of the capillaries in the traumatized tissue is increased, causing loss of plasma proteins together with fluid to the interstitial space. In each of these settings, the loss of fluid from the vascular compartment causes extracellular volume receptors to respond as if total extracellular volume were decreased, resulting in the retention of Na⁺ and water and a further expansion of ECF volume in an attempt to restore plasma volume. However, as will be discussed in the next Clinical Note, salt and water retention are also driven by other mechanisms in disease states such as these and can persist even when both the plasma and interstitial fluid compartments are expanded.

---

and the distal segments of the nephron. Figure 1 shows schematically the major locations of Na⁺ reabsorption and the relation to amounts filtered and excreted on a daily basis. In this example, the glomerular filtration rate (GFR) is 180 L/day and the plasma Na⁺ concentration is 139 mmol/L. This gives a daily rate of Na⁺ filtration of $(139 \times 180) = 25{,}000$ mmol/day. To maintain Na⁺ balance, the kidney must excrete only as much Na⁺ as ingested, or 150 to 250 mmol/day. Thus the amount of Na⁺ excreted is ranges at most from 0.01–5% of the filtered load, meaning that 95–99.99% is reabsorbed. With normal dietary intake of Na⁺, the reabsorption

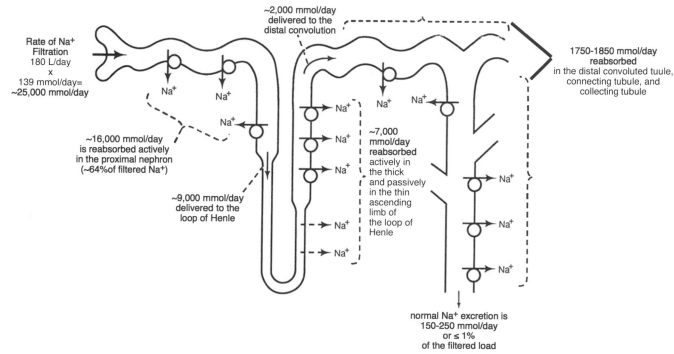

FIGURE 1   Location and amount of Na⁺ reabsorption along the nephron. More than 99% of the filtered Na⁺ is normally reabsorbed and returned to the plasma. *Note:* For Na⁺, 1 mmol is equal to 1 mEq.

of Na^+ must be finely regulated in the range of 99–99.99%. The processes involved in this delicate balancing process are considered in the subsequent sections of this chapter.

The Na^+ concentration in the urine reveals little about the rate of Na^+ excretion. As for any solute, the rate of excretion is equal to the product of the urine flow rate and the solute concentration, $UF \cdot U_{Na}$. Consequently, the Na^+ concentration in the urine will depend on the urine osmolality and can vary from 5–200 mmol/L, depending on the rate of Na^+ excretion relative to water excretion.

When considered on a day-to-day basis, the renal Na^+ excretory system responds relatively slowly to changes in Na^+ input in the diet. This is illustrated in Fig. 2, which shows the responses of urinary Na^+ output and body weight to changes in the dietary intake of Na^+. The human subject for whom the data are given in Fig. 2 has been ingesting an extremely low Na^+ diet of about 15 mmol/day. If that rate of Na^+ intake is abruptly increased 10-fold to 150 mmol/day, Na^+ excretion also begins to rise, but not as rapidly as the sudden increase in Na^+ intake. Because Na^+ excretion does not rise as rapidly as Na^+ intake, a transient period of positive sodium balance results. In other words, the total amount of Na^+ in the body increases until a new steady state is established in which Na^+ output is equal to Na^+ input. As a consequence of the positive Na^+ balance, this subject experienced an increase in ECF

volume manifested by an increase in body weight. When the Na^+ intake rate is abruptly returned to 15 mmol/day, Na^+ output remains temporarily higher than Na^+ intake until a new steady state is established. During this transient period, Na^+ excretion by the kidneys exceeds Na^+ intake, resulting in a negative Na^+ balance and a loss of ECF volume that is manifested by a decrease in body weight. Just the opposite happens with the decrease in daily Na^+ intake from the normal level of 150–250 mmol/day. There is a transient period of Na^+ loss resulting in a contracted ECF volume and decreased body weight. This is reversed when Na^+ intake is returned to normal.

The slower response of the Na^+ excretory system can be contrasted with the relatively rapid response of the thirst and the vasopressin mechanisms to changes in plasma osmolality. As discussed in Chapter 28, an increase in water intake is rapidly followed by a diuresis that results in excretion of the excess water in 1–2 hr. In contrast, if one ingests or infuses the equivalent of 1 L of isotonic saline, it takes the body about 2–4 days to excrete this excess load of both salt and water. Because of this slower response of the Na^+ regulatory mechanism, the regular daily intake of Na^+ can be an important determinant of the basal ECF volume. Individuals who by habit have a higher daily Na^+ intake may have a slightly expanded ECF volume, whereas those individuals on a low-salt diet have a lower ECF volume in proportion to their body size.

Because plasma volume is proportional to ECF volume and is an important determinant of the mean circulatory filling pressure, ECF volume plays a role in the maintenance of a normal blood pressure. A higher dietary Na^+ intake disposes the individual to a higher ECF volume and a higher blood pressure, whereas a lower Na^+ intake is associated with a decreased ECF volume and a lower blood pressure. Many individuals maintain a relatively constant blood pressure despite wide variations in daily Na^+ intake. On the other hand, other individuals seem to be quite sensitive to changes in Na^+ intake. Some patients with hypertension respond with a fall in blood pressure when they are placed on a low-salt diet with no drug therapy. Others may show a fall in blood pressure only when salt restriction is combined with diuretics. Individuals such as these are referred to as *salt-sensitive hypertensive patients,* and their hypertension can sometimes be controlled with Na^+ restriction and diuretics. It would appear that these individuals have a greater propensity to retain Na^+ in the presence of a high Na^+ intake and, thus, expand their ECF volume and raise their basal blood pressure. Alternatively, in a salt-sensitive hypertensive patient the elevated blood pressure may be necessary to produce sufficient Na^+ excretion via pressure natriuresis to

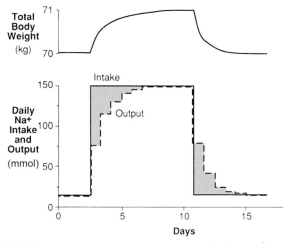

FIGURE 2 Change in total body output of Na^+ when Na^+ intake is suddenly increased and then decreased. Note that the changes in output always lag behind the changes in intake, causing a net accumulation or loss of Na^+ during a transient period. The increase in ECF volume is indicated by the increase in body weight with increased intake and by the decrease in the body weight with decreased intake. (Modified from Reineck HJ, Stein JH. Sodium metabolism. In Maxwell MH, Kleeman CR, Narins RG, eds. *Clinical disorders of fluid and electrolyte metabolism,* 4th ed. New York: McGraw-Hill, 1987, pp 33–59.)

maintain body Na$^+$ balance (see later discussion). Therefore, daily Na$^+$ intake and the efficiency with which the kidney excretes Na$^+$ are important determinants of total ECF volume and blood pressure.

The next sections consider how Na$^+$ excretion by the kidneys is regulated in response to changes in dietary intake. To understand this feedback regulation of body Na$^+$ content, it is necessary to identify the receptors that sense changes in ECF volume, the effectors that mediate changes in Na$^+$ reabsorption by the kidney, and the final effects on different nephron segments that reabsorb Na$^+$.

SENSING OF EXTRACELLULAR FLUID VOLUME

The large veins and atria possess receptors that sense relative vascular filling and, thus, plasma volume. These receptors are neural *stretch receptors* that increase their rate of firing as the wall of the vessel is expanded in much the same way as the stretch receptors of the carotid sinus (see Chapter 15). The venous stretch receptors are probably the volume receptors that modulate vasopressin release from the posterior pituitary and that can override the normal signals of plasma osmolality in regulating vasopressin release. As discussed later, the atria can also release a peptide (atrial natriuretic peptide) that may regulate renal Na$^+$ excretion directly.

Distension of the large veins in the thorax or distension of the atria results in a relatively rapid response of natriuresis and diuresis. Arterial baroreceptors also seem to be important in regulating this response. It is likely that there are arterial stretch receptors similar to those located in the carotid sinus that respond to an increase in arterial blood pressure or pulse pressure with an increase in firing. With decreased stretch of these receptors, which would occur with a falling cardiac output or a diminished plasma volume, the renal excretion of Na$^+$ and water falls rapidly. The afferent neural arcs that carry the information from stretch receptors in the atria and large veins run primarily with parasympathetic fibers in the vagus nerve. These fibers impinge on the central nervous system in a variety of different centers that may be associated with the regulation of vasopressin secretion, sympathetic firing to the kidneys, and cardiovascular centers among others.

What factors normally stimulate these volume receptors? First, an actual decrease in extracellular volume due to dehydration leads to a decrease in plasma volume with a fall in central venous pressure and atrial filling by venous return. This decreases stretch in the walls of both the venous and arterial volume receptors, eventually leading to an appropriate retention of Na$^+$. Second, there could be a decrease in plasma volume with a normal or increased ECF volume (see Clinical Note on edema earlier in this chapter). The decreases in venous and arterial filling lead to retention of Na$^+$ by the kidneys. Finally, under conditions of normal or expanded ECF volume when cardiac output is decreased or effective perfusion of the tissues is decreased because of heart failure, there is a strong stimulus for Na$^+$ retention despite the expanded ECF volume. Thus, for example, heart failure is often accompanied by extracellular volume expansion, resulting in a progression from peripheral edema to pulmonary edema and eventual decompensation of cardiac contraction due to excessive ventricular filling. This general condition is referred to as *functional hypovolemia* (see following Clinical Note)—the body is reacting to the poor tissue perfusion as if it were due to a low circulating blood volume, whereas the blood volume may, in fact, be dangerously expanded.

In the remainder of the chapter, the emphasis is on how these signals indicating a change in ECF volume result in changes in the rate of Na$^+$ excretion by the kidney. This is obviously one of the most important functions of the kidney, and it is vital to the maintenance of normal cardiac output and blood pressure and, thus, to survival. Multiple mechanisms regulate Na$^+$ reabsorption in various segments of the nephron, but the apparent redundancy of these mechanisms may have a survival advantage in that the failure of one will not lead to catastrophe. Nevertheless, it is clear that a better understanding of the interrelation among the various mechanisms must be achieved before we can fully understand the regulation of Na$^+$ excretion under normal physiologic circumstances, as well as in pathologic derangements that lead to such conditions as hypertension and congestive heart failure.

THREE MECHANISMS REGULATE Na$^+$ EXCRETION

The regulation of Na$^+$ excretion by the kidney is accomplished by three basic mechanisms, each of which is also influenced by other signaling systems. In general, a reduction in GFR is associated with Na$^+$ retention, and an increase in GFR enhances Na$^+$ excretion. The second mechanism, aldosterone, controls Na$^+$ reabsorption in the CNT and principal cells of the connecting tubule and the collecting duct, i.e., the aldosterone-responsive distal nephron (ARDN). The third mechanism also depends primarily on changing the rate of Na$^+$ reabsorption by the nephron through the effects of natriuretic hormones on transport.

Clinical Note

Functional Hypovolemia

Congestive heart failure, liver disease with ascites (ECF accumulation in the peritoneal cavity), and the nephrotic syndrome are all characterized by avid renal retention of NaCl and water despite the presence of often-dramatic edema. The preceding Clinical Note presented situations in which edema could be present but in which the plasma volume is diminished. In these situations, decreased vascular filling is an appropriate signal for salt and water retention. However, in the three conditions considered here, most typically both the interstitial and vascular components of the ECF volume are increased as evidenced by edema and venous congestion (neck vein distension, increased central venous pressure).

In congestive heart failure, the underlying problem is usually a reduced cardiac output that provides inadequate perfusion of the tissues, including the kidneys. However, heart failure can also occur in a situation of normal or increased cardiac output due to an arteriovenous (AV) anastomosis such as a fistula, in which case most of the cardiac output is shunted to the venous side without perfusing the organs.

In liver disease, AV anastomoses are also the primary cause of poor tissue perfusion. These anastomoses occur in the liver, possibly because of hypertension in the portal circulation, but they also occur for unknown reasons in the skin and other organs. In the skin, the anastomoses form spider-like angiomas characteristic of liver disease. The hypoproteinemia of liver disease can contribute to the edema as a result of the decreased colloid osmotic pressure of the plasma as discussed in the earlier Clinical Note on edema, but this effect is neutralized by increasing tissue hydrostatic pressure as the tissues are distended by edema.

In the nephrotic syndrome, NaCl and water retention appear to be an intrinsic response of the kidney to the disease process and directly result in the expansion of ECF volume and edema. Again one might reason that the edema occurs as a consequence of the loss of plasma proteins in the urine, but it is found that if steroid administration is effective in countering the autoimmune response, salt and water excretion increase dramatically before plasma proteins are restored.

In each of the three diseases, the characteristic sign is the presence of low urine Na^+ concentration (<25 mmol/L) in a hypertonic urine, together with marked edema. In each situation, the mechanisms regulating Na^+ and water excretion are responding as if the patient were volume depleted (both Na^+ and water are being retained) despite the edema and vascular congestion. For this reason they are referred to collectively as examples of *functional* hypovolemia.

Changes in Glomerular Filtration Rate

Changes in the GFR result in a proportional change in the filtered load of Na^+. Normally, renal blood flow (RBF) and GFR are maintained relatively constant by autoregulation and by the tubuloglomerular feedback mechanism. However, as discussed in Chapter 24, renal blood flow and GFR can be reduced by increased sympathetic nerve input to the kidney, circulating catecholamines, and angiotensin II. It is important to recognize that small changes in GFR that would not normally be detectable can result in marked changes in the rate of Na^+ excretion.

When GFR changes, the proximal tubule reabsorbs a constant *fraction* of the filtered load of salt and water. In other words, the rate of proximal tubule salt and water reabsorption increases with increasing GFR so that the fraction of the filtered load that is reabsorbed remains constant. This response is referred to as *glomerulotubular balance* (GT balance). Under normal conditions, 60–66% of the filtered load of Na^+ is reabsorbed in the proximal tubule. However, we need to consider the rate at which salt and water are passed on to the loop of Henle and distal nephron when the GFR changes. Because the proximal tubule acts like a mass reabsorber that adjusts its rate of reabsorption to match the rate of filtration, it delivers a constant fraction of the filtered Na^+ to the loop of Henle. However, the absolute *amount* of this filtered salt and water that leaves the proximal tubule increases as the filtered amount increases. Considering Fig. 1, let us assume that the GFR increases by 10%. This would increase the filtered load of Na^+ from 25,000 to 27,500 mmol/day. The proximal tubule would increase its rate of reabsorption so that a constant fraction of about 64% would be reabsorbed. Thus, reabsorption would increase from 16,000 to 17,600 mmol/day. However, this would leave an unreabsorbed Na^+ flow of (27,500 − 17,600) = 9900 mmol/day that would be passed on to the loop of Henle. Thus, despite the increased Na$^+$ reabsorption by the proximal tubule in response to the increased rate of filtration, a larger amount of Na$^+$ flows out of the proximal tubule.

The thick ascending limb of the loop of Henle responds to the increased load by increasing its rate of Na$^+$ reabsorption, but it too does not completely compensate. Consequently, increased amounts of salt and water are passed on to the distal nephron, which has a fixed rate of reabsorption unless this is altered by aldosterone. Therefore, the extra amount of filtered Na$^+$ that is passed on to the distal nephron by a small increase in GFR can represent a large amount of Na$^+$ in comparison with the usual rate of excretion of 25–250 mmol/day. For this reason, subtle changes in GFR can produce quite marked changes in Na$^+$ excretion merely by delivering more Na$^+$ to the distal nephron. Unfortunately, it is experimentally extremely difficult to reliably detect GFR changes of less than 10%, even though such changes could result in significant changes in Na$^+$ excretion. Other than the changes with eating described next, it is thought that the GFR is relatively constant in humans. Nevertheless, changes that would not normally be detectable could regulate Na$^+$ excretion.

GFR in humans increases with the ingestion of food, particularly a high-protein meal. GFR increases begin 30–60 minutes after the meal and reach a peak 20–50% above baseline within 1–2 hr. This increase in GFR can last for 3–6 hr, depending on the rate of gastric emptying and the size of the meal. The reason for this postprandial increase in GFR is not known. It could result from the effects of intestinal hormones on the renal circulation, the direct effect of an increased concentration of circulating amino acids, or from neurally mediated changes in afferent and efferent arterial resistance. Nevertheless, this increase in GFR can produce natriuresis and diuresis.

Natriuresis and diuresis are also known to be associated with increases in systemic blood pressure, a phenomenon referred to as *pressure natriuresis*. A 50% increase in systolic blood pressure can lead to a three- to fivefold increase in both urine flow and Na$^+$ excretion. Nevertheless, as discussed in Chapter 24, GFR and RBF are not observed to change as blood pressure rises because of the phenomenon of autoregulation. However, it is difficult to state with certainty that a GFR increase could not explain the pressure natriuresis. As noted earlier, even relatively subtle changes in the GFR lead to substantial changes in the rate of Na$^+$ excretion. Alternatively, blood pressure may affect the relative influence of other local hormones on the rate of Na$^+$ reabsorption in the nephron.

Renin, Angiotensin, and Aldosterone

In Chapter 27, we considered the mechanisms involved in regulating ion transport in the connecting tubule and the collecting duct by aldosterone. Aldosterone is known to increase Na$^+$ reabsorption by increasing Na$^+$ entry across the luminal membrane of CNT and principal cells, by increasing the number of Na$^+$,K$^+$-ATPase pumps, and by increasing metabolism in these cells. However, aldosterone normally regulates the excretion of Na$^+$ only in the range of 0.1–2.0% of the filtered load. As noted in Chapter 27, even in the complete absence of aldosterone, only about 2% of the filtered Na$^+$ load is excreted.

Aldosterone also has a relatively slow effect on Na$^+$ reabsorption by the connecting tubule and collecting duct. It normally takes at least 2 hr to see an effect of a change in the plasma aldosterone level, and the full effect is not experienced for 12–24 hr. Consequently, it is unlikely that aldosterone is involved in the rapid regulation of Na$^+$ excretion as occurs, for example, when a large bolus of isotonic saline is infused intravenously or after a hemorrhage. In these circumstances, the rate of Na$^+$ excretion is altered rapidly and dramatically with immediate effects that cannot be attributed to aldosterone. Thus, it appears that aldosterone is more important in regulating the day-to-day excretion of Na$^+$ to match variations in Na$^+$ intake in the diet.

In Chapter 27, the various stimuli that increase aldosterone secretion by the adrenal cortex were presented in Table 1. Which of these factors would be involved in increasing aldosterone production in the face of a falling extracellular volume? As discussed earlier, a decrease in ECF volume is not usually accompanied by a change in either the plasma Na$^+$ or K$^+$ concentration because the vasopressin mechanism maintains a constant plasma osmolality and, thus, constant concentrations of the constituent solutes. In most cases of clinical importance, changes in aldosterone secretion by the adrenal cortex are produced by changes in the circulating angiotensin II levels.

Figure 3 shows the feedback mechanism that regulates renin secretion by the juxtaglomerular (JG) cells in the afferent arterioles, as discussed previously in Chapter 24. Renin secretion by the JG cells is increased by a variety of factors, but the common denominator for all of them is decreased vascular volume, blood pressure, or poor tissue perfusion (e.g., functional hypovolemia). As expected, this feedback mechanism responds to a decrease in ECF volume by decreasing renal Na$^+$ and water excretion. Conversely, expansion of ECF volume diminishes renin secretion. The increased plasma angiotensin II that is produced by renin secretion increases Na$^+$ reabsorption in both the proximal and distal

TABLE 1   Factors That Alter Renin Release from Juxtaglomerular Cells

	Mechanism
Factors promoting renin release	
↓ BP	Intrinsic renal baroreceptor function of JG cells in afferent arteriole
↓ Blood volume or ↓ ECF volume	Systemic baroreceptors and venous stretch receptors, via renal efferent nerves and β-adrenergic effect of circulating catecholamines on JG cells to increase intracellular cAMP
"Functional hypovolemia" (edematous states: congestive heart failure, nephrotic syndrome, cirrhosis of the liver)	Above mechanisms augmented by decreased renal blood flow, and generalized poor tissue perfusion
PGE_2 and PGI_2	Released by the kidney and act locally as paracrine and autocrine regulators of JG cells
↓ NaCl delivery to the loop of Henle and distal nephron	Macula densa cells sense decrease NaCl delivery; release ATP and/or adenosine to signal JG cells to release renin
Factors inhibiting renin release	
↑ BP	Direct effect on JG cells in afferent arteriole to decrease renin release
↑ Blood volume or ↑ ECF volume *(even without ↑ BP)*	Decreased firing by systemic volume receptors such as stretch receptors in the veins and atria of the heart
Atrial natriuretic peptide (ANP; see next section of text)	Circulating hormone from the atria of the heart directly inhibits renin release by JG cells
Cyclooxygenase inhibitors (NSAIDs)	↓ Renal prostaglandin production, especially important in the setting of ECF volume depletion, stress, or trauma
↑ NaCl delivery to the loop of Henle and distal nephron	Macula densa cells sense increased NaCl delivery and decrease ATP and/or adenosine release diminishing signal to JG cells
$β_2$-Adrenergic agonists	Direct effect on JG cells to ↓ cAMP

segments of the nephron. In the proximal tubule, angiotensin acts directly to augment the activity of the Na^+-H^+ exchanger. Angiotensin II indirectly stimulates Na^+ reabsorption in the CNT and principal cells of the ARDN by stimulating zona glomerulus cells in the adrenal cortex to secrete aldosterone. As discussed in Chapter 24, angiotensin II also decreases RBF and GFR, which, in turn, decrease Na^+ and water excretion for the reasons discussed earlier. Finally, angiotensin II also acts in the central nervous system to stimulate thirst.

Table 1 presents some of the most important factors that augment or diminish renin release as shown in Fig. 3. JG cells in the afferent arterioles function as stretch receptors and their intracellular Ca^{2+} concentration rises with increased wall stretch. When blood pressure falls, therefore, the intracellular Ca^{2+} concentration in the JG cells falls, resulting in increased renin secretion. Although renin secretion is the expected response to a decrease in blood pressure, the intracellular signaling system is unique because in most other secretory cells (e.g., the vasopressin secreting cells in the posterior pituitary) secretion is stimulated by a *rise* in intracellular Ca^{2+}.

Decreased blood pressure and decreased vascular volume are sensed systemically by arterial baroreceptors and venous stretch receptors. Through central nervous system reflex mechanisms, the decreased blood pressure results in increased renal nerve activity resulting in β-adrenergic stimulation of the JG cells. β-Adrenoceptor stimulation increases cAMP levels in the JG cells, which also augments renin release. The JG cells also have β-

adrenergic receptors, and $β_2$-adrenergic agonists inhibit intracellular cAMP production and thus renin release.

Renin release by JG cells also responds to changes in NaCl delivery to the macula densa such that decreased NaCl delivery stimulates renin release and vice versa. As discussed in Chapter 24, the effect of NaCl delivery to the distal nephron on renin release is confusing in the context of the tubuloglomerular feedback mechanism. This feedback involves an increase in single nephron GFR in response to decreased NaCl delivery to the macula densa, which is opposite to what one would expect if angiotensin II levels were raised by renin secretion. However, it is important to recognize that angiotensin II is not the mediator of the tubuloglomerular feedback mechanism. Furthermore, tubuloglomerular feedback operates on a local level to control the single nephron GFR, whereas the change in plasma renin levels in response to changes in NaCl and water flow to the macula densa occur as a generalized phenomenon in all nephrons. Although the macula densa cells are in proximity to the JG cells, no direct anatomic connection exists between the two. Recent evidence indicates that macula densa cells release ATP and/or adenosine, which stimulates JG cells to increase renin secretion as shown in Fig. 3.

The renin-angiotensin-aldosterone system is an important target of emerging drug therapy for hypertension because of its central role in the regulation of ECF volume. These drugs include inhibitors of the enzyme that converts angiotensin I to angiotensin II—so-called ACE inhibitors, as well as angiotensin II receptor blockers. Because there are at least two angiotensin II

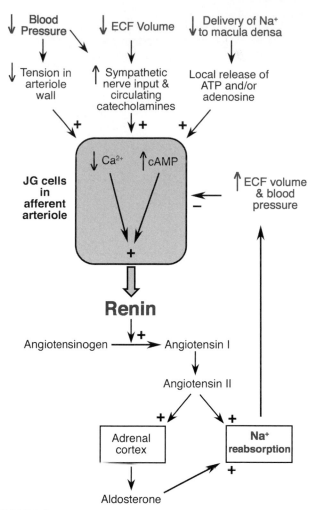

FIGURE 3 Renin secretion and the regulation of ECF volume and blood pressure. Renin is secreted by JG cells in the afferent arteriole in response to reductions in blood pressure, ECF volume, and GFR. Blood pressure is sensed directly by stretch receptors in the afferent arteriole. Decreased stretch results in a fall in intracellular Ca^{2+} and increased renin release. Renin secretion is also activated by increases in cAMP produced by increased sympathetic nerve input and circulating catecholamines, and by ATP and/or adenosine from macula densa cells. Elevated plasma angiotensin II produced by the release of renin acts to restore ECF volume and blood pressure by increasing Na^+ reabsorption through two mechanisms. Angiotensin II has a direct effect on proximal tubular cells to increase Na^+ reabsorption and stimulates the release of aldosterone from the adrenal cortex, which in turn increases Na^+ reabsorption by the aldosterone-responsive distal nephron.

receptor isoforms, drug development is focusing on specific receptor blockers that may differentiate, for example, between vascular and adrenal cortical angiotensin II receptors.

Natriuretic Hormones

The idea that there might be a circulating natriuretic hormone came from classical animal experiments in which both the GFR and plasma aldosterone concentration were kept constant during the infusion of isotonic saline. Despite the absence of changes in GFR or aldosterone levels, the infusion resulted in a natriuresis. In parallel experiments, it was found that a small volume of plasma from a volume-expanded animal produced natriuresis in a second animal that had not been volume expanded and in which GFR and aldosterone were held constant. Thus, it was proposed that a third factor (i.e., in addition to changes in GFR and aldosterone) must be involved in the regulation of Na^+ excretion—a natriuretic hormone that was released in response to volume expansion and caused decreased reabsorption of salt and water. It now appears that there are two classes of natriuretic factors, peptide hormones and ouabain-like factors.

After two decades of trial and error, evidence for the first natriuretic factor, which was called *atrial natriuretic peptide* (ANP), was published in 1980. It was subsequently found that the circulating form of ANP is a 28-amino-acid peptide with a molecular weight of about 3000 Da, and radioimmunoassays indicate a normal circulating concentration in the human of 3–5 pM. ANP is stored in granules in the myocytes of the left and right atria and is released in response to increased atrial filling, which occurs with extracellular volume expansion. Similar peptides are also produced by the brain, the heart, and the kidney, the latter being called *urodilatin*. When ANP is given in pharmacologic doses, rapid and marked natriuresis is observed, but the physiologic role of ANP has been questioned. It appears that physiologic concentrations of ANP can produce a natriuresis only in the presence of true ECF volume expansion, and then produce only a three- to fivefold increase in Na^+ excretion. Furthermore, the permissive volume expansion must be sufficient to elevate the blood pressure and/or renal blood flow. It is for this reason that ANP is ineffective in counteracting the volume expansion in functional hypovolemia.

The effects of ANP on the kidney are due to a small increase in GFR that, for the reasons discussed earlier, would automatically increase Na^+ excretion and inhibit Na^+ reabsorption in the inner medullary collecting duct. The increase in GFR may be due to a slight increase in the ultrafiltration coefficient (K_f), accompanied by an increase in efferent arteriolar resistance with or without an increase in afferent arteriolar resistance. Receptors for ANP are also present in many areas of the brain, and they may have other effects through neural efferent arcs. ANP also causes a decrease in renin release, which promotes natriuresis.

There appears to be a second type of natriuretic factor that is biochemically almost identical to ouabain, which is a member of the class of drugs referred to as *cardiac glycosides*. There is evidence that this *ouabain-like factor* (OLF) is synthesized and released by the brain and/or by the adrenal cortex in response to expansion of the ECF volume. (Because the cardiac glycosides are structurally quite similar to steroids, the latter site seems plausible.) It is proposed that OLF, as all cardiac glycosides, is an Na^+,K^+-ATPase inhibitor and thereby decreases Na^+ reabsorption, possibly in all segments of the nephron.

CHANGES IN GT BALANCE

The three systemic mechanisms for regulating Na^+ excretion just discussed (changes in GFR, plasma aldosterone levels, and circulating natriuretic factors) may not be responsible for the rapid changes in sodium excretion in response to sudden alterations in ECF volume. Changes in GFR hold promise as potential modulators of Na^+ excretion, but they are extremely difficult to demonstrate. Aldosterone appears to be an appropriate hormone for the day-to-day regulation of Na^+ excretion over a relatively modest range, but it cannot account for the rapid natriuresis observed with volume expansion. ANP at physiologic concentrations has a rather modest effect on Na^+ excretion, and the physiologic role of OLF remains to be demonstrated.

As discussed earlier, the proximal tubule normally reabsorbs a constant fraction (60–66%) of the filtered load of salt and water. However, this GT balance can be disturbed dramatically under pathologic conditions. In nephrotic syndrome or liver disease (in which large amounts of plasma volume are lost to the interstitial fluid), severe dehydration, or cardiac failure, the proximal tubule can increase its rate of Na^+ and water reabsorption to 90% of the filtered load. On the other hand, when the ECF volume is expanded by NaCl or plasma infusion, the proximal tubule can decrease its rate of Na^+ and water reabsorption to <50% of the filtered load. These changes occur rapidly and result in dramatic changes in Na^+ delivery to the distal nephron and, thus, in dramatic changes in Na^+ excretion.

The hormones known to influence proximal tubular reabsorption are listed in Table 2. However, all of these hormones have primary actions in other organ systems, and their effects on proximal tubular reabsorption may be relatively slow in onset and not directly related to changes in ECF balance. The main regulators of proximal tubular volume reabsorption appear to be the renal nerves, the renin-angiotensin system, and the Starling forces that are responsible for the uptake of reabsorbed fluid into the peritubular capillaries.

Renal Nerves

As discussed in Chapter 24, strong firing of the renal sympathetic nerves produces a fall in GFR and a corresponding decrease in Na^+ excretion. However, these levels of sympathetic firing are usually found under relatively nonphysiologic circumstances. Thus, this mechanism may be important in severe volume depletion, as occurs with hemorrhage or under conditions of decreased cardiac output due to heart failure. Under these circumstances, RBF and GFR fall, resulting in a decrease in Na^+ excretion and in expansion of the ECF volume.

The renal nerves also have effects on renal Na^+ reabsorption. These effects are exerted directly on the nephron rather than indirectly by changes in GFR. Acute denervation of the kidneys results in decreased Na^+ reabsorption. On the other hand, low-level stimulation of the renal nerves results in Na^+ and water retention because of increased proximal tubule reabsorption. Adrenergic fibers have nerve endings that release catecholamines in the proximal tubule and portions of the distal nephron. Norepinephrine increases Na^+ reabsorption by stimulating the Na^+-H^+ antiporter in the proximal tubule. Thereby, even low-level sympathetic stimulation decreases Na^+ excretion by enhancing its reabsorption in the proximal tubule.

As discussed earlier (see Fig. 3), sympathetic firing to the kidney, produced, for example, by ECF volume contraction, increases renin release from JG cells by activating β-adrenergic receptors on these cells. The resulting renin release, acting via angiotensin II, also enhances aldosterone production, which increases Na^+ reabsorption by CNT and principal cells in the ARDN. Angiotensin II also acts like norepinephrine to increase proximal tubular Na^+ reabsorption by stimulating the Na^+-H^+ antiporter in proximal tubules.

Peritubular Factors

The rate of salt and water reabsorption by the proximal tubule may also be affected by the efficiency with which the reabsorbed fluid can be taken up into peritubular capillaries. As discussed in Chapter 24, the

TABLE 2 Hormones Affecting Volume Reabsorption in the Proximal Tubule

↑ Proximal volume reabsorption	↓ Proximal volume reabsorption
Norepinephrine	Parathyroid hormone (PTH)
Angiotensin II	Dopamine
Thyroid hormone	
Insulin	
Glucagon	

uptake of fluid from the cortical interstitial space into peritubular capillaries is favored by the Starling forces present there. As shown in Fig. 4, normal renal interstitial pressure is relatively high and the hydrostatic pressure in the peritubular capillaries is low because it has been reduced by both the afferent and efferent arterioles. Furthermore, the colloid osmotic pressure of the capillaries is increased because of the filtration of fluid at the glomerulus. All of these factors produce a net driving force that is normally quite high in favor of absorption of fluid from the interstitium.

The peritubular capillary pressure depends on the resistances of the afferent and efferent arterioles. When these resistances increase, peritubular capillary pressure decreases, enhancing uptake of fluid into the peritubular capillaries, as shown in Fig. 4. On the other hand, dilation of these resistance vessels leads to an increased capillary pressure and decreased uptake.

Glomerular dynamics also influence the colloid osmotic pressure in peritubular capillaries and, thus, the force favoring fluid uptake. When the filtration fraction is increased, for example, by increased GFR at a constant renal blood flow, the increase in the filtration fraction elevates the colloid osmotic pressure in the peritubular capillaries, which increases their uptake of fluid. On the other hand, when the filtration fraction is low, peritubular colloid osmotic pressure is decreased.

The peritubular capillary and interstitial colloid osmotic pressures and the capillary and interstitial hydrostatic pressures are commonly referred to as *peritubular factors*. As discussed in Chapter 26 (see Fig. 7 of that chapter), the net driving force for uptake into the capillary may also change the permeability characteristics of the junctional complexes between cells. When peritubular capillary uptake is diminished, interstitial fluid pressure may rise, which may, in turn, distort junctions between the cells and increase the backflow of fluid from the interstitial space to the tubular lumen. This would decrease the net reabsorption of salt and water by the proximal tubule. These mechanisms could be responsible for the rapid natriuresis and diuresis observed when saline solutions are administered.

Prostaglandins, Bradykinin, and Dopamine

Cells within the kidney produce prostaglandins and bradykinin. In the plasma, bradykinin is also produced by the action of plasma kallikrein. Increased plasma kallikrein levels and renal kallikrein excretion is associated with natriuresis and diuresis. Infusion of prostaglandin E_2 or I_2, or of bradykinin, results in natriuresis and diuresis. These actions of the prostaglandins and bradykinin appear to be due to their hemodynamic effects. They increase RBF with a lesser increase in the GFR. Thus, the filtration fraction falls and a lower colloid osmotic pressure favoring fluid uptake into the peritubular capillaries could lead to natriuresis and diuresis by the mechanisms discussed.

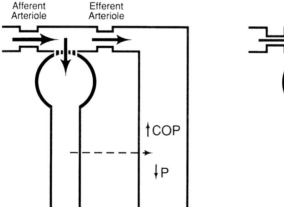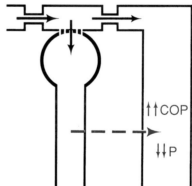

FIGURE 4   Effect of afferent and efferent arteriolar resistances on the Starling forces for fluid uptake in peritubular capillaries. With normal afferent and efferent arteriolar resistances (left panel), approximately 20% of the renal plasma flow is filtered at the glomerulus. The ultrafiltration of the protein-free fluid raises the average colloid osmotic pressure (COP) of the plasma and the peritubular capillaries. The resistance afforded by the two arterioles in series produces a lower hydrostatic pressure (P) in the peritubular capillaries. Both the increased colloid osmotic pressure and decreased hydrostatic pressure favor absorption of fluid from interstitial fluid space. With increasing afferent and efferent arteriolar resistances (right panel), total renal blood flow is reduced. A higher filtration fraction results in a higher average colloid osmotic pressure in the peritubular capillaries. Because of the higher resistances of the arterioles, hydrostatic pressure in the peritubular capillaries is reduced. The greater increase in colloid osmotic pressure and the greater decrease in hydrostatic pressure produce an increased driving force for fluid absorption.

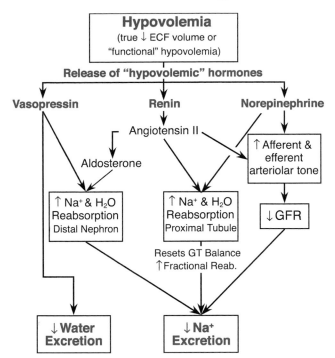

FIGURE 5 Primary mechanisms that conserve Na⁺ and ECF volume in true or functional hypovolemia.

Dopamine is synthesized in the kidney and has local actions that are mediated by receptors in both the proximal tubule and the distal nephron. Dopamine synthesis and excretion is elevated in natriuresis, and dopamine infusion causes natriuresis and diuresis. Dopamine has been found to inhibit Na⁺ reabsorption by both the proximal tubule and the collecting duct, and it inhibits the action of vasopressin in the collecting duct. Dopamine is sometimes used clinically in conjunction with diuretics to treat edema.

INTEGRATED CONTROL OF Na⁺ EXCRETION: A SUMMARY

As noted in the introduction, the factors that regulate Na⁺ excretion are multiple and overlapping. It is easy, therefore, to lose track of the primary feedback mechanisms that preserve Na⁺ balance on a day-to-day basis. Figures 5 and 6 summarize the interactions of these mechanisms that regulate Na⁺ excretion and thus extracellular volume, in hypovolemia and hypervolemia, respectively.

As shown in Fig. 5, the effects of hypovolemia can be produced either by a true deficit in ECF volume or in response to a perceived deficit in ECF volume resulting from low cardiac output, poor tissue perfusion, or renal disease, that is, *functional* hypovolemia. The primary

response to hypovolemia is the systemic release of the three hypovolemic hormones: vasopressin, renin, and norepinephrine. Angiotensin II generation in response to renin has three effects that decrease Na⁺ excretion. It acts directly on proximal tubule cells to increase Na⁺ reabsorption. Angiotensin II also increases the secretion of aldosterone, which in turn increases Na⁺ reabsorption by CNT and principal cells in the ARDN. Finally, angiotensin II also decreases the GFR, which, by decreasing Na⁺ and water delivery to the loop of Henle, decreases Na⁺ excretion.

Norepinephrine released by nerve endings and epinephrine, which is released by the adrenal medulla as a stress hormone in response to volume contraction, have two primary effects to conserve Na⁺. They decrease the GFR by their action on the afferent and efferent arterioles, and they directly stimulate Na⁺ reabsorption by the proximal tubule. As discussed in the previous chapter, a reduction of the ECF volume by 5% or more produces high plasma vasopressin levels regardless of the plasma osmolality, and the resulting water retention can lead to hyponatremia. Vasopressin also acts directly on the CNT and principal cells of the ARDN to increase Na⁺ reabsorption.

In hypervolemia (Fig. 6), the release of the three hypovolemia hormones is suppressed. Natriuretic hormones such as ANP and possibly OLF are released, and they increase GFR while inhibiting Na⁺ reabsorption. Volume expansion may also be accompanied by

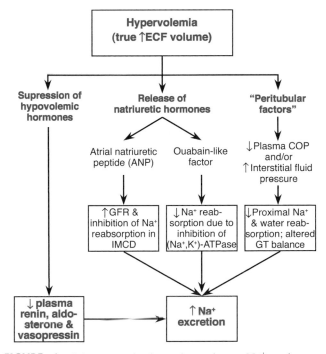

FIGURE 6 Primary mechanisms that enhance Na⁺ and water excretion in hypervolemia.

dilution of plasma proteins, resulting in a fall of the plasma colloid osmotic pressure. There is also an increase in renal interstitial fluid pressure. Both of the latter effects diminish the net uptake of fluid into the peritubular capillaries and thus decrease Na$^+$ and water reabsorption by the proximal tubule.

Suggested Readings

Humphreys MH, Valentin J-P. Natriuretic humoral agents. In Seldin DW, Giebisch G, eds. *The kidney: Physiology and pathophysiology,* Vol 1, 3rd ed. Philadelphia: Lippincott Williams & Wilkins, 2000, pp 1371–1410.

Kirchner KA, Stein JH. Sodium metabolism. In Narins RG, ed. *Maxwell & Kleeman's clinical disorders of fluid and electrolyte metabolism,* 5th ed. New York: McGraw-Hill, 1994, pp 45–80.

Navar LG, Inscho EW, Majid DSA, Imig JD, Harrison-Bernard WM, Mitchell KD. Paracrine regulation of the renal microcirculation. *Physiol Rev* 1996;76:425–536.

Palmer BF, Alpern RJ, Seldin RW. Physiology and pathophysiology of sodium retention. In Seldin DW, Giebisch G, eds. *The kidney: Physiology and pathophysiology,* Vol 1, 3rd ed. Philadelphia: Lippincott Williams & Wilkins, 2000, pp 1473–1518.

Rose BD, Post TW. *Clinical physiology of acid–base and electrolyte disorders,* 5th ed. New York: McGraw-Hill, 2001, pp 239–284.

30

Renal Regulation of Potassium, Calcium, and Magnesium

JAMES A. SCHAFER

KEY POINTS

- Relatively small changes in plasma concentrations of K^+, Ca^{2+}, and Mg^{2+} can have important effects on neuromuscular excitability and cardiac rhythm. Balance of these cations is maintained largely by regulation of their excretion in the urine.

- Most of the K^+ in the body is present in cells, so the only ~2% of the total body K^+ is located in the extracellular fluid compartment. Thus, the balance between the uptake of K^+ into cells and its loss from cells is the most important determinant of acute changes in the plasma K^+ concentration.

- Changes in intracellular K^+ generally parallel changes in extracellular K^+. Thus, the transfer of K^+ into or out of cells helps to buffer the effects of K^+ intake or loss on the plasma K^+ concentration.

- Alkalemia, β-adrenergic agonists, insulin, and aldosterone enhance K^+ uptake into cells. K^+ is lost from cells in response to acidemia, hyperosmotic states, α-adrenergic agonists,

and with exercise. Cell damage can cause the release of large amounts of K^+ from cells.

- K^+ is reabsorbed passively in the proximal tubule, and is reabsorbed passively and actively in the loop of Henle. The regulation of K^+ secretion that occurs in the connecting tubule and collecting duct, i.e., the aldosterone-responsive distal nephron (ARDN), determines the rate of urinary K^+ excretion.

- More than 99% of the total body Ca^{2+} is in bone and this portion does not exchange readily with free Ca^{2+} in the extracellular fluid.

- Calcium is reabsorbed passively in the proximal tubule at a rate that follows closely the rates of Na^+ and water reabsorption. Ca^{2+} reabsorption in the loop of Henle occurs primarily in the ascending limb by passive diffusion via the paracellular pathway, and it is driven by a favorable concentration gradient and the lumen-positive voltage.

- Calcium is actively reabsorbed via Ca^{2+}-selective channels in CNT cells in the connecting tubule,

---

KEY POINTS (*continued*)

and this reabsorptive process is stimulated by PTH and 1,25-dihydroxy-cholecalciferol [1,25(OH)$_2$D$_3$].

- To maintain Mg^{2+} balance, the kidneys must excrete daily the amount of Mg^{2+} absorbed by the intestines, which is about 120 mg. This amount is ~5% of the filtered load of Mg^{2+}, so substantial reabsorption of Mg^{2+} must occur.

- Because of its low Mg^{2+} permeability, the proximal tubule reabsorbs less than 20% of the filtered load of Mg^{2+}.

- More than 60% of the filtered Mg^{2+} is reabsorbed in the loop of Henle, primarily in the thick ascending limb by paracellular diffusion that is facilitated by paracellin-1 in the junctional complexes.

K$^+$, Ca^{2+}, and Mg^{2+} constitute the majority of the cations in the body. As is the case for Na$^+$, homeostasis for each of these cations is maintained largely by regulation of its rate of renal excretion. Disturbances in the total body content and plasma concentrations of any of these cations can produce a wide variety of clinical manifestations. Of particular importance is the effect of the plasma concentration of these cations on *neuromuscular excitability* and cardiac rhythm. In general, an increase in the plasma K$^+$ concentration, or a decrease in the plasma Ca^{2+} or Mg^{2+} concentration, causes neuromuscular hyperexcitability. Some extreme examples are the potentially fatal *cardiac arrhythmias* produced by *hyperkalemia* and the *tetany* observed with severe *hypocalcemia* (low plasma Ca^{2+} concentration). Conversely, hypokalemia, hypercalcemia, or hypermagnesemia are associated with depression of neuromuscular excitability.

Much is known about the normal regulation of K$^+$ balance by the kidney and about hormones that regulate the distribution of K$^+$ between the intra- and extracellular compartments. Most of the body stores of K$^+$ are intracellular, and total body K$^+$ is an important determinant of normal cell volume and composition. Potassium deficiency is associated with muscle wasting, weakness, fatigue, and, in children, with a general failure to thrive.

Only in the past few years have researchers identified the most important transport mechanisms for Ca^{2+} and Mg^{2+} reabsorption and their regulation. Normal body stores of Ca^{2+} and Mg^{2+} are, of course, essential for bone development, and Mg^{2+} is an important cofactor for many enzyme systems, including mitochondrial adenosine triphosphatase (ATPase) and other transport ATPases.

POTASSIUM BALANCE

The extracellular fluid (ECF) K$^+$ concentration is normally 3.5–5.0 mmol/L, and deviations outside these limits are associated with the changes in neuromuscular excitability just noted. However, when these changes occur over a relatively longer time, greater deviations may be tolerated. As illustrated in Fig. 1, more than 98% of the total body store of K$^+$ (about 3700 mmol for a 70-kg human) is located inside cells, and the ECF contains only ~65 mmol. Thus, the typical daily intake and output of K$^+$, 50–150 mmol, exceeds the total amount in the extracellular compartment. The intake of K$^+$ is almost as high as that of Na$^+$ because it is present in large amounts in the plant and animal cells in the diet.

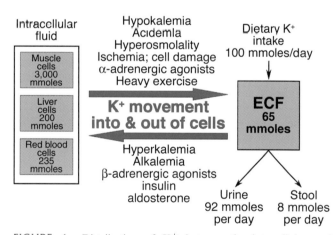

FIGURE 1   Distribution of K$^+$ between the intracellular and extracellular (ECF) fluid compartments. Only ~2% of the total body K$^+$ is present in the ECF, with most of the remainder in intracellular fluid, here represented by three of the largest cellular compartments and the approximate millimoles of K$^+$ in each. The distribution represents a balance between active uptake into cells by the Na$^+$,K$^+$-ATPase and the passive leak of K$^+$ out of cells. This balance is influenced by the K$^+$ concentration of the ECF as well as by the factors listed above and below the two arrows. A typical daily K$^+$ intake of 100 mmol is greater than the total amount of ECF K$^+$ and is matched by the sum of a small excretion in the stool and the regulated excretion of K$^+$ by the kidneys. (Values of K$^+$ content and portions of figure adapted from Giebisch G, Malnic G, Berliner RW. Control of renal potassium excretion. In Brenner MB, ed. *The kidney*, 5th ed. Philadelphia: WB Saunders, 1996.)

A small amount of this ingested K^+ is excreted in the feces, and the reabsorption of K^+ from the gastrointestinal tract is regulated in response to some of the same stimuli that affect the kidney. The colon responds to aldosterone with an increase in K^+ secretion and elimination in the feces; however, the regulation of colonic K^+ transport is not as important for K^+ balance as the renal transport, nor is it as precise.

Distribution of K^+ between Intracellular and Extracellular Fluids

The distribution of K^+ between the intracellular and extracellular fluid compartments is determined by the balance of bidirectional K^+ fluxes into and out of cells. Uptake of K^+ into cells is dependent on the activity of Na^+,K^+-ATPase, whereas efflux from cells is passive and depends on the number and type of K^+ channels in various cells. Figure 1 illustrates its importance in maintaining the critical extracellular K^+ concentration. Because the intake of K^+ in 1 day is more than the total content of the ECF compartment, the amount of K^+ in even a single meal could cause lethal hyperkalemia. To be sure the kidneys respond by increasing K^+ secretion after an increase in intake, but the response is relatively slow, requiring several hours. In the interim, the intracellular compartment forms an important buffer to prevent large changes in ECF K^+ concentration. As ECF K^+ rises, its uptake into cells increases, and when ECF K^+ falls it shifts out of cells. The mechanisms responsible for these movements are both passive and active. K^+ diffuses into or out of cells through ion channels in accordance with its electrochemical potential gradient, which is directly related to the ECF K^+ concentration. The Na^+,K^+-ATPase, which actively transports K^+ into cells, is stimulated by an elevation and inhibited by a decrease in the extracellular K^+ concentration.

Changes in acid–base status are also important determinants of K^+ distribution, as is also discussed in Chapter 31. In acidemia (low plasma pH), the resulting rise in intracellular H^+ concentration titrates some of the anionic intracellular proteins and organic phosphates. The resulting fall in fixed anionic charge in the cells obligates less K^+ to remain in the cells for electroneutrality, and it is lost to the ECF. Conversely, the shift of H^+ out of cells in *alkalemia* produces a corresponding shift of K^+ into cells. Plasma *hyperosmolality* also causes a shift of K^+ out of cells when transporters that allow KCl to diffuse out of cells are activated to mitigate cell shrinkage.

As shown in Fig. 1, insulin and β-adrenergic agonists are also important agents that shift K^+ into cells and produce a fall in ECF K^+. Independent of its effect on glucose entry into cells, insulin increases the activity of

Na^+,K^+-ATPase, which is also stimulated by β-adrenergic agonists (such as propranolol) acting primarily through β_2 receptors. On the other hand, α-adrenergic agonists (such as phenylephrine) inhibit Na^+,K^+-ATPase and cause a rise in ECF K^+. Aldosterone acts on K^+ transport in epithelia besides the kidney, including the colon and sweat and salivary glands, but there is some controversial evidence that it also causes a K^+ shift into other cells.

Exercise can also cause K^+ loss from cells and hyperkalemia. K^+ is lost from muscle cells during the action potentials producing muscle contraction, but most of this K^+ enters the T-tubules in the muscle and is reaccumulated by the Na^+,K^+-ATPase. However, with strenuous exercise a significant loss of muscle K^+ to the circulation occurs, which can lead to increased urinary excretion. In marathon runners, for example, plasma K^+ concentrations as high as 10 mmol/L have been recorded immediately after a race. The urinary loss of K^+ that results from its increased filtration at the high plasma concentration, together with the K^+ lost in the sweat, can produce a significant decrease in total body K^+. When the exercise is stopped and cells reaccumulate K^+, the hyperkalemia quickly changes to an often severe hypokalemia, which contributes to the weakness and dizziness the marathon runner experiences at the end of a race.

Mechanisms of Renal K^+ Transport

Figure 2 shows schematically those regions of the nephron that transport K^+ and the relative amounts handled by each. At a plasma concentration of 4 mmol/L and a glomerular filtration of 180 L/day, about 720 mmol/day are filtered at the glomerulus. Of this, about 7–20% is finally excreted. K^+ excretion is generally not regulated by changes in the glomerular filtration rate (GFR); however, when the GFR is significantly reduced (as in renal disease), hyperkalemia often develops due to the inability to filter K^+ at rates sufficient to match a daily intake at the higher end of the normal range.

Although the rate of K^+ excretion is typically 50–150 mmol/day as shown in Fig. 2, the rate can vary considerably in response to dietary intake. With very low K^+ intake, the kidney can respond by excreting extremely small amounts of K^+. On the other hand, with excessive K^+ intake, the amount of K^+ excreted can actually *exceed* the filtered load by 20–100%. The fact that the K^+ excretion rate can exceed its rate of filtration provides definitive evidence that K^+ is actively secreted at some point in the nephron. As discussed later, this secretion of K^+ is primarily a function of the aldosterone-responsive distal nephron (ARDN), i.e., the connecting tubule and collecting duct.

Clinical Note

Hyperkalemia and Its Treatment

Disturbances in K^+ balance are seen frequently. Changes in plasma K^+ concentration have effects on neuromuscular excitability because of the importance of the K^+ concentration gradient across cell membranes in determining the resting membrane potential. For example, an acute increase in the ECF K^+ concentration leads to depolarization of the resting potential, which, by itself, should increase neuromuscular excitability; however, with chronic depolarization of the resting potential, the voltage-gated Na^+ channels that are responsible for the rising phase of the action potential are inactivated, leading to diminished excitability. Chronic hypokalemia has the opposite effect—the hyperpolarization of the membrane potential activates the Na^+ channels. Whereas hypokalemia can be treated easily by dietary K^+ supplements or, if severe, by administering intravenous K^+, the proper treatment of hyperkalemia requires a better understanding of its origin.

Hyperkalemia results either from a shift of K^+ out of cells to the ECF or from reduced K^+ excretion in the urine. Metabolic alkalosis, uncontrolled diabetes mellitus, β-adrenergic blocking drugs, and strenuous exercise are all causes of hyperkalemia. In addition, tissue damage, such as occurs with traumatic injury or cancer chemotherapy, can produce massive K^+ loss from cells. Hyperkalemia due to inadequate excretion is found in advanced renal failure or when the blood supply to the kidney is inadequate, as in low output heart failure. It can also be due to low plasma aldosterone concentrations. Plasma aldosterone is low not only in Addison's disease, but also in the presence of drugs such as converting enzyme inhibitors or angiotensin II receptor blockers, which interfere with the stimulation of aldosterone secretion by angiotensin II. Plasma aldosterone levels can also decrease with nonsteroidal anti-inflammatory drugs (NSAIDs) because they inhibit prostaglandin production, which is one of the stimuli to renin release and angiotensin II production. Finally, potassium-sparing diuretics, by blocking Na^+ reabsorption in the collecting duct, indirectly decrease K^+ secretion and can produce hyperkalemia.

In treating hyperkalemia, it is necessary first to establish that it is not due to acidosis because, as will be discussed in Chapter 31, total body K^+ depletion often accompanies hyperkalemia in chronic acidosis. In this case, merely correcting the cause of the acidosis usually will restore K^+ balance. It should also be established that the hyperkalemia is not due to one of the drugs mentioned here. Mild hyperkalemia from other causes can be treated by giving a loop diuretic, which augments K^+ secretion in the connecting tubule and collecting duct, for the reasons discussed later and in Chapter 27. More severe hyperkalemia must be treated more rapidly. A cation exchange resin that binds K^+ in the intestinal tract can be given orally or by enema, and it can reduce plasma K^+ within hours. An even more rapid response to dangerously high plasma K^+ concentration can be achieved by administering insulin plus glucose, a β_2-adrenergic agonist, or $NaHCO_3$ to create metabolic alkalosis. All of these procedures operate within 30–60 min by shifting K^+ into cells, as illustrated in Fig. 1. In life-threatening cases of hyperkalemia, when signs of cardiac arrhythmias are already present, Ca^{2+} is administered intravenously in the form of calcium gluconate because, by mechanisms that are not yet understood, it rapidly but transiently reverses the inactivation of Na^+ channels.

The proximal tubule reabsorbs $\sim$60% of the filtered load of K^+ (about 430 mmol/day). This reabsorption is primarily passive and occurs because of the rise in the tubular fluid K^+ concentration as fluid is reabsorbed. Diuretics that act proximally (such as acetazolamide or osmotic diuretics) increase K^+ excretion by decreasing the amount of fluid and, hence, K^+ reabsorbed in the proximal tubule.

Somewhat in excess of 30% of the filtered K^+ ($\sim$230 mmol/day) is reabsorbed in the loop of Henle. Most of this reabsorption occurs in the thick ascending limb of the loop of Henle. As discussed in Chapter 27, K^+ is transported by the same luminal membrane transport system that carries Na^+ and Cl^-, i.e., the Na^+-K^+-$2Cl^-$ cotransporter. Although much of the K^+ that is transported into the thick ascending limb cell by this mechanism diffuses back into the lumen via the K^+ channel in the luminal membrane, substantial amounts move from the cell toward the plasma, resulting in net K^+ reabsorption.

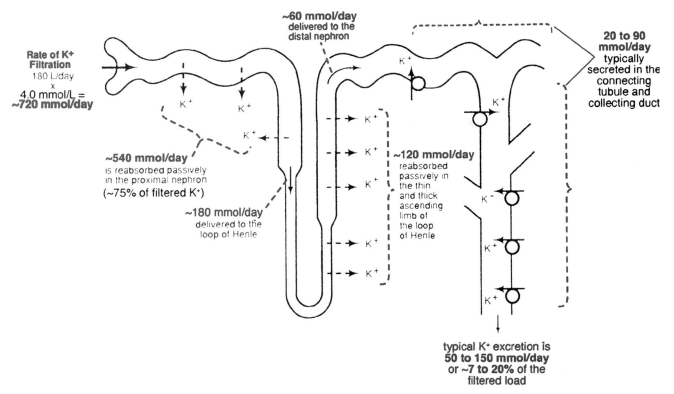

FIGURE 2 Sites of K^+ reabsorption and secretion along the nephron. K^+ is passively reabsorbed in the proximal tubule. Although there is passive potassium secretion by diffusion into the descending limb, overall there is net reabsorption in the loop of Henle due to greater reabsorption in the ascending limb, so that less than 10% of the filtered load is delivered to the distal convolution. Active secretion of K^+ by the CNT and principal cells in the connecting tubule and collecting duct adds to the amount delivered so that the daily excretion of K^+ ranges from 7–20% of the filtered load.

Reabsorption in the proximal tubule and the loop of Henle reduces the K^+ delivered to the distal convolution to less than 10% of the filtered load. Because the amount of K^+ normally excreted exceeds this amount, K^+ must be added to the distal tubular urine by secretion. Normally, the connecting tubule and collecting duct secrete 20–90 mmol/day, and this amount is adjusted appropriately to match the daily excretion to the daily intake.

Not only are there marked changes in the rate of K^+ secretion in response to changes in dietary K^+ intake, but these functional changes are also associated with morphologic changes. Increased dietary K^+ intake is associated with increased K^+ secretion by the connecting tubule and collecting duct, an increase in the basolateral membrane surface area, and the density of Na^+,K^+-ATPase. These changes are mediated by increase in aldosterone release by the adrenal cortex that accompanies hyperkalemia. In addition, aldosterone as well as vasopressin increases the activity of K^+ channels in the luminal membrane of the CNT and principal cells in the ARDN.

On the other hand, with a K^+-deficient diet K^+ secretion is drastically reduced, and the ARDN can even

exhibit net K^+ reabsorption and reduce the K^+ excretion virtually to zero. This net reabsorption of K^+ is produced primarily by the intercalated cells of the cortical collecting duct. In chronic, severe K^+ deficiency, these cells express a second type of H^+ pump, the H^+,K^+-ATPase (see Chapter 31). This pump actively secretes H^+ in exchange for an uptake of K^+ into the cells that mediate its active reabsorption.

The plasma K^+ concentration is an important regulator of K^+ secretion by the connecting tubule and collecting duct, i.e., the ARDN. As shown in Fig. 3A, a rise in plasma K^+ concentration stimulates secretion even in the presence of a constant aldosterone concentration. The mechanism for this direct effect of K^+ concentration is not well understood, but it may relate to a corresponding increase in cell K^+ that is available for secretion as the extracellular K^+ concentration rises.

As shown in Fig. 3B, the daily regulation of K^+ secretion by the ARDN is regulated by aldosterone. As discussed in Chapter 27, aldosterone acts to increase K^+ secretion by CNT and principal cells in association with increased Na^+ reabsorption. Extracellular volume depletion, in which increased plasma aldosterone

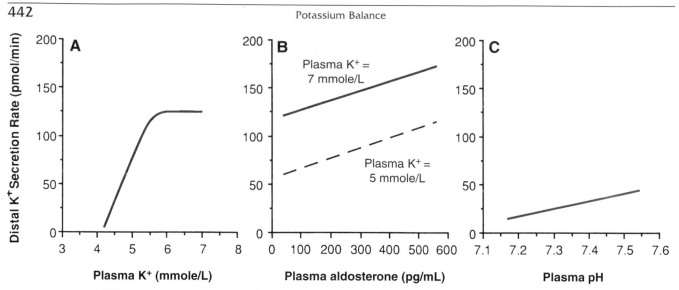

FIGURE 3   Effects of plasma K$^+$ concentration, plasma aldosterone level, and plasma pH on K$^+$ in the ARDN. The rate of K$^+$ secretion for a *single* connecting tubule and collecting duct is given. (A) The K$^+$ secretion rate increases sharply with increases in plasma K$^+$ in the range of 4–5.5 mmol/L. (B) Increasing plasma aldosterone concentration also accelerates K$^+$ secretion, and this effect is additive to the effect of a high plasma K$^+$ concentration. (C) When basal K$^+$ secretion is low because of low plasma K$^+$ and aldosterone concentrations, the rate of secretion can be decreased by acidemia and increased by alkalemia. (Modified from Wright FS, Giebisch G. Regulation of K$^+$ excretion. In Seldin DW, Giebisch G, eds. *The kidney: Physiology and pathophysiology*. New York: Raven Press, 1985, pp 1223–1249.)

augments Na$^+$ reabsorption, also leads to increased K$^+$ secretion. Therefore, dehydration is often associated with increased K$^+$ excretion that can lead to significant depletion of body K$^+$ stores.

Acute respiratory and metabolic acid–base disturbances also affect renal K$^+$ excretion. As shown in Fig. 3C, alkalosis is associated with increased K$^+$ secretion, whereas acidosis is associated with decreased K$^+$ secretion by CNT and principal cells of the ARDN. Part of this effect may be due to the change in intracellular K$^+$ concentration with changes in plasma pH. The shift of K$^+$ into cells in alkalosis could increase the concentration driving force for K$^+$ secretion, with the reverse occurring in acidosis. However, there is also a direct effect of pH on the K$^+$ channels in the luminal membrane of the principal cell—alkalosis stimulates and acidosis inhibits them. Chronic acid–base disturbances produce complex changes in K$^+$ secretion, which will be discussed in Chapter 31.

Factors Affecting K$^+$ Secretion by the Aldosterone-Responsive Distal Nephron

As discussed in Chapter 27 and in the Clinical Note on the effects of diuretics in that chapter, K$^+$ secretion is also considerably enhanced when the delivery of Na$^+$ or fluid flow through the connecting tubule and collecting duct (the ARDN) is increased. Because of the clinical importance of this effect, it is worthwhile to consider again the mechanism responsible for K$^+$ secretion in the

ARDN, which is illustrated in Fig. 4. (This figure has been adapted from Fig. 5 in Chapter 27 to highlight the K$^+$ secretory mechanism.) K$^+$ is secreted from CNT and principal cells into the tubular fluid by moving passively down its electrochemical potential gradient through K$^+$-selective channels in the luminal membrane. There is a driving force for K$^+$ to leave the cell where its concentration is higher than in the tubular fluid. Furthermore, because of the Na$^+$ permeability of the luminal membrane, the voltage across the membrane is less negative from cell to lumen (−30 mV) than across the basolateral membrane (−80 mV). This lower opposing negative voltage gives a larger net electrochemical potential driving force for K$^+$ to leave the cell across the luminal membrane. Thus, three factors influence the rate of K$^+$ secretion: the intracellular K$^+$ concentration, the luminal K$^+$ concentration, and the voltage across the luminal membrane. When intracellular K$^+$ is reduced, secretion decreases because the concentration difference across the luminal membrane decreases. With increased luminal flow rate, K$^+$ secretion increases because the K$^+$ concentration in the lumen remains lower (due to the larger volume of fluid) despite a higher K$^+$ secretion rate. Although the luminal K$^+$ concentration remains lower, the final excretion of K$^+$ is greater because the flow rate is greater. Increased Na$^+$ delivery to the distal tubule also tends to enhance K$^+$ secretion. With an increased Na$^+$ concentration in the lumen, greater depolarization of the luminal membrane is seen, resulting in a lower opposing voltage and, thus, a greater

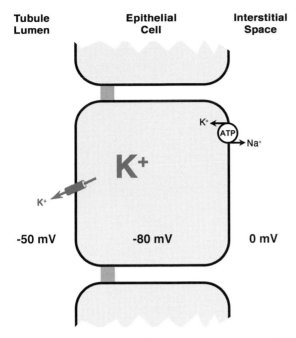

Tubule Lumen   **Epithelial Cell**   **Interstitial Space**

-50 mV   -80 mV   0 mV

FIGURE 4   Driving forces for K$^+$ secretion by CNT and principal cells of the ARDN. A high intracellular K$^+$ concentration is maintained by the Na$^+$,K$^+$-ATPase, which provides a favorable concentration gradient for K$^+$ to diffuse through K$^+$ channels in the luminal membrane. The voltage difference across the luminal membrane (−30 mV, cell compared to lumen) is also less than that across the basolateral cell membrane (−80 mV, cell compared to interstitial fluid) because it is depolarized by the Na$^+$ channels (not shown in this diagram). Thus, there is less of an electrical force retarding K$^+$ movement into the lumen. The net result is a favorable electrochemical potential gradient for K$^+$ to diffuse from the cells into the lumen resulting in secretion.

K$^+$ secretion. In summary, diuretics enhance K$^+$ secretion in the ARDN by providing a larger volume of fluid into which the K$^+$ is secreted, as well as by producing a more negative luminal voltage, both of which enhance the electrochemical potential gradient down which K$^+$ leaves the cell.

Finally, the rate of delivery of anions other than Cl$^-$ to the distal tubule is also an important determinant of K$^+$ secretion. K$^+$ secretion is enhanced when poorly reabsorbed anions are present in the tubular fluid, probably because less salt and water are reabsorbed. The loss of K$^+$ via this route can become quite acute, for example, in diabetic ketoacidosis, in which large amounts of acetoacetate and β-hydroxybutyrate are delivered to the distal tubule and collecting duct.

CALCIUM BALANCE

Calcium is the most abundant ion in the body but 99% of it is in the skeleton and does not participate in the regulation of the plasma Ca^{2+} concentration.

Typical adults ingest about 600 to 1500 mg of calcium daily; however, the gastrointestinal tract absorbs only 150–200 mg/day. This rate of absorption is regulated largely by the most active metabolite of vitamin D$_3$, *1,25-dihydroxycholecalciferol* [1,25(OH)$_2$D$_3$; see Chapter 43], in response to the demands for the maintenance of the normal ECF calcium concentration and the remodeling of bone. The plasma Ca^{2+} concentration is also closely regulated by PTH, which controls the release of Ca^{2+} from bone, as discussed in Chapter 43. Nevertheless, the kidney is essential in regulating the final output of Ca^{2+} to match the daily intake and to maintain a constant total body Ca^{2+}.

Although the normal plasma Ca^{2+} concentration is about 8–10 mg/dL (2.1–2.6 mmol/L), ~40% of this Ca^{2+} is bound to plasma proteins and cannot be filtered at the glomerulus. Thus, the rate of Ca^{2+} filtration is one-half the usual product: GFR · P$_{Ca}$, giving a daily filtration rate of ~9 g of Ca^{2+} per day for a GFR of 180 L/day. However, typically only 0.5–1.0% of this amount is excreted, indicating substantial reabsorption along the nephron.

The major regions of the nephron responsible for reabsorbing Ca^{2+} are shown schematically in Fig. 5. Reabsorption in the proximal tubule is mostly passive, although some active transport may be involved. In this region, the rate of Ca^{2+} reabsorption is proportional to that of Na$^+$. Therefore, a proximal diuretic such as acetazolamide and osmotic diuresis increases Ca^{2+} excretion. Approximately 30% of the filtered load of Ca^{2+} is reabsorbed in the loop of Henle. This reabsorption occurs primarily in the thick ascending limb of the loop of Henle and, again, is proportional to the rate of Na$^+$ reabsorption. Most if not all of this reabsorption is passive and is driven by the lumen-positive voltage.

Approximately 8% of the filtered load of Ca^{2+} is reabsorbed in the late distal convoluted tubule and the connecting tubule. Active Ca^{2+} reabsorption by these regions of the nephron is the primary target of factors that regulate the final excretion of Ca^{2+}. The most important factor that changes the Ca^{2+} reabsorptive rate is the plasma Ca^{2+} concentration. As it rises, reabsorption decreases and vice versa. The late distal convoluted tubule and the connecting tubule also express receptors for the hormones PTH, 1,25(OH)$_2$D$_3$, and calcitonin. Parathyroid hormone and 1,25(OH)$_2$D$_3$ increase Ca^{2+} reabsorption in this region of the nephron. High plasma levels of calcitonin inhibit Ca^{2+} reabsorption, but the regulatory significance of calcitonin in Ca^{2+} reabsorption is doubtful (see also Chapter 43).

Ca^{2+} reabsorption is stimulated in metabolic alkalosis and by volume contraction, whereas it is inhibited by phosphate depletion, metabolic acidosis, or ECF

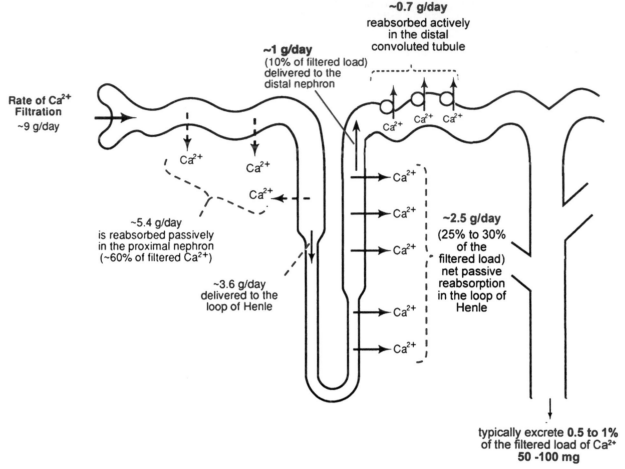

FIGURE 5 Calcium reabsorption along the nephron. Reabsorption in the proximal tubule and the loop of Henle is passive, and proportional to the reabsorption of Na$^+$ and water. Reabsorption by CNT cells in the late distal convoluted tubule and connecting tubule is active and is hormonally regulated.

volume expansion. Diuretics also increase Ca^{2+} excretion, presumably because of increased distal tubular flow and a decreased opportunity to reabsorb Ca^{2+} because of that more rapid flow.

The reabsorption of Ca^{2+} in the proximal tubule and the loop of Henle is passive and involves the paracellular diffusion of Ca^{2+} down its electrochemical potential gradient from the lumen to the interstitial fluid. A favorable concentration gradient for Ca^{2+} diffusion is produced as water reabsorption in the proximal tubule and the descending limb of the loop of Henle causes the Ca^{2+} concentration in the tubular fluid to rise. The lumen-positive transepithelial voltage in the late proximal tubule and in the thick ascending limb of the loop of Henle also contributes to a favorable electrochemical potential gradient for Ca^{2+} reabsorption. Nevertheless, the passive reabsorption of Ca^{2+} also requires that the junctional complexes be quite permeable to Ca^{2+}. Because of a high permeability, the factional reabsorption of Ca^{2+} in the proximal tubule (~60% of the

amount filtered) and the loop of Henle (~30%) approximates that for Na$^+$. In addition, because the favorable electrochemical potential gradient for passive Ca^{2+} reabsorption is dependent on Na$^+$ and water reabsorption in the proximal tubule and the loop of Henle, those factors that alter Na$^+$ reabsorption in these segments produce parallel changes in Ca^{2+} reabsorption.

Ca^{2+} is actively reabsorbed by CNT cells in the late distal convoluted tubule and connecting tubule. Specific Ca^{2+} channels mediate Ca^{2+} transport across the luminal membrane of these cells as shown in Fig. 6. Because the intracellular Ca^{2+} concentration in these cells is very low (on the order of 100 nmol/L, or 100 · 10^{-6} mmol/L), Ca^{2+} entering from the lumen must be transported out of the cell across the basolateral membrane at the same rate to maintain that low intracellular concentration. The mechanisms responsible for active Ca^{2+} transport across the basolateral membrane (Fig. 6), include an active Ca^{2+}-ATPase

Tubule Lumen   **Epithelial Cell**   **Interstitial Space**

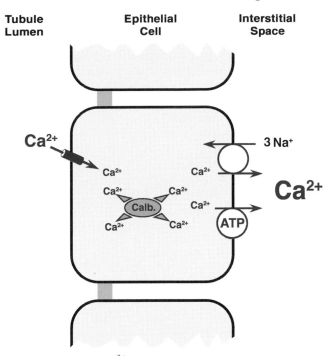

FIGURE 6   Active Ca^{2+} reabsorption by CNT cells in the late distal convoluted tubule and connecting tubule. Ca^{2+} enters CNT cells passively down its electrochemical potential gradient through Ca^{2+}-selective channels. The very low intracellular concentration of ionized Ca^{2+} is maintained by active transport out of the cell across the basolateral membrane. There are two active transport mechanisms: a Na^{+}-Ca^{2+} exchanger by which the active transport of 1 Ca^{2+} out of the cell is energetically coupled to the entry of 3 Na^{+}, and a Ca^{2+}-ATPase. Despite the high transcellular flux of Ca^{2+} in these cells, the intracellular Ca^{2+} concentration remains low due to the presence of the Ca^{2+}-binding protein calbindin-D_{28K} (labeled *Calb.* in the illustration), which also facilitates the diffusion of Ca^{2+} across the cell.

that uses energy from ATP hydrolysis to extrude Ca^{2+} from the cell and an electrogenic Na^{+}-Ca^{2+} exchange mechanism that uses the energy available from the entry of three Na^{+} ions for each Ca^{2+} ion driven out of the cell. CNT cells also express a Ca^{2+} binding protein called calbindin-D_{28K}, which also helps to maintain a low intracellular Ca^{2+} concentration and enables more rapid diffusional equilibration of Ca^{2+} within the cell. The CNT cells appear to be specifically differentiated to control the reabsorption and hence the final excretion of Ca^{2+}. They are the only cells along the nephron that express the Ca^{2+}-specific channel, and they also express both active Ca^{2+} transporters, the three related hormone receptors, and calbindin.

MAGNESIUM BALANCE

As is the case for calcium, most of the body's magnesium resides in bone. The normal intake of magnesium is on the order of 300 mg/day, of which the gastrointestinal tract reabsorbs less than half. A greater fraction of the ingested amount is reabsorbed in magnesium depletion; a lesser fraction is reabsorbed when body magnesium stores are excessive. However, the mechanisms involved in this regulation are not fully understood.

The kidney must excrete an amount equivalent to the daily intake of Mg^{2+} by absorption from the intestine, or about 120 mg/day. Normally 5–20% of the filtered load of Mg^{2+} is excreted, but the amount excreted can

---

Clinical Note

Calcium and Phosphate Balance in Renal Failure
Renal failure patients, among their other problems, also suffer from hyperparathyroidism and a resulting reabsorption of bone that leads to osteoporosis and increased risk of fractures. The reasons for the hyperparathyroidism are not completely known. According to one hypothesis, the decreased clearance of phosphate as GFR falls leads to a small increase in its plasma concentration, causing a fall in the plasma Ca^{2+} concentration and the secretion of PTH. Elevated levels of PTH help the kidneys to maintain phosphate balance by decreasing phosphate reabsorption in the proximal tubule (see Chapter 26), but they also cause the bone pathology.

An additional or alternate possibility is that phosphate retention as well as the reduction

in renal mass would diminish $1,25(OH)_2D_3$ production. (The kidney is the primary site where dietary vitamin D is converted to the highly active calcitriol form; see Chapter 43.) The fall in calcitriol decreases its normal tonic inhibition of PTH secretion, leading to increased PTH secretion. It has been found that when patients on hemodialysis are given calcitriol intravenously, plasma PTH levels are markedly reduced.

The usual mode of therapy for the phosphatemia and hyperparathyroidism of renal failure is to have the patient eat a low-phosphate diet and to administer phosphate binding agents such as calcium carbonate to reduce intestinal absorption of phosphate.

approach the filtered load with excessive magnesium intake, and it can even exceed the filtered load in the extreme. Alternatively, filtered Mg^{2+} can be reabsorbed almost completely in states of Mg^{2+} depletion. Presumably, these changes in renal handling occur because of hormonal influences, but these mechanisms are at present unknown.

The proximal tubule passively reabsorbs a far smaller fraction of Mg^{2+} (only 20–30%) than of Na^+, K^+, or Ca^{2+}. Although the Mg^{2+} concentration in the tubular fluid of the proximal tubule and the descending limb of the loop of Henle rises considerably as fluid is reabsorbed, its reabsorption is limited due to the low Mg^{2+} permeability of the junctional complexes in these segments of the nephron. However, about two-thirds of the filtered Mg^{2+} is reabsorbed passively in the thick ascending limb of the loop of Henle. The elevated luminal concentration of Mg^{2+} and the lumen-positive voltage in these segments provide a large driving force for this passive reabsorption, but it would not occur if it were not for the presence of a protein called *paracellin* in the junctional complexes of the thick ascending limb. It appears that paracellin mediates the facilitated diffusion of Mg^{2+}, and to a lesser extent Ca^{2+}, and it is the first known example of such a transporter in the junctional complexes. Paracellin was discovered serendipitously through research on a rare congenital defect that causes hypomagnesemia due to excessive urinary Mg^{2+} excretion. Genomic analysis of a kindred that exhibited this trait revealed a linked gene mutation. Antibodies that were raised to the protein coded by the normal gene revealed that it was localized exclusively to the junctional complexes between thick ascending limb cells, leading to

its name paracellin. There seems to be some competition between Mg^{2+} and Ca^{2+} for this transporter, but because of its limited reabsorption in the more proximal regions of the nephron, the absence of paracellin has a greater effect on Mg^{2+} than Ca^{2+} excretion. A small amount of the filtered Mg^{2+} is also reabsorbed in the distal convolution, the connecting tubule and collecting duct by mechanisms that have not been identified but appear to be passive.

Suggested Readings

Friedman PA. Renal calcium metabolism. In Seldin DW, Giebisch G, eds. *The kidney: Physiology and pathophysiology*, Vol 2, 3rd ed. Philadelphia: Lippincott Williams & Wilkins, 2000, pp 1749–1790.

Hoenderop JG, Nilius B, Bindels RGM. Molecular mechanism of active Ca^{2+} reabsorption in the distal nephron. *Annu Rev Physiol* 2002;64:529–549.

Malnic G, Muto S, Giebisch G. Regulation of renal potassium excretion. In Seldin DW, Giebisch G, eds. *The kidney: Physiology and pathophysiology*, Vol 2, 3rd ed. Philadelphia: Lippincott Williams & Wilkins, 2000, pp 1575–1614.

Quamme GA, DeRouffignac CP. Renal magnesium handling. In Seldin DW, Giebisch G, eds. *The kidney: Physiology and pathophysiology*, Vol 2, 3rd ed. Philadelphia: Lippincott Williams & Wilkins, 2000, pp 1711–1730.

Rosa, RM, Epstein FH. Extrarenal potassium metabolism. In Seldin DW, Giebisch G, eds. *The kidney: Physiology and pathophysiology*, Vol 2, 3rd ed. Philadelphia: Lippincott Williams & Wilkins, 2000, pp 1551–1574.

Rose BD, Post TW. *Clinical physiology of acid–base and electrolyte disorders*, 5th ed. New York: McGraw-Hill, 2001, pp 372–403.

Simon DB, Lu Y, Choate KA, Velazquez H, Al-Sabban E, Praga M, Casari G, Bettinelli A, Colussi G, Rodriguez-Soriano J, McCredle D, Milford D, Sanjad S, Lifton RP. Paracellin-1, a renal tight junction protein required for paracellular Mg^{2+} reabsorption. *Science* 1999;285:103–106.

The Role of the Kidney in Acid–Base Balance

JAMES A. SCHAFER

KEY POINTS

- The body maintains a balance of acid production, intake, and excretion so that the plasma pH varies little despite daily variations in the production of acid by the metabolism of ingested foods.
- The kidney is responsible for maintaining the plasma HCO_3^- concentration at normal levels, typically by reabsorbing all the HCO_3^- filtered at the glomerulus and generating new HCO_3^- by excreting titratable acidity and ammonium ions.
- Acute changes in H^+ intake or loss from the plasma can be partially compensated by buffering from plasma and cellular buffers including proteins, phosphate, and other weak acids and bases.
- The HCO_3^-/CO_2 buffer system is particularly important because of the rapidity with which

changes in ventilation can change plasma pH, but this buffer system is limited by the availability of HCO_3^-.
- In both the proximal and distal segments of the nephron, H^+ is secreted actively into the lumen, while HCO_3^- returns to the plasma. These transport processes result in the reabsorption of 90–95% of the filtered HCO_3^- in the proximal tubule and reduction of the urine pH in the distal segments.
- On a typical Western diet, HCO_3^- reabsorption continues in the distal nephron segments; however, these segments are also able to produce "new" HCO_3^- by excreting titratable acidity and ammonium, thus reclaiming HCO_3^- that has been used to buffer fixed acids in the diet.

Essential Medical Physiology, Third Edition

KEY POINTS (*continued*)

- Filtered phosphate that has not been reabsorbed in the proximal tubule and loop of Henle is normally the primary buffer anion in the urine; that is, it is responsible for most of the titratable acidity.

- Ammonium is an important route by which H^+ is excreted from the body. By excreting nitrogen in the form of ammonium rather than urea, the bicarbonate that would have been consumed to produce urea is spared.

The pH of extracellular fluids is normally maintained very near to 7.40 with minor deviations due to diet and fluctuations in respiratory minute volume. Under pathologic conditions, plasma pH may be as low as 7.0 or as high as 7.7, which represents a wide range of hydrogen ion concentration from 100 nmol/L ($100 \cdot 10^{-9}$ mol/L) to 20 nmol/L, respectively. In the Western hemisphere, the normal diet contains an excess of fixed acids because of the metabolic conversion of sulfur-containing amino acids to sulfuric acid and of nucleic acids and phospholipids to phosphoric acid. This is equivalent to a daily proton (H^+) production of about 1 mmol/kg body weight per day or, for convenience, on the order of 100 mmol/day in a normal adult.

As with all the solutes taken in and produced in the body, homeostasis demands a rate of excretion equal to this daily rate of intake and production of H^+. The kidney is responsible for excreting this acid load. It does so by conserving all of the bicarbonate filtered at the glomerulus and secreting hydrogen ions into the urine while generating "new" bicarbonate in the renal tubules. However, the kidney can also compensate for a net base input or acid loss from the body by failing to reabsorb all of the filtered bicarbonate. The loss of bicarbonate in the urine is equivalent to the loss of base from the body and causes a fall in the pH of the extracellular fluid.

BUFFERS IN THE MAINTENANCE OF ACID–BASE BALANCE

As substrates are metabolized to end products such as phosphoric and sulfuric acid, phosphate anions and proteins both in the plasma and in cells buffer the protons from these strong acids. Proteins contain multiple negative charges because of dissociated carboxyl groups on the acidic amino acids, and they can be represented as polyvalent anions, P^{n-}. (The valence n− indicates that there are n negative charges.) On the average, plasma proteins have a valence of −11 to −12. In the following reactions, which are examples of the buffering of H^+ that normally occurs in the plasma and within cells, a protein with a valence of −12 is used

as an example:

$$H^+ + HPO_4{}^{2-} \Leftrightarrow H_2PO_4{}^-$$

$$H^+ + P^{-12} \Leftrightarrow HP^{-11}$$

This action of extra- and intracellular buffers prevents rapid and wide fluctuations in the pH of body fluids. However, the most important buffer is the HCO_3^-/CO_2 system because the end product, CO_2, is rapidly regulated by the rate of respiration. The equilibrium reaction between HCO_3^-, H^+, carbonic acid (H_2CO_3), and CO_2 was discussed in Chapters 20 and 26 and is repeated here because it forms the basis for understanding the role of this buffer system in renal acid–base regulation.

$$H^+ + HCO_3{}^- \Leftrightarrow H_2CO_3 \xrightleftharpoons[\text{carbonic anhydrase}]{} CO_2 + H_2O$$

$$[1]$$

In this reaction scheme, the double-headed arrows indicate the equilibrium relation between reactants and products. The equilibrium between H_2CO_3 and $CO_2 + H_2O$ is catalyzed by carbonic anhydrase. CO_2 levels in the plasma can be regulated rapidly by respiration, but the finite body stores of HCO_3^- limit the buffering capacity. When the protons from the fixed acids ingested and formed by metabolism are buffered by HCO_3^-, the HCO_3^- concentration falls as it reacts with the protons to form H_2CO_3, as shown by the change in the equilibrium represented by a directional arrow ($\rightarrow$ or $\rightarrow$) and the concentration changes by up or down arrows ($\uparrow$ or $\rightarrow$):

$$H^+ + HCO_3{}^- \rightarrow H_2CO_3 \rightarrow CO_2 + H_2O \qquad [2]$$

As the hydrogen ions are buffered, carbonic acid and CO_2 concentrations rise, but this is countered by an increase in respiration to maintain a normal CO_2 tension (PCO2) in the plasma. As the reaction proceeds constantly to the right, the amount of HCO_3^- in the extracellular fluid is reduced and would eventually be exhausted. To maintain homeostasis, the HCO_3^- consumed in buffering reactions must be restored. This is accomplished by the kidneys, which excrete H^+ in the urine while returning HCO_3^- to the plasma. Therefore, the kidneys essentially reverse the reaction in Eq. [3] and return the plasma HCO_3^- concentration to normal.

Not all individuals are faced with a net dietary acid load. On a vegetarian diet, there is an excess intake of the salts of weak acids. The anions such as oxalate, citrate, butyrate, lactate, and others represent a base load (proton acceptors) that is buffered by all body buffers, including the HCO_3^-/CO_2 buffer system. The HCO_3^-/CO_2 buffer system is also effective in preventing alkaline shifts in plasma pH because of ingestion of strong bases or the loss of hydrogen ions. For example, an excess intake of hydroxyl ions, as occurs in poisoning with lye, is buffered by CO_2:

$$OH^- + CO_2 \underset{\substack{\text{carbonic}\\\text{anhydrase}}}{\overset{OH^-}{\longleftrightarrow}} HCO_3^- \qquad [3]$$

This reaction is also catalyzed by carbonic anhydrase so that it is in equilibrium, but the excess HCO_3^- formed by titrating the alkali must be eliminated from the body by the kidneys.

The relation between the concentrations of the base (A^-) and the associated acid forms (HA) of an acid is given by the *Henderson-Hasselbalch equation:*

$$pH = pK_A + \log\frac{[A^-]}{[HA]} \qquad [4]$$

For the HCO_3^-/CO_2 buffer system, the Henderson-Hasselbalch equation can be written as:

$$pH = 6.1 + \log\frac{[HCO_3^-]}{[CO_2]} \qquad [5]$$

In this equation, the concentration of CO_2 has been substituted for the weak acid H_2CO_3 because it can be determined from the arterial P_{CO_2} reported by the clinical laboratory (see later discussion). In the pH range of 7.0–7.7, more than 99% of the total CO_2 in the plasma is in the form of dissolved CO_2 rather than H_2CO_3. The pK_a value defines the pH at which the ratio $[A^-]/[HA]$ (or $[HCO_3^-]/[CO_2]$) equals 1.0. At this pH, the buffering capacity is maximal. However, the physiologic pH of the extracellular fluid (7.4) is far from 6.1 (the pK_a value of the HCO_3^-/CO_2 buffer system) and, in fact, represents a $[HCO_3^-]/[CO_2]$ ratio of 20:1. Although the normal extracellular pH lies far from the pK_a of this buffer system, its unique effectiveness in reducing fluctuations

in pH derives from the fact that $[CO_2]$ is rapidly regulated by pulmonary ventilation.

Practically speaking, the concentrations of CO_2 or H_2CO_3 are not reported directly from the laboratory but must be computed from the measured partial pressure of CO_2 in arterial blood (P_{CO_2}). Under physiologic conditions, the solubility of CO_2 in plasma is a linear function of the arterial P_{CO_2} such that:

$$[CO_2](\text{in mmol/L})$$
$$= \alpha(\text{mmol/L per mm Hg}) \cdot P_{CO_2}(\text{mm Hg}) \qquad [6]$$

where α is the solubility coefficient for CO_2 (0.03 mmol/L per mm Hg at body temperature). Thus, as discussed in Chapter 20, the working form of the Henderson-Hasselbalch equation becomes:

$$pH = 6.1 + \log\frac{[HCO_3^-]}{0.03 \cdot P_{CO_2}} \qquad [7]$$

A simpler formulation of the relationship among the variables pH (or H^+ concentration), P_{CO_2}, and $[HCO_3^-]$ is the *Henderson equation,* which does not involve the logarithm function. The Henderson equation expresses the hydrogen ion concentration in nanomoles per liter (i.e., 10^{-9} mol/L) rather than as the pH. This requires one to think either in terms of the absolute H^+ concentration ($[H^+]$) or to convert $[H^+]$ to pH, remembering that $pH = -\log[H^+]$. The Henderson equation is:

$$[H^+] = \frac{24 \cdot P_{CO_2}}{[HCO_3^-]} \qquad [8]$$

where the constant 24 includes the pK_a of the HCO_3^-/CO_2 system and the solubility coefficient of CO_2 (α) that were separate terms in the Henderson-Hasselbalch formulation. Equation [8] shows that for a normal P_{CO_2} of 40 mm Hg and a plasma $[HCO_3^-]$ of 24 mmol/L the H^+ concentration is 40 nmol/L, which is equivalent to a pH of 7.4. The equilibrium concentrations for normal acid–base conditions in the human body are as follows:

$$\underset{\substack{40\,\text{mmol/L}\\(pH=7.4)}}{H^+} + \underset{24\,\text{mmol/L}}{HCO_3^-} \Leftrightarrow H_2CO_3 \Leftrightarrow \underset{\substack{1.2\,\text{mmol/L}\\(P_{CO_2}=40\,\text{mm Hg})}}{CO_2} + H_2O$$

$$[9]$$

Any one of the above three concentrations (or pH) can be derived from the other two using either the Henderson-Hasselbalch equation (Eq. [7]) or the equivalent Henderson equation (Eq. [8]).

Table 1 gives the equivalent hydrogen ion concentration in nM for various pH values in the physiologic

TABLE 1   Relation between pH and H$^+$ Concentration

pH	[H+] (nM)	[H+] (nM) approximated as
7.00	100	$\cong 79 \times 1.25$
7.10	79	$\cong 100 \times 0.8$, or $\cong 63 \times 1.25$
7.20	63	$\cong 79 \times 0.8$, or $\cong 50 \times 1.25$
7.30	50	$\cong 63 \times 0.8$, or $\cong 40 \times 1.25$
7.40	**40**	**(normal reference values)**
7.50	32	$\cong 40 \times 0.8$, or $\cong 25 \times 1.25$
7.60	25	$\cong 32 \times 0.8$, or $\cong 20 \times 1.25$
7.70	20	$\cong 25 \times 0.8$

range. One can easily convert one to the other from one remembered value in this table. For an increase in pH of 0.1 unit, the H$^+$ concentration falls to 0.8 times the previous H$^+$ concentration. For a decrease in pH of 0.1 unit, the H$^+$ concentration is 1.25 times the previous value. For example, remembering that a pH of 7.4 is 40 nmol/L H$^+$, one can compute the H$^+$ concentration corresponding to a pH of 7.3 as (1.25 · 40) 50 nmol/L, or for pH 7.5 as (0.8 · 40) 32 nmol/L H$^+$. Use of the Henderson equation and this scheme has the distinct advantage of eliminating the use of logarithms in the solution of clinical problems, and it enables one to focus more directly on the important parameter of interest: the hydrogen ion concentration.

Both the Henderson-Hasselbalch and Henderson equations show that the HCO$_3^-$/CO$_2$ buffer system can be manipulated to change the plasma pH by altering the ratio of the plasma HCO$_3^-$ concentration to Pco$_2$. Changes in the plasma HCO$_3^-$ concentration are produced by alterations in the renal retention or excretion of HCO$_3^-$, and changes in Pco$_2$ are brought about by changes in the rate of respiration. The physiologic regulation of plasma pH by these mechanisms is considered in the next sections.

PULMONARY REGULATION OF Pco$_2$

As shown earlier, when a strong acid is buffered in the plasma, the concentration of bicarbonate falls, whereas buffering of a base has the opposite effect. In both situations, changes in the pulmonary excretion of CO$_2$ will serve to return the pH toward normal. As noted earlier, the pulmonary regulation of CO$_2$ is the major biologic advantage of the HCO$_3^-$/CO$_2$ buffer system.

In metabolic acidosis (discussed later), the HCO$_3^-$ concentration falls because it buffers the excess hydrogen ions. In response to the acidosis, hyperventilation lowers Pco$_2$ and maintains plasma pH nearer to normal. This is referred to as *respiratory compensation* for the metabolic acidosis. Clinically, the increased ventilation

in the severely acidotic patient is discernible to the physician as a deep, unhastened breathing referred to as *Kussmaul breathing*. In this respiration pattern, the acidotic patient increases his or her tidal volume to dilute the concentration of CO$_2$ in alveolar air and, thus, in the arterial plasma. This slow, deep breathing, which is characteristic of acidosis, can be easily distinguished from the shallow, rapid breathing that occurs in pulmonary congestion because of pulmonary edema, infarction, or infection.

In contrast to the respiratory response to metabolic acidosis, in metabolic alkalosis hypoventilation occurs. This increases the concentration of alveolar CO$_2$ and, hence, its partial pressure in body fluids. The increase in the concentration of H$_2$CO$_3$ in the body fluids results in increased formation of both H$^+$ and HCO$_3^-$, which counteract the alkalosis. However, the hypoventilatory effect in alkalosis is not as effective as that of hyperventilation in acidosis. In metabolic alkalosis, respiratory compensation requires a *decreased* minute volume of ventilation, which also decreases arterial Po$_2$, and which would be opposed by normal regulatory feedback mechanisms that counteract hypoxia.

RENAL REGULATION OF PLASMA BICARBONATE CONCENTRATION

Figure 1 illustrates the primary mechanisms of H$^+$ and HCO$_3^-$ transport that were described in detail in Chapters 26 and 27. The basic mechanism of moving acid and base from blood to urine or vice versa is the same throughout the nephron. Under normal acid–base conditions, H$^+$ is secreted into the lumen and HCO$_3^-$ moves from the epithelial cells to the blood in both the proximal and distal nephron segments. These regions of the nephron differ only in the mechanisms used to transport H$^+$ and HCO$_3^-$ and the effects of H$^+$ secretion into the lumen. In the proximal tubule, H$^+$ secretion occurs against a small concentration gradient and serves primarily to drive HCO$_3^-$ reabsorption. H$^+$ secreted into the lumen of the proximal tubule is titrated by HCO$_3^-$ to form CO$_2$, and the HCO$_3^-$ generated in the cell by H$^+$ secretion leaves the cell across the basolateral membrane and enters the plasma. In the distal tubule segments that have *intercalated cells,* H$^+$ secretion produced by the H$^+$-ATPase (adenosine triphosphatase) drives urine pH lower, resulting in the excretion of titratable acidity and ammonium and the generation of new bicarbonate ions that are returned to the blood (described in more detail later). Intercalated cells are found from the distal convoluted tubule to the medullary collecting duct, especially in the late distal convoluted tubule, the connecting tubule, and the cortical and outer medullary

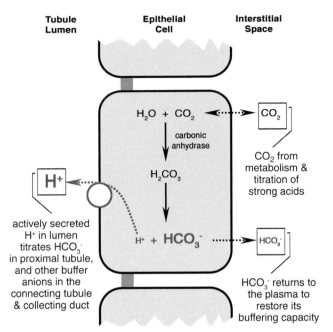

Tubule Lumen	Epithelial Cell	Interstitial Space

FIGURE 1   General mechanism for H^+ secretion and return of bicarbonate to the plasma in renal tubular cells. As H^+ is actively secreted into the lumen, the H^+ concentration within the cells falls and mass balance favors the increased production of HCO_3^- in the cells. The rising cell HCO_3^- concentration produces an electrochemical potential gradient that drives the movement of HCO_3^- out of the cell, which occurs only across the basolateral membrane. This process is present in proximal tubule cells and in intercalated cells of the distal nephron. Only the details of the luminal and basolateral membrane transporters differ.

regions of the collecting duct. For convenience, in the remainder of this chapter, these segments will be referred to collectively as the *distal nephron.*

Bicarbonate Reabsorption

Bicarbonate reabsorption occurs in the proximal tubule, and to a more limited extent in the distal nephron segments, because of the active secretion of H^+, which titrates luminal HCO_3^- to CO_2 and water. The proximal tubular mechanism is a high-capacity system that accounts for the absorption of 90–95% of the filtered load of HCO_3^- so normally little HCO_3^- flows on to the distal nephron segments. In the proximal tubule, this process is catalyzed by luminal *carbonic anhydrase,* so the introduction of an inhibitor of carbonic anhydrase depresses the reabsorption of HCO_3^- by the proximal tubule and results in the excretion of HCO_3^- together with Na^+. This is the basis of the mild diuretic and natriuretic action of carbonic anhydrase inhibitors such as acetazolamide (Diamox).

Despite its high HCO_3^- reabsorptive capacity, the proximal tubule is limited by its ability to generate a

sufficiently acid luminal fluid to titrate all the filtered HCO_3^-. The distal tubular mechanism for reabsorbing HCO_3^- is similar to the proximal mechanism except that it is capable of reclaiming the remaining 5–10% of the filtered HCO_3^- from the tubular fluid. To accomplish this, however, the distal mechanism must be able to secrete H^+ against a steep concentration gradient that in the collecting ducts approaches 1000:1 (blood pH 7.4, or $40 \cdot 10^{-9}$ M versus a urine pH 4.5, or $32 \cdot 10^{-6}$ M). For example, if the luminal CO_2 partial pressure is the same as in the plasma (40 mm Hg), the Henderson-Hasselbalch equation shows that the HCO_3^- concentration is less than 1.0 mmol/L at a urine pH of 6.0. At a urine pH of 5, the HCO_3^- concentration is ~0.1 mmol/L. The collecting duct is able to develop the greater H^+ gradient because active H^+ secretion is coupled to ATP hydrolysis by the H^+-ATPase located in the luminal membrane (see Fig. 8 in Chapter 27).

Under normal conditions, with a plasma HCO_3^- concentration of 24 mmol/L and a glomerular filtration rate (GFR) of 130 mL/min, 3.1 mmol of HCO_3^- are filtered and 3.1 mmol are reabsorbed every minute, which is equivalent to about 4.5 mol of HCO_3^- filtered and returned to the plasma by reabsorption every day. Normally, the urine contains virtually no HCO_3^-, however, as shown in Fig. 2, when the plasma HCO_3^- concentration increases above 25 mmol/L, the excess filtered HCO_3^- appears in the urine. Increases in the plasma HCO_3^- may occur as a consequence of the ingestion of HCO_3^- or other bases, the metabolism of salts of organic acids, or the loss of acid as occurs with vomiting. The plasma level at which HCO_3^- appears in the urine is referred to as the *plasma threshold* for

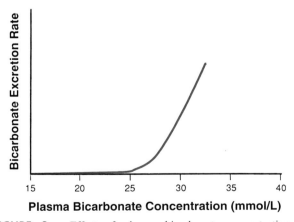

FIGURE 2   Effect of plasma bicarbonate concentration on bicarbonate excretion rate. Virtually no bicarbonate is excreted at plasma bicarbonate concentrations below 25 mmol/L, which is often referred to as the plasma threshold for bicarbonate. Above this plasma concentration, bicarbonate excretion increases with increasing plasma bicarbonate concentration.

HCO_3^-. Thus, the basic mechanism for excreting excess HCO_3^- is by an "overflow" of filtered HCO_3^- into the urine whenever the plasma concentration rises above normal levels, unless there is another signal that alters HCO_3^- reabsorption.

Regulation of HCO_3^- Reabsorption

Table 2 lists the factors that alter H^+ secretion and, thus, HCO_3^- reabsorption by the proximal tubule. First, as would be expected, one of the most important factors is the systemic acid–base status, as reflected by the plasma pH. Acidemia (plasma pH < 7.4) is a direct stimulus to H^+ secretion, whereas alkalemia (plasma pH > 7.4) is an inhibitor. It is found that chronic acidosis causes an increase in the activity of the Na^+-H^+ exchanger in the proximal tubule and the H^+-ATPase in the intercalated cells, whereas alkalosis produces the opposite effect. Second, increases in plasma Pco_2 stimulate HCO_3^- reabsorption, whereas a fall in Pco_2 depresses it. This effect of Pco_2 on HCO_3^- reabsorption is evidenced as a shift in the curve relating HCO_3^- excretion to the plasma HCO_3^- concentration, as shown in Fig. 3. With an elevation in Pco_2, the threshold for HCO_3^- excretion is shifted to higher plasma HCO_3^- concentrations, and the opposite is observed with a fall in Pco_2. This mechanism is part of the renal adaptation to respiratory acidosis and alkalosis. For example, as described later, when hyperventilation decreases the plasma Pco_2 and, thus, produces alkalosis, HCO_3^- reabsorption is decreased in response to the fall in Pco_2 and the resulting urinary loss of HCO_3^- counteracts the alkalosis.

The third factor in Table 2 is the plasma K^+ concentration. HCO_3^- reabsorption is increased in hypokalemia and decreased in hyperkalemia. It is likely that the intracellular K^+ concentration rather than the plasma K^+ concentration is the important variable, but the intracellular concentration tends to rise or fall in proportion to the extracellular or plasma concentration. Depletion of intracellular K^+ augments the renal tubular capacity for reabsorbing HCO_3^- and results in an increase in the plasma HCO_3^- concentration above normal levels with no urinary loss.

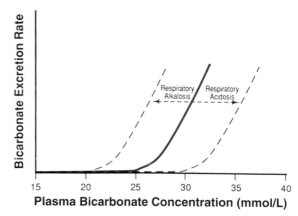

FIGURE 3   Effect of respiratory alkalosis and acidosis on the plasma bicarbonate threshold.

Finally, HCO_3^- reabsorption is augmented by decreases in the plasma Cl^- concentration and extracellular fluid (ECF) volume. When the body is depleted of Cl^- and a hypovolemic status develops, the plasma HCO_3^- concentration increases and the tubular reabsorptive capacity for HCO_3^- increases, which shifts the HCO_3^- excretion curve (see Fig. 2) toward the right. Conversely, expansion of the body ECF depresses HCO_3^- reabsorption. The reason for this reciprocal relationship is not entirely clear, but relates in part to the need to conserve Na^+ after volume depletion. In the presence of ECF volume depletion, Na^+ is avidly conserved by the proximal tubule, as discussed in Chapter 29. If Cl^- is not available for reabsorption with Na^+, protons are secreted in exchange for reabsorbed Na^+, resulting in the reabsorption of HCO_3^-. However, the loss of Cl^- appears to be an important factor because merely restoring the extracellular fluid volume without replacing Cl^- does not correct the alkalosis resulting from HCO_3^- retention.

Bicarbonate Generation by the Distal Nephron

In addition to reabsorbing filtered HCO_3^- in the proximal tubule, the kidney must also regenerate HCO_3^- to replenish the HCO_3^- stores depleted during the buffering of strong acids produced by metabolism. The renal excretion of H^+ allows new HCO_3^- generation, as shown in Fig. 1. However, merely excreting acid urine does not account for the necessary daily excretion of H^+. Even at the minimum urine pH of 4.5, the excretion of 1–2 L of unbuffered urine would eliminate less than 1 mmol of acid per day, compared with the daily intake and production of about 100 mmol. Thus, the H^+ must be excreted in a form other than as a dissociated

TABLE 2   Factors That Alter H^+ Secretion and HCO_3^- Reabsorption by the Proximal Tubule

Increase	Decrease
Acidosis	Alkalosis
↑ Pco_2 (as in respiratory acidosis)	↓ Pco_2 (as in respiratory alkalosis)
Hypokalemia	Hyperkalemia
↓ ECF volume with Cl^- depletion	↑ ECF volume

acid, and this is accomplished by the formation of *titratable acidity*, i.e., H^+ combined with urine buffers such as HPO_4^{2-} and SO_4^{2-}.

All buffer anions that are filtered at the glomerulus can augment the excreted titratable acidity, and these anions buffer secreted H^+ ions even in the lumen of the proximal tubule. However, the formation of titratable acidity in the proximal tubule is limited by the luminal pH that can be achieved. Maximal titration of urine buffers to their associated acid forms requires the production of more acid urine, which is accomplished primarily by those regions of the nephron that have interacted cells. These regions include the late distal convoluted tubule, the connecting tubule, and the cortical and outer medullary collecting duct that are referred to here collectively as the distal nephron. The effect of urine pH on the rate of excretion of titratable acidity can be best illustrated by phosphate, which is the primary urinary buffer. The pK_a of the phosphate buffer system is 6.8, so from the Henderson-Hasselbalch equation, one can compute that at a normal plasma pH of 7.4 the ratio of HPO_4^{2-} to $H_2PO_4^-$ is 4:1. When the pH falls, as it does in the urine, HPO_4^{2-} is titrated to $H_2PO_4^-$ and the ratio falls. At the lowest urinary pH of 4.5, virtually all the HPO_4^{2-} has been converted to $H_2PO_4^-$:

$$pH = 6.8 + \log \frac{[HPO_4^{2-}]}{[H_2PO_4^-]}$$

$$\text{at pH 7.4,} \quad \frac{[HPO_4^{2-}]}{[H_2PO_4^-]} = 4.0$$

$$\text{at pH 4.5,} \quad \frac{[HPO_4^{2-}]}{[H_2PO_4^-]} = 0.0013 \qquad [10]$$

Other weak acids present in the urine are also titrated to their undissociated acid form by urinary acidification in the distal nephron segments. Although phosphate is usually the major urinary buffer, other weak acids such as creatinine become increasingly important as the urinary pH falls. In diabetic ketoacidosis, β-hydroxybutyrate and acetoacetate can be important urinary buffers at low urine pH.

TABLE 3 Factors That Alter H^+ Secretion and Formation of Titratable Acidity by the Distal Tubule and Collecting Duct

Increase	Decrease
Acidosis	Alkalosis
↑ P_{CO_2} (as in respiratory acidosis)	↓ P_{CO_2} (as in respiratory alkalosis)
Hypokalemia	Hyperkalemia
Mineralocorticoid excess (e.g., Cushing's syndrome)	Mineralocorticoid deficit (e.g., Addison's disease)

Table 3 presents the factors that alter H^+ secretion in the distal tubule and collecting duct and, thus, alter the rate of excretion of titratable acidity by changing the urine pH. As in the proximal tubule, H^+ secretion by intercalated cells in the distal portions of the nephron is stimulated by (1) acidosis, (2) elevation of plasma P_{CO_2}, and (3) hypokalemia, whereas the opposite changes decrease H^+ secretion. In contrast to the proximal tubule, ECF volume does not affect distal H^+ secretion; however, another factor, variation in plasma mineralocorticoid hormones, operates exclusively in the aldosterone-responsive distal nephron (ARDN) and, thus, has a greater effect on the formation of a maximally acid urine than on HCO_3^- reabsorption. Increased endogenous or exogenous levels of mineralocorticoids and/or the 17-OH glucocorticoids will increase the renal tubular capacity for H^+ secretion. Thus, alkalosis is a common accompaniment of Cushing's syndrome (excess production of adrenal corticosteroids; see Chapter 40).

The intercalated cells in the distal nephron can respond to alkalosis not only by increasing the urine pH and, thus, preventing the titration of urinary buffers (i.e., reducing or eliminating titratable acidity), but also by actually secreting HCO_3^- and thus increasing the excretion of this buffer base. As discussed in Chapter 27 and illustrated in Fig. 4, the normal orientation of the H^+-ATPase in the luminal membrane and the HCO_3^-/Cl^- exchange mechanism in the basolateral membrane can be reversed so that protons move toward the plasma and HCO_3^- is secreted into the lumen. However, the proximal tubule, merely by failing to reabsorb all of the filtered HCO_3^-, is a much more important contributor to HCO_3^- excretion in alkalosis.

In response to acidosis, the amounts of buffer anions in the urine and the minimum urinary pH set a limit on the quantity of H^+ that can be excreted as titratable acidity and, thus, on the ability to restore plasma HCO_3^-. But the kidney has a highly efficient means of overcoming this limitation. It does so by generating increasing amounts of ammonium ion (NH_4^+) that are excreted in the urine. NH_4^+ excretion by the kidney helps to raise body HCO_3^- stores by decreasing the consumption of HCO_3^- by hepatic ureagenesis.

As illustrated schematically in Fig. 5, the metabolism of amino acids in the liver produces NH_4^+ and HCO_3^- plus small quantities of SO_4^{2-} from the sulfur-containing amino acids. The NH_4^+ and HCO_3^- are converted primarily into urea, which is effectively excreted by the kidney, and glutamine. Glutamine in the circulation can be metabolized in the kidney to NH_4^+ and α-ketoglutarate. Further metabolism of the α-ketoglutarate results in the production of HCO_3^-,

A.
H⁺ Secretion

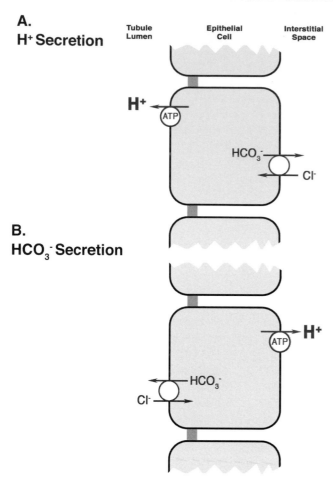

B.
HCO₃⁻ Secretion

FIGURE 4 H^+ and HCO_3^- transport in the collecting duct. (A) Normally, the collecting duct actively secretes H^+ and returns HCO_3^- to the plasma. (B) In alkalosis, the orientation of the transporters can be reversed so that HCO_3^- is secreted into the lumen.

which leaves renal cells only across their basolateral membranes. Therefore, the excretion of urea by the kidney causes the loss of both NH_4^+ and HCO_3^-, whereas the metabolism of glutamine results in the excretion of NH_4^+ and the return of HCO_3^- to the plasma. In other words, when the kidney metabolizes glutamine it essentially restores body HCO_3^- that would otherwise have been lost by the excretion of urea. The effect is to generate additional HCO_3^- that can compensate for the HCO_3^- consumed in the titration of phosphoric and sulfuric acids produced by metabolism, as shown on the right-hand side of Fig. 5.

The kidney is one of the few organs capable of gluconeogenesis, which involves the deamination and deamidation of amino acids and the production of NH_4^+ especially in proximal tubule cells. In chronic acidosis, the kidney increases the synthesis of the enzymes for gluconeogencsis, particularly glutaminase, and thus increases its capability for *ammoniagenesis,* that is, the production of NH_4^+. As shown in Table 4, the rate of

ammonium excretion can increase by 10-fold in diabetic ketoacidosis. Under normal acid–base conditions, the kidney excretes roughly equal amounts of titratable acidity and NH_4^+. In chronic acidosis, although both increase, NH_4^+ becomes the dominant contributor to H^+ excretion and, thus, it is the primary contributor to the generation of new HCO_3^- or what might be regarded as sparing HCO_3^- from consumption by urea-genesis. In alkalosis, the excretion of titratable acidity diminishes primarily because H^+ secretion is decreased and, thus, the urine pH rises. Glutamine metabolism also decreases, causing less NH_4^+ to be excreted, and thus, less HCO_3^- to be spared from consumption in ureagenesis.

CLINICAL DISTURBANCES OF ACID–BASE BALANCE

In understanding the physiologic mechanisms that normally regulate acid–base balance with dietary intake and metabolic production of acids and bases, it is helpful to examine the common acid–base disturbances. These conditions show the various ways in which acid–base imbalances can develop and how physiologic control mechanisms operate to restore the plasma H^+ concentration as close to normal as possible. In considering these examples, it should also be realized that normal acid–base balance does not merely mean the presence of a normal plasma pH. As can be seen from the Henderson equation (Eq. [8]), an H^+ concentration of 40 nmol/L could occur with an unlimited number of Pco_2 and HCO_3^- values, as long as the ratio of Pco_2 to HCO_3^- remains 40:24 = 1.67. For example, a patient might have metabolic alkalosis because of severe vomiting, which increases his or her plasma HCO_3^- concentration to 36 mmol/L, and a simultaneous respiratory acidosis with a Pco_2 of 60 mm Hg caused by inadequate respiratory ventilation (e.g., because of the simultaneous presence of pneumonia). The calculated H^+ concentration of 40 nmol/L (pH 7.4) suggests that the patient has no acid–base disturbance, whereas in fact he or she has coexisting metabolic alkalosis and respiratory acidosis. For this reason, it is important to make the distinction between an elevated plasma hydrogen ion concentration (which is referred to as *acidemia*) and acidosis, which may or may not involve acidemia. Similarly, *alkalemia* should be distinguished from alkalosis.

Acid–base changes can be discussed most easily in terms of their effect on the mass balance between H^+, HCO_3^-, and CO_2 in the equilibrium reactions for the HCO_3^-/CO_2 buffer system (Eq. [1]). In other words, when the concentration of one of the reactants or

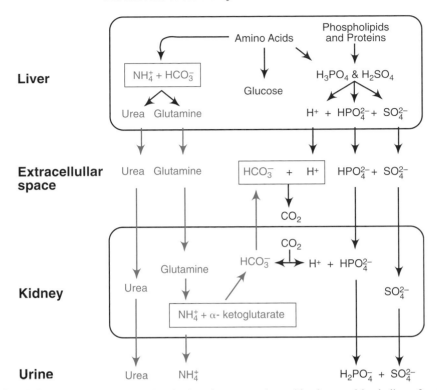

FIGURE 5   Excretion of ammonium in the urine spares plasma bicarbonate. Metabolism of amino acids in the liver results in the production of NH_4^+ and HCO_3^-. These products can be used to form either glutamine or urea, which is excreted by the kidney. When glutamine is utilized by gluconeogenic pathways in the kidney, NH_4^+ and α-ketoglutarate are produced. The NH_4^+ is excreted in the urine, whereas HCO_3^- is returned to the plasma to buffer hydrogen ions produced by metabolism of amino acids and lipids. Thus, the utilization of glutamine by the kidneys allows excretion of NH_4^+ from amino acid metabolism and spares plasma HCO_3^- that would be otherwise consumed in ureagenesis. The normal metabolism of phospholipids and proteins (right side of figure) leading to the production of strong fixed acids is indicated. The anions of these strong acids are excreted in the urine and contribute to the titratable acidity.

TABLE 4   Excretion of Titratable Acidity and Ammonium

	Total acid excreted (mmol/day)	Ratio of NH_4^+ to titratable acidity
Normal diet:		
ammonium	30–50	
titratable acidity	10–30	1–2.5
Diabetic ketoacidosis:		
ammonium	300–500	
titratable acidity	75–250	2–5
Severe chronic renal failure:		
ammonium	0.5–15	
titratable acidity	2–20	0.2–1.5

products is altered, what happens to the concentrations of the other reactants or products as a new equilibrium is achieved? In the equations that follows, the larger, bold arrow indicates the change that initiates a shift in acid–base balance. The resulting direction of change in the equilibrium reaction due to the perturbation is indicated by the arrow symbol ($\rightarrow$ or $\leftarrow$).

Respiratory Acidosis

Rapid changes in plasma pH can occur because of alterations in pulmonary ventilation. For example, CO_2 retention due to chronic lung disease, or drugs or diseases that affect the nervous system and depress respiration, will cause an increase in the arterial P_{CO_2} and a decrease in the ratio of HCO_3^- to CO_2 concentrations, with concomitant acidosis. In pulmonary arrest, acidosis can be as threatening to life as hypoxia. In respiratory acidosis, the primary event is the increase in plasma P_{CO_2} resulting in a shift in the equilibrium reactions by mass balance:

$$CO_2 + H_2O \rightarrow H_2CO_3 \rightarrow H^+ + HCO_3^- \qquad [11]$$

From the equilibrium reactions, we can see that the hallmark of primary respiratory acidosis is a rise in both H^+ and HCO_3^- concentrations. In acute cases, we find that the HCO_3^- concentration rises about 1 mmol/L per each 10 mm Hg rise in the P_{CO_2}. The renal response to hypercapnia (the high P_{CO_2}) is to increase the rate of H^+ secretion in both the proximal and distal tubules. This results both in an increase in proximal HCO_3^- reabsorption, despite the elevation in the plasma HCO_3^- concentration (the bicarbonate excretion curve moves toward the right in Fig. 3), and in increased HCO_3^- generation in the distal tubule and collecting duct. Both mechanisms tend to counteract the acidosis by further increasing the plasma HCO_3^- concentration; however, this renal adaptation to hypercapnia begins slowly and requires at least 24–48 hr. Thus, the initial defense against hypercapnia falls on the plasma and tissue buffers.

As the kidney responds to the acidosis, it creates an increase in the plasma HCO_3^- concentration, which exceeds that due to the hypercapnia alone. This chronic renal compensation will raise the HCO_3^- concentration by about 3.5 mmol/L for each mm Hg that the P_{CO_2} has risen above the normal level of 40 mm Hg. The increased plasma HCO_3^- buffers the increase in H^+, but this correction is never complete, so the acidosis persists even with the higher HCO_3^- concentration. Thus, both acute respiratory acidosis as well as chronic respiratory acidosis with metabolic compensation produce an increase in the concentration of both H^+ and HCO_3^- in the plasma. The latter state is sometimes referred to as a "compensated" respiratory acidosis.

Respiratory Alkalosis

Respiratory alkalosis is the converse of respiratory acidosis. It develops secondary to hyperventilation, which occurs in several settings in which ventilation exceeds the metabolic demands of the body. The light-headedness one feels after blowing up a few balloons is a manifestation of respiratory alkalosis. Perhaps the most common setting is in the patient on a respirator when the minute volume has been set too high. Respiratory alkalosis also occurs after hyperventilation in response to low ambient O_2, as at high altitude. Mountain climbers are in a chronic state of respiratory alkalosis that is compensated only partially by the renal excretion of HCO_3^-. Inappropriate hyperventilation may occur because of abnormal stimuli to the central nervous system respiratory centers by tumors, infections, or poisoning with drugs such as aspirin. The hyperventilation lowers P_{CO_2} and increases the ratio of plasma HCO_3^- to P_{CO_2} as can be seen from the mass balance relationship:

$$CO_2 + H_2O \leftarrow H_2CO_3 \leftarrow H^+ + HCO_3^- \qquad [12]$$
$$\downarrow \qquad\qquad \downarrow \qquad\quad \downarrow \qquad \downarrow$$

Again, as in respiratory acidosis, the H^+ and HCO_3^- concentrations move in the same direction, but in respiratory alkalosis, they fall rather than rise. The appropriate renal compensation for the acute hypocapnia is to reduce the secretion of H^+, thereby excreting HCO_3^- and reducing its plasma concentration. The primary mechanism involved is a decrease in proximal bicarbonate reabsorption that is reflected in a shift to the left of the HCO_3^- excretion curve in Fig. 3. This renal response is much more rapid than the increase in plasma HCO_3^- concentration seen in respiratory acidosis because the kidney can rapidly "dump" bicarbonate into the urine merely by failing to reabsorb the filtered load completely. Because of the rapid loss of HCO_3^- by the kidneys, even with the acute development of respiratory alkalosis the HCO_3^- concentration falls by about 2.5 mmol/L per each 10 mm Hg fall in the P_{CO_2} below the normal 40 mm Hg. With chronic respiratory alkalosis, the plasma HCO_3^- concentration falls by about 5 mmol/L per each 10 mm Hg fall in the P_{CO_2}. Thus, the hallmark of respiratory alkalosis even with renal compensation is a fall in *both* H^+ and HCO_3^-.

Metabolic Acidosis

The primary event responsible for metabolic acidosis may fall into one of four broad categories: (1) increased acid intake, (2) increased metabolic production of acid, (3) decreased acid excretion by the kidneys, and (4) increased loss of alkali from the body. In the first three categories, the *primary* event is an increase in the total H^+ concentration in the plasma, resulting in the following shift in the buffer equilibrium:

$$H^+ + HCO_3^- \rightarrow H_2CO_3 \qquad [13]$$
$$\downarrow \qquad\quad \downarrow$$

In other words, the excess H^+ is buffered by plasma HCO_3^- which, by mass action, results in a fall in the plasma HCO_3^- concentration. The excess carbonic acid generated is in equilibrium with CO_2, but the excess CO_2 can be rapidly disposed of by pulmonary ventilation.

In the case of excessive alkali loss from the body, which may occur from the gastrointestinal tract or the kidney, the primary event is a fall in the plasma HCO_3^- concentration with the following consequences to the equilibrium:

$$H^+ + HCO_3^- \leftarrow H_2CO_3 \qquad [14]$$
$$\uparrow \qquad\qquad \downarrow$$

The HCO_3^- losses are partially replaced by dissociation of H_2CO_3 resulting in the acidemia. But, in the cases of both a primary H^+ excess as well as a HCO_3^- loss, note that the hallmark of metabolic acidosis is acidemia with a fall in HCO_3^-, exactly the opposite pattern from that observed in respiratory acidosis. To maintain electroneutrality, this *fall* in HCO_3^- concentration must be compensated by a corresponding rise in the plasma concentration of Cl^- or of the anionic weak base form of the ingested or metabolically produced acid, for example, lactate. Examination of the relative changes in Cl^- and HCO_3^- concentrations provides a useful tool in establishing the cause of a metabolic acidosis (see next section on the anion gap).

The pulmonary system responds to a fall in plasma pH by an increase in minute ventilation, and the resulting drop in P_{CO_2} normally compensates partially for the metabolic acidosis. Because of the rapidity with which an increase in minute ventilation can change the plasma P_{CO_2}, respiratory compensation is usually found to coexist with the metabolic acidosis, a state also referred to as *compensated metabolic acidosis*. There is a decrease of approximately 1.25 mm Hg in the P_{CO_2} for each 1 mmol/L fall in the HCO_3^- concentration (below the normal 24 mmol/L) produced by the metabolic acidosis. However, note what happens to the mass balance relations as the P_{CO_2} falls with hyperventilation. As would be desired, the H^+ concentration falls, although not completely back to normal, but the HCO_3^- concentration also falls. Therefore, with the respiratory compensation the plasma is less acid, but the HCO_3^- concentration falls even lower. Thus, in metabolic acidosis, whether compensated or not, acidemia coexists with a low plasma HCO_3^- concentration.

The eventual correction of a metabolic acidosis depends on the loss of the excess acid or a net gain of HCO_3^-. The kidney can return the lost HCO_3^- by conserving all the filtered HCO_3^- (as it normally does), and by increasing the rate at which it generates new HCO_3^- by the excretion of titratable acidity and ammonium. These processes are usually adequate to reverse completely the tendency toward metabolic acidosis from normal metabolism, but in disease states the kidney may be incapable of keeping up with the demand for HCO_3^- production and chronic acidosis will persist until the original cause is corrected.

Anion Gap

The anion gap is a means of approximating the total concentrations of anions other than Cl^- and HCO_3^- in the plasma. It is calculated as the difference between the plasma Na^+ concentration and the sum of the Cl^- and HCO_3^- concentrations: $([Na^+] + [K^+]) - ([Cl^-] + [HCO_3^-])$. This difference is normally about 12 mmol/L (normal range 8–16 mmol/L). Because there must be equal concentrations of anions and cations in the plasma for electroneutrality, the anion gap indicates that about 12 mmol/L of anions are not measured in

TABLE 5 Causes of Metabolic Acidosis

With an increase in unmeasured anions (anion gap > ~16)	*Without* an increase in unmeasured anions (anion gap < ~16)
Diabetic ketoacidosis	Diarrhea
Renal failure	Renal tubular acidosis
Starvation	Hypoaldosteronism
Salicylate poisoning	Hyperparathyroidism
Alcoholic ketoacidosis	Carbonic anhydrase inhibitors
Ethylene glycol poisoning	Ammonium chloride ingestion
Methyl alcohol poisoning	Drainage of pancreatic juice
Paraldehyde poisoning	Ureterosigmoidostomy
Lactic acidosis	

routine blood chemistries. These anions normally include sulfate, phosphate, lactate, urate, oxalate, pyruvate, and small concentrations of several other organic acids.

In analyzing the cause of a metabolic acidosis, calculation of the anion gap helps to focus on the possible etiology, as summarized in Table 5. The anion gap in patients with normal renal function and no acid–base disturbance will be 12 mmol/L or less. An anion gap greater than 12, signals the presence of increased concentrations of acid anions usually associated with the conditions in the left-hand column. In the presence of renal failure, a large anion gap indicates accumulation of anions such as HPO_4^- and SO_4^- because of the decrease in glomerular filtration. Metabolic acidosis not associated with an increase in immeasurable anions in a patient with normal renal function will immediately focus attention on the eight possibilities listed in the right-hand column of Table 5.

Metabolic Alkalosis

Metabolic alkalosis results when there is an excessive loss of H^+ from the body or excessive HCO_3^- retention. It may also occur, but more rarely, when there is an excessive production of HCO_3^-. When H^+ is lost from the body, as in vomiting, the HCO_3^- concentration rises:

$$H^+ + HCO_3^- \leftarrow H_2CO_3 \qquad [15]$$

On the other hand, if HCO_3^- intake or production is increased, mass action causes the H^+ concentration to fall:

$$H^+ + HCO_3^- \rightarrow H_2CO_3 \qquad [16]$$

Note that in either case metabolic alkalosis is characterized by a rise in plasma HCO_3^-. Respiratory compensation occurs by hypoventilation to increase P_{CO_2}, which partially corrects the alkalemia but results in a further

Clinical Note

Examples of Metabolic Acidosis

Because dangerous amounts of acid would rarely be ingested in normal foodstuffs, metabolic acidosis because of increased acid intake is usually observed only in poisoning. Typical causes of such poisoning also involve the simultaneous ingestion of an acid anion so that the anion gap increases. For example, among common acidic poisons, the metabolism of methyl alcohol results in formic acid, ethylene glycol in oxalic acid, paraldehyde in acetic acid, and aspirin in acetylsalicylic acid. However, ingestion of either ammonium chloride or hydrochloric acid will result in acidosis with a normal anion gap because the accompanying anion is Cl^-.

Increased metabolic production of acid is commonly observed in diabetic ketoacidosis, in which acetoacetic and β-hydroxybutyric acid are produced. In severe illnesses with generalized poor tissue perfusion, toxic reactions, septicemia, or massive catabolism, high rates of lactic acid production may occur. In both diabetic ketoacidosis and lactic acidosis, the anion gap can be markedly elevated.

Chronic renal failure can also be associated with a buildup of the anions of weak acids that are not being excreted at high enough rates by the kidneys. Decreased glomerular filtration leads to a decreased excretion of titratable acidity, and the general decrease in renal mass decreases the ability of the kidney to form ammonium. Both of these effects compromise the ability of the kidney to generate HCO_3^-.

Loss of bicarbonate leading to metabolic acidosis can occur from either the gastrointestinal tract or the urine. The most common cause is the loss of HCO_3^- and other ions in diarrhea, as occurs, for example, in inflammatory and infectious bowel diseases. Loss of HCO_3^- by this route may also be iatrogenic because of the postoperative drainage of pancreatic juice or because of the routing of urine excretion through loops of bowel after surgical procedures in the lower urinary tract (ureterosigmoidostomy).

A less common cause of metabolic acidosis produced by excessive HCO_3^- losses or decreased excretion of acid is *renal tubular acidosis* (RTA). RTA is due to a defect in the renal excretion of H^+ or in the reabsorption of HCO_3^- or both without a proportional impairment of GFR. In general, there are two forms of RTA: a proximal form characterized by impairment of proximal tubular HCO_3^- reabsorption, resulting in bicarbonaturia (often the consequence of drugs that inhibit carbonic anhydrase), and a distal form characterized by an inability of the distal H^+ transport mechanism to establish a normal gradient, resulting in the excretion of a relatively alkaline urine. Consequently, inadequate amounts of titratable acidity and NH_4^+ are excreted and positive acid balance develops. RTA is suspected when acidosis exists in the absence of an abnormal anion gap and there is no evidence of extrarenal alkali loss.

---

rise in HCO_3^-. However, respiratory compensation is limited by the fact the hypoventilation compromises the arterial P_{O_2}. Thus, the compensatory change in P_{CO_2} is much less than with respiratory compensation for metabolic acidosis. With metabolic alkalosis, P_{CO_2} rises by approximately 0.6 mm Hg for each 1 mmol/L rise in the HCO_3^- concentration above the normal 24 mmol/L.

COMMON ACID–BASE DISTURBANCES: A SUMMARY

Figures 6 through 9 present a systematic approach to the analysis of common acid–base disturbances and their compensatory counterparts based on the information given in the preceding sections. Given a set of arterial blood gas determinations, the first step (Fig. 6) is to determine whether acidemia (pH < 7.37) or alkalemia (pH > 7.43) exists. However, even if the pH does not fall outside the normal range of 7.37–7.43, there may still be a mixture of acid–base disorders that fortuitously give a plasma pH in the normal range. For example, a metabolic alkalosis due to vomiting can be complicated by coexisting hypercapnia due to emphysema. For this reason, the other blood gas parameters (HCO_3^- and P_{CO_2}) should also be examined to determine if they fall in the normal ranges.

In step 2 of the analysis (Fig. 7), the P_{CO_2} and HCO_3^- values are examined, remembering that in respiratory

---

Clinical Note

Examples of Metabolic Alkalosis

It is difficult to induce alkalosis in a normal individual by administering large quantities of alkali because of the ease of increasing the renal excretion of HCO_3^-. Ordinarily, the factor responsible for the generation of the elevated plasma HCO_3^- is easily identified. With *vomiting* or gastric aspiration, the loss of hydrochloric acid together with NaCl, K^+, and volume results in a rise in the HCO_3^- concentration, as in the first equilibrium reaction above. *Loop diuretics* and *thiazide diuretics* increase H^+ excretion. Augmented renal excretion of H^+ also accounts for the generation of HCO_3^- in conditions in which there is an excess of *adrenal steroids*, which directly stimulate H^+ secretion in the distal tubule and collecting duct.

When alkalosis occurs because of gastric volume losses, excessive HCO_3^- reabsorption often continues even when renal function is normal. This excessive reabsorption can be corrected only when the ECF volume and Cl^- deficit is replaced, and thus this condition is referred to as *contraction alkalosis*. The explanation for the increased reabsorption of HCO_3^- after such volume losses may involve many factors including Cl^- and K^+ depletion. Loss of Na^+ and the associated contraction of extracellular volume restrict HCO_3^- excretion by increasing the avidity with which the proximal tubule reabsorbs Na^+. The increased Na^+-H^+ antiport in the proximal tubule also increases HCO_3^- reabsorption, but Cl^- depletion also plays an important role. For reasons that are not clear, volume replacement must include Cl^- for the alkalosis to be corrected.

Step 1

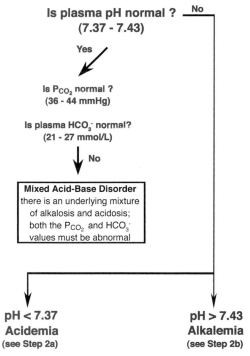

FIGURE 6   Step 1 in a systematic analysis of acid–base status. When examining the values obtained from an arterial blood gas analysis, the first step is to determine if the plasma pH is normal. A plasma pH below 7.37 is called *acidemia* and it suggests the presence of an acidosis as the primary disorder. A plasma pH above 7.43 is called *alkalemia* and it suggests the presence of an alkalosis as the primary disorder. However, the plasma pH may also be in the normal range if there are offsetting primary acid–base disorders.

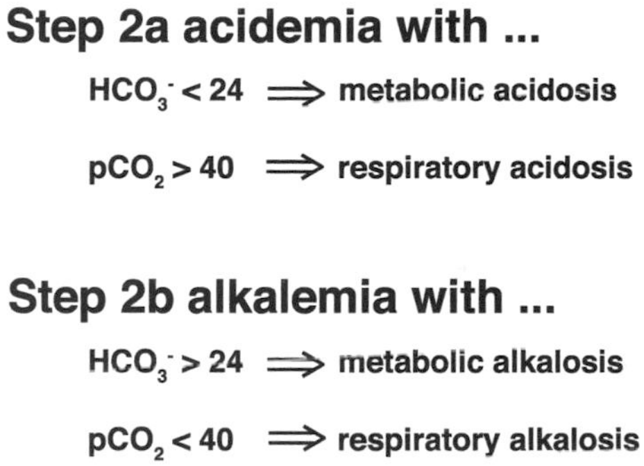

Step 2a acidemia with ...

$HCO_3^- < 24 \implies$ **metabolic acidosis**

$pCO_2 > 40 \implies$ **respiratory acidosis**

Step 2b alkalemia with ...

$HCO_3^- > 24 \implies$ **metabolic alkalosis**

$pCO_2 < 40 \implies$ **respiratory alkalosis**

FIGURE 7   Step 2 in a systematic analysis of acid–base status. Based on whether acidemia or alkalemia is found in step 1, the analysis proceeds to determine the primary acid–base disorder based on the reported values for the concentration of HCO_3^- and P_{CO_2} in the arterial plasma. (Note that in each substep only one of the two conditions can be true if the values for HCO_3^- and P_{CO_2} are consistent with the plasma pH according to the Henderson-Hasselbalch equation.) (Adapted with alterations from the scheme of analysis presented in Andreoli TE, Carpenter CCG, Griggs RC, Loscalzo J, eds. *Cecil essentials of medicine*. Philadelphia: WB Saunders, 2001, p. 249.)

disorders the HCO_3^- concentration varies from normal in the same direction as the H^+ concentration, whereas in metabolic disorders they change in the opposite direction. Thus, for example, an acidemia (high H^+ concentration) is accompanied by an elevated HCO_3^- if the primary disorder is a respiratory acidosis, but a decreased HCO_3^- if it is metabolic acidosis. Alternatively, one can focus on the P_{CO_2}, expecting it to be

Step 3. Is compensation appropriate?

<u>Metabolic acidosis</u> - P_{CO_2} decreases by ~1.25 mm Hg for each 1 mmol/L fall in HCO_3^- below 24

<u>Metabolic alkalosis</u> - P_{CO_2} increases by ~0.6 mm Hg for each 1 mmol/L rise in HCO_3^- above 24

<u>Respiratory Acidosis</u>

<u>Acute:</u> HCO_3^- increases by ~1 mmol/L for each 10 mm Hg rise in P_{CO_2} above 40 mm Hg

<u>Chronic:</u> HCO_3^- increases by ~3.5 mmol/L for each 10 mm Hg rise in P_{CO_2} above 40 mm Hg

<u>Respiratory Alkalosis</u>

<u>Acute:</u> HCO_3^- decreases by ~2 mmol/L for each 10 mm Hg decrease in P_{CO_2} below 40 mm Hg

<u>Chronic:</u> HCO_3^- decreases by ~5 mmol/L for each 10 mm Hg decrease in P_{CO_2} below 40 mm Hg

FIGURE 8   Step 3 in a systematic analysis of acid–base status. In metabolic acid–base disorders, the respiratory center changes the minute rate of ventilation to raise the plasma P_{CO_2} in alkalosis and lower it in acidosis. The change in P_{CO_2} for a normal regulatory response is indicated. In respiratory acid–base disorders, the kidney compensates by changing the rate of HCO_3^- excretion and thus its plasma concentration. This renal compensation occurs over a period of hours and days in contrast to the extremely rapid changes in P_{CO_2} that can occur with a change in ventilation. Thus the expected change in plasma HCO_3^- is greater with chronic than with acute respiratory acidosis or alkalosis. (Adapted with alterations from the scheme of analysis presented in Andreoli TE. Carpenter CCG, Griggs RC, Loscalzo J, eds. *Cecil essentials of medicine*. Philadelphia: WB Saunders, 2001, p. 249.)

elevated above the normal 40 mm Hg if, for example, the acidemia is due to respiratory acidosis.

With any primary acid–base disturbance, the alteration in bicarbonate or P_{CO_2} is expected to be accompanied by an offsetting change in the other parameter due to normal physiologic compensation. Step 3 of the analysis (Fig. 8) is designed to determine if the change in the other parameter is of the magnitude expected for normal compensation. Because renal compensation occurs more slowly, one should also take note from the history of the patient if the symptoms of a primary respiratory disorder have occurred acutely or chronically. Changes outside of the predicted range for the second parameter are an indication that the primary

Step 4. Examine the anion gap

Anion gap is the difference in plasma ion concentrations given by:

$Na^+ - (Cl^- + HCO_3^-)$, which is normally ≤ 12 mmol/L

Only a metabolic acidosis produced by accumulation of an acid with an unmeasured anion causes an anion gap >20 mmol/L

FIGURE 9   Step 4 in a systematic analysis of acid–base status. The final step in the assessment of an acid–base disorder is to calculate the anion gap. A value greater than 20 mmol/L it is a definitive indication of the metabolic acidosis produced by some acid with an accompanying anion other than Cl^- or HCO_3^-, for example, the production of formic acid from ingested ethanol causes a metabolic acidosis. Although an anion gap is expected only in metabolic acidosis, it should always be calculated because of the possibility of multiple coexisting acid–base disorders.

acid–base disorder is accompanied by a secondary one. In other words, that there is a mixed acid–base disorder. Note that the equations used for this analysis have been developed empirically by analysis of blood gas data from many patients. Nevertheless, these guidelines should not be regarded as absolute, and care should be taken to be sure that the diagnosis of the blood gas disorder is consistent with the history and physical observations for the patient.

The terminology used to describe acid base disturbances can sometimes be confusing. It is best to reserve the terms *acidosis* and *alkalosis* for primary pathologic disturbances and not for physiologically appropriate compensatory mechanisms. Consider, for example, a patient with diabetic ketoacidosis who has a bicarbonate level of 14 mmol/L and the expected drop in P_{CO_2} of 1.25 mm Hg for each 1 mmol/L of bicarbonate below 24 mmol/L (P_{CO_2} = 28 mm Hg). This patient should be described as having a metabolic acidosis with respiratory compensation. From this perspective, it is *not* appropriate to refer to the respiratory response as a "compensatory respiratory alkalosis" because a normal physiologic response rather than a pathologic process produces it. On the other hand, patients often have a second pathologic process that produces what can be identified as an inappropriate compensation. Using the preceding example, if the patient had emphysema in addition to his or her diabetic ketoacidosis, the bicarbonate concentration of 14 mmol/L might be accompanied by a P_{CO_2} of 38 mm Hg because the ventilatory defect prevents the expected decrease in P_{CO_2} in response to the metabolic acidosis. Thus, the patient would be described as having both a metabolic and a respiratory acidosis. Despite the fact that the P_{CO_2} of 38 is not significantly different from normal, it is much higher than would be expected in the presence of the metabolic acidosis.

The most difficult discrimination is often the etiology of a metabolic acidosis, because it can result from excessive acid intake or production or from excessive HCO_3^- loss. In these cases, an abnormal anion gap can point to the presence of excessive intake or production of an acid with accompanying anion as described in step 4 (Fig. 9). Although metabolic alkalosis is a less common disorder, its correction often is more dependent on restoring normal ECF volume and Cl^- than on the acid loss or base excess itself.

Suggested Readings

Alpern RJ. Renal acidification mechanisms. In Brenner MB, ed. *Brenner & Rectors the kidney,* 6th ed. Philadelphia: WB Saunders, 2000, pp 455–519.

Andreoli TE, Cecil RL, Carpenter CCJ, Griggs RC, Loscalzo J. *Cecil essentials of medicine,* 5th ed. Philadelphia: WB Saunders, 2001, pp 245–252.

Halperin ML, Kamel KS, Ethier JM, Stinebaugh BJ, Jungas RL. Biochemistry and physiology of ammonium excretion. In Seldin DW, Giebisch G, eds. *The kidney: Physiology and pathophysiology,* Vol 2, 2nd ed. New York: Raven Press, 1992, pp 2645–2680.

Hamm L, Alpern RJ. Cellular mechanisms of renal tubular acidification. In Seldin DW, Giebisch G, eds. *The kidney: Physiology and pathophysiology,* Vol 2, 2nd ed. New York: Raven Press, 1992, pp 2581–2626.

Rose BD, Post TW. *Clinical physiology of acid–base and electrolyte disorders,* 5th ed. New York: McGraw-Hill, 2001, pp 647–695.

PART V

GASTROINTESTINAL PHYSIOLOGY

C H A P T E R

32

Regulation

LEONARD R. JOHNSON

KEY POINTS

- The functions of the gastrointestinal tract are regulated by *chemical mediators* including (1) hormones, which are delivered in the blood; (2) paracrines, which diffuse through the interstitial fluid; and (3) neurocrines, which are released from neurons.
- Neural regulation of the gastrointestinal (GI) tract is both extrinsic and intrinsic.
- Parasympathetic and sympathetic nerves relay information to and from the GI tract.
- The *enteric nervous system* within the gut receives and relays information to and from the extrinsic nerves and conducts signals along the gut.
- Two families of related *peptides* regulate many of the functions of the GI tract: (1) gastrin and

cholecystokinin and (2) secretin, vasoactive intestinal peptide (VIP), gastric inhibitory peptide (GIF), and glucagon.
- *Gastrointestinal hormones* are found in endocrine cells scattered over large areas of mucosa. They are released by chemicals found in food, neural activity, or physical distension and may stimulate or inhibit several processes.
- There are five established *gastrointestinal tract hormones*: gastrin, cholecystokinin, secretin, GIF, and motilin.
- Somatostatin and histamine are two important *paracrine agents*.
- Three peptides function as *neurocrines*: VIP, bombesin (GRP), and enkephalins.

In the vertebrate body plan, the gastrointestinal (GI) tract is a tube highly specialized for processing ingesta and absorbing nutrients into the body. As such, the inside of the tube or lumen is frequently considered to be outside the body itself. Often called the digestive system, the GI system has four major activities, of which digestion is only one. These are motility, secretion, digestion, and absorption.

Motility refers to the contractions of the muscles of the GI tract that result in the orderly movement of ingested material from the mouth to the anus. The contractions of the GI tract serve not only to propel material but also to mix and reduce the size of its particles. The movement of material through the tract is regulated to optimize the time needed for both digestion and absorption.

The second major activity of the GI tract is *secretion*. Secretions consist primarily of water, electrolytes, enzymes, and mucus and are provided both by the mucosal lining of the tract and by accessory glands. The salivary glands, pancreas, and liver produce highly specialized and essential secretions that enter the lumen of the GI tract through ducts. The secretory activities of these glands and those present within the mucosa itself are finely regulated.

Digestion refers to the chemical breakdown of food into molecules able to cross the mucosa and gain entry to the blood. Most digestion is accomplished by enzymes either secreted into the lumen by the various glands or inserted into the surface of the small intestinal mucosa as integral parts of the plasma membranes of the enterocytes. Gastric acid accounts for some digestion as well. Digestion per se is not regulated, but the availability of the digestive enzymes depends in part on the regulation of secretion.

The final function of the GI system is the *absorption* of nutrients, vitamins, minerals, water, and electrolytes.

Although there is little evidence that absorptive processes are regulated over the short term, motility and secretion are closely regulated to optimize digestion and, in turn, absorption.

The regulation and integration of motility and secretion are mediated, in all known instances, by the action of chemicals on receptors of target cells within the digestive tract. These chemicals are delivered either by nerves and are termed *neurocrines,* by the bloodstream and termed *hormones,* or by diffusion through the interstitial fluid and termed *paracrines*. In the last two instances, the chemicals are peptides synthesized by endocrine cells within the epithelial layer of the mucosa.

This chapter presents an overview of the neural and chemical regulation of GI function. In subsequent chapters in this section, regulation of the motility and secretion of the specific organs is covered in detail. The material in this chapter is a framework for that discussion.

GENERAL STRUCTURE

The two surfaces of the GI tract are referred to as the *mucosa* and the *serosa*. Although the outer surface or serosa is a thin layer of connective tissue covered with a single layer of squamous epithelial cells, the inner or mucosal surface is quite complex. The mucosa actually consists of three components (Fig. 1): a single layer of epithelial cells called the *epithelium,* the *lamina propria,* and the *muscularis mucosae*. Although the structure of the GI tract varies considerably from region to region, most of that variation is within the mucosa. The epithelial cells themselves are the primary source of variation and are specifically modified to carry out the functions of

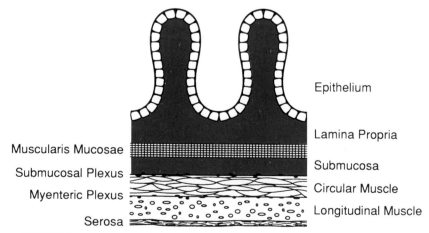

FIGURE 1   The general histology of the wall of the GI tract, seen in transverse cross section.

each region of the tract. Epithelial cells may secrete fluid and electrolytes, enzymes, or mucus into the lumen or chemical messengers into the intracellular spaces. Other epithelial cells are modified specifically for the absorption of nutrients or water and electrolytes. Blood vessels and lymph nodes are found within the lamina propria, which consists primarily of connective tissue. Contractions of the muscularis mucosae change the shape of the mucosa, producing ridges and valleys that alter the absorptive or secretory area. The muscularis is two thin layers of smooth muscle oriented so that the inner layer is circular and the outer layer runs longitudinally.

The submucosa consists primarily of collagen and elastin loosely woven into a connective tissue that contains some glands and the larger blood vessels of the wall of the tract. Also within the submucosa, at its juncture with the circular muscle layer, is a network of nerves called the *submucosal plexus* or Meissner's plexus.

The smooth muscle responsible for motility of the GI tract is found in an inner, circular layer and an outer, longitudinal layer. The thickness of these layers varies over the different regions of the tract. In general, contractions of the circular layer narrow the diameter of the lumen, whereas contractions of the longitudinal layer shorten a particular segment of the tract. Between the circular and longitudinal muscle layers is the second nerve network, called the *myenteric plexus*. The myenteric plexus and the submucosal plexus are the main components of the enteric nervous system and contain numerous interconnecting cell bodies and nerve processes.

INNERVATION OF THE GASTROINTESTINAL TRACT

Extrinsic innervation of the GI tract is provided by both the parasympathetic and sympathetic systems. Together with the enteric or intrinsic nervous system, they make up the autonomic nervous system. The innervation of the GI tract is referred to as *autonomic* because we are unaware of its activities and have no conscious control over the functions it regulates.

Parasympathetic Innervation

Down to the level of the transverse colon, parasympathetic innervation to the GI tract is supplied by the vagus nerve. The pelvic nerve innervates the descending colon, sigmoid colon, rectum, and anal canal (Fig. 2). The striated muscle of the upper third of the esophagus and the external anal sphincter receive cholinergic innervation from the vagus and pelvic nerves, respectively. Preganglionic fibers arise from cell bodies within

the medulla of the brain (vagus) and the sacral region of the spinal cord (pelvic). These predominantly cholinergic fibers synapse with ganglion cells located in the enteric nervous system (Fig. 3). Thus, vagal activity can affect secretion, motility, or the release of hormones, as indicated in Fig. 3. The mediator at the target cells is generally acetylcholine, but this is not always the case. In many cases, the chemical mediator has not been identified.

It is also important to realize that the vagus is a mixed nerve in which approximately 75% of the fibers are afferent. Receptors in the mucosa and smooth muscle relay information back to higher centers via the vagus and pelvic nerves (Fig. 3). This afferent information may trigger a reflex whose efferent limb is also present in the vagus nerve. Reflexes of this type are called *vagovagal reflexes*.

Sympathetic Innervation

Unlike the parasympathetic system, preganglionic fibers from the sympathetic system synapse outside of the GI tract in prevertebral ganglia (Fig. 2). Preganglionic, cholinergic efferent fibers from the cord synapse in four major ganglia: the celiac, superior mesenteric, inferior mesenteric, and hypogastric. Postganglionic adrenergic fibers from these ganglia innervate the cells of the myenteric and submucosal plexuses. Elements from the enteric system then innervate smooth muscle, secretory, and endocrine cells. Some blood vessels, certain smooth muscle cells, and the muscularis mucosae receive direct postganglionic sympathetic innervation (Fig. 3). Approximately 50% of the fibers present in sympathetic nerves to the gut are afferent. Thus, information is also relayed from the gut to the spinal cord.

Intrinsic Innervation

Intrinsic innervation of the gut is provided by the enteric nervous system consisting primarily of the networks formed by the submucosal and myenteric plexuses. The elements of the enteric nervous system not only relay information to and from the gut via the parasympathetic and sympathetic systems, but also relay information along the gut. In other words, local or intramural (contained within the wall) reflexes can produce a response in one part of the tract after a stimulus to another part even in the absence of all extrinsic innervation (Fig. 3).

Many different chemicals act as neurocrines within the enteric nervous system. Acetylcholine is released by most of the extrinsic preganglionic fibers and acts on neurons within the prevertebral ganglia and enteric nervous system. Most of the endings of the postganglionic

A Parasympathetic   B Sympathetic

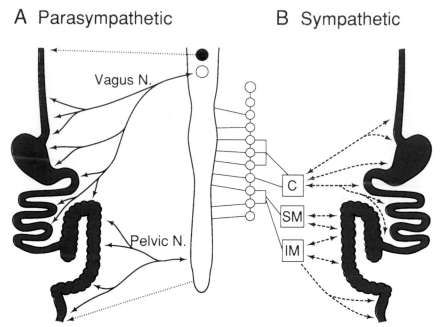

FIGURE 2   The extrinsic branches of the autonomic nervous system. (A)
Parasympathetic. Dotted lines indicate cholinergic innervation of the striated muscle
in the esophagus and external anal sphincter. Solid lines indicate afferent and
preganglionic innervation of the remaining GI tract. (B) Sympathetic. Solid lines
denote the afferent and preganglionic efferent pathways between the spinal cord and
the prevertebral ganglia. C, celiac; SM, superior mesenteric; IM, inferior mesenteric.
Dashed lines indicate the afferent and postganglionic efferent innervation.

sympathetic neurons release norepinephrine, which also acts on neurons within the enteric system. The enteric nervous system itself contains many different mediators. These include acetylcholine, nitric oxide, vasoactive intestinal peptide (VIP), enkephalins, serotonin, and substance P. Our knowledge of the localization of neurocrines is far from complete, and in many instances a single neuron may release more than one mediator.

CHARACTERISTICS OF GASTROINTESTINAL PEPTIDES

The GI peptides regulate many different functions: water and electrolyte secretion from the stomach, pancreas, liver, and gut; enzyme secretion of the stomach and pancreas; and contraction and relaxation of the smooth muscle of the stomach, small and large bowel, various sphincters, and gall bladder. These peptides also regulate the release of GI peptides and some other endocrines, such as insulin, glucagon, and calcitonin. Some GI peptides have trophic effects, regulating the growth of the exocrine pancreas and the mucosa of the stomach, small and large intestine, and gall bladder. Many peptides have identical effects on an end organ. Others may produce opposite effects.

Thus, the end-organ response is an integration of the actions of the various peptides reaching its receptors. Fortunately, many of these effects are pharmacologic and are produced by doses of peptide greater than those present as a result of normal release. The important actions of the GI peptides are the physiologic ones—those that occur with amounts of hormone released as a result of the stimulus provided by the ingestion and digestion of a meal. The actions of the GI peptides also vary in intensity and direction among species. Wherever the data are available, the remainder of this section concerns the actions that occur in humans.

Although some GI peptides are located in nerves, many others are found in mucosal endocrine cells. These cells are found not in discrete isolated glands but as single cells scattered over a wide area of mucosa. This more generalized distribution ensures that endocrine release is regulated by the events taking place in a relatively large part of the gut. Thus, the release of these peptides is an integrated response in time as well as area. Because the gut endocrine cells are scattered, one cannot surgically remove a GI "endocrine gland" to examine the effects of the absence of these peptides—a standard technique of endocrinologists to determine the physiologically significant actions of hormones. This distribution also means that studies of the release of these

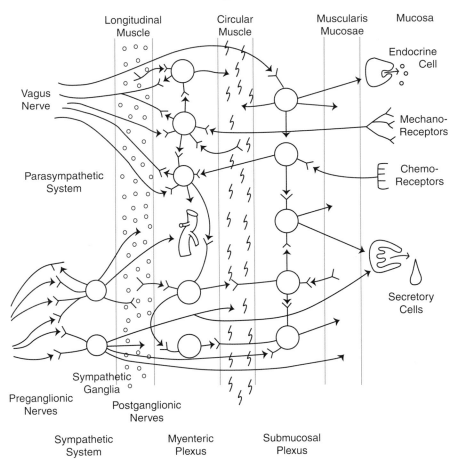

FIGURE 3   The integration of the extrinsic (parasympathetic and sympathetic) nervous system with the enteric (myenteric and submucosal plexuses) nervous system. The preganglionic fibers of the parasympathetic synapse with ganglion cells located in the enteric nervous system. Their cell bodies, in turn, send signals to smooth muscle, secretory, and endocrine cells. They also receive information from receptors located in the mucosa and in the smooth muscle, which is relayed to higher centers via vagal afferents. This may result in vagovagal (long) reflexes. Postganglionic efferent fibers from the sympathetic ganglia innervate the elements of the enteric system, but they also innervate smooth muscle, blood vessels, and secretory cells directly. The enteric nervous system relays information up and down the length of the GI tract, which may result in short or intrinsic reflexes.

hormones are difficult, because *in vitro* experiments are impossible and there is often no venous blood supply to isolate and sample.

At first it was believed that all GI peptides were hormones. Although all GI hormones are peptides, we now know that the GI peptides can be classified as endocrine hormones, paracrines, or neurocrines (Fig. 4).

Hormones are released into the portal circulation, pass through the liver, and are then circulated to all tissues of the body (unless excluded from the brain by the blood–brain barrier). Target tissues are those that have specific receptors for the hormones in their plasma membranes. Thus, specificity is a property of the target tissue. *Paracrines* are released from endocrine cells within the mucosa and diffuse through the extracellular spaces or are carried short distances by capillaries to their target cells. Here, specificity depends not only on the presence of receptors but also on the proximity of the target cell to the endocrine cell. Thus, the effects of paracrines are limited by diffusion, but because the releasing cells may be spread over a large area of mucosa it is still possible for paracrines to act as major mediators of function. Histamine, a derivative of the amino acid histidine, is an important agent in the stimulation of gastric acid secretion that acts as a paracrine.

Neurocrines are synthesized in the cell bodies of neurons and migrate to the axonal ending, where they can be released by an action potential. Once released, neurocrines diffuse the short distance across the synaptic cleft to their target cells. Acetylcholine, although not a peptide, is the best known neuroregulator in the GI tract.

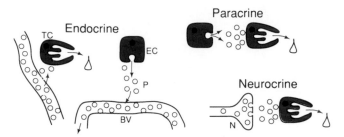

FIGURE 4   Different methods of delivery of peptides from cells of the GI tract. BV, blood vessel; EC, endocrine cell; N, neuron; P, peptide; TC, target cell.

HORMONES

Four criteria must be satisfied to prove the existence of a GI hormone: (1) A physiologic stimulus, such as the presence of food, applied to one part of the digestive tract produces a response in another part; (2) the response must be able to be elicited after nervous connections between the two portions of the tract are severed; (3) an extract of mucosa taken from the site where the stimulus was applied reproduces the response when injected into the blood; and (4) the substance must be isolated, purified, identified chemically, and synthesized.

Five GI peptides have satisfied these criteria and are considered hormones. They are, in the order of their discovery, secretin, gastrin, cholecystokinin (CCK), gastric inhibitory peptide, and motilin. There is also an extensive list of peptides, found in endocrine cells of the GI tract, whose members for one reason or another have failed to satisfy one of these criteria. These substances are called *candidate hormones*.

The science of endocrinology actually began in 1902 with the discovery of *secretin* (meaning, literally, "to stimulate secretion") by Bayliss and Starling. They demonstrated that hydrochloric acid placed on the duodenal mucosa stimulated pancreatic water and bicarbonate secretion. The effect persisted after denervation, and an acid extract of canine duodenal mucosa stimulated pancreatic secretion when injected into the blood of a second dog. Secretin was isolated and its amino acid sequence determined by Jorpes and Mutt in 1966. It was synthesized by Bodanszky and coworkers later in the same year.

Edkins discovered *gastrin* (which means, literally, "to stimulate the stomach") in 1905, describing it to the Royal Society by saying "in the process of the absorption of digested food in the stomach a substance may be separated from the cells of the mucous membrane which, passing into the blood or lymph, later stimulates the secretory cells of the stomach to functional activity." The existence of gastrin remained controversial for 43 years, primarily because histamine, found in large quantities in the gastric mucosa, had been shown by Popielski also to stimulate gastric acid secretion. In 1938, Komarov convincingly showed that gastrin was a peptide separate from histamine. In 1964, Gregory and his colleagues finished the extraction and isolation of hog antral gastrin. Kenner synthesized gastrin the same year, making it the first GI peptide to satisfy all of the criteria and establishing it as a hormone. The word *hormone* (from the Greek for "to set in motion" or "to spur on") was actually coined by W. B. Hardy and used by Starling in 1905 to describe secretin and gastrin and to convey the concept of bloodborne messengers.

In 1928, Ivy and Oldberg demonstrated that fat in the duodenum stimulated gall bladder contraction. They named the humoral mediator *cholecystokinin* (*cholecysto* refers to the gall bladder, *kin* means to move, and *in* means to stimulate). In 1943, Harper and Raper described a hormone released from the mucosa of the small intestine that stimulated pancreatic enzyme secretion. They named it *pancreozymin*. With the purification of these two substances by Jorpes and Mutt in 1968, it became obvious that the ability to stimulate both gall bladder contraction and pancreatic enzyme secretion resided in the same peptide, but we still call this hormone CCK because this was its first action to be described.

In 1969, Brown and his coworkers described a potent inhibitor of gastric acid secretion extracted from intestinal mucosa. After its isolation and purification in 1971, it was named *gastric inhibitory peptide* (GIF) for its ability to inhibit gastric acid secretion. GIF is released from the intestinal mucosa by glucose and also stimulates insulin release. The release of insulin was shown to be a physiologic phenomenon, making GIF the fourth GI hormone. Because the insulinotropic effect of GIF requires elevated serum glucose and because the inhibitory effects on the stomach require pharmacologic doses of hormone, it has been suggested that its name be changed to (G)lucose-dependent (I)nsulinotropic (P)eptide. In either case, GIF is still its common name.

Motilin was also purified by Brown and his associates in early 1970. Motilin is a linear 22-amino-acid peptide found primarily in the proximal small intestine. It is released approximately every 90 min in fasting humans and, as its name implies, stimulates GI motility. Release is under neural control, and this peptide is responsible for the interdigestive migrating myoelectric complex.

Gastrin

Gastrin is a 17-amino-acid straight chain peptide (Fig. 5). Most peptide hormones are heterogeneous, occurring in two or more molecular forms. Gastrin was

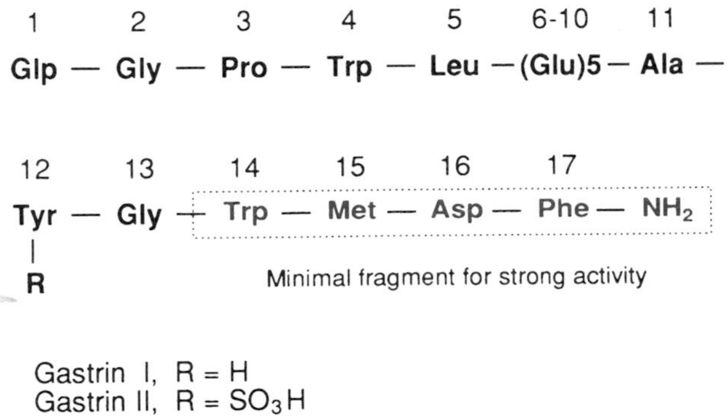

```
 1      2      3      4      5     6-10    11
Glp — Gly — Pro — Trp — Leu —(Glu)5— Ala —


 12     13     14     15     16     17
Tyr — Gly —┊ Trp — Met — Asp — Phe — NH₂ ┊
 |         ┊                              ┊
 R           Minimal fragment for strong activity
```

Gastrin I, R = H
Gastrin II, R = SO₃H

FIGURE 5   Structure of human "little" gastrin (HG-17).

originally isolated from hog antral mucosa as a hepta-decapeptide called G-17 or "little" gastrin. The form shown in Fig. 5 is human G-17 and differs from the gastrins of other species only by one or two amino acid substitutions in the middle of the molecule. Little gastrin accounts for about 90% of the gastrin found in antral mucosa. The major component of gastrin in serum of fasting subjects, however, is a larger molecule called "big" gastrin. On isolation, big gastrin was found to contain 34 amino acids, and is therefore called G-34. Trypsin splits G-34 to yield G-17 plus another inactive heptadecapeptide. Therefore, G-34 is not a dimer of G-17. Current evidence indicates that G-17 is not produced from G-34 or vice versa, because both pro G-17 and pro G-34 are produced. Between meals, referred to as the *interdigestive state,* most human serum gastrin is G-34. The human duodenal mucosa also contains significant amounts of gastrin, which is predominantly G-34. In response to a meal, a large amount of antral G-17 is released along with smaller amounts of G-34 from the antral mucosa. G-34 may be released from duodenal mucosa under special conditions. G-17 and G-34 are equipotent, although the half-life of G-34 is 38 min and the half-life of G-17 is approximately 7 min.

Radioimmunoassay of plasma from normal fasting humans gives values between 50 and 100 pg/mL (pg or picogram $= 10^{-12}$ g). During a meal, these values usually increase 50–100%.

In Fig. 5, note that position 12 is a tyrosyl residue. If the tyrosine is sulfated, the molecule is called *gastrin II;* unsulfated, it is called *gastrin I.* Both forms occur with equal frequency in nature and are equipotent for most effects. The N terminus of gastrin is a pyroglutamyl residue (Glp). The –NH₂ group that follows the C-terminal Phe indicates that this is the amidated form, phenylalamide. The pyroglutamyl and phenylala-mide residues protect gastrin from amino peptidases and carboxypeptidases.

The most interesting structural feature of gastrin is that all the biologic activity is contained in the four C-terminal amino acids. This tetrapeptide is the minimal fragment of gastrin necessary for strong activity and is about one-sixth as potent as the whole hormone.

The physiologically significant actions of gastrin are the stimulation of gastric acid secretion and its trophic activity (Table 1). The receptor for gastrin in the so-called CCK-B or CCK-2 receptor, which also binds CCK, but its affinity for CCK is much lower than its affinity for gastrin. Gastrin stimulates acid secretion by acting directly on the acid-secreting parietal cell and by releasing histamine from the enterochromaffin-like cell (ECL cell). Gastrin also stimulates histamine synthesis in, and growth of, the ECL cells. On a molar basis, gastrin is 1500 times more potent than histamine for stimulating acid secretion. Gastrin stimulates the growth of the oxyntic gland (acid-secreting portion) mucosa of the stomach and also the colonic mucosa. If most endogenous gastrin is surgically removed by resecting the gastric antrum, these tissues atrophy. Patients having high serum levels of gastrin because of continual release of the hormone from a tumor have hyperplasia and hypertrophy of the acid-secreting portion of the stomach. The trophic actions of gastrin and the other GI hormones are direct effects and are specific for GI tissues. There is considerable interest in the role of gastrin in promoting the growth of colon cancers. Many colon tumors have receptors for gastrin and some also secrete the hormone, which may act as an autocrine agent. Interestingly, progastrin and glycine extended G-17 and G-34, steps in the synthesis of the hormone, are sometimes secreted and have trophic effects.

Gastrin is physiologically released from G cells of the antral mucosa, and to a lesser extent the duodenal mucosa, in response to a meal. The most important releasers are protein digestion products, small peptides, and amino acids (Table 1). Phenylalanine and trypto-phan are the amino acids with the greatest gastrin-releasing activity. The release of gastrin requires luminal contact with the antral mucosa because intravenous amino acids are ineffective. Activation of the vagus and local cholinergic reflexes also release gastrin. The vagal mediator of gastrin release is bombesin (also called gastrin-releasing peptide, GRP). Gastrin is also released by physical distension of the wall of the stomach. For example, inflating a balloon in the antrum initiates a reflex release of gastrin. This response is triggered during a meal by the pressure of ingested food.

Release of gastrin is inhibited by acidification of gastric luminal contents below pH 3. This is an important feedback inhibition of gastrin release and ensures that additional gastrin is not released when the stomach is already acidified. It is important to realize that raising the pH by neutralizing gastric contents does

TABLE 1   Gastrointestinal Hormones

Hormone	Action		Site of release	Releaser
Gastrin	Stimulates		Antrum (duodenum)[a]	Peptides
		Peptides		Amino acids
		Gastric acid secretion		Distention Vagal stimulation
		Growth of gastric oxyntic giant mucosa		
CCK	Stimulates		Duodenum	Peptides
		Gallbladder contraction	Jejunum	amino acids
		Pancreatic enzyme secretion		fatty acids > 8c
		Pancreatic bicarbonate secretion		(Acid)[a]
		Growth of exocrine pancreas		
	Inhibits			
		Gastric emptying		
Secretin	Stimulates		Duodenum	Acid
		Pancreatic bicarbonate secretion		(Fat)[a]
		Biliary bicarbonate secretion		
		Growth of exocrine pancreas		
		Pepsin secretion		
	Inhibits			
		Gastric acid secretion		
		Trophic effect of gastrin		
GIP	Stimulates		Duodenum	Glucose
		Insulin release	Jejunum	Amino acids
	Inhibits			Fatty acids
		Gastric acid secretion		
Motilin	Stimulates		Duodenum	Nerves
		Gastric motility	Jejunum	
		Intestinal motility		

[a] Parentheses indicate an item of lesser importance. CCK, cholecystokinin; GIP, gastric inhibitory peptide.

not by itself stimulate gastrin release. It allows the other stimuli to be effective.

Cholecystokinin

Cholecystokinin (CCK) is structurally related to gastrin. As shown in Fig. 6 the C-terminal 5 amino acids of CCK are identical to those of gastrin. CCK was originally isolated as a 33-amino-acid peptide (Fig. 6). In addition to the 33-amino-acid form (CCK-33), CCK-39 and CCK-8 have also been isolated. The structural feature that determines whether a peptide of the gastrin-CCK family behaves like gastrin (stimulates acid secretion) or CCK (stimulates gall bladder contraction and pancreatic enzyme secretion) is the location of the tyrosyl residue. In peptides with gastrin-like activity, the tyrosyl residue is in position 6 from the C terminus, and they bind to the CCK-B receptor. Sulfation of the tyrosyl residue does not alter the potency of gastrin peptides. In CCK peptides, the tyrosyl residue is in position 7 from the C terminus, and it is always sulfated. These peptides bind to the CCK-B receptor and their pattern of activity to that of gastrin. Thus, in order to act like CCK, one of these peptides must have a sulfated tyrosyl residue in position 7 from the C terminus.

All others, including the C-terminal tetra-, penta-, and hexapeptides have gastrin-like activity. Given high enough doses, each of these two hormones can produce all the effects of the other. The difference between the two is determined by which effects occur at physiologic doses.

CCK physiologically stimulates gall bladder contraction and relaxes the sphincter of Oddi to facilitate the emptying of the gall bladder. CCK is a potent stimulator of pancreatic enzyme secretion, but only a weak stimulator of pancreatic water and bicarbonate. However, if administered with secretin, low doses of CCK greatly potentiate the secretion of bicarbonate stimulated by secretin. This important interaction is a physiologically significant effect of CCK. CCK is an important inhibitor of gastric emptying and is the reason fatty meals empty more slowly than nonfat meals. CCK also has trophic effects, stimulating the growth of the exocrine pancreas and the mucosa of the gall bladder (Table 1).

CCK is released from the I-cells of the duodenal and jejunal mucosa by peptides and single amino acids. Fatty acids or their monoglycerides containing eight or more carbon atoms are the most potent stimuli for CCK release. Fat must be in an absorbable form to release CCK; therefore, triglycerides are ineffective. Acid is a weak releaser of CCK, but may do so in high doses.

FIGURE 6   Structure of porcine cholecystokinin.

Secretin

Secretin is a member of a family of peptides chemically homologous to glucagon (Fig. 7). Secretin contains 27 amino acids, 14 of which are identical in kind and position to those of glucagon. Members of this family of peptides have no active fragments, all amino acids being required for activity. The active forms of these peptides appear to be α-helices, and all amino acids must be in position to form the active tertiary structure.

The primary effect of secretin is the stimulation of pancreatic bicarbonate and water secretion. It also has an identical effect on the liver and is the most potent choleretic of the GI peptides. Like CCK, secretin stimulates the growth of the exocrine pancreas, an effect that is greatly potentiated in the presence of both hormones. Secretin inhibits gastric acid secretion and the trophic effect of gastrin on the acid-producing portion of the stomach. All of the actions of secretin discussed earlier serve to reduce the amount of acid in the duodenum (see Table 1). For this reason, secretin has been referred to as "nature's antacid." The only other action of secretin that one should be aware of is that it stimulates pepsin secretion from the stomach.

Secretin is released from the S cells of the duodenal mucosa. By far the most important stimulus for secretin release is acid. Secretin release occurs when the pH of the duodenal contents falls below 4.5. To a lesser extent, secretin is also released by fatty acids. Because the duodenum often contains large amounts of dietary fat, this may be a physiologically significant mechanism of release.

Gastric Inhibitory Peptide

GIF is a member of the secretin-glucagon family of peptides (see Fig. 7). It has nine amino acids identical to those of secretin. In pharmacologic doses, GIF has most of the actions of secretin and glucagon.

The only action of GIF that is known to be physiologically significant is the stimulation of insulin release in the presence of glucose. This hormone is responsible for the finding that an oral glucose load is cleared from the blood more rapidly than an intravenous glucose load of the same magnitude. As its name implies, GIF inhibits gastric acid secretion.

GIF is the only hormone released by all three major foodstuffs. Fat must be in the hydrolyzed form in order to release GIF. Amino acids such as arginine, histidine, leucine, lysine, and others, which are not potent CCK releasers, release GIF. GIF is released by oral but not intravenous glucose. Almost all GIF release occurs from the duodenum and proximal jejunum.

Motilin

Motilin is a linear 22-amino-acid peptide whose structure is unrelated to that of either gastrin or secretin. Motilin is released approximately every 90 min during fasting. Release is prevented by atropine, indicating that it is regulated by a cholinergic pathway, and by the ingestion of a mixed meal. Serum levels of motilin increase just before and during a wave of GI motility, termed the *interdigestive migrating myoelectric complex* (see Chapter 33). If the increase in serum motilin is prevented, this complex fails to occur. Initiation of this pattern of motility is the only known physiologic action of motilin.

Candidate Hormones

Some peptides have been isolated and for one reason or another are not considered to be hormones. These are called *putative* or *candidate hormones*. Those listed in Table 2 may turn out to have physiologic significance.

Pancreatic polypeptide is a 36-amino-acid peptide released from the pancreas by all three foodstuffs. A protein meal, however, is by far the most important stimulus for release. Pancreatic polypeptide inhibits both pancreatic enzyme and bicarbonate secretion. The significance of the inhibition of pancreatic secretion itself, however, is doubtful, for the antiserum to pancreatic polypeptide was found to have no effect on pancreatic secretion.

Peptide YY has 36 amino acid residues, 18 of which are identical to those of pancreatic polypeptide. It is named for the fact that both its amino and carboxyl amino acid residues are tyrosines. Most peptide YY is found in ileal and colonic mucosa and is released by

		1	2	3	4	5	6	7	8	9	10	11	12	13	14	15
Secretin	(27)	His	Ser	Asp	Gly	Thr	Phe	Thr	Ser	Glu	Leu	Ser	Arg	Leu	Arg	Asp
VIP	(28)	His	Ser	Asp	Ala	Val	Phe	Thr	Asp	Asn	Tyr	Thr	Arg	Leu	Arg	Lys
GIP	(42)	Tyr	Ala	Glu	Gly	Thr	Phe	Ile	Ser	Asp	Tyr	Ser	Ile	Ala	Met	Asp
Glucagon	(29)	His	Ser	Gln	Gly	Thr	Phe	Thr	Ser	Asp	Tyr	Ser	Lys	Tyr	Leu	Asp

		16	17	18	19	20	21	22	23	24	25	26	27	28	29	30-42
Secretin	(27)	Ser	Ala	Arg	Leu	Gln	Arg	Leu	Leu	Gln	Gly	Leu	Val	NH₂		
VIP	(28)	Gln	Met	Ala	Val	Lys	Lys	Tyr	Leu	Asn	Ser	Ile	Leu	Asn	NH₂	
GIP	(42)	Lys	Ile	Arg	Gln	Gln	Asp	Phe	Val	Asn	Trp	Leu	Leu	Ala	Gln	...13 more
Glucagon	(29)	Ser	Arg	Arg	Ala	Gln	Asp	Phe	Val	Gln	Trp	Leu	Met	Asp	Thr	

FIGURE 7   Structures of members of the secretin family of peptides. Shaded residues are identical to secretin. Circled numbers are the total amino acid residues.

meals, and especially fat. Serum levels may be sufficient to inhibit gastric secretion and emptying, and it has been proposed as an *enterogastrone* (substance from the small intestine, *entero-*, that inhibits, *-one*, the stomach, *gastr-*).

Enteroglucagon is "glucagon-like immunoactivity" shown to be present in the distal small intestine and released into the bloodstream. The enteroglucagons are products of the same gene present in the pancreatic alpha cell, processed to form glucagon, and the intestinal L cell, which makes three forms of glucagon. One of these, GLP-1 (glucagon-like peptide 1), may function as an insulin releaser and also to inhibit gastric secretion and emptying. The significance of these effects has not been established.

PARACRINES

Paracrines are synthesized in and released from endocrine cells in the same fashion as hormones. However, the amounts released do not reach the general circulation in concentrations sufficient to produce effects. Instead, paracrines act on cells in their immediate vicinity, reaching them by simple diffusion or perhaps movement through capillaries before being diluted in the general circulation. One can assess the biologic significance of a hormone by correlating its plasma levels with its effects. Similar studies are not possible with paracrines or neurocrines. Current studies make use of specific pharmacologic blocking agents or antisera directed toward these substances. Some investigators have perfused organs *in vitro* and successfully measured changes in the release of paracrine agents in the small volumes of venous effluent.

At this time, *somatostatin is* the only GI peptide we know of that functions physiologically as a paracrine (Table 3). It was first isolated from the hypothalamus as growth hormone release inhibitory factor (GHRIF),

and it appears to inhibit the release of all peptide hormones. It is now known to exist throughout the GI mucosa and to inhibit the release of all gut hormones. Somatostatin is released by luminal acid and mediates the inhibition of the release of gastrin when the gastric contents are acidified. It also directly inhibits acid secretion from the parietal cells. Vagal activation prevents somatostatin release. It may also act physiologically as a paracrine in the pancreatic islets to regulate insulin and glucagon release.

The only other known significant paracrine agent in gut physiology is histamine. Histamine is not a peptide, but is found throughout the mucosa in endocrine-like cells called ECL cells (enterochromaffin-like). Gastrin stimulates the release of histamine and its synthesis by activating the enzyme histidine decarboxylase present in the ECL cells. Gastrin also stimulates the growth of these cells. Released histamine then stimulates the parietal cells to secrete acid. Histamine also potentiates the effects of gastrin and acetylcholine on acid secretion. This is why drugs like Tagamet (cimetidine) and Zantac (ranitidine) that block histamine H_2-receptors are such potent inhibitors of acid secretion.

NEUROCRINES

At first, all GI peptides were believed to originate from endocrine cells. Powerful cytochemical techniques for the localization of peptides, however, have shown that many were located in nerves throughout the mucosa and smooth muscle of the gut.

Vasoactive Intestinal Peptide

Currently, three peptides are known to function physiologically in the gut as neurocrines (Table 4).

TABLE 2   Candidate Hormones

Peptide	Action	Site of release	Releaser
Pancreatic polypeptide	Inhibits Pancreatic bicarbonate secretion Pancreatic enzyme secretion	Pancreas	Protein (fat)[a] (Glucose)
Peptide YY	Inhibits Gastric secretion Gastric emptying	Ileum Colon	Fat
Glucagon-like-peptide-1 GLP-1	Stimulates Insulin release Inhibits Gastric secretion Gastric emptying	Ileum Colon	Fat Glucose

[a] Parentheses indicate an item of lesser importance.

TABLE 3   Paracrines

Substance	Action	Site of release	Releaser

Somatostatin	Inhibits
	  Gastrin release
	  Other peptide hormone release
	  Gastric acid secretion
		GI mucosa
		Pancreatic islets
			Acid
			Vagus inhibits release

Histamine	Stimulates Gastric acid secretion	Oxyntic gland mucosa ECL cell	Gastrin

GI, gastrointestinal; ECL, enterochromaffin-like.

Vasoactive intestinal peptide (VIP) is chemically a member of the secretin-glucagon family (see Fig. 7). It is found only in nerves and appears to mediate the relaxation of GI smooth muscle. VIP may also relax vascular smooth muscle to physiologically mediate vasodilation. Many of the effects of VIP on relaxing smooth muscle may be mediated by *nitric oxide,* the synthesis of which is stimulated by VIP. When injected into the bloodstream, VIP has many of the effects of its relatives, secretin, GIP, and glucagon. It stimulates pancreatic fluid and bicarbonate secretion and water and electrolyte secretion from the intestinal mucosa. It also inhibits gastric secretion.

VIP appears to be the agent responsible for pancreatic cholera or watery diarrhea syndrome. The symptoms of this frequently lethal disease result from the secretion of a peptide from a pancreatic islet cell tumor. The peptide has been identified as VIP in both tumor tissue and plasma samples from these patients. Because VIP stimulates the secretion of cholera-like fluid from intestinal mucosa, it is probably the causative agent.

Bombesin or Gastrin-Releasing Peptide

A variety of biologically active peptides has been isolated from amphibian skin. Many of these have mammalian counterparts with physiologic functions. Bombesin is one of these. It was first isolated from the skin of a frog, for which it is named. The mammalian counterpart of bombesin is GRP, which is found in nerves of the gastric mucosa. GRP is released by vagal stimulation and appears to mediate vagally stimulated gastrin release.

Enkephalins

Two pentapeptides isolated from brain activate opiate receptors and have been found in nerves of the mucosa and smooth muscle of the GI tract. They are identical except for the C-terminal amino acid, which is methionine in one (met-enkephalin) and leucine in the other (leu-enkephalin). The enkephalins contract smooth muscle and appear to mediate the contraction of the lower esophageal, pyloric, and ileocecal sphincters. They may also function physiologically in peristalsis.

Opiates have been used for years to treat diarrhea. Their effectiveness is attributed to the fact that they slow intestinal transit. The enkephalins are also potent inhibitors of intestinal fluid secretion. Thus, the antidiarrheal property of opiates is probably due to this action as well as the motility effects.

INTEGRATION OF REGULATORY MECHANISMS

This chapter outlined peripheral and intrinsic nervous regulatory mechanisms and hormonal, paracrine, and neurocrine mechanisms. It is essential to understand that each does not occur as an isolated event but that all may integrate into one control mechanism. Extrinsic nerves may alter hormone release, which in turn may

TABLE 4   Neurocrines

Peptide	Action	Site of release
VIP	Relaxes sphincters	Mucosa and smooth muscle of GI tract
	Relaxes gut circular muscle	
	Stimulates intestinal secretion	
	Stimulates pancreatic secretion	
Bombesin or GRP	Stimulates gastrin release	Gastric mucosa
Enkephalins	Stimulates smooth muscle contraction	Mucosa and smooth muscle of GI tract
	Inhibits intestinal secretion	

VIP, vasoactive inhibitory peptide; GRP, gastrin-releasing peptide; GI, gastrointestinal.

be affected by paracrine agents. This is the case with gastrin. Gastric acid secretion (see Chapter 34) is regulated by extrinsic nerves, local enteric reflexes, paracrines, and hormones acting in an integrated manner. This variety of mechanisms ensures that events outside the GI tract, seeing or smelling food for example, can be integrated with those occurring within the tract, and that one portion of the tract can influence digestive processes occurring in other, more remote, portions.

Suggested Readings

Dockray GJ. Physiology of enteric neuropeptides. In Johnson LR, ed. *Physiology of the GI tract,* 3rd ed. New York: Raven Press, 1994, pp 169–210.

Dockray GJ, Varro A, Dimaline R, Wang T. The gastrins: Their production and biological activities. *Annu Rev Physiol* 2001;63:119–139.

Walsh JH. GI hormones. In Johnson LR, ed. *Physiology of the GI tract,* 3rd ed. New York: Raven Press, 1994, pp 1–129.

Wood JD. Physiology of the enteric nervous system. In Johnson LR, ed. *Physiology of the GI tract,* 3rd ed. New York: Raven Press, 1994, pp 423–482.

Motility

LEONARD R. JOHNSON

KEY POINTS

- Gastrointestinal smooth muscle cells contract as a unit because of anatomic and electrical coupling.
- Smooth muscles may contract and relax over a few seconds (*phasic*), or contractions may last from minutes to hours (*tonic*).
- Material moves through the esophagus and entire gastrointestinal (GI) tract from regions of higher to regions of lower intraluminal pressure.
- *Primary peristaltic contractions* are initiated in the esophagus by swallowing and are responsible for moving most material through the esophagus; *secondary peristaltic contractions* initiated by distension and local reflexes remove any "leftover" material.

- The principal motility function of the orad (proximal) stomach is *receptive relaxation*, which is mediated by a vagovagal reflex and allows the stomach to store ingested material. The principal activity of the caudal (distal) stomach is mixing and emptying.
- Gastric contractions are triggered by regularly (3–5 min) occurring depolarizations called *slow waves*.
- Small intestinal motility is characterized by brief, irregular contractions that are interrupted during fasting approximately every 90 min by a wave of intense contractions that sweeps the entire length of the small intestine. After a meal, these

Essential Medical Physiology, Third Edition

GASTROINTESTINAL SMOOTH MUSCLE

The smooth muscle cells in each part of the gastrointestinal (GI) tract have structural and functional differences. However, certain basic properties are common to all of these cells. Smooth muscle cells make up all of the contractile tissue of the GI tract with the exception of the pharynx, the upper one-third to one-half of the esophagus, and the external anal sphincter, which are striated muscle.

Structure of Smooth Muscle Cells

Smooth muscle cells are smaller than skeletal muscle cells and are long, narrow, and spindle shaped. Most are 4–10 μm wide and 50–200 μm long. Cells are loosely packed with relatively large intracellular spaces that contain bands of collagen, elastin, and other connective tissue. This network allows contractile forces to be transmitted from one cell to others nearby. Cells are arranged in bundles that branch and anastomose with one another. These bundles, or fasciae, are surrounded by connective tissue and are the functional units of gut smooth muscle.

The cells belonging to a bundle are functionally coupled so that contractions of all cells are synchronous. Smooth muscle tissue is therefore classified as *unitary* (Fig. 1). Coupling occurs by actual fusion of cell membranes to form areas of low electrical resistance termed *gap junctions* or *nexuses*. Only a few smooth muscle cells in each bundle are actually innervated. The nerve axons run through the bundles releasing neurotransmitters from swellings along their length. The swellings are actually removed from the muscle cells so that no neuromuscular junctions exist. The neurotransmitters either excite or inhibit only a few cells in each bundle, and the effect of the transmitter on membrane potential is spread directly from one cell to another.

Contractile elements of smooth muscle cells are not arranged in the orderly fashion found in skeletal muscle cells. Sarcomeres and, thus, striations are absent. Instead, the contractile proteins (actin and myosin} are arranged in myofilaments crossing from one side of the cell to the other at oblique angles. Thin filaments consist of actin and tropomyosin, and thick filaments are made up of myosin. Smooth muscle contains much more actin and less myosin and troponin than skeletal muscle. The ratio of thin to thick filaments in smooth muscle is about 15:1, compared with 2:1 for skeletal muscle. Smooth muscle cells also contain a network of intermediate filaments that form a type of internal skeleton. The contractile filaments may anchor to this network, thus transmitting their contractile force over much of the cell.

Smooth Muscle Contraction

Depolarization of circular muscle results in the rapid conduction of the depolarization around the gut so that a ring of smooth muscle contracts. The depolarization moves more slowly longitudinally from this ring of depolarized cells. The opposite occurs when longitudinal muscle is depolarized. In this case, the depolarizing current is transmitted much more rapidly in the longitudinal direction and spreads slowly around the gut.

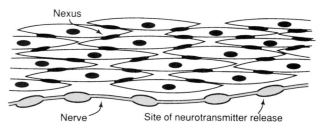

FIGURE 1   Structural characteristics of unitary, smooth muscle. Cells are coupled anatomically and electrically by areas of fusion of their cell membranes. These areas are called *nexuses*.

In both cases, however, the depolarization moves rapidly with the long axes of the smooth muscle cells. This pattern is produced simply because the current spreads much more rapidly through the low resistance of the cytoplasm of the cells than across the higher resistance of the cell membranes. Because smooth muscle cells are 10–20 times longer than they are wide, the resistance to current per unit distance is much less over the long axis of the cells compared with the short axis.

Depolarization results in different responses from smooth muscle cells in different portions of the GI tract. Muscles of the esophagus, the distal one-third of the stomach (antrum), and the small intestine contract and relax rapidly, in a matter of seconds. These contractions are called *phasic*. On the other hand, the smooth muscle of the lower esophageal sphincter, the orad stomach, the ileocecal sphincter, and the internal anal sphincter sustain contractions that may last from minutes to hours. These contractions are called *tonic*. The type of contraction depends on the smooth muscle cell itself and is adapted to carry out the motility function of the organ involved.

The contractile pattern of the smooth muscle in most parts of the GI tract is determined by the basic electrical rhythm of depolarizations of the resting membrane potential. These oscillations in resting membrane potential are referred to as *slow waves* and have different frequencies in different parts of the tract (3–12 cycles/min). The slow waves determine the pattern of spike or action potentials produced. The activity of extrinsic nerves and the presence of hormones and paracrines modulate this activity, determining the strength of the contractions and whether or not variations in membrane potentials result in action potentials and subsequent contractions. Wave frequency is extremely constant and is virtually unaffected by neural and hormonal activity. Slow waves are inherent in the smooth muscle and are influenced only by body temperature and metabolic activity. As these increase, the frequency of slow waves also increases. The origin of slow waves is in cells called the *interstitial cells of Cajal*. These specialized cells receive a large amount of neural input and form gap junctions with smooth muscle cells.

Contractions of smooth muscle cells depend on levels of free intracellular calcium, as do contractions of other muscle cells. The threshold for interaction of the contractile proteins occurs at approximately 10^{-7} M calcium, and maximal contraction occurs at about 10^{-5} M. Points of disagreement exist regarding the actual steps involved, but the most likely summary of events is shown in Fig. 2. Calcium first binds to calmodulin, one of the calcium-binding proteins. The complex then activates a myosin light chain kinase, which in turn splits adenosine triphosphate (ATP) to phosphorylate

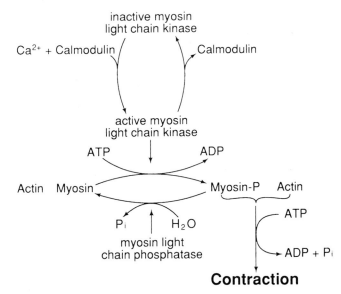

FIGURE 2 Biochemical steps in the contraction of smooth muscle. Increased Ca^{2+} activates myosin light chain kinase, phosphorylating myosin. Myosin-P then binds to actin, resulting in contraction. As Ca^{2+} levels decrease, the kinase is inactivated, the phosphatase removes the phosphate from myosin-P, and relaxation occurs. (Modified from Johnson LR, ed., *Gastrointestinal physiology*, 6th ed. St. Louis: CV Mosby, 2001.)

one of the components of myosin. The phosphorylated form of myosin now interacts with actin to produce a contraction. The contraction is supported by the additional release of energy from ATP. As intracellular calcium levels decrease, myosin is dephosphorylated by a specific phosphatase. The interaction between myosin and actin ceases, and relaxation occurs.

The remainder of this chapter focuses on how these contractile events are produced in each part of the GI tract and how they are modified to result in the patterns of motility specific to each region.

SWALLOWING

Swallowing consists of chewing, a pharyngeal phase, movement of material through the esophagus, and the relaxation of the stomach to receive the ingested material.

Chewing

Chewing has three major functions: (1) It facilitates swallowing and prevents injury to the lining of the pharynx and esophagus by reducing the size of the ingested particles and mixing them with saliva; (2) starches in food are exposed to salivary amylase, the enzyme that begins their chemical breakdown (digestion); and (3) it increases the surface area of ingested material, speeding up the rate at which it can be acted on by digestive enzymes. Particles are usually reduced to a few cubic millimeters in humans, but other animals such as the dog and cat almost bolt their food, chewing it only into lumps small enough to pass into the pharynx. The act of chewing is both voluntary and involuntary, and most of the time proceeds by reflex actions void of conscious input.

The *chewing reflex* is initiated by food in the mouth that inhibits muscles of mastication, and the jaw drops. A subsequent stretch reflex of the jaw muscles produces a contraction that automatically raises the jaw, closing the teeth on the food bolus. Compression of the bolus against the mucosal surface of the mouth inhibits the jaw muscles to repeat the process.

Pharyngeal Phase

Normally, liquids are propelled immediately from the mouth to the oropharynx and swallowed. Swallowing begins with a posterior movement of the tongue, forcing the liquid ahead of it. At the start of a swallow of solid material, the tip of the tongue separates a bolus of material between the tongue and the hard and soft palates (Fig. 3A). The tongue then raises against the hard palate and sweeps backward, forcing the bolus into the oropharynx (Fig. 3B). At this time, the nasopharynx is closed by a downward movement of the soft palate and contraction of the superior constrictor muscles of the pharynx. Simultaneously, respiration is inhibited and the laryngeal muscles contract to close the glottis and elevate the larynx. A peristaltic contraction now begins in the superior constrictor muscles (Fig. 3B) and proceeds through the middle and inferior constrictors, propelling the bolus through the pharynx (Fig. 3C). The upper esophageal sphincter relaxes, allowing these contractions to move the material into the esophagus (Fig. 3D).

Although swallowing can be initiated voluntarily, these efforts fail unless there is something, at least a small amount of saliva, to trigger the swallowing reflex. A voluntary movement of the tongue begins the swallow by forcing material into contact with a large number of receptors lining the pharynx. These receptors initiate impulses via vagal and glossopharyngeal afferents to the *swallowing center* in the medulla (Fig. 4). Although the oral and pharyngeal phases of swallowing are brief, lasting less than a second, the afferent impulses arriving at the swallowing center evoke a coordinated sequential output of efferent activity lasting as long as 9 sec. This efferent activity occurs via a variety of nuclei including the nucleus ambiguus and those of the trigeminal, facial, and hypoglossal nerves, and sequentially activates the muscles of the pharynx and esophagus.

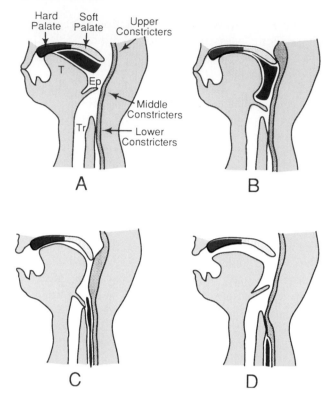

FIGURE 3   Oral and pharyngeal stages of a swallow. The dark area represents the bolus to be swallowed. See text for detailed explanation. T, tongue; Ep, epiglottis; Tr, trachea. (Modified from Johnson LR, ed., *Gastrointestinal physiology*, 6th ed. St. Louis: CV Mosby, 2001.)

Receptors present in the pharynx and esophagus send feedback information to the swallowing center via interneurons and vagal afferents, leading to further coordination of the muscular contractions involved. The sequential nature of the firing of efferents from the swallowing center and the overall coordination result in the peristaltic nature of the contractions (Fig. 4).

Because the contractions of the muscles of the pharynx are under control by extrinsic nerves and the swallowing center, a variety of neurologic lesions result in swallowing problems. Cerebrovascular accidents involving the medulla and swallowing center result in a loss of the pharyngeal phase of swallowing. Aspiration is often the result of such lesions, because the movement of material through the pharynx and upper esophageal sphincter is no longer coordinated.

MOVEMENT OF MATERIAL THROUGH THE ESOPHAGUS

The obvious function of the esophagus is to serve as a conduit for and to propel swallowed material to

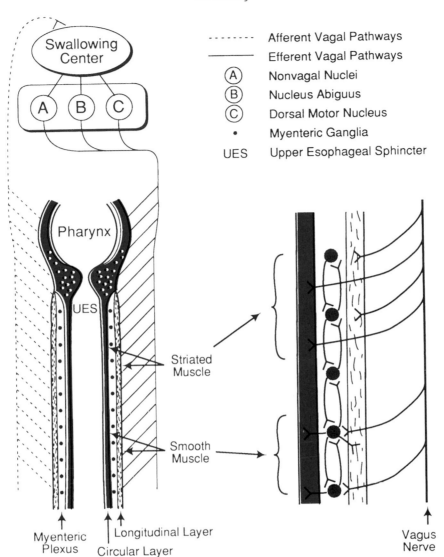

FIGURE 4   Neural pathways involved in the regulation of pharyngeal and esophageal peristalsis. Vagal sensory input is relayed to the swallowing center in the medulla, where output to muscles is coordinated with respiration. Muscles of the pharynx and the striated esophagus are innervated via the nucleus ambiguus; smooth muscles of the esophagus are innervated via the dorsal motor nucleus. Enlarged area details the direct innervation of striated muscle by vagal fibers and indirect innervation of smooth muscle via the interneurons. The sequential nature of the innervation, which makes peristalsis possible, is also illustrated here. (Adapted from Johnson LR, ed., *Gastrointestinal physiology*, 6th ed. St. Louis: CV Mosby, 2001.)

the stomach. The esophagus, however, also has two important barrier functions: (1) It prevents the entry of air at the upper end and (2) the entry or reflux of gastric contents at the lower end. The barrier functions are carried out by two sphincters, one at each end. The propulsive function is accomplished by the coordinated contractions of the two muscle layers of the esophageal wall.

The anatomy of the esophagus follows the general pattern outlined in the previous chapter with an inner layer of circular muscle and an outer layer of longitudinal muscle. The major difference is that in the human the upper one-third to one-half of this muscle is striated. The lower one-half of the esophagus is smooth muscle, and there is a brief transition zone between the two types. The *upper esophageal* (or pharyngoesophageal) *sphincter* consists of a thickening of the circular layer of striated muscle, which is anatomically identified as the cricopharyngeal muscle. The *lower esophageal* (or gastroesophageal) *sphincter* cannot be identified anatomically, but consists of a 1- to 2-cm zone of increased pressure that is maintained at a level greater than the pressure in the stomach.

Material moves through the GI tract from regions of higher intraluminal pressure to regions of lower pressure. These pressures can be monitored by pressure-sensing devices placed (usually in swallowed tubes of known length) at various levels in the GI tract. Figure 5 depicts data from six such devices placed sequentially along the esophagus. Note that resting intraesophageal pressures are below atmospheric, because most of the esophagus is located in the thorax. In fact, intrathoracic pressure is determined by measuring intraesophageal pressure. The pressure at the upper esophageal sphincter is above atmospheric. The pressure becomes equal to atmospheric pressure as the esophagus passes through the diaphragm into the abdomen and then increases considerably at the lower esophageal sphincter. The upper esophageal sphincter produces pressures of approximately 60 mm Hg over a 1- to 2-cm zone. The body of the esophagus is a flaccid tube that directly reflects the pressures within the thorax and abdomen. Hence, within the thorax, the intraluminal pressure drops during inspiration and rises with expiration (Fig. 5A). After passing the diaphragm, the excursions reverse to reflect intra-abdominal pressures. The lower esophageal sphincter consists of a zone of increased pressure approximately 20–40 mm Hg higher than the pressures on either side of it. This zone may occur over distances of from a few millimeters to several centimeters.

During a swallow, the upper esophageal sphincter relaxes immediately before the lower pharyngeal muscles contract. This decreases the pressure in the area of the sphincter (Fig. 5B) and allows the bolus to enter the esophagus. The sphincter then contracts to prevent reflux and entry of air. Pressure at the sphincter returns to the basal level as the muscle assumes its resting tone. A *primary peristaltic contraction* of the esophageal body now begins just below the upper esophageal sphincter. The coordinated and sequential nature of the contractions produces a zone of increased pressure that moves down the esophagus with the bolus in front of it. As the peristaltic contraction passes, the esophageal muscle returns to its resting tone and becomes flaccid again. As the bolus approaches the lower esophageal sphincter, the sphincter relaxes (Fig. 5B), allowing the bolus to enter the stomach. The sphincter then contracts back to its resting tone, which results in a pressure higher than that in the stomach, thereby preventing reflux.

Esophageal peristalsis is slow, moving down the esophagus at velocities ranging from 2–6 cm/sec, and may take a total of 10 sec. However, the time it takes for

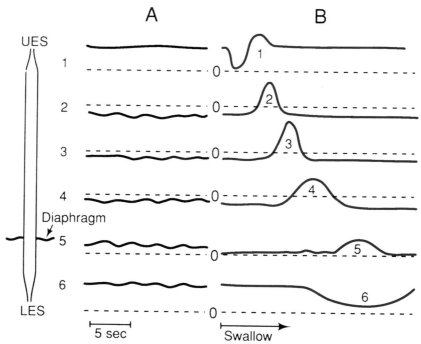

FIGURE 5   Intraluminal esophageal pressures recorded (A) between swallows and (B) during a swallow. X axis, time; UES, upper esophageal sphincter; LES, lower esophageal sphincter. (Adapted from Johnson LR, ed., *Gastrointestinal physiology*, 6th ed. St. Louis: CV Mosby, 2001.)

a bolus to reach the stomach also depends greatly on gravity and the physical nature of the material swallowed. Liquids swallowed in an upright position will reach the stomach before the peristaltic contraction. Therefore, although both esophageal sphincters must relax, peristalsis per se is not always necessary for esophageal transport. For most material, however, peristalsis is essential for transport to the stomach. The efficiency of esophageal peristalsis can be demonstrated readily by drinking a liquid while standing on ones head.

Frequently, however, the esophagus is not totally emptied by the original peristaltic contraction initiated by the swallow. The resulting distension, induced by the residual material, initiates another peristaltic contraction beginning just above the area of distension. This contraction, which occurs in the absence of a swallow or the pharyngeal phase, is called *secondary peristalsis* and is involuntary and not usually sensed. Secondary peristaltic contractions serve to "sweep" the esophagus clean of material left from a swallow or refluxed from the stomach. At times, it may take several secondary contractions to remove all material from the esophagus.

Esophageal motility is regulated by both central and peripheral mechanisms, all of which are not completely understood. Closure of the upper esophageal sphincter depends on the tone and elasticity of the cricopharyngeal muscle. Relaxation of this muscle during a swallow is coordinated through the swallowing center, as previously discussed. The body of the esophagus receives innervation through the vagus nerves (see Fig. 4). Somatic motor nerves from the nucleus ambiguus synapse directly with the striated muscle in the upper portion of the esophagus. Visceral motor nerves, arising from the dorsal motor nucleus, synapse with neurons of the myenteric plexus (between the circular and longitudinal layers of smooth muscle). These interneurons in turn innervate the smooth muscle layers (see Fig. 4). The interneurons also communicate signals with each other—in other words, up and down the esophagus—and relay input to the central nervous system (CNS) via afferent nerves in the vagus. Afferent input, for example, initiates secondary peristaltic contractions and alters the intensity of esophageal contractions.

Although primary esophageal peristalsis is initiated during a swallow via sequentially fired nerves from the swallowing center, the CNS is not necessary for the entire peristaltic contraction. Interruption of central influences by bilateral vagotomy produces abnormalities of peristalsis in the striated muscle portion of the esophagus, but peristalsis occurs normally in the smooth muscle portions. Peristalsis can also be induced in excised esophagi placed in physiologic salt solutions.

This indicates that the interneurons or smooth muscle cells themselves are able to coordinate the contractions.

Resting tone of the lower esophageal sphincter appears to be largely myogenic. Passive stretching of this portion of the esophagus results in smooth muscle contraction independent of neural and hormonal input. This resting tone, however, is modified by various agents. Gastrin increases tone, and numerous descriptions of lower esophageal sphincter regulation state that this is a physiologically important control mechanism. Current information, however, indicates that in physiologic amounts, gastrin does not play a normal role in the regulation of tone or contractions of the sphincter. Acetylcholine and related compounds increase resting tone, and prostaglandin E and isoproterenol decrease tone.

Relaxation of the sphincter as a peristaltic wave approaches is mediated neurally. Vagal stimulation decreases sphincter tone and pressure. The mediator of this response is vasoactive intestinal peptide (VIP) (see Chapter 32), which initiates the synthesis of nitric oxide (NO). NO is a potent relaxer of smooth muscle and directly causes the lower esophageal sphincter to relax.

In the condition known as *achalasia,* the lower esophageal sphincter fails to relax during swallowing and the primary peristaltic contractions are weak and nonpropulsive. Thus, the esophagus becomes functionally obstructed and swallowed material builds up, the lower esophagus dilates, and aspiration may occur. Treatment involves stretching the sphincter with a balloon or surgically weakening the sphincter muscle.

Motor abnormalities of the esophagus also occur in diffuse esophageal spasm, resulting in simultaneous strong contractions of long duration. These contractions are largely nonpropulsive. Obstruction of the esophagus often occurs in esophageal cancer. Scleroderma frequently begins as nerve degeneration in the esophagus and progresses to the replacement of smooth muscle with noncontractile fibrous tissues. This patient experiences heartburn and dysphagia (difficulty swallowing).

Receptive Relaxation of the Stomach

Swallowing also involves the stomach. The pressure within the orad portion of the stomach is essentially equal to intra-abdominal pressure, which is slightly higher than atmospheric. During a swallow, the orad stomach relaxes before the arrival of the bolus at the same time that the lower esophageal sphincter relaxes. After the bolus enters the stomach, pressure returns slowly to basal levels. In this manner, the stomach can accommodate volume changes of as much as 1500 mL with negligible increases in pressure. Accommodation is made possible by the active relaxation of the gastric smooth muscle, a process referred to as *receptive relaxation*.

Receptive relaxation is mediated by a *vagovagal reflex,* which means that afferent information from the stomach is relayed to the CNS via the vagus and that the signal from the CNS resulting in relaxation also reaches the gastric smooth muscle via vagal efferents. Receptive relaxation can be elicited in the absence of a swallow by distending or stretching the stomach. Vagotomy abolishes this reflex. The neurotransmitter has not been identified but nitric oxide appears to be involved. The hormone cholecystokinin (CCK) functions physiologically to make the orad stomach more distensible, thus facilitating this process.

GASTRIC MOTILITY

Activity of the smooth muscle of the stomach carries out three principal functions: (1) The muscle relaxes to accommodate large volumes ingested during a meal; (2) contractions mix ingested material with gastric juice, facilitating digestion, solubilizing some constituents, and reducing the size of the particles; and (3) contractions propel material into the duodenum at a rate regulated to provide optimal time for intestinal digestion and absorption.

Structure and Innervation of the Stomach

Unlike the rest of the GI tract, the stomach has three layers of smooth muscle cells: an outer longitudinal layer, a middle circular layer, and an inner oblique layer. The longitudinal layer is incomplete, being absent from the anterior and posterior surfaces, whereas the circular layer is the most prominent. The oblique layer is formed from two bands of muscle that radiate from the lower esophageal sphincter over the anterior and posterior surfaces to fuse with the circular muscle in the caudad portion of the stomach. The thickness of the circular and longitudinal layers increases as one moves distally toward the pylorus (Fig. 6).

The innervation of the stomach is similar to that described in the previous chapter for the GI tract in general. Although both plexuses are present, the myenteric is the most prominent and receives extrinsic innervation from the vagus and from fibers originating in the celiac plexus of the sympathetic nervous system. The myenteric plexus also receives innervation from other intrinsic plexuses. Axons from neurons of the myenteric plexus synapse with smooth muscle cells and with secretory and endocrine cells of the mucosa.

As shown in Fig. 6, the stomach can be divided into two regions based on patterns of motility: the *orad* or *proximal region* is responsible for accommodating an

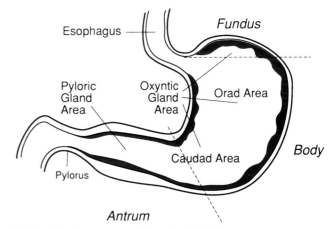

FIGURE 6 Anatomic and functional divisions of the stomach. To consider motility, the stomach is divided into an orad (proximal) region and a caudad (distal) region. Fundus, body, and antrum are anatomic designations for regions of the stomach.

ingested meal, and the *caudad* or *distal region* is responsible for the contractions that lead to mixing and propulsion into the duodenum.

Motility of the Orad Area

As indicated earlier, the most important motor activity of the orad region of the stomach is accommodation to ingested material. The musculature of the orad stomach is thin, and during the remainder of the digestive process, contractions are weak and infrequent. Therefore, ingested material is deposited in layers and may remain relatively unmixed for up to an hour. This allows salivary amylase, which is denatured when mixed with gastric acid, to digest a significant proportion of the carbohydrates present. The weak contractions of the orad region reduce the size of the stomach as its contents empty. It is uncertain whether the musculature contracts tonically to maintain pressure on the remaining contents or whether contractions propel the contents into the caudad stomach. Little is known about the regulation of these contractions. In addition to facilitating accommodation, CCK inhibits these contractions, making the stomach more distensible and ultimately decreasing the rate of gastric emptying.

Motility of the Caudad Area

In the fasted state, for the greater part of the time, the stomach is quiescent. However, vigorous contractions may occur at the same time in the caudad stomach and the upper part of the small intestine. In humans these contractions occur at intervals of approximately 90 min and are referred to as the *migrating motility complex* or *migrating myoelectric complex*. The contractions begin in the midportion of the stomach and move distally over the caudad area. These periods of cyclical contractions, which last 3–5 minutes, propel any residue from the previous meal and accumulated mucus into the duodenum. Eating abolishes the migrating motility complex. After a meal, contractions are more frequent and of varying strength, depending on the nature of the material ingested. These contractions are peristaltic and originate in the midstomach. As they move distally toward the pylorus, both velocity and force of contraction increase (Fig. 7). In humans, each contraction lasts from 2–20 seconds, and contractions may occur at rates of 3–5 per minute. This is the primary form of motility of the caudad stomach and both mixes food with digestive juices and propels it into the duodenum. As the peristaltic contraction approaches the pylorus, its velocity actually increases so that it overtakes most of the gastric contents moving in front of it. Some of the contents are "squirted" into the duodenum, but most

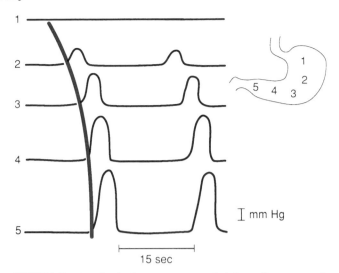

FIGURE 7 Intraluminal pressures recorded from five sensors in the stomach. Note that there is little activity in the orad region and that peristaltic contractions increase in force and velocity as they near the pylorus. (Modified from Johnson LR, ed., *Gastrointestinal physiology*, 6th ed. St. Louis: CV Mosby, 2001.)

are propelled back into the stomach because the wave of contraction has closed the distal antrum. This "retropulsion" back into the stomach thoroughly mixes the contents and reduces the size of solid particles.

Contractions of the caudad area of the stomach are initiated by intrinsic electrical activity generated by pacemaker cells known as the interstitial cells of Cajal, which have unstable resting membrane potentials. The pacemaker conductance is regulated by periodic releases of Ca^{2+}, resulting in a rhythmic depolarization and repolarization. These cyclic fluctuations are called *slow waves*, and if they reach a threshold level of depolarization, they trigger a contraction. These electrical changes are often referred to as *pacesetter potentials* or the *basic electrical rhythm* of the stomach. These latter designations refer to the fact that the slow waves determine the timing and pattern of contractions. Slow waves are always present even if not of sufficient magnitude to produce contractions. In humans, their frequency is constant and ranges between three and five cycles per minute. Thus, as previously noted, the frequency of contractions is also three to five per minute.

Whether an individual slow wave results in a contraction and the amplitudes of contractions are determined by hormonal and neural events dependent on the nature of the gastric contents and the digestive state. Slow waves originate from an area in the middle of the stomach, near the border of the orad and caudad areas. This is often referred to as the *gastric pacemaker*. The slow wave then spreads toward the gastroduodenal junction, increasing in both velocity and amplitude as it nears the pylorus. The increases in velocity and amplitude account for the similar changes in the gastric peristaltic wave as it moves toward the duodenum.

The interstitial cells of Cajal are intercalated between the intramural neurons and the smooth muscle cells and possess receptors for neurotransmitters and some hormones including CCK. Additional evidence indicates that the interstitial cells of Cajal are coupled with smooth muscle cells via gap junctions. Similar arrangements of these cells also exist in the small and large bowels and are responsible for generating the electrical activity resulting in the movements of these tissues as well.

Three gastric slow waves and their resulting contractile events are depicted in Fig. 8. The slow wave consists of an initial rise (depolarization) called the upstroke potential and a relatively flat plateau potential (Fig. 8A). For contraction to occur, the plateau potential must exceed threshold, as in Fig. 8B. At times, the plateau potential may have a number of spike potentials superimposed on it (Fig. 8C). These lead to increased strength of contraction and may initiate and prolong contractions. Spike potentials occur most frequently in the muscle of the distal portion of the antrum. Once threshold is reached, the greater the amplitude of the slow wave, the greater the force of contraction.

Although digestive events and the resulting neural and hormonal input to the stomach are not necessary for slow waves to occur, they markedly affect the amplitude of the slow waves and the degree of spiking on the plateau potential and, thus, the frequency of the contractions. These events have important regulatory effects on gastric motility. In general, vagal stimulation increases the force and frequency of contractions, whereas sympathetic stimulation decreases both of these parameters. Gastrin and motilin stimulates contractions, and secretin and gastric inhibitory peptide (GIP) decrease contractions. Motilin is responsible for the periodic contractions of the migrating motor complex during fasting. The physiologic significance of the gastric motor effects of gastrin, secretin, and GIP is doubtful.

Gastric Emptying

After a normal mixed meal, the stomach may contain 1500 mL of solids, ingested liquids, and gastric juice that take approximately 3 hr to empty into the duodenum. Gastric emptying is regulated by a variety of mechanisms to ensure that it occurs at a rate optimal for the digestion and absorption of nutrients and the neutralization of gastric contents. In general, the greater the volume, the more rapidly the contents empty. Liquids empty more rapidly than solids, and solids must be reduced in size to particles of 2 mm^3 or less for emptying to occur. The regulation of emptying based on volume and particle size is, for the most part, intrinsic to the gastric smooth muscle itself.

The most rapidly emptying substance is isotonic saline. Both hypo- and hypertonic saline empty more slowly. The addition of calories, especially in the form of lipids, or acid further slows emptying. The receptors for these responses are located in the duodenal mucosa and are sensitive to changes in osmolarity, pH, or lipid content. Receptor activation triggers several neural and hormonally mediated mechanisms that inhibit gastric emptying. For example, fats release CCK, which physiologically inhibits emptying. Acid, when placed in the duodenum, inhibits motility and gastric emptying with a latent period as short as 20–40 sec, indicating that the inhibition is due to a neural reflex. This reflex appears to be entirely intrinsic, with information from the duodenal receptors to the gastric smooth muscle carried by the neurons of the intramural plexuses. Other hormones such as secretin and GIP also inhibit emptying, but do not appear to do so in physiologic concentrations. Much of the regulation of gastric emptying is mediated by as yet undefined pathways.

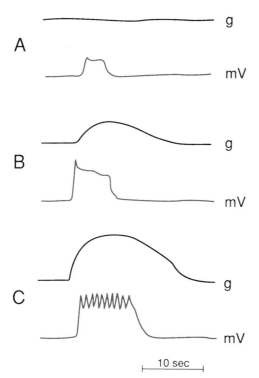

A   g / mV

B   g / mV

C   g / mV

⊢ 10 sec ⊣

FIGURE 8   Three sets of mechanical (g) and electrical (mV) tracings from the caudad region of the stomach. (A) Slow-wave depolarization of insufficient magnitude to cause contraction. (B) Increased depolarization results in contraction. (C) Electrical slow wave with multiple spike potentials and extended plateau produces a vigorous and extended contraction.

Clinical Note

Most disorders of gastric motility result in abnormalities of gastric emptying. Impaired emptying produces symptoms of fullness, loss of appetite, nausea, and sometimes vomiting. Reduced emptying may be caused by obstruction of the gastroduodenal canal by peptic ulcers or cancer. Vagotomy, which may be performed to reduce acid secretion in patients with peptic ulcer disease, delays gastric emptying. This is usually prevented by a procedure called *pyloroplasty*, in which the surgeon cuts and weakens the muscle in the pyloric area. A rapid rate of gastric emptying may result in diarrhea due to the osmotic load placed in the small intestine during a given time period. Increased rates of gastric emptying are also associated with duodenal ulcer disease, indicating that the acid entering the intestine cannot be neutralized before it damages the duodenal mucosa.

Inhibition of gastric emptying results from a variety of changes in gastric and duodenal motility. These include increased distensibility of the orad stomach (this is the effect of CCK, for example), decreased frequency and force of peristaltic contractions in the caudad stomach, decreased diameter of the pylorus, and increased tone and, hence, pressure of the proximal duodenum. Although noticeable differences exist between the motility of the distal stomach and proximal duodenum, it is unclear whether a true sphincter exists at the pylorus. Some investigators have demonstrated a zone of increased pressure (indicative of a sphincter) between the human stomach and duodenum, but others have failed to do so. It is also uncertain whether a distinct anatomic ring of circular muscle exists between the two. However, recent studies have shown that the pylorus can contract independently and produce a large effect on gastric emptying.

SMALL INTESTINAL MOTILITY

The primary functions of the small intestine are the digestion and absorption of nutrients and the absorption of fluid and electrolytes. Almost all nutrient absorption occurs in the duodenum and jejunum. The motility patterns of the small intestine are organized to optimize these functions. These include (1) mixing of contents with digestive enzymes and other secretions; (2) further reduction of particle size and solubilization; (3) circulation of contents to achieve optimal exposure to the membranes of intestinal cells, which contain additional digestive enzymes and are the absorbing surfaces; and (4) net propulsion of contents through the small intestine into the large bowel.

Structure and Innervation of the Small Intestine

The small intestine conforms to the general anatomic pattern already discussed. Both circular and longitudinal layers of smooth muscle are present and complete throughout its length. The circular muscle is somewhat thicker than the longitudinal, and the thickness of each decreases distally toward the ileocecal junction. The major part of digestion and absorption occurs in the duodenum and jejunum. The duodenum is only approximately 20 cm long in humans, whereas the jejunum and ileum together are 7–8 m long. Extrinsic innervation to the small intestine is supplied by the vagus nerve and by sympathetic fibers from the celiac and superior mesenteric ganglia. Although many vagal fibers are cholinergic and many sympathetic fibers are adrenergic, other neuromediators are also involved that have not been identified. Extrinsic nerves innervate elements of a well-developed enteric nervous system, which contains neurons, receptors, and nerve endings arranged in networks to form several plexuses.

Movements of the Small Intestine

In the fasting human, contractions of the small intestine are spaced unevenly over time. At any one point, there are cycles of few contractions and no contractions. Most contractions involve short (1- to 5-cm) lengths of bowel. Approximately every 90 min there are periods of intense contractions that move from the duodenum to the ileocecal sphincter. These are the migrating motility complexes similar to those previously described in the stomach. It takes about 90 min for one group of contractions to sweep the entire length of the small intestine, and then the complex is repeated. This complex clears undigested residue from the bowel.

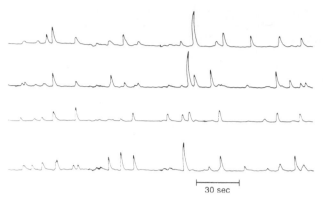

FIGURE 9   Intraluminal pressure changes recorded from the human duodenum following a meal. Recordings are from four sensors placed 1 cm apart. Note that a large contraction at one site is not necessarily propagated to the next. (Modified from Johnson LR, ed., *Gastrointestinal physiology*, 6th ed. St. Louis: CV Mosby, 2001.)

After a meal, the migrating motility complex disappears and contractions are spread more uniformly over time and are present about 30% of the time at any one locus. Most of these contractions are isolated, occurring over short (1- to 5-cm) lengths of bowel, and produce intraluminal pressure waves that are relatively uniform in shape and last approximately 5 sec (Fig. 9). The effect that any one of these contractions has on the intestinal content depends on the contractions occurring in adjoining sections of bowel. If a contraction is isolated and not coordinated with movement above or below, it separates the bowel into segments, propelling contents in both orad and caudad directions. When the contracting area relaxes, the contents flow back into the original segment with the result that mixing has occurred without net propulsion. This is the predominant motor activity in the fed individual and divides the bowel into segments, which accounts for the name *segmentation* given to this pattern. However, if contractions of adjacent segments are coordinated in a proximal to distal manner, a short peristaltic contraction occurs, resulting in net propulsion of contents. In normal individuals, peristaltic or proximal to distal contractions occur only over short distances of bowel. Rapid movement of chyme from the duodenum to the ileum would be disadvantageous because the time for digestion and absorption would be insufficient.

Control of Small Intestinal Motility

As in the stomach, interstitial cells of Cajal are responsible for a pacemaker potential that results in smooth muscle cell membrane depolarizations and repolarizations of 5–15 mV (Fig. 10). This slow wave activity or basic electrical rhythm is similar to that in the stomach in that it is always present whether or not

contractions occur and sets the maximal frequency at which contractions can occur in any one part of the intestine. The slow waves of the intestine differ from those of the stomach in that they do not trigger contractions themselves. Contractions do not occur unless one or more spike potentials are present on the plateau portion of the slow wave (Fig. 10). Spike potentials are qualitatively similar to action potentials, but for a variety of reasons, some of which are historical, they are referred to as spike potentials in discussions of the small intestine. The slow waves occur after time delays in descending portions of the intestine; thus, they have the appearance of being propagated (Fig. 10). Distally, the slow waves become broader (longer duration); thus, the frequency of slow waves decreases distally.

Although the slow wave frequency decreases distally, the frequency at any one point is constant, ranging from 11 to 12 cycles per minute in the duodenum to 8 to 9 in the terminal ileum. The slow waves, therefore, determine the maximal frequency at which contractions can occur. In the proximal duodenum, this frequency is 12 cycles per minute. Even if spike potentials are not present on every slow wave, the slow waves determine the intervals between contractions. Again, in the proximal duodenum most contractions occur at time intervals divisible by 5 sec (12/min).

Whether spike potentials and, hence, contractions occur depends on neural and hormonal input to the smooth muscle. This of course depends, in turn, on the nature and volume of luminal contents and the digestive

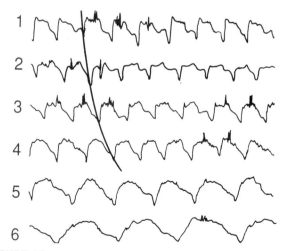

FIGURE 10   Six tracings of slow waves and spike potentials from successive and progressively distal areas of small intestine. The solid curved tracing shows apparent propagation of slow waves. Note that slow-wave frequency decreases distally. Rapid transient depolarizations on some slow waves are spike potentials and result in contractions. (Modified from Johnson LR, ed., *Gastrointestinal physiology*, 6th ed. St. Louis: CV Mosby, 2001.)

state of the individual. In general, the excitability and frequency of contractions of intestinal smooth muscle are enhanced by parasympathetic and decreased by sympathetic stimulation. There is evidence that VIP and nitric oxide act together to mediate neurally controlled relaxation, and that the opioid peptides may mediate some of the contractile responses. Serotonin may also be a neurotransmitter involved in peristalsis. Motilin is released into the bloodstream before each migrating motor complex and mediates this response. In addition, gastrin, CCK, and insulin stimulate contractions, whereas secretin and glucagon inhibit them, but the physiologic significance of these effects is doubtful.

Intestinal motility is also regulated by a number of neural reflexes that depend on either or both the intrinsic and extrinsic nervous systems. When the intestinal wall is distended by the placement of a bolus within the lumen, the muscle above the bolus contracts and that below it relaxes, resulting in a peristaltic contraction. This contraction normally propels the bolus a short distance, but may move it the entire length of the gut. This is known as the peristaltic reflex or, as described originally by Bayliss and Starling, the *law of the intestines*. This reflex depends on the intrinsic or enteric nervous system. Another reflex prevents the movement of additional material into an already severely distended section of bowel. This *intestinointestinal reflex* inhibits contractile activity in the remaining bowel after severe distension of one portion. This reflex depends on the extrinsic nervous system and is abolished when the extrinsic nerves are sectioned. Increased motor activity of the ileum accompanies gastric secretion and emptying. This results in the movement of ileal contents into the large intestine after a meal and is called the *gastroileal reflex*. Motility of both the small and large intestines is also influenced by the higher centers of the nervous system. The fact that emotions alter bowel functions is appreciated by most individuals.

Vomiting

Vomiting (emesis) is the forceful expulsion of intestinal and gastric contents through the mouth. General discharge of the autonomic nervous system precedes and accompanies vomiting, resulting in copious salivation, sweating, rapid breathing, and an irregular heartbeat. In humans, vomiting is usually, but not necessarily, associated with the feeling of nausea. Vomiting is normally preceded in humans by retching, which is the pattern of activity that overcomes the normal antireflux mechanisms of the GI tract.

A wave of reverse peristalsis begins in the distal small intestine moving intestinal contents orad. Retching begins as this wave moves through the duodenum.

Reverse peristalsis has not been described in humans simply because the experiment has not been done with the proper recording methods, but the phenomenon has been adequately described in dogs and cats using surgically implanted recording devices. A retch begins with a deep inspiration against a closed glottis and a strong contraction of the abdominal muscles. This increases intra-abdominal pressure and decreases intrathoracic pressure, so that the pressure gradient within the portions of the GI tract located in the two regions may become as much as 200 mm Hg. With each retch, the abdominal portion of the esophagus and a portion of the stomach actually slide through the hiatus in the diaphragm and move into the thorax. A contraction of the antrum and continued reverse peristalsis force the gastric contents through a relaxed lower esophageal sphincter into the flaccid esophagus. As the retch subsides, the stomach moves back into the relaxed abdomen and most of the contents drain from the esophagus back into the stomach. The cycle may be repeated several times, during which the upper esophageal sphincter has remained closed, preventing gastric contents from entering the mouth or pharynx.

Vomiting itself occurs after a sequence of stronger or developed retches. A sudden strong contraction of the abdominal muscles raises the diaphragm high into the thorax, resulting in an increase in intrathoracic pressure that may equal 100 mm Hg. The larynx and hyoid bone are then reflexly drawn forward, and the increased intrathoracic pressure forces the contents of the esophagus past the upper esophageal sphincter and out of the mouth. Material remaining in the esophagus empties into the stomach after the abdominal muscles relax. If the stomach still contains a significant volume, the cycle may be repeated until almost all the contents have been expelled.

Vomiting is reflexively controlled by the vomiting center located in the medulla. Electrical stimulation of this area results in immediate vomiting without retching. Stimulation of another medullary area can result in retching without vomiting. In the normal situation, however, the areas interact with each other and their activities are closely correlated. The vomiting center is activated by afferent impulses triggered by diverse stimuli from many parts of the body. These include tickling the back of the throat, distension of the stomach or duodenum, dizziness, unequal vestibular stimulation (seasickness), pain from the urogenital system, and other painful injuries. Various chemicals stimulate vomiting by acting on either central or peripheral receptors. A group of receptors located in the floor of the fourth ventricle of the brain constitutes a "chemoreceptor trigger zone," which is activated by emetics in the blood or cerebrospinal fluid. Stimulation of this zone also occurs during vomiting arising from radiation or motion sickness. Chemically sensitive receptors also occur peripherally, and primarily in the GI tract. Most of these are located in the duodenum and respond to emetics such as ipecac.

In general, vomiting is a protective mechanism to rid the body of noxious or toxic substances. Prolonged vomiting, however, can cause severe problems in fluid and electrolyte balance, especially in children. Because gastric juice contains relatively (to plasma) high concentrations of H^+ and K^+, these individuals may develop metabolic alkalosis and hypokalemia.

LARGE INTESTINAL MOTILITY

Approximately 7–10 L of ingested or secreted water enters the small intestine during a 24-hr period. Of this amount, about 600 mL reach the colon. The motility patterns of the large intestine are organized so that all but about 100 mL are absorbed. The remaining fecal material is then stored until it can be evacuated conveniently.

Structure and Innervation of the Large Intestine

Beginning from the ileocecal junction, the large intestine is anatomically divided into the cecum; the ascending, transverse, descending, and sigmoid colon; the rectum; and the anal canal (Fig. 11). The inner, circular layer of smooth muscle fibers is continuous from the cecum to the anal canal. Within the rectum and anal canal, the circular layer thickens to form the internal anal sphincter. The outer, longitudinal layer of smooth muscle fibers is discontinuous, consisting of three bands that run the entire length of the large

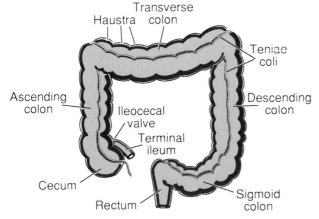

FIGURE 11   The anatomy of the large intestine.

intestine. These bands are called the *teniae coli* and fan out over the rectum to become relatively continuous. Layers of striated muscle distal to and overlapping the internal anal sphincter form the external anal sphincter.

The human large intestine is considerably thicker and shorter than the small intestine, and the teniae coli are easily visible. In addition, the colon is divided into segments, called *haustra* or haustrations, that give the appearance of a chain of small sacs. Haustra disappear during, and reappear after, contractions of a specific segment. They may also reform at other loci, so their locations are not fixed. Nevertheless, they probably have a structural basis as well as a functional one. Functionally, they may depend on the contractile state of the smooth muscle fibers; structurally, they may be influenced by areas of concentration of muscular tissue and by folds in the mucosa.

The myenteric plexus in the large intestine is concentrated underneath the teniae, that is, between the layers of longitudinal and circular muscle. The myenteric plexus receives input from local receptors and from both the parasympathetic and sympathetic systems. Parasympathetic fibers from the vagus innervate the cecum and ascending and transverse colons, and the descending and sigmoid colons and the rectum and anus are innervated by the pelvic nerves from the sacral region of the spinal cord. The pelvic nerves enter the colon near the rectosigmoid junction and project anteriorly and posteriorly along the myenteric plexus innervating it as well. Sympathetic innervation is supplied to the proximal colon via the superior mesenteric ganglion, whereas the inferior mesenteric ganglion provides fibers to the distal colon. The rectum and anal canal receive sympathetic innervation from the hypogastric plexus. Within the autonomic system acetylcholine, substance P, and tachykinins mediate contraction, whereas VIP and nitric oxide mediate relaxation. The voluntary external anal sphincter is innervated by the somatic pudendal nerves, where transmission is mediated by acetylcholine.

Motility of the Cecum and Proximal Colon

The pressure recorded by a sensor placed in the ileocecal sphincter is several millimeters of mercury greater than that in the ileum or colon. The pressure, however, is not constant. When the ileum is distended, the sphincter relaxes, allowing contents to flow into the colon. When the colon is distended the sphincter contracts, preventing reflux into the ileum. Thus, material moves intermittently from the ileum to the colon.

Most of the movements of the proximal colon are segmental, serving to mix contents back and forth, and

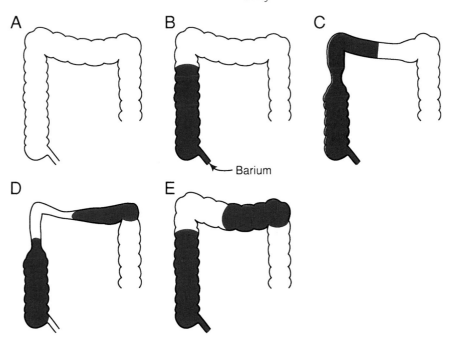

FIGURE 12 Mass movement. (A) Colon before entry of barium sulfate. (B) Barium enters proximal ascending colon, showing haustra. (C) As more barium enters, the haustra disappear from a portion of the ascending and transverse colons, and a contraction begins in this area. (D) The contraction has moved a portion of the barium into the caudad transverse colon. (E) Haustra return.

exposing them to absorptive surfaces. These contractions last from 12–60 seconds and generate intraluminal pressures between 10 and 50 mm Hg. These segmental contractions are believed to be partly responsible for the formation of haustra.

Approximately one to three times a day, peristaltic contractions move significant amounts of material caudally from one region of the colon to another. At the start of a so-called *mass movement,* segmentation ceases and haustrations disappear from the segment of bowel involved (Fig. 12). Contents are moved in a narrow tubular length of bowel at least 20 cm long. One mass movement may transport contents from the transverse to the sigmoid colon or rectum. After the mass movement, haustrations reappear and segmentation begins.

Motility of the Descending and Sigmoid Colon

Most absorption of water and electrolytes occurs in the proximal colon, so that material present in the distal colon is in a semisolid state. Even so, the primary form of motility is nonpropulsive segmentation. This activity produces a certain amount of resistance, which retards movement of material into the rectum. Net caudad movement of contents in this region also occurs by mass movements that propel fecal contents into the rectum.

Motility of the Rectum and Anal Canal

The rectum has more segmental contractions than the sigmoid colon. Therefore, unless filled by a mass movement, it is kept empty or nearly so by retrograde movement of contents into the sigmoid colon. This accounts for the retention and subsequent absorption of suppositories. As fecal material is forced into the rectum, the rectum contracts and the internal anal sphincter relaxes (Fig. 13). Normally, the anal canal is closed by contraction of the internal sphincter. The relaxation of the internal anal sphincter after contraction of the rectum is the *rectosphincteric reflex*. Filling the rectum to about 25% of capacity produces an urge to defecate. Defecation is prevented by the external anal sphincter, which is normally in a state of tonic contraction maintained by reflex activation through dorsal roots in the sacral segments. In paraplegics lacking this tonic contraction, the rectosphincteric reflex results in defecation. In the normal individual, if defecation is not convenient, the external sphincter remains closed, the receptors in the rectum accommodate to the distension stimulus, the internal sphincter regains its tone (Fig. 13), and the urge to defecate subsides.

If the rectosphincteric reflex occurs under convenient circumstances, defecation may occur through a series of acts that are both involuntary and voluntary. The external anal sphincter is relaxed voluntarily, and the

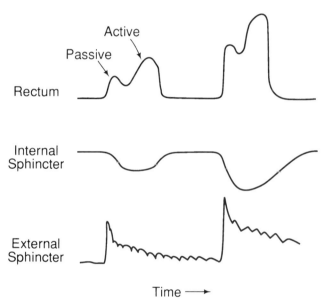

FIGURE 13 Pressures recorded within the rectum and the internal and external anal sphincters. As contents distend the rectum, the pressure increases passively; if sufficient, active contraction increases pressure further. This is accompanied by relaxation of the internal anal sphincter and contraction of the external anal sphincter. As contents continue to enter the rectum, the pressure in the internal anal sphincter decreases and the pressure in the external sphincter increases.

longitudinal muscles of the rectum and distal colon contract. This shortens the rectum and straightens the angle between the rectum and sigmoid colon, facilitating the passage of feces. This plus the resulting increase in pressure is often sufficient for evacuation. The process, however, is enhanced by inspiration and contraction of the chest muscles against a closed glottis and full lungs. This raises both intrathoracic and intra-abdominal pressures. Contraction of the abdominal muscles increases pressure further. Intra-abdominal pressures may increase to as much as 100–200 mm Hg. The pelvic floor—which supports the abdominal contents—relaxes during defecation, allowing the increased pressure to force the floor downward, which straightens the rectum.

The hemodynamic consequences of the pressure changes involved in defecation are substantial. There is an abrupt rise in arterial pressure as the increased intrathoracic pressure is transmitted across the wall of the heart and aorta. The large veins in the thorax collapse, stopping venous return and causing peripheral venous pressure to rise. Together, these events result in decreased stroke volume and cardiac output and a resultant drop in systemic arterial pressure. In a normal individual, these events are of no consequence, but death can result from cerebral vascular accidents caused by increased intracranial pressures produced while straining to defecate.

Control of Large Intestinal Motility

The resting tone of the ileocecal sphincter is primarily myogenous, that is, a property of the smooth muscle cells themselves. Relaxation of the sphincter in response to distension of the ileum and its contraction after distension of the proximal colon are neural reflexes. These are probably mediated entirely by the enteric nerves. Shortly after a meal, the ileum contracts and the sphincter relaxes, resulting in the propulsion of ileal contents into the colon. This is referred to as the *gastroileal reflex,* and appears to be mediated by the extrinsic autonomic nerves. There is some evidence that digestive hormones may also be involved.

As is the case with the stomach and small intestine, the membrane potential of the smooth muscle cells of

Clinical Note

The major advance, during the past 10 years or so, in our understanding of the control of gastrointestinal motility has been the appreciation of the role played by the interstitial cells of Cajal (ICCs). However, in addition to being responsible for generating the pacemaker potentials of the various tissues (as outlined in this chapter), strong evidence is accumulating that defects in these cells may be the cause of a variety of human motility disorders. ICCs express a novel protein, and antibodies to this protein have allowed clinicians to evaluate ICC function in a variety of conditions. These studies have shown that ICCs are reduced in number in pseudo-obstruction, achalasia, ulcerative colitis, Chaga's disease, diabetes, and slow transit constipation. Investigators are using animal models of various genetic mutations to determine whether changes in ICCs are the causes or the results of these conditions.

the colon and rectum fluctuates. Again, this fluctuation is initiated in the interstitial cells of Cajal. Spike potentials have also been recorded from the slow, wavelike depolarizations. Although these events initiate contractions, they are more irregular than those of the small intestine, and their correlation with contractile events is not totally understood.

Enteric or intrinsic nerves play a major role in colonic motility, for haustrations and mass movements occur in the absence of extrinsic innervation. The major effect of the enteric nerves is inhibitory, because interruption of their influence results in tonic contraction. This inhibition of contraction is probably mediated by VIP and nitric oxide. In *megacolon* or Hirschsprung's disease, the ganglion cells are absent from a segment of colon and the tissue concentration of VIP is extremely low. This results in constriction and loss of coordinated movements in the affected segment. Colonic contents accumulate proximal to the constriction, and the colon becomes grossly distended and hyperplastic. This condition is therefore the colonic correlate of esophageal achalasia. Surgical removal of the aganglionic segment usually restores

normal function. Defecation is controlled by both extrinsic and intrinsic nerves. Activity within the spinal cord reinforces the rectosphincteric reflex, which is primarily under control of the intrinsic nerves. The sensation of distension and the voluntary control of the external anal sphincter are mediated by nerves within the spinal cord to the cerebral cortex.

Suggested Readings

Conklin JL, Christensen J. Motor functions of the pharynx and esophagus. In Johnson LR, ed., *Physiology of the GI tract,* 3rd ed. New York: Raven Press, 1994, pp 903–928.

Christensen J. Motility of the colon. In Johnson LR, ed., *Physiology of the GI tract,* 3rd ed. New York: Raven Press, 1994, pp 991–1024.

Makhlouf GM. Neuromuscular function of the small intestine. In Johnson LR, ed., *Physiology of the GI tract,* 3rd ed. New York: Raven Press, 1994, pp 977–990.

Mayer EA. The physiology of gastric storage and emptying. In Johnson, LR, ed., *Physiology of the GI tract,* 3rd ed. New York: Raven Press, 1994, pp 977–990.

Sanders KM, Ördög T, Ward SM. Physiology and pathophysiology of the interstitial cells of Cajal: From bench to bedside. *Am J Physiol Gastrointest Liver Physiol* 2002; 282:G747–G756.

Secretion

LEONARD R. JOHNSON

KEY POINTS

- Saliva is produced in large volumes relative to the weight of the glands, contains relatively high concentrations of K^+, is hypotonic, and contains specialized organic substances.
- Acini produce the *primary saliva*, containing water and electrolytes in concentrations approximately equal to those of plasma.
- *Gastric acid* is secreted in concentrations as high as 150 mmol/L by the parietal cells, which contain a proton pump (H^+,K^+-ATPase) in their apical membranes. Acid converts inactive pepsinogen to pepsin, solubilizes some food, and kills bacteria.
- The major stimulants of acid secretion are the hormone gastrin, the vagal mediator acetylcholine, and histamine, which is released from the enterochromaffin-like (ECL) cells by gastrin and acetylcholine and acts as a paracrine to stimulate the parietal cells.
- The *cephalic phase of gastric secretion* is mediated by the vagus nerve, which stimulates the parietal cell via acetylcholine and releases

Essential Medical Physiology, Third Edition

INTRODUCTION

Three major types of secretory products are elaborated by the cells and organs of the digestive tract. These are water and electrolytes, mucus, and specialized organic molecules.

Throughout the entire length of the GI tract, epithelial cells secrete water and electrolytes. In addition, the salivary glands, pancreas, and liver produce copious secretions of fluid and ions. These secretions function in a variety of ways. Water liquifies the luminal contents, dissolves a significant portion of nutrients, and provides a medium for the chemical reactions of digestion to occur. Hydrogen kills some bacteria, takes part in the digestion of protein, and acidifies the gastric contents. Bicarbonate ion protects the mucosa from acid injury and increases the pH of the intestinal contents into the range for optimal enzyme activity.

The primary functions of mucus are derived from its glycoproteins. Although the chemical nature of these may vary slightly in different parts of the GI tract, its functions to lubricate and protect the mucosa are the same throughout. Mucus is produced and secreted by specialized cells found along the entire length of the tract. Chewing mixes food with mucus secreted by the salivary glands so that it can be swallowed easily. In the colon, mucus causes fecal particles to adhere and coats them so they can be excreted readily. Throughout the tract mucus lines the digestive cavity, preventing abrasions and keeping the cells from coming in contact with acid.

Digestive enzymes constitute the major group of specialized organic compounds secreted into the GI tract. Important steps in the digestion of all three classes of nutrients occur with the lumen of the gut. In addition, some organic compounds such as bile acids and intrinsic factor are essential for optimal digestion and/or absorption.

This chapter describes these secretions and their sources and how the various specialized cells of the mucosa and glands produce and release these secretory products. Considerable space is given to describing the sometimes involved regulatory processes which ensure that the secretions are elaborated in the quantities and sequence such that optimal digestion and absorption are achieved.

SALIVARY SECRETION

The functions of saliva fall into three general categories: digestion, lubrication, and protection. Saliva is produced in large volumes, relative to the weight of the salivary glands, by an active process. Unlike the process in the other gastrointestinal glands, the secretion of saliva is almost totally under neural control. Both branches of the autonomic nervous system stimulate salivary secretion, although the parasympathetic system provides a much stronger input. The healthy adult secretes approximately 1 L of saliva per day.

Functions of Saliva

The digestive actions of saliva are the results of two enzymes, one directed toward carbohydrates and the other toward fat. Saliva contains an α-amylase, called *ptyalin,* that cleaves internal α-1,4-glycosidic bonds. The enzyme is identical to pancreatic amylase, and the products of exhaustive digestion are maltose, maltotriose, and α-limit dextrins, which contain the α-1,6 branch points of the starch molecule. The pH optimum for the enzyme is 7, and it is rapidly denatured when exposed to gastric juice at a pH of less than 4. However, because a large portion of a meal often remains unmixed in the orad stomach, salivary amylase is able to act within this mass of ingested material and digest up to 75% of the starch before being mixed with acid. Without salivary amylase, there is no deficiency in carbohydrate digestion, because the pancreatic enzyme is secreted in sufficient amounts to digest all of the starch present in the small intestine.

Lingual lipase is secreted by salivary glands of the tongue and begins the digestion of triglycerides. Its acidic pH optimum allows it to remain active throughout the stomach and into the proximal duodenum.

Although this is not strictly a digestive function, saliva dissolves many dietary constituents. This process of solubilization increases the sensitivity of the taste buds. Saliva also washes food particles from the taste buds, so that subsequently ingested material can be tasted.

The lubricating properties of saliva are due primarily to its mucus content. Chewing mixes saliva with ingested material, thoroughly lubricating it and facilitating the swallowing process. The lubricated bolus moves more easily down the esophagus. Lubrication provided by saliva is also necessary for speech.

Saliva protects the mouth by diluting and buffering harmful substances. Hot solutions of tea, coffee, or soup, for example, are diluted by saliva and their temperatures lowered. Foul-tasting substances can eventually be washed out of the mouth by copious salivation. Before vomiting, salivation is stimulated strongly. This saliva dilutes and neutralizes corrosive gastric juice, preventing damage to the mouth and esophagus. Dry mouth, or *xerostomia,* is associated with chronic infections of the buccal mucosa and with dental caries. Saliva dissolves and washes food particles from between the teeth. Saliva also contains a number of organic substances that are bacteriocidal. These include a *lysozyme,* which attacks bacterial cell walls; the *binding glycoprotein for immunoglobulin A (IgA),* which together with IgA forms secretory IgA, which in turn is immunologically active against bacteria and viruses; and *lactoferrin,* which chelates iron, preventing access by organisms that require iron for growth. Various compounds such as fluoride and calcium phosphate are taken up by the salivary glands and secreted in the saliva in concentrated amounts. These in turn may be incorporated into the teeth.

Anatomy and Innervation of the Salivary Glands

The major salivary glands are three paired structures that deliver their secretions into the mouth through ducts. The largest of these are the parotid glands, located between the angle of the jaw and the ear; the submaxillary glands are located below the angle of the jaw; and the sublingual glands, as their name implies, are found below the tongue. The parotid glands are made up only of serous cells and secrete a watery fluid. The other two pairs are mixed glands containing cells that secrete mucin glycoprotein as well. Smaller salivary glands occur within the mucosa of the tongue, lips, palate, and other areas of the buccal cavity.

The microscopic structure of the salivary glands is similar to that of the pancreas and analogous to a bunch of grapes. A single grape corresponds to the *acinus,* which is the blind end of a branching duct system and is made up of a group of cells called *acinar cells* (Fig. 1). The acinar cells secrete the initial salivary fluid, consisting of electrolytes, mucus, and enzymes. From the acinus, saliva passes relatively unchanged through a short, *intercalated duct* and into the *striated duct.* The striated duct is lined by columnar epithelial cells that function like renal tubule cells to modify the inorganic composition of saliva. The combination of the acinus, intercalated duct, and striated duct represent the secretory unit or *salivon* of the salivary gland. The basement membranes of the acini and intercalated ducts are covered in part by specialized contractile cells called *myoepithelial cells.* These cells are shaped somewhat like stars, and the motile extensions contain actin and myosin. Contraction of the myoepithelial cells occurs

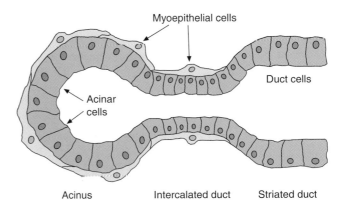

FIGURE 1 Schematic diagram of the functional histology of the salivon, the secretory unit of the salivary glands.

when salivary secretion is stimulated, and results in the rapid expulsion of saliva into the mouth. Contraction of these cells expels saliva from the acinus, shortens and widens the intercalated duct, and prevents distension of the acinus. Similar cells can be found in the mammary gland and pancreas.

The salivary glands receive a high blood flow proportional to their weight. The direction of the arterial flow is opposite to the flow of saliva through the salivon. Separate capillary beds, the vessels of which appear to be parallel to each other and the salivons, supply the ductules and the acini. These capillaries are extremely permeable, allowing rapid movement of water and other molecules across their basement membranes. The rate of blood flow through resting salivary tissue is approximately 20 times that through muscle. This, in part, accounts for the ability of these glands to produce prodigious amounts of saliva relative to their weight.

The salivary glands are innervated by both components of the autonomic nervous system. Parasympathetic innervation is delivered by the facial and glossopharyngeal nerves, whereas sympathetic innervation is from thoracic spinal nerves via the superior cervical ganglion. The autonomic nervous system regulates the secretion, blood flow, and growth of the salivary glands.

Composition of Saliva

Saliva is primarily a mixture of water, electrolytes, and some organic compounds. The *major characteristics of saliva* are (1) its large volume relative to the mass of the salivary glands, (2) its high potassium concentration, (3) its low osmolarity, and (4) the specialized organic materials it contains.

Inorganic Composition

Saliva can be elaborated in large volumes compared with the secretions of similar organs. Maximal secretory rates may be as high as 1 mL per gram of gland per minute, which is comparable to the secretory rate for the entire pancreas. On the basis of tissue weight, this amounts to a 50-fold higher rate of secretion.

At all secretory rates except the highest, saliva is significantly hypotonic to plasma. As the rate of secretion increases, the osmolarity of saliva increases and approaches isotonicity at maximal rates.

The concentrations of electrolytes in saliva vary with the rate of secretion (Fig. 2). The K^+ concentration is always greater than that found in plasma, indicating that the salivary glands secrete K^+ against its electrochemical gradient. As the flow rate of saliva increases, the K^+ concentration in the fluid decreases, plateaus, and remains relatively constant at higher flow rates. In

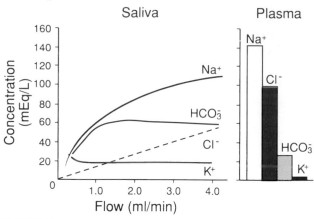

FIGURE 2 Concentrations of major ions in saliva as a function of the secretory rate. Note that, at all rates, Na^+ and Cl^- concentrations are lower and that K^+ and HCO_3^- are higher than their respective concentrations in plasma.

most species, the concentration of Na^+ in saliva is always less than that of plasma, and as the rate of secretion increases, the concentration of Na^+ also increases. In general, Cl^- concentrations parallel those of Na^+. These findings indicate that Na^+ and Cl^- are secreted and then reabsorbed as the saliva passes through the ducts. The concentration of HCO_3^- in saliva is higher than that found in plasma except at low flow rates, which accounts for changes in the pH of saliva as well. At unstimulated rates of flow, the pH is slightly acidic, but with stimulation it rapidly rises to around pH 8. The relationships between flow rates and ion concentrations shown in Fig. 2 vary somewhat depending on the nature of the stimulus.

Two basic types of studies indicate how the final saliva is produced and explain the relationship between ion concentrations and flow rates shown in Fig. 2. First, analysis of fluid collected by micropuncture of the intercalated ducts shows that Na^+, Cl^-, K^+, and HCO_3^- are present in concentrations approximately equal to their concentrations in plasma. This fluid is also isotonic to plasma. Second, if one perfuses a salivary gland duct with a solution containing Na^+, Cl^-, K^+, and HCO_3^- at concentrations similar to those present in plasma, the fluid collected at the duct opening has lower concentrations of Na^+ and Cl^- and higher concentrations of K^+ and HCO_3^-. In addition, the fluid becomes hypotonic. The longer one allows the fluid to remain in the duct, the greater the changes. These data indicate (1) that the acini produce a fluid having ion concentrations similar to those of plasma and (2) that as the fluid moves down the ducts, Na^+ and Cl^- are reabsorbed and K^+ and HCO_3^- are secreted into the saliva. The higher the flow of saliva, the less time available for these changes to take place. Thus, at high secretory rates the ionic composition of saliva more closely resembles that

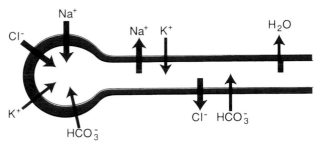

FIGURE 3 Fluxes of the primary ions and water across the salivon. The fluid leaving the acinus is isotonic to plasma. Na$^+$ and Cl$^-$ leave the duct, and K$^+$ and HCO$_3^-$ enter. The thickness of the arrows indicates that more Na$^+$ and Cl$^-$ leave the duct than K$^+$ and HCO$_3^-$ enter. Because the membrane is relatively impermeable to water, the saliva becomes hypotonic.

of plasma (Fig. 2). At low rates of flow, there is considerably more K$^+$ and considerably less Na$^+$ and Cl$^-$. The HCO$_3^-$ concentration remains fairly high even at high rates of secretion, because HCO$_3^-$ secretion is stimulated by most salivary gland agonists. Some K$^+$ and HCO$_3^-$ enter in exchange for Na$^+$ and Cl$^-$, but much more Na$^+$ and Cl$^-$ leave the ducts, making the saliva hypotonic. Because the epithelia of the salivary gland ducts are relatively impermeable to water, the saliva remains hypotonic. These processes are summarized in Fig. 3.

Current evidence indicates that Cl$^-$ is the major ion species actively secreted by the acinar cells (Fig. 4A). There is no evidence for direct active secretion of Na$^+$. The secretory mechanism for Cl$^-$ is inhibited by ouabain, indicating that it depends on the Na$^+$-K$^+$ pump in the basolateral membrane. Na$^+$ moves across the apical membrane of the acinar cell into the lumen to preserve electroneutrality, and water follows down its osmotic gradient. K$^+$ and HCO$_3^-$ enter saliva passively as well, but there is evidence for an active component for each. Within the ducts, Na$^+$ is actively absorbed and K$^+$ actively secreted (Fig. 4B). Some K$^+$ is secreted in exchange for H$^+$. In addition, HCO$_3^-$ is actively secreted in exchange for Cl$^-$. The net result is a decrease in Na$^+$ and Cl$^-$ concentrations and an increase in K$^+$ and HCO$_3^-$ concentrations and pH. The active absorption of Na$^+$ and secretion of K$^+$ is dependent on the (Na$^+$,K$^+$)-ATPase in the basolateral membrane. The ductule epithelium is relatively impermeable to water compared with that of the acini. Thus, there is a decrease in Na$^+$ and Cl$^-$ concentrations and an increase in K$^+$ and HCO$_3^-$ concentrations and pH as saliva moves through the ducts. Because more ions leave than enter, the saliva becomes hypotonic. Agents that stimulate salivary secretion increase the activity of these channels and transport processes. Aldosterone acts at the apical membrane to increase the absorption of Na$^+$ and secretion of K$^+$.

Organic Composition

Several of the organic materials synthesized and secreted by the salivary glands are discussed in the section describing the functions of saliva. These include the α-amylase ptyalin, lingual lipase, mucus, lysozymes, glycoproteins, and lactoferrin. Another enzyme produced by salivary glands is *kallikrein,* which does not have a digestive function, but converts a plasma protein into the potent vasodilator *bradykinin.* Kallikrein is released during increased metabolic activity of the salivary gland cells and results in increased blood flow to the secreting glands. Saliva also contains the blood group substances A, B, AB, and O.

The synthesis of salivary gland enzymes, their storage, and their release are similar to the same processes in the pancreas and are discussed later in this chapter. The total protein concentration of saliva is approximately one-tenth the concentration of proteins in the plasma.

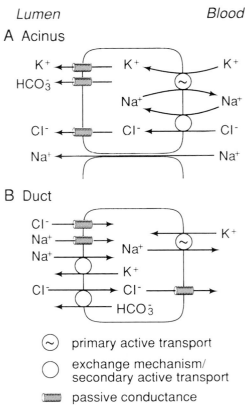

FIGURE 4 Transport mechanisms proposed to explain salivary secretion. (A) In the acinus, Cl$^-$ enters the cell as Na$^+$ enters down the concentration gradient created by the (Na$^+$,K$^+$)-ATPase. Cl$^-$ then diffuses into the lumen and Na$^+$ enters to maintain electrical neutrality. K$^+$ and HCO$_3^-$ are present in amounts equal to their concentrations in plasma. (B) In the ducts Na$^+$ leaves the lumen as it moves down the gradient created by the (Na$^+$,K$^+$)-ATPase. Some Na$^+$ leaves in exchange for K$^+$. Cl$^-$ leaves with Na$^+$ to preserve neutrality. Some Cl$^-$ leaves in exchange for HCO$_3^-$. In both cell types, the driving force is the basolateral (Na$^+$,K$^+$)-ATPase.

Regulation of Salivary Gland Secretion

For all practical purposes, salivation is under total control of the autonomic nervous system. Aldosterone and antidiuretic hormone (ADH) modify the ionic content of saliva by decreasing the Na^+ and increasing the K^+ concentrations, but they do not regulate the flow of saliva. The regulation of salivary gland secretion differs in this respect from the control of the secretions of the other digestive glands. The gastrointestinal hormones have major roles in regulating the secretions of the stomach, pancreas, and liver. The control of salivation is unusual in one other respect, because although the parasympathetic system exerts far greater influence than the sympathetic, activation of either of these systems stimulates secretion.

Stimulation of the parasympathetic nerves to the salivary glands initiates and maintains salivary secretion. Increased secretion is due to the activation of transport processes in both acinar and duct cells. Secretion is facilitated by the contraction of the myoepithelial cells, which are directly innervated by the parasympathetic nerves. Parasympathetic fibers also innervate the surrounding blood vessels, stimulating vasodilation and increasing the supply of blood to the secreting cells. Increased cellular activity in response to parasympathetic stimulation is accompanied by an increased consumption of glucose and oxygen and the production of metabolites, which also increase blood flow through vasodilation. In addition, kallikrein is released and the vasodilator bradykinin produced. Eventually, increased cellular activity results in growth of the salivary glands. Section of the parasympathetic nerves to the salivary glands causes the glands to atrophy, whereas sympathetic section has little effect. These processes are outlined in Fig. 5.

Sympathetic stimulation also increases secretion, metabolism, and growth of the salivary glands, although these responses are less pronounced and of shorter duration than those produced by the parasympathetics. The myoepithelial cells also contract in response to sympathetic stimulation. Stimulation via the sympathetic nerves produces a biphasic change in blood flow to the salivary glands. The earliest response is a decrease in blood flow caused by activation of α-adrenergic receptors and vasoconstriction. However, as vasodilator metabolites are produced, blood flow increases over resting levels. The effects of sympathetic stimulation are also summarized in Fig. 5.

Salivary glands contain receptors to many different mediators, but the most important physiologically are the muscarinic cholinergic and β-adrenergic receptors. The parasympathetic mediator is acetylcholine, which acts on muscarinic receptors. This results in the formation of inositol triphosphate and the subsequent release of Ca^{2+} from intracellular stores. Ca^{2+} may also enter the cell from the plasma. VIP and substance P are also released from neurons within the salivary glands, and they stimulate Ca^{2+} release. The primary sympathetic mediator is norepinephrine, which binds to α-adrenergic receptors, resulting in the formation of adenosine 3′,5′-cyclic monophosphate (cAMP). Formation of these second messengers leads to protein phosphorylation and enzyme activation, resulting eventually in increased salivary gland function. Agonists that increase cAMP usually increase the enzyme and mucus content of saliva, whereas those that increase Ca^{2+} have a greater effect on the volume of secretion from the acinar cells.

The common stimuli and inhibitors of salivary gland activity are shown in Fig. 5. Salivation is stimulated by the presence of food in the mouth. These stimuli include taste, smell, and the physical sensations produced by chewing and the pressure of the food. Salivation can also be initiated through cortical centers by simply thinking of appetizing food and by conditioned reflexes. Sour-tasting foods and certain chemicals present in spicy foods are also potent stimulators of saliva flow. As mentioned in Chapter 33, nausea leads to an intense production of mucus-rich saliva. Inhibition of salivary gland activity occurs during sleep and periods of dehydration. Other external events inhibiting salivary flow are fatigue and fear.

GASTRIC SECRETION

Four components of gastric juice—*hydrochloric acid, pepsin, mucus,* and *intrinsic factor*—have physiologic functions. Acid is necessary for the conversion of

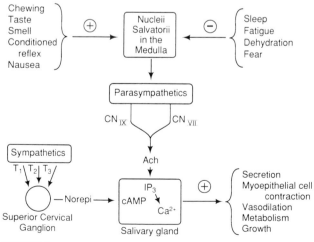

FIGURE 5 Regulation of the salivary glands by the central nervous system. (Modified from Johnson LR, ed., *Gastrointestinal physiology,* 6th ed. St. Louis: CV Mosby, 2001.)

inactive *pepsinogen* to the active enzyme pepsin. Acid and pepsin begin the digestion of dietary protein. However, in the absence of the stomach, pancreatic enzymes hydrolyze all ingested proteins, so that they are totally absorbed. Gastric acid is bacteriostatic, and most bacteria entering the gastrointestinal tract with ingested food are killed in the stomach. Without gastric acid or in cases in which its secretion is severely reduced, there is a higher incidence of bacterial infections of the intestines. Mucus is secreted as a protective coating for the stomach and serves as a physical lubricant and barrier between the cells and ingested material. Mucus and bicarbonate trapped in the mucus layer also maintain the mucosal surface at a near -neutral pH. This is part of the so-called gastric mucosal barrier that protects the stomach from acid and pepsin. Intrinsic factor binds vitamin B_{12} and is necessary for its absorption in the ileum. This is the only indispensable substance in gastric juice. Patients who have undergone a total gastrectomy must take injections of vitamin B_{12}.

Functional Anatomy of the Stomach

The stomach is divided into two regions according to secretory function. The proximal 80% consists of acid-secreting *oxyntic gland mucosa*. The remaining distal 20% does not secrete acid but contains endocrine cells that synthesize and release the hormone gastrin. The mucosa in this region is referred to as the *pyloric gland mucosa*, and the region itself is often called the *antrum* (Fig. 6).

The gastric mucosa contains glands that open into pits in the mucosal surface (Fig. 7). The pits of the oxyntic gland area are shallow and lined with mucous or surface epithelial cells, which line the surface as well. At the bases of the pits are the openings of the glands.

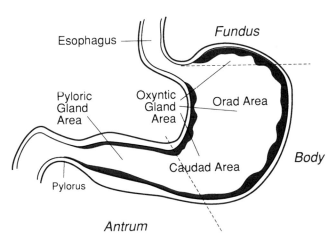

FIGURE 6   Anatomic and functional divisions of the stomach. In discussions of secretion, the stomach is divided into the acid-secreting oxyntic gland area and the gastrin-producing pyloric gland area.

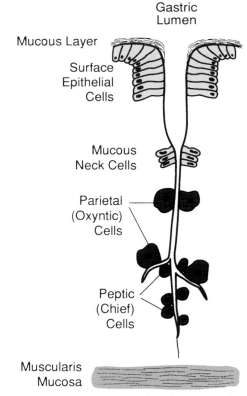

FIGURE 7   Oxyntic gland and surface pit, showing the positions of the various cell types.

The glands project into the mucosa toward the serosa and account for approximately three-fourths of the total mucosal thickness. The oxyntic glands contain acid-producing, *parietal*, or oxyntic cells and the *chief* or *peptic cells,* which synthesize and secrete the enzyme precursor pepsinogen (Fig. 7). Pyloric glands contain the hormone-producing *gastrin* (G) *cells* and *mucous cells,* which also secrete pepsinogen. The oxyntic glands contain *mucous neck cells,* located primarily where the glands open into the pits. These cells secrete soluble mucus, which is thinner than the visible mucus produced by the *surface epithelial cells.* Each gland also contains a stem cell located in the area of the mucous neck cells. These are proliferative cells: After division, one daughter cell remains anchored as the new stem cell. The other undergoes a number of divisions and the resulting cells may migrate to the surface, where they differentiate into surface epithelial cells, or they may migrate down into the glands, where they become parietal cells. The pits of the pyloric gland area are much deeper than those of the oxyntic gland area, occupying about two-thirds of the mucosa. Stem cells are found both in the neck regions and in the glands themselves. In the pyloric gland area, daughter cells differentiate into mucous cells or into G cells. Peptic cells normally arise by mitosis, but during times when damaged mucosa is being repaired, they may also differentiate from stem cells.

Parietal cells secrete hydrochloric acid at concentrations ranging from 150–160 mmol/L in amounts of 1–2 L/day. The human stomach contains approximately 1 billion parietal cells, and the number of parietal cells determines the secretory capacity of the stomach. Because the pH of this solution is less than 1 and that of the blood is 7.4, it means that the parietal cells concentrate H^+ several million times. The energy required for this process comes from adenosine triphosphate (ATP) produced by the numerous mitochondria within the cells (Fig. 8A). In humans, parietal cells also secrete intrinsic factor. In some species, intrinsic factor is produced by the chief cells.

The ultrastructure of the parietal cell is unique and reflects its function. The cytoplasm of the nonsecreting parietal cell (Fig. 8A) contains a branching *intracellular* or *secretory canaliculus* that is closed to the lumen of the gland. The canaliculi are lined by short microvilli, which are not prominent. The cytoplasm of the resting cell also contains an elaborate system of tubular and vesicular membranes called *tubulovesicles*. These are usually concentrated in the apical region of the cell. Within a few minutes after the stimulation of secretion, the secretory canaliculus (the apical membrane) begins to expand and becomes open to the lumen. The microvilli increase greatly in number and length, actually becoming filamentous, so that the surface area of the apical membrane may increase 6- to 10-fold (Fig. 8B). This expansion is matched by a decrease in the surface area of the membranes of the tubulovesicles. Removal of the secretory stimulus leads to a collapse of the canaliculi and reappearance of the tubulovesicles. This morphologic transformation is a complex process, and several theories have been proposed to account for it. Most evidence favors a membrane recycling process that proposes that the tubulovesicles fuse with the apical plasma membrane when the cell is stimulated, increasing its surface area. After stimulation, the surface membrane is believed to be reincorporated into tubulovesicles by a process of endocytosis.

The active transport mechanism responsible for the secretion of H^+ is the (H^+, K^+)-ATPase enzyme, which in the resting cell is located in the membranes of the tubulovesicles. Thus, during the stimulation of secretion, the H^+ pump is relocated to the apical or secretory membrane. The activities of (H^+, K^+)-ATPase and carbonic anhydrase, another enzyme involved in acid secretion, increase after the stimulation of secretion. Acid secretion begins within 10 min of administering a stimulant. The lag time is believed to be due to the morphologic conversion described and the activation of these enzymes.

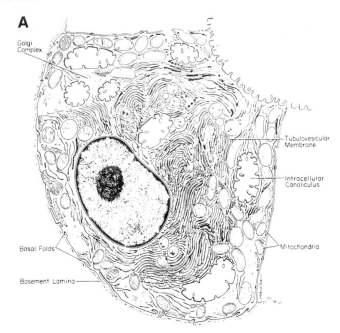

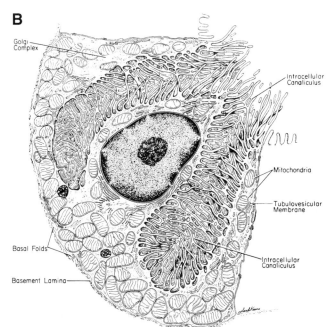

FIGURE 8 (A) Nonsecreting parietal cell is characterized by numerous tubulovesicles and an internalized distended intracellular canaliculus, that has few microvilli. (B) In the secreting parietal cell, the tubulovesicles have fused with the membrane of intracellular canaliculus, which now opens into the lumen of the gland. The fusion has produced abundant, long microvilli within the canaliculus. (Modified from Ito S., Functional gastric morphology. In Johnson LR, ed., *Physiology of the gastrointestinal tract*. New York: Raven Press, 1981, p 531.)

The surface epithelial or mucous cells are cuboidal and contain large numbers of mucous granules at their apical surfaces. During secretion, the membranes of the granules fuse with the plasma membrane, releasing mucus into the lumen.

Like other cells that synthesize and secrete enzymes, the peptic or chief cells have a highly developed endoplasmic reticulum. After synthesis by the rough endoplasmic reticulum, the enzyme precursor pepsinogen molecules are transported to the numerous Golgi structures and packaged into zymogen granules. The granules migrate to the apical region of the cell, where they are stored in the cytoplasm. A secretory stimulus triggers membrane fusion and exocytosis of pepsinogen. This entire process of enzyme synthesis, packaging, and secretion is discussed later in this chapter in the section concerning the pancreas.

Electrolytes of Gastric Juice

At all rates of secretion, gastric juice is essentially isotonic to plasma. The ionic composition of gastric juice, however, varies with the rate of secretion (Fig. 9). At basal (unstimulated) rates of secretion, gastric juice is primarily a solution of NaCl with small amounts of H^+ and K^+. As the rate of secretion increases, the concentration of Na^+ decreases and the concentration of H^+ increases, so that at peak rates gastric juice is primarily HCl with small amounts of K^+ and Na^+. At all rates of secretion, H^+ and Cl are secreted against their electrochemical gradients. It is important to realize that even at basal secretory rates, gastric juice is extremely acidic. The concentration of H^+ may range from 10–20 mmol/L basally up to 130–150 mmol/L at peak rates. In other words, the pH will range from pH 2 to pH < 1.

The changes in ionic composition with the rate of gastric secretion are due to the manner in which the juice is produced. There are actually two separate secretions. A *nonparietal secretion* contains primarily Na^+ and Cl^- with K^+ and HCO_3^- present in amounts approximately equal to their concentrations in plasma. In the absence of all H^+ secretion, HCO_3^- may be detected in gastric juice in concentrations up to 30 mmol/L. This nonparietal component is produced continually at a constant low rate. The *parietal component,* which is a solution of 150 mmol/L HCl plus 10–20 mmol/L KCl, is secreted against this background at rates depending on the degree of stimulation. Therefore, at all rates of secretion, gastric juice is a mixture of these two components. At low rates, the nonparietal component predominates. As the secretory rate increases, and because the increase is due only to the parietal cell component, gastric juice begins to resemble pure parietal cell secretion.

This so-called *two-component theory* of gastric secretion is probably an oversimplification. There is no doubt exchange of H^+ for Na^+ as the parietal secretion moves up the gland into the lumen. Although such changes are minimal at high rates of secretion, they contribute significantly to the final composition of the juice at lower rates, when more time is available for exchange.

Knowledge of the composition of gastric juice is important in treating a patient who has lost significant volumes of gastric juice by aspiration or chronic vomiting. Replacement of only NaCl will result in *hypokalemic metabolic alkalosis.*

Acid Secretory Process

The cellular processes that best explain the secretion of HCl are shown in Fig. 10. The exact metabolic steps in the production of H^+ are not known. However, the reaction is summarized by

$$HOH \rightarrow OH^- + H^+, \qquad [1]$$
$$OH^- + CO_2 \rightarrow HCO_3^-. \qquad [2]$$

The H^+ is actively pumped across the apical membrane in exchange for luminal K^+ by the (H^+, K^+)-ATPase already discussed. K^+ is accumulated within the cell by the (Na^+, K^+)-ATPase in the basolateral membrane. The accumulated K^+ moves down its electrochemical gradient, leaking through conductive pathways in both membranes. The K^+ entering the luminal space is, therefore, being recycled by the (H^+, K^+)-ATPase. Cl^- enters the cell across the basolateral membrane in exchange for HCO_3^-. The HCO_3^- is formed from CO_2 and OH^-, which is accumulated within the cell as H^+ is pumped out. *Carbonic anhydrase* (Ca) catalyzes the formation of HCO_3^- from OH^- and CO_2. The HCCV is carried away by the blood. The pH of the venous blood from an actively secreting stomach may actually be higher than the pH of the arterial blood, because of this so-called *alkaline tide.* In addition, the $(Na^+ - K^+)$ pump maintains a low intracellular Na^+ concentration.

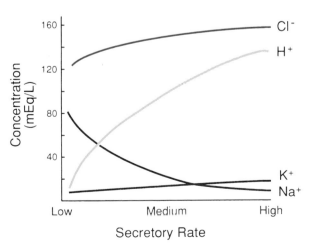

FIGURE 9 Relationships between the concentrations of the principal ions in gastric juice and the rate of secretion.

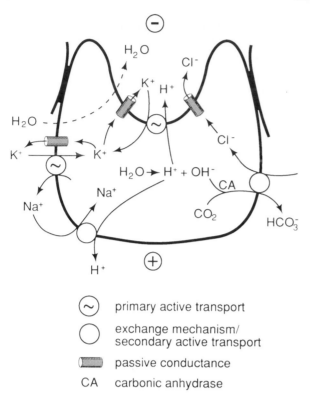

~   primary active transport

◯   exchange mechanism/
     secondary active transport

▨   passive conductance

CA   carbonic anhydrase

FIGURE 10   Cellular processes that best account for the production of HCl by the gastric parietal cell.

Some Na^+ moves down its concentration gradient, entering the cell across the basolateral membrane in exchange for H^+. This is a secondary active transport of H^+ out of the cell. This Na^+ is recycled by the sodium pump. The Na^+-H^+ exchange also increases the amount of base in the cell, facilitating the entrance of Cl^-. Therefore, the movement of Cl^- from the blood to the lumen against its electrochemical gradient is achieved by virtue of excess OH^- in the cell after H^+ has been pumped out. HCO_3^- is produced from CO_2 and OH^- and diffuses down its concentration gradient in exchange for Cl^- entering the cell. Water moves into the lumen by osmosis in response to the secretion of ions.

The potential difference across the resting gastric mucosa is approximately -70 to -80 mV lumen negative with respect to the blood. This significant charge difference is due to the active secretion of Cl^- by surface epithelial cells as well as parietal cells (see Chapter 36 for a discussion of the Cl^- secretory mechanism). When acid secretion is stimulated, the potential difference decreases to -30 or -40 mV because of the transport of positively charged H^+ in the same direction as Cl^-. Thus, H^+ is actually secreted down an electrical gradient, facilitating its transport against the concentration gradient of several million-fold.

The potential differences across the mucosa of the various regions of the gastrointestinal tract vary

according to location, and this information is frequently used by physicians in locating the tips of catheters. For example, as a catheter with an attached electrode moves from the esophagus (-15 mV) into the stomach, the potential difference will increase dramatically. It will then decrease from approximately -80 to -5 mV as the tip passes the pylorus and enters the duodenum.

The secretion of H^+ against such a great concentration gradient and the maintenance of an electrical gradient require minimal leakage of ions and acid back into the mucosa. The absence of significant leakage across the gastric mucosa is attributed to the so-called gastric mucosal barrier. When this barrier is disrupted by bile acids, alcohol, aspirin, or any of a number of other agents that damage the mucosa, the potential difference decreases considerably as ions leak down their electrochemical gradients. H^+ then enters the mucosa and damage occurs. The anatomic nature of the barrier is not totally defined; its properties and role in mucosal damage are more fully discussed in the section on the pathophysiology of ulcer disease.

Stimulants of Acid Secretion

Only a few substances directly stimulate acid secretion from the parietal cells. These include the antral hormone gastrin, the parasympathetic mediator acetylcholine, and the paracrine agent histamine. The amounts of gastrin and acetylcholine reaching the stomach vary in response to the digestive state and are responsible for the regulation of acid secretion. In addition to these three main stimulants, an unknown hormone of intestinal origin stimulates acid secretion. This peptide has been named *enterooxyntin* to denote both its origin and action. In humans, an additional amount of stimulation is caused by circulating amino acids after their absorption from the small intestine.

Histamine is released from the *enterochromaffin-like cells* (ECLs) within the lamina propria and diffuses through the extracellular fluid to the parietal cells. Histamine acts as a paracrine to stimulate acid secretion. Figure 11 illustrates the relationships between gastrin, acetylcholine, and histamine. The ECL cell contains receptors for gastrin and acetylcholine. Gastrin is the major regulator of the ECL cells. Acutely gastrin stimulates the release of histamine from the ECL cells and the synthesis of new histamine by increasing the activity of the enzyme, histidine decarboxylase. Over the longer term, gastrin stimulates the proliferation of the ECL cells. Although histamine release and synthesis are also increased by acetylcholine, its effects are not as pronounced as those of gastrin.

The parietal cell membrane has separate receptors for histamine, gastrin, and acetylcholine (Fig. 11).

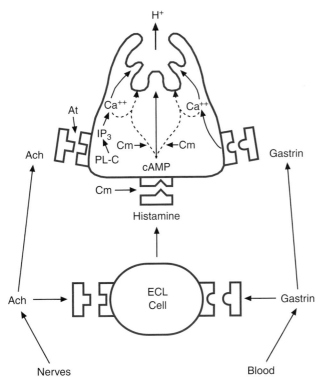

FIGURE 11 Interactions of histamine, gastrin, and acetylcholine (Ach) on the parietal cell. This model accounts for the potentiation between stimuli and the inhibitory effects of atropine (At) and cimetidine (Cra).

The final rate of secretion, however, depends on the interactions of these secretagogues. At this point, it is important to understand the concept of potentiation. *Potentiation* occurs between two stimulants when the response to their simultaneous administration exceeds the sums of the responses to each agent administered separately. Another convenient definition is that the maximal response to the two agents acting together exceeds the maximal response to either agent alone. Potentiation allows small amounts of endogenous stimuli to produce near maximal effects, and within the gastrointestinal tract it is a common physiologic event.

At the parietal cell, histamine potentiates the actions of gastrin and acetylcholine. In a similar manner, potentiation also exists between gastrin and acetylcholine. These interactions between stimuli are one reason why specific antihistamines that block acid secretion stimulated by histamine (H$_2$-receptor blockers such as cimetidine) also block secretory responses to acetylcholine and gastrin. As one would predict, atropine, the specific antagonist of the muscarinic actions of acetylcholine, also blocks secretion stimulated by histamine and gastrin. Thus, the effects of the H$_2$ blockers on gastrin- or acetylcholine-stimulated secretion are due to the inhibition of the portion of the secretory response resulting from histamine potentiation as well as to the inhibition of histamine released by their actions on the ECL cells. Similarly, the inhibition of histamine- and gastrin-stimulated acid secretion by atropine is caused by removal of the potentiating interactions with acetylcholine. The points at which atropine and cimetidine affect the secretory process are indicated in Fig. 11. No specific and potent receptor blocker for gastrin is available for similar studies.

The exact intracellular events leading to potentiation are unknown. However, potentiation occurs only between those agents that act through different second messenger systems after binding to their receptors. Acetylcholine binds to the muscarinic receptor, resulting in the formation of inositol trisphosphate and the subsequent increase in intracellular Ca^{2+}. Histamine binding activates adenylate cyclase, resulting in the formation of cAMP. The messenger system for gastrin has not been worked out, but appears to involve Ca^{2+} rather than cAMP and to be somewhat different from the cholinergic system.

Stimulation of Acid Secretion

Basal or *interdigestive secretion* is that which occurs in the absence of all gastrointestinal stimulation. Basal secretion is equal to about 10% of the maximal response to a meal. In humans, basal secretion shows a circadian rhythm, with the highest acid output in the evening and the lowest in the morning. The cause of this variation is unknown, but it is not matched by changes in plasma gastrin levels. In both dogs and humans, basal acid secretion is reduced by vagotomy, further reduced by antrectomy (removal of the source of gastrin), and inhibited by histamine H$_2$ antagonists. These results indicate that the presence of background amounts of acetylcholine, gastrin, and histamine account for at least part of basal secretion.

Between meals, therefore, the emptied stomach contains a relatively small volume of gastric juice. The pH of this fluid, however, is usually less than 2.0, and the mucosa is acidified. Acidification of the antral mucosa prevents gastrin release, and the acidification of the oxyntic gland mucosa has an inhibitory effect on acid secretion. Both of these mechanisms appear to involve the acid-mediated release of *somatostatin* and the paracrine action of somatostatin on the G cells in the antrum and the parietal cells in the oxyntic gland area.

The stimulation of acid secretion is divided into the cephalic, gastric, and intestinal phases for convenience in understanding the different mechanisms involved. This division is based on the location of the receptors initiating the secretory responses. The division is somewhat artificial, for during most of the response to a meal stimulation is initiated simultaneously from more than one area. The stomach of a normal 70-kg man has the capacity to secrete about 20 mmol of acid per hour.

Cephalic Phase

Stimulation during the cephalic phase accounts for about 30% of the response to a meal. The presence of food in the mouth stimulates mechanoreceptors (pressure) and chemoreceptors (smell and taste) located in the tongue and the buccal and nasal cavities. Central pathways can also be activated by the thought of an appetizing meal or events triggering conditioned reflexes. In each case, afferent impulses are relayed to the vagal nucleus, and vagal efferent nerves carry impulses to the stomach (Fig. 12). The entire cephalic phase is therefore blocked by vagotomy.

The cephalic phase can be studied by a procedure known as *sham feeding*. A subject is given an appetizing meal and allowed to feed himself and chew but not to swallow the food. The cephalic phase can also be activated by direct vagal stimulation. In humans, this is mimicked by inducing hypoglycemia, which activates hypothalamic centers that stimulate secretion via vagal pathways. Hypoglycemia can be induced with insulin or tolbutamide or by giving glucose analogues, such as 3-methylglucose or 2-deoxyglucose, which interfere with glucose metabolism.

The vagus increases acid secretion through two mechanisms: It stimulates the parietal cells directly and stimulates the release of gastrin. The mediator at the parietal cell is acetylcholine, and the mediator at the G cell is *bombesin,* also called *gastrin-releasing peptide* (GRP). Atropine blocks the direct effects on the

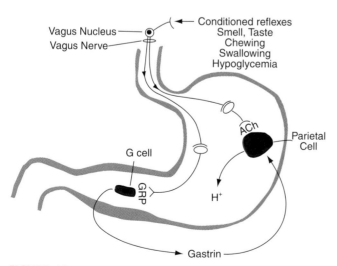

FIGURE 12 Mechanisms stimulating gastric acid secretion during the cephalic phase. (Modified from Johnson LR, ed., *Gastrointestinal physiology,* 6th ed. St. Louis: CV Mosby, 2001.)

parietal cell, but does not block the release of gastrin. In humans, the direct effect on the parietal cell is more important, because selective vagotomy of the acid-secreting portion of the stomach abolishes the response to sham feeding, whereas antrectomy only moderately reduces it. During sham feeding experiments or other studies designed to examine the cephalic phase, gastrin release will not occur unless steps are taken to neutralize the acid present in the stomach. The mechanisms involved in the cephalic phase are outlined in Fig. 12.

Gastric Phase

As swallowed food enters the stomach, it mixes with the small volume of acid present. Buffers (primarily proteins) in the food neutralize the acid, raising the intragastric pH from around 2 to as high as 6. Because gastrin release is inhibited at antral pHs below 3, little or no gastrin is released from a stomach that does not contain food. When the pH increases above 3, vagal stimulation from the cephalic phase initiates gastrin release. Gastrin release is maintained during the gastric phase by both neural and chemical mechanisms. The gastric phase accounts for approximately 60% of the acid response to a meal.

Distension of the gastric wall activates mechano-receptors initiating *extramural* or *vagovagal reflexes* and *intramural reflexes* that stimulate both gastrin release and acid secretion. Distension of the oxyntic gland area stimulates the parietal cells directly via a local or intramural reflex mediated by acetylcholine. Extramural reflexes whose afferent and efferent paths are contained in the vagus nerve are also triggered. This vagovagal reflex results in effects identical to those of the cephalic phase, namely, parietal cell stimulation by acetylcholine and gastrin release mediated by GRP. Distension of the antrum causes gastrin release via a local, intramural reflex that appears to be mediated by acetylcholine. Antral distension also elicits a vagovagal reflex that results in both gastrin release via GRP and stimulation of the parietal cells directly via acetylcholine. All gastrin release stimulated by distension is blocked by acidifying the antral mucosa. Distension with acidified solutions, however, will still result in increased acid secretion through local and vagovagal reflexes, directly stimulating the parietal cells via acetylcholine. The reflexes are summarized in Fig. 13.

The only major nutrient that stimulates gastric acid secretion chemically is protein. To be effective, the protein must be partially digested to peptides and amino acids. The entire stimulation of acid secretion by protein digestion products appears to be due to the direct chemical release of gastrin. This release of gastrin is not blocked by vagotomy or atropine, but it is blocked by

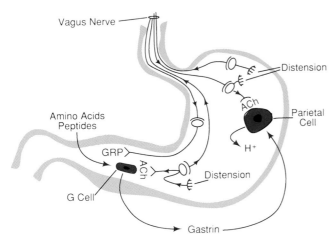

FIGURE 13 Mechanisms stimulating gastric acid secretion during the gastric phase. (Modified from Johnson LR, ed., *Gastrointestinal physiology*, 6th ed. St. Louis: CV Mosby, 2001.)

acidifying the antral mucosa below pH 3. The most effective amino acids for releasing gastrin in humans are phenylalanine and tryptophan.

The only other commonly ingested agents known to stimulate acid secretion chemically are calcium, alcohol, and caffeine. In the human, calcium ions are a strong stimulus for gastrin release and acid secretion. High concentrations of ethanol release gastrin, and intravenous ethanol increases acid secretion by an unknown mechanism. In general, solutions of ethanol up to 15% have little or no effect on gastric secretion. Caffeine is a direct stimulant of acid secretion in humans. It probably acts by inhibiting the phosphodiesterase that breaks down cAMP. Caffeine, however, is not the major stimulant present in coffee, because decaffeinated coffee is as strong a stimulant of secretion as regular coffee.

Intestinal Phase

Acid secretion as a result of stimuli acting from the intestine is minor, accounting for only 10% of the total response to a meal. Protein digestion products in the duodenum stimulate acid secretion by at least two mechanisms. One seems to involve the release of an unidentified hormone by distension. This hormone has been named *enterooxyntin to* denote both its location and effect. Intravenously infused amino acids also stimulate acid secretion. Therefore, part of the intestinal phase of stimulation is due to the circulation of absorbed amino acids to the parietal cells. Intestinal mucosa contains gastrin, but it is not released under normal conditions in humans. The factors stimulating acid secretion during the entire response to a meal are summarized in Fig. 14. Note that histamine is present to stimulate the parietal cell and potentiate the other stimuli during all phases.

Regulation of Gastrin Release

A summary of the pathways believed to be involved in the regulation of gastrin release is depicted in Fig. 15. Endocrine cells containing somatostatin are located close to the gastrin-containing G cells. Somatostatin, acting as a paracrine, inhibits gastrin release. Vagal stimulation increases gastrin release via two mechanisms. First, some neurons release GRP (bombesin) at the G cells, stimulating them directly to release gastrin. Second, other inhibitory neurons go to the somatostatin cell, where they release acetylcholine, which inhibits somatostatin release. This dual pathway mechanism accounts for the observations that vagus-stimulated gastrin release is not blocked by atropine and that vagal stimulation decreases somatostatin release into mucosal perfusates. Digested protein and other chemicals act directly on the G cell to stimulate gastrin release. It is believed that the apical surface of the G cell contains specific receptors to recognize these stimulants. These receptors, however, have not been identified. Acid in the lumen resulting in a pH < 3 acts directly on the somatostatin cell to stimulate somatostatin release. When somatostatin is released, it blocks the effect of all stimulants of gastrin release. Within the oxyntic gland area, somatostatin acts directly on the parietal cell to inhibit acid secretion.

Inhibition of Acid Secretion

The duration of the acid secretory response to a meal is determined primarily by the intragastric pH and the nature of the chyme entering the duodenum. As shown in Fig. 16, at the start of a meal the stomach contains a small volume of acidic gastric juice, which acidifies the antral mucosa. Somatostatin is released and acts through a paracrine mechanism to inhibit gastrin release and to inhibit directly the parietal cells as well. As a result, the

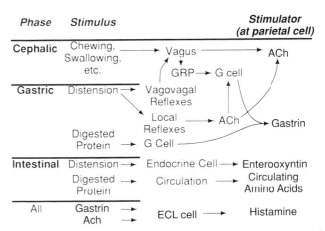

FIGURE 14 Summary of the mechanisms for stimulating gastric acid secretion.

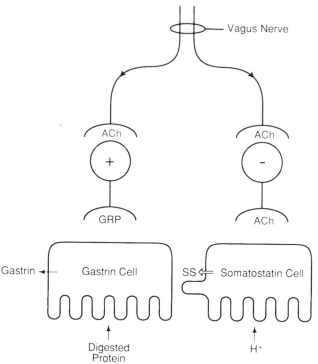

FIGURE 15   Mechanism accounting for the regulation ot gastrin release.

and the pH of the contents has begun to decrease. As the pH of the contents continues to fall, gastrin release is inhibited, removing a significant factor for the stimulation of acid secretion, and somatostatin inhibits the parietal cells. This negative feedback mechanism is extremely important in the regulation of acid secretion.

As chyme enters the duodenum, it initiates a number of processes that inhibit acid secretion at the level of the parietal cell. These mechanisms are both neural and humoral and are triggered by the pH, osmolarity, and fat content of the chyme. Thus, these processes are similar, and some may be identical to those that inhibit gastric emptying. Large amounts of acid may release sufficient secretin from the duodenal mucosa to inhibit the parietal cells. Acid also triggers an inhibitory intramural neuroreflex. Hyperosmotic solutions and fatty acids release an as yet unidentified hormone or hormones that inhibit acid secretion. These substances are called *enterogastrones,* denoting their location and inhibitory effect on the stomach. Gastric inhibitory peptide (GIF) may also take part in the inhibition of both acid secretion and gastrin release. None of the hormonal mediators of these responses is firmly established. The important thing for the student to understand is that both neural and humoral mechanisms exist for the inhibition of acid secretion, and that these are triggered by chyme in the duodenum. Our knowledge of the inhibition of acid secretion is summarized in Fig. 17.

Pepsin

Pepsin is stored and secreted as the inactive precursor pepsinogen, which has a molecular weight of 42,500. At intragastric pH lower than 5, pepsinogen is split to form the active enzyme pepsin, which has a molecular weight of 35,000. Pepsin can catalyze the formation of

rate of secretion is low (basal). As the meal is ingested, the small volume of acid is rapidly neutralized by buffers present in the food, intragastric pH increases to 5 or 6, and acid secretion begins. Secretion is initiated by the direct vagal component of the cephalic response, but as the intragastric pH rises above 3, gastrin release is triggered by stimuli of the cephalic and gastric phases. An hour after the meal, the rate of acid secretion is maximal, the buffering capacity of the meal is saturated, a significant portion of the meal has emptied from the stomach,

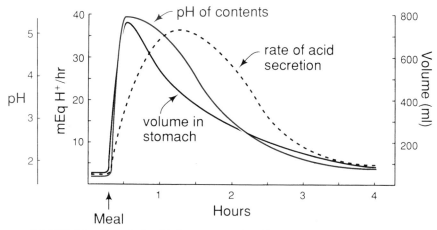

FIGURE 16   The relationship between gastric acid secretory rate, intragastric pH, and the volume of gastric contents during a meal.

Region	Stimulus	Mediation	Inhibits gastrin release	Inhibits acid secretion
Oxyntic gland area	Acid (pH < 3.0)	Somatostatin		YES
Antrum			YES	
Duodenum	Acid	Secretin	YES	YES
		Nervous reflex		YES
	Hyper-osmotic solutions	unidentified enterogasterone		YES
Duodenum and Jejunum	fatty acids	GIP	YES	YES
		unidentified enterogasterone		YES

FIGURE 17   Mechanisms inhibiting gastric acid secretion. Blanks indicate actions that either do not occur or are not significant physiologically.

additional enzyme from pepsinogen. At an intragastric pH of 2 or less, the conversion of pepsinogen to pepsin is almost instantaneous. Pepsin begins the digestion of protein by splitting interior peptide bonds.

There are two main groups of pepsinogens. Group I pepsinogens are secreted by peptic (chief) and mucous cells of the oxyntic glands. Group II pepsinogens come from mucous cells of the pyloric gland area and the duodenum as well as the oxyntic gland mucosa. Pepsinogens can be measured in the blood, where their quantities correlate with the presence of duodenal ulcer disease.

Acetylcholine is the strongest and most important stimulator of pepsin secretion. Vagal stimulation during both the cephalic and gastric phases results in the majority of the pepsin response to a meal. Acid is necessary not only for the activation of pepsin but also because it initiates at least two other mechanisms stimulating pepsin secretion. First, acid triggers a local cholinergic reflex that stimulates the chief cells to secrete. This mechanism is sensitive to atropine and local anesthetics and is mediated by the enteric nervous system. Second, acid releases secretin from the duodenum, and secretin stimulates pepsin secretion. There is also evidence that cholinergic stimulation results in more pepsin secretion when the intragastric pH is low. These mechanisms ensure that large amounts of pepsinogen are not secreted unless sufficient acid is present to convert it to the active enzyme.

Mucus

The cells of the gastric mucosa secrete two types of mucus. Vagal nerve stimulation and acetylcholine stimulate soluble mucus secretion from the mucous neck cells.

Soluble mucus mixes with the other secretions of the glands and lubricates the gastric chyme. Surface mucous cells secrete visible or insoluble mucus in response to chemical or physical irritation. Chemicals, such as ethanol, and friction with ingested material stimulate the release of visible mucus, which is secreted as a gel forming an unstirred layer over the mucosa. This layer traps dead cells as they are sloughed from the mucosa and forms a protective lubricating coat. In addition, bicarbonate ions from the alkaline component of secretion are trapped within this layer and maintain the pH at the surface of the stomach near neutrality despite luminal pHs of less than 2. During the digestion of a meal, this physical and alkaline barrier prevents damage from friction and keeps the mucosal cells from coming into contact with pepsin and high concentrations of acid.

Intrinsic Factor

Intrinsic factor is secreted by the parietal cells in humans as a 55,000-molecular-weight mucoprotein. It combines with dietary vitamin B_{12}, forming a complex necessary for the absorption of the vitamin by an active process that is located in ileal mucosa. The absence of intrinsic factor results in the condition known as pernicious anemia, a disease associated with achlorhydria and the absence of parietal cells. The development of this disease is poorly understood because the liver stores vitamin B_{12} in amounts sufficient for several years. Therefore, the developmental stages of the disease occur a number of years before the condition is recognized. The secretion of intrinsic factor is the only reason the stomach is necessary for life, and patients

TABLE 1   Basal and Maxim (Histamine-Stimulated) Acid Outputs From Normal Humans and Those With Conditions Affecting Acid Secretion

Condition	Acid output (mmol/hr)	
	Basal	Maximal
Normal	1–5	6–40
Pernicious anemia	0	0
Gastric cancer	0–5	0–40
Gastric ulcer	0–3	1–20
Duodenal ulcer	2–10	15–60
Gastrinoma	10–30	30–80

with pernicious anemia or after gastrectomy must take injections of vitamin B_{12}.

Acid Secretion and Serum Gastrin

Table 1 shows the basal and maximal rates of gastric acid secretion for normal humans and for patients having clinical conditions in which acid secretion is usually altered. Basal acid output is that which occurs during the interdigestive period when the stomach is empty. Maximal acid output is measured in response to a maximal dose of histamine. It is important to note the degree of overlap between acid outputs in the various groups. Thus, although these values indicate that, in general, a patient with duodenal ulcer secretes more acid than a normal individual, such measurements have little meaning in the diagnosis of individual cases.

Because of the feedback mechanism whereby acid inhibits the release of gastrin, serum gastrin values are, in general, inversely related to acid secretory rates. In other words, patients with gastric cancer or gastric ulcer disease usually have higher than normal serum gastrin levels. Patients with pernicious anemia, who secrete no acid, often have astronomically high serum gastrins. The obvious exceptions are gastrinoma (Zollinger-Ellison syndrome) patients, who have both high rates of acid production and extremely high gastrin levels. In this group of patients, however (as discussed in Chapter 32), the gastrin is being released from a tumor and is not subject to acid feedback inhibition.

Clinical Note

Peptic Ulcer

Gastric and *duodenal ulcers* form when the respective mucosal linings of the gastrointestinal tract are digested by acid and pepsin. The presence of acid and pepsin are normally required for ulcer formation and, as such, both diseases are classified as *peptic ulcers*. Despite being classified under one heading, however, the etiologies of the two conditions are quite different. Put simply, an ulcer forms when damaging factors such as acid and pepsin overcome the mucosal protective factors such as bicarbonate secretion, mucus, and cell renewal. In the case of duodenal ulcer, there is good evidence that the defect is increased amounts of acid and pepsin. In the case of gastric ulcer, the defect appears to be in the mucosa itself, so that its defense mechanisms have been weakened. This analysis is oversimplified and generalized, and both factors no doubt play some role in most cases of ulcer.

Increased acid and pepsin secretion have been implicated in duodenal ulcer disease. As shown in Table 1, as a group, these patients have higher than normal rates of secretion. In general, duodenal ulcer disease patients have higher than normal serum gastrin levels in response to a meal and about double the normal number of gastric parietal cells. Whether the increased number of parietal cells is due to the trophic effect of gastrin is unknown. Increased serum gastrin levels are due in part to a defective mechanism for the inhibition of gastrin release by mucosal acidification. There is also evidence that parietal cells in patients with duodenal ulcer are actually more sensitive to gastrin as well. In addition, the secretion of pepsin, as determined from serum pepsinogen levels, is almost doubled in the duodenal ulcer group. The ability of increased acid secretion to produce duodenal ulcer disease is dramatically demonstrated in patients with gastrinoma. These patients always develop duodenal ulcers, never gastric ulcers.

The lower than normal rate of acid secretion in gastric ulcer is believed to be due in part to the inability to collect acid that has been secreted and then has leaked back into the damaged mucosa. The normal gastric mucosa is relatively impermeable to acid, but a number of events may cause this so-called gastric mucosal barrier to become weakened. These include abnormalities

Clinical Note (continued)

in mucosal blood flow, altered rates of cell renewal, decreased mucus secretion, bacterial infection, and damaging agents such as alcohol, bile acids, and aspirin.

The exact nature of the barrier is unknown, and mucosal resistance to acid probably includes physiologic processes as well as morphologic entities. Cell membranes and junctional complexes prevent normal back-diffusion of H^+. Diffused H^+ is normally transported back into the lumen. Processes that no doubt have some role in maintaining the barrier also include cell renewal and migration, blood flow, and mucous and bicarbonate secretion. Chemical factors such as gastrin, epidermal growth factor, and prostaglandin have all been shown to decrease the severity and promote the healing of ulcers.

It has now been established that the major acquired causative factor in the genesis of both gastric and duodenal ulcer is infection with the bacteria *Helicobacter pylori*. Virtually 100% of gastric ulcer patients, excluding those whose ulcers were caused by chronic aspirin or other nonsteroidal anti-inflammatory drugs (NSAIDs), are infected, whereas 95% of duodenal ulcer patients are *H. pylori* positive. The major characteristic of *H. pylori* is high urease activity and the production of NH_3 from urea, which allows the bacteria to survive and colonize in the acid environment of the gastric mucosa. NH_3 directly damages the epithelial cells, breaking the gastric mucosal barrier and allowing H^+ to diffuse back into cells. Although NH_3 is the major cytotoxic agent, the bacterium also releases a variety of factors and cytokines that damage cells and contribute to gastric ulcer formation.

Recent evidence also suggests that *H. pylori* is responsible for the increased acid secretion found in duodenal ulcer patients. After eradication of the bacteria from a group of duodenal ulcer patients, their increased basal acid output, increased GRP-stimulated acid output, increased ratio of basal to gastrin-stimulated maximal acid output, and increased ratio of GRP-stimulated maximal acid output to gastrin-stimulated maximal acid output returned to normal. Only the increased maximal acid output in response to gastrin failed to return to normal. Because the maximal acid output is probably due to an increased number of parietal cells from the trophic action of the gastrin, this too may return to normal after serum gastrin levels have been reduced. Increased acid secretory and serum gastrin responses appear to be related in part to a decreased inhibition of gastrin release and parietal cell secretion by somatostatin in *H. pylori*-infected individuals. There is also evidence that NH_3^+ directly stimulates gastrin release. All of the effects of *H. pylori* on the mucosa have not been elucidated, but the treatment of gastric and duodenal ulcer diseases involves the eradication of the infection.

Why *H. pylori* infection causes gastric ulcer in one person and duodenal ulcer in another has not been firmly established. However, strong evidence indicates that *H. pylori* infection of the corpus of the stomach results in gastric ulcer, while a primarily antral infection causes duodenal ulcer. In the corpus, gastritis results in decreased acid secretion, damage, gastric ulcer, and the risk of neoplasia. In the antrum, gastritis inhibits somatostatin release, increases gastrin release, increases acid secretion, and results in a duodenal ulcer.

The treatment of peptic ulcer disease is based entirely on its pathophysiology. Medical treatment usually consists of administering antisecretory drugs such as a histamine H_2-receptor blocking agent. The most potent antisecretory drugs are those like omeprazole that inhibit the (H^+,K^+)-ATPase. These compounds are substituted benzimidazoles that accumulate in acid spaces and are activated at low pH. Once activated, these drugs bind irreversibly to sulfhydryl groups present on the active site of the (H^+,K^+)-ATPase, inhibiting the enzyme. Omeprazole effectively treats peptic ulcer, even duodenal ulcers caused by gastrinoma (Zollinger-Ellison syndrome). Medical treatment includes the eradication of *H. pylori*, which is extremely effective in preventing recurrence of the ulcer. Effective medical treatment has made surgery for peptic ulcer disease all but disappear. When done, however, surgery usually consisted of vagotomy and/or antrectomy. These procedures decrease acid secretion by 60–80% by removing one or both major stimulants of acid secretion and their potentiating interactions with histamine.

PANCREATIC SECRETION

Pancreatic exocrine secretion consists of an *aqueous* or *bicarbonate component* and an *enzymatic component.* The aqueous component consists primarily of water and sodium bicarbonate and is produced by the cells lining the pancreatic ducts. The aqueous component neutralizes duodenal contents, preventing injury to the duodenal mucosa and bringing the contents within the pH range necessary for optimal enzymatic digestion of nutrients. The enzymatic or protein component is a low-volume secretion from the pancreatic acinar cells that contains enzymes for the digestion of all major foodstuffs. Unlike the enzymes secreted into saliva and gastric juice, the pancreatic enzymes are essential for normal digestion and absorption.

Functional Anatomy of the Pancreas

The structure of the exocrine pancreas resembles a cluster of grapes and its functional units are similar to the salivons of the salivary glands. Pyramidal *acinar cells* are oriented with the apices toward a lumen to form an *acinus* (Fig. 18). Groups of acini form lobules separated from each other by areolar tissue. The lumen of each spherical acinus is drained by a *ductule* whose epithelium extends into the acinus in the form of centroacinar cells. Within each lobule, ductules join to form intralobular ducts. These in turn drain into extralobular ducts, which join to form the major pancreatic collecting duct draining the gland.

The acinar cells comprise approximately 80% of the pancreas by volume and secrete a small volume of juice containing the pancreatic enzymes. *Ductule cells,* which comprise about 4% of the gland, together with the centroacinar cells, secrete the aqueous component, a large volume secretion of water and $NaHCO_3$. The endocrine cells of the pancreas account for only 2% of its mass and are found in the islets of Langerhans distributed throughout the pancreatic parenchyma. The islets contain both the insulin-secreting β cells and the

glucagon-secreting α cells. The islets also contain large amounts of somatostatin, which may act as a paracrine to inhibit the release of both insulin and glucagon. The pancreas also produces pancreatic polypeptide, a candidate hormone that inhibits pancreatic exocrine secretion (see Chapter 32).

Efferent nerves from both the sympathetic and parasympathetic systems influence pancreatic secretion. Sympathetic innervation is provided by postganglionic fibers from the celiac and superior mesenteric plexuses. Sympathetic fibers enter the pancreas along with the arteries to the organ. Parasympathetic preganglionic fibers to the pancreas are contained in branches of the vagus nerves. These fibers course down the surface of the stomach, entering the pancreas from the antral-duodenal region. Within the pancreas, the vagal fibers terminate either at acini and islets or on other intrinsic cholinergic nerves of the pancreas. Parasympathetic nerves stimulate pancreatic exocrine secretion, whereas sympathetic nerves are largely inhibitory.

Aqueous Component of Pancreatic Secretion

The 100-g human pancreas secretes approximately 1 L of fluid per day, which is sufficient to neutralize most of the acid entering the duodenum. At all rates of secretion, pancreatic juice is essentially isotonic with plasma. At low rates of secretion, pancreatic juice is primarily a solution of Na^+ and Cl^-, whereas at high rates, Na^+ and HCO_3^- predominate. The concentration of Na^+ in pancreatic juice approximately equals its concentration in plasma, and at all rates of secretion K^+ is also found in concentrations equal to its plasma levels (Fig. 19).

As in gastric juice, the ionic composition of pancreatic juice varies with the rate of secretion (Fig. 19). Just as the concentrations of the cations Na^+ and H^+ were

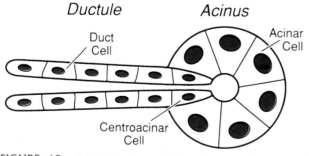

FIGURE 18   Schematic diagram illustrating the histology of a functional unit of the pancreas.

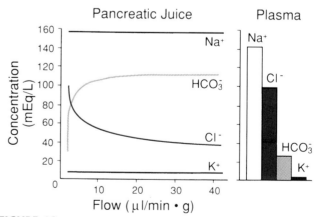

FIGURE 19   Relationship between the principal ions in pancreatic juice and secretory rate. Concentrations of the ions in plasma are shown for comparison.

reciprocally related in gastric juice, the concentrations of Cl^- and HCO_3^- vary inversely in pancreatic juice. As one might expect, theories analogous to those proposed to explain the ionic composition of gastric juice also explain the variation in pancreatic juice composition. It is believed that one cell type, perhaps the acinar cell, secretes a small volume of juice that primarily contains Na^+ and Cl^-. This fluid is the basal secretion and is equal to only about 2% of the maximal secretory rate. In response to stimulation, other cell types—the ductule cells and the centroacinar cells—secrete large volumes of fluid containing primarily Na^+ and HCO_3^-. As the rate of secretion increases, the small amount of Cl^- is diluted by the HCO_3^-, and pancreatic juice becomes a solution of Na^+ and HCO_3^- with small amounts of Cl^-. This concept is analogous to the two-component hypothesis used to explain the composition of gastric juice. Evidence also exists to indicate that as HCO_3^- moves through the duct system of the pancreas, it moves down its concentration gradient, leaving the ducts in exchange for Cl^-. Thus, the higher the secretory rate, the less time for exchange to occur, and the more likely that pancreatic juice will contain primarily HCO_3^-. At low secretory rates, this exchange moves toward completeness and Cl^- becomes the predominant anion. In reality, the final composition of pancreatic juice is probably due to both of these processes.

The aqueous component is secreted by the ductule and centroacinar cells and may contain HCO_3^- at a concentration equal to several times its concentration in plasma. The lumen of the pancreatic ducts is 5–9 mV negative with respect to the blood. Therefore, HCO_3^- is secreted against both its chemical and electrical gradients. This has been interpreted as evidence that HCO_3^- is actively transported across the apical surface of the ductule cell. Although the exact mechanism has not been elucidated, evidence has accumulated for the HCO_3^- secretory mechanism illustrated in Fig. 20. In this model the (Na^+, K^+)-ATPase of the basolateral membrane creates an electrochemical gradient for Na^+ to move into the cell in exchange for H^+, which leaves the cell against its concentration gradient. There is some evidence that a H^+-ATPase may actually pump H^+ out of the cell. In either case, CO_2 diffuses readily into the alkalinized cell, combining with water to form HCO_3^-. This latter step is catalyzed by carbonic anhydrase. The continued movement of H^+ across the basolateral membrane leads to a buildup of HCCV and the movement of HCCV across the apical membrane in exchange for Cl^-. The Cl^--HCO_3^- exchange mechanism is found in many secretory cells of the gastrointestinal tract. As H^+ leaves the cell, it combines with HCO_3^- in the plasma to produce more CO_2, which is free to diffuse into the cell. Because of this secretory process, venous

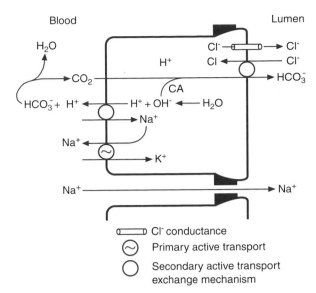

FIGURE 20   Cellular processes that best account for the secretion of HCO_3^- by the pancreatic ductule cells.

blood from the actively secreting pancreas has a lower pH than blood from a nonsecreting gland. As HCO_3^- is secreted into the lumen, Na^+ also moves across the epithelium, preserving electrical neutrality. Much of the Na^+ diffuses through intercellular (paracellular) pathways. Water then moves into the lumen along its osmotic gradient. The rate of HCO_3^- secretion depends on luminal Cl^- and the activity of the Cl^- channel in the apical membrane. This conductance is activated by cAMP in response to stimulation by secretin and is present in ductule but not acinar cells. This channel occurs in many types of epithelial cells and is defective in cystic fibrosis.

This model accounts for evidence indicating that (1) HCO_3^- secretion is ouabain sensitive, (2) secretion involves Na^+-H^+ and Cl-HCO_3^- exchange, (3) secretion of HCO_3^- occurs against an electrochemical gradient and is an active process, (4) most of the HCO_3^- in pancreatic juice is derived from plasma, and (5) in the absence of extracellular Cl^-, HCO_3^- secretion decreases significantly.

Enzymatic Component of Pancreatic Secretion

The pancreas supplies the principal enzymes for the digestion of all major foodstuffs. All pancreatic enzymes are synthesized and secreted by the acinar cells. Pancreatic lipase and amylase are secreted as active enzymes whereas, like pepsin, the pancreatic proteases are secreted as inactive precursors that are converted to active enzymes in the lumen of the small intestine. The individual enzymes and the nature of their specific activities are discussed in Chapter 35, which is concerned with the digestion and absorption of nutrients.

Between meals, pancreatic enzymes are stored in zymogen granules that have migrated to an area near the apical membrane of the acinar cell. A secretory stimulus results in the fusion of the granule membrane with the apical membrane of the cell and the secretion of enzyme contents into the lumen. This process of exocytosis is the only step in the synthesis of the proteins, formation of granules, and secretion of the enzymes known to be controlled by neural and hormonal input.

The steps involved in the synthesis and secretion of pancreatic enzymes are outlined in Fig. 21. In general, these same steps are involved in the production and secretion of salivary enzymes and of pepsin from the

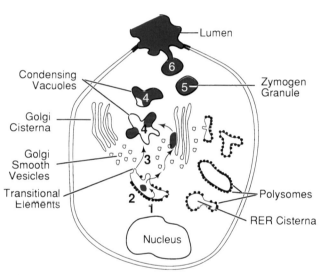

FIGURE 21 Schematic diagram of a pancreatic acinar cell illustrating the primary steps in the synthesis and secretion of enzymes.

stomach. Step 1 is the actual synthesis of the enzyme protein on the polysomes attached to the cisternae of the rough endoplasmic reticulum. As the protein is being synthesized, it elongates and, directed by a leader sequence of hydrophobic amino acids, enters the cisternae of the rough endoplasmic reticulum, where it is collected and where it may undergo some post-translational modification (step 2). Once within the lumen, the enzymes remain membrane bound until they are secreted from the cell. In step 3, the enzymes are transferred to the Golgi complex. This transfer may involve the budding off of enzyme-containing transition elements from the rough endoplasmic reticulum. After association with the Golgi vesicles, the enzymes are transported to condensing vacuoles (step 4), where they are concentrated to form zymogen granules (step 5). The enzymes are then stored in the zymogen granules until a secretory stimulus triggers their expulsion into the lumen of the acinus (step 6). Energy is required for the transport of the enzymes through the endoplasmic reticulum and Golgi vesicles to the condensing granules. Steps 1 through 5 occur without a secretory stimulus.

The process just outlined is accepted by most authorities in this field. However, acini that have been depleted of granules by constant stimulation are still capable of secreting enzymes. The interrelationship of these two mechanisms has not been explained experimentally.

Normally, the pancreatic proteases are not activated until they are secreted into the duodenum. There, enterokinase, an enzyme present in the apical membranes of the enterocytes, converts inactive trypsinogen to the active enzyme trypsin. Trypsin then catalyzes the conversion of the other protease proenzymes to active enzymes. Numerous measures are in place to ensure that proteases are not activated in the cytoplasm of the acinar cell where

Clinical Note

Pancreatitis

Pancreatitis, or inflammation of the pancreas, occurs when activated pancreatic proteases digest pancreatic tissue itself. The most common causes of this condition are excessive consumption of alcohol and blockage of the pancreatic or common duct. Duct blockage is usually due to gallstones and frequently occurs at the ampulla of Vater. Pancreatic secretions build up behind the obstruction, trypsin accumulates, and activates the other pancreatic proteases as well as additional trypsin. Eventually the normal

defense mechanisms are overwhelmed and digestion of pancreatic tissue begins.

Approximately 10% of pancreatitis is hereditary. Mutations of the genes associated with the normal defense mechanisms have been identified as accounting for most of these cases. Two mutations occur in the trypsinogen gene itself. One (R122H) enhances trypsin activity by impairing its autodigestion. The other (N29I) leads to an increased rate of autoactivation. In addition, mutations in the predominant native trypsin inhibitor have been shown to increase the risk of disease.

they can digest the pancreas itself. First, and foremost, the proteases are synthesized as inactive precursors. Second, pancreatic enzymes are membrane bound from the time of synthesis until they are secreted from zymogen granules. Third, acinar cells contain a trypsin inhibitor which destroys active trypsin. Fourth, trypsin itself is capable of autodigestion.

Regulation of Secretion

The secretion of the aqueous component (fluid and bicarbonate) is regulated by the amount of acid entering the duodenum. Because the function of this component of pancreatic secretion is to neutralize the intestinal contents, this operates as another negative feedback system for the regulation of a gastrointestinal secretion. Similarly, the secretion of the enzymatic component is regulated by the amount of fat and protein in the intestinal contents. Most stimulation of pancreatic secretion is initiated during the intestinal phase, although stimuli from both the cephalic and gastric phases also contribute. The primary agents involved in stimulating the pancreatic cells are secretin, CCK, and acetylcholine acting via vagovagal reflexes.

In humans, there is little basal pancreatic secretion. The aqueous component is secreted at rates of 2–3% of maximal, and the enzymatic component at 10–15% of maximal. The stimuli involved in basal pancreatic secretion are unknown, and basal secretion may be an intrinsic property of the organ.

Cephalic Phase

Truncal vagotomy decreases the pancreatic secretory response to a meal by approximately 60%. Most of this reduction, however, is due to the interruption of vagovagal reflexes initiated during the intestinal phase.

The stimuli for the cephalic phase of pancreatic secretion are identical to those discussed previously for gastric secretion and contribute approximately 20% of the pancreatic response. Vagal efferents to the pancreas release acetylcholine at both the ductule and acinar cells (Fig. 22). Stimulation has a greater effect on the enzymatic component (acinar cells) than on the aqueous component. This results in the production of a low volume of secretion with a high enzymatic content. In humans, gastrin plays little or no role in the cephalic phase of pancreatic secretion.

Gastric Phase

Distension of either the proximal or distal regions of the stomach stimulates pancreatic secretion by initiating vagovagal reflexes. Because these are mediated by

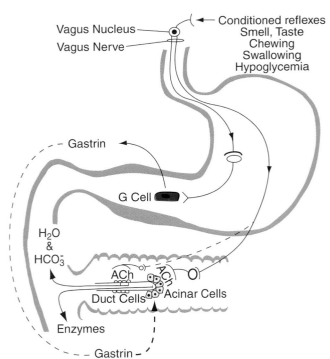

FIGURE 22   Mechanisms involved in the stimulation of pancreatic secretion during the cephalic phase. Dashed lines represent mechanisms of little importance in humans. (Modified from Johnson LR, ed., *Gastrointestinal physiology*, 6th ed. St. Louis: CV Mosby, 2001.)

acetylcholine, secretion is primarily the low-volume enzymatic component. There is no strong evidence that gastrin contributes to the human pancreatic response. The percentage of pancreatic secretion due to the gastric phase is small, amounting to only 5–10% of the total.

Intestinal Phase

Hydrogen ion and fat and protein digestion products in the lumen of the small intestine account for 70–80% of the human pancreatic secretory response to a meal. Acid releases secretin from the S cells of duodenal mucosa, and secretin is the principal stimulant of the aqueous component. Fat and protein digestion products release CCK from duodenal I cells. CCK is the primary humoral stimulant of the enzymatic component. The enzymatic component is also stimulated by acetylcholine via vagovagal reflexes initiated by acid, fatty acids, and peptides and amino acids acting on receptors present in the duodenal mucosa. Neither CCK nor acetylcholine has much effect on the ductule cells in the absence of secretin. However, both potentiate the effects of secretin, allowing the pancreas to secrete large amounts of water and bicarbonate. The mechanisms involved in the stimulation of pancreatic secretion during the intestinal phase are summarized in Fig. 23.

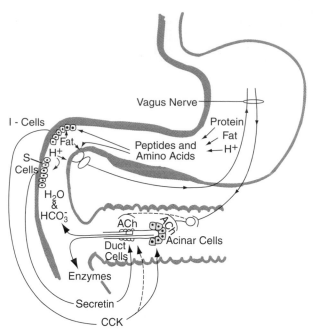

FIGURE 23 Mechanisms involved in the stimulation of pancreatic secretion during the intestinal phase. Dashed lines indicate potentiative interactions with secretin. (Modified from Johnson LR, ed., *Gastrointestinal physiology,* 6th ed. St. Louis: CV Mosby, 2001.)

Secretin is released when the duodenal mucosa is acidified to pH 4.5 or lower (Fig. 24). As the pH of the duodenal contents is lowered to 3.0, secretin release increases linearly. Below pH 3.0, secretin release does not depend on the pH, but depends only on the amount of titratable acid entering the duodenum. As more acid enters the duodenum, more S cells are acidified to release secretin. Therefore, below pH 3.0, secretin release depends on only the area of mucosa acidified. Secretin can also be released by high concentrations of fatty acids. This mechanism, however, is not nearly as important physiologically as release by hydrogen ion.

During the response to a normal meal, the pH of duodenal contents rarely drops below 3.5 or 4, and only the duodenal bulb and proximal duodenum are sufficiently acidified to release secretin. Therefore, even though secretin is present in relatively equal amounts throughout the duodenal and jejunal mucosa, little is released. In fact, the amount released is, by itself, sufficient to account for only a small fraction of the bicarbonate and water secreted by the pancreas. However, in the presence of fat and protein digestion products, lowering the duodenal pH even slightly below 4.5 leads to near maximal increases in water and bicarbonate secretion. This magnified response is due to the potentiation of small amounts of secretin by CCK and acetylcholine. The potentiation of secretin effects is an important physiologic interaction either between two hormones or between a hormone and a neurocrine.

Potentiation allows maximal effects to be produced with small amounts of hormones, leading in turn to a conservation of hormone supplies. Because only small amounts of hormone circulate, potentiation also ensures that "pharmacologic" effects of the peptides do not occur.

CCK is released by L-isomers of amino acids and by fatty acids containing eight or more carbon atoms. Not all amino acids are strong releasers of CCK. In humans, phenylalanine, methionine, and tryptophan appear to be the most potent. Three peptides, all containing glycine, are also effective releasers of CCK. These are glycylphenylalanine, glycyltryptophan, and phenylalanylglycine. Interestingly, dipeptides of glycine and glycine itself are not effective stimuli for CCK release. There is evidence that additional peptides, some containing at least four amino acids, also release CCK. Fatty acids longer than eight carbon atoms release CCK. Lauric, palmitic, stearic, and oleic acids are potent and equally effective. All of these protein and fat digestion products, as well as hydrogen ion, also initiate vagovagal reflexes. CCK is present in equal concentrations throughout the mucosa of the duodenum and jejunum. Digestion products are equally effective at stimulating pancreatic enzyme secretion when applied to any segment of this part of the small intestine.

Response to a Meal

As digestion proceeds in the stomach, much of the gastric acid is buffered by proteins. As mixing, digestion, and secretion continue, the buffers become saturated and the pH drops to around 2. As the stomach empties, the maximal load of titratable acid delivered to the duodenum may be as great as 20–30 mmol/hr. This includes both free H^+ and that bound to buffers. This

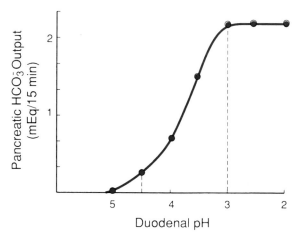

FIGURE 24 Pancreatic bicarbonate output in response to the acidification of a fixed length of duodenum. Bicarbonate output is used as an index of secretin release.

Clinical Note

A number of conditions in which the pancreas is directly involved decrease pancreatic secretion. These include pancreatitis, cystic fibrosis, and tumors of the pancreas. Pancreatic secretion is reduced by severe protein deficiency as in kwashiorkor. Chronic pancreatitis usually results in decreased volume and bicarbonate output, whereas individuals with cystic fibrosis or kwashiorkor exhibit decreases in both components. Cystic fibrosis is a relatively common, lethal, autosomal recessive disorder that causes viscid mucous secretion by pancreatic, airway and intestinal epithelia as well as a generalized disorder of fluid and electrolyte transport. In the pancreas the result is generalized pancreatic exocrine insufficiency. The critical abnormality is defective regulation of apical membrane Cl^- channels (see Fig. 20). In cystic fibrosis the gene product is apparently normal, however, Cl^- conductance is not inserted into the apical membrane. Thus, the problem is in protein trafficking. This defect results in the inability of cAMP to stimulate Cl^- secretion and a severe decrease in secretion of the aqueous component. As a result proteinacious acinar secretions become concentrated and finally block the lumen of the duct, eventually destroying the gland. Pancreatic exocrine function is determined by collecting the duodenal content with a double-lumen nasogastric tube. One lumen is used to evacuate potentially contaminating gastric juice and the other to collect duodenal content, which is presumed to be primarily pancreatic in origin. Symptomatically, pancreatic insufficiency may show up as steatorrhea, the presence of fat in the stool. Pancreatic enzyme secretion must be reduced to less than 20% of normal before enough fat is left undigested to appear in the stool. Thus, pancreatic enzymes are secreted in great excess. Steatorrhea can be the result of a number of problems beside insufficient pancreatic lipase. To determine whether the condition is pancreatic in origin, it is necessary only to measure the concentration of any pancreatic enzyme in the jejunal content after a meal.

rate of delivery approximates the maximal acid secretory rate of the stomach, which in turn equals the maximal rate of bicarbonate secretion by the pancreas. In the first part of the duodenum, both the free acid and bound acid are rapidly neutralized by bicarbonate and the pH is raised from 2 at the pylorus to around 4 beyond the duodenal bulb. Within the remainder of the proximal duodenum, the pH of chyme is quickly increased to near neutrality. Most of the neutralization is accomplished by pancreatic bicarbonate, but an additional amount of bicarbonate is secreted by the duodenal mucosa and the liver. A small amount of hydrogen ion is absorbed directly by the mucosa.

Pancreatic enzyme secretion increases abruptly shortly after chyme enters the stomach, and peaks within 30 min to levels equal to 70–80% of those attainable with maximal stimulation by CCK and cholinergic reflexes. Enzyme secretion continues at this rate as chyme enters the duodenum. The enzyme response may be kept below the maximal possible rate by inhibitors such as pancreatic polypeptide.

Changes in the proportion of the nutrients in the diet change the proportion of enzymes in pancreatic secretion. For example, ingestion of a high-protein, low-carbohydrate diet for several days results in an increase in the proportion of proteases and a decrease in the proportion of amylase in pancreatic secretion. Adaptation of pancreatic juice to the diet is now known to be mediated by hormones at the level of gene expression. CCK increases the expression of the protease genes. GIF and secretin increase the expression of the lipase gene. In diabetics, insulin regulates amylase gene expression, but we do not know how amylase expression is regulated normally.

Molecular Basis for Potentiation

After the delivery of a secretory stimulus, the entire chain of intracellular events leading to secretion has not been identified. However, the concept of potentiation requires that the stimuli bind to different receptors and trigger different chains of events. Some of the steps involved have been elucidated using isolated rodent acini and are diagrammed schematically in Fig. 25. Secretin triggers an increase in cAMP formation after it binds to its membrane receptor. Acetylcholine binds to a receptor different from that of secretin and increases the formation of inositol trisphosphate, which in turn mobilizes Ca^{2+} from intracellular stores. CCK binds to a third distinct receptor and also increases intracellular

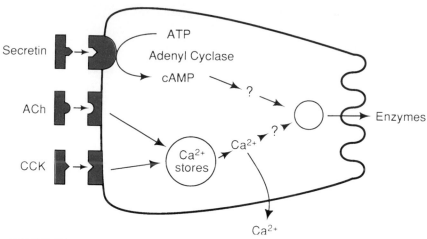

FIGURE 25   Diagram of the mechanisms leading to potentiation of enzyme secretion from the rodent pancreatic acinar cell. Similar potentiative interactions occur in human ductule cells but not in human acinar cells. (Modified from Johnson LR, *Gastrointestinal physiology*, 6th ed. St. Louis: CV Mosby, 2001.)

free Ca^{2+} via the inositol trisphosphate pathway. Interactions between secretin and acetylcholine or secretin and CCK, therefore, result in potentiation. The effects of combining acetylcholine and CCK, which trigger identical mechanisms, are only additive. Obviously, at some point the different intracellular mechanisms must interact to produce the potentiated response. The point of this interaction and the actual events producing potentiation are not understood. VIP and glucagon also increase cAMP levels, and substance P, GRP, and gastrin increase Ca^{2+} levels in acinar cells. There is no evidence, however, that any of these agents is released in sufficient quantities to be a physiologically significant regulator of pancreatic secretion.

As previously mentioned, these interactions have been elucidated for isolated rodent acini. However, a similar situation is likely to exist in the human ductule cell for the potentiation of the effect of secretin by CCK and acetylcholine. Why secretin does not potentiate CCK and acetylcholine effects on human acinar cells is not known.

BILE SECRETION AND GALL BLADDER FUNCTION

Bile is responsible for the principal digestive functions of the liver. The presence of bile in the small intestine is necessary for the digestion and absorption of lipids. The problem of the insolubility of fats in water is solved by the constituents of bile. The bile salts and other organic components of bile are responsible in part for emulsifying fat so that it can be digested by pancreatic lipase. The bile acids also take part in solubilizing the digestion products into micelles. Micellar formation is essential for the optimal absorption of fat digestion products. Bile also serves as a vehicle for the elimination of a variety of substances from the body. These include endogenous products like cholesterol and bile pigments, as well as some drugs and heavy metals.

The liver also carries out many additional nondigestive functions, most of which are treated in chapters relating to the pertinent organ system or in a biochemistry textbook. Because in this text there is no single chapter on the liver in which all of its functions are discussed, the more important ones are briefly summarized here.

The liver plays important roles in carbohydrate, lipid, and protein metabolism. Blood glucose levels are held relatively constant by hormones acting on the liver to regulate glycogenolysis (release of glucose from glycogen stored in the liver) and gluconeogenesis (synthesis of glucose from noncarbohydrate precursors). Hormones also regulate the formation of ketone bodies by β-oxidation of fatty acids. This process occurs in the liver, and the acetoacetate, acetone, and β-hydroxybutyrate produced are released into the blood to serve as calorie sources during periods of starvation. The liver is responsible for the synthesis of all nonessential amino acids and for the production of urea from ammonia.

The liver also synthesizes a number of proteins with specific functions. These include all plasma proteins with the exception of the immunoglobulins. Albumins and globulins are responsible for most of the plasma oncotic pressure and for transporting many specialized molecules that are bound to them. The fibrinogens and other proteins are involved in blood clotting. The liver synthesizes very-low-density lipoprotein, which is

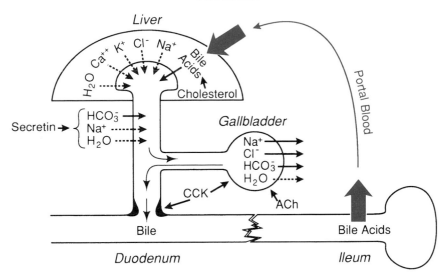

FIGURE 26   Overview of the biliary system and the enterohepatic circulation of bile acids. Solid arrows indicate active transport processes.

converted into the other lipoproteins. The lipoproteins transport cholesterol and triglycerides throughout the body.

The liver degrades many drugs and toxic products as well as a wide variety of hormones. Some are excreted in the bile. Others are made water soluble by conjugation to substances like glucuronic acid or by chemical transformation, and are then readily excreted by the kidneys.

The liver also serves as a storehouse for a number of substances whose availability to the body may be sporadic. These include iron and vitamins B_{12}, A, and D.

Overview of the Biliary System

Fig. 26 is a schematic illustration of the biliary system and the circulation of bile acids between the intestines and the liver. *Bile* is continuously produced by the hepatocytes. The principal organic constituents of bile are the *bile acids,* which are synthesized by the hepatocytes. The secretion of the bile acids carries water and electrolytes into the bile by osmotic filtration. Additional water and electrolytes, primarily $NaHCO_3$, are added by cells lining the ducts. This latter secretory component is stimulated by secretin and is essentially identical to the aqueous component of pancreatic secretion. The secretion of bile increases pressure in the hepatic ducts, causing the gall bladder to fill. Within the gall bladder, the bile is stored and concentrated by the absorption of water and electrolytes. When a meal is eaten, the gall bladder is stimulated to contract by CCK and vagal nerve stimulation. Within the lumen of the intestine, bile participates in the emulsification,

hydrolysis, and absorption of lipids. Most bile acids are reabsorbed either passively throughout the intestine or by an active process in the ileum. Those that are lost in the feces are replaced by resynthesis in the hepatocytes. The absorbed bile acids are returned to the liver via the portal circulation, where they are actively extracted from the blood by the hepatocytes. Together with newly synthesized bile acids, the returning bile acids are secreted into the bile canaliculi. Canalicular bile is produced in response to the osmotic effects of anions that are secreted by ATP-dependent transporters present in the biliary canaliculus. In humans, almost all bile formation is driven by bile acids and is, therefore, referred to as being *bile acid dependent*. The portion of bile stimulated by secretin and contributed by the ducts is termed *ductular secretion*.

Organic Constituents of Bile

Bile is a complex mixture of inorganic and organic components, the physical properties of which account for the ability of bile to solubilize normally insoluble fat digestion products. In fact, bile itself contains molecules that are insoluble in water but that are solubilized in bile because of the interactions of its various organic constituents.

Bile Acids

Bile acids account for about 50% of the organic components of bile. They are synthesized in the liver from cholesterol and contain the steroid nucleus with a branched side chain of three to nine carbon atoms

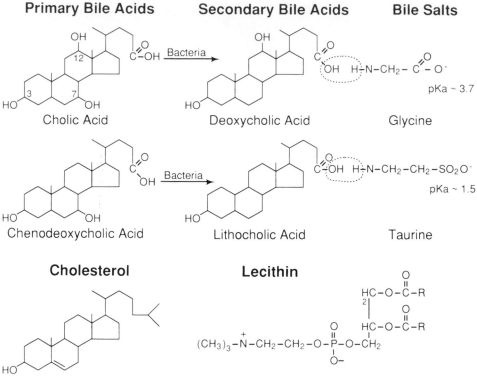

FIGURE 27   Principal organic constituents of bile. The two primary bile acids may be converted to secondary bile acids in the intestine. Each of the four bile acids may be conjugated to either glycine or taurine to form bile salts. The R-groups of lecithin represent fatty acids.

ending in a carboxyl group (Fig. 27). Chemically, therefore, they are carboxylic acids. Four bile acids are present in bile; there are also trace amounts of others that are chemical modifications of the four. The liver synthesizes two bile acids, *cholic* and *chenodeoxycholic acid*. These are the primary bile acids. Within the lumen of the gut, a fraction of each is dehydroxylated by bacteria to form *deoxycholic* and *lithocholic acids* (Fig. 27). These are called *secondary bile acids*. All four are returned to the liver in the portal blood and secreted into bile. The relative amounts of the bile acids in bile are approximately four cholic to two chenodeoxycholic to one deoxycholic to only small amounts of lithocholic.

The solubility of the bile acids depends on the number of hydroxyl groups present and the state of the terminal carboxyl group. Cholic acid, with three hydroxyl groups, is the most soluble, whereas lithocholic, a monohydroxy acid, is the least soluble. The pKs of the bile acids are near the pH of duodenal contents so there are relatively equal amounts in the protonated (insoluble) form and the ion (soluble). The liver, however, conjugates the bile acids to the amino acids glycine or taurine with pKs of 3.7 and 1.5, respectively. Thus, at the pH of intestinal contents the bile acids are largely ionized and water soluble. These conjugated bile

acids exist as salts of various cations, primarily sodium, and are referred to as *bile salts* (Fig. 27).

The unique properties of the bile salts are due to the fact that they are *amphipathic* molecules—that is, they have both hydrophilic and hydrophobic portions. The molecules are planar and all of the hydrophilic groups project in the same direction out from the hydrophobic sterol nucleus. The hydrophilic groups are the hydroxyls of the cholesterol nucleus, the peptide bond of the side chain, and either the carbonyl or sulfate group of glycine or taurine (Fig. 28). The bile salts align themselves at air–water or oil–water interfaces with the hydrophilic portions in water. In the lumen of the duodenum, the bile salts arrange themselves around droplets of lipid, keeping them dispersed into a suspension called an *emulsion*. An emulsion consists of droplets approximately 0.5–1 μm in diameter. In this form, the fat has an increased surface area exposed to the action of pancreatic lipase.

Above a certain concentration, called the *critical micellar concentration,* bile salts form molecular aggregates called micelles (Fig. 28). *Micelles* are cylindrical, having the bile salts on the outside with their hydrophilic portions oriented outward. The inside of the micelle is made up of various molecules that are insoluble in water. Micelles, 40–70 Å in diameter, are in true solution; hence, the bile salts serve to solubilize

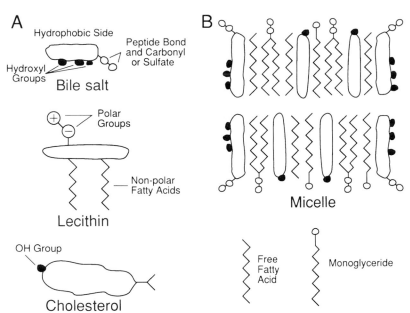

FIGURE 28   Structures of components of micelles and micelles themselves. (A) Representations of bile salt, lecithin, and cholesterol molecules, illustrating the separation of polar and nonpolar surfaces. (B) Cross section of a micelle, showing the arrangement of these molecules plus the principal products of fat digestion. Micelles are cylindrical disks whose outer curved surfaces are composed of bile salts.

these other molecules. Within bile itself, the bile salts are always present in amounts above the critical micellar concentration, and the micelles also contain phospholipids and cholesterol. The osmolarity of bile drawn from the cystic duct is always near 300 mOsm. However, the cation concentration, primarily Na^+, may be as high as 300 mmol/L. The apparent discrepancy is resolved by the fact that a single micelle osmotically inactivates many molecules and sequesters a significant number of the inorganic cations. Within the lumen of the proximal small bowel, micelles contain fatty acids, monoglycerides, and fat-soluble vitamins in addition to the bile salts, cholesterol, and phospholipids.

Phospholipids

Phospholipids, primarily lecithins (see Fig. 27), are the second most abundant organic component of bile and account for approximately 30–40% of the solids present. Phospholipids are also amphipathic, and the hydrophilic phosphatidylcholine group is oriented outward, whereas the hydrophobic fatty acid carbon chains are buried in the core of the micelle (see Fig. 28). As mentioned previously, the phospholipids themselves are insoluble and are solubilized in micelles. Phospholipids, however, are extremely important because they increase the ability of bile salts to form micelles and solubilize cholesterol. Approximately 2 mol of lecithin are solubilized per mole of bile salt.

Cholesterol

Bile is the primary excretory pathway for cholesterol. The insoluble cholesterol makes up roughly 4% of the organic material in bile and is solubilized in the core of the micelle (see Fig. 28). If more cholesterol is present than can be solubilized, crystals of cholesterol form in the bile. These crystals may serve as the nidus for gallstone formation.

Bile Pigments

The fourth significant group of organic compounds found in bile is the bile pigments and it accounts for approximately 2% of the solids. The principal bile pigment is *bilirubin,* produced from hemoglobin by cells of the reticuloendothelial system. Chemically, the bile pigments are tetrapyrroles, which are derived from porphyrins. Bilirubin is insoluble in water, but within the liver it is made soluble by conjugation to glucuronic acid. It is secreted as the soluble salt *bilirubin glucuronide* and, therefore, is not found within the micelles.

Secretion of Bile

Functional Histology of the Liver

The functional organization of the liver is shown schematically in Fig. 29. The liver is divided into lobules organized around a central vein that receives blood

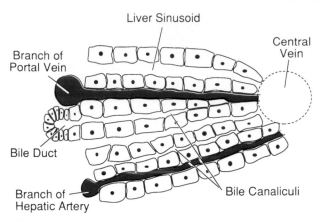

FIGURE 29 Schematic diagram of the relationship between blood vessels, hepatocytes, and bile canaliculi in the liver. Each hepatocyte is exposed to blood at one membrane surface and a bile canaliculus at the other.

through separations surrounded by plates of hepato-cytes. The separations are called *sinusoids,* and they in turn are supplied by blood from both portal vein and the hepatic artery. The plates of hepatocytes are no more than two cells thick, so every hepatocyte is exposed to the blood. Openings between the plates ensure that the blood is exposed to a large surface area. The hepatocytes remove substances from the blood and secrete them into the biliary canaliculi lying between the adjacent hepatocytes. The bile in the canaliculi flows toward the periphery, countercurrent to the flow of blood, and drains into peripheral bile ducts. The countercurrent relationship between bile flow and blood flow within the lobule minimizes the concentration gradients between substances in the blood and bile and contri-butes to the liver's efficiency in extracting substances from the blood.

Bile Acids and the Enterohepatic Circulation

The total bile acid pool in the human body is approximately 2.5 g and includes the bile acids present in the liver, ducts, gall bladder, bowel, and blood, that is, within the entire *enterohepatic circulation* (see Fig. 26). The bile acid pool may circulate through the entero-hepatic circulation several times during the digestion of a meal, so that 15–30 g bile acids may enter the duode-num during a 24-hr period.

Bile acids are secreted continuously by the liver. The secreted bile acids include those extracted from the portal blood and those newly synthesized. Approxi-mately 90–95% of bile acids in portal blood are carried bound to plasma proteins. Within the liver, the hepa-tocytes extract the bile acids from the portal blood via a secondary active transport mechanism (Fig. 30). The basolateral (blood side) membrane of the hepatocyte

contains a transport protein that binds the bile acid and Na^+, transporting them both to the cytoplasm. The (Na^+, K^+)-ATPase pumps the Na^+ out of the cell, providing the concentration gradient for the entry of the bile acids. This extraction of bile acids from the portal blood is nearly 100% efficient.

The rate of synthesis of new bile acids is inversely correlated with the return of bile acids via the portal blood. Under normal conditions, approximately 600 mg of bile acids are synthesized per day. This amount replaces that lost into the stool every day. Bile acids extracted from the portal blood act to feedback-inhibit the 7-α-hydroxylase, the rate-limiting enzyme for bile acid synthesis from cholesterol. If the enterohepatic circulation is interrupted (e.g., by a biliary fistula draining bile to the outside), the rate of synthesis of bile acids becomes maximal. In humans, this is equal to 3–5 g per day. In patients incapable of reabsorbing bile acids, the maximal rate of synthesis equals the total amount of bile acids secreted by the liver. Although this is greater than the bile acid pool, because of the recirculation of bile acids, it is only a fraction of the 15–30 g normally secreted by the liver per day.

Although bile acids returning to the liver in the portal blood inhibit new bile acid synthesis, they are a potent stimulus of bile secretion and greatly increase the rate of bile production. Such a substance is called a *choleretic.* The volume of bile produced in response to the osmotic gradient caused by the secretion of bile acids, both returning and newly synthesized, is known as the bile acid-dependent fraction of the overall secretion of bile.

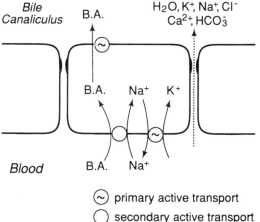

FIGURE 30 Mechanism for the secretion of bile by the hepatocytes. Bile acids (B.A.) are reabsorbed from the blood by a secondary active transport mechanism dependent on the Na^+ gradient established by the (Na^+, K^+)-ATPase. They then enter the bile canaliculus via a facilitated diffusion pathway. Water and electrolytes enter the bile through paracellular pathways along an osmotic gradient. This process is known as *osmotic filtration.*

The process by which returning and newly synthesized bile salts are secreted across the apical membrane of the hepatocyte into the bile canaliculus depends on ATP. Two ATP-dependent transporters have been defined in the canalicular membrane. One has specificity for monovalent bile salts and the second for divalent bile salt conjugates.

Bile acids are absorbed throughout the entire small intestine by passive diffusion and in the terminal ileum by an active transport mechanism. This mechanism requires Na^+, and is secondary to the Na^+ gradient created by the (Na^+,K^+)-ATPase. More than 95% of the secreted bile acids are normally absorbed and returned to the liver. Because bile acids are large molecules, their passive absorption depends on their lipid solubility. When bile acids enter the duodenum they are 100% conjugated, almost totally dissociated and, therefore, lipid insoluble. Because the pK of glycine conjugates is higher than that of taurine conjugates, bile acids conjugated to glycine are likely to be absorbed earlier than those conjugated to taurine. Within the lumen bacteria deconjugate the bile salts, increasing their lipid solubility and increasing their passive absorption approximately ninefold. Bacteria also dehydroxylate bile acids, which increases passive absorption about fourfold. In humans, approximately 50% of the secreted bile acids are actively absorbed. The active mechanism absorbs primarily conjugated bile acids, although a small amount of unconjugated acids also may be absorbed. The location of the active absorptive mechanism in the terminal ileum ensures the presence of adequate amounts of bile acids for micelle formation until all of the fat digestion products are absorbed. The absorption of fat is usually completed by the end of the jejunum. Some bile acids pass into the colon, where they are further deconjugated and modified by bacteria. Some of these may be absorbed passively. The remainder is excreted in the stool. After extraction from the portal blood by the hepatocytes, both primary and secondary bile acids are reconjugated to glycine or taurine and some secondary bile acids are rehydroxylated.

Loss of excessive bile acids into the colon produces diarrhea. Bile acids or their degradation products inhibit colonic absorption of sodium and H_2O. This results in diarrhea without loss of fat in the stool, if the resynthesis of bile acids can keep up with the loss. In cases of extreme loss, such as after ileal resection, the digestion and absorption of long-chain triglycerides is also impaired and steatorrhea results as well. The steatorrhea can be cured by substituting medium-chain triglycerides for long-chain triglycerides in the diet. Medium-chain triglycerides do not require micelle formation for absorption.

Phospholipids and Cholesterol

Phospholipids and cholesterol are secreted by the hepatocytes into the bile. The exact mechanism of secretion is unknown, but the quantity secreted is directly related to the secretion of bile acids. Within the small intestine, phospholipids and cholesterol secreted with the bile are handled in the same manner as those ingested (see Chapter 35).

Bilirubin

Cells of the reticuloendothelial system degrade hemoglobin from worn-out red blood cells. The porphyrin is converted into the yellow pigment bilirubin. The insoluble bilirubin is released into the blood, where it is carried to the liver tightly bound to plasma albumin (Fig. 31).

Hepatocytes extract bilirubin from the blood and conjugate it to glucuronic acid to form water-soluble bilirubin glucuronide. Uptake from the blood is mediated by an active anion transport system different from the one that extracts bile acids. This system, however, is also virtually 100% efficient. Bilirubin glucuronide is secreted into the bile and is partially responsible for its golden color. Because it is water soluble, it does not take part in micelle formation. The failure to clear sufficient bilirubin from the blood in patients with liver damage may result in the yellow appearance of the skin and eyeballs, the condition known as *jaundice*.

Bilirubin glucuronide is not absorbed from the intestine in appreciable amounts, and it is excreted in the feces. A fraction, however, is deconjugated and reduced to *urobilinogen* by bacteria of the distal small bowel and intestine. Some urobilinogen is excreted in the feces, but some is also absorbed and transported to the liver via the portal circulation. The liver extracts most of the urobilinogen via an active mechanism and secretes it into the bile. The capacity of the liver to extract urobilinogen is low, so that a measurable fraction passes into the systemic circulation, where it is filtered by the kidneys and excreted in the urine. If the liver becomes damaged, the amount of urobilinogen in the urine may increase severalfold. Urobilinogen is colorless; however, in the urine and feces, it is exposed to oxygen and oxidized to urobilin and stercobilin, respectively. These pigments are partially responsible for the color of the excretory products.

Water and Electrolytes

As bile acids are secreted into the bile canaliculi, water and electrolytes enter the bile by the process of osmotic filtration (see Fig. 30). Canalicular bile is, thus, primarily

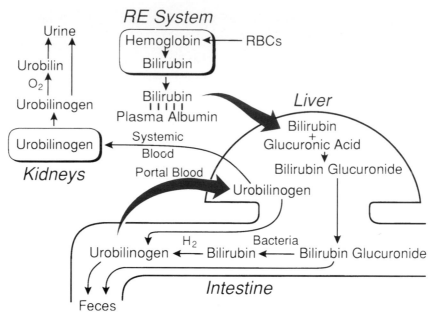

FIGURE 31   Bilirubin excretory pathways. Heavy arrows represent active processes.

an ultrafiltrate of plasma as far as the concentrations of water and electrolytes are concerned. In some species, there is evidence for the active secretion of Na^+ into the bile canaliculi. However, this has not been established in humans. The higher the rate of return of bile acids to the liver, the faster they are secreted and the greater the volume of bile. This component is therefore referred to as the bile acid-dependent fraction.

The contribution of the bile ductules and duct to bile production is identical to that of the pancreatic ducts to pancreatic juice. Secretin stimulates the secretion of HCO_3^- and water from the ductule cells (see Fig. 26). This results in significant increases in bile volume, HCO_3^- concentration, and pH, and a decrease in the concentration of bile salts. The mechanism of HCO_3^- secretion by the ducts of the liver involves active transport and is also similar to the one employed by the pancreas. When stimulated by secretin, the HCO_3^- concentration of the bile may increase two- or threefold over that of the plasma.

Gall Bladder Function

Filling

The liver secretes bile continuously. Because the hepatic end of the system is blind (closed), secretion results in the generation of hydrostatic pressure within the ducts. In humans, the secretory pressure normally ranges between 10 and 20 mm Hg. The maximal secretory pressure, above which bile cannot be produced, has been measured at 23 mm Hg.

During the interdigestive period, the gall bladder is flaccid and the sphincter of Oddi at the opening of the duct into the duodenum is closed. This causes bile to flow into the gall bladder. The capacity of the human gall bladder ranges between 20 and 60 mL. The volume of bile produced by the liver before the gall bladder empties, however, may be several times this amount.

Concentration of the Bile

The gall bladder epithelium concentrates the bile by actively reabsorbing Na^+, Cl^-, and HCO_3^-. Water follows down the osmotic gradient created. The active transport mechanism employed varies between species. In some species, NaCl and $NaHCO_3$ are absorbed via a coupled transport system that is not electrogenic. In humans, an electrical potential difference, inside approximately -8 mV, is created, indicating that the system is not coupled, that Na^+ may be the only ion actively transported, and that Cl^- and HCO_3^- follow passively. Hydrogen ion is also secreted, which neutralizes HCO_3^-, and the resulting CO_2 is absorbed across the gall bladder epithelium. In any case, the net result is a decrease in the pH, total amounts of Na^+, Cl^-, HCO_3^-, and water in the bile and a large increase in the concentration of the organic constituents of bile (Table 2).

Bile remains isosmotic to plasma even though the organic constituents are concentrated greatly. The bile salts, cholesterol, and phospholipids are present in osmotically inactive micelles. Because bile salts are anions, many of the inorganic cations are bound within the micelles and are also osmotically inactive. This

Clinical Note

Gallstones

Approximately 10% of the white population over 29 years of age in the United States is estimated to have gallstones. There are basically two types of gallstones, cholesterol stones and pigment stones, although both have a mixed composition.

Cholesterol stones are 50–75% cholesterol and most contain a pigment center (bilirubin) that probably acted as a nidus for stone formation. During the course of a day, about 50% of bile secreted by nonstone individuals is supersaturated with cholesterol. All bile produced by patients with cholesterol stones is supersaturated with cholesterol. The solubility of cholesterol in bile depends on micelle formation, which in turn depends on the relative amounts of bile salts, lipids, and cholesterol in the bile. In nonobese cholesterol stone patients, the amount of lipids is decreased. In obese patients, the amount of cholesterol is increased. In both groups, the bile acid pool is reduced for some reason to about 50% of normal. Supersaturation results in crystal formation and the growth of the crystals into stones.

Pigment stones are produced when unconjugated bilirubin precipitates with calcium to form a stone. In normal bile, bilirubin is solubilized by conjugation to glucuronic acid, and approximately only 1% remains unconjugated. Gall bladder bile from patients with pigment stones, however, is saturated with unconjugated bilirubin. The enzyme β-glucuronidase is responsible for the deconjugation of bilirubin within the gall bladder. The enzyme is released from the wall of the damaged gall bladder, and it is present in high concentrations in a variety of bacteria, including *Escherichia coli*, which may infect the gall bladder.

The treatment for gallstones is surgical removal of the gall bladder, known as cholecystectomy. Normal digestion and absorption are unaffected in individuals after cholecystectomy. Without the gall bladder, bile is released continuously into the duodenum as it is produced by the liver. Amounts produced are higher than normal during interdigestive periods, because bile salts are not being stored in the gall bladder. They therefore return to the liver continuously to stimulate bile production. The large increases of bile secretion into the duodenum at the beginning of a meal, however, are absent in the cholecystectomized patient, and the overall secretion of bile in response to a meal is somewhat reduced. The gall bladder is naturally absent from some species of animals, including the rat and the horse.

also explains the great excess of inorganic cations over inorganic anions in concentrated bile. The mechanism of water absorption across the gall bladder epithelium is similar to that across other tight epithelia (see Chapter 36). The layer of epithelial cells has large lateral intercellular spaces. These are closed by tight junctions at the apical ends, but are relatively permeable to water and solutes at the basal ends. As Na^+ is pumped into the intercellular space near the apical end, Cl^- follows, creating a hypertonic solution, and water enters passively. The movement of water molecules creates an osmotic gradient, hypertonic at the apical end and isotonic at the basal end, within the lateral spaces. This is known as a *standing osmotic gradient*. During absorption, the spaces swell because of the increased hydrostatic pressure created by the entering water. This hydrostatic pressure produces a gradient for the absorption of an isotonic solution by the nearby capillaries. In this manner an isotonic solution is absorbed and carried away by the blood.

Expulsion of Bile

The gall bladder begins to contract rhythmically and expel bile into the duodenum within 30 min of a meal. The principal stimulus is hormonal, although contrac-

TABLE 2   Approximate Values for Major Components of Liver and Gall Bladder Bile

Component	Liver Bile	Gall bladder bile
Na^+ (mmol/L)	150	300
K^+ (mmol/L)	4.5	10
Ca^{2+} (mmol/L)	4	20
Cl^- (mmol/L)	80	5
Bile salts (mmol/L)	30	315
pH	7.4	6.5
Cholesterol (mg/100 mL)	110	600
Bilirubin (mg/100 mL)	100	1,000

TABLE 3   Regulatory Mechanisms in Biliary Physiology

Stimulus	Target	Effect
Bile acids	Liver	Stimulate bile flow
(in portal circulation)	Hepatocytes	Inhibit bile acid synthesis
Secretin	Liver	Stimulates water and
	Bile ducts	HCO_3^- secretion
		into the bile
CCK	Gall bladder	Stimulates contraction
		(emptying)
	Sphincter	Causes relaxation
Vagus	Gall bladder	Stimulates contraction

CCK, cholecystokinin

tion is also stimulated somewhat by vagal activity during all phases of digestion. The major stimulus for gall bladder contraction is CCK released by fat and protein digestion products within the lumen of the duodenum. CCK has two actions that result in bile expulsion: It contracts the smooth muscle of the gall bladder and relaxes the sphincter of Oddi (see Fig. 26). The sphincter of Oddi is a thickening of the circular muscle of the bile duct at its entrance into the duodenum. The sphincter is embedded in the duodenal wall, but is distinct from it. However, bile flow through the sphincter is markedly influenced by the contractile activity of the duodenum. Bile enters the duodenum in spurts when the muscle of the duodenum relaxes; bile does not enter during duodenal contraction.

Summary of Control Mechanisms

Table 3 summarizes the physiologically important mechanisms resulting in the secretion of bile and its expulsion into the duodenum. Bile acids returning to the liver via the portal circulation inhibit bile acid synthesis and stimulate bile production by the hepatocytes. Secretin stimulates water and HCO_3^- secretion from the cells lining the bile ducts. There is little or no effect of neural stimulation on bile production. Other hormones such as CCK and gastrin do not have significant physiologic effects on bile secretion.

CCK stimulates the contraction of the gall bladder and relaxes the sphincter of Oddi. Vagal activity during all three phases of the response to a meal also aids in contracting the gall bladder. Contractile activity of the duodenal wall influences the opening and closing of the sphincter of Oddi.

Suggested Readings

Argent BE, Case RM. Pancreatic ducts: Cellular mechanism and control of bicarbonate secretion. In Johnson LR, ed. *Physiology of the gastrointestinal tract*, 3rd ed. New York: Raven Press, 1994, pp 1473–1498.

Cook DI, van Lennep EW, Roberts M, Young JA. Secretion by the major salivary glands. In Johnson LR, ed. *Physiology of the gastrointestinal tract*, 3rd ed. New York: Raven Press, 1994, pp 1061–1118.

Hersey SJ. Gastric secretion of pepsins. In Johnson LR, ed. *Physiology of the gastrointestinal tract*, 3rd ed. New York: Raven Press, 1994, pp 1227–1238.

Hoffman AF. Biliary secretion and excretion: The hepatobiliary component of the enterohepatic circulation of bile acids. In Johnson LR, ed. *Physiology of the gastrointestinal tract*, 3rd ed. New York: Raven Press, 1994, pp 1555–1576.

Lloyd K,CK, Debas H. Peripheral regulation of gastric acid secretion. In Johnson LR, ed. *Physiology of the gastrointestinal tract*, 3rd ed. New York: Raven Press, 1994, pp 1185–1226.

Modlin IM, Sachs G. *Acid related diseases*. Koustauz: Schuetztor-Verlag, 1998.

Scheele GA. Extracellular and intracellular messengers in diet-induced regulation of the pancreatic gene expression. In Johnson LR, ed. *Physiology of the gastrointestinal tract*, 3rd ed. New York: Raven Press, 1994, pp 1543–1554.

Silen W. Gastric mucosal defense and repair. In Johnson LR, ed. *Physiology of the gastrointestinal tract*, 2nd ed. New York: Raven Press, 1987, pp 1055–1070.

Solomon TE. Control of exocrine pancreatic secretion. In Johnson LR, ed. *Physiology of the gastrointestinal tract*, 3rd ed. New York: Raven Press, 1994, pp 1499–1530.

Suprenaut A. Control of the gastrointestinal tract by enteric neurons. *Annu Rev Physiol* 1994;56:117–165.

Sweeney JT, Ulrich CD. Genetics of pancreatic disease. *Clin Persp Gastroenterol* 2002;5:110–116.

Yule DI, Williams JA. Stimulus-secretion coupling in the pancreatic acinus. In Johnson LR, ed. *Physiology of the gastrointestinal tract*, 3rd ed. New York: Raven Press, 1994, pp 1447–1472.

Digestion and Absorption

LEONARD R. JOHNSON

KEY POINTS

- Almost all physiologically significant digestion and absorption occurs in the small intestine.
- *Digestion* is the chemical breakdown of food by enzymes secreted into the lumen or associated with the brush border membrane of enterocytes.
- Amylase catalyzes the luminal digestion of carbohydrates; the *brush border membrane* contains specific enzymes to digest disaccharides and remaining polymers of glucose.
- Glucose and galactose are absorbed actively by a Na^+-dependent carrier.
- The major enzymes that digest protein within the lumen of the gut are secreted in inactive

forms by the pancreas and then activated by trypsin, which is itself activated by enterokinase on the brush border.
- A membrane-bound peptidase aids in the digestion of large peptides to di- and tripeptides, which are absorbed across the brush border membrane along with free amino acids by a variety of specific transporters.
- *Within the cytoplasm* most peptides are digested to free amino acids that exit the basolateral membrane via a number of carriers with different specificities.
- Fat *digestion* is primarily carried out by the pancreatic enzymes, lipase, phospholipase A_2, and cholesterol ester hydrolase.

KEY POINTS (*continued*)

- Aided by colipase, pancreatic lipase digests triglycerides suspended in an emulsion of fat droplets to free fatty acids and 2-monoglycerides, which are solubilized in micelles by bile salts and other amphipathic molecules.
- Micelles are able to diffuse through the unstirred water layer at the surface of the enterocyte, allowing the fat digestion products to contact and diffuse through the brush border membrane.

- Within the enterocyte, triglycerides and phospholipids are resynthesized in the smooth endoplasmic reticulum and then packaged into chylomicrons, which contain an apoprotein on their surfaces.
- Chylomicrons leave the enterocyte by exocytosis and enter the lacteals of the lymphatics because they are too large to pass through the capillary walls.

Chapters 33 and 34 described gastrointestinal motility and secretion and their regulation. Each of these processes is regulated to optimize the digestion and absorption of nutrients. From a teleologic viewpoint, secretion is controlled to provide enzymes, fluids, and electrolytes in the proper locations, at the right times, and in the required quantities to digest foodstuffs. Motility patterns ensure that the ingesta move through the alimentary canal at a rate that provides optimal time for digestion and subsequent absorption. This chapter explains how food is broken down into absorbable molecules and how these by-products are absorbed into the body.

STRUCTURAL CONSIDERATIONS

Most digestion and, for practical purposes, all absorption of nutrients take place in the small intestine. The macroscopic surface area of the human small intestine is approximately 2.5 m², but structural modifications increase the overall surface area manyfold so that the exposure of nutrients to absorptive cells and membrane-bound digestive enzymes is greatly enhanced. The surface of the small intestine is thrown up into longitudinal folds, called *folds of Kerckring*. Finger-like villi project from these folds 0.5–1.5 mm into the lumen (Fig. 1). The villi are longest in the duodenum and shortest in the distal ileum. The surface of the villus is a layer of columnar epithelial cells, called *enterocytes,* interspersed with mucus-secreting *goblet cells.* The apical surface of an enterocyte is covered by hundreds of tiny processes called *microvilli* (Fig. 1). Owing to its appearance, the microvillous surface is called the *brush border* and greatly increases the exposure of luminal contents to the enterocyte membrane. Collectively, these structural modifications result in a total small intestinal absorptive area of approximately 400 m². Tube-like *crypts* project down into the surface at the

base of each villus. There are three crypts per villus, and these are lined with cells that are continuous with those of the villi. Crypts are approximately 0.3–0.5 mm deep, with the deepest found in the duodenum and the shallowest in the terminal ileum.

The various cell types along the crypt–villus unit have different functions (Fig. 1). The crypt cells themselves are the proliferative cells of the intestine. Crypts are monoclonal, meaning that there is only one stem cell per crypt. After division, one daughter cell remains anchored as the stem cell and the other divides several times while migrating up the crypt. Crypt cells are also capable of secreting fluid and electrolytes. As the crypt cells divide, the daughter cells migrate out of the crypts and up on to a villus. Some daughter cells differentiate into mucus-secreting goblet cells. The function of intestinal mucus is not totally clear, but it no doubt plays a role in physical, chemical, and perhaps immunologic protection. Most daughter cells differentiate into enterocytes, which are the principal digestive and absorptive cells of the small bowel. As new cells migrate onto a villus, old cells are extruded from the tip. During the course of its migration up the villus, an enterocyte

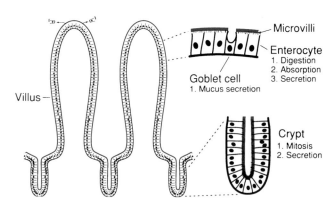

FIGURE 1   Schematic diagram of intestinal villi and crypt. Cell types and functions of individual cells are noted.

matures, the brush border becomes highly developed, the activities of its membrane-bound enzymes increase, and specific transport processes and membrane carriers differentiate. Thus, cells near the villus tip are much more capable of digestion and absorption than those at the base. Additional villous cells differentiate into hormone-producing enteroendocrine cells. These cells, for example, are responsible for the synthesis and release of secretin and cholecystokinin. A few cells migrate to the bottom of the crypts and become *Paneth cells,* whose functions are not entirely understood. Paneth cells contain lysozyme, a bacteriolytic enzyme, as well as immunoglobulin. This coupled with findings of degenerating bacteria and protozoa within their lysosomal elements suggest that Paneth cells may regulate the microbiological flora in the gut.

The entire epithelium of the small intestine is replaced every 3–6 days, making it one of the fastest growing tissues of the body. The rates of cell division and turnover are influenced by luminal contents, including nutrients and growth factors, and by gastrointestinal hormones. In general, increased cell loss results in higher rates of cell proliferation. The mechanisms balancing cell loss and renewal are poorly understood, but result in rapid repair of damaged mucosa. The high mitotic rate of the gastrointestinal mucosa makes it susceptible to x-radiation and cancer chemotherapy with resulting diarrhea and other problems.

DIGESTION

Digestion is the chemical breakdown of ingested food into absorbable molecules and is due to enzymes secreted into the lumen or bound to the apical membranes of the enterocytes. This suggests two major types of enzymes and digestion: *Luminal* or *cavital digestion* takes place in the lumen of the gastrointestinal tract and is carried out by enzymes secreted by the salivary glands, stomach, and pancreas. *Membrane digestion,* also referred to as *contact digestion,* is carried out by hydrolytic enzymes synthesized by the enterocytes and inserted in their brush borders as integral parts of the microvillar membrane. Enterocytes continuously synthesize these enzymes to replace those that are broken down or lost into the lumen. As the enterocyte migrates up the villus, the total number of enzymes present and, hence, its capacity to hydrolyze nutrients increase. The important digestive enzymes and their sources are shown in Table 1.

A significant surplus of digestive enzymes is normally present. The enzymes of the salivary glands and stomach are not essential. All luminal digestion can easily be accomplished by the pancreatic enzymes. The pancreatic

TABLE 1 Digestive Enzymes According to Their Sources

Source	Enzyme
Salivary glands	Amylase
	Lingual lipase
Stomach	Pepsin
Pancreas	Amylase
	Trypsin
	Chymotrypsin
	Carboxypeptidase
	Elastase
	Lipase-colipase
	Phospholipase A_2
	Cholesterol esterase (nonspecific)
Intestinal mucosa	Enterokinase
	Disaccharidases
	Sucrase
	Maltase
	Trehalase
	α-Dextrinase (isomaltase)
	Peptidases
	Aminooligopeptidase
	Dipeptidase

enzymes are also present in excess, and maldigestion and malabsorption do not occur unless secretion is reduced to a small fraction of normal. *Steatorrhea* (dietary fat in the stool), for example, does not occur unless lipase secretion is reduced by at least 80%. Hydrolysis by membrane-bound enzymes is also extremely rapid. Digestion and absorption of essentially all major dietary components are complete before the ingesta enter the ileum.

Developmental Changes

In some species, notably the rat (in which most studies have been performed), levels of disaccharidases except lactase are extremely low at birth. At the time of weaning, the third week of life, activities of sucrase, maltase, trehalase, and isomaltase increase dramatically, and lactase activity decreases to near zero. This shift in the pattern of enzyme activities to accommodate the change in diet is genetically programmed and is not dependent on the change in diet. The enzyme shift can be triggered by corticosterone, but it also occurs in adrenalectomized rats.

Humans do not undergo these changes in enzyme activities. All disaccharidases are present from birth at adult levels. Lactase levels are actually higher in the newborn than in the adult. In fact, in many adults—blacks and Asians in particular—lactase levels are sufficiently depressed so that these individuals cannot digest milk and other lactose-containing dairy products. This is not surprising if one stops to think that man is the only mammal that consumes milk after weaning. Interestingly, adult human populations showing the smallest degree of lactose intolerance are those originating in northern Europe, where herd animals were first domesticated.

Dietary Regulation

The pattern of enzymes produced by the pancreas shifts in response to chronic changes in the diet. This type of change on the part of an organism that allows it to cope with an alteration in its environment is called an *adaptation*. Substitution of protein for a significant portion of the carbohydrates present in the diet results in a decrease in the amount of amylase secreted and an increase in the amounts of proteases. Similarly, the total amount of lipase secreted is increased, compared with the other enzymes, after exposure to a high-fat diet. These changes are produced at the level of enzyme synthesis, for, in general, the enzymes are secreted in the same proportion as they are synthesized. The mechanism of this feedback regulation involves effects of cholecystokinin (CCK), secretin, and insulin on the expression of messenger RNAs (mRNAs) for the various enzymes (see Chapter 34).

ABSORPTION

Absorption normally refers to the movement of nutrients, water, and electrolytes through the epithelial cell lining into the blood or lymph. *Secretion* refers to

the movement of water and electrolytes in the opposite direction.

Absorptive Pathway

The pathways taken by nutrients on their journey from the lumen of the intestine to the blood or lymph are shown in Fig. 2. They actually consist of eight different barriers:

1. An unstirred layer of fluid that is crossed by diffusion.
2. The glycocalyx, a filamentous glycoprotein surface coat, whose function is not entirely understood.
3. The apical membrane of the cell, which contains protein carriers for the specific transport of some nutrients.
4. The cytoplasm of the enterocyte.
5. The basolateral cell membrane.
6. The intercellular space.
7. The basement membrane.
8. The wall of the capillary or lymph vessel.

The last three of these are normally crossed by simple diffusion; specific carrier-mediated transport processes exist in the basolateral membrane for most electrolytes and nutrients.

Blood is supplied to each villus by one or more central arterioles, which break up into capillaries that form a network beneath the bases of the absorbing cells. Thus, blood flow ascends centrally and descends peripherally, and absorbed substances enter the descending network. Eventually, this blood drains into the portal vein. Each villus contains a central lymphatic

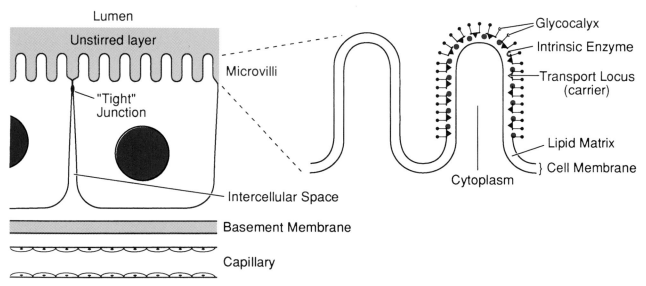

FIGURE 2   The pathway traversed by absorbed solutes. Beginning in the lumen, barriers consist of an unstirred layer of fluid, the glycocalyx, the apical cell membrane, the cytoplasm of the enterocyte, the basolateral cell membrane, the intercellular space, the basement membrane, and finally, the wall of the capillary or lymphatic vessel. (Modified from Johnson LR, ed., *Gastrointestinal physiology,* 6th ed. St Louis: Mosby, 2001.)

vessel called a *lacteal*. The products of fat digestion enter the lacteals and, eventually, the bloodstream via the left thoracic duct.

Transport Processes

The apical membrane of the enterocyte is capable of transporting solutes by a number of different mechanisms. Many solutes cross the apical membrane by the process of simple or passive diffusion. *Simple diffusion* takes place through pores of the membrane or, in the case of lipid-soluble molecules, through the lipid domain of the membrane. It is governed only by the concentration and electrical gradient of the molecule across the membrane. Many nutrients are taken into the cell by *active transport*. This requires the combination of the absorbed molecule with a specific protein carrier. The carrier often also requires the binding of Na^+. These processes are termed *secondary active transport* mechanisms because they make use of a Na^+ gradient derived from the energy of the (Na^+,K^+)-ATPase in the basolateral membrane. Active transport results in the accumulation of substances against their electrochemical gradients. Some molecules combine with specific membrane carriers in order to be transported across the membrane, but they cannot be accumulated against a concentration gradient, and no energy is involved. This type of transport also depends on the concentration gradient of the molecule being transported and is called *facilitated diffusion*. Finally, some large molecules may be transported by the process of *pinocytosis,* in which the plasma membrane invaginates around the molecule, "swallowing" it. This process usually takes place at the bases of the microvilli. Some movement of water and small electrolytes takes place through the tight junctions directly into the intercellular spaces. This is termed *paracellular transport*.

Transport processes are not all evenly distributed over the entire length of the small intestine. In general, the absorptive capacity for most nutrients is greater proximally and decreases distally. Active mechanisms for the absorption of vitamin B_{12} and bile salts, however, are present only in the ileum. The absorptive capacity of the small intestine greatly exceeds the load placed on it. As much as 60–70% of the bowel may be removed without significantly decreasing the absorption of protein and carbohydrate digestion products. However, if the ileum is removed, the ability to absorb vitamin B_{12} is lost, and steatorrhea occurs because of the inability to actively reabsorb bile salts. The liver cannot synthesize sufficient new bile acids to replace those necessary for fat digestion and absorption.

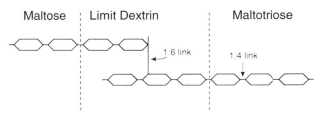

FIGURE 3   Products of exhaustive hydrolysis of starch by α-amylase.

CARBOHYDRATES

Carbohydrates constitute approximately 50% of the "typical" American diet. The average daily intake is 300–500 g, which amounts to an average of 1600 calories per day. The major dietary form of carbohydrate is starch, accounting for 50% of the total intake. Most dietary starch is in the form of the polysaccharide amylopectin or in the form of plant starch and consists of chains of glucose molecules linked at the α-1,6 carbons (Fig. 3). Animal starch or amylose is also present in the diet. This polysaccharide is a straight chain polymer of glucose, linked α-1,4. The disaccharides sucrose (glucose-fructose), lactose (glucose-galactose), and maltose (glucose-glucose) make up approximately 30%, 6%, and 2% of the intake, respectively. The remainder is accounted for by small amounts of glucose, fructose, and trehalose. The polysaccharide cellulose is also ingested, but its β-1,4 linkages are not attacked by mammalian enzymes. Approximately 80% of the cellulose, however, is digested by the microbial flora of the lower bowel. Cellulose is a primary constituent of "fiber," and it is unclear how much of that digested by the flora is absorbed and how much is used by the flora.

Digestion

To be absorbed, dietary carbohydrates must be broken down into monosaccharides. Luminal digestion of starch begins in the mouth with the action of salivary α-amylase and ends in the small intestine following the action of pancreatic α-amylase. These identical enzymes have a pH optimum near 7.0, are activated by Cl^-, and produce identical digestion products. Most luminal starch digestion is carried out by the pancreatic enzyme, and no problems result if salivary amylase is absent. In fact, most salivary amylase is inactivated by gastric acid. However, if a starchy meal is eaten and remains unmixed in the orad stomach for a time, salivary amylase may digest more than 50% of the starch present. Amylase hydrolyzes only the interior 1,4-glycosidic bonds of amylase. Therefore, exhaustive digestion of starch by amylase yields *maltose, maltotriose,* and so-called α-limit

Carbohydrates

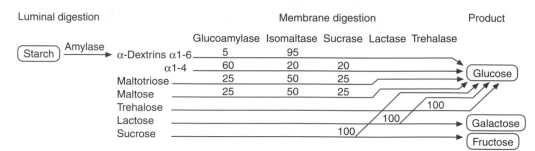

FIGURE 4   Substrate specificities of the enzymes involved in carbohydrate digestion and the resulting products. Numbers are percentages of each substrate broken down by a particular enzyme. Note that only three final products are formed.

dextrins, which contain the α-1,6 linkage branch points (Fig. 3) and α-1,4 linkages adjacent to the branch points. These 1,4 linkages are resistant to amylase. Free glucose is never produced by the action of amylase.

The digestion of the oligosaccharides produced by the action of amylase on starch and the digestion of other forms of dietary carbohydrates are carried out by enzymes located in the brush border membrane of the enterocytes (Fig. 4). Many of these enzymes hydrolyze more than one substrate. As shown in Fig. 4, *glucoamylase* cleaves the α-1,4 bonds of the limit dextrins, freeing glucose molecules sequentially from the nonreducing ends. *Sucrase* and *isomaltase* (also called α-dextrinase) break these bonds as well. When the α-1,6 branch point is reached, it is hydrolyzed by isomaltase. Isomaltase also accounts for approximately half the digestion of maltotriose and maltose. In addition to hydrolyzing some of the α-dextrins, glucoamylase hydrolyzes maltotriose and maltose. Sucrase aids in the digestion of maltose and maltotriose and accounts for 100% of the breakdown of sucrose. Two disaccharidases have specificity for only one substrate. Lactose hydrolyzes all of the lactose present, and *trehalase* is the only enzyme that digests trehalose, an α-1,1-linked dimer of glucose.

The rate-limiting step in carbohydrate assimilation is absorption. Both the luminal and membrane-bound enzymes are present in excess, so that digestion is normally complete in the proximal jejunum. The human digestive tract is incapable of digesting the β-glucose bonds found in cellulose and hemicellulose. As a result, these carbohydrates make up undigestible fiber in the diet and are eliminated in the feces.

Absorption

Carbohydrate digestion produces the monosaccharides glucose, galactose, and fructose. These three hexoses are the only dietary sugars of any consequence that are absorbed. In general, the hexoses are too large to pass through the aqueous channels between the enterocytes

or through the pores in the apical cell membranes. Thus, only a small fraction of sugar absorption takes place by passive diffusion. The amount passively absorbed is somewhat variable, influenced by bulk flow. The large majority of hexose uptake is by mediated transport.

Fructose is absorbed by facilitated diffusion. The carrier involved is GLUT-5, which is located in the apical membrane of the enterocyte (Fig. 5). Hence, it is not absorbed against a concentration gradient. However, it is taken up more rapidly than can be explained by simple diffusion.

Glucose and galactose are absorbed by an active transport process requiring the presence of Na^+ in the lumen. This Na^+-dependent carrier (SGLT-1) is identical for both sugars because each competes with the other for transport. The uptake of glucose and galactose depends on the electrochemical Na^+ gradient generated by the (Na^+, K^+)-ATPase in the basolateral membrane, and is therefore a secondary active transport system. A model illustrating the basic principles of this system is

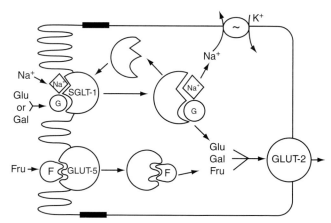

FIGURE 5   Model of the carriers involved in the absorption of glucose (Glu), galactose (Gal), and fructose (Fru). SGLT-1 is the Na^+-dependent carrier for the active absorption of glucose and galactose. GLUT-5 is the facilitated diffusion carrier for fructose. All three sugars exit the cell via the facilitated process, GLUT-2, in the basolateral membrane.

Clinical Note

Abnormalities of Carbohydrate Assimilation

Although predominant forms of dietary carbohydrate are polysaccharides and oligosaccharides, it is the monosaccharide products that are absorbed. Most problems of carbohydrate assimilation are due to enzyme defects and the failure to reduce one of the dietary forms to monosaccharides. Undigested carbohydrates remaining in the lumen increase the osmotic pressure of the luminal contents, preventing the absorption of a comparable amount of water. In addition, these carbohydrates are fermented by bacteria in the distal ileum and colon, adding to this osmotic effect. The retention of water results in *diarrhea*. Symptoms of diarrhea include cramps and abdominal distension.

Lactose intolerance is the most common problem of carbohydrate digestion and is caused by the absence of the brush border enzyme lactase. The incidence and distribution of this condition were discussed earlier in this chapter in the section on developmental changes. The condition is diagnosed by feeding lactose and monitoring the appearance of glucose in the blood. If blood glucose levels do not increase, it is evidence of lactase deficiency. However, one must also show

that the individual is capable of absorbing glucose when it is fed, because the absence of the carrier for glucose and galactose has been documented in rare instances. Lactase-deficient individuals do fine if they avoid lactose in their diets. Those deficient in the glucose carrier are, of course, diagnosed as infants, and will thrive if fed fructose. In this condition, enzyme activities are normal.

Maltose intolerance has also been described, but this is extremely rare. Because three different brush border enzymes have specificity for maltose, it is not likely that a genetic defect will eliminate all maltase activity.

Sucrase and isomaltase are subunits of a single protein, and an inherited absence occurs in about 0.2% of North Americans. These individuals cannot digest sucrose, but do well on diets low in this sugar.

Assimilation of dietary sugars may also be impaired by diseases that involve a large area of the intestinal mucosal surface. Conditions leading to general structural impairment of the mucosa and decreased brush border enzyme activities include celiac disease and some bacterial and parasitic infections. These diseases often alter the assimilation of all nutrients.

---

shown in Fig. 5. Within the cell, the Na^+ concentration is maintained low by the (Na^+, K^+)-ATPase. Thus, inside the cell, the equilibrium is such that Na^+ leaves the carrier. Glucose and galactose within the cell exit across the basolateral membrane and are carried away by the circulation; therefore, there is also a concentration gradient favoring the removal of the sugars from the carrier. Sugars leave the cell by an Na^+-independent facilitated diffusion process (GLUT-2). The exit of sugars from the cell is rapid, and there is little intracellular accumulation. The carrier is then free to combine with more Na^+ and glucose or galactose in the luminal compartment. Because of this mechanism, the addition of glucose or galactose to the lumen will increase Na^+ absorption.

PROTEIN

Dietary protein intake varies greatly among individuals of like background and among groups with different geographic, cultural, and economic foundations.

Adult humans require 0.5–0.7 g of protein per kilogram of body weight per day to maintain nitrogen balance. Young children may require 4 g/kg per day.

In addition to dietary intake, protein is present in gastrointestinal secretions and the cells shed into the lumen of the digestive tract. This protein is handled in the same manner as dietary protein and accounts for an additional 35–55 g per day (10–30 g from secretions and 25 g from cells).

Essentially all ingested protein is assimilated. The capacity to digest and absorb protein greatly exceeds the dietary load, and normally dietary protein is absorbed completely before the end of the jejunum. The 10% of the total present in the tract that is excreted in the stool is primarily in the form of desquamated cells, bacteria, and mucoproteins, much of which is derived from distal regions of the tract.

Digestion

Two general classes of proteases are secreted into the lumen (Table 2). *Endopeptidases* hydrolyze interior

TABLE 2   Functions of Principal Luminal Proteases

Enzyme	Primary action
Endopeptidases	*Hydrolyze(s) interior peptide bonds containing*
Pepsin	Aromatic amino acids
Trypsin	Basic amino acids; produces peptides with the C-terminal basic amino acids
Chymotrypsin	Aromatic amino acids, leucine, methionine, and glutamine; produces peptides with these amino acids at the C-terminus
Elastase	Neutral aliphatic amino acids; produces peptides with these at the C-terminus
Exopeptidases	*Hydrolyze(s) external peptide bonds containing*
Carboxypeptidase A	Aromatic and neutral Aliphatic amino acids at the C-terminus
Carboxypeptidase B	Basic amino acids at the C-terminus

peptide bonds, whereas the *exopeptidases* (both are carboxypeptidases) remove one amino acid at a time from the C-terminal ends of proteins and peptides.

Protein digestion begins in the stomach with the action of pepsin. The chief cells secrete the precursor pepsinogen, which is converted to pepsin by gastric acid (see Chapter 34). Three isozymes of pepsin have been identified, and each has a pH optimum between pH 1 and 3. All are irreversibly denatured after exposure to pH 5 or higher. Pepsin is inactivated when the chyme enters the duodenum and is neutralized by bicarbonate. Under optimal conditions, pepsin may result in the digestion of 10% of the protein present in the stomach. The contribution of pepsin to the digestion of a meal is dispensable, because patients with pernicious anemia, who secrete no acid, assimilate normal amounts of ingested protein. The pancreatic proteases are able to carry out all luminal digestion of protein.

The five major proteases of the pancreas are secreted as inactive precursors. *Trypsinogen* is converted to active *trypsin* by the brush border enzyme *enterokinase* (Fig. 6). Enterokinase cleaves a hexapeptide from the N-terminal end of *trypsinogen* to produce *trypsin,* an enzyme containing 223 amino acids. Trypsin is autocatalytic, and once enterokinase initiates the process, most of the remaining trypsinogen is activated by trypsin. Trypsin also converts chymotrypsinogen, *proelastase,* and the *procarboxypeptidases* to their active counterparts (Fig. 6). The principal actions of the pancreatic proteases are shown in Table 2. Therefore, luminal digestion of protein results in free amino acids plus small peptides of different lengths.

The pancreatic proteases also digest each other, so these enzymes are rapidly inactivated within the lumen.

These secreted proteins are digested and absorbed in the same manner as dietary proteins.

Absorption

Free amino acids produced by luminal digestion are absorbed across the enterocyte apical membrane. However, unlike carbohydrate absorption (in which only monosaccharides are absorbed), di- and tripeptides are absorbed intact. Larger peptides are absorbed poorly or not at all, and these are reduced to amino acids and di- and tripeptides by peptidases located in the brush border membrane. The digestion products of the brush border hydrolases are rapidly absorbed. These processes are summarized in Fig. 7. The absorption of whole proteins and large peptides is so minor as to be insignificant nutritionally. Absorption of whole proteins, however, is significant immunologically, and in some species other than humans the neonatal intestine has a high capacity to absorb whole proteins and especially immune globulins from colostrum. Absorption takes place by a pinocytotic mechanism and decreases with age.

Although a small amount of amino acids is absorbed by passive diffusion, L-isomers of most amino acids are absorbed by carrier-mediated secondary active transport systems requiring Na^+. This mechanism is analogous to that previously described for the absorption of glucose and galactose (see Fig. 5). Studies of competition between different amino acids have identified separate carrier systems requiring Na^+ (Table 3). Note that there is some overlap in amino acid specificity

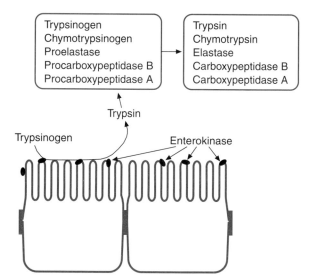

FIGURE 6   Means of activating the pancreatic proteases that are secreted as proenzymes. Trypsinogen is activated by enterokinase. The resulting trypsin converts the other proenzymes plus additional trypsinogen.

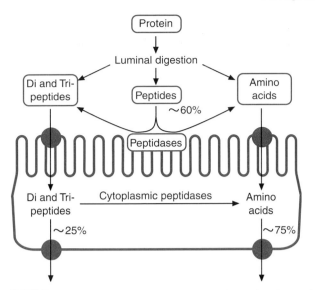

FIGURE 7   Summary of protein digestion and absorption. Approximately 40% of ingested amino acids are absorbed as free amino acids or di- and tripeptides after luminal digestion. The remaining 60% are absorbed after being broken down further by membrane-bound peptidases. Free amino acids leave the cell by facilitated diffusion and by simple diffusion.

when presented as an equal amount of free amino acids. Furthermore, amino acids presented to the gut as di- or tripeptides do not compete for transport carriers with identical free amino acids. These data indicate that carriers different from those for amino acids mediate the uptake of small peptides. The different peptide systems have not been well defined, but at least some appear to be secondary active transport systems dependent on Na^+. After a meal, most of the protein is absorbed in the form of di-and tripeptides.

Within the enterocytes, most di- and tripeptides are hydrolyzed to amino acids by cytoplasmic peptidases. Free amino acids diffuse across the basolateral membrane or cross it by carrier-mediated processes. A number of carriers exist in the basolateral membrane. Some of these require Na^+ and others that are identical to those in the apical membrane do not. In general, the more hydrophobic an amino acid, the greater the proportion that leaves the cell by diffusion. A small percentage of peptides is not broken down and enters the extracellular space intact. This entire scheme for protein digestion and absorption is summarized in Fig. 7.

among these carriers, and that the energy for the active accumulation of the amino acids comes ultimately from the (Na^+, K^+)-ATPase of the brush border membrane. Other amino acids and some of those transported actively can be absorbed by facilitative processes that do not require Na^+.

Although the uptake of free amino acids is significant, the transport of equimolar amounts of amino acids in the form of di- or tripeptides is more rapid and more efficient. Amino acids appear more rapidly in the portal blood when fed as di- or tripeptides than as free amino acids. A higher proportion of amino acids is assimilated when presented to the gut as a protein hydrolysate than

LIPIDS

The daily dietary intake of fat varies from below 25 g, or less than 12% of caloric intake (found among rice-eating Asian cultures), to as much as 140–160 g, or 40% of the caloric intake (found among some individuals in Western meat-eating cultures). The difference is due primarily to the increased intake, in the latter group, of saturated fatty acids found in red meat and dairy products. The intake of unsaturated fatty acids is approximately the same in both groups.

Among those substances discussed in this section are dietary triglycerides, phospholipids, and cholesterol. Fats in general are an important source of calories, but in addition, they are major structural components of the body. Because fat is insoluble in water, it is the major component of cell membranes, and each of the three compounds listed above plays an integral role in membrane structure and function. In addition, cholesterol is used by the body to synthesize the bile acids and a variety of steroid hormones.

Because it is insoluble in water, the assimilation of dietary fat by the body is a complicated process. Compared with the digestion of carbohydrates and proteins, numerous steps are involved before dietary fat reaches the bloodstream for distribution. A problem at any one of these steps can result in an abnormality in lipid assimilation.

TABLE 3   Amino Acid Carrier Systems

Transport system	Amino acids	Na^+ dependence	
Neutral	Neutral, aromatic, aliphatic	Yes	
PHE	Phenylalanine, Methionine	Yes	
Acidic	Aspartate, glutamate	Yes	Brush border membrane
Imino	Proline, hydroxyproline	Yes	
y^+	Basic	No	
L	Hydrophobic neutral	No	
A	Neutral	Yes	
ASC	3 or 4 carbon neutral	Yes	Basolateral membrane
L	Hydrophobic neutral	No	
y^+	Basic	No	

ASC, alanine serine cysteine plus other 3 or 4 carbon neutral acids.

Clinical Note

Abnormalities of Protein Assimilation

Decreases in the amounts (or absence) of proteolytic enzymes occur in cases of pancreatic insufficiency as might develop in patients with pancreatitis or cystic fibrosis. Such decreases in the amounts of active enzymes may impair protein digestion to the point at which absorption is decreased and nitrogen is lost in the stool. The congenital absence of trypsin alone can also result in protein malabsorption. Trypsin is important for adequate protein assimilation not only because it digests dietary protein but because it is also necessary for the conversion of chymotrypsinogen, proelastase, and procarboxypeptidases to the active enzymes. Thus, patients lacking trypsin also lack active forms of the other proteolytic enzymes.

Several conditions have been described in which the uptake of amino acids is impaired because of the genetic lack of one of the transport systems for amino acids. These carrier proteins appear to be identical in the gut and kidney, because the same amino acids that cannot be absorbed in the gut are also lost in the urine. *Cystinuria* is a condition in which the dibasic amino acids cystine, lysine, arginine, and ornithine are not absorbed by the gut or reabsorbed by the proximal renal tubule. Another similar condition, called *Hartnup's disease,* affects the uptake of neutral amino acids. These patients, however, do not become amino acid deficient because the same amino acids can be absorbed as di- and tripeptides. This is excellent evidence that the carriers for amino acids are distinct from those for peptides. There is also a certain amount of overlap in the affinities of the transport carriers, so that a small amount of amino acid may be absorbed in the absence of its primary carrier.

Digestion

The major component of dietary fat consists of *triglycerides,* three fatty acids esterified to glycerol. The principal saturated fatty acids are palmitic (C16) and stearic (C18). The unsaturated ones are oleic (CIS, one double bond) and linoleic (C18, two double bonds). All fatty acids contain an even number of carbon atoms. *Phospholipids* are present in the diet in small amounts and consist of glycerol esterified to two fatty acids and to either choline or inositol at the 3 position of glycerol. *Cholesterol* is ingested in the form of cholesterol esters with the fatty acid esterified to the 3-OH group of the sterol nucleus. These three types of compounds are shown in Fig. 8.

Within the Stomach

The churning and mixing that occur in the distal portion of the stomach break lipids into small droplets, greatly increasing the total surface area available to digestive enzymes. These droplets are kept apart by *emulsifying agents.* The latter are amphipathic compounds (see Chapter 34) such as bile salts, phospholipids, fatty acids, monoglycerides, and proteins. The major emulsifying agents in the stomach are some of the dietary proteins. However, the overall process of emulsification is inhibited by low pH. Approximately 10% of dietary triglycerides are hydrolyzed to glycerol and free fatty acids within the stomach by gastric lipase secreted by cells in the fundus. In humans, lingual lipase is much less important. These enzymes have pH optima between 3 and 6. The resulting lipase breakdown products aid in the emulsification process.

The rate of gastric emptying is carefully controlled to deliver amounts of fat into the duodenum that can be digested easily and absorbed without overloading the system. Duodenal receptors respond to fat by releasing CCK and enterogastrones and by triggering neural reflexes that inhibit the rate of gastric emptying.

Within the Duodenum

As chyme enters the duodenum, the fat is emulsified further by the constituents of bile. Bile acids themselves are poor emulsifying agents, but bile also contains lecithin. The combination of lecithin with bile salts and the polar lipids (lysolecithin and monoglycerides) produced by digestion results in a mixture with a greatly enhanced emulsifying power. As a result, fats are totally dispersed into a fine suspension of droplets ranging from 0.5–1.0 μm in diameter. This is called an *emulsion,* and each droplet is covered with negatively charged emulsifying agents that prevent them from coalescing. The final result is a significant increase in the surface area available for enzyme action. The pancreas secretes three enzymes and one additional protein that play major roles in luminal fat digestion. The most important

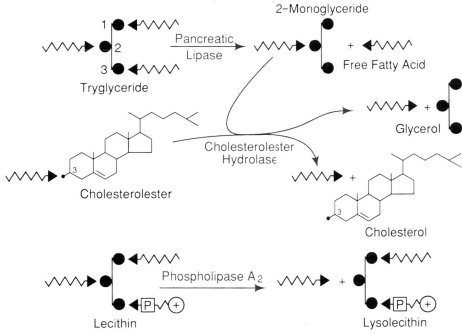

FIGURE 8 Three major classes of dietary fat and their digestion products. The primary enzymes involved in digestion are also illustrated.

of these is *pancreatic lipase,* also called glycerol ester hydrolase. It is secreted as an active enzyme with a pH optimum of 8. It functions at the oil–water interface to hydrolyze the ester linkages at the 1 and 3 positions of glycerol. This results in the production of fatty acids and 2-monoglycerides (see Fig. 8). In the presence of bile salts, the pH optimum of pancreatic lipase is reduced to 6.0, which is the approximate pH of the duodenal contents. The exact mechanism for this shift is unknown. Bile salt molecules, however, occupy the oil–water interface and inhibit the activity of pancreatic lipase by displacing the enzyme molecules from the surface of the emulsion droplets. This inactivation is prevented by a protein called *colipase,* which is secreted by the pancreas as inactive procolipase. Procolipase is activated by proteolytic enzymes (mainly trypsin), which cleave a peptide fragment, reducing it from 105 amino acids to 94 amino acids. Colipase has no lipolytic activity itself, but it binds to lipase in a 1:1 ratio and displaces a bile salt molecule from the interface. It thus serves as an anchor for lipase, allowing it to digest triglycerides. One molecule of colipase also binds one bile salt micelle, thus keeping the micelle near the site of hydrolysis and facilitating the removal and solubilization of the by-products of lipolysis. Pancreatic lipase is secreted in great excess, and the amount of active enzyme must be reduced by approximately 80% before dietary fat appears in the stool (steatorrhea). Duodenal pH of less than 3 inactivates lipase, but this does not

occur in individuals with normal rates of gastric and pancreatic bicarbonate secretion.

The second noteworthy pancreatic lipolytic enzyme is *cholesterol ester hydrolase,* which is probably identical to *nonspecific esterase.* Approximately 15% of dietary cholesterol consists of cholesterol esterified to fatty acids. This linkage is cleaved by cholesterol ester hydrolase giving rise to free cholesterol and fatty acids (see Fig. 8). Cholesterol ester hydrolase lacks the specificity of pancreatic lipase and also hydrolyzes ester linkages at all 3 positions of triglycerides as well as those of vitamins A and D. The action of this enzyme, therefore, results in the production of free glycerol.

The most important enzyme for the digestion of phospholipids is pancreatic *phospholipase A₂,* which is secreted as the proenzyme. The approximately 2 g of daily dietary phospholipids are augmented by the 12 g contained in biliary secretions and sloughed cells. As shown in Fig. 8, phospholipase A₂ releases fatty acids from the 2 position to yield lysophospholipids (1-acylglycerophosphatides). Pancreatic lipase can also hydrolyze phospholipids at the 1 position, but only at a low rate. Therefore, almost all phospholipids are absorbed as lysophosphatides. Phospholipase A₂ requires bile salts for its activity and has a pH optimum of 7.5.

Some foods also contain enzymes that digest dietary lipids. One of these of importance to humans is *milk lipase* found in breast milk. The activity of this enzyme is increased by low concentrations of bile salts. It is sometimes referred to as bile salt-sensitive lipase and is not active until it reaches the duodenum. This enzyme is important to newborns for the digestion of milk and aids pancreatic lipase in the hydrolysis of triglycerides, for it hydrolyzes all three ester linkages.

Absorption

The products of lipid digestion undergo an interesting journey before they reach the bloodstream. First, lipolytic by-products are solubilized in micelles within the intestinal lumen. (See Chapter 34 for a discussion of micelle formation.) They then diffuse from the micelle across the enterocyte brush border membrane and enter the cytoplasm of the cell. Within the cell, triglycerides are resynthesized. The resynthesized lipids are combined with p-lipoprotein to form particles called chylomicrons. The chylomicrons leave the cell by exocytosis and enter the lymph via the lacteals present in the villi. Finally, they reach the bloodstream via the thoracic duct.

Transport into the Enterocytes

Lipolytic products are solubilized in mixed micelles within the intestinal lumen. Mixed micelles are cylindrical disks 30–70 Å in diameter consisting of an outer layer of bile salts surrounding the hydrophobic products of fat digestion (Fig. 9; also see Chapter 34, Fig. 28). The bile salts are oriented with their hydrophilic portions outward in the aqueous phase and hydrophobic sides toward the center. In this fashion, the hydrophobic products of lipid digestion are made water soluble. In addition to bile salts, mixed micelles contain fatty acids, monoglycerides, phospholipids, cholesterol, and fat-soluble vitamins. Glycerol is water soluble and is not a constituent of micelles.

All fat digestion products were believed, until recently, to be absorbed by simple diffusion. However, evidence now indicates that some carrier-dependent mechanisms are involved. In some cases transport shows a specificity that cannot be explained by simple diffusion. Transport maxima or saturation has been shown to occur for some products of digestion such as linoleate.

Micelles diffuse through the unstirred water layer, which is a major diffusion barrier to the absorption of lipids, and bring the products of lipolysis into contact with the brush border membrane. An equilibrium is established between lipids in the micellar and aqueous phases, with lipids moving in and out of micelles and striking the brush border membrane. When they contact the brush border membrane, because of their lipid solubilities, they are able to diffuse through the membrane and enter the cells (see Fig. 9). Thus, the lipolytic products diffuse into the cells in direct proportion to their concentration in solution. The concentrations of lipolytic products in solution are, in turn, dependent on their concentration in the micelles. A low pH microclimate exists within the unstirred layer at the surface of the intestine, and fatty acids become protonated, reducing their micellar solubility. This facilitates absorption by shifting the equilibrium so that more lipolytic products leave the micelles and are free to enter the cell membranes. Short-chain and medium-chain triglycerides are sufficiently water soluble in that they can diffuse through the unstirred layer without the aid of micelles. Therefore, their uptake, like that of glycerol, is independent of micelle formation.

The evidence for the absorption of lipids from micelles is several-fold. First, lipid digestion products

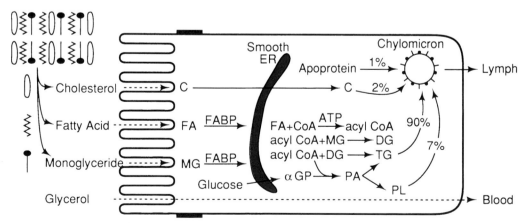

FIGURE 9   Summary of lipid absorption, resynthesis, and chylomicron formation. Percentages refer to relative amounts present in chylomicrons. Chylomicrons leave the cell by the process of exocytosis. FABP, fatty acid-binding proteins.

form mixed micelles with bile salts within the lumen of the intestine. Second, long-chain fatty acids and monoglycerides are absorbed more rapidly from micellar solution than from emulsions. Third, lipolytic products are absorbed at different rates, with cholesterol absorbed more slowly than the other constituents of micelles. As micelles move distally through the gut, their concentration of cholesterol increases. These findings also provide evidence that the micelles themselves are not absorbed. Finally, most ingested fat is absorbed by midjejunum, but the bile salts themselves are not actively reabsorbed until the ileum. Thus, bile salts are present in sufficient concentrations to ensure that lipid absorption is complete. The separate mechanism for active bile salt absorption is another strong indication that lipids are not absorbed as part of intact micelles.

Intracellular Processing

Within the enterocytes the products of lipolysis are reesterified (see Fig. 9). The resynthesized triglycerides and phospholipids, together with cholesterol, are combined with apoprotein to form chylomicrons, which are taken up by the lymph vessels in the villi.

Fatty Acid-Binding Protein

Lipolytic products partition into the lipid matrix of the cell membrane rather than the cytosol. The cytosol, however, contains specific proteins with high affinities for fatty acids. These proteins are called *fatty acid-binding proteins* (FABPs) (see Fig. 9). They have a high specificity for long-chain fatty acids and their monoglycerides, and low affinity for medium- and short-chain fatty acids. FABPs preferentially bind unsaturated fatty acids but transport all long-chain fatty acids from the membrane to the smooth endoplasmic reticulum, where resynthesis takes place. This explains the findings that resynthesized triglycerides contain only long-chain fatty acids and that medium-chain fatty acids are absorbed directly into the bloodstream without resynthesis into triglyceride.

Monoglyceride Acylation Pathway

The most important resynthesis pathway makes use of 1- and 2-monoglycerides, although 2-monoglycerides are preferred and the most abundant. Fatty acids are activated to acyl coenzyme A (CoA), a reaction catalyzed by acyl CoA synthetase and making use of coenzyme A, ATP, and Mg^{2+}. In the presence of mono- and

diglyceride transferases, the reaction then proceeds to form triglycerides (see Fig. 9):

1. Acyl CoA + monoglyceride → diglyceride.
2. Acyl CoA + diglyceride → triglyceride.

All of these reactions take place within the smooth endoplasmic reticulum, which becomes engorged with lipid after a meal containing fat.

Phosphatidic Acid Pathway

In the presence of monoglyceride, the phosphatidic acid pathway is the minor route of triglyceride resynthesis. During fasting, however, it becomes the major mechanism. The overall reaction combines three molecules of acyl CoA with α-glycerophosphate, derived from hexose metabolism, to form one molecule of triglyceride. Less than 4% of absorbed glycerol is used for triglyceride resynthesis. The intermediate product of this reaction is phosphatidic acid, which can be used for triglyceride or phospholipid synthesis (see Fig. 9):

$$2\text{Acyl CoA} + \alpha\text{-glycerophosphate} \rightarrow \text{phosphatidic acid} \qquad [1]$$

In the production of triglycerides, phosphatidic acid is dephosphorylated to yield a 1,2-diglyceride, but this pool of diglyceride remains separate from that produced by the monoglyceride acylation pathway. Thus, the two pathways function independently.

Phospholipids can be produced independently of the phosphatidic acid pathway by direct acylation of absorbed lysophospholipids. Thus, the acylation of lysolecithin produces lecithin (phosphatidylcholine), which is an important constituent of chylomicrons.

Cholesterol

Cholesterol is absorbed in free form, but a significant portion is re-esterified with fatty acids within the enterocytes. The ratio of free to esterified cholesterol leaving the enterocyte depends on the amount of dietary cholesterol. As dietary cholesterol decreases, the proportion of free cholesterol in the lymph increases.

Chylomicron Formation

Within the enterocytes, resynthesized triglycerides, cholesterol and cholesterol esters, phospholipids, and apoproteins form *chylomicrons*, lipid-carrying particles. Chylomicrons are the largest of a variety of lipid-carrying lipoproteins found in lymph. Chylomicrons

Clinical Note

Abnormalities of Lipid Assimilation

Compared with carbohydrate and protein assimilation, lipid assimilation is much more complicated, and malabsorption of lipids occurs more frequently. Malabsorption of fats can occur because of a defect at any one of the steps involved in the process of lipid assimilation. Fat malabsorption is defined as the excretion of more than 7 g of fat per day in the stool. This condition is called *steatorrhea*. Normal individuals are able to absorb as much as 150–200 g of fat per day, and, of the 3–5 g fat excreted in the stool each day, approximately one-half is of dietary origin. The remainder comes from desquamated cells and colonic flora added to the lumen distal to the regions where pancreatic lipase is active and bile salts are present.

Luminal digestion of triglycerides is inadequate in conditions of pancreatic deficiency. In the absence of pancreatic lipase, two-thirds of dietary fat is not absorbed and appears in the stool as undigested triglyceride. Conditions resulting in decreased pancreatic lipase activity include pancreatitis, cystic fibrosis, and other pancreatic diseases. Increased acidity of the duodenal contents resulting from gastric hypersecretion (Zollinger-Ellison syndrome) or inadequate bicarbonate secretion may lower the pH significantly below the optimum for pancreatic lipase or actually denature the enzyme. Pancreatic lipase, however, is secreted in great excess, and digestion is unimpaired in the presence of only 15–20% of the normal amount of enzyme. In the absence of bile salts, approximately 50% of the dietary fat appears in the stool. Although bile salts are important in the emulsification process, the defect is not in hydrolysis, but is a failure to absorb the products of lipolysis because almost all of the stool fat is in the form of free fatty acids. All of the steps before micelle formation can occur in the absence of bile salts because fat is emulsified by protein, lecithin, lysolecithin, and, once hydrolysis begins, by monoglycerides. Once the concentration of bile salts drops below the critical micellar concentration, however, fat malabsorption occurs. Decreases in the concentration of effective bile salts may be caused by (1) liver disease; (2) interruption of the enterohepatic circulation, such as occurs after ileal resection; (3) bacterial overgrowth of the small intestine and subsequent bile salt deconjugation by bacteria; or (4) increased duodenal acidity, which causes the bile salts to become protonated and un-ionized, less soluble, and unable to form micelles.

Absorption of fat is also decreased by a number of conditions that affect or decrease the number of absorbing cells. These include tropical sprue and gluten enteropathy.

No disease has been associated with triglyceride resynthesis. However, as mentioned earlier, failure to synthesize apoprotein B prevents chylomicron formation and leads to a buildup of fat within the enterocyte. This condition is known as abetalipoproteinemia.

Some aspects of lipid malabsorption can be treated by altering the diet. Substitution of medium-chain triglycerides for long-chain triglycerides will eliminate steatorrhea. Medium-chain triglycerides are more water soluble and can be hydrolyzed more rapidly than long-chain triglycerides. The resulting medium-chain fatty acids are water soluble and are absorbed directly without depending on micelle formation. Medium-chain fatty acids also pass through the enterocyte without being resynthesized into triglycerides. They do not take part in chylomicron formation and are absorbed directly into the portal blood. Substitution of medium-chain triglycerides for long-chain triglycerides will not prevent malabsorption of other lipids, such as sterols, whose uptake depends on micelle formation.

range in size from 750–6000 Å, with a mean diameter of 1200 Å, and are made up of 90% triglycerides, 2% cholesterol (about equally divided between free and esterified), 7% phospholipids, and 1% protein (see Fig. 9). The core of the chylomicron contains most of the triglycerides and cholesterol but no phospholipids. The surface membrane of the chylomicron is a monolayer composed of phospholipids, apoproteins, free cholesterol, and a small amount of saturated triglycerides. Phospholipids cover 80–90% of the surface area, whereas *apoproteins* are present in amounts sufficient to cover only 10–20%. The exact mechanism of chylomicron packaging is not known; however, apoproteins are essential for transporting the lipids out of the cell. Inhibition or lack of apoprotein synthesis leads to an accumulation of lipid within the enterocytes.

The many types of apoproteins are divided into A, B, C, and E classes. Intestinal cells have been shown to synthesize Apo AI, Apo AIV, Apo B, and Apo CII. The failure to synthesize Apo B results in abetalipoproteinemia and the inability to transport chylomicrons out of the intestinal cells. Some apoproteins are also synthesized by the liver, and the intestinal cells can make use of circulating lipoprotein remnants. Other lipoproteins exist in addition to chylomicrons and include very-low-density lipoprotein (VLDL), low-density lipoprotein (LDL), and high-density lipoprotein (HDL). VLDL is smaller than chylomicrons and is the major lipoprotein synthesized during fasting. The intestine synthesizes 11–40% of VLDL during fasting, and the mechanisms for synthesis are independent of those for chylomicron synthesis. LDL is believed to be derived from the hepatic breakdown of plasma VLDL. HDL can be synthesized directly by the small intestine and the liver or derived from the catabolism of chylomicrons or VLDL. The lipoproteins not only differ in size, but their apoprotein and lipid compositions differ as well.

Although the exact steps in chylomicron formation are not entirely known, their path of intracellular transport is clear. After triglyceride synthesis within the smooth endoplasmic reticulum, chylomicrons begin to appear in the Golgi apparatus. Accumulation of chylomicrons in the Golgi leads to the formation of secretory vesicles. The vesicles containing the chylomicrons migrate to the basolateral membrane. The vesicle membrane fuses with the cell membrane and the chylomicrons are secreted by the process of exocytosis. The chylomicrons traverse the basement membrane and enter the lacteals by moving through the gaps between the endothelial cells lining the lymphatics. Chylomicrons are too large to enter the capillaries and eventually reach the bloodstream via the thoracic duct.

VITAMINS AND MINERALS

Vitamins are organic compounds required in small amounts by the body as cofactors or coenzymes for a wide variety of metabolic reactions. They cannot be synthesized and must be obtained from the diet. They are included in this chapter on the digestion and absorption of nutrients because the mechanisms governing their absorption are similar to and based on the same principles as those for nutrients. Vitamins are easily (and conveniently, for our purposes) divided according to whether they are soluble in water or fat (Table 4). In terms of quantity, the primary minerals absorbed by the small intestine are calcium and iron. Both of these are absorbed by specific mechanisms similar in many ways to those for nutrients.

TABLE 4 Vitamins

Vitamin	Water soulble	Fat soluble
A (retinol)		+
B$_1$ (thiamine)	+	
B$_2$ (riboflavin)	+	
B$_6$ (pyridoxine)	+	
B$_{12}$ (cobalamin)	+	
C (ascorbic acid)	+	
D (calciferol)		+
E (alpha, beta, and gamma tocopherol)		+
K (phytonadione, menaquinone, and menadione)		+
H Biotin	+	
M Folic (pteroyglutamic) acid	+	
Niacin (nicotinic acid)	+	
Pantothenic acid	+	

Water-Soluble Vitamins

For years it was believed that the primary mechanism for the absorption of water-soluble vitamins was simple diffusion. This was the result, in part, of using vitamins in doses that saturated the carriers. We now know that carrier-mediated mechanisms similar to those for sugars and amino acids play important roles in water-soluble vitamin absorption as well.

Vitamin B$_1$ (Thiamine)

Thiamine is absorbed by an Na$^+$-dependent active transport process in the jejunum.

Vitamin B$_2$ (Riboflavin)

Riboflavin is absorbed in the proximal human small intestine by facilitated transport. Absorption is increased by bile salts, which may act to increase vitamin solubility.

Vitamin B$_6$ (Pyridoxine)

The only known mechanism for the absorption of pyridoxine is simple diffusion.

Vitamin B$_{12}$ (Cobalamin)

Vitamin B$_{12}$ is required for the maturation of red blood cells. Its deficiency results in the condition known as *pernicious anemia*. The liver stores B$_{12}$ in amounts (2–5 mg) sufficient to supply the body for 3–6 years. Thus, pernicious anemia develops long after the body ceases to assimilate B$_{12}$. In the stomach, digestion by pepsin and acid releases cobalamin from foods. It is then bound to glycoproteins, known as R proteins, which are secreted into the saliva and gastric juice. These transcobalamins are more important for carrying B_{12} in plasma than in the lumen of the digestive tract. Within the duodenum, pancreatic enzymes begin to break down the R proteins, so that B_{12} is transferred to *intrinsic factor,* which normally binds with less affinity than the R proteins. Intrinsic factor is a glycoprotein secreted by the gastric parietal cell (see Chapter 34) and the intrinsic factor–B_{12} complex is resistant to digestion by pancreatic enzymes. The intrinsic factor–B_{12} complex travels to the ileum, where it binds to highly specific receptors on the enterocyte brush border membrane. The complex probably enters the cell intact, but the mechanism is unclear. The receptors on the ileal enterocytes bind only to intrinsic factor and not to R proteins or free vitamin. Thus, in the absence of pancreatic enzymes, B_{12} remains bound to R proteins and cannot be absorbed. Binding of the complex to the cell membrane also requires divalent cations and a pH greater than 5.5. Approximately 1–2% of oral B_{12} can be absorbed without binding to intrinsic factor. Therefore, in the absence of intrinsic factor, massive doses (1 mg/day) of vitamin B_{12} will prevent pernicious anemia. Pernicious anemia occurs most frequently after atrophy of the gastric parietal cells.

Vitamin C (Ascorbic Acid)

Most animal species synthesize ascorbic acid from glucose in their liver. These species absorb vitamin C slowly across their intestinal mucosa by simple diffusion. A few species, including humans, have lost the ability to synthesize the vitamin and have developed a specific active mechanism for the uptake of ascorbic acid. In humans, the active transport process is located in the ileum and depends on the Na^+ gradient.

Biotin

In all species studied, biotin is taken up by an active transport mechanism in the proximal portion of the small bowel. The mechanism requires Na^+ in the lumen.

Folk Acid

Despite many studies, the uptake of folk acid (pteroylglutamic acid) is poorly understood. It is absorbed by a carrier-mediated transport system in the jejunum. This mechanism may be active.

Nicotinic Acid (Niacin)

The details for the absorption of nicotinic acid are not clear. However, the vitamin is taken up by an Na^+-dependent saturable mechanism located in the jejunum.

Pantothenic Acid

Pantothenic acid, the precursor of coenzyme A, is absorbed by a Na^+-dependent active transport mechanism present in the jejunum.

Fat-Soluble Vitamins

The four fat-soluble vitamins are classified as polar, water-insoluble, nonswelling lipids. As such, they all partition into micelles and are dependent on micelles for delivery to the absorbing surface of the enterocyte. Their rates of absorption, however, are influenced by the polar nature of their different side chains and, as such, may be altered by luminal pH. The largest proportion of the fat soluble vitamins is absorbed into the lymph in chylomicrons.

Vitamin A

The majority of vitamin A or retinol is absorbed as its precursor, carotene. Absorption is independent of bile salt concentration and is enhanced by decreasing luminal pH. The uptake mechanism is concentration dependent and passive. Vitamin A is stored in the liver in amounts sufficient for approximately 1 year.

Vitamin D

Vitamin D absorption is similar to that of vitamin A. The liver stores vitamin D sufficient for 2–6 months.

Vitamin E

Vitamin E is absorbed into the enterocyte by a passive mechanism and enters the lymph without modification.

Vitamin K

Vitamin K_1 is the form present in the diet and is absorbed by an energy-requiring, carrier-mediated process. Vitamin K_2, which is derived from bacterial sources and has a different side chain from vitamin K_1, is absorbed by a passive process.

Calcium

All portions of the small intestine actively absorb calcium ions. The amount of calcium absorbed far exceeds the amount of other divalent ions absorbed by the gut mucosa. Calcium is stored in the body primarily

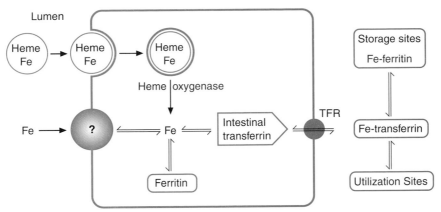

FIGURE 10   Model for the absorption of iron by the small intestine. TFR, transferrin receptor.

Iron is necessary for the synthesis of hemoglobin, myoglobin, the cytochromes, catalase, and peroxidase. Thus, it plays an essential role in oxygen storage, transfer, and metabolism.

in the form of bone, and calcium ions are essential as second messengers and metabolic regulators as well as for the excitation process of skeletal and cardiac muscle.

Calcium absorption is regulated by *vitamin D* and *parathyroid hormone* and is matched to the dietary intake, urinary excretion, and plasma levels of the ion. The regulation of calcium homeostasis and the mechanisms involved are covered in detail in Chapter 43.

Iron

Iron is necessary for the synthesis of hemoglobin, myoglobin, the cytochromes, catalase, and peroxidase. Thus, it plays an essential role in oxygen storage, transfer, and metabolism.

The total body content of iron in the normal adult is only about 4 g, of which approximately 65% is in the form of hemoglobin, 5% is in the form of myoglobin, and 1% is incorporated in enzymes; the remainder is stored in the forms of ferritin and hemosiderin, primarily in the liver. The dietary intake of iron in developed nations is approximately 20 mg/day, derived primarily from the ingestion of meat. Only a small fraction of this daily intake is absorbed, and this appears to be determined, in part, by the bodily needs for this mineral. Thus, the normal adult male absorbs approximately 1 mg/day, and this is sufficient to balance the losses of iron resulting from desquamation of epidermal and intestinal epithelial cells. Growing children and premenopausal adult women must absorb approximately 2 mg/day to meet their bodily needs. Iron deficiency resulting from inadequate dietary intake or impaired absorption is one of the most common causes of anemia in the world today.

Iron is absorbed primarily in the duodenum and upper jejunum. As the result of digestive processes, the iron in the luminal contents of these segments is in two forms. The first is iron bound to hemoglobin and

myoglobin referred to as "heme iron." The second is free, ionized iron in the ferric (Fe^{3+}) or ferrous (Fe^{2+}) forms.

The cellular mechanisms responsible for the absorption of these two forms are incompletely understood, but a reasonable model is presented in Fig. 10. Heme iron is relatively well absorbed, probably by binding to a putative heme receptor. Some heme may cross the membrane directly because of its lipophilic nature. Within the cell, heme is bound to heme oxygenase and free iron is released. The iron then associates with ferritin and is transported to the basolateral surface. The mechanism by which free iron, primarily in the ferrous (Fe^{2+}) form, enters the cell across the apical membrane is conjectural, but undoubtedly involves binding to a specific protein or carrier. Almost all absorption of nonheme iron takes place in the duodenum.

Free or ionized iron is cytotoxic. Thus, most of the iron in the enterocyte is bound either to a storage protein called apoferritin to form ferritin or to a protein that is responsible for transporting iron through the cytoplasm from the entry step at the apical membrane to the exit step at the basolateral membrane. This protein is often referred to as *intestinal transferrin*. After exiting the cell, by a mechanism that is not well defined but probably involves recognition by a transferrin receptor, iron is found in the plasma bound to a β-globulin called *transferrin*, which, as the name implies, is responsible for transporting or transferring iron from the small intestine to storage sites and from storage sites to sites of use such as the bone marrow. This protein is different from intestinal transferrin. Iron exists within the storage sites, mainly the liver, in the form of ferritin, which, as in the small intestine, appears to be in dynamic equilibrium with iron bound to transferrin.

The prevalent view regarding the regulation of iron absorption by the small intestine is the notion of "mucosal block." According to this notion, iron uptake across the apical membrane is limited by the ability of intestinal transferrin to bind it. When apoferritin in the storage organs and transferrin in the plasma are fully saturated, exit of iron across the basolateral membrane is impeded, leading to saturation of apoferritin and intestinal transferrin in the enterocyte and, in turn, inhibition of further iron uptake across the apical membrane. Conversely, when body iron stores are reduced (e.g., after hemorrhage), apoferritins and trans-ferrins are unsaturated and iron uptake can proceed until these stores are replete.

Suggested Readings

Alpers DH. Digestion and absorption of carbohydrates and proteins. In Johnson LR, ed. *Physiology of the gastrointestinal tract,* 3rd ed. New York: Raven Press, 1994, pp 1723–1750.

Ganapathy V, Brandsch M, Leibach FH. Intestinal transport of amino acids and peptides. In Johnson LR, ed. *Physiology of the gastrointestinal tract,* 3rd ed. New York: Raven Press, 1994, pp 1773–1794.

Madara L, Trier JS. Functional morphology of the mucosa of the small intestine. In Johnson LR, ed. *Physiology of the gastro-intestinal tract,* 3rd ed. New York: Raven Press, 1994, pp 1577–1622.

Rose RC. Intestinal absorption of water-soluble vitamins. In Johnson LR, ed. *Physiology of the gastrointestinal tract,* 2nd ed. New York: Raven Press, 1987, pp 1581–1596.

Tso P. Intestinal lipid absorption. In Johnson LR, ed. *Physiology of the gastrointestinal tract,* 3rd ed. New York: Raven Press, 1994, pp 1867–1908.

Wright EM, Hirayama BA, Loo BBF, *et al.* Intestinal sugar transport. In Johnson LR, ed. *Physiology of the gastrointestinal tract,* 3rd ed. New York: Raven Press, 1994, pp 1751–1772.

Wright RM, Turk E, Martin MG. Molecular basis for glucose-galactose malabsorption. *Cell Biochem Biophys* 2002;36:115–121.

Intestinal Electrolyte and Water Transport

STANLEY G. SCHULTZ

KEY POINTS

- About 85% of the 9 L of water and 30 g of NaCl presented to the gastrointestinal (GI) tract daily is absorbed by the small intestine. All but 100–200 mL of the remaining fluid is reabsorbed by the colon.
- Solutes and water may cross the epithelial cell lining of the GI tract through transcellular and paracellular routes. The *transcellular route* involves movements across the apical and basolateral membranes arranged in series. The *paracellular route* involves movements through the *zonula occludens,* or tight junctions, and the underlying intracellular space.
- The permeability of the paracellular pathway is highest in the duodenum and decreases progressively in the aboral direction.
- The *Na$^+$-K$^+$ pump,* which extrudes three Na$^+$ ions in exchange for two K$^+$ ions per adenosine triphosphate (ATP) hydrolyzed, is the principal

mechanism responsible for Na$^+$ movement across the basolateral membranes of the epithelial cells. This mechanism is responsible for the low intracellular Na$^+$ and high intracellular K$^+$ activities of these cells.
- Some chloride is absorbed in the intestinal tract by diffusion through paracellular pathways driven in part by the transepithelial electrical potential difference established by Na$^+$ absorption.
- Some cells possess a *neutral chloride entry mechanism* that mediates the movement of 1 Na$^+$, 1 K$^+$, and 2 Cl$^-$ ions across the apical membrane, energized by the Na$^+$ gradient across that barrier.
- *Potassium absorption* by the small intestines is by means of diffusion through paracellular pathways. K$^+$ absorption by the large intestine is an active process mediated by a K$^+$,H$^+$-ATPase

Essential Medical Physiology, Third Edition

OVERVIEW OF THE DAILY TASK: INPUT VERSUS OUTPUT

In addition to efficiently digesting and absorbing the many grams of nutrients (i.e., carbohydrates, proteins, and fats) ingested daily, as described in Chapter 38, the intestinal tract faces a prodigious task in absorbing ions and water.

Normally, the average human ingests approximately 5–10 g of NaCl per day and consumes about 2 L of water. But approximately 7 L of water and 25 g of Na^+

are added to this ingested load by salivary, gastric, biliary, pancreatic, and intestinal secretions (Fig. 1). These secretions represent approximately 20% of our total body water (see Chapter 1) and 15% of our total body Na^+ content. Clearly, failure to reabsorb this impressive load would rapidly lead to serious dehydration and electrolyte imbalance.

Approximately, 7.5 L or 85% of the total load of water and Na^+ presented to the small intestine are absorbed by that organ; this represents only about 40% of its maximum absorptive capacity. Approximately 1.5 L of

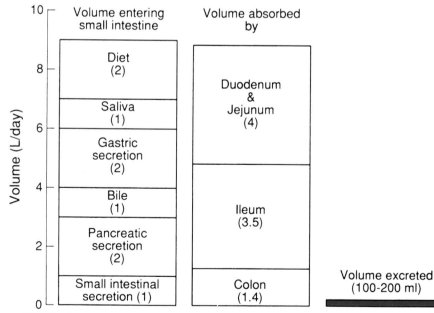

FIGURE 1   Approximate volumes of fluid entering the small intestine daily, absorbed by the small and large intestines and excreted in the feces. (From Sellin JH. The pathophysiology of diarrhea. In: Schultz SG, Andreoli TE, Brown AM, Fambrough DM, Welsh MJ, eds. *Molecular biology of membrane transport disorders*. New York: Plenum Press, 1996; 541-563.)

fluid cross the ileocecal valve daily and enter the colon, and more than 95% of this load is absorbed. The maximum absorptive capacity of the large intestine is estimated to be 4–6 L per day. Thus, there is a large *absorptive reserve capacity* that serves to protect the individual from excessive losses of fluid and electrolytes when small intestine function is impaired. However, the colon is incapable of absorbing most nutrients (i.e., the products of protein, fat, and carbohydrate digestion), so any of these that escape absorption by the small intestine will be lost in the stool or broken down by colonic bacteria.

One of the products of bacterial catabolism of undigested and unabsorbed carbohydrates that pass through the ileocecal valve are short-chain free fatty acids such as acetate, propionate, and butyrate. Because these are readily absorbed by the large intestine, this organ is capable of salvaging some nutrients that escape from the small intestine. In herbivores, short-chain free fatty acids absorbed by the colon contribute significantly to the nutritional needs of the animal. In humans and other omnivores, the extent to which they contribute to the caloric intake is considerably less under normal conditions, but it can become significant if absorptive capacity of the small intestine is decreased.

The daily fecal excreta contains approximately 100 mL of water and 25–50 g of solids. The solids are for the most part bacteria (30%), undigested dietary fiber (30%), lipids (10–20%), and inorganic matter (10–20%). Although luminal contents that enter the colon from the ileum are relatively rich in Na^+ and Cl^-, the stool is normally relatively rich in K^+ and HCO_3^-.

The remainder of this chapter focuses on how water and electrolytes are transported by the small and large intestines and how impairment of these normally highly efficient processes may result in life-threatening diarrheas (derived from the Greek, "to run through").

TRANSCELLULAR AND PARACELLULAR ROUTES OF ABSORPTION

The epithelial cells that line the small and large intestines are bound together at their luminal-facing ends by a band of specialized proteins referred to as the *zonulae occludens* or tight junctions (Fig. 2). These junctions not only preserve the integrity of this sheet of cells, but also form the boundary between their apical (or luminal-facing) surfaces and their basolateral membranes, which face the serosa. These two membranes possess markedly different properties. For example, as discussed in Chapter 38, the apical membranes of small intestinal cells possess enzymes responsible for the surface or contact digestion of disaccharides and

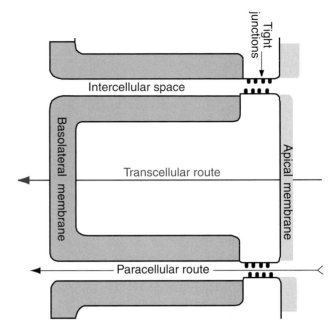

FIGURE 2   Functional anatomy of the small intestinal absorptive cells, illustrating the transcellular and paracellular routes for absorption.

peptides; these enzymes are not present in the basolateral membranes. At the same time, the basolateral membranes of all intestinal epithelial cells contain the Na^+,K^+-ATPase (adenosine triphosphatase), or Na^+-K^+ pump; this enzyme pump is not present in their apical membranes. It appears that, in addition to binding these epithelial cells, the junctional complexes prevent intermingling of these membrane-bound proteins and thus preserve the asymmetric properties (or polarity) of these cells, which, as we will see, are essential for absorption and secretion.

The name *zonulae occludens,* or occluding zones, stems from the fact that these intercellular barriers are essentially impermeable to macromolecules such as proteins and polysaccharides. On the other hand, their permeabilities to water and small ions are variable. Some epithelia are characterized by junctional complexes that are highly permeable to water and small ions and for this reason are often referred to as leaky or low-resistance epithelia. Examples of leaky epithelia in the gastrointestinal (GI) tract are the gall bladder and the small intestine. Other epithelia are characterized by junctional complexes that are essentially impermeable to water and electrolytes and they are referred to as tight or high-resistance epithelia. Between these two extremes are epithelia such as mammalian colon, whose junctional complexes are moderately leaky to water and small ions.

The leakiness or tightness of the junctional complexes determines the magnitude of paracellular movements of water and small ions. Because paracellular movements are the results of diffusion or, in the case of water movements, osmosis, they tend to dissipate differences in ion concentrations and osmolarity across the epithelium. The more leaky the epithelium, the lower the differences in ionic concentrations and total osmolarity between the luminal contents and the plasma. Thus, the ionic composition and total osmolarity of the luminal contents in the very leaky small intestine do not differ markedly from those of the plasma; like the renal proximal tubule (see Chapter 29), the small intestine carries out isosmotic absorption of fluid and electrolytes. On the other hand, the colon, being a much tighter epithelium than the small intestine, can sustain relatively large transepithelial gradients. Thus, the 100–200 mL of fluid contained in the daily excreta may be hypertonic with respect to plasma and have a Na^+ concentration of only 25–50 mmol/L and a K^+ concentration as high as 90 mmol/L; these asymmetries would not be possible if the junctional complexes were highly permeable to water and ions.

In short, there are in general two routes for the transepithelial transport of water and small ions. One is *transcellular*, involving movements across both the apical and basolateral membranes. Substances absorbed via this route must enter the cell across the apical membrane and then leave across the basolateral membrane, whereas the opposite sequence pertains to substances that are secreted. This double membrane model of transcellular transport is universally accepted. The other route is *paracellular*, in which movements of water and small ions circumvent the two limiting membranes of the epithelial cell and are driven solely by passive forces. The rate-limiting step that determines the magnitudes of these paracellular flows is the permeability of the tight junctions.

SODIUM ABSORPTION BY THE SMALL AND LARGE INTESTINES

Four mechanisms have been identified that are responsible for the movement of Na^+ from the luminal solution across the apical membranes of different Na^+-absorbing epithelial cells. As illustrated in Fig. 3, the first and simplest (a) is restricted diffusion of Na^+ through water-filled, highly selective channels that are often inhibited by the diuretic agent amiloride; the second (b) is Na^+ entry coupled to the entry of organic solutes such as sugars and amino acids (i.e., cotransport); the third (c) is Na^+ entry coupled to the entry of Cl^- most often mediated by a so-called "tritransporter" that brings about the entry of 1 Na^+ ion and 1 K^+ ion coupled to the entry of 2 Cl^- ions (another example of

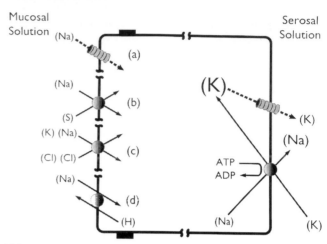

FIGURE 3   A composite model illustrating the four different mechanisms responsible for the movement of Na^+ from the luminal solution across the apical membranes of small and large intestinal cells. S, organic solute.

cotransport); and the fourth (d) is Na^+ entry coupled to the countertransport of H^+. In every instance, Na^+ entry is a "downhill" or "passive" process. In the first (a), it is driven entirely by its electrochemical potential difference across the apical membrane, and, in the second (b), Na^+ entry is energized by the combined electrochemical potential differences for Na^+ and those of the cotransported solutes. Mechanisms (c) and (d) are electrically neutral so that Na^+ entry by these routes is driven by the combined chemical potential differences of Na^+ and those of the cotransported or countertransported solutes and is not *directly* affected by the electrical potential difference across the apical membrane.

The Na^+ that enters these absorptive cells by any of these four different mechanisms is then extruded from the cell across the basolateral membrane by the Na^+-K^+ pump. This pump is responsible for maintaining the low intracellular Na^+ concentration and, because Na^+ extrusion is coupled to K^+ uptake, the high intracellular K^+ concentration characteristic of all these cells. Most, if not all, of the K^+ pumped into the cell in exchange for Na^+ recycles across the basolateral membrane through K^+ leaks or channels.

The distribution of the four Na^+ entry mechanisms illustrated in Fig. 3 differs along the intestinal tract. Na^+ entry across the apical membranes of small intestinal absorptive (villus) cells is mediated predominantly by the carrier mechanisms b, c, and d.

On the other hand, Na^+ entry across the apical membranes of colonic absorptive cells, particularly in the distal colon, is the result for the most part of restricted diffusion through highly selective channels (mechanism a). Further, the number of active channels in the apical membranes of colonic absorptive cells appears to be regulated by mineralocorticoids such as aldosterone. Any stimulus that brings about an increase in plasma aldosterone levels (e.g., sodium deprivation, hypovolemia) results in an increase in the number of active channels, an increase in the rate of Na^+ entry into these cells and, in turn, an increase in the rate of Na^+ absorption.

Thus, in many respects, Na^+ absorption by small intestine cells resembles that of renal proximal tubule cells, whereas Na^+ absorption by large intestine cells resembles that of the distal nephron (see Chapters 29 and 30).

CHLORIDE ABSORPTION AND BICARBONATE SECRETION BY THE INTESTINAL TRACT

Chloride absorption by the small and large intestines involves both paracellular and transcellular routes.

The active absorption of Na^+ by mechanisms a and b illustrated in Fig. 3 establishes an electrical potential difference across the intestinal epithelium oriented such that the serosal solution or plasma is electrically positive with respect to the mucosal solution or lumen. The magnitude of this electrical potential difference is determined by the rate of transcellular active Na^+ absorption (i.e., the "positive current" directed from the mucosal to the serosal solution) and the resistance of the paracellular pathways to the flow of ions. In the very leaky or low-resistance small intestine, the transepithelial electrical potential difference is small, generally between 2 and 5 mV. In the much less leaky, higher resistance large intestine, the transepithelial electrical potential difference may exceed 20 mV. These electrical potential differences provide a driving force for the absorption of Cl^- by diffusion through the paracellular pathways.

Transcellular Cl^- absorption involves two mechanisms by which Cl^- gains entry into the absorptive cells. The first is mechanism (c) illustrated in Fig. 3, in which Cl^- entry is coupled to the entry (cotransport) of Na^+ and, often, K^+; this mechanism can be blocked by so-called "loop diuretics" such as furosemide. The second is a countertransport mechanism that brings about Cl^- entry across the apical membrane in exchange for HCO_3^-, which is produced within the cell by the hydration of CO_2 catalyzed by carbonic anhydrase (Fig. 4). Both entry mechanisms are present in small intestinal absorptive cells, but only the HCO_3^- exchange mechanism is found in the colon, where it is responsible for the alkaline pH of fecal water. Both processes are capable of driving Cl^- into the cell against an electrochemical potential difference; that is, they are examples

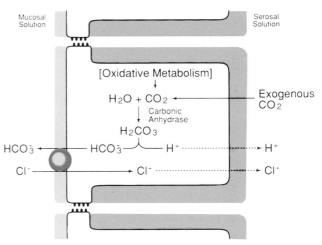

FIGURE 4   Cellular model for Cl^- absorption coupled to HCO_3^- secretion. The mechanisms responsible for H^+ and Cl^- exit from the cell across the basolateral membrane are unclear.

of secondary active transport. In the first instance, the energy for the uphill movement of Cl^- is derived from coupling to the downhill movement of Na^+ into the cell; in the second instance, it is derived from coupling to the downhill movement of HCO_3^- out of the cell.

The mechanisms responsible for Cl^- exit from the cells across the basolateral membranes are unclear. But, inasmuch as intracellular Cl^- is at a higher electrochemical potential than that in the serosal fluid or plasma, some of the Cl^- may exit the cell by diffusion.

POTASSIUM TRANSPORT BY THE INTESTINES

K^+ is both absorbed and secreted by the intestines. The bulk of the K^+ that enters the intestinal tract, derived from salivary, gastric, pancreatic, and biliary secretions as well as dietary intake, is reabsorbed in the small intestine. The primary mechanism responsible is probably diffusion through paracellular pathways. Thus, the reabsorption of water, secondary to the active absorption of Na^+ and other solutes (see below), will tend to concentrate K^+ in the chyme. But, as noted earlier, because the junctional complexes in the small intestine are very leaky to small ions, large concentration differences between the chyme and plasma cannot be sustained. Thus, the reabsorption of water establishes a concentration difference for 5^+ between chyme and plasma that serves as the driving force for K^+ diffusion through paracellular pathways from the former to the latter. In contrast to the situation in the small intestine, K^+ is also absorbed by the colon by means of a K^+,H^+-ATPase located in the apical membrane of the surface epithelial cell. K^+ that is pumped into the cell in exchange for H^+ then exits across K^+ channels

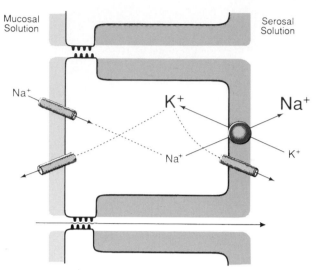

FIGURE 5   Cellular model for K^+ secretion by colonic epithelial cells.

located in the basolateral membrane. The activity of the K^+,H^+-ATPase appears to be upregulated by K^+ depletion and acidosis.

K^+ is also actively secreted by colonic epithelial cells by the mechanism illustrated in Fig. 5. The apical membranes of these cells appear to be permeable to K^+. Thus, some of the K^+ pumped into the cell by the Na^+-K^+ pump at the basolateral membranes does not recycle across those barriers but, instead, exits across the apical membranes. This is responsible for the high K^+ concentration in fecal water. But, inasmuch as the total fecal water is normally only 100–200 mL/day, the actual amount of K^+ normally lost in the excreta is small. Finally, as noted earlier, aldosterone stimulates Na^+ absorption by colonic cells and, by virtue of the mechanism illustrated in Fig. 5, also increases K^+ secretion by these cells. Thus, the effects of aldosterone in the colon closely resemble those in the distal nephron (see Chapter 32).

WATER ABSORPTION

It is well established that water absorption by epithelial tissues is dependent on and proportional to total solute absorption. The nature of this proportionality is illustrated in Fig. 6. Thus, in leaky epithelia such as gall bladder and small intestine, the absorbed fluid is isotonic with that in the lumen and plasma; in short, approximately 1 L of water accompanies the absorption of 300 mOsm of solute. In other tighter epithelia, less than 1 L of water accompanies the absorption of 300 mOsm so that the absorbed fluid or absorbate is hypertonic with respect to plasma, and the fluid in the lumen becomes increasingly hypotonic.

These observations gave rise to the notions that water absorption is secondary to solute absorption and is a passive process driven solely by osmotic forces, and that the osmolarity of the absorbate is determined by the hydraulic permeability or filtration coefficient (see Chapter 2) of the transcellular and paracellular pathways. The models that have emerged to explain the observations illustrated in Fig. 6 are shown in Fig. 7.

Figure 7A is a model of a leaky epithelium, such as the small intestine, where the apical and basolateral membranes are permeable to water and the junctional complexes are highly permeable to both water and small ions, such as Na^+, K^+, and Cl^-. In this model, the apical and basolateral membrane mechanisms responsible for the *transcellular* absorption of Na^+, Cl^-, sugars, amino acids, and other solutes deposit the solutes in the confined regions that surround the basolateral membrane: the intercellular spaces and the subepithelial spaces. This increases the osmolarity of those regions and establishes the driving force for water absorption. Inasmuch as the pathways for water flow are highly permeable to water, very small differences in osmolarity amounting to only a few mOsm/L are sufficient to keep the flow of water close to the flow of solutes in nearly isotonic proportions. Thus, as in the renal proximal tubule (see Chapter 29), fluid absorption by the very leaky small intestine is essentially isosmotic; in other works the osmolarity of the luminal contents does not differ from that of the plasma by more than a few mOsm/L. Further, as mentioned earlier, because of the leakiness of the junctions, the concentrations of Na^+, Cl^-, and K^+ in the luminal fluid do not differ markedly from those in the plasma. Finally, because the junctions are leaky to these small ions, the flow of water through these junctions entraps and drags these small ions along; this phenomenon is referred to as *solvent drag*, and it serves to augment ion absorption by the small intestine.

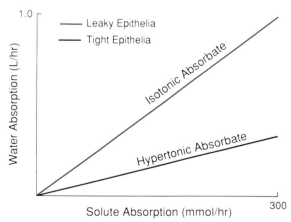

FIGURE 6   Relation between water (or volume) and solute absorption observed in GI epithelia.

Lumen   Plasma

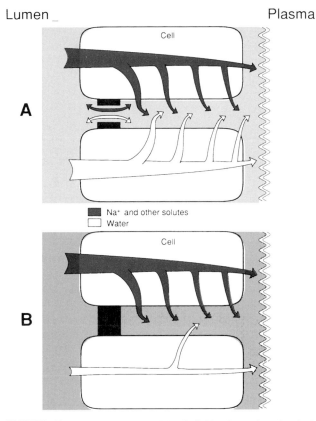

A

■ Na⁺ and other solutes
□ Water

B

FIGURE 7 (A) Cellular model of fluid absorption by leakey epithelia. (B) Cellular model of fluid absorption by tight epithelia.

In tight epithelia, the situation is generally quite different, as illustrated in Fig. 7B. Whereas the basolateral membranes are highly permeable to water, the junctions and apical membrane are, generally, far less so. Thus, although solute absorption still generates an osmotic driving force, water has difficulty keeping up with the rate of solute absorption. The result is the absorption of a hypertonic solution from the lumen that may render the latter hypotonic. In the distal colon, however, this is offset by the secretions of K^+ and HCO_3^- and the accumulation of osmotically active breakdown products of bacterial digestion. Consequently, the fecal water may be hypertonic to plasma.

In summary, it is universally accepted that water absorption is a passive process that is always secondary to and completely dependent on solute absorption. Although several models have been suggested to explain this observation, the simplest and most widely held is that described earlier in which the mechanism is simple osmosis into subepithelial regions that are rendered hypertonic by solute absorption. The osmolarity of the absorbate is determined by the ability of water absorption to keep pace with solute absorption, and this is largely determined by the permeabilities of the apical membranes and junctional pathways to water.

Intestinal Secretion

The small and large intestines also have the ability to secrete water and electrolytes. This secretory capacity appears to reside mainly in cells located in their crypts, whereas absorption by these organs is primarily carried out by cells located in the upper two-thirds of the villi.

The mechanism of secretion is illustrated in Fig. 8. The basolateral membranes of these secretory cells possess a $NaKCl_2$ cotransport ("tritransport") mechanism that mediates the neutral influx of Na^+, K^+, and Cl^- and is capable of driving Cl^- into the cell against an electrochemical potential difference; thus, the intracellular concentration of Cl^- is much greater than that predicted by the Nernst equation (see Chapter 2) for a passive distribution. The energy needed to achieve this is derived from the coupling of the uphill movement of Cl^- to the downhill movement of Na^+ (i.e., the Na^+ gradient).

The apical membranes of these secretory cells possess two types of Cl^- channels. One type is activated by elevations in cell Ca^{2+}. The other type is called the *cystic fibrosis transmembrane conductance regulator* and is identical to Cl^- channels found in the apical membranes of secretory cells in the airway epithelia (i.e., bronchioles) and pancreatic acinar cells. These channels are activated by phosphorylation mediated by (cyclic adenosine monophosphate) cAMP-dependent protein kinase A. Defects in these channels may result in impaired secretory ability by these cells, which is responsible for the pulmonary, pancreatic, and intestinal

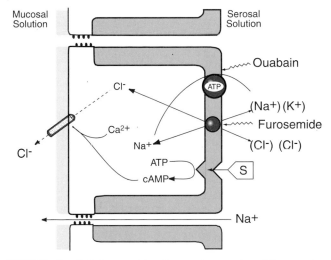

FIGURE 8 Cellular model of secondary active Cl^- secretion accompanied by passive Na^+ secretion by secretory cells in the crypts of the small and large intestines. S, secretagogue. (Note: Although only one Cl^- is depicted, as discussed in the text, there are at least two families of apical membrane Cl^- channels: one activated by Ca^{2+} and the other by cAMP.)

abnormalities characteristic of the inherited congenital disease cystic fibrosis.

Under resting or unstimulated conditions, the Cl^- channels are inactive, and these secretory cells are poised to secrete Cl^-. Stimuli that bring about an increase in cell cAMP and/or Ca^{2+} open these channels and permit Cl^- to diffuse from the cell into the lumen. This negative current, directed from the serosal fluid (plasma) into the lumen, generates a crypt-negative (lumen-negative) electrical potential difference that provides the driving force for the movement of Na^+ through paracellular pathways. The end result is the secretion of NaCl into the crypts. This establishes an osmotic pressure difference that draws water into the crypts. The fluid that emerges from the crypts into the lumen is essentially isotonic saline.

Intestinal secretion is an important physiologic event as well as a potentially life-threatening pathologic event. Secretion is stimulated by many of the GI hormones and neurotransmitters activated after a meal. These secretagogues interact with receptors on the basolateral membranes of the secretory cells and initiate the cascade of events leading to an elevation of cell cAMP and/or Ca^{2+}. For example, acetylcholine, which is released by the enteric nervous system during a meal, brings about an increase in cell Ca^+, whereas the GI hormones secretin and VIP result in an elevation of cell cAMP. The resulting secretion undoubtedly assists in the

Clinical Note

Diarrhea

Diarrhea is the main cause of death or disability in the world today, afflicting individuals of all ages. Death is the direct or indirect result of hypovolemia and circulatory collapse compounded by hypokalemia and metabolic acidosis.

The etiology of diarrheas can be subdivided into three broad categories:

1. *Diarrhea due to impaired absorption by the small or large intestines.* This can result from infection, inflammation, or anything that leads to a reduction of total effective absorptive surface.

2. *Osmotic diarrhea due to the accumulation within the small intestine of nonreabsorbable solutes.* As discussed earlier, because the small intestine is very leaky to water, the luminal contents are always isotonic with the plasma. It follows that the accumulation of nonreabsorbable, osmotically active solutes in the lumen will generate an osmotic pressure difference that draws fluid into the lumen. For example, in the condition of glucose-galactose malabsorption mentioned in Chapter 39, the accumulation of these solutes in the small intestine results in an osmotic diarrhea. This problem may be compounded by the intraluminal accumulation of osmotically active breakdown products of bacterial metabolism.

3. *Secretory diarrheas resulting from excessive stimulation of the secretory cells present in the crypts of the small intestine and colon.* These are among the most prevalent afflictions of mankind.

In some instances, they are caused by excessively high blood levels of normal secretagogues that may result, for example, from functional tumors of endocrine cells. But by far the greatest cause of secretory diarrheas is infestation of the small intestine with enteropathic bacteria such as *Vibrio cholerae* or certain strains of *Escherichia coli.* These organisms secrete enterotoxins that bind to receptors on the apical membranes of the crypt cells and result in the stimulation of the adenylate cyclase activity located on their basolateral membranes. Adenylate cyclase activity is maximally and permanently stimulated for the life of the cell, causing maximal secretion by these cells that may result in salt and water losses of many liters per day despite the considerable absorptive reserve capacity of the colon. Secretory diarrheas can also be induced by excessive secretion of endogenous secretagogues. For example, prostaglandins, leukotrienes, histamine, and bradykinin released in immune and/or inflammatory reactions interact with receptors on the basolateral membranes and result in elevations in cell cAMP and/or Ca^{2+} and, in turn, activation of apical membrane Cl^- channels. Also, certain tumors called *carcinoids* secrete serotonin, which brings about an elevation of cell Ca^{2+} and may result in watery diarrhea.

In short, whereas intestinal secretion is a normal and essential process, overstimulation of the underlying physiologic mechanisms can result in severe disability and death.

processes of digestion and absorption by maintaining the liquidity of the small intestine chyme. These small intestinal secretions, which may be as much as 1500 mL/day, are normally reabsorbed by the large intestine. Thus, under physiologic conditions there is a careful balance between absorption by the small intestinal villus cells and secretion by their crypt cells. This concerted effort results in the efficient digestion and absorption of nutrients without compromising the liquidity of the chyme.

BALANCE BETWEEN ABSORPTION AND SECRETION: ORAL REHYDRATION THERAPY

The amount of fluid that flows per unit of time from the small intestine through the ileocecal valve is determined by the balance between absorptive processes carried out by cells located on the upper half of the villus and by secretory processes apparently mediated by cells located in the crypts (Fig. 9). Under physiologic conditions, these oppositely directed fluid movements are balanced to achieve the highly efficient digestion and absorption of nutrients and at the same time maintain the liquidity of the luminal contents. The fluid that enters the colon is relatively devoid of nutrients, and its volume is normally well below the maximum absorptive capacity of that organ. Secretory diarrheas result when secretion by the crypt cells so markedly exceeds the villus absorptive processes that the volume of fluid entering the colon exceeds the maximum absorptive capacity of that organ.

The presence of sugars or amino acids in the lumen greatly increases the rate of total solute absorption by the small intestinal villous cells via the Na^+-coupled entry mechanisms illustrated in Fig. 3B. This finding suggested that fluid loss from the small intestine because of secretion might be diminished by stimulating fluid absorption by that organ. Clearly, because the absorption of

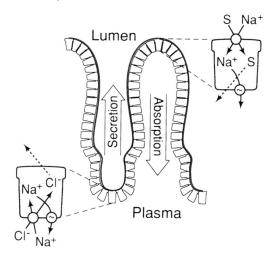

FIGURE 9 Illustration of the relation between absorptive and secretory flows that, in the final analysis, determine net fluid movments. S, solute.

sugars and amino acids is coupled to Na^+, every molecule of sugar or amino acid absorbed will be accompanied by one Na^+ ion and one anion, primarily Cl^-, a total of three osmotically active solutes. This reasoning led to the use of glucose- and amino acid-containing saline solutions that, when administered orally, markedly reduce salt and water losses resulting from secretory diarrheas. These oral replacement (rehydration) therapies have markedly reduced deaths and disabilities attributable to secretory diarrheas worldwide.

Suggested Readings

Schultz SG. Cellular models of epithelial ion transport. In Andreoli TE, Huffman JF, Fanestil DD, Schultz SG, eds. *Physiology of membrane disorders,* 2nd ed. New York: Plenum Press, 1986, pp 519–534.

Sellin JH. Intestinal electrolyte absorption and secretion. In Feldman M, Friedman LS, Sleisenger H, eds. *Sleisenger Fordtran's gastrointestinal and liver disease,* 7th ed. Philadelphia: WB Saunders, 2002, pp 1693–1714.

PART VI

ENDOCRINE PHYSIOLOGY

C H A P T E R

37

Introduction to Endocrine Physiology

H. MAURICE GOODMAN

KEY POINTS

- *Peptide* and *protein hormones* are usually stored in membrane-bound vesicles and secreted by exocytosis triggered by an increase in cytosolic calcium.
- *Steroid hormones* are derivatives of cholesterol and are generally not stored, but are secreted as they are synthesized.
- After secretion, hormones distribute throughout extracellular fluid and are rapidly cleared from the blood.
- Many hormones circulate bound specifically and reversibly to large carrier proteins. *Protein binding* protects against loss of hormone by the kidney, slows the rate of hormone degradation by decreasing cellular uptake, buffers changes in free hormone concentrations, and in certain cases may facilitate delivery to target cells.

- Hormones are rapidly degraded in the liver and kidney as well as by their target tissues.
- Hormone levels in blood are not static, but vary widely in accordance with physiologic demand. Hormone secretion may be episodic, pulsatile, or follow a daily rhythm.
- Measurement of hormones, although difficult because of their low concentrations, is essential for diagnostic, therapeutic, and research purposes. *Immunoassays* take advantage of the exquisite sensitivity and specificity of antibody antigen reactions to recognize and quantitate hormones in biologic fluids.
- *Modulation of responding systems:* Hormones regulate both the sensitivity and the capacity of target tissues to respond either to themselves or to other hormones. Sensitivity is the acuity of a cell's ability to recognize and respond to a

KEY POINTS (*continued*)

signal. The capacity to respond depends on the abundance of competent cells and their level of development.

- *Reinforcement:* Complementary actions of hormones reinforce each other and increase both the magnitude and rapidity of the overall response.

- *Redundancy:* More than one mechanism is often present to regulate critical functions.
- *Push–pull mechanisms:* Rapid change in a process may be achieved by simultaneous stimulation by one agent and relief of inhibition by another.

THE ENDOCRINE SYSTEM

Chapter 2 considered the general issue of chemical communication between cells. This section focuses on the classic endocrine glands and their hormones. A *hormone* is a chemical substance that is released into the blood in small amounts and that, after delivery by the circulation, elicits a typical physiologic response in other cells. An *endocrine gland* is a group of cells that produce and secrete a hormone. Endocrine glands are also called *ductless glands* to distinguish them from exocrine glands, which deliver their products through ducts to the outside of the body or to the lumen of the gastrointestinal tract. Endocrine glands typically are highly vascular, and their capillaries are usually fenestrated. Both of these features presumably facilitate two-way trafficking of signals to the gland and hormones to the blood. Except for these vascular features, endocrine glands have no special distinguishing characteristics. They are usually, but not always, bilateral structures, and they may arise during development from any of the primitive germ layers. Although the discussion in this text is limited largely to the classic endocrine glands listed in Table 1, it is abundantly clear that many other organs, including brain, kidney, heart, and even adipose tissue, also behave as endocrine glands. Unfortunately, endocrine physiology can sometimes involve a bewildering array of facts, not all of which can be derived from basic principles. The outline of goals and objectives shown in Fig. 1 should be useful for organizing and digesting this necessarily large volume of material.

Table 1　Classical Endocrine Glands

Pituitary gland
Thyroid gland
Parathyroid glands
Islets of Langerhans
Adrenal glands
Gonads (ovaries and testes)
Placenta

SYNTHESIS, SECRETION, AND METABOLISM OF HORMONES

The classic hormones fall into three categories: (1) derivatives of the amino acid tyrosine; (2) steroids, which are derivatives of cholesterol; and (3) peptides and proteins, which comprise the most abundant and diverse class of hormones. Table 2 lists some examples of each category. Endocrine glands synthesize their secretory products from simple precursors such as amino acids and acetate or transform complex precursors taken up from the blood. General steps in protein biosynthesis, storage, and secretion common to all protein and peptide hormones are reviewed in Chapter 2. These hormones are stored, sometimes in large amounts, in membrane-bound secretory vesicles, and can be released

Goals and Objectives

A. The student should be familiar with

   1. Essential features of feedback regulation

   2. Essentials of competitive binding assays

B. For each hormone, the student should know:

   1. Its cell of origin

   2. Its chemical nature, including
     a. Distinctive features of its chemical composition
     b. Biosynthesis
     c. Whether it circulates free or bound to plasma proteins
     d. How it is degraded and removed from the body

   3. Its principal physiological actions
     a. At the whole body level
     b. At the tissue level
     c. At the cellular level
     d. At the molecular level
     e. Consequences of inadequate or excess secretion

   4. What signals or perturbations in the internal or external environment evoke or suppress its secretion
     a. How those signals are transmitted
     b. How that secretion is controlled
     c. What factors modulate the secretory response
     d. How rapidly the hormone acts
     e. How long it acts
     f. What factors modulate its action

FIGURE 1　　Goals and objectives in endocrine physiology.

Table 2 Chemical Nature of the Classical Hormones

Tyrosine derivatives	Steroids	Peptides (< 20 amino acids)	Proteins (> 20 amino acids)
Epinephrine	Testosterone	Oxytocin	Insulin
Norepinephrine	Estradiol	Vasopressin	Glucagon
Dopamine	Progesterone	Angiotensin	Adrenocorticotropic hormone
Triiodothyronine	Cortisol	Melanocyte- stimulating hormone	Thyroid-stimulating hormone
			Follicle-stimulating hormone
Thyroxine	Aldosterone	Somatostatin	Luteinizing hormone
	Vitamin D	Thyrotropin-releasing hormone	Growth hormone
			Prolactin
		Gonadotropin-releasing hormone	Growth hormone-releasing hormone
			Parathyroid hormone
			Calcitonin
			Chorionic gonadotropin
			Corticotropin-releasing hormone
			Placental lactogen

rapidly. In general, the same stimuli that provoke secretion of peptide hormones also activate transcription of the genes that encode them so that new synthesis replenishes the storage pools. In contrast, steroid hormones are not stored in significant amounts and must be synthesized *de novo*. Secretion of these hormones therefore increases slowly. Relevant details of synthesis and storage of the amino acid and steroid hormones are presented in the discussion of each gland.

Protein and peptide hormones are synthesized on ribosomes as larger molecules (*prohormones* and *prepro-hormones*) than the final secretory product and undergo a variety of postsynthetic steps of transformation into the final secretory product. Postsynthetic processing to the final biologically active form is not limited to peptide hormones or to the time prior to secretion. *Postsecretory transformations* to more active forms may occur in liver, kidney, fat, or blood, as well as in the target tissues themselves. For example, thyroxine, the major secretory product of the thyroid gland, is converted extrathyroidally to triiodothyronine, which is the biologically active form of the hormone. Testosterone, the male hormone, is converted to a more potent form, dihydrotestosterone, within some target tissues and is converted to the female hormone, estrogen, in other tissues. Peripheral transformations add another level of complexity that must be considered when evaluating causes of endocrine disease.

HORMONES IN BLOOD

Hormones are secreted into extracellular fluid and readily enter the blood by passive diffusion driven by steep concentration gradients. Diffusion through pores in capillary endothelium also largely accounts for delivery of hormones to the extracellular fluid that bathes both target and nontarget cells. Receptor-mediated transfer across capillary endothelial cells may facilitate delivery of insulin, and perhaps other hormones, to target cells, but the importance of this mechanism of hormone delivery has not been established. In general, hormones distribute rapidly throughout the extracellular fluid and are not preferentially directed toward their target tissues.

Most hormones are cleared from the blood soon after secretion and have a half-life in blood of less than 30 min. The *half-life of a hormone in blood* is defined as that period of time needed for its concentration to be reduced by half. Clearance of a hormone from the blood depends on the rapidity with which it can escape from the circulation to equilibrate with extravascular fluids as well as on its rate of degradation. Some hormones, for example, epinephrine, have half-lives of the order of seconds, whereas thyroid hormones have half-lives of the order of days.

The half-life of a hormone in blood must be distinguished from the duration of its hormonal effect. Some hormones produce effects virtually instantaneously, and the effects may disappear as rapidly as the hormone is cleared from the blood. Other hormones produce effects only after a lag time that may last minutes or even hours, and the time the maximum effect is seen may bear little relation to the time of maximum hormone concentration in the blood. The time for decay of a hormonal effect is also highly variable; it may be only a few seconds, or it may require several days. Some responses persist well after hormonal concentrations have returned to basal levels. Understanding the time course of a hormone's survival in blood as well as the onset and duration of its action is obviously important for understanding normal physiology, endocrine disease, and the limitations of hormone therapy.

Protein Binding

Most hormones are quite soluble, and they circulate completely dissolved in plasma water. Steroid hormones and thyroid hormones, whose solubility in water is limited, circulate bound specifically to large carrier proteins with only a small fraction, sometimes less than 1%, present in free solution. Some peptide hormones also bind to specific proteins in plasma. Protein binding is reversible; free and bound hormone are in equilibrium. However, only free hormone can cross the capillary endothelium and reach its receptors in target cells. Protein binding protects against loss of hormone by the kidney, slows the rate of hormone degradation by decreasing cellular uptake, and buffers changes in free hormone concentrations. In some cases hormone binding may facilitate or impede hormone delivery to target cells.

Hormone Degradation

Implicit in any regulatory system involving hormones or any other signal is the necessity for the signal to disappear once the appropriate information has been conveyed. Recall that neurotransmitters are either rapidly destroyed in the synaptic cleft or taken up by nerve endings. Little hormone is destroyed in association with production of its biologic effects, and the remainder must therefore be inactivated and excreted. Degradation of hormones and their subsequent excretion are processes that are just as important as secretion. In general, rates of hormone degradation are characteristic for each hormone and follow first-order kinetics. That is, a constant percentage of hormone present in blood is destroyed per unit time by processes that are not usually subject to regulation. Inactivation of hormones occurs enzymatically in blood or intercellular spaces, in liver or kidney cells, or in target cells. Inactivation may involve complete metabolism so that no recognizable product appears in urine, or it may be limited to some simple one- or two-step process such as addition of a methyl group or glucuronic acid. In the latter cases, recognizable degradation products are found in urine and can be measured to obtain a crude index of the rate of hormone production.

Changing Levels of Hormones in Blood

Hormone concentrations in plasma fluctuate from minute to minute and may vary widely in the normal individual over the course of a day. Because rates of degradation usually do not vary, changes in concentration reflect changes in secretion rates. Hormone secretion may be episodic, pulsatile, or follow a daily rhythm (Fig. 2). In most cases, it is necessary to make multiple

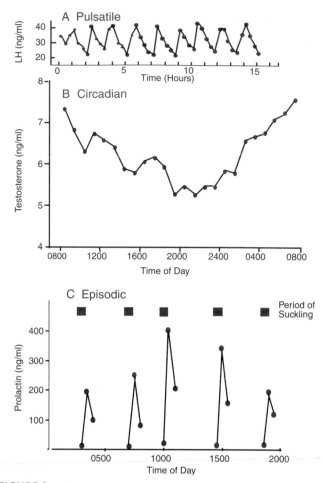

FIGURE 2 Changes in hormone concentrations in blood may follow different patterns. (A) Hourly rhythm of LH secretion. (From Yamaji *et al.*, *Endocrinology* 1972;90:771.) (B) Daily rhythm in testosterone secretion. (From Bremer *et al.*, *J Clin Endocrinol Metab* 1983;56:1278.) (C) Episodic secretion of prolactin. (From Hwang *et al.*, *Proc Natl Acad Sci USA*, 1971; 68:1902.)

serial measurements of hormones before a diagnosis of a hyper- or hypofunctional state can be confirmed. Endocrine disease occurs when the concentration of hormone in blood is inappropriate for the physiologic situation rather than because the absolute amounts in blood are high or low.

Measurement of Hormones

Whether for the purpose of diagnosing a patient's disease, monitoring therapy, or research to gain understanding of normal physiology, it is often necessary to measure how much hormone is present in some biologic fluid. Chemical detection of hormones in blood is difficult. With the exception of the thyroid hormones, which contain large amounts of iodine, there is no unique chemistry that sets hormones apart from other bodily constituents. Furthermore, hormones circulate in blood at minute concentrations, usually in the nanomolar range, which further complicates the problem of their detection. Consequently, the earliest methods developed for measuring hormones were bioassays that depended on the ability of a hormone to produce a characteristic biologic response. For example, induction of ovulation in the rabbit in response to an injection of urine from a pregnant woman is an indication of the presence of the placental hormone chorionic gonadotropin and is the basis for the rabbit test that was used for many years as an indicator of early pregnancy. Before hormones were identified chemically, they were quantitated in *units* of the biologic responses they produced. For example, a unit of insulin is defined as one-third of the amount needed to lower blood sugar in a 2-kg rabbit to convulsive levels within 3 hr. Although bioassays are now seldom used, some hormones, including insulin, are still standardized in terms of biologic units.

Immunoassays

As knowledge of hormone structure increased, it became evident that peptide hormones were not identical in all species. Small differences in amino acid sequence, which may not affect the biologic activity of a hormone, were found to produce antibody reactions after prolonged administration. This finding led to the realization that antibody–antigen reactions might be applied to the measurement of hormones in biologic fluids. Antibodies can recognize and bind with high avidity to tiny amounts of the material (antigens) that evoked their production, even in the presence of large amounts of other substances that may be similar or different. Reaction of a hormone with an antibody results in a complex with altered physical properties as compared with either free hormone or free antibody. Ingenious techniques have been devised to isolate, detect, and quantify hormone antibody complexes. Although antibodies are commonly produced with protein hormones as antigens, production of specific antibodies to nonprotein hormones can also be induced by first attaching these compounds to some protein, for example, serum albumin.

Radioimmunoassays

The earliest and still widely used application of antibodies to the measurement of hormone concentrations in biologic fluids is the *radioimmunoassay*. Radioimmunoassays depend on competition between a radioactively labeled hormone prepared in the laboratory and unlabeled hormone present in a biologic sample for binding to a limited amount of antibody. The sensitivity needed for detection and quantitation of tiny amounts of hormone is obtained by incorporating

iodine of high specific radioactivity into tyrosine residues of peptides. For hormones that lack a site capable of incorporating iodine such as the steroids, radioactive carbon or tritium can be used or a chemical tail containing tyrosine can be added. To perform a radioimmunoassay, a sample of plasma containing an unknown amount of hormone is mixed in a test tube with a small amount of antibody and a known amount of radioactive hormone.

The unlabeled hormone in the biological sample competes with the labeled hormone for a limited number of antibody binding sites. The more hormone present in the plasma sample, the less radioactive hormone can bind to the antibody (Fig. 3A). The antibody-bound radioactivity is then separated from unbound radioactivity by a variety of physicochemical means and counted. The amount of hormone present in the plasma sample is inversely related to the amount of radioactivity recovered in the antibody complex and can be quantitated precisely by comparison with a standard curve constructed with known amounts of unlabeled hormone (Fig. 3B).

Sandwich-Type Assays

Once the amino acid sequence of a hormone is known, small peptides corresponding to specific regions of the hormone can be synthesized and used for producing antibodies that bind to specific portions of the hormone. If the hormone is large enough so that two different antibodies can bind to it without interfering with each other, we can obtain a "sandwich" in which the hormone becomes the "filling" linking two antibodies (Fig. 4). One antibody can be linked to a solid support such as a plastic dish with multiple wells. The plasma sample containing the hormone to be measured is then added and allowed to bind to the immobilized antibodies. A second "reporter" antibody linked to an enzyme such as a peroxidase or a phosphatase is then added, allowed to bind, and the excess washed away. A reaction is then run to quantitate the amount of reporter antibody retained by measuring a colored product. This scheme is one variation of the *enzyme-linked immunosorbent assay* (ELISA). Many other variations of this approach are also in use, some relying on a fluorescent moiety rather than an enzyme attached to the reporter antibody. Quantitation of hormone in the biologic sample is obtained by comparison with a standard curve constructed with known amounts of the hormone.

The major limitation of immunoassays is that *immunologic* rather than *biologic* activity is measured, because the portions of the hormone molecule recognized by the antibodies are unlikely to be the same as the portion recognized by the hormone receptor. Thus, a protein hormone that may be biologically inactive may

A

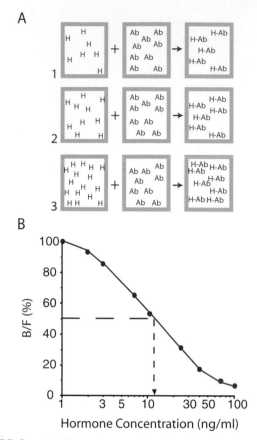

FIGURE 3   (A) Competing reactions that form the basis of the radioimmunoassay. Labeled hormone (H) shown in blue competes with hormone in biological sample shown in black for a limiting amount of antibodies (Ab). As the concentration of hormone in the biological sample rises (rows 1, 2, and 3), decreasing amounts of the labeled hormone appear in the hormone–antibody (H-Ab) complex and the ratio of bound-to-free labeled hormone (B/F) decreases. (B) A typical standard curve used to estimate the amount of hormone in the biological sample. A B/F ratio of 50% corresponds to 12 ng/mL in this example.

retain its immunologic activity. For example, the biologically active portion of parathyroid hormone resides in the amino-terminal one-third of the molecule, but the carboxy-terminal portion formed by partial degradation of the hormone has a long half-life in blood and accounts for nearly 80% of the immunoreactive parathyroid hormone in human plasma. Similarly, biologically inactive prohormones may be detected. By and large, discrepancies between biologic activity and immunoactivity have not presented insurmountable difficulties and in several cases have even led to increased understanding.

REGULATION OF HORMONE SECRETION

For hormones to function as carriers of critical information, their secretion must be turned on and off at

precisely the right times. The organism must have some way of knowing when there is a need for a hormone to be secreted and when that need has passed. The necessary components of endocrine regulatory systems include the following:

1. Detector of an actual or threatened homeostatic imbalance.
2. Coupling mechanism to activate the secretory apparatus.
3. Secretory apparatus.
4. Hormone.
5. End organ capable of responding to the hormone.
6. Detector to recognize that the hormonal effect has occurred, and that the hormonal signal can now be shut off, usually the same as component 1.
7. Mechanism for removing the hormone from target cells and blood.
8. Synthetic apparatus to replenish hormone in the secretory cell.

As we discuss hormonal control, it is important to identify and understand the components of the regulation of each hormonal secretion because (1) derangements in any of the components are the bases of endocrine disease and (2) manipulation of any component provides an opportunity for therapeutic intervention.

Negative Feedback

Secretion of most hormones is regulated by negative feedback. *Negative feedback* means that some

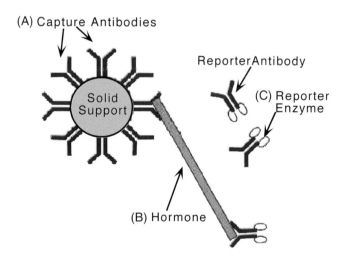

FIGURE 4   Sandwich-type assay. (A) The first (capture) antibody linked to a solid support such as an agarose bead. (B) The hormone to be measured. (C) The second (reporter) antibody linked to an enzyme, which upon reacting with a test substrate gives a colored product. In this model, the amount of reporter antibody captured is directly proportional to the amount of hormone in the sample being tested.

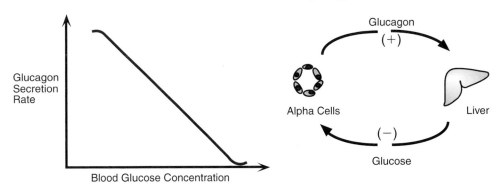

FIGURE 5 Negative feedback of glucose production by glucagon. (−) = inhibits, (+) = stimulates.

consequence of hormone secretion acts directly or indirectly on the secretory cell in a negative way to inhibit further secretion. A simple example from everyday experience is the thermostat. When the temperature in a room falls below some preset level, the thermostat signals the furnace to produce heat. When room temperature rises to the preset level, the signal from the thermostat to the furnace is shut off, and heat production ceases until the temperature again falls. This is a simple closed-loop feedback system and is analogous to the regulation of glucagon secretion. A fall in blood glucose detected by the alpha cells of the islets of Langerhans causes them to release glucagon, which stimulates the liver to release glucose and thereby increase blood glucose concentrations (Fig. 5). With restoration of blood glucose to the set-point level, further secretion of glucagon is inhibited. This simple example involves only secreting cells and responding cells. Other systems may be considerably more complex and involve one or more intermediary events, but the essence of negative feedback regulation remains the same: *Hormones produce biologic effects that directly or indirectly inhibit their further secretion.*

A problem that emerges with this system of control is that the thermostat maintains room temperature constant only if the natural tendency of the temperature is to fall. If the temperature were to rise, it could not be controlled by simply turning off the furnace. This problem is at least partially resolved in hormonal systems, because at physiologic set points the basal rate of secretion usually is not zero. In the preceding example, when blood glucose concentration rises, glucagon secretion can be decreased and therefore diminish the impetus on the liver to release glucose. Some regulation above and below the set point can therefore be accomplished with just one feedback loop; this mechanism is seen in some endocrine control systems. Regulation is more efficient, however, with a second, opposing loop, which is activated when the controlled variable deviates in the opposite direction. With regulation of blood glucose, for example, that second loop is provided by insulin. Insulin inhibits glucose production by the liver and is secreted in response to an elevated blood glucose level (Fig. 6). Protection against deviation in either direction is often achieved in biologic systems by the opposing actions of antagonistic control systems.

Closed-loop negative feedback control as described earlier can maintain conditions only at a state of

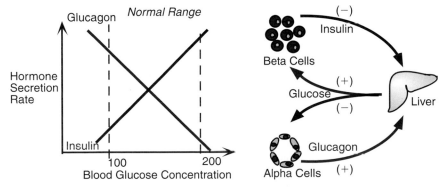

FIGURE 6 Negative feedback regulation of blood glucose concentration by insulin and glucagon. (−) = inhibits, (+) = stimulates.

constancy. Such systems are effective in guarding against upward or downward deviations from some predetermined set point, but changing environmental demands often require temporary deviation from constancy. This is accomplished in some cases by adjusting the set point, and in other cases, by a signal that overrides the set point. For example, epinephrine secreted by the adrenal medulla in response to some emergency inhibits insulin secretion and increases glucagon secretion, even though the concentration of glucose in the blood may already be high. Whether the set point is changed or overridden, deviation from constancy is achieved by the intervention of some additional signal from outside the negative feedback system. In most cases that additional signal originates with the nervous system.

Hormones also initiate or regulate processes that are not limited to steady or constant conditions. Virtually all of these processes are self-limiting, and their control resembles negative feedback, but of the open-loop type. For example, oxytocin is a hormone secreted by hypothalamic nerve cells whose axons terminate in the posterior pituitary gland. Its secretion is necessary for extrusion of milk from the lumen of the mammary alveolus into secretory ducts so that the infant suckling at the nipple can receive milk. In this case, sensory nerve endings in the nipple detect the signal and convey afferent information to the central nervous system, which in turn signals release of oxytocin from axon terminals in the pituitary gland. Oxytocin causes myoepithelial cells to contract, resulting in delivery of milk to the infant. When the infant is satisfied, the suckling stimulus at the nipple ceases.

Positive Feedback

Positive feedback means that some consequence of hormonal secretion acts on the secretory cells to stimulate further secretion. Rather than being self-limiting, as with negative feedback, the drive for secretion becomes progressively more intense. Positive feedback systems are unusual in biology, because they terminate with some cataclysmic, explosive event. A good example of a positive feedback system involves oxytocin and its other effect: causing contraction of uterine muscle during childbirth (Fig. 7). In this case the stimulus for oxytocin secretion is dilation of the uterine cervix. Upon receipt of this information through sensory nerves, the brain signals the release of oxytocin from nerve endings in the posterior pituitary gland. Enhanced uterine contraction in response to oxytocin results in greater dilation of the cervix, which strengthens the signal for oxytocin release and so on until the infant is expelled from the uterine cavity.

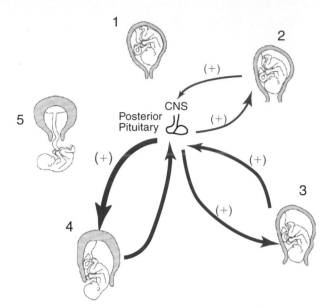

FIGURE 7   Positive feedback regulation of oxytocin secretion. (1) Uterine contractions at the onset of parturition apply mild stretch to the cervix. (2) In response to sensory input from the cervix, oxytocin is secreted from the posterior pituitary gland, and stimulates (+) further contraction of the uterus, which, in turn stimulates secretion of more oxytocin (3) leading to further stretching of the cervix, and even more oxytocin secretion (4), until the fetus is expelled (5).

ACTIONS OF HORMONES

Regulation of bodily functions by hormones is achieved by regulating the activities of individual cells. Hormones signal cells to start or stop secreting, contracting, dividing, or differentiating. They may also accelerate or slow the rates of these processes, or they may modify responses to other hormones. All of these cellular actions summate to produce the biologic responses we observe at the level of tissues, organs, and the whole body. The factors that determine the magnitude of cellular responses to a hormone are listed in Table 3. Although hormones are distributed throughout the blood and extracellular fluid, only

Table 3   Determinants of the Magnitude of a
Hormonal Response

1. Concentration of Hormone at Target Cell Surface
   a. rate of secretion
   b. rate of delivery by the circulation
   c. rate of hormone degradation
2. Sensitivity of Target Cells
   a. number of functional receptors per cell
   b. receptor affinity for the hormone
   c. postreceptor amplification capacity
   d. abundance of available effector molecules
3. Number of Functional Target Cells

certain target cells respond to any given hormone. Most cells are targets for more than one hormone and many hormones target more than a single cell type. The molecular mechanisms that underlie signal reception and transduction are discussed in Chapter 2 and are considered in subsequent chapters in relation to the actions of specific hormones.

Whether or not a cell responds to a hormone depends on whether or not it has receptors for that hormone (see Chapter 2). However, it is important to recognize that the nature of the specific response elicited in a given target cell is determined by that cell rather than by the hormone. In fact, different cell types may respond to the same hormone in different ways. For example, the adrenal hormone cortisol stimulates net protein breakdown in skeletal muscle and net protein synthesis in liver. Cortisol activates the same receptor in muscle and liver to produce these divergent actions, but other hormones may produce their effects through different receptors expressed in different cells. Vascular smooth muscle cells and epithelial cells in the collecting ducts of renal tubules are targets for the posterior pituitary hormone vasopressin. When stimulated by vasopressin, arterioles contract, whereas collecting ducts increase their permeability to water. The V_1 receptor in vascular smooth muscle and the V_2 receptor in renal tubules couple to different intracellular signaling pathways as well as to different effector molecules (see Chapter 28).

Modulation of Responding Systems

Not all aspects of hormonal control are determined simply by how much hormone is secreted or even when a hormone is secreted. Receptivity of target tissues to hormonal stimulation is not constant and can be changed under a variety of circumstances. Receptivity of target tissues to hormonal stimulation can be expressed in terms of two separate, but related, aspects: *sensitivity* to stimulation and the *capacity* to respond. Sensitivity describes the acuity of a cell's ability to recognize a signal and to respond in proportion to the intensity of that signal. Sensitivity is often described in terms of the concentration of hormone needed to produce a half-maximal response. The relationship between the magnitude of a hormonal response and the concentration of hormone that produces the response can be described by a sigmoidal curve (Fig. 8). An increase in sensitivity lowers the concentration of hormone needed to elicit a half-maximal response and produces a leftward shift of the dose–response curve. Conversely, a decrease in sensitivity increases the concentration of hormone needed to evoke the same response and thus results in a rightward shift in the dose–response curve. In the example shown in Fig. 8, we can assume that curve B

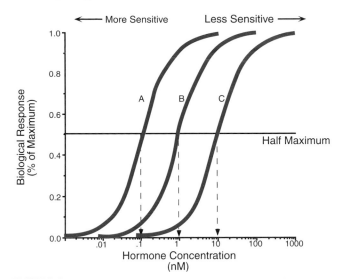

FIGURE 8   The relationship between concentration and response. Three levels of sensitivity are shown. Arrows indicate the concentration of hormone that produces a half-maximal response for each level of sensitivity. Note that hormone concentrations are expressed on a logarithmic scale.

represents the basal sensitivity that may be increased (curve A) or decreased (curve C) in different physiologic or pathologic conditions. Changes in capacity to respond are illustrated in Fig. 9. In this case the maximum response may be increased (curve A) or decreased (curve C), but the sensitivity (i.e., the concentration of hormone needed to produce the half-maximal response) remains unchanged at 1 ng/mL.

One mechanism by which hormones adjust the sensitivity of target cells is by regulating the availability of hormone receptors. These changes are referred to as *up-regulation* or *down-regulation*. Hormone–receptor interactions depend on the likelihood that a molecule

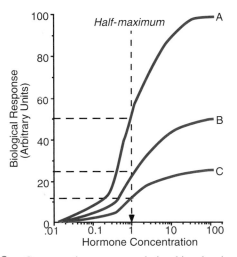

FIGURE 9   Concentration response relationships showing different capacities to respond. Note that the concentration needed to produce the half-maximal response is identical for all three response capacities.

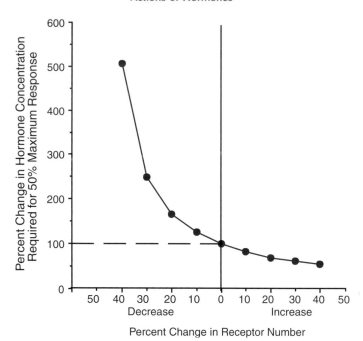

FIGURE 10   The effects of up- or down-regulation of receptor number on sensitivity to hormonal stimulation.

of hormone will encounter a molecule of receptor. If there are no hormones or no receptors, there can be no response. The higher the concentration of hormone, the more likely it is to interact with its receptors. Similarly, the more receptors available to interact with any particular concentration of hormone, the greater the likelihood of a response. In other words, the probability that a molecule of hormone will encounter a molecule of receptor is related to the abundance of both the hormone and the receptor. The effects of changing receptor abundance on hormone sensitivity are shown in Fig. 10. Note that the relationship is not linear, and that increasing the receptor number by 40% decreases the hormone concentration needed to produce a half-maximal response by almost a factor of 2, while decreasing the number of receptors by 40% increases the needed concentration of hormone by a factor of 5. Down-regulation is frequently encountered and may result from inactivation of receptors by covalent modification, from an increased rate of sequestration of internalized membrane receptors, or from a change in the rates of receptor degradation or synthesis.

The other determinant of sensitivity is the affinity of the receptor for the hormone. Affinity reflects the "tightness" of binding or the likelihood that an encounter between a hormone and its receptor will result in binding. Affinity is usually defined in terms of the concentration of hormone needed to occupy half of the available receptors. Although the affinity of a receptor for its hormone may be adjusted by covalent

modifications such as phosphorylation or dephosphorylation, in general, it appears the number of receptors is regulated rather than their affinity.

Biological responses do not necessarily parallel hormone binding and, therefore, are not limited by the affinity of the receptor for the hormone. Because they depend on many postreceptor events, responses to some hormones may be at a maximum at concentrations of hormone that do not saturate all of the receptors (Fig. 11). When fewer than 100% of the receptors need to be occupied to obtain a maximum response, cells are said to express *spare receptors*. For example, glucose uptake by fat cells is stimulated by insulin in a dose-dependent manner, but the response reaches a maximum when only a small percent of available receptors are occupied by insulin. Consequently, a half-maximal response to insulin is achieved at a concentration that is considerably lower than that required to occupy half of the receptors and, hence, the sensitivity of the cell is considerably greater than the affinity of the receptor. Recall that sensitivity is measured in terms of a biological response, which is the physiologically meaningful parameter, whereas affinity is independent of the postreceptor events that produce the biological response. The magnitude of a cellular response to a hormone is determined by the summation of the signal generated by each of the occupied receptors, and therefore is related to the number of receptors that are activated rather than the fraction of the total receptor pool that is bound to hormone. However, because the

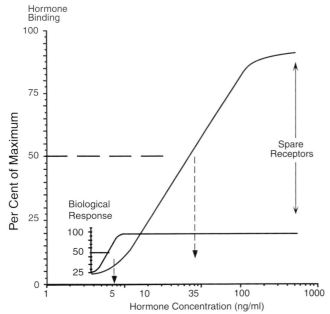

FIGURE 11   Spare receptors. Note that the concentration of hormone needed to produce a half-maximal response is considerably lower than that needed to occupy half of the receptors.

percentage of available receptors that bind to hormone is determined by the hormone concentration, the number of activated receptors needed to a produce a half-maximal response will be equivalent to a smaller and smaller fraction as the total number of receptors increases. In the example shown in Fig. 11 expression of five times more receptors than needed for a maximum response increases the sensitivity sevenfold.

Another consequence of spare membrane receptors relates to the rapidity with which hormone can be cleared from the blood. Degradation of peptide hormones depends in part on receptor-mediated internalization of the hormone and hence access to degrading enzymes. Some membrane receptors such as the C-type receptors for the atrial natriuretic hormone (see Chapter 29) lack the biochemical components needed for signal transduction, and have a role only in hormone degradation. Spare receptors thus may blunt potentially harmful overresponses to rapid changes in peptide hormone concentrations.

Sensitivity to hormonal stimulation can also be modulated in ways that do not involve changes in receptor number or affinity. Postreceptor modulation may affect any of the steps in the biological pathway through which hormonal effects are produced. Up- or down-regulation of effector molecules such as enzymes, ion channels, and contractile proteins may amplify or dampen responses and, hence, change the relationship between receptor occupancy and magnitude of response. For example, the activity of cAMP phosphodiesterase

increases in adipocytes in the absence of pituitary hormones. Recall from Chapter 2 that this enzyme catalyzes the degradation of cAMP, and when its activity is increased, less cAMP can accumulate after stimulation of adenylyl cyclase by a hormone such as epinephrine. Therefore, if all other things were equal, a higher concentration of epinephrine would be needed to produce a given amount of lipolysis than might be necessary in the presence of normal amounts of pituitary hormones; hence, sensitivity to epinephrine appears reduced.

At the tissue, organ, or whole-body level, the response to a hormone is the aggregate of the contributions of all of the stimulated cells, so that the magnitude of the response is determined both by the number of responsive cells and their competence. For example, the pituitary hormone ACTH stimulates the adrenal glands to secrete their hormone, cortisol, in a dose-related manner. However, immediately after removal of one adrenal gland, changes in the concentration of cortisol in response to ACTH administration would be only half as large as seen when both glands are present. Therefore, a much higher dose of ACTH will be needed to achieve the same change as was produced preoperatively.

INTERACTIONS OF RESPONDING SYSTEMS

Maintaining the integrity of the internal environment or successfully meeting an external challenge typically involves the coordinated interplay of several physiologic systems and the integration of multiple hormonal signals. Solutions to physiologic problems require integration of a large variety of simultaneous events that together may produce results that are greater or less than the simple algebraic sum of the individual hormonal responses. In the following section, we consider some of the ways in which endocrine regulatory systems may interact.

Reinforcement

Although a hormone may trigger an overall cellular response by affecting some fundamental rate-determining reaction, several cellular processes may be affected simultaneously. Hormonal effects exerted at several locales within a single cell *reinforce each other* and sum to produce the overall response. Let us consider, for example, just some of the ways insulin acts on the fat cell to promote storage of triglycerides:

- It acts at the cell membrane to increase availability of substrate for triglyceride synthesis.
- It activates several cytosolic and mitochondrial enzymes critical for fatty acid synthesis.

- It inhibits breakdown of already formed triglycerides.
- It induces synthesis of the extracellular enzyme lipoprotein lipase needed to take up lipids from the circulation.

Any one of these actions might accomplish the end of increasing fat storage, but collectively, these different effects make possible an enormously broader range of response in a shorter time frame. These effects of insulin will be considered in detail in Chapter 41.

Reinforcement can also be observed at the level of the whole organism, where a hormone may act in different ways in different tissues to produce complementary effects. A good example of this is the action of cortisol to promote hepatic synthesis of glucose. It acts in extrahepatic tissues to mobilize substrate and in the liver to increase conversion of these substrates to glucose (see Chapter 40). Either the extrahepatic action or the hepatic action would increase glucose synthesis, but together, these complementary actions reinforce each other and increase both the magnitude and rapidity of the overall response.

Redundancy

Fail-safe mechanisms govern crucial functions. Just as each organ system has built-in excess capacity, giving it the potential to function at levels beyond the usual day-to-day demands, so too is excess regulatory capacity provided in the form of seemingly duplicative or overlapping controls. Simply put, the body has more than one way to achieve a given end. For example (Fig. 12), conversion of liver glycogen to blood glucose can be signaled by at least two hormones, glucagon from the alpha cells of the pancreas and epinephrine from the adrenal medulla (see Chapters 40 and 41). Both of these hormones increase cAMP production in the liver, and

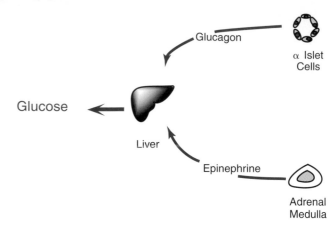

FIGURE 12   Redundant mechanisms to stimulate hepatic glucose production.

thereby activate the enzyme phosphorylase, which catalyzes glycogenolysis. Two hormones secreted from two different tissues, sometimes in response to different conditions, thus produce the same end result.

Redundancy can also be seen at the molecular level. Using the same example of conversion of liver glycogen to blood glucose, there are even two ways that epinephrine can activate phosphorylase (Fig. 13). By stimulating β-adrenergic receptors, epinephrine increases cAMP formation as already mentioned. By stimulating α-adrenergic receptors, epinephrine also activates phosphorylase, but these receptors operate through the agency of increased intracellular calcium concentrations produced by the release of inositol trisphosphate (see Chapter 2).

Redundant mechanisms not only assure that a critical process will take place, but they also offer opportunity for flexibility and subtle fine-tuning of a process. Although redundant in the respect that two different

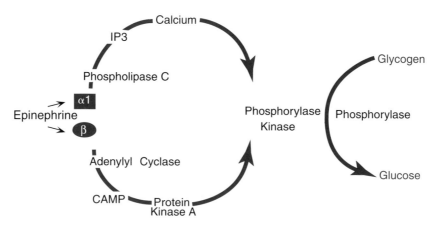

FIGURE 13   Redundant mechanisms to activate glycogen phosphorylase by a single hormone, epinephrine, acting through both α and β receptors. IP3 = Inositol trisphate; CAMP = cyclic adenosine monophosphate.

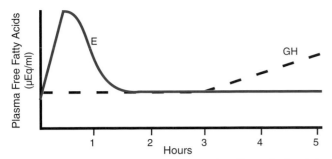

FIGURE 14   Idealized representation of the effects of epinephrine and growth hormone on plasma concentrations of free fatty acids.

hormones may have some overlapping effects, the actions of the two hormones are usually not identical in all respects. Within the physiologic range of its concentrations in blood, glucagon's action is restricted to the liver. In contrast, epinephrine produces a variety of other responses in many extrahepatic tissues while increasing glycogenolysis in the liver. Variations in the relative input from both hormones allows for a wide spectrum of changes in blood glucose concentrations relative to such other effects of epinephrine as increased heart rate.

Two hormones that produce common effects may differ not only in their range of actions, but also in their time constants (Fig. 14). One may have a rapid onset and short duration of action, whereas another may have a longer duration of action, but a slower onset. For example, epinephrine increases blood concentrations of free fatty acids (FFA) within seconds or minutes, and this effect dissipates as rapidly when epinephrine secretion is stopped. Growth hormone also increases blood concentrations of FFA, but its effects are seen only after a lag period of 2–3 hr and persist for many hours. A hormone like epinephrine may therefore be used to meet short-term needs, and another, like growth hormone, may answer sustained needs.

One of the implications of redundancy for the understanding both of normal physiology and endocrine disease is that partial, or perhaps even complete, failure of one mechanism can be compensated by increased reliance on a redundant mechanism. Thus, functional deficiencies may be evident only in subtle ways and may not show up readily as overt disease. Some deficiencies may become apparent only after appropriate provocation or perturbation of the system. Conversely, strategies for therapeutic interventions designed to increase or decrease the rate of a process must take into account the redundant inputs that regulate that process. Merely accelerating or blocking one regulatory input may not produce the desired effect because independent adjustments in redundant pathways may completely compensate for the intervention.

Synergy and Attenuation

The net effect of two or more redundant mechanisms acting simultaneously may be equal to, greater than, or less than the sum of their individual effects. *Synergy* or *potentiation* are the terms used when the combined effects of two or more hormones are greater than the algebraic sum of their individual effects. An example of this phenomenon is shown in Fig. 15. Both growth hormone and adrenal cortical steroids stimulate the hydrolysis of triglycerides in adipocytes and promote the release of glycerol and fatty acids. Each has a small effect, but together they produce about twice as great an increase in glycerol as would be expected from simply adding their individual effects. Synergy occurs when one of two hormones that act at different sites in a reaction sequence increase a reaction that would otherwise limit the actions of the other agent. For example, if hormone A stimulates cAMP production and hormone B inhibits cAMP degradation, each would somewhat increase the cellular concentration of cyclic AMP, but their combined actions would produce a very large increase. Following the same rationale, when two hormones act through a common step in a reaction sequence, the sum of their individual effects may be greater than their combined effects because some subsequent step in the reaction sequence may become limiting. Such *attenuation* is less commonly seen than synergy and usually occurs only at very high levels of stimulation or under pathologic conditions.

A special case of synergy that is related to the preceding examples has been called *permissive action*. A hormone acts permissively when its presence is necessary for, or permits, a biological response to occur, even though the hormone does not initiate or participate in the response. Permissive effects were originally described

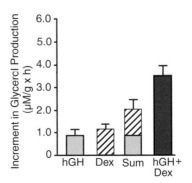

FIGURE 15   Synergistic effects of human growth hormone (hGH) and the synthetic glucocorticoid dexamethasone (Dex) on lipolysis as measured by the increase in glycerol release from rat adipocytes. Both hGH and Dex were effective when added individually, but when added together their effect was greater than the sum of their individual effects. (From Gorin *et al., Endocrinology* 1990;126:2973.)

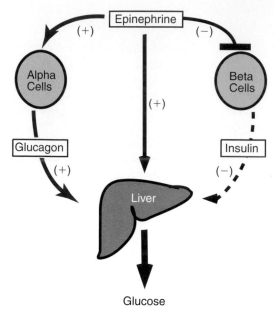

FIGURE 16   Push–pull mechanism. Epinephrine inhibits insulin secretion while promoting glucagon secretion. This combination of effects on the liver stimulates glucose production while simultaneously relieving an inhibitory influence.

for the adrenal cortical hormones, but appear to occur for other hormones as well. Permissive actions are not limited to responses to hormones, but pertain to any cellular response to any signal and presumably reflect some ongoing action of a hormone to maintain some basal cellular functions.

Push–Pull Mechanisms

Many critical processes are under dual control by agents that act antagonistically either to stimulate or to inhibit. Such dual control allows for more precise regulation through negative feedback. The example cited earlier was hepatic production of glucose, which is increased by glucagon and inhibited by insulin. In emergency situations or during exercise, epinephrine and norepinephrine released from the adrenal medulla and norepinephrine released from sympathetic nerve endings override both negative feedback systems by inhibiting insulin secretion and stimulating glucagon secretion (Fig. 16). The effect

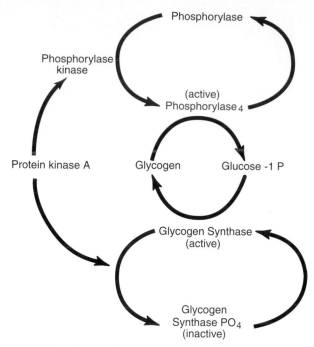

FIGURE 17   Push–pull mechanism to activate glycogen phosphorylase while simultaneously inhibiting glycogen synthase.

of adding a stimulatory influence and simultaneously removing an inhibitory influence is a rapid and large response, more rapid and larger than could be achieved by simply affecting either hormone alone, or than could be accomplished by the direct glycogenolytic effect of epinephrine or norepinephrine.

Another type of push–pull mechanism can be seen at the molecular level. Net synthesis of glycogen from glucose depends on the activity of two enzymes: glycogen synthase, which catalyzes the formation of glycogen from glucose, and glycogen phosphorylase, which catalyzes glycogen breakdown (Fig. 17). The net reaction rate is determined by the balance of the activity of the two enzymes. The activity of both enzymes is subject to regulation by phosphorylation, but in opposite directions: Addition of a phosphate group activates phosphorylase, but inactivates synthase. In this case, a single agent, cAMP, which activates protein kinase A, increases the activity of phosphorylase and simultaneously inhibits synthase.

Pituitary Gland

H. MAURICE GOODMAN

KEY POINTS

- The *pituitary gland* consists of an *anterior lobe* derived from the primitive gut and a *neural lobe* derived from the brain stem.
- The neural lobe receives its blood supply directly through the superior and inferior hypophysial arteries.
- The anterior lobe receives its blood supply through a system of capillary beds arranged in series such that it is supplied with blood that first perfused the median eminence of the hypothalamus.
- The anterior lobe synthesizes and secretes six protein hormones: thyroid-stimulating hormone (TSH), follicle-stimulating hormone (FSH), luteinizing hormone (LH), adrenocorticotropic hormone (ACTH), growth hormone (GH), and prolactin (PRL).
- Secretion of the anterior pituitary hormones is controlled by hypothalamic neurons that secrete neurohormones into the hypophysial portal circulation and by hormones of their target glands.
- Secretion of posterior pituitary hormones is controlled by neural signals arising either within the brain or at peripheral sites.

Essential Medical Physiology, Third Edition

OVERVIEW

The pituitary gland has usually been thought of as the "master gland" because its hormone secretions control the growth and activity of three other endocrine glands: the thyroid, adrenals, and gonads. Because the secretory activity of the master gland is itself controlled by hormones that originate in either the brain or the target glands, it is perhaps better to think of the pituitary gland as the relay between the control centers in the central nervous system and the peripheral endocrine organs. The pituitary hormones are not limited in their activity to regulation of endocrine target glands; they also act directly on nonendocrine target tissues. Secretion of all of these hormones is under the control of signals arising in both the brain and the periphery.

MORPHOLOGY

The pituitary gland is located in a small depression in the sphenoid bone, the *sella turcica,* just beneath the hypothalamus, and is connected to the hypothalamus by a thin stalk called the *infundibulum*. It is a compound organ consisting of a neural or posterior lobe derived embryologically from the brain stem, and a larger anterior portion, the *adenohypophysis,* which derives embryologically from the primitive foregut. The cells located at the junction of the two lobes comprise the *intermediate lobe,* which is not readily identifiable as an anatomic entity in humans (Fig. 1).

Histologically, the anterior lobe consists of large polygonal cells arranged in cords and surrounded by a sinusoidal capillary system. Most of the cells contain secretory granules, although some are only sparsely granulated. Based on their characteristic staining with standard histochemical dyes and immunofluorescent stains, it is possible to identify the cells that secrete each of the pituitary hormones. It was once thought that there was a unique cell type for each of the pituitary hormones, but it is now recognized that some cells

may produce more than one hormone. Although particular cell types tend to cluster in central or peripheral regions of the gland, the functional significance, if any, of their arrangement within the anterior lobe is not known.

The posterior lobe consists of two major portions: the infundibulum, or stalk, and the infundibular process, or neural lobe. The posterior lobe is richly endowed with nonmyelinated nerve fibers that contain electron-dense secretory granules. The cell bodies from which these fibers arise are located in the bilaterally paired *supraoptic* and *paraventricular nuclei* of the hypothalamus. These cells are characteristically large compared to other hypothalamic neurons and hence are referred to as *magnocellular*. Secretory material synthesized in cell bodies in the hypothalamus is transported down the axons and stored in bulbous nerve endings within the posterior lobe. Dilated terminals of these fibers lie in proximity to the rich capillary network whose fenestrated endothelium allows secretory products to enter the circulation readily.

The vascular supply and innervation of the two lobes reflect their different embryological origins and provide important clues that ultimately led to an understanding of their physiologic regulation. The anterior lobe is sparsely innervated and lacks any secretomotor nerves. This fact might argue against a role for the pituitary as a relay between the central nervous system and peripheral endocrine organs, except that communication between the anterior pituitary and the brain is through vascular, rather than neural, channels.

The anterior lobe is linked to the brain stem by the *hypothalamo-hypophysial portal system* through which it receives most of its blood supply (Fig. 2). The superior hypophysial arteries deliver blood to an intricate network of capillaries, the *primary plexus,* in the median eminence of the hypothalamus. Capillaries of the primary plexus converge to form long hypophyseal portal vessels, which course down the infundibular stalk to deliver their blood to capillary sinusoids interspersed among the secretory cells of the anterior lobe. The inferior hypophysial arteries supply a similar

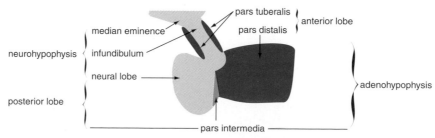

FIGURE 1   Midsagittal section of the human pituitary gland indicating the nomenclature of its various parts.

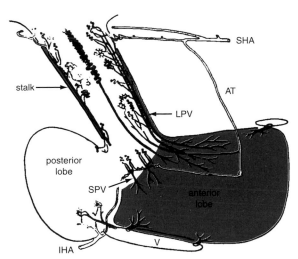

FIGURE 2   Vascular supply of the human pituitary gland. Note the origin of long portal vessels (LPV) from the primary capillary bed and the origin of short portal vessels (SPV) from the capillary bed in the lower part of the stalk. Both sets of portal vessels break up into sinusoidal capillaries in the anterior lobe. SHA and IHA, superior and inferior hypophysial arteries, respectively; AT, trabecular artery that forms an anastomotic pathway between SHA and IHA; V, venous sinuses. (Redrawn from Daniel PM, Prichard MML. Observations on the vascular anatomy of the pituitary gland and its importance in pituitary function. *Am Heart J* 1966;72:147.)

capillary plexus in the lower portion of the infundibular stem. These capillaries drain into short portal vessels, which supply a second sinusoidal capillary network within the anterior lobe. Nearly all of the blood that reaches the anterior lobe is carried in the long and short portal vessels. The anterior lobe receives only a small portion of its blood supply directly from the paired trabecular arteries, which branch off the superior hypophysial arteries. In contrast, the circulation in the posterior pituitary is unremarkable. It is supplied with blood by the inferior hypophysial arteries. Venous blood drains from both lobes through a number of short veins into the nearby cavernous sinuses.

The portal arrangement of blood flow is important because blood that supplies the secretory cells of the anterior lobe first drains the hypothalamus. Portal blood can thus pick up chemical signals released by neurons of the central nervous system and deliver them to secretory cells of the anterior pituitary. As might be anticipated, because hypophysial portal blood flow represents only a tiny fraction of the cardiac output, only minute amounts of neural secretions are needed to achieve biologically effective concentrations in pituitary sinusoidal blood when delivered in this way. More than 1000 times more secretory material would be needed if it were dissolved in the entire blood volume and delivered through the arterial circulation. This arrangement also provides a measure of specificity to hypothalamic

secretion, because pituitary cells are the only ones exposed to concentrations that are high enough to be physiologically effective.

PHYSIOLOGY OF THE ANTERIOR PITUITARY GLAND

There are six anterior pituitary hormones whose physiologic importance is clearly established. They include the hormones that govern the function of the thyroid and adrenal glands, the gonads, the mammary glands, and bodily growth. They have been called *trophic* or *tropic* from the Greek *trophos,* meaning "to nourish," or *tropic,* meaning "to turn toward." Both terms are generally accepted. We thus have, for example, thyrotrophin, or thyrotropin, which is also more accurately called thyroid-stimulating hormone (TSH). Because its effects are exerted throughout the body or *soma* in Greek, growth hormone (GH) has also been called the somatotropic hormone (STH), or somatotropin. Table 1 lists the anterior pituitary hormones and their various synonyms. The various anterior pituitary cells are named for the hormones they contain. Thus we have thyrotropes, corticotropes, somatotropes, and lactotropes. Because a substantial number of growth hormone-producing cells also secrete prolactin, they are called *somatomammotropes.* Some evidence suggests that somatomammotropes are an intermediate stage in the interconversion of somatotropes and lactotropes. The two gonadotropins are found in a single cell type, called the *gonadotrope.*

All of the anterior pituitary hormones are proteins or glycoproteins. They are synthesized on ribosomes and translocated through various cellular compartments where they undergo post-translational processing. They are packaged in membrane-bound secretory granules and secreted by exocytosis. The pituitary gland stores relatively large amounts of hormone, sufficient to meet physiologic demands for many days. Over the course of many decades, these hormones were extracted, purified, and characterized. Now their three-dimensional configurations and the structure of their genes are known, and we can group the anterior pituitary hormones by families.

Glycoprotein Hormones

The glycoprotein hormone family includes TSH, whose only known physiologic role is to stimulate secretion of thyroid hormone, and the two gonadotropins, follicle-stimulating hormone (FSH) and luteinizing hormone (LH). Although named for their function in women, both gonadotropic hormones are crucial for the

Table 1 Hormones of the Anterior Pituitary Gland

Hormone	Target	Major actions in humans
Glycoprotein family		
Thyroid-stimulating hormone (TSH), also called thyrotropin	Thyroid gland	Stimulates synthesis and secretion of thyroid hormones
Follicle-stimulating hormone (FSH)	Ovary	Stimulates growth of follicles and estrogen secretion
	Testis	Acts on Sertoli cells to promote maturation of sperm
Luteinizing hormone (LH)	Ovary	Stimulates ovulation of ripe follicle and formation of corpus luteum; stimulates estrogen and progesterone synthesis by corpus luteum
	Testis	Stimulates interstitial cells of Leydig to synthesize and secrete testosterone
Growth hormone/prolactin family		
Growth hormone (GH), also called somatotropic hormone (STH)	Most tissues	Promotes growth in stature and mass; stimulates production of IGF-I; stimulates protein synthesis; usually inhibits glucose utilization and promotes fat utilization
Prolactin	Mammary glands	Promotes milk secretion and mammary growth
Pro-opiomelanocortin family (POMC)		
Adrenocorticotropic hormone (ACTH), also known as adrenocorticotropin or corticotropin	Adrenal cortex	Promotes synthesis and secretion of adrenal cortical hormones
β-Lipotropin	Fat	Physiologic role not established

function of testes as well as ovaries. In women FSH promotes growth of ovarian follicles and in men it promotes formation of spermatozoa by the germinal epithelium of the testis. In women LH induces ovulation of the ripe follicle and formation of the corpus luteum from remaining glomerulosa cells in the collapsed, ruptured follicle. It also stimulates synthesis and secretion of the ovarian hormones estrogen and progesterone. In men LH stimulates secretion of the male hormone, testosterone, by interstitial cells of the testis. The actions of these hormones are discussed in detail in Chapters 45 and 46.

The three glycoprotein hormones are synthesized and stored in pituitary basophils and, as their name implies, each contains sugar moieties covalently linked to asparagine residues in the polypeptide chains. All three are comprised of two peptide subunits, designated α and β, which, though tightly coupled, are not covalently linked. The α-subunit is common to all three hormones, and is the product of a single gene located on chromosome 6. The β-subunits of each are somewhat larger than the α-subunit and confer physiologic specificity. Both α- and β-subunits contribute to receptor binding and both must be present in the receptor binding pocket to produce a biological response.

β-Subunits are encoded in separate genes located on different chromosomes: TSH β on chromosome 1, FSH β on chromosome 11, and LH β on chromosome 19, but there is a great deal of homology in their amino acid sequences. Both subunits contain carbohydrate moieties that are considerably less constant in their composition than are their peptide chains. α Subunits are synthesized in excess over β-subunits, and hence it is synthesis of β-subunits that appears to be rate limiting for production of each of the glycoprotein hormones. Pairing of the two subunits begins in the rough endoplasmic reticulum and continues in the Golgi apparatus, where processing of carbohydrate components of the subunits is completed. The loosely paired complex then undergoes spontaneous refolding in secretory granules into a stable, active hormone. Control of expression of the α- and β-subunit genes is not perfectly coordinated, and the free α- and the β-subunits of all three hormones may be found in blood plasma.

The placental hormone, human chorionic gonadotropin (hCG), is closely related chemically and functionally to the pituitary gonadotropic hormones. It, too, is a glycoprotein and consists of an α and a β chain. The α chain is a product of the same gene as the α chain of pituitary glycoprotein hormones. The peptide sequence of the β chain is identical to that of LH except that it is longer by 32 amino acids at its carboxyl terminus. Curiously, although there is only a single gene for each β-subunit of the pituitary glycoprotein hormones, the human genome contains 7 copies of the hCG β gene, all located on chromosome 19 in proximity to the LH β gene. Not surprisingly, hCG has biological actions that are similar to those of LH (Chapter 47).

Growth Hormone and Prolactin

Growth hormone (GH) is required for attainment of normal adult stature (see Chapter 44) and produces metabolic effects that may not be directly related to its growth-promoting actions. Metabolic effects include mobilization of free fatty acids from adipose tissue and inhibition of glucose metabolism in muscle and adipose tissue. The role of GH in energy balance is discussed in Chapter 42. Somatotropes are by far the most abundant anterior pituitary cells, and they account for about half of the cells of the adenohypophysis. GH, which is secreted throughout life, is the most abundant of the pituitary hormones. The human pituitary gland stores between 5 and 10 mg of GH, an amount that is 20–100 times greater than other anterior pituitary hormones. Structurally, GH is closely related to another pituitary hormone, *prolactin* (PRL), which is required for milk production in postpartum women (see Chapter 47). The functions of PRL in men or nonlactating women are not firmly established, but a growing body of evidence suggests that it may stimulate cells of the immune system. These pituitary hormones are closely related to the placental hormone *human chorionic somatomammotropin* (hCS), which has both growth-promoting and milk-producing activity in some experimental systems. Because of this property, hCS is also called *human placental lactogen* (hPL). Although the physiologic function of this placental hormone has not been established with certainty, it may regulate maternal metabolism during pregnancy and prepare the mammary glands for lactation (see Chapter 47).

Growth hormone, PRL, and hCS appear to have evolved from a single ancestral gene that duplicated several times; the GH and PRL genes before the emergence of the vertebrates, and GH and the hCS genes after the divergence of the primates from other mammalian groups. The human haploid genome contains two GH and three hCS genes all located on the long arm of chromosome 17, and a single PRL gene located on chromosome 6. These genes are similar in the arrangement of their transcribed and nontranscribed portions as well as their nucleotide sequences. GH and

hCS have about 80% of their amino acids in common, and a region 146 amino acids long is similar in hGH and PRL. Only one of the GH genes (hGH N) is expressed in the pituitary, but because an alternative mode of splicing of the RNA transcript is possible, two GH isoforms are produced and secreted by the somatotropes. The larger form is the 22-kDa molecule (22K GH), which is about 10 times more abundant than the smaller, 20-kDa molecule (20K GH), which lacks amino acids 32 to 46. The other GH gene (hGH V) appears to be expressed only in the placenta and is the predominant form of GH in the blood of pregnant women. It encodes a protein that appears to have the same biological actions as the pituitary hormone although it differs from the pituitary hormone in 13 amino acids and also in that it may be glycosylated.

Considering the similarities in their structures, it is not surprising that GH shares some of the lactogenic activity of PRL and hCS. However, human GH also has about two-thirds of its amino acids in common with GH molecules of cattle and rats, but humans are completely insensitive to cattle or rat GH and respond only to the GH produced by humans or monkeys. This requirement of primates for primate GH is an example of *species specificity* and largely results from the change of a single amino acid in GH and a corresponding change of a single amino acid in the binding site in the GH receptor. Because of species specificity, human GH was in short supply for treatment of GH-deficient children until the advent of recombinant DNA technology, which made possible an almost limitless supply.

Adrenocorticotropin Family

Portions of the cortex of the adrenal glands are controlled physiologically by *adrenal corticotropic hormone* (ACTH), which is also called *corticotropin* or *adrenocorticotropin*. This family of pituitary peptides includes α- and β-*melanocyte-stimulating hormones* (MSH), β- and α-*lipotropin* (LPH), and β-*endorphin*. Of these, ACTH is the only peptide whose physiologic role in humans is established. The MSHs, which disperse melanin pigment in melanocytes in the skin of lower vertebrates, have little importance in this regard in humans and are not secreted in significant amounts. β-LPH is named for its stimulatory effect on mobilization of lipids from adipose tissue in rabbits, but the physiologic importance of this action is uncertain. The 91-amino-acid chain of β-LPH contains at its carboxyl end the complete amino acid sequence of β-endorphin (from *endogenous morphine*), which reacts with the same receptors as morphine.

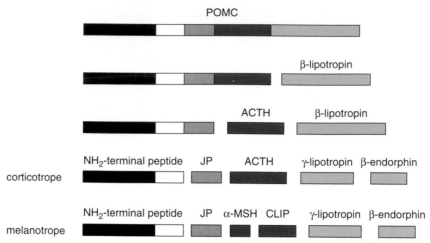

FIGURE 3   Proteolytic processing of pro-opiomelanocortin (POMC). POMC after removal of the signal peptide is shown on the first line. The first cleavage by prohormone convertase 1 releases β-lipotropin. The second cleavage releases ACTH. A third cleavage releases the joining peptide (JP) to produce the principal secretory products of the corticotropes of the anterior pituitary gland. A third and fourth cleavage takes place in the melanotropes of the intermediate lobe and splits ACTH into α-melanocyte stimulating hormone (α-MSH) and the corticotropin-like intermediate lobe peptide (CLIP) and divides β-lipotropin into γ-lipotropin and β-endorphin. Some cleavage of β-lipotropin also takes place in the corticotrope. Additional post-translational processing (not shown) includes removal of the carboxyl-terminal amino acid from each of the peptides and glycosylation and phosphorylation of some of the peptide fragments. In neural tissue the NH₂-terminal peptide shown by the clear area is also released to produce γ_3-MSH.

The ACTH-related peptides constitute a family because (1) they contain regions of homologous amino acid sequences that may have arisen through exon duplication and (2) because they all arise from the transcription and translation of the same gene (Fig. 3). The gene product is called *pro-opiomelanocortin* (POMC), which consists of 239 amino acids after removal of the signal peptide. The molecule contains 10 doublets of basic amino acids (arginine and lysine in various combinations), which are potential sites for cleavage by trypsin-like endopeptidases, called *prohormone convertases*. POMC is expressed by cells in the anterior lobe of the pituitary, the intermediate lobe, and various cells in the central nervous system; but tissue-specific differences in the way the molecule is processed after translation give rise to differences in the final secretory products. More than seven different enzymes carry out these post-translational modifications. The predominant products of human corticotropes are ACTH and β-LPH. Because final processing of POMC occurs in the secretory granule, β-LPH is secreted along with ACTH. Cleavage of β-LPH also occurs to some extent in human corticotropes, so that some β-endorphin may also be released, particularly when ACTH secretion is brisk. The intermediate lobe in some animals gives rise principally to α- and β-MSH. Because the intermediate lobe of the pituitary gland of

humans is thought to be nonfunctional except perhaps in fetal life, it is not discussed further. Some of the POMC peptides produced in hypothalamic neurons may play an important role in regulating food intake (see Chapter 42) and in coordinating overall responses to stress.

REGULATION OF ANTERIOR PITUITARY FUNCTION

Secretion of the anterior pituitary hormones is regulated by the central nervous system and hormones produced in peripheral target glands. Input from the central nervous system provides the primary drive for secretion and peripheral input plays a secondary, though vital, role in modulating secretory rates. Secretion of all of the anterior pituitary hormones except PRL declines severely in the absence of stimulation from the hypothalamus as can be produced, for example, when the pituitary gland is removed surgically from its natural location and reimplanted at a site remote from the hypothalamus. In contrast, PRL secretion is dramatically increased. The persistent high rate of secretion of PRL under these circumstances indicates not only that the pituitary glands can revascularize and survive in a new location but also that PRL secretion is normally under tonic inhibitory control by the hypothalamus.

Secretion of each of the anterior pituitary hormones follows a characteristic diurnal pattern entrained by activity, sleep, or light–dark cycles. Secretion of each of these hormones also occurs in a pulsatile manner probably reflecting periodic pulses of hypothalamic neurohormone release into hypophysial portal capillaries. Pulse frequency varies widely from about two pulses per hour for ACTH to one pulse every 3–4 hr for TSH, GH, and PRL. Modulation of secretion in response to changes in the internal or external environment may be reflected as changes in the amplitude or frequency of secretory pulses, or by episodic bursts of secretion. In this chapter we discuss only general aspects of the regulation of anterior pituitary function. A detailed description of the control of the secretory activity of each hormone is given in subsequent chapters in conjunction with a discussion of its role in regulating physiologic processes.

Hypophysiotropic Hormones

As already mentioned, the central nervous system communicates with the anterior pituitary gland by means of neurosecretions released into the hypothalamo-hypophysial portal system. These neurosecretions are called *hypophysiotropic hormones* and are listed in Table 2. The hypothalamic nuclei that house hypophysiotropic hormone producing neurons are shown in Fig. 4. The fact that only small amounts of the hypophysiotropic hormones are synthesized, stored, and secreted frustrated efforts to isolate and identify them for nearly 25 years. Their abundance in the

hypothalamus is less than 0.1% of that of even the scarcest pituitary hormone in the anterior lobe.

Thyrotropin-releasing hormone (TRH), the first of the hypothalamic neurohormones to be characterized, was found to be a tripeptide. It was isolated, identified, and synthesized almost simultaneously in the laboratories of Roger Guillemin and Andrew Schally, who were subsequently recognized for this monumental achievement with the award of a Nobel Prize. Guillemin's laboratory processed 25 kg of sheep hypothalami to obtain 1 mg of TRH. Schally's laboratory extracted 245,000 pig hypothalami to yield only 8.2 mg of this tripeptide. The human TRH gene encodes a 242-residue preprohormone molecule that contains six copies of TRH. TRH is synthesized primarily in parvocellular (small) neurons in the paraventricular nuclei of the hypothalamus and is stored in nerve terminals in the median eminence of the hypothalamus. TRH is also expressed in neurons that are widely dispersed throughout the central nervous system and probably acts as a neurotransmitter that mediates a variety of other responses. Actions of TRH that regulate TSH secretion and thyroid function are discussed further in Chapter 39.

Gonadotropin-releasing hormone (GnRH) was the next hypophysiotropic hormone to be isolated and characterized. Hypothalamic control over secretion of FSH and LH is exerted by this decapeptide. Endocrinologists originally had some difficulty accepting the idea that secretion of both gonadotropins is controlled by a single hypothalamic releasing hormone because FSH and LH appear to be secreted independently under certain circumstances. We now know that other factors can account for partial independence of LH and

Table 2 Hypophysiotropic Hormones

Hormone	Amino acids	Hypothalamic source	Physiologic actions on the pituitary
Corticotropin releasing hormone (CRH)	41	Parvoneurons of the paraventricular nuclei	Stimulates secretion of ACTH and β-lipotropin
Gonadotropin releasing hormone (GnRH), originally called luteinizing hormone-releasing hormone (LHRH)	10	Arcuate nuclei	Stimulates secretion of FSH and LH
Growth hormone-releasing hormone (GHRH)	44	Arcuate nuclei	Stimulates GH secretion
Growth hormone releasing peptide (ghrelin)	28		Increases response to GHRH and may directly stimulate GH secretion
Somatotropin release-inhibiting factor (SRIF); somatostatin	14 or 28	Anterior hypothalamic periventricular system	Inhibits secretion of GH
Prolactin-stimulating factor (?)	?	?	Stimulates prolactin secretion (?)
Prolactin inhibiting factor (PIF)	Dopamine secretion	Tuberohypophysial neurons	Inhibits prolactin
Thyrotropin-releasing hormone (TRH)	3	Parvoneurons of the paraventricular nuclei	Stimulates secretion of TSH and prolactin
Arginine vasopressin (AVP)	9	Parvoneurons of the paraventricular nuclei	Acts in concert with CRH to stimulate secretion of ACTH

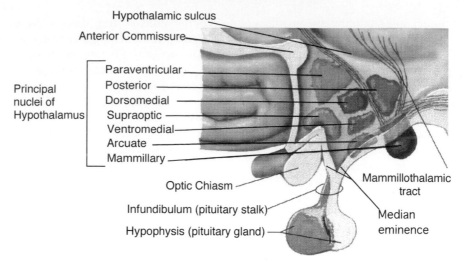

FIGURE 4   Principal hypothalamic nuclei (in midsagittal section) and their relationship to the pituitary gland. (From Netter, F.H. *Atlas of Human Anatomy*, 2nd ed., Summit, NJ: Novartis, 1989. Reprinted with permission from Icon Learning Systems, LLC, a subsidiary of Multimedia USA, Inc.)

FSH secretion. The frequency of pulses of GnRH release determines the ratio of FSH and LH secreted. In addition, target glands secrete hormones that selectively inhibit secretion of either FSH or LH. These complex events are discussed in detail in Chapters 45 and 46.

The GnRH gene encodes a 92-amino-acid preprohormone that contains the 10-amino-acid GnRH peptide and an adjacent 56-amino-acid GnRH associated peptide (GAP), which may also have some biological activity. GAP is found along with GnRH in nerve terminals and may be cosecreted with GnRH. Cell bodies of the neurons that release GnRH into the hypophysial portal circulation reside primarily in the arcuate nucleus in the anterior hypothalamus, but GnRH-containing neurons are also found in the preoptic area and project to extrahypothalamic regions where GnRH release may be related to various aspects of reproductive behavior. GnRH is also expressed in the placenta. Curiously, humans and some other species have a second GnRH gene that is expressed elsewhere in the brain and appears to have no role in gonadotropin release.

Growth hormone secretion is controlled by the *growth hormone-releasing hormone* (GHRH) and a GH release-inhibiting hormone, *somatostatin,* which is also called *somatotropin release-inhibiting factor* (SRIF). In addition, a newly discovered peptide called *ghrelin* may act both on the somatotropes to increase GH secretion by augmenting the actions of GHRH and on the hypothalamus to increase secretion of GHRH. The physiologic role of this novel peptide, which is synthesized both in the hypothalamus and the stomach, has

not been established. GHRH is a member of a family of gastrointestinal and neurohormones that includes vasoactive intestinal peptide (VIP), glucagon (see Chapter 41), and the probable ancestral peptide in this family, called PACAP (pituitary adenylate cyclase activating peptide). GHRH-containing neurons are found predominantly in the arcuate nuclei, and to a lesser extent in the ventromedial nuclei of the hypothalamus. Curiously, GHRH was originally isolated from a pancreatic tumor and is normally expressed in the pancreas, the intestinal tract, and other tissues, but the physiologic role of extrahypothalamically produced GHRH is unknown.

Somatostatin was originally isolated from hypothalamic extracts based on its ability to inhibit GH secretion. The somatostatin gene codes for a 118-amino-acid preprohormone from which either a 14-amino-acid or a 28-amino-acid form of somatostatin is released by proteolytic cleavage. Both forms are similarly active. The remarkable conservation of the amino acid sequence of the somatostatin precursor and the presence of processed fragments that accompany somatostatin in hypothalamic nerve terminals have suggested to some investigators that additional physiologically active peptides may be derived from the somatostatin gene. The somatostatin gene is widely expressed in neuronal tissue as well as in the pancreas (see Chapter 41) and the gastrointestinal tract. The somatostatin that regulates GH secretion originates in neurons present in the preoptic, periventricular, and paraventricular nuclei. It appears that somatostatin is secreted nearly continuously and that it restrains GH secretion except during periodic brief episodes that coincide with increases in GHRH secretion. Coordinated episodes of decreased somatostatin release and increased GHRH secretion produce a pulsatile pattern of GH secretion.

Corticotropin releasing hormone is a 41-amino-acid polypeptide derived from a preprohormone of 192 amino acids. CRH is present in greatest abundance in the parvocellular neurons in the paraventricular nuclei whose axons project to the median eminence. About half of these cells also express *arginine vasopressin* (AVP), which also acts as a corticotropin releasing hormone. AVP has other important physiologic functions and is a hormone of the posterior pituitary gland (see later discussion). The wide distribution of CRH-containing neurons in the central nervous system suggests that it has other actions besides regulation of ACTH secretion.

The simple monoamine neurotransmitter *dopamine* appears to satisfy most of the criteria for a *PRL inhibitory factor* whose existence was suggested by the persistent high rate of PRL secretion by pituitary glands transplanted outside the sella turcica. It is possible that there is also a PRL releasing hormone, but although several candidates have been proposed, general agreement on its nature or even its existence is still lacking.

Although, in general, the hypophysiotropic hormones affect the secretion of one or another pituitary hormone specifically, TRH can increase the secretion of PRL at least as well as it increases the secretion of TSH. The physiologic meaning of this experimental finding is not understood. Under normal physiologic conditions, PRL and TSH appear to be secreted independently, and increased PRL secretion is not necessarily seen in circumstances that call for increased TSH secretion. However, in laboratory rats and possibly human beings as well, suckling at the breast increases both PRL and TSH secretion in a manner suggestive of increased TRH secretion. In the normal individual, somatostatin may inhibit secretion of other pituitary hormones in addition to GH, but again the physiologic significance of this action is not understood. In disease states, specificity of responses of various pituitary cells for their own hypophysiotropic hormones may break down, or cells might even begin to secrete their hormones autonomously.

The neurons that secrete the hypophysiotropic hormones are not autonomous. They receive input from many structures within the brain as well as from circulating hormones. Neurons that are directly or indirectly excited by actual or impending changes in the internal or external environment, from emotional changes, and from generators of rhythmic activity signal to hypophysiotropic neurons by means of classical neurotransmitters as well as neuropeptides. In addition, their activity is modulated by hormonal changes in the general circulation. Integration of responses to all of these signals may take place in the hypophysiotropic

neurons themselves or information may be processed elsewhere in the brain and relayed to the hypophysiotropic neurons. Conversely, hypophysiotropic neurons or neurons that release hypophysiotropic peptides as their neurotransmitters communicate with other neurons dispersed throughout the central nervous system to produce responses that presumably are relevant to the physiologic circumstances that call forth pituitary hormone secretion.

Hypophysiotropic hormones increase both secretion and synthesis of pituitary hormones. All appear to act through stimulation of G-protein-coupled receptors (see Chapter 2) on the surfaces of anterior pituitary cells to increase the formation of cyclic AMP or inositol-trisphosphate-diacylglyceride second messenger systems. Release of hormone almost certainly is the result of an influx of calcium, which triggers and sustains the process of exocytosis. The actions of hypophysiotropic hormones on their target cells in the pituitary are considered further in later chapters.

Feedback Control of Anterior Pituitary Function

We have already indicated that the primary drive for secretion of all of the anterior pituitary hormones except PRL is stimulation by the hypothalamic releasing hormones. In the absence of the hormones of their target glands, secretion of TSH, ACTH, and the gonadotropins gradually increases manyfold. Secretion of these pituitary hormones is subject to negative feedback inhibition by secretions of their target glands. Regulation of secretion of anterior pituitary hormones in the normal individual is achieved through the interplay of stimulatory effects of releasing hormones and inhibitory effects of target gland hormones (Fig. 5). Regulation of the secretion of pituitary hormones by hormones of target glands could be achieved equally well if negative feedback signals acted at the level of (1) the hypothalamus to inhibit secretion of hypophysiotropic hormones or (2) the pituitary gland to blunt the response to hypophysiotropic stimulation. Actually some combination of these two mechanisms applies to all of the anterior pituitary hormones except PRL.

In experimental animals it appears that secretion of GnRH is variable and highly sensitive to environmental influences, e.g., day length, or even the act of mating. In humans and other primates, secretion of GnRH after puberty appears to be somewhat less influenced by changes in the internal and external environment, but there is ample evidence that GnRH secretion is modulated by factors in both the internal and external environment. It has been shown experimentally in

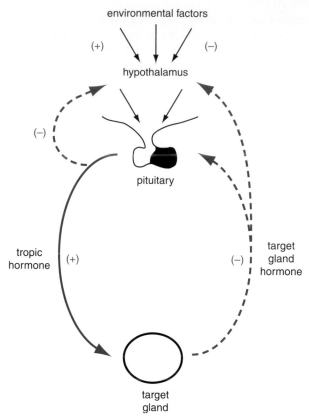

FIGURE 5 Regulation of anterior pituitary hormone secretion. Environmental factors may increase or decrease pituitary activity by increasing or decreasing hypophysiotropic hormone secretion. Pituitary secretions increase the secretion of target gland hormones, which may inhibit further secretion by acting at either the hypothalamus or the pituitary. Pituitary hormones may also inhibit their own secretion by a short feedback loop. (From Goodman HM. The pituitary gland. In Mountcastle VB, ed., *Medical physiology*, 14th ed. St. Louis: Mosby, 1980.)

rhesus monkeys and human subjects that all of the complex changes in the rates of FSH and LH secretion characteristic of the normal menstrual cycle can occur when the pituitary gland is stimulated by pulses of GnRH delivered at an invariant frequency and amplitude. For such changes in pituitary secretion to occur, changes in secretion of target gland hormones that accompany ripening of the follicle, ovulation, and luteinization must modulate the responses of gonadotropes to GnRH (see Chapter 46). However, in normal humans it is evident that target gland hormones also act at the level of the hypothalamus to regulate both the amplitude and frequency of GnRH secretory bursts.

Adrenal cortical hormones exert a negative feedback effect both on pituitary corticotropes where they decrease transcription of POMC and on hypothalamic neurons where they decrease CRH synthesis and secretion (see Chapter 40). The rate of CRH secretion is also profoundly affected by changes in both the internal and external environment. Physiologically, CRH is secreted in increased amounts in response to nonspecific stress. This effect is seen even in the absence of the adrenal glands, and hence the inhibitory effects of its hormones, indicating that CRH secretion must be controlled by positive inputs as well as the negative effects of adrenal hormones.

Control of GH secretion is more complex because it is under the influence of a releasing hormone, GHRH, probably a release-enhancing hormone, ghrelin, and a release-inhibiting hormone, somatostatin. In addition, GH secretion is under negative feedback control by products of its actions in peripheral tissues. As discussed in detail in Chapter 44, GH evokes production of a peptide called insulin-like growth factor-I (IGF-I), which mediates the growth-promoting actions of GH. IGF-I exerts powerful inhibitory effects on GH secretion by decreasing the sensitivity of somatotropes to GHRH. It also acts on the somatostatin-secreting neurons to increase the release of somatostatin and to inhibit the release of GHRH. GH is also a metabolic regulator, and products of its metabolic activity such as increased glucose or free fatty acid concentrations in blood may also inhibit its secretion.

Modulating effects of target gland hormones on the pituitary gland are not limited to inhibiting secretion of their own provocative hormones. Target gland hormones may modulate pituitary function by increasing the sensitivity of other pituitary cells to their releasing factors or by increasing the synthesis of other pituitary hormones. Hormones of the thyroid and adrenal glands are required for normal responses of the somatotropes to GHRH. Similarly, estrogen secreted by the ovary in response to FSH and LH increases PRL synthesis and secretion.

In addition to feedback inhibition exerted by target gland hormones, there is evidence that pituitary hormones may inhibit their own secretion. In this so-called *short-loop feedback system,* pituitary cells respond to increased concentrations of their own hormones by decreasing further secretion. The physiologic importance of short-loop feedback systems has not been established, nor has that of the postulated *ultrashort-loop feedback* in which high concentrations of hypophysiotropic hormones may inhibit their own release.

PHYSIOLOGY OF THE POSTERIOR PITUITARY

The posterior pituitary gland secretes two hormones. They are *oxytocin,* which means "rapid birth" in reference to its action to increase uterine contractions during parturition, and *vasopressin,* in reference to its ability to contract vascular smooth muscle and thus raise blood pressure. Because the human hormone has an arginine in position 8 instead of the lysine found in the hormone that was originally isolated from pigs, it is called *arginine vasopressin* (AVP). Both oxytocin and AVP are nonapeptides and differ from each other by only two amino acid residues. Similarities in the structures of their genes and in their post-translational processing make it virtually certain that these hormones evolved from a single ancestral gene. The genes that encode them occupy adjacent loci on chromosome 20, but in opposite transcriptional orientation.

Each of the posterior pituitary hormones has other actions in addition to those for which it was named. Oxytocin also causes contraction of the myoepithelial cells that envelop the secretory alveoli of the mammary glands and thus enables the suckling infant to receive milk. AVP is also called *antidiuretic hormone* (ADH) for its action to promote reabsorption of "free water" by renal tubules (see Chapter 28). These two effects are mediated by different heptihelical receptors that are coupled to different G-protein-dependent second messenger systems. V_1 receptors signal vascular muscle contraction by means of the inositol-trisphosphate-diacylglycerol pathway (Chapter 2), whereas V_2 receptors utilize the cyclic AMP system to produce the antidiuretic effect in renal tubules. Oxytocin acts through a single class of G-protein receptors that signals through the inositol-trisphosphate-diacylglycerol pathway. Physiologic actions of these hormones are considered further in Chapters 28 and 47.

Oxytocin and AVP are stored in and secreted by the neurohypophysis, but are synthesized in magnocellular neurons whose cell bodies are present in both the *supraoptic* and *paraventricular nuclei* of the hypothalamus. Cells in the supraoptic nuclei appear to be the major source of neurohypophysial AVP, whereas cells in the paraventricular nuclei may be the principal source of oxytocin. After transfer to the Golgi apparatus, the oxytocin and AVP prohormones are packaged in secretory vesicles along with the enzymes that cleave them into the final secreted products. The secretory vesicles are then transported down the axons to the nerve terminals in the posterior gland where they are stored in relatively large amounts. It has been estimated that sufficient AVP is stored in the neurohypophysis to provide for 30–50 days of secretion at basal rates or 5–7 days at maximal rates of secretion. Oxytocin and AVP are stored as 1:1 complexes with 93–95 residue peptides called *neurophysins,* which are actually adjacent segments of their prohormone molecules. The neurophysins are co-secreted with AVP or oxytocin, but have no known hormonal actions. The neurophysins, however, play an essential role in the post-translational processing of the neurohypophysial hormones. The

amino acid sequence of the central portion of the neurophysins is highly conserved across many vertebrate species, and mutations in this region of the preprohormone are responsible for hereditary deficiencies in AVP that produce the disease *diabetes insipidus* (see Chapter 28) even though expression of the AVP portion of the preprohormone is normal.

As already discussed, AVP is also synthesized in small cells of the paraventricular nuclei and is delivered by the hypophysial portal capillaries to the anterior pituitary gland where it plays a role in regulating ACTH secretion. AVP is produced in considerably larger amounts in the magnocellular neurons and is carried directly into the general circulation by the veins that drain the posterior lobe. It is unlikely that AVP that originates in magnocellular neurons acts as a hypophysiotropic hormone, but it can reach the corticotropes and stimulate ACTH secretion when its concentration in the general circulation increases sufficiently. Oxytocin, like AVP, may also be synthesized in parvocellular neurons at other sites in the nervous system and be released from axon terminals that project to a wide range of sites within the central nervous system. Oxytocin may also be produced in some reproductive tissues where it acts as a paracrine factor.

REGULATION OF POSTERIOR PITUITARY FUNCTION

Because the hormones of the posterior pituitary gland are synthesized and stored in nerve cells, it should not be surprising that their secretion is controlled in the same way as that of more conventional neurotransmitters. Action potentials that arise from synaptic input to the cell bodies within the hypothalamus course down the axons in the pituitary stalk, trigger an influx of calcium into nerve terminals, and release the contents of neurosecretory granules. Vasopressin and oxytocin are released along with their respective neurophysins, other segments of the precursor molecule, and presumably the enzymes responsible for cleavage of the precursor. Tight binding of AVP or oxytocin to their respective neurophysins is favored by the acidic pH of the secretory granule, but upon secretion, the higher pH of the extracellular environment allows the hormones to dissociate from their neurophysins and circulate in an unbound form. Oxytocin and vasopressin are rapidly cleared from the blood with a half-life of about 2 min.

As discussed in Chapter 37, signals for the secretion of oxytocin originate in the periphery and are transmitted to the brain by sensory neurons. After appropriate processing in higher centers, cells in the supraoptic and paraventricular nuclei are signaled to release their

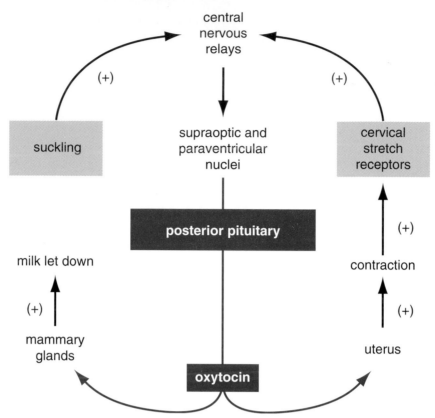

FIGURE 6   Regulation of oxytocin secretion showing a positive feedback arrangement. Oxytocin stimulates the uterus to contract and causes the cervix to stretch. Increased cervical stretch is sensed by neurons in the cervix and transmitted to the hypothalamus, which signals more oxytocin secretion. Oxytocin secreted in response to suckling forms an open-loop feedback system in which positive input is interrupted when the infant is satisfied and stops suckling. Further details are given in Chapter 47.

hormone from nerve terminals in the posterior pituitary gland (Fig. 6). The importance of neural input to oxytocin secretion is underscored by the observation that it may also be secreted in a conditioned reflex. A nursing mother sometimes releases oxytocin in response to cries of her baby even before the infant begins to suckle. Oxytocin is also secreted at the same low basal rate in both men and women, but no physiologic role for oxytocin in men and nonlactating women has been established.

Signals for AVP secretion in response to increased osmolality of the blood are thought to originate in hypothalamic neurons and possibly in the AVP secretory cells themselves. Osmoreceptor cells are exquisitely sensitive to small increases in osmolality and signal cells in the supraoptic and paraventricular nuclei to secrete AVP. AVP is also secreted in response to decreased blood volume. Although the specific cells responsible for monitoring blood volume have not been identified, volume monitors appear to be located within the thorax and relay their information to the central nervous system in afferent neurons in the vagus

nerves. The control of AVP secretion is shown in Fig. 7 and is discussed more fully in Chapter 28. Under basal conditions blood levels of AVP fluctuate in a diurnal rhythmic pattern that closely resembles that of ACTH.

Suggested Reading

Andersen B, Rosenfeld MG. POU domain factors in the neuroendocrine system: Lessons from developmental biology provide insights into human disease. *Endocr Rev* 2001;22:2–35.

Eipper BA, Mains RE. Structure and biosynthesis of Pro-ACTH/endorphin and related peptides. *Endocr Rev* 1980;1:1–27.

Gharib SD, Wierman ME, Shupnik, MA, Chin WW. Molecular biology of the pituitary gonadotropins. *Endocr Rev* 1990;11: 177–199.

Ling N, Zeytin F, Böhlen P, Esch F, Brazeau P, Wehrenberg WB, Baird A, Guillemin R. Growth hormone releasing factors. *Annu Rev Biochem* 1985;54:404–424.

Miller WL, Eberhardt NL. Structure and evolution of the growth hormone gene family. *Endocr Rev* 1983;4:97–130.

Osorio M, Kopp P, Marui S, Latronico AC, Mendonca BB, Arnhold IJ. Combined pituitary hormone deficiency caused by a novel mutation

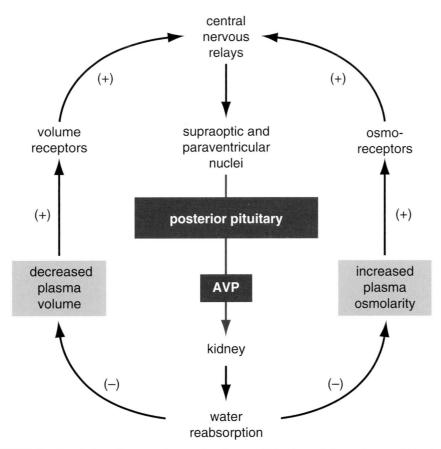

FIGURE 7   Regulation of vasopressin secretion. Increased blood osmolality or decreased blood volume are sensed in the brain or thorax, respectively, and increase vasopressin secretion. Vasopressin, acting principally on the kidney produces changes that restore osmolality and volume, thereby shutting down further secretion in a negative feedback arrangement.

of a highly conserved residue (F88S) in the homeodomain of PROP-1. *J Clin Endocrinol Metab* 2000;85:2779–2785.

Palkovits M. Interconnections between the neuroendocrine hypothalamus and the central autonomic system. *Frontiers Neuroendocrinol* 1999;20:270–295.

Shupnik MA, Ridgway EC, Chin WW. Molecular biology of thyrotropin. *Endocr Rev* 1989;10:459–475.

Vale W, Rivier C, Brown MR, Spiess J, Koob G, Swanson L, Bilezikjian L, Bloom F, Rivier J. Chemical and biological characterization of corticotropin releasing factor. *Rec Prog Horm Res* 1984;40:245–270.

Vitt UA, Hsu SY, Hsueh AJ. Evolution and classification of cystine knot-containing hormones and related extracellular signaling molecules. *Mol Endocrinol.* 2001;15:681–694.

39

Thyroid Gland

H. MAURICE GOODMAN

KEY POINTS

- The thyroid hormones thyroxine (T_4) and triiodothyronine (T_3) are α-amino acids derived from the condensation of two tyrosine molecules and contain two atoms of iodine on the inner ring and one (T_3) or two (T_4) iodine atoms on the outer ring.
- *Biosynthetic reactions* include uptake, peroxidation, and incorporation of iodine into tyrosine residues that are in peptide linkage within the large glycoprotein, thyroglobulin, followed by peroxidase-catalyzed coupling of nearby tyrosine residues.
- *Secretion* involves endocytosis of thyroglobulin from the follicular lumen followed by proteolysis in lysosomes to release the T_4 and T_3. Synthesis and secretion are regulated by the

thyroid-stimulating hormone (TSH) of the pituitary gland.
- Thyroid hormones circulate bound to plasma proteins and have long half-lives.
- T_3 and T_4 promote growth largely through interactions with growth hormone. Both are required for normal development of the central nervous system (CNS) during the perinatal period, and both increase sensitivity to sympathetic stimulation.
- T_3 and T_4 increase oxygen consumption and the basal metabolic rate and accelerate virtually all aspects of carbohydrate, lipid, and protein metabolism.
- In the absence of T_3 and T_4, defective cholesterol excretion leads to hypercholesterolemia, and some defect in protein metabolism leads

OVERVIEW

In the adult human, normal operation of a wide variety of physiologic processes affecting virtually every organ system requires appropriate amounts of thyroid hormone. Governing all of these processes, thyroid hormone acts as a modulator, or gain control, rather than an all-or-none signal that turns the process on or off. In the immature individual, thyroid hormone plays an indispensable role in growth and development. Its presence in optimal amounts at a critical time is an absolute requirement for normal development of the nervous system. In its role in growth and development too, its presence seems to be required for the normal unfolding of processes whose course it modulates but does not initiate. Because thyroid hormone affects virtually every system in the body in this way, it is difficult to give a simple, concise answer to the naive but profound question: What does thyroid hormone do? The response of most endocrinologists would be couched in terms of consequences of hormone excess or deficiency. Indeed, deranged function of the thyroid gland is among the most prevalent of endocrine diseases and may affect as many as 4–5% of the population in the United States. In regions of the world where the trace element iodine is scarce, the incidence of deranged thyroid function may be even higher.

MORPHOLOGY

The human thyroid gland is located at the base of the neck and wraps around the trachea just below the cricoid cartilage (Fig. 1). The two large lateral lobes that comprise the bulk of the gland lie on either side of the trachea and are connected by a thin isthmus. A third structure, the pyramidal lobe, which may be a remnant of the embryonic thyroglossal duct, is sometimes also seen as a finger-like projection extending headward from the isthmus. The thyroid gland in the normal human being weighs about 20 g but is capable of enormous growth, sometimes achieving a weight of several hundred grams when stimulated intensely over a long period of time. Such enlargement of the thyroid gland, which may be grossly obvious, is called a *goiter* and is one of the most common manifestations of thyroid disease.

The thyroid gland receives its blood supply through the inferior and superior thyroid arteries, which arise from the external carotid and subclavian arteries. Relative to its weight, the thyroid gland receives a greater flow of blood than most other tissues of the body. Venous drainage is through the paired superior, middle, and inferior thyroid veins into the internal jugular and innominate veins. The gland is also endowed with a rich lymphatic system that may play an important role in delivery of hormone to the general circulation. The thyroid gland also has an abundant supply of sympathetic and parasympathetic nerves. Some studies suggest that sympathetic stimulation or infusion of epinephrine or norepinephrine may increase secretion of thyroid hormone, but it is probably only of minor importance in the overall regulation of thyroid function.

The functional unit of the thyroid gland is the *follicle,* which is composed of epithelial cells arranged as hollow vesicles of various shapes ranging in size from 0.02–0.3 mm in diameter; it is filled with a glycoprotein colloid called *thyroglobulin* (Fig. 2). The adult human thyroid gland has about three million follicles. Epithelial cells lining each follicle may be cuboidal or columnar, depending on their functional state, with the height of the epithelium being greatest when its activity is highest. Each follicle is surrounded by a dense capillary network separated from epithelial cells by a well-defined basement membrane. Groups of densely packed follicles are bound together by connective tissue septa

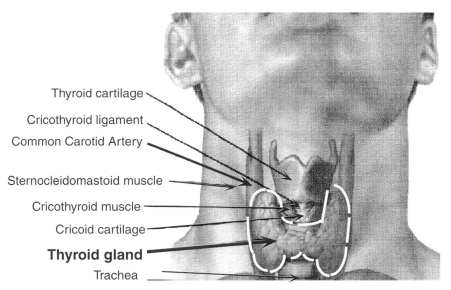

Thyroid cartilage

Cricothyroid ligament

Common Carotid Artery

Sternocleidomastoid muscle

Cricothyroid muscle

Cricoid cartilage

Thyroid gland

Trachea

FIGURE 1   Gross anatomy of the thyroid gland. (From Netter FH, *Atlas of human anatomy*, 2nd ed., Summit, N.J.: Novartis, 1989. Reprinted with permission from ICON Learning Systems, LLC, a subsidiary of Multimedia USA, Inc.)

to form lobules that receive their blood supply from a single small artery. The functional state of one lobule may differ widely from that of an adjacent lobule. Secretory cells of the thyroid gland are derived embryologically and phylogenetically from two sources. Follicular cells, which produce the classical thyroid hormones, thyroxine and triiodothyronine, arise from endoderm of the primitive pharynx. Parafollicular, or C cells, are located between the follicles and produce the polypeptide hormone calcitonin, which is discussed in Chapter 43.

THYROID HORMONES

The thyroid hormones are α-amino acid derivatives of tyrosine (Fig. 3). The thyronine nucleus consists of two benzene rings in ether linkage, with an alanine side chain

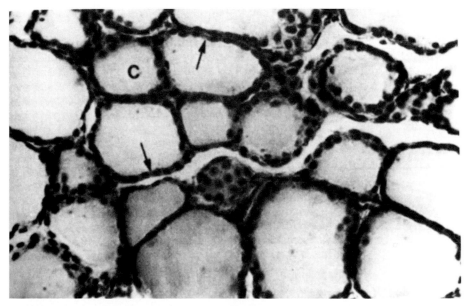

FIGURE 2   Histology of the human thyroid. Simple cuboidal cells (arrows) make up the follicles. C, thyroid colloid (thyroglobulin), which fills the follicles. (From Borysenko M, Beringer T. *Functional histology*, Boston: Little, Brown, 1979, p. 312.)

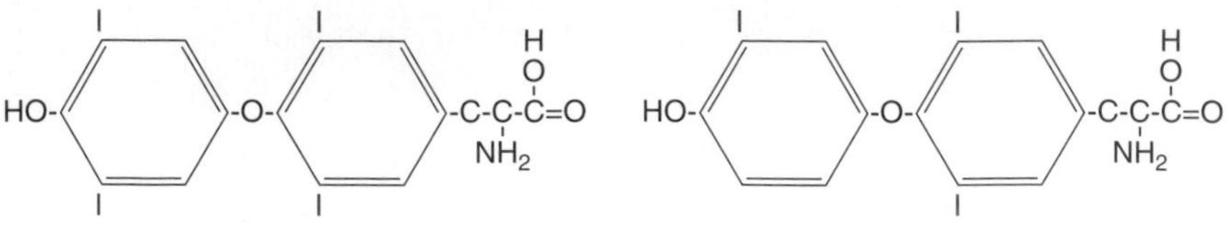

tetraiodothyronine
thyroxine
T4

triiodothyronine
T3

FIGURE 3   Thyroid hormones.

in the para position on the inner or tyrosyl ring and a hydroxyl group in the para position in the outer or phenolic ring. *Thyroxine* was the first thyroid hormone to be isolated and characterized. Its name derives from *thyr*oid *oxyin*dole, which describes the chemical structure erroneously assigned to it in 1914. *Triiodothyronine, a* considerably less abundant, but three times more potent hormone than thyroxine in most assay systems, was not discovered until 1953. Both hormone molecules are exceptionally rich in iodine, which comprises more than half of their molecular weight. Thyroxine contains four atoms of iodine and is abbreviated as T_4 while triiodothyronine, which has three atoms of iodine is abbreviated as T_3.

Biosynthesis

Several aspects of the production of thyroid hormone are unusual: (1) Thyroid hormones contain large amounts of iodine. Biosynthesis of active hormone requires adequate amounts of this scarce element. This need is met by an efficient energy-dependent mechanism that allows thyroid cells to take up and concentrate iodide. The thyroid gland is also the principal site of storage of this rare dietary constituent. (2) Thyroid hormones are partially synthesized extracellularly at the luminal surface of follicular cells and stored in an extracellular compartment, the follicular lumen. (3) The hormone therefore is doubly secreted, in that the precursor molecule, *thyroglobulin,* is released from apical surfaces of follicular cells into the follicular lumen, only to be taken up again by follicular cells and degraded to release T_4 and T_3, which are then secreted into the blood from the basal surfaces of follicular cells. (4) Thyroxine, the major secretory product, is not the biologically active form of the hormone, but must be transformed to T_3 at extrathyroidal sites.

Biosynthesis of thyroid hormones can be considered as the sum of several discrete processes (Fig. 4), all of which depend on the products of three genes that are expressed predominantly, if not exclusively, in thyroid follicle cells: the sodium iodide symporter (NIS), thyroglobulin, and thyroid peroxidase.

Iodine Trapping

Under normal circumstances iodide is about 25–50 times more concentrated in the cytosol of thyroid follicular cells than in blood plasma, and during periods of active stimulation, it may be as high as 250 times that of plasma. Iodine is accumulated against a steep concentration gradient by the action of an electrogenic "iodide pump" located in the basolateral membranes. The pump is actually a sodium iodide symporter that couples the transfer of two ions of sodium with each ion of iodide. Iodide is thus transported against its concentration driven by the favorable electrochemical gradient for sodium. Energy is expended by the sodium potassium ATPase (the sodium pump), which then exchanges two ions of sodium for three ions of potassium to maintain the electrochemical gradient for sodium. Outward diffusion of potassium maintains the membrane potential. Like other transporters, the sodium iodide symporter has a finite capacity and can be saturated. Consequently, other anions, e.g., perchlorate, pertechnetate, and thiocyanate, that compete for binding sites on the sodium iodide symporter can block the uptake of iodide. This property can be exploited for diagnostic or therapeutic purposes.

Thyroglobulin Synthesis

Thyroglobulin is the other major component needed for synthesis of thyroxine and triiodothyronine. Thyroglobulin is the matrix for thyroid hormone synthesis and is the form in which hormone is stored in the gland. It is a large glycoprotein that forms a stable dimer with a molecular mass of about 660,000 Da. Like other secretory proteins, thyroglobulin is synthesized on ribosomes, glycosylated in the cisternae of the endoplasmic reticulum, translocated to the Golgi apparatus, and packaged in secretory vesicles that discharge it from the apical surface into the lumen. Because thyroglobulin secretion into the lumen is coupled with its synthesis, follicular cells do not have the extensive accumulation of secretory granules that is characteristic of protein-secreting cells. Iodination to form mature thyroglobulin

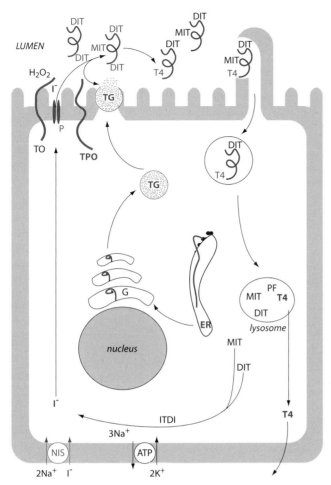

FIGURE 4   Thyroid hormone biosynthesis and secretion. Iodide(I^-) is transported into the thyroid follicular cell by the sodium iodide symporter (NIS) in the basal membrane and diffuses passively into the lumen through the iodide channel called pendrin (P). Thyroglobulin (TG) is synthesized by the rough endoplasmic reticulum (ER), processed in the ER and the Golgi (G), where it is packaged into secretory granules and released into the follicular lumen. In the presence of hydrogen peroxide (H_2O_2) produced in the lumenal membrane by thyroid oxidase (TO), the thyroid peroxidase (TPO) oxidizes iodide, which reacts with tyrosine residues in TG to produce monoiodotyrosyl (MIT) and diiodotyrosyl (DIT) residues within the TG. The TPO reaction also catalyzes the coupling of iodotyrosines to form thyroxine (T_4) and some triiodothyronine (T_3, not shown) residues within the TG. Secretion of T_4 begins with phagocytosis of TG, fusion of TG-laden endosomes with lysosomes, and proteolytic digestion to peptide fragments (PF), MIT, DIT, and T_4, which leaves the cell at the basal membrane. MIT and DIT are deiodinated by iodotyrosine deiodinase (ITDI) and recycled.

does not take place until after the thyroglobulin is discharged into the lumen.

Incorporation of Iodine

Iodide that enters at the basolateral surfaces of the follicular cell must be delivered to the follicular lumen

where hormone biosynthesis takes place. Iodide diffuses throughout the follicular cell and exits from the apical membrane by way of a sodium-independent iodide transporter called *pendrin* that is also expressed in brain and kidney. In order for iodide to be incorporated into tyrosine residues in thyroglobulin, it must first be converted to some higher oxidized state. This step is catalyzed by the thyroid-specific *thyroperoxidase* in the presence of hydrogen peroxide, whose formation may be rate limiting. Hydrogen peroxide is generated by the catalytic activity of a calcium-dependent NADPH oxidase that is present in the brush border. Thyroperoxidase is the key enzyme in thyroid hormone formation and is thought to catalyze the iodination and coupling reactions described later as well as the activation of iodide. Thyroperoxidase spans the brush border membrane on the apical surface of follicular cells and is oriented such that its catalytic domain faces the follicular lumen.

Addition of iodine molecules to tyrosine residues in thyroglobulin is called *organification*. Thyroglobulin is iodinated at the apical surface of follicular cells as it is extruded into the follicular lumen. Iodide acceptor sites in thyroglobulin are in sufficient excess over the availability of iodide that no free iodide accumulates in the follicular lumen. Although post-translational conformational changes orchestrated by endoplasmic reticular proteins organize the configuration of thyroglobulin to increase its ability to be iodinated, iodination and hormone formation do not appear to be particularly efficient. Tyrosine is not especially abundant in thyroglobulin and comprises only about 1 in 40 residues of the peptide chain. Only about 10% of the 132 tyrosine residues in each thyroglobulin dimer appear to be in positions favorable for iodination. The initial products formed are *monoiodotyrosine* (MIT) and *diiodotyrosine* (DIT), and they remain in peptide linkage within the thyroglobulin molecules. Normally more DIT is formed than MIT, but when iodine is scarce there is less iodination and the ratio of MIT to DIT is reversed.

Coupling

The final stage of thyroxine biosynthesis is the coupling of two molecules of DIT to form T_4 within the peptide chain. This reaction is also catalyzed by thyroperoxidase. Only about 20% of iodinated tyrosine residues undergo coupling, with the rest remaining as MIT and DIT. After coupling is complete, each thyroglobulin molecule normally contains one to three molecules of T_4. T_3 is considerably scarcer, with one molecule being present in only 20–30% of thyroglobulin molecules. T_3 may be formed by deiodination of T_4 or coupling of one residue of DIT with one of MIT.

Exactly how coupling is achieved is not known. One possible mechanism involves joining two iodo-tyrosine residues that are in proximity to each other on either two separate strands of thyroglobulin or adjacent folds of the same strand. Free radicals formed by the action of thyroperoxidase react to form the ether linkage at the heart of the thyronine nucleus, leaving behind in one of the peptide chains the serine or alanine residue that was once attached to the phenyl group that now comprises the outer ring of T_4 (Fig. 5). An alternative mechanism involves coupling a free diiodophenyl-pyruvate (deaminated DIT) with a molecule of DIT in peptide linkage within the thyroglobulin molecule by a similar reaction sequence. Regardless of which model proves correct, it is sufficient to recognize the central importance of thyroperoxidase for formation of the thyronine nucleus as well as iodination of tyrosine residues. In addition, the mature hormone is formed while in peptide linkage within the thyroglobulin molecule, and remains a part of that large storage molecule until lysosomal enzymes set it free during the secretory process.

Hormone Storage

The thyroid is unique among endocrine glands in that it stores its product extracellularly in follicular lumens as large precursor molecules. In the normal individual, approximately 30% of the mass of the thyroid gland is thyroglobulin, which corresponds to about 2–3 months' supply of hormone. Mature thyroglobulin is a high molecular weight (660,000 Da) molecule, probably a dimer of the thyroglobulin precursor peptide, and contains about 10% carbohydrate and about 0.5% iodine. The tyrosine residues that are situated just a few amino acids away from the C and N termini are the principal sites of iodothyronine formation. MIT and DIT at other sites in thyroglobulin comprise an important reservoir for iodine and constitute about 90% of the total pool of iodine in the body.

Secretion

Thyroglobulin stored within follicular lumens is separated from extracellular fluid and the capillary endothelium by a virtually impenetrable layer of follicular cells. In order for secretion to occur, thyroglobulin must be brought back into follicular cells by a process of endocytosis. On acute stimulation with TSH, long strands of protoplasm (pseudopodia) reach out from the apical surfaces of follicular cells to surround chunks of thyroglobulin, which are taken up in endocytic

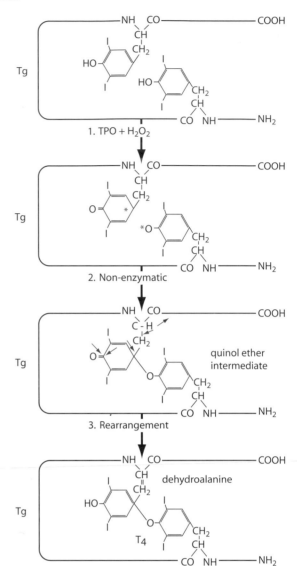

FIGURE 5   Hypothetical coupling scheme for intramolecular formation of T_4 based on model reaction with purified thyroid peroxidase. (From Taurog A. in Braverman LE, Utiger RD, eds., *Werner's the thyroid*, 8th ed., Philadelphia: Lippincott Williams and Wilkins, p. 71.)

vesicles (Fig. 6). In chronic situations uptake is probably less dramatic than that shown in Fig. 6, but nevertheless requires an ongoing endocytic process. The endocytic vesicles migrate toward the basal portion of the cells and fuse with lysosomes, which simultaneously migrate from the basal to the apical region of the cells to meet the incoming endocytic vesicles. As fused lysoendosomes migrate toward the basement membrane, thyroglobulin is degraded to peptide fragments and free amino acids, including T_4, T_3, MIT, and DIT. Of these, only T_4 and T_3 are released into the bloodstream, in a ratio of about 20:1, perhaps by a process of simple diffusion down a concentration gradient.

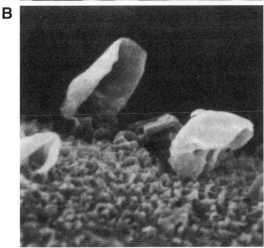

FIGURE 6 Scanning electron micrographs of the luminal microvilli of dog thyroid follicular cells. (A) TSH secretion suppressed by feeding thyroid hormone. (B) At 1 hr after TSH. A, 36,000×; B, 16,500×. (From Balasse PD, Rodesch FR, Neve PE, et al. Observations en microscopie a balayage de la surface apicale y de cellules folliculares thyroidiennes chez le chien. *C Roy Acad Sci [D] (Paris),* 1972;274:2332.)

Monoiodotyrosine and DIT cannot be utilized for synthesis of thyroglobulin and are rapidly deiodinated by a specific microsomal deiodinase. Virtually all of the iodide released from iodotyrosines is recycled into thyroglobulin. Deiodination of iodotyrosine provides about twice as much iodide for hormone synthesis as the iodide pump and is therefore of great significance in hormone biosynthesis. Patients who are genetically deficient in thyroidal tyrosine deiodinase readily suffer symptoms of iodine deficiency and excrete MIT and DIT in their urine. Normally, virtually no MIT or DIT escape from the gland.

Synthesis of thyroglobulin and its export in vesicles into the follicular lumen is an ongoing process that takes place simultaneously with uptake of thyroglobulin in other vesicles moving in the opposite direction. These opposite processes, involving vesicles laden with thyroglobulin moving into and out of the cells, are somehow regulated so that under normal circumstances thyroglobulin neither accumulates in the cells or follicular lumens nor is depleted. The physiologic mechanisms for such traffic control are not yet understood.

REGULATION OF THYROID FUNCTION

Effects of Thyroid-Stimulating Hormone

Although the thyroid gland can carry out all of the steps of hormone biosynthesis, storage, and secretion in the absence of any external signals, autonomous function is far too sluggish to meet bodily needs for thyroid hormone. The principal regulator of thyroid function is the thyroid-stimulating hormone (TSH), which is secreted by thyrotropes in the pituitary gland (see Chapter 38). It may be recalled that TSH consists of two glycosylated peptide subunits including the same β-subunit that is also found in FSH, and LH (Chap. 38). The β-subunit is the part of the hormone that confers thyroid-specific stimulating activity, but free β-subunits are inactive, and stimulate the thyroid only when linked to α-subunits in a complex three dimensional configuration.

Thyroid-stimulating hormone binds to a single class of heptihelical G protein-coupled receptors (see Chapter 2) in the basolateral surface membranes of thyroid follicular cells. The TSH receptor is the product of a single gene, but it is comprised of two subunits held together by a disulfide bond. It appears that after the molecule has been properly folded and its disulfide bonds formed, a loop of about 50 amino acids is excised proteolytically from the extracellular portion of the receptor. The α-subunit includes about 300 residues at the amino terminus and contains most of the TSH binding surfaces. The β-subunit contains the seven-membrane-spanning α-helices and the short carboxyl-terminal tail in the cytoplasm. Reduction of the disulfide bond may lead to release of the α-subunit into the extracellular fluid, and may have important implications for the development of antibodies to the TSH receptor and thyroid disease (see below). Binding of TSH to the receptor results in activation of both adenylyl cyclase through $G\alpha_s$ and phospholipase C through $G\alpha_q$ and leads to increases in both the cyclic AMP and diacylglycerol/IP$_3$ second messenger pathways (see Chapter 2). Activation of the cyclic AMP pathway appears to be the more important transduction mechanism because all of the known effects of TSH can be duplicated by cyclic AMP. Because TSH increases cyclic AMP production at much lower concentrations than are needed to increase phospholipid turnover, it is likely that IP$_3$ and DAG are redundant mediators that reinforce the effects of cyclic AMP at times of intense stimulation, but it is also possible that these second messengers signal some unique responses. Increased turnover of phospholipid is associated with release of arachidonic acid and the consequent increased production of prostaglandins that also follows TSH stimulation of the thyroid.

In addition to regulating all aspects of hormone biosynthesis and secretion, TSH increases blood flow to the thyroid, and with prolonged stimulation TSH also increases the height of the follicular epithelium (hypertrophy) and can stimulate division of follicular cells (hyperplasia). Stimulation of thyroid follicular cells by TSH is a good example of a *pleiotropic* effect of a hormone in which multiple separate but complementary actions sum to produce an overall response. Each step of hormone biosynthesis, storage, and secretion appears to be directly stimulated by a cyclic AMP-dependent process that is accelerated independently of the preceding or following steps in the pathway. Thus, even when increased iodide transport is blocked with a drug that specifically affects the iodide pump, TSH nevertheless accelerates the remaining steps in the synthetic and secretory process. Similarly, when iodination of tyrosine is blocked by a drug specific for the organification process, TSH still stimulates iodide transport and thyroglobulin synthesis.

Most of the responses to TSH depend on activation of protein kinase A and the resultant phosphorylation of proteins including transcription factors such as CREB (cyclic AMP response element-binding protein; see Chapter 2). TSH increases expression of the genes for the sodium iodide symporter, thyroglobulin, thyroid oxidase, and thyroid peroxidase. These effects are exerted through cooperative interactions of TSH-activated nuclear proteins with thyroid-specific transcription factors whose expression is also enhanced by TSH. TSH appears to increase blood flow by activating the gene for nitric oxide synthase, which increases production of the potent vasodilator, nitric oxide, and by inducing expression of paracrine factors that promote capillary growth (angiogenesis). Precisely how TSH increases thyroid growth is not understood, but it is apparent that synthesis and secretion of a variety of local growth factors is induced.

Autoregulation of Thyroid Hormone Synthesis

Although production of thyroid hormones is severely impaired when too little iodide is available, iodide uptake and hormone biosynthesis are temporarily blocked when the concentration of iodide in blood plasma becomes too high. This effect of iodide has been exploited clinically to produce short-term suppression of thyroid hormone secretion. This inhibitory effect of iodide apparently depends on its being incorporated into some organic molecule and is thought to represent an autoregulatory phenomenon that protects against overproduction of thyroxine. Blockade of thyroid hormone production is short lived, and the gland eventually "escapes" from the inhibitory effects of iodide by mechanisms that include down-regulation of the sodium iodide symporter.

Biosynthetic activity of the thyroid gland may also be regulated by the thyroglobulin that accumulates in the follicular lumen. Evidence has been presented that thyroglobulin, acting through a receptor on the apical

---

Clinical Note

Effects of the Thyroid-Stimulating Immunoglobulins Overproduction of thyroid hormone, *hyperthyroidism,* which is also known as Graves' disease, is usually accompanied by extremely low concentrations of TSH in blood plasma, yet the thyroid gland gives every indication of being under intense stimulation. This paradox was resolved when it was found that blood plasma of affected individuals contains a substance that stimulates the thyroid gland to produce and secrete thyroid hormone. This substance is an immunoglobulin secreted by lymphocytes and is almost certainly an antibody to the TSH receptor. *Thyroid-stimulating immunoglobulin* (TSI) can be found in the serum of virtually all patients with Graves' disease, suggesting an autoimmune etiology to this disorder. It is of interest to note that when reacting with the TSH receptor, antibodies trigger the same sequence of responses that are produced when TSH interacts with the receptor. This fact indicates that all information needed to produce the characteristic cellular response to TSH resides in the receptor rather than the hormone. The role of the hormone therefore must be limited to activation of the receptor. Similar effects have also been seen with antibodies to receptors for other hormones.

surface of follicular cells, decreases the expression of thyroid-specific transcription factors and thereby decreases expression of the genes for thyroglobulin, the thyroid peroxidase, the sodium iodide symporter, and the TSH receptor. Further effects of thyroglobulin include increased transcription of pendrin, which delivers iodide from the follicular cell to the lumen. Thus thyroglobulin may have significant effects in regulating its own synthesis and may temper the stimulatory effects of TSH, which remains the primary—and most important—regulator of thyroid function.

THYROID HORMONES IN BLOOD

More than 99% of thyroid hormone circulating in blood is firmly bound to three plasma proteins. They are *thyroxine-binding globulin* (TBG), *transthyretin* (TTR), and albumin. Of these, TBG is quantitatively the most important and accounts for more than 70% of the total protein-bound hormone (both T_4 and T). About 10–15% of circulating T_4 and 10% of circulating T_3 is bound to TTR and nearly equal amounts are bound to albumin. TBG carries the bulk of the hormone even though its concentration in plasma is only 6% of that of TTR and less than 0.1% of that of albumin because its affinity for both T_4 and T_3 is so much higher than that of the other proteins. All three thyroid hormone binding proteins bind T_4 at least 10 times more avidly than T_3. All are large enough to escape filtration by the renal glomerular membranes and little crosses the capillary endothelium. The less than 1% of hormone present in free solution is in equilibrium with bound hormone and is the only hormone that can escape from capillaries to produce biological activity or be acted on by tissue enzymes.

The total amount of thyroid hormone bound to plasma proteins represents about three times as much hormone as is secreted and degraded in the course of a single day. Thus plasma proteins provide a substantial reservoir of extrathyroidal hormone. We should therefore not expect acute increases or decreases in the rate of secretion of thyroid hormones to bring about large or rapid changes in circulating concentrations of thyroid hormones. For example, if the rate of thyroxine secretion were doubled for 1 day, we could expect its concentration in blood to increase by no more than 30% even if there were no accompanying increase in the rate of hormone degradation. A 10-fold increase in the rate of secretion lasting for 60 min would only give a 12% increase in total circulating thyroxine, and if thyroxine secretion stopped completely for 1 hr, its concentration would decrease by only 1%. Furthermore, because the binding capacity of plasma proteins for thyroid hormones is far from saturated, even a massive increase in secretion rate would have little effect on the percentage of hormone that is unbound. These considerations seem to rule out changes in thyroid hormone secretion as effectors of minute-to-minute regulation of any homeostatic process. On the other hand, because so much of the circulating hormone is bound to plasma binding proteins, we might expect that the total amount of T_4 and T_3 in the circulation would be affected significantly by decreases in the concentration of plasma binding proteins, as might occur with liver or kidney disease.

METABOLISM OF THYROID HORMONES

Because T_4 is bound much more tightly by plasma proteins then T_3, a greater fraction of T_3 is free to diffuse out of the vascular compartment and into cells where it can produce its biological effects or be degraded. Consequently, it is not surprising that the half-time for disappearance of an administered dose of ^{125}I-labeled T_3 is only one-sixth of that for T_4, or that the lag time needed to observe effects of T_3 is considerably shorter than that needed for T_4. However, because of the binding proteins, both T_4 and T_3 have unusually long half-lives in plasma, measured in days rather than seconds or minutes (Fig. 7). It is noteworthy that the half-lives of T_3 and T_4 are increased with thyroid deficiency and shortened with hyperthyroidism.

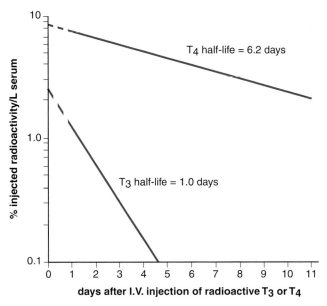

FIGURE 7   Rate of loss of serum radioactivity after injection of labeled thyroxine or triiodothyronine into human subjects. (Plotted from data of Nicoloff JD, Low JC, Dussault JH, et al. Simultaneous measurement of thyroxine and triiodothyronine peripheral turnover kinetics in man. *J Clin Invest* 1972;51:473.)

Although the main secretory product of the thyroid gland and the major form of thyroid hormone present in the circulating plasma reservoir is T_4, abundant evidence indicates that it is T_3 and not T_4 that binds to the thyroid hormone receptor (see later discussion). In fact, T_4 can be considered to be a prohormone that serves as the precursor for extrathyroidal formation of T_3.

Observations in human subjects confirm that T_3 is actually formed extrathyroidally and can account for most of the biological activity of the thyroid gland. Thyroidectomized subjects given pure T_4 in physiologic amounts have normal amounts of T_3 in their circulation. Furthermore, the rate of metabolism of T_3 in normal subjects is such that about 30 μg of T_3 is replaced daily, even though the thyroid gland secretes only 5 μg each day. Thus, nearly 85% of the T_3 that turns over each day must be formed by deiodination of T_4 in extrathyroidal

tissues. This extrathyroidal formation of T_3 consumes about 35% of the T_4 secreted each day. The remainder is degraded to inactive metabolites.

Extrathyroidal metabolism of T_4 centers around selective and sequential removal of iodine from the thyronine nucleus catalyzed by three different enzymes called deiodinases (Fig. 8). The type I deiodinase is expressed mainly in the liver and kidney, but is also found in the central nervous system, the anterior pituitary gland, and the thyroid gland itself. The type I deiodinase is a membrane-bound enzyme with its catalytic domain oriented to face the cytoplasm. Despite its intracellular location, however, T_3 formed by deiodination, especially in the liver and kidney, readily escapes into the circulation and accounts for about 80% of the T_3 in blood. The type I deiodinase can remove an iodine molecule either from the outer (phenolic) ring of

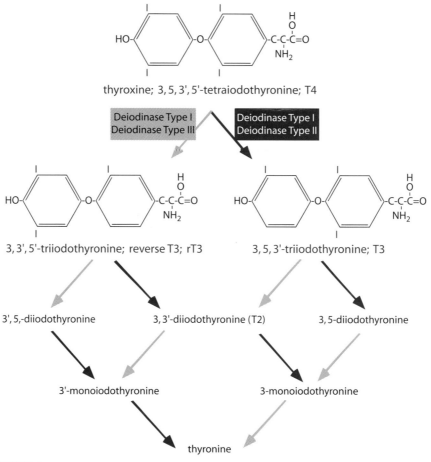

FIGURE 8   Metabolism of thyroxine. About 90% of thyroxine is metabolized by sequential deiodination catalyzed by deiodinases (types I, II and III); the first step removes an iodine from either the phenolic or tyrosyl ring producing an active (T_3) or an inactive (rT_3) compound. Subsequent deiodinations continue until all of the iodine is recovered from the thyronine nucleus. Dark blue arrows designate deiodination of the phenolic ring and light blue arrows indicate deiodination of the tyrosyl ring. Less than 10% of thyroxine is metabolized by shortening the alanine side chain prior to deiodination.

T_4, or from the inner (tyrosyl) ring. Iodines in the phenolic ring are designated $3'$ and $5'$, whereas iodines in the inner ring are designated simply 3 and 5. The 3 and 5 positions on either ring are chemically equivalent, but there are profound functional consequences of removing an iodine from the inner or outer rings of thyroxine. Removing an iodine from the outer ring produces $3',3,5$ triiodothyronine, usually designated as T_3, and converts thyroxine to the form that binds to the thyroid hormone receptor. Removal of an iodine from the inner ring produces $3',5',3$ triiodothyronine, also called *reverse T_3* (rT_3), which cannot bind to the thyroid hormone receptor and can only be further deiodinated.

The type II deiodinase is absent from the liver, but is found in many extrahepatic tissues including the brain and pituitary gland where it is thought to produce T_3 to meet local tissue demands independently of circulating T_3, although these tissues can also take up T_3 from the blood. Expression of the type II deiodinase is regulated by other hormones; its expression is highest when blood concentrations of T_4 are low. In addition, hormones that act through the cyclic AMP second messenger system (Chapter 2) and growth factors stimulate type II deiodinase expression. These characteristics support the idea that this enzyme may provide T_3 to meet local demands.

The type III deiodinase removes an iodine from the tyrosyl ring of T_4 or T_3, and hence its function is solely degradative. It is widely expressed by many tissues throughout the body. Reverse T_3, a product of both the type I and type III deiodinases, may be further deiodinated by the type III deiodinase by removal of the second iodide from inner ring (Fig. 8). Reverse T_3 is also a favored substrate for the type I deiodinase, and although it is formed at a similar rate as T_3 it is degraded much faster than T_3. Some rT_3 escapes into the bloodstream where it is avidly bound to TBG and TTR.

All three deiodinases can catalyze the oxidative removal of iodine from partially deiodinated hormone metabolites, and through their joint actions the thyronine nucleus can be completely stripped of iodine. The liberated iodide is then available to be taken up by the thyroid and recycled into hormone. A quantitatively less important route for degradation of thyroid hormones includes shortening of the alanine side chain to produce tetraiodothyroacetic acid (Tetrac) and its subsequent deiodination products. Thyroid hormones are also conjugated with glucuronic acid and excreted intact in the bile. Bacteria in the intestine can split the glucuronide bond, and some of the thyroxine liberated can be taken up from the intestine and be returned to the general circulation. This cycle of excretion in bile and absorption from the intestine is called the *enterohepatic circulation* and may be of importance in maintaining normal thyroid economy when thyroid function is marginal or dietary iodide is scarce. Thyroxine is one of the few naturally occurring hormones that is sufficiently resistant to intestinal and hepatic destruction that it can readily be given by mouth.

PHYSIOLOGIC EFFECTS OF THYROID HORMONES

Growth and Maturation

Skeletal System

One of the most striking effects of thyroid hormones is on bodily growth (see Chapter 44). Although fetal growth appears to be independent of the thyroid, growth of the neonate and attainment of normal adult stature require optimal amounts of thyroid hormone. Because stature or height is determined by the length of the skeleton, we might anticipate an effect of thyroid hormone on growth of bone. However, there is no evidence that T_3 acts directly on cartilage or bone cells to signal increased bone formation. Rather, at the level of bone formation, thyroid hormones appear to act permissively or synergistically with growth hormone, insulin-like growth factor (see Chapter 44), and other growth factors that promote bone formation. Thyroid hormones also promote bone growth indirectly by actions on the pituitary gland and hypothalamus. Thyroid hormone is required for normal growth hormone synthesis and secretion.

Central Nervous System

The importance of the thyroid hormones for normal development of the nervous system is well established. Thyroid hormones and their receptors are present early in the development of the fetal brain, well before the fetal thyroid gland becomes functional. T_4 and T_3 present in the fetal brain at this time probably arise in the mother and readily cross the placenta to the fetus. Some evidence suggests that maternal hypothyroidism may lead to deficiencies in postnatal neural development, but direct effects of thyroid deficiency on the fetal brain have not been established. However, failure of thyroid gland development in babies born to mothers with normal thyroid function have normal brain development if properly treated with thyroid hormones after birth. Maturation of the nervous system during the perinatal period has an absolute dependence on thyroid hormone. During this critical period thyroid hormone must be present for normal development of the brain. In rats made hypothyroid at birth, cerebral and cerebellar growth and nerve myelination are severely delayed. Overall size of the brain is reduced along with its vascularity, particularly at the capillary level. The decrease in size may be partially accounted for by a decrease in axonal density and dendritic branching. Thyroid hormone deficiency also leads to specific defects in cell migration and differentiation.

Autonomic Nervous System

Interactions between thyroid hormones and the autonomic nervous system, particularly the sympathetic branch, are important throughout life. Increased secretion of thyroid hormone exaggerates many of the responses that are mediated by the neurotransmitters norepinephrine and epinephrine released from sympathetic neurons and the adrenal medulla (see Chapter 40). In fact, many symptoms of hyperthyroidism, including tachycardia (rapid heart rate) and increased cardiac output, resemble increased activity of the sympathetic nervous system. Thyroid hormones increase the number of receptors for epinephrine and norepinephrine (β-adrenergic receptors) in the myocardium and some other tissues. Thyroid hormones may also increase the stimulatory G-protein (G_s) associated with adrenergic receptors and down-regulate the inhibitory G-protein (G_i), either of which results in greater production of cyclic AMP. Furthermore, sympathetic stimulation acting by way of cyclic AMP activates the type II deiodinase, which accelerates local conversion of T_4 to T_3. Because thyroid hormones exaggerate a variety of responses mediated by β-adrenergic receptors, pharmacologic blockade of these receptors is useful for reducing some of the symptoms of hyperthyroidism. Conversely, the diverse functions of the sympathetic nervous system are compromised in hypothyroid states.

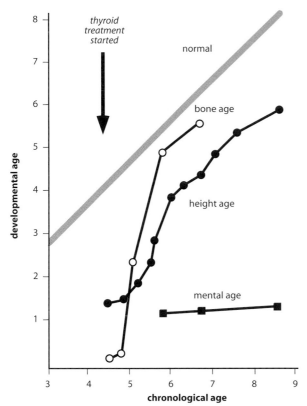

FIGURE 9 Effects of thyroid therapy on growth and development of a child with no functional thyroid tissue. Daily treatment with thyroid began at 4.5 years of age (vertical arrow). Bone age rapidly returned toward normal, and the rate of growth (height age) paralleled the normal curve. Mental development, however, remained infantile. (From Wilkins L. *The diagnosis and treatment of endocrine disorders in childhood and adolescence,* Springfield, IL: Charles C Thomas, 1965.)

Metabolism

Oxidative Metabolism and Thermogenesis

More than a century has passed since it was recognized that the thyroid gland exerts profound effects on oxidative metabolism in humans. The so-called *basal metabolic rate* (BMR), which is a measure of oxygen consumption under defined resting conditions, is highly sensitive to thyroid status. A decrease in oxygen consumption results from a deficiency of thyroid hormones, and excessive thyroid hormone increases BMR. Oxygen consumption in all tissues except brain, testis, and spleen is sensitive to the thyroid status and increases in response to thyroid hormone (Fig. 10). Even though the dose of thyroid hormone given to hypothyroid animals in the experiment shown in Fig. 10 was large, there was a delay of many hours before effects were observable. In fact, the rate of oxygen consumption in the whole animal did not reach its maximum until 4 days after a single dose of hormone. The underlying mechanisms for increased oxygen consumption are not well understood.

Oxygen consumption ultimately reflects activity of mitochondria and is coupled with formation of high-energy bonds in ATP. Physiologically, oxygen consumption is proportional to energy utilization. Thus if consumption of oxygen increase, there must be increased utilization of energy or the efficiency of coupling ATP production with oxygen consumption must be altered. T_3 appears to accelerate ATP-dependent processes, including activity of the sodium potassium ATPase that maintains ionic integrity of all cells and to decrease efficiency of oxygen utilization. In normal individuals activity of the sodium potassium ATPase is thought to account for about 20% of the resting oxygen consumption. Activity of this enzyme is decreased in hypothyroid individuals, and its synthesis is accelerated by thyroid hormone. A variety of other metabolic reactions are also accelerated by T_3, and the accompanying increased turnover of ATP contributes to the increase in oxygen consumption.

ATP synthesis is limited by availability of ADP in the mitochondria. ADP is regenerated from breakdown of ATP at various extramitochondrial sites and must be returned to the interior of the mitochondria. Transport of ADP across the inner mitochondrial membrane and phosphorylation to ATP are driven by the proton gradient created by the electron transport system, which thus couples ATP synthesis with oxygen consumption. Leakage of protons across the inner mitochondrial membrane "uncouples" oxygen consumption from ATP production by dissipating the gradient. As a result, electron transport to oxygen proceeds at a rate that is unlimited by availability of ADP, and the energy derived is dissipated as heat rather than as formation of high-energy phosphate bonds. Leakage of protons into the mitochondria depends on the presence of special uncoupling proteins (UCP) in the inner mitochondrial membrane. To date three proteins thought to have uncoupling activity have been identified in mitochondrial membranes of various tissues. All three appear to be up-regulated by T_3. Although the physiologic importance of UCP-1 seems firmly established (see later discussion), the physiologic roles of UCP-2 and UCP-3 remain controversial.

Thyroid Hormone and Temperature Regulation

Splitting of ATP not only energizes cellular processes but also results in heat production. Thyroid hormones are said to be *calorigenic* because they promote heat production. It is therefore not surprising that one of the classical signs of hypothyroidism is decreased tolerance to cold, whereas excessive heat production and sweating are seen in hyperthyroidism. Effects of thyroid hormone on oxidative metabolism are seen only in animals

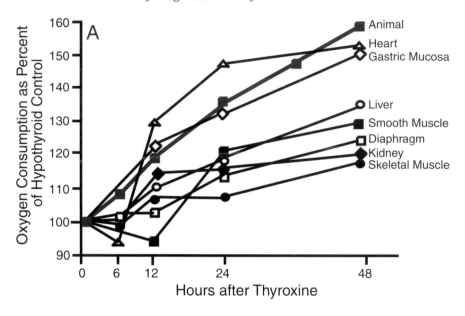

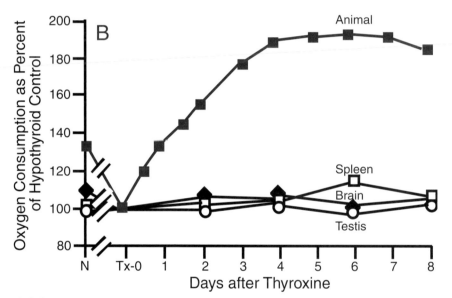

FIGURE 10   Effects of thyroxine on oxygen consumption by various tissues of thyroidectomized rats. Note that in (A) the abscissa is in units of hours and in (B) the units are days. N, normal; Tx-0, thyroidectomized just prior to thyroxine. (Redrawn from Barker SB, Klitgaard HM. Metabolism of tissues excised from thyroxine-injected rats. *Am J Physiol* 1952;170:81.)

that maintain a constant body temperature, consistent with the idea that calorigenic effects may be related to thermoregulation. Thyroidectomized animals have severely reduced ability to survive cold temperature. T_3 contributes to both heat production and heat conservation.

Individuals exposed to a cold environment maintain constant body temperature by increasing heat production by at least two mechanisms: (1) shivering, which is a rapid increase in involuntary activity of skeletal muscle,

and (2) the so-called nonshivering thermogenesis seen in cold-acclimated individuals. Details of the underlying mechanisms for each of these responses are still not understood. As we have seen, the metabolic effects of T_3 have a long lag time and hence increased production of T_3 cannot be of much use for making rapid adjustments to cold temperatures. The role of T_3 in the shivering response is probably limited to maintenance of tissue sensitivity to sympathetic stimulation. In this context, the importance of T_3 derives from actions that were established before exposure to cold temperature. Maintenance of sensitivity to sympathetic stimulation permits efficient mobilization of stored carbohydrate and fat needed to fuel the shivering response and to make circulatory adjustments for increased activity of skeletal muscle. It may be also recalled that the sympathetic nervous system regulates heat conservation by decreasing blood flow through the skin. Piloerection in animals increases the thickness of the insulating layer of fur. These responses are likely to be of importance in both acute and chronic responses to cold exposure.

Chronic nonshivering thermogenesis appears to require increased production of T_3, which acts in concert with the sympathetic nervous system to increase heat production and conservation. Some data indicate that norepinephrine may increase permeability of brown fat and skeletal muscle cells to sodium. Increased activity of the sodium pump could account for increased oxygen consumption and heat production in the cold acclimatized individual. In muscles of cold-acclimated rats, activity of the sodium-potassium ATPase is increased in a manner that appears to depend on thyroid hormone. Some experimental results support a similar effect on calcium pumps.

Brown fat is an important source of heat in newborn humans and throughout life in small mammals. This form of adipose tissue is especially rich in mitochondria, which give it its unique brown color. Mitochondria in this tissue contain UCP 1, sometimes called *thermogenin,* which allows them to oxidize relatively large amounts of fatty acids and produce heat unfettered by limitations in availability of ADP. Although both T_3 and the sympathetic neurotransmitter norepinephrine can each induce the synthesis of UCP-1, their cooperative interaction results in production of 3–4 times as much of this mitochondrial protein as the sum of their independent actions. In addition, T_3 increases the efficacy of norepinephrine to release fatty acids from stored triglycerides and thus provides fuel for heat production. Brown adipose tissue increases synthesis of the type II deiodinase in response to sympathetic stimulation, and produces abundant T_3 locally to meet its needs. Adult humans have little brown fat, but may increase heat production through similar effects of UCP-2 and UCP-3 in white fat and muscle, but supporting evidence for this possibility is not available.

In rodents and other experimental animals, exposure to cold temperatures is an important stimulus for increased TSH secretion from the pituitary and the resultant increase in T_4 and T_3 secretion from the thyroid gland. Cold exposure does not increase TSH secretion in humans except in the newborn. In humans and experimental animals, however, exposure to cold temperatures increases conversion of T_4 to T_3 probably

as a result of increased sympathetic nervous activity, which leads to increased cyclic AMP production in various tissues. Recall that expression of the type II deiodinase is activated by cyclic AMP.

Carbohydrate Metabolism

T_3 accelerates virtually all aspects of metabolism including carbohydrate utilization. It increases glucose absorption from the digestive tract, glycogenolysis and gluconeogenesis in hepatocytes, and glucose oxidation in liver, fat, and muscle cells. No single or unique reaction in any pathway of carbohydrate metabolism has been identified as the rate-determining target of T_3 action. Rather, carbohydrate degradation appears to be driven by other factors, such as increased demand for ATP, the content of carbohydrate in the diet, or the nutritional state. Although T_3 may induce synthesis of specific enzymes of carbohydrate and lipid metabolism, e.g., the malic enzyme, glucose-6-phosphate dehydrogenase, and 6-phosphogluconate dehydrogenase, it appears to behave principally as an amplifier or gain control working in conjunction with other signals (Fig. 11). In the example shown, induction of the malic enzyme in hepatocytes was dependent both on the concentration of glucose in the culture medium and on the concentration of T_3. T_3 had little effect on enzyme induction when there was no glucose but amplified the effectiveness of glucose as an inducer of gene expression. This experiment provides a good example of how T_3 can amplify readout of genetic information.

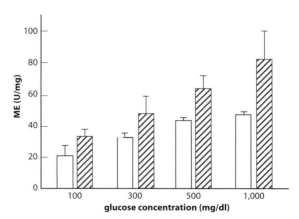

FIGURE 11 Effects of glucose and T_3 on the induction of malic enzyme (ME) in isolated hepatocyte cultures. Note that the amount of enzyme present in tissues was increased by growing cells in higher and higher concentrations of glucose. Open bars show effects of glucose in the presence of a low (10^{-10} M) concentration of T_3. Solid bars indicate that the effects of glucose were exaggerated when cells were grown in a high concentration of T_3 (10^{-8} M). (From Mariash GN, Oppenheimer JH. Thyroid hormone-carbohydrate interaction at the hepatic nuclear level. *Fed Proc* 1982;41:2674.)

Clinical Note

Increased blood cholesterol (*hypercholesterolemia*) is typically found in hypothyroidism. Thyroid hormones reduce cholesterol in the plasma of normal subjects and restore blood concentrations of cholesterol to normal in hypothyroid subjects. Hypercholesterolemia in hypothyroid subjects results from decreased ability to excrete cholesterol in bile rather than overproduction of cholesterol. In fact, cholesterol synthesis is impaired in the hypothyroid individual. T_3 may facilitate hepatic excretion of cholesterol by increasing the abundance of low-density lipoprotein (LDL) receptors in hepatocyte membranes, thereby enhancing uptake of cholesterol from the blood.

Lipid Metabolism

Because glucose is the major precursor for fatty acid synthesis in both liver and fat cells, it should not be surprising that optimal amounts of thyroid hormone are necessary for lipogenesis in these cells. Once again the primary determinant of lipogenesis is not T_3, but, rather, the amount of available carbohydrate or insulin (see Chapter 41), with thyroid hormone acting as a gain control. Similarly, mobilization of fatty acids from storage depots in adipocytes is compromised in the thyroid-deficient subject and increased above normal when thyroid hormones are present in excess. Once again, T_3 amplifies physiologic signals for fat mobilization without itself acting as such a signal.

Nitrogen Metabolism

Body proteins are constantly being degraded and resynthesized. Both synthesis and degradation of protein are slowed in the absence of thyroid hormones; conversely, both are accelerated by thyroid hormone. In the presence of excess T_4 or T_3, the effects of degradation predominate, and severe catabolism of muscle often results. In hyperthyroid subjects body protein mass decreases despite increased appetite and ingestion of dietary proteins. With thyroid deficiency there is a characteristic accumulation of a mucous-like material consisting of protein complexed with hyaluronic acid and chondroitin sulfate in extracellular spaces, particularly in the skin. Because of its osmotic effect, this material causes water to accumulate in these spaces, giving rise to the edema typically seen in hypothyroid individuals and to the name *myxedema* for hypothyroidism.

REGULATION OF THYROID HORMONE SECRETION

As already indicated, secretion of thyroid hormones depends on stimulation of thyroid follicular cells by TSH, which bears the primary responsibility for integrating thyroid function with bodily needs (Chapter 38). In the absence of TSH, thyroid cells are quiescent and atrophy, and, as we have seen, administration of TSH increases both synthesis and secretion of T_4 and T_3. Secretion of TSH by the pituitary gland is governed by positive input from the hypothalamic hormone thyrotropin-releasing hormone (TRH) and negative input from thyroid hormones. Little TSH is produced by the pituitary gland when it is removed from contact with the hypothalamus and transplanted to some extrahypothalamic site, and disruption of the TRH gene reduces the TSH content of mouse pituitaries to less than half that of wild-type litter mate controls. Positive input for thyroid hormone secretion thus originates in the central nervous system by way of TRH and the anterior pituitary gland. TRH increases the expression of the genes for both the α- and β-subunits of TSH and increases the post-translational incorporation of carbohydrate that is required for normal potency of TSH, but these processes can occur at a reduced level in the absence of TRH. Blood levels of thyroid hormones in mice lacking a functional TRH gene are less than half of normal, but the mice grow, develop, and reproduce almost normally, indicating that their hypothyroidism is relatively mild.

Maintaining constant levels of thyroid hormones in blood depends on the negative feedback effects of T_4 and T_3, which inhibit synthesis and secretion of TSH (Fig. 12). The contribution of free T_4 in blood is quite significant in this regard. Because thyrotropes are rich in type II deiodinase, they can convert this more abundant form of thyroid hormone to T_3 and thereby monitor the overall amount of free hormone in blood. High concentrations of thyroid hormones may shut off TSH secretion completely and, when maintained over time, produce atrophy of the thyroid gland. Measurement of relative concentrations of TSH and thyroid hormones in the blood provide critically important information for diagnosing thyroid disease. For instance, low blood concentrations of free T_3 and T_4 in the presence of elevated levels of TSH signal a primary defect in the thyroid gland, whereas high concentrations of free T_3

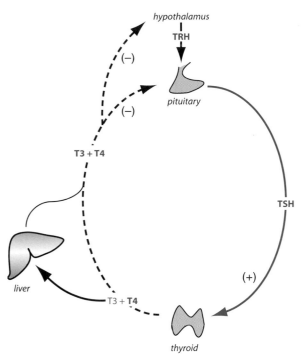

FIGURE 12   Feedback regulation of thyroid hormone secretion. (+), stimulation; (−), inhibition.

transcription of the genes for both the α- and β-subunits of TSH and for TRH receptors. In addition, T_3 inhibits release of stored hormone and accelerates TRH receptor degradation. The net consequence of these actions of T_3 is a reduction in the sensitivity of the thyrotropes to TRH (Fig. 14).

MECHANISM OF THYROID HORMONE ACTION

As must already be obvious, virtually all cells appear to require optimal amounts of thyroid hormone for normal operation, even though different aspects of cellular function may be affected in different cells. Thyroid hormones are quite hydrophobic and may either diffuse across the cell membrane or enter target cells by a carrier-mediated transport process. T_3 formed within the target cell by deiodination of T_4 appears to mix freely with T_3 taken up from the plasma and to enter the nucleus, where it binds to specific receptors (see Chapter 2). Thyroid hormone receptors are members of the large family of nuclear hormone receptors and bind to specific nucleotide sequences (thyroid response elements or TREs) in the genes they regulate. Unlike most other nuclear receptors, thyroid hormone receptors bind to their response elements in the absence of hormone. They bind as monomers or as homodimers composed either of two thyroid hormone receptors, or they may form heterodimers with other nuclear receptor family members, usually the receptor for an isomer of retinoic acid. In the absence of T_3, the unoccupied receptor, in conjunction with a corepressor protein, inhibits T_3-dependent gene expression by maintaining the DNA in a tightly coiled configuration that bars access of transcription activators or RNA polymerase. Upon binding T_3, the configuration of the receptor is modified in a way that causes it to release the corepressor and bind instead to a coactivator. Although T_3 acts in an analogous way to suppress expression of some genes, the underlying mechanism for negative control of gene expression is not understood.

Nuclear receptors for T_3 are encoded in two genes, designated TRα and TRβ. The TRα gene resides on chromosome 17 and gives rise to two isoforms, $TR\alpha_1$ and $TR\alpha_2$, as a result of alternate splicing that deletes the T_3 binding site from the $TR\alpha_2$ isoform. The $TR\alpha_2$ isoform, therefore, cannot act as a hormone receptor, but it nevertheless plays a vital physiologic role (see later discussion). The TRβ gene maps to chromosome 3 and also gives rise to two alternately spliced products, $TR\beta_1$ and $TR\beta_2$. $TR\alpha_1$ and $TR\beta_1$ are widely distributed throughout the body and are present in different ratios in the nuclei of all target tissues examined, but $TR\beta_2$

and T_4 accompanied by high concentrations of TSH reflect a defect in the pituitary or hypothalamus. As already noted, the high concentrations of T_4 and T_3 seen in Graves' disease are accompanied by very low concentrations of TSH in blood as a result of negative feedback inhibition of TSH secretion.

Negative feedback inhibition of TSH secretion results from actions of thyroid hormones exerted both on TRH neurons in the paraventricular nuclei of the hypothalamus and on thyrotropes in the pituitary. Results of animal studies indicate that T_3 and T_4 inhibit TRH synthesis and secretion. Events thought to occur within the thyrotropes are illustrated in Fig. 13. TRH binds its G-protein-coupled heptihelical receptors (Chapter 2) on the surface of thyrotropes. The resulting activation of phospholipase C generates the second messengers inositol trisphosphate (IP_3) and diacylglycerol (DAG). IP_3 promotes calcium mobilization, and DAG activates protein kinase C, both of which rapidly stimulate release of stored hormone. This effect is augmented by influx of extracellular calcium following activation of membrane calcium channels. In addition, transcription of genes for both subunits of TSH is increased. TRH also promotes processing of the carbohydrate components of TSH necessary for maximum biological activity. Meanwhile, both T_4 and T_3 enter the cell at a rate determined by their free concentrations in blood plasma, and T_4 is deiodinated to T_3 in the cytoplasm. T_3 enters the nucleus, binds to its receptors, and down-regulates

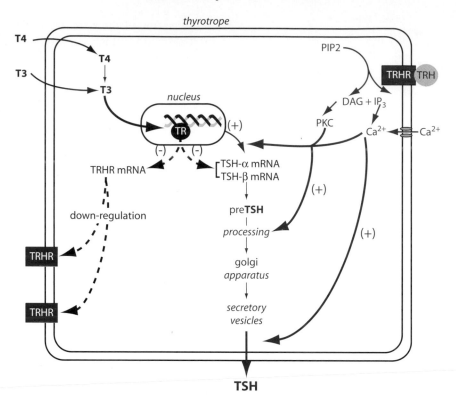

FIGURE 13   Effects of TRH, T_3, and T_4 on the thyrotrope. T_3 down-regulates expression of genes for TRH receptors and both subunits of TSH (dashed arrows). TRH effects (shown in blue) include up-regulation of TSH gene expression, enhanced TSH glycosylation, and accelerated secretion. ($+$), increase; ($-$), decrease.

appears to be expressed primarily in the anterior pituitary gland and the brain.

Efforts to determine which T_3 responses are mediated by each form of the T_3 receptor have been greatly advanced by the advent of technology that permits disruption or "knock out" of individual genes in mouse embryos. Mice lacking both $TR\beta$ isoforms have no development deficiencies, are fertile, and exhibit no obvious behavioral abnormalities. However, these animals have abnormally high rates of TSH, T_4, and T_3 secretion presumably because the $TR\beta_2$ mediates the negative feedback action of T_3. These symptoms are remarkably similar to those seen in a rare genetic disease that is characterized by resistance to thyroid hormone. Like the knockout mice, patients exhibit few abnormalities but have increased circulating levels of TSH, T_4, and T_3. They have enlarged thyroid glands (goiter) stemming from increased TSH levels, but exhibit none of the consequences of T_4 hypersecretion. This disease typically results from mutations in the $TR\beta$ gene.

No effects on life span or fertility result from manipulation of $TR\alpha$ gene so that it encodes only the $TR\alpha_2$ isoform that cannot bind T_3. However, animals that lack the $TR\alpha_1$ isoform have low heart rates and low body temperature. When the $TR\alpha$ gene was knocked out

so that neither the α_1 nor α_2 isoform could be expressed, the animals stopped growing after about 2 weeks and died shortly after weaning with apparent failure of intestinal development. Thus although few symptoms of hypothyroidism result from knock out of any of the thyroid hormone receptors that are capable of binding T_3, loss of the α_2 isoform produced devastating effects, suggesting that it plays a critical, though perhaps T_3-independent, role in gene transcription. The combined absence of $TR\alpha_1$, $TR\beta_1$, and $TR\beta_2$ produces more symptoms of hypothyroidism than lack of either $TR\alpha_1$ or $TR\beta$, suggesting that these receptors have redundant or overlapping functions. However, the hypothyroid symptoms are mild compared to those seen when the complement of TRs is normal but thyroid hormone is absent. Unoccupied TRs that repress gene expression may therefore produce harmful effects. Consistent with this idea, a mutation of the $TR\beta$ gene that prevents it from binding T_3 produced severe defects in neurologic development similar to those seen in hypothyroid mice even though T_3 was abundant. Thus at least one of the physiologic roles of T_3 may be to counteract the consequences of T_3 receptor silencing of some genes.

Although extensive evidence indicates that T_3 and T_4 produce the majority of their actions through nuclear

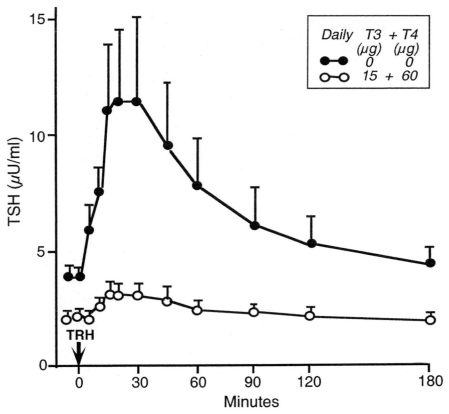

FIGURE 14   Effect of treatment with thyroid hormones for 3–4 weeks on TSH secretion in normal young men in response to an intravenous injection of TRH. Six normal subjects received 25 mg of TRH indicated by the arrow. Values are expressed as means ± SEM. (From Snyder PJ, Utiger RD. Inhibition of thyrotropin response to thyrotropin-releasing hormone by small quantities of thyroid hormones. *J Clin Invest* 1972;52:2077.)

receptors, extranuclear specific binding proteins for thyroid hormones have also been found in the cytosol and mitochondria. The function, if any, of these proteins is not known. In addition, some rapid effects of T_3 and T_4 that may not involve the genome have also been described. It is highly likely that T_3 and T_4 have physiologically important actions that are not dependent on nuclear events, but detailed understanding will require further research.

Suggested Reading

Braverman LE, Utiger RD, eds. *Werner and Ingbar's the thyroid,* 8th ed. Philadelphia: Lippincott Williams and Wilkins, 2000. (This book provides excellent coverage of a broad range of basic and clinical topics.)

De La Vieja A, Dohan O, Levy O, Carrasco N. Molecular analysis of the sodium/iodide symporter: Impact on thyroid and extrathyroid pathophysiology. *Physiol Rev* 2000;80:1083–1105.

Gershengorn MC, Osman R. Molecular and cellular biology of thyrotropin-releasing hormone receptors. *Physiol Rev* 1996;76:175–191.

Köhrle J. Local activation and inactivation of thyroid hormones: The deiodinase family. *Mol Cell Endocrinol* 1999;151:103–119.

Rapoport B, Chazenbalk D, Jaume JC, McLachlan, SM. The thyrotropin (TSH) receptor: Interactions with TSH and autoantibodies. *Endocr Rev* 1998;19:673–716.

Vassart G, Dumont J. The thyrotropin receptor and the regulation of thyrocyte function and growth. *Endocr Rev* 1992;13:596–611.

Yen PM. Physiologic and molecular basis of thyroid hormone action. *Physiol Rev* 2001;81:1097–1142.

C H A P T E R

40

Adrenal Glands

H. MAURICE GOODMAN

KEY POINTS

- The adrenal cortex secretes three classes of steroid hormones: *mineralocorticoids, gluco-corticoids,* and *androgens.*
- Adrenal cortical hormones are derived from cholesterol. Relatively minor differences in functional groups on the steroid nucleus profoundly affect biological activity.
- Adrenal steroid hormones circulate in blood bound to plasma proteins and are degraded chiefly in the liver.
- The principal mineralocorticoid is *aldosterone* whose primary function is to maintain the vascular volume by regulating sodium retention by the kidney. Aldosterone also regulates potassium excretion by the kidney, sodium

absorption in the colon, and sodium and potassium secretion in sweat and saliva.
- Angiotensin II is the primary stimulus for aldosterone secretion, but increased plasma concentrations of potassium also increase aldosterone secretion.
- The kidney responds to a decrease in vascular volume by secreting the enzyme *renin,* which hydrolyzes angiotensinogen in the blood to release angiotensin I, which is then converted to angiotensin II.
- *Cortisol* is the principal glucocorticoid. It preserves carbohydrate reserves by promoting gluconeogenesis in the liver, and by promoting protein degradation in muscle and lymphoid

KEY POINTS (*continued*)

tissues and triglycerides in adipose tissue to provide substrates for gluconeogenesis. Cortisol also decreases sensitivity of tissues to insulin and decreases triglyceride synthesis in adipose tissue and glucose metabolism in muscle.

- Glucocorticoids protect against injury.
- Glucocorticoid secretion is controlled by adrenocorticotropic hormone (ACTH) in a negative feedback manner in which glucocorticoids act both at the hypothalamic and pituitary levels to

inhibit secretion of corticotropin-releasing hormone (CRH) and ACTH.
- The adrenal medullae are modified sympathetic ganglia, and as components of the sympathetic nervous system participate in emergency responses characterized by the "fright, fight, or flight" reactions.
- The principal hormones of the adrenal medullae are *norepinephrine* and *epinephrine*, which are derivatives of tyrosine.

OVERVIEW

The adrenal glands are complex polyfunctional organs whose secretions are required for maintenance of life. Without adrenal hormones, deranged electrolyte or carbohydrate metabolism leads to circulatory collapse or hypoglycemic coma and death. The hormones of the outer region, or cortex, are steroids and act at the level of the genome to regulate the expression of genes that govern the operation of fundamental processes in virtually all cells. The inner region, the adrenal medulla, is actually a component of the sympathetic nervous system and participates in the wide array of regulatory responses that are characteristic of that branch of the nervous system. The three major categories of adrenal steroid hormones are *mineralocorticoids,* whose actions defend the body content of sodium and potassium; *glucocorticoids,* whose actions affect body fuel metabolism, responses to injury, and general cell function; and *androgens,* whose actions are similar to the hormone of the male gonad. We focus on actions of these hormones on the limited number of processes that are most thoroughly studied, but the reader should keep in mind that adrenal cortical hormones directly or indirectly affect almost every physiologic process and hence are central to the maintenance of homeostasis. Secretion of mineralocorticoids is primarily controlled by the kidney through secretion of *renin* and the consequent production of *angiotensin.* Secretion of glucocorticoids and androgens is controlled by the anterior pituitary gland through secretion of ACTH.

The adrenal cortex and the medulla often behave as a functional unit and together confer a remarkable capacity to cope with changes in the internal or external environment. Fast-acting medullary hormones are signals for physiologic adjustments, and slower acting cortical hormones maintain or increase sensitivity of

tissues to medullary hormones and other signals; they also maintain or enhance the capacity of tissues to respond to such signals. The cortical hormones thus tend to be modulators rather than initiators of responses.

MORPHOLOGY

The adrenal glands are bilateral structures situated above the kidneys. They are comprised of an outer region or cortex, which normally makes up more than three-quarters of the adrenal mass, and an inner region or medulla (Fig. 1). The medulla is a modified sympathetic ganglion that, in response to signals reaching it through cholinergic, preganglionic fibers, releases either or both of its two hormones, epinephrine and norepinephrine, into adrenal venous blood. The cortex arises from mesodermal tissue and produces a class of lipid-soluble hormones derived from cholesterol and called *steroids.* The cortex is subdivided histologically into three zones. Cells in the outer region, or *zona glomerulosa,* are arranged in clusters (glomeruli) and produce the hormone *aldosterone.* In the *zona fasciculata,* which comprises the bulk of the cortex, rows of lipid-laden cells are arranged radially in bundles of parallel cords (*fasces*). The inner region of the cortex consists of a tangled network of cells and is called the *zona reticularis.* The fasciculata and reticularis, which produce both *cortisol* and the adrenal androgens, are functionally separate from the zona glomerulosa.

The adrenal glands receive their blood supply from numerous small arteries that branch off the renal arteries or the lumbar portion of the aorta and its various major branches. These arteries penetrate the adrenal capsules and divide to form the subcapsular plexus from which small arterial branches pass centripetally toward the medulla. The subcapsular plexuses also

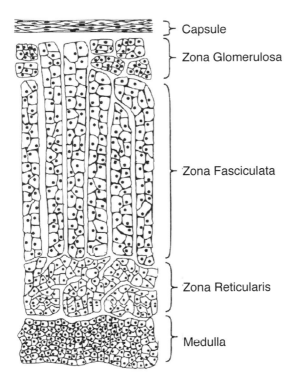

FIGURE 1 Histology of the adrenal gland.

give rise to long loops of capillaries that pass between the cords of fascicular cells and empty into sinusoids in the reticularis and medulla. Sinusoidal blood collects through venules into a single large central vein in each adrenal and drains into either the renal vein or the inferior vena cava.

ADRENAL CORTEX

In all species thus far studied, the adrenal cortex is essential for maintenance of life. Insufficiency of adrenal cortical hormones (Addison's disease) produced by pathologic destruction or surgical removal of the adrenal cortices results in death within 1–2 weeks unless replacement therapy is instituted. Virtually every organ system goes awry with adrenal cortical insufficiency, but the most likely cause of death appears to be circulatory collapse secondary to sodium depletion. When food intake is inadequate, death may result instead from insufficient amounts of glucose in the blood (hypoglycemia).

Adrenal cortical hormones have been divided into two categories based on their ability to protect against these two causes of death. The so-called *mineralocorticoids* are necessary for maintenance of sodium and potassium balance. Aldosterone is the physiologically important mineralocorticoid, although some deoxycorticosterone, another potent mineralocorticoid, is also produced by the normal adrenal gland (Fig. 2). Cortisol

and, to a lesser extent, corticosterone are the physiologically important *glucocorticoids* and are so named for their ability to maintain carbohydrate reserves. Glucocorticoids have a variety of other effects as well. At high concentrations, aldosterone may exert glucocorticoid-like activity and, conversely, cortisol and corticosterone may exert some mineralocorticoid activity (see later discussion). The adrenal cortex also produces *androgens*, which as their name implies have biological effects similar to those of the male gonadal hormones (see Chapter 45). Adrenal androgens mediate some of the changes that occur at puberty. Adrenal steroid hormones are closely related to steroid hormones produced by the testis and ovary and are synthesized from common precursors. In some abnormal states the adrenals may secrete any of the gonadal steroids.

Adrenocortical Hormones

All of the adrenal steroids are derivatives of the polycyclic phenanthrene nucleus, which is also present in cholesterol, ovarian and testicular steroids, bile acids, and precursors of vitamin D. Use of some of the standard conventions for naming the rings and the carbons

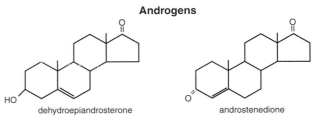

FIGURE 2 Principal adrenal steroid hormones.

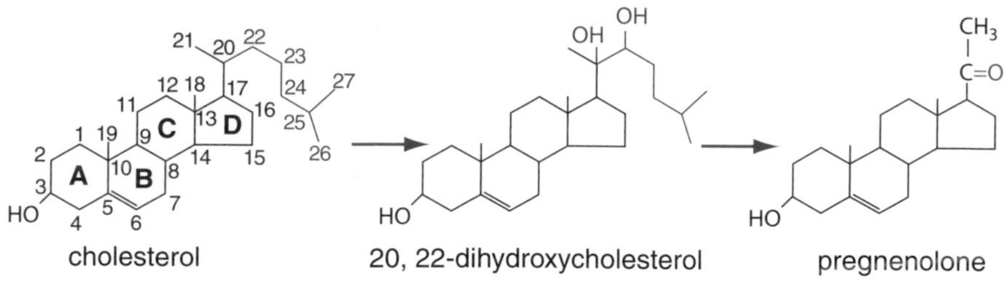

FIGURE 3   Conversion of cholesterol to pregnenolone. Carbons 20 and 22 are sequentially oxidized (in either order) followed by oxidative cleavage of the bond between them. All three reactions are catalyzed by cytochrome P450scc.

facilitates discussion of the biosynthesis and metabolism of the steroid hormones. When drawing structures of steroid hormones, carbon atoms are indicated by junctions of lines that represent chemical bonds. The carbons are numbered and the rings lettered as shown in Fig. 3. Remember that steroid hormones have complex three-dimensional structures; they are not flat, two-dimensional molecules as we depict them for simplicity. Substituents on the steroid nucleus that project toward the reader are usually designated by the prefix β. Those that project away from the reader are designated by α and are shown diagrammatically with dashed lines. The fully saturated 21-carbon molecule is called *pregnane*. When a double bond is present in any of the rings, the -*ane* in the ending is changed to -*ene*, or to -*diene* when there are two double bonds, i.e., pregn*ene* or pregna*diene*. The location of the double bond is designated by the Greek letter Δ followed by one or more superscripts to indicate the location. The presence of a hydroxyl group (OH) is indicated by the ending -*ol*, and the presence of a keto group (O) by the ending -*one*. Thus the important intermediate in the biosynthetic pathway for steroid hormones shown in Fig. 3 has a double bond in the B ring, a keto group on carbon 20, and a hydroxyl group on carbon 3, and hence is called Δ^5 pregnenolone.

The starting material for steroid hormone biosynthesis is cholesterol, most of which arrives at the adrenal cortex in the form of low-density lipoproteins (LDL), which are avidly taken up from blood by a process of receptor-mediated endocytosis. Adrenal cortical cells also synthesize cholesterol from carbohydrate or fatty acid precursors. Substantial amounts of cholesterol are stored in steroid hormone-producing cells in the form of fatty acid esters.

Key reactions in the biosynthesis of the adrenal hormones are catalyzed by a particular class of oxidizing enzymes, the cytochromes P450, that includes a large number of hepatic detoxifying enzymes called *mixed function oxidases*. They contain a heme group covalently linked through a sulfur-iron bond and absorb light in the visible range. The name P450 derives from the

property of these Pigments to absorb light at 450 nm when reduced by carbon monoxide. The P450 enzymes utilize molecular oxygen and electrons donated from $NADPH^+$ to oxidize their substrates. Although they have a single substrate-binding site, some of the P450 enzymes catalyze more than one oxidative step in steroid hormone synthesis.

The rate-limiting step in the biosynthesis of all of the steroid hormones is cleavage of the side chain to convert the 27-carbon cholesterol molecule to the 21-carbon pregnenolone molecule (Fig. 3). The enzyme that catalyzes this complex reaction, is called P450scc (for side-chain cleavage) and is located on the inner mitochondrial membrane. P450scc catalyzes the oxidation of carbons 20 and 22 and then cleaves the bond between them to shorten the side chain. This initial step in hormone biosynthesis requires a complicated series of molecular events. Cholesterol must first be released from its esterified storage form by the action of an esterase. The free, but water-insoluble cholesterol must then be transferred to the mitochondria perhaps through the agency of a cholesterol binding protein and participation of cytoskeletal elements. Cholesterol must then enter the mitochondria to gain access to P450scc whose activity is limited primarily by availability of its substrate. The steroid acute regulatory (StAR) protein plays an indispensable role in presenting cholesterol to P450scc through mechanisms that are still incompletely understood. The StAR protein has a very short half-life, and stimulation of its synthesis appears to be the critical regulated step in steroid hormone biosynthesis. Unlike cholesterol, 21-carbon steroids apparently pass through the mitochondrial membrane rather freely.

Pregnenolone is the common precursor of all steroid hormones produced by the adrenals or the gonads. An early step in hormone biosynthesis is oxidation of the hydroxyl group at carbon 3 to a keto group. This reaction is catalyzed by the enzyme 3-β-hydroxysteroid dehydrogenase (3βHSD) and initiates a rearrangement that shifts the double bond from the B ring to the A ring. *A ketone group at carbon 3 is found in all biologically important*

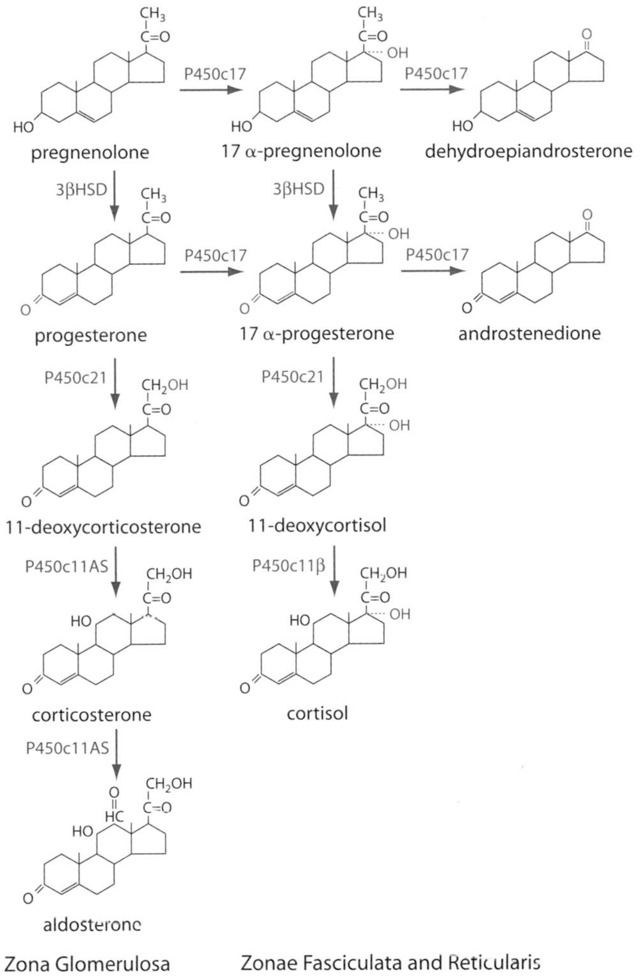

FIGURE 4 Biosynthesis of adrenal cortical hormones. The pathway on the left represents that found in the zona glomerulosa. The pathways on the right are followed in the zonae fasciculata and reticularis. Hydroxylation at carbon 17 usually precedes hydroxylation at carbon 11. Note that changes produced in each reaction are shown in blue.

adrenal steroids and appears necessary for physiologic activity. Biosynthesis of the various steroid hormones involves oxygenation of carbons 21, 17, 11, and 18, as depicted in Fig. 4. The exact sequence of hydroxylations may vary, and some of the reactions may take place in a different order than that presented in the figure. The specific hormone that is ultimately secreted once the cholesterol–pregnenolone road block has been passed is determined by the enzymatic makeup of the particular cells involved. For example, two different P450 enzymes catalyze the hydroxylation of carbon 11. One of these is found exclusively in cells of the zona glomerulosa (P450c11AS) and catalyzes the oxidation of both carbon 11 and carbon 18. The other enzyme (P450c11β) is found in cells of the zonae fasciculata and reticularis and can oxidize only carbon 11. At the same time, cells of the zonae fasciculata and reticularis, but not of the zona glomerulosa, express the enzyme P450c17, also called

P450 17α-hydroxylase/lyase, which catalyzes the oxidation of the carbon at position 17. Hence, glomerulosa cells can produce corticosterone, aldosterone, and deoxycorticosterone, but not cortisol, while cells of the zonae fasciculata and reticularis can form cortisol and 17α-hydroxyprogesterone. When reducing equivalents are delivered to P450c17 rapidly enough, the reaction continues beyond 17α-progesterone hydroxylation to cleavage of the carbon 17-carbon 20 bond. Removal of carbons 20 and 21 produces the 19-carbon androgens. Hence androgens can also be produced by these cells but not by glomerulosa cells.

As is probably already apparent, steroid chemistry is complex and can be bewildering; but because these compounds are so important physiologically and therapeutically, some familiarity with their structures is required. We can simplify the task somewhat by noting that steroid hormones can be placed into three major categories: those that contain 21 carbon atoms, those that contain 19 carbon atoms, and those that contain 18 carbon atoms. In addition, there are relatively few sites where modification of the steroid nucleus determines its physiologic activity.

The physiologically important steroid hormones of the *21-carbon series* are as follows:

1. *Progesterone,* which has the simplest structure, can serve as a precursor molecule for all of the other steroid hormones. Note that the only modifications to the basic carbon skeleton of the 21-carbon steroid nucleus are keto groups at positions 3 and 20. Normal adrenal cortical cells convert progesterone to other products so rapidly that none escapes into adrenal venous blood. Progesterone is a major secretory product of the ovaries and the placenta.

2. *Addition of a hydroxyl group to carbon 21 of progesterone is the minimal change required for adrenal corticoid activity.* This addition produces *11-deoxycorticosterone,* a potent mineralocorticoid that is virtually devoid of glucocorticoid activity. Deoxycorticosterone is only a minor secretory product of the normal adrenal gland but may become important in some disease states.

3. *A hydroxyl group at carbon 11 is found in all glucocorticoids.* Adding the hydroxyl group at carbon 11 confers glucocorticoid activity to deoxycorticosterone and reduces its mineralocorticoid activity 10-fold. This compound is *corticosterone* and can be produced in cells of all three zones of the adrenal cortex. Corticosterone is the major glucocorticoid in the rat but is of only secondary importance in humans.

4. Corticosterone is a precursor of *aldosterone,* which is produced in cells of the zona glomerulosa by oxidation of carbon 18 to an aldehyde. The oxygen at

carbon 18 increases the mineralocorticoid potency of corticosterone by a factor of 200 and only slightly decreases glucocorticoid activity.

5. *Cortisol* differs from corticosterone only by the presence of a hydroxyl group at carbon 17. Cortisol is the most potent of the naturally occurring glucocorticoids. It has 10 times as much glucocorticoid activity as aldosterone, but less than 0.25% of aldosterone's mineralocorticoid activity in normal human subjects. Synthetic glucocorticoids with even greater potency than cortisol are available for therapeutic use.

6. *Cortisone* differs from cortisol only in that the substituent on carbon 11 is a keto group rather than a hydroxyl group. Cortisone is produced from cortisol at extraadrenal sites by oxidation of the hydroxyl group on carbon 11 and circulates in blood at about one-fifth the concentration of cortisol. Oxidation of cortisol to cortisone profoundly lowers its affinity for adrenal steroid hormone receptors and hence inactivates it. Cortisol is oxidized to cortisone in mineralocorticoid target cells, and cortisone can be reduced to cortisol in the liver and other glucocorticoid target tissues. This so-called cortisol-cortisone shuttle is catalyzed by two enzymes, 11-hydroxysteroid dehydrogenase (HSD)-I and HSD-II, which are products of different genes and are expressed in different tissues. HSD-I can catalyze the reaction in either direction, and hence may activate or inactivate the hormone. HSD-II, which is expressed in all mineralocorticoid target tissues, catalyzes only the oxidation of cortisol to cortisone (Fig. 5).

Steroids in the *19-carbon series* usually have androgenic activity and are precursors of the estrogens (female hormones). Hydroxylation of either pregnenolone or progesterone at carbon 17 is the critical prerequisite for cleavage of the $C_{20,21}$ side chain to yield the adrenal androgens *dehydroepiandrosterone* or *androstenedione* (see Fig. 4). The principal testicular androgen is *testosterone,* which has a hydroxyl group rather than a keto group at carbon 17. Although 19 carbon androgens are

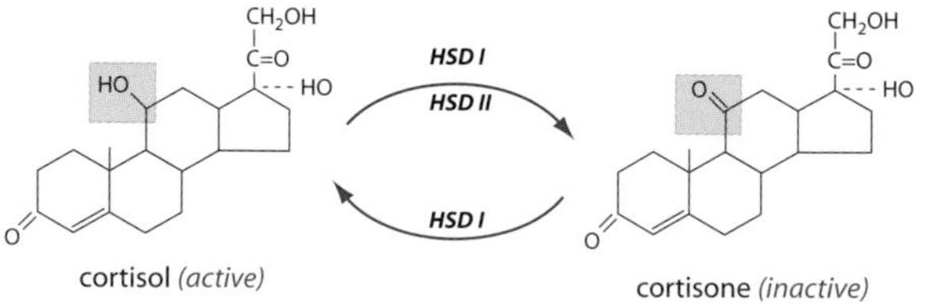

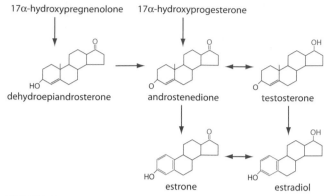

FIGURE 6   Biosynthetic pathway for androgens and estrogens. In the adrenal the sequence does not usually proceed all the way to testosterone and the estrogens, which are the gonadal hormones. Because the cells of the zona glomerulosa lack 17α-hydroxylase, these reactions can occur only in the inner zones.

products of the same enzyme that catalyzes 17α-hydroxylation in the adrenals and the gonads, cleavage of the bond linking carbons 17 and 20 in the adrenals normally occurs to a significant extent only after puberty, and then is confined largely to cells of the zona reticularis.

Steroids of the *18-carbon series* usually have estrogenic activity. Estrogens characteristically have an unsaturated A ring. Oxidation of the A ring (a process called *aromatization*) results in loss of the methyl carbon at position 19 (Fig. 6). This reaction, which takes place principally in ovaries and placenta, can also occur in a variety of nonendocrine tissues where aromatization of the A ring of either testicular or adrenal androgens comprises the principal source of estrogens in men and postmenopausal women.

Regulation of Adrenal Cortical Hormone Synthesis

Effects of ACTH

Adrenocorticotropic hormone maintains normal secretory activity of the inner zones of the adrenal cortex. After removal of the pituitary gland, little or no steroidogenesis occurs in the zona fasciculata or reticularis, but the zona glomerulosa continues to function. In cells of all three zones, ACTH interacts with a specific G-protein-coupled membrane receptor and triggers production of cyclic AMP by activating adenylyl cyclase (see Chapter 2). Cyclic AMP activates protein kinase A, which catalyzes the phosphorylation of a variety of proteins and thereby modifies their activity. In the zonae fasciculata and reticularis this results in accelerated deesterification of cholesterol esters, increased transport of cholesterol to the mitochondria, and increased synthesis

FIGURE 5   The cortisol-cortisone shuttle. Two enzymes, 11-hyrdoxysteroid dehydrogenase (HSD-I and HSD-II), catalyze the inactivating conversion of cortisol to cortisone. HSD-I can also catalyze conversion of the inactive cortisone to cortisol.

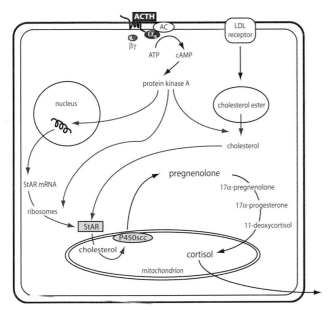

FIGURE 7 Stimulation of steroidogenesis in a zona fasciculata cell of the adrenal cortex by ACTH. The rate-determining reaction is the conversion of cholesterol to pregnenolone, which requires mobilization of cholesterol from its storage droplet and transfer to the P450scc (side-chain cleavage enzyme) on the inner mitochondrial membrane. See text for discussion. ACTH may also increase cholesterol uptake by increasing the number or affinity of low density lipoprotein (LDL) receptors. α_s, stimulatory α-subunit of the guanine nucleotide-binding protein; AC, adenylyl cyclase, $\beta\gamma$, $\beta\gamma$-subunits of the guanine nucleotide-binding protein; StAR, steroid acute regulatory protein.

of StAR protein. Early studies of how ACTH increases steroidogenesis revealed a puzzling requirement for protein synthesis that ultimately led to the discovery of the StAR protein. It appears that ACTH increases StAR protein by stimulating its synthesis on preexisting mRNA templates and by promoting gene transcription (Fig. 7). Thus the immediate actions of ACTH accelerate the delivery of cholesterol to P450scc to form pregnenolone. Once pregnenolone is formed, remaining steps in steroid biosynthesis can proceed without further intervention from ACTH, although some evidence suggests that ACTH may also speed up some later reactions in the biosynthetic sequence. With continued stimulation, ACTH, acting through cyclic AMP and protein kinase A, also stimulates transcription of the genes encoding the P450 enzymes (P450scc, P450c21, P450c17, and P450c11β) and the LDL receptor responsible for uptake of cholesterol.

ACTH is the only hormone known to control synthesis of the adrenal androgens. These 19-carbon steroids are produced primarily in the zona reticularis. Their production is limited first by the rate of conversion of cholesterol to pregnenolone and subsequently by cleavage of the carbon 17-20 bond. As already mentioned, P450c17, the same enzyme that catalyzes

α-hydroxylation of carbon 17 of cortisol also catalyzes the second oxidative reaction at carbon 17 (17,20-lyase) that removes the $C_{20,21}$ side chain. Some evidence suggests that the lyase reaction is increased by phosphorylation of the P450c17, and other studies suggest that androgen production is driven by the capacity of reticularis cells to deliver reducing equivalents to the reaction. Little or no androgen is produced in young children whose adrenal glands contain only a rudimentary zona reticularis. The reticularis with its unique complement of enzymes develops shortly before puberty. The arrival of puberty is heralded by a dramatic increase in production of the adrenal androgens, principally dehydroepiandrosterone sulfate (DHEAS), which are responsible for growth of pubic and axillary hair. Secretion of DHEAS gradually rises to reach a maximum by age 20–25, and thereafter declines. This pattern of androgen secretion is quite different from the pattern of cortisol secretion and therefore appears to be governed by other factors than simply the ACTH-dependent rate of pregnenolone formation. These findings have led some investigators to propose separate control of androgen production, possibly by another, as yet unidentified pituitary hormone, but to date no such hormone has been found. It is important to emphasize that increased stimulation of both the fasciculata and the reticularis by ACTH can profoundly increase adrenal androgen production.

Effects of ACTH on the adrenal cortex are not limited to accelerating the rate-determining step in steroid hormone production. ACTH either directly or indirectly also increases blood flow to the adrenal glands and thereby provides not only the needed oxygen and metabolic fuels but also increased capacity to deliver newly secreted hormone to the general circulation. ACTH maintains the functional integrity of the inner zones of the adrenal cortex: Absence of ACTH leads to atrophy of these two zones, while chronic stimulation increases their mass.

Stimulation with ACTH increases steroid hormone secretion within 1–2 min, and peak rates of secretion are seen in about 15 min. Unlike other glandular cells, steroid-producing cells do not store hormones, and hence biosynthesis and secretion are components of a single process regulated at the step of cholesterol conversion to pregnenolone. Because steroid hormones are lipid soluble they can diffuse through the plasma membrane and enter the circulation through simple diffusion down a concentration gradient. Even under basal conditions, cortisol concentrations are more than 100 times higher in fasciculata cells than in plasma. It is not surprising, therefore, that biosynthetic intermediates may escape into the circulation during intense stimulation. Human adrenal glands normally produce about

20 mg of cortisol, about 2 mg of corticosterone, 10–15 mg of androgens, and about 150 μg of aldosterone each day, but with sustained stimulation they can increase this output manyfold.

Regulation of Aldosterone Synthesis

The regulation of aldosterone synthesis is considerably more complex than that of the glucocorticoids and is not completely understood. Although cells of the zona glomerulosa express ACTH receptors and ACTH is required for optimal secretion, ACTH is not an important regulator of aldosterone production in most species. Angiotensin II, an octapeptide whose production is regulated by the kidney (see later discussion and Chapter 29) is the hormonal signal for increased production of aldosterone (Fig. 8). The cellular events entrained by angiotensin II are not as well established as those described for ACTH. Like ACTH, angiotensin II reacts with specific G-protein-coupled receptors, but angiotensin II does not activate adenylyl cyclase or use cyclic AMP as its second messenger. Instead, it acts through inositol trisphosphate (IP$_3$) and calcium to promote the formation of pregnenolone from cholesterol. The ligand-bound angiotensin receptor associates with Gα_q and activates phospholipase C to release IP$_3$ (see Chapter 2)

and diacylglycerol (DAG) from membrane phosphatidylinositol bisphosphate. Gα_q may also interact directly with potassium channels and cause them to close. The resulting depolarization of the membrane opens voltage-gated calcium channels and allows calcium to enter. Simultaneously, the $\beta\gamma$-subunits directly activate calcium channels, thus further promoting calcium entry. Intracellular calcium concentrations are also increased by interaction of IP$_3$ with its receptor in the endoplasmic reticulum to release stored calcium. Increased intracellular calcium activates a calmodulin-dependent protein kinase that promotes transfer of cholesterol into the mitochondria by increasing synthesis of the StAR protein in the same manner as described for protein kinase A in fasciculata cells. The increase in cytosolic calcium in turn raises the intramitochondrial calcium concentration and stimulates P450c11AS, which catalyzes the critical final reactions in aldosterone synthesis. The DAG that is released by activation of phospholipase C activates protein kinase C, but the role of this enzyme is uncertain. It may augment synthesis of StAR protein, and it may participate in calcium channel activation. Protein kinase C may also play an important role in mediating the hypertrophy of the zona glomerulosa seen after prolonged stimulation of the adrenal glands with angiotensin II.

Cells of the zona glomerulosa are exquisitely sensitive to changes in concentration of potassium in the extracellular fluid and adjust aldosterone synthesis accordingly. An increase of as little as 0.1 mM in the concentration of potassium, which corresponds to a change of only about 2–3%, may increase aldosterone production by as much as 35%. Increased extracellular potassium depolarizes the plasma membrane and activates voltage-gated calcium channels. The resulting increase in intracellular calcium stimulates aldosterone synthesis as already described. The rate of aldosterone secretion can also be affected by the concentration of sodium in the extracellular fluid, but relatively large changes are required. A decline in sodium concentration increases sensitivity to angiotensin II and potassium, but direct effects of sodium on glomerulosa cells are relatively unimportant except in extreme cases of sodium depletion. However, sodium profoundly affects aldosterone synthesis indirectly through its influence on production of angiotensin II as described later and in Chapter 29.

Synthesis and secretion of aldosterone are also negatively regulated by the *atrial natriuretic factor* (ANF) secreted primarily by the cardiac atria This hormone and its role in normal physiology are discussed in Chapter 29. ANF receptors have intrinsic guanylyl cyclase activity and, when bound to ANF, catalyze the conversion of guanosine triphosphate to cyclic

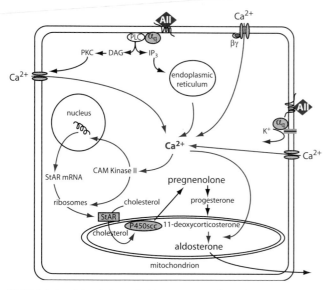

FIGURE 8 Stimulation of aldosterone synthesis by angiotensin II (AII). AII accelerates the conversion of cholesterol to pregnenolone and 11-deoxycorticosterone to aldosterone. α_q, $\beta\gamma$, subunits of the guanine nucleotide-binding protein; PLC, phospholipase C; DAG, diacylglycerol; IP$_3$, inositol trisphosphate; PKC, protein kinase C; CAM kinase II, calcium-calmodulin-dependent protein kinase II; StAR, steroid acute regulatory protein. High concentrations of potassium (K$^+$) in blood also increase aldosterone synthesis and secretion by increasing intracellular calcium secondary to partial membrane depolarization.

guanosine monophosphate (cyclic GMP). Precisely how an increase in cyclic GMP antagonizes the actions of angiotensin and potassium on aldosterone synthesis has not been established. Cyclic GMP is known to activate the enzyme cyclic AMP phosphodiesterase, and may thereby lower basal levels of cyclic AMP, or it may act through stimulating cyclic GMP-dependent protein kinase.

Adrenal Steroid Hormones in Blood

Adrenal cortical hormones are transported in blood bound to a specific plasma protein, *corticosteroid binding globulin* (CBG), which is also called *transcortin,* and to a lesser extent to albumin. Like albumin, CBG is synthesized and secreted by the liver, but its concentration of ~1 μM in plasma is only about 1000th that of albumin. CBG is a glycoprotein with a molecular weight of about 58,000, and is a member of the serine proteinase inhibitor (SERPIN) superfamily of proteins. It has a single steroid hormone binding site whose affinity for glucocorticoids is nearly 20 times higher than for aldosterone. About 95% of the glucocorticoids and about 60% of the aldosterone in blood are bound to protein. Under normal circumstances the concentration of free or unbound cortisol in plasma is about 100 times that of aldosterone. Probably because they circulate bound to plasma proteins, adrenal steroids have a relatively long half-life in blood: 1.5–2 hr for cortisol, and about 15 min for aldosterone.

Metabolism and Excretion of Adrenal Cortical Hormones

Because mammals cannot degrade the steroid nucleus, elimination of steroid hormones is achieved by inactivation through metabolic changes that make them unrecognizable to their receptors. Inactivation of glucocorticoids occurs mainly in liver and is achieved primarily by reduction of the A ring and its keto group at position 3. Conjugation of the resulting hydroxyl group on carbon 3 with glucuronic acid or sulfate increases water solubility and decreases binding to CBG so the steroid can now pass through renal glomerular capillaries and be excreted in the urine. The major urinary products of steroid hormone degradation are glucuronide esters of 17-hydroxycorticosteroids (17-OHCS) derived from cortisol, and 17-ketosteroids (17-KS) derived from glucocorticoids and androgens. Because recognizable hormonal products can be identified in the urine, it is possible to estimate the daily secretory rate of steroid hormones by the noninvasive technique of analyzing urinary excretory products.

PHYSIOLOGY OF THE MINERALOCORTICOIDS

Although several naturally occurring adrenal cortical hormones, including glucocorticoids, can produce mineralocorticoid effects, aldosterone is by far the most important mineralocorticoid physiologically. In its absence there is a progressive loss of sodium by the kidney, which results secondarily in a loss of extracellular fluid (see Chapter 29). Recall that the kidney adjusts the composition of the extracellular fluid by processes that involve formation of an ultrafiltrate of plasma followed by secretion or selective reabsorption of solutes and water. With severe loss of blood volume (*hypovolemia*), water is retained in an effort to restore volume, and the concentration of sodium in blood plasma may gradually fall (*hyponatremia*) from the normal value of 140 mEq/liter to 120 mEq/liter or even lower in extreme cases. With the decrease in concentration of sodium, the principal cation of extracellular fluid, there is a net transfer of water from extracellular to intracellular space, further aggravating hypovolemia. Diarrhea is frequently seen and it, too, worsens hypovolemia. Loss of plasma volume increases the hematocrit and the viscosity of blood (*hemoconcentration*). Simultaneous with the loss of sodium, the ability to excrete potassium is impaired, and with continued dietary intake, plasma concentrations of potassium may increase from the normal value of 4 mEq/liter to 8–10 mEq/liter (*hyperkalemia*). Increased concentrations of potassium in blood, and therefore in extracellular fluid, result in partial depolarization of plasma membranes of all cells, leading to cardiac arrhythmia and weakness of muscles including the heart. Blood pressure falls from the combined effects of decreased vascular volume, decreased cardiac contractility, and decreased responsiveness of vascular smooth muscle to vasoconstrictor agents caused by hyponatremia. Mild acidosis is seen with mineralocorticoid deficiency, partly as a result of deranged potassium balance and partly from lack of the direct effects of aldosterone on hydrogen ion excretion.

All of these life-threatening changes can be reversed by administration of aldosterone and can be traced to the ability of aldosterone to promote inward transport of sodium across epithelial cells of kidney tubules and the outward transport of potassium and hydrogen ions into the urine. It has been estimated that aldosterone is required for the reabsorption of only about 2% of the sodium filtered at the renal glomeruli; even in its absence, about 98% of the filtered sodium is reabsorbed. However, 2% of the sodium filtered each day corresponds to the amount present in about 3.5 L of extracellular fluid. Aldosterone also promotes sodium and potassium transport by the sweat glands, the colon, and the salivary glands. Of these target tissues, the kidney is by far the most important.

Effects of Aldosterone on the Kidney

Initial insights into the action of aldosterone on the kidney were obtained from observations of the effects of hormone deprivation or administration on the composition of the urine. Mineralocorticoids decrease the ratio of urinary sodium to potassium concentrations; in the absence of mineralocorticoids, the ratio increases. However, although aldosterone promotes both sodium conservation and potassium excretion, the two effects are not tightly coupled, and sodium is not simply exchanged for potassium. Indeed, the same amount of aldosterone that increased both sodium retention and potassium excretion when given to adrenalectomized dogs stimulated only potassium excretion in normal, sodium-replete dogs. Similarly, when normal human subjects were given aldosterone for 25 days, the sodium-retaining effects lasted only for the first 15 days, but increased excretion of potassium persisted for as long as the hormone was given (Fig. 9). Renal handling of sodium and

potassium is complex, and compensatory mechanisms exerted at aldosterone insensitive loci within the kidney can offset sustained effects of aldosterone on sodium absorption when measured in the otherwise normal subject.

Renal tubular epithelial cells are polarized. Permeability properties of the membrane that faces the lumen are different from those of the basolateral membranes that face the interstitium. Reabsorption of sodium depends on entry through channels in the luminal membrane followed by extrusion by the sodium-potassium-dependent ATPase in the basolateral membranes. This enzyme is energized by cleavage of ATP and exchanges three sodium ions for two potassium ions. Potassium, which would otherwise accumulate within the cells, can then passively diffuse through channels located in both the luminal and basolateral membranes (Fig. 10). Consequently, movement of sodium from the lumen to the interstitium is not necessarily accompanied by equivalent movement of potassium in the opposite direction. The proportion of potassium that back-diffuses into the interstitium depends on the relative strengths of the electrochemical gradients across the luminal and basolateral

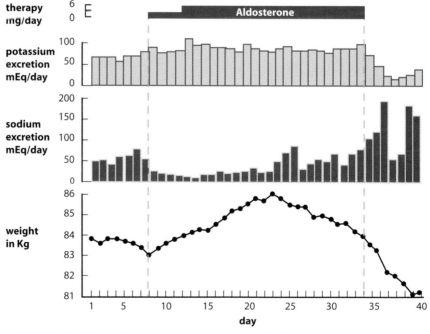

FIGURE 9   Effects of continuous administration of aldosterone to a normal man. Aldosterone (3–6 mg/day) increased potassium excretion and sodium retention, represented here as a decrease in urinary sodium. The increased retention of sodium, which continued for 2 weeks, caused fluid retention and hence an increase in body weight. The subject "escaped" from the sodium-retaining effects but continued to excrete increased amounts of potassium for as long as aldosterone was given. (From August JT, Nelson DH, Thorn GW. Response of normal subjects to large amounts of aldosterone. *J Clin Invest* 1958;37:1549–1559.)

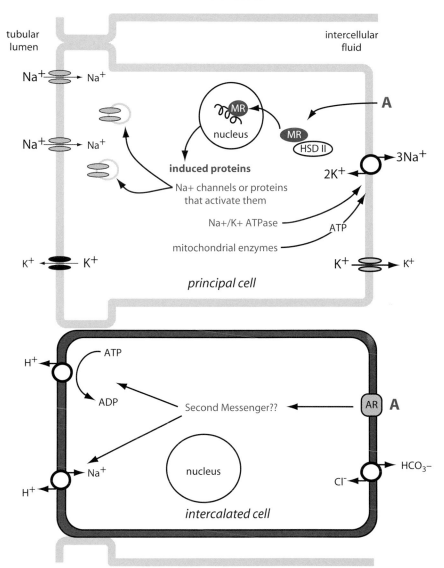

FIGURE 10 Proposed mechanisms for renal actions of aldosterone. In the principal cells of the cortical collecting ducts aldosterone (A) induces transcription and translation of new proteins. Note that entry of sodium into the principal cell and increased activity of the sodium potassium ATPase increase the electrochemical gradient that provides the driving force for outflow of potassium into the lumen or back into the interstitium. MR, mineralocorticoid receptor; HSD-II, 11-hydroxysteroid dehydrogenase II. In the intercalated cells, aldosterone promotes the secretion of protons by a mechanism that bypasses the nucleus and probably involves an aldosterone receptor on the cell surface (AR) and activation of some second messenger.

membranes, and these are in turn determined by the ionic composition of the interstitial fluid and the urine.

The principal cells in the connecting segments and the cortical collecting ducts are the most important targets for aldosterone. Aldosterone increases the sodium permeability of the luminal membrane. This action of aldosterone requires a lag period of at least 30 min, is sensitive to inhibitors of RNA and protein synthesis, and is mediated by transcriptional events

initiated by nuclear receptors (see Chapter 2). After binding aldosterone, the mineralocorticoid receptor in conjunction with other nuclear regulatory proteins regulates transcription of certain genes. Aldosterone induces transcription of sodium channel genes and may increase expression of proteins that transform nonconductive sodium channels to a functionally active state and perhaps insert sequestered channels into the luminal membrane. It also increases expression of sodium-potassium ATPase in the basolateral membrane by inducing its synthesis and membrane insertion. At the same time, aldosterone increases the capacity for ATP generation by promoting synthesis of some enzymes of the citric acid cycle in mitochondria, such as isocitrate dehydrogenase.

Aldosterone promotes hydrogen ion excretion through actions exerted on the intercalated cells of the distal tubule and collecting duct. The underlying mechanisms have not been firmly established, but aldosterone appears to stimulate an electrogenic proton pump (H^+-ATPase) in the luminal membrane. Proton secretion is accompanied by exchange of chloride for bicarbonate in the basolateral membrane. There is increasing evidence that aldosterone may also activate sodium:proton exchange. This effect is too rapid to depend on transcription and may be mediated by membrane-bound rather than nuclear receptors. Because sodium ions are almost a million times more abundant than hydrogen ions in extracellular fluid, secretion of protons in exchange for sodium accounts for only a minuscule fraction of the sodium that is retained. It is possible that transcriptional changes in the intercalated cells also contribute to the effects of aldosterone on acid excretion.

Similar responses to aldosterone are seen in extrarenal tissues. Aldosterone also promotes the absorption of sodium and secretion of potassium in the colon and decreases the ratio of sodium to potassium concentrations in sweat and salivary secretions. Formation of sweat and saliva is analogous to formation of urine. Initial secretions are ultrafiltrates of blood plasma whose ionic composition is modified by epithelial cells of the duct that carries the fluid from its site of generation to its site of release. Under the influence of aldosterone, sodium is reabsorbed, in exchange for potassium probably by the same cellular mechanism as described for the cells of the cortical collecting duct (Fig. 10). The effect of aldosterone on these secretions is not subject to the escape phenomenon. Because perspiration can be an important avenue for sodium loss, minimizing sodium loss in sweat is physiologically significant. Persons suffering from adrenal insufficiency are especially sensitive to

extended exposure to a hot environment and may become severely dehydrated.

The Cortisol/Cortisone Shuttle and the Mechanism of Mineralocorticoid Specificity

Receptors for adrenal steroids were originally classified based on their affinity and selectivity for mineralocorticoids or glucocorticoids. The mineralocorticoid receptors, also called type I receptors, have a high and nearly equal affinity for aldosterone and cortisol. The type II, or glucocorticoid receptors, have a considerably greater affinity for cortisol than for aldosterone. Expression of mineralocorticoid receptors is confined to aldosterone target tissues and the brain, whereas glucocorticoid receptor expression is widely disseminated. Both receptors reside in the cytosol bound to other proteins in the unstimulated state. Upon binding hormone, they release their associated proteins and migrate as dimers to the nucleus where they activate gene expression.

Because the mineralocorticoid receptor binds aldosterone and cortisol with equal affinity, it cannot distinguish between the two classes of steroid hormones. Nevertheless, even though the concentration of cortisol in blood is about 1000 times higher than that of aldosterone, mineralocorticoid responses normally reflect only the availability of aldosterone. This is due in part to differences in plasma protein binding; only 3–4% of cortisol is in free solution compared to nearly 40% of the aldosterone. Although hormone binding lowers the discrepancy in the available hormone concentrations by 10-fold, free cortisol is 100 times as abundant as free aldosterone, and it readily diffuses into mineralocorticoid target cells. Access to mineralocorticoid receptors, however, is guarded by the enzyme HSD-II, which colocalizes with the receptors and defends mineralocorticoid specificity. The high efficiency of this enzyme inactivates cortisol by converting it to cortisone, which is released into the blood (see Fig. 5). Consequently, the kidneys, which are the major targets for aldosterone, are the major source of circulating cortisone.

Clinical Note

Persons with a genetic defect in HSD-II suffer from symptoms of mineralocorticoid excess (hypertension and hypokalemia) as a result of constant saturation of the mineralocorticoid receptor by cortisol. An acquired form of the same ailment is seen after ingestion of excessive amounts of licorice, which contains the potent inhibitors of HSD, glycyrrhizic acid and its metabolite glycyrrhetinic acid.

Regulation of Aldosterone Secretion

Angiotensin II is the primary stimulus for aldosterone secretion, although ACTH and high concentrations of potassium are also potent stimuli. Angiotensin II is formed in blood by a two-step process that depends on proteolytic cleavage of the plasma protein, angiotensinogen, by the enzyme renin, to release the inactive decapeptide angiotensin I. Angiotensin I is then converted to angiotensin II by the ubiquitous angiotensin-converting enzyme. Control of angiotensin II production is achieved by regulating the secretion of renin from smooth muscle cells of the afferent glomerular arterioles. The principal stimulus for renin secretion is a decrease in the vascular volume, and the principal physiologic role of aldosterone is to defend the vascular volume. Aldosterone secretion is regulated by negative feedback, with vascular volume, not the concentration of aldosterone, as the controlled variable. Reabsorption of sodium is accompanied by a proportionate reabsorption of water, and because sodium remains extracellular, sodium retention expands the extracellular volume and hence blood volume. Expansion of the blood volume, which is the ultimate result of sodium retention, provides the negative feedback signal for regulation of renin and aldosterone secretion (Fig. 11). Although preservation of body sodium is central to aldosterone action, the concentration of sodium in blood does not appear to be monitored directly, and fluctuations in plasma concentrations have little direct effect on the secretion of renin.

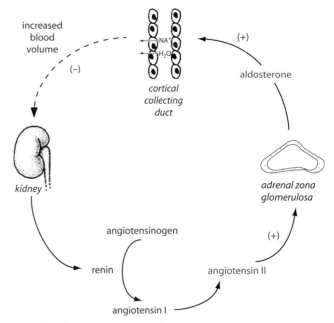

FIGURE 11 Negative feedback control of aldosterone secretion. The monitored variable is blood volume. (+), stimulates; (−), inhibits.

PHYSIOLOGY OF THE GLUCOCORTICOIDS

Although named for their critical role in maintaining carbohydrate reserves, glucocorticoids produce diverse physiologic actions, many of which are still not well understood and therefore can be considered only phenomenologically. Virtually every tissue of the body is affected by an excess or deficiency of glucocorticoids (Table 1). If any simple phrase could be used to describe the role of glucocorticoids, it would be *"coping with adversity."*

Even if sodium balance could be preserved and carbohydrate intake were adequate to meet energy needs, individuals suffering from adrenal insufficiency would still teeter on the brink of disaster when faced with a threatening environment. We consider here only the most thoroughly studied actions of glucocorticoids.

Effects on Energy Metabolism

Ability to maintain and draw on metabolic fuel reserves is ensured by the actions and interactions of many hormones and is critically dependent on normal function of the adrenal cortex (Fig. 12). Although we speak of maintaining carbohydrate reserves as the hallmark of glucocorticoid activity, it must be understood that metabolism of carbohydrate, protein, and lipid are inseparable components of overall energy balance. This complex topic is considered further in Chapter 42.

Clinical Note

In the absence of adrenal function, even relatively short periods of fasting may produce a catastrophic decrease in blood sugar (*hypoglycemia*) accompanied by depletion of muscle and liver glycogen. A drastically compromised ability to produce sugar from nonglucose precursors (*gluconeogenesis*) forces these individuals to rely almost exclusively on dietary sugars to meet their carbohydrate needs. Their metabolic problems are further complicated by decreased ability to utilize alternate substrates such as fatty acids and protein.

TABLE 1 Some Effects of Glucocorticoids

Tissue	Effects
Central nervous system	Taste, hearing, and smell ↑ in acuity with adrenal cortical insufficiency and ↓ in Cushing's disease ↓ Corticotropin releasing hormone (see text) ↓ ADH secretion
Cardiovascular system	Maintain sensitivity to epinephrine and norepinephrine ↑ Sensitivity to vasoconstrictor agents Maintain microcirculation
Gastrointestinal tract	↑ Gastric acid secretion ↓ Gastric mucosal cell proliferation
Liver	↑ Gluconeogenesis
Lungs	↑ Maturation and surfactant production during fetal development
Pituitary	↓ ACTH secretion and synthesis
Kidney	↑ GFR Needed to excrete dilute urine
Bone	↑ Resorption ↓ Formation
Muscle	↓ Fatigue (probably secondary to cardiovascular actions) ↑ Protein catabolism ↓ Glucose oxidation ↓ Insulin sensitivity ↓ Protein synthesis
Immune system (see text)	↓ Mass of thymus and lymph nodes ↓ Blood concentrations of eosinophils, basophils, and lymphocytes ↓ Cellular immunity
Connective tissue	↓ Activity of fibroblasts ↓ Collagen synthesis

Glucocorticoids promote gluconeogenesis by complementary mechanisms:

1. *Extrahepatic actions provide substrate.* Glucocorticoids promote proteolysis and inhibit protein synthesis in muscle and lymphoid tissues, thereby causing amino acids to be released into the blood. In addition, they increase blood glycerol concentrations by acting with other hormones to increase lipolysis in adipose tissue.

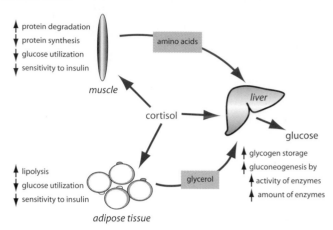

FIGURE 12   Principal effects of glucocorticoids on glucose production and metabolism of body fuels. ↑, increased; ↓, decreased.

2. *Hepatic actions enhance the flow of glucose precursors through existing enzymatic machinery and induce the synthesis of additional gluconeogenic and glycogen-forming enzymes along with enzymes needed to convert amino acids to usable precursors of carbohydrate.*

Nitrogen excretion during fasting is lower than normal with adrenal insufficiency, reflecting decreased conversion of amino acids to glucose. High concentrations of glucocorticoids, as seen in states of adrenal hyperfunction, inhibit protein synthesis and promote rapid breakdown of muscle and lymphoid tissues that serve as repositories for stored protein. These effects result in increased blood urea nitrogen (BUN) and enhanced nitrogen excretion.

Glucocorticoids defend against hypoglycemia in yet another way. In experimental animals glucocorticoids decrease utilization of glucose by muscle and adipose tissue and lower the responsiveness of these tissues to insulin. Prolonged exposure to high levels of glucocorticoids often leads to diabetes mellitus (Chapter 41); about 20% of patients with Cushing's disease are also diabetic, and virtually all of the remainder have some milder impairment of glucose metabolism. Despite the relative decrease in insulin sensitivity and increased tendency for fat mobilization seen in experimental animals, patients with Cushing's disease paradoxically accumulate fat in the face (moon face), between the shoulders (buffalo

---

Clinical Note

Individuals with hyperfunction of the adrenal cortex (Cushing's disease) characteristically have spindly arms and legs, reflecting increased breakdown of their muscle protein. Protein wasting in these patients may also extend to skin and connective tissue, and it contributes to their propensity to bruise easily.

hump), and in the abdomen (truncal obesity). As yet there is no explanation for this redistribution of body fat.

Water Balance

In the absence of the adrenal glands, renal plasma flow and glomerular filtration are reduced and it is difficult to produce either concentrated or dilute urine. One of the diagnostic tests for adrenal cortical insufficiency is the rapidity with which a water load can be excreted. Glucocorticoids facilitate excretion of free water and are more important in this regard than mineralocorticoids. The mechanism for this effect is still debated. It has been suggested that glucocorticoids may maintain normal rates of glomerular filtration by acting directly on glomeruli or glomerular blood flow, or indirectly by facilitating the action or production of the atrial natriuretic hormone (see Chapter 29). In addition, in the absence of glucocorticoids, antidiuretic hormone (vasopressin) secretion is increased.

Lung Development

One of the dramatic physiologic changes that must be accommodated in the newborn infant is the shift from the placenta to the lungs as site of oxygen and carbon dioxide exchange. Preparation of fetal lungs for this transition includes thinning of the alveolar walls to diminish the diffusion barrier and the production of surfactant. Surfactant consists largely of phospholipids associated with specific proteins. It reduces alveolar surface tension, increases lung compliance, and allows even distribution of inspired air. Glucocorticoids play a crucial role in maturation of alveoli and production of surfactant. A major problem of preterm delivery is a condition known as *respiratory distress syndrome,* which is caused by impaired pulmonary mechanics resulting from incomplete development of pulmonary alveoli and production of surfactant. Although the fetal adrenal gland is capable of secreting some glucocorticoids by about the 24th week of pregnancy, the major secretory products of the fetal adrenal gland are androgens that serve as precursors for placental estrogen synthesis (see Chapter 47). Administration of large doses of glucocorticoid to mothers carrying fetuses at risk for premature delivery diminishes the incidence of respiratory distress syndrome. Steady-state levels of mRNA for surfactant proteins may occur within about 15 hr after exposure of fetuses to glucocorticoids.

Glucocorticoids and Responses to Injury

One of the most remarkable effects of glucocorticoids was discovered almost by chance during the late 1940s when it was observed that glucocorticoids dramatically reduced the severity of disease in patients suffering from rheumatoid arthritis. This observation called attention to the *anti-inflammatory effects* of glucocorticoids. Because supraphysiologic concentrations are needed to produce these effects therapeutically, some investigators consider them to be pharmacological side effects. As understanding of the anti-inflammatory actions has increased, however, it is apparent that glucocorticoids are physiologic modulators of the inflammatory response through actions mediated by the activated glucocorticoid receptor. It is likely that free cortisol concentrations are higher at local sites of tissue injury than in the general circulation. Partial degradation of CBG by the proteolytic enzyme elastase, which is secreted by activated mononuclear leukocytes, decreases its affinity for cortisol and raises the concentration of free cortisol at the site of inflammation. In addition, inflammatory mediators up-regulate 11 βHSD-I in some cells and cause localized conversion of circulating cortisone to cortisol. As might be anticipated, glucocorticoids and related compounds devised by the pharmaceutical industry are exceedingly important therapeutic agents for treating such diverse conditions as poison ivy, asthma, a host of inflammatory conditions, and various autoimmune diseases. The latter reflects their related ability to diminish the immune response.

Anti-Inflammatory Effects

Inflammation is the term used to encompass the various responses of tissues to injury. It is characterized by redness, heat, swelling, pain, and loss of function. Redness and heat are manifestations of increased blood flow and result from vasodilation. Swelling is due to formation of a protein-rich exudate that collects because capillaries and venules become leaky to proteins. Pain is caused by chemical products of cellular injury and sometimes by mechanical injury to nerve endings. Loss of function may be a direct consequence of injury or secondary to the pain and swelling that injury evokes. An intimately related component of the early response to tissue injury is the migration of white blood cells to the injured area and the subsequent recruitment of the immune system.

The initial pattern of the inflammatory response is independent of the injurious agent or causal event. This response is presumably defensive and may be a necessary antecedent of the repair process. Increased blood flow accelerates delivery of the white blood cells that combat invading foreign substances or organisms and clean up the debris of injured and dead cells. Increased blood flow also facilitates dissemination of chemoat-tractants to white blood cells and promotes their migration to the site of injury. In addition, increased blood flow provides more oxygen and nutrients to cells at the site of damage and facilitates removal of toxins and wastes. Increased permeability of the microvascu-lature allows fluid to accumulate in the extravascular space in the vicinity of the injury and thus dilute noxious agents.

Although we are accustomed to thinking of physiolo-gic responses as having beneficial effects, it is apparent that some aspects of inflammation may actually cause or magnify tissue damage. Lysosomal hydrolases released during phagocytosis of cellular debris or invad-ing organisms may damage nearby cells that were not harmed by the initial insult. Loss of fluid from the microvasculature at the site of injury may increase viscosity of the blood, slowing its flow, and even leaving some capillaries clogged with stagnant red blood cells. Decreased perfusion may cause further cell damage. In addition, massive disseminated fluid loss into the extra-vascular space sometimes compromises cardiovascular function. Consequently long-term survival demands that checks and balances be in place to prevent the defensive and positive aspects of the inflammatory response from becoming destructive. We may regard the physiologic role of the glucocorticoids to modulate inflammatory responses as a major component of such checks and balances. Exaggeration of such physiologic modulation with supraphysiologic amounts of glucocorticoids upsets the balance in favor of suppression of inflammation and provides the therapeutic efficacy of pharmacologic treatment.

Inflammation is initiated, sustained, and amplified by the release of a large number of chemical mediators derived from multiple sources. Cytokines are a diverse group of peptides that range in size from about 8–40 kDa and are produced mainly by cells of the hemato-poietic and immune systems, but many can be synthe-sized and secreted by virtually any cell. Cytokines may promote or antagonize development of inflammation or have a mixture of pro- and anti-inflammatory effects depending on the particular cells involved. Prostaglan-dins and leukotrienes are released principally from vascular endothelial cells and macrophages, but virtually all cell types can produce and release them. They may also produce either pro- or anti-inflammatory effects depending on the particular compound formed and the cells acted on. Histamine and serotonin are released from mast cells and platelets. Enzymes and superoxides released from dead or dying cells or from cells that remove debris by phagocytosis contribute directly and indirectly to the spread of inflammation by activating other mediators (e.g., bradykinin) and leukocyte attractants that arise from humoral precursors associ-ated with the immune and clotting systems.

Glucocorticoids and the Metabolites of Arachidonic Acid

Prostaglandins and the closely related leukotrienes are derived from the polyunsaturated essential fatty acid arachidonic acid (Fig. 13). Because of their 20-carbon backbone they are also sometimes referred to collectively as *eicosanoids*. These compounds play a central role in the inflammatory response. They generally act locally on cells in the immediate vicinity of their production, including the cells that produced them, but some also survive in blood long enough to act on distant tissues. Prostaglan-dins act directly on blood vessels to cause vasodilation and indirectly increase vascular permeability by poten-tiating the actions of histamine and bradykinin. Prosta-glandins sensitize nerve endings of pain fibers to other mediators of inflammation such as histamine, serotonin, bradykinin, and substance P, thereby producing increased sensitivity to touch (*hyperalgesia*). The leukotrienes stimulate production of cytokines and act directly on the microvasculature to increase permeability. Leuko-trienes also attract white blood cells to the site of injury and increase their stickiness to vascular endothelium. The physiology of arachidonate metabolites is complex, and a thorough discussion is not possible here. There are a large number of these compounds with different biological activities. Although some eicosanoids have anti-inflam-matory actions that may limit the overall inflammatory response, arachidonic acid derivatives are major contri-butors to inflammation.

Arachidonic acid is released from membrane phos-pholipids by phospholipase A_2 (PLA_2; see Chapter 2), which is activated by injury, phagocytosis, or a variety of other stimuli in responsive cells. Activation is mediated by a cytosolic PLA_2 activating protein that closely resem-bles a protein in bee venom called *mellitin*. In addition, PLA_2 activity also increases as a result of increased enzyme synthesis. The first step in the production of prostaglandins from arachidonate is catalyzed by a cytosolic enzyme, cyxlooxygenase. One isoform of this enzyme, called COX 1, is constitutively expressed. A second form, COX 2, is induced by the inflammatory response. Glucocorticoids suppress the formation of prostaglandins by inhibiting synthesis of COX 2 and probably also by inducing expression of a protein that inhibits PLA_2. Nonsteroidal anti-inflammatory drugs such as indomethacin and aspirin also block the cyclooxy-genase reaction catalyzed by both COX 1 and COX 2. Some of the newer anti-inflammatory drugs specifically block COX 2 and hence may target inflammation more specifically.

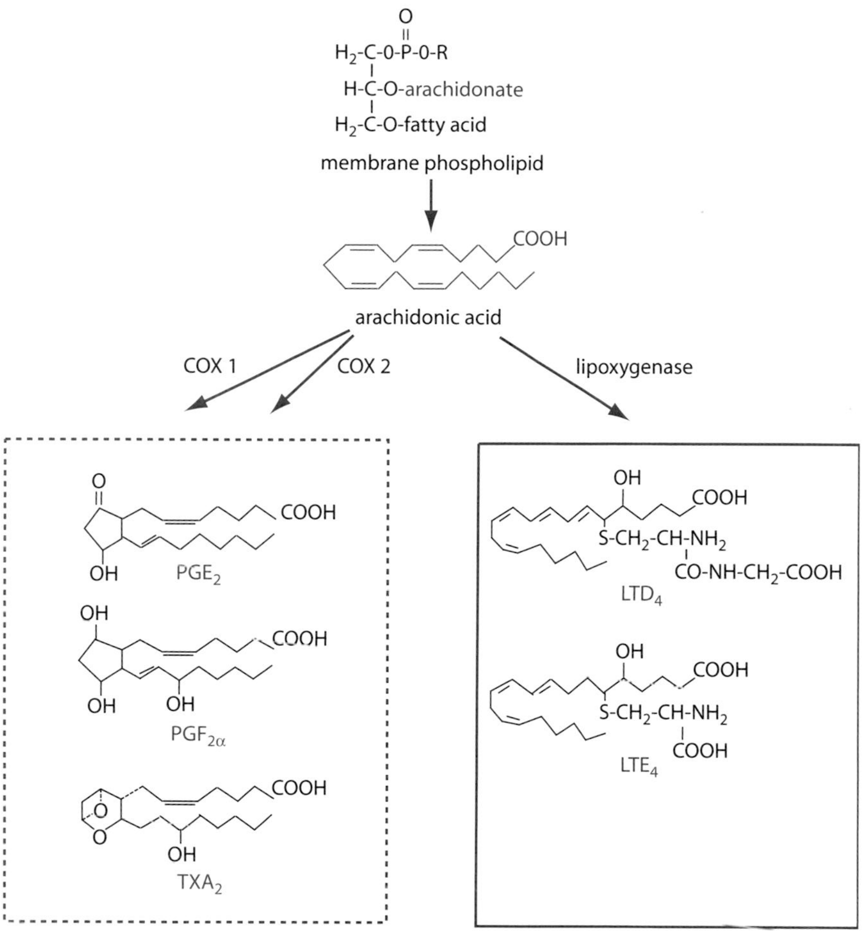

FIGURE 13 Synthesis and structures of some arachidonic acid metabolites. R may be choline, inositol, serine or ethanolamine. PG, prostaglandin; LT, leukotriene. The designations E_2 or $F_{2\alpha}$ refer to substituents on the ring structure of the PG. The designations D_4 and E_4 refer to glutathione derivatives in thioester linkage at carbon 6 of LT.

Glucocorticoids and Cytokines

The large number of compounds designated as cytokines include one or more isoforms of the interleukins (IL-1 through IL-18), tumor necrosis factor (TNF), the interferons (IFN-α, -β, -γ), colony stimulating factor (CSF), granulocyte-macrophage colony-stimulating factor (GM-CSF), transforming growth factor (TGF), leukemia inhibiting factor, oncostatin, and a variety of cell- or tissue-specific growth factors. It is not clear just how many of these hormone-like molecules are produced, and not all have a role in inflammation. Two of these factors, IL-1β and TNFα are particularly important in the development of inflammation. The intracellular signaling pathways and biological actions of these two cytokines are remarkably similar. They enhance each other's actions in the inflammatory response and differ only in the respect that TNFα may promote cell death (apoptosis), whereas IL-1 does not.

Interleukin-1 is produced primarily by macrophages and to a lesser extent by other connective tissue elements, skin, and endothelial cells. Its release from macrophages is stimulated by interaction with immune complexes, activated lymphocytes, and metabolites of arachidonic acid, especially leukotrienes. IL-1 is not stored in its cells of origin but is synthesized and secreted within hours of stimulation in a response mediated by increased intracellular calcium and protein kinase C. IL-1 acts on many cells to produce a variety of responses, all of which are components of the inflammatory/immune response. They are illustrated in Fig. 14. Many of the consequences of these actions can be recognized from personal experience as nonspecific symptoms of viral infection. TNFα is also produced in macrophages and other cells in response to injury and immune complexes, and it can act on many cells including those that secrete it. Secretion of both IL-1 and TNFα and

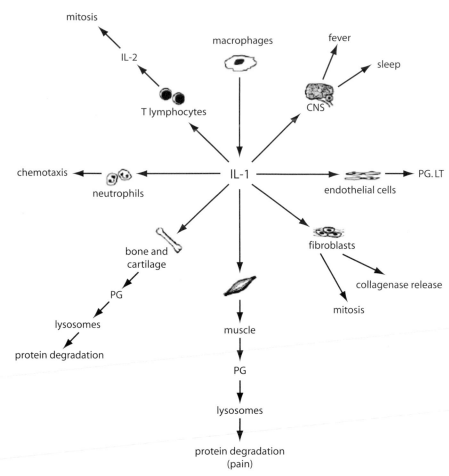

FIGURE 14   Effects of IL-1. PG, prostaglandin; LT, leukotriene.

their receptors are increased by some of the cytokines and other mediators of inflammation whose production they increase, so that an amplifying positive feedback cascade is set in motion. Some products of these cytokines also feed back on their own production in a negative way to modulate the inflammatory response. Glucocorticoids play an important role as negative modulators of IL-1 and TNFα by (1) inhibiting their production, (2) interfering with signaling pathways, and (3) inhibiting the actions of their products. Glucocorticoids also interfere with the production and release of other pro-inflammatory cytokines as well, including IFN-γ, IL-2, IL-6, and IL-8.

Production of IL-1 and TNFα and many of their effects on target cells are mediated by activation of genes by the transcription factor, nuclear factor kappa B (NF-κB). In the unactivated state NF-κB resides in the cytoplasm bound to the NF-κB inhibitor (I-κB). Activation of the signaling cascade by some tissue insult or the binding of IL-1 and TNFα to their respective receptors is initiated by activation of a kinase (IκK) that phosphorylates I-κB, causing it to dissociate from NF-κB and to be degraded. Free NF-κB is then able to translocate to the nucleus where it binds to response elements in genes that it regulates, including genes for the cytokines IL-1, TNFα,, IL-6, and IL-8 and for such enzymes as PLA$_2$, COX$_2$, and nitric oxide synthase (Fig. 15). IL-6 is an important pro-inflammatory cytokine that acts on the hypothalamus, liver, and other tissues, and IL-8 plays an important role as a leukocyte attractant. Nitric oxide is important as a vasodilator and may have other effects as well. Glucocorticoids interfere with the actions of IL-1 and TNFα by promoting the synthesis of I-κB, which traps NF-κB in the cytosol, and by interfering with the ability of NF-κB that enters the nucleus to activate target genes. The mechanism for interference with gene activation is thought to involve protein–protein interaction between the liganded glucocorticoid receptor and NF-κB. Glucocorticoids also appear to interfere with IL-1 or TNFα-dependent activation of other genes by the AP-1 transcription complex. In addition, cortisol induces expression of a protein that inhibits PLA$_2$ and destabilizes the mRNA for COX$_2$. It is noteworthy that many of the responses attributed to IL-1 may be mediated by prostaglandins or other arachidonate metabolites. For example, IL-1, which is identical with what was once

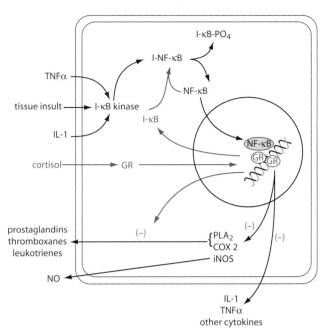

FIGURE 15   Anti-inflammatory actions of cortisol. Cortisol induces the formation of the inhibitor of nuclear factor κB (I-κB), which binds to nuclear factor κB (NF-κB) and prevents it from entering the nucleus and activating target genes. The activated glucocorticoid receptor (GR) also interferes with NF-κB binding to its response elements in DNA, thus preventing induction of phospholipase A_2 (PLA$_2$), cyclooxygenase 2 (COX$_2$), and inducible nitric oxide synthase (iNOS). TNFα, tumor necrosis factor-α; IL-1, interleukin-1; NO, nitric oxide.

called *endogenous pyrogen*, may cause fever by inducing the formation of prostaglandins in the thermoregulatory center of the hypothalamus. Glucocorticoids might therefore exert their antipyretic effect at two levels: at the level of the macrophage by inhibiting IL-1 production, and at the level of the hypothalamus by interfering with prostaglandin synthesis.

Glucocorticoids and the Release of Other Inflammatory Mediators

Granulocytes, mast cells, and macrophages contain vesicles filled with serotonin, histamine, or degradative enzymes, all of which contribute to the inflammatory response. These mediators and lysosomal enzymes are released in response to arachidonate metabolites, cellular injury, reaction with antibodies, or during phagocytosis of invading pathogens. Glucocorticoids protect against the release of all of these compounds by inhibiting cellular degranulation. It has also been suggested that glucocorticoids inhibit histamine formation and stabilize lysosomal membranes, but the molecular mechanisms for these effects are unknown.

Glucocorticoids and the Immune Response

The immune system, whose function is destruction and elimination of foreign substances or organisms, has two major components: the B lymphocytes which are formed in bone marrow and develop in liver or spleen, and the thymus-derived T lymphocytes. Humoral immunity is the province of B lymphocytes, which, on differentiation into plasma cells, are responsible for production of antibodies. Large numbers of B lymphocytes circulate in blood or reside in lymph nodes. Reaction with a foreign substance (antigen) stimulates B cells to divide and produce a clone of cells capable of recognizing the antigen and producing antibodies to it. Such proliferation depends on cytokines released from the macrophages and helper T cells. Antibodies, which are circulating immunoglobulins, bind to foreign substances and thus mark them for destruction. Glucocorticoids inhibit cytokine production by macrophages and T cells and thus decrease normal proliferation of B cells and reduce circulating concentrations of immunoglobulins. At high concentrations glucocorticoids may also act directly on B cells to inhibit antibody synthesis and may even kill B cells by activating apoptosis (programmed cell death).

T cells are responsible for cellular immunity, and they participate in destruction of invading pathogens or cells that express foreign surface antigens as might follow viral infection or transformation into tumor cells. IL-1 stimulates T lymphocytes to produce IL-2, which promotes proliferation of T lymphocytes that have been activated by coming in contact with antigens. Antigenic stimulation triggers the expression of IL-2 receptors only in those T cells that recognize the antigen. Consequently, only certain clones of T cells are stimulated to divide because there are no receptors for IL-2 on the surface membranes of T lymphocytes until they interact with their specific antigens. Glucocorticoids block the production of IL-2, but probably not the response to it and thereby inhibit proliferation of T lymphocytes. IL-2 also stimulates T lymphocytes to produce IFN-γ, which participates in destruction of virus-infected or tumor cells and also stimulates macrophages to produce IL-1. Macrophages, T lymphocytes, and secretory products are thus arranged in a positive feedback relationship and produce a self-amplifying cascade of responses. Glucocorticoids restrain the cycle by suppressing production of each of the mediators. Glucocorticoids also activate apoptosis in some T lymphocytes.

The physiologic implications of the suppressive effects of glucocorticoids on humoral and cellular immunity are incompletely understood. It has been suggested that suppression of the immune response might prevent development of autoimmunity that might otherwise follow from the release of fragments of injured cells. However, it must be pointed out that much of the immunosuppression by glucocorticoids requires concentrations that may never be reached under physiologic conditions. High doses of glucocorticoids can so impair immune responses that relatively innocuous infections with some organisms can become overwhelming and cause death. Thus, excessive anti-immune or anti-inflammatory influences are just as damaging as unchecked immune or inflammatory responses. Under normal physiologically relevant circumstances, these influences are balanced and protective.

Other Effects of Glucocorticoids on Lymphoid Tissues

Sustained high concentrations of glucocorticoids produce a dramatic reduction in the mass of all lymphoid tissues including thymus, spleen, and lymph nodes. The thymus contains germinal centers for lymphocytes, and large numbers of T lymphocytes are formed and mature within it. Lymph nodes contain large numbers of both T and B lymphocytes. Immature lymphocytes of both lineages have glucocorticoid receptors and respond to hormonal stimulation by the same series of events as is seen in other steroid-responsive cells except that the DNA transcribed expresses the program for apoptosis. Loss in mass of thymus and lymph nodes can be accounted for by the destruction of lymphocytes rather than the stromal or supporting elements. Mature lymphocytes and germinal centers seem to be unresponsive to this action of glucocorticoids.

Glucocorticoids also decrease circulating levels of lymphocytes and particularly a class of white blood cells known as *eosinophils* (for their cytological staining properties). This decrease is partly due to apoptosis and partly to sequestration in the spleen and lungs. Curiously, the total white blood cell count does not decrease because glucocorticoids also induce a substantial mobilization of neutrophils from bone marrow.

Maintenance of Vascular Responsiveness to Catecholamines

A final action of glucocorticoids relevant to inflammation and the response to injury is maintenance of sensitivity of vascular smooth muscle to vasoconstrictor effects of norepinephrine released from autonomic nerve endings or the adrenal medulla. By counteracting local vasodilator effects of inflammatory mediators, norepinephrine decreases blood flow and limits the availability of fluid to form the inflammatory exudate. In addition, arteriolar constriction decreases capillary and venular pressure and favors reabsorption of extracellular fluid, thereby reducing swelling. The vasoconstrictor action of norepinephrine is compromised in the absence of glucocorticoids. The mechanism for this action is not known, but at high concentrations glucocorticoids may block inactivation of norepinephrine.

Adrenal Cortical Function during Stress

During the mid-1930s the Canadian endocrinologist Hans Selye observed that animals respond to a variety of seemingly unrelated threatening or noxious circumstances with a characteristic pattern of changes that include an increase in size of the adrenal glands, involution of the thymus, and a decrease in the mass of all lymphoid tissues. He inferred that the adrenal glands are stimulated whenever an animal is exposed to any unfavorable circumstance, which he called *stress*. Stress does not directly affect adrenal cortical function but, rather, increases the output of ACTH from the pituitary gland (see later discussion). In fact, stress is now defined operationally by endocrinologists as any of the variety of conditions that increase ACTH secretion.

Although it is clear that relatively benign changes in the internal or external environment may become lethal in the absence of the adrenal glands, we understand little more than Selye did about what cortisol might be doing to protect against stress. The favored experimental model used to investigate this problem was the adrenalectomized animal, which might have further complicated an already complex experimental question. It appears that many cellular functions require glucocorticoids either directly or indirectly for their maintenance, suggesting that these steroid hormones govern some process that is fundamental to normal operation of most cells. Consequently, without replacement therapy many systems are functioning only marginally even before the imposition of stress. Any insult may therefore prove overwhelming. It further became apparent that gluco-corticoids are required for normal responses to other hormones or to drugs, even though these steroid hormones themselves do not initiate similar responses in the absence of these agents.

Treatment of adrenalectomized animals with a constant basal amount of glucocorticoid prior to and during a stressful incident prevented the devastating effects of stress and permitted expression of expected responses to stimuli. This finding introduced the idea that glucocorticoids act in a normalizing, or *permissive,* way. That is, by maintaining normal operation of cells, gluco-corticoids permit normal regulatory mechanisms to act. Because it was not necessary to increase the amounts of adrenal corticoids to ensure survival of stressed adrena-lectomized animals, it was concluded that increased secretion of glucocorticoids was not required to combat stress. However, this conclusion is not consistent with clinical experience. Persons suffering from pituitary insufficiency or who have undergone hypophysectomy have severe difficulty withstanding stressful situations even though at other times they get along reasonably well on the small amounts of glucocorticoids produced by their adrenals in the absence of ACTH. Patients suffering from adrenal insufficiency are routinely given increased doses of glucocorticoids before undergoing surgery or other stressful procedures. We have already seen that glucocorticoids suppress the inflammatory response. It is also known that these hormones increase the sensitivity of various tissues to epinephrine and norepinephrine, which are also secreted in response to stress (see later discussion). Although we still do not understand the role of increased concentrations of glucocorticoids in the physiologic response to stress, it appears likely that they are beneficial. The question remains open, however, and will not be resolved until a better understanding of glucocorticoid actions is obtained.

Mechanism of Action of Glucocorticoids

With few exceptions, the physiologic actions of cortisol at the molecular level fit the general pattern of steroid hormone action described in Chapter 2. The gene for the glucocorticoid receptor gives rise to two isoforms as a result of alternate splicing of RNA. The alpha isoform binds glucocorticoids, sheds its associated proteins and migrates to the nucleus where it can form homodimers that bind to response elements in target genes. The beta isoform cannot bind hormone, is constitutively located in the nucleus, and apparently cannot bind to DNA. The beta isoform, however, can dimerize with the alpha isoform and diminish or block the ability of the alpha isoform to activate transcription.

Some evidence suggests that formation of the beta isoform may be a regulated process that modulates glucocorticoid responsiveness.

Glucocorticoids act on a great variety of cells and produce a wide range of effects that depend on activating or suppressing transcription of specific genes. The ability to regulate different genes in different tissues presumably reflects differing accessibility of the activated glucocorticoid receptor to glucocorticoid-responsive genes in each differentiated cell type, and presumably reflects the presence or absence of different coactivators and corepressors. Glucocorticoids also inhibit expression of some genes that lack glucocorticoid response elements. Such inhibitory effects are thought to be the result of protein–protein interactions between the glucocorticoid receptor and other transcription factors to modify their ability to activate gene transcription. The mechanisms for such interference are the subject of active research. The glucocorticoid receptor can be phosphorylated to various degrees on serine residues. Phosphorylation may modulate the affinity of the receptor for hormone, or DNA, or may modify its ability to interact with other proteins.

Regulation of Glucocorticoid Secretion

Secretion of glucocorticoids is regulated by the anterior pituitary gland through the hormone ACTH, whose effects on the inner zones of the adrenal cortex have already been described (see earlier discussion). In the absence of ACTH, the concentration of cortisol in blood decreases to very low values, and the inner zones of the adrenal cortex atrophy. Regulation of ACTH secretion requires vascular contact between the hypothalamus and the anterior lobe of the pituitary gland, and it is driven primarily by corticotropin-releasing hormone (CRH). CRH-containing neurons are widely distributed in the forebrain and brain stem but are heavily concentrated in the paraventricular nuclei in close association with vasopressin-secreting neurons. They stimulate the pituitary to secrete ACTH by releasing CRH into the hypophysial-portal capillaries (Chapter 38). Vasopressin (AVP) also exerts an important influence on ACTH secretion by augmenting the response to CRH. AVP is co-secreted with CRH, particularly in response to stress. Note that the AVP that is secreted into the hypophysial-portal vessels along with CRH arises in a different population of paraventricular neurons from those that produce the AVP that is secreted by the posterior lobe of the pituitary in response to changes in blood osmolality or volume (Chapter 38).

CRH binds to G-protein-coupled receptors in the corticotrope membrane and activates adenylyl cyclase.

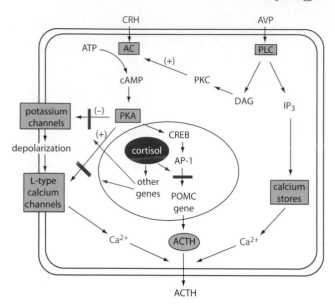

FIGURE 16 Hormonal interactions that regulate ACTH secretion by pituitary corticotropes. CRH, corticotropin releasing hormone; AVP, arginine vasopressin; AC, adenylyl cyclase; PLC, phospholipase C; ATP, adenosine triphosphate; cAMP, cyclic adenosine monophosphate; PKC, protein kinase C; DAG, diacylglycerol; IP$_3$, inositol trisphosphate; PKA, protein kinase A; CREB, cyclic AMP response element binding protein; AP-1, activator protein-1; POMC, proopiomelanocortin.

The resulting increase in cyclic AMP activates protein kinase A, which directly or indirectly inhibits potassium outflow through at least two classes of potassium channels. Buildup of positive charge within the corticotrope decreases the membrane potential and results in calcium influx through activation of voltage-sensitive calcium channels. Direct phosphorylation of these channels may enhance calcium entry by lowering their threshold for activation. Increased intracellular calcium and perhaps additional effects of protein kinase A on secretory vesicle trafficking trigger ACTH secretion. Protein kinase A also phosphorylates CREB, which initiates production of the AP-1 nuclear factor that activates POMC transcription. AVP binds to its G-protein-coupled receptor and activates phospholipase C to cause the release of DAG and IP$_3$. This action of AVP has little effect on CRH secretion in the absence of CRH, but in its presence amplifies the effects of CRH on ACTH secretion without affecting synthesis. As described in Chapter 2, IP$_3$ stimulates release of calcium from intracellular stores, and DAG activates protein kinase C, although the role of this enzyme in ACTH secretion is unknown. These effects are summarized in Fig. 16.

Upon stimulation with ACTH, the adrenal cortex secretes cortisol, which inhibits further secretion of ACTH in a typical negative feedback arrangement (Fig. 17). Cortisol exerts its inhibitory effects both on CRH neurons in the hypothalamus and on corticotropes

in the anterior pituitary. These effects are mediated by the glucocorticoid receptor. The negative feedback effects on secretion depend on transcription of genes that code for proteins that either activate potassium channels or block the effects of PKA-catalyzed phosphorylation on these channels and may also act at the level of secretory vesicle trafficking. Initial actions of glucocorticoids suppress secretion of CRH and ACTH from storage granules. Subsequent actions of glucocorticoids result from inhibition of transcription of the genes for CRH and POMC in hypothalamic neurons and corticotropes perhaps by direct interaction of the glucocorticoid receptor with transcription factors that regulate synthesis of CRH and POMC. The inhibitory effect of cortisol on the corticotrope are shown in Fig. 16. This feedback system closely resembles the one described earlier for regulation of thyroid hormone secretion even though the adrenal-ACTH system is much more dynamic and subject to episodic changes.

The relative importance of the pituitary and the CRH-producing neurons of the paraventricular nucleus for negative feedback regulation of ACTH secretion has been explored in mice that were made deficient in CRH by disruption of the CRH gene. These CRH knockout mice secrete normal basal levels of ACTH and glucocorticoid,

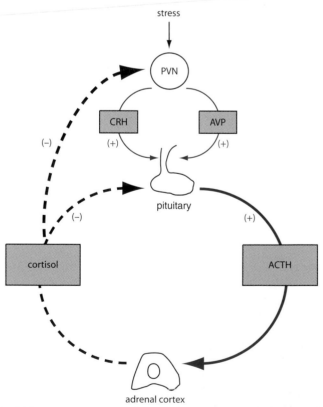

FIGURE 17 Negative feedback control of glucocorticoid secretion. PVN, paraventricular nuclei; CRH, corticotropin releasing hormone; AVP, arginine vasopressin; (+), stimulates; (−), inhibits.

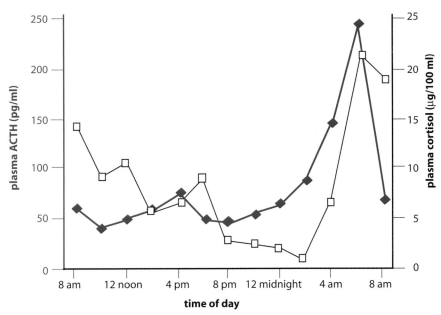

FIGURE 18 Variations in plasma concentrations of ACTH and cortisol at different times of day. (From Matsukura S, West CD, Ichikawa Y, Jubiz W, Harada G, Tyler FH. A new phenomenon of usefulness in the radioimmunoassay of plasma adrenocortico-tropic hormone. *J Lab Clin Med* 1971;77:490–500.)

and their corticotropes express normal levels of mRNA for POMC. In normal mice, disruption of negative feedback by surgical removal of the adrenal glands results in a prompt increase both in POMC gene expression and ACTH secretion. Adrenalectomy of CRH knockout mice produces no increase in ACTH secretion, although POMC mRNA increases normally. These animals also suffer a severe impairment, but not total lack of ACTH secretion in response to stress. Thus it seems that basal function of the pituitary/adrenal negative feedback system does not require CRH, but that CRH is crucial for increasing ACTH secretion above basal levels. Further, it appears that transcription of the POMC gene is inhibited by glucocorticoids even under basal conditions.

It was pointed out earlier that negative feedback systems ensure constancy of the controlled variable. However, even in the absence of stress, ACTH and cortisol concentrations in blood plasma are not constant but oscillate with a 24-hr periodicity. This so-called *circadian rhythm* is sensitive to the daily pattern of physical activity. For all but those who work the night shift, hormone levels are highest in the early morning hours just before arousal and lowest in the evening (Fig. 18). This rhythmic pattern of ACTH secretion is consistent with the negative feedback model shown in Fig. 17 and is sensitive to glucocorticoid input throughout the day. In the negative feedback system, the positive limb (CRH and ACTH secretion) is inhibited when the negative limb (cortisol concentration in blood) reaches

some set point. For basal ACTH secretion, the set point of the corticotropes and the CRH-secreting cells is thought to vary in its sensitivity to cortisol at different times of day. Decreased sensitivity to inhibitory effects of cortisol in the early morning results in increased output of CRH, ACTH, and cortisol. As the day progresses, sensitivity to cortisol increases, and there is a decrease in the output of CRH and consequently of ACTH and cortisol. The cellular mechanisms underlying the periodic changes in set point are not understood, however, *although they vary with time of day, cortisol concentrations in blood are precisely controlled throughout the day.*

Negative feedback also governs the response of the pituitary-adrenal axis to most stressful stimuli. Different mechanisms appear to apply at different stages of the response. With the imposition of a stressful stimulus, a sharp increase in ACTH secretion occurs that is driven by CRH and AVP. The rate of ACTH secretion is determined by both the intensity of the stimulus to CRH-secreting neurons and the negative feedback influence of cortisol. In the initial moments of the stress response, pituitary corticotropes and CRH neurons monitor the rate of change rather than the absolute concentration of cortisol and decrease their output accordingly. After about 2 hr, negative feedback seems to be proportional to the total amount of cortisol secreted during the stressful episode. With chronic stress a new steady state is reached, and the negative feedback system again seems to monitor the concentration of cortisol in blood but with the set point readjusted at a higher level.

Each phase of negative feedback involves different cellular mechanisms. During the first few minutes, the inhibitory effects of cortisol occur without a lag period and are expressed too rapidly to be mediated by altered gene expression. Indeed the rapid inhibitory action of cortisol is unaffected by inhibitors of protein synthesis. Its molecular basis is unknown, but it may be mediated by nongenomic responses of receptors in neuronal membranes. The negative feedback effect of cortisol in the subsequent interval occurs after a lag period and seems to require RNA and protein synthesis typical of the steroid actions discussed earlier. In this phase cortisol restrains secretion of CRH and ACTH but not their synthesis. At this time, corticotropes are less sensitive to CRH. With chronic administration of glucocorticoids or with chronic stress, negative feedback is also exerted at the level of POMC gene transcription and translation.

Major features of the regulation of ACTH secretion include the following:

1. Basal secretion of ACTH follows a diurnal rhythm driven by CRH and perhaps by intrinsic rhythmicity of the corticotropes.
2. Stress increases CRH and AVP secretion through neural pathways.
3. ACTH secretion is subject to negative feedback control under basal conditions and during the response to most stressful stimuli.
4. Cortisol inhibits secretion of both CRH and ACTH.

Some observations suggest that cytokines produced by cells of the immune system may directly affect secretion by the hypothalamic-pituitary-adrenal axis. In particular, IL-1, IL-2, and IL-6 stimulate CRH secretion, and may also act directly on the pituitary to increase ACTH secretion. IL-2 and IL-6 may also stimulate cortisol secretion by a direct action on the adrenal gland. In addition, lymphocytes express ACTH and related products of the POMC gene and are responsive to the stimulatory effects of CRH and the inhibitory effects of glucocorticoids. Because glucocorticoids inhibit cytokine production, there is another negative feedback relationship between the immune system and the adrenals (Fig. 19). It has been suggested that this communication between the endocrine and immune systems provides a mechanism to alert the body to the presence of invading organisms or antigens.

In our discussion of the regulation of cortisol and ACTH secretion, we have ignored other members of the ACTH family that reside in the same secretory granule and are released along with ACTH. Endocrinologists have focused their attention on the physiologic implications of increased secretion of ACTH and glucocorticoids in response to stress. Recent observations suggest that other peptides such as β-endorphin and α-MSH, whose concentration in blood increases in parallel with ACTH may exert anti-inflammatory actions.

ADRENAL MEDULLA

The adrenal medulla accounts for about 10% of the mass of the adrenal gland and is embryologically and physiologically distinct from the cortex, although cortical and medullary hormones often act in a complementary

---

Clinical Note

Understanding of the negative feedback relation between the adrenal and pituitary glands has important diagnostic and therapeutic applications. Normal adrenocortical function can be suppressed by injection of large doses of glucocorticoids. For these tests a potent synthetic glucocorticoid, usually dexamethasone, is administered, and at a predetermined time later the natural steroids or their metabolites are measured in blood or urine. If the hypothalamo-pituitary-adrenal system is intact, production of cortisol is suppressed and its concentrations in blood is low. If on the other hand, cortisol concentrations remain high, an autonomous adrenal or ACTH-producing tumor may be present.

Another clinical application is treatment of the adrenogenital syndrome. As pointed out earlier, adrenal glands produce androgenic steroids by extension of the synthetic pathway for glucocorticoids (see Fig. 4). Defects in production of glucocorticoids, particularly in enzymes responsible for hydroxylation of carbons 21 or 11, may lead to increased production of adrenal androgens. Overproduction of androgens in female patients leads to masculinization, which is manifest, for example, by enlargement of the clitoris, increased muscular development, and growth of facial hair. Severe defects may lead to masculinization of the genitalia of female infants and in male babies produce the supermasculinized

manner. Cells of the adrenal medulla have an affinity for chromium salts in histological preparations and hence are called *chromaffin cells*. They arise from neuroectoderm and are innervated by neurons whose cell bodies lie in the intermediolateral cell column in the thoracolumbar portion of the spinal cord. Axons of these cells pass through the paravertebral sympathetic ganglia to form the splanchnic nerves. Chromaffin cells are thus modified postganglionic neurons. Their principal secretory products, epinephrine and norepinephrine, are derivatives of the amino acid tyrosine and belong to a class of compounds called *catecholamines*. About 5–6 mg of catecholamines are stored in membrane-bound granules within chromaffin cells. Epinephrine is about five times as abundant in the human adrenal medulla as norepinephrine, but only norepinephrine is found in postganglionic sympathetic neurons and extra-adrenal chromaffin tissue. Although medullary hormones affect virtually every tissue of the body and play a crucial role in the acute response to stress, the adrenal medulla is not required for survival as long as the rest of the sympathetic nervous system is intact.

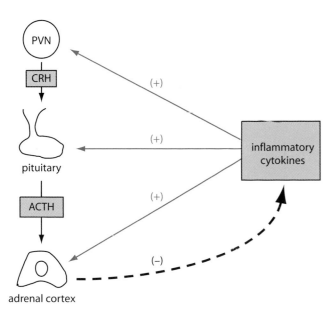

FIGURE 19 Negative feedback regulation of the hypothalamic-pituitary-adrenal axis by inflammatory cytokines. PVN, paraventricular nuclei; CRH, corticotropin releasing hormone.

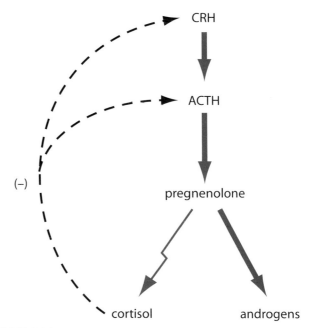

FIGURE 20 Consequences of a partial block of cortisol production by defects in either 11- or 21-hydroxylase. Pregnenolone is diverted to androgens, which exert no feedback activity on ACTH secretion. The thickness of the arrows connotes relative amounts. The broken arrows indicate impairment in the inhibitory limb of the feedback system. Administration of glucocorticoids shuts down androgen production by inhibiting ACTH secretion.

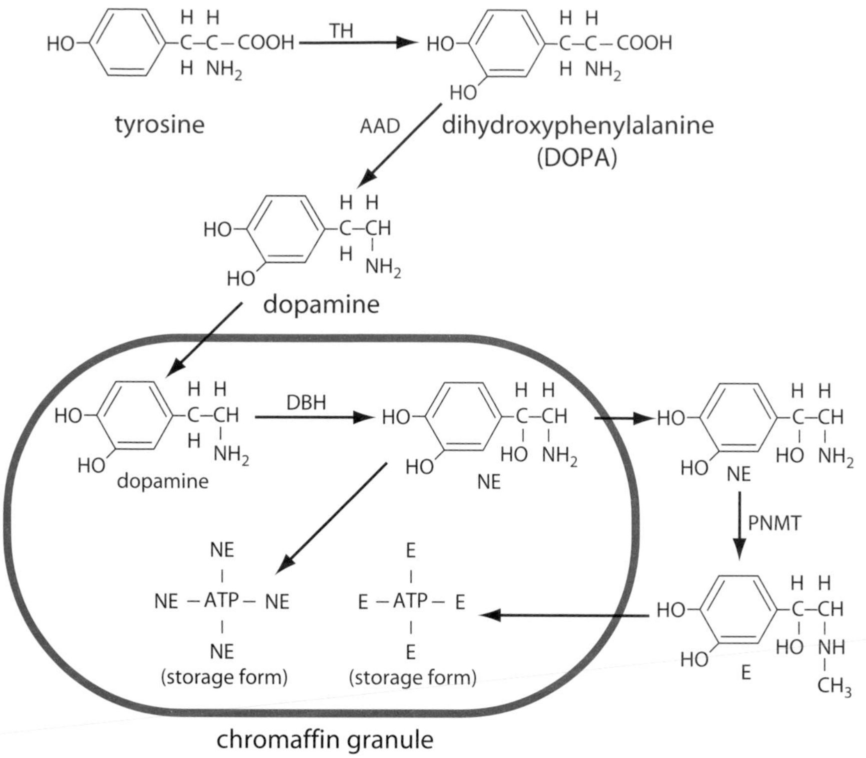

FIGURE 21   Biosynthetic sequence for epinephrine (E) and norepinephrine (N) in adrenal medullary cells. TH, tyrosine hydroxylase; AAD, aromatic L-amino acid decarboxylase (also called DOPA decarboxylase); DBH, dopamine beta-hydroxylase; PNMT, phenyletha-nolamine-N-methyltransferase,

Biosynthesis of the Medullary Hormones

The biosynthetic pathway for epinephrine and norepinephrine is shown in Fig. 21. Hydroxylation of tyrosine to form dihydroxyphenylalanine (DOPA) is the rate-determining reaction and is catalyzed by the enzyme tyrosine hydroxylase. Activity of this enzyme is inhibited by catecholamines (product inhibition) and stimulated by phosphorylation. In this way, regulatory adjustments are made rapidly and are closely tied to bursts of secretion. A protracted increase in secretory activity induces synthesis of additional enzyme after a lag time of about 12 hr.

Tyrosine hydroxylase and DOPA decarboxylase are cytosolic enzymes, but the enzyme that catalyzes the β-hydroxylation of dopamine to form norepinephrine resides within the secretory granule. Dopamine is pumped into the granule by an energy-dependent, stereospecific process. For sympathetic nerve endings and those adreno-medullary cells that produce norepinephrine, synthesis is complete with the formation of norepinephrine, and the hormone remains in the granule until it is secreted. Synthesis of epinephrine, however, requires that norepinephrine reenter the cytosol for the final methylation reaction. The enzyme required for this reaction, phenylethanolamine-N-methyltransferase (PNMT), is

at least partly inducible by glucocorticoids. Induction requires concentrations of cortisol that are considerably higher than those found in peripheral blood. The vascular arrangement in the adrenals is such that interstitial fluid surrounding cells of the medulla can equilibrate with venous blood that drains the cortex and therefore has a much higher content of glucocorticoids than arterial blood. Glucocorticoids may thus determine the ratio of epinephrine to norepinephrine production. Once methylated, epinephrine is pumped back into the storage granule, whose membrane protects stored catecholamines from oxidation by cytosolic enzymes.

Storage, Release, and Metabolism

Catecholamines are stored in secretory granules in close association with ATP and at a molar ratio of 4:1, suggesting some hydrostatic interaction between the positively charged amines and the four negative charges on ATP. Some opioid peptides, including the enkephalins, β endorphin, and their precursors, are also found in these granules. Acetylcholine released during neuronal stimulation increases sodium conductance of the chromaffin cell membrane. The resulting influx of sodium ions depolarizes the plasma membrane, leading to an influx of calcium through voltage-sensitive channels. Calcium is required for catecholamine secretion. Increased cytosolic concentrations of calcium promote phosphorylation of microtubules and the consequent translocation of secretory granules to the cell surface. Secretion occurs when membranes of the chromaffin granules fuse with plasma membranes and the granular contents are extruded into the extracellular space. Fusion of the granular membrane with the cell membrane may also require calcium. ATP, opioid peptides, and other contents of the granules are released along with epinephrine and norepinephrine. As yet, the physiologic significance of opioid secretion by the adrenals is not known, but it has been suggested that the analgesic effects of these compounds may be of importance in the stress response.

All of the epinephrine in blood originates in the adrenal glands. However, norepinephrine may reach the blood by either adrenal secretion or diffusion from sympathetic synapses. The half-lives of medullary hormones in the peripheral circulation have been estimated to be less than 10 sec for epinephrine and less than 15 sec for norepinephrine. Up to 90% of the catecholamines are removed in a single passage through most capillary beds. Clearance from the blood requires uptake by both neuronal and non-neuronal tissues. Significant amounts of norepinephrine are taken up by sympathetic nerve endings and incorporated into secretory granules for release at a later time. Epinephrine and norepinephrine that are taken up in excess of storage capacity are degraded in neuronal cytosol principally by the enzyme monoamine oxidase (MAO). This enzyme catalyzes oxidative deamination of epinephrine, norepinephrine, and other biologically important amines (Fig. 22). Catecholamines taken up by endothelium, heart, liver, and other tissues are also inactivated enzymatically, principally by catecholamine-O-methyl-transferase (COMT), which catalyzes transfer of a methyl group from S-adenosyl methionine to one of the hydroxyl groups. Both of these enzymes are widely distributed and can act sequentially in either order on both epinephrine and norepinephrine. A number of pharmaceutical agents have been developed to modify the actions of these enzymes and thus modify sympathetic responses. Inactivated catecholamines, chiefly vanillylmandelic acid (VMA) and 3-methoxy-4-hydroxyphenylglycol (MHPG), are conjugated with sulfate or glucuronide and excreted in urine. As with steroid hormones, measurement of urinary metabolites of catecholamines is a useful, noninvasive source of diagnostic information.

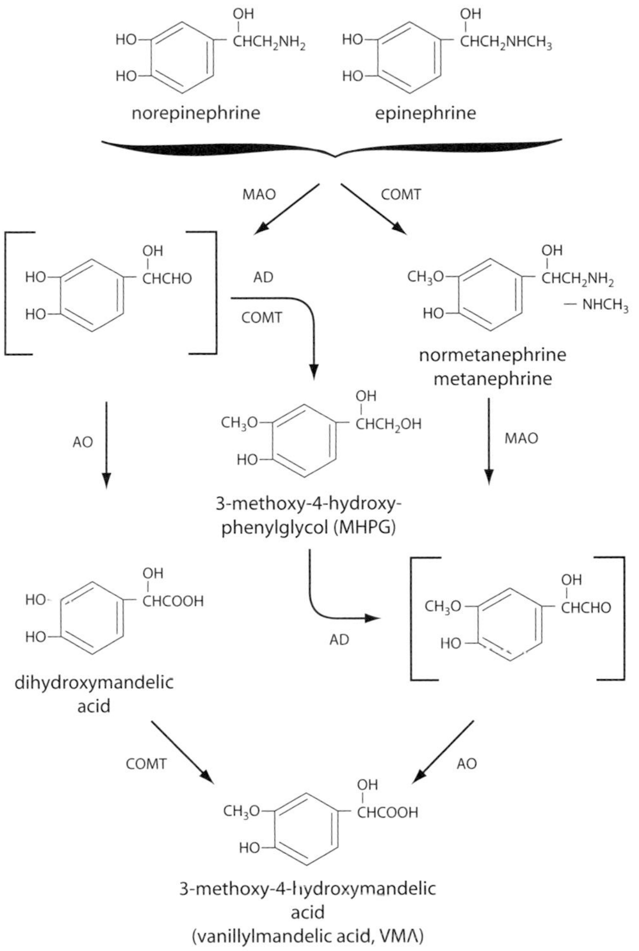

Physiologic Actions of Medullary Hormones

The sympathetic nervous system and adrenal medullary hormones, like the cortical hormones, act on a wide

FIGURE 22 Catecholamine degradation. MAO, monoamine oxidase; COMT, catechol-O-methyltransferase; AD, alcohol dehydrogenase; AO, aldehyde oxidase. (From Cryer PE, In Felig P, Baxter JD, Frohman LA, eds., *Endocrinology and metabolism*, 3rd ed., McGraw-Hill, New York, 1995, p 716.)

variety of tissues to maintain the integrity of the internal environment both at rest and in the face of internal and external challenges. Catecholamines enable us to cope with emergencies and equip us for "fright, fight, or flight." Responsive tissues make no distinctions between blood-borne catecholamines and those released locally from nerve endings. In contrast to adrenal cortical hormones, effects of catecholamines are expressed within seconds and dissipate as rapidly when the hormone is removed. Medullary hormones are thus ideally suited for making the rapid short-term adjustments demanded by a changing environment, whereas cortical hormones, which act only after a lag period of at least 30 min, are of little use at the onset of stress. The cortex and medulla together, however, provide an effective "one–two punch," with cortical hormones maintaining and even amplifying the effectiveness of medullary hormones.

TABLE 2 Typical Responses to Stimulation of the Adrenal Medulla

Target	Responses
CARDIOVASCULAR SYSTEM	
Heart	↑ Force and rate of contraction
	↑ Conduction
	↑ Blood flow (dilation of coronary arterioles)
	↑ Glycogenolysis
Arterioles	
Skin	Constriction
Mucosae	Constriction
Skeletal muscle	Constriction
	Dilation
METABOLISM	
Fat	↑ Lipolysis
	↑ Blood FFA and glycerol
Liver	↑ Glycogenolysis and gluconeogenesis
	↑ Blood sugar
Muscle	↑ Glycogenolysis
	↑ Lactate and pyruvate release
BRONCHIAL MUSCLE	Relaxation
STOMACH AND INTESTINES	↑ Motility
	↑ Sphincter contraction
URINARY BLADDER	↑ Sphincter contraction
SKIN	↑ Sweating
EYES	Contraction of radial muscle of the iris
SALIVARY GLAND	↑ Amylase secretion
	↑ Watery secretion
KIDNEY	↑ Renin secretion
SKELETAL MUSCLE	↑ Tension generation
	↑ Neuromuscular transmission (defatiguing effect)

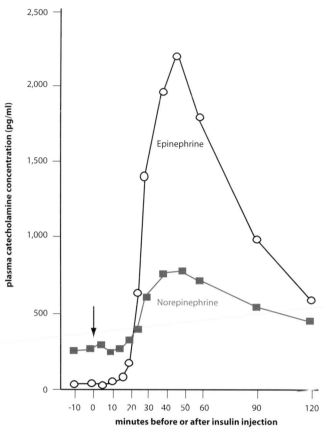

FIGURE 23   Changes in blood concentrations of epinephrine and norepinephrine in response to hypoglycemia. Insulin, which produces hypoglycemia, was injected at the time indicated by the arrow. (From Garber AJ, Bier DM, Cryer PE, Pagliara AS. Hypoglycemia in compensated chronic renal insufficiency. Substrate limitation of gluconeogenesis. *J Clin Invest* 1976;58:7–15.)

Cells in virtually all tissues of the body express G-protein-coupled receptors for epinephrine and norepinephrine on their surface membranes. These so-called adrenergic receptors were originally divided into two categories, α and β based on their activation or inhibition by various drugs. Subsequently, the α and β receptors were further subdivided into α_1, α_2, β_1, β_2, and β_3 receptors. All of these receptors recognize both epinephrine and norepinephrine at least to some extent, and a given cell may have more than one class of adrenergic receptor.

Biochemical mechanisms of signal transduction follow the pharmacologic subdivisions of the adrenergic receptors. All of the β-adrenergic receptors communicate with adenylyl cyclase through the stimulatory guanosine binding protein ($G\alpha_s$; see Chapter 2) and activate adenylyl cyclase, but subtle differences distinguish them. β_3 Receptors may couple to both G_s and G_i heterotrimeric proteins, and hence give a less robust response than β_1 and β_2 receptors. From a physiologic perspective, the only difference between β_1 and β_2 receptors is the low sensitivity of the β_2 receptors to norepinephrine. Stimulation of α_2 receptors inhibits adenylyl cyclase and may block the increase in cyclic AMP produced by other agents. For α_2 effects, the receptor communicates with adenylyl cyclase through G_i. Responses initiated by the α_1 receptor, which couples with $G\alpha_q$, are mediated by IP_3/DAG (see Chapter 2).

Some of the physiologic effects of catecholamines are listed in Table 2. Although these actions may seem diverse, in actuality they constitute a magnificently coordinated set of responses. When producing their effects, catecholamines maximize the contributions of each of the various tissues to resolve the challenges to survival. On the whole, cardiovascular effects maximize cardiac output and ensure perfusion of the brain and working muscles. Metabolic effects ensure an adequate supply of energy-rich substrate. Relaxation of bronchial muscles facilitates pulmonary ventilation. Ocular effects increase visual acuity. Effects on skeletal muscle and transmitter release from motor neurons increase muscular performance, and quiescence of the gut permits diversion of blood flow, oxygen, and fuel to reinforce these effects.

Regulation of Adrenal Medullary Function

The sympathetic nervous system, including its adrenal medullary component, is activated by any actual or threatened change in the internal or external environment. It responds to physical changes, emotional inputs, and anticipation of increased physical activity. Input reaches the adrenal medulla through its sympathetic innervation. Signals arising in the hypothalamus and other integrating centers activate both the neural and hormonal components of the sympathetic nervous system but not necessarily in an all-or-none fashion. Activation may be general or selectively limited to discrete targets. The adrenals can be preferentially stimulated, and it is even possible that norepinephrine- or epinephrine-secreting cells may be selectively activated, as shown in Fig. 23. In response to hypoglycemia detected by glucose-monitoring cells in the central nervous system, the concentration of norepinephrine in blood increased 3-fold, whereas that of epinephrine, which tends to be a more effective hyperglycemic agent, increased 50-fold. Metabolic actions of epinephrine are discussed further in Chapter 42.

Suggested Reading

Blalock JE. Proopiomelanocortin and the immune-neuroendocrine connection. *Ann NY Acad Sci* 1999;885:161–172.

Clark AJL, Weber A. Adrenocorticotropin insensitivity syndromes. *Endocr Rev* 1998;19:828–843.

Dallman MF, Bhatnagar S. Chronic stress and energy balance: Role of the hypothalamo-pituitary-adrenal axis. In McEwen BS, ed., *Handbook of physiology, Section 7, Endocrinology,* Vol IV, *Coping with the environment: Neural and endocrine mechanisms,* 2001, New York: Oxford University Press, pp 179–210.

Funder JW. Steroids, receptors, and response elements: the limits of signal specificity. *Rec Prog Horm Res* 1991;47:191–210.

Keller-Wood ME, Dallman MF. Corticosteroid inhibition of ACTH secretion. *Endocr Rev* 1984;5:1–24.

McKay LI, Cidlowski JA. Molecular control of immune/inflammatory responses: Interactions between nuclear factor-κB and steroid receptor-signaling pathways. *Endocr Rev* 1999;20:435–459.

Needleman P, Turk J, Jakschik BA, Morrison AR, Lefkowith JB. Arachidonic acid metabolism. *Annu Rev Biochem* 1986;55:69–102.

Orth DN. Corticotropin-releasing hormone in humans. *Endocr Rev* 1992;13:164–191.

Rogerson FM, Fuller PJ. Mineralocorticoid action. *Steroids* 2000;65:61–73.

Sapolsky RM, Romero LM, Munck AU. How do glucocorticoids influence stress responses? Integrating permissive, suppressive, stimulatory, and preparative actions. *Endocr Rev* 2000;21:55–89.

Stocco DM. StAR protein and the regulation of steroid hormone biosynthesis. *Annu Rev Physiol* 2001;63:193–213.

Ungar A, Phillips JH. Regulation of the adrenal medulla. *Physiol Rev* 1983;63:787–843.

Weninger SC, Majzoub JA. Regulation and actions of corticotropin releasing hormone. In McEwen BS, ed., *Handbook of physiology, Section 7, Endocrinology,* Vol IV, *Coping with the environment: Neural and endocrine mechanisms,* 2001, New York: Oxford University Press, pp 103–124.

White PC, Mune T, Agarwal AK. 11β-Hydroxysteroid dehydrogenase and the syndrome of apparent mineralocorticoid excess. *Endocr Rev* 1997;18:135–156.

Young JB, Landsberg L. Synthesis, storage and secretion of adrenal medullary hormones: physiology and pathophysiology. In McEwen BS, ed., *Handbook of physiology, Section 7, Endocrinology,* Vol IV, *Coping with the environment: Neural and endocrine mechanisms,* 2001, New York: Oxford University Press, pp 3–20.

41

The Pancreatic Islets

H. MAURICE GOODMAN

KEY POINTS

- One to two million highly vascularized and richly innervated islets comprise the endocrine pancreas. These islets contain insulin-secreting beta cells, glucagon-secreting alpha cells, and somatostatin-secreting delta cells.
- *Glucagon* is a small, single-chain peptide hormone that is cleared rapidly from the circulation. Its physiologic role is to stimulate hepatic production of metabolic fuels and increase blood concentrations of glucose and ketone bodies.
- *Glucagon secretion:* (1) Alpha cells respond directly to blood glucose concentrations. Glucagon secretion is maximal at low blood glucose concentrations and inhibited at high glucose concentrations; (2) hypoglycemia sensed by the central nervous system activates parasympathetic and sympathetic (including epinephrine and norepinephrine from the adrenal medulla) stimulation of glucagon secretion; (3) increased

concentrations of amino acids in blood and the gut hormone cholecystokinin stimulate glucagon secretion; and (4) somatostatin, insulin, ketone bodies, and free fatty acids inhibit glucagon secretion.
- *Insulin* is a small, two-chain peptide hormone that is cleared rapidly from the circulation. The principal physiologic action of insulin is to promote storage of metabolic fuels.
- Insulin increases transport of glucose in *adipose tissue,* thereby promoting fatty acid storage as triglycerides, inhibition of free fatty acid release, and increased triglyceride synthesis.
- Insulin increases glucose transport and glycogen synthesis *in muscle* and inhibits the production of glucose and ketone bodies *in the liver.*
- Insulin binds to a surface receptor and activates an intrinsic receptor *tyrosine kinase.* Tyrosine phosphorylation of insulin receptor substrate

Essential Medical Physiology, Third Edition

Clinical Note (*continued*)

provides docking sites for association of cytoplasmic proteins and initiates cascades of protein phosphorylation that ultimately express insulin responses.

- *Insulin secretion:* (1) Beta cells increase their rates of insulin secretion in response to increases in blood glucose concentration; (2) gastrointestinal hormones secreted in response to nutrients in the intestinal lumen stimulate insulin secretion; (3) insulin secretion is inhibited by somatostatin and by norepinephrine released from sympathetic innervation or epinephrine and norepinephrine secreted by the adrenal medulla; and (4) increased concentrations of blood glucose are sensed by the beta cells through the increased flux of glucose metabolites that increase adenosine triphosphate production.

OVERVIEW

The principal pancreatic hormones are *insulin* and *glucagon,* whose opposing effects on the liver regulate hepatic storage, production, and release of energy-rich fuels. Insulin is an anabolic hormone that promotes sequestration of carbohydrate, fat, and protein in storage depots throughout the body. Its powerful actions are exerted principally on skeletal muscle, liver, and adipose tissue, whereas those of glucagon are restricted to the liver, which responds by forming and secreting energy-rich water-soluble fuels: glucose, acetoacetic acid, and β-hydroxybutyric acid. Interplay of these two hormones contributes to constancy in the availability of metabolic fuels to all cells. *Somatostatin* is a third islet hormone, but a physiologic role for pancreatic somatostatin has not been established. A fourth substance, *pancreatic polypeptide,* is even less understood. Glucagon acts in concert with other fuel-mobilizing hormones to counterbalance the fuel-storing effects of insulin. Because compensatory changes in secretion of all of these hormones are readily made, states of glucagon excess or deficiency rarely lead to overt human disease. Insulin, on the other hand, acts alone, and prolonged survival is not possible in its absence. Inadequacy of insulin due either to insufficient production [*diabetes mellitus type I,* also called insulin-dependent diabetes mellitus (IDDM)] or end-organ unresponsiveness [*diabetes mellitus type II,* also called noninsulin-dependent diabetes mellitus (NIDDM)] results in one of the most common of the endocrine diseases and affects more than 3% of the American population.

MORPHOLOGY OF THE ENDOCRINE PANCREAS

The 1 to 2 million islets of the human pancreas range in size from about 50–500 mm in diameter and contain from 50–300 endocrine cells. Collectively the islets comprise only 1–2% of the pancreatic mass. They are highly vascular, with each cell seemingly in direct contact with a capillary. Blood is supplied by the pancreatic artery and drains into the portal vein, which thus delivers the entire output of pancreatic hormones to the liver. The islets are also richly innervated with both sympathetic and parasympathetic fibers that terminate on or near the secretory cells.

Histologically, the islets consist of three cell types. *Beta cells,* which synthesize and secrete insulin, make up about 60–75% of a typical islet. *Alpha cells* are the source of glucagon and comprise perhaps as much as 20% of islet tissue. *Delta cells,* which are considerably less abundant, produce somatostatin. An additional but rarer cell type, the F cell, may also appear in the exocrine part of the pancreas. It contains and secretes a compound of called *pancreatic polypeptide,* which inhibits pancreatic exocrine cell functions.

Beta cells occupy the central region of the islet or microlobules within islets, whereas alpha cells occupy the outer rim. Delta cells are interposed between them and are thus in contact with both types (Fig. 1). Gap junctions link alpha cells to each other, beta cells to each other, and alpha cells to beta cells. Although experimental proof is lacking, this arrangement may account for synchronous secretory activity. There are also tight junctions between various islet cells. These sites of close apposition or actual fusion of plasma membranes of adjacent cells may affect diffusion of substances into or out of intercellular spaces. This arrangement could either facilitate or hinder paracrine communication between alpha, beta, and delta cells. Blood flows through an anastomosing network of capillaries from the center of the islet toward the periphery. This arrangement favors intra-islet delivery of insulin from the centrally located beta cells to the peripherally located alpha cells. The physiologic consequences of these complex anatomic specializations are not understood.

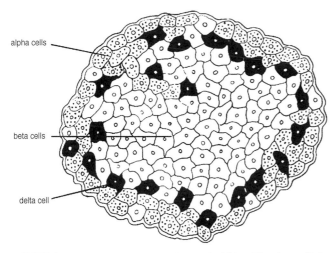

FIGURE 1   Arrangement of cells in a typical islet. The clear cells in the center of the islet are the beta (insulin secreting) cells. The stippled cells in the periphery are the alpha (glucagon secreting) cells, and the solid black cells are the delta (somatostatin secreting) cells. (From Orci L, Unger R., Functional subdivision of islets of Langerhans and possible role of D cells. *Lancet*, 1975;2:1243–1244.)

GLUCAGON

Biosynthesis, Secretion, and Metabolism

Glucagon is a simple unbranched peptide chain that consists of 29 amino acids and has a molecular weight of about 3500. Its amino acid sequence has been remarkably preserved throughout evolution of the vertebrates. The glucagon gene, which is located on chromosome 2, is expressed primarily in the alpha cells, L cells of the intestinal epithelium, and discrete brain areas. It encodes a large, 158-amino-acid preproglucagon protein that is processed in a tissue-specific manner to give rise to at least six biologically active peptides that contain similar amino acid sequences. In alpha cells the preproglucagon molecule is enzymatically cleaved to release glucagon and the major proglucagon fragment (Fig. 2).

In the intestine and the hypothalamus, the principal cleavage products are the *glucagon-like peptide 1* (GLP-1), which has important effects on islet function (see later discussion), and *glucagon-like peptide 2* (GLP-2), whose major actions are exerted on the intestine. GLP-1 may also regulate the rate of gastric emptying and feeding behavior. Proglucagon is a member of a superfamily of genes that encodes gastrointestinal hormones and neuropeptides including secretin, vasoactive inhibitory peptide (VIP), pituitary adenylyl cyclase activating peptide (PACAP), glucose-dependent insulinotropic peptide (GIP), and the growth hormone-releasing hormone, GHRH. Glucagon is packaged, stored in membrane-bound granules, and secreted by exocytosis like other peptide hormones.

Glucagon circulates without binding to carrier proteins and has a half-life in blood of about 5 min. Its concentrations in peripheral blood are considerably lower than in portal venous blood. This difference reflects not only greater dilution in the general circulation but also the fact that about 25% of the secreted glucagon is destroyed during passage through the liver. The kidney is another important site of degradation, and a considerable fraction of circulating glucagon is destroyed by plasma peptidases.

Physiologic Actions of Glucagon

The physiologic role of glucagon is to stimulate hepatic production and secretion of glucose and, to a lesser extent, ketone bodies, which are derived from fatty acids. Under normal circumstances, liver and possibly pancreatic beta cells are the only targets of glucagon action. A number of other tissues including fat and heart express glucagon receptors, and can respond to glucagon experimentally, but considerably higher concentrations of glucagon are needed than are normally found in peripheral blood. Glucagon stimulates the liver to release glucose and produces a prompt increase in

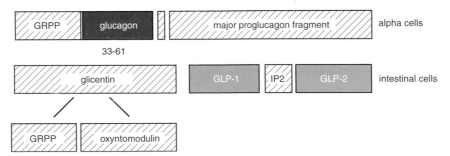

FIGURE 2   Cell-specific post-translational processing of preproglucagon. GRPP, glicentin related pancreatic peptide; GLP-1, glucagon-like peptide-1; GLP-2, glucagon-like peptide-2; IP2, intervening peptide 2. Intervening peptide-1 is the small fragment between glucagon and the major glucagon fragment at the top of the figure.

blood glucose concentration. Glucose that is released from the liver is obtained from breakdown of stored glycogen (*glycogenolysis*) and new synthesis (*gluconeogenesis*). Because the principal precursors for gluconeogenesis are amino acids, especially alanine, glucagon also increases hepatic production of urea (*ureogenesis*) from the amino groups. Glucagon also increases production of ketone bodies (*ketogenesis*) by directing metabolism of long-chain fatty acids toward oxidation and away from esterification and export as lipoproteins. Concomitantly, glucagon may also promote breakdown of hepatic triglycerides to yield long-chain fatty acids, which, along with fatty acids that reach the liver from peripheral fat depots, provide the substrate for ketogenesis.

All of the effects of glucagon appear to be mediated by cyclic AMP (see Chapter 2). In fact, it was studies of the glycogenolytic action of glucagon that led to the discovery of cyclic AMP and its role as a second messenger. Activation of protein kinase A by cyclic AMP results in phosphorylation of enzymes, which increases or decreases their activity, or phosphorylation of the transcription factor CREB, which usually increases transcription of target genes. Glucagon may also increase intracellular concentrations of calcium by a mechanism that depends on activation of protein kinase A, and the increased calcium may reinforce some actions of glucagon, particularly on glycogenolysis.

Glucose Production

To understand how glucagon stimulates the hepatocyte to release glucose, we must first consider some of the biochemical reactions that govern glucose metabolism in the liver. Biochemical pathways that link these reactions are illustrated in Fig. 3. It is important to recognize that not all enzymatic reactions are freely reversible under conditions that prevail in living cells. Phosphorylation and dephosphorylation of substrate usually require separate enzymes. This sets up substrate cycles, which would spin futilely in the absence of some regulatory influence exerted on either or both opposing reactions. These reactions are often strategically situated at or near branch points in metabolic pathways and can therefore direct flow of substrates toward one fate or another. Regulation is achieved both by modulating the activity of enzymes already present in cells and by increasing or decreasing rates of enzyme synthesis and therefore amounts of enzyme molecules. Enzyme activity can be regulated allosterically by changes in conformation produced by substrates or cofactors, or covalently by phosphorylation and dephosphorylation of regulatory sites in the enzymes themselves. Changing the activity of an enzyme requires only seconds, whereas

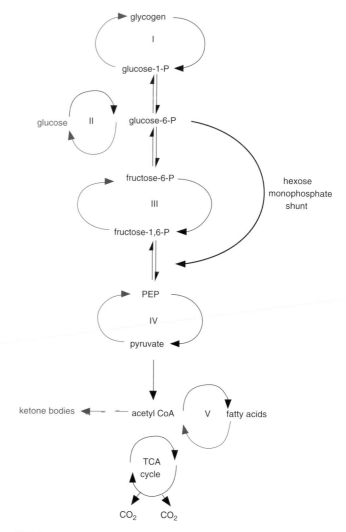

FIGURE 3   Biochemical pathways of glucose metabolism in hepatocytes. Reactions that are accelerated in the presence of glucagon are shown in blue.

many minutes or even hours are needed to change the amount of an enzyme.

Glycogenolysis

Cyclic AMP formed in response to the interaction of glucagon with its G-protein-coupled receptors on the surface of hepatocytes activates protein kinase A, which catalyzes phosphorylation, and hence activation, of an enzyme called *phosphorylase kinase* (Fig. 4). This enzyme, in turn, catalyzes phosphorylation of another enzyme, *glycogen phosphorylase*, which cleaves glycogen stepwise to release glucose-1-phosphate. Glucose-1-phosphate is the substrate for *glycogen synthase*, which catalyzes the incorporation of glucose into glycogen. Glycogen synthase is also a substrate for protein kinase A and is inactivated when phosphorylated. Thus by increasing the formation of cyclic AMP, glucagon

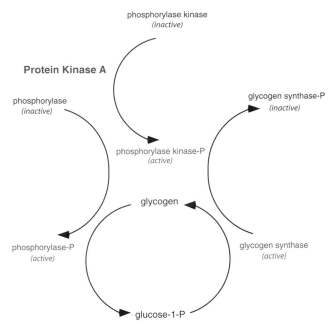

Protein Kinase A

phosphorylase kinase
(inactive)

phosphorylase kinase-P
(active)

glycogen synthase-P
(inactive)

phosphorylase
(inactive)

glycogen

phosphorylase-P
(active)

glycogen synthase
(active)

glucose-1-P

FIGURE 4   Role of protein kinase A (cyclic AMP-dependent protein kinase) in glycogen metabolism.

simultaneously promotes glycogen breakdown and prevents recycling of glucose to glycogen. Cyclic AMP-dependent phosphorylation of enzymes that regulate the glycolytic pathway at the level of phosphofructokinase and acetyl coenzyme A (CoA) carboxylase (see later discussion) prevents consumption of glucose-6-phosphate by the hepatocyte itself, leaving dephosphorylation and diffusion into the blood as the only pathway open to newly depolymerized glucose.

Gluconeogenesis

Precursors of glucose enter the gluconeogenic pathway as 3- or 4-carbon compounds. Glucagon directs their conversion to glucose by accelerating their condensation to fructose phosphate while simultaneously blocking their escape from the gluconeogenic pathway (cycles III and IV in Fig. 3). Cyclic AMP controls production of a potent allosteric regulator of metabolism called *fructose-2,6-bisphosphate*. This compound, when present even in tiny amounts, activates *phosphofructokinase* and inhibits *fructose-1,6-bisphosphatase*, thereby directing flow of substrate toward glucose breakdown rather than glucose formation (Fig. 5). Fructose-2,6-bisphosphate, which should not be confused with fructose-1,6-bisphosphate, is formed from fructose-6-phosphate by the action of an unusual bifunctional enzyme that catalyzes either phosphorylation of fructose-6-phosphate to fructose-2,6-bisphosphate or dephosphorylation of fructose-2,6-bisphosphate to fructose-6-phosphate, depending on its own state of phosphorylation. This enzyme is a substrate

for protein kinase A and behaves as a phosphatase when it is phosphorylated. Its activity in the presence of cyclic AMP rapidly depletes the hepatocyte of fructose-2,6-bisphosphate, and substrate therefore flows toward glucose production.

The other important regulatory step in gluconeogenesis is phosphorylation and dephosphorylation of pyruvate (cycle IV in Fig. 3). It is here that the 3- and 4-carbon fragments enter or escape from the gluconeogenic pathway. The cytosolic enzyme that catalyzes dephosphorylation of phosphoenol pyruvate (PEP) was inappropriately named *pyruvate kinase* before it was recognized that direct phosphorylation of pyruvate does not occur under physiologic conditions and that this enzyme acts only in the direction of dephosphorylation (Fig. 6). Regulation of this enzyme is complex. Pyruvate kinase is another substrate for protein kinase A and is powerfully inhibited when phosphorylated, but the inhibition can be overcome by fructose-1,6-bisphosphate. Thus, activation of protein kinase A has the duel effect of decreasing pyruvate kinase activity directly and of decreasing the abundance of its activator, fructose-1,6-bisphosphate, by reactions shown in Fig. 5. Inhibiting pyruvate kinase may be the single most important effect of glucagon on the gluconeogenic pathway. On a longer time scale, glucagon inhibits the synthesis of pyruvate kinase. Phosphorylation of pyruvate requires a complex series of reactions in which pyruvate must first enter mitochondria where it is carboxylated to form oxaloacetate. Entry of pyruvate across the mitochondrial membrane is accelerated by glucagon, but the mechanism for this effect is not known. Oxaloacetate is converted to cytosolic PEP by the catalytic activity of *PEP carboxykinase*. Synthesis of this enzyme is

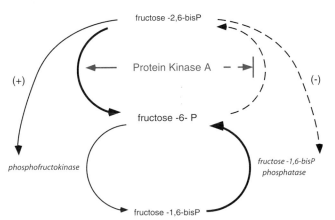

fructose -2,6-bisP

Protein Kinase A

(+)

(-)

fructose -6- P

phosphofructokinase

fructose -1,6-bisP
phosphatase

fructose -1,6-bisP

FIGURE 5   Regulation of fructose-1,6-bisphosphate metabolism by protein kinase A (cyclic AMP-dependent protein kinase) and fructose-2,6-bisphosphate. Fructose-2,6-bisphosphate, whose formation depends on protein kinase A, activates (+) phosphofructokinase and inhibits (−) fructose-1,6-bisphosphatase.

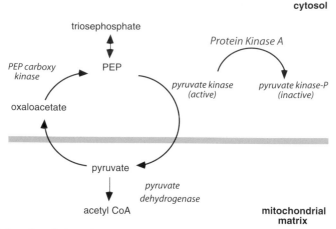

FIGURE 6   Regulation of PEP formation by protein kinase A (cyclic AMP-dependent protein kinase). Protein kinase A catalyzes the phosphorylation and, hence, inactivation of pyruvate kinase whose activity limits the conversion of PEP to pyruvate.

accelerated by cyclic AMP-dependent phosphorylation of CREB (see Chapter 2).

Lipogenesis and Ketogenesis

An alternate fate of pyruvate in mitochondria is decarboxylation to form acetyl CoA (Fig. 7). This 2-carbon acetyl unit is the building block of fatty acids and eventually finds its way back to the cytosol where fatty acid synthesis (*lipogenesis*) takes place. Lipogenesis is the principal competitor of gluconeogenesis for 3-carbon precursors. The first committed step in fatty acid synthesis is the carboxylation of acetyl CoA to form malonyl CoA. *Acetyl CoA carboxylase*, the enzyme that catalyzes this reaction, is yet another substrate for protein kinase A and is powerfully inhibited when phosphorylated. Inhibition of fatty acid synthesis not only preserves substrate for gluconeogenesis but also prevents oxidation of glucose by the hexose monophosphate shunt pathway (Fig. 3). NADP, which is required for shunt activity, is reduced in the initial reactions of this pathway and can be regenerated only by transferring protons to the elongating fatty acid chain.

Fatty acid synthesis and oxidation constitute another substrate cycle and another regulatory site for cyclic AMP action. The same reaction that inhibits fatty acid synthesis promotes fatty acid oxidation and consequently *ketogenesis* (ketone body formation) (Fig. 7). Long-chain fatty acid molecules that reach the liver can either be oxidized or esterified and exported to adipose tissue as the triglyceride component of low-density lipoproteins. To be esterified, fatty acids must remain in the cytosol, and to be oxidized they must enter the mitochondria. Long-chain fatty acids can cross the mitochondrial membrane only when linked to carnitine.

Carnitine acyl transferase, the enzyme that catalyzes this linkage, is powerfully inhibited by malonyl CoA. Thus when malonyl CoA concentrations are high, coincident with fatty acid synthesis, fatty acid oxidation is inhibited. Conversely, when the formation of malonyl CoA is blocked, fatty acids readily enter mitochondria and are oxidized to acetyl CoA. Because long-chain fatty acids typically contain 16 and 18 carbons, each molecule that is oxidized yields eight or nine molecules of acetyl CoA. The ketone bodies β-hydroxybutyrate and acetoacetate are formed from condensation of two molecules of acetyl CoA and the subsequent removal of the CoA moiety.

By reducing the concentration of malonyl CoA, glucagon sets the stage for ketogenesis, but the actual

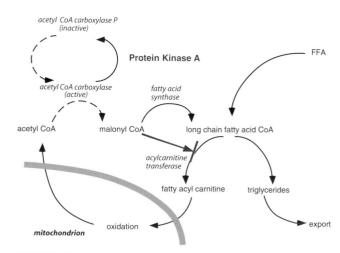

FIGURE 7   Protein kinase A (cyclic AMP-dependent protein kinase) indirectly stimulates ketogenesis by decreasing the formation of malonyl CoA, thus removing a restriction on accessibility of fatty acids to intramitochondrial oxidative enzymes.

amount of ketone production is determined by the amount of long-chain fatty acids available for oxidation. Most fatty acids oxidized in liver originate in adipose tissue, but glucagon, through cyclic AMP and protein kinase A, may also activate a lipase in liver and thereby provide fatty acids from breakdown of hepatic triglycerides.

Ureogenesis

Whenever carbon chains of amino acids are used as substrate for gluconeogenesis, amino groups must be disposed of in the form of urea, which thus becomes a by-product of gluconeogenesis. By promoting gluconeogenesis, therefore, glucagon also increases the formation of urea (*ureogenesis*). Carbon skeletons of most amino acids can be converted to glucose, but because of peculiarities of peripheral metabolism, alanine is quantitatively the most important glucogenic amino acid. By accelerating conversion of pyruvate to glucose (see earlier discussion), glucagon indirectly accelerates transamination of alanine to pyruvate. Glucagon also accelerates ureogenesis by increasing transport of amino acids across hepatocyte plasma membranes by an action that requires synthesis of new RNA and protein. In addition, glucagon also promotes the synthesis of some urea cycle enzymes.

Regulation of Glucagon Secretion

The concentration of glucose in blood is the most important determinant of glucagon secretion in normal individuals. When the plasma glucose concentration exceeds 200 mg/dL, glucagon secretion is maximally inhibited. Inhibitory effects of glucose are proportionately less at lower concentrations and disappear when its concentration falls below 50 mg/dL. Except immediately after a meal rich in carbohydrate, the blood glucose concentration remains constant at around 90 mg/dL. The set point for glucose concentration thus falls well within the range over which glucagon secretion is regulated, and alpha cells can respond to changes in blood glucose with either an increase or a decrease in glucagon output. The alpha cells appear to respond directly to changes in glucose concentration, but we do not yet understand how they monitor blood glucose concentration and translate that information to an appropriate rate of glucagon secretion. Little is understood of the intracellular molecular events that bring about an increase or decrease of glucagon secretion.

Low blood glucose (*hypoglycemia*) not only relieves inhibition of glucagon secretion, but this life-threatening circumstance stimulates the central nervous system to signal both parasympathetic and sympathetic nerve endings within the islet to release their neurotransmitters,

acetylcholine and VIP (vasoactive intestinal peptide), from parasympathetic endings and norepinephrine and NPY (neuropeptide Y) from sympathetic endings. Alpha cells express receptors for these neurotransmitters, and they secrete glucagon in response to both parasympathetic and sympathetic stimulation. The sympathetic response to hypoglycemia also involves secretion of epinephrine and norepinephrine from the adrenal medulla (see Chapter 40). Adrenomedullary hormones further stimulate alpha cells to secrete glucagon.

Glucagon secretion is evoked by a meal rich in amino acids. Alpha cells respond directly to increased blood levels of certain amino acids, particularly arginine. In addition, digestion of protein-rich foods triggers the release of cholecystokinin from cells in the duodenal mucosa (see Chapter 32). This gastrointestinal hormone is a secretagogue for islet hormones as well as pancreatic enzymes and may alert alpha cells to an impending influx of amino acids. Increased secretion of glucagon in response to a protein meal not only prepares the liver to dispose of excess amino acids by gluconeogenesis but also signals the liver to release glucose and thus counteracts the hypoglycemic effects of insulin, whose secretion is simultaneously increased by amino acids (see later discussion).

Glucose is not the only physiologic inhibitor of glucagon secretion. Insulin, somatostatin, GLP-1, glucose-dependent insulinotropic peptide (GIP), and free fatty acids (FFA) also exert inhibitory influences on glucagon secretion (Fig. 8). Insulin, which may reach alpha cells by either the endocrine or paracrine route, directly inhibits glucagon secretion and is required for expression of inhibitory effects of glucose. In fact, it has been suggested that glucose may inhibit glucagon secretion indirectly through increased secretion of insulin. In persons suffering from insulin deficiency

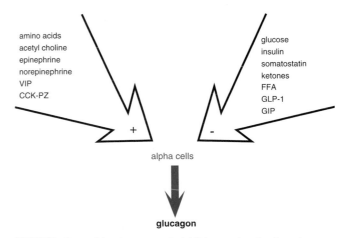

amino acids
acetyl choline
epinephrine
norepinephrine
VIP
CCK-PZ

glucose
insulin
somatostatin
ketones
FFA
GLP-1
GIP

alpha cells

glucagon

FIGURE 8   Stimulatory and inhibitory signals for glucagon secretion.

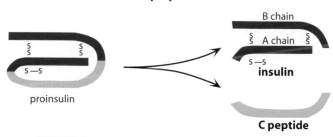

FIGURE 9   Post-translational processing of preproinsulin.

INSULIN

Biosynthesis, Secretion, and Metabolism

Insulin is composed of two unbranched peptide chains joined together by two disulfide bridges (Fig. 9). The single gene that encodes the preproinsulin molecule consists of three exons and two introns and is located on chromosome 11. The two chains of insulin and their disulfide cross bridges are derived from the single-chain proinsulin molecule, from which a 31-residue peptide, called the *connecting peptide* (C peptide), is excised by stepwise actions of two trypsin-like enzymes called *prohormone convertases*. Conversion of proinsulin to insulin takes place slowly within storage granules. The C peptide therefore accumulates within granules in equimolar amounts with insulin. When insulin is secreted, the entire contents of secretory vesicles are disgorged into extracellular fluid. Consequently, the C peptide and any remaining proinsulin and processing intermediates are released into the circulation. When secretion is rapid, proinsulin may comprise as much as 20% of the circulating peptides detected by insulin antibodies, but it contributes little biological activity. Although several biological actions of the C peptide have been described, no physiologic role for the C peptide has yet been established.

The insulin storage granule contains a variety of proteins that are also released whenever insulin is secreted. Most of these proteins are thought to maintain optimal conditions for storage and processing of insulin, but some may also have biological activity. Their fate and actions, if any, are largely unknown. One such protein, however, called *amylin,* may contribute to the amyloid that accumulates in and around beta cells in states of insulin hypersecretion and may contribute to

islet pathology. A wide variety of biological actions of amylin have been described including antagonism to the actions of insulin in various tissues, suppression of appetite, and delaying of gastric emptying, but a physiologic role for amylin remains to be established and is the subject of some controversy.

Insulin is cleared rapidly from the circulation with a half-life of 4–6 min and is destroyed by a specific enzyme, called insulinase or insulin degrading enzyme, that is present in liver, muscle, kidney, and other tissues. The first step in insulin degradation is receptor-mediated internalization through an endosomal mechanism. Degradation may take place within endosomes or after fusion of endosomes with lysosomes. The liver is the principal site of insulin degradation and inactivates about 30–70% of the insulin that reaches it in hepatic portal blood. Insulin degradation in the liver appears to be a regulated process governed by changes in availability of metabolic fuels and changing physiologic circumstances. The liver may thus regulate the amount of insulin that enters the systemic circulation. The kidneys destroy about half of the insulin that reaches the general circulation following receptor-mediated uptake both from the glomerular filtrate and from postglomerular blood plasma. Normally, little or no insulin is found in urine. Muscle and other insulin-sensitive tissues throughout the body apparently account for destruction of the remainder. Proinsulin has a half-life that is at least twice as long as insulin and is not converted to insulin outside the pancreas. The kidney is the principal site of degradation of proinsulin and the C peptide. Because little degradation of the C peptide occurs in the liver, its concentration in blood is useful for estimating the rate of insulin secretion and evaluation of beta-cell function in patients who are receiving injections of insulin.

Physiologic Actions of Insulin

The physiologic role of insulin is to promote storage of metabolic fuel. Insulin has many effects on different

Clinical Note

In many areas of endocrinology, basic insights into the physiologic role of a hormone can be gained from examining the consequences of its absence. In such studies, secondary or tertiary effects may overshadow the primary cellular lesion but nevertheless ultimately broaden our understanding of cellular responses in the context of the whole organism. Insights into the physiology of insulin and the physiologic processes it affects directly and indirectly were originally gained from clinical observation. Consideration of some of the classic signs of this disease therefore provides a good starting point for discussing the physiology of insulin.

Hyperglycemia: In the normal individual the concentration of glucose in blood is maintained at around 90 mg/dL of plasma. Blood glucose in diabetics may be 300–400 mg/dL and even reach 1000 mg/dL on occasion. Diabetics have particular difficulty removing excess glucose from their blood. Normally, after ingestion of a meal rich in carbohydrate, there is only a small and transient increase in the concentration of blood glucose, and excess glucose disappears rapidly from plasma. The diabetic, however, is "intolerant" of glucose, and the ability to remove it from plasma is severely impaired.

Oral glucose tolerance tests, which assess the ability to dispose of a glucose load, are used diagnostically to evaluate existing or impending diabetic conditions. A standard load of glucose is given by mouth and the blood glucose concentration is measured periodically over the course of the subsequent 4 hr. In normal subjects, blood glucose concentrations return to baseline values within 2 hr, and the peak value does not rise above 180 mg/dL. In the diabetic or "prediabetic," blood glucose values rise much higher and take a longer time to return to basal levels (Fig. 10).

Glycosuria: Normally the renal tubule has adequate capacity to transport and reabsorb all of the glucose filtered at the glomerulus so that little or none escapes in the urine. Because of hyperglycemia, however, the concentration of glucose in the glomerular filtrate is so high that it exceeds the capacity for reabsorption and "spills" into the urine, causing *glycosuria* (excretion of glucose in urine).

Polyuria is defined as excessive production of urine. Because more glucose is present in the glomerular filtrate than can be reabsorbed by proximal tubules, it remains in the tubular lumen and exerts an osmotic hindrance to water and salt reabsorption in this portion of the nephron, which normally reabsorbs about two-thirds of the glomerular filtrate (see Chapter 26). The abnormally high volume of fluid that remains cannot be reabsorbed by more distal portions of the nephron, with the result that water excretion is increased (*osmotic diuresis*). Increased flow through the nephron increases urinary loss of sodium and potassium as well.

Polydipsia: Dehydration results from the copious flow of urine and stimulates thirst, a condition called *polydipsia,* or excessive drinking. The untreated diabetic is characteristically thirsty and consumes large volumes of water to compensate for water lost in urine. Polydipsia is often the first symptom that is noticed by the patient or parents of a diabetic child.

Polyphagia: By mechanisms that are not yet understood, appetite is increased in what seems to be an effort to compensate for urinary loss of glucose. The condition is called *polyphagia* (excessive food consumption).

Weight loss: Despite increased appetite and food intake, however, insulin deficiency reduces all anabolic processes and accelerates catabolic processes. Accelerated protein degradation, particularly in muscle, provides substrate for gluconeogenesis. Increased mobilization and utilization of stored fats indirectly leads to increased triglyceride concentration in plasma and often results in *lipemia* (high concentration of lipids in blood). Fatty acid oxidation by the liver results in increased production of the ketone bodies (*ketosis*), which are released into the blood and cause *ketonemia*. Because ketone bodies are small, readily filtrable molecules that are actively reabsorbed by a renal mechanism of limited capacity, high blood levels may result in loss of ketone bodies in the urine (*ketonuria*). Ketone bodies are organic acids and produce acidosis, which may be aggravated by excessive washout of sodium and potassium in the urine. Plasma pH may become so low that acidotic coma and death may follow unless insulin therapy is instituted.

The hyperglycemia that causes this whole sequence of events arises from an "underutilization" of glucose by muscle and adipose tissue cells. Even within a single cell it produces multiple effects that are both complementary and reinforcing. Insulin acts on adipose tissue, skeletal muscle, and liver to defend and expand reserves of triglyceride, glycogen, and protein. Within a few minutes after intravenous injection of insulin, there is a striking decrease in the plasma concentrations of glucose, amino acids, FFA, ketone bodies, and potassium. If the dose of insulin is large enough, blood glucose may fall too low to meet the needs of the central nervous system, and *hypoglycemic coma* may occur. Insulin lowers blood glucose in two ways: It increases uptake by muscle and adipose tissue and decreases output by liver. It lowers the concentration of amino acids by stimulating their uptake by muscle and reducing their release. Insulin lowers the concentration of FFA by blocking their release from adipocytes, and this action in turn lowers the blood ketone level. The decrease in potassium results from stimulation of the sodium/potassium ATPase (sodium pump) in the plasma membranes of muscle, liver, and fat cells. The physiologic significance of this response to insulin is not understood.

Effects on Adipose Tissue

Storage of fat in adipose tissue depends on multiple insulin-sensitive reactions, including (1) synthesis of long-chain fatty acids from glucose; (2) synthesis of triglycerides from fatty acids and glycerol (*esterification*); (3) breakdown of triglycerides to release glycerol and long-chain fatty acids (*lipolysis*); and (4) uptake of fatty acids from the lipoproteins of blood. The relevant biochemical pathways are shown in Fig. 11.

Lipolysis and esterification are central events in the physiology of the adipocyte. The rate of lipolysis depends on the activity of triglyceride lipase. Lipolysis proceeds at a basal rate in the absence of hormonal stimulation but increases dramatically when cyclic AMP is increased. *Hormone-sensitive lipase* catalyzes the breakdown of triglycerides into fatty acids and glycerol. Fatty acids can either escape from the adipocyte and become the FFA of blood or be re-esterified to triglyceride. Fatty acid esterification requires a source of glycerol that is phosphorylated in its α carbon; *free glycerol cannot be used.* Because adipose tissue lacks the enzyme α-glycerol kinase, all of the free glycerol that is produced by lipolysis escapes into the blood. The only source of α-glycerol phosphate available for esterification of fatty acids is derived from phosphorylated 3-carbon intermediates formed from oxidation of glucose.

As its name implies, hormone-sensitive lipase is activated by lipolytic hormones, which stimulate the formation of cyclic AMP and thereby promote its phosphorylation by protein kinase A. Insulin accelerates the degradation of cyclic AMP by activating the enzyme

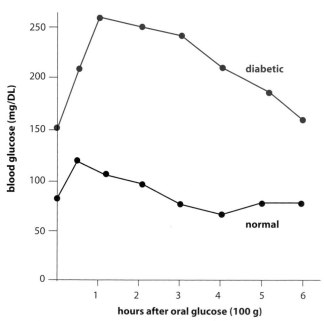

FIGURE 10   Idealized glucose tolerance tests in normal and diabetic subjects.

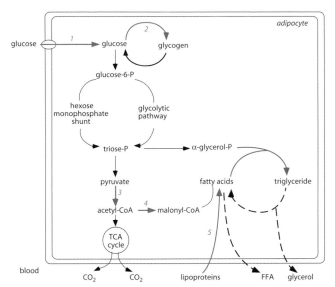

FIGURE 11 Carbohydrate and lipid metabolism in adipose tissue. Reactions enhanced by insulin (blue arrows) are as follows: (1) transport of glucose into adipose cell; (2) conversion of excess glucose to glycogen; (3) decarboxylation of pyruvate; (4) initiation of fatty acid synthesis; and (5) uptake of fatty acids from circulating lipoproteins. Breakdown of triglycerides is inhibited by insulin (broken arrow). Esterification of fatty acids to triglycerides follows from availability of α-glycerol phosphate.

cyclic AMP phosphodiesterase and thus interferes with activation of hormone-sensitive lipase. Simultaneously, insulin increases the rate of fatty acid esterification by increasing the availability of α-glycerol phosphate. The net result of these actions is preservation of triglyceride stores at the expense of plasma FFA, whose concentration in blood plasma promptly falls. Decreases in FFA concentrations are seen with doses of insulin that are too low to affect blood glucose and appear to be the most sensitive response to insulin.

Because glucose does not readily diffuse across the plasma membrane, its entry into adipocytes and most other cells depends on carrier-mediated transport. Insulin increases cellular uptake and metabolism of glucose by accelerating transmembrane transport of glucose and structurally related sugars. This action depends on the availability of glucose transporters in the plasma membrane. Glucose transporters (abbreviated GLUT) are large proteins that weave in and out of the membrane 12 times to form stereospecific channels through which glucose can diffuse down its concentration gradient. At least five isoforms of GLUT are expressed in various cell types. In addition to GLUT 1, which is present in the plasma membrane of most cells, insulin-sensitive cells such as adipocytes contain pools of intracellular membranous vesicles that are rich in GLUT 4. Insulin increases the number of glucose transporters on the adipocyte surface by stimulating

the translocation of GLUT 4-containing vesicles toward the cell surface and fusion of their membranes with the adipocyte plasma membrane (Fig. 12).

Insulin may accelerate synthesis of fatty acids by increasing the uptake of glucose and by activating at least two enzymes that direct the flow of glucose carbons into fatty acids. Insulin increases conversion of pyruvate to acetyl CoA, which provides the building blocks for long-chain fatty acid synthesis, and stimulates carboxylation of acetyl CoA to malonyl CoA, which is the initial and rate-determining reaction in fatty acid synthesis. In humans, adipose tissue is not an important site of fatty acid synthesis, particularly in Western cultures where the diet is rich in fat. Fat stored in adipose tissue is derived mainly from dietary fat and triglycerides synthesized in the liver. Fat destined for storage reaches adipose tissue

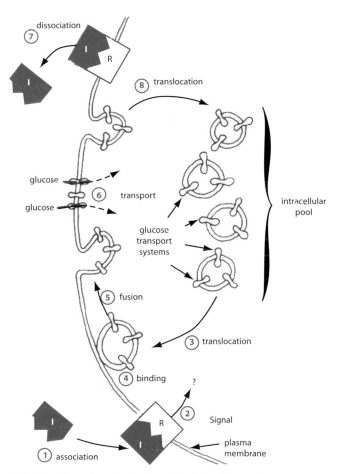

FIGURE 12 Hypothetical model of insulin's action on glucose transport. Upon associating with its receptors (R) in the cell membrane, insulin (I) signals the translocation of glucose transport systems to the plasma membrane. The stepwise sequence of events is indicated by the blue circled numbers. (From Karnieli E, Zarnowski MJ, Hissin RJ, Salans LB, Cushman SW, Insulin-stimulated translocation of glucose transport systems in the isolated rat adipose cell. Time course, reversal, insulin concentration dependency, and relationship to glucose transport activity. *J Biol Chem* 1981;256:4772–4777).

in the form of low-density lipoproteins and chylomi-crons. Uptake of fat from lipoproteins depends on cleavage of ester bonds in triglycerides by the enzyme *lipoprotein lipase* to release fatty acids. Lipoprotein lipase is synthesized and secreted by adipocytes and adheres to the endothelium of adjacent capillaries. Insulin promotes synthesis of lipoprotein lipase and thus facilitates the transfer of fatty acids from lipoproteins to triglyceride storage droplets in adipocytes.

Effects on Muscle

Insulin increases uptake of glucose by muscle and directs its intracellular metabolism toward the formation of glycogen (Fig. 13). Because muscle comprises nearly 50% of body mass, uptake by muscle accounts for the majority of the glucose that disappears from blood after injection of insulin. As in adipocytes, glucose utilization in muscle is limited by permeability of the plasma membrane. Insulin accelerates entry of glucose into muscle by mobilizing GLUT 4-containing vesicles by the same mechanism that is operative in adipocytes.

Metabolism of glucose begins with conversion to glucose-6-phosphate catalyzed by either of the two isoforms of the enzyme *hexokinase* that are present in muscle. Insulin not only increases the synthesis of hexokinase II, but it also appears to enhance the efficiency of hexokinase II activity by promoting its association with the outer membrane of mitochondria, which optimizes access to ATP. In the basal state, glucose is phosphorylated almost as rapidly as it enters the cell, and hence the intracellular concentration of free glucose is only about one-tenth to one-third that of extracellular fluid.

Glucose-6-P is an allosteric inhibitor of hexokinase and an allosteric activator of glycogen synthase. Stimulation of glycogen synthesis by insulin and glucose-6-phosphate protects hexokinase from the inhibitory effect of glucose-6-phosphate when entry of glucose into the muscle cell is rapid. Glycogen synthase activity is low when the enzyme is phosphorylated and increased when it is dephosphorylated. The degree of phosphorylation of glycogen synthase is determined by the balance of kinase and phosphatase activities. Insulin shifts the balance in

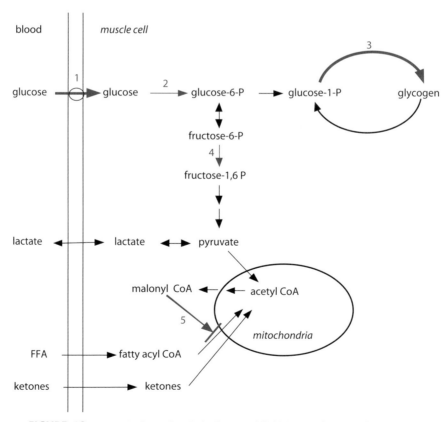

FIGURE 13   Metabolism of carbohydrate and lipid in muscle. Rate-limiting reactions accelerated by insulin (blue arrows) are as follows: (1) transport of glucose into muscle cells; (2) phosphorylation of glucose by hexokinase; (3) storage of glucose as glycogen; (4) addition of the second phosphate by phosphofructokinase; and (5) inhibition of fatty acid entry into mitochondria by malonyl CoA.

favor of dephosphorylation in part by inhibiting the enzyme *glycogen synthase kinase 3* (GSK-3) and in part by activating a phosphatase. Dephosphorylation of glycogen synthase not only increases its activity directly, but also increases its responsiveness to stimulation by its substrate, glucose-6-P. Hence the powerful effects of insulin on muscle glycogen synthesis are achieved by the complementary effects of increased glucose transport, increased glucose phosphorylation, and increased glycogen synthase activity.

The alternative fate of glucose-6-P, metabolism to pyruvate in the glycolytic pathway, is also increased by insulin. Access to the glycolytic pathway is guarded by phosphofructokinase, whose activity is precisely regulated by a combination of allosteric effectors including ATP, ADP, and fructose-2,6-bisphosphate. This complex enzyme behaves differently in intact cells and in the broken cell preparations typically used by biochemists to study enzyme regulation. Because conflicting findings have been obtained under a variety of experimental circumstances, no general agreement has been reached on how insulin increases phosphofructokinase activity. In contrast to the liver, the isoform of the enzyme that forms fructose-2,6-bisphosphate in muscle is not regulated by cyclic AMP. The effects of insulin are likely to be indirect.

Note that oxidation of fat profoundly affects the metabolism of glucose in muscle and that insulin also increases all aspects of glucose metabolism in muscle as an indirect consequence of its action on adipose tissue to decrease FFA production. When insulin concentrations are low, increased oxidation of fatty acids decreases oxidation of glucose by inhibiting the decarboxylation of pyruvate and the transport of glucose across the muscle cell membrane. In addition, products of fatty acid oxidation appear also to inhibit hexokinase, but recent studies have called into question the relevance of earlier findings that fatty acid oxidation may inhibit phosphofructokinase. Insulin not only limits the availability of fatty acids, but also inhibits their oxidation. Insulin increases the formation of malonyl CoA, which blocks entry of long chain fatty acids into the mitochondria as described for liver (Fig. 7). These effects are discussed in Chapter 42.

Protein synthesis and degradation are ongoing processes in all tissues and in the nongrowing individual are completely balanced so that on average there is no net increase or decrease in body protein (Fig. 14). In the absence of insulin there is net degradation of muscle protein and muscle becomes an exporter of amino acids, which serve as substrate for gluconeogenesis and ureogenesis in the liver. As with its effects on carbohydrate and fat metabolism, insulin intercedes in protein synthesis at several levels, and has both rapidly apparent and delayed effects. Insulin increases uptake of amino

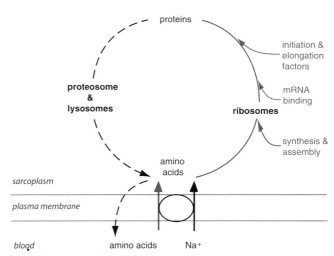

FIGURE 14   Effects of insulin on protein turnover in muscle. Reactions stimulated by insulin are shown in blue. The dashed arrows indicate inhibition.

acids from blood by stimulating their transport across the plasma membrane. Insulin increases protein synthesis by promoting phosphorylation of the initiation factors (e.g., eIF-2 for *eukaryotic initiation factor-2*) that govern translation of mRNA. Under the influence of insulin, attachment of mRNA to ribosomes is enhanced, as reflected by the higher content of polysomes compared to monosomes. This effect of insulin appears to be selective for mRNAs for specific proteins. On a longer timescale, insulin increases total RNA in muscle by increasing synthesis of RNA and protein components of ribosomes. Understanding of how insulin decreases protein degradation is incomplete, but it appears that insulin decreases ATP-dependent protein degradation both by decreasing expression of various elements of the proteasomal protein degrading apparatus and by modulating the protease activity of its components.

Effects on Liver

Insulin reduces outflow of glucose from the liver and promotes storage of glycogen. It inhibits glycogenolysis, gluconeogenesis, ureogenesis, and ketogenesis, and it stimulates the synthesis of fatty acids and proteins. These effects are accomplished by a combination of actions that change the activity of some hepatic enzymes and rates of synthesis of other enzymes. Hence not all the effects of insulin occur on the same timescale. Although we use the terms *block* and *inhibit* to describe the actions of insulin, it is important to remember that these verbs are used in the relative and not the absolute sense. Rarely would inhibition of an enzymatic transformation be absolute. In addition, all of the hepatic effects of insulin are reinforced indirectly by actions of insulin on muscle and fat to reduce the influx of substrates for gluconeogenesis and ketogenesis. The actions of insulin on hepatic metabolism are always superimposed on a background of other regulatory influences exerted by metabolites, glucagon, and a variety of other regulatory agents. The magnitude of any change produced by insulin is thus determined not only by the concentration of insulin, but also by the strength of the opposing or cooperative actions of these other influences. Rates of secretion of both insulin and glucagon are dictated by physiologic demand. Because of their antagonistic influences on hepatic function, however, it is the ratio, rather than the absolute concentrations, of these two hormones that determines the overall hepatic response.

Glucose Production:

In general, liver takes up glucose when the circulating concentration is high and releases it when the blood level is low. Glucose transport into or out of hepatocytes depends on a high-capacity insulin-insensitive isoform of the glucose transporter GLUT 2. Because the movement of glucose is passive, net uptake or release depends on whether the concentration of free glucose is higher in extracellular or intracellular fluid. The intracellular concentration of free glucose depends on the balance between phosphorylation and dephosphorylation of glucose (Fig. 15, cycle II). The two enzymes that catalyze phosphorylation are *hexokinase*, which has a high affinity for glucose and other 6-carbon sugars, and *glucokinase*, which is specific for glucose. The kinetic properties of glucokinase are such that phosphorylation increases proportionately with glucose concentration over the entire physiologic range. In addition, glucokinase activity is regulated by glucose. When glucose concentrations are low, much of the glucokinase is bound to an inhibitory protein that sequesters it within the nucleus. An increase in glucose concentration releases glucokinase from its inhibitor and allows it to move into the cytosol where glucose phosphorylation can take place.

Phosphorylated glucose cannot pass across the hepatocyte membrane. Dephosphorylation of glucose requires the activity of *glucose-6-phosphatase*. Insulin suppresses synthesis of glucose-6-phosphatase and increases synthesis of glucokinase, thereby decreasing net output of glucose while promoting net uptake. This response to insulin is relatively sluggish and contributes to long-term adaptation rather than to minute-to-minute regulation. The rapid effects of insulin to suppress glucose release are exerted indirectly through decreasing the availability of glucose-6-phosphate, hence starving the phosphatase of substrate. Uptake and phosphorylation by glucokinase is only one source of glucose-6-P. Glucose-6-P is also produced by gluconeogenesis and glycogenolysis. Insulin not only inhibits these processes, but it also drives them in the opposite direction.

Most of the hepatic actions of insulin are opposite to those of glucagon, discussed earlier, and can be traced to inhibition of cyclic AMP accumulation. Rapid actions of insulin largely depend on changes in the phosphorylation state of enzymes already present in hepatocytes. Insulin decreases hepatic concentrations of cyclic AMP by accelerating its degradation by cyclic AMP phosphodiesterase, and it may also interfere with cAMP formation and, perhaps, activation protein kinase A. The immediate consequences can be seen in Fig. 15 and are in sharp contrast to the changes in glucose metabolism produced by glucagon, as shown in Fig. 3. Insulin promotes glycogen synthesis and inhibits glycogen breakdown. These effects are accomplished by the combination of interference with cyclic AMP-dependent processes that drive these reactions in the opposite direction (see Fig. 4); by inhibition of glycogen synthase

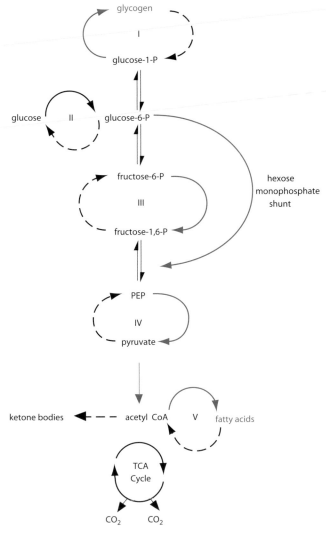

FIGURE 15   Effects of insulin on glucose metabolism in hepatocytes. Blue arrows indicate reactions that are increased; broken arrows indicate reactions that are decreased.

kinase, which, like protein kinase A, inactivates glycogen synthase; and by activation of the phosphatase that dephosphorylates both glycogen synthase and phosphorylase. The net effect is that glucose-6-P is incorporated into glycogen.

By lowering cAMP concentrations, insulin decreases the breakdown and increases the formation of fructose-2,6-phosphate, which potently stimulates phosphofructokinase and promotes the conversion of glucose to pyruvate. Insulin affects several enzymes in the PEP substrate cycle (see Fig. 15, cycle IV) and in so doing directs substrate flow away from gluconeogenesis and toward lipogenesis (Fig. 16). With relief of inhibition of pyruvate kinase, PEP can be converted to pyruvate, which then enters mitochondria. Insulin activates the mitochondrial enzyme that catalyzes decarboxylation of pyruvate to acetyl CoA and indirectly accelerates this reaction by decreasing the inhibition imposed by fatty

acid oxidation. Decarboxylation of pyruvate to acetyl coenzyme A irreversibly removes these carbons from the gluconeogenic pathway and makes them available for fatty acid synthesis. The roundabout process that transfers acetyl carbons across the mitochondrial membrane to the cytoplasm, where lipogenesis occurs, requires condensation with oxaloacetate to form citrate. Citrate is transported to the cytosol and cleaved to release acetyl CoA and oxaloacetate. Recall from earlier discussion that oxaloacetate is a crucial intermediate in gluconeogenesis and is converted to PEP by PEP carboxykinase. Insulin bars the flow of this lipogenic substrate into the gluconeogenic pool by inhibiting synthesis of *PEP carboxykinase*. The only fate left to cytosolic oxaloacetate is decarboxylation to pyruvate.

Finally, insulin increases the activity of *acetyl CoA carboxylase*, which catalyzes the rate-determining reaction in fatty acid synthesis. Activation is accomplished in part by relieving cyclic AMP-dependent inhibition and in part by promoting the polymerization of inactive subunits of the enzyme into an active complex. The resulting malonyl CoA not only condenses to form long-chain fatty acids but also prevents oxidation of newly formed fatty acids by blocking their entry into mitochondria (see Fig. 7). On a longer timescale, insulin increases the synthesis of acetyl CoA carboxylase.

Note that hepatic oxidation of either glucose or fatty acids increases delivery of acetyl CoA to the cytosol, but ketogenesis results only from oxidation of fatty acids. The primary reason is that lipogenesis usually accompanies glucose utilization and provides an alternate pathway for disposal of acetyl CoA. There is also a quantitative difference in the rate of acetyl CoA production from the two substrates: 1 mol of glucose yields only 2 mol of acetyl CoA compared to 8 or 9 mol for each mole of fatty acids.

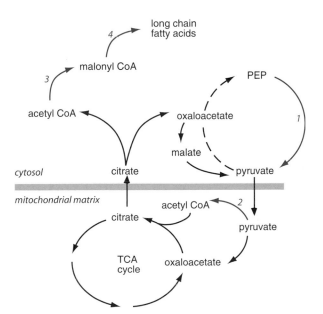

FIGURE 16   Effects of insulin on lipogenesis in hepatocytes. Blue arrows indicate reactions that are increased, and broken arrow indicates reaction that is decreased. (1) Pyruvate kinase, (2) pyruvate dehydrogenase, (3) acetyl CoA carboxylase, and (4) fatty acid synthase.

Mechanism of Insulin Action

The many changes that insulin produces at the molecular level—membrane transport, enzyme activation, gene transcription, and protein synthesis—have been described. The molecular events that link these changes with the interaction of insulin and its receptor are still incompletely understood but are the subjects of intense investigation. Many of the intermediate steps in the action of insulin have been uncovered, but others remain to be identified. It is clear that transduction of the insulin signal is not accomplished by a linear series of biochemical changes, but rather that multiple intracellular signaling pathways are activated simultaneously and may intersect at one or more points before the final result is expressed (Fig. 17).

The insulin receptor is a tetramer composed of two α and two β glycoprotein subunits that are held together by disulfide bonds that link the α-subunits to the β-subunits and the α-subunits to each other (Fig. 18). The α- and β-subunits of insulin are encoded in a single gene. The α-subunits are completely extracellular and contain the insulin-binding domain. The β-subunits span the plasma membrane and contain *tyrosine kinase* activity in the cytosolic domain. Binding to insulin is thought to produce a conformational change that relieves

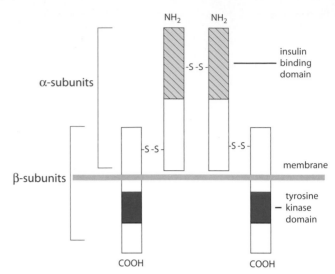

FIGURE 18   Model of the insulin receptor.

the β-subunit from the inhibitory effects of the α-subunit allowing it to phosphorylate itself and other proteins on tyrosine residues. Autophosphorylation of the kinase domain is required for full activation. Tyrosine phosphorylation of the receptor also provides docking sites for other proteins that participate in transducing the hormonal signal. Docking on the phosphorylated receptor may position proteins optimally for phosphorylation by the receptor kinase.

Among the proteins that are phosphorylated on tyrosine residues by the insulin receptor kinase are four cytosolic proteins called *insulin receptor substrates* (IRS-1, IRS-2, IRS-3, and IRS-4). These relatively large proteins contain multiple tyrosine phosphorylation sites and act as scaffolds on which other proteins are assembled to form large signaling complexes. IRS-1 and IRS-2 appear to be present in all insulin target cells, whereas IRS-3 and IRS- 4 have more limited distribution. Despite their names, the IRS proteins are not functionally limited to transduction of the insulin signal, but are also important for expression of effects of other hormones and growth factors. Moreover, they are not the only substrates for the insulin receptor kinase. A variety of other proteins that are tyrosine phosphorylated by the insulin receptor kinase have also been identified. Proteins recruited to the insulin receptor and IRS proteins may have enzymatic activity or they may in turn recruit other proteins by providing sites for protein–protein interactions. The assemblage of proteins initiates signaling cascades that ultimately express the various actions of insulin described earlier. One of the most important of the proteins that is activated is *phosphatidylinositol-3 (PI3) kinase*. PI3 kinase plays a critical role in activating many downstream effector

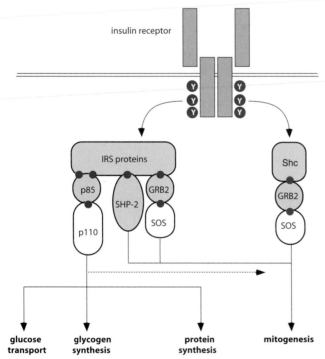

FIGURE 17   Current model of the insulin receptor signaling. Phosphorylated tyrosine residues (Y) on the insulin receptor serve as anchoring sites for cytosolic proteins (IRS proteins and Shc), which in turn are phosphorylated on tyrosines (dark blue circles) and dock with other proteins. IRS, insulin receptor substrate; Shc, Src homology containing protein; p85 and p110 are subunits of phosphoinositol-3-kinase; SHP-2, protein tyrosine phosphatase-2; GrB-2, growth factor receptor binding protein-2; SOS, son of sevenless (a GTPase activating protein). (From Virkamki A, Ueki K, Kahn CR. Protein–protein interaction in insulin signaling and the molecular mechanisms of insulin resistance. *J Clin Invest* 1999;103:931).

molecules including protein kinase B, which is thought to mediate the effects of insulin on glycogen synthesis and GLUT 4 translocation. PI3 kinase, however, is also activated by a variety of other hormones, cytokines, and growth factors whose actions do not necessarily mimic those of insulin. The uniqueness of the response to insulin probably reflects the unique combination of biochemical consequences produced by the simultaneous activity of multiple signaling pathways and the particular set of effector molecules expressed in insulin target cells. Although insulin is known to regulate expression of more than 150 genes, few of the nuclear regulatory proteins that are activated by insulin are known, and precisely how the insulin receptor communicates with these regulatory proteins is unknown. A more detailed discussion of the complex molecular events that govern insulin action can be found in articles listed at the end of this chapter.

Regulation of Insulin Secretion

As might be expected of a hormone whose physiologic role is promotion of fuel storage, insulin secretion is greatest immediately after eating and decreases during between-meal periods (Fig. 19). Coordination of insulin secretion with nutritional state as well as with fluctuating demands for energy production is achieved through stimulation of beta cells by metabolites, hormones, and neural signals. Because insulin plays the primary role in regulating storage and mobilization of metabolic fuels, the beta cells must be constantly apprised of bodily needs, not only with regard to feeding and fasting, but also to the changing demands of the environment. Energy needs differ widely when an individual is at peace with the surroundings and when struggling for survival. Maintaining constancy of the internal environment is achieved through direct monitoring of circulating metabolites by beta cells themselves. This input can be overridden or enhanced by hormonal or neural signals that prepare the individual for rapid storage of an influx of food or for massive mobilization of fuel reserves to permit a suitable response to environmental demands.

Glucose

Glucose is the most important regulator of insulin secretion. In normal individuals its concentration in blood is maintained within the narrow range of about 70 or 80 mg/dL after an overnight fast to about 150 mg/dL immediately after a glucose-rich meal. When blood glucose increases above a threshold value of about 100 mg/dL, insulin secretion increases proportionately. At lower concentrations adjustments in insulin secretion

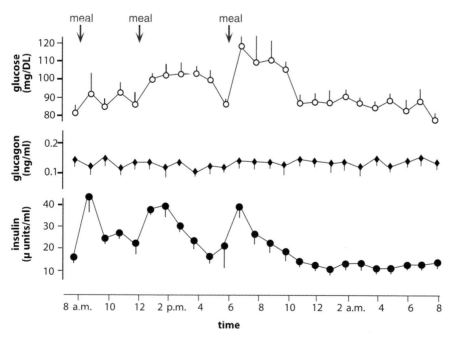

FIGURE 19   Changes in the concentrations of plasma glucose, glucagon, and insulin throughout the day. Values are the mean ± SEM (*n*, 4). (From Tasaka Y, Sekine M, Wakatsuki M, Ohgawara H, Shizume K, Levels of pancreatic glucagon, insulin and glucose during twenty-four hours of the day in normal subjects. *Horm Metab Res* 1975;7:205–206.)

are largely governed by other stimuli (see later discussion) that act as amplifiers or inhibitors of the effects of glucose. The effectiveness of these agents therefore decreases as glucose concentration decreases.

Other Circulating Metabolites

Amino acids are important stimuli for insulin secretion. The transient increase in plasma amino acids after a protein-rich meal is accompanied by increased secretion of insulin. Arginine, lysine, and leucine are the most potent amino acid stimulators of insulin secretion. Insulin secreted at this time may facilitate storage of dietary amino acids as protein and prevents their diversion to gluconeogenesis. Amino acids are effective signals for insulin release only when blood glucose concentrations are adequate. Failure to increase insulin secretion when glucose is in short supply prevents hypoglycemia that might otherwise occur after a protein meal that contains little carbohydrate. Fatty acids and ketone bodies may also increase insulin secretion, but only when they are present at rather high concentrations. Because fatty acid mobilization and ketogenesis are inhibited by insulin, their ability to stimulate insulin secretion provides a feedback mechanism to protect against excessive mobilization of fatty acids and ketosis.

Hormonal and Neural Control

In response to carbohydrate in the lumen, the intestinal mucosa secretes one or more factors, called *incretins*, that reach the pancreas through the general circulation and stimulate the beta cells to release insulin even though the increase in blood glucose is still quite small. Incretins are thought to act by amplifying the stimulatory effects of glucose. This anticipatory secretion of insulin prepares tissues to cope with the coming influx of glucose and dampens what might otherwise be a large increase in blood sugar. Various gastrointestinal hormones including gastrin, secretin, cholecystokinin, glucagon-like peptide (GLP-1), and glucose-dependent insulinotropic peptide (GIP), can evoke insulin secretion when tested experimentally, but of these hormones, only GLP-1 and GIP appear to be physiologically important incretins.

Secretion of insulin in response to food intake is also mediated by a neural pathway. The taste or smell of food or the expectation of eating may increase insulin secretion during this so-called *cephalic phase* of feeding. Parasympathetic fibers in the vagus nerve stimulate beta cells by releasing acetylcholine or the neuropeptide VIP. Activation of this pathway is initiated by integrative centers in the brain and involves input from sensory endings in the mouth, stomach, small intestine, and

portal vein. An increase in the concentration of glucose in portal blood is detected by glucose sensors in the wall of the portal vein and the information is relayed to the brain via vagal afferent nerves. In response, vagal efferent nerves stimulate the pancreas to secrete insulin and the liver to take up glucose.

Insulin secretion by the human pancreas is virtually shut off by epinephrine or norepinephrine delivered to beta cells by either the circulation or sympathetic neurons. This inhibitory effect is seen not only as a response to low blood glucose, but may occur even when the blood glucose level is high. It is mediated through α_2-adrenergic receptors on the surface of beta cells. Physiologic circumstances that activate the sympathetic nervous system thus can shut down insulin secretion and thereby remove the major restraint on mobilization of metabolic fuels needed to cope with an emergency.

Secretory activity of beta cells is also enhanced by growth hormone and cortisol by mechanisms that are not yet understood. Although they do not directly evoke a secretory response, basal insulin secretion is increased when these hormones are present in excess, and beta cells become hyperresponsive to signals for insulin secretion. Conversely, insulin secretion is reduced when either is deficient. Excessive growth hormone or cortisol decreases tissue sensitivity to insulin and can produce diabetes (see Chapter 42). The factors that regulate insulin secretion are shown in Fig. 20.

Cellular Events

Beta cells increase their rates of insulin secretion within 30 sec of exposure to increased concentrations of glucose and can shut down secretion as rapidly. The question of how the concentration of glucose is monitored and translated into a rate of insulin secretion has not been answered completely, but many of the

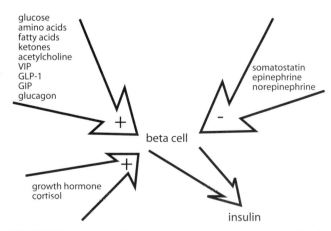

FIGURE 20   Metabolic, hormonal, and neural influences on insulin secretion.

important steps are known. The beta cell has specific receptors for glucagon, acetylcholine, GLP-1, and other compounds that increase insulin secretion by promoting the formation of cyclic AMP or IP_3 and DAG, but it does not appear to have specific receptors for glucose. To effect insulin secretion, glucose must be metabolized by the beta cell, indicating that some consequence of glucose oxidation, rather than glucose itself, is the critical determinant. Beta-cell membranes contain the glucose transporter GLUT 2, which has a high capacity, but relatively low affinity, for glucose. Consequently, as glucose concentrations increase above about 100 mg/dl, glucose enters the beta cell at a rate that is limited by its concentration and not by availability of transporters. It is likely that glucokinase, which is specific for glucose and catalyzes the rate-determining reaction for glucose metabolism in beta cells, has the requisite kinetic characteristics to behave as a glucose sensor. Mutations that affect the function of this enzyme result in decreased insulin secretion in response to glucose that may be severe enough to cause a form of diabetes.

Secretion of insulin, like secretion of other peptide hormones, requires increased cytosolic calcium. Perhaps through the agency of a calmodulin-activated protein kinase, calcium promotes movement of secretory granules to the periphery of the beta cell, fusion of the granular membrane with the plasma membrane, and the consequent extrusion of granular contents into the extracellular space. To increase insulin secretion, increased metabolism of glucose must somehow bring about an increase in intracellular calcium concentration. Linkage between glucose metabolism and intracellular calcium concentration appears to be achieved by their mutual relationship to cellular concentrations of ATP and ADP.

In resting pancreatic beta cells, efflux of potassium through open potassium channels maintains the membrane potential at about −70 mV. Some potassium channels in these cells are sensitive to ATP, which inhibits (closes) them, and to ADP, which activates (opens) them. When blood glucose concentrations are low, the effects of ADP predominate even though its concentration in beta cell cytoplasm is about 1000 times lower than that of ATP. Because glucose transport is not rate-limiting in beta cells, increased concentrations in blood accelerate glucose oxidation and promote ATP formation at the expense of ADP. As a result, ADP levels become insufficient to counter the inhibitory effects of ATP, and potassium channels close. The consequent buildup of positive charge within the beta cell causes the membrane to depolarize, which activates voltage-sensitive calcium channels. When the depolarizing membrane potential reaches about −50 mV, calcium channels open. An influx of positively charged calcium reverses the membrane potential. Electrical recording of these events produces a

pattern of voltage changes that resembles an action potential. The frequency and duration of electrical discharges in beta cells increase as glucose concentrations increase. In addition to triggering insulin secretion, elevated intracellular calcium inhibits voltage-sensitive calcium channels, and activates calcium-sensitive potassium channels allowing potassium to exit and the cell to repolarize (Fig. 21).

Although entry of calcium triggers insulin secretion, it appears that glucose and the various hormonal modulators may stimulate secretion by acting at additional regulatory sites downstream from calcium. Hormones and neurotransmitters that increase insulin

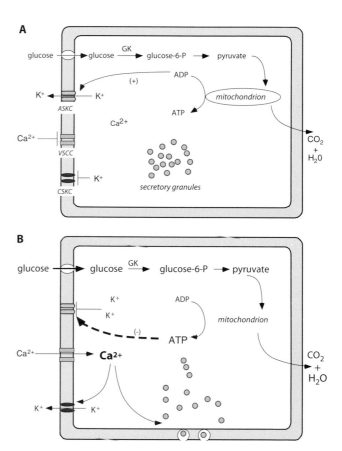

FIGURE 21   Regulation of insulin secretion by glucose. (A) "Resting" beta cell (blood glucose < 100 mg/dl). ADP/ATP ratio is high enough so that ATP-sensitive potassium channels (ASKC) are open, and the membrane potential is about −70 mV. Voltage-sensitive calcium channels (VSCC) and calcium-sensitive potassium channels (CSKC) are closed. (B) Beta-cell response to increased blood glucose. In response to increased glucose entry and metabolism, the ratio of ADP/ATP decreases, and ATP-sensitive potassium channels close. Voltage-sensitive calcium channels are activated; calcium enters and stimulates insulin secretion. Increased cytosolic calcium inhibits voltage-sensitive calcium channels and activates calcium-sensitive potassium channels, thereby allowing the cell membrane to repolarize and calcium channels to close. Persistence of high glucose results in repeated spiking of electrical discharges and oscillation of intracellular calcium concentrations.

secretion act through either the cyclic AMP or DAG/IP$_3$ second messenger pathways to enhance the stimulatory effect of glucose. Some evidence suggests that voltage-sensitive calcium channels may be substrates for protein kinase A, and that phosphorylation may lower their threshold for activation. Additional actions of protein kinase A appear to enhance later steps in the secretory pathway and further increase insulin secretion under conditions when the rate of calcium influx is maximal. This might explain how glucagon and other hormones that activate adenylyl cyclase increase insulin secretion. Agents like acetylcholine increase IP$_3$ and may thus stimulate release of calcium from intracellular storage sites. In addition, activation of protein kinase C enhances aspects of the secretory process that are independent of calcium. Norepinephrine and somatostatin block insulin secretion by way of the inhibitory guanine nucleotide binding protein (G$_i$), which may directly inhibit voltage-sensitive calcium channels as well as adenylyl cyclase.

In addition to serving as the principal signal for insulin secretion, glucose appears to be the most important stimulator of insulin synthesis. Both glucose and cyclic AMP increase transcription of the insulin gene. The mRNA template for insulin turns over slowly and has a half-life of about 30 hr. Glucose also appears to regulate its stability. Hyperglycemia prolongs its half-life more than twofold while hypoglycemia accelerates its degradation. In addition, glucose increases translation of the preproinsulin mRNA by stimulating both the initiation and elongation reactions. Concurrently, glucose also up-regulates production of the enzymes needed to process preproinsulin to insulin.

SOMATOSTATIN

Biosynthesis, Secretion, and Metabolism

Somatostatin was originally isolated from hypothalamic extracts that inhibited the secretion of growth hormone. Somatostatin is widely distributed in many neural tissues where it presumably functions as a neurotransmitter. It is found in many secretory cells (delta cells) outside of the pancreatic islets, particularly in the lining of the gastrointestinal tract. Somatostatin is stored in membrane-bound vesicles and secreted by exocytosis. Measurable increases in the somatostatin concentration can be found in peripheral blood after ingestion of a meal rich in fat or protein, with the vast majority secreted by intestinal cells rather than islet cells. It is cleared rapidly from the blood and has a half-life of only about 3 min.

Physiologic Actions

The physiologic importance of pancreatic somatostatin is not understood. Because it can inhibit secretion of both insulin and glucagon, it has been suggested that somatostatin, by acting in a paracrine fashion, may contribute to the regulation of glucagon and insulin secretion. However, anatomic relationships and the direction of blood flow in the microcirculation in the islets are inconsistent with such a role. Somatostatin also inhibits secretion of various gastrointestinal hormones and decreases acid secretion by the gastric mucosa and enzyme secretion by the acinar portion of the pancreas. In addition, somatostatin decreases intestinal motility and may slow the rate of absorption of nutrients from the digestive tract. Increased fecal excretion of fat is a prominent feature in patients suffering from somatostatin-secreting tumors. At the cellular level the inhibitory effects of somatostatin are mediated by G-protein-coupled receptors that signal through Gα_i to inhibit adenylyl cyclase, and through $\beta\gamma$-subunits that activate potassium channels and hyperpolarize cell membranes.

Regulation of Secretion

Increased concentrations of glucose or amino acids in blood stimulate somatostatin secretion by intestinal delta cells. In addition, glucose or fat in the gastrointestinal tract elicits a secretory response by pancreatic delta cells, mediated perhaps by glucagon or gastrointestinal hormones. Somatostatin secretion is also increased by norepinephrine and inhibited by acetylcholine.

Suggested Reading

Becker AB, Roth RA. Insulin receptor structure and function in normal and pathological conditions. *Ann Rev Med* 1990;41:99–116.

Burant CF, Sivitz WI, Fukumoto H, Kayano T, Nagamatsu S, Seino S, Pessin JE, Bell GI. Mammalian glucose transporters: Structure and molecular regulation. *Rec Prog Horm Res* 1991; 47:349–387.

Cheatham B, Kahn CR. Insulin action and the insulin signaling network. *Endocr Rev* 1995;16:117–142.

Jefferson LS, Cherrington AD., eds. *Handbook of physiology, Section 7, Endocrinology, Vol II. The endocrine pancreas and regulation of metabolism*, 2001, New York: Oxford University Press. (This volume covers a wide range of topics relevant to items discussed in this chapter.)

Kieffer TJ, Habener JF. The glucagon-like peptides. *Endocr Rev* 1999;20:876–913.

Kimball SR, Vary TC, Jefferson LS. Regulation of protein synthesis by insulin. *Ann Rev Physiol* 1994;56:321–348.

Miller RE. Pancreatic neuroendocrinology: Peripheral neural mechanisms in the regulation of the islets of Langerhans. *Endocr Rev* 1981;2:471–494.

Pilkis SJ, Granner DK. Molecular physiology of the regulation of hepatic gluconeogenesis and glycolysis. *Ann Rev Physiol* 1992;54:885–909.

Rajan AS, Aguilar-Bryan L, Nelson DA, Yaney GC, Hsu WH, Kunze DL, Boyd AE, III. Ion channels and insulin secretion. *Diabetes Care* 1990;13:340–363.

Taylor SI, Cama A, Accili D, Barbetti F, Quon MJ, de la Luz Sierra M, Suzuki Y, Koller E, Levy-Toledano R, Wertheimer E, Moncada VY, Kadowaki H, Kadowaki T. Mutations in the insulin receptor gene. *Endocr Rev* 1992;13:566–595.

Unger RH, Orci L. Physiology and pathophysiology of glucagon. *Physiol Rev* 1976;56:778–838.

CHAPTER

42

Hormonal Regulation of Fuel Metabolism

H. MAURICE GOODMAN

KEY POINTS

- Glucose can be used by all cells, but *free glucose* is available for only about 1 hour's worth of fuel needs.
- *Glycogen stores* provide little more than half a day's fuel requirement.
- *Proteins* can be drawn on to meet energy needs.
- *Fat (triglyceride)* is the most efficient storage form of fuel. The normal human has enough fat to meet 30–40 days worth of energy demands.
- Most cells can derive their energy from oxidation of glucose, fat, or amino acids, but some

cells, including brain, have an absolute requirement for glucose.
- *Problems inherent in the use of glucose and fat as metabolic fuels:* (1) Conversion of carbohydrate to fat consumes energy, and fat is largely not reconvertable to carbohydrate and, hence, unavailable to meet the energy needs of brain and some other tissues; (2) limited water solubility of fat requires protein carriers for interorgan transport or conversion to water-soluble ketone bodies for access to brain; and

---

Essential Medical Physiology, Third Edition

659

KEY POINTS (*continued*)

(3) fat also cannot be used anaerobically or by cells that lack mitochondria.

- *The glucose-fatty acid cycle:* When in ample supply, glucose limits oxidation of fat, and when glucose is scarce, oxidation of fat limits glucose consumption. Hormonal regulation of metabolism largely operates through the glucose-fatty acid cycle.
- Glucose is preserved for the brain and other glucose-dependent cells during food deprivation: Insulin secretion decreases, whereas secretion of glucagon and growth hormone (GH) increases, and T_3 production from T_4 decreases.
- Substrate utilization during short-term maximal effort depends largely on anaerobic degradation of glycogen stores.
- Plasma free fatty acids provide 50–60% of the fuel during sustained exercise, and plasma glucose and muscle glycogen provide the remainder.
- Long-term regulation of fuel stores depends on signals arising in the adipose tissue and the hypothalamus to regulate food intake and energy expenditure.

OVERVIEW

Mammalian survival in a cold, hostile environment demands an uninterrupted supply of metabolic fuels to maintain body temperature, to escape from danger, and to grow and reproduce. A constant supply of glucose and other energy-rich metabolic fuels to the brain and other vital organs must be available at all times despite wide fluctuations in food intake and energy expenditure. Constant availability of metabolic fuel is achieved by storing excess carbohydrate, fat, and protein principally in liver, adipose tissue, and muscle and drawing on those reserves when needed. We consider here how fuel homeostasis is maintained in a minute-to-minute, day-to-day, and year-to-year manner by regulating fuel storage and mobilization, the mixture of fuels consumed, and food intake. Homeostatic regulation is provided by the endocrine system and the autonomic nervous system. The strategy of hormonal regulation of metabolism during starvation or exercise is to provide sufficient substrate to working muscles while maintaining an adequate concentration of glucose in blood to satisfy the needs of brain and other glucose-dependent cells. When dietary or stored carbohydrate is inadequate, availability of glucose is ensured by (1) gluconeogenesis from lactate, glycerol, and alanine; and (2) inhibition of glucose utilization by those tissues that can satisfy their energy needs with other substrates, notably fatty acids and ketone bodies. The principal hormones that govern fuel homeostasis are insulin, glucagon, epinephrine, cortisol, growth hormone (GH), thyroxine (T_4) and the newly discovered adipocyte hormone, leptin. The principal target organs for these hormones are adipose tissue, liver, and skeletal muscle.

GENERAL FEATURES OF ENERGY METABOLISM

In discussing how hormones regulate fuel metabolism, we consider first the characteristics of metabolic fuels and the intrinsic biochemical regulatory mechanisms on which hormonal control is superimposed.

Body Fuels

Glucose

Glucose is readily oxidized by all cells. One gram yields about 4 Calories. The average 70-kg man requires approximately 2000 Calories per day and therefore would require a reserve supply of approximately 500 g of glucose to ensure sufficient substrate to survive 1 day of food deprivation. If glucose were stored as an isosmolar solution, approximately 10 L of water (10 kg) would be needed to accommodate a single day's energy needs, and the 70-kg man would have to carry around a storage depot equal to his own weight if he were to survive only 1 week of starvation. Actually, only about 20 g of free glucose is dissolved in extracellular fluids, or enough to provide energy for about 1 hr.

Glycogen

Polymerizing glucose to glycogen eliminates the osmotic requirement for large volumes of water. To meet a single day's energy needs, only about 1.8 kg of "wet" glycogen is required; that is, 500 g of glycogen obligates only about 1.3 L of water. Glycogen stores in the well fed 70-kg man are only enough to meet part of a day's energy needs—about 100 g in the liver and about 200 g in muscle.

Protein

Calories can also be stored in somewhat more concentrated form as protein. Storage of protein, however, also obligates storage of some water, and oxidation of protein creates unique by-products: ammonia, which must be detoxified to form urea at metabolic expense, and sulfur-containing acids. The body of a normal 70-kg man in nitrogen balance contains about 10–12 kg of protein, most of which is in skeletal muscle. Little or no protein is stored as an inert fuel depot, so that mobilization of protein for energy necessarily produces some functional deficits. Under conditions of prolonged starvation, as much as one-half of the body protein may be consumed for energy before death ensues, usually from failure of respiratory muscles.

Fat

Triglycerides are by far the most concentrated storage form of high-energy fuel (9 Calories/g), and they can be stored essentially without water. One day's energy needs can be met by less than 250 g of triglyceride. Thus a 70-kg man carrying 10 kg of fat maintains an adequate depot of fuel to meet energy needs for more than 40 days. Most fat is stored in adipose tissue, but other tissues such as muscle also contain small reserves of triglycerides.

Problems Inherent in the Use of Glucose and Fat as Metabolic Fuels

1. Fat is the most abundant and efficient energy reserve, but efficiency has its price. When converting dietary carbohydrate to fat, about 25% of the energy is dissipated as heat. More importantly, synthesis of fatty acids from glucose is an irreversible process. Once the carbons of glucose are converted to fatty acids, virtually none can be reconverted to glucose. The glycerol portion of triglycerides remains convertible to glucose, but glycerol represents only about 10% of the mass of triglyceride.
2. Limited water solubility of fat complicates transport between tissues. Triglycerides are "packaged" as very-low-density lipoproteins or as chylomicrons for transport in blood to storage sites. Uptake by cells follows breakdown to fatty acids by lipoprotein lipase at the external surface or within capillaries of muscle or fat cells. Mobilization of stored triglycerides also requires breakdown to fatty acids, which leave adipocytes in the form of free fatty acids (FFA). FFA are not very soluble in water and are transported in blood firmly bound to albumin. Because they are bound to albumin, FFA have limited access to tissues such as brain; they can be processed to water-soluble forms in the liver, however, which converts them to 4-carbon ketoacids (*ketone bodies*), which can cross the blood–brain barrier.
3. Energy can be derived from glucose without simultaneous consumption of oxygen, but oxygen is required for degradation of fat. Therefore, glucose must be constantly available in the blood to satisfy the needs of red blood cells, which lack mitochondria, and cells in the renal medullae, which function under low oxygen tension. Under basal conditions these cells consume about 50 g of glucose each day and release an equivalent amount of lactate into the blood. Because lactate is readily reconverted to glucose in the liver, however, these tissues do not act as a drain on carbohydrate reserves.
4. In a well-nourished person the brain relies almost exclusively on glucose to meet its energy needs and consumes nearly 150 g per day. Brain does not derive energy from oxidation of FFA or amino acids. Ketone bodies are the only alternative substrates to glucose, but studies in experimental animals indicate that only certain regions of the brain can substitute ketone bodies for glucose. Total fasting for 4–5 days is required before the concentrations of ketone bodies in blood are high enough to provide a significant fraction of the brain's energy needs. Even after several weeks of total starvation, the brain continues to satisfy about one-third of its energy needs with glucose. The brain stores little glycogen and hence must depend on the circulation to meet its minute-to-minute fuel requirements. The rate of glucose delivery depends on its concentration in arterial blood, the rate of blood flow, and the efficiency of extraction. Although an increased flow rate might compensate for decreased glucose concentration, the mechanisms that regulate blood flow in brain are responsive to oxygen and carbon dioxide, rather than glucose. Under basal conditions the concentration of glucose in arterial blood is about 5 mM (90 mg/dL), of which the brain extracts about 10%. The fraction extracted can double, or perhaps even triple, when the concentration of glucose is low; but when the blood glucose falls below about 30 mg/dL, metabolism and function are compromised. Thus the brain is exceedingly vulnerable to hypoglycemia, which can quickly produce coma or death.

Fuel Consumption

The amount of metabolic fuel consumed in a day varies widely and normally is balanced by variations in food intake, but the adipose tissue reservoir of triglycerides can shrink or expand to accommodate imbalances in fuel intake and expenditure. Muscle comprises about

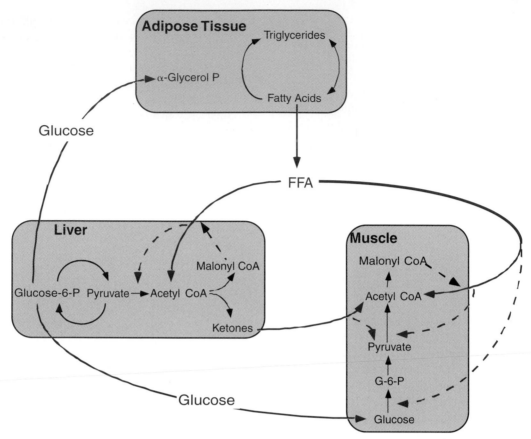

FIGURE 1 Intraorgan flow of substrate and the competitive regulatory effects of glucose and fatty acids that comprise the glucose-fatty acid cycle. Dashed arrows denote inhibition. See text for details.

50% of body mass and is by far the major consumer of metabolic fuel. Even at rest muscle metabolism accounts for about 30% of the oxygen consumed. Although normally a 56-kg woman or a 70-kg man or consumes about 1600 or 2000 Calories in a typical day, daily caloric requirements may range from about 1000 with complete bed rest to as much as 6000 with prolonged physical activity. For example, marathon running may consume 3000 Calories in only 3 hr. Under basal conditions an individual on a typical mixed diet derives about half of the daily energy needs from the oxidation of glucose, a small fraction from consumption of protein, and the remainder from fat. With starvation or with prolonged exercise, limited carbohydrate reserves are quickly exhausted unless some restriction is placed on carbohydrate consumption by muscle, whose fuel needs far exceed those of any other tissue and can be met by increased utilization of fat. In fact, simply providing muscle with fat restricts its ability to consume carbohydrate. Hormonal regulation of energy balance is largely accomplished through adjusting the flux of energy-rich fatty acids and their derivatives to muscle, and the consequent sparing of carbohydrate and protein.

GLUCOSE-FATTY ACID CYCLE

The self-regulating interplay between glucose and fatty acid metabolism is called the *glucose-fatty acid cycle*. This cycle constitutes an important biochemical mechanism for limiting glucose utilization when alternative substrate is available, and conversely limiting the consumption of stored fat when glucose is available. Fatty acids that are produced in adipose tissue in an ongoing cycle of lipolysis and reesterification may either escape from fat cells to become the free fatty acids, or they may be retained as triglycerides, depending on the availability of α-glycerol phosphate (Fig. 1). The only source of α-glycerol phosphate for reesterification of fatty acids is the pool of triose phosphates derived from glucose oxidation, because adipose tissue is deficient in the enzyme required to phosphorylate and hence reuse glycerol released from triglycerides. Consequently, when glucose is abundant, α-glycerol phosphate is readily available, the rate of reesterification is high relative to lipolysis, and the rate of release of FFA is low. Conversely, when glucose is scarce, more fatty acids escape and plasma concentrations of FFA increase.

Exposure of muscle to elevated levels of FFA for several hours decreases transport of glucose across the plasma membrane and phosphorylation to glucose-6-phosphate. The resulting decrease of glucose-6-phosphate, which is both a substrate and an allosteric activator of glycogen synthase, results in decreased glycogen formation as well as decreased glucose oxidation by glycolysis. Glycolysis may be further curtailed by inhibition of phosphofructokinase. Oxidation of fatty acids or ketone bodies also limits the oxidation of pyruvate to acetyl CoA. Recall from Chapter 41 that long-chain fatty acids must be linked to carnitine to gain entry into mitochondria where they are oxidized. Activity of acylcarnitine transferase is increased by long-chain fatty acid coenzyme A (CoA) and inhibited by malonyl CoA whose formation is accelerated when glucose is plentiful. Oxidation of long-chain fatty acids or ketone bodies to acetyl CoA reduces the cofactor nicotinamide-adenine dinucleotide (NAD) to NADH at a rate that exceeds oxidative regeneration in the nonworking muscle. The resulting scarcity of NAD and free CoA limits the breakdown of pyruvate directly, and also activates the mitochondrial enzyme pyruvate dehydrogenase (PDH) kinase that inactivates a key enzyme of pyruvate oxidation. The activity of PDH kinase, in turn, is inhibited by pyruvate.

Influx of fatty acids to the liver promotes ketogenesis and gluconeogenesis largely by the same mechanisms that diminish glucose metabolism in muscle. Metabolism of long-chain fatty acids inhibits the intramitochondrial oxidation of pyruvate to acetyl CoA. Gluconeogenic precursors arriving at the liver in the form of pyruvate, lactate, alanine, or glycerol are thus spared oxidation in the tricarboxylic acid cycle and instead are converted to phosphoenol pyruvate (PEP). Conversely, when glucose is abundant, the concentration of glucose-6-phosphate increases, and gluconeogenesis is inhibited both at the

level of fructose-1,6-bisphosphate formation and at the level of pyruvate kinase (see Fig. 5 in Chapter 41). Under these circumstances malonyl CoA formation is increased and fatty acids are restrained from entering the mitochondria and subsequent degradation.

Through the reciprocal effects of glucose and fatty acids, glucose indirectly regulates its own rate of utilization by a negative feedback process that increases gluconeogenesis. Because of the intrinsic allosteric regulatory properties of the glucose-fatty acid cycle, hormones may regulate metabolic processes not only by altering the activities or amounts of enzymes, but also by influencing the flow of metabolites. The glucose-fatty acid cycle operates in normal physiology even though the concentration of glucose in blood remains nearly constant. In fact, the contribution of some hormones, notably glucocorticoids and GH, to the maintenance of blood glucose and muscle glycogen depends in part on the glucose-fatty acid cycle. Conversely, in addition to stimulating glucose entry into muscle, insulin indirectly increases glucose metabolism by decreasing FFA mobilization from adipose tissue, thereby shutting down the inhibitory influence of the glucose-fatty acid cycle. This effect may be accelerated by a further effect of insulin to increase the formation of malonyl CoA in liver and muscle, thereby diminishing access of fatty acids to the mitochondrial oxidative apparatus.

OVERALL REGULATION OF BLOOD GLUCOSE CONCENTRATION

Despite vagaries in dietary input and large fluctuations in food consumption, the concentration of glucose in blood remains remarkably constant. Its concentration at any time is determined by the rate of input and the rate of removal by the various body tissues (Fig. 2).

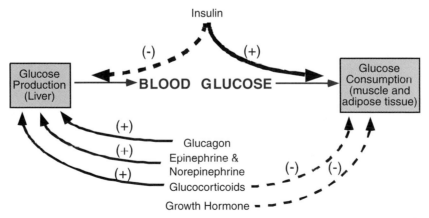

FIGURE 2   Interaction of hormones to maintain the blood glucose concentration. Solid arrows (+) denote increase; dashed arrows (−) denote decrease.

The rate of glucose removal from the blood varies over a wide range depending on physical activity and environmental temperature. The rate of input largely reflects activity of the liver, since even immediately after eating, glucose absorbed from the intestines must pass through the liver before entering the circulation. Liver glycogen is the immediate source of blood glucose under most circumstances. Hepatic gluconeogenesis may contribute to blood glucose directly but is more important for replenishing glycogen stores. The kidneys are also capable of gluconeogenesis, but their role as providers of blood glucose has not been studied to any great extent, except in acidosis when glucose production from glutamate accompanies renal production and excretion of ammonia. Recent studies in patients undergoing liver transplantation, however, indicate that glucose production by the kidneys immediately after removal of the liver can be substantial, at least for a short time.

Minute-to-minute regulation of blood glucose depends on (1) insulin, which, in promoting fuel storage, drives glucose concentrations down; and (2) glucagon, and to a lesser extent catecholamines, which, in mobilizing fuel reserves, drive glucose concentrations up. Effects of these hormones are evident within seconds or minutes and dissipate as quickly. Insulin acts at the level of the liver to inhibit glucose output, and on muscle and fat to increase glucose uptake. Liver is more responsive to insulin than muscle and fat, and because of its anatomic location, is exposed to higher hormone concentrations. Smaller increments in insulin concentration are needed to inhibit glucose production than to promote glucose uptake. Glucagon and catecholamines act on hepatocytes to promote glycogenolysis and gluconeogenesis. They have no direct effects on glucose uptake by peripheral tissues, but epinephrine and norepinephrine may decrease the demand for blood glucose by mobilizing alternative fuels, glycogen and fat, within muscle and adipose tissue. Increased blood glucose is perceived directly by pancreatic beta cells, which respond by secreting insulin. Hypoglycemia is perceived not only by the glucagon-secreting alpha cells of pancreatic islets, but also by the central nervous system, which activates sympathetic outflow to the islets and the adrenal medullae. Sympathetic stimulation of pancreatic islets increases secretion of glucagon and inhibits secretion of insulin. In addition, hypoglycemia evokes secretion of the hypothalamic releasing hormones that stimulate ACTH and GH secretion from the pituitary gland (Fig. 3). Cortisol, secreted in response to ACTH, and GH act only after a substantial delay and hence are unlikely to contribute to rapid restoration of blood glucose. However, they are important for withstanding a sustained hypoglycemic challenge.

Long-term regulation, operative on a timescale of hours or perhaps days, depends on direct and indirect

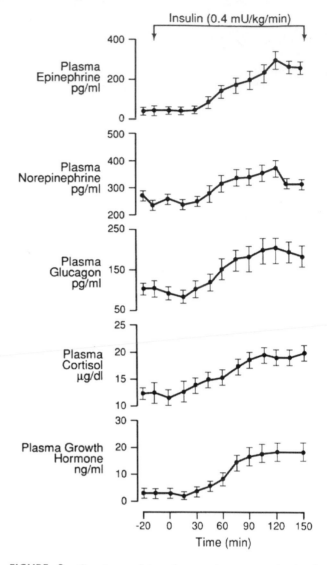

FIGURE 3 Counter-regulatory hormonal responses to insulin-induced hypoglycemia. The infusion of insulin reduced plasma glucose concentration to 50–55 mg/dL. (From Sacca L, Sherwin R, Hendler R, Felig P. Influence of continuous physiologic hyperinsulinemia on glucose kinetics and counterregulatory hormones in normal and diabetic humans. *J Clin Invest* 1979;63:849–857.)

actions of many hormones and ultimately ensures (1) that the peripheral drain on glucose reserves is minimized and (2) that liver contains an adequate reservoir of glycogen to satisfy the minute-to-minute needs of glucose-dependent cells. To achieve these ends, peripheral tissues, mainly muscle, must be provided with alternate substrate and limit their consumption of glucose. At the same time, gluconeogenesis must be stimulated and supplied with adequate precursors to provide the 150–200 g of glucose needed each day by the brain and other glucose-dependent tissues. Long-term regulation includes all of the responses that govern glucose utilization as well as all those reactions

Clinical Note

Hypoglycemia may be the presenting symptom in patients suffering from a variety of ailments and may be the principal symptom of hypopituitarism that results in the combined deficiency of GH and ACTH, or of either of these hormones alone. Similarly, hypoglycemia is a prominent symptom of adrenal cortical failure (Addison's disease). Hypoglycemia due to a glucagon deficiency is not seen, presumably because of redundancy with sympathetic mechanisms for rapidly increasing glucose production. Not surprisingly, severe hepatic damage caused by hepatitis or a variety of hepatic toxins may produce hypoglycemia in patients with inadequate food intake. Carnitine deficiency, which results in impaired fatty acid transport into mitochondria also results in hypoglycemia during fasting. Plasma concentrations of insulin are appropriately low in all of these cases of hypoglycemia. Because the restraining effect of insulin on fatty acid mobilization is lacking, ketonemia is a prominent feature of all of these hypoglycemias except that resulting from carnitine deficiency.

that govern storage of fuel as glycogen, protein, or triglycerides.

INTEGRATED ACTIONS OF METABOLIC HORMONES

Metabolic fuels absorbed from the intestine are largely converted to storage forms in liver, adipocytes and muscle. It is fair to state that storage is virtually the exclusive province of insulin, which stimulates biochemical reactions that convert simple compounds to more complex storage forms and inhibits fuel mobilization. Hormones that mobilize fuel and defend the glucose concentration of the blood are called *counter-regulatory* and include glucagon, epinephrine, norepinephrine, cortisol, and GH. Secretion of most or all of these hormones is increased whenever there is increased demand for energy. These hormones act synergistically and together produce effects that are greater than the sum of their individual actions. In the example shown in Fig. 4, glucagon and epinephrine raised the blood glucose level primarily by increasing hepatic production. When cortisol was given simultaneously, these effects were magnified, even though cortisol had little effect when given alone. Triiodothyronine (T_3) must also be considered in this context, as its actions increase the rate of fuel consumption and the sensitivity of target cells to insulin and counter-regulatory hormones. Before examining the interactions of these hormones in the whole body, it is useful to summarize their effects on individual tissues.

Adipose Tissue

The central event in adipose tissue metabolism is the cycle of fatty acid esterification and triglyceride lipolysis (Fig. 5). Although reesterification of fatty acids can regulate FFA output from fat cells, regulation of lipolysis and hence the rate at which the cycle spins provides a wider range of control. It has been estimated that under basal conditions 20% of the fatty acids released in lipolysis are reesterified to triglycerides, and that reesterification may decrease to 9–10% during active fuel consumption. Under the same conditions,

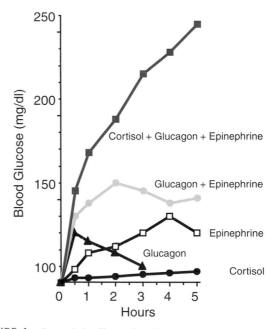

FIGURE 4 Synergistic effects of cortisol, glucagon, and epinephrine on increasing the plasma glucose level. Note that the hyperglycemic response to the triple hormone infusion is far greater than the additive response of all three hormones given singly. (Redrawn from data of Eigler N, Sacca L, Sherwin RS. Synergistic interactions of physiologic increments of glucagon, epinephrine, and cortisol in the dog. *J Clin Invest* 1979;63:114–123.)

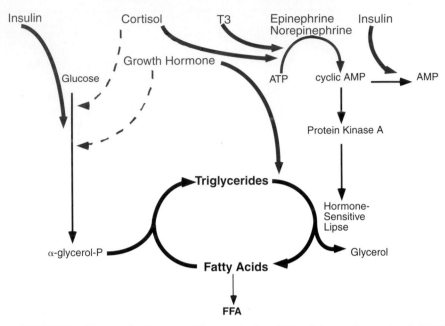

FIGURE 5 Hormonal effects on FFA production. Epinephrine and norepinephrine stimulate hormone-sensitive lipase through a cyclic AMP-mediated process. Insulin antagonizes this effect by stimulating cyclic AMP degradation. T_3, cortisol, and growth hormone increase the response of adipocytes to epinephrine and norepinephrine. Growth hormone also directly stimulates lipolysis. Insulin indirectly antagonizes the production of FFA by increasing reesterification. Growth hormone and cortisol increase FFA release by inhibiting (dashed arrows) reesterification.

lipolysis may be varied over a 10-fold range. Catecholamines and insulin, through their antagonistic effects on cyclic AMP metabolism, increase or decrease the activity of hormone-sensitive lipase. Responses to these hormones are expressed within minutes. Other hormones, especially cortisol, T_3, and GH, modulate the sensitivity of adipocytes to insulin and catecholamines. Modulation is not a reflection of abrupt changes in hormone concentrations but, rather, stems from long-term tuning of metabolic machinery. Finally, GH produces a sustained increase in lipolysis after a delay of about 2 hr. Growth hormone and cortisol also decrease fatty acid esterification by inhibiting glucose metabolism both directly and by decreasing responsiveness to insulin. These hormonal effects on adipose tissue are summarized in Table 1 .

Muscle

By inhibiting FFA mobilization, insulin promptly decreases plasma FFA concentrations and thus removes a deterrent of glucose utilization in muscle at the same time that it promotes transport of glucose into myocytes. The response to insulin can be divided into two components: Stimulation of glucose transport and glycogen synthesis are direct effects and are seen within minutes. Increased oxidation of glucose that results from release

of inhibition requires several hours. Epinephrine and norepinephrine promptly increase cyclic AMP production and glycogenolysis. When the rate of glucose production from glycogen exceeds the need for ATP production, muscle cells release pyruvate and lactate, which can be reconverted to glucose in liver. Growth hormone and cortisol directly inhibit glucose uptake by muscle and indirectly decrease glucose metabolism in myocytes through the agency of the glucose-fatty acid cycle. By indirectly inhibiting glucose metabolism, GH and cortisol decrease glycogen breakdown. The resulting

TABLE 1   Hormonal Effects on Metabolism in Adipocytes

Hormone	Glucose uptake	Lipolysis	Reesterification	Rapid (R) or slow (S)
Insulin	↑	↓	↑	R
Epinephrine and norepinephrine	↑[a]	↑ ↑	↑	R
Growth hormone	↓	↑	↓	S
Cortisol	↓	↑[b]	↓	S
T_3	↑	↑[b]	↑	S

[a]Increased glucose uptake and reesterification are driven by the increase in lipolysis.

[b]Permissive effects.

TABLE 2   Hormonal Effects of Glucose Metabolism in Muscle

	Glucose uptake	Glucose phosphorylation	Glycolysis	Glycogen storage
Insulin	↑ (D)	↑ (I)	↑ (I) (D)	↑(D)
Epinephrine, norepinephrine	↓[a] (I)	↓[a] (I)	↑	↓ (D)
Growth hormone	↓ (D,I)	↓ (I)	↓ (I)	↑ (I)
Cortisol	↓ (D,I)	↓ (I)	↓ (I)	↑ (I)
T3	↑	↑	↑	↑ or ↓[b]

Key: D, direct effect; I, indirect effect via the glucose-fatty acid cycle.
[a]Immediate effect is secondary to glycogenolysis; later effect is secondary to the glucose-fatty acid cycle
[b]Dependent on the dose; high rates of oxygen consumption may decrease glycogen.

preservation of muscle glycogen has been called the *glycostatic effect* of GH, and is part of the overall effect of cortisol that gives rise to the term *glucocorticoid*. Cortisol also inhibits the uptake of amino acids and their incorporation into proteins and simultaneously promotes degradation of muscle protein. As a result, muscle becomes a net exporter of amino acids, which provide substrate for gluconeogenesis in liver. These events are summarized in Table 2 .

Liver

The antagonistic effects of insulin and glucagon on gluconeogenesis, ketogenesis, and glycogen metabolism in hepatocytes are described in Chapter 41. Epinephrine and norepinephrine, by virtue of their effects on cyclic AMP metabolism, share all the actions of glucagon. In addition, these medullary hormones activate α_1-adrenergic receptors and reinforce these effects through the agency of the diacylglycerol (DAG)–inositol trisphosphate (IP$_3$)–calcium system (see Chapter 2). Cortisol is indispensable as a permissive agent for the actions of glucagon and catecholamines on gluconeogenesis and glycogenolysis. In addition, cortisol induces synthesis of a variety of enzymes responsible for gluconeogenesis and glycogen storage. By virtue of its actions on protein degradation in muscle, cortisol is also indispensable for

providing substrate for gluconeogenesis. T$_3$ promotes glucose utilization in liver by promoting synthesis of enzymes required for glucose metabolism and lipid formation. Growth hormone is thought to increase hepatic glucose production, probably as a result of increased FFA mobilization, and it also increases ketogenesis largely by increasing mobilization of FFA. These hormonal influences on hepatic metabolism are summarized in Table 3 .

Pancreatic Islets

Alpha and beta cells of pancreatic islets are targets for metabolic hormones as well as producers of glucagon and insulin. Glucagon can stimulate insulin secretion but the physiologic significance of such an action is not understood. Insulin inhibits glucagon secretion, and in its absence, responsiveness of alpha cells to glucose is severely impaired. Conversely, insulin apparently also exerts autocrine effects on the beta cells and is required to maintain the normal secretory response to increased glucose concentrations. Epinephrine and norepinephrine inhibit insulin secretion and stimulate glucagon secretion. Growth hormone, cortisol, and T$_3$ are required for normal secretory activity of beta cells, whose capacity for insulin secretion is reduced in their absence. The effects of

TABLE 3   Hormonal Regulation of Metabolism in Liver

Hormone	Glucose output	Glycogen synthesis (S) or breakdown (B)	Gluconeogenesis	Ketogenesis	Ureogenesis	Lipogenesis
Insulin	↓ (D)	S	↓ (D)	↓ (I, D)	↓ (I, D)	↑ (D)
Glucagon	↑ (D)	B	↑ (D)	↑ (D)	↑ (D)	↓ (D)
Epinephrine, norepinephrine	↑ (D)	B	↑ (D)	↑ (I, D)	↑ (D)	↓ (D)
GH	↑ (I)	S	↑ (I)	↑ (I)	↓[a] (I)	↓ (I)
Cortisol	↑ (I, D)	S	↑ (I, D)	↑ (I)	↑ (I)	↓ (I)
T$_3$	↑ (I)	B	↑ (I)	↑ (I)	↑ (I)	↑ (D)

Key: D, direct effect; I, indirect effect.
[a]Growth hormone promotes protein synthesis and hence decreases availability of amino acids for ureogenesis.

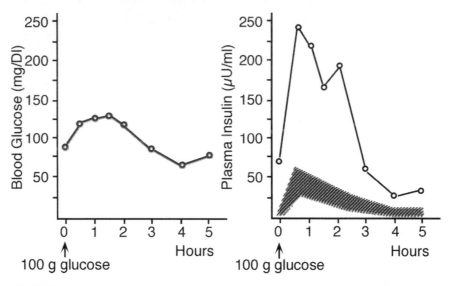

FIGURE 6   Normal glucose tolerance following ingestion of 100 g of glucose in a subject with a growth hormone-secreting pituitary tumor is accompanied by an exaggerated increase in plasma insulin indicative of decreased insulin sensitivity. The shaded area in the right-hand panel represents the plasma insulin response of 43 normal subjects who showed the same changes in glucose concentration after ingestion of 100 g of glucose. (From Daughaday WH, Kipnis DM. The growth promoting and anti-insulin actions of somatotropin. *Rec Prog Horm Res* 1966;22:49–99.)

GH and cortisol on insulin secretion are somewhat paradoxical. Although their effects in adipose tissue, muscle, and liver are opposite to those of insulin, GH and cortisol nevertheless increase the sensitivity of beta cells to signals for insulin secretion and exaggerate responses to hyperglycemia (Fig. 6). When cortisol or growth hormone is present in excess, higher than normal concentrations of insulin are required to maintain blood glucose in the normal range. Higher concentrations of insulin itself may contribute to decreased sensitivity by down-regulating insulin receptors in fat and muscle. Hormonal effects on insulin secretion and sensitivity of tissues to insulin are summarized in Table 4.

REGULATION OF METABOLISM DURING FEEDING AND FASTING

Postprandial Period

Immediately after eating, metabolic activity is directed toward the processing and sequestration of energy-rich substrates that are absorbed by the intestines. This phase is dominated by insulin, which is secreted in response to three inputs to the beta cells. The *cephalic,* or psychological aspect of eating, stimulates insulin secretion through acetylcholine and vasoactive inhibitory peptide (VIP) released from vagal fibers that innervate islet cells. Food in the small intestine stimulates secretion of

Clinical Note

When either growth hormone or glucocorticoids are present in excess for prolonged periods, diabetes mellitus often results. Approximately 30% of patients suffering from excess GH (acromegaly) and a similar percentage of persons suffering from Cushing's disease (excess glucocorticoids) experience diabetes mellitus as a complication of their disease. Most of the others have some decrease in their ability to dispose of a glucose load (decreased glucose tolerance). In the early stages diabetes is reversible and disappears when the excess pituitary or adrenal secretion is corrected. Later, however, diabetes may become irreversible, and the islet cells may be destroyed. This so-called diabetogenic effect is an important consideration with chronic glucocorticoid therapy and argues against use of GH to build muscle mass in athletes.

TABLE 4   Hormonal Effects on Insulin Secretion and Sensitivity of Target Cells to Insulin

	Insulin secretion by beta cells	Sensitivity of target cells to insulin
Insulin		↓[a]
Glucagon	↑	↓[b]
Epinephrine, norepinephrine	↓	↓[b]
Growth hormone	↑ (I)	↓
Cortisol	↑ (I)	↓
T$_3$	↑ (I)	↑

(*I*) Indirect effect; increases sensitivity to direct stimuli.
[a]Down-regulation of receptors.
[b]Stimulates opposite effects in liver.

intestinal hormones, especially glucagon-like peptide-1 (GLP-1) and glucose-dependent insulinotropic peptide (GIP), which are potent secretagogues for insulin. Finally, the beta cells respond directly to increased glucose and amino acids in arterial blood (see Chapter 41). During the postprandial period the concentration of insulin in peripheral blood may rise from a resting value of about 10 μUnits/mL to perhaps as much as 50 μUnits/mL. Glucagon secretion may also increase at this time in response to amino acids in arterial blood. Dietary amino acids may also stimulate growth hormone secretion. Characteristically, the sympathetic nervous system is relatively quiet during the postprandial period, and there is little secretory activity of the adrenal medulla or cortex at this time. Under the dominant influence of insulin, dietary carbohydrates and lipids are transferred to storage depots in liver, adipose tissue, and muscle; and amino acids are converted to proteins in various tissues. Extrahepatic tissues use dietary glucose and fat to meet their needs instead of glucose derived from hepatic glycogen or fatty acids mobilized from adipose tissue. Hepatic glycogen increases by an amount equivalent to about half of the ingested carbohydrate. Fatty acid mobilization is inhibited by the high concentrations of insulin and glucose in blood. Of course, the composition of the diet profoundly affects postprandial responses. Obviously, a diet rich in carbohydrate elicits quantitatively different responses from one that is mainly composed of fat.

Postabsorptive Period

Several hours after eating, when metabolic fuels have largely been absorbed from the intestine, the body begins to draw on fuels that were stored during the postprandial period. During this period insulin secretion returns to relatively low basal rates and is governed principally by the concentration of glucose in blood, which has returned to about 5 mM (90 mg/dL). About 75% of the glucose

secreted by the liver derives from glycogen, and the remainder comes from gluconeogenesis, driven principally by glucagon. Although the rate of glucagon secretion is relatively low at this time, the decline in insulin enables the actions of glucagon to prevail. Growth hormone and cortisol are also secreted at relatively low basal rates in the postabsorptive period. About 75% of the glucose consumed by extrahepatic tissues during this period is taken up by brain, blood cells, and other tissues whose consumption of fuels is independent of insulin. Muscle and adipose tissue, which are highly dependent on insulin account for the remaining 25%. FFA gradually increase as adipose tissue is progressively relieved of the restraint imposed by high levels of insulin during the postprandial period. Blood glucose remains constant during this period, but glucose metabolism in muscle decreases as the restrictive effects of the glucose-fatty acid cycle become operative. Liver gradually depletes its glycogen stores and begins to rely more heavily on gluconeogenesis from amino acids and glycerol to replace glucose consumed by extrahepatic tissues.

Fasting

More than 24 hr after the last meal, the individual can be considered to be fasting. At this time, circulating insulin concentrations decrease further, and glucagon and GH increase. Cortisol secretion follows its basal diurnal rhythmic pattern (see Chapter 40) unaffected by fasting at this early stage; but basal concentrations of cortisol play their essential permissive role in allowing gluconeogenesis and lipolysis to proceed. Glucocorticoids and GH also exert a restraining influence on glucose metabolism in muscle and adipose tissue. With the further decrease in insulin concentration, any remaining restraint on lipolysis is removed. The lipolytic cycle speeds up, fatty acid esterification decreases, and FFA mobilization is accelerated. This effect is supported and accelerated by GH and cortisol. Decreased insulin permits net breakdown of muscle protein; and the amino acids that consequently leave muscle, mainly as alanine, provide the substrate for gluconeogenesis. Fuel consumption after 24 hr of fasting is shown in Fig. 7.

With prolonged fasting of 3 days or more, increased GH and decreased insulin concentrations in blood result in even greater mobilization of FFA. Ketogenesis becomes significant, driven by the almost unopposed action of glucagon. By about the third day of starvation, ketone bodies in blood reach concentrations of 2–3 mM and begin to provide for an appreciable fraction of the brain's metabolic needs. Urinary nitrogen excretion decreases to the postabsorptive level or below as the rapidly turning over pool of proteins diminishes. During subsequent weeks of total starvation, nitrogen excretion

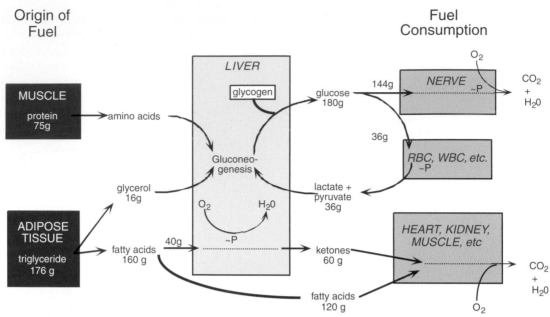

FIGURE 7   Quantitative turnover of substrates in a hypothetical person in the basal state after fasting for 24 hr (−1800 Calories). (From Cahill GF Jr. Starvation in man. *N Engl J Med* 1970;282:668–675.)

continues at a low but steady rate with carbon skeletons from amino acids providing substrate for gluconeogenesis and the intermediates needed to maintain the tricarboxylic acid cycle. Glycerol liberated from triglycerides provides the other major substrate for gluconeogenesis. Renal gluconeogenesis from glutamate accompanies production of ammonia stimulated by ketoacidosis. Virtually all other energy needs are met by oxidation of fatty acids and ketones until triglyceride reserves are depleted. In the terminal stages of starvation, proteins may become the only remaining substrate and are rapidly broken down to amino acids. Gluconeogenesis briefly increases once again until cumulative protein loss precludes continued survival. Curiously, continued slow loss of protein during starvation of the extremely obese individual may result in death from protein depletion even before fat depots are depleted.

Figure 8 shows some representative values for hormone concentrations in blood in the transition from the fed to the fasting state. Values for cortisol remain unchanged or might even decrease somewhat until late in starvation. Concentrations of cortisol shown in the figure represent morning values and change with the time of day in a diurnal rhythmic pattern that is not altered by fasting (see Chapter 40). Even though its concentration does not increase during fasting, cortisol nevertheless is an essential component of the survival mechanism. In its absence, mechanisms for producing and sparing carbohydrates are virtually inoperative, and death from hypoglycemia is inevitable. The role of glucocorticoids in fasting is a good example of permissive action, in which a hormone maintains the instruments of metabolic

adjustments so that other agents can manipulate those instruments effectively. Hypoglycemia or perhaps nonspecific stress may account for increased cortisol in the terminal stages of starvation.

The decrease in plasma concentrations of T_3 are not indicative of decreased secretion of TSH or thyroid hormone, but rather reflect decreased conversion of plasma T_4 to T_3. At least during the first few days of fasting, T_4 concentrations in plasma remain constant. The slight decline in T_4 seen with more prolonged fasting probably reflects a decrease in plasma binding proteins. Recall that T_3, which is formed mostly in extrathyroidal tissue, is the biologically active form of the hormone (see Chapter 39). Deiodination of thyroxine can lead to the formation of T_3 or the inactive metabolite rT_3. With starvation, the concentration of rT_3 in plasma increases, suggesting that metabolism of thyroxine shifted from the formation of the active to the inactive metabolite. Some of this increase may also be accounted for by a somewhat slower rate of degradation of rT_3. Decreased production of T_3 results in an overall decrease in metabolic rate and can be viewed as an adaptive mechanism for conservation of metabolic fuels.

Secretion of GH follows a pulsatile pattern that is exaggerated during starvation (see Chapter 44). Fasting increases both the frequency of secretory pulses and their amplitude. The values for GH shown in Fig. 8 represent concentrations present in a mixed sample of blood that was continuously drawn at a very slow rate over a 24-hr period. The metabolic changes produced by an increase in GH secretion are similar to those that result from a decrease in insulin secretion. Growth hormone increases

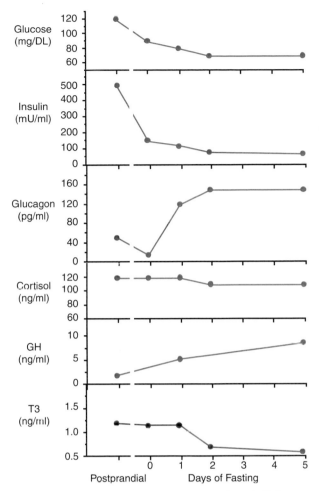

FIGURE 8 Changes in plasma concentrations of metabolic hormones during fasting.

lipolysis, decreases glucose utilization in muscle and fat, and increases glucose production by the liver. These effects of GH are relatively small compared to the effects of diminished insulin secretion. However, persons suffering from a deficiency of GH may become hypoglycemic during fasting, while treatment with GH helps to maintain their blood glucose (Fig. 9). In the nonfasting individual, GH stimulates the liver and other tissues to secrete insulin-like growth factor-1 (IGF-1), which stimulates protein synthesis. The liver becomes insensitive to this effect of GH during fasting, and plasma concentrations of IGF-1 fall dramatically. This too may be an adaptive mechanism that maximizes availability of amino acids for gluconeogenesis and turnover of critical proteins.

HORMONAL INTERACTIONS DURING EXERCISE

During exercise, overall oxygen consumption may increase 10–15 times in a well-trained young athlete. The requirements for fuel are met by mobilization of reserves within muscle cells and from extramuscular fuel depots. Rapid uptake of glucose from blood can potentially deplete, or at least dangerously lower, glucose concentrations and hence jeopardize the brain unless some physiologic controls are operative. We can consider two forms of exercise: short-term maximal effort, characterized by sprinting for a few seconds, and sustained aerobic work, characterized by marathon running.

Short-Term Maximal Effort

For the few seconds of the 100-yard dash, endogenous ATP reserves in muscle, creatine phosphate and glycogen, are the chief sources of energy. For short-term maximal effort, energy must be released from fuel before circulatory adjustments can provide the required oxygen. Breakdown of glycogen to lactate provides the needed ATP and is activated in part through intrinsic biochemical mechanisms that activate glycogen phosphorylase and phosphofructokinase. For example, calcium released from the sarcoplasmic reticulum in response to neural stimulation not only triggers muscle contraction but also activates glycogen phosphorylase. These intrinsic mechanisms are reinforced by epinephrine and norepinephrine released from the adrenal medullae and sympathetic nerve endings in response to central activation of the sympathetic nervous system.

The endocrine system is important primarily for maintaining or replenishing fuel reserves in muscle. Through the actions of hormones and the glucose-fatty acid cycle already discussed, glycogen reserves in muscle are sustained at or near capacity, so that muscle is always prepared to respond to demands for maximal effort. During the recovery phase lactate released from working muscles is converted to glucose in liver and can be exported back to muscle in the classic Cori cycle. Insulin secreted in response to increased dietary intake of glucose or amino acids promotes reformation of glycogen.

Sustained Aerobic Exercise

Glucose taken up from the blood or derived from muscle glycogen is also the most important fuel in the early stages of moderately intense exercise, but with continued effort dependence on fatty acids increases. Although fat is a more efficient fuel than glucose from a storage point of view, glucose is more efficient than fatty acids from the perspective of oxygen consumption and yields about 5% more energy per liter of oxygen. Table 5 shows the changes in fuel consumption with time in subjects exercising at 30% of their maximal oxygen consumption. For reasons that are not fully understood,

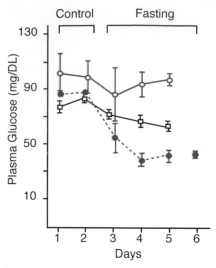

FIGURE 9 Concentrations of glucose in the plasma of normal subjects (Control) and patients suffering from isolated deficiency of GH (shown in blue) while eating normally and while fasting. Some GH-deficient patients were untreated while others were given 5 mg of human GH per day (treated). Fasting began after collection of blood on day 2. From Merimee, TJ, Felig, P, Marliss, E, Fineberg, SE, Cahill, GF Jr. (1971). Glucose and lipid homeostasis in absence of human growth hormone. *J. Clin. Invest.* 50:574–582. With permission.

working muscles, even in the trained athlete, cannot derive more than about 70% of their energy from oxidation of fat. Hypoglycemia and exhaustion occur when muscle glycogen is depleted. With sustained exercise, the decline in insulin and the increase in all of the counter-regulatory hormones contribute to supplying fat to the working muscles and maximizing gluconeogenesis (Fig. 10).

Anticipation of exercise may be sufficient to activate the sympathetic nervous system, which is of critical importance not only for supplying the fuel for the working muscles but for making the cardiovascular adjustments that maintain blood flow to carry fuel and oxygen to muscle, gluconeogenic precursors to liver, and heat to sites of dissipation. Insulin secretion is shut down by sympathetic activity. This removes the major inhibitory influence on production of glucose by the liver,

glycogen breakdown in muscle, and FFA release from adipocytes. At first glance decreasing insulin secretion might seem deleterious for glucose consumption in muscle. However, the decrease in insulin concentration only decreases glucose uptake by nonworking muscles. Mobilization of GLUT4 and transport of glucose across the sarcolemma is stimulated by muscular contractions independently of insulin. Glucose metabolism in working muscles is therefore not limited by membrane transport but, rather, by phosphofructokinase, which in turn is responsive to a variety of intracellular metabolites that coordinate its activity with energy demand.

Increased hepatic glucose production results primarily from the combined effects of the fall in insulin secretion and the rise in glucagon secretion with some contribution from catecholamines. The contributions of the increased

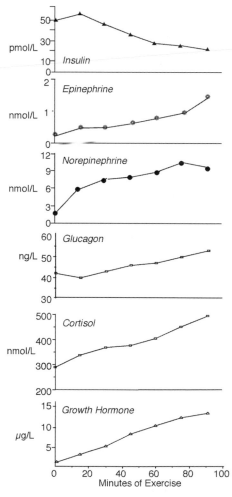

FIGURE 10 Changes in concentration of insulin and counterregulatory hormones during prolonged moderate exercise. Values shown are the means obtained for eight young men exercising on a bicycle ergometer at ∼50% of maximum oxygen consumption. (Drawn from data of Davis SN, Galassetti P, Wasserman DH, Tate D. Effects of gender on neuroendocrine and metabolic counterregulatory responses to exercise in normal man. *J Clin Endocrinol Metab* 2000;85:224–230).

TABLE 5   Fuels Consumed by Leg Muscles of Man during Mild Prolonged Exercise

	%Contribution to oxygen uptake		
Period of exercise	Plasma glucose	Plasma free fatty acids	Muscle glycogen
40	27	37	36
90	41	37	22
180	36	50	14
240	30	62	8

Data from Ahlborg et al. *J Clin Invest* 1974;53:1080–1090.

secretion of GH and cortisol to this effect are unlikely to be important initially, but with sustained exercise the contributions of both are likely to increase. Actions of both hormones on adipocytes increase the output of FFA and glycerol and decrease glucose utilization by adipocytes and nonworking muscles. Additionally, the increased cortisol would be expected to increase the expression of gluconeogenic enzymes in the liver.

Glycogen reserves of nonworking muscles may provide an important source of carbohydrate for working muscles during sustained exercise and for restoring muscle glycogen after exercise. Epinephrine and norepinephrine stimulate glycogenolysis in nonworking as well as working muscles. Glucose phosphate produced from glycogen can be completely broken down to carbon dioxide and water in working muscles, but nonworking muscles convert it to pyruvate and lactate, which escape into the blood. Liver then reconverts these 3-carbon acids to glucose, which is returned to the circulation and selectively taken up by the working muscles (Fig. 11).

LONG-TERM REGULATION OF FUEL STORAGE

Adipose tissue diffusely scattered throughout the body has an almost limitless capacity for fuel storage.

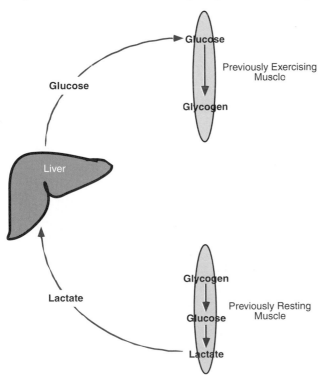

FIGURE 11 Postulated interaction between previously exercising muscle and previously resting muscle via the Cori cycle during recovery from prolonged arm exercise. (From Ahlborg G, Wahren J, Felig P. Splanchnic and peripheral glucose and lactate metabolism during and after prolonged arm exercise. *J Clin Invest* 1986;77:690–699).

Increasing or decreasing the total amount of fat stored is achieved primarily by changes in the volume of the fat droplets within each adipocyte by the mechanisms of fat deposition and mobilization already discussed. In addition, the number of adipocytes is not fixed and may increase throughout life by multiplication and differentiation of precursor cells. Conversely, fat cells may become depleted of their triglyceride stores and undergo dedifferentiation and apoptosis. The process of adipogenesis depends on locally produced growth factors and cytokines as well as hormones. Insulin and cortisol promote differentiation of human preadipocytes and tend to favor accumulation of body fat, whereas GH inhibits both differentiation and storage. These chronic actions of insulin and GH are consistent with short-term actions already discussed, but the effects of cortisol are unexpected in light of its short-term actions to promote lipolysis and decrease fatty acid reesterification. However, removal of the adrenal glands in experimental animals prevents or reverses all forms of genetic or experimentally induced obesity. Additionally, chronic excess production of glucocorticoids in humans is associated with increased body fat, especially in the torso (truncal obesity), the face (moon face), and between the scapulae (buffalo hump).

For most people, body fat reserves are maintained at a nearly constant level throughout adult life despite enormous variations in daily food consumption and energy expenditure. Figure 12 summarizes the findings of five independent studies of changes in body weight and fat mass with aging in about 12,000 individuals. Although total body fat increased with increasing age, the increase averaged less than a gram per day when averaged over a period of 50 years and corresponded to a daily positive energy balance of about 6 Calories. Assuming that daily energy consumption averages about 2000 Calories, the intake of fuel in a mixed diet matched the rate of energy utilization with an error of 0.3%. However, the affluence, ready access to high-calorie foods, and technology that fosters sedentary activities in contemporary society have so distorted the balance between caloric intake and energy expenditure that 30% of the American population is now classified as obese.

An understanding of the mechanisms that govern long-term fuel storage requires an understanding of the regulation of energy expenditure as well as food intake. Physical activity accounts for only about 30% of daily energy expenditure. Sixty percent of energy consumption is expended at rest for maintenance of ion gradients, renewal of cellular constituents, neuronal activity, and to support cardiopulmonary work. The remaining 10% is dissipated as the thermogenic effect of feeding and the consequent processes of assimilation.

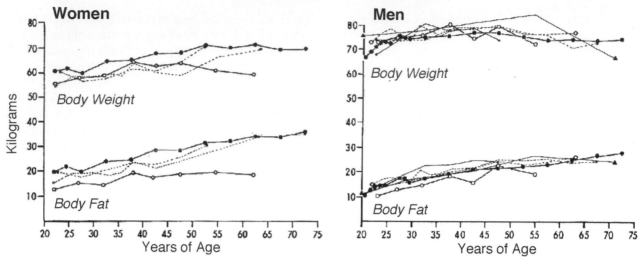

FIGURE 12   Cross-sectional data showing changes in body weight and fat content with aging obtained in five independent studies. (From Forbes GB, Reina JC. Adult lean body mass declines with age: Some longitudinal observations. Metabolism 1970;19:653–663).

Neither resting nor thermogenic energy expenditure are fixed, but are adjustable in a manner that tends to keep body fat reserves constant. In the experiment illustrated in Fig. 13, normal human subjects were overfed or underfed in order to increase or decrease body weight by 10%. They were then given just enough food each day to maintain their new weight at a constant level. Energy utilization increased disproportionally in the overfed subjects and decreased disproportionally in the underfed subjects. Such compensatory changes in energy expenditure opposed maintenance of the change from initial body fat content. How such changes in energy expenditure are brought about is not understood. One possibility is that metabolic efficiency may be regulated by varying expression of genes that encode proteins that uncouple ATP generation from oxygen consumption (see Chapter 39).

Hypothalamic Control of Appetite and Food Intake

Studies such as those illustrated in Figs. 12 and 13 and many older observations gave rise to the idea that the mass of the fat storage depot is monitored and maintained at a nearly constant set point through feedback mechanisms that regulate food consumption and energy expenditure (Fig. 14). Clinical observations and studies in experimental animals established that the hypothalamus coordinates the drive for food intake with such energy-consuming processes as temperature regulation, growth, and reproduction. Various injuries to the hypothalamus can produce either insatiable eating behavior accompanied by severe obesity, or food

avoidance and lethal starvation. A complex neural network interconnects "satiety centers" in the medial hypothalamus and "hunger centers" in the lateral hypothalamus with each other, with autonomic integrating centers in the hypothalamus and brain stem, and

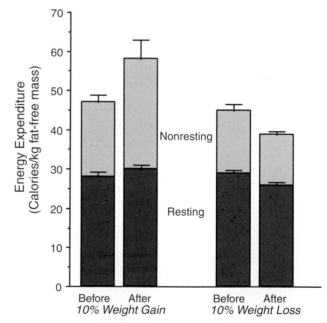

FIGURE 13   Changes in energy expenditure after increase or decrease of body weight. Thirteen normal subjects were overfed a defined diet until their body weight increased by 10%. Eleven normal subjects were underfed until their body weight decreased by 10%. Both groups were then fed just enough to maintain their new weights for 2 weeks at which time energy expenditure and lean body mass were measured. (Drawn from data of Leibel RL, Rosenbaum M, Hirsch J. Changes in energy expenditure resulting from altered body weight. *N Engl J Med* 1995;332:673–674).

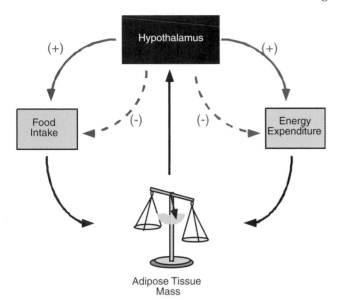

FIGURE 14 Hypothetical regulatory system for maintaining constancy of adipose mass in which the mass of total stored fat is monitored. Adjustments in energy intake and expenditure are made to maintain constancy. Solid lines (+) denote increase; dashed arrows (−) denote decrease.

with neurons in the arcuate and paraventricular nuclei that regulate secretion of hypophysiotropic hormones (see Chapter 38). Important advances have been made in our understanding of the complex process of regulating food intake in recent years including discovery of the adipocyte hormone, leptin, and some of the neuropeptide transmitters associated with appetite control and their receptors.

Leptin

Leptin, which means "thin," is expressed primarily, but not exclusively, in adipocytes. Inactivating mutations of the gene that encodes leptin or its receptor results in hyperphagia (excess food consumption), obesity, diabetes, impaired temperature regulation, and infertility in mice. In the very rare cases that have been reported in humans, mutation of the genes that code either for leptin or its receptor results in hyperphagia, morbid obesity, and impaired sexual development. When administered to obese, leptin-deficient mice, leptin decreases body weight by reducing food intake and increasing energy utilization (Fig. 15).

Leptin concentrations in blood correlate positively with body fat content (Fig. 16), suggesting that leptin might provide a means for monitoring fat stores. Blood levels of leptin also reflect changes in nutritional state. Within hours after initiation of fasting, leptin concentrations decrease sharply (Fig. 17) and, conversely, sustained overfeeding increases plasma levels. Because leptin concentrations change to a far greater extent in

response to nutritional status than to changes in adipose mass, it has been suggested that a fall in blood leptin concentration acts as a starvation signal to increase food intake and initiate energy conservation.

Leptin is encoded as a 167-amino-acid prohormone that is the product of the *ob* gene. Its tetrahelical structure resembles that of the class of cytokines and hormones that includes GH and prolactin. Because mRNA levels correlate with circulating leptin concentrations and with low levels of the hormone present in adipocytes, secretion of leptin is thought to be regulated at the level of gene transcription. Little is known of the cellular events that are associated with leptin secretion or of the mechanisms that regulate synthesis. Studies of isolated adipocytes indicate that enlargement of the lipid storage droplet increases leptin mRNA production, But the cellular mechanisms that are activated by fat cell enlargement are not understood. Insulin and cortisol act synergistically to increase leptin synthesis and secretion, whereas norepinephrine or increased activity of the sympathetic nervous system decreases leptin production. Plasma concentrations of leptin appear to follow a circadian pattern with highest levels found at night. Frequent spikes in leptin concentration in blood are indicative of synchronized pulsatile secretion, but how secretion by diffusely distributed adipocytes is coordinated is not understood. More than 40% of the leptin in blood is bound to protein. Leptin is cleared from the blood primarily by the kidney.

The leptin receptor, like receptors for GH and prolactin, belongs to the class of transmembrane cytokine

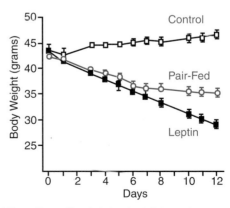

FIGURE 15 Effects of leptin in leptin-deficient mice. Body weights of female obese mice treated with saline (control) or 270 pg of leptin/day were compared to body weights of obese mice treated with saline, but fed an amount of food equal to that consumed by the leptin-treated mice (pair-fed). Note that the loss of body weight produced by leptin was not accounted for simply by decreased food intake. (From Levin N, Nelson C, Gurney A, Vandlen R, de Sauvage F. Pair-feeding studies provide compelling evidence that the ob protein exerts adipose-reducing effects in excess of those induced by reductions in food intake. *Proc Natl Acad Sci USA* 1996;93:1726–1730.)

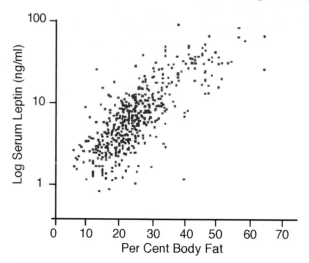

FIGURE 16 Leptin concentrations in blood plasma correlate with body fat content in 500 human subjects. (From Caro JF, Sinha MK, Kolaczynski JW, Zhang PL, Considine RV. Leptin: The tale of an obesity gene. *Diabetes* 1996;45:1455–1462.)

receptors that signals by activation of the cytosolic tyrosine kinase JAK2 (see Chapter 2). Multiple splice variants of leptin receptor mRNA give rise to different isoforms, including one that circulates as a soluble protein. Other isoforms have truncated cytoplasmic domains, but only the form with the full-length cytoplasmic tail appears to be capable of signaling. Truncated forms, which are expressed in vascular endothelium and the choroid plexus, may serve a transport function to facilitate passage of leptin across blood vessels in the brain to breech the blood–brain barrier and thus deliver leptin to target cells in the central nervous system.

The principal targets for leptin are neurons in the arcuate nuclei of the hypothalamus, but leptin receptors are also found in neurons of the paraventricular, ventromedial, and dorsomedial nuclei and neurons in the lateral hypothalamus. Neuropeptide transmitters of some of these neurons have been identified along with their receptors, and the sites of their expression have been located. These neurons project to hypothalamic and brain-stem autonomic integrating centers. Neurons from the arcuate and paraventricular nuclei project to the median eminence of the hypothalamus where they release hypophysiotropic hormones into the hypophysial-portal circulation (see Chapter 38). Through these neural connections, leptin integrates nutritional status with adrenal, gonadal, and thyroidal function and with GH and prolactin secretion. Some of the relevant neuropeptides are as follows:

- Neuropeptide Y (NPY) is a 36-amino-acid peptide that is abundantly expressed in neurons of the arcuate nuclei of the hypothalamus whose axons project to the paraventricular nuclei and the lateral hypothalamic area. Its expression is increased during fasting. When administered into the hypothalamus of rodents, NPY stimulates food intake, lowers energy expenditure, and with continued administration may produce obesity. Expression of NPY is increased in leptin deficiency; production, and transcription and release of NPY are suppressed by leptin administration.
- Pro-opiomelanocortin (POMC), the precursor of ACTH in anterior pituitary cells, is also expressed in the arcuate nucleus of the hypothalamus where

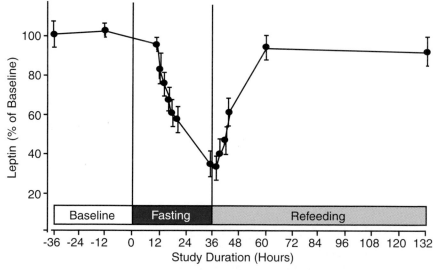

FIGURE 17 Effect of fasting and refeeding on leptin concentrations in the plasma of normal human subjects. (From Kolaczynski JW, Considine RV, Ohannesian J, Marco C, Opentanova I, Nyce MR, Myint M, Caro JF. Responses of leptin to short-term fasting and refeeding in humans: A link with ketogenesis but not ketones themselves. *Diabetes* 1996;45:1511–1515).

post-translational processing (see Fig. 3 in Chapter 38) gives rise to α-melanocyte stimulating hormone (α-MSH). As a neuropeptide, α-MSH is a potent negative regulator of food intake and activates melanocortin receptors in neurons in the dorsomedial and paraventricular nuclei of the hypothalamus and the lateral hypothalamic area. Expression of α-MSH in rodents is increased by leptin and suppressed by overfeeding. Pharmacological blockade or genetic depletion of brain melanocortin receptors results in obesity.

- In the skin, α-MSH increases expression of a black pigment in hair follicles. A protein, called *agouti,* competes with α-MSH for the melanocortin receptor, and under its influence, a yellow pigment is expressed. The observation that a mutation that results in ubiquitous inappropriate expression of the agouti gene in mice also produces obesity led to the discovery that neurons in the arcuate nucleus express a similar protein, the agouti-related protein (AGRP), which competes with α-MSH for binding to the melanocortin receptor in the brain. AGRP and neuropeptide Y are coexpressed in arcuate neurons, and their combined actions provide a strong drive for food intake. Expression of AGRP is increased in leptin deficiency and decreased by leptin treatment.

- Melanin concentrating hormone (MCH) was originally described as the factor that opposes the melanophore dispersing activity of MSH in fish, but MCH is also expressed in neurons in the lateral hypothalamus of mammals. When administered to rodents, MCH stimulates feeding behavior and antagonizes the inhibitory effects of α-MSH. Unlike AGRP, MCH does not bind to the same receptor as α-MSH. Fasting increases MCH expression, and absence of the MCH gene in mice results in reduction in body fat content.

- Another appetite-suppressing peptide is the cocaine and amphetamine regulated transcript (CART) whose discovery arose out of studies of drugs of abuse. It is widely expressed in the brain and other endocrine tissues including some neurons in the arcuate nuclei that also express α-MSH. Administration of CART inhibits the feeding response to NPY. Mice that are deficient in CART become obese.

Other neuropeptides and amine transmitters that originate in neurons in various brain loci and the gastrointestinal tract also participate in the complex regulation of feeding behavior. A complete "wiring diagram" cannot yet be drawn, but some of the relationships are shown in Fig. 18.

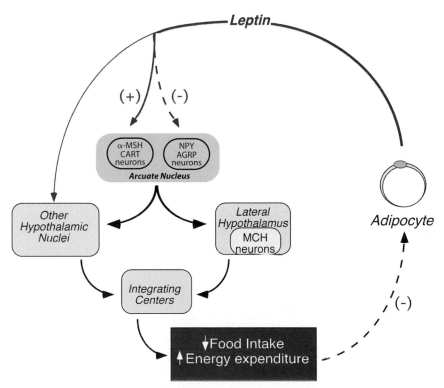

FIGURE 18   Proposed negative feedback control of leptin secretion, α-MSH, α melanocyte stimulating hormone; CART, cocaine and amphetamine regulated transcript; NPY, neuropeptide Y; AGRP, agouti-related peptide; MCH, melanin concentrating hormone.

Other Effects of Leptin

Through its actions on neurons associated with autonomic and anterior pituitary functions, leptin affects temperature regulation, reproduction, and adrenal cortical function. Leptin receptors are also found in many cells outside of the central nervous system. Adipocytes express leptin receptors and respond to leptin in an autocrine manner with an increase in lipolysis. Leptin acts directly on pancreatic beta cells to inhibit insulin secretion and thus forms one arm of a negative feedback arrangement between beta cells and adipocytes in which stimulation of leptin secretion by insulin leads to inhibition of insulin secretion by leptin.

The presence of leptin receptors in the gonads suggests that peripheral actions of leptin may complement the fertility-promoting effects exerted at the hypothalamic level. Other peripheral effects of leptin include actions in bone marrow to promote hematopoiesis and actions in capillary endothelium to increase angiogenesis (blood vessel formation). Leptin is also produced by the placenta and may play a role in fetal development.

Role of Insulin

Although insulin promotes the storage of excess metabolic fuels in adipose tissue, it may also serve as an indicator of total body fat and signal to the hypothalamus to limit food intake and fat storage. Insulin sensitivity decreases as fat storage depots increase. Therefore, higher concentrations of insulin are required to maintain blood glucose within normal limits, and hence insulin levels in blood increase or decrease in parallel with changes in body fat. Insulin receptors are present in neurons in the arcuate nucleus where insulin acts to inhibit NPY synthesis and produce other leptin-like effects. Insulin also stimulates leptin synthesis in adipocytes and in typical negative feedback fashion, leptin inhibits insulin secretion. While understanding of the interplay between insulin and leptin is incomplete, it is likely that both hormones play a major roles in long term regulation of fat storage.

Role of Gastrointestinal Hormones

Secretions from the intestinal tract also influence appetite and satiety, and influence both meal size and the frequency of eating. Although discovered for its actions to increase GH secretion, ghrelin (see Chapters 38 and 44), secreted by endocrine cells in the mucosal lining of the stomach may also play an important role in signaling hunger. Its concentrations in blood increase with fasting and are promptly reduced by food ingestion. Ghrelin stimulates cells in the arcuate nuclei to secrete NPY and AGRP. The duodenal hormone cholecystokinin (CCK) which is secreted by the intestinal mucosa upon ingestion of proteins and fat (see Chapter 32) acts as a satiety signal to limit meal size. Genetic defects in the CCK receptor produce abesity in rodents. Leptin and insulin may increase the sensitivity to CCK.

Suggested Reading

Ahima RS, Flier JS. Leptin. *Ann Rev Physiol* 2000;62:413–437.

Felig P, Sherwin RS, Soman V, Wahren J, Hendler R, Sacca L, Eigler N, Goldberg D, Walesky M. Hormonal interactions in the regulation of blood glucose. *Rec Prog Horm Res* 1979;35:501–528.

Jefferson LS, Cherrington AD. The endocrine pancreas and regulation of metabolism, in *Handbook of physiology, Section 7, Vol II, The endocrine pancreas and regulation of metabolism*, Oxford University Press, New York, 2001.

Randle PJ, Kerbey AL, Espinal J. Mechanisms decreasing glucose oxidation in diabetes and starvation: Role of lipid fuels and hormones. *Diabetes/Metab Rev* 1988;4:623–638.

Ruderman NB, Saha AK, Vavvas D, Witters LA. Malonyl-CoA, fuel sensing, and insulin resistance. *Am J Physiol* 1999;276:E1–E18.

Schwartz, M, Woods, SC, Porte, D, Jr., Seeley, RJ, Baskin, DG, Central nervous system control of food intake. *Nature* 2000;404:661–671

Wasserman DH, Cherrington AD. Regulation of extramuscular fuel sources during exercise. In Rowell LB, Shepherd JT, eds., *Exercise: Regulation and integration of multiple systems, Handbook of physiology, Section 12*, Oxford University Press, New York, 1996, pp 1036–1074.)

CHAPTER

43

Hormonal Regulation of Calcium Metabolism

H. MAURICE GOODMAN

KEY POINTS

- The concentration of ionized calcium in blood is regulated within narrow limits.
- Maintenance of the concentration of calcium in extracellular fluid depends on the rates of calcium absorption from the intestine, excretion in the urine, and exchange with bone.
- *Parathyroid hormone* (PTH) increases blood calcium by stimulating calcium mobilization from bone and calcium reabsorption from the glomerular filtrate. It also indirectly stimulates calcium absorption from the gut by increasing the synthesis of the active form of vitamin D.
- Parathyroid hormone lowers blood phosphate by decreasing the reabsorption of phosphate in the proximal tubules of the kidney.

- Secretory cells of the parathyroid glands directly monitor blood calcium concentrations and increase their rates of PTH secretion when calcium levels decline. Conversely, high concentrations of blood calcium inhibit PTH secretion.
- Calcitonin is secreted by the C cells of the thyroid gland in response to increasing concentrations of blood calcium. Its principal physiologic effect is to inhibit the activity of osteoclasts in bone.
- Ultraviolet light catalyzes the conversion of 7-dehydrocholesterol to vitamin D_3 in the skin. Successive hydroxylations in the liver at carbon 25 and in the kidney at carbon 1 result in the active form, $1,25(OH)_2D_3$.

KEY POINTS (*continued*)

- 1,25(OH)$_2$D$_3$ increases calcium absorption in the intestine and the kidney and promotes calcium mobilization from bone.
- In *hypocalcemia*, increased secretion of PTH increases the formation of 1,25(OH)$_2$D$_3$. These two hormones cooperate to restore blood calcium by increasing mobilization from bone, decreasing loss by the kidney, and increasing absorption of dietary calcium.

- In response to *hypercalcemia*, shutdown of PTH secretion and 1,25(OH)$_2$D$_3$ synthesis allows calcium levels to decline slowly, whereas increased secretion of calcitonin promptly inhibits the bone resorbing activity of osteoclasts.
- Deficiency of estrogen or excessive thyroid hormone or glucocorticoids decrease skeletal mass.

OVERVIEW

Adequate amounts of calcium in its ionized form, Ca^{2+}, are needed for normal function of all cells. Calcium ion regulates a wide range of biological processes and is one of the principal constituents of bone. In terrestrial vertebrates, including humans, maintenance of adequate concentrations calcium ion[1] in the extracellular fluid requires the activity of two hormones, *parathyroid hormone* (PTH) and a derivative of vitamin D called *1α,25-dihydroxycholecalciferol* (1,25(OH)$_2$D$_3$) or *calcitriol*. In more primitive vertebrates living in a marine environment, guarding against excessively high concentrations of calcium requires another hormone, *calcitonin,* which appears to have only vestigial activity in humans.

Body calcium ultimately is derived from the diet, and daily intake is usually offset by urinary loss. The skeleton acts as a major reservoir of calcium and can buffer the concentration of calcium in extracellular fluid by taking up or releasing calcium phosphate. PTH promotes the transfer of calcium from bone, the glomerular filtrate, and intestinal contents into the extracellular fluid. It acts on bone cells to promote calcium mobilization and on renal tubules to reabsorb calcium and excrete phosphate. It promotes intestinal transport of calcium and phosphate indirectly by increasing the formation of 1,25(OH)$_2$D$_3$ required for calcium uptake by intestinal cells. This vitamin D metabolite also promotes calcium mobilization from bone and reinforces the actions of PTH on this process. In addition, 1,25(OH)$_2$D$_3$ promotes reabsorption of calcium and phosphate by renal tubules. The rate of PTH secretion is inversely related to the concentration of blood calcium, which directly inhibits secretion by the chief cells of the parathyroid glands. Calcitonin inhibits the activity of bone-resorbing cells, and thus blocks inflow of calcium to the extracellular

fluid compartment. Its secretion is stimulated by high concentrations of blood calcium.

GENERAL FEATURES OF CALCIUM BALANCE

Calcium enters into a wide range of cellular and molecular processes. Changes of its concentration within cells regulate enzymatic activities and such fundamental cellular events as muscular contraction, secretion, and cell division. As already discussed (see Chapter 2), calcium and calmodulin also act as intracellular mediators of hormone action. In the extracellular compartment, calcium is vital for blood clotting and maintenance of normal membrane function. Calcium is the basic mineral of bones and teeth and thus plays a structural as well as a regulatory role. Not surprisingly, its concentration in extracellular fluid must be maintained within narrow limits. Deviations in either direction are not readily tolerated and, if severe, may be life threatening.

Distribution of Calcium in the Body

The adult human body contains approximately 1000 g of calcium, about 99% of which is sequestered in bone, primarily in the form of hydroxyapatite crystals ($Ca_{10}(PO_4)_6(OH)_2$). In addition to providing structural support, bone serves as an enormous reservoir for calcium salts. Each day about 600 mg of calcium is exchanged between bone mineral and the extracellular fluid. Much of this exchange reflects resorption and reformation of bone as the skeleton undergoes constant remodeling, but some also occurs by exchange with a labile calcium pool in bone.

Most of the calcium that is not in bone crystals is found in cells of soft tissues bound to proteins within the sarcoplasmic reticulum, mitochondria, and other organelles. Energy-dependent transport of calcium by these organelles and the cell membrane maintains the resting

---

[1]Calcium is present in several forms within the body, but only the ionized form, Ca^{2+}, is monitored and regulated. In this discussion, calcium refers to the ionized form except when otherwise specified.

Clinical Note

Electrical excitability of cell membranes increases when the extracellular concentration of calcium is low, and the threshold for triggering action potentials may be lowered almost to the resting potential, which results in spontaneous, asynchronous, and involuntary contractions of skeletal muscle called *tetany*. A typical attack of tetany involves muscular spasms in the face and characteristic contortions of the arms and hands. Laryngeal spasm and contraction of respiratory muscles may compromise breathing. Pronounced *hypocalcemia* (low blood calcium) may produce more generalized muscular contractions and convulsions.

Increased concentrations of calcium in blood (*hypercalcemia*) may cause calcium salts to precipitate out of solution because of their low solubility at physiologic pH. "Stones" form, especially in the kidney, where they may produce severe painful damage (renal colic), which may lead to renal failure and hypertension.

concentration of free calcium in cytosol at low levels: 0.1–1 μM. Cytosolic calcium can increase 10-fold or more, however, with just a brief change in membrane permeability or affinity of intracellular binding proteins. The rapidity and magnitude of changes in cytosolic calcium are consistent with its role as a biological signal.

The concentration of calcium in interstitial fluid is about 1.5 mM. Interstitial calcium consists mainly of free, ionized calcium, but about 10% is complexed with such anions as citrate, lactate, or phosphate. Ionized and complexed calcium pass freely through capillary membranes and equilibrate with calcium in blood plasma. The total calcium concentration in blood is nearly twice that of interstitial fluid because calcium is avidly bound by albumin and other proteins. Total calcium in blood plasma is normally about 10 mg/dL (5 mEq/L or 2.5 mM), but only the ionized component appears to be monitored and regulated. Because so large a fraction of blood calcium is bound in protein, diseases that produce substantial changes in albumin concentrations may produce striking abnormalities in total plasma calcium content, even though the concentration of ionized calcium may be normal.

Calcium Balance

Normally, adults are in calcium balance; that is, on average, daily intake equals daily loss in urine and feces. Except for lactation and pregnancy, deviations from balance reflect changes in the metabolism of bone. Immobilization of a limb, bed rest, weightlessness, and malignant disease are examples of circumstances that produce negative calcium balance, whereas growth of the skeleton produces positive calcium balance. Dietary intake of calcium in the United States typically varies between 500 and 1500 mg/day, primarily in the form of dairy products. For example, an 8-oz. glass of milk contains about 290 mg of calcium. Calcium absorbed from the gut exchanges with the various body pools and ultimately is lost in the urine so that there is no net gain or loss of calcium in the extracellular pool in young adults. These relations are illustrated in Fig. 1. It is noteworthy that the entire extracellular calcium pool turns over many times in the course of a day. Hence, even small changes in any of these calcium fluxes can have profound effects.

Intestinal Absorption

Calcium is taken up along the entire length of the small intestine, but uptake is greatest in the ileum and jejunum. Secretions of the gastrointestinal tract are rich in calcium and add to the minimum load that must be absorbed to maintain balance. Net uptake is usually in the range of 100–200 mg/day. Absorption of calcium requires metabolic energy and the activity of specific carrier molecules in the luminal membrane (brush border) of intestinal cells. Although detailed understanding is not yet at hand, it appears that carrier-mediated transport across the brush border determines the overall rate. Calcium is carried down its concentration gradient into the cytosol of intestinal epithelial cells and is extruded from the basolateral surfaces in exchange for sodium, which must then be pumped out at metabolic expense. Overall transfer of calcium from the intestinal lumen to interstitial fluid proceeds against a concentration gradient and is largely dependent on $1,25(OH)_2D_3$ (see later discussion). Although some calcium is taken up passively, simple diffusion is not adequate to meet body needs even when the concentration of calcium in the intestinal lumen is high.

Bone

Understanding regulation of calcium balance requires at least a rudimentary understanding of the physiology of bone. Metabolic activity in bone must satisfy two needs. The skeleton must attend to its own structural integrity through continuous remodeling and renewal, and it must

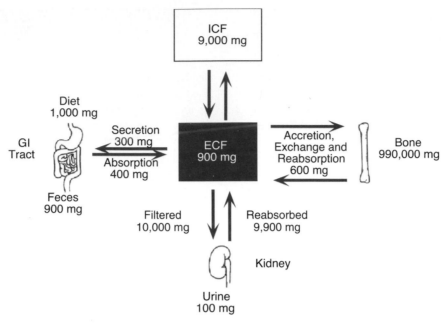

FIGURE 1   Daily calcium balance in a typical adult.

respond to systemic needs for adequate amounts of calcium in the extracellular fluid. By and large, maintenance of adequate concentrations of calcium in blood takes precedence over maintenance of structural integrity of bone. However, the student must recognize that these two homeostatic functions, though driven by different forces, are not completely independent. Diseases of bone that disrupt skeletal homeostasis may have consequences for overall calcium balance; and, conversely, inadequacies of calcium balance lead to inadequate mineralization of bone.

The *extracellular matrix* is the predominant component of bone. One-third of the bony matrix is organic, and two-thirds is comprised of highly ordered mineral crystals. The organic component, called *osteoid,* is composed primarily of collagen and provides the framework on which bone mineral is deposited. Collagen molecules in osteoid aggregate and cross-link to form fibrils of precise structure. Spaces between the ends of collagen molecules within fibrils provide initiation sites for crystal formation. Most calcium phosphate crystals are found within collagen fibrils and have their long axes oriented in parallel with the fibrils.

Cortical (compact) bone is the most prevalent form and is found in the shafts of long bones and on the surfaces of the pelvis, skull, and other flat bones. The basic unit of cortical bone is called an *osteon* and consists of concentric layers, or lamellae, of bone arranged around a central channel (*haversian canal*), which contains the capillary blood supply. Other canals, which run roughly perpendicularly, penetrate the osteons and

form an anastomosing array of channels through which blood vessels in haversian canals connect with vessels in the *periosteum.* The entire bone is surrounded on its outer surface by the periosteum and is separated from the marrow by the *endosteum.*

Cancellous (trabecular) bone is found at the ends of the long bones, in the vertebrae, and in the internal portions of the pelvis, skull, and other flat bones. It is also called spongy bone, a term that describes well its appearance in section (Fig. 2). Although only about 20% of the skeleton is comprised of trabecular bone, its sponge-like organization provides at least five times as much surface area for metabolic exchange as compact bone. The trabeculae of spongy bone are not penetrated by blood vessels, but the spaces between them are filled with blood sinusoids or highly vascular marrow. The trabeculae are completely surrounded by endosteum. Distributed throughout the lamellae of both forms of bone are tiny chambers, or lacunae, each of which houses an *osteocyte.* The lacunae are interconnected by an extensive network of canaliculi, which extend to the endosteal and periosteal surfaces. Osteocytes receive nourishment and biological signals by way of cytoplasmic processes that extend through the canaliculi to form gap junctions with each other and cells of the endosteum or periosteum (Fig. 3).

It is important to recognize that the mineralized matrix in both forms of bone and the bone extracellular fluid are separated from the extracellular compartment of the rest of the body by a continuous layer of cells sometimes called the *bone membrane.* This layer of cells

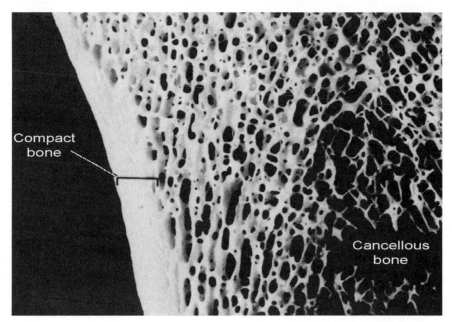

FIGURE 2   A thick ground section of the tibia illustrating cortical compact bone and the lattice of trabeculae of cancellous bone. (From Fawcett DW. *A textbook of histology,* 11th ed. Saunders, Philadelphia, 1986, p 201.)

is comprised of the endosteum, periosteum, osteocytes, and cells that line the haversian canals. Crystallization or solubilization of bone mineral is determined by physicochemical equilibria related to the concentrations of calcium, phosphate, hydrogen, and other constituents in bone water. The fluxes of calcium and phosphate into or out of the bone extracellular fluid involve active participation of the cells of the bone membrane.

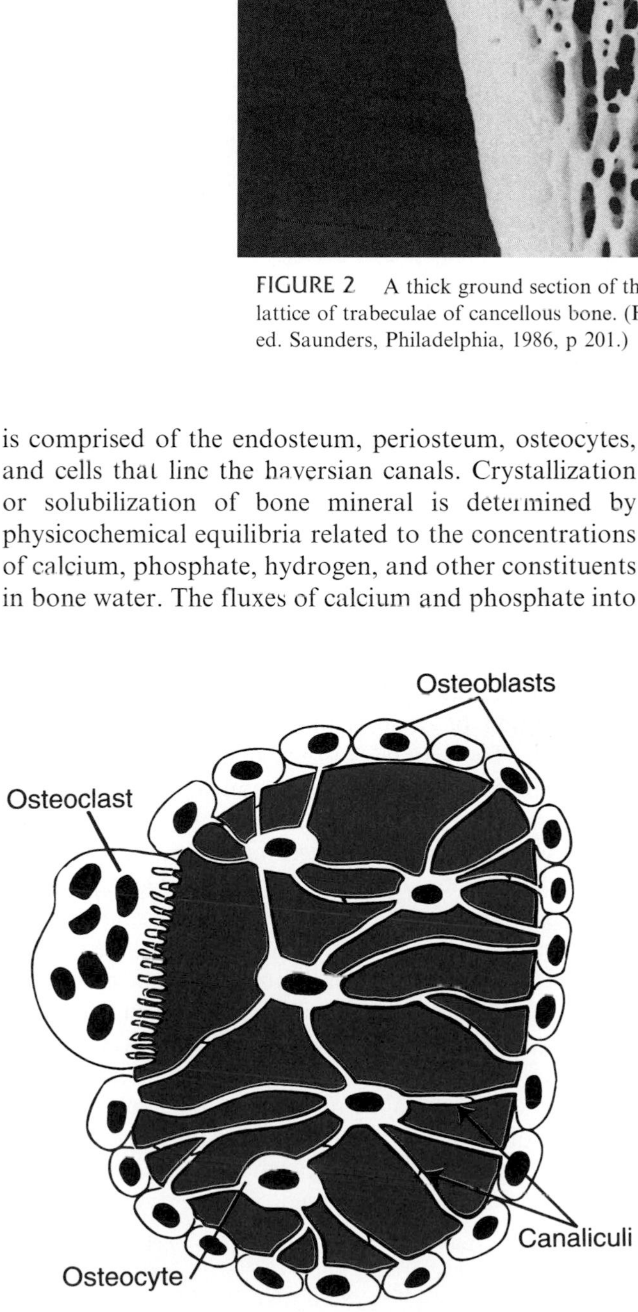

Osteoblasts are the cells responsible for formation of bone. They arise from progenitors in connective tissue and marrow stroma and form a continuous sheet on the surface of newly forming bone. When actively laying down bone, osteoblasts are cuboidal or low columnar in shape. They have a dense rough endoplasmic reticulum consistent with synthesis and secretion of collagen and other proteins of bone matrix. Osteoblasts probably also promote mineralization, but their role in this regard and details of the mineralization process are somewhat controversial. Under physiologic conditions, calcium and phosphate are in metastable solution. That is, their concentrations in extracellular fluid would be sufficiently high for them to precipitate out of solution were it not for other constituents, particularly pyrophosphate, which stabilize the solution. During mineralization osteoblasts secrete alkaline phosphatase, which cleaves pyrophosphate and thus removes a stabilizing influence, and at the same time increases local concentrations of phosphate, which promotes crystallization. In addition, during bone growth and perhaps during remodeling of mature bone, osteoblasts secrete calcium-rich vesicles into the calcifying osteoid.

During growth or remodeling of bone, some osteoblasts become entrapped in matrix and differentiate into osteocytes. Osteocytes are the most abundant cells in bone and are about 10 times as abundant in human bone as osteoblasts. On completion of growth or remodeling,

FIGURE 3   Cross section through a bony trabecula. The shaded area indicates mineralized matrix.

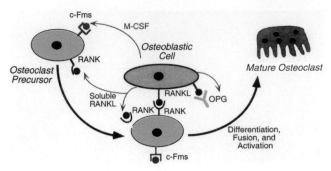

FIGURE 4   Differentiation and activation of osteoclasts. c-FMS, receptor for macrophage colony stimulating factor; M-CSF, macrophage colony stimulating factor; RANK, receptor activators of NF-κB; RANKL, RANK ligand; OPG, osteoprotegerin. (Modified from Khosla S. Minireview: The OPG/RANKL/RANK system. *Endocrinology* 1002;124:5050–5055.)

surface osteoblasts dedifferentiate to become the flattened, spindle-shaped cells of the endosteum that lines most of the surface of bone. These cells may be reactivated in response to stimuli for bone formation. Thus osteoblasts, osteocytes, and quiescent lining cells represent three stages of the same cellular lineage and together comprise most, or perhaps all, of the bone membrane. In subsequent discussion, these cells along with their stromal and periosteal precursors are referred to as *osteoblastic* cells.

Osteoclasts are responsible for bone resorption (Fig. 4). They are large cells that arise by fusion of mononucleated hematopoietic cells; they may have as many as 20–40 nuclei. Precursors of osteoclasts originate in bone marrow and migrate through the circulation from thymus and other reticuloendothelial tissues to sites of bone destined for resorption. Differentiation and activation of osteoclasts require direct physical contact with osteoblastic cells that govern these processes by producing at least two indispensable cytokines. Osteoclasts arise from the cellular lineage that also gives rise to macrophages and that expresses receptors for the *macrophage colony stimulating factor* (m-CSF) on their surface membranes. Osteoblastic cells secrete m-CSF. Osteoclasts and their precursors also express receptor activators of NF-κB (*RANK*) on their surfaces. Nuclear factor-κB is a transcription factor that translocates from the cytosol to the nucleus on activation (see Chapter 40). These receptors belong to the *tumor necrosis factor-α* (TNF-α) family of cytokine receptors. Osteoblastic cells express a membrane-bound cytokine called RANK ligand (*RANKL*). This cytokine is a member of the TNF-α family, and binds to RANK on the surface of osteoclasts or their precursors when cells of the two lineages come in physical contact. Steps in the differentiation of osteoclasts are illustrated in Fig. 4. Another member of the TNF receptor family exerts negative control over osteoclast formation and activity.

This compound, called *osteoprotegerin* (OPG), can bind RANK with high affinity, but unlike typical members of the TNF-α receptor family, OPG lacks a transmembrane domain and is a secreted soluble protein. Osteoprotegerin, which is produced by many different cells, competes with RANK for RANKL on the surface of osteoblastic cells and thereby limits osteoclast formation and activity. An unusual and confusing aspect of osteoclast regulation is the seemingly inverted participation of a membrane-bound ligand (RANK) and a soluble receptor (OPG). Although the factors that regulate expression of RANK and m-CSF are known (see later discussion), the factors that regulate OPG expression have not been identified.

In histological sections, osteoclasts are usually found on the surface of bone in pits created by their erosive action. Integrins on the surface of osteoclasts form tight bonds with osteocalcin, osteopontin, and other proteins of the bone matrix and create a sealed off region of extracellular space between the osteoclasts and the surface of the bony matrix. The specialized part of the osteoclast that faces the bony surface bone is thrown into many folds called the *ruffled border*. The ruffled border sweeps over the surface of bone, continuously changing its configuration as it releases acids and hydrolytic enzymes that dissolve the mineral crystals and the protein matrix. Small bone crystals are often seen in phagocytic vesicles deep in its folds. On completion of resorption, osteoclasts are inactivated and lose some of their nuclei. Complete inactivation involves fission of the giant, polynucleate cell back to mononuclear cells, which may undergo apoptosis, but some multinuclear cells remain quiescent on the bone surface interspersed among the lining cells.

Resorption of bone is precisely coupled with bone formation. The pattern of events in bone remodeling typically begins with differentiation and activation of osteoclasts followed sequentially by bone resorption, osteoblast activation and migration to the site of bone resorption, and finally bone formation. Details of the signaling mechanisms that couple bone formation with bone resorption are not yet known, but it appears that osteoblasts secrete a variety of autocrine and paracrine growth factors that are trapped and stored in the bone matrix during osteogenesis. Resorption of the matrix by osteoclastic activity appears to release these factors, which in turn may activate quiescent osteoblasts and promote differentiation of new osteoblasts from progenitor cells that may be attracted to the site by protein fragments of partially degraded osteoid.

Kidney

Ionized and complexed calcium pass freely through glomerular membranes. Normally 98–99% of the

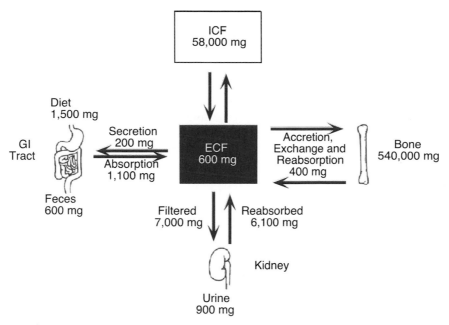

FIGURE 5   Daily phosphorous balance in a typical adult.

10,000 mg of calcium filtered by the glomeruli each day is reabsorbed by the renal tubules. About two-thirds of the reabsorption occurs in the proximal tubule, tightly coupled to sodium reabsorption and, for the most part, dragged passively along with water. Much of the remaining calcium is resorbed in the loop of Henle and is also tightly linked to sodium reabsorption (see Chapter 30). Normally only about 10% of the filtered calcium reaches the distal nephron. Reabsorption of calcium in the vicinity of the junction of the distal convoluted tubules and the collecting ducts is governed by an active, saturable process that is independent of sodium reabsorption. Active transport of calcium in this region is hormonally regulated (see later discussion).

Phosphorous Balance

Because of their intimate relationship, the fate of calcium cannot be discussed without also considering phosphorus. Calcium is usually absorbed in the intestines accompanied by phosphorus, and deposition and mobilization of calcium in bones always occurs in conjunction with phosphorus. Phosphorus is as ubiquitous in its distribution and physiologic role as calcium. The high-energy phosphate bond of ATP and other metabolites is the coinage of biological energetics. Phosphorus is indispensable for biological information transfer. It is a component of nucleic acids and second messengers such as cyclic AMP and IP_3, and as the addend that increases or decreases enzymatic activities or guides protein–protein interactions.

About 90% of the 500–800 g of phosphorus in the adult human is deposited in the skeleton. Much of the remainder is incorporated into organic phosphates distributed throughout soft tissues in the form of phospholipids, nucleic acids, and soluble metabolites. Daily intake of phosphorus is in the range of 1000–1500 mg, mainly in dairy products. Organic phosphorus is digested to inorganic phosphate before it is absorbed in the small intestine by both active and passive processes. Net absorption is linearly related to intake and appears not to saturate. The concentration of inorganic phosphate in blood is about 3.5 mg/dL. About 55% is present as free ions, about 35% is complexed with calcium or other cations, and 10% is protein bound. Phosphate concentrations are not tightly controlled and may vary widely under such influences as diet, age, and sex. Ionized and complexed phosphate pass freely across glomerular and other capillary membranes. Phosphate in the glomerular filtrate is actively reabsorbed by a sodium-coupled cotransport process in the proximal tubule. These relations in daily phosphorous balance are shown in Fig. 5.

PARATHYROID GLANDS AND PARATHYROID HORMONE

The parathyroid glands arose relatively recently in vertebrate evolution, coincident with the emergence of ancestral forms onto dry land. They are not found in fish and are seen in amphibians such as the salamander only after metamorphosis to the land-dwelling form. The importance of the parathyroids in normal calcium economy was established during the latter part of the 19th century when it was found that parathyroidectomy resulted in lethal tetany. Diseases resulting from overproduction or underproduction of parathyroid hormone (PTH) are relatively uncommon.

Human beings typically have four parathyroid glands, but as few as two and as many as eight have been observed. Each gland is a flattened ellipsoid measuring about 6 mm in its longest diameter. The aggregate mass of the adult parathyroid glands is about 120 mg in men and about 140 mg in women. These glands adhere to the posterior surface of the thyroid gland or occasionally are embedded within thyroid tissue. They are well vascularized and derive their blood supply mainly from the inferior thyroid arteries. Parathyroid glands are comprised of two cell types (Fig. 6). The *chief cells* predominate and are arranged in clusters or cords. They are the source of PTH and have all of the cytological characteristics of cells that produce protein hormones: rough endoplasmic reticulum, prominent Golgi apparatus, and some membrane-bound storage granules. The *oxyphil cells,* which appear singly or in small groups, are larger than chief cells and contain a remarkable number of mitochondria. Oxyphil cells have no known function and are thought by some to be degenerated chief cells. Their cytological properties are not characteristic of secretory cells. Few oxyphil cells are seen before puberty, but their number increases thereafter with age.

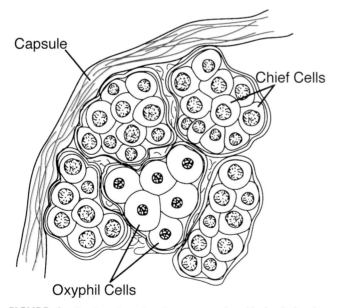

FIGURE 6   Section through a human parathyroid gland showing small chief cells and larger oxyphil cells. The cells are arranged in cords surrounded by loose connective tissue. (From Borysenko, M., and Beringer, T. *Functional histology,* 2nd ed. Little, Brown, Boston, 1984, p 316.)

Biosynthesis, Storage, and Secretion of PTH

The secreted form of PTH is a simple straight-chain peptide of 84 amino acids. There are no disulfide bridges. As many as 50 amino acids can be removed from the carboxyl terminus without compromising biological potency, but removal of just the serine at the amino terminus virtually inactivates the hormone. All of the known biological effects of PTH can be reproduced with a peptide corresponding to amino acids 1–34. PTH is expressed as a larger "preprohormone" and is the product of a single-copy gene located on chromosome 11. Sequential cleavage forms first a 90-amino-acid prohormone and then the mature hormone. The larger, transient forms have little or no biological activity and are not released into the circulation even at times of intense secretory activity. Synthesis of PTH is regulated at both transcriptional and at post-transcriptional sites. Cyclic AMP, acting through the cyclic AMP response element binding protein (CREB) up-regulates PTH gene transcription, while vitamin D down-regulates transcription. Low plasma calcium concentrations prolong the survival time of the mRNA transcript, while low concentrations of plasma phosphate accelerate message degradation. PTH is synthesized continuously, but the glands store little hormone; only enough to sustain maximal secretion rates for about 90 min.

Parathyroid cells are unusual in the respect that hormone degradation as well as synthesis is adjusted according to physiologic demand for secretion. As much as 90% of the hormone synthesized may be destroyed within the chief cells, which break down PTH at an accelerated rate when plasma calcium concentrations are high. Proteolytic enzymes incorporated into secretory granules cleave the PTH molecules so that fragments are cosecreted with the intact 84-amino-acid PTH. Similar fragments are produced by degradation of PTH in liver and kidney and also enter the bloodstream. PTH fragments are cleared from blood by filtration at the glomeruli, but remain in the blood hours longer than intact hormone, which has a half-life of only about 2–4 min. The intact 84-amino-acid peptide is the only biologically active form of PTH in the blood.

A substance closely related to PTH, *parathyroid hormone related peptide* (PTHrP), is found in the plasma of patients suffering from certain malignancies and accounts for the accompanying hypercalcemia. The gene for PTHrP was isolated from some tumors and found to encode a peptide whose first 13 amino acid residues are remarkably similar to the first 13 amino acids of PTH; thereafter, the structures of the two molecules diverge. The similarities of the N-terminal primary sequence and presumably the secondary structure of subsequent segments allow PTHrP to bind with high affinity to the PTH receptor and therefore produce the same biological effects as PTH. PTH and PTHrP are immunologically distinct and do not cross-react in immunoassays. PTHrP is synthesized in a wide range of tissues and acts locally as a paracrine factor to regulate a variety of processes that are unrelated to regulation of calcium concentrations in extracellular fluid. Little or no PTHrP is found in blood plasma of normal individuals.

Mechanisms of PTH Actions

Binding of PTH to G-protein-coupled receptors on the surfaces of target cells increases the formation of cyclic AMP and of inositol trisphosphate (IP_3) and diacylglycerol (DAG) (see Chapter 2). The PTH receptor is coupled to adenylyl cyclase through a stimulatory G protein (G_s) and to phospholipase C through G_q. Consequently, protein kinases A and C are also activated and intracellular calcium is increased. Rapid responses almost certainly result from protein phosphorylation, while delayed responses result from altered expression of genes regulated by CREB. It is likely that the two second messenger pathways activated by PTH are redundant and reinforce each other (see Clinical Note above).

Physiologic Actions of PTH

Parathyroid hormone is the principal regulator of the extracellular calcium pool. It increases the calcium concentration and decreases the phosphate concentration in blood by various direct and indirect actions on bone, kidney, and intestine. In its absence, the concentration of calcium in blood, and hence interstitial fluid, decreases dramatically over a period of several hours while the concentration of phosphate increases.

Actions on Bone

Increases in PTH concentration in blood result in mobilization of calcium phosphate from the bone matrix due to increased osteoclastic activity. The initial phase is seen within 1–2 hr and results from activation of preformed osteoclasts already present on the bone surface. A later and more pronounced phase of the response to PTH becomes evident after about 12 hr and is characterized by widespread resorption of both mineral and organic components of bone matrix, particularly in trabecular bone. Evidence of osteoclastic activity is reflected not only by calcium phosphate mobilization, but also by increased urinary excretion of hydroxyproline and other products of collagen breakdown.

Although activity of all bone cell types is affected by PTH, only cells of osteoblastic lineage express receptors for PTH. Osteoclasts are thus not direct targets for PTH. Activation, differentiation, and recruitment of osteoclasts in response to PTH depends on increased expression of at least two cytokines by cells of osteoblastic lineage within the bone marrow (Fig. 7). PTH induces these cells to synthesize and secrete m-CSF and to express RANKL on their surfaces (see earlier section on osteoclasts). These cytokines increase the differentiation and activity of osteoclasts and protect them from apoptosis. In addition, PTH may decrease osteoblastic expression of OPG, the competitive inhibitor of RANK

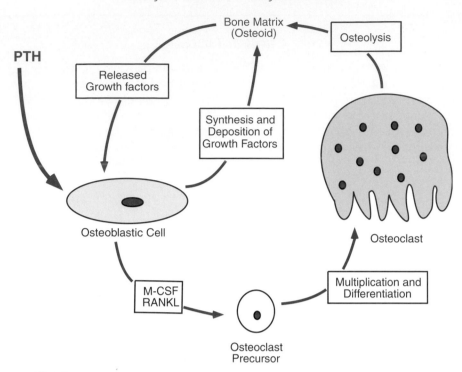

FIGURE 7 Effects of PTH in the activation of osteoclasts and osteoblasts. M-CSF, macrophage colony stimulating factor; RANKL, receptor activator of N-κB ligand. PTH stimulates osteoblastic cells in the bone marrow to secrete M-CSF and to express RANKL on their surface. Contact with osteoclast precursors initiates their transformation to osteoclasts, which digest bone matrix and release previously sequestered growth factors that stimulate osteoblasts to lay down new bone.

for binding RANKL. PTH also induces retraction of osteoblastic lining cells and thus exposes bony surfaces to which osteoclasts can bind.

Synthesis and secretion of collagen and other matrix proteins by osteoblasts are inhibited in the early phases of PTH action, but are reactivated subsequently as a result of the biological coupling discussed earlier, and new bone is laid down. At this time PTH stimulates osteoblasts to synthesize and secrete growth factors including insulin-like growth factor I (IGF-I) and transforming growth factor β (TGFβ), which are sequestered in the bone matrix and also act in an autocrine or paracrine manner to stimulate osteoblast progenitor cells to divide and differentiate. At least part of the stimulus for osteoblastic activity that follows osteoclastic activity may come from liberation of growth factors that were sequestered in the bony matrix when the bone was laid down. With prolonged continuous exposure to high concentrations of PTH, as seen with hyperparathyroidism, osteoclastic activity is greater than osteoblastic activity, and bone resorption predominates. However, intermittent stimulation with PTH leads to net formation of bone.

In addition to its critical role in maintaining blood calcium concentrations, PTH is also important for skeletal homeostasis. As already mentioned, bone remodeling continues throughout life. Remodeling of bone not only ensures renewal and maintenance of strength, but also adjusts bone structure and strength to accommodate the various stresses and strains of changing demands of daily living. Increased stress leads to bone formation strengthening the affected area, while weightlessness or limb immobilization leads to mineral loss. Osteocytes entrapped in the bony matrix are thought to function as mechanosensors that signal remodeling through release of prostaglandins, nitric oxide, and growth factors. In animal studies these actions are facilitated and enhanced by PTH.

Actions on Kidney

In the kidney, PTH produces three distinct effects, each of which contributes to the maintenance of calcium homeostasis. In the distal nephron it promotes the reabsorption of calcium, and in the proximal tubule it inhibits reabsorption of phosphate and promotes hydroxylation and, hence, activation of vitamin D (see later discussion). In producing these effects, PTH binds to G-protein-coupled receptors in both the proximal and distal tubules and stimulates the production of cyclic

AMP, DAG, and IP$_3$. Some cyclic AMP escapes from renal tubular cells and appears in the urine. About one-half of the cyclic AMP found in urine is attributable to renal actions of PTH.

Calcium Reabsorption

The kidney reacts quickly to changes in PTH concentrations in blood and is responsible for minute-to-minute adjustments in blood calcium. PTH acts directly on the distal portion of the nephron to decrease urinary excretion of calcium well before significant amounts of calcium can be mobilized from bone. About 90% of the filtered calcium is reabsorbed in the proximal tubule and the loop of Henle independently of PTH. Therefore, because its actions are limited to the distal reabsorptive mechanism, PTH can provide only fine-tuning of calcium excretion. Even small changes in the fraction of calcium reabsorbed from the glomerular filtrate, however, can be of great significance. Hypoparathyroid patients whose blood calcium is maintained in the normal range excrete about three times as much urinary calcium as normal subjects.

The cellular mechanisms that account for increased calcium reabsorption in response to PTH are shown in Fig. 8. Activation of G-protein-coupled receptors on the basolateral surface of cells in the distal convoluted tubules causes intracellular vesicles that harbor calcium channels to migrate to the luminal surface and fuse with the luminal membrane. Calcium ions in tubular fluid flow passively down their concentration gradient into the cells.

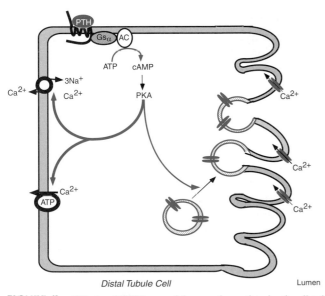

FIGURE 8   Effects of PTH on calcium reabsorption in the distal nephron. Gsα, α-subunit of the stimulatory G protein; AC, adenylyl cyclase; cAMP, cyclic adenosine monophosphate; PKA, protein kinase A.

In the basolateral membrane PTH activates sodium calcium antiporters that exchange three ions of extracellular sodium for one ion of intracellular calcium, a calcium ATPase that pumps calcium across the membrane, and the sodium potassium ATPase that maintains the electrochemical gradient across the membrane. Even when maximally stimulated, this PTH-sensitive mechanism has a low capacity that saturates when the filtered load of calcium is high. The filtered load of calcium is determined in part by the plasma calcium concentration and, hence, may increase with continued increased secretion of PTH. Because reabsorption of calcium in the proximal portions of the nephron is proportional to sodium and water reabsorption, a nearly constant fraction of the filtered load reaches the distal nephron. Consequently when the filtered load is high, more calcium may reach the distal nephron than the reabsorbing mechanism can handle. This circumstance accounts for the paradoxical increase in urinary calcium seen in later phases of PTH action. Regardless of the absolute amount excreted, however, PTH decreases the fraction of filtered calcium that escapes in the urine.

Phosphate Excretion

Parathyroid hormone powerfully inhibits tubular reabsorption of phosphate and thus increases the amount excreted in urine. This effect is seen within minutes after injection of PTH and is exerted in the proximal tubules, where the bulk of phosphate reabsorption occurs. Decreased reabsorption of phosphate results from decreased capacity for sodium-phosphate cotransport across the luminal membrane of tubular cells. In a manner analogous—but opposite—to its effects on calcium reabsorption, PTH decreases the abundance of sodium-phosphate cotransporters in the brush border of proximal tubule cells by stimulating their translocation to intracellular vesicles (Fig. 9).

Effects on Intestinal Absorption

Calcium balance ultimately depends on intestinal absorption of dietary calcium. Calcium absorption is severely reduced in hypoparathyroid patients and dramatically increased in those with hyperparathyroidism. Within a day or two after treatment of hypoparathyroid subjects with PTH, calcium absorption increases. Intestinal uptake of calcium is stimulated by an active metabolite of vitamin D. PTH stimulates the renal enzyme that converts vitamin D to its active form (see Fig. 9 and later discussion), but has no direct effects on intestinal transport of either calcium or phosphate.

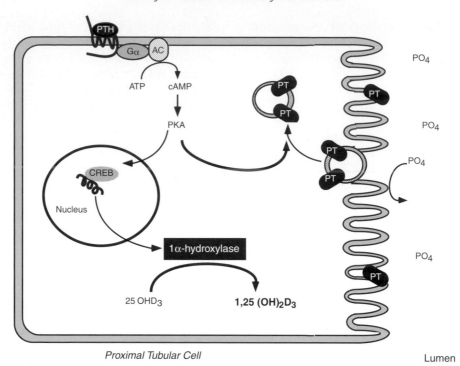

FIGURE 9 Effects of PTH on proximal tubule cells. PT, sodium phosphate cotransporter. $25OHD_3$ and $1,25(OH)_2D_3$ are metabolic forms of vitamin D (see Fig. 15). Gsα, α-subunit of the stimulatory G protein; AC, adenylyl cyclase; cAMP, cyclic adenosine monophosphate; PKA, protein kinase A; CREB, cyclic AMP response element binding protein.

Regulation of PTH Secretion

Chief cells of the parathyroid glands are exquisitely sensitive to changes in extracellular calcium and rapidly adjust their rates of PTH secretion in a manner that is inversely related to the concentration of ionized calcium (Fig. 10). The resulting increases or decreases in blood levels of PTH produce either positive or negative changes in the plasma calcium concentration and thereby provide negative feedback signals for regulation of PTH secretion. The activated form of vitamin D, whose synthesis depends on PTH, is also a negative feedback inhibitor of PTH synthesis (see later discussion). Although blood levels of phosphate are also affected by PTH, high phosphate appears to exert little or no effect on the secretion of PTH, but may exert some effects on hormone synthesis. Under experimental conditions, high concentrations of magnesium in plasma may also inhibit PTH secretion, but the concentration range in which magnesium affects secretion is well beyond that seen physiologically. A decrease in ionized calcium in blood appears to be the only physiologically relevant signal for PTH secretion.

Chief cells are programmed to secrete PTH unless inhibited by extracellular calcium, but secretion is not totally suppressed even when plasma concentrations are

very high. Through mechanisms that are not understood, normal individuals secrete PTH throughout the day in pulses with frequencies of one to three pulses per hour. Blood levels of PTH also follow a diurnal pattern with peak values seen shortly after midnight and minimal values seen in late morning. Diurnal fluctuations appear to arise from endogenous events in the chief cells and may promote anabolic responses of bone to PTH.

The cellular mechanisms by which extracellular calcium regulates PTH secretion are poorly understood. These cells are equipped with calcium-sensing receptors in their plasma membranes and can adjust secretion in response to as little as a 2–3% change in extracellular calcium concentration. Calcium-sensing receptors are members of the G-protein-coupled receptor superfamily (see Chapter 2) and bind calcium in proportion to its concentration in extracellular fluid. Because they appear to be coupled to adenylyl cyclase through G_i, and to phospholipase C, and perhaps membrane calcium channels probably through G_q, several second messengers appear to be involved in governing PTH secretion. Increased extracellular calcium increases production of DAG and IP_3 and decreases production of cyclic AMP. Cytosolic calcium increases as a result of IP_3-mediated release from intracellular stores followed by influx

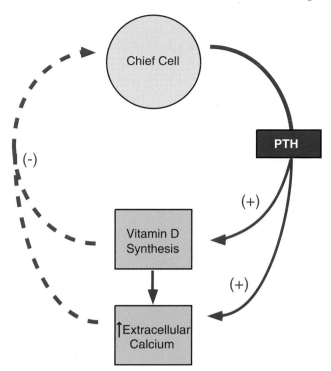

FIGURE 10   Regulation of PTH secretion. (−), decrease; (+), increase.

disulfide ring at the amino terminus, calcitonin has no remarkable structural features. Immunoreactive circulating forms are heterogeneous in size, reflecting the presence of precursors, partially degraded hormone, and disulfide-linked dimers and polymers. The active hormone has a half-life in plasma of about 5–10 min and is cleared from the blood primarily by the kidney. The gene that encodes calcitonin also encodes a neuropeptide called calcitonin gene related peptide (CGRP). This gene contains six exons, but only the first four are represented in the mRNA transcript that codes for the precursor of calcitonin. In the mRNA that codes for CGRP, the portion corresponding to exon 4 is deleted and replaced by exons 5 and 6. Because the first exon codes for an untranslated region, and the peptide corresponding to exons 2 and 3 is removed in post-translational processing,

through activated membrane channels. Paradoxically, in chief cells, unlike most other secretory cells, increased calcium is associated with inhibition rather than stimulation of hormone secretion. It is not yet understand how these changes in intracellular messages combine to inhibit PTH secretion. These events are summarized in Fig. 11.

CALCITONIN

Cells of Origin

Calcitonin is sometimes also called *thyrocalcitonin* to describe its origin in the *parafollicular cells* of the thyroid gland. These cells, which are also called *C cells,* occur singly or in clusters in or between thyroid follicles. They are larger and stain less densely than follicular cells in routine preparations (Fig. 12). Like other peptide hormone-secreting cells, they contain membrane-bound storage granules. Parafollicular cells arise embryologically from neuroectodermal cells that migrate to the last branchial pouch, and, in submammalian vertebrates, give rise to the ultimobranchial glands.

Biosynthesis, Secretion, and Metabolism

Calcitonin consists of 32 amino acids and has a molecular weight of about 3400. Except for a seven-member

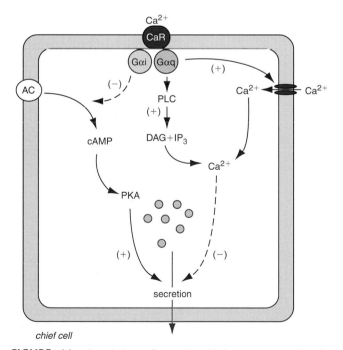

FIGURE 11   Regulation of parathyroid hormone secretion by calcium (Ca^{2+}). Heptihelical calcium receptors (CaR) on the surface of chief cells communicate with Ca^{2+} channels, adenylyl cyclase (AC), and phospholipase C (PLC) by way of guanosine nucleotide binding proteins (G_i and G_q). The resulting increase in Ca^{2+} and inhibition of adenylyl cyclase lowers cyclic AMP (cAMP) and interferes with protein kinase A (PKA) mediated events that lead to secretion. DAG, diacyl glycerol; IP_3; inositol trisphosphate.

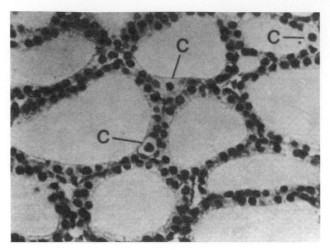

FIGURE 12 Low-power photomicrograph of a portion of the thyroid gland of a normal dog. Parafollicular (C) cells are indicated in the walls of the follicles. (From Ham AW, Cormack DH. *Histology*, 8th ed., Lippincott, Philadelphia, 1979, p 802.)

the mature products have no common amino acid sequences (Fig. 13).

Physiologic Actions of Calcitonin

Calcitonin escaped discovery for many years because no obvious derangement in calcium balance or other homeostatic function results from deficient or excessive production. Thyroidectomy does not produce a tendency toward hypercalcemia, and thyroid tumors that secrete massive amounts of calcitonin do not cause hypocalcemia. Attention was drawn to the possible existence of a calcium-lowering hormone by the experimental finding that direct injection of a concentrated solution of calcium into the thyroid artery in dogs caused a more rapid fall in

blood calcium than parathyroidectomy. Indeed, calcitonin promptly and dramatically lowers the blood calcium concentration in many experimental animals. Calcitonin is not a major factor in calcium homeostasis in humans, however, and does not participate in minute-to-minute regulation of blood calcium concentrations. Rather, the importance of calcitonin may be limited to protection against excessive bone resorption.

Actions on Bone

Calcitonin lowers blood calcium and phosphate primarily, and perhaps exclusively, by inhibiting osteoclastic activity. The decrease in blood calcium produced by calcitonin is greatest when osteoclastic bone resorption is most intense and is least evident when osteoclastic activity is minimal. Binding of calcitonin by G-protein-coupled receptors on the osteoclast surface promptly increases cyclic AMP formation, and within minutes the expanse and activity of the ruffled border diminishes. Osteoclasts pull away from the bone surface and begin to dedifferentiate. Synthesis and secretion of lysosomal enzymes is inhibited. In less than an hour, fewer osteoclasts are present, and those that remain have decreased bone-resorbing activity.

Osteoclasts are the principal, and probably only, target cells for calcitonin in bone. Osteoblasts do not have receptors for calcitonin and are not directly affected by it. Curiously, although they are uniquely expressed in either osteoblasts or osteoclasts, receptors for PTH and calcitonin are closely related and have about one-third of their amino acid sequences in common, suggesting they evolved from a common ancestral molecule. Because of the coupling phenomenon in the cycle of bone resorption and bone formation

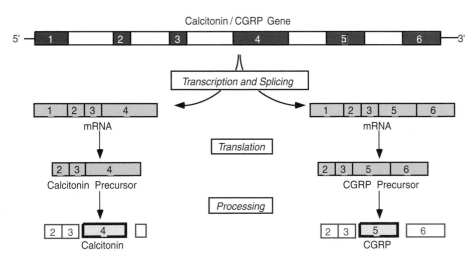

FIGURE 13 Alternate splicing of calcitonin/calcitonin gene related peptide (CGRP) mRNA gives rise to either calcitonin or CGRP, with no shared sequence of amino acids.

discussed earlier, inhibition of osteoclastic activity by calcitonin eventually decreases osteoblastic activity as well. All cell types appear quiescent in histological sections of bone that were chronically exposed to high concentrations of calcitonin. Although osteoclasts express very high numbers of receptors for calcitonin, they quickly become insensitive to the hormone because continued stimulation results in massive down-regulation of the receptors.

Actions on Kidney

At high concentrations calcitonin may increase urinary excretion of calcium and phosphorus, probably by acting on the proximal tubules. In humans these effects are small, last only a short while, and are not physiologically important for lowering blood calcium. Renal handling of calcium is not disrupted in patients with thyroid tumors that secrete large amounts of calcitonin. Kidney cells "escape" from prolonged stimulation with calcitonin and become refractory to it, probably as a result of down-regulation of receptors.

Regulation of Secretion

Circulating concentrations of calcitonin are low when blood calcium is in the normal range or below, but they increase proportionately when calcium rises above about 9 mg/dL (Fig. 14). Parafollicular cells respond directly to ionized calcium in blood and express the same G-protein-coupled calcium sensing receptor in their surface membranes as the parathyroid chief cells. Both cell types respond to extracellular calcium over the

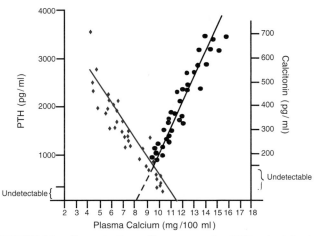

FIGURE 14 Concentrations of immunoreactive PTH and calcitonin in pig plasma as a function of plasma calcium. (From Arnaud CD, Littledike T, Tsao HS. Simultaneous measurements of calcitonin and parathyroid hormone in the pig. In Taylor S, Foster GV, eds., *Proceedings of the symposium on calcitonin and C cells*. Heinemann, London, 1969.)

same concentration range, but their secretory responses are opposite, presumably because of differences in the transduction and secretory apparatus. Although little information is available concerning intracellular mechanisms in parafollicular cells, the relation between stimulation and secretion is more typical of endocrine cells than the regulation of PTH secretion.

In addition to responding directly to high concentrations of calcium, calcitonin secretion may also increase after eating. Gastrin, a hormone produced by gastric mucosal cells, stimulates parafollicular cells to secrete calcitonin. Other gastrointestinal hormones, including cholecystokinin, glucagon, and secretin, have similar effects, but gastrin is the most potent among these agents. Secretion of calcitonin in anticipation of an influx of calcium from the intestine is a feed-forward mechanism that may guard against excessive concentrations of plasma calcium by decreasing osteoclastic activity. This phenomenon is analogous to the anticipatory secretion of insulin after a carbohydrate-rich meal (see Chapter 41). Although the importance of this response in humans is not established, sensitivity of parafollicular cells to gastrin has been exploited clinically as a provocative test for diagnosing medullary carcinoma of the thyroid.

VITAMIN D-ENDOCRINE SYSTEM

A derivative of vitamin D_3, 1,25-dihydroxycholecalciferol (1,25$(OH)_2D_3$), is indispensable for maintaining adequate concentrations of calcium in the extracellular fluid and adequate mineralization of bone matrix. Vitamin D deficiency leads to inadequate calcification of bone matrix and severe softening of the skeleton, called *osteomalacia*, and may result in bone deformities and fractures. Osteomalacia in children is called *rickets* and may produce permanent deformities of the weight-bearing bones (bowed legs). Although vitamin D is now often called the vitamin D-endocrine system, when it was discovered as the factor in fish oil that prevents rickets, its hormone-like nature was not suspected. One important distinction between hormones and vitamins is that hormones are synthesized within the body from simple precursors, but vitamins must be provided in the diet. Actually, vitamin D_3, can be synthesized endogenously in humans, but the rate is limited by a non-enzymatic reaction that requires radiant energy in the form of light in the near-ultraviolet range, hence the name *sunshine vitamin*. The immediate precursor of vitamin D_3, 7-dehydrocholesterol, is synthesized from acetyl coenzyme A (CoA) and is stored in skin. Conversion of 7-dehydrocholesterol to vitamin D_3 proceeds spontaneously in the presence of sunlight that penetrates the epidermis to the outer layers of the dermis. Vitamin D deficiency became a significant public health problem as a by-product of industrialization. Urban living, smog, and increased indoor activity limit exposure of the populace to sunshine and hence endogenous production of vitamin D_3. This problem is readily addressed by adding vitamin D to foods, particularly milk.

1,25-Dihydroxy-vitamin D_3 also fits the description of a hormone in the respect that it travels through the blood in small amounts from its site of production to affect cells at distant sites. Another major difference between a vitamin and a hormone is that vitamins usually are cofactors in metabolic reactions, whereas hormones behave as regulators and interact with specific receptors. $1,25(OH)_2D_3$ produces many of its biological effects in a manner characteristic of steroid hormones (see Chapter 2). It binds to a specific nuclear receptor that is a member of the same superfamily as the receptors for steroid and thyroid hormones.

Synthesis and Metabolism

The form of vitamin D produced in mammals is called *cholecalciferol,* or vitamin D_3; it differs from vitamin D_2 (ergosterol), which is produced in plants, only in the length of the side chain. Irradiation of the skin results in photolysis of the bond that links carbons 9 and 10 in 7-dehydrocholesterol, and thus opens the B ring of the steroid nucleus (Fig. 15). The resultant cholecalciferol is biologically inert but, unlike its precursor, has a high affinity for a vitamin D-binding protein in plasma. Vitamin D_3 is transported by the blood to the liver, where it is oxidized to form 25-hydroxycholecalciferol ($25OH-D_3$) by the same P450 mitochondrial enzyme that oxidizes cholesterol on carbons 26 and 27 in the pathway leading to formation of bile acids. This reaction appears to be controlled only by the availability of substrate. $25OH-D_3$ has high affinity for the vitamin D-binding protein and is the major circulating form of vitamin D. It has little biological activity. In the proximal tubules of the kidney, a second hydroxyl group is added at carbon 1 by another P450 enzyme to yield the compound, $1,25(OH)_2D_3$, which is about 1000 times as active as $25 OH-D_3$, and probably accounts for all of the biological activity of vitamin D. $1,25(OH)_2D_3$ is considerably less abundant in blood than its 25 hydroxylated precursor and binds less tightly to vitamin D-binding globulin than its precursor $25 OH-D_3$. Consequently, $1,25(OH)_2D_3$ has a half-life in blood of 15 hr compared to 15 days for $25 OH-D_3$.

Physiologic Actions of $1,25(OH)_2D_3$

Overall, the principal physiologic actions of $1,25(OH)_2D_3$ increase calcium and phosphate concentrations in extracellular fluid. These effects are exerted primarily on intestine and bone and, to a lesser extent, on kidney. Vitamin D receptors are widely distributed, however, and a variety of other actions that are not obviously related to calcium balance have been described or postulated. Because these latter effects are neither well understood nor germane to regulation of calcium balance, they are not discussed further.

Actions on Intestine

Uptake of dietary calcium and phosphate depends on active transport by epithelial cells lining the small intestine. Deficiency of vitamin D severely impairs intestinal transport of both calcium and phosphorus. Although calcium uptake is usually accompanied by phosphate uptake, the two ions are transported by independent mechanisms, both of which are stimulated by $1,25(OH)_2D_3$. Increased uptake of calcium is seen about 2 hr after $1,25(OH)_2D_3$ is given to deficient subjects and is maximal within 4 hr. A much longer time is required when vitamin D is given, presumably because of the time needed for sequential hydroxylations in liver and kidney.

Calcium uptake by duodenal epithelial cells is illustrated in Fig. 16. Calcium enters passively down its electrochemical gradient through two novel channels, the epithelial calcium channel (ECaC) and calcium transporter 1 (CaT1). Upon entry into the cytosol calcium is bound virtually instantenously by calcium binding proteins called calbindins and carried through the cytosol to the basolateral membrane where it is extruded into the interstitium by calcium ATPase (calcium pump) and sodium/calcium cytosolic calcium concentration low and thus maintain a gradient favorable for calcium influx while affording protection from deleterious effects of high concentrations of free calcium. It appears the abundance of ECaC and CaTI in the luminal membrane and at least one of the calbindins in the cytosol depends on $1,25(OH)_2D_3$ through regulation of gene transcription. Similarly, $1,25(OH)_2D_3$ is thought to regulate expression of sodium phosphate transporters in the luminal membrane.

Some evidence obtained in experimental animals and in cultured cells suggests that $1,25,(OH)_2D_3$ may also produce some rapid actions that are not mediated by altered genomic expression. Among these are rapid transport of calcium across the intestinal epithelium by a process that may involve both the IP_3-DAG and the cyclic AMP second messenger systems (see Chapter 2) and the activation of membrane calcium channels. The physiologic importance of these rapid actions of $1,25,(OH)_2D_3$ and the nature of the receptor that signals them are not known.

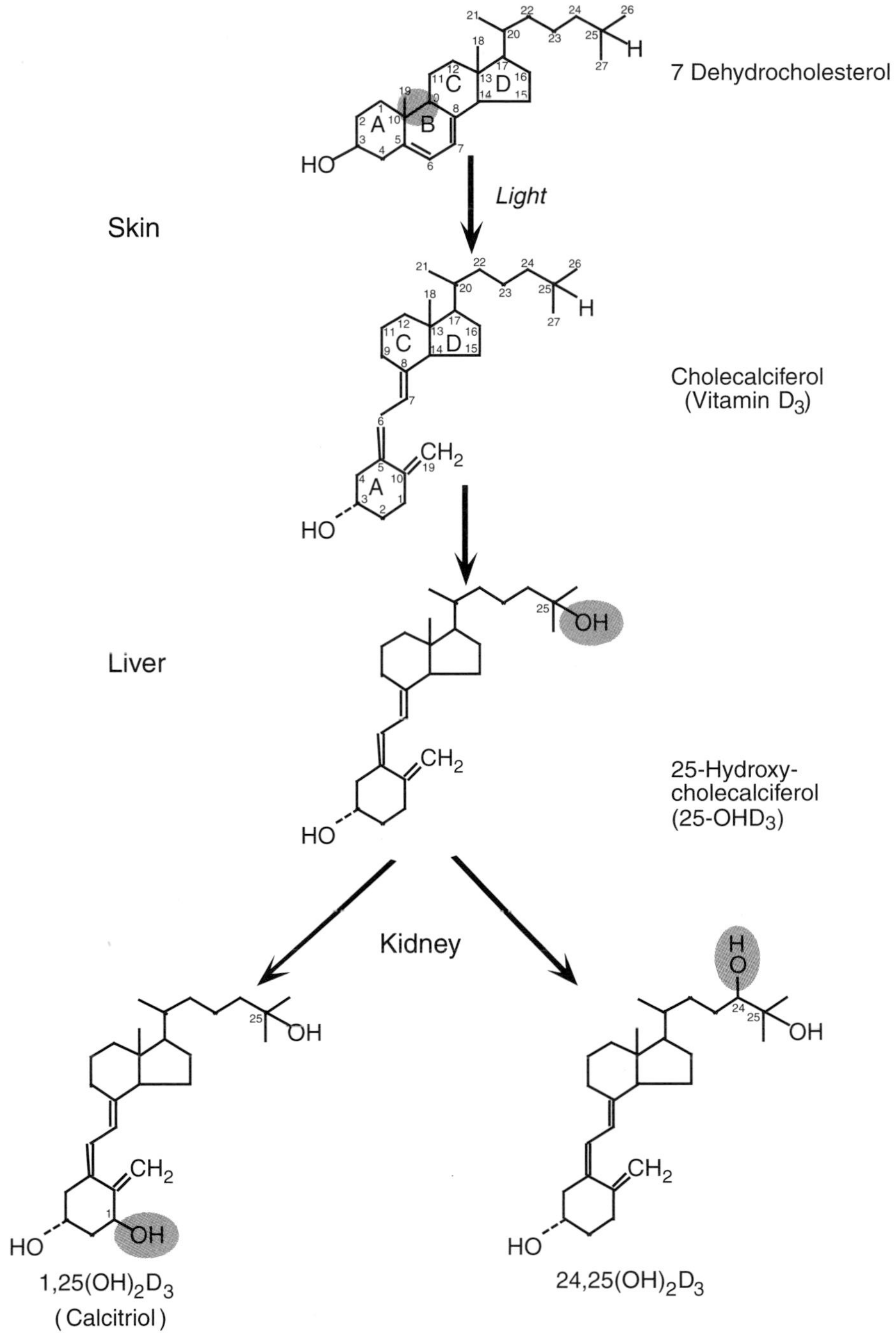

FIGURE 15   Biosynthesis of 1α,25 dihydroxycholecalciferol.

Actions on Bone

Although the most obvious consequence of vitamin D deficiency is decreased mineralization of bone, $1,25(OH)_2D_3$ apparently does not directly increase bone formation or calcium phosphate deposition in osteoid. Rather, mineralization of osteoid occurs spontaneously when adequate amounts of these ions are available. Ultimately, increased bone mineralization is made possible by increased intestinal absorption of calcium and phosphate. Paradoxically, perhaps, $1,25(OH)_2D_3$ acts on bone to promote resorption in a manner that resembles the effects of PTH. Like PTH, $1,25(OH)_2D_3$ increases both the number and activity of osteoclasts. As seen for PTH, osteoblasts, rather than mature osteoclasts, have receptors for $1,25(OH)_2D_3$. Like PTH, $1,25(OH)_2D_3$ stimulates osteoblastic cells to express m-CSF and RANK ligand as well as a variety of other proteins. Sensitivity of bone to PTH decreases with vitamin D deficiency; conversely, in the absence of PTH, 30–100

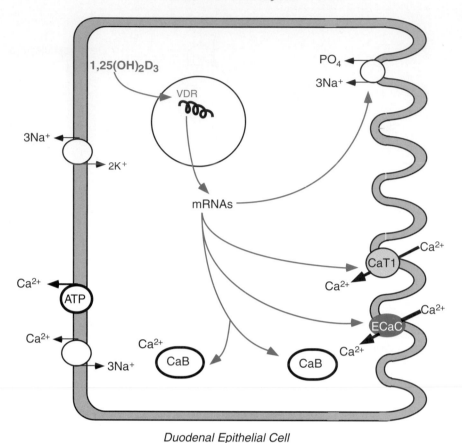

Duodenal Epithelial Cell

FIGURE 16   Actions of $1,25(OH)_2D_3$ on intestinal transport of calcium. VDR, vitamin D receptor; CaT1, calcium transporter 1; ECaC, epithelial calcium transporter; CaB, calbindin.

times as much $1,25(OH)_2D_3$ is needed to mobilize calcium and phosphate. The molecular site(s) of cooperative interaction of these two hormones in osteoblasts is(are) not known.

Actions on Kidney

When given to vitamin D-deficient subjects, $1,25(OH)_2D_3$ increases reabsorption of both calcium and phosphate. The effects on phosphate reabsorption are probably indirect. PTH secretion is increased in vitamin D deficiency (see later discussion), and hence tubular reabsorption of phosphate is restricted. Replenishment of $1,25(OH)_2D_3$ decreases the secretion of PTH and thus allows proximal tubular reabsorption of phosphate to increase. Effects of $1,25(OH)_2D_3$ on calcium reabsorption are probably direct. Specific receptors for $1,25(OH)_2D_3$ are found in the distal nephron, probably in the same cells in which PTH stimulates calcium uptake. These cells also express the same vitamin D-dependent calcium-binding protein, calbindin, that is found in intestinal cells, and calbindin is likely to play the same role in renal tubular cells as in intestinal

epithelium. It is unlikely that $1,25(OH)_2D_3$ regulates calcium balance on a minute-to-minute basis. Instead, it may act in a permissive way to support the actions of PTH, which is the primary regulator. The molecular basis for this interaction has not been elucidated.

Actions on the Parathyroid Glands

The chief cells of the parathyroid glands are physiologic targets for $1,25(OH)_2D_3$ and respond to it in a manner that is characteristic of negative feedback. In this case, negative feedback is exerted at the level of synthesis rather than secretion. The promoter region of the PTH gene contains a vitamin D response element. Binding of the liganded receptor suppresses transcription of the gene and leads to a rapid decline in the preproPTH mRNA. Because the chief cells store relatively little hormone, decreased synthesis rapidly leads to decreased secretion. In a second negative feedback action, $1,25(OH)_2D_3$ indirectly decreases PTH secretion by virtue of its actions to increase plasma calcium concentration. Consistent with the crucial role of calcium in regulating PTH secretion, the negative feedback effects of $1,25(OH)_2D_3$

on PTH synthesis are modulated by the plasma calcium concentration. Nuclear receptors for $1,25(OH)_2D_3$ are down-regulated when the plasma calcium concentration is low and up-regulated when it is high.

Regulation of $1,25(OH)_2D_3$ Production

As true of any hormone, the concentration of $1,25(OH)_2D_3$ in blood must be appropriate for prevailing physiologic circumstances if it is to exercise its proper role in maintaining homeostasis. Production of $1,25(OH)_2D_3$ is subject to feedback regulation in a fashion quite similar to that of other hormones. The most important regulatory step is the hydroxylation of carbon 1 by cells in the proximal tubules of the kidney. The rate of this reaction is determined by the availability of the requisite P450 enzyme, which has a half-life of only about 2–4 hr. Regulation of synthesis of $1,25(OH)D_3$ is achieved by regulating transcription of the gene that codes for the 1α-hydroxylase.rate of enzyme. Several cyclic AMP response elements (CRE) are present in its promoter region. PTH activates transcription of the 1α-hydroxylase gene through increasing production of cyclic AMP, activation of protein kinase A, and phosphorylation of CREB. Activation of protein kinase C through the IP_3-DAG second messenger system also appears to play some role in up-regulation of this enzyme. In the absence of PTH, the concentration of 1α-hydroxylase in renal cells quickly falls. A major component of the negative feedback system is the powerful inhibitory effect of $1,25(OH)_2D_3$ to suppress PTH gene expression in the parathyroid chief cells.

Through a "short" feedback loop, $1,25(OH)_2D_3$ acts as a negative feedback inhibitor of its own production by rapidly down-regulating 1α-hydroxylase expression. At the same time, $1,25(OH)D_3$ up-regulates the enzyme that hydroxylates vitamin D metabolites on carbon 24 to produce $24,25(OH)_2D_3$ or $1,24,25(OH)_3D_3$. Hydroxylation at carbon 24 is the initial reaction in the degradative pathway that culminates in the production of calcitroic acid, the principal biliary excretory product of vitamin D. Up-regulation of the 24 hydroxylase by $1,25(OH)D_3$ is not confined to the kidney, but is also seen in all $1,25(OH)D_3$ target cells.

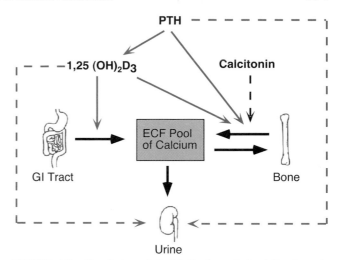

FIGURE 17   Regulation of $1\alpha,25$-dihydroxycholecalciferol synthesis. Solid arrows indicate stimulation; dashed arrows represent inhibition.

Finally, the results of its actions, increased calcium and phosphate concentrations in blood, directly or indirectly silence the two activators of $1,25(OH)_2D_3$ production, PTH, and low phosphate. The regulation of $1,25(OH)_2D_3$ production is summarized in Fig. 17.

INTEGRATED ACTIONS OF CALCITROPIC HORMONES

Response to a Hypocalcemic Challenge

Because some calcium is always lost in urine, even a short period of total fasting can produce a mild hypocalcemic challenge. More severe challenges are produced by a diet deficient in calcium or anything that might interfere with calcium absorption by renal tubules or the intestine. The parathyroid glands are exquisitely sensitive to even a small decrease in ionized calcium and promptly increase PTH secretion (Fig. 18). Effects of PTH on calcium reabsorption from the glomerular filtrate coupled with some calcium mobilization from bone are evident after about an hour, providing the first line of defense against a hypocalcemic challenge. These actions are adequate only to compensate for a mild or

---

Clinical Note

The importance of the negative feedback effects of $1,25(OH)_2D_3$ on PTH secretion is highlighted by the condition known as *renal osteodystrophy* that is found in patients with chronic renal failure.

This disorder is characterized by abnormalities in bone mineralization due largely to secondary hyperparathyroidism that results from inability to synthesize $1,25(OH)_2D_3$.

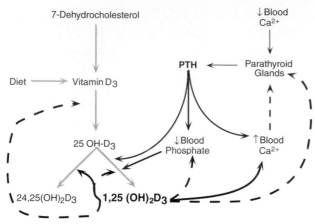

FIGURE 18 Overall regulation of calcium balance by PTH, calcitonin, and 1,25(OH)$_2$D$_3$. Solid arrows indicate stimulation; dashed arrows represent inhibition.

brief challenge. When the hypocalcemic challenge is large and sustained, additional, delayed responses to PTH are needed. After about 12–24 hr, increased formation of 1,25(OH)$_2$D$_3$ increases the efficiency of calcium absorption from the gut. Osteoclastic bone resorption in response to both PTH and 1,25(OH)$_2$D$_3$ taps the almost inexhaustible reserves of calcium in the skeleton. If calcium intake remains inadequate, skeletal integrity may be sacrificed in favor of maintaining blood calcium concentrations.

Response to a Hypercalcemic Challenge

Hypercalcemia is rarely seen under normal physiologic circumstances, but it may be a complication of a variety of pathologic conditions usually accompanied by increased blood concentrations of PTH or PTHrP. An example of hypercalcemia that might arise under physiologic circumstances is the case when a person who has been living for some time on a low-calcium diet ingests calcium-rich food. Under the influence of high concentrations of PTH and 1,25(OH)$_2$D$_3$ that would result from calcium insufficiency, osteoclastic activity transfers bone mineral to the extracellular fluid. In addition, calcium absorptive mechanisms in the intestine and renal tubules are stimulated to their maximal efficiency. Consequently the calcium that enters the gut is absorbed efficiently and blood calcium is increased by a few tenths of a milligram per deciliter. Calcitonin secretion is promptly increased and would provide some benefit through suppression of osteoclastic activity. Although PTH secretion promptly decreases, and its effects on calcium and phosphate transport in renal tubules quickly diminish, several hours pass before hydroxylation of 25 OH-D$_3$ and osteoclastic

bone resorption diminish. Even after its production is shut down, many hours are required for responses to 1,25(OH)$_2$D$_3$ to decrease. Although some calcium phosphate may crystallize in demineralized osteoid, renal loss of calcium is the principal means of lowering blood calcium. The rate of renal loss, however, is limited to only 10% of the calcium present in the glomerular filtrate, or about 40 mg/hr, even after complete shutdown of PTH-sensitive transport.

OTHER HORMONES AFFECTING CALCIUM BALANCE

In addition to the primary endocrine regulators of calcium balance discussed earlier, it is apparent that many other endocrine and paracrine factors influence calcium balance. Bone growth and remodeling involve a still incompletely understood interplay of local and circulating cytokines, growth factors and hormones including insulin-like growth factor-1, growth hormone (see Chapter 44), the cytokines interleukin 1 (see Chapter 40), interleukin 6, interleukin 11, TNFα, TGFβ, and doubtless many others. The prostaglandins (see Chapter 40) also have calcium mobilizing activity and stimulate bone lysis. Production of prostaglandins and interleukins is increased in a variety of inflammatory conditions and can lead to systemic or localized destruction of bone.

Many of the systemic hormones directly or indirectly have an impact on calcium balance. Obviously, special demands are imposed on overall calcium balance during growth, pregnancy, and lactation. All of the hormones that govern growth, namely, growth hormone, the insulin-like growth factors, and thyroidal and gonadal hormones (see Chapter 44), directly or indirectly influence the activity of bone cells and calcium balance. The gonadal hormones, particularly estrogens, play a critical role in maintaining bone mass, which decreases in their absence, leading to *osteoporosis*. This condition is common in postmenopausal women. Osteoblastic cells express receptors for estrogens, which stimulate proliferation of osteoblast progenitors and inhibit production of cytokines such as interleukin-6 that activate osteoclasts. Consequently in their absence osteoclastic activity is increased and osteoblastic activity is decreased, and net loss of bone results.

Defects in calcium metabolism are also seen in hyperthyroidism and in conditions of excess or deficiency of adrenal cortical hormones. Excessive thyroid hormone accelerates activity of both the osteoclasts and osteoblasts, which may result in net bone resorption and a decrease in bone mass. This action may produce a mild hypercalcemia and secondarily suppress PTH secretion and hence $1,25(OH)_2D_3$ production. These hormonal changes result in increased urinary loss of calcium and decreased intestinal absorption. Excessive glucocorticoid concentrations also decrease skeletal mass. Although glucocorticoids stimulate the differentiation of osteoclast progenitors, they decrease proliferation of these progenitor cells, which ultimately leads to a decrease in active osteoblasts. Glucocorticoids also antagonize the actions and formation of $1,25(OH)_2D_3$ by some unknown mechanism, and directly inhibit calcium uptake in the intestine. These changes may increase PTH secretion and stimulate osteoclasts. Conversely, adrenal insufficiency may lead to hypercalcemia due largely to decreased renal excretion of calcium.

CONCLUDING COMMENTS

The regulation of calcium homeostasis provides good examples of some of the principles of hormonal integration discussed in Chapter 37. The principle of cooperativity is illustrated by the multiple actions of PTH on different tissues to produce a variety of effects that contribute to the central theme of increasing calcium in the extracellular compartment. Overlapping effects of PTH and $1,25(OH)_2D_3$ illustrate redundancy. The opposing actions of PTH and calcitonin on osteoclasts are typical of a push–pull system. The

thoughtful student will undoubtedly find other examples in this exquisitely coordinated system.

Suggested Reading

Brommage R, DeLuca HF. Evidence that 1,25-dihydroxyvitamin D_3 is the physiologically active metabolite of vitamin D_3. *Endocr Rev* 1985;6:491–511.

Brown EM, Pollak M, Seidman CE, Seidman JG, Chou YH, Riccardi D, Hebert SC. Calcium ion-sensing cell-surface receptors. *New Engl J Med* 1995;333:234–240.

Diaz R, Fuleihan GEH, Brown EM. Parathyroid hormone and polyhormones: Production and export. In Fray, JCS, ed., *Endocrine regulation of water and electrolyte balance, Vol 3, Handbook of physiology, Section 7, The endocrine system*. Oxford University Press, New York, 2000, pp 607–662.

Jones G, Strugnall SA, DeLuca H. Current understanding of the molecular actions of vitamin D. *Physiol Rev* 1998;78:1193–1231.

Malloy PJ, Pike JW, Feldman D. The vitamin D receptor and the syndrome of hereditary 1,25-dihydroxyvitamin D-resistant rickets. *Endocr Rev* 1999;20:156–188.

Mannstadt M, Jüppner H, Gardella TJ. Receptors for PTH and PTHrP: Their biological importance and functional properties. *Am J Physiol* 1999;277:F665–F675.

Muff R, Fischer JA. Parathyroid hormone receptors in control of proximal tubular function. *Ann Rev Physiol* 1992;54:67–79.

Nijweide PJ, Burger EH, Feyen JHM. Cells of bone: Proliferation, differentiation, and hormonal regulation. *Physiol Rev* 1986; 66:855–886.

Suda T, Takahashi N, Udagawa N, Jimi E, Gillespie MT, Martin TJ. Modulation of osteoclast differentiation and function by new members of the tumor necrosis factor receptor and ligand families. *Endocr Rev* 1999;20:245–397.

Hormone Control of Growth

H. MAURICE GOODMAN

KEY POINTS

- Growth hormone (GH) is indispensable for the attainment of adult stature.
- Growth hormone stimulates growth of the long bones by stimulating production of insulin-like growth factor I (IGF-I).
- Insulin-like growth factor I produced locally and in the liver stimulates growth and maturation of chondrocytes in epiphysial plates.
- Growth hormone increases lean body mass, decreases body fat, promotes fatty acid utilization, and limits carbohydrate utilization.
- Growth hormone is secreted in discrete pulses about every 3 hours, with the largest pulse occurring in the early phases of nocturnal sleep. GH is secreted throughout life; the highest

blood levels are seen during adolescence and early adulthood and decline with increasing age.
- Growth hormone secretion is controlled by the hypothalamic GH-releasing hormone (GHRH), by the GH-release-inhibiting hormone somatostatin, and by IGF-I.
- Thyroid hormone is required by the somatotrope for normal GH synthesis and secretion. Thyroid hormones maintain tissue sensitivity to GH and IGF-I and hence are required for normal expression of GH actions.
- Optimal concentrations of insulin in blood are required to maintain normal growth during postnatal life.

Essential Medical Physiology, Third Edition

OVERVIEW

The simple word *growth* describes a variety of processes, both living and nonliving, that share the common feature of increase in mass. For purposes of this chapter, we limit the definition of growth to mean the *organized addition of new tissue that occurs normally in development from infancy to adulthood.* This process is complex and depends on the interplay of genetic, nutritional, and environmental influences as well as actions of the endocrine system. Growth of an individual or an organ involves increases both in cell number and cell size, differentiation of cells to perform highly specialized functions, and tissue remodeling that may require apoptosis as well as new cell formation. Most of these processes depend on locally produced growth factors that operate through paracrine or autocrine mechanisms. Many of these processes continue to operate throughout life, providing not only for cell renewal but also for adaptations to meet physiological demands. Dozens of growth factors have been described and an unknown number of others await discovery. Regulation of growth by the endocrine system can be viewed as coordination of local growth processes with overall development of the individual and with external environmental influences. This chapter describes the hormones that play important roles in growth and their interactions at critical times in development.

Growth is most rapid during prenatal life. In only 9 months, the body length of a human increases from just a few micrometers to almost 30% of final adult height. The growth rate decelerates after birth but during the first year of life is rapid enough that the infant increases half again in height to about 45% of final adult stature. Thereafter growth decelerates and continues at a slower rate, about 2 inches per year, until puberty. Steady growth during this juvenile period contributes the largest fraction, about 40%, to final adult height. With the onset of sexual development, growth accelerates to about twice the juvenile rate and contributes about 15 to 18% of final

adult height before stopping altogether (Figs. 1 and 2). Our understanding of hormonal influences on growth is limited largely to the juvenile and adolescent periods, but emerging information is providing insight into regulation of prenatal growth, which is largely independent of the classical hormones. During the juvenile period, the influence of growth hormone is preeminent, but appropriate secretion of thyroid and adrenal hormones and insulin is essential for optimal growth. The adolescent growth spurt reflects the added input of androgens and estrogens, which speed up growth and the maturation of bone that brings it to a halt.

GROWTH HORMONE

Growth hormone (GH), which is also called somatotropin (STH), is the single most important hormone for

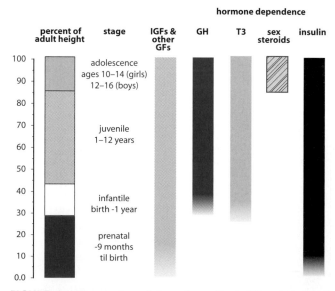

FIGURE 1 Hormonal regulation of growth at different stages of life. IGFs, insulin-like growth factors; GH, growth hormone; T3, triiodothyronine.

Clinical Note

Pituitary Dwarfism

Pituitary dwarfism is the failure of growth that results from lack of GH during childhood. Pituitary dwarfs typically are of normal weight and length at birth and grow rapidly and nearly normally during early infancy. Before the end of the first year, however, growth is noticeably below the normal rate and continues slowly for many years. Left untreated, they may reach heights of around 4 feet. Typically, the pituitary dwarf retains a juvenile appearance because of the retention of "baby fat" and the disproportionately small size of maxillary and mandibular bones. Pituitary dwarfism is not a single entity and may encompass a range of defects. The deficiency in GH may be accompanied by deficiencies in several or all other anterior pituitary hormones (*panhypopituitarism*) as might result from defects in pituitary development (see Chapter 38). Alternatively, traumatic injury to the pituitary gland or a tumor that either destroys pituitary cells themselves or their connections to the hypothalamus might also interfere with normal pituitary function. These individuals do not mature sexually and suffer from inadequacies of thyroid and adrenal glands as well. The lack of GH might also be an isolated inherited defect, with no abnormalities in other pituitary hormones. Aside from their diminutive height, such individuals are normal in all respects and can reproduce normally. Causes of *isolated GH deficiency* are multiple, and include derangements in synthesis, secretion, and end-organ responsiveness.

Overproduction of GH may occur either as a result of some derangement in mechanisms that control secretion by normal pituitary cells or from tumor cells that secrete GH autonomously. Overproduction of GH during childhood results in *gigantism*, in which an adult height in excess of 8 feet has occasionally been reported. Overproduction of GH during adulthood, after the growth plates of long bones have fused (see below), produces growth only by stimulation of responsive osteoblastic progenitor cells in the periosteum. There is thickening of the cranium and mandible, as well as enlargement of some facial bones and bones in the hands and feet. Growth and deformities in these acral parts give rise to the name *acromegaly* to describe this condition. Persistence of responsive cartilage progenitor cells in the costochrondral junctions leads to elongation of the ribs to give a typical barrel-chested appearance. Thickening of the skin and disproportionate growth of some soft tissues, including spleen and liver, are also observed in acromegalic patients.

normal growth. Attainment of adult size is absolutely dependent on GH; in its absence, growth is severely limited, and when it is present in excess, growth is excessive.

When thinking about giants and dwarfs, it is important to keep in mind the limitations of GH action. The pediatric literature makes frequent use of the term *genetic potential* in discussions of diagnosis and treatment of disorders of growth. Predictions of how tall a child will be as an adult are usually based on the average of parental height plus 2.5 inches for boys or minus 2.5 inches for girls. We can think of GH as the facilitator of expression of genetic potential for growth rather than as the primary determinant. The entire range over which GH can influence adult stature is only about ≈30% of genetic potential. A person destined by his genetic makeup to attain a final height of 6 feet will attain a height of about 4 feet even in the absence of GH and is unlikely to exceed 8 feet in height even with massive overproduction of GH from birth. We do not understand what determines the genetic potential for growth; but it is clear that, although both arise from a single cell, a hypopituitary elephant is enormously larger than a giant mouse. Within the same species, something other than aberrations in GH secretion accounts for the large differences in size of miniature and standard poodles or Chihuahuas and Great Danes.

Synthesis, Secretion, and Metabolism

Although the anterior pituitary gland produces at least six hormones, more than one-third of its cells synthesize and secrete GH. Of the GH produced by somatotropes, 90% is comprised of 191 amino acids and has a molecular weight of about 22,000; the remaining 10% (called 20K GH) has a molecular weight of 20,000 and lacks the 15-amino-acid sequence corresponding to residues 32 to 46. Both forms are products of the same gene and result from alternate splicing of the RNA transcript. Both forms of hormone are secreted and have

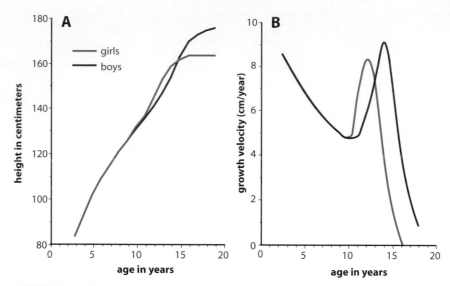

FIGURE 2   (A) Typical growth curves for boys and girls. Note that growth is not strictly linear and that it proceeds at the same rate in juvenile boys and girls. At puberty, which begins earlier in girls than in boys, a spurt in growth immediately precedes growth arrest. (B) Nonlinearity of growth is more clearly evident when plotted as the changes in growth rate over time. Note that growth, which is vary rapid in the newborn, slows during the juvenile period and accelerates at puberty.

similar growth-promoting activity, although metabolic effects of the 20K form are reduced.

About half of the GH in blood circulates bound to a protein that has the same amino acid sequence as the extracellular domain of the GH receptor (see below). In fact, the plasma GH binding protein is a product of the same gene that encodes the GH receptor and originates by proteolytic cleavage of the receptor at the outer surface of target cells. It is thought that the binding protein serves as a reservoir of GH that buffers changes in free hormone concentration and prolongs the half-life of the hormone. The monomeric form of free GH can readily cross capillary membranes, but bound hormone presumably cannot. The half-life of GH in blood is about 20 min. GH that crosses the glomerular membrane is reabsorbed and destroyed in the kidney, which is the major site of GH degradation. Less than 0.01% of the hormone secreted each day reaches the urine in recognizable form. GH is also degraded in its various target cells following uptake by receptor mediated endocytosis.

Mode of Action

Like other peptide and protein hormones, GH binds to its receptor on the surface of target cells. The GH receptor is a glycoprotein that has a single membrane-spanning region and a relatively long intracellular tail; it neither has catalytic activity nor interacts with G proteins. The GH receptor binds to a cytosolic enzyme called *Janus kinase 2* (JAK-2), which catalyzes the phosphorylation of the receptor and other proteins on tyrosine residues (see Chapter 2). Growth hormone activates a signaling cascade by binding sequentially to two GH receptor molecules to form a receptor dimer that sandwiches the hormone between the two receptor molecules. Such dimerization of receptors is also seen for other hormone and cytokine receptors of the super-family to which the GH receptor belongs. Dimerization of receptors brings the bound JAK-2 enzymes into favorable alignment to promote tyrosine phosphorylation and activation of their catalytic sites. In addition, dimerization may also recruit JAK-2 molecules to unoccupied binding sites on the receptors. Tyrosine phosphorylation provides docking sites for other proteins and facilitates their phosphorylation. One group of target proteins, called STATs (signal transducers and activators of transcription), migrate to the nucleus, and activate gene transcription (see Chapter 2). Another target group, the mitogen-activated protein (MAP) kinases, are also thought to have a role in promoting gene transcription. Activation of the GH receptor also results in an influx of extracellular calcium through voltage-regulated channels which may further promote transcription of target genes. All in all, GH produces its effects in various cells by stimulating the transcription of specific genes.

Effects on Skeletal Growth

The ultimate height attained by an individual is determined by the length of the skeleton and, in

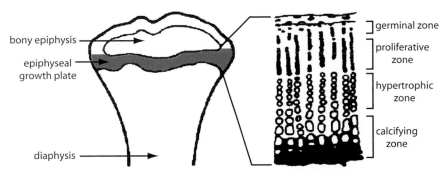

FIGURE 3   Schematic representation of the tibial epiphyseal growth plate. (From Ohlsson C, Isgaard J, Törnell J, Nilsson A, Isaksson OGP, Lindahl A., *Acta Paediatr* 1993; 381(suppl.):33–40. With permission.)

particular, the vertebral column and long bones of the legs. Growth of these bones occurs by a process called *endochondral ossification*, in which proliferating cartilage is replaced by bone. The ends of long bones are called *epiphyses* and arise from ossification centers that are separate from those responsible for ossification of the diaphysis, or shaft. In the growing individual, the epiphyses are separated from the diaphysis by cartilaginous regions called *epiphyseal plates*, in which continuous production of chondrocytes provides the impetus for diaphyseal elongation. Chondrocytes in epiphyseal growth plates are arranged in orderly columns in parallel with the long axis of the bone (Fig. 3). Frequent division of small, flattened cells in the germinal zone at the distal end of the growth plate provides for continual elongation of columns of chondrocytes. As they grow and mature, chondrocytes produce the mucopolysaccharides and collagen that constitute the cartilage matrix. Cartilage cells hypertrophy, become heavily vacuolated, and degenerate as the surrounding matrix becomes calcified. Ingrowth of blood vessels and migration of osteoblast progenitors from the marrow result in replacement of calcified cartilage with true bone. Proliferation of chondrocytes at the epiphyseal border of the growth plate is balanced by cellular degeneration at the diaphyseal end, so in the normally growing individual the thickness of the growth plate remains constant as the epiphyses are pushed further and further outward by the elongating shaft of bone. Eventually, progenitors of chondrocytes are either exhausted or lose their capacity to divide. As remaining chondrocytes go through their cycle of growth and degeneration, the epiphyseal plate becomes progressively narrower and is ultimately obliterated when diaphyseal bone fuses with the bony epiphyses. At this time, the epiphyseal plates are said to be closed, and the capacity for further growth is lost. In the absence of GH there is severe atrophy of the epiphyseal plates, which become narrow as proliferation of cartilage progenitor cells slows markedly. Conversely, after GH is given to a hypopituitary subject, resumption of

cellular proliferation causes columns of chondrocytes to elongate and epiphyseal plates to widen.

Growth of bone requires that diameter as well as length increase. Thickening of long bones is accomplished by proliferation of osteoblastic progenitors from the connective tissue sheath (*periosteum*) that surrounds the diaphysis. As it grows, bone is also subject to continual reabsorption and reorganization, with the incorporation of new cells that originate in both the periosteal and endosteal regions. Remodeling, which is an intrinsic property of skeletal growth, is accompanied by destruction and replacement of calcified matrix, as described in Chapter 43. Treatment with GH often produces a transient increase in urinary excretion of calcium and phosphorus, reflecting bone remodeling. Increased urinary hydroxyproline derives from breakdown and replacement of collagen in bone matrix.

The Somatomedin Hypothesis

The epiphyseal growth plates are obviously stimulated after GH is given to hypophysectomized animals, but little or no stimulation of cell division, protein synthesis, or incorporation of radioactive sulfur into mucopolysaccharides of cartilage matrix was observed in early experiments in which epiphyseal cartilage was taken from hypophysectomized rats and incubated with GH. In contrast, when cartilage taken from the same rats was incubated with blood plasma from hypophysectomized rats that had been treated with GH, there was a sharp increase in matrix formation, protein synthesis, and DNA synthesis. Blood plasma from normal rats produced similar effects, but plasma from hypophysectomized rats that had not been given GH had little effect. These experiments gave rise to the hypothesis that GH may not act directly to promote growth but, instead, stimulates the liver to produce an intermediate, bloodborne substance that activates chondrogenesis and perhaps other GH-dependent growth processes in other tissues. This substance was later named *somatomedin* (somatotropin mediator) and, upon subsequent purification, was found to consist of two closely related substances that also produce the insulin-like activity that persists in plasma after all the authentic insulin is removed by immunoprecipitation. These substances are now called *insulin-like growth factors*, or IGF-I and IGF-II. Of the two, IGF-I appears to be the more important mediator of the actions of GH and has been studied more thoroughly. Although some aspects of the original somatomedin hypothesis have been discarded (see below), the crucial role of IGF-I as an intermediary in the growth promoting action of GH is now firmly established.

In general, plasma concentrations of IGF-I reflect the availability of GH or the rate of growth. They are higher than normal in blood of persons suffering from acromegaly and are very low in GH-deficient individuals. Children whose growth rate is higher than average have higher than average concentrations of IGF-I, while children at the lower extreme of normal growth have lower values. When GH is injected into GH-deficient patients or experimental animals, IGF-I concentrations increase after a delay of about 6 to 8 hours and remain elevated for more than a day. Children or adults who are resistant to GH because of a receptor defect have low concentrations of IGF-I in their blood despite high concentrations of GH. Growth of these children is restored to nearly normal rates following daily administration of IGF-I (Fig. 4). Disruption of the IGF-I gene in mice causes severe growth retardation despite high concentrations of GH in their blood. Daily treatment with even large doses of GH does not accelerate their growth. Similarly, a child with a homozygous deletion of the IGF-I gene suffered severe pre- and postnatal growth retardation that was partially corrected by daily treatment with IGF-I.

While overwhelming evidence indicates that IGF-I stimulates cell division in cartilage and many other tissues and accounts for much and perhaps all of the growth-promoting action of GH, the somatomedin hypothesis as originally formulated is inconsistent with some experimental findings. Production of IGF-I is not limited to the liver and may be increased by GH in many tissues, including cells in the epiphyseal growth plate. Direct infusion of small amounts of GH into epiphyseal cartilage of the proximal tibia in one leg of hypophysectomized rats was found to stimulate tibial growth of that limb, but not of the contralateral limb. Only a direct action of GH on osteogenesis can explain such localized stimulation of growth, because IGF-I that arises in the liver is equally available in the blood supply to both hind limbs. It is now apparent that GH stimulates prechondrocytes and other cells in the epiphyseal plates to synthesize and secrete IGFs that act locally in an autocrine or paracrine manner to stimulate cell division, chondrocyte maturation, and bone growth. Evidence to support this conclusion includes findings of receptors for both GH and IGF-I in cells in the epiphyseal plates along with the GH-dependent increase in mRNA for IGFs. Thus, growth of the long bones might be stimulated by IGF-I that reaches them either through the circulation or by diffusion from local sites of production, or some combination of the two. The failure of the original experiments to demonstrate these actions may be attributable to deficiencies in the culture conditions.

A genetic engineering approach was adopted to evaluate the relative importance of locally produced and bloodborne IGF-I. A line of mice was developed in which the IGF-I gene was selectively disrupted only in hepatocytes. Concentrations of IGF-I in the blood of these animals were severely reduced, but their growth and body proportions were no different from those of control animals that produced normal amounts of IGF-I in their livers; however, even though the lengths of their limb bones were normal, the diameters of these bones were smaller than normal. These findings indicate that locally produced IGF-I is sufficient to account for normal growth at the epiphyses, but that IGF-I in the circulation is important for circumferential growth, primarily of cortical bone. In addition, the average concentration of GH in the blood of these genetically altered mice was considerably increased. This finding is consistent with findings that IGF-I exerts a negative feedback effect on GH secretion. The current view of the

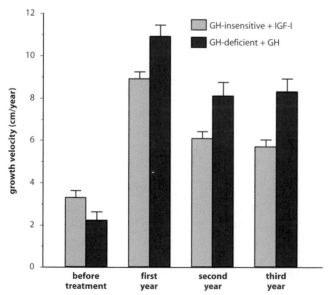

FIGURE 4 Insulin-like growth factor-I (IGF-I) treatment of children with growth hormone (GH) receptor deficiency compared to GH treatment of children with GH deficiency. (Plotted from data of Guevara-Aguirre J, Rosenbloom AL, Vasconez O, Martinez V, Gargosky S, Allen L, Rosenfeld R, *J Clin Endocrinol Metab* 1997; 82:629–633.)

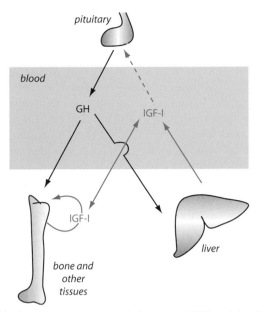

FIGURE 5 The roles of growth hormone (GH) and insulin-like growth factor I (IGF-I) in promoting growth. GH stimulates IGF-I production in liver and epiphyseal growth plates. Epiphyseal growth is stimulated primarily by autocrine/paracrine actions of IGF-I. Hepatic production of IGF-I stimulates circumferential growth of bone and acts primarily as a negative feedback regulator of GH secretion. Liver is the principal source of IGF-I in blood, but other GH target organs may also contribute a small amount to the circulating pool. Dashed arrow indicates inhibition.

relationship between GH and IGF-I is summarized in Fig. 5. GH acts directly on both the liver and its peripheral target tissues to promote IGF-I production. Liver is the principal source of IGF in blood, but target tissues also make some contributions. Stimulation of growth at the epiphyses is provided primarily by locally produced IGF-I acting in an autocrine/paracrine manner, while IGF-I produced in liver contributes to periosteal growth. The additional role of bloodborne IGF is regulation of GH secretion.

Properties of the Insulin-Like Growth Factors

Insulin-like growth factor I and II are small unbranched peptides that have molecular masses of about 7500 Da. They are encoded in separate genes and are expressed in a wide variety of cells. Their protein structures are very similar to each other and to proinsulin (see Chapter 41) both in terms of amino acid sequence and in the arrangement of disulfide bonds. The IGFs share about 50% amino acid identity with insulin. In contrast to insulin, however, the region corresponding to the connecting peptide is retained in the mature form of the IGFs which also have a C-terminal extension. Both IGF-I and IGF-II are present in blood throughout life but their concentrations differ at different stages of life.

Two receptors for the IGFs have been identified. The IGF-I receptor, which binds IGF-I with greater affinity than IGF-II, is remarkably similar to the insulin receptor and signals in a similar manner. The IGF-II receptor is structurally unrelated to the IGF-I receptor and binds IGF-II with a very much higher affinity than IGF-I. It consists of a single membrane-spanning protein with a short cytosolic domain that is thought to lack signaling capabilities. Curiously, the IGF-II receptor is identical to the mannose-6-phosphate receptor that binds mannose-6-phosphate groups on newly synthesized lysosomal enzymes and transfers them from the trans-Golgi vesicles to the endosomes and then to lysosomes. It may also transfer mannose-6-phosphate-containing glycoproteins from the extracellular fluid to the lysosomes by an endocytotic process. The IGF-II receptor likely plays an important role in clearing IGF-II from extracellular fluids.

The IGFs circulate in blood tightly bound to IGF binding proteins (IGFBPs). Six different IGFBPs, each the product of a separate gene, are found in mammalian plasma and extracellular fluids. Their affinities for both IGF-I and IGF-II are considerably higher than the affinity of the IGF-I receptor for either IGF-I or IGF-II. The combined binding capacity of all the IGFBPs in plasma is about twice the concentration of the IGFs, so that less than 1% of the IGFs are free. IGFBP-3, the synthesis of which is stimulated by GH, IGF-I, and insulin, is the most abundant form and is complexed with most of the IGF-I and IGF-II in plasma. The IGFBP-3 and its cargo of IGFs form a large 150-kDa ternary complex with a third protein, the so-called *acid-labile subunit* (ALS), the synthesis of which is also stimulated by GH. Consequently, the concentrations of both proteins are quite low in the blood of GH-deficient subjects and increase upon treatment with GH. The remainder of the IGFs in plasma are distributed among the other IGFBPs that do not bind to ALS and hence form complexes of about 50-kDa which are small enough to escape across the capillary endothelium. Of these, IGFBP-2 is the most important quantitatively. Its concentration in blood is increased in plasma of GH-deficient patients and is decreased by GH but rises dramatically after administration of IGF-I.

Normally, the binding capacity of IGFBP-3 is saturated, while the other IGFBPs have free binding sites. Consequently, the IGFs do not readily escape from the vascular compartment and have a half-life in blood of about 15 hours. Proteolytic "clipping" of IGFBP-3 by proteases present in plasma lowers its binding affinity and releases IGF-I, which may then form a lower molecular weight complex with other IGFBPs, which may then deliver it to the extracellular fluid. The major functions of the IGF binding proteins in blood are to provide a plasma reservoir of IGF-I and IGF-II, to slow their degradation, and to regulate their bioavailability.

The IGFBPs are synthesized locally in conjunction with IGF in a wide variety of cells and are widely distributed in extracellular fluid. Their biology is complex and not completely understood. It may be recalled that the IGFs mediate localized growth in response to a variety of signals in addition to GH. A wide variety of cells both produce and respond to IGF-I, which is a small and readily diffusible molecule. The IGFBPs may provide a means of restricting cell growth to the extent and location dictated by physiological demand. Because their affinity for both IGFs is so much greater than the affinity of the IGF-I receptor, the IGFBPs can successfully compete with the IGF-I receptor for binding free IGF and thus restrict their bioavailability. Conversely, the IGFBPs have also been found to enhance the actions of IGF-I. Some of the IGFBPs bind to extracellular matrices, where they may provide a local reservoir of growth factors that might be released by proteolytic modification of the binding or matrix proteins. Binding to the cell surface lowers the affinity of some of the binding proteins for the growth factor and thus provides a means of targeted delivery of IGF-I to receptive cells. In this way, the IGFBPs provide a means of specific tissue or cellular localization of the IGFs. Some evidence also suggests that IGFBPs may produce biological effects that are independent of the IGFs.

Effects of GH/IGF-I on Body Composition

Growth-hormone-deficient animals and human subjects have a relatively high proportion of fat, compared to water and protein, in their bodies. Treatment with GH changes the proportion of these bodily constituents to resemble the normal juvenile distribution. Body protein stores increase, particularly in muscle, accompanied by a relative decrease in fat. Despite their relatively higher fat content, subjects who are congenitally deficient in GH or are unresponsive to it actually have fewer total adipocytes than normal individuals. Their adiposity is due to an increase in the amount of fat stored in each cell. GH increases the proliferation of fat cell precursors through autocrine stimulation by IGF-I secretion by adipocyte precursor cells and can restore normal cellularity. Curiously, however, GH also restrains the differentiation of fat cell precursors into mature adipocytes. The overall decrease in body fat produced by GH results from decreased deposition of fat and accelerated fat mobilization and increased reliance of fat for energy production (see Chapter 42).

Most internal organs grow in proportion to body size, except liver and spleen, which may be disproportionately

enlarged by prolonged treatment with GH. The heart may also be enlarged in acromegalic subjects in part from hypertension, which is frequently seen in these individuals, and in part from stimulation of cardiac myocyte growth by GH-induced IGF-I production. Conversely, GH deficiency beginning in childhood is associated with decreased myocardial mass due to decreased thickness of the ventricular walls. Treatment of these individuals leads to increased myocardial mass and performance. Skin and the underlying connective tissue also increase in mass, but GH does not appear to influence growth of the thyroid, gonads, or reproductive organs.

Changes in body composition and organ growth have been monitored by studying changes in the biochemical balance of body constituents (Fig. 6). When human subjects or experimental animals are given GH repeatedly for several days, there is net retention of nitrogen, reflecting increased protein synthesis. Urinary nitrogen is decreased, as is the concentration of urea in blood. Net synthesis of protein is increased without an accompanying change in the net rates of protein degradation. Increased retention of potassium reflects the increase in intracellular water that results from increased size and number of cells. An increase in sodium retention and the consequent expansion of extracellular volume are characteristic of GH replacement and may result from activation of sodium channels in the distal portions of the nephron. Increased phosphate retention reflects expansion of the cellular and skeletal mass and is brought about in part by activation of sodium phosphate cotransporters in the proximal tubules and activation of the 1α-hydroxylase that catalyzes production of calcitriol (Chapter 43).

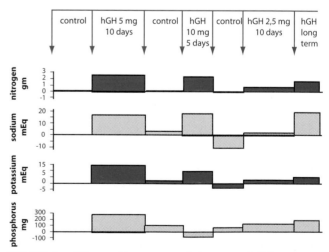

FIGURE 6   Effects of human growth hormone (hGH) treatment on nitrogen, sodium, potassium, and phosphorus balances in an 11.5-year-old girl with pituitary dwarfism. Changes above the control baseline represent retention of the substance; below the line, they represent loss. (From Hutchings JJ, Escamilla RF, Deamer WC and Li, CH, *J Clin Endocrinol Metab* 1959; 19:759–764. With permission.)

Patterns of Growth Hormone Secretion

Growth is a slow, continuous process that takes place over more than a decade. It might be expected, therefore, that concentrations of GH in blood would be fairly static. In contrast to such expectations, however, frequent measurements of GH concentrations in blood plasma throughout the day reveal wide fluctuations indicative of multiple episodes of secretion rate. Because metabolism of GH is thought to be invariant, changes in plasma concentration imply changes in secretion. In male rats, GH is secreted in regular pulses every 3.0 to 3.5 hours in what has been called an *ultradian rhythm*. In humans, GH secretion is also pulsatile, but the pattern of changes in blood concentrations is usually less obvious than in rats. Frequent bursts of secretion occur throughout the day, with the largest being associated with the early hours of sleep (Fig. 7). In addition, stressful changes in the internal and external environment can produce brief episodes of hormone secretion. Little information or diagnostic insight, therefore, can be obtained from a single random measurement of the GH concentration in blood. Because secretory episodes last only a short while, multiple, frequent measurements are necessary to evaluate functional status or to relate GH secretion to physiological events. Alternatively, it is possible to withdraw small amounts of blood continuously over the course of a day and, by measuring GH in the pooled sample, to obtain a 24-hr integrated concentration of GH in blood.

The possible physiological significance of intermittent as compared with constant secretion of GH has received much attention experimentally. Pulsatile administration of GH to hypophysectomized rats is more effective in stimulating growth than continuous infusion of the same daily dose. However, similar findings have not been made in human subjects, whose rate of growth, like that of experimental animals, can be restored to normal or near normal with single daily or every other day injections of GH. While expression of some hepatic genes appears to be sensitive to the pattern of changes in plasma GH concentrations in rodents, there is neither evidence for comparable effects in humans nor an obvious relationship of the affected genes to growth of rodents. In normal human adults, the same amount of GH given either as a constant infusion or in eight equally spaced brief infusions over 24 hours increased IGF-I and IGFBP-3 to the same extent.

Effects of Age

Using the continuous sampling method, it was found that GH secretion, though most active during the adolescent growth spurt, persists throughout life long after the epiphyses have fused and growth has stopped (Fig. 8). In mid-adolescence the pituitary secretes between 1 and 2 mg of GH per day. Between ages 20 and 40 years, the daily rate of secretion gradually decreases in both men and women, but remarkably, even during middle age, the pituitary continues to secrete about 0.1 mg of GH every day. Low rates of GH secretion in the elderly may be related to loss of lean body mass in later life. Changes in GH secretion with age primarily reflect changes in the magnitude of secretory pulses.

Regulation of Growth Hormone Secretion

In addition to spontaneous pulses, secretory episodes are induced by such metabolic signals as a rapid fall in blood glucose concentration or an increase in certain amino acids, particularly arginine and leucine. The

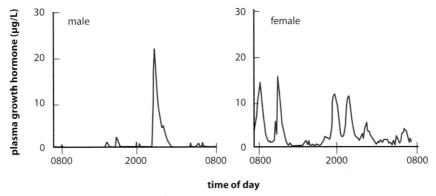

FIGURE 7 Growth hormone concentrations in blood sampled at 10-min intervals over a 24-hr period in a normal adult male and a normal adult female subject. The large pulse in the male coincides with the early hours of sleep. Note that the pulses of secretion are more frequent and of greater amplitude in the female. (From Asplin CM, Faria CA, Carlsen EC *et al.*, *J Clin Endocrinol Metab* 1989; 69:239–245. With permission.)

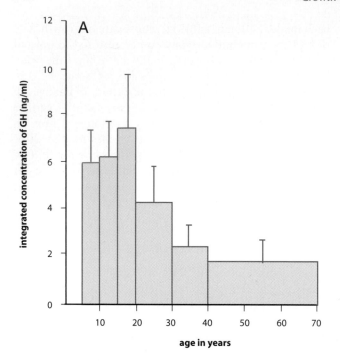

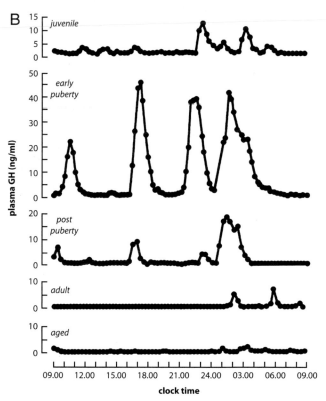

FIGURE 8 (A) Relation between the integrated concentration of growth hormone and age in 173 normal male and female subjects. (From Zadik Z, Chalew SA, McCarter RJ, Jr, Meistas M, Kowarski AA, *J Clin Endocrinol Metab* 1985; 60:513–516. With permission.) (B) Changing patterns of GH secretion with age. (Modified from Robinson ICAF, Hindmarsh PC. In: Kostyo JL, Ed., *Handbook of physiology*, Section 7: *The endocrine system*, Vol. V: *Hormonal Control of Growth*, New York: Oxford University Press, 1999, pp. 329–396.)

physiological significance of these changes in GH secretion is not understood, but provocative tests using these signals are helpful for judging the competence of the GH secretory apparatus (Fig. 9). Traumatic and psychogenic stresses are also powerful inducers of GH secretion in humans and monkeys, but whether increased secretion of GH is beneficial for coping with stress is not established and is not universally seen in mammals. In rats, for example, GH secretion is inhibited by the same signals that increase it in humans. However, regardless of their significance, these observations indicate that GH secretion is under minute-to-minute control by the nervous system. That control is expressed through the hypothalamo–hypophysial portal circulation, which delivers two hypothalamic neuropeptides to the somatotropes: *GH-releasing hormone* (GHRH) and *somatostatin*. It is possible that a third hormone, *ghrelin* (see Chapter 38), also plays a role in this regard, but data to support this premise are not yet in hand. GHRH provides the primary drive for GH synthesis and secretion. In its absence, or when a lesion interrupts hypophysial portal blood flow, secretion of GH ceases. Somatostatin reduces or blocks the response of the pituitary to GHRH on GH secretion but has little or no influence on GH synthesis. Somatostatin and GHRH also exert reciprocal inhibitory influences on GHRH and somatostatin neurons (Fig. 10).

In addition to neuroendocrine mechanisms that adjust secretion in response to changes in the internal or external environment, secretion of GH is also under negative feedback control. As for other negative feedback systems, products of GH action, principally IGF-I, act as inhibitory signals (Fig. 11). IGF-I acts primarily at the pituitary level, where it decreases GH secretion in response to GHRH. IGF-I may also increase somatostatin secretion. Increased concentrations of FFA or glucose, which are also related to GH action (Chapter 42), also exert inhibitory effects, probably through increased somatostatin secretion, but fasting, which also leads to increased FFA, appears to inhibit somatostatin secretion. Growth hormone exerts a short loop feedback effect by inhibiting the secretion of GHRH and increasing the secretion of somatostatin.

Negative feedback control sets the overall level of GH secretion by regulating the amounts of GH secreted in each pulse. The phenomenon of pulsatility and the circadian variation that increases the magnitude of the secretory pulses at night are entrained by neural mechanisms. Pulsatility appears to be the result of reciprocal intermittent secretion of both GHRH and somatostatin. It appears that bursts of GHRH secretion are timed to coincide with interruptions in somatostatin secretion. Experimental evidence obtained in rodents indicates that GHRH-secreting neurons in the arcuate nuclei

Clinical Note

Causes of Growth Hormone Deficiency
Defective hypothalamic production or secretion of GHRH may be a more common cause of GH deficiency than defects in the pituitary gland. GH concentrations in plasma are restored to normal in many GH-deficient individuals after treatment with GHRH, suggesting that their somatotropes are competent but not adequately stimulated. Defects in somatostatin synthesis or secretion are not known to be responsible for disease states, but long-acting analogs of somatostatin can be used to decrease GH secretion for treatment of acromegaly.

communicate with somatostatin-secreting neurons in the periventricular nuclei, either directly or through interneurons; conversely, somatostatinergic neurons communicate with GHRH neurons. However, understanding of how reciprocal changes in secretion of these two neurohormones are brought about is still incomplete. GHRH neurons express receptors for ghrelin, but the role of ghrelin in the regulation of GH secretion remains to be uncovered.

Actions of GHRH, Somatostatin, and IGF-I on the Somatotrope

Secretion of GH is under the direct control of at least three hormones: GHRH, somatostatin, and IGF-I. Receptors for all three hormones are present on the surface of somatotropes. The complex interplay of these hormones is illustrated in Fig. 12. Receptors for GHRH and somatostatin are coupled to several G proteins and express their antagonistic effects on GH secretion in part through their opposing influences on cAMP production, cAMP action, and cytosolic calcium concentrations. GHRH activates adenylyl cyclase through a typical G_s-linked mechanism (see Chapter 2). Cyclic AMP activates protein kinase A, some of which migrates to the nucleus and phosphorylates the cAMP response element binding (CREB) protein. Activation of CREB

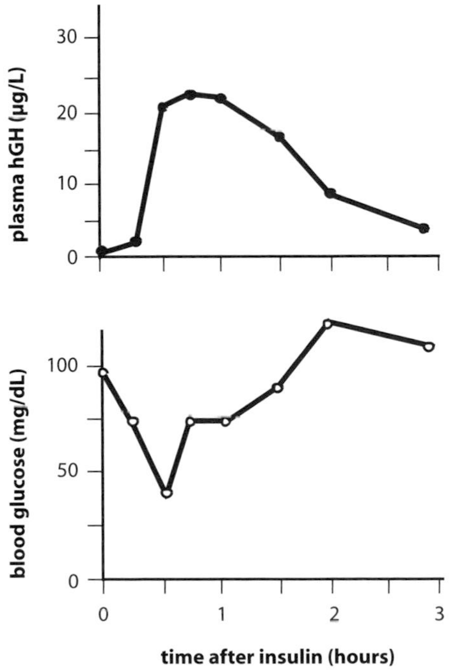

FIGURE 9 Acute changes in plasma growth hormone concentration (upper panel) in response to insulin-induced hypoglycemia (lower panel). (From Roth J, Glick SM, Yalow RS, and Berson S, *Science* 1963; 140:987–989. With permission.)

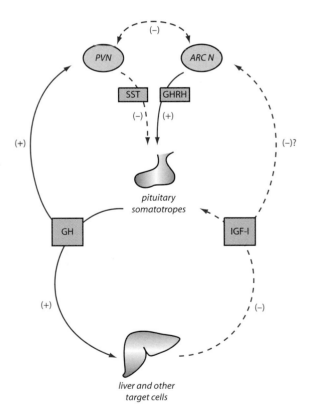

FIGURE 10 Regulation of growth hormone (GH) secretion. PVN, periventricular nuclei; ARCN, arcuate nuclei; SST, somatostatin; GHRH, growth hormone releasing hormone; IGF-I, insulin-like growth factor-I; +, stimulation; –, inhibition.

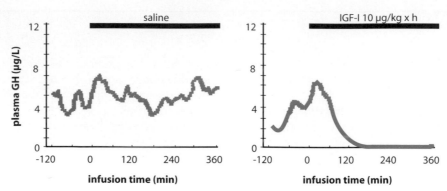

FIGURE 11   Effects of insulin-like growth factor I (IGF-I) on growth hormone (GH) secretion in normal fasted men. Values shown are averages for the same 10 men given infusions of either physiological saline (control) or IGF-I for the periods indicated. *Note:* IGF-I completely blocked GH secretion after a lag period of 1 hour. (Redrawn from Hartman ML, Clayton PE, Johnson ML, Celniker A, Perlman AJ, Alberti KG, Thorner MO, *J Clin Invest* 1993; 91:2453–2462.)

promotes the expression of the transcription factor, Pit 1, which in turn increases transcription of the GH gene and the GHRH receptor gene. In addition, cAMP-dependent phosphorylation of voltage-sensitive calcium channels is thought to lower their threshold and increase their likelihood of opening. Voltage-sensitive calcium channels are also activated by a G-protein-dependent mechanism that depolarizes the somatotrope membrane by activating sodium channels and inhibiting potassium channels. The resulting increase in cytosolic calcium

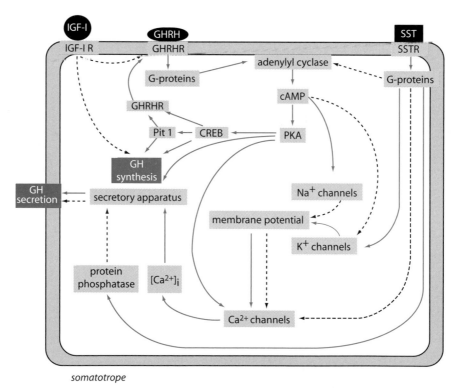

FIGURE 12   Effects of growth-hormone-releasing hormone (GHRH), insulin-like growth factor I (IGF-I), and somatostatin (SST) on the somatotrope. IGF-I R, GHRHR, and SSTR are receptors for IGF-I, GHRH, and SST, respectively; cAMP, cyclic adenosine monophosphate; CREB, cAMP response element binding protein; PKA, protein kinase A; PIT-1, pituitary-specific transcription factor; $[Ca^{2+}]_i$, intracellular free calcium concentration. Solid blue arrows indicate increase or stimulation; dashed black arrows indicate decrease or inhibition. See text for discussion.

concentration ($[Ca^{2+}]_i$) triggers exocytosis of GH. Increased $[Ca^{2+}]_i$ also limits the secretory event by inhibiting voltage-sensitive calcium channels and restores membrane polarity by activating potassium channels. Somatostatin acts through the inhibitory guanine nucleotide binding protein (G_i) to antagonize activation of adenylyl cyclase. Somatostatin receptors also inhibit calcium channels and activate potassium channels through a G-protein-mediated mechanism. Activation of potassium channels hyperpolarizes the plasma membrane and thereby prevents activation of voltage-sensitive calcium channels. Somatostatin also activates a protein phosphatase through a G-protein-dependent mechanism and thereby antagonizes activation of the secretory apparatus by protein kinase A.

The negative feedback effects of IGF-I are slower in onset than the G-protein-mediated effects of GHRH and somatostatin and require tyrosine-phosphorylation-initiated changes in gene expression that downregulate GHRH receptors and GH synthesis. Somatotropes also express G-protein-coupled receptors for ghrelin (Chapter 38). In cultured cells, the activated ghrelin receptor signals through the IP$_3$-DAG second messenger system. Release of calcium from intracellular stores in response to IP$_3$ and DAG-dependent activation of protein kinase C complement the actions of GHRH and enhance GH secretion. Because the physiology of ghrelin remains to be elaborated, it not included in Fig. 12.

THYROID HORMONES

As already mentioned in Chapter 39, growth is stunted in children suffering from unremediated deficiency of thyroid hormones. Treatment of hypothyroid children with thyroid hormone results in rapid catch-up growth rates, which are accompanied by accelerated maturation of bone. Conversely, hyperthyroidism in childhood increases the rate of growth, but, because of early closure of the epiphyses, the maximum height attained is not increased. Thyroidectomy of juvenile experimental animals produces nearly as drastic an inhibition of growth as hypophysectomy, and restoration of normal amounts of triiodothyronine (T3) or thyroxine (T4) promptly reinitiates growth. Young mice grow somewhat faster than normal after treatment with thyroxine, but adult size, although attained earlier, is no greater than normal.

The effects of thyroid hormones on growth are intimately entwined with GH. T3 and T4 have little if any growth-promoting effect in the absence of GH. Plasma concentrations of both GH and IGF-I are reduced in hypothyroid children and adults and restored by treatment with thyroid hormone (Fig. 13).

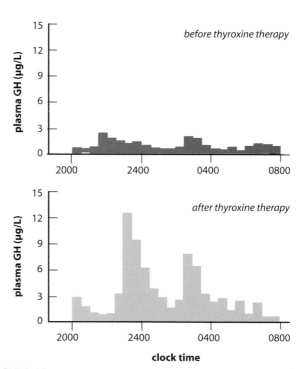

FIGURE 13 Nocturnal growth hormone secretion in a hypothyroid boy in the early stages of puberty before and 2 months after treatment with thyroxine. Each bar represents the average plasma GH concentration during a 30-min period of continuous withdrawal of blood. Plasma IGF-I was increased more than fourfold during treatment. (From Chernausek SD, Turner R. *J. Pediatr* 1989; 114:968–972. With permission.)

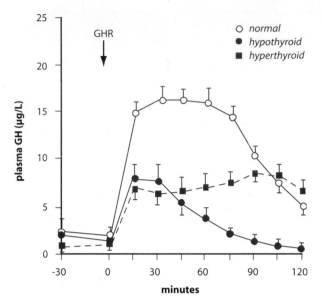

FIGURE 14 . Decreased growth hormone (GH) secretion in response to a test dose of growth-hormone-releasing hormone (GHRH) in hypothyroid and hyperthyroids individuals. Data shown represent average responses from 30 normal, 25 hypothyroid, and 38 hyperthyroid adult patients. (From Valcavi R, Zini M, Portioli M, *J Endocrinol Invest* 1992; 15:313–330. With permission.)

This decrease is due to decreased amplitude of secretory pulses and possibly also to a decrease in frequency consistent with impairments at both the hypothalamic and pituitary levels. Insulin-induced hypoglycemia and other stimuli for GH secretion produce abnormally small increases in the concentration of GH in plasma of hypothyroid subjects. Such blunted responses to provocative signals probably reflect decreased sensitivity to GHRH as well as depletion of GH stores. Curiously, GH secretion in response to a test dose of GHRH is decreased when there is either a deficiency or an excess of thyroid hormone (Fig. 14).

Dependence of Growth Hormone Synthesis and Secretion on T3

The promoter region of the rodent GH gene contains a thyroid hormone response element, and its transcriptional activity is enhanced by T3. Furthermore, T3 increases the stability of the GH messenger RNA transcripts. GH synthesis comes to an almost complete halt and the somatotropes become severely depleted of GH only a few days after thyroidectomy. The human GH gene lacks the thyroid hormone response element, and its transcription is not directly activated by T3. However, thyroid hormones affect synthesis of human GH indirectly via their effects on the expression of the

GHRH receptor. Blunted responses to GHRH in hypothyroid children and adults probably result from decreased expression of GHRH receptors by somatotropes. Thyroidectomy also decreases the abundance of GHRH receptors in rodent somatotropes, and hormone replacement restores both the number of receptors and receptor mRNA.

Importance of T3 for Expression of Growth Hormone Actions

Failure of growth in thyroid-deficient individuals is largely due to a deficiency of GH, which may be compounded by a decrease in sensitivity to GH. Treatment of thyroidectomized animals with GH alone can reinitiate growth, but even large amounts cannot sustain a normal rate of growth unless some thyroid hormone is also given. In rats that were both hypophysectomized and thyroidectomized, T4 decreased the amount of GH needed to stimulate growth (increased sensitivity) and exaggerated the magnitude of the response (increased efficacy). Thyroid hormones increase expression of GH receptors in rodent tissues. T3 and T4 also potentiate effects of GH on the growth of long bones and increase its effects on protein synthesis in muscle and liver. IGF-I concentrations are reduced in the blood of hypothyroid individuals partly because of decreased circulating GH and partly because of decreased hepatic responsiveness to GH. In addition, tissues isolated from thyroidectomized animals are less responsive to IGF-I.

INSULIN

Although neither GH nor T4 appears to be an important determinant of fetal growth, insulin may act as a growth-promoting hormone during the fetal period. Infants born of diabetic mothers are often larger than normal, especially when the diabetes is poorly controlled. Because glucose readily crosses the placenta, high concentrations of glucose in maternal blood increase fetal blood glucose and stimulate the fetal pancreas to secrete insulin. In the rare cases of congenital deficiency of insulin that have been reported, fetal size is below normal. Structurally, insulin is closely related to IGF-I and IGF-II and, when present in adequate concentrations, can react with IGF-I receptors, which are closely related structurally to the insulin receptor. We do not know if the effects of insulin on fetal growth are mediated by the insulin receptor or IGF-I receptors.

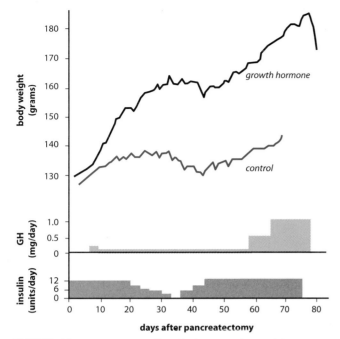

FIGURE 15 Requirement of insulin for normal growth in response to GH. All rats were pancreatectomized 3 to 7 weeks before the experiment was begun. Each rat was fed 7 g of food per day. The treated group was injected with the indicated amounts of growth hormone daily. Note the failure to respond to GH in the period between 20 and 44 days, coincident with the decrease in daily insulin dose, and the resumption of growth when the daily dose of insulin was restored. (From Scow RO, Wagner EM, and Ronov E, *Endocrinology* 1958; 62:593–604. With permission.)

Optimal concentrations of insulin in blood are required to maintain normal growth during postnatal life, but it has been difficult to obtain a precise definition of the role of insulin. Because life cannot be maintained for long without insulin, dramatic effects of sustained deficiency on final adult size are not seen; however, growth is often retarded in insulin-dependent diabetic children, particularly in the months leading up to the diagnosis of full-blown disease. Studies in pancreatectomized rats indicate a direct relation between the effectiveness of GH and the dose of insulin administered. Treatment with GH sustained a rapid rate of growth as long as the daily dose of insulin was adequate, but growth progressively decreased as the dose of insulin was reduced (Fig. 15). Conversely, insulin cannot sustain a normal rate of growth in the absence of GH.

The effects of insulin on postnatal growth cannot be attributed to changes in GH secretion, which, if anything, is increased in human diabetics. Although insulin was sometimes used diagnostically to provoke GH secretion, it is the resulting hypoglycemia, rather than insulin *per se*, that stimulates GH release. Expression of IGF-I mRNA in liver and other tissues is decreased in diabetes and in low insulin states such as fasting or caloric restriction,

consistent with the possibility that insulin is permissive for growth. Insulin stimulates protein synthesis and inhibits protein degradation; in its absence, protein breakdown is severe. Consequently, without insulin, normal responses to GH are not seen; the anabolic effects of GH on body protein either cannot be expressed or are masked by simultaneous, unchecked catabolic processes.

GONADAL HORMONES

Awakening of the gonads at the onset of sexual maturation is accompanied by a dramatic acceleration of growth. The adolescent growth spurt, like other changes at puberty, is attributable to steroid hormones of the gonads and perhaps the adrenals. Because the development of pubic and axillary hair at the onset of puberty is a response to increased secretion of adrenal androgens, this initial stage of sexual maturation is called *adrenarche*. The physiological mechanisms that trigger increased secretion of adrenal androgens and the awakening of the gonadotropic secretory apparatus are poorly understood; they are considered further in Chapter 45. At the same time that gonadal steroids promote linear growth, they accelerate closure of the epiphyses and therefore limit the final height that can be attained. Children who undergo early puberty and hence experience their growth spurt while their contemporaries continue to grow at the slower prepubertal rate are likely to be the tallest and most physically developed in grade school or junior high but among the shortest in their high school graduating class. Deficiency of gonadal hormones, if left untreated, delays epiphyseal closure, and, despite the absence of a pubertal growth spurt, such hypogonadal individuals tend to be tall and have unusually long arms and legs.

In considering the relationship of the gonadal hormones to the pubertal growth spurt it is important to understand that:

- Androgens and estrogens are produced and secreted by both the ovaries and the testes.
- Androgens are precursors of estrogens and are aromatized to estrogens by a reaction catalyzed by P450 aromatase in the gonads and extragonadal tissues.
- Estrogens produce their biological effects at concentrations that are more than 1000 times lower than the concentrations at which androgens produce their effects.

Until recently it was generally accepted that androgens produce the adolescent growth spurt in both sexes. This idea was rooted in the observations that even at the

Clinical Note

In one case report, a man with a homozygous disruption of the α estrogen receptor had normal sexual development but failed to experience an adolescent growth spurt. At age 28, when he was diagnosed, he had attained a height of 6 feet, 8 inches, and his epiphyses had still not closed. In contrast, genetic males with nonfunctional androgen receptors experience a normal pubertal growth spurt and their epiphyses close at the normal time.

relatively low doses used therapeutically in women, administration of estrogens inhibits growth, while administration of androgens stimulates growth. Some "experiments of nature" that have come to light in recent years have challenged this idea and led to the opposite conclusion. Girls with ovarian agenesis (congenital absence of ovaries) have short stature and do not experience an adolescent growth surge. Their growth is increased by treatment with doses of estrogen that are below the threshold needed to cause breast development. In normal girls, the adolescent growth spurt usually occurs before estrogen secretion is sufficient to initiate growth of breasts.

Patients of either sex who have a homozygous disruption of the aromatase gene do not experience a pubertal growth spurt despite supranormal levels of androgens and continue to grow at the juvenile rate well beyond the time of normal epiphyseal fusion unless estrogens are given. Although sexual development cannot occur in girls with this defect, affected boys develop normally. These observations established that estrogen rather than androgen is responsible for both acceleration of growth at puberty and maturation of the epiphyseal plates.

It is noteworthy that estrogen levels increase in the plasma of both boys and girls early in puberty and reach similar concentrations at the onset of the growth spurt. The well-established growth-promoting effect of androgens administered to children whose epiphyses are not yet fused is likely attributable to their conversion to estrogen. Synthetic androgens that are chemically modified in ways that prevent aromatization are ineffective in promoting growth, even though other aspects of androgen activity (Chapter 45) are fully evident. Nevertheless, the maximal rate of growth achieved in adolescence is greater in males than females, and a supportive growth-promoting role for androgens is not ruled out. In addition, androgens stimulate growth of muscle, particularly in the upper body. Androgen secretion during puberty in boys produces a doubling of muscle mass by increasing the size and number of muscle cells. Such growth of muscle can occur in the absence of GH or thyroid hormones and is mediated by the same

androgen receptors that are expressed in other androgen-sensitive tissues (see Chapter 45). Stimulation of muscle growth by androgens is most pronounced in androgen-deficient or hypopituitary subjects, and only small effects, if any, are seen in men with normal testicular function, except perhaps when very large doses of so-called anabolic steroids are used.

Most if not all of the increase in height stimulated by estrogens or androgens at puberty is due to increased secretion of GH (Fig. 16). During the pubertal growth spurt or when androgens are given to prepubertal children, there is an increase in both frequency and

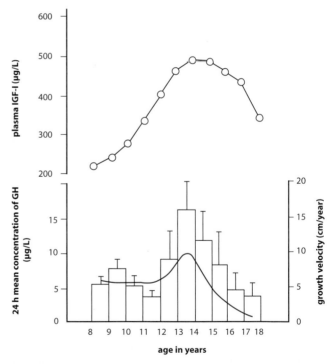

FIGURE 16 Changes in plasma IGF-I and growth hormone concentrations in the peripubertal period in normal boys. Bars in the lower panel indicate 24-hr integrated concentrations of GH. The blue curve is the idealized growth velocity curve for North American boys. (Upper panel from Juul A, Dalgaard P, Blum WF, Bang P, Hall K, Michaelsen KF, Muller J, Skakkebaek NE, *J Clin Endocrinol Metab* 1995; 80:2534–2542; lower panel from Martha PM, Jr, Rogol AD, Veldhuis JD, Kerrigan JR, Goodman DW, Blizzard RM, *J Clin Endocrinol Metab* 1989; 69:563–570. With permission.)

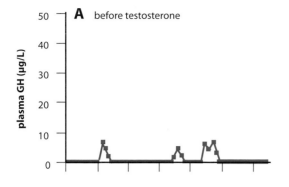

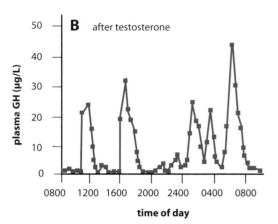

FIGURE 17   Effects of testosterone in a boy with short stature and delayed puberty. (A) before testosterone, (B) during therapy with long acting testosterone. Note the increase in frequency and amplitude of growth hormone secretory episodes in the treated subjects. (From Link K, Blizzard RM, Evans WS, Kaiser DL, Parker MW, Rogol AD, *J Clin Endocrinol Metab* 1986; 62:159–164. With permission.)

amplitude of secretory pulses of GH (Fig. 17), and GHRH concentrations are increased in peripheral blood of boys and girls during puberty. The concentration of IGF-I in blood also increases during the pubertal growth spurt or after androgens are given to prepubertal children. This increase is probably a consequence of increased secretion of GH.

We still do not understand the basis for the either stimulatory or inhibitory effects of estrogen on linear growth. At the same concentrations that inhibit growth, estrogens increase GH secretion, and, as we have seen in Fig. 7, plasma concentrations of GH are higher in women than in men. In addition, the GH secretory apparatus tends to be more sensitive to environmental influences in women than in men, and the circulating concentrations of GH tend to rise more readily in women in response to provocative stimuli. Inhibitory effects on growth appear to result from interference with the action of GH at the level of its target cells. Estrogens act directly on the epiphyseal plates, causing them to lose their capacity to replenish cartilage progenitor cells. Estrogens, which are not catabolic, also antagonize effects of GH on nitrogen retention and minimize the increase in IGF-I in blood of hypophysectomized or hypopituitary individuals treated with GH. Neither estrogens nor androgens affect the ability of IGF-I to act.

GLUCOCORTICOIDS

Normal growth requires secretion of glucocorticoids, the widespread effects of which promote optimal function of a variety of organ systems (see Chapter 40), a sense of health and well-being, and normal appetite. Glucocorticoids are required for synthesis of GH and have complex effects on GH secretion. When given acutely, they may enhance GH gene transcription and increase responsiveness of somatotropes to GHRH. However, GH secretion is reduced by excessive glucocorticoids, probably as a result of increased somatostatin production. Consistent with their catabolic effects in muscle and lymphoid tissues, glucocorticoids also antagonize the actions of GH. Hypophysectomized rats grew less in response to GH when cortisone was given simultaneously (Fig. 18). Glucocorticoids similarly blunt the response to GH administered to hypopituitary children. The cellular mechanisms for this antagonism are not yet understand. IGF-I production may be reduced by treatment with glucocorticoids,

---

Clinical Note

Glucocorticoids and Growth

Children suffering from overproduction of glucocorticoids (Cushing's disease) experience some stunting of their growth. In a recently reported case involving identical twins, the unaffected twin was 21 cm taller at age 15 than her sister, whose disease was untreated until around the time of puberty. Similar impairment of growth is seen in children treated chronically with high doses of glucocorticoids to control asthma or inflammatory disorders.

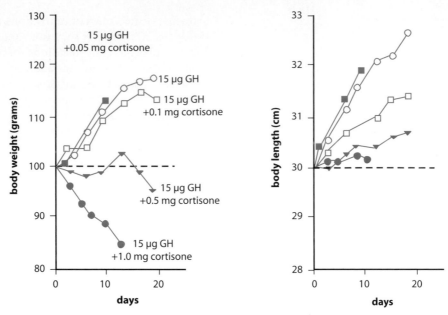

FIGURE 18   Effects of cortisone on growth in hypophysectomized rats given growth hormone replacement. The growth-promoting response to GH, measured as a change in either body weight or length, decreased progressively as the dose of cortisone was increased from 0.1 to 1.0 mg per day. The decrease in body weight seen when 1.0 mg per day of cortisone was given probably resulted from net breakdown of muscle mass (see Chapter 40). (From Soyka LF, Crawford JD, *J Clin Endocrinol Metab* 1965; 25:469–475. With permission.)

but we do not know if this is a cause or an effect of the decreased action of GH.

Suggested Readings

Baumann G. Growth hormone heterogeneity: genes, isohormones, variants and binding proteins. *Endocrine Rev* 1991; 12:424–449.

Clemmons DR. Use of mutagenesis to probe IGF-binding protein structure/function relationships. *Endocrine Rev* 2001; 22:800–817.

Giustina A, Veldhuis JD, Pathophysiology of the neuroregulation of growth hormone secretion in experimental animals and the human. *Endocrine Rev* 1998; 19:717–797.

Jones JI, Clemmons DR. Insulin-like growth factors and their binding proteins: biological actions. *Endocrine Rev* 1995; 16:3–34.

LeRoith DF, Bondy C, Yakar S, Liu JL, Butler A. The somatomedin hypothesis: 2001. *Endocrine Rev* 2001; 22:53–74.

Ohlsson C, Bengtsson B-Å, Isaksson OGP, Andreassen TT, Slootweg MC. Growth hormone and bone. *Endocrine Rev* 1998; 19:55–79

Rosenfeld RG, Rosenbloom AL, Guevara-Aguirre J. Growth hormone (GH) insensitivity due to primary GH receptor deficiency. *Endocrine Rev* 1994; 15:369–390.

Smit L, Meyer DJ, Argetsinger LS, Schwartz J, Carter S.-C. Molecular events in growth hormone–receptor interaction and signaling. In: Kostyo JL, Ed., *Handbook of physiology*, Section 7: *The endocrine system,* Vol. V: *Hormonal control of growth,* New York: Oxford University Press, 1999, pp. 445–480.

Hormonal Control of Reproduction in the Male

H. MAURICE GOODMAN

KEY POINTS

- The testes serve the dual function of producing sperm and hormones. Sperm are produced in the seminiferous tubules. Testosterone is produced in the interstitial cells of Leydig.
- Both *follicle-stimulating hormone* (FSH) and *luteinizing hormone* (LH) are required for normal spermatogenesis.
- Follicle-stimulating hormone acts on Sertoli cells and stimulates growth of the seminiferous tubules and the initiation of spermatogenesis.
- Luteinizing hormone stimulates Leydig cells to secrete testosterone and indirectly stimulates spermatogenesis through the indispensable actions of testosterone on the Sertoli cells.
- Testosterone promotes growth, differentiation, and function of accessory sex organs in the

male genital tract and maintains their normal function. It also promotes development and maintenance of nongenital aspects of the male phenotype, including actions on hair growth, skeletal and muscle growth, and deepening of the voice.
- The sexually indifferent gonad differentiates into a testis under the influence of the product of the sex-determining gene (SRY) on the Y chromosome.
- The early embryo develops tubular structures that can differentiate into the internal genitalia of males (the *wolffian ducts*) or females (*Müllerian ducts*).
- Development of the male phenotype requires the production and secretion of testicular

OVERVIEW

The testes serve the dual function of producing sperm and hormones. The principal testicular hormone is the steroid *testosterone*, which has an intratesticular role in sperm production and an extratesticular role in promoting delivery of sperm to the female genital tract. In this respect, testosterone promotes development and maintenance of accessory sexual structures responsible for nurturing gametes and ejecting them from the body, development of secondary sexual characteristics that make the man attractive to women, and those behavioral characteristics that promote successful procreation. Testicular function is driven by the pituitary through the secretion of two gonadotropic hormones: follicle-stimulating hormone (FSH) and luteinizing hormone (LH). Secretion of these pituitary hormones is controlled by: (1) the central nervous system, through intermittent secretion of the hypothalamic hormone gonadotropin-releasing hormone (GnRH); and (2) the testes, through the secretion of testosterone and inhibin. Testosterone, its potent metabolite 5α-dihydrotestosterone, and an additional testicular secretion called *anti-müllerian hormone*, also function as determinants of sexual differentiation during fetal life.

THE TESTES

The testes are paired ovoid organs located in the scrotal sac outside the body cavity. The extra-abdominal location, coupled with vascular countercurrent heat exchangers and muscular reflexes that retract the testes to the abdomen, permits testicular temperature to be maintained constant at about 2°C below body temperature. For reasons that are not understood, this small reduction in temperature is crucial for normal *spermatogenesis* (sperm production). Failure of the testes to descend into the scrotum results in failure of spermatogenesis, although production of testosterone may be maintained. The two principal functions of the testis—sperm production and steroid hormone synthesis—are carried out in morphologically distinct compartments. Sperm are formed and develop within *seminiferous tubules*, which comprise the bulk of testicular mass. Testosterone is produced by the *interstitial cells of Leydig*, which lie between the seminiferous tubules (Fig. 1). The entire testis is encased in an inelastic fibrous capsule consisting of three layers of dense connective tissue and some smooth muscle.

Blood reaches the testes primarily through paired spermatic arteries and is first cooled by heat exchange with returning venous blood in the *pampiniform plexus*. This complex tangle of blood vessels is formed by the highly tortuous and convoluted artery intermingling with equally tortuous venous branches that converge to form the spermatic vein. This arrangement provides a large surface area for warm arterial blood to transfer heat to cooler venous blood across thin vascular walls. Rewarmed venous blood returns to the systemic circulation primarily through the internal spermatic veins.

Leydig cells are embedded in loose connective tissue that fills the spaces between seminiferous tubules. They are large polyhedral cells with an extensive smooth endoplasmic reticulum characteristic of steroid-secreting cells. Although extensive at birth, Leydig cells virtually disappear after the first 6 months of postnatal life, only to reappear more than a decade later with the onset of puberty.

Seminiferous tubules are highly convoluted loops that range from about 120 to 300 μm in diameter and from 30 to 70 cm in length. They are arranged in lobules bounded by fibrous connective tissue. Each testis has hundreds of such tubules that are connected at both ends to the *rete testis* (Fig. 2). It has been estimated that, if laid end to end, the seminiferous tubules of the human testes would extend more than 250 m. The seminiferous epithelium that lines the tubules consists of three types of cells: *spermatogonia*, which are stem cells; *spermatocytes*, which are in the process of becoming sperm; and *Sertoli cells*, which nurture developing sperm and secrete a variety of products into the blood and the lumens of seminiferous tubules. Seminiferous tubules are surrounded by a thin coating of peritubular epithelial cells, which

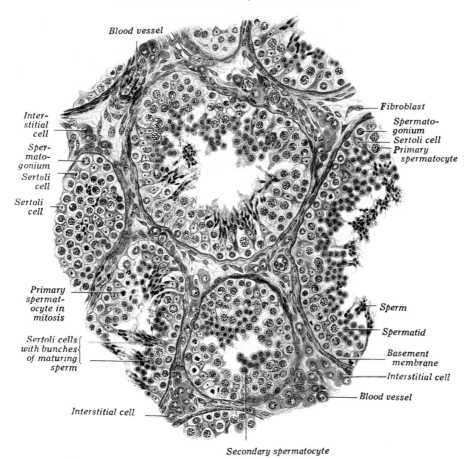

Blood vessel

Inter-
stitial
cell

Sper-
mato-
gonium

Sertoli
cell

Sertoli
cell

Primary
spermat-
ocyte in
mitosis

Sertoli cells
with bunches
of maturing
sperm

Interstitial cell

Fibroblast

Spermato-
gonium

Sertoli cell

Primary
spermatocyte

Sperm

Spermatid

Basement
membrane

Interstitial cell

Blood vessel

Secondary spermatocyte

FIGURE 1   Histological section of human testis. The transected tubules show various stages of spermatogenesis. (From Fawcett DW, *A Textbook of histology*, 11th ed., Philadelphia: W.B. Saunders, 1986, p. 804. With permission.)

in some species are contractile and help propel the non-motile sperm through the tubules toward the rete testis.

Spermatogenesis goes on continuously from puberty to senescence along the entire length of the seminiferous tubules. Though a continuous process, spermatogenesis can be divided into three discrete phases: (1) *mitotic divisions*, which replenish the spermatogonia and provide the cells destined to become mature sperm; (2) *meiotic divisions*, which reduce the chromosome number and produce a cluster of haploid spermatids; and (3) transformation of spermatids into mature sperm (*spermiogenesis*), a process involving the loss of most of the cytoplasm and the development of flagella. These events occur along the length of the seminiferous tubules in a definite temporal and spatial pattern. A *spermatogenic cycle* includes all of the transformations from spermatogonium to spermatozoan and requires about 64 days. As the cycle progresses, germ cells move from the basal portion of the germinal epithelium toward the lumen. Successive cycles begin before the previous one has been completed, so that at any given point along a tubule different stages of the cycle are seen at different depths

of the epithelium (Fig. 3). Spermatogenic cycles are synchronized in adjacent groups of cells, but the cycles are slightly advanced in similar groups of cells located immediately upstream, so that cells at any given stage of the spermatogenic cycle are spaced at regular intervals along the length of the tubules. This complex series of events ensures that mature spermatozoa are produced continuously. About 2 million spermatogonia, each giving rise to 64 sperm cells, begin this process in each testis every day. Hundreds of millions of spermatozoa are thus produced daily throughout six or more decades of reproductive life.

Sertoli cells are remarkable polyfunctional cells whose activities are intimately related to many aspects of the formation and maturation of spermatozoa. They extend through the entire thickness of the germinal epithelium from basement membrane to lumen, and in the adult take on exceedingly irregular shapes determined by the changing conformation of the developing sperm cells embedded in their cytoplasm (Fig. 4). Differentiating sperm cells are isolated from the bloodstream and must rely on Sertoli cells for their sustenance. Adjacent Sertoli

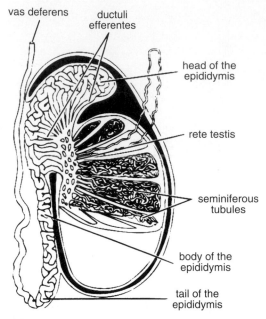

FIGURE 2   Cutaway diagram of the architecture of the human testis. (From Fawcett DW, *A Textbook of histology*, 11th ed., Philadelphia: W.B. Saunders, 1986, p. 797; modified from Hamilton, *Textbook of human anatomy*, London: Macmillan, 1957. With permission.)

ejaculation, sperm are expelled into the *vas deferens* and ultimately through the *urethra*. An accessory storage area for sperm lies in the ampulla of the vas deferens, posterior to the *seminal vesicles*. These elongated, hollow evaginations of the deferential ducts secrete a fluid rich in citric acid and fructose that provides nourishment for the sperm after ejaculation. Metabolism of fructose provides the energy for sperm motility. Additional citrate and a variety of enzymes are added to the ejaculate by the *prostate*, which is the largest of the accessory secretory glands. Sperm and the combined secretions of the accessory glands make up the *semen*, of which less than 10% is sperm.

CONTROL OF TESTICULAR FUNCTION

Physiological activity of the testis is governed by two pituitary gonadotropic hormones: follicle-stimulating hormone (FSH) and luteinizing hormone (LH) (see Chapter 38). The same gonadotropic hormones are

cells arch above the clusters of spermatogonia that nestle between them at the level of the basement membrane, and they form a series of tight junctions that limit passage of physiologically relevant molecules into or out of seminiferous tubules. This so-called *blood–testis barrier* actually has selective permeability that allows rapid entry of testosterone, for example, but virtually completely excludes cholesterol. The physiological significance of the blood–testis barrier has not been established, but it is probably of some importance that spermatogonia are located on the blood side of the barrier, whereas developing spermatids are restricted to the luminal side. In addition to harboring developing sperm, Sertoli cells secrete a watery fluid that transports spermatozoa through the seminiferous tubules and into the *epididymis*, where 99% of the fluid is reabsorbed.

The remaining portion of the male reproductive tract consists of modified excretory ducts that ultimately deliver sperm to the exterior along with secretions of accessory glands that promote sperm survival and fertility. Sperm leave the testis through multiple *ductuli efferentes*, whose ciliated epithelium facilitates passage from the rete testis into the highly convoluted and tortuous duct of the epididymis. The epididymis is the primary area for maturation and storage of sperm, which remain viable within its confines for months.

Sperm are advanced through the epididymis, particularly during sexual arousal, by rhythmic contractions of circular smooth muscle surrounding the duct. At

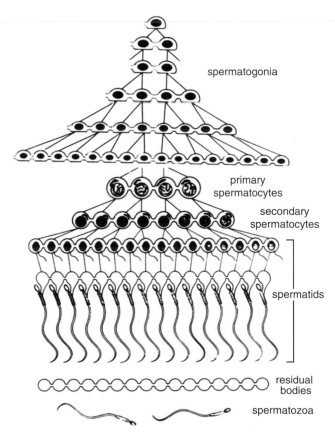

FIGURE 3   The formation of mammalian germ cells. Each primary spermatogonia ultimately gives rise to 64 sperm cells. Cytokinesis is incomplete in all but the earliest spermatogonial divisions, resulting in expanding clones of germ cells that remain joined by intercellular bridges. (From Fawcett DW, *A Textbook of histology*, 11th ed., Philadelphia: W.B. Saunders, 1986, p. 815. With permission.)

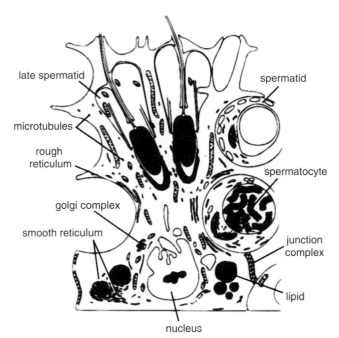

late spermatid

spermatid

microtubules

rough
reticulum

spermatocyte

golgi complex

smooth reticulum

junction
complex

lipid

nucleus

FIGURE 4 Ultrastructure of the Sertoli cell and its relation to the germ cells. The spermatocytes and early spermatids occupy niches in the sides of the columnar supporting cell, whereas late spermatids reside in deep recesses in its apex. (From Fawcett DW, *A Textbook of histology*, 11th ed., Philadelphia: W.B. Saunders, 1986, p. 834. With permission.)

produced in pituitary glands of men and women, but, because the physiology of these hormones has been studied more extensively in women, the names that have been adopted for them describe their activity in the ovary (see Chapters 46 and 47). FSH and LH are closely related glycoprotein hormones that consist of a common α subunit and unique β subunits that confer FSH or LH specificity. The three subunits are the products of three genes that are regulated independently. Both gonadotropins are synthesized and secreted by a single class of pituitary cells, the gonadotropes. Their sites of stimulation of testicular function, however, are discrete, as LH acts on the Leydig cells and FSH acts on the Sertoli cells in the germinal epithelium.

Leydig Cells

The principal role of Leydig cells is synthesis and secretion of testosterone in response to stimulation by LH. Testosterone is an important paracrine regulator of intratesticular functions as well as a hormonal regulator of a variety of extratesticular cells. In addition to stimulating steroidogenesis, LH controls the availability of its own receptors (downregulation) and governs growth and differentiation of Leydig cells. After hypophysectomy of experimental animals, Leydig cells atrophy and lose their extensive smooth endoplasmic reticulum where

the bulk of testosterone synthesis takes place. LH restores them to normal and can produce frank hypertrophy if given in excess. Leydig cells, which are abundant in newborn baby boys, regress and die shortly after birth. Secretion of LH at the onset of puberty causes dormant Leydig cell precursors to proliferate and differentiate into mature steroidogenic cells. In the fetus, growth and development of Leydig cells depend initially on the placental hormone, *chorionic gonadotropin*, which is present in high concentrations and which stimulates LH receptors, and later on LH secreted by the fetal pituitary gland.

The LH receptor is a member of the superfamily of receptors that are coupled to heterotrimeric G-proteins. LH stimulates the formation of cyclic adenosine monophosphate (cAMP) (see Chapter 2), which in turn activates protein kinase A and the subsequent phosphorylation of proteins that promote steroidogenesis, gene transcription, and other cellular functions. As with the adrenal cortex (see Chapter 40), the initial step in the synthesis of testosterone is the conversion of cholesterol to pregnenolone. This reaction requires mobilization of cholesterol from storage droplets and its translocation from cytosol to the intramitochondrial compartment, where cleavage of the side chain occurs. Access of cholesterol to the P450scc enzyme in the mitochondria is governed by a yet to be defined action of the steroid acute regulatory protein (StAR) the expression and phosphorylation of which are accelerated by cAMP. Stored cholesterol may derive either from *de novo* synthesis within the Leydig cell or from circulating cholesterol, which enters the cell by receptor-mediated endocytosis of low-density lipoproteins. The biochemical pathway for testosterone biosynthesis is shown in Fig. 5. Neither LH nor cAMP appears to accelerate the activity of any of the four enzymes responsible for conversion of the 21-carbon pregnenolone to testosterone. It may be recalled that the rate of steroid hormone secretion, as well as synthesis, is determined by the rate of conversion of cholesterol to pregnenolone. However, in maintaining the functional integrity of the Leydig cells, LH maintains the levels of all of the steroid-transforming enzymes. Transcription of the gene that encodes P450c17, the enzyme responsible for the two-step conversion of 21-carbon steroids to 19-carbon steroids, appears to be especially sensitive to cAMP. Testosterone released from Leydig cells may diffuse into nearby capillaries for transport in the general circulation or may diffuse to nearby seminiferous tubules, where it performs its essential role in spermatogenesis.

The testes also secrete small amounts of estradiol and some androstenedione, which serves as a precursor for extratesticular synthesis of estrogens. The Leydig cell is the chief source of testicular estrogens, but immature Sertoli cells have the capacity to convert testosterone to estradiol. In addition, developing sperm cells express the

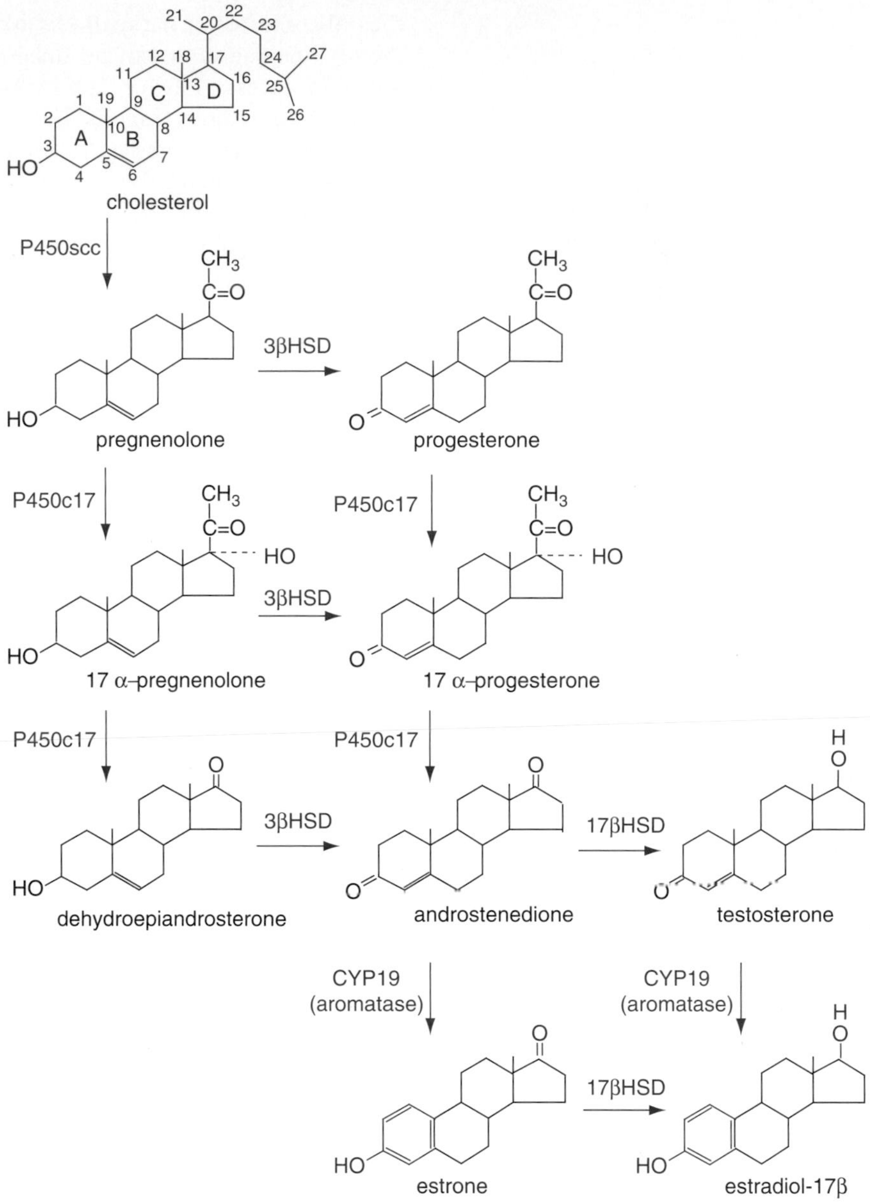

FIGURE 5 Biosynthesis of testicular steroids. Catalyzed changes at each step are shown in blue. Testosterone comprises more than 99% of testicular steroid hormone production.

enzyme P450arom (aromatase) and may also convert androgens to estrogens. Estradiol is present in seminal fluid and may be essential for reabsorption of seminal tubular fluid in the rete testis. The presence of estrogen receptors in the epididymis and several testicular cells, including Leydig cells, suggests that estradiol may have other important actions in normal sperm formation and maturation.

Germinal Epithelium

The function of the germinal epithelium is to produce large numbers of sperm that are capable of fertilization.

The Sertoli cells, which are interposed between the developing sperm and the vasculature, harbor and nurture sperm as they mature. Sertoli cells are the only cells known to express FSH receptors in human males and therefore are the only targets of FSH. In the immature testis, FSH increases Sertoli cell proliferation and differentiation and probably maintains their functional state throughout life; in its absence, testicular size is severely reduced and sperm production, which is limited by Sertoli cell availability, is severely restricted. It has been known for many years that FSH, LH, and testosterone all play vital roles in spermatogenesis. It is likely that FSH indirectly regulates development of spermatogonia by stimulating Sertoli cells to produce both growth and survival factors that prevent germ cell apoptosis. Withdrawal of FSH and LH arrests spermatogonial development, which is the major rate-limiting step in spermatogenesis. Once formed, spermatocytes progress through meiosis normally in the absence of gonadotropic support. Although FSH and testosterone are required for initiation of normal rates of spermatogenesis, sperm formation can be maintained indefinitely with very high doses of testosterone alone or with sufficient LH to stimulate testosterone production.

Sertoli cells lack receptors for LH but are richly endowed with androgen receptors, indicating that the actions of LH on Sertoli cell function are indirect and are mediated by testosterone, which reaches them in high concentration by diffusion from adjacent Leydig cells, and perhaps also by peptide factors produced by Leydig cells. Testosterone readily passes through the blood–testis barrier and is found in high concentrations in seminiferous fluid. However, the absence of androgen receptors in developing human sperm cells indicates that support of sperm cell development by testosterone is also exerted indirectly by way of the Sertoli cells. Although testosterone is critically important for spermatogenesis, it is ineffective in this regard when administered in amounts sufficient to restore normal blood concentrations. For reasons that are not understood, the intratesticular concentration needed to support

spermatogenesis is many times higher than that necessary to saturate androgen receptors. The concentration of testosterone in testicular venous blood is 40 to 50 times that found in peripheral blood, and its concentration in aspirates of human testicular fluid is more than 100 times higher than the concentration found in blood plasma.

The FSH receptor is closely related to the LH receptor, and, when stimulated, activates adenylyl cyclase through the agency of the stimulatory alpha G protein ($G\alpha_s$; see Chapter 2). The resulting activation of protein kinase A catalyzes phosphorylation of proteins that regulate the cytoskeletal elements that maintain the tortuous shape of these cells, production of the membrane glycoproteins that govern adherence to developing sperm, and expression of specific genes that code for proteins that directly and indirectly regulate germ cell development (Fig. 6). Some of the proteins secreted by Sertoli cells into the seminiferous tubules are thought to facilitate germ cell maturation in the epididymis and perhaps more distal portions of the reproductive tract. Upon stimulation by FSH, Sertoli cells may also secrete paracrine factors that enhance Leydig cell responses to LH.

FSH and testosterone have overlapping actions on Sertoli cells and act synergistically, but the precise actions of each remain unknown. Recently described "experiments of nature" have shed some light on the relative importance of FSH and testosterone in spermatogenesis. Inactivating mutations of the β subunit of LH in humans

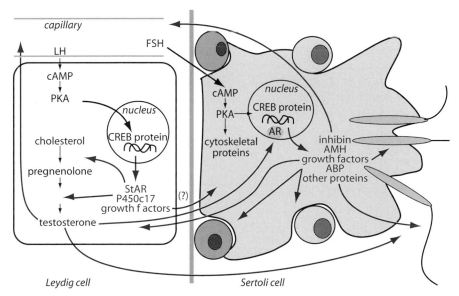

FIGURE 6   Actions of FSH and LH on the testis. FSH acts directly only on Sertoli cells, while LH acts directly solely on Leydig cells. Paracrine cross-talk between mediated by growth factors likely takes place between Sertoli and Leydig cells and between Sertoli cells and germ cells. cAMP, cyclic adenosine monophosphate; PKA, protein kinase A; CREB, cAMP response element binding protein; StAR, steroid acute regulatory protein; P450c17, 17α hydroxylase/lyase; AR, androgen receptor; AMH, anti-müllerian hormone; ABP, androgen-binding protein.

or rodents results in total failure of spermatogenesis, despite normal prenatal sexual development (see below), suggesting that testosterone is indispensable. In contrast, inactivating mutations of the gene for the FSH receptor did not prevent affected men or mice from fathering offspring. Thus, FSH apparently is not absolutely required for spermatogenesis, but the patients and rodents with inactive FSH receptors had small testes, low sperm counts, and a preponderance of defective sperm. Stimulation by FSH at some period of life, then, is required for production of a normal quantity and quality of sperm.

TESTOSTERONE

Secretion and Metabolism

Testosterone is the principal androgen secreted by the mature testis. The normal young man produces about 7 mg each day, of which less than 5% is derived from adrenal secretions. This amount decreases somewhat with age, so that by the seventh decade and beyond testosterone production may have decreased to 4 mg per day; however, in the absence of illness or injury, men do not experience a sharp drop in testosterone production akin to the abrupt cessation of estrogen production in the postmenopausal woman. As with the other steroid hormones, testosterone in blood is largely bound to plasma protein, with only about 2 to 3% present as free hormone. About 50% is bound to albumin and about 45% to sex-hormone-binding globulin (SHBG), which is also called testosterone–estradiol-binding globulin (TeBG). This glycoprotein binds both estrogen and testosterone, but its single binding site has a higher affinity for testosterone. Its concentration in plasma is decreased by androgens. Consequently, SHBG is more than twice as abundant in the circulation of women than men. In addition to its functions as a carrier protein, SHBG may also act as an enhancer of hormone action. Persuasive evidence has been amassed to indicate that SHBG binds to specific receptors on cell membranes and increases the formation of cAMP when its steroid binding site is occupied. The nature of the receptor has not been characterized, nor has the physiological importance of this action been established. Although testosterone decreases expression of SHBG in hepatocytes, both testosterone and FSH increase transcription of the same gene in Sertoli cells, where its protein product was given the name androgen-binding protein (ABP) before its identity with SHBG was known. ABP is secreted into the lumens of seminiferous tubules.

Testosterone that is not bound to plasma proteins diffuses out of capillaries and enters nontarget as well as target cells. In some respects, testosterone can be considered to be a prohormone, because it is converted in extratesticular tissues to other biologically active steroids. Testosterone may be reduced to the more potent androgen, 5α-dihydrotestosterone, in the liver in a reaction catalyzed by the enzyme 5α-reductase type I and returned to the blood. This enzyme is a component of the steroid hormone degradative pathway and also reduces 21-carbon adrenal steroids. Testosterone is also reduced to dihydrotestosterone in the cytoplasm of its target cells mainly through the catalytic activity of 5α-reductase type II, whose abundance in these cells is increased by the actions of testosterone. Dihydrotestosterone is only about 5% as abundant in blood as testosterone and is derived primarily from extratesticular metabolism. Some testosterone is also metabolized to estradiol (Fig. 7) in both androgen target and nontarget tissues. A variety of cells, including some in brain, breast, and adipose tissue, can convert testosterone and androstenedione to estradiol and estrone, which produce cellular effects that are different from, and sometimes opposite to, those of testosterone. The concentration of estrogens in blood of normal men is similar to that of women in the early follicular phase of the menstrual cycle (see Chapter 46). About two-thirds of these estrogens are formed from androgen outside of the testis. Although less than 1% of the peripheral pool of testosterone is converted to estrogens, it is important to recognize that estradiol produces its biological effects at concentrations that are far below those required for androgens to produce their effects. In other tissues, including liver, reduction catalyzed by 5β-reductase destroys androgenic potency. The liver is the principal site of degradation of testosterone and releases water-soluble sulfate or glucuronide conjugates into blood for excretion in the urine.

Mechanism of Action

Like other steroid hormones, testosterone penetrates the target cells whose growth and function it stimulates. Androgen target cells generally convert testosterone to 5α-dihydrotestosterone before it binds to the androgen receptor. The androgen receptor is a ligand-dependent transcription factor that belongs to the nuclear receptor superfamily (see Chapter 2). It binds both testosterone and dihydrotestosterone, but the dihydro- form dissociates from the receptor much more slowly than testosterone and therefore is the predominant androgen associated with DNA. It is likely that the higher affinity of dihydrotestosterone for the androgen receptor accounts for its greater biological potency compared to testosterone. Upon binding testosterone or dihydrotestosterone, the liganded receptor complex binds to androgen response elements in specific target genes and, along with a cell-specific array of transcription factors and coactivators,

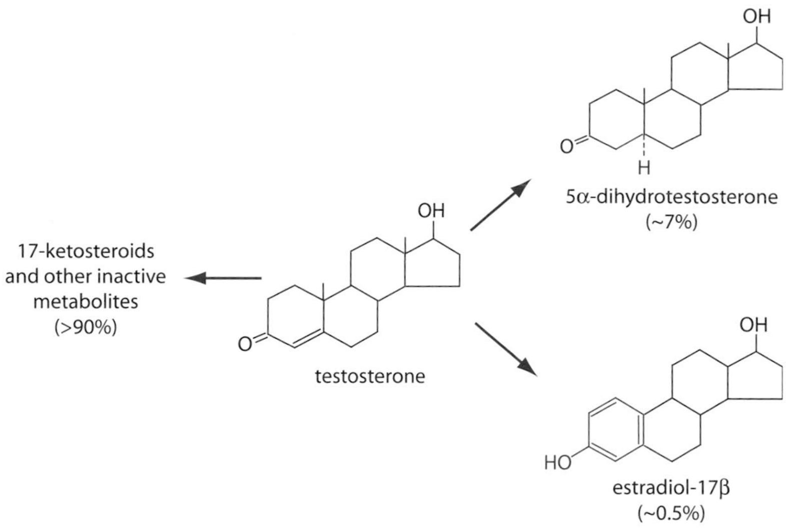

FIGURE 7   Metabolism of testosterone. Most of the testosterone secreted each day is degraded by the liver and other tissues by reduction of the A ring, oxidation of the 17 hydroxyl group, and conjugation with polar substituents. Conversion to 5α-dihydrotestosterone takes place in target cells catalyzed mainly by the type II dehydrogenase and in nontarget cells mainly but not exclusively by the type I dehydrogenase. Aromatization of testosterone to estradiol may occur directly or after conversion to androstenedione. Note that 5-α dihydrotestosterone cannot be aromatized or reconverted to testosterone.

regulates expression of a cadre of genes that are characteristic of each particular target cell. These events are summarized in Fig. 8. Some nongenomic actions of testosterone have also been described and include rapid increases in intracellular calcium. We do not understand the cellular mechanisms of these effects or their physiological importance.

Effects on the Male Genital Tract

Testosterone promotes growth, differentiation, and function of accessory organs of reproduction. Its effects on growth of the genital tract begin early in embryonic life and are not completed until adolescence, after an interruption of more than a decade. Maintenance of normal reproductive function in the adult also depends on continued testosterone secretion. The secretory epithelia of the seminal vesicles and prostate atrophy after castration but can be restored with injections of androgen.

Effects on Secondary Sexual Characteristics

In addition to its effect on organs directly related to transport and delivery of sperm, testosterone affects a variety of other tissues and thus contributes to the morphological and psychological components of masculinity. These characteristics are clearly an integral part of reproduction, for they are related to the attractiveness of the male to the female. During early adolescence, androgens that arise from the adrenals and later from the testes stimulate growth of pubic hair. Growth of chest, axillary, and facial hair is also stimulated, but scalp hair is affected in the opposite manner. Recession of hair at the temples is a typical response to androgen, and adequate amounts allow expression of genes for baldness. Growth

Clinical Note

Prostate-Specific Antigen (PSA)
One of the proteins whose expression in the prostate is stimulated by dihydrotestosterone is the so-called *prostate-specific antigen* (PSA), which is found in blood in high concentrations in patients afflicted with prostate cancer. Its abundance in plasma is now widely used diagnostically as a marker of prostate cancer. PSA is a serine protease synthesized in the columnar cells of the glandular epithelium and secreted into the semen. Cleavage of *seminogelin* by PSA causes liquefaction of the ejaculate and is thought to increase sperm mobility.

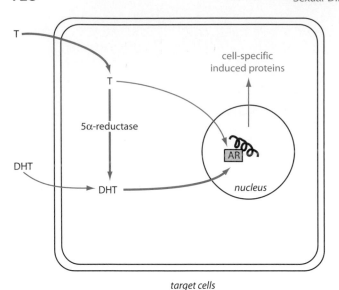

FIGURE 8 Action of testosterone. Testosterone (T) enters its target cell and binds to its nuclear androgen receptor (AR) either directly or after it is converted to dihydrotestosterone (DHT). The thickness of the arrows reflects the quantitative importance of each reaction. The hormone–receptor complex binds to DNA along with a variety of cell specific nuclear regulatory proteins to induce formation of the RNA that encodes the proteins that express effects of the hormone. *Not shown:* Testosterone may also bind to membrane receptors and initiate rapid ionic changes that may reinforce its genomic effects. Testosterone may also produce rapid changes in cAMP through the binding of the sex-hormone-binding globulin (SHBG) to surface receptors.

and secretion of sebaceous glands in the skin are also stimulated, a phenomenon undoubtedly related to the acne of adolescence. Dihydrotestosterone is the important androgen for recession of scalp hair and stimulation of the sebaceous glands.

Androgen secretion at puberty stimulates growth of the larynx and thickening of the vocal chords, thus lowering the pitch of the voice. Also at this time, the characteristic adolescent growth spurt results from an interplay of testosterone and growth hormone (see Chapter 44) that promotes growth of the vertebrae and long bones. Development of the shoulder girdle is pronounced. This growth is self-limiting, as androgens, after extragonadal conversion to estrogens, accelerate epiphyseal closure (see Chapter 44). Androgens promote growth of muscle, especially in the upper torso. Indeed, men have almost half again as much muscle mass as women. In some animals, the temporal and masseter muscles are particularly sensitive to androgenic stimulation. Growth and nitrogen retention, of course, are also related to stimulation of appetite and increased food intake. Accordingly, androgens bring about increased physical vigor and a feeling of well-being. Testosterone also stimulates red blood cell production by direct effects on bone marrow and by stimulating secretion of the hormone

erythropoietin from the kidney. This action of androgens accounts for the higher hematocrit in men than women. In both men and women, androgens increase sexual drive (*libido*).

SEXUAL DIFFERENTIATION

Primordial components of both male and female reproductive tracts are present in early embryos of both sexes, and their development is either stimulated or suppressed by humoral factors arising in the testes. The indifferent gonads present in the early embryo differentiate into testes under the influence of the product of the sex determining gene SRY (for sex-determining region of the Y chromosome). By about the seventh week of embryonic life, the medulla of the primitive gonad becomes distinguishable as a testis with the appearance of cords of cells that give rise to seminiferous tubules. Leydig cells appear about 10 days later and undergo rapid proliferation for the next 6 to 8 weeks in response to chorionic gonadotropin, which is produced by the placenta in large amounts at this time, and perhaps also to LH secreted by the fetal pituitary gland. Fetal Leydig cells secrete sufficient testosterone to raise blood concentrations to the same levels as those seen in adult men. Testosterone accumulation is enhanced by an additional effect of the SRY gene product, which blocks expression of aromatase and thus prevents conversion of testosterone to estrogens.

Development of Internal Reproductive Ducts and Their Derivatives

Regardless of its genetic sex, the embryo has the potential to develop phenotypically either as male or female. The pattern for female development is expressed unless overridden by secretions of the fetal testis. The early embryo develops two sets of ducts that are the precursors of either male or female internal genitalia (Fig. 9). Seminal vesicles, epididymes, and vasa deferentia arise from primitive mesonephric, or *wolffian*, ducts. Internal genitalia of the female, including the uterus, fallopian tubes, and upper vagina, develop from paramesonephric, or *Müllerian*, ducts. When stimulated by testosterone, the wolffian ducts differentiate into male reproductive structures, but in the absence of androgen they regress and disappear. In contrast, Müllerian ducts develop into female reproductive structures unless actively suppressed. Testosterone does not stimulate müllerian regression. Under the influence of the SRY gene product and specific transcription factors, Sertoli cells in newly differentiated seminiferous tubules secrete a glycoprotein called

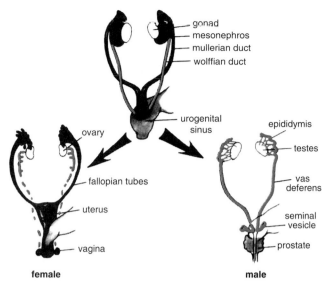

FIGURE 9 Development of the male and female internal genitalia. (From Jaffe RB, In: Yen SC, Jaffe RB, Eds., *Reproductive endocrinology*, 2nd ed., Philadelphia: W.B. Saunders, 1986, p. 283. With permission.)

anti-müllerian hormone (AMH), which causes apoptosis of tubular epithelial cells and reabsorption of the Müllerian ducts. In experiments in which only one testis was removed from embryonic rabbits, the Müllerian duct regressed only on the side with the remaining gonad, indicating that anti-müllerian hormone must act locally as a paracrine factor. The wolffian duct regressed on the opposite side, suggesting that testosterone, too, must act locally to sustain the adjacent wolffian duct, as the amounts that reached the contralateral duct through the circulation were inadequate to prevent its regression (Fig. 10).

Sertoli cell production of AMH is not limited to the embryonic period but continues into adulthood. AMH is present in adult blood serum and in seminal plasma, where it binds to sperm and may increase their motility. Plasma concentrations of AMH are highest in the prepubertal period and fall as testosterone concentrations rise. Its secretion is stimulated by FSH and strongly inhibited by testosterone. In the testis, AMH inhibits Leydig cell differentiation and expression of steroidogenic enzymes, particularly P450c17. AMH is also expressed in the adult ovary and is found in the plasma of women as well as men. No extragonadal role for AMH in adults has yet been established.

AMH is a member of the transforming growth factor β (TGF-β) family of growth factors. As for other members of the TGF-β family, AMH signals by way of membrane receptors that have intrinsic enzymatic activity that catalyzes phosphorylation of proteins on serine and threonine residues. The AMH receptor consists of two non-identical subunits, each of which has a single

membrane-spanning region and an intracellular kinase domain. Binding of AMH to its specific primary receptor causes it to complex with and phosphorylate a secondary signal-transducing subunit that may also be a component of receptors for other agonists of the TGF-β family. The activated receptor complex associates with and phosphorylates cytosolic proteins called *Smads*, which enter the nucleus and activate transcription of specific genes (Fig. 11).

Development of the External Genitalia

The urogenital sinus and genital tubercle are the primitive structures that give rise to the external genitalia in both sexes. Masculinization of these structures to form the penis, scrotum, and prostate gland depends on secretion of testosterone by the fetal testis. Unless stimulated by androgen, these structures develop into female external genitalia. When there is insufficient

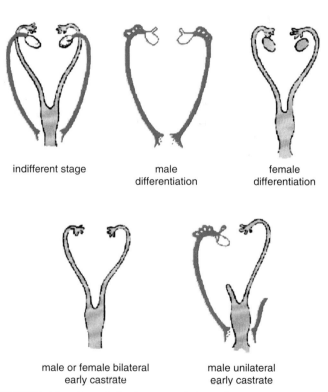

FIGURE 10 Normal development of the male and female reproductive tracts. Tissues destined to form the male tract are shown in blue; tissues that develop into the female tract are shown in gray. Bilateral castration of either male or female embryos results in development of the female pattern. Early unilateral castration of male embryos results in development of the normal male duct system on the side with the remaining gonad, but female development on the contralateral side. This pattern develops because both testosterone and anti-müllerian hormone act as paracrine factors. (Modified from Jost A, In: Jones HW, Scott WW, Eds., *Hermaphroditism, genital anomalies and related endocrine disorders*, 2nd ed., Baltimore, MD: Williams & Wilkins, 1971, p. 16.)

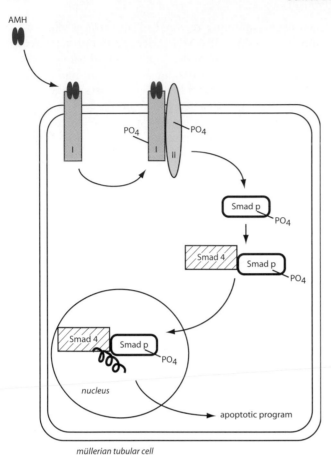

AMH

PO$_4$

PO$_4$

Smad p

PO$_4$

Smad 4

Smad p

PO$_4$

Smad 4

Smad p

PO$_4$

nucleus

apoptotic program

müllerian tubular cell

FIGURE 11 Anti-mullerian hormone (AMH) signaling pathway. AMH binds to its specific primary receptor (I) which then forms a heterodimer with and phosphorylates the secondary signal transducing subunit (II). The activated receptor complex then catalyzes phosphorylation of Smad proteins on serine and threonine residues, causing them to bind Smad 4, which carries them into the nucleus, where transcription of specific genes results in expression of an apoptotic program and resorption of the Müllerian duct cells.

androgen in male embryos, or too much androgen in female embryos, differentiation is incomplete and the external genitalia are ambiguous. Differentiation of the masculine external genitalia depends on dihydrotestosterone rather than testosterone. The 5α-reductase type II responsible for conversion of testosterone to dihydrotestosterone is present in tissues destined to become external genitalia even before the testis starts to secrete testosterone. In contrast, this enzyme does not appear in tissues derived from the wolffian ducts until after they differentiate, indicating that testosterone rather than dihydrotestosterone was the signal for differentiation of the wolffian derivatives.

The importance of androgen action in sexual development is highlighted by a fascinating human syndrome called *testicular feminization*, which can be traced to an inherited defect in the single gene on the X chromosome that encodes the androgen receptor. Afflicted individuals

have the normal female phenotype but have sparse pubic and axillary hair and no menstrual cycles. Genetically, they are male and have intra-abdominal testes and circulating concentrations of testosterone and estradiol that are within the range found in normal men, but their tissues are totally unresponsive to androgens. Their external genitalia are female because, as already mentioned, the primordial tissues develop in the female pattern unless stimulated by androgen. Because AMH production and responsiveness are normal and their wolffian ducts are unable to respond to androgen, both of these duct systems regress and neither male nor female internal genitalia develop. Secondary sexual characteristics including breast development appear at puberty in response to unopposed action of estrogens formed extragonadally from testosterone.

Postnatal Development

Aside from a brief surge in androgen production during the immediate neonatal period, testicular function enters a period of quiescence, and further development of the male genital tract is arrested until the onset of puberty. Increased production of testosterone at puberty promotes growth of the penis and scrotum and increases pigmentation of the genitalia as well as the depth of rugal folds in scrotal skin. Further growth of the prostate, seminal vesicles, and epididymes also occurs at this time. Although differentiation of the epididymes and seminal vesicles was independent of dihydrotestosterone during the early fetal period, later acquisition of 5α-reductase type II makes this more active androgen the dominant form stimulating growth and secretory activity during the pubertal period. Increased secretion of FSH at puberty stimulates multiplication of Sertoli cells and growth of the seminiferous tubules, which constitute the bulk of the testicular mass.

The importance of some of the foregoing information is highlighted by another interesting genetic disorder that has been described as "penis at twelve." Affected individuals have a deletion or inactivating mutation in the gene that codes for 5α-reductase type II and cannot convert testosterone to dihydrotestosterone in derivatives of the genital tubercle. Although testes and wolffian derivatives develop normally, the prostate gland is absent, and external genitalia at birth are ambiguous or overtly feminine. Affected children have been raised as females. With the onset of puberty, testosterone production increases, and there is an increase in the expression of 5α-reductase type I in the skin. Significant growth of the penis occurs at this time in response to the 5α-dihydrotestosterone produced in the liver and skin by the catalytic activity of 5α-reductase type I.

REGULATION OF TESTICULAR FUNCTION

Testicular function, as we have seen, depends on stimulation by two pituitary hormones, FSH and LH. Without them, the testes lose spermatogenic and steroidogenic capacities and either atrophy or fail to develop. Secretion of these hormones by the pituitary gland is driven by the central nervous system through its secretion of the gonadotropin-releasing hormone (GnRH), which reaches the pituitary by way of the hypophysial portal blood vessels (see Chapter 38). Separation of the pituitary gland from its vascular linkage to the hypothalamus results in total cessation of gonadotropin secretion and testicular atrophy. The central nervous system and the pituitary gland are kept apprised of testicular activity by signals related to each of the testicular functions: steroidogenesis and gametogenesis. Characteristic of a negative feedback, signals from the testis are inhibitory. Castration results in a prompt increase in secretion of both FSH and LH. The central nervous system also receives and integrates other information from the internal and external environments and modifies GnRH secretion accordingly.

Gonadotropin-Releasing Hormone and the Hypothalamic Pulse Generator

Gonadotropin-releasing hormone is a decapeptide produced by a diffuse network of about 2000 neurons whose perikarya are located primarily in the arcuate nuclei in the medial basal hypothalamus and whose axons terminate in the median eminence in the vicinity of the hypophysial portal capillaries. GnRH-secreting neurons also project to other parts of the brain and may mediate some aspects of sexual behavior. GnRH is released into the hypophysial portal circulation in discrete pulses at regular intervals ranging from about one every hour to one every 3 hours or longer. Each pulse lasts only a few minutes, and the secreted GnRH disappears rapidly with a half-life of about 4 minutes. GnRH secretion is difficult to monitor directly because hypophysial portal blood is inaccessible and because its concentration in peripheral blood is too low to measure even with the most sensitive assays. The pulsatile nature of GnRH secretion has been inferred from results of frequent measurements of LH concentrations in peripheral blood (Fig. 12). FSH concentrations tend to fluctuate much less, largely because FSH has a longer half-life than LH, 2 to 3 hours compared to 20 to 30 minutes.

Pulsatile secretion requires synchronous firing of many neurons, which therefore must be in communication with each other and with a common pulse generator. Because pulsatile secretion of GnRH continues even after

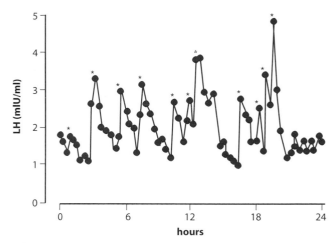

FIGURE 12 Luteinizing hormone (LH) secretory pattern observed in a normal 36-year-old man; ★, statistically significant discrete pulse. (From Crowley WF, Jr, In: Krieger DT, Bardin CW, Eds., *Current topics in endocrinology and metabolism*, New York: Marcel Decker, 1985, p. 157. With permission.)

experimental disconnection of the medial basal hypothalamus from the rest of the central nervous system, the pulse generator must be located within this small portion of the hypothalamus. Pulsatile secretion of GnRH by neurons maintained in tissue culture indicate that episodic secretion is an intrinsic property of GnRH neurons. There is good correspondence between electrical activity in the arcuate nuclei and LH concentrations in blood, as determined in rhesus monkeys fitted with permanently implanted electrodes. The frequency and amplitude of secretory pulses and corresponding electrical activity can be modified experimentally (Fig. 13) and are regulated physiologically by gonadal steroids and probably by other information processed within the central nervous system.

The significance of the pulsatile nature of GnRH secretion became evident in studies of reproductive function in rhesus monkeys whose arcuate nuclei had been destroyed and whose secretion of LH and FSH therefore came to a halt. When GnRH was given as a constant infusion, gonadotropin secretion was restored only for a short while. FSH and LH secretion soon decreased and stopped even though the infusion of GnRH continued. Only when GnRH was administered intermittently for a few minutes of each hour was it possible to sustain normal gonadotropin secretion in these monkeys. Similar results have been obtained in human patients and applied therapeutically.

The cellular mechanisms that account for the complex effects of GnRH on gonadotropes are not fully understood. The GnRH receptor is a G-protein coupled heptihelical receptor that activates phospholipase C through $G\alpha_q$ (Chapter 2). The resulting formation of inositol trisphosphate (IP_3) and diacylglycerol (DAG) results in mobilization of intracellular calcium and activation of protein kinase C. Transcription of genes for FSHβ, LHβ, and the common α subunit depends upon increased cytosolic calcium and several protein kinases for which the activation pathways are not understood. Secretion of gonadotropins depends upon the increase in intracellular calcium achieved by mobilizing calcium from intracellular stores and by activating membrane calcium channels. Desensitization of gonadotropes after prolonged uninterrupted exposure to GnRH appears to result from the combined effects of downregulation of GnRH receptors, downregulation of calcium channels associated with secretion, and a decrease in the releasable storage pool of gonadotropin.

Negative Feedback Regulators

The hormones FSH and LH originate in the same pituitary cell whose secretory activity is stimulated by the same hypothalamic hormone, GnRH. Nevertheless, secretion of FSH is controlled independently of LH secretion by negative feedback signals that relate to the separate functions of the two gonadotropins. Although castration is followed by increased secretion of both FSH and LH, only LH is restored to normal when physiological amounts of testosterone are given. Failure of testicular descent into the scrotum (*cryptorchidism*) may result in destruction of the germinal epithelium without affecting Leydig cells. With this condition, blood levels of testosterone and LH are normal, but FSH is elevated. Thus, testosterone, which is secreted in response to LH, acts as a feedback regulator of LH and hence of its own secretion. By this reasoning, we would expect that spermatogenesis, which is stimulated by FSH, might be associated with secretion of a substance that reflects gamete production. Indeed, FSH stimulates the Sertoli cells to synthesize and secrete a glycoprotein called *inhibin*, which acts as a feedback inhibitor of FSH.

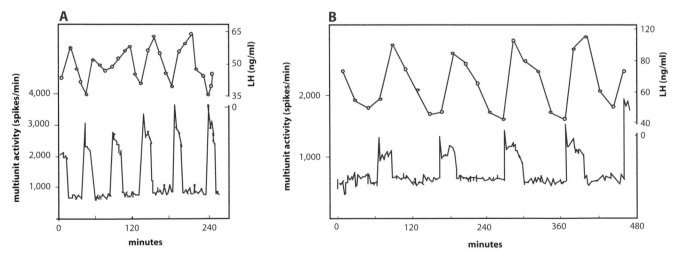

FIGURE 13   Recording of multiple unit activity (MUA) in the arcuate nuclei of conscious (A) and anesthetized (B) monkeys fitted with permanently implanted electrodes. Simultaneous measurements of LH in peripheral blood are shown in the upper tracings. (From Wilson RC, Kesner JS, Kaufman JN *et al.*, *Neuroendocrinology* 1984; 39:256. With permission.)

Inhibin, which was originally purified from follicular fluid of the pig ovary, is a disulfide-linked heterodimer comprised of an α subunit and either of two forms of a β subunit, β_A or β_B. The physiologically important form of inhibin secreted by the human testis is the $\alpha\beta_B$ dimer called *inhibin B*. Its concentration in blood plasma is reflective of the number of functioning Sertoli cells and spermatogenesis. Both inhibin A and inhibin B are produced by the ovary (see Chapter 46). Little is known about the significance of alternate β subunits or the factors that determine when each form is produced. All three subunits are encoded in separate genes and presumably are regulated independently. They are members of the same family of growth factors that includes AMH and TGF-β. Of additional interest is the finding that dimers formed from two β subunits produce effects that are opposite those of the $\alpha\beta$ dimer and stimulate FSH release from gonadotropes maintained in tissue culture. These compounds are called *activins* and function in a paracrine mode in the testis and many other tissues. While the production of the α subunit is largely confined to male and female gonads, β subunits are produced in many extragonadal tissues where activins mediate a variety of functions. Activins are produced in the pituitary and appear to play a supportive role in FSH production. The pituitary and other tissues also produce an unrelated protein called *follistatin*, which binds activins and blocks their actions.

The feedback relations that fit best with our current understanding of the regulation of testicular function in the adult male are shown in Fig. 14. Pulses of GnRH originating in the arcuate nuclei evoke secretion of both FSH and LH by the anterior pituitary. FSH and LH are positive effectors of testicular function and stimulate release of inhibin and testosterone, respectively. Testosterone has an intratesticular action that reinforces the effects of FSH. It also travels through the circulation to the hypothalamus, where it exerts its negative feedback effect primarily by slowing the frequency of GnRH pulses. Because secretion of LH is more sensitive to frequency of stimulation than is secretion of FSH, decreases in GnRH pulse frequency lower the ratio of LH to FSH in the gonadotropic output. In the castrate monkey, the hypothalamic pulse generator discharges once per hour and slows to once every 2 hours after testosterone is replaced. This rate is about the same as that seen in normal men. The higher frequency in the castrate triggers more frequent bursts of gonadotropin secretion, resulting in higher blood levels of both FSH and LH. Testosterone may also decrease the amplitude of the GnRH pulses somewhat and may also exert some direct restraint on LH release from gonadotropes. In high enough concentrations, testosterone may inhibit GnRH release sufficiently to shut off secretion of both

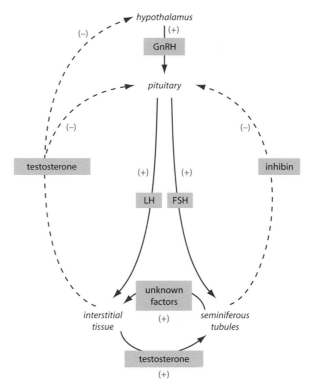

FIGURE 14 Negative feedback regulation of testicular function; +, stimulation; –, inhibition. Direct effects of testosterone on the pituitary gland are still uncertain.

gonadotropic hormones. The negative feedback effect of inhibin appears to be exerted exclusively on gonadotropes to inhibit FSHβ transcription and FSH secretion in response to GnRH. Some evidence indicates that inhibin may also exert local effects on Leydig cells to enhance testosterone production.

Prepuberty

Testicular function is critical for development of the normal masculine phenotype early in the prenatal period. All of the elements of the control system are present in the early embryo. GnRH and gonadotropins are detectable at about the time that testosterone begins stimulating wolffian duct development. The hypothalamic GnRH pulse generator and its negative feedback control are functional in the newborn. Both the frequency and amplitude of GnRH and LH pulses are similar to those observed in the adult. After about the sixth month of postnatal life and for the remainder of the juvenile period, the GnRH pulse generator is restrained and gonadotropin secretion is low. The amplitude and frequency of GnRH pulses decline but do not disappear, and responsiveness of the gonadotropes to GnRH diminishes. It is evident that negative feedback regulation remains operative, however, because blood levels of

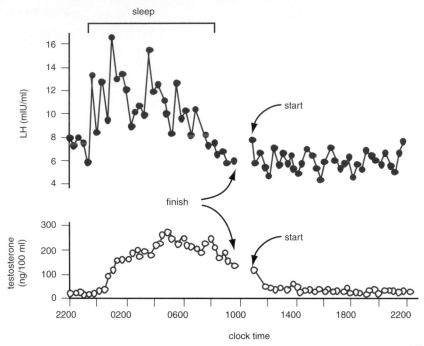

FIGURE 15 Plasma LH and testosterone measured every 20 minutes reveal nocturnal pulsatile secretion of GnRH in a pubertal 14-year-old boy. (From Boyer RM, Rosenfeld RS, Kapen S *et al.*, *J Clin Invest* 1974; 54:609. With permission.)

gonadotropins increase after gonadectomy in prepubertal subjects and fall with gonadal hormone administration. The system is extremely sensitive to feedback inhibition during this time, but suppression of the pulse generator cannot be explained simply as a change in the set point for feedback inhibition. The plasma concentration of gonadotropins is high in juvenile subjects whose testes have failed to develop and who consequently lack testosterone, but rise even higher when these subjects reach the age when puberty would normally occur. Thus, restraint of the GnRH pulse generator imposed by the central nervous system diminishes at the onset of puberty.

Early stages of puberty are characterized by the appearance of high-amplitude pulses of LH during sleep (Fig. 15). Testosterone concentrations in plasma follow the gonadotropins, and there is a distinct day/night pattern. As puberty progresses, high-amplitude pulses are distributed throughout the day at the adult frequency of about one every 2 hours. Sensitivity of the pituitary gland to GnRH increases during puberty, possibly as a result of a self-priming effect of GnRH on gonadotropes. GnRH increases the amount of releasable FSH and LH in the gonadotropes and may also increase the number of its receptors on the gonadotrope surface. The underlying neural mechanisms for suppression of the GnRH pulse generator in the juvenile period are not understood. Increased inhibitory input from neuropeptide Y (NPY)- and γ-aminobutyric acid (GABA)-secreting neurons has been observed, but neither the factors

that produce nor terminate such input are understood. Clearly, the onset of reproductive capacity is influenced by, and must be coordinated with, metabolic factors and attainment of physical size. In this regard, as we have seen (Chapter 44) puberty is intimately related to growth. Onset of puberty, especially in girls, has long been associated with adequacy of body fat stores, and it appears that adequate circulating concentrations of leptin (Chapter 43) are permissive for the onset of puberty, but available evidence indicates that leptin is not the trigger. It is likely that some confluence of genetic, developmental, and nutritional factors signals readiness for reproductive development and function.

Suggested Readings

Crowley WF, Jr, Filicori M, Spratt DI, Santoro NF. The physiology of gonadotropin-releasing hormone (GnRH) in men and women. *Rec Prog Horm Res* 1985; 41:473–526.

Crowley WF, Jr, Whitcomb RW, Jameson JL, Weiss J, Finkelstein JS, O'Dea LSL. Neuroendocrine control of human reproduction in the male. *Recent Prog Horm Res* 1991; 47:349–387.

George FW, Wilson JD. Hormonal control of sexual development. *Vitamins Horm* 1986; 43:145–196.

Hayes FJ, Hall JE, Boepple PA, Crowley WF, Jr. Clinical review 96—differential control of gonadotropin secretion in the human: endocrine role of inhibin. *J Clin Endocrinol Metab* 1998; 83: 1835–1841.

Huhtaniemi I. Mutations of gonadotrophin and gonadotrophin receptor genes: what do they teach us about reproductive physiology? *J Reprod Fertil* 2000; 119:173–186.

Kierszenbaum AL. Mammalian spermatogenesis *in vivo* and *in vitro*: a partnership of spermatogenic and somatic cell lineages. *Endocrine Rev* 1994; 15:116–134.

Mooradian AD, Morley JE, Korenman, SG. Biological actions of androgens. *Endocrine Rev* 1987; 8:1–28.

Naor Z, Harris D, Shacham S. Mechanism of GnRH receptor signaling: combinatorial cross-talk of Ca^{2+} and protein kinase C. *Frontiers Neuroendocrinol* 1998; 19:1–19.

Payne AH, Youngblood GL. Regulation of expression of steroidogenic enzymes in Leydig cells. *Biol Reprod* 1995; 52:217–225.

Plant TM, Marshall GR. The functional significance of FSH in spermatogenesis and control of its secretion in male primates. *Endocrine Rev* 2001; 22:764–786.

Rosner WDJ, Hryb Khan MS, Nakhla AM, Romas NA. Sex hormone-binding globulin mediates steroid hormone signal transduction at the plasma membrane. *J Steroid Biochem Molec Biol* 1999; 69:481–485.

Teixeira JS, Donahoe, PK. Müllerian inhibiting substance: an instructive developmental hormone with diagnostic and possible therapeutic applications. *Endocrine Rev* 2001; 22:657–674.

Wilson JD, Androgen abuse by athletes. *Endocrine Rev* 1988; 9:181–199.

Wilson JD, Griffin JE, Russell DW. Steroid 5α-reductase 2 deficiency. *Endocrine Rev* 1993; 14:577–593.

Ying S-Y. Inhibins activins and folliculostatins: gonadal proteins modulating the secretion of follicle-stimulating hormone. *Endocrine Rev* 1988; 9:267–293.

Hormonal Control of Reproduction in the Female: The Menstrual Cycle

H. MAURICE GOODMAN

KEY POINTS

- The ovaries have the dual function of producing gametes and hormones.
- Ova are produced in the ovarian follicles, which also produce estradiol and inhibin.
- After release of the ovum, the follicle is reorganized to form the corpus luteum, which secretes progesterone, estradiol, and the peptide hormones inhibin and relaxin.
- Estrogen and progesterone are derived from cholesterol. Their concentrations in blood vary widely and they circulate bound to plasma proteins. They are inactivated by the liver and excreted by the kidneys.
- The ovarian cycle lasts about 28 days and consists of a follicular phase (12–14 days), in which the follicle grows to maturity; ovulation (1 day); and a luteal phase (12–14 days), in which the corpus luteum functions for its programmed life-span.
- A sharp increase in secretion of luteinizing hormone (LH) and to a lesser extent follicle-stimulating hormone (FSH) triggers ovulation.

KEY POINTS (*continued*)

- Luteinizing hormone causes granulosa cells to differentiate to form the corpus luteum and stimulates the corpus luteum to secrete hormones.
- Estrogens promote growth and function of the reproductive tract and stimulate proliferation of the endometrial lining. Estrogens also promote growth of the mammary glands.
- Progesterone promotes endometrial differentiation and, in conjunction with estrodiol, promotes mammary growth.
- Loss of estrogen and progesterone when the corpus luteum regresses results in shedding of the endometrial lining that was built up and sustained by these hormones (menstruation).
- Secretion of FSH and LH is stimulated by intermittent release of gonadotropin-releasing hormone (GnRH) by neurons in the arcuate nuclei of the hypothalamus.
- Secretion of FSH and LH during the follicular and luteal phases of the cycle is restrained by the negative feedback effects of ovarian hormones (negative feedback).
- Growth in the number and secretory capacity of granulosa cells during the follicular phase results in increasing blood concentrations of estradiol, which, upon reaching sustained high levels, triggers release of the ovulatory burst of LH and FSH (positive feedback).
- Negative feedback effects of progesterone on the hypothalamus slow the frequency of GnRH pulses and, on the pituitary, block the estrogen-induced surge of gonadotropin secretion.
- Both negative and positive feedback effects of estradiol are exerted primarily on the pituitary and can occur without a change in the pattern of GnRH secretion, but GnRH pulses increase in magnitude coincident with release of the ovulatory burst of gonadotropin secretion.
- Cues for timing critical events in the menstrual cycle arise in the ovary.

OVERVIEW

The ovaries serve the dual function of producing eggs and the hormones that support reproductive functions. Unlike men, in whom large numbers of gametes are produced continuously from stem cells, women release only one gamete at a time from a limited pool of preformed gametes in a process that is repeated at regular monthly intervals. Each interval encompasses the time needed for the ovum to develop, for preparation of the reproductive tract to receive the fertilized ovum, and for that ovum to become fertilized and pregnancy to be established. If the ovary does not receive a signal that an embryo has begun to develop, the process of gamete maturation begins anew. The principal ovarian hormones are the steroids *estradiol* and *progesterone* and the peptide *inhibin*, which orchestrate the cyclic series of events that unfold in the ovary, pituitary, and reproductive tract. As the ovum develops within its follicle, estradiol stimulates growth of the structures of the reproductive tract that receive the sperm, facilitate fertilization, and ultimately house the developing embryo. Estradiol along with a variety of peptide growth factors produced in the developing follicle acts within the follicle to stimulate proliferation and secretory functions of granulosa cells and thereby enhances its own production. Progesterone is produced by the corpus luteum that develops from the follicle after the egg is shed. It prepares the uterus for successful implantation and growth of the embryo and is absolutely required for the maintenance of pregnancy. Ovarian function is driven by the two pituitary gonadotropins follicle-stimulating hormone (FSH) and luteinizing hormone (LH), which stimulate ovarian steroid production, growth of the follicle, ovulation, and development of the corpus luteum. Secretion of these hormones depends on stimulatory input from the hypothalamus through the gonadotropin-releasing hormone (GnRH) and complex inhibitory and stimulatory input from ovarian steroid and peptide hormones.

FEMALE REPRODUCTIVE TRACT

Ovaries

The adult human ovaries are paired, flattened ellipsoid structures that measure about 5 cm in their longest dimension. They lie within the pelvic area of the abdominal cavity attached to the broad ligaments that extend from either side of the uterus by peritoneal folds called the *mesovaria*. Both the gamete-producing and hormone-producing functions of the ovary take place in the outer or cortical portion. It is within the ovarian cortex that the precursors of the female gametes, the *oocytes*, are stored and develop into *ova* (eggs). The functional unit is the *ovarian follicle*, which initially consists of a single oocyte surrounded by a layer of *granulosa cells* enclosed within a basement membrane, the *basal lamina*, which separates the follicle from cortical stroma. When they emerge from the resting stage, follicles become ensheathed in a layer of specialized cells called the *theca folliculi*. Follicles in many stages of development are found in the cortex of the adult ovary along with structures that form when the mature ovum is released by the process of *ovulation*. Ovarian follicles, in which the ova develop and the corpora lutea derived from them, are also the sites of ovarian hormone production. The inner portion of the ovary, the *medulla*, consists chiefly of vascular elements that arise from anastomoses of the uterine and ovarian arteries. A rich supply of unmyelinated nerve fibers also enters the medulla along with blood vessels (Fig. 1).

Folliculogenesis

In contrast to the testis, which produces hundreds of millions of sperm each day, the ovary normally produces a single mature ovum about once each month. The testis must continuously renew its pool of germ cell precursors

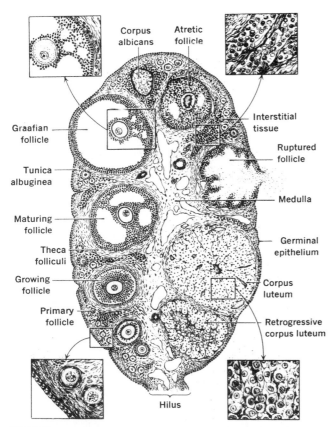

FIGURE 1 Mammalian ovary showing the various stages of follicular and luteal development. Obviously, events depicted occur sequentially and are not all present in any section of a human ovary. (From Turner CD, Bagnara JT. *General endocrinology*, 6th ed., Philadelphia: W.B. Saunders, 1976, p. 453. With permission.)

throughout reproductive life in order to sustain this rate of sperm production, while the ovary must draw upon its initial endowment of primordial oocytes to provide the approximately 500 mature ova ovulated during the four decades of a woman's reproductive life. Although ovulation, the hallmark of ovarian activity, occurs episodically at 28-day intervals, examination of the ovary at any time during childhood or the reproductive life of a mature woman reveals continuous activity with multiple follicles at various stages in their life cycle.

Folliculogenesis begins in fetal life. Primordial germ cells multiply by mitosis. They begin to differentiate into primary oocytes and enter meiosis between the 11th and 20th weeks after fertilization. Primary oocytes remain arrested in prophase of the first meiotic division until meiosis resumes at the time of ovulation, which may be more than four decades later for some oocytes. Meiosis is not completed until the second polar body is extruded at the time of fertilization. Around the 20th week of fetal life, about 6 to 7 million oocytes are available to form primordial follicles, but the human female is born with about only 300,000 to 400,000 primordial follicles in each ovary. Oocytes that fail to form into primordial follicles are lost by apoptosis. The vast majority of primordial follicles remain in a resting state for many years. In a seemingly random process, some follicles enter into a growth phase and begin the long journey toward ovulation, but the vast majority of developing follicles become atretic. This process begins during the fetal period and continues until *menopause* at around age 50, when all of the primordial follicles are exhausted.

As primordial follicles emerge from the resting stage, the oocyte grows from a diameter of about 20 μm to about 100 μm, and a layer of extracellular mucopolysaccharides and proteins called the *zona pellucida* forms around it (Fig. 2). Growth of primary follicles is accompanied by migration and differentiation of mesenchymal cells to form the *theca folliculi*. Its inner layer, the *theca interna*, is composed of secretory cells with an extensive smooth endoplasmic reticulum characteristic of steroidogenic cells. The *theca externa* is formed by reorganization of surrounding stromal cells. At this time, also, a dense capillary network develops around the follicle. The oocyte completes its growth by accumulating stored nutrients and the messenger RNA and protein synthesizing apparatus that will be activated upon fertilization. As the follicle continues to grow, granulosa cells increase in number and begin to form multiple layers. The innermost granulosa cells are in intimate contact with the oocyte through cellular processes that penetrate the zona pellucida and form gap junctions with its plasma membrane. Granulosa cells also form gap junctions with each other and function as nurse cells supplying nutrients to the oocyte, which is

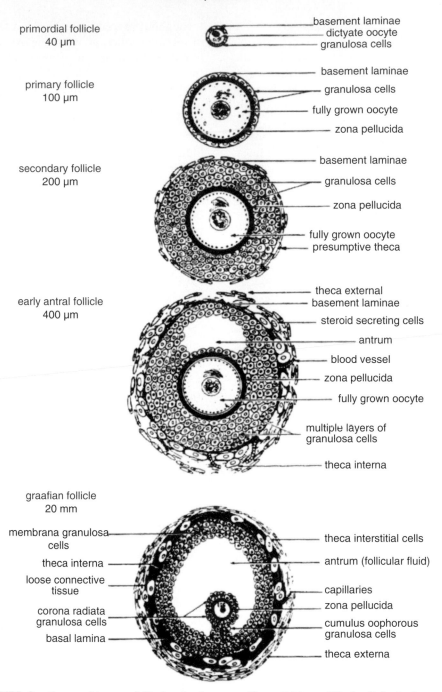

primordial follicle
40 μm
— basement laminae
— dictyate oocyte
— granulosa cells

primary follicle
100 μm
— basement laminae
— granulosa cells
— fully grown oocyte
— zona pellucida

secondary follicle
200 μm
— basement laminae
— granulosa cells
— zona pellucida
— fully grown oocyte
— presumptive theca

early antral follicle
400 μm
— theca external
— basement laminae
— steroid secreting cells
— antrum
— blood vessel
— zona pellucida
— fully grown oocyte
— multiple layers of granulosa cells
— theca interna

graafian follicle
20 mm
membrana granulosa cells
theca interna
loose connective tissue
corona radiata granulosa cells
basal lamina
— theca interstitial cells
— antrum (follicular fluid)
— capillaries
— zona pellucida
— cumulus oophorous granulosa cells
— theca externa

FIGURE 2　Stages of human follicular development. (From Erickson GF. In: Felig P, Baxter JJ, Frohman LA, Eds. *Endocrinology and metabolism,* 3rd ed., New York: McGraw-Hill, 1995, pp. 973–1015. With permission.)

separated from direct contact with capillaries by the basal lamina and the granulosa cells.

Follicular development continues with further proliferation of granulosa cells and gradual elaboration of fluid within the follicle. Follicular fluid is derived from blood plasma and contains plasma proteins, including hormones, and various proteins and steroids secreted by the granulosa cells and the ovum. Accumulation of

follicular fluid brings about further enlargement of the follicle and the formation of a central fluid-filled cavity called the *antrum*. Follicular growth up to this stage is independent of pituitary hormones, but, without support from follicle-stimulating hormone (FSH; see Chapters 38 and 45), further development is not possible and the follicles become atretic. Any follicle can be arrested at any stage of its development and undergo the

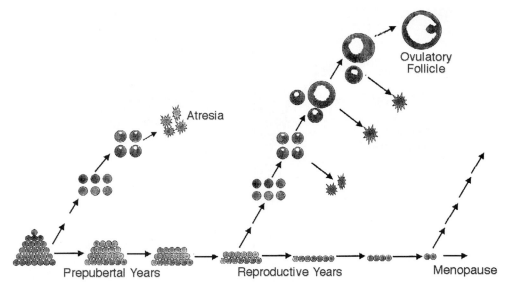

FIGURE 3   Follicular development through the life of a woman. Note the gradual depletion of the pool of primary follicles. (Adapted from McGee EA, Hsueh AJW. *Endocrine Rev* 2000; 21:200–214.)

degenerative changes of atresia. Atresia is the fate of all of the follicles that enter the growth phase before puberty, and more than 99% of the 200,00 to 400,000 remaining at puberty. The physiological mechanisms that control this seemingly wasteful process are poorly understood.

In the presence of FSH, antral follicles continue to develop slowly for about 2 months until they reach a critical size. About 20 days before ovulation, a group or cohort of 6 to 12 of these follicles enters into the final rapid growth phase, but in each cycle normally only one survives and ovulates, while the others become atretic and die (Fig. 3). The surviving follicle has been called the dominant follicle because it may contribute to the demise of other developing follicles. As the dominant follicle matures, the fluid content in the antrum increases rapidly, possibly in response to increased colloid osmotic pressure created by partial hydrolysis of dissolved mucopolysaccharides. The ripe, preovulatory follicle reaches a diameter of 20 to 30 mm and bulges into the peritoneal cavity. At this time it consists of about 60 million granulosa cells arranged in multiple layers around the periphery. The ovum and its surrounding layers of granulosa cells, the *corona radiata*, are suspended by a narrow bridge of granulosa cells (the *cumulus oophorous*) in a pool of more than 6 ml of follicular fluid. At ovulation, a point opposite the ovum in the follicle wall, called the *stigma*, erodes and the ovum with its corona of granulosa cells is extruded into the peritoneal cavity in a bolus of follicular fluid (Fig. 4).

Ovulation is followed by ingrowth and differentiation of the remaining mural granulosa cells, thecal cells, and some stromal cells, which fill the cavity of the collapsed follicle to form a new endocrine structure, the *corpus*

luteum. The process by which granulosa and thecal cell are converted to luteal cells is called *luteinization* (meaning yellowing) and is the morphological reflection of the accumulation of lipid. Luteinization also involves biochemical changes that enable the corpus luteum to become the most active of all steroid-producing tissues per unit weight. The corpus luteum consists of large polygonal cells containing smooth endoplasmic reticulum and a rich supply of fenestrated capillaries. Unless

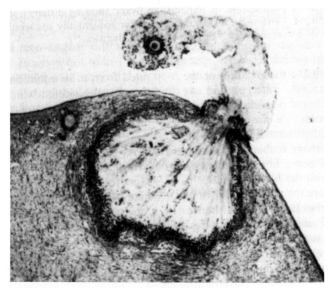

FIGURE 4   Ovulation in a rabbit. Follicular fluid, granulosa cells, some blood, and cellular debris continue to ooze out of the follicle even after the egg mass has been extruded. (From Hafez ESE, Blandau RJ. In: Hafez ESE, Blandau RJ, Eds., *The mammalian oviduct*, Chicago: University of Chicago Press, 1969. With permission.)

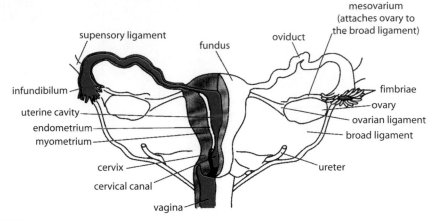

FIGURE 5 Uterus and associated female reproductive structures. The left side of the figure has been sectioned to show the internal structures. (From Tortora GJ, Anagnostakos NP. *Principles of anatomy and physiology,* 3rd ed., New York: Harper & Row, 1981, p. 721. With permission.)

pregnancy ensues, the corpus luteum regresses after 2 weeks, leaving a scar on the surface of the ovary.

Oviducts and Uterus

The primitive müllerian ducts that develop during early embryonic life give rise to the duct system that in primitive organisms provides the route for ova to escape to the outside (Fig. 5). In mammals, these tubes are adapted to provide a site for fertilization of ova and nurture of embryos. In female embryos, the müllerian ducts are not subjected to the destructive effects of the anti-müllerian hormone (see Chapter 45) and, instead, develop into the oviducts, uterus, and upper portion of the vagina. Unlike the development of the sexual duct system in the male fetus, this differentiation is independent of gonadal hormones.

The paired oviducts (*fallopian tubes*) are a conduit for transfer of the ovum to the uterus (see Chapter 47). The ovarian end comes in close contact with the ovary and has a funnel-shaped opening, the *infundibulum*, surrounded by finger-like projections called *fimbriae*. The oviduct, particularly the infundibulum, is lined with ciliated cells, the synchronous beating of which plays an important role in egg transport. The lining of the oviduct also contains secretory cells whose products provide nourishment for the zygote in its 3- to 4-day journey to the uterus. The walls of the oviducts contain layers of smooth muscle cells oriented either longitudinally or circumferentially.

Distal portions of the müllerian ducts fuse to give rise to the uterus. In the nonpregnant woman, the uterus is a small, pear-shaped structure extending about 6 to 7 cm in its longest dimension. It is capable of enormous expansion, partly by passive stretching and partly by growth, so that at the end of pregnancy it may reach 35 cm or more

in its longest dimension. Its thick walls consist mainly of smooth muscle and are called the *myometrium*. The secretory epithelial lining is called the *endometrium* and varies in thickness with changes in the hormonal environment, as discussed below. The oviducts join the uterus at the upper, rounded end. The caudal end constricts to a narrow cylinder called the *uterine cervix*, whose thick wall is composed largely of dense connective tissue rich in collagen fibers and some smooth muscle. The cervical canal is lined with mucus-producing cells and is usually filled with mucus. The cervix bulges into the upper reaches of the vagina, which forms the final link to the outside. The lower portion of the vagina, which communicates with the exterior, is formed from the embryonic urogenital sinus.

OVARIAN HORMONES

The principal hormones secreted by the ovary are estrogens (*estradiol-17β* and *estrone*) and *progesterone*. These hormones are steroids and are derived from cholesterol by the series of reactions depicted in Fig. 6. Their biosynthesis is intricately interwoven with the events of the ovarian cycle. In addition, the ovary produces a large number of biologically active peptides, most of which act within the ovary as paracrine growth factors, but at least two, *inhibin* and *relaxin*, are produced in sufficient amounts to enter the blood and produce effects in distant cells.

Estrogens

Unlike humans, who, it has been said, "eat when they are not hungry, drink when they are not thirsty, and make love at all seasons of the year," most vertebrate animals mate only at times of maximum fertility of the female. This period of sexual receptivity is called *estrus*, derived from the Greek word for vehement desire. Estrogens are compounds that promote estrus and were originally isolated from the follicular fluid of sow ovaries. Characteristic of steroid-secreting tissues, little hormone is stored within the secretory cells themselves. Estrogens circulate in blood loosely bound to albumin and tightly bound to the *testosterone–estradiol-binding globulin* (TeBG), which is also called the *sex-hormone-binding globulin* (SHBG; see Chapter 45). Plasma concentrations of estrogens are considerably lower than those of other gonadal steroids and vary over an almost 20-fold range during the cycle.

Liver is the principal site of metabolic destruction of the estrogens. Estradiol and estrone are completely cleared from the blood by a single passage through the liver and are inactivated by hydroxylation and conjugation with sulfate and glucuronide. About half the protein-bound estrogen in blood is conjugated with sulfate or glucuronide. Conjugated estrogens excreted in the bile are reabsorbed in the lower gut and returned to the liver in portal blood in a typical enterohepatic circulatory pattern. The kidney is the chief route of excretion of estrogenic metabolites.

Progesterone

Pregnancy, or *gestation*, requires the presence of another ovarian steroid hormone, progesterone. In the nonpregnant woman, progesterone secretion is largely confined to cells of the corpus luteum, but, because it is an intermediate in the biosynthesis of all steroid hormones, small amounts may also be released from the adrenal cortex. Some progesterone is also produced by granulosa cells just before ovulation. The rate of progesterone production varies widely. Its concentration in blood ranges from virtually nil during the early preovulatory part of the ovarian cycle to as much as 2 mg/dL after the corpus luteum has formed. Progesterone circulates in blood in association with plasma proteins and has a high affinity for the corticosteroid-binding globulin (CBG). Liver is the principal site of progesterone inactivation, which is achieved by reduction of the A ring and the keto groups at carbons 3 and 20 to give *pregnanediol*, which is the chief metabolite found in urine. Considerable degradation also occurs in the uterus.

Inhibin

As discussed in Chapter 45, inhibin is a 32-kDa disulfide-linked dimer of an α subunit and either of two β subunits, β_A or β_B and enters the circulation as either inhibin A (α/β_A) or inhibin B (α/β_B). Expression of the β_A subunit is greatest in luteal cells, while expression of the β_B subunit is a product of granulosa cells. Consequently, blood levels of inhibin B are highest during the periods of preovulatory growth and expansion of granulosa cells, while blood levels of inhibin A are highest during peak luteal cell function. In addition to serving as a circulating hormone, inhibin probably exerts paracrine actions in the ovary, and activin formed by dimerization of two β subunits also exerts important intraovarian paracrine actions. The activin-binding peptide *follistatin*, which blocks activin action, also plays an important intraovarian role. Although some activin is found in the circulation, its concentrations do not change during the ovarian cycle, and its source is primarily extra-ovarian.

Relaxin

The corpus luteum also secretes a second peptide hormone called *relaxin*, which was named for its ability to relax the pubic ligament of the pregnant guinea pig. In other species, including humans, it also relaxes the myometrium and plays an important role in parturition by causing softening of the uterine cervix. Its peptide structure, particularly the organization of its disulfide bonds places it in the same family as insulin and the insulin-like growth factors (IGFs). A physiological role for relaxin in the nonpregnant woman has not been established.

EFFECTS OF FOLLICLE STIMULATING HORMONE AND LUTEINIZING HORMONE

Follicular development beyond the antral stage depends on two gonadotropic hormones secreted by the anterior pituitary gland: FSH and LH. In addition to follicular growth, gonadotropins are required for ovulation, luteinization, and steroid hormone formation by both the follicle and the corpus luteum. The relevant molecular and biochemical characteristics of these glycoprotein hormones are described in Chapters 38 and 45. Follicular growth and function also depend on paracrine effects of estrogens, androgens, and possibly progesterone, as well as peptide paracrine factors, including IGF-II, activin, members of the transforming growth factor-β family, and others. The sequence of rapid follicular growth, ovulation, and the subsequent formation and degeneration of the corpus luteum is repeated about every 28 days and constitutes the ovarian cycle. The part of the cycle devoted to the final rapid growth of the ovulatory follicle lasts about 14 days and is called the *follicular phase*. The remainder of the cycle is dominated

Clinical Note

During the first few days of the follicular phase, multiple follicles have the potential to ovulate and can be rescued from atresia by treatment with supraphysiological amounts of FSH. This accounts for the high frequency of multiple births following therapies for infertility that involve administration of gonadotropins or agents such as clomiphene that stimulate endogenous gonadotropin secretion. Production of multiple ova (superovulation) induced in this way is used for harvesting eggs for *in vitro* fertilization technologies.

by the corpus luteum and is called the *luteal phase*. It also lasts about 14 days. Ovulation occurs at midcycle and requires only about a day. These events are orchestrated by a complex pattern of pituitary and ovarian hormonal changes. Even as the ripening follicle is prepared for ovulation, preparations are underway in the background to set the stage for the next cycle in the event that fertilization does not occur.

Early growth of follicles from the primordial to the preantral stage is independent of pituitary hormones and is likely governed by paracrine factors produced by the ovum itself as well as by granulosa and thecal cells. Follicular sensitivity to gonadotropins becomes evident at the early antral stage and gradually increases as individual follicles in each cohort slowly increase from about 0.2 to 2 mm in diameter, at which time they are capable of undergoing the rapid growth and development that lead to ovulation. Approximately 85 days elapse between entry of a cohort of follicles into the gonadotropin-responsive stage and ovulation of one of them. During this time all follicles in both ovaries are exposed to wide swings in gonadotropin concentrations, but their capacity to respond is limited by their degree of development. Ovulation occurs at the midpoint of the fourth cycle after a cohort of follicles reaches the antral stage and has been growing in response to gonadotropins; consequently, the next three cohorts are already being prepared to ovulate. The number of follicles with the potential to ovulate in each cohort is reduced by atresia at all stages of development.

Under normal circumstances, only one follicle ovulates in each cycle. The ovulatory follicle is randomly located on either the right or left ovary. Usually 6 to 12 follicles are mature enough to enter into the final preovulatory growth period near the end of the luteal phase of the preceding cycle and begin to grow rapidly in response to the increase in FSH that occurs at that time. The ovulatory or "dominant" follicle is selected from this group early in the follicular phase that leads to its ovulation. The physiological mechanisms for selection of a single dominant follicle are not understood. Recruitment of the next cohort of follicles does not begin as long as the dominant follicle or its resultant corpus luteum is present and functional. Experimental destruction of either the dominant follicle or the corpus luteum is promptly followed by selection and development of a new ovulatory follicle from the next cohort.

Estradiol Production

Granulosa cells in antral follicles are the only targets for FSH. No other ovarian cells are known to have FSH receptors. Granulosa cells of the ovulatory follicle are the major and virtually only source of estradiol in the follicular phase of the ovarian cycle and secrete estrogens in response to FSH. Until about the middle of the follicular phase, LH receptors are found only in cells of the theca interna and the stroma. LH stimulates thecal cells to produce androstenedione. Follicular synthesis of estrogen depends on complex interactions between the two gonadotropins and between theca and granulosa cells. Although isolated theca and granulosa cells may be able to synthesize some estrogens, cooperative interaction of both cell types is required for physiologically relevant rates of estradiol production in the follicular phase. Neither granulosa nor theca cells express the full complement of enzymes needed for synthesis of estradiol. Granulosa cells are limited in their capacity to produce pregnenolone because they have little access to cholesterol delivered by the circulation in the form of low-density lipoproteins (LDLs), express few LDL receptors, and have minimal levels of the P450scc necessary to convert cholesterol to pregnenolone (Fig. 6). Theca cells, which have a direct capillary blood supply, express high levels of LDL receptors, as well as high levels of P450scc and P450c17. Theca cells thus can metabolize the 21-carbon pregnenolone to the 19-carbon androstenedione but lack aromatase and hence cannot synthesize estrogens. Granulosa cells, on the other hand, express ample aromatase but cannot produce its 19-carbon substrate because they lack P450c17; however, granulosa cells readily aromatize androgens provided by diffusion from the theca interna. This two-cell interaction is illustrated in Fig. 7. The participation of two different

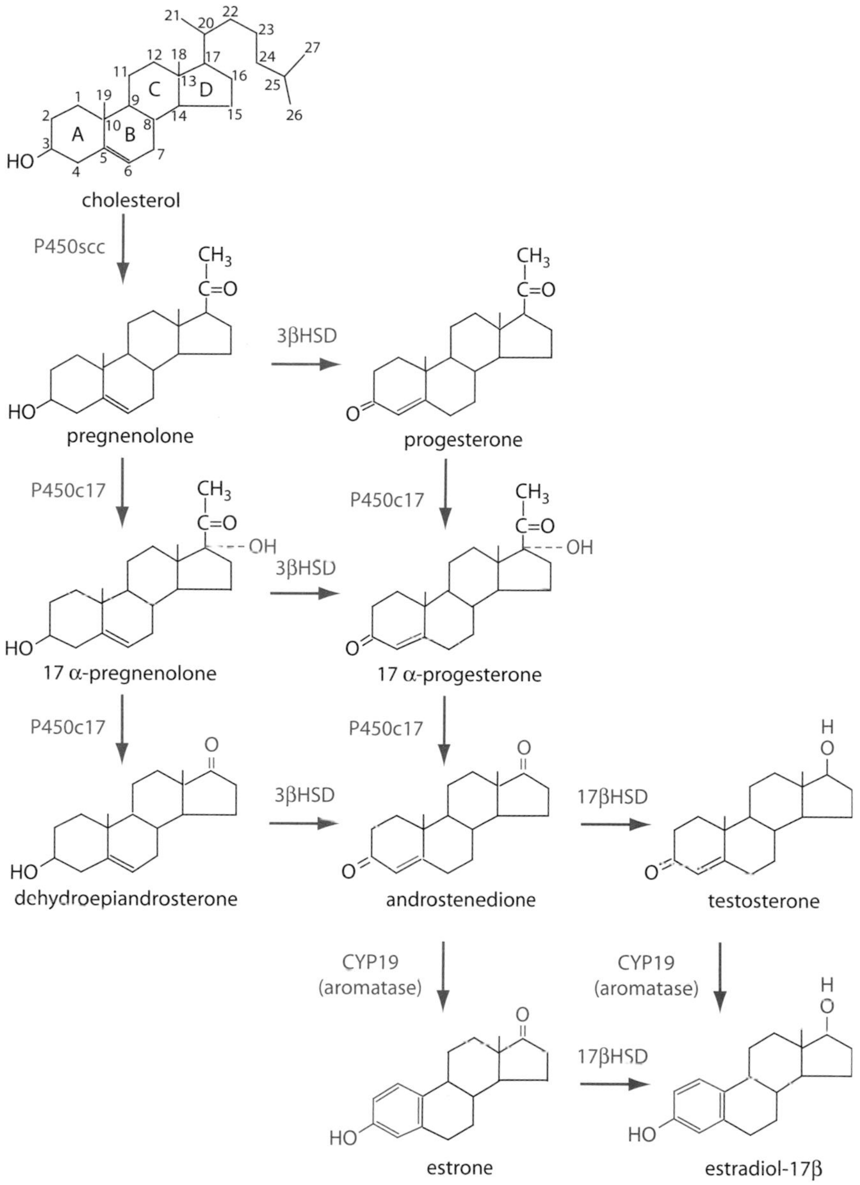

FIGURE 6 Biosynthesis of ovarian hormones. Cleavage of the cholesterol side chain by P450scc between carbons 21 and 22 gives rise to 21-carbon progestins. Removal of carbons 20 and 21 by the two-step reaction catalyzed by P450c17 (17α hydroxylase/lyase) produces the 19-carbon androgen series. Aromatization of ring A catalyzed by P450cyp19 (aromatase) eliminates carbon 19 and yields 18-carbon estrogens. 3βHSD, 3β hydroxysteroid dehydrogenase; 17βHSD, 17β hydroxysteroid dehydrogenase.

cells, each stimulated by its own gonadotropin, accounts for the requirement of both pituitary hormones for adequate estrogen production and hence for follicular development.

Follicular Development

Under the influence of FSH, granulosa cells in the follicle destined to ovulate increase by more than 100-fold and the follicle expands about 10-fold in diameter, mainly because of the increase in follicular fluid. Differentiation of granulosa cells leads to increased expression of a number of genes including that for the LH receptor. In early antral follicles, granulosa cells have few if any receptors for LH and are unresponsive to it. By about the middle of the follicular phase, granulosa cell begin to express increasing amounts of LH receptor (Fig. 8), which becomes quite abundant just prior to ovulation. Acquisition of LH receptors in response to FSH enables granulosa cells to respond to both FSH and LH.

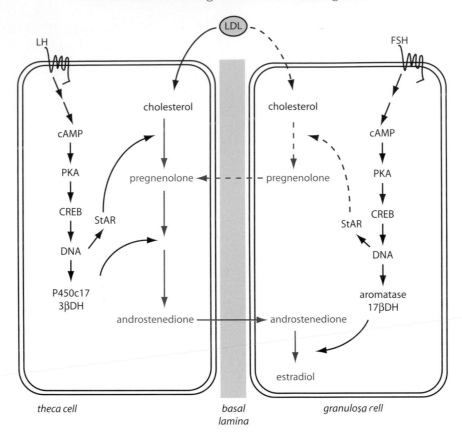

FIGURE 7 Theca and granulosa cell cooperation in estrogen synthesis. Theca cells produce androgens in response to LH. Granulosa cells respond to FSH by aromatizing androgens to estrogens. LDL, low density lipoproteins; cAMP, cyclic adenosine monophosphate; PKA, protein kinase A; DNA, deoxyribonucleic acid; CREB, cAMP response element binding protein; StAR, steroid acute regulatory protein; P450c17, 17α dehydrogenase/lyase; 3βDH, 3β hydroxysteroid dehydrogenase; 17βDH, 17β hydroxysteroid dehydrogenase. Dashed arrows indicate minor or questionable effect.

Induction of LH receptors and other actions of FSH on follicular development are amplified by paracrine actions of peptide growth factors and steroid hormones. In response to FSH, granulosa cells secrete insulin-like growth factor II (IGF-II), inhibin, activin, and other growth factors, including vascular endothelial growth factor (VEGF), which greatly increases vascularization around the theca. IGF-II not only stimulates the growth and secretory capacity of granulosa cells but also acts synergistically with LH to increase synthesis of androgens by cells of the theca interna. Similarly, inhibin produced by granulosa cells in response to FSH also stimulates thecal cell production of androgens, while activin enhances FSH-induced expression of P450 aromatase in the granulosa cells. Thus, even though theca cells lack FSH receptors, they nevertheless respond indirectly to FSH with increased production of the androgens required by the granulosa cells for estrogen secretion. Similarly, by stimulating thecal cells to produce androgens, LH augments growth and development of granulosa cells through the paracrine actions of both estrogen and androgens, the receptors of which are abundantly expressed in these cells. These events constitute a local positive feedback circuit that gives the follicle progressively greater capacity to produce estradiol and makes it increasingly sensitive to FSH and LH as it matures.

Cellular Actions

Follicle stimulating hormone and LH each bind to specific G-protein-coupled receptors on the surface of granulosa or theca cells and activate adenylyl cyclase in the manner already described (Chapters 2 and 45). Increased concentrations of cAMP activate protein kinase A, which catalyzes phosphorylation of cAMP response element binding (CREB) protein and other nuclear and cytoplasmic proteins that ultimately lead to increased transcription of genes that encode growth factors and other proteins critical for cell growth and steroid hormone production. As described for adrenal cortical (Chapter 40) and Leydig cells (Chapter 45), the rate-limiting step in steroid hormone synthesis requires synthesis of the steroid acute regulatory protein (StAR) to deliver cholesterol to the intramitochondrial enzyme P450scc, which converts it to pregnenolone (Fig. 6). In addition, activation of protein kinase A results in increased expression of P450c17 in theca cells and P450 aromatase in granulosa cells. Along with increased expression of growth factors, the ripening follicle also expresses increased amounts of a protease that specifically cleaves IGF-4 binding protein (IGFBP-4) which is present in follicular fluid and thereby increases availability of free IGF-II.

Ovulation

Luteinizing hormone is the physiological signal for ovulation. Its concentration in blood rises sharply and reaches a peak about 16 hours before ovulation (see below). Blood levels of FSH also increase at this time, and, although large amounts of FSH can also cause ovulation, the required concentrations are not achieved during the normal reproductive cycle. The events that lead to follicular rupture and expulsion of the ovum are not fully understood, but the process is known to be initiated by increased production of cAMP in theca and granulosa cells in response to LH and the consequent release of paracrine factors and enzymes.

As the follicle approaches ovulation, it accumulates follicular fluid, but, despite the preovulatory swelling, intrafollicular pressure does not increase. The follicular wall becomes increasingly distensible due to activity of proteolytic enzymes that digest the collagen framework and other proteins of the intercellular matrix. One of these enzymes, plasmin, accumulates in follicular fluid in the form of its inactive precursor, plasminogen. Granulosa cells secrete *plasminogen activator* in response to hormonal stimulation. Because of their newly acquired receptors, granulosa cells of the preovulatory follicle respond to LH by secreting progesterone, which is thought to induce the formation of prostaglandins. The finding that pharmacological blockade of either prostaglandin or progesterone synthesis prevents ovulation indicates that these agents play essential roles in the ovulatory process. Prostaglandins appear to activate release of lysosomal proteases in a discrete region of the follicle wall called the *stigma*. Breakdown of the extracellular matrix of the theca and the surface epithelium of the ovary facilitates extrusion of the ovum into the abdominal cavity.

Although little or no progesterone is produced throughout most of the follicular phase, granulosa cells of the preovulatory follicle acquire the capacity for progesterone production. Stimulation by LH evokes expression of LDL receptors, P450scc, and doubtless other relevant proteins, enabling them to take up cholesterol and convert it to pregnenolone. Because the capacity to remove the side chain at carbon 17 remains limited, 21-carbon steroids are formed faster than they can be processed to estradiol and hence are secreted as progesterone. Furthermore, as granulosa cells acquire the ability to respond to LH, they also begin to lose aromatase activity (Fig. 8). This is reflected in the abrupt decline in estrogen production that just precedes ovulation.

Corpus Luteum Formation

Luteinizing hormone was named for its ability to induce formation of the corpus luteum after ovulation; however, as already mentioned, luteinization may actually begin before the follicle ruptures. Granulosa cells removed from mature follicles complete their luteinization in tissue culture without further stimulation by gonadotropin. Nevertheless, luteinization within the ovary depends on LH and is accelerated by the increased concentration of LH that precedes ovulation. Occasionally, luteinization occurs in the absence of ovulation and results in the syndrome of luteinized unruptured follicles, which may be a cause of infertility in some women whose reproductive cycles seem otherwise normal.

Development of a vascular supply is critical for development of the corpus luteum and its function. Although granulosa cells of the preovulatory follicle are avascular, the corpus luteum, is highly vascular, and when fully developed each steroidogenic cell appears to be in contact with at least one capillary. After extrusion of the ovum infolding of the collapsing follicle causes the highly vascular theca interna to interdigitate with layers of granulosa cells that line the follicular wall. Under the influence of LH, granulosa cells express high levels of VEGF, which stimulates growth and differentiation of capillary endothelial cells. It has been estimated that vascular endothelial cells make up fully half of the cells of the mature corpus luteum.

Oocyte Maturation

Granulosa cells not only provide nutrients to the ovum but may also prevent it from completing its meiotic division until the time of ovulation. Granulosa cells are thought to secrete a substance called *oocyte maturation inhibitor* (OMI) into follicular fluid. LH triggers resumption of meiosis at the time of ovulation, perhaps by blocking production of this factor or interfering with its action.

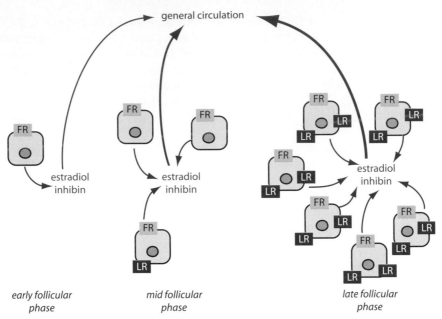

FIGURE 8 Proliferation and differentiation of granulosa cells during follicular development. Initially, granulosa cells are few and have receptors only for FSH on their surface. In response to continued stimulation with FSH, granulosa cells proliferate and by the midfollicular phase LH receptors begin to appear. By late in the follicular phase a large number of granulosa cells are present and they are responsive to both LH and FSH. They are now competent to secrete sufficient estradiol to trigger the ovulatory surge of gonadotropins. FR, FSH receptor; LR, LH receptor.

Corpus Luteal Function

Maintenance of steroid production by the corpus luteum depends on continued stimulation with LH. Decreased production of progesterone and premature demise of the corpus luteum is seen in women whose secretion of LH is blocked pharmacologically. In this respect, LH is said to be *luteotropic*. The corpus luteum has a finite life-span, however, and about a week after ovulation becomes progressively less sensitive to LH and finally regresses despite continued stimulation with LH. Estradiol and prostaglandin $F_{2\alpha}$, which are produced by the corpus luteum, can hasten luteolysis and may be responsible for its demise. We do not understand the mechanisms that limit the functional life-span of the human corpus luteum.

Ovarian Blood Flow

Luteinizing hormone also increases blood flow to the ovary and produces ovarian hyperemia throughout the cycle. In addition to increased angiogenesis, blood flow may be enhanced by release of histamine or perhaps prostaglandins. Increased ovarian blood flow increases the opportunity for delivery of steroid hormones to the general circulation and for delivery to the ovary of cholesterol-laden LDL needed to support high rates of steroidogenesis. Increased blood flow to the developing follicle may also be important for preovulatory swelling of the follicle, which depends on increased elaboration of fluid from blood plasma.

PHYSIOLOGICAL ACTIONS OF OVARIAN STEROID HORMONES

As described above, intraovarian actions of estradiol and progesterone are intimately connected to ovulation and formation of the corpus luteum. In general, extra-ovarian actions of these hormones ensure that the ovum reaches its potential to develop into a new individual. Ovarian steroids act on the reproductive tract to prepare it for fulfilling its role in fertilization, implantation, and development of the embryo, and they induce changes elsewhere that equip the female physically and behaviorally for conceiving, giving birth, and rearing the child. Although estrogens, perhaps in concert with progesterone, drive females of subprimate species to mate, androgens, rather than estrogens, are responsible for libido in humans of either sex. Estrogens and progesterone tend to act in concert and sometimes enhance or antagonize each other's actions. Estrogen secretion usually precedes progesterone secretion and primes the target tissues to respond to progesterone. Estrogens induce the synthesis of progesterone receptors. Without estrogen priming,

TABLE 1   Effects of Estrogen and Progesterone on the Reproductive Tract

Organ	Estrogen	Progesterone
Oviducts		
Lining	↑ Cilia formation and activity	↑ Secretion
Muscular wall	↑ Contractility	↓ Contractility
Uterus		
Endometrium	↑ Proliferation	↑ Differentiation and secretion
Myometrium	↑ Growth and contractility	↓ Contractility
Cervical glands	Watery secretion	Dense, viscous secretion
Vagina		
	↑ Epithelial proliferation	↑ Differentiation
	↑ Glycogen deposition	↓ Proliferation

progesterone has little biological effect. Conversely, progesterone downregulates its own receptors and estrogen receptors in some tissues and thereby decreases responses to estrogens.

Effects on the Reproductive Tract

At puberty estrogens promote growth and development of the oviducts, uterus, vagina, and external genitalia. Estrogens stimulate cellular proliferation in the mucosal linings as well as in the muscular coats of these structures. Even after they have matured, maintenance of size and function of internal reproductive organs requires continued stimulation by estrogen and progesterone. Prolonged deprivation after ovariectomy results in severe involution of both muscular and mucosal portions. Dramatic changes are also evident, especially in the mucosal linings of these structures, as steroid hormones wax and wane during the reproductive cycle. These effects of estrogen and progesterone are summarized in Table 1.

Menstruation

Nowhere are the effects of estrogen and progesterone more obvious than in the endometrium. Estrogens secreted by developing follicles increase the thickness of the endometrium by stimulating growth of epithelial cells in terms of both number and height. Endometrial glands form and elongate. Endometrial growth is accompanied by increased blood flow, especially through the spiral arteries, which grow rapidly under the influence of estrogens. This stage of the uterine cycle is known as the *proliferative phase* and coincides with the follicular phase of the ovarian cycle. Progesterone secreted by the corpus luteum causes the newly proliferated endometrial lining to differentiate and become secretory. This action is consistent with its role of preparing the uterus for nurture and implantation of the newly fertilized ovum if successful mating has occurred.

The so-called uterine milk secreted by the endometrium is thought to nourish the blastocyst until it can implant. This portion of the uterine cycle is called the *secretory phase* and coincides with the luteal phase of the ovarian cycle (Fig. 9).

Maintaining the thickened endometrium depends on the continued presence of the ovarian steroid hormones. After the regressing corpus luteum loses its ability to produce adequate amounts of estradiol and progesterone, the outer portion of the endometrium degenerates and is sloughed into the uterine cavity. The mechanism for shedding the uterine lining is incompletely understood, although prostaglandin $F_{2\alpha}$ appears to play an important role, perhaps in producing vascular spasm and ischemia and in stimulating release of lysosomal proteases. Loss of the proliferated endometrium is accompanied by bleeding. This monthly vaginal discharge of blood is known as *menstruation*. The typical menstrual period lasts 3 to 5 days and the total flow of blood seldom exceeds 50 ml. The first menstrual bleeding, called *menarche*, usually occurs at about age 13. Menstruation continues at monthly intervals until menopause, normally interrupted only by periods of pregnancy.

In the myometrium, estrogen increases expression of contractile proteins, gap junction formation, and spontaneous contractile activity. In its absence, uterine muscle is insensitive to stretch or other stimuli for contraction. Further estrogen increases the irritability of uterine smooth muscle and, in particular, increases its sensitivity to oxytocin, in part as a consequence of inducing uterine receptors for oxytocin (see Chapter 47). The latter phenomenon may be of significance during parturition. Progesterone counteracts these effects and decreases both the amplitude and frequency of spontaneous contractions. Withdrawal of progesterone prior to menstruation is accompanied by increased myometrial prostaglandin formation. Myometrial contractions in response to prostaglandins are thought to account for the discomfort that precedes menstruation.

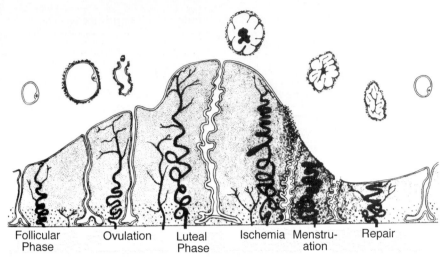

FIGURE 9   Endometrial changes during a typical menstrual cycle. Simultaneous events in the ovary are also indicated. The endometrium thickens during the follicular phase; uterine glands elongate, and spiral arteries grow to supply the thickened endometrium. During the early luteal phase there is further thickening of the endometrium, marked growth of the coiled arteries, and increased complexity of the uterine glands. As the corpus luteum wanes, the endometrial thickness is reduced by loss of ground substance. Increased coiling of spiral arteries causes ischemia and finally sloughing of endometrium. (From Bartelmez GW. *Am J Obstet Gynecol* 1957; 74:931. With permission.)

Effects on the Mammary Glands

Development of the breasts begins early in puberty and is due primarily to estrogen, which promotes development of the duct system and growth and pigmentation of the nipples and areolar portions of the breast. In cooperation with progesterone, estrogen may also increase the lobuloalveolar portions of the glands, but alveolar development also requires the pituitary hormone prolactin (see Chapter 47). Secretory components, however, account for only about 20% of the mass of the adult breast. The remainder is stromal tissue and fat. Estrogen also stimulates stromal proliferation and fat deposition.

Other Effects of Ovarian Hormones

Estrogen also acts on the body in ways that are not necessarily related to reproduction. As already indicated (see Chapter 44), it contributes to the pubertal growth spurt and its termination by directly stimulating growth of cartilage progenitors, increased growth hormone secretion, and closure of the epiphyseal plates. In adolescent females and males, estrogens increase bone density, and in the adult they contribute to the maintenance of bone density by stimulating osteoblastic activity and inhibiting bone resorption. Estrogen can also cause selective changes in bone structure, especially widening of the pelvis, which facilitates passage of the infant through the birth canal. It promotes deposition of subcutaneous fat, particularly in the thighs and buttocks, and increases hepatic synthesis of steroid- and thyroid-hormone-binding proteins. Based on epidemiological evidence, estrogens are considered to have cardiovascular protective effects. Indeed, estrogen receptors are expressed in vascular smooth muscle and cardiac myocytes, and estrogens have been found to alter plasma lipid profiles. Estradiol also acts on various cells in the central nervous system and is responsible for some behavioral patterns, especially in lower animals.

Clinical Note

Treatment of Neoplastic Breast Disease

Responsiveness of breast tissue elements to growth-promoting effects of estrogen is of significance in treatment of neoplastic breast disease. Some forms of breast cancer continue to express estrogen receptors and remain partially or completely dependent on estrogens for growth. These cancers are sensitive to treatment with estrogen antagonists, such as tamoxifen, or with a newer class of drugs (aromatase inhibitors) that block the activity of the enzyme P450arom and therefore starve these estrogen-dependent tumors of estrogens.

Clinical Note

Progesterone and Body Temperature
Progesterone has a mild thermogenic effect and may increase basal body temperature by as much as 1°F. Because the appearance of progesterone indicates the presence of a corpus luteum, a woman can readily determine when ovulation has

occurred, and hence the time of maximum fertility, by monitoring her temperature daily. This simple, noninvasive procedure has been helpful for couples seeking to conceive a child or who are practicing the "rhythm method" of contraception.

More recently evidence has been brought forth that estrogens may protect cognitive functions.

Progesterone also acts on the central nervous system, and, in addition to its effects on regulation of gonadotropin secretion (see below), it may produce changes in behavior or mood. It is curious that more dramatic effects may result from withdrawal of progesterone than from administering it. Thus, withdrawal of progesterone may trigger menstruation, lactation, parturition, and the postpartum psychic depression experienced by many women.

Mechanism of Action

Estrogens and progesterone, like other steroid hormones, readily penetrate cell membranes and bind to receptors that are members of the nuclear receptor superfamily of transcription factors (see Chapter 2). In addition, various nongenomic effects of both hormones have been reported but the molecular processes involved in these actions are incompletely understood. Two separate estrogen receptors, designated ERα and ERβ, are products of different genes and are expressed in many different cells in both reproductive and nonreproductive tissues of both men and women. Their DNA binding domains are almost identical. While some differences are present in their ligand binding domains, ERα and ERβ differ mainly in the amino acid sequences of their activation function (AF) domains; therefore, they interact with different transcriptional coactivator and other nuclear regulatory proteins to control expression of different sets of genes. Upon binding ligand, ERα and ERβ may form homo- or heterodimers before binding to DNA in estrogen-sensitive genes. The resulting synthesis of new messenger RNA is followed, in turn, by the formation of a variety of proteins that modify cellular activity. Both receptors can also be activated by phosphorylation in the absence of ligand, and both receptors can modulate the transcription-activating properties of other nuclear regulatory proteins without directly binding to DNA. Finally, estrogen receptors have the interesting property of assuming different conformational changes depending upon the particular ligand that

is bound. Because the conformation of the liganded receptor profoundly affects its ability to interact with other transcription regulators for which expression is cell type specific, some compounds produced in plants (*phytoestrogens*) or pharmaceutically manufactured "antiestrogens" may block the effects of estrogen in some tissues while acting as estrogen agonists in other tissues. Additionally, some compounds may bind to activate ERα but antagonize actions of ERβ. These properties have given rise to a very important category of drugs called *selective estrogen receptor modulators* (SERMS), which, for example, may block the undesirable proliferative or neoplastic actions of estrogens on the breast and uterus while mimicking desirable effects on maintenance of bone density in postmenopausal women. One of these compounds, tamoxifen, is routinely used in treatment of breast cancer.

Two isoforms of the progesterone receptor, PRA and PRB, are expressed by progesterone-responsive cells. They are products of the same gene whose mRNA transcript contains two alternate translation start sites. PRA and PRB differ by the presence of an additional sequence of amino acids at the amino terminus of PRB that provides an additional region for interacting with nuclear regulatory proteins. Liganded PRs bind to the DNA of target genes as homodimers or heterodimers and in different combinations may activate different subsets of genes. When expressed together in some cells, liganded PRA can repress the activity of PRB. Both PRA and PRB expression are induced by prior exposure of cells to estrogens and can repress the activity or expression of ERα. Studies under way in mice in which expression of specific progesterone or estrogen receptors is disrupted by genetic manipulation are determining which particular responses to ovarian steroid hormones are mediated by each receptor.

REGULATION OF THE REPRODUCTIVE CYCLE

The central event of each ovarian cycle is ovulation, which is triggered by a massive increase in blood LH concentration. This surge of LH secretion must be timed to occur when the ovum and its follicle are ready. The corpus luteum must secrete its hormones to optimize the opportunity for fertilization and establishment of pregnancy. The period after ovulation during which the ovum can be fertilized is brief and lasts less than 24 hours. If fertilization does not occur, a new follicle must be prepared. Coordination of these events requires two-way communication between the pituitary and the ovaries and between the ovaries and the reproductive tract. Examination of the changing pattern of hormones in blood throughout the ovarian cycle provides some insight into these communications.

Pattern of Hormones in Blood During Ovarian Cycle

Figure 10 illustrates daily changes in the concentrations of major hormones in a typical cycle extending from one menstrual period to the next. The only remarkable feature of the profile of gonadotropin concentrations is the dramatic peak in LH and FSH that precedes ovulation. Except for the 2 to 3 days of the midcycle peak, LH concentrations remain at nearly constant low levels throughout the follicular and luteal phases. The concentration of FSH is also low during both phases of the cycle, but a secondary peak is evident early in the follicular phase and diminishes as ovulation approaches. Blood levels of FSH remain low throughout most of the luteal phase but begin to rise 1 or 2 days before the onset of menstruation.

The ovarian hormones follow a different pattern. Early in the follicular phase, the concentration of estradiol is low. It then gradually increases at an increasing rate until it reaches its zenith about 12 hours before the peak in LH; thereafter, estradiol levels fall abruptly and reach a nadir just after the LH peak. During the luteal phase, there is a secondary rise in estradiol concentration, which then falls to the early follicular level a few days before the onset of menstruation. Progesterone is barely or not at all detectable throughout most of the follicular phase and then begins to rise along with LH at the onset of the ovulatory peak. Progesterone continues to rise and reaches its maximum concentration several days after the LH peak has ended. Progesterone levels remain high for about 7 days and then gradually fall and reach almost undetectable levels a day or two before the onset of menstruation. Inhibin B concentrations are low early in the follicular phase and then rise and fall in parallel with FSH. The apparent peak that coincides with ovulation is thought to result from the absorption of inhibin B already present in high concentration in the expelled follicular fluid rather than concurrent secretion by

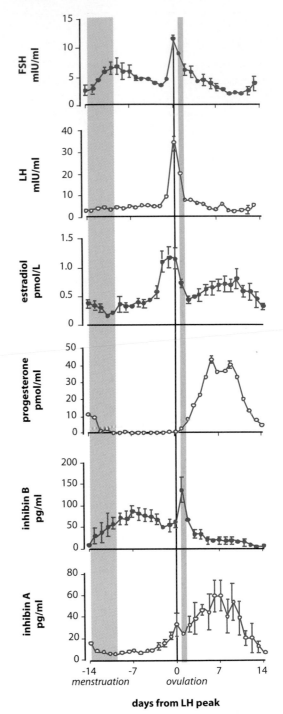

FIGURE 10   Mean values of LH, FSH, progesterone, estradiol, and inhibin in daily serum samples of women during ovulatory menstrual cycles. Data from various cycles are combined, using the midcycle peak of LH as the reference point (day 0). Vertical bars indicate standard error of mean. (Adapted from Groome NJ, Illingworth PJ, O'Brien M, Pal R, Faye ER, Mather JP, McNeilly AS. *J Clin Endocrinol Metab* 1996; 81:1401–1405.)

granulosa cells. Concentrations of inhibin A reach their highest levels in the luteal phase before declining in parallel with progesterone.

Regulation of Follicle-Stimulating Hormone and Luteinizing Hormone Secretion

At first glance, this pattern of hormone concentrations is unlike anything seen for other anterior pituitary hormones and the secretions of their target glands. Indeed, there are unique aspects, but during most of the cycle, gonadotropin secretion is under negative feedback control similar to that seen for TSH (Chapter 39), ACTH (Chapter 40), and the gonadotropins in men (Chapter 45). The ovulatory burst of FSH and LH secretion is brought about by a positive feedback mechanism unlike any we have considered. Secretion of FSH and LH is also controlled by GnRH, which is secreted in synchronized pulsatile bursts by neurons whose cell bodies reside in the arcuate nuclei and the medial preoptic area of the hypothalamus (see Chapter 45).

Negative Feedback Aspects

As we have seen, FSH and LH stimulate production of ovarian hormones (Fig. 11, follicular phase). In the absence of ovarian hormones after ovariectomy or menopause, concentrations of FSH and LH in blood may increase as much as five- to tenfold. Treatment with low doses of estrogen lowers circulating concentrations of gonadotropins to levels seen during the follicular phase. When low doses of estradiol are given to subjects whose ovaries are intact, inhibition of gonadotropin secretion results in failure of follicular development. Progesterone alone, unlike estrogen, is ineffective in lowering high levels of FSH and LH in the blood of postmenopausal women, but it can synergize with estrogen to suppress gonadotropin secretion. These findings exemplify classical

negative feedback. Inhibin may also provide some feedback inhibition of FSH secretion during the follicular phase and may contribute the low level of FSH during the luteal phase, but its effects are probably small. In ovariectomized rhesus monkeys, the normal pattern of gonadotropin concentrations can be reproduced by treating with only estradiol and progesterone. The rise in FSH concentration follows the fall in estrogen, progesterone, and inhibin secretion at the end of the luteal phase and stimulates growth of the cohort of follicles that will produce the ovum in the next cycle. Throughout the follicular and luteal phases of the cycle, steroid concentrations appear to be sufficient to suppress LH secretion.

Although the ovarian steroids and inhibin suppress FSH and LH secretion, estrogen and progesterone concentrations change during the cycle in ways that seem independent of gonadotropin concentrations. For example, estrogen rises dramatically as the follicular phase progresses, even though LH remains constant and FSH is diminishing. The mechanism for this increase in estradiol is implicit in what has already been presented and is consistent with negative feedback. Estrogen production by the maturing follicle increases without a preceding increase in gonadotropin concentration because the mass of responsive theca and granulosa cells increases as does their sensitivity to gonadotropins. In fact, the decrease in FSH during the transition from early to late follicular phase probably results from feedback inhibition by the increasing concentration of estrogen, perhaps in conjunction with inhibin. Although luteinizing cells do not divide, progesterone and estrogen concentrations continue to increase during the early luteal phase, well after FSH and LH have returned to basal levels. Increasing steroid hormone production at this time reflects completion of

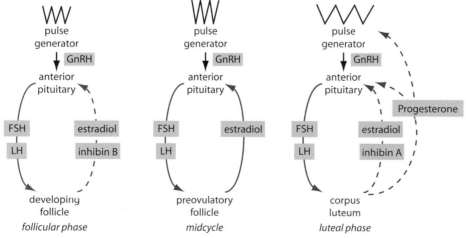

FIGURE 11   Ovarian–pituitary interactions at different phases of the menstrual cycle. Solid arrows, stimulation; dashed arrows, inhibition. Note that the *frequency* of the GnRH pulse generator slows in the luteal phase, and that the *amplitude* may increase at midcycle.

the luteinization process. Conversely, the gradual loss of sensitivity of luteal cells to LH accounts for the decrease in progesterone and estrogen secretion during the latter part of the luteal phase. Thus, one of the unique features of the female reproductive cycle is that changes in steroid hormone production result more from changes in the number or sensitivity of competent target cells than from changes in gonadotropin concentrations (see Chapter 37).

Positive Feedback Aspects

Rising estrogen levels in the late follicular phase trigger the massive burst of LH secretion that just precedes ovulation. This LH surge can be duplicated experimentally in monkeys and women given sufficient estrogen to raise their blood levels above a critical threshold level for 2 to 3 days. This compelling evidence implicates increased estrogen secretion by the ripening follicle as the causal event that triggers the massive release of LH and FSH from the pituitary (Fig. 11, midcycle). It can be considered positive feedback because LH stimulates estrogen secretion, which in turn stimulates more LH secretion in a self-generating explosive pattern.

Progesterone concentrations begin to rise about 6 hours before ovulation. This change is probably a response to the increase in LH rather than its cause. It is significant that large doses of progesterone given experimentally block the estrogen-induced surge of LH in women, which may account for the absence of repeated LH surges during the luteal phase, when the concentrations of estrogen might be high enough to trigger the positive feedback effect (Fig. 11, luteal phase). This action of progesterone, which may contribute to the decline in the LH surge, also contributes to its effectiveness as an oral contraceptive agent. In this regard, progesterone also inhibits follicular growth.

Neural Control of Gonadotropin Secretion

It is clear that secretion of gonadotropins is influenced to a large measure by ovarian steroid hormones. It is equally clear that secretion of these pituitary hormones is controlled by the central nervous system. Gonadotropin secretion ceases after the vascular connection between the anterior pituitary gland and the hypothalamus is interrupted or after the arcuate nuclei of the medial basal hypothalamus are destroyed. Less drastic environmental inputs, including rapid travel across time zones, stress, anxiety, and other emotional changes, can also affect the reproductive function in women, presumably through neural input to the medial basal hypothalamus. As discussed in Chapter 45, secretion of gonadotropins requires the operation of a hypothalamic pulse generator that produces intermittent stimulation of the pituitary gland by GnRH.

Sites of Feedback Control

The ovarian steroids might produce their positive or negative feedback effects by acting at the level of the hypothalamus or the anterior pituitary gland, or both. The GnRH pulse generator in the medial basal hypothalamus drives gonadotropin secretion regardless of whether negative or positive feedback prevails. Gonadotropin secretion falls to zero after bilateral destruction of the arcuate nuclei in rhesus monkeys and cannot be increased by either ovariectomy or treatment with the same amount of estradiol that evokes a surge of FSH and LH in normal animals. When such animals are fitted with a pump that delivers a constant amount of GnRH in brief pulses every hour, the normal cyclic pattern of gonadotropin is restored and the animals ovulate each month. Identical results have been obtained in women suffering from Kallman's syndrome, in which there is a developmental deficiency in GnRH production by the hypothalamus (Fig. 12). In both cases, administration of GnRH in pulses of constant amplitude and frequency was sufficient to produce normal ovulatory cycles. Because both positive and negative feedback aspects of gonadotropin secretion can be produced even when hypothalamic input is "clamped" at constant frequency and amplitude, these effects of estradiol must be exerted at the level of the pituitary.

Although changes in amplitude and frequency of GnRH pulses are not necessary for the normal pattern of gonadotropin secretion during an experimental or therapeutic regimen, variations in frequency and amplitude nevertheless may occur physiologically in a way that complements and reinforces the intrinsic pattern already described. During the normal reproductive cycle, GnRH pulses are considerably less frequent in the luteal phase than in the follicular phase. Complementing its negative feedback effects, estradiol may decrease the amplitude of GnRH pulses, and progesterone slows their frequency, perhaps by stimulating hypothalamic production of endogenous opioids. It appears that an increase in amplitude of GnRH pulses precedes the LH surge, and there is good evidence that progesterone acts at the level of the hypothalamus to block the estradiol-induced LH surge. Thus, feedback effects of estradiol appear to be exerted primarily, but not exclusively, on the pituitary and those of progesterone primarily, but probably not exclusively, on the hypothalamus.

We do not yet understand the intrapituitary mechanisms responsible for the negative and positive feedback effects of estradiol. As seen with the ovary, changes in hormone secretion may be brought about by changes in

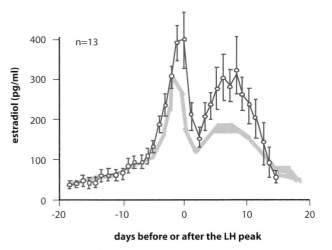
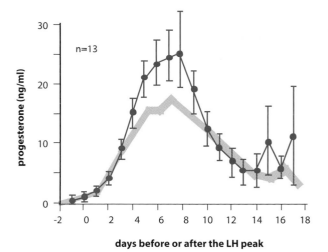

FIGURE 12   Results of ovulation induction employing a physiological frequency of GnRH administration to hypogonadotropic hypogonadal women upon ovarian steroid secretion. Normal values are
represented by the shaded areas. (From Crowley WF, Jr, Filicori M, Spratt DJ *et al. Rec Prog Horm Res*
1985; 41:473. With permission.)

the sensitivity of target cells, as well as by changes in
concentration of a stimulatory hormone. Women and
experimental animals are more responsive to a test dose
of GnRH at midcycle than at any other time. An increase
in the number of receptors for GnRH at this time has
been reported, and there is evidence that the high
concentrations of estradiol that precede the LH surge
upregulate GnRH receptors and stimulate LH synthesis
and storage.

Timing of Reproductive Cycles

Although the pacemaker for rhythmic release of
GnRH resides in the hypothalamus, the timekeeper for
the slower monthly rhythm of the ovarian cycle resides in
the ovary. As already indicated, the corpus luteum has a
built-in life-span of about 12 days and involutes despite
continued stimulation with LH. A new cohort of follicles
cannot arise as long as the corpus luteum remains
functional. Its demise appears to relieve inhibition of
follicular growth and secretion of FSH, which increases
sufficiently in blood to stimulate growth of the next
cohort of follicles. Thus, the interval between the LH
surge and the emergence of the new cohort of follicles is
determined by the ovary. The principal event around
which the menstrual cycle revolves is ovulation, which
depends on an ovulatory surge of LH. The length of the
follicular phase may be somewhat variable and may be
influenced by extraovarian events, but the timing of the
LH surge resides in the ovary. It is only when the
developing follicle signals its readiness to ovulate with
increasing blood levels of estradiol that the pituitary
secretes the ovulatory spike of gonadotropin. Hence
throughout the cycle it is the ovary that notifies the
pituitary and hypothalamus of its readiness to proceed to
the next stage.

Clinical Note

Gonadotropin Secretion and Antifertility Drugs
The negative feedback effects of estrogens and
progesterone on gonadotropin secretion have
been exploited to produce antifertility drugs.
Chemical modifications that prolong their half-
lives and enable them to escape destruction by the
liver has made possible their administration by
mouth. Preparations of estrogens alone, in combination with progestational compounds, or administered sequentially prevent follicular development
by blocking gonadotropin secretion, thus preventing ovulation. Sequential administration of estrogens for 2 weeks followed by progestins and then
withdrawal for a few days reproduces normal
endometrial events, including menstruation.
Depot injections of long-acting progestins can not
only block ovulation by preventing the ovulatory
surge of gonadotropin secretion but also stimulate
production of dense cervical mucus, which
prevents passage of sperm into the uterine cavity.

The beginning and end of cyclic ovarian activity (*i.e.*, menarche and menopause) occur on a longer time scale. The events associated with the onset of puberty were considered in Chapter 45. Although we still do not know what biological phenomena signal readiness for reproductive development and the end of the juvenile period, it appears that the timekeeper for this process resides in the central nervous system, which initiates sexual development and function by activating the GnRH pulse generator. Termination of cyclic ovarian activity coincides with the disappearance or exhaustion of primordial follicles, but during the final decade of a woman's reproductive life there is a paradoxical doubling of the rate of loss of follicles by atresia. Aging of the GnRH pacemaker may be a factor in this acceleration of follicular loss, as studies in normally cycling women indicate that both the amplitude of LH pulses and the interval intervening between pulses increases with increased age.

Suggested Readings

Ackland JF, Schwartz NB, Mayo KE, Dodson RE. Nonsteroidal signals originating in the gonads. *Physiol Revs* 1992; 72:731–788.

Dorrington JH, Armstrong DT. Effects of FSH on gonadal function. *Recent Prog Horm Res* 1979; 35:301–332.

Giudice LC. Insulin-like growth factors and ovarian follicular development. *Endocrine Rev* 1992; 13:641–669.

Gougeon A. Regulation of ovarian follicular development in primates: facts and hypotheses. *Endocrine Rev* 1996; 17:121–155.

Graham D, Clarke CL. Physiological action of progesterone in target tissues. *Endocrine Rev* 1997; 18: 502-519.

Hayes FJ, Crowley WF, Jr. Gonadotropin pulsations across development. *Horm Res* 1998; 49:163–168.

Knobil E, Hotchkiss J. The menstrual cycle and its neuroendocrine control, in: Knobil E, Neill JD, Eds., *The physiology of reproduction*, Vol. 2, 2nd ed., New York, Raven Press, 1994, pp. 711–749.

Marshall JC, Dalkin AC, Haisleder DJ, Paul SJ, Ortolano GA, Kelch RP. Gonadotropin-releasing hormone pulses: regulators of gonadotropin synthesis and ovarian cycles. *Recent Prog Horm Res* 1991; 47:155–187.

Matzuk MM. Revelations of ovarian follicle biology from gene knockout mice. *Mol Cell Endocrinol* 2000; 163:61–66.

McGee EA, Hsueh AJW. Initial and cyclic recruitment of ovarian follicles. *Endocrine Rev* 2000; 21:200–214.

Nilsson S, Makela S, Treuter E, Tujague M, Thomsen J, Andersson G, Enmark E, Pettersson K, Warner M, Gustafsson JA. Mechanisms of estrogen action. *Physiol Rev* 2001; 81:1535–1565.

Richards JS, Jahnsen T, Hedin L, Lifka J, Ratoosh S, Durica JM, Goldring NB. Ovarian follicular development: from physiology to molecular biology. *Rec Prog Horm Res* 1987; 43:231–270.

Hormonal Control of Reproduction in the Female: Pregnancy and Lactation

H. MAURICE GOODMAN

KEY POINTS

- Estrogen secretion during the follicular phase prepares the female reproductive tract for efficient sperm transport to the site of fertilization in the ampulla of the oviducts.
- Ciliary action and muscular contractions of the oviduct propel the ovum and later the blastocyst toward the uterus.
- Implantation occurs about 5 days after ovulation and involves invasion of the endometrium and differentiation of the outer portion of the blastocyst into the primitive placenta.

- The placenta is a highly vascular organ specialized for bidirectional transport of metabolites and gases and for hormone synthesis.
- Secretion of human chorionic gonadotropin (hCG) by the early trophoblast prolongs the life of the corpus luteum and maintains its capacity for progesterone production for most of the first trimester.
- Human chorionic somatomammotropin (hCS) is secreted by the placenta in large quantities, particularly in the latter half of pregnancy, and

KEY POINTS (*continued*)

has both growth-hormone-like and prolactin-like activities.

- The placenta constitutively synthesizes large amounts of progesterone from cholesterol taken up from the maternal circulation.
- Although the placenta cannot synthesize 19-carbon steroids, it produces large amounts of estrogens by aromatizing androgens secreted by the maternal and fetal adrenal glands.
- Corticotropin-releasing hormone (CRH) produced by the placenta stimulates the fetal pituitary to secrete ACTH and also acts directly on the fetal adrenal cortex to stimulate production of dehydroepiandrosterone-sulfate which serves as a precursor for placental estrogen production.
- Estrogens drive their own production and the production of progesterone by increasing placental uptake of cholesterol and conversion to pregnenolone, which serves as a precursor for placental production of progesterone and fetal adrenal production of dehydroepiandrosterone-sulfate.

- Parturition is initiated by convergence of self-amplifying effects of CRH, cortisol, prostaglandins, mechanical stretching of the myometrium, and perhaps other factors that are not fully understood.
- Oxytocin promotes uterine contractions and speeds up parturition but is not secreted until parturition has been initiated.
- Estrogen and progesterone promote growth and branching of the mammary duct system.
- Prolactin and hCS in the presence of estrogen and progesterone during pregnancy stimulate growth of the mammary alveolar tissue.
- Tactile stimulation of the nipples or sensory input related to nursing signal release of oxytocin.
- Oxytocin stimulates myoepithelial cells surrounding the mammary alveoli and ducts to contract and deliver pressurized milk to the lactiferous ducts.
- Suckling stimulates prolactin secretion.
- Prolactin is secreted in response to suckling and stimulates milk synthesis in preparation for the next episode of suckling.

OVERVIEW

Successful reproduction depends not only on the union of eggs and sperm but also on survival of adequate numbers of the new generation to reach reproductive age and begin the cycle again. In some species, parental involvement in the reproductive process ends with fertilization of the ova; thousands or even millions of embryos may result from a single mating, with just a few surviving long enough to procreate. Higher mammals, particularly humans, have adopted the alternative strategy of producing only few or a single fertilized ovum at a time. Prolonged parental care during the embryonic and neonatal periods substitutes for huge numbers of unattended offspring as the means for increasing the likelihood of survival. Estrogen and progesterone prepare the maternal body for successful internal fertilization and hospitable acceptance of the embryo. The conceptus then takes charge. After lodging firmly within the uterine lining and gaining access to the maternal circulation, it secretes protein and steroid hormones that ensure continued maternal acceptance, and it directs maternal functions to provide for its development. Simultaneously, the conceptus withdraws whatever nutrients it needs from the maternal circulation. At the appropriate time, the fetus signals its readiness to depart the uterus and initiates the birth process. While in utero, placental hormones prepare the mammary glands to produce the milk needed for nurture after birth. Finally, suckling stimulates continued milk production.

FERTILIZATION AND IMPLANTATION

Gamete Transport

Fertilization takes place in a distal portion of the oviduct called the *ampulla*, far from the site of sperm deposition in the vagina. To reach the ovum, sperm must swim through the cervical canal, cross the entire length of the uterine cavity, and then travel up through the muscular isthmus of the oviduct. Even with the aid of contractions of the female reproductive tract, the journey is formidable. Only about one of every million sperm deposited in the vagina reach the ampulla; here, if they arrive first, they await the arrival of the ovum. Sperm usually remain fertile within the female reproductive tract for 1 to 2 days, but as long as 4 days is possible. Access to

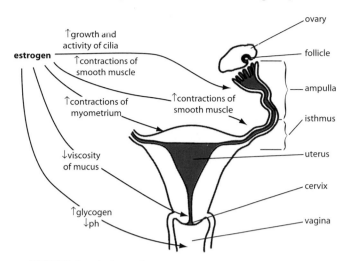

FIGURE 1   Actions of estrogen to promote sperm transport.

the upper reaches of the reproductive tract is heavily influenced by ovarian steroid hormones.

Estrogen is secreted in abundance late in the follicular phase of the ovarian cycle and prepares the reproductive tract for efficient sperm transport (Fig. 1). Glycogen deposited in the vaginal mucosa under its influence provides substrate for the production of lactate, which lowers the pH of vaginal fluid. An acidic environment increases the motility of sperm that is essential for their passage through the cervical canal. In addition, the copious watery secretion produced by cells lining the cervical canal under the influence of estrogen increases access to the uterine cavity. When estrogen is absent or when its effects are opposed by progesterone, the cervical canal is filled with a viscous mucus that resists sperm penetration. Vigorous contractions of the uterus propel sperm toward the oviducts, where they may appear anywhere from 5 to 60 minutes after ejaculation. Prostaglandins present in seminal plasma and oxytocin released from the pituitary in response to intercourse may stimulate contraction of the highly responsive estrogen-dominated myometrium.

Role of the Oviducts

The oviducts are uniquely adapted for facilitating transport of sperm toward the ovary and transporting the ovum in the opposite direction toward its rendezvous with sperm. It is also within the oviducts that sperm undergo a process called *capacitation*, which prepares them for a successful encounter with the ovum and its adherent mass of granulosa cells, the *cumulus oophorus*. Capacitation is an activation process that involves both enhancement of flagellar activity and the biochemical

and structural changes in the plasma membrane of the sperm head that prepare sperm to undergo the *acrosomal reaction*. The acrosome is a membranous vesicle that is positioned at the tip of the sperm head. It is filled with several hydrolytic enzymes. Contact of the sperm head with the *zona pellucida* that surrounds the ovum initiates fusion of the acrosomal membrane with the plasma membrane and the exocytotic release of enzymes that digest a path through the zona pellucida, clearing a path for sperm to reach the ovum. The acrosome reaction and the events that produce it are highly reminiscent of the sequelae of hormone receptor interactions and involve activation of membrane calcium channels, tyrosine phosphorylation, phospholipase C, and other intracellular signaling mechanisms. Initiation of the acrosomal reaction is facilitated by progesterone secreted by the mass of cumulus cells that surround the ovum. The sperm plasma membrane contains progesterone receptors that trigger an influx of calcium within seconds. These membrane-associated receptors differ from the classical nuclear receptors.

In response to estrogens or perhaps other local signals associated with impending ovulation, muscular activity in the distal portion of the oviduct brings the infundibulum into close contact with the surface of the ovary. At ovulation, the ovum, together with its surrounding granulosa cells, is released into the peritoneal cavity and is swept into the ostium of the oviduct by the vigorous, synchronous beating of cilia on the infundibular surface. Development of cilia in the epithelial lining and their synchronized rhythmic activity are consequences of earlier exposure to estrogens. Movement of the egg mass through the ampulla toward the site of fertilization near the ampullar–isthmic junction depends principally on currents set up in tubal fluid by the beating of cilia and to a lesser extent by contractile activity of the ampullar wall to produce a churning motion.

Propulsion of sperm through the isthmus toward the ampulla is accomplished largely by muscular contractions of the tubal wall. Circular smooth muscle of the isthmus is innervated with sympathetic fibers and has both α-adrenergic receptors, which mediate contraction, and β-adrenergic receptors, which mediate relaxation. Under the influence of estrogen, the α receptors predominate. Subsequently, as estrogenic effects are opposed by progesterone, the β receptors prevail, and isthmic smooth muscles relax. This reversal in the response to adrenergic stimulation may account for the ability of the oviduct to facilitate sperm transport through the isthmus toward the ovary and subsequently to promote passage of the embryo in the opposite direction toward the uterus.

After fertilization, the oviduct retains the embryo for about 3 days and nourishes it with secreted nutrients before facilitating its entry into the uterine cavity. These complex events, orchestrated by the interplay of estrogen, progesterone, and autonomic innervation, require participation of the smooth muscle of the walls of the oviduct as well as secretory and ciliary activity of the epithelial lining. As crucial as these mechanical actions may be, however, the oviduct does not contribute in an indispensable way to fertility of the ovum or sperm or to their union, as modern techniques of *in vitro* fertilization bypass it with no ill effects.

The period of fertility is short; from the time the ovum is shed until it can no longer be fertilized is only about 6 to 24 hours. As soon as a sperm penetrates the ovum, the second polar body is extruded, and the fertilized ovum begins to divide. By the time the fertilized egg enters the uterine cavity, it has reached the blastocyst stage and consists of about 100 cells. Timing of the arrival of the blastocyst in the uterine cavity is determined by the balance between antagonistic effects of estrogen and progesterone on the contractility of the oviductal wall. Under the influence of estrogen, circularly oriented smooth muscle of the isthmus is contracted and bars passage of the embryo to the uterus. As the corpus luteum organizes and increases its capacity to secrete progesterone, β-adrenergic receptors gain ascendancy, muscles of the isthmus relax, and the embryonic mass is allowed to pass into the uterine cavity. Ovarian steroids can thus "lock" the ovum or embryo in the oviduct or cause its delivery prematurely into the uterine cavity.

Implantation and Formation of the Placenta

The blastocyst floats freely in the uterine cavity for about a day before it implants, normally on about the fifth day after ovulation. Experience with *in vitro* fertilization indicates that there is about a 3-day period of uterine receptivity in which implantation leads to full-term pregnancy. It should be recalled that this period of endometrial sensitivity coincides with the period of maximal progesterone output by the corpus luteum (Fig. 2). In the late luteal phase of the menstrual cycle, the outer layer of the endometrium differentiates to form the *decidua*. Decidualized stromal cells enlarge and transform from an elongated spindle shape to a rounded morphology with accumulation of glycogen. Decidualization requires high concentrations of progesterone and may be enhanced by activity of cytokines and relaxin. Decidual cells express several proteins that may facilitate implantation, but the precise roles of these proteins either in implantation or pregnancy have not been determined definitively. One such protein is the hormone prolactin, which continues to be secreted throughout pregnancy. Another is the IGF-1 binding protein (IGFBP-1).

At the time of implantation, the blastocyst consists of an inner mass of cells destined to become the fetus and an outer rim of cells called the *trophoblast*. It is the trophoblast that forms the attachment to maternal decidual tissue and gives rise to the fetal membranes and the definitive placenta (Fig. 3). Cells of the trophoblast proliferate and form the multinucleated *syncytial trophoblast*, the specialized functions of which enable it to destroy adjacent decidual cells and allow the blastocyst to penetrate deep into the uterine endometrium. Killed decidual cells are phagocytosed by the trophoblast as the embryo penetrates the subepithelial connective tissue and eventually becomes completely enclosed within the endometrium. Products released from degenerating decidual cells produce hyperemia and increased capillary permeability. Local extravasation of blood from damaged capillaries forms small pools of blood that are in direct contact with the trophoblast and provide nourishment to the embryo until the definitive placenta forms. From the time the ovum is shed until the blastocyst implants, metabolic needs are met by secretions of the oviduct and the endometrium.

The syncytial trophoblast and an inner *cytotrophoblast* layer of cells soon completely surround the inner cell mass and send out solid columns of cells that further erode the endometrium and anchor the embryo. These columns of cells differentiate into the *placental villi*. As they digest the endometrium, pools of extravasated maternal blood become more extensive and fuse into a complex labyrinth that drains into venous sinuses in the endometrium. These pools expand and eventually receive an abundant supply of arterial blood. By the third week, the villi are invaded by fetal blood vessels as the primitive circulatory system begins to function (see Chapter 66).

Although much uncertainty remains regarding details of implantation in humans, it is perfectly clear that progesterone secreted by the ovary at the height of luteal

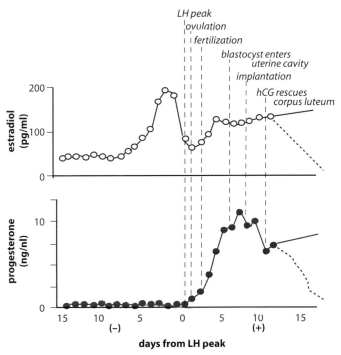

FIGURE 2   Relation between events of early pregnancy and steroid hormone concentrations in maternal blood (estradiol and progesterone concentrations are redrawn from data given in Fig. 10 of Chapter 46). By the 10th day after the LH peak, there is sufficient hCG to maintain and increase estrogen and progesterone production, which would otherwise decrease (dotted lines) at this time.

function is indispensable for all of these events to occur. Removal of the corpus luteum at this time or blockade of progesterone secretion or progesterone receptors prevents implantation. Progesterone is indispensable for maintenance of decidual cells, quiescence of the myometrium, and formation of the dense, viscous cervical mucus that essentially seals off the uterine cavity from the outside. It is noteworthy that the implanting trophoblast and the fetus are genetically distinct from the mother, yet the maternal immune system does not reject the implanted embryo as a foreign body. Progesterone plays a decisive role in immunological acceptance of the embryo. It promotes tolerance by regulating accumulation of lymphocyte types in the uterine cavity, by suppressing lymphocyte toxicity, and by inhibiting the production of cytolytic cytokines. The importance of progesterone for implantation and retention of the blastocyst is underscored by the development of a progesterone antagonist (RU486) that prevents implantation or causes an already implanted conceptus to be shed along with the uterine lining.

PLACENTAL HORMONES

The placenta is a complex, primarily vascular organ adapted to optimize the exchange of gases, nutrients, and electrolytes between maternal and fetal circulations. In humans, the placenta is also a major endocrine gland capable of producing large amounts of both steroid and peptide hormones and neurohormones. The placenta is the most recently evolved of all mammalian organs, and its endocrine function is highly developed in primates. It is unique among endocrine glands in that, as far as is known, its secretory activity is autonomous and not subject to regulation by maternal or fetal signals. In experimental animals such as the rat, pregnancy is terminated if the pituitary gland is removed during the first half of gestation or if the ovaries, and consequently the corpora lutea, are removed at any time. In primates, the pituitary gland and ovaries are essential only for a brief period after fertilization. After about 7 weeks, the placenta produces enough progesterone to maintain pregnancy. In addition, it also produces large amounts of estrogen, human chorionic gonadotropin (hCG), and human chorionic somatomammotropin (hCS), which is also called human placental lactogen (hPL). It can also secrete growth hormone (GH), thyroid-stimulating hormone (TSH), adrenocorticotropic hormone (ACTH), gonadotropin-releasing hormone (GnRH), corticotropin-releasing hormone (CRH), and a long list of other biologically active peptides. During pregnancy there is the unique situation of hormones secreted by one individual, the fetus, regulating the physiology of

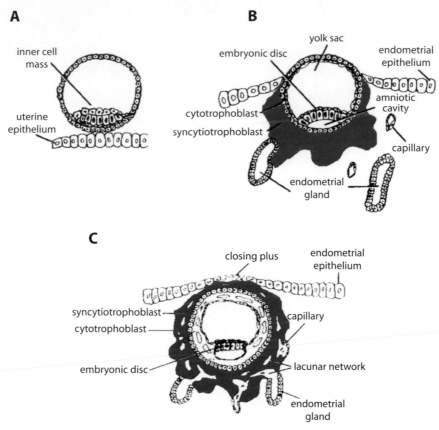

FIGURE 3 (A) Six-day-old blastocyst settles on endometrial surface. (B) By the 8th day, the blastocyst has begun to penetrate the endometrium; expanding syncytiotrophoblast (blue) invades and destroys decidualized endometrial cells. (C) By 12 days the blastocyst has completely embedded itself in the decidualized endometrium, and a clot or plug has formed to cover the site of entry. The trophoblast has continued to invade the endometrium and has eroded uterine capillaries and glands. A network of pools of extravasated blood (lacunar network) has begun to form. (Adapted from Khong TY, Pearce JM. In: Lavery JP, Ed., *The human placenta: clinical perspectives*, Rockville MD: Aspen Publishers, 1987, p. 26.)

another, the mother. By extracting needed nutrients and adding hormones to the maternal circulation, the placenta redirects some aspects of maternal function to accommodate the growing fetus (see Chapter 66).

Human Chorionic Gonadotropin

As already discussed (see Chapter 46), the functional life of the corpus luteum in infertile cycles ends by the 12th day after ovulation. About 2 days later, the endometrium is shed, and menstruation begins. For pregnancy to develop, the endometrium must be maintained; therefore, the ovary must be notified that fertilization has occurred. The signal to the ovary in humans is a luteotropic substance secreted by the conceptus: hCG. Human chorionic gonadotropin rescues the corpus luteum (*i.e.*, extends its life-span) and stimulates it to continue secreting progesterone and estrogens, which in turn maintain the endometrium in a state favorable for implantation

and placentation (Fig. 4). Continued secretion of luteal steroids and inhibin notifies the pituitary gland that pregnancy has begun and inhibits secretion of the gonadotropins that would otherwise stimulate development of the next cohort of follicles. Pituitary gonadotropins remain virtually undetectable in maternal blood throughout pregnancy as a result of the negative feedback effects of high circulating concentrations of estrogens and progesterone. Relaxin secretion by the corpus luteum increases in early pregnancy, becomes maximum at around the end of the first trimester, and then declines somewhat but continues throughout pregnancy. Relaxin may synergize with progesterone in early pregnancy to suppress contractile activity of uterine smooth muscle.

Human chorionic gonadotropin is a glycoprotein that is closely related to the pituitary glycoprotein hormones (see Chapter 38). Although there are wide variations in the carbohydrate components, the peptide backbones of the glycoprotein hormones are closely related and consist

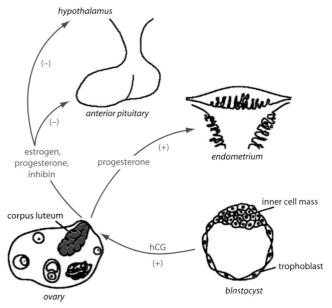

FIGURE 4   Maternal responses to hCG.

of a common α subunit and activity-specific β subunits. The α subunits of follicle-stimulating hormone (FSH), luteinizing hormone (LH), TSH, and hCG have the same amino acid sequence and are encoded in the same gene. In humans, seven genes or pseudogenes code for hCG β, but only two or three of them are expressed. The β subunit of hCG is almost identical to the β subunit of LH, differing only by a 32-amino-acid extension at the carboxyl

terminus of hCG. It is not surprising, therefore, that hCG and LH act through a common receptor and that hCG has LH-like bioactivity. Human chorionic gonadotropin contains considerably more carbohydrate, particularly sialic acid residues, than its pituitary counterparts, which accounts for its extraordinary stability in blood. The half-life of hCG is more than 30 hours, as compared to just a few minutes for the pituitary glycoprotein hormones. The long half-life facilitates rapid build-up of adequate concentrations of this vital signal produced by a few vulnerable cells.

Trophoblast cells of the developing placenta begin to secrete hCG early, with detectable amounts already present in blood by about the eighth day after ovulation, when luteal function under the influence of LH is still at its height. Production of hCG increases dramatically during the early weeks of pregnancy (Fig. 5). Blood levels continue to rise and during the third month of pregnancy reach peak values that are perhaps 200 to 1000 times that of LH at the height of the ovulatory surge. Presumably because of its high concentration, hCG is able to prolong the functional life of the corpus luteum while LH, at the prevailing concentrations in the luteal phase of an infertile cycle, cannot. High concentrations of hCG at this early stage of fetal development are also critical for male sexual differentiation, which occurs before the fetal pituitary can produce adequate amounts of LH to stimulate testosterone synthesis by the developing testis. Secretion of testosterone by the fetal testes is crucial for

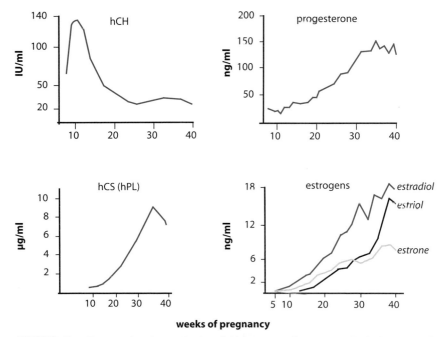

weeks of pregnancy

FIGURE 5   Changes in plasma levels of "hormones of pregnancy" during normal gestation. (From Freinkel N, Metzger BE. In: Wilson JD, Foster DW, Eds., *Williams textbook of endocrinology*, 8th ed., Philadelphia: W.B. Saunders, 1992. With permission.)

survival of the wolffian duct system and formation of the male internal genitalia (see Chapter 45). Human chorionic gonadotropin stimulation of the fetal adrenal gland may augment estrogen production later in pregnancy (see below). Finally, it is the appearance of hCG in large amounts in urine that is used as a diagnostic test for pregnancy. Because its biological activity is like that of LH, urine containing hCG induces ovulation when injected into estrous rabbits in the classical rabbit test. Now hCG can be measured with a simple sensitive immunological test, and pregnancy can be detected even before the next expected menstrual period.

Secretion of significant amounts of progesterone by the corpus luteum diminishes after about the eighth week of pregnancy despite continued stimulation by hCG. Measurements of progesterone in human ovarian venous blood indicate that the corpus luteum remains functional throughout most of the first trimester, and, although some capacity to produce progesterone persists throughout pregnancy, continued presence of the ovary is not required for a successful outcome. Well before the decline in luteal steroidogenesis, placental production of progesterone becomes adequate to maintain pregnancy.

Human Chorionic Somatomammotropin

The other placental protein hormone that is secreted in large amounts is hCS. Like hCG, hCS is produced by the syncytial trophoblast and becomes detectable in maternal plasma early in pregnancy. Its concentration in maternal plasma increases steadily from about the third week after fertilization and reaches a plateau by the last month of pregnancy (Fig. 5) when the placenta produces about 1 g of hCS each day. The concentration of hCG in maternal blood at this time is about 100 times higher than that normally seen for other protein hormones in women or men. Human chorionic somatomammotropin has a short half-life and, despite its high concentration at parturition, is undetectable in plasma after the first postpartum day.

Despite its abundance and its ability to produce a variety of biological actions in the laboratory, the physiological role of hCS has not been established definitively. Human chorionic somatomammotropin has strong prolactin-like activity and can induce lactation in test animals, but lactation normally does not begin until long enough after parturition for hCS to be cleared from maternal blood. However, it is likely that hCS promotes mammary growth in preparation for lactation. It is also likely that hCS contributes to the availability of nutrients for the developing fetus by operating like GH to mobilize maternal fat and decrease maternal glucose consumption (see Chapter 42). In this context, hCS

may be responsible for the decreased glucose tolerance (so-called *gestational diabetes*) experienced by many women during pregnancy. Although secretion of hCS is directed predominantly into maternal blood, appreciable concentrations are also found in fetal blood in midgestation. Receptors for hCS are present in human fetal fibroblasts and myoblasts, and these cells release IGF-II when stimulated by hCS. As already discussed (see Chapter 44), fetal growth is independent of GH, but the role of hCS in this regard is unknown.

Despite these observations, evidence from genetic studies makes it unlikely that hCS is indispensable for the successful outcome of pregnancy. Human chorionic somatomammotropin is a member of the growth hormone–prolactin family (see Chapter 38) and shares large regions of structural homology with both of these pituitary hormones. Five genes of this family are clustered on chromosome 17, including three that encode hCS and two that encode GH. Two of the hCS genes are expressed and code for identical secretory products. The third hCS gene appears to be a pseudogene, the transcription of which does not produce fully processed mRNA. No adverse consequences for pregnancy, parturition, or early postnatal development were seen in three cases in which a stretch of DNA that contains both hCS genes and one hGH gene was missing from both chromosomes. No immunoassayable hCS was present in maternal plasma, but it is possible that the remaining hCS pseudogene was expressed under these circumstances or that recombination of the remaining fragments of these genes produced a chimeric protein with hCS-like activity. Regardless of whether or not hCS is indispensable for normal gestation, important functions are often governed by redundant mechanisms, and it is likely that hCS contributes in some way to a successful outcome of pregnancy.

Progesterone

As progesterone secretion by the corpus luteum declines, the trophoblast becomes the major producer of progesterone. Placental production of progesterone increases as pregnancy progresses, so that during the final months upward of 250 mg may be secreted per day. This huge amount is more than ten times the daily production by the corpus luteum at the height of its activity and may be even greater in women bearing more than one fetus. Because the placenta cannot synthesize cholesterol, it imports cholesterol in the form of low-density lipoproteins (LDLs) from the maternal circulation. In late pregnancy, progesterone production consumes an amount of cholesterol equivalent to about 25% of the daily turnover in a normal nonpregnant woman.

Production of progesterone by the placenta is not subject to regulation by any known extraplacental factors other than availability of substrate. As in the adrenals and gonads, the rate of conversion of cholesterol to pregnenolone by P450scc determines the rate of progesterone production. In the adrenals and gonads, ACTH and LH stimulate synthesis of the steroid acute regulatory protein (StAR), which is required for transfer of cholesterol from cytosol to the mitochondrial matrix where P450scc resides (Chapter 40). The placenta does not express StAR. Access of cholesterol to the interior of mitochondria is thought to be provided by a similar protein that is constitutively expressed in the trophoblast. Consequently, placental conversion of cholesterol to pregnenolone bypasses the step that is regulated in all other steroid hormone-producing tissues. Ample expression of 3β-hydroxysteroid dehydrogenase allows rapid conversion of pregnenolone to progesterone. All of the pregnenolone produced is either secreted as progesterone or exported to the fetal adrenal glands to serve as substrate for adrenal steroidogenesis (Fig. 6).

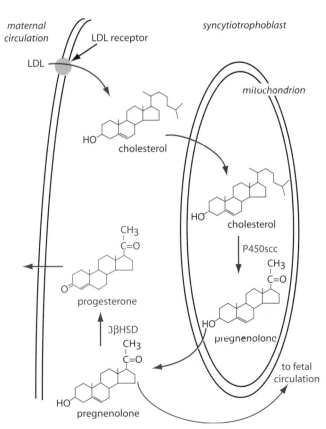

FIGURE 6   Progesterone synthesis by the trophoblast. Cholesterol is taken up via low density lipoprotein (LDL) receptors and transferred to the inner mitochondrial matrix by constitutively expressed protein(s), where its C22–C27 side chain is removed by P450scc. Pregnenolone exits the mitochondria and is oxidized to progesterone by 3β-hydroxysteroid dehydrogenase (3βHSD).

Estrogens

The human placenta is virtually the only site of estrogen production after the corpus luteum declines. However, the placenta cannot synthesize estrogens from cholesterol or use progesterone or pregnenolone as substrate for estrogen synthesis. The placenta does not express P450c17, which cleaves the carbon-20,21 side chain to produce the requisite 19-carbon androgen precursor. Reminiscent of the dependence of granulosa cells on thecal cell production of androgens in ovarian follicles (see Chapter 46), estrogen synthesis by the trophoblast depends on import of 19-carbon androgen substrates which are secreted by the adrenal glands of the fetus and, to a lesser extent, the mother (Fig. 7). The trophoblast expresses an abundance of P450aromatase, the activity of which is sufficient to aromatize all of the available substrate. The cooperative interaction between the fetal adrenal and the placenta has given rise to the term *fetoplacental unit* as the source of estrogen production in pregnancy. The placental estrogens are *estradiol*, *estrone*, and *estriol*, which differs from estradiol by the presence of an additional hydroxyl group on carbon-16. Of these, estriol is by far the major estrogenic product. Its rate of synthesis may exceed 45 mg/day by the end of pregnancy.

Despite its high rate of production, however, concentrations of unconjugated estriol in blood are lower than those of estradiol (Fig. 5) due to its high rate of metabolism and excretion. Although estriol can bind to estrogen receptors, it contributes little to overall estrogenic bioactivity, as it is only about 1% as potent as estradiol and 10% as potent as estrone in most assays; however, estriol is almost as potent as estradiol in stimulating uterine blood flow. It is possible that the fetus uses this elaborate mechanism of estriol production to ensure that uterine blood flow remains adequate for its survival.

The Role of the Fetal Adrenal Cortex

The fetal adrenal glands play a central role in placental steroidogenesis and hence maintenance of pregnancy and may also have a role in provoking the onset of labor at the end of pregnancy. The adrenal glands of the fetus differ significantly in both morphology and function from the adrenal glands of the adult. They are bounded by a thin outer region, called the *definitive cortex*, which will become the zona glomerulosa, and a huge, inner *fetal zone*, which regresses and disappears shortly after birth. The transitional zone at the interface of the two zones gives rise to the fasciculata and reticularis of the adult (see Chapter 40). In midpregnancy, the fetal adrenals are large—larger in fact, than the kidneys—and the fetal zone constitutes 80% of its mass. Growth, differentiation,

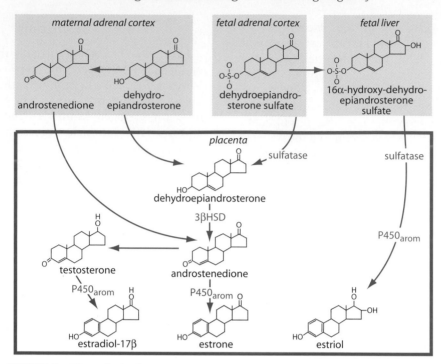

FIGURE 7   Biosynthesis of estrogens during pregnancy. Note that androgens formed in either the fetal or maternal adrenals are the precursors for all three estrogens and that the placenta cannot convert progesterone to androgens. Hydroxylation of dehydroepiandrosterone sulfate on carbon 16 by the fetal liver gives rise to estriol, which is derived almost exclusively from fetal sources. Fetal androgens are secreted as sulfate esters and must first be converted to free androgens by the abundant placental sulfatase before conversion to estrogens by the enzyme P450aromatase (P450arom). 3βHSD, 3β-hydroxysteroid dehydrogenase.

and secretory activity of the fetal adrenals are controlled by ACTH, the actions of which are augmented by a variety of fetal growth factors including insulin-like growth factor II (IGF-II). The fetal pituitary is the main source of ACTH, but the placenta also secretes some ACTH. In addition, the placenta secretes corticotropin-releasing hormone (CRH), which not only stimulates the fetal pituitary to secrete ACTH but also directly stimulates steroidogenesis by the fetal adrenal glands.

The chief product of the fetal zone is the 19-carbon androgen *dehydroepiandrosterone* (DHEA), which is secreted as the biologically inert sulfate ester (DHEA-S). Sulfation protects against masculinization of the genitalia in female fetuses and prevents aromatization in extra-gonadal fetal tissues. The fetal zone produces DHEA-S at an increasing rate that becomes detectable by about the eighth week of pregnancy, well before cortisol and aldosterone are produced by the definitive and transitional zones. At term, secretion of DHEA-S may reach 200 mg per day. The cholesterol substrate for DHEA-S production is synthesized in both the fetal liver and fetal adrenals. It is likely that pregnenolone released into the fetal circulation by the placenta also provides substrate.

Much of the DHEA-S in the fetal circulation is oxidized at carbon-16 in the fetal liver to form 16α-DHEA-S

and then exported to the placenta. The placenta is highly efficient at extracting 19 carbon steroids from both maternal and fetal blood. It is rich in sulfatase and converts 16α-DHEA-S to 16α-DHEA prior to aromatization to form estriol. Because synthesis of estriol reflects the combined activities of the fetal adrenals, the fetal liver, and the placenta, its rate of production, as reflected in maternal estriol concentrations, has been used as an indicator of fetal well-being. DHEA-S that escapes 16α-hydroxylation in the fetal liver is converted to androstenedione or testosterone in the placenta after hydrolysis of the sulfate bond and then aromatized to form estrone and estradiol.

ROLES OF PROGESTERONE AND ESTROGENS IN SUSTAINING PREGNANCY

As its name implies, progesterone is essential for maintaining all stages of pregnancy, and pharmacologic blockade of its actions at any time terminates the pregnancy. Progesterone sustains pregnancy by opposing the forces that conspire to increase uterine contractility and expel the fetus. One of these forces is physical stretch of the myometrium by the growing fetus. Stretch or

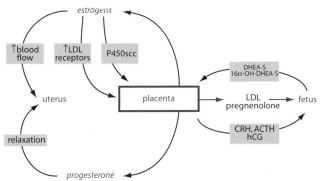

FIGURE 8 Effects of estrogen on production of placental steroid hormones. By increasing uterine blood flow and inducing low-density lipoprotein (LDL) and the P450 side-chain cleavage (P450scc) enzyme, estrogens increase placental production of pregnenolone, which is used as substrate for androgen production in the fetal adrenals. Uptake of LDL from the maternal circulation may also transfer cholesterol to the fetal circulation. DHEA-S, dehydroepiandrosterone sulfate; 16α-DHEA-S, 16α hydroxy-dehydroepiandrosterone sulfate; CRH, corticotropin releasing hormone; ACTH, adrenocorticotropic hormone; hCG, human chorionic gonadotropin.

tension coupled with estrogens and progesterone promotes myometrial growth and hypertrophy in parallel with growth of the conceptus. Estrogens promote expression of genes that code for contractile proteins, gap junction proteins that electrically couple myometrial cells, oxytocin receptors, receptors for prostaglandins, ion channel proteins, and doubtlessly many others that directly or indirectly tend to increase contractility. Throughout pregnancy, estrogens act through a positive feedback mechanism not only to increase their own production but also to increase synthesis of progesterone, which suppresses their excitatory effects (Fig. 8). Estrogens accelerate progesterone synthesis by increasing the delivery of its precursor substrate, cholesterol, to the trophoblast and by upregulating P450scc. Estrogens increase uterine blood flow by stimulating endothelial cells to produce the potent vasodilator nitric oxide and promote uterine formation of prostaglandin I, which is also a vasodilator. In addition, they increase receptor-mediated cholesterol uptake by stimulating expression of LDL receptors. Pregnenolone and LDLs that cross the placenta and enter the fetal circulation serve as substrates for adrenal production of DHEA-S. DHEA-S is then converted by the placenta to estrogens in what amounts to a positive feedback system that progressively increases estrogen production in parallel with progesterone production.

PARTURITION

Pregnancy in the human lasts about 40 weeks. The process of birth, or *parturition*, is the expulsion of a viable baby from the uterus at the end of pregnancy and is the culmination of all the processes discussed in this and the previous two chapters. Studying the phenomenon of parturition has revealed a surprising array of strategies that have been adopted by different species, regulate parturition. Humans and the great apes have evolved mechanisms that appear to be unique, and the scarcity of experimental models that employ strategies similar to those of humans has hampered efforts to study underlying mechanisms of timing and initiating parturition in humans. Consequently, our understanding of the processes that bring about this climactic event in human reproductive physiology is still incomplete.

Successful delivery of the baby can take place only after the myometrium acquires the capacity for forceful, coordinated contractions and the cervix softens and becomes distensible (called ripening) so that uterine contractions can drive the baby through the cervical canal. These changes reflect the triumph of the excitatory effects of estrogens over the suppressive effects of progesterone that had prevailed thus far. Indeed, in most animals parturition is heralded by a decline in progesterone production coincident with an increase in estrogen production. Humans and higher primates are unique in the respect that there is neither an abrupt increase in plasma concentrations of estrogens nor a fall in progesterone at the onset of parturition. It is highly likely that multiple gradual changes, gaining momentum over days or weeks, tip the precarious estrogen/progesterone balance in favor of estrogen dominance.

In theory, signals to terminate pregnancy could originate with either the mother or the fetus. Most investigators favor the idea that the fetus, which has essentially controlled events during the rest of pregnancy, signals its readiness to be born. In sheep, the triggering event for parturition is an ACTH-dependent increase in cortisol production by the fetal adrenals. In this species, cortisol stimulates expression of P450c17 in the placenta and thereby shifts production of steroid hormones away from progesterone and toward estrogen. While there is neither a stimulation of P450c17 expression in the human placenta nor a fall in progesterone in humans, the human fetal adrenal may nevertheless have an essential role in orchestrating the events that lead up to parturition.

Role of Corticotropin-Releasing Hormone

Although the ability to secrete 19-carbon androgens is acquired by the fetal zone of the adrenals early in gestation, the definitive and transitional zones mature much later. The capacity to produce significant amounts of cortisol is not acquired until about the 30th week of

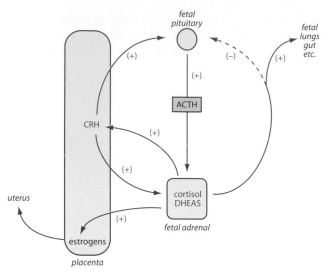

FIGURE 9 Effects of cortisol production in late pregnancy. By stimulating placental secretion of CRH (corticotropin-releasing hormone), cortisol initiates direct and indirect (via the fetal pituitary) positive feedback loops that enhance its own secretion and increases secretion of DHEA-S (dehydroepiandrosterone sulfate). In this way cortisol-induced maturation of the fetus occurs simultaneously with increased production of estrogens, which prepare the uterus for parturition. ACTH, adrenocorticotropic hormone.

gestation. An abundant supply of cortisol in the final weeks of pregnancy is indispensable for maturation of the lungs, the gastrointestinal tract, and other systems to prepare the fetus for extra-uterine life. Cortisol also antagonizes the suppressive effects of progesterone on CRH production in the placenta and hence increases transcription of the CRH gene. This paradoxical effect of cortisol on placental expression of CRH is opposite to its negative feedback effects on CRH production in the hypothalamus. Instead of suppressing CRH production, fetal production of cortisol initiates a positive feedback loop (Fig. 9). It may be recalled that CRH not only stimulates the fetal pituitary to secrete ACTH but also directly stimulates steroidogenesis in the fetal adrenal cortex; consequently, there is an increasing drive to the adrenal to increase production of cortisol and DHEA-S. Accelerating secretion of DHEA-S accounts for the increasingly steep rise in estrogen concentrations in maternal blood in the last weeks of pregnancy (as shown in Figure 5).

Prostaglandin production is also increased in fetal membranes and the uterus in late pregnancy. Prostaglandins $F_{2\alpha}$ and E participate in or initiate events that lead to rupture of the fetal membranes, softening of the uterine cervix, and contraction of the myometrium. CRH stimulates their formation by the fetal membranes. These prostaglandins, in turn, stimulate placental production of CRH and establish a second positive feedback loop. We might expect cortisol to oppose prostaglandin formation

in the fetus as it does in extrauterine tissues (Chapter 40); however, in fetal membranes, cortisol paradoxically increases expression of the prostaglandin synthesizing enzyme COX2 and inhibits formation of the principal prostaglandin-degrading enzyme. Prostaglandins also stimulate CRH secretion by the fetal hypothalamus, increasing ACTH secretion and providing further drive for cortisol secretion and consequent stimulation of CRH secretion.

Concentrations of CRH in maternal plasma increase exponentially as pregnancy progresses, but there is only a slight rise in ACTH and free cortisol. Discordance between CRH plasma concentrations and pituitary and adrenal secretory activity is due largely to the presence of a CRH binding protein (CRH-BP) that is present in plasma of pregnant as well as nonpregnant women. Additionally, responsiveness of the maternal pituitary to CRH is decreased during pregnancy possibly because of downregulation of CRH receptors in corticotropes. Despite the somewhat blunted sensitivity to CRH, however, maternal ACTH secretion follows the normal diurnal rhythmic pattern and increases appropriately in response to stress. Until about three weeks before parturition, concentrations of CRH-BP in maternal plasma vastly exceed those of CRH and there is little or no free CRH. For reasons that are not understood, CRH-BP concentrations fall dramatically at the same time that placental production of CRH is increasing most rapidly and exceeds the capacity of CRH-BP. Free CRH in maternal plasma stimulates prostaglandin production in the myometrium and cervix, causing increased contractility and cervical ripening.

In addition to CRH-related positive feedback loops, a large number of genes that encode gap junction proteins, ion channels, oxytocin receptors, prostaglandin receptors, proteases that breakdown cervical collagen fibers, and a host of other proteins are activated to an increasing extent by stretch and probably paracrine and autocrine factors that arise in the placenta or decidua. There is also evidence that progesterone-inactivating enzymes that are induced in the myometrium, the placenta, and the cervix in the final weeks may lower tissue concentrations of progesterone and hence its effectiveness. While there appears to be no single event that precipitates parturition, the various processes that are set in motion weeks earlier gradually build up to overwhelm progesterone dominance and unleash excitatory forces that expel the fetus. CRH and the factors that regulate its production appear to play a crucial but not exclusive, role (Fig. 10).

Role of Oxytocin

Oxytocin is a neurohormone secreted by nerve endings in the posterior lobe of the pituitary gland in

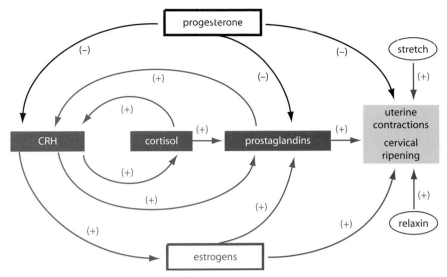

FIGURE 10   Positive feedback cycles that contribute to initiation of parturition. CRH, corticotropin-releasing hormone; +, stimulate; −, inhibit. See text for details.

response to neural stimuli received by cell bodies in the paraventricular and supraoptic nuclei of the hypothalamus (Chapter 38). It produces powerful synchronized contractions of the myometrium at the end of pregnancy, when uterine muscle is highly sensitive to it. Oxytocin is sometimes used clinically to induce labor. As parturition approaches, responsiveness to oxytocin increases in parallel with estrogen-induced increases in oxytocin receptors in both the endometrium and myometrium. Oxytocin is not the physiological trigger for parturition, however, as its concentration in maternal blood normally does not increase until after labor has begun. Oxytocin is secreted in response to stretching of the uterine cervix and hastens expulsion of the fetus and the placenta (as described in Chapter 37), but probably has little to do in initiating parturition. As a consequence of its action on myometrial contraction, oxytocin protects against hemorrhage after expulsion of the placenta. Just prior to delivery, the uterus receives nearly 25% of the cardiac output, most of which flows through the low resistance pathways of the maternal portion of the placenta. Intense contraction of the newly emptied uterus acts as a natural tourniquet to control loss of blood from the massive wound left when the placenta is torn away from the uterine lining.

LACTATION

The mammary glands are specialized secretory structures derived from the skin. As the name implies, they are unique to mammals. The secretory portion of the mammary glands is arranged in lobules consisting of branched tubules, the *lobuloalveolar ducts*, from which multiple evaginations or alveoli emerge in an arrangement resembling a bunch of grapes. The alveoli consist of a single layer of secretory epithelial cells surrounded by a meshwork of contractile *myoepithelial cells* (Fig. 11) Many lobuloalveolar ducts converge to form a *lactiferous duct*, which carries the milk to the nipple. Each mammary gland consists of 10 to 15 lobules, each with its own lactiferous duct opening separately to the outside. In the

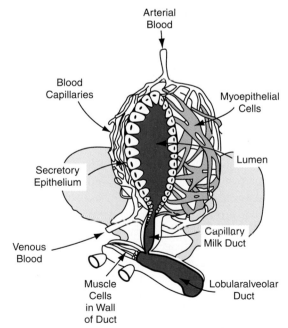

FIGURE 11   Mammary alveolus consisting of milk-producing cells surrounded by a meshwork of contractile myoepithelial cells. Milk-producing cells are targets for prolactin, while myoepithelial cells are targets for oxytocin. (From Turner CW. *Harvesting your milk crop*, Oakbrook, IL: Babson Bros., 1969, p. 17. With permission.)

inactive, nonlactating gland, alveoli are present only in rudimentary form, with the entire glandular portion consisting almost exclusively of lobuloalveolar ducts. The mammary glands have an abundant vascular supply and are innervated with sympathetic nerve fibers and a rich supply of sensory fibers to the nipple and areola.

Milk secreted by the mammary glands provides nourishment and immunoglobulins to offspring during the immediate postnatal period and for varying times thereafter, depending on culture and custom. Milk provides all of the basic nourishment, vitamins, minerals, fats, carbohydrates, and proteins needed by the infant until the teeth erupt. In addition, milk contains maternal immunoglobulins that are absorbed intact by the immature intestine and provide passive immunity to common pathogens. The extraordinarily versatile cells of the mammary alveoli simultaneously synthesize large amounts of protein, fat, and lactose and secrete these constituents, by various mechanisms, along with a large volume of aqueous medium for which the ionic composition differs substantially from blood plasma. Human milk consists of about 1% protein (principally in the form of casein and lactalbumin), about 4% fat, and about 7% lactose. Each liter of milk also contains about 300 mg of calcium. After lactation is established, the well-nourished woman suckling a single infant may produce about a liter of milk per day and as much as 3 liters per day if suckling twins. It should be apparent, therefore, that, in addition to hormonal regulation at the level of the mammary glands, milk production requires extramammary regulation by all those hormones responsible for compensatory adjustments in intermediary metabolism (see Chapters 41 and 42), calcium balance (see Chapter 43), and salt and water balance (see Chapter 29).

Growth and Development of the Mammary Glands

Prenatal growth and development of the mammary glands are independent of sex hormones and genetic sex. Until the onset of puberty, there are no differences in the male and female breast. With the onset of puberty, the duct system grows and branches under the influence of estrogen. Surrounding stromal and fat tissue also proliferate in response to estrogens. Progesterone, in combination with estrogen, promotes growth and branching of the lobuloalveolar tissue, but for these steroids to be effective prolactin, growth hormone, IGF-1, and cortisol must also be present. Lobuloalveolar growth and regression occur to some degree during each ovarian cycle. Growth, differentiation, and proliferation of mammary alveoli are pronounced during pregnancy, when estrogens, progesterone, prolactin, and hCS circulate in high concentrations.

Milk Production

Once the secretory apparatus has developed, production of milk depends primarily on continued episodic stimulation with high concentrations of prolactin, but adrenal glucocorticoids and insulin are also important in a permissive sense that needs to be defined more precisely. All of these hormones and hCS are present in abundance during late stages of pregnancy, yet lactation does not begin until after parturition. High concentrations of estrogen and particularly progesterone in maternal blood inhibit lactation by interfering with the action of prolactin on mammary epithelium. With parturition, the precipitous fall in estrogen and progesterone levels relieves this inhibition, and prolactin receptors in alveolar epithelium may increase as much as 20-fold. Development of full secretory capacity, however, takes some time. Initially, the mammary glands secrete only a watery fluid called *colostrum*, which is rich in protein but poor in lactose and fat. It takes about 2 to 5 days for the mammary glands to secrete mature milk with a full complement of nutrients. It is not clear whether this delay reflects a slow acquisition of secretory capacity or a regulated sequence of events timed to coincide with the infant's capacity to utilize nutrients.

Mechanism of Prolactin Action

Prolactin acts on alveolar epithelial cells to stimulate expression of genes for milk proteins such as casein and lactalbumin, enzymes needed for synthesis of lactose and triglycerides, and the proteins that govern the various steps in the secretory process. The prolactin receptor is a large peptide with a single membrane-spanning domain. It is closely related to the GH receptor and transmits its signal by activating tyrosine phosphorylation of intracellular proteins as described for the GH receptor (see Chapter 44). Binding of prolactin causes two receptors to dimerize and activate the cytosolic enzyme, Janus kinase 2 (JAK-2). Some of the proteins thus phosphorylated belong to the STAT family (for signal transduction and activation of transcription) and dimerize and migrate to the nucleus, where they activate transcription of specific genes. Prolactin may also signal through activation of a tyrosine kinase related to the *src* oncogene and by activating membrane ion channels. The signaling cascades set in motion the various events that accompany growth of the secretory alveoli as well as synthesis and secretion of milk.

Neuroendocrine Mechanisms

Continued lactation requires more than just the correct complement of hormones. Milk must also be removed regularly by suckling. Failure to empty the mammary alveoli causes lactation to stop within about a week and the lobuloalveolar structures to involute. Involution results not only from prolactin withdrawal, but also from the presence of inhibitory factors in milk that block secretion if allowed to remain in alveolar lumens. Suckling triggers two neuroendocrine reflexes critical for the maintenance of lactation: the so-called milk letdown reflex and surges of prolactin secretion.

Milk Letdown Reflex

Because each lactiferous duct has only a single opening to the outside and alveoli are not readily collapsible, application of negative pressure at the nipple does not cause milk to flow. The milk letdown reflex, also called the *milk ejection reflex*, permits the suckling infant to obtain milk. This neuroendocrine reflex involves the hormone oxytocin, which is secreted in response to suckling. Oxytocin stimulates contraction of myoepithelial cells that surround each alveolus, creating positive pressure of about 10 to 20 mm Hg in the alveoli and the communicating duct system. Suckling merely distorts the valve-like folds of tissue in the nipple and allows the pressurized milk to be ejected into the infant's mouth. Sensory input from nerve endings in the nipple is transmitted to the hypothalamus by way of thoracic nerves and the spinal cord and stimulates neurons in the supraoptic and paraventricular nuclei to release oxytocin from terminals in the posterior lobe (Fig. 12). These neurons can also be activated by higher brain centers, so that the mere sight of the baby or hearing it cry is often

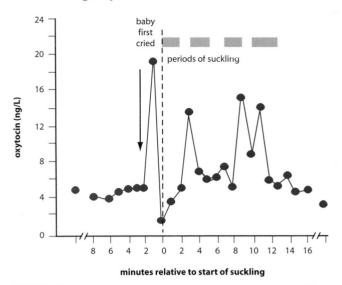

FIGURE 13   Relation of blood oxytocin concentrations to suckling. Note that the initial rise in oxytocin preceded the initial period of suckling. (From McNeilly AS, Robinson ICA, Houston MJ *et al. Br Med. J* 1983; 286:257. With permission.)

sufficient to produce milk letdown (Fig. 13). Conversely, stressful conditions may inhibit oxytocin secretion, preventing the suckling infant from obtaining milk even though the breast is full.

Cellular Actions of Oxytocin

The oxytocin receptor is expressed principally in uterine smooth muscle and myoepithelial cells that surround mammary alveoli. It is a typical heptahelical receptor that is coupled through G_q to phospholipase Cβ. When activated, it signals through inositol trisphosphate (IP_3) and diacylglycerol (DAG) released from phosphatidylinositol in the membrane. Inositol trisphosphate increases intracellular calcium by signaling its release from intracellular stores. Diacylglycerol activates protein kinase C, which may phosphorylate membrane calcium channels and further increase intracellular calcium. Increased intracellular calcium binds to calmodulin and activates myosin light-chain kinase, which initiates contraction of myoepithelial or myometrial cells.

Control of Prolactin Secretion

Suckling is also an important stimulus for secretion of prolactin. During suckling, the prolactin concentration in blood may increase by tenfold or more within just a few minutes (Fig. 14). Although suckling evokes secretion of both oxytocin and prolactin, the two secretory reflexes are processed separately in the central nervous system, and the two hormones are secreted independently.

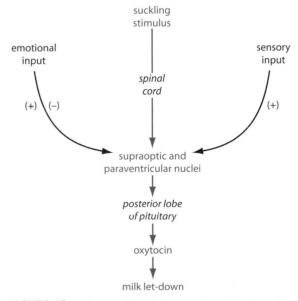

FIGURE 12   Control of oxytocin secretion during lactation.

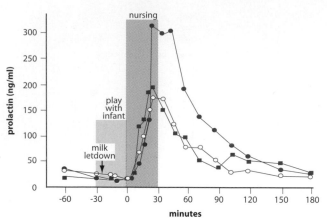

FIGURE 14 Plasma prolactin concentrations in three women during nursing and anticipation of nursing. Note that, although anticipation of nursing apparently resulted in oxytocin secretion, increased prolactin secretion did not occur until well after suckling began. (From Noel GL, Suh HK, Franz AG. *J Clin Endocrinol Metab* 1974; 38:413. With permission.)

Emotional signals that release oxytocin and produce milk letdown are not followed by prolactin secretion. It is unlikely that prolactin secreted during suckling can act quickly enough to increase milk production to meet current demands. Rather, such episodes of secretion are important for producing the milk needed for subsequent feedings. Milk production is thus related to frequency of suckling, which gives the newborn some control over its nutritional supply and is an extension into the postnatal period of the self-serving control over maternal function that the fetus exercised *in utero*.

Increased secretion of prolactin and even milk production do not require a preceding pregnancy. Repeated stimulation of the nipples can induce lactation in some women who have never borne a child. In some cultures, postmenopausal women act as wet nurses for infants whose mothers produce inadequate milk. This fact underscores the lack of involvement of the ovarian steroids in lactation once the glandular apparatus has been formed.

Prolactin is unique among the anterior pituitary hormones in the respect that its secretion is increased rather than decreased when the vascular connection between the pituitary gland and the hypothalamus is interrupted. Prolactin secretion is controlled primarily by an inhibitory hypophysiotropic hormone, dopamine. Dopamine is synthesized by sequential hydroxylation and decarboxylation of tyrosine (see Fig. 21 in Chapter 40). Surgical transection of the human pituitary stalk increases plasma prolactin concentrations in peripheral blood (*hyperprolactinemia*) and may lead to the onset of lactation. Stimulation of prolactin secretion by suckling results from inhibition of dopamine secretion into the hypophysial portal circulation by dopaminergic neurons whose cell bodies are located in the arcuate nuclei. It has been found experimentally that abrupt relief from dopamine inhibition results in a surge of prolactin secretion. It is possible that prolactin secretion is also under positive control by way of a yet to be identified prolactin-releasing factor. Experimentally, prolactin secretion is increased by neuropeptides such as thyrotropin-releasing hormone (TRH) and vasoactive inhibitory peptide (VIP). In spite of its potency as a prolactin-releasing agent, it is unlikely that TRH is a physiological regulator of prolactin secretion. Normally, TSH and prolactin are secreted independently. TSH secretion does not increase during lactation. The physiological importance of VIP as a prolactin-releasing hormone has not been established.

Lactotropes express estrogen receptors and increase their production of prolactin mRNA and protein in response to estrogens. Estradiol, which stimulates proliferation and hypertrophy of lactotropes, is probably responsible for the increased number of lactotropes in the pituitary and their prolactin content during pregnancy. Estradiol may therefore increase prolactin secretion by increasing its availability. In addition, although it does not act directly as a prolactin-releasing factor, estradiol decreases the sensitivity of lactotropes to dopamine. Paradoxically, however, estradiol also increases dopamine synthesis and concentration in the hypothalamus and may therefore increase dopamine secretion (Fig. 15).

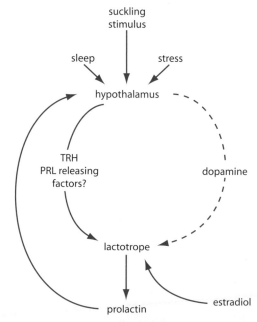

FIGURE 15 Control of prolactin secretion. Dashed line indicates inhibition. A physiological role for TRH (thyrotropin-releasing hormone) and other postulated releasing hormones has not been established. Estradiol stimulates secretion and may interfere with the inhibitory action of dopamine.

To date, there is no known product of prolactin action that feeds back to regulate prolactin secretion. The effects of suckling and estrogen on prolactin secretion are open loops. Experiments in animals suggest that prolactin itself may act as its own "short-loop" feedback inhibitor by stimulating dopaminergic neurons in the arcuate nucleus. It is not certain that such an effect is applicable to humans. If prolactin is a negative effector of its own secretion, it is not clear what mechanisms override feedback inhibition to allow prolactin to rise to high levels during pregnancy.

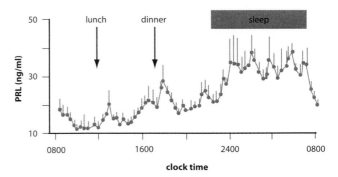

FIGURE 17 Around-the-clock prolactin concentrations in eight normal women. Acute elevation of prolactin level occurs shortly after onset of sleep and begins to decrease shortly before awakening. (From Yen SC, Jaffe RB. In: Yen SC, Jaffe R, Eds., *Reproductive physiology*, 4th ed., Philadelphia: Saunders, 1999, p. 268. With permission.)

Cellular Regulation of Prolactin Secretion

As in most other endocrine cells, the secretory activity of lactotropes is enhanced by increased cytosolic concentrations of calcium and cyclic adenosine monophosphate (cAMP) (Fig. 16). Dopamine acting through G-protein-coupled receptors, inhibits prolactin secretion

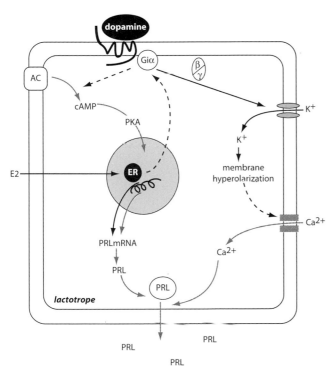

FIGURE 16 Cellular events in the regulation of prolactin secretion. Steps in basal synthesis and secretion are indicated by the pale blue arrows. Effects of dopamine are shown by the dark blue arrows. Effects of estradiol are shown in black. Dashed lines indicate inhibition. $G_{i\alpha}$ and $\beta\gamma$ are the subunits of the heterotrimeric inhibitory G protein. AC, adenylate cyclase; cAMP, cyclic adenosine monophosphate; PKA, protein kinase A; E2, estradiol; ER, estrogen receptor; PRL, prolactin. *Not shown*: TRH increases prolactin secretion through activation of its heptihelical receptor, which is coupled through $G_{\alpha q}$ to phospholipase C_β and causes release of inositol trisphosphate, diacylglyceride, and increased intracellular calcium.

through several temporally distinct mechanisms. Initial inhibitory effects are detectable within seconds and result from membrane hyperpolarization, which deactivates voltage-sensitive calcium channels and lowers intracellular calcium. This effect appears to result from direct stimulation of potassium influx by G-protein-gated channels. Minutes later there is a decrease in cAMP which leads to decreased transcription of the prolactin gene. Estrogens are thought to decrease responsiveness to dopamine by uncoupling dopamine receptors from G proteins. On a longer time scale, dopamine antagonizes the proliferative effects of estrogen by mechanisms that are not understood.

Prolactin in Blood

Prolactin is secreted continuously at low basal rates throughout life, regardless of sex. Its concentration in blood increases during nocturnal sleep in a diurnal rhythmic pattern. Basal values are somewhat higher in women than in men and prepubertal children, presumably reflecting the effects of estrogens. Episodic increases in response to eating and stress are superimposed on this basal pattern (Fig. 17). Prolactin concentrations rise steadily in maternal blood throughout pregnancy to about 20 times the nonpregnant value (Fig. 18). After delivery, prolactin concentrations remain elevated, even in the absence of suckling, and slowly return to the prepregnancy range, usually within less than 2 weeks. Prolactin also increases in fetal blood as pregnancy progresses, and during the final weeks reaches levels that are higher than those seen in maternal plasma. The fetal kidney apparently excretes prolactin into the amniotic fluid where at midpregnancy the prolactin concentration is five to ten times higher than that of either maternal or fetal blood. Although some of the prolactin in maternal

Clinical Note

Hyperprolactinemia

Hyperprolactinemia, often resulting from a small prolactin-secreting pituitary tumor (micro adenoma), is now recognized as a common cause of infertility and abnormal or absent menstrual cycles. Treatment with bromocriptine, a drug that activates dopamine receptors, suppresses prolactin secretion and restores normal reproductive function.

blood is produced by uterine decidual cells, prolactin in fetal blood originates in the fetal pituitary and does not cross the placental barrier. The high prolactin concentration seen in the newborn decreases to the low levels of childhood within the first week after birth. The physiological importance of any of these changes in prolactin concentration in either prenatal or postnatal life are not understood.

suckling is high. Ovarian activity is limited to varying degrees of incomplete follicular development, and even in those women who ovulate, luteal function is deficient. The delay in resumption of cyclicity results from decreased amplitude and frequency of GnRH release by the hypothalamic GnRH pulse generator (Chapter 45). Pulsatile administration of GnRH to lactating women promptly restores ovulation and normal corpus luteal function.

Lactation and Resumption of Ovarian Cycles

Menstrual cycles resume as early as 6 to 8 weeks after delivery in women who do not nurse their babies. With breast-feeding, however, the reappearance of normal ovarian cycles may be delayed for many months. This delay, called *lactational amenorrhea*, serves as a natural, but unreliable, form of birth control. Lactational amenorrhea is related to high plasma concentrations of prolactin. Delayed resumption of fertile cycles therefore is most pronounced when breast milk is not supplemented with other foods and consequently the frequency of

Suggested Readings

Ben-Jonathan N, Hnasko H. Dopamine as a prolactin (PRL) inhibitor. Endocrine Rev 2001; 22:724–763.

Challis JRG, Matthews SG, Gibb W, Lye SJ, Endocrine and paracrine regulation of birth at term and preterm. Endocrine Rev 2000; 21:514–550.

Gimpl G, Fahrenholz F. The oxytocin receptor system: structure, function, and regulation. Physiol Rev 2001; 81:630–683.

Hodgen GD. Hormonal regulation in in vitro fertilization. Vitamins Horm 1986; 43:251–282.

Jansen RPS. Endocrine response in the fallopian tube. Endocrine Rev 1984; 5:525–552.

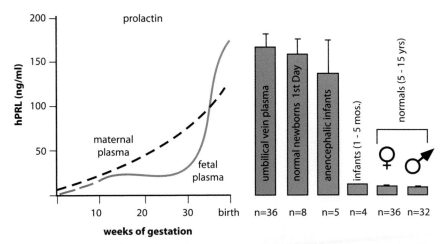

FIGURE 18   (Left) Comparison of the pattern of change of prolactin concentrations in fetal and maternal plasma during gestation. (Right) Plasma levels in normal and anencephalic newborns are compared with those of normal infants and adults. The high concentrations seen in anencephalic babies presumably reflect prolactin secretion by an anterior pituitary gland uninhibited by influences of the brain. (From Aubert ML, Grumbach MM, Kaplan SL. *J Clin Invest* 1975; 56:155. With permission.)

Mesiano S, Jaffe RB. Developmental and functional biology of the primate fetal adrenal cortex. Endocrine Rev 1997; 18: 378–403.

Neill JD, Nagy GM. Prolactin secretion and its control, in: Knobil E, Neill JD, Eds., The physiology of reproduction, Vol. 1, 2nd ed., New York: Raven Press, 1994, pp. 1833–1860.

Pepe GJ, Albrecht ED. Actions of placental and fetal adrenal steroid hormones in primate pregnancy. Endocrine Rev 1995; 16:608–648.

Tucker HA. Lactation and its hormonal control, in: Knobil E, Neill JD, Eds., The physiology of reproduction, Vol. 2, 2nd ed., New York: Raven Press, 1994, pp. 1065–1098.

PART VII

CENTRAL NERVOUS SYSTEM PHYSIOLOGY

Organization of the Nervous System

DIANNA A. JOHNSON

KEY POINTS

- The nervous system consists of more than 20 billion neurons linked together in motor and sensory pathways with distinctive anatomic, biochemical, and electrophysiologic properties.
- The general anatomic organization of the nervous system evolves at early developmental stages, preceding that of most other organs and providing a mechanism for neuronal control over subsequent developmental events.
- The lumen of the neural tube expands during development to form a series of four interconnected *ventricles* that circulate *cerebrospinal fluid* through the internal regions of the brain.
- The *peripheral nervous system* is composed of groups of cell bodies organized within two types of ganglia: (1) *sensory ganglia*, located near the

- spinal cord or brain stem, and (2) *autonomic motor ganglia*, some of which are located near the muscle or gland to be innervated.
- Sensory pathways share common organizational motifs, with primary sensory cell bodies in sensory ganglia sending central processes through spinal cord/brain stem tracts to reach primary receiving areas in the cerebral cortex.
- Voluntary motor pathways from the cortex control movement of skeletal muscle; sympathetic and parasympathetic nuclei in the brain stem and spinal cord provide autonomic control of smooth muscles and glands; and short interconnected sensory/motor pathways in the spinal cord and brain stem coordinate subconscious reflexes.

The central nervous system, consisting of brain and spinal cord, represents the most complex of all human organs. It resides in a highly protected environment, surrounded by a cushion of cerebrospinal fluid (CSF) and encased within the bony calvarium and vertebral column. Ten million input neurons feed sensory information to it; a half million output neurons extend from it to provide neuronal control over all the body's muscles and glands. However, the overwhelming majority of nerve cells in the body, roughly 20 billion, reside entirely within the brain proper and provide computational links between sensory and motor pathways. Together these neuronal circuits constantly monitor the internal and external environment, store and recall motor and sensory data, and send out a continual flow of messages that allows the body to maintain vegetative functions and respond to changes in the environment.

Unlike computer circuits, which contain only a handful of different elements, the human nervous system operates with a highly diverse set of elements represented by 50 million distinct types of nerve cells. The identity of each neuronal cell type is established by its specific pre- and postsynaptic partners and its relative position within a specified neuronal pathway; its molecular makeup, including the nature of the neurotransmitters it releases and to which it responds; and its electrophysiologic properties. For this reason, the study of the nervous system and the treatment of neuronal disorders requires a highly interdisciplinary approach, including anatomic, biochemical, and electrophysiologic techniques. An understanding of neuroanatomy provides the basis for correlating a patient's functional loss with the specific location of the causative lesion within the affected neuronal pathway. Likewise, knowledge of neurotransmitter metabolism is a prerequisite for developing neuropharmacologic treatments for disease states such as Parkinsonism and Alzheimer's disease. Electrophysiologic analyses, such as the *electroencephalogram* (EEC), help establish the parameters for determining appropriate treatment for epilepsy or for sleep disorders.

EARLY DEVELOPMENTAL STAGES OF THE NERVOUS SYSTEM

The dominant position of the nervous system within the body is established early during development. It is the first tissue to differentiate, it maintains the fastest rate of growth of any embryonic tissue, and it is the largest organ during the gestational period of development. Embryonic neurons and their precursors synthesize and release *trophic factors* that affect the viability of surrounding cells, as well as *tropic factors*, which influence the direction of neural growth, thus influencing subsequent maturation

of neuronal pathways as well as their targets, including muscles and glands.

The steps of neuronal development include (1) formation of the *neural plate*, (2) folding of the plate to form the *neural tube*, and (3) bulging and bending of the tube to form a curved configuration of five brain vesicles attached to a straight tubular spinal cord (Figs. 1 and 2). The neural plate, from which the nervous system develops, is formed by a thickening of the ectoderm during the third week of gestation in humans. Through a process called *neural induction*, regions of the neural plate become genetically programmed to form particular regions of the nervous system. Growth of the neural plate is accelerated along its lateral edges, causing the edges to curve toward each other, eventually fuse, and form an open neural tube with both cranial and caudal openings called *neuropores*. Closure of both neuropores is normally complete by the fourth week.

Preferential growth at three nodes along the cranial portion of the sealed neural tube causes intermittent bulging of tissue, formation of three brain vesicles (*prosencephalon*, *mesencephalon*, and *rhombencephalon*), and bending of the tube at flexion points generally located between the vesicles. Subsequent growth and division of the three vesicles leads to the formation of five major brain regions: (1) *telencephalon*, (2) *diencephalon* (regions 1 and 2 are collectively called the *forebrain*), (3) *mesencephalon* (*midbrain*), (4) the *metencephalon* (*pons* and the overlying *cerebellum*), and (5) *myelencephalon* (*medulla*). The spinal cord retains its original configuration as a relatively uniform tubular structure with the generation of slight enlargements at cervical and lumbar levels.

DEVELOPMENT OF BRAIN VESICLES

The closed lumen of the neural tube is altered in shape as a result of enlargement of the three brain vesicles, particularly the telencephalon. As the cerebral hemispheres of the telencephalon increase in size, two lateral expansions of the lumen form the first and second (*lateral*) ventricles, connected to each other and to the third ventricle by a thin, Y-shaped midline channel called the *foramen of Monro*. Continued posterior and inferior expansion of the hemispheres eventually forces the cortex and the underlying lateral ventricles into a C shape. Tissue surrounding the lateral ventricles develops into: (1) *primary motor* and *sensory cortex*, which provides conscious motor control and sensory perception; (2) *limbic structures*, which establish mood and emotion; and (3) *basal ganglia*, which contain relay nuclei for motor pathways. Two structures develop from the diencephalon and surround the third ventricle: the *thalamus,* which

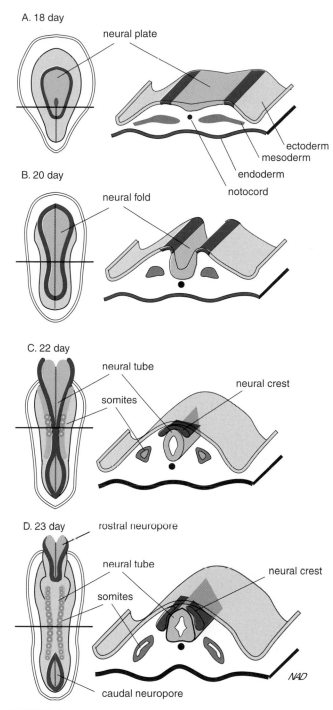

FIGURE 1 Formation of neural plate, neural crest, and neural tube. Human embryos at approximately (A) 18 days, (B) 20 days, (C) 22 days, and (D) 23 days. Left side, dorsal view; right side, cross section at the level indicated on the dorsal view. Note that the neural crest closes into a neural tube in the center first, and this closure moves rostrally and caudally, leaving a neuropore at each end that closes at about 4 weeks.

contains sensory relay nuclei, and the *hypothalamus,* which provides neurohormones controlling reproductive and other behaviors. A narrow cerebral aqueduct extends rostrally from the third ventricle through brain stem

structures (midbrain, pons, and medulla), areas that contain ascending, descending, and reflex pathways for motor and sensory circuits, as well as nuclei controlling vegetative functions. The fourth ventricle emerges from the *cerebral aqueduct* and is flattened between the caudal portion of the brain stem and the overlying cerebellum, an area involved in fine motor control (Fig. 3).

Cerebrospinal fluid fills the ventricles, providing a cushion of support for delicate neuronal tissue. It is generated by the *choroid plexus,* which is composed of a capillary network surrounded by cuboidal or columnar epithelium lining the ventricular walls, in particular the roof of the lateral and fourth ventricles. Structurally similar to the distal and collecting tubules of the kidney, the choroid plexus maintains the chemical stability of the CSF. It has directional capabilities in that it continually produces CSF and actively transpors metabolites out of the ventricles. Cerebrospinal fluid circulates through the ventricles, exits through openings in the fourth ventricle to the *subarachnoid space* surrounding the outer surface of the brain, is reabsorbed by the arachnoid granulations, and eventually collects in the venous system of the meningeal covering of the cortex. This circulation serves an important function in maintaining the appropriate ionic milieu necessary for neuronal activity. Blockage of CSF flow by insufficient absorption, by obstructions such as brain tumors or genetic malformations, or by internal bleeding caused by trauma or stroke can cause increases in intracranial pressure, enlargement of ventricles, and development of hydrocephalus.

Samples of CSF are often removed from the spinal cord region for analysis in order to screen for possible infections or electrolyte imbalances. CSF bathes the full extent of the spinal cord within its surrounding dural sack, the most caudal region of which contains only groups of exiting spinal nerves called the *cauda equina*. This affords a reasonably safe region from which CSF samples can be removed with little danger of injury to brain or spinal cord. In a procedure called a *spinal tap,* a small amount of CSF is removed by inserting a needle through an intervertebral space in the lumbosacral region and piercing the underlying dura.

OVERVIEW OF THE PERIPHERAL NERVOUS SYSTEM

The peripheral nervous system arises from specialized cells (*neural crest cells*) located along the lateral edges of the neural plate (see Fig. 1). These cells bud off and become separated from the neural tube as it forms. Some groups of neural crest cells migrate to form a linear array of *dorsal root ganglia* located adjacent to the lateral borders of the spinal cord. These cells function as sensory

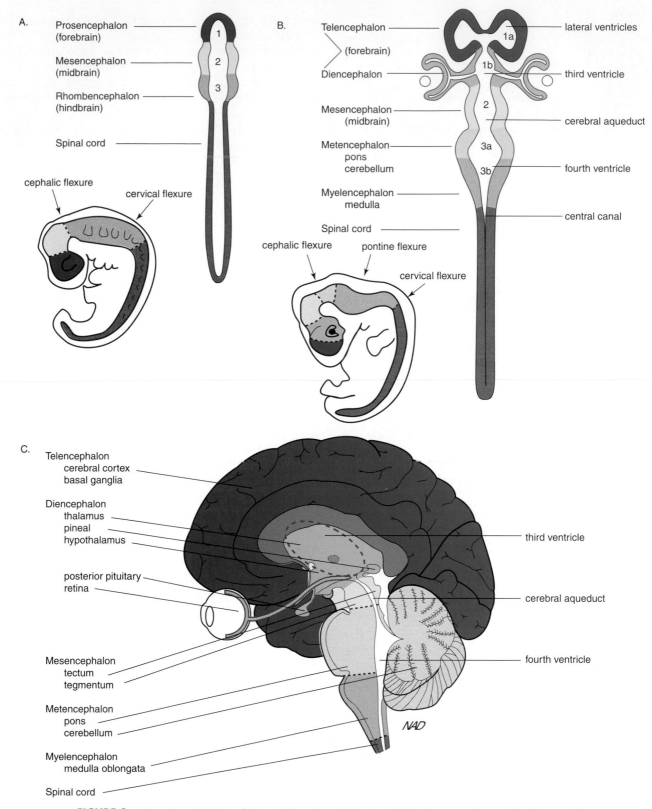

FIGURE 2   Growth and folding of the rostral portion of the neural tube. Human embryos at the three-vesicle stage at about 28 days (A) and the five-vesicle stage at about 38 days (B). Dorsal views of the unfolded embryos show the expansion of the neural tissue during development. Lateral views of the embryos show the three bends or flexures that occur as the embryo develops. The organization of the adult brain (C) reveals the ultimate fate of the three embryonic areas shown in differing shades of blue and gray. Brains are not drawn to scale.

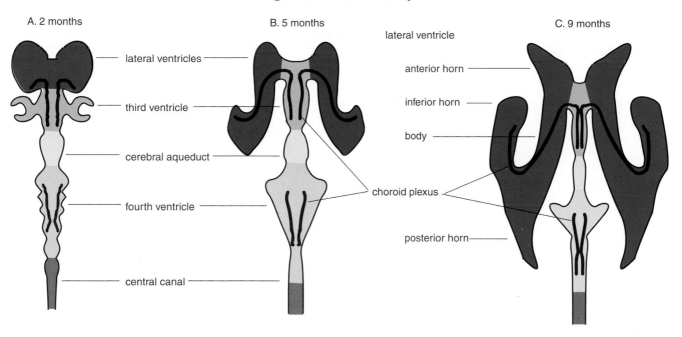

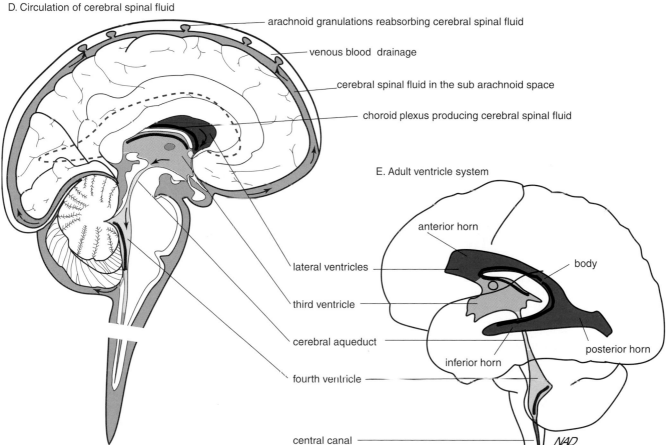

FIGURE 3 Development of the ventricles, choroid plexus, and circulation of the cerebral spinal fluid, as well as the ventricles and choroid plexus. Shades of blue and gray of the spaces correspond to the shades of the surrounding neural tissue in Fig. 1A. Cerebral spinal fluid is formed by the choroid plexus located in the two lateral, third, and fourth ventricles. It flows through the cerebral aqueduct and out three holes in the fourth ventricle into the subarachnoid space, bathing the entire brain and spinal cord. Cerebral spinal fluid is reabsorbed by the arachnoid granulations and returned to the venous blood.

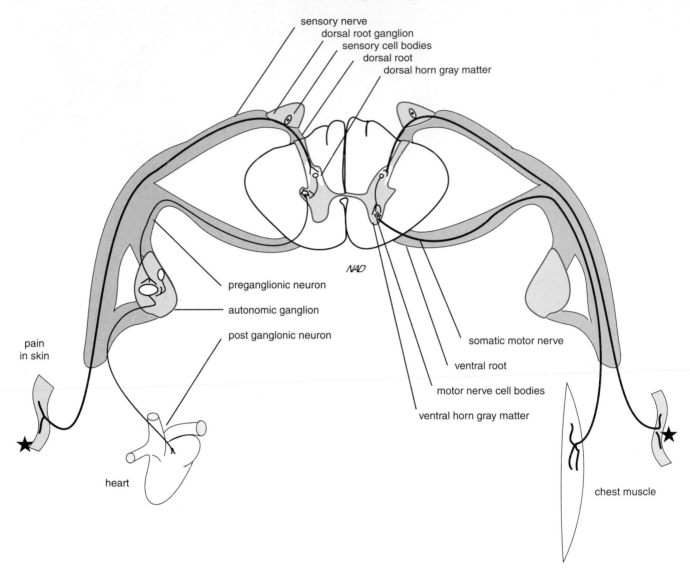

FIGURE 4   Organization of the peripheral nervous system. Sensory information (afferent neurons) from all parts of the body enters the dorsal part of the spinal cord; motor information (efferent neurons) exits the ventral part of the spinal cord. Primary sensory cell bodies are located in the dorsal root ganglia, and their axons project via the dorsal root to the dorsal horn gray matter (and on to higher centers in brain, not shown). Cell bodies of somatic motor neurons are located in the gray matter of the anterior horn, and their axons pass via the ventral root directly to somatic muscles. Autonomic motor neurons also exit the ventral root but synapse in autonomic ganglia, and postganglionic neurons then contact target tissue. The example on the left illustrates the autonomic (sympathetic) neurons that speed the heart. Cell bodies and synaptic connections between neurons are located in areas indicated in blue.

neurons. Their cell bodies remain in the spinal ganglia, a distal process projects to the tissue to be innervated, and a proximal process projects into the appropriate level of the dorsal aspect of the spinal cord. Other groups of neural crest cells migrate to equivalent positions along the medulla and pons, where they form *sensory ganglia* associated with the cranial nerves, and still other groups form *motor ganglia* associated with both sympathetic and parasympathetic components of the autonomic nervous system (Fig. 4).

OVERVIEW OF SENSORY PATHWAYS

It is apparent from the descriptions given earlier that specific areas of brain, spinal cord, or peripheral nervous tissue are assigned specific functional roles based on their relative position with the nervous system (Fig. 5). A general rule for neurons might be stated, "Where you are is what you are." However, the nervous system should not be viewed as a series of isolated modules with specified duties; rather, it more correctly resembles

A. Cross sections of the nervous system   B. Coronal view of the nervous system   C. Lateral view of the nervous system

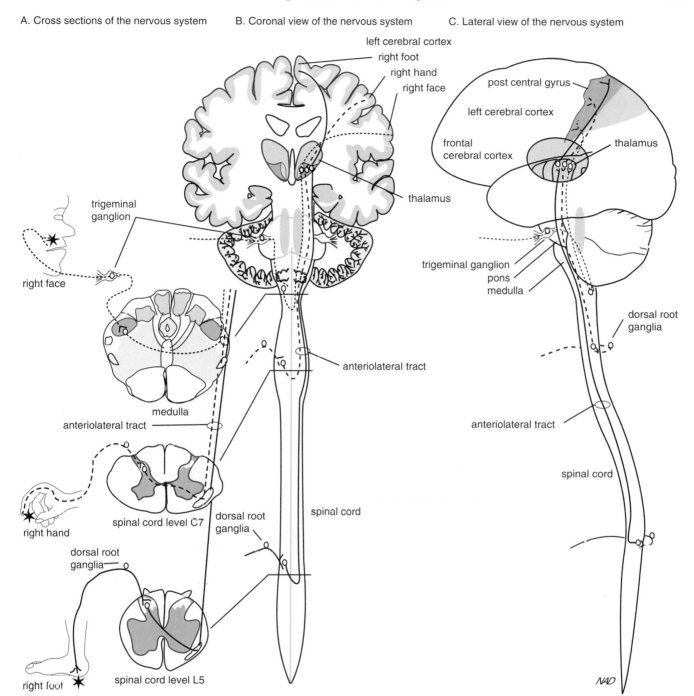

FIGURE 5   Sensory axis of the nervous system. Primary sensory neurons carrying pain and temperature information project to the dorsal part of the spinal cord; their cell bodies are located in the dorsal root ganglia. Neurons carrying pain and temperature information from the face project through the trigeminal nerve to the lateral part of the pons; their cell bodies are in the trigeminal ganglion. Pain information is relayed through synaptic connections to second-order and on to higher order neurons in areas shown in blue. Pain relay areas are present in both the spinal cord and thalamus, within pathways that project to the post central gyrus of the cerebral cortex. Note that information from the right side of the body projects to the left side of the brain.

a series of interconnected circuits, each consisting of sensory receptor neurons, relay and processing nuclei, primary receiving areas, memory banks, connecting links to other pathways, feedback loops, and final output pathways.

The major input portion of this overall scheme is the sensory axis of the nervous system. At the spinal level of the sensory axis, somatosensory cell bodies reside in dorsal root ganglia, and their fibers exit through spinal nerves to innervate skin and skeletal muscles of the body. Through projections of the cranial nerves, sensory cell bodies residing in cranial nerve ganglia innervate analogous structures in the head and neck region plus organs of the special senses of vision, taste, hearing, balance, and smell. The only areas not innervated by sensory fibers are bone and nervous tissue itself.

After entering appropriate regions of the spinal cord or brain stem, proximal sensory fibers are classified as sensory tracts, and they ultimately project to primary receiving areas located in posterior and temporal regions of the cortex after a relay in the thalamus. Many sensory tracts cross the midbrain of the neural axis and project to the contralateral cerebral hemisphere. Only after

activation of a primary cortical receiving area will a sensation be consciously perceived.

Each individual sensory pathway maintains strict spatial or *somatotopic organization*. For example, pain or touch input from a certain area of the body is localized to specified regions of the cortex within the primary receiving area. Likewise, specific modality attributes (*e.g.*, wavelength of light, actual pitch of a sound) are associated with given subregions of the primary sensory cortex.

OVERVIEW OF MOTOR PATHWAYS

The output or motor axis of the nervous system has two major divisions: One is involved in conscious control of skeletal muscle contraction and involves the corticospinal tract; the other operates subconsciously either through basic motor reflexes of skeletal muscle orchestrated by subcortical motor areas such as the cerebellum, basal ganglia, and brain stem nuclei, or through autonomic nervous system control of smooth muscle. Cortical and subcortical motor regions are somatotopically organized, providing finely tuned control,

---

Clinical Note

Spina Bifida, Epilepsy, and Hydrocephalus

Abnormalities during early developmental stages of the nervous system can cause profound neuronal birth defects. For example, alterations in the production and differentiation rates of neural plate neurons may result in abnormalities in the closure of the neural pores. The cranial and caudal neuropores usually close at 25 and 27 days of gestation, respectively. Failure of the caudal neuropore to close in a timely fashion disrupts the functional development of the lower parts of the spinal cord that normally control movement of the muscles of the legs as well as smooth muscles and glands of the lower body cavity. This severely crippling condition is called *spina bifida*. Even more devastating is the condition called *anencephaly*, which results from failure of the rostral or cranial neuropore to close. Gross abnormalities in brain structures can result, leading to severe mental retardation or death of the infant.

Abnormalities in neuronal migration during embryogenesis can also result in clinical disorders, many of which may be so severe that they are incompatible with life. Even if the abnormality

is relatively minor, neurologic symptoms may persist through adulthood. The most common condition in these cases is *epilepsy*, caused by abnormal, synchronous activity in large groups of neurons that results in convulsive seizures. Even small patches of abnormally placed cells in the cerebral cortex can sufficiently disrupt normal physiology to result in epilepsy. Although other types of epilepsy may be responsive to drug therapy, seizures associated with neuronal migration disorders characteristically are not.

Occasionally, infants are born with an obstruction that blocks the flow of CSF through the ventricles. Fluid collects in the ventricles, causing increased pressure and swelling. Because the skull is soft and incompletely formed, the head itself will expand, sparing the brain from damage. This condition, called *hydrocephalus,* is treated by inserting a drainage tube or ventricular shunt into the swollen ventricle. In adults, hydrocephalus is more problematic. Because the adult skull does not expand, increased pressure caused by a ventricular blockage from intracranial bleeding, tumor, and so on will compress and damage brain tissue if not treated quickly.

A. Cross sections
   of the spinal cord

B. Coronal view of the nervous system

C. Lateral view of the nervous system

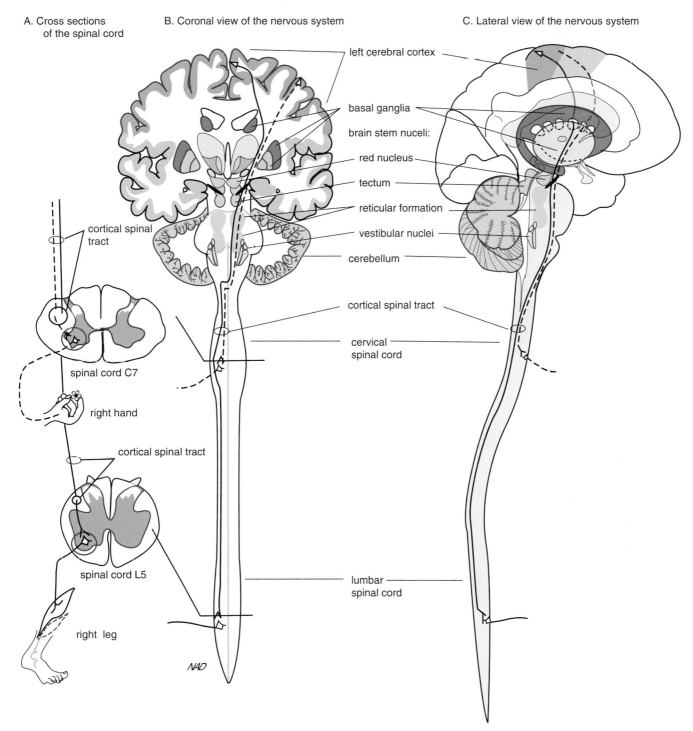

left cerebral cortex

basal ganglia

brain stem nuceli:

red nucleus

tectum

reticular formation

vestibular nuclei

cerebellum

cortical spinal tract

cervical spinal cord

lumbar spinal cord

cortical spinal tract

spinal cord C7

right hand

cortical spinal tract

spinal cord L5

right leg

NAD

FIGURE 6　Motor axis of the nervous system. Blue indicates the many areas of the brain contributing to motor control. The corticospinal tract from the cerebral cortex to the spinal cord is shown in black. In addition to commands delivered directly to motor neurons in the brain stem and spinal cord, the cerebral cortex communicates to other motor areas, calling forth the movements specialized by each. For example, the reticular formation and cerebellum program the static and dynamic motor adjustments in space, respectively. Note that, except for the cerebellum, the right brain controls the left hand and leg.

particularly in the extremities (e.g., the digits) (Fig. 6). In contrast, autonomic innervation is more diffuse and movements are highly stereotyped. Many common lesions of the nervous system are relatively localized. Stroke, tumor, or infection many affect only a small portion of the motor or sensory axis. Because of the somatotopic organization of somatic sensory and motor systems, resulting deficits may also be highly specific. For example, a patient experiencing a stroke within a small area of the left cerebral cortex may lose only the ability to move the right hand because of the loss of appropriate input through the corticospinal tract.

Suggested Readings

Conn PM. *Neuroscience in medicine*. Philadelphia: JB Lippincott, 1995.

Cowan WM. The development of the brain. *Sci Am* 1979; 241(3): 112–133.

Martin JH. *Neuroanatorny*. Stamford: Appleton and Lang, 1996.

Purves D, Lichtman JW. *Principles of neural development*. Sunderland, MA: Sinuaer, 1985.

49

Sensory Receptors of the Somatosensory System

DIANNA A. JOHNSON

KEY POINTS

- Separate sensory pathways carry information about each of the special senses of vision, taste, olfaction, and equilibrium and about the general somatosensory sensations of touch, proprioception, pain, and temperature.
- *Somatosensory mechanoreceptors* display distinct physiologic properties based on receptive field size, ability to adapt to the stimulus with time, stimulus threshold, and response time.
- The propagation rate of action potentials varies with the diameter of the nerve fiber and the degree of myelination.
- *Pacinian corpuscles* are rapidly adapting mechanoreceptors that act as rate detectors with large receptive fields.
- *Ruffini's endings* are mechanoreceptors with large receptive fields and sustained responses.
- *Meissner's corpuscles* and *Merkel's disks* are mechanoreceptors with small receptive fields. The former is a rapidly adapting cell; the latter is slowly adapting.
- *Nociceptors* contribute to the perception of pain by transmitting information about high

intensities of mechanical, thermal, or chemical stimulation.
- *Nociceptors* and central nervous system neurons that comprise central pain pathways may both exhibit *hyperalgesia* after exposure to a painful stimulus.
- *Thermoreceptors* in the skin are rapidly adapting and have a limited response range.
- Thermoreceptors within the central nervous system exhibit acute sensitivity and allow the body to maintain the temperature of the nervous system within its narrow functional range.
- *Proprioceptors* in muscles and joints send information about body position and movement to the cerebellum and other motor centers in the brain in order to maintain posture and aid in muscle reflexes.
- *Visceral sensory fibers* provide general information about mechanical, thermal, and chemical stimuli, but the sensation is not well localized and is often referred, or perceived as emanating from some unstimulated part of the body.

GENERAL PROPERTIES OF SENSORY RECEPTORS

Sensory information from the body reaches the nervous system piecemeal through a series of separate sensory pathways associated with specific sensory modalities. These pathways consist of sensory receptors and their projections to designated receiving areas in the cortex. A unified perception of our physical world emerges from the coordinated response of these primary receiving areas and secondary association areas.

Separate pathways for the special senses of vision, hearing, taste, and olfaction and for the vestibular sense of equilibrium begin with single-modality receptors that are concentrated at relatively small, specialized locations within the head (see Chapters 51 to 54). The other senses comprise the somatosensory system, whose diverse receptor types are found throughout the body and provide four general sensations:

1. *Touch*, in response to mechanical contact with the skin and pressure or stretching of internal organs.
2. *Proprioceptive information*, in response to changes in position and movement of muscles and joints.
3. *Pain*, in response to noxious (tissue-damaging) stimuli.
4. *Thermal sensations*, in response to changes in temperature.

Although each of these elemental modalities is transmitted by a designated class of receptors (Fig. 1; see following sections), a much wider range of sensations or submodalities is discernible through the selective combinations of inputs. For example, superficial touch is distinct from deep pressure; limb proprioception involves both static position and dynamic kinesthesia information; and wetness is a combination of touch and thermal sensations.

Most somatosensory receptors are *mechanoreceptors* that respond to physical distortion such as bending or stretching. They have been extensively studied, and their general attributes serve as a model for many other receptor types for both the general and special senses. They are discussed in detail here in order to illustrate the general properties of sensory receptors.

These receptors share common transduction mechanisms that convert movement to electrical signals in the form of generator currents and action potentials. Each of the six major classes of mechanoreceptors shown in Fig.1 has a distinctive morphology, and each exhibits different physiologic properties based to a large extent on morphologic properties.

Sensory fibers sending information to the central nervous system (CNS) are classified as *afferent fibers*. In contrast, motor fibers, which project from the CNS to specified muscle groups, are classified as *efferent fibers*. Afferent and efferent fibers may be bundled together within a single, "mixed" nerve such as the sciatic nerve. Different types of sensory or motor fibers have different propagation velocities based on the diameter of the fiber and the degree to which the fiber is sheathed by supportive Schwann cells. Sheaths provide insulation for the nerve fiber and are composed of overlapping layers of Schwann cell membrane, called *myelin*. The fastest conduction rates are achieved by fibers designated as type I_a or A-alpha, which have large diameters and heavy myelin sheaths. Intermediate rates are achieved by myelinated fibers of medium (type II or A-beta) or small (type III or A-gamma) diameters. Slowest rates are sustained by unmyelinated fibers (type IV or C). Virtually all mechanoreceptors have type I or II fibers and are capable of conveying touch information from the extremities to the CNS within 4 ms. Thermoreceptors and nociceptors are type III (A-gamma) and IV (C) fibers, which take 2 s or more to transmit information about pain and temperature. Shown in Fig. 2 is an example of conduction velocities of different fibers observed in experimental recordings from the sciatic nerve.

TOUCH RECEPTORS

The five major classes of touch receptors illustrated in Fig. 1 include three (Pacinian corpuscles, Merkel's disks, and Ruffini's corpuscles) that are common to both hairy and glabrous (nonhairy) skin. Hair follicle receptors are unique to hairy skin, and Meissner's corpuscles are found primarily in glabrous skin. All of these types of somatosensory mechanoreceptors have fibers designated as type II, indicating that they have a similar diameter, a similar degree of myelination, and a similar conduction rate.

Pacinian corpuscles and *Ruffini's corpuscles* are found deep within the dermal layers. Both have relatively large, specialized end organs that are widely spaced. These attributes permit each sensory fiber to respond to stimuli over a wide expanse of skin, thus producing a relatively large receptive field for each cell. A difference is noted, however, in the ability of these two receptor types to sustain their responses to long-lasting stimuli. Pacinian corpuscles respond rapidly to a change in pressure on the skin, then quickly adapt and stop firing even though the stimulus continues. Ruffini's endings are relatively slow to adapt to a sustained stimulus and continue to respond during long stimulus periods.

These different adaptation characteristics reflect the cellular organization of the end organ rather than electrophysiological properties of the individual nerves. An experimental demonstration of this fact is illustrated in Fig. 3. The end organ of the Pacinian corpuscle is encapsulated by 20 to 70 concentric layers of connective tissue arranged like layers of an onion. In response to

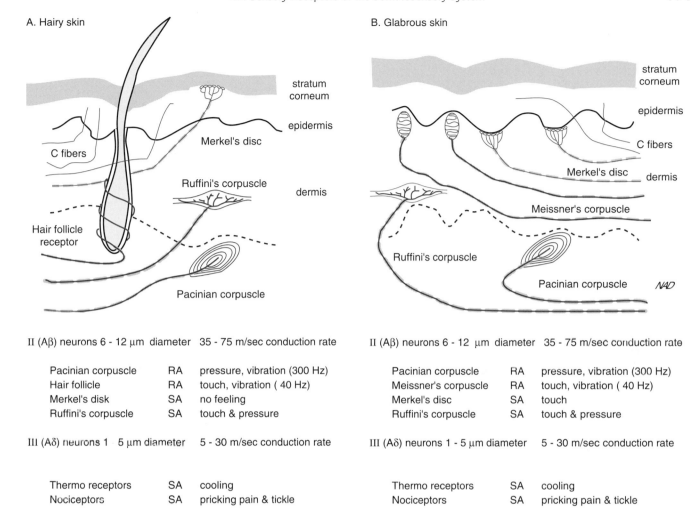

A. Hairy skin

stratum corneum

epidermis

Merkel's disc

C fibers

Ruffini's corpuscle

dermis

Hair follicle receptor

Pacinian corpuscle

B. Glabrous skin

stratum corneum

epidermis

C fibers

Merkel's disc   dermis

Meissner's corpuscle

Ruffini's corpuscle

Pacinian corpuscle   NAD

II (Aβ) neurons 6 - 12 μm diameter		35 - 75 m/sec conduction rate
Pacinian corpuscle	RA	pressure, vibration (300 Hz)
Hair follicle	RA	touch, vibration (40 Hz)
Merkel's disk	SA	no feeling
Ruffini's corpuscle	SA	touch & pressure

III (Aδ) neurons 1 5 μm diameter		5 - 30 m/sec conduction rate
Thermo receptors	SA	cooling
Nociceptors	SA	pricking pain & tickle

IV (C) neurons 0.2 - 1.5 μm diameter		0.5 - 2 m/sec conduction rate
Thermo receptors	SA	heat
Nociceptors	SA	slow burning pain

II (Aβ) neurons 6 - 12 μm diameter		35 - 75 m/sec conduction rate
Pacinian corpuscle	RA	pressure, vibration (300 Hz)
Meissner's corpuscle	RA	touch, vibration (40 Hz)
Merkel's disc	SA	touch
Ruffini's corpuscle	SA	touch & pressure

III (Aδ) neurons 1 - 5 μm diameter		5 - 30 m/sec conduction rate
Thermo receptors	SA	cooling
Nociceptors	SA	pricking pain & tickle

IV (C) neurons 0.2 - 1.5 μm diameter		0.5 - 2 m/sec conduction rate
Thermo receptors	SA	heat
Nociceptors	SA	slow burning pain

FIGURE 1   Specialized somatic sensory nerve endings in the skin: (A) hairy skin such as the arm, and (B) glabrous skin such as the palm of the hand. Fibers are classified (type II, III, or IV) according to conduction velocity. All of the axons shown are covered with an intermittent, insulating wrapping of myelin except for the type IV (C) fibers. Rapidly adapting receptors (RAs) respond during the beginning and end of a movement but not during a steady stimulus. Slowly adapting receptors (SAs) respond throughout a stimulus. Not shown are the sensory neurons transmitting proprioceptive information from muscles and joints. These large (13–20 μm in diameter), heavily myelinated, and fast (80–120 m/s) axons are classified as type I fibers.

pressure on the surface of the skin, layers are compressed and the compression is transmitted to the underlying nerve terminal membrane, where mechanosensitive channels are triggered to open. Increasing the conductance of the membrane to sodium or to sodium and potassium leads to small depolarizations called *generator potentials*. These channels are not the type of voltage-gated sodium channels that directly produce an action potential; however, action potentials will result from a given stimulus if the generator potentials produced are able to summate and bring the cell to the threshold level for activating voltage-gated channels. The number of action potentials produced will be determined by the strength of the generator potential, which in turn reflects the degree of mechanical distortion.

The response of the Pacinian corpuscle is not sustained, however, because the encapsulating layers are covered with a viscous liquid that allows them to slip past each other and dissipate the pressure to surrounding tissues. When the pressure is relieved, slippage of the layers back

A. Nerve bundle containing thousands of axons

B. Compound action potential

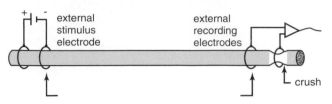

120 mm between the stimulus and recording electrodes

FIGURE 2   The compound action potential from the sciatic nerve. The different conduction velocities of sensory fibers from a single nerve are seen in recordings from external electrodes (A). The distal nerve is crushed to prevent the action potentials from reaching the second (reference) recording electrode. Action potentials from the fastest sensory neurons (classified as type I or I_a) travel the 120 mm to the recording electrode in only 1 ms. Smaller fibers are slower, and their action potentials take longer to travel the distance.

to their original position also changes pressure on the neuronal membrane, and action potentials may again result. During a period of rapid change in pressure, either applied or relieved, layers have less time to compensate and more action potentials will be generated. The physical configuration of the end organ determines what rates of change provide the most potent stimuli. Accordingly, Pacinian corpuscles are most sensitive to rates of change (vibrations) of 200 to 300 Hz.

These simple mechanisms permit the Pacinian corpuscle to act as a *rate detector* as well as a *pressure detector*. Rate detectors are critically important in allowing the brain to make predictions about ongoing changes in body position. Pacinian corpuscles and other receptors located in or near the joint capsules help detect rates of movement of the different parts of the body and allow the brain to predict where feet or arms will be during any precise moment during the ongoing movement. Appropriate motor signals can then be sent to make anticipatory corrections.

Ruffini's endings are mechanoreceptors with large receptive fields and sustained responses. They have pressure transduction mechanisms similar to those of Pacinian corpuscles, but they lack a layered capsule (Fig. 3). Without sliding layers to facilitate dissipation of the pressure, they adapt relatively slowly to a sustained stimulus and thus can provide information about the degree of sustained mechanical distortion of the skin. A Pacinian corpuscle can be made to respond like a Ruffini's ending by experimentally removing its connective tissue capsule.

Two other important mechanoreceptor types are typically found in more superficial layers of glabrous skin. *Meissner's corpuscles* are about one tenth the size of Pacinian corpuscles and are located in the ridges of glabrous skin, such as the raised parts of the fingerprints. *Merkel's disks* are located within the epidermis and consist of a nerve terminal and a flattened non-neural epithelial cell. Merkel's disks are slowly adapting, whereas Meissner's corpuscles are rapidly adapting and are most

sensitive to low-frequency stimuli of around 50 Hz. Both of these receptors have relatively small receptive fields compared to Pacinian corpuscles or Ruffini's endings; however, absolute receptive field size for any single class of receptor varies from one area of the body to another. This holds true for all types of mechanoreceptors in the skin. The size of the receptive field is inversely proportional to the spatial resolution of the touch sensation, so that the smaller the receptive field, the greater the ability to discriminate. A simple test of spatial resolution is the two-point discrimination test, which measures the minimum distance necessary to differentiate between two simultaneous stimuli. The value varies 20-fold across the body surface, with fingertips showing highest resolution (2 mm) and the back showing the lowest (40 mm).

Hairy skin is innervated by *hair follicle receptors*, which may be either slowly adapting or rapidly adapting. Each hair follicle is innervated by a single free nerve ending that varies according to hair type. Bending of the hair deforms the associated nerve ending membrane, changes its conductance, and leads to changes in generator potentials and firing rates of the neuron.

THERMORECEPTORS

Changes in ambient temperature are monitored by separate cold and warm *thermoreceptors* in the skin. Each receptor has a receptive field size of approximately 1 mm in diameter with little overlap between adjacent receptive fields. These individual domains can be mapped experimentally to demonstrate that small, discrete areas of skin can produce either warm or cold sensations, but not both. Both thermoreceptor subtypes are most sensitive to abrupt changes in temperature and adapt to a new stimulus level within several seconds. They are also limited in their response range. *Heat receptors* are active at temperatures of 30 to 45°C, at which point they become silent and nociceptive thermoreceptors

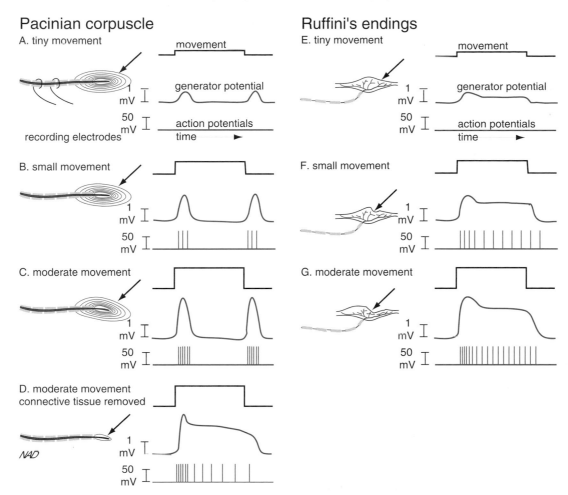

FIGURE 3   Characteristics of Pacinian corpuscles and Ruffini's endings: (A) to (D) experimental analysis of rapidly adapting Pacinian corpuscles, and (E) to (G) slowly adapting Ruffini's endings. Each example shows the receptor on the left and three recordings on the right: the movement (top), generator potentials (middle), and action potentials (bottom). In (A) and (E), a tiny movement produces small generator potentials, but these are too small to start action potentials. In (B) and (F), a larger movement produces larger generator potentials. This causes a burst of action potentials in the Pacinian corpuscle and a continuous stream of action potentials in the Ruffini's endings. With a still larger movement, the generator potentials in (C) and (G) are even greater, and the action potentials occur at higher frequency. In (D), with most of the corpuscle removed, the Pacinian receptor responds in a manner similar to that for a Ruffini's ending.

begin firing (see below). Heat receptors have small, unmyelinated fibers with relatively slow conduction rates that are designated as type IV or C fibers (Fig. 1). *Cold receptors* respond preferentially to temperatures of 18 to 30°C and cease firing entirely below 10°C, which is one of the reasons why cold temperature acts as an effective anesthetic. Cold receptors have fibers of medium-sized diameter and limited layers of myelin and are thus designated as type III fibers (Fig.1).

Besides the conscious perception of temperature changes provided by thermoreceptors in the skin, other thermoreceptors clustered in the hypothalamus and spinal cord monitor internal temperatures. Changes in core body temperature lead to subconscious reflexes that regulate temperature but provide no conscious sensations.

These receptors show exquisite sensitivity, effectively signaling changes as small as 0.01°C. A stable core temperature close to 37°C is particularly important for maintaining normal brain function. Temperatures above 40.5°C cause serious dysfunction; brain temperatures below 30°C depress all neuronal activity, including that necessary for maintaining minimal respiratory rate.

Pain Receptors

Receptors that convey information about noxious or potentially noxious stimuli are called *nociceptors* (from the Latin *nocere*, "to hurt"). Nociceptors are divided into four major subtypes based on the types of painful stimuli to which they respond. Three of the subtypes—*mechanoreceptors*, *thermal receptors*, and *chemoreceptors*—respond to a single modality, whereas the fourth subtype is *polymodal* and is capable of responding to mechanical, thermal, and chemical features of the noxious stimuli.

Although other receptor types respond to similar modalities, nociceptors are most sensitive to a higher range of stimulus intensities, specifically those that could be expected to cause tissue damage. For example, general thermal receptors in the skin normally respond to temperatures below 45°C, whereas nociceptive thermal receptors respond preferentially to temperatures above that level. Similarly, general mechanoreceptors respond to most types of pressure, but nociceptive mechanoreceptors are preferentially sensitive to pressure from sharp objects. It is interesting to note that as nociceptive pathways become activated, perceptions change accordingly. The sensation of "hot" changes to that of "scalding" at approximately 45°C. Likewise, moderate pressure from a smooth object can feel pleasant, but it feels uncomfortable when the same pressure is applied from a sharp object.

A number of different general chemical receptors are found throughout the body, including sensors for specific compounds, such as carbon dioxide in the blood. Nociceptive chemoreceptors are particularly sensitive to potential irritants—namely, high levels of potassium, extremes in tissue pH, bradykinins, and histamines.

All nociceptors have either *free nerve endings* or simple, nonencapsulated end organs. They are located throughout the body except for brain and bone tissue. In general, all nociceptors subtypes have fibers that fall into two categories; the first includes type III fibers that generate the sensation of pricking pain or tickle, and the second includes type IV (C) fibers that generate the sensation of slow burning pain (see Fig. 1).

Nociceptors are unique among sensory receptors in that their sensitivity to stimuli increases after tissue in the area near the nerve ending has been damaged. This

---

Clinical Note

Regulation of Pain

It should be recognized that the perception of pain is not synonymous with activation of nociceptors. Pain is the feeling that normally accompanies irritating, aching, or burning sensations transmitted by nociceptive pathways; however, other factors such as stress, fright, or other strong emotional states may trigger the perception of pain in the absence of nociceptive activity. Similarly, nociception does not always lead to the perception of pain. Signals from nociceptors may be modified by two major mechanisms that can shunt or "gate" pain information so that it fails to reach conscious centers of the brain. The first component of the *pain-gating mechanism* is localized in the spinal cord and consists of inhibitory interneurons that link large-diameter mechanoreceptor axons with small-diameter nociceptive axons as they enter the dorsal horn of the cord. When the mechanoreceptor pathway is active, it causes the interneurons to inhibit nearby nociceptive axons, thus suppressing painful information before it projects up the spinal cord. This pathway presumably accounts for the fact that it feels less painful if you rub your toe after you stub it. It also forms the basis of electrical treatment for some kinds of chronic, intractable pain. Wires taped to the patient's skin can be used to electrically stimulate large-diameter sensory axons and thus suppress transmission of nociceptive information.

The second component of the pain-gating mechanism originates in an area of the brain stem called the *periaqueductal gray matter* (*PAG*), so named because it encircles the cerebral aqueduct. The PAG receives information from several brain regions regarding emotional status and then sends that information through descending projections to the dorsal horn of the spinal cord, where they effectively depress the activity of nociceptive neurons. Through this pathway, strong emotion, stress, or even stoic resolve can blunt the perception of pain. One type of neurotransmitter substance used within this pathway has been identified as an endogenous morphine-like substance called *endorphin*. Endorphin is manufactured in cells of the PAG, is released in the dorsal horn, and interacts with inhibitory *endorphin receptors* on nociceptive neurons. A number of drugs—notably, opium and similar compounds such as morphine, codeine, and heroin—are all capable of binding directly to the endorphin receptor, also called the *opioid receptor*. In this way, they produce profound analgesia when taken systemically. Through interactions with opioid receptors elsewhere in the brain, these potentially addictive drugs also cause mood changes, drowsiness, mental clouding, nausea, vomiting, and constipation.

process, termed *primary hyperalgesia*, occurs at the site of injury and is a result of a change within the receptor neuron itself, usually manifested as a decrease in the threshold response. In some cases, there is also a change from unimodal to bimodal responses, with more nerves becoming responsive to both mechanical and thermal stimuli. Similar changes may also occur in CNS neurons located within stimulated sensory pathways in brain and spinal cord. These higher order neurons can also exhibit decreased thresholds and in addition may show an increase in receptive field size. This condition is called *secondary hyperalgesia* (Fig. 4). Both primary and secondary hyperalgesia are linked to a greater than normal release of neuroactive substances that sensitizes nearby sensory receptors or central neurons. Bradykinin, histamine, serotonin, leukotrienes, prostaglandins, substance P, and cytokines have been implicated in primary hyperalgesia. Substance P, calcitonin-gene-related peptide (CGRP), and excitatory amino acids such as glutamate released from activated neurons in the brain and spinal cord may be involved in secondary hyperalgesia.

SENSORY RECEPTORS IN MUSCLES, JOINTS, AND VISCERAL ORGANS

Proprioceptors (from the Latin *proprius,* "one's own") are so named because they provide information about body position and movement in space. Signals generated from proprioceptors generally do not reach cortical levels and thus do not contribute significantly to conscious perception. However, proprioceptive information is critically important in coordinating voluntary and reflex movement of skeletal muscle. Two major classes of muscle proprioceptors are involved in this process: (1) *muscle spindles,* which monitor length and rate of stretch of the innervated muscle; and (2) *Golgi tendon organs*, which gauge muscle force by monitoring muscle tension. Additional proprioceptive information is provided by a variety of mechanoreceptors in the connective tissue of joints. Some are rapidly adapting and provide information about limb movement; others are slowly adapting and continuously monitor body posture. These various sources of information work in an integrated manner so that loss of one input is compensated for by those remaining. For this reason, patients who have hip replacements and thus are devoid of proprioceptors in the affected joint still receive sufficient proprioceptive information from other sources to walk in a coordinated fashion.

Sensations from smooth muscle and from internal organs are much less discrete than those from somatic structures, such as skin and skeletal muscle. *Visceral mechanoreceptors*, *thermoreceptors*, and *chemoreceptors* are broadly distributed throughout the body and provide sensory information that activates local involuntary

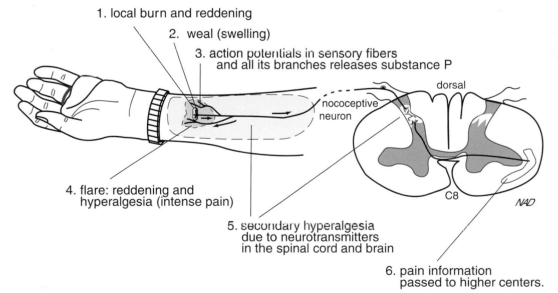

1. local burn and reddening
2. weal (swelling)
3. action potentials in sensory fibers and all its branches releases substance P
dorsal
nococeptive neuron
4. flare: reddening and hyperalgesia (intense pain)
C8
NAD
5. secondary hyperalgesia due to neurotransmitters in the spinal cord and brain
6. pain information passed to higher centers.

FIGURE 4   Hyperalgesia: increased sensitivity to a superficial burn on the forearm is shown here. (1) Injury of local tissue causes release of bradykinin and other pain-producing chemicals that also cause vasodilation and reddening. (2) Dilated capillaries leak fluid (edema), which produces a weal (swelling.) (3) Action potentials spread to all branches of the injured neuron and release neurotransmitters into the skin and spinal cord. (4) Substance P releases histamine from mast cells, causing additional vasodilation (reddening) and primary hyperalgesia (lower threshold). (5) Secondary hyperalgesia in the surrounding area is caused by sensitization of neurons in the CNS. (6) Pain information is transmitted through the spinal cord and passed to higher centers.

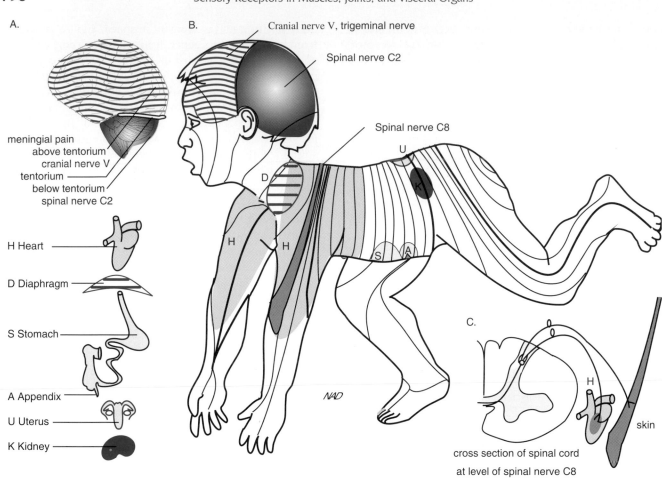

A.

meningial pain
above tentorium
cranial nerve V
tentorium
below tentorium
spinal nerve C2

H Heart

D Diaphragm

S Stomach

A Appendix

U Uterus

K Kidney

B.   Cranial nerve V, trigeminal nerve

Spinal nerve C2

Spinal nerve C8

NAD

C.

cross section of spinal cord

at level of spinal nerve C8

skin

FIGURE 5   Instances of referred pain. Painful irritation of a given internal organ (A) is perceived as pain in a specific area of skin, called a *dermatome* (see Chapter 50), as shown on the right (B). This occurs because the nociceptive neurons from the two areas converge on the same secondary neurons in the spinal cord. For example, (C) pain afferents from the heart converge on the same spinal cord neurons as those from the skin of the left shoulder and arm (the C8 dermatome). Note that intracranial head pain arises from irritation of blood vessels and meninges (A). Pain does not occur from irritation of the neuronal tissue of the brain.

reflexes and contributes to feelings of hunger and thirst as well as to an overall feeling of malaise or well-being. These receptors provide relatively few centrally project- ing fibers, accounting for less than 10% of the sensory fibers found in the spinal cord. They synapse on some of the same spinal neurons that receive somatic sensory input. For this reason, visceral pain is often *referred*, providing a perception that the sensation (in many cases, discomfort) actually originates from skin or muscle that may be some distance from the painful organ (Fig. 5). For example, angina pectoris (heart pain) is usually

referred to the left chest, shoulder, and upper arm. Visceral sensory neurons also have broad overlapping receptive fields, which further decrease the ability to localize visceral pain precisely.

Suggested Readings

Bear MF, Connors BW, Paradiso MA. *Neuroscience: exploring the brain.* Baltimore: Williams & Wilkins, 1996.
Kandel ER, Schwartz JH, Jessell TM. *Principles of neuroscience.* New York: Elsevier, 1991.
Martin JH. *Neuroanatomy.* Stamford: Appleton and Lang, 1996.

Central Projections of the General Somatosensory System

DIANNA A. JOHNSON

KEY POINTS

- Transmission of sensory information to the central nervous system is triggered when receptors reach the threshold required to initiate action potentials; the number of action potentials reflect the strength of the triggering stimulus.

- *Somatic sensory* neurons release glutamate at their axon terminals in order to stimulate postsynaptic neurons within the central somatosensory pathway.

- Different modalities and submodalities of somatosensory sensation project in discrete pathways within the dorsal and lateral portions of the spinal cord.

- Centrally projecting somatosensory fibers are *somatotopically* arranged, with cells in each dorsal root ganglion responsible for sensations from one area of the body, defined as a *dermatome*.

- Neuronal cell bodies occupy the central core of the spinal cord and are surrounded by ascending (sensory) fibers in the dorsal and

lateral portions of the cord and by descending (motor) fibers in the ventral regions.

- The sensory pathway carrying information about discriminatory touch and proprioception ascends in the dorsal columns of the spinal cord, crosses the midline in the lower medulla, and projects to the thalamus before reaching the *somatosensory cortex.*

- Information about pain and crude touch is transmitted through fibers that decussate in the spinal cord, ascend in the lateral columns of the spinal cord, and synapse in the thalamus before joining other somatosensory projections in the primary sensory cortex.

- Somatosensory information from the head region is carried by *cranial nerve V.*

- Subcortical regions that receive somatosensory information include: (1) the cerebellum, which uses proprioceptive information to coordinate movements and posture; and (2) motor nuclei in the spinal cord and brain stem, which control

Essential Medical Physiology, Third Edition

GENERAL PROPERTIES OF CENTRAL SENSORY PATHWAYS

Sensory receptors encode information about stimulus modality, location, intensity, and novelty. As described earlier, this is achieved through the specific transducer elements within individual receptor endings that convert the energy from chemical, thermal, or mechanical stimuli into bioelectrical signals in the form of generator potentials and action potentials. Generator potentials are localized responses to sensory stimuli that vary in intensity based on the intensity of the stimulus. Generator potentials of sufficient size will initiate action potentials; even larger generator potentials will generate multiple action potentials. Action potentials are not localized responses; rather, they are propagated along the central projection fibers of the neuron. In this way, sensory information passes from peripheral to central elements of the nervous system. Specifics of the transmitted information are encoded as follows: Stimulus *modality* is encoded by the preferential response of a given sensory neuron to its appropriate stimulus; *location* is encoded by the receptive field properties of the activated neuron; and *intensity* is encoded by the frequency of action potentials. Rapidly adapting neurons fire only at stimulus onset (or offset) and thus transmit information about *stimulus change* or *novelty*.

Cell bodies of somatic sensory neurons are located in the *dorsal root ganglia* and are designated as *first-order sensory afferents* because they provide the first link within the chain of neurons constituting the *primary sensory pathways*. Their central fibers project within short *dorsal spinal rootlets* to their synaptic targets on *second-order sensory neurons* in designated areas of the spinal cord and brain stem. Action potentials, generated by stimulation of receptor end organs, travel back to the cell body and along the central process. This wave of depolarization stimulates the release of a variety of neuroactive substances (including substance P, calcitonin-gene-related product [CGRP], and glutamate) that act as chemical neurotransmitters to excite or inhibit second-order neurons. Sensory information is thus transmitted along individual cells by action potentials and between neurons within the pathway by chemical transmission. This basic plan establishes a polarity for neurons, with input arriving on peripheral dendritic processes and output being provided by central axonal processes with nerve terminals containing stores of neurotransmitters positioned for release. Most sensory and motor pathways operate in this manner.

An interesting exception is represented by certain dorsal root ganglia cells, which apparently release neuroactive substances at dendritic endings as well as central axon terminals, providing direct feedback as well as feedforward responses. As noted in the previous chapter, peripheral release of transmitter substances from nociceptive sensory neurons is a trigger for primary hyperalgesia.

Cell bodies from all types of somatic sensory nerves are grouped together within the dorsal root ganglia; however, their central projections maintain distinctly separate positions within the spinal cord. Two major parallel sensory pathways exist: one transmitting information about touch and proprioception, and the other transmitting information about pain and temperature. Subdivisions are observed within the main pathways with specific locations designated for each modality or even submodality (e.g., light-touch and crude-touch fibers are grouped separately). Relative position within the main pathway also reflects the location of the receptive field that is represented. Thus, each pathway has within it a *somatotopic arrangement* of its fibers, with the lower extremities layered into the tract first, followed by the trunk, arms, and head.

ORGANIZATION OF THE SPINAL CORD

To better understand somatotopic and modality-specific arrangements, it is helpful to review the general organization of the spinal cord (Figs. 1 and 2). *Dorsal roots* carrying sensory information and *ventral roots* transmitting motor signals occur in left and right pairs attached at 30 regular intervals along the spinal cord. Each point of attachment establishes a separate *spinal segment*. Rootlets from individual segments merge just distal to the dorsal root ganglia to form *spinal nerves*, which must pass through the vertebral column at notches between the vertebrae. Vertebrae and associated spinal

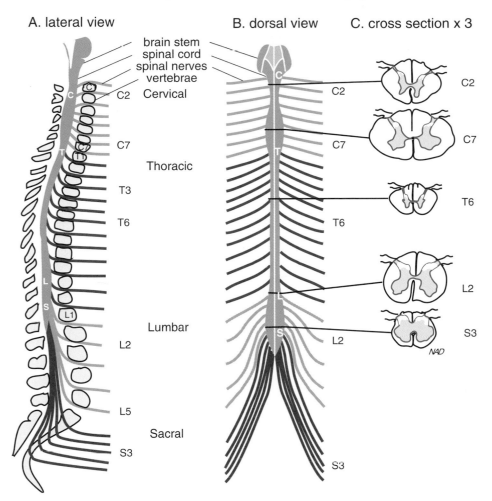

FIGURE 1   Organization and internal structure of the spinal cord. Pairs of spinal nerves leave the vertebral
column through spaces between the vertebra. These nerves are classified by the region of the body they innervate:
cervical, thoracic, lumbar, or sacral. (A) Lateral and (B) dorsal views of the entire spinal cord. The spinal cord is
shown in blue, and spinal neurons are in black. Note that the spinal cord is shorter than the vertebral column,
leaving a space at the caudal end of the vertebral column that contains only spinal nerves. (C) Cross sections of
the spinal cord at different levels. Note that the white matter fiber tracts (shown in white) increase in size from
caudal to rostral levels. Also, the areas of gray matter, containing cell bodies and many synaptic connections
(shown in blue), are larger in the cervical and lumbar regions because of the many motor and sensory fibers that
innervate arms and legs respectively.

segments are divided into four groups: *cervical* (segments
C1–C8); *thoracic* (segments T1–T12); *lumbar* (segments
L1–L5); and *sacral* (segments S1–S5). At higher levels of
the cord, nerves and vertebral notches are fairly well
aligned, whereas at lower levels the alignment is lost
because the cord is considerably shorter than the vertebral
column. Long projections of spinal nerves called the
cauda equina (Latin for "horse's tail") fill the vertebral
column at lumbar and sacral levels, providing clinicians
with a relatively safe area from which to extract cerebrospinal fluid, as described in Chapter 48.

Cell bodies within each dorsal root ganglion collectively innervate one area of the body designated as a
dermatome (Table 1). Although not traditionally included

in illustrations of dermatome distributions (including
Figs. 2 and 3), central projections from each ganglion
usually overlap two or three spinal cord levels, resulting
in significant overlap between adjacent dermatomes.
Thus, sensory sensation can be reasonably maintained
if only a single ganglion is lesioned.

Once the spinal nerves leave the bony protection of
the vertebrae or skull, they reorganize according to their
peripheral destinations. For example, the majority of the
neurons from cervical nerves C4 to C8 and thoracic
nerves T1 and T2 coalesce to form the brachial plexus.
They then redivide into anterior and posterior branches
and finally segregate into individual peripheral nerves
such as the radial nerve, composed of both sensory and

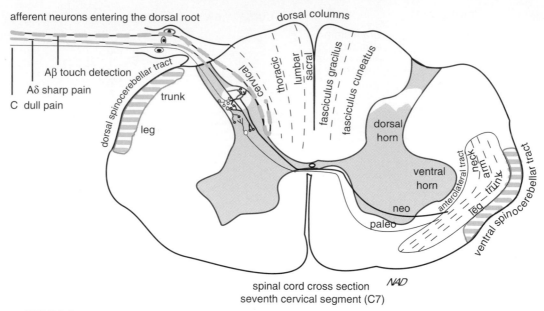

FIGURE 2   Internal structure of the spinal cord. Primary neurons carrying information about proprioception and discriminatory touch enter the spinal cord and ascend uncrossed in the dorsal columns to the gracilus or cuneate nuclei in the brain stem. Fibers are somatopically organized, with cervical and thoracic fibers projecting in the fasciculus cuneatus, and lumbar and sacral fibers projecting in the fasciculus gracilis. Primary neurons carrying information about pain, temperature, and crude touch enter the spinal cord and make one or more synaptic connections in the dorsal horn before crossing the midline and ascending in the anteriolateral tract or the ventral spinocerebellar tract.

motor fibers. This major nerve collects sensory information from the back of the arm and carries motor commands to extensor–supinator muscles of the arm.

Within the internal structure of the spinal cord, there are no apparent markers of segmental boundaries. Cell bodies within the spinal cord are concentrated in a continuous central core of gray matter that bulges anteriorly to form the *ventral horns* associated with motor systems and posteriorly to form the *dorsal horns* associated with sensory systems. White matter, consisting of afferent and efferent fibers, fills in the contours of the gray matter and helps accentuate the butterfly pattern of gray matter seen in cross sections of the cord. From a three-dimensional perspective, the white matter extends as a series of

TABLE 1   Examples of Dermatomal Distributions

Dermatome	Area of Body Represented
C5	Lateral shoulder
C6	Thumb
C7	Index and middle fingers
C8, T1	Ring, little fingers
T4	Nipple
T10	Navel
L3, L4	Anterior thigh
L5	Dorsal foot
S1	Lateral foot, sole

columns within the spinal cord. Ascending sensory fibers are found in the *dorsal* and *lateral columns*; descending motor fibers are concentrated in *lateral* and *anterior columns*. Although there are no anatomic boundaries between segments, variations in internal organization reflect functional differences among the four levels of the spinal cord. For example, cervical and lumbar segments are slightly enlarged overall and have pronounced ventral horns because of the large numbers of motor neurons required for innervation of muscles in the extremities, whereas the thoracic cord has very narrow ventral horns. Likewise, the number of efferent fibers ascending within the spinal cord increases at higher levels. The cervical cord contains sensory projections from the entire body and thus has the largest proportion of white matter, whereas the sacral level contains only fibers from sacral dermatomes and has correspondingly less white matter.

DORSAL COLUMN/LATERAL LEMNISCAL PATHWAY

Sensory fibers within the dorsal rootlets project into the cord in a spatially precise and modality-specific fashion. A shallow groove, the *dorsal sulcus*, marks the point of attachment of the spinal nerves on the dorsal surface of the cord. Sensory fibers enter the underlying

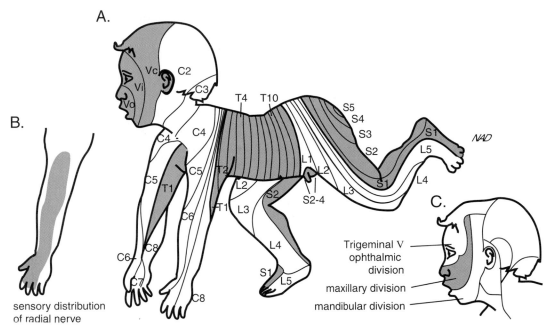

FIGURE 3 Dermatomes map the orderly distribution of sensory projects from the skin to the spinal cord. (A) Cell bodies within each dorsal root ganglion collectively innervate one region of skin called a *dermatome*; the name of each dermatome (from C2 through S5) denotes the ganglion that innervates it, Dermatomes of the face are named to reflect innervation by the three divisions (oral, V_O; interposed, V_I; caudal, V_C) of the trigeminal ganglion. Each dorsal root ganglion sends central fibers primarily to one level of the spinal cord, with some degree of overlap (not shown). However, the fibers that project from each ganglion to the skin for the most part do not travel as separate nerves; rather, they are combined with sensory nerve fibers from other ganglia and also with motor nerves projecting to muscles in the same region. Two examples are shown. (B) The peripheral projections of the radial nerve combine fibers from portions of C6, C7, C8, and T1 dorsal root ganglia. (C) Peripheral projections of the trigeminal ganglion are rearranged into ophthalmic, maxillary, and mandibular divisions of the trigeminal nerve.

cord and segregate into two major pathways. The first pathway, consisting of fibers carrying information about light touch and proprioception, ascends in the dorsal columns. Axons innervating the lower body assume a position along the posterior midline; fibers from the upper body are added sequentially in a mediolateral fashion. In the midthoracic level of the cord, the dorsal column of fibers is divided into two *fasciculi*. The more medial area, called the *gracile fasciculus*, contains fibers from sacral, lumbar, and lower thoracic dermatomes. The more lateral area, called the *cuneate fasciculus*, carries fibers from upper thoracic and cervical dermatomes. The first synaptic contact for these first-order sensory neurons is in the brain stem, at the level of the medulla in the *gracile* and *cuneate nuclei*, respectively. Fibers in the pathway do not cross the midline of the spinal cord as they ascend. Thus, a neuronal lesion affecting only one side of the spinal cord will interrupt information about light touch and proprioception from the ipsilateral side of the body below the level of the lesion. The patient will lose the ability to identify objects by touch and will have difficulty with balanced

movement on the affected side. Note that this primary sensory pathway does not make intermediate synaptic contacts with neurons within the spinal cord.

Cells in the gracilis and cuneate nuclei represent second-order neurons within this pathway, and their axons cross the midline to ascend in the *medial lemniscal tract* to the *ventral posterolateral nucleus* (VPL) of the thalamus. These thalamic neurons then project to specific regions of the primary somatosensory cortex. The entire four-neuron chain comprises the dorsal column–medial lemniscal pathway (Fig. 4).

Points of synaptic contact within a pathway do not simply pass information along unchanged. Each relay serves as a computing center that allows the information to be processed, amplified, sharpened, suppressed, or otherwise altered. For example, the function of the dorsal column nuclei is to shape tactile information in time and space. These nuclei make edges seem sharper and abrupt pressure changes seem more sudden than they really are. These enhancements of textures are produced by lateral inhibition in much the same way as visual information is altered, as will be discussed in Chapter 51.

ANTEROLATERAL PATHWAY

The second major ascending sensory system is called the *spinothalamic* or *anterolateral pathway*, which carries information about pain and temperature from the body. This pathway is best understood as two subdivided tracts. The first is called the *neospinothalamic tract*, which carries information about sharp pain within A-delta (small myelinated) fibers. Unlike the dorsal column–lateral lemniscal pathway, the first-order neurons in this pathway synapse with neurons within the dorsal horn of the spinal cord. Projection fibers then cross the midline of the cord and form the lateral portion of the anterolateral tract as they ascend the cord to the VPL nucleus of the thalamus.

The *paleospinothalamic subdivision* carries information about dull, aversive pain; temperature; and crude touch within small, unmyelinated C fibers. There are two synaptic relays with cells in the dorsal horn of the spinal cord. Axons of projection neurons then cross the cord and join the anterolateral tract, terminating in the intralaminar nucleus of the thalamus. Unlike other tracts in the somatosensory tract, the paleospinothalamic tract is not somatotopically organized (Fig. 5).

Because of multiple synapses in the spinal cord, pain information can spread to adjacent neurons in the dorsal horn, causing a widespread hyperalgesia. Pain information can also be suppressed through synaptic input at the spinal cord level so that neither local spinal reflexes nor conscious centers are informed of the injury. As discussed in Chapter 49, the first synapse in this path can be inhibited, both pre- and postsynaptically, by endorphins, and this neurotransmitter can be mimicked by drugs such as morphine. This gating or interruption of nociceptive information allows an organism to cope with an emergency or stress without the distraction of pain.

SOMATOSENSORY PATHWAYS IN THE HEAD REGION

Somatic sensory cranial nerve pathways are homologous in many ways to the medial lemniscal and spinothalamic pathways of spinal nerves. The *trigeminal nerve*, or *cranial nerve V*, is the primary sensory nerve involved in transmitting touch and pain information from the face and neck. It contains fibers from mechanoreceptors, thermoreceptors, and nociceptors for the face, oral mucosa, distal part of the tongue, portions of the dura mater covering of the cerebral cortex, and tooth pulp. First-order neuronal cell bodies are located in the *trigeminal ganglia* adjacent to the brain stem. Central fibers enter the pons and separate into two main tracts. Axons that carry information about light touch and proprioception synapse in the *principal* or *main sensory nucleus* of cranial nerve V, which in turn sends fibers across the midline to ascend to the *ventral posterior medial* nucleus (VPM) of the thalamus. This pathway is analogous to the medial lemniscal pathway of the spinal cord. First-order neurons carrying information about pain and temperature send fibers to the *spinal trigeminal nucleus* located in the lower medulla. Second-order axons cross the midline and ascend to the VPM and intralaminar thalamic nuclei in a manner similar to the spinothalamic pathway of spinal nerves.

Clinical Note

Shingles and Herpes Zoster Virus

Herpes zoster virus, commonly known as chicken pox, preferentially infects neurons of the peripheral nervous system, particularly dorsal root ganglion cells. Individuals infected with the virus during childhood usually display red, itchy spots on the skin for approximately 1 week and are symptom free thereafter. However, the virus may remain dormant, usually residing in a single dorsal root ganglion, and can become reactivated in some individuals decades later to produce a condition known as *shingles*. The revived virus increases the excitability of sensory cells in the ganglion so that sensory nerves have lower thresholds as well as spontaneous activity. This activity triggers burning or stabbing sensations that are agonizingly painful. The reactivation phase of the infection may last months or years, during which time the skin first becomes inflamed, then blisters, and finally appears scaly. Because the infection is restricted to a single dorsal root ganglion, the affected areas of the skin reflect the dermatomal distribution of the affected ganglion. Thoracic and facial areas are most commonly involved, although instances of infection of every dermatome have been reported. In fact, the observations of many shingles patients and their infected areas helped to provide the currently used map of dermatomal distributions of individual dorsal root ganglia. Antiviral agents and steroids are used as treatment for this condition.

A. Cross sections of the nervous system

B. Coronal view of the nervous system

C. Lateral view of the nervous system

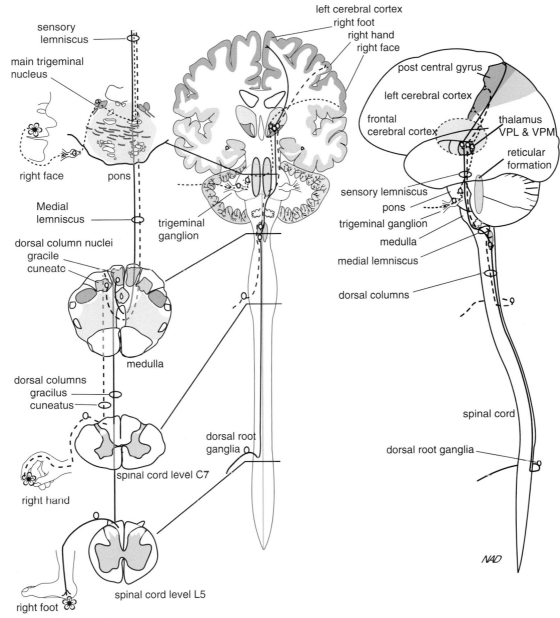

FIGURE 4 Dorsal column–lateral lemniscal pathway; neuronal pathways of proprioception and light, discriminatory touch: (A) cross section, (B) coronal view, and (C) lateral view of spinal cord and brain stem with associated body part. Sensory fibers enter the spinal cord and ascend in the dorsal columns to the dorsal column nuclei. Secondary neurons project from the nuclei, cross the midline, and ascend contralaterally in the medial lemniscus to the ventral posterior lateral nucleus of the thalamus. Tertiary neurons project to the post-central gyrus of the cerebral cortex. Tactile information from the face enters the trigeminal nerve (V) and synapses in the trigeminal main sensory nucleus. Axons from secondary neurons cross the midline, form the trigeminal lemniscus, join the medial lemniscus to form the sensory lemniscus, and ascend to the ventral posterior medial nucleus of the thalamus. Tertiary neurons project to the lower part of the post-central gyrus.

A. Cross sections of the nervous system

B. Coronal view of the nervous system

C. Lateral view of the nervous system

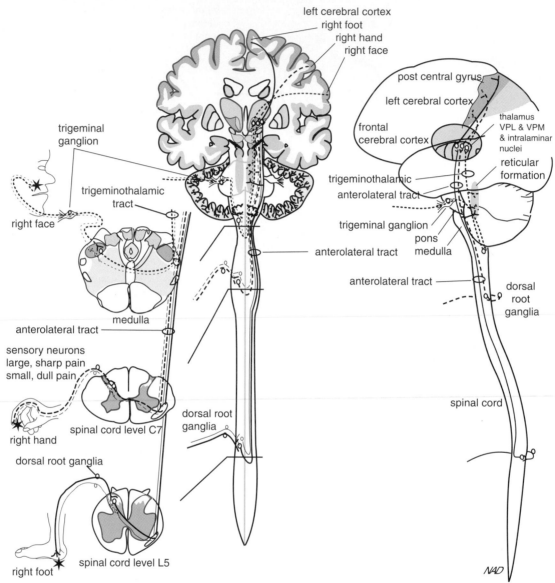

left cerebral cortex
right foot
right hand
right face

post central gyrus

left cerebral cortex

thalamus VPL & VPM & intralaminar nuclei

frontal cerebral cortex

reticular formation

trigeminal ganglion

trigeminothalamic tract

trigeminothalamic

anterolateral tract

right face

trigeminal ganglion

pons
medulla

anterolateral tract

medulla

anterolateral tract

anterolateral tract

dorsal root ganglia

sensory neurons
large, sharp pain
small, dull pain

dorsal root ganglia

right hand

spinal cord level C7

spinal cord

dorsal root ganglia

right foot

spinal cord level L5

NAD

FIGURE 5   Anterolateral path to the brain; neuronal pathways of pain: (A) cross section, (B) coronal view, and (C) lateral view of spinal cord and brain stem with associated body parts. Nociceptive information is carried by two paths: the neo-anterolateral path (heavy lines), which carries information about sharp, pricking pain; and the paleo-anterolateral path (light lines), which carries information about dull aversive pain. Both enter the spinal cord and make synaptic connections in the dorsal horn of the spinal cord. Projection neurons from the dorsal horn ascend contralaterally in the anterolateral tract to the thalamus. These sensory neurons from the face have cell bodies in the trigeminal ganglia; their fibers enter the brain in the middle of the pons and descend to the spinal trigeminal nucleus. After synapsing, the projection neurons cross and ascend contralaterally in the trigeminal lemniscus to the thalamus. The neo-tracts project to the ventral posterior lateral nuclei (body) and ventral posterior medial nuclei (face) of the thalamus and from there to the post central gyrus of the cerebral cortex. The paleo-tracts terminate in the intralaminar nuclei of the thalamus. Both systems send information to the reticular formation. Temperature and crude touch follow similar paths to the cerebral cortex.

A. Cerebral cortex

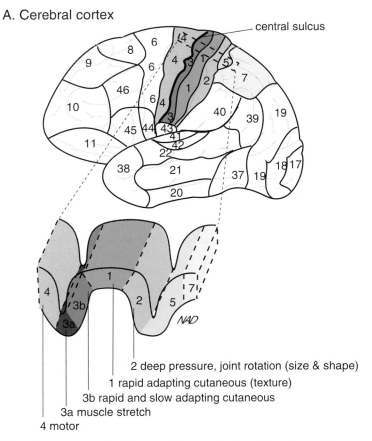

2 deep pressure, joint rotation (size & shape)
1 rapid adapting cutaneous (texture)
3b rapid and slow adapting cutaneous
3a muscle stretch
4 motor

B. column organization and layers in the 3b region of the cerebral cortex

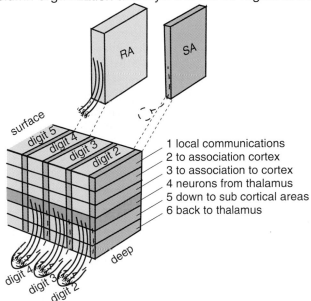

1 local communications
2 to association cortex
3 to association to cortex
4 neurons from thalamus
5 down to sub cortical areas
6 back to thalamus

FIGURE 6   Modular organization of the somatosensory cortex. (A) Organization of the primary receptive areas (darker blue; Brodmann's areas 1, 2, and 3, also called collectively S1) on the post-central gyrus of the cerebral cortex. Note that area 3 (both 3a and 3b) lies on the posterior wall of the central sulcus. Each area receives specific types (submodalities) of touch information from designated areas of the body in a somatotopic manner. (B) Detail of the 3b cortical area receiving information from mechanical receptors in three left-hand fingers. Projection neurons from the thalamus synapse in layer 4, within columns devoted to rapidly adapting (RA) and slowly adapting (SA) receptors. Fibers that project from the cortex segregate into five layers based on their destination.

THALAMIC PROJECTIONS TO SUBCORTICAL REGIONS AND SOMATOSENSORY CORTEX

Two anatomically smaller but nonetheless important elements of the ascending sensory system do not project to cortical regions and therefore do not contribute to conscious perception. First, many fibers carrying proprioceptive information do not travel with the dorsal column–medial lemniscal pathway but rather project in the lateral columns directly to the ipsilateral cerebellum. The cerebellum uses proprioceptive information to modulate movement of muscles in the extremities.

Another important ascending system forms the afferent arm of protective reflex pathways that operate on the subconscious level. As they ascend to the thalamus, all spinal and cranial sensory tracts send collateral branches to the *brain stem reticular formation*, a diffuse network of neurons within the core of the brain stem. Various nuclear groups within the reticular formation function as coordinating centers that extract relevant sensory information and in turn activate appropriate muscle groups through their motor axons. Different regions of the reticular formation have been identified that elicit the *cough reflex*, *blink reflex*, and *gag reflex*. Respiratory and cardiovascular centers within the reticular formation regulate breathing rate and heart rate.

Sensations are not perceived until sensory information reaches the primary sensory cortex. The *post-central gyrus* (area SI) acts as the primary receiving area for somatosensory projections from the thalamus, and, like most cortical regions, it has an elaborate organizational framework with several major components:

1. Areas that receive input from one submodality are arranged in four succeeding cortical strips that run parallel to the central sulcus: *area 3a* (most anterior), muscle stretch receptors; *area 3b*, cutaneous receptors; *area 1*, rapidly adapting cutaneous receptors; *area 2* (most posterior), deep pressure receptors.

2. Information from different parts of the body project to different parts of each cortical strip. Thus, each strip contains an orderly representation of the body called a *homunculus* (Latin for "little man") that is derived from the location of the receptive fields of the input neurons. The homunculus is not true to scale, however, and certain areas, such as lips and fingertips, are grossly overrepresented. The high density of sensory receptors in these locations requires proportionally large cortical receiving areas. Experiments in primates have recently demonstrated that the homunculus is not invariant, and its boundaries shift slightly in response to the level of activity within each set of sensory fibers. Concert pianists might be expected to have disproportionately large fingertip receiving areas.

3. Each homuncular area is subdivided further into functional modules or columns that lie perpendicular to the cortical surface. All the cells within a vertical column have similar function; for example, columns of slowly adapting cells are separated from columns of rapidly adapting cells.

4. The cortex is layered like a cake, with each layer containing functional aggregates of cells and fibers. Input fibers from the thalamus synapse on cells in layer 4. Layer 6 contains fibers that feed back to the thalamus. Fibers projecting to other subcortical structures are present in layer 5. More superficial layers contain fibers that project to other association areas of the cortex.

The modular organization of the primary somatosensory cortex exemplifies one of the major organizing principles of the brain (Fig. 6). Neurons that have similar functions occupy similar positions. Principles of modular organization apply to most sensory association areas, including *secondary somatosensory cortex* (SII) located at the lateral tip of SI and even higher association areas found in the lower part of the parietal lobe. However, functional attributes of the cells within the modules change in succeeding stages of information processing. Through convergence of input from SI cells, the neurons in SII display larger receptive fields and more complex response properties. Input from several different modalities also converges on a single module at higher levels, so that separate aspects of a sensory stimulus can come together as a unified perception of an object.

Suggested Readings

Dubner R, Bennett GJ. Spinal and trigeminal mechanisms of nociception. *Annu Rev Neurosci* 1983; 6:381–418.

Garner EP, Hamalainen HA, Palmer CI, Warren S. Touching the outside world: representation of the motion and direction within primary somatosensory cortex, In: Lund JS, Ed., *Sensory processing in the mammalian brain: neuron substrates and experimental strategies*. New York: Oxford University Press, 1989, pp. 49–66.

Willis WD, Coggeshall R. *Sensory mechanisms of the spinal cord*, 2nd ed., New York: Plenum Press, 1991.

51

The Visual System

DIANNA A. JOHNSON

KEY POINTS

- Only one small region of retina, the *fovea*, provides high visual acuity because the foveal arrangement of photoreceptors places the light-sensitive cellular elements in direct alignment with focused, incoming light.
- *Cone photoreceptors* are specialized for high-acuity color vision under bright-light (photopic) conditions, whereas *rod photoreceptors* have more sensitivity but less acuity and are specialized for dim-light (scotopic) vision.
- The *retinal pigment epithelium* provides a number of crucially important support roles for photoreceptors, including removal of cellular

debris and recycling of molecular substrates in the visual transduction cascade.
- The absorption of light by photopigments in rods and cones triggers a *phototransduction cascade* that ultimately determines the rate at which glutamate is released from photoreceptor terminals.
- A three-neuron chain consisting of *photoreceptor*, *bipolar*, and *ganglion cells* transmits visual information through the retina, the optic nerve and track, and on to higher visual centers.
- The *receptive field* of any given neuron in the visual pathway is defined as the portion

KEY POINTS (*continued*)

- of the visual field to which that neuron responds.
- Information about location of an object in visual space is transmitted through *magnocellular* (M) pathway ganglion cells; information about color and shape of an object is transmitted through *parvocellular* (P) pathway ganglion cells.
- Images that fall on the two retinas are merged in stages along the visual pathway, culminating in the visual cortex with an integrated, unified perception of visual space.
- Because all parts of the primary visual pathway maintain a *retinotopic organization*, lesions at specific points in the pathway result in specific visual field deficits.

- Cells within *ocular dominance layers* of the *lateral geniculate* project to cells clustered *in ocular dominance columns* in the *visual cortex*.
- Functional clustering of other cortical cells establishes cortical modules containing *orientation-specific columns* and *color-specific blobs* within the visual cortex.
- Perception of the color of an object or light source results from a comparative assessment of the *hue*, *saturation*, and *brightness* of the direct or reflected light.
- Specialized arrangements of extraocular muscles continually align the eyeballs, allowing the fovea to capture a focused image of the object of interest.

Humans are highly visual animals. The sense of sight is usually acknowledged as the most important of all the senses based on the vast amount of usable information it provides about the environment. The relative amount of brain tissue devoted to visual information processing is proportionally large, accounting for more than half of the total brain mass. Pathways linking the retina and the primary visual cortex in the occipital lobe provide conscious perception of visual input; storage of visual memory occupies significant regions of the parietal and temporal lobes; visual reflex pathways within the brain stem and spinal cord control eye movement and provide protective reflexes; and light-driven circadian rhythms regulate general metabolic rates, control hormonal function, and influence mood through pathways involving the retina, pineal gland, and portions of the diencephalon. The extensive coverage given to the the visual system in this chapter reflects its dominance within the structure of the brain as well as its significance in the clinical setting. Visual testing is included in most standard physical exams of the patient because it can be easily administered and it can be used to pinpoint possible lesions in the many different visual areas of the brain as well as to evaluate other functions such as cognitive status (visual memory), motor reflexes (postural reflex), autonomic function (pupil dilation), and even inner ear problems (eye movement). Visual testing in the experimental setting can also be accomplished with relative ease. The retina, in particular, has been intensively studied as one of the most approachable parts of the brain, and it has served well as a widely used model for understanding many aspects of nervous system function.

STRUCTURE OF THE EYE

The primary visual receptive tissue is the retina, a thin sheet of neurons with the consistency of wet cellophane. The position of the retina within the eyeball is somewhat analogous to the position of film in a camera (Fig. 1). Because light must enter the eye through the pupillary opening, the image that falls on the retina is reversed and inverted.

One class of retinal neurons, the *photoreceptor cell*, is primarily responsible for specific absorption of photons, and paradoxically these cells occupy the retinal layer furthermost from the incoming light (Fig. 2). In general, light must first penetrate through all retinal layers before encountering molecules of light-absorbing photopigment located in the outer segments of photoreceptors. As one might expect, this arrangement results in some distortion of the visual image. Regional specialization in the arrangement of retinal layers compensates for this distortion. The portion of the retina in line with the *visual axis* (a theoretical line drawn horizontally through the center of the lens) is modified so that all cellular layers of the retina are displaced radially except for the outer segments of photoreceptors. Even blood vessels are absent in this region. This allows incident light to project more directly on the photopigment-rich regions with minimal distortion. The cellular displacement results in a depression (or *foveal pit*) about 1.5 mm in diameter in the retinal surface. The fovea and the area immediately surrounding it are collectively referred to as the *macula*. The fovea is responsible for central, high-acuity vision. With functional loss of the fovea, an individual is considered legally

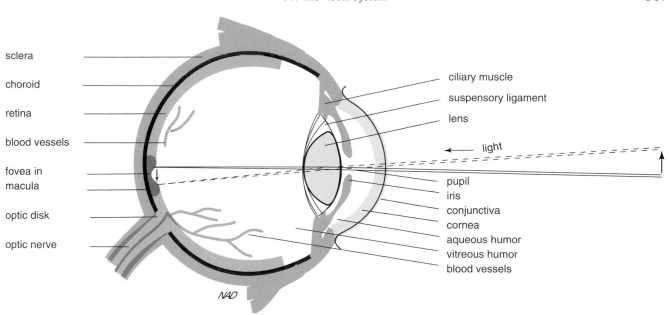

sclera

choroid

retina

blood vessels

fovea in
macula

optic disk

optic nerve

ciliary muscle

suspensory ligament

lens

light

pupil

iris

conjunctiva

cornea

aqueous humor

vitreous humor

blood vessels

NAD

FIGURE 1 Horizontal section through the left human eye ball. Neurons of the retina are shown in blue; retinal thickness is highly magnified. Retinal ganglion cell axons coalesce to form the head of the optic nerve and exit through the retina, resulting in a blind spot at this location. Blood vessels travel along the optic nerve and spread over the surface of the retina, except over the fovea. Structures of the anterior eye are discussed in subsequent sections of this chapter.

blind, even if peripheral vision served by the rest of the retina is retained. The oculomotor system, which will be described later, can be considered as a foveal adaptation designed to maintain the targeted image in register with the fovea. More peripheral regions of the retina do not provide high-acuity images but instead provide very sensitive responses to moving objects in low light.

SPECIALIZED FEATURES OF ROD AND CONE PHOTORECEPTORS

The two types of retinal photoreceptor cells, *rods* and *cones*, are not evenly distributed across the retina (Fig. 3). Cones are concentrated in the foveal region, whereas rods are absent. Several of the cellular characteristics of cones make them particularly suited for high-acuity vision. Because of the tightly packed arrangement of foveal cones, the foveal region is capable of high-resolution (fine-grained) responses. In addition, the *outer segment*, which contains most of the visual pigment of the cone, is relatively short and is most sensitive to direct axial rays of light. The photochemical cascade in cones is very fast, allowing the foveal region to have a very rapid response time.

Cones are adapted for color vision. Three types of cones have been described: one containing visual pigment that is most sensitive to blue light, one that is sensitive to green light, and one that is most sensitive to long

wavelengths, including red and yellow light (referred to as the *red-sensitive cone*). As will be discussed later, the fovea is comprised primarily of red- and green-sensitive cones. The perception of color from a given light source is derived from a comparison of the relative amount of light activation that is generated in red- versus green- versus. blue-sensitive cones.

Cones are also adapted primarily for daytime (*photopic*) vision. They saturate only in intense light and are generally less sensitive to light than are rod photoreceptors. During low-light (*scotopic*) conditions, such as those that exist during dawn or dusk, cones are not responsive, and we have no color or high-acuity vision at these times; visual images are achromatic, and we have difficulty distinguishing small shapes. Under these low-light conditions, vision is primarily mediated by rods. The same visual pigment, *rhodopsin*, is found in all rods. It maximally absorbs blue–green light, but because the perception of color is derived from a comparison of activation resulting from different photopigments, we do not perceive scotopic vision as having any specific color content.

Rods have much more photopigment than cones. Their outer segments, which contain the rhodopsin, are much longer than cones; thus, a single rod can respond to a much broader angle of incident light. Rods also respond more slowly than cones, which allows them to summate more light responses. These design features are responsible in part for the high sensitivity of individual rods which, in humans, has reached the theoretical limit of

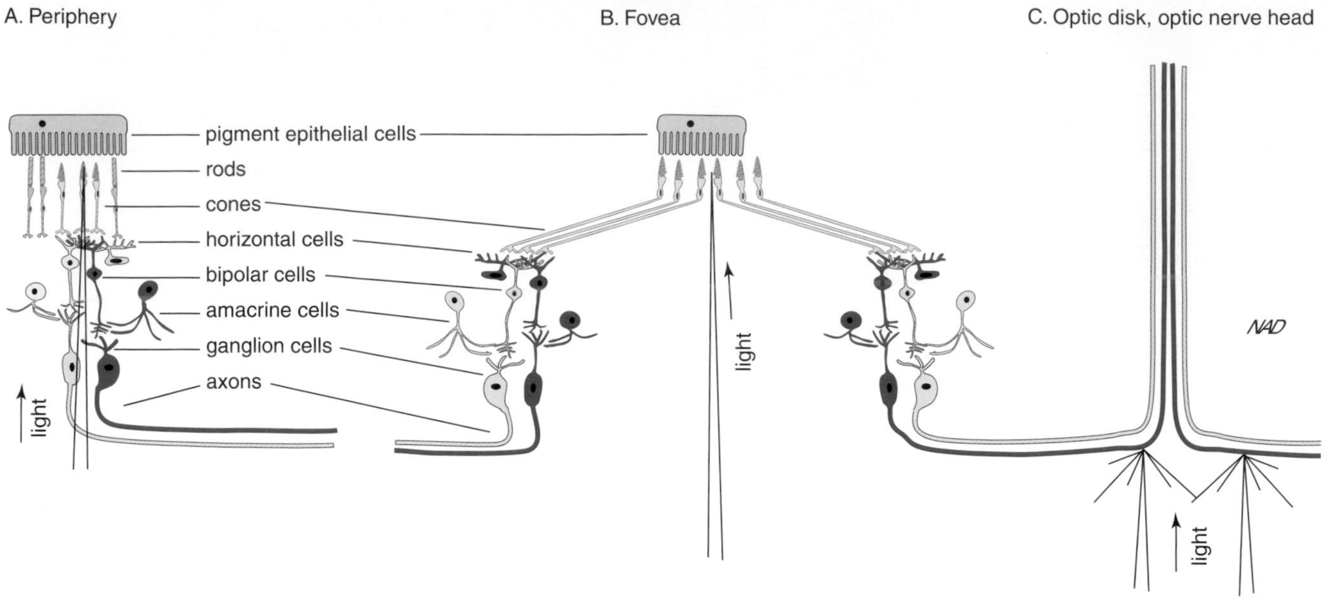

A. Periphery   B. Fovea   C. Optic disk, optic nerve head

FIGURE 2   Cellular organization of different regions of the retina. (A) The peripheral retina contains mostly rod photoreceptors, whereas the fovea contains no rods, only cones; other neurons in the retinal visual pathway are similar in the two regions. (B) In the fovea, all cellular elements except the outer segments are displaced radially out of the path of light. (C) The optic nerve head is comprised of ganglion cell axons and thus is devoid of other retinal neurons.

detection. Under ideal conditions, the human retina can detect a single photon of light.

The outer segments of rods and cones are in close contact with a monolayer of pigmented epithelial cells (*retinal pigment epithelium*, or RPE), which serves important metabolic and supportive functions for photoreceptors. The pigment granules within the RPE help absorb stray light and thus reduce glare. RPE cells produce a sticky extracellular matrix material that helps keep the outer segments straight and aligned. They participate in the breakdown of bleached photopigment and recycle the molecular byproducts to photoreceptors for resynthesis of pigments. They also ingest the outermost tips of the outer segments after they are shed by photoreceptors and thus aid in the continual renewal of outer segments.

PHOTOTRANSDUCTION CASCADE

Rhodopsin consists of a chromophore, *11-cis retinal* (the aldehyde form of vitamin A), and a protein, *opsin* (Fig. 4). Absorption of a photon of light changes the chromophore to the all-*trans* form of retinal. This conformational change produces a short-lived intermediate, *meta-rhodopsin II*, which activates a G-protein called *transducin*. Transducin activates a phosphodiesterase enzyme, leading to a reduction in the intracellular levels of cyclic guanosine monophosphate (cGMP). As cGMP

levels are lowered, the conductance of sodium channels in the rod plasma membranes is decreased, and the membrane becomes hyperpolarized in proportion to the amount of light initially absorbed.

It may initially seem counterintuitive for rods to become hyperpolarized in response to increased light exposure; however, it seems less so in light of the fact that rods do not have action potentials. Sodium channels of photoreceptors are not voltage dependent, so their conductances are regulated almost entirely through the cGMP regulatory site. Rods are capable of responding to both increases and decreases in light by producing graded potentials that reflect an increase or decrease in the levels of cGMP. The final output of the rod is in the form of an increase or decrease in the amount of transmitter (glutamate) released at the photoreceptor synapse.

The sodium channels of rods also gate calcium ions that provide a feedback loop to regulate cGMP synthesis. In low levels of light, cGMP levels are high, channels are open, and sodium and calcium can enter the cell. The inward sodium current (called the *dark current*) keeps the cell relatively depolarized; in addition, the influx of calcium inhibits guanylate cyclase activity. Because both synthetic and degradative enzymes for cGMP are relatively inactive under these conditions, cGMP levels remain fairly constant. An increase in light decreases cGMP levels and causes a rapid decrease in the conductance of both sodium and calcium.

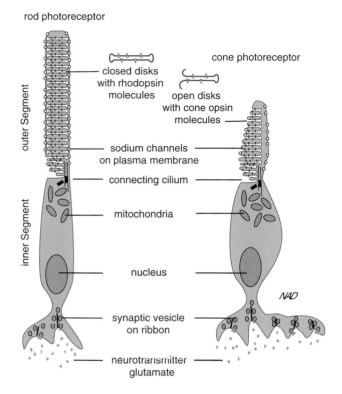

Rods

Long outer segment
Big synaptic terminal

High sensitivity
   over 1,000 closed disks
   more photopigment
   about 40,000 rhodopsin/disk
   scotopic night vision

High amplification
   one photon stops the entry
   of about 10^7 Na^+ ions

Saturates in daylight

Low temporal resolution
   slow response
   long integration time

Sensitive to scattered light
   poor spatial resolution

Rod system

Low acuity
   rods absent in central fovea
   highly convergent pathways

Achromatic
   one type of rhodopsin

Cones

Short outer segment
Huge synaptic terminal

Low sensitivity
   many fewer open disks
   less photopigment
   less per disk
   photopic day vision

Less amplification

Saturates only in intense light

High temporal resolution
   fast response
   short integration time

Sensitive to direct axial light
   good spatial resolution

Cone system

High acuity
   cones concentrated in fovea
   slight divergence in pathways

Chromatic (color vision)
   three types of cones each
   with different photopigment

FIGURE 3 Comparison of the cellular features of rod and cone photoreceptors. Structural and chemical differences of rods and cones make them suitable for their specialized functions: rods for scotopic vision in dim light and cones for photopic vision in bright light conditions. Humans have one class of rods and three classes of cones producing black and white vision at night and color vision in the day, respectively. Other vertebrates have different combinations; for example, frogs have two of each.

Reopening of the channel requires synthesis of new cGMP. cGMP production is enhanced under increased light conditions because the calcium inhibition of guanylate cyclase has been removed and as a result sodium conductance begins to increase. This is one of the major mechanisms responsible for the phenomenon of *light adaptation*. The adaptation response is slower than the initial hyperpolarization response because synthesis of cGMP is slower than degradation.

Reset mechanisms also operate to maintain the necessary levels of photopigment. As with cGMP, the degradation of rhodopsin is much faster than its resynthesis. The light-stimulated conversion of rhodopsin into dissociated opsin protein and chromophore (all-*trans* retinal) is aided (speeded) by two other processes. A protein called *arrestin* binds to the opsin; in addition, opsin is phosphorylated by *opsin kinase*. These steps also help prevent any reassociation of opsin with the chromophore. Resynthesis of 11-*cis* retinal requires an isomerase enzyme found in the RPE. All-trans retinal must be released from the rod, taken up by the RPE, reisomerized, and sent back to the photoreceptor to reassociate with opsin.

It was recently discovered that opsin has a molecular structure similar to that of the family of membrane receptors that activates G-proteins. For example, it has seven hydrophobic transmembrane spanning regions and significant sequence homology with another member of this protein family, namely, the beta-adrenergic receptor. Thus, an analogy can be made between phototransduction and chemical transmission. The RPE secretes a ligand (the chromophore, 11-*cis* retinal) that binds to a membrane receptor (opsin) located on rod photoreceptors. The photoisomerization and removal of the ligand (by light) leads to the activation of a G-protein that regulates membrane conductance and determines neurotransmitter output of the rod to second-order neurons. From this perspective, the cellular mechanisms used in photo-transduction are similar to those commonly used in other signaling systems, with the one unique feature being the light-sensitive properties of the ligand-receptor interaction.

Although phototransduction mechanisms in rods and cones appear to be quite similar, several significant differences are recognized. The chromophore involved (11-*cis* retinal) is the same in cones as in rods; however, the primary structure of the three different cone opsins differs slightly from each other and from rod opsin. These small differences in opsin structure influence the conformation of the chromophore in its bound state, which in turn establishes the specific spectral absorption properties of the four classes of photoreceptor cells.

A. Light / dark responses in rods   B. Steps in the visual transduction cascade

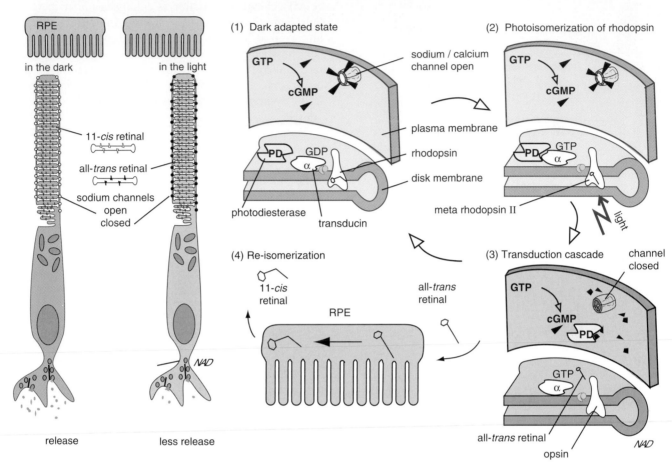

FIGURE 4 Steps in the phototransduction cascade of rods. (A) A rod photoreceptor in the depolarized state (blue) and in the hyperpolarized state (gray). Light changes 11-*cis* retinal to all-*trans* retinal, closes sodium channels, hyperpolarizes the rod, and decreases glutamate release. (B) Arrangement of molecules and components in the outer segment; four steps in the process of phototransduction are depicted: (1) In the dark-adapted state, cyclic guanidine monophosphate (cGMP) binds the sodium/calcium channel, allowing a continuous influx of sodium to keep the cell depolarized. (2) Light photoisomerizes the 11-*cis* retinal component of rhodopsin to all-*trans* retinal and allows rhodopsin to become meta-rhodopsin II. (3) Meta-rhodopsin II activates the G-protein transducin, which activates phosphodiesterase, which metabolizes cGMP and allows the sodium-calcium channel to close; closing the channel hyperpolarizes the rod (gray). (4) All-*trans* retinal diffuses away from the opsin and is carried by a series of retinal binding proteins to the retinal pigment epithelium (RPE), where isomerase converts it back to 11-*cis* retinal, which is carried back to the rod on retinal binding proteins where it binds to opsin and forms light-sensitive rhodopsin.

A second difference is that recycling of cone photopigments occurs within the cone and does not require involvement of the RPE.

RETINAL PROCESSING

Rods and cones form a regular mosaic arrangement across the outer or external surface of the retina. Photoreceptors are the primary sensory neurons of the visual pathway, with *bipolar cells* and *ganglion cells* representing the secondary and tertiary sensory neurons, respectively (Fig. 5). Synaptic connections between photoreceptors and bipolar cells are localized in the *outer plexiform layer*; synapses between bipolar and ganglion cells are found in the *inner plexiform layer*. Three types of interneurons are present in the retina: *horizontal cells*, which provide feedback and feedforward interactions in the outer plexiform layer; *amacrine cells*, which influence synaptic activity in the inner plexiform layer; and *interplexiform cells*,

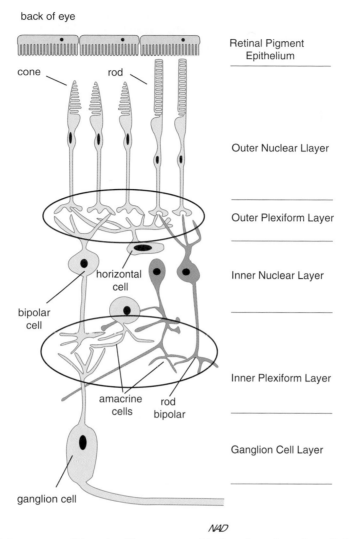

back of eye

cone   rod

Retinal Pigment Epithelium

Outer Nuclear Llayer

Outer Plexiform Layer

horizontal cell

bipolar cell

Inner Nuclear Layer

amacrine cells   rod bipolar

Inner Plexiform Layer

Ganglion Cell Layer

ganglion cell

NAD

FIGURE 5   Cellular anatomy of the retina. Photoreceptors, bipolar cells, and ganglion cells form the "straight-through" retinal pathway with synaptic relays for processing visual information in the outer and inner plexiform layers. Horizontal and amacrine cells provide lateral interactions within outer and inner plexiform layers, respectively. Interplexiform cells (not shown) feed back information from the inner to the outer plexiform layer.

which provide feedback from inner to outer plexiform layers.

The visual pathway not only must transmit information about the intensity of light present in the environment and its spectral properties but must also convey information about the location of light sources. The anatomic arrangement of the visual system must be such that spatial information about the visual field can be encoded and transmitted to higher cortical centers for processing. Thus, retinal circuitry is highly ordered along vertical and horizontal planes. In the vertical plane, cell bodies and synapse are arranged in distinct layers. Superimposed on this laminar structure are horizontally repeating modular arrangements of cells that form functional columns or channels (not to be confused with ion channels in membranes)—for example, a color-specific channel consisting of a cone or small group of cones and their associated bipolar and ganglion cells.

ON and OFF Channels of Visual Information

Two types of functional channels established by retinal circuitry are designated as *ON* and *OFF channels* (Fig. 6). An OFF channel is formed by a column of cells that are all depolarized when the light is decreased. All photoreceptors are depolarized in the dark and therefore release more glutamate when the light is OFF. Many bipolar cells exhibit OFF responses, as well, because they are depolarized by the increased levels of glutamate released by photoreceptors in the dark. Other bipolar cells have the opposite (ON) response and are hyperpolarized when light levels decrease. The reason for this

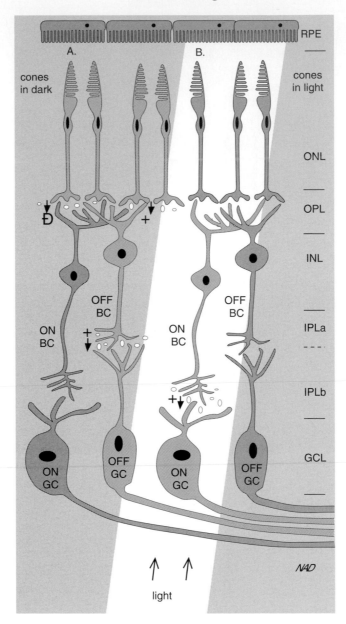

FIGURE 6   Synaptic interactions of ON and OFF channels in the retina. (A) When light levels are decreased, cone photoreceptors are depolarized (blue) and release more glutamate, which interacts with excitatory glutamate receptors (+) on OFF bipolar cells (OFF BCs) and inhibitory (metabotropic) glutamate receptors (−) on ON bipolar cells (ON BCs). OFF and ON bipolar cells in turn excite OFF and ON ganglion cells (GCs), respectively. (B) When light levels are increased, the cone photoreceptors hyperpolarize (gray) and release less glutamate, resulting in less excitation of the OFF BCs and less inhibition of the ON BCs. This signal is passed on to the GCs. Thus, OFF cells collectively form an OFF channel whose component cells are active when the light is off, and ON cells form an ON channel whose components are active when the light is on.

sign-inverting response was only recently explained with the discovery of a special class of hyperpolarizing (metabotropic) glutamate receptors localized to bipolar cells with ON responses. This arrangement leads to the establishment of ON and OFF channels at the level of the photoreceptor-to-bipolar cell synapse in the outer plexiform layer. Separate ON and OFF channel responses are maintained throughout much of the visual pathway, including ganglion cells, as well as cells in the lateral geniculate and in the visual cortex. Both channels respond to increases and decreases in light levels but in opposite directions. From an engineering perspective, this doubling of the visual signal by creating a mirror image provides an internal comparison mechanism that could help in decreasing noise and maintaining fidelity of the visual signal as it is sent to the visual cortex.

Center Versus Surround Receptive Fields

These ON and OFF responses result from the direct synaptic activation of first-order to second-order to third-order neurons that form the straight-through pathway within a given module. In addition to these vertical responses, horizontal interactions between adjacent modules are mediated by retinal interneurons: *horizontal cells* at the level of the outer plexiform layer and *amacrine cells* in the inner plexiform layer. Horizontal cells are depolarized along with OFF bipolar cells in response to glutamate released from photoreceptors (Fig. 7). Horizontal cells are inhibitory interneurons so they in effect produce an inhibitory influence that spreads from any given active module to adjacent modules. Thus, the dark activation of an OFF bipolar cell via the straight-through pathway from a photoreceptor may be counterbalanced by inhibitory input from horizontal cells if surrounding modules are also being activated by darkness at the same time. In other words, if a central module and its surrounding modules are exposed to uniform darkness, then OFF bipolar cells will not be active; the same holds true for ON channels. Likewise, both ON and OFF channels are essentially silent when a central module and its surrounding modules are exposed to uniform levels of light.

The functional classification of ON and OFF responses denotes the response of a bipolar cell to stimulation via the straight-through pathway—namely, the central portion of the module. Because of inhibitory horizontal cells, a bipolar cell is responsive not only to input from the straight-through (central) pathway, but also to adjacent or surrounding pathways as well, and the response to central output is opposite that from the surround under similar lighting conditions. The arrangement is called *center versus surround inhibition*. Each bipolar cell receives central input from a small circular region of retina containing one or a small group of cones and surround input from a larger area encircling the central region. These regions together comprise the area of the retina from which a given bipolar cell will receive input, and they define the *receptive field* for that particular cell. Each bipolar cell has a distinctive receptive field. If a receptive field map of all bipolar cells were displayed on the retina, the entire surface area would be covered by overlapping circles, with those near the fovea being very small and those in the peripheral retina being much larger.

Because of center versus surround inhibition, an OFF bipolar cell will be maximally stimulated when the portion of the retina representing the center of the receptive field of the cell is relatively dark and the surrounding area is light. Under these conditions, the bipolar cell will be receiving direct dark stimulation from the central pathway but no inhibitory input from surrounding OFF pathways. Similarly, OFF bipolar cells will be maximally inhibited when a small spot of light surrounded by a dark annulus is centered over its receptive field. The opposite is true for ON bipolar cells. They are maximally stimulated when their receptive fields are illuminated by a small spot of light surrounded by a dark annulus and maximally inhibited by a small dark spot surrounded by a bright annulus. Conditions for maximal stimulation or inhibition can be artificially produced under laboratory conditions, but they are rarely encountered in the real world. During normal visual experiences, individual receptive fields may be uniformly dark or light (in which case there is no bipolar response), or they may be nonuniformly illuminated such as in the case when an edge of light falls across the receptive field. As long as the receptive field is nonuniformly illuminated, the bipolar cell will respond by being either partially depolarized or hyperpolarized. Thus, bipolar cells act as *local edge detectors*.

The receptive field properties of bipolar cells represent a series of templates that selectively encode information about boundaries of light and dark in the visual field. As this coded information is transmitted through neurons in the primary visual pathway, other processing steps are performed; however, the basic center versus surround property is retained through many synaptic relays. Bipolar cells, ganglion cells, lateral geniculate cells, and several types of cells in the visual cortex all have circular ON or OFF center versus surround receptive fields.

Magnocellular and Parvocellular Pathways

There is a large variation in the receptive field sizes of ganglion cells. One type of ganglion cell has a relatively small receptive field size, roughly equivalent to the receptive field size of one bipolar cell or a small group of bipolar cells. These cells can encode finely detailed information about the shape and texture of objects in the visual field and constitute the *parvocellular (P) pathway*. Another type of ganglion cell has a large receptive field equivalent to the summed receptive field sizes of hundreds of bipolar cells. This highly convergent pathway, called the *magnocellular (M) pathway*, is not capable of distinguishing fine detail; it is better suited for the task of mapping the general location of objects within the visual field. A third type of ganglion cell, not as well understood, has more complex receptive field properties and responds specifically to the movement of objects.

CENTRAL VISUAL PATHWAYS

Information about the visual world is simultaneously encoded by both retinas separately. These two images must be merged and integrated within the central visual pathway in order to provide a single unified perception

OFF centered channel

ON centered channel

light change

A.

cones

HC

OFF BC

OFF GC

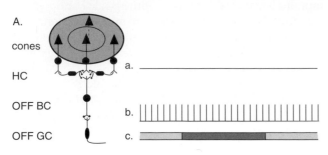

a.

b.

c.

A.

cones

HC

ON BC

ON GC

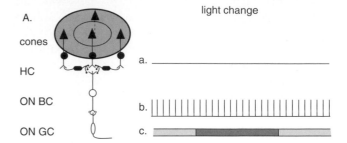

a.

b.

c.

B.

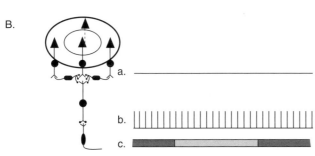

a.

b.

c.

B.

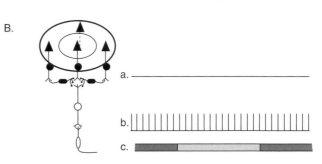

a.

b.

c.

C.

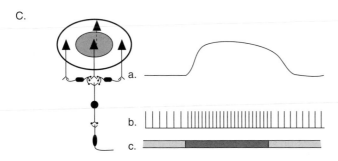

a.

b.

c.

C.

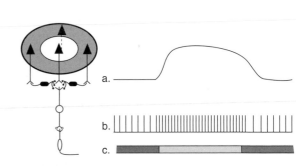

a.

b.

c.

D.

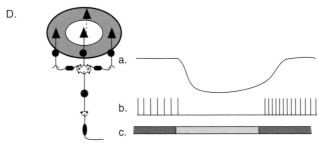

a.

b.

c.

D.

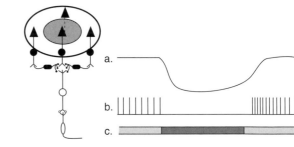

a.

b.

c.

E.

NAD

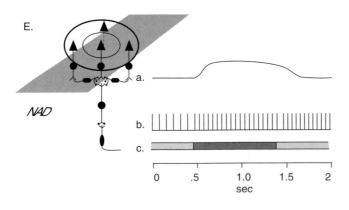

a.

b.

c.

0   .5   1.0   1.5   2
sec

E.

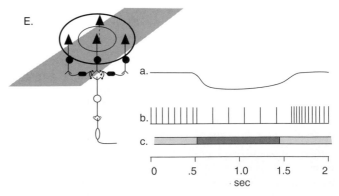

a.

b.

c.

0   .5   1.0   1.5   2
sec

of visual space. Merging occurs in stages at different points within the visual pathway. As a result, a unilateral lesion at a more distal point within the pathway will produce different visual deficits from one at a more proximal point because the degree of merging will be different. With sufficient knowledge of central visual pathways, the location of a lesion can be deduced from the type of visual deficit present (for details, see below).

Each retina can be divided into quadrants based on superior, inferior, nasal, and temporal coordinates (Fig. 8). The fovea is located at the posterior pole or intersection point of the quadrants. When light rays enter the pupillary opening, the image projected on the retina is reversed and inverted. Thus, the image of an object located within a given region of the peripheral visual field is projected to the opposite position in the peripheral retinal field. Objects located in the center of the visual field can be focused on foveal regions of both eyes simultaneously. Visual fields are usually tested separately for each eye, with the patient looking straight ahead and the eye immobile. Regions of the visual field are designated as right, left, upper, and lower quadrants.

It is important to note that visual field deficits are always defined in terms of visual field loss, not retinal field location. In the 180° visual field viewed with both eyes, a central zone of about 90° (the *binocular zone*) can be viewed with both eyes (Figs. 9 and 10). The more peripheral regions of the visual field are visible only by the ipsilateral eye (because the nose gets in the way), creating two temporal, crescent, *monocular zones*. A *blind spot* in each retina is created by the optic nerve head or optic disk; here, the exiting fibers of ganglion cells coalesce and become invested with myelin. The optic disc is located along the horizontal meridian and displaced in the nasal direction from the fovea. Even with one eye closed, there is no conscious perception of a blind spot, just as there is no perception of the shadows that are cast by retinal blood vessels. This occurs because the visual cortex is able to compensate for these normal interruptions within the visual field; however, both can be demonstrated during visual exams.

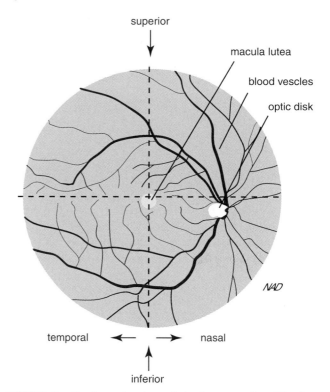

FIGURE 8 Fundoscopic view of the human retina. The inner (vitreal) surface of the retina of the right eye is illustrated. Axons of ganglion cells from the entire retina converge and exit the eye at the optic disk, where they penetrate the retina and produce an area with no neuronal elements other than the optic nerve, thus creating a blind spot. Blood vessels enter the eye at the optic disk and spread over the surface of the retina except in the region of the fovea located in the macula lutea. Two hypothetical lines crossing at right angles through the fovea delineate the four quadrants of the retina. The optic disk is located in the inferior nasal quadrant.

Left and Right Visual Hemifields

Optic nerve fibers exit the eye and project to the *optic chiasm*, where partial decussation occurs (see Fig. 10). Fibers originating from the nasal portions of both retinas cross the midline and merge with uncrossed fibers from the temporal retina of the opposite eye. Fibers proximal

---

FIGURE 7 Receptive field properties of bipolar and ganglion cells. Responses in both OFF (left panel) and ON (right panel) channels are shown. Figures (A) to (E) in each panel illustrate the connections among cones, horizontal cells (HCs), bipolar cells (BCs), and ganglion cells (GCs). The specific pattern of light/dark stimulation being experimentally applied to the receptive fields of these cells is indicated by the shading (darkness = dark gray; light = white). To the right of each figure is a set of recordings showing the response of the cells to the light/dark stimulation: the graded electrical response of the bipolar cells (a) and the action potentials generated in ganglion cells (b). The duration of the stimulation is shown in (c). OFF bipolar cells (left panel) are stimulated by direct input from cone photoreceptors located in the center of its receptive field but are indirectly inhibited by cones in the surrounding area through stimulation of inhibitory horizontal cells. Exposure to uniform dark (A) or light (B) produces no change in BC membrane potentials or GC action potentials, because the center effects are exactly counteracted by the surround effects. A very small centralized dark spot (C) or light spot (D) restricted to the center of the receptive filed produces maximal excitatory and inhibitory responses, respectively. Any nonuniform illumination of the receptive field (E) may produce a detectable response because center excitation will no longer be completely counterbalanced by the inhibitory surround. Similar mechanisms underlie electrical responses and receptive field properties of all cells in the ON channel (right panel).

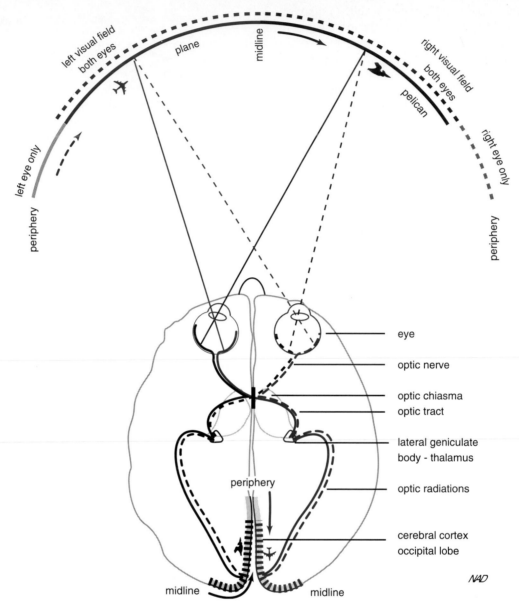

FIGURE 9 Visual pathways to the cerebral cortex. The left visual hemifield is shown in blue; the right visual hemifield, in black. Left eye projections are solid lines; right eye projections are dashed lines. Each retina receives information from both visual hemifields (except for the far periphery (not shown), because the nose blocks this input); however, outgoing information from each retina is split. Fibers from the nasal portions of each retina cross to the contralateral side at the optic chiasm, whereas fibers from the temporal regions remain ipsilateral. As a result, information from a single visual hemifield (transmitted from both eyes) is combined and sent to the contralateral lateral geniculate body and cerebral cortex. Hence, the image of an airplane seen on the left is received in the right visual cortex, and the image of a pelican seen on the right is received by the left cortex.

to this point, designated as the *optic tract*, project to the *lateral geniculate body* and from there to the primary visual cortex. As a result of the partial decussation, each optic tract carries information solely about the contralateral visual field. There are two views of the contralateral visual hemifield: one provided by temporal fibers from the ipsilateral retina and the other provided by nasal fibers from the contralateral retina. Thus, all information from the right visual field is transferred to the left brain, regardless of which eye it enters.

Binocularly responsive neurons in the cortex receive retinotopic information from both eyes. By comparing the slight disparities in the positional information from the two eyes, these binocular cells provide the major

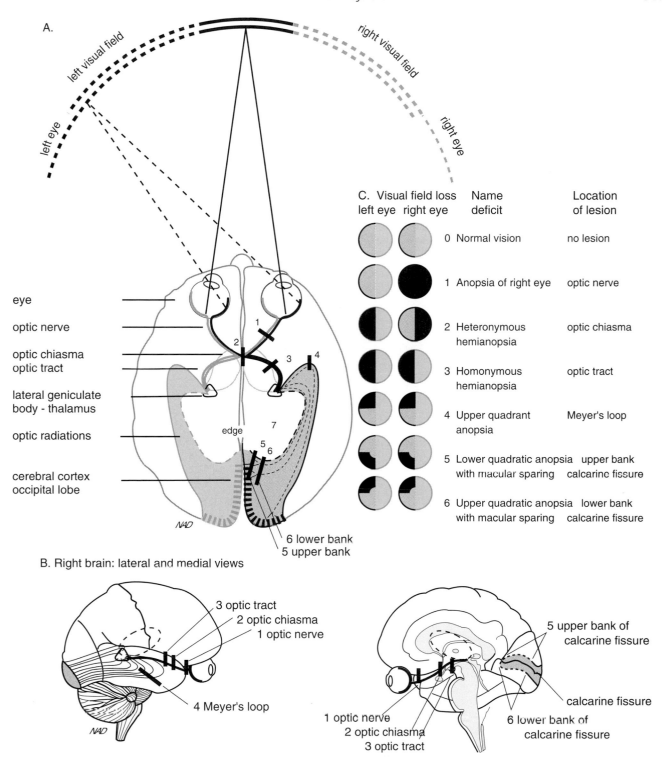

A.

left visual field

right visual field

left eye

right eye

C. Visual field loss
left eye   right eye

Name
deficit

Location
of lesion

eye

optic nerve

optic chiasma
optic tract

lateral geniculate
body - thalamus

optic radiations

cerebral cortex
occipital lobe

edge

NAD

6 lower bank
5 upper bank

0 Normal vision   no lesion

1 Anopsia of right eye   optic nerve

2 Heteronymous   optic chiasma
  hemianopsia

3 Homonymous   optic tract
  hemianopsia

4 Upper quadrant   Meyer's loop
  anopsia

5 Lower quadratic anopsia   upper bank
  with macular sparing   calcarine fissure

6 Upper quadratic anopsia   lower bank
  with macular sparing   calcarine fissure

B. Right brain: lateral and medial views

3 optic tract
2 optic chiasma
1 optic nerve

5 upper bank of
calcarine fissure

4 Meyer's loop

NAD

1 optic nerve
2 optic chiasma
3 optic tract

calcarine fissure

6 lower bank of
calcarine fissure

FIGURE 10   Visual field deficits. Discrete lesions in the primary visual pathway lead to at least six characteristic visual deficits. The left visual field is shown in blue; the right visual field, in gray. Central visual field (macula) projections are shown as solid lines; peripheral visual field projections are shown as dashed lines. (A) View of the visual fields, visual pathways, and locations of six possible lesions. (B) Lateral view (left) and medial view (right) of the human brain showing the locations of the same lesions. Cerebellum is removed in the medial view. (C) Visual field deficits (black) that result from each injury. Note that information from both eyes is interrupted in every case, except for the optic nerve injury, and that the deficit is in the left visual field when the right brain is injured. Also note the unique pattern of loss when the crossing fibers of the optic chiasm are destroyed. This can happen when a pituitary tumor expands.

mechanism (called *stereopsis*) for creating a three-dimensional representation of objects in space and a means to achieve depth perception.

There is no conscious perception of the split of the visual field between the left and right visual occipital cortex. The re-merging of visual information into a single precept about an object with both left and right parts does not occur until the signals reach visual association areas in the parietal cortex. Decussating fibers in the *corpus callosum* provide cross-talk between left and right cortical areas to reestablish the complete visual field.

The projection of retinal and lateral geniculate fibers occurs in a highly ordered, *retinotopic* manner. Foveal representation of the contralateral visual hemifield covers a large area of the posterior pole of the *calcarine fissure*, whereas peripheral vision has a more modest representation in the anterior areas.

Retinal Projections to the Lateral Geniculate Nucleus

The lateral geniculate is a C-shaped structure consisting of six cellular layers, each receiving input from distinct sets of ganglion cells (Fig. 11). Ganglion cells with large receptive field sizes project to large neurons in layers 1 and 2, thus constituting the magnocellular (M) pathway. As described earlier, this pathway carries information about the spatial position of objects in the visual field. Ganglion cells with small receptive field sizes project to the small neurons found in layers 3 to 6, thus forming the parvocellular (P) pathway. This pathway carries information about the detailed shape of visual objects. The separation of left/right ocular input is retained so that layers 1, 4, and 6 receive contralateral projections, and layers 2, 3, and 5 receive ipsilateral projections. Some ganglion cells bypass the lateral geniculate, project directly to the brain stem, and participate in visual reflexes.

THE VISUAL CORTEX

Cortical Ocular Dominance Columns

Cells within the six *ocular dominance layers* of the lateral geniculate project to distinct *ocular dominance stripes* or *columns* within Brodmann's area 17 (also called area VI) of the primary visual cortex (see Fig. 11). Pairs of adjacent ocular dominance columns—one with ipsilateral projections from the *lateral geniculate nucleus* (LGN), and the other with contralateral projections—receive information about the same point in visual space. Adjacent pairs are collectively designated as a *hypercolumn*. Hypercolumns form the basic structural modules of the visual cortex, and each displays at least three

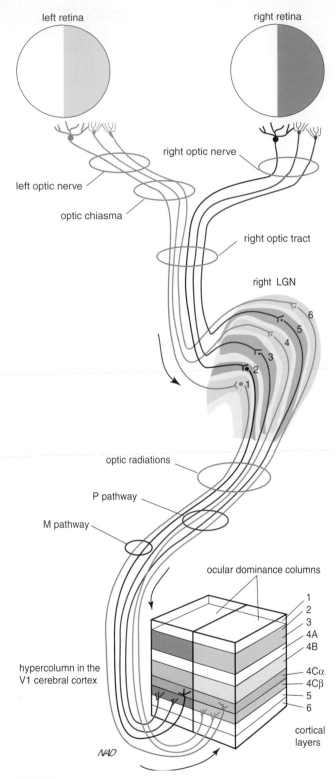

FIGURE 11 Projections of the M and P pathways from retina to the lateral geniculate nucleus and visual cortex. Projections from the left eye are shown in light shades; projections from the right eye are dark shades. Two channels of information leave the eye: the magnocellular (M) channel (gray), starting with large ganglion cells that have large cell bodies and large dendritic fields, and the parvocellular (P) channel (blue), starting with the more numerous and smaller ganglion cells. Both

different functional subdivisions in addition to ocular dominance columns. These include *input/output layers*, *orientation-specific columns*, and *color-specific blobs*. Layer 4 functions as the major input layer within a hypercolumn. It contains *stellate cells* that receive information from axons of lateral geniculate cells. In turn, stellate cells synapse on *pyramidal cells* in other cortical layers. Pyramidal cells in layers 2, 3, and 4 project to higher centers in the visual association cortex: Brodmann's area 18 (also called areas V2 and V3) and Brodmann's area 19 (also called areas V4 and V5). Pyramidal cells in layer 5 project to visual reflex areas in the upper brain stem; cells in layer 6 project back to the lateral geniculate.

Orientation-Specific Columns and Color-Specific Blobs

Like most retinal and LGN neurons, the stellate cells of the visual cortex have circular center-surround receptive fields (Fig. 12). Several stellate cells with overlapping receptive fields synapse on each simple pyramidal cell, and the summation of this input establishes a bar-shaped, center-surround receptive field for the pyramidal cell. For example, an OFF center simple pyramidal cell would be maximally stimulated by a dark bar on a light background positioned on the appropriate region of the retina. In addition, different pyramidal cells have different preferred orientations so that the degree of stimulation is dependent on the orientation of the bar. Cortical cells with the same orientation selectivity are grouped together in *orientation columns*. Present in each ocular dominance column is a complete set of orientation columns. Each column of cells is preferentially responsive to a given axis of orientation ±5°. LGN cells in the M pathway project to stellate cells and pyramidal cells in orientation columns. LGN cells in the P pathway have a split projection (Fig. 13). Some synapse in orientation columns; others, designated as the *K channel*, carry color-specific information to color-specific blob regions in the cortex that are not orientation selective.

types of ganglion cells project to the lateral geniculate nucleus (LGN) by way of the optic nerve, chiasm, and tract. The LGN has six layers. The M channel projects to layers 1 and 2 (gray) and the P channel to layers 3, 4, 5, and 6 (blue). Contralateral neurons go to layers 1, 4, and 6 (light shade), and ipsilateral neurons go to layers 2, 3, and 5. Neurons leave these layers as the optic radiations and enter the primary visual area (V1) of the cerebral cortex. A column of cortex, about 1 mm × 1 mm, extending from the surface (layer 1) down about 2 mm to the deepest layer (layer 6), is designated as a hypercolumn. The major inputs of most LGN neurons are to layer 4C. The M channel synapses in 4Ca and the P channel synapses in 4Cb. The input from the two eyes remains segregated, with the contralateral (light color) and ipsilateral (dark) information forming ocular dominance columns.

Higher Visual Processing

Cells of the M pathway continue their projection from area VI to association areas designated as V2, V3, V4, V5 (Brodmann's areas 17, 18, and 19) and finally to other regions in the parietal lobe. This parietal stream carries spatial and movement information important for localizing objects in space (Fig. 14). Cells from both components of the P pathway project to association areas V2 and V4 and finally to inferior regions of the temporal lobe. At each stage of processing, receptive field properties change from spots (stellate cells) to bars (simple pyramidal cells), angles, and so on, in higher order neurons. This pathway is concerned with object recognition. About 10% of the cells in the *inferotemporal cortex* have very large and complex receptive fields and are selective for very specific stimuli, such as the hand or a familiar face. A condition known as *prosopagnosia* has been reported in patients with bilateral lesions along the inferior surface of the occipital and temporal lobes. In this condition, patients have impaired recognition of familiar faces. Although there is considerable crossover and mixing of the information at several points along the M and P pathways, it does appear that specific visual processing tasks (determining *where* an object is versus *what* an object is) are generally assigned to different regions of visual association cortex (see Fig. 14). Other higher order visual tasks may likewise be localized to specific cortical regions.

COLOR VISION

The subjective experience of color involves three components:

1. *Hue* (normally thought of as *color*) is defined as the proportion in which the three cone systems are activated; 200 gradations can be recognized.
2. *Saturation* (the degree of dilution by grayness) is defined as the degree to which all three cone systems are stimulated based on their maximal response; 20 steps can be recognized.
3. *Brightness* is defined as the total effect of all three cone mechanisms; 500 steps can be recognized.

Color vision thus has $200 \times 20 \times 500 = 1$ million gradations. Achromatic vision has only 500.

Photopigments of Cones

Individual cones contain only one of three pigments (Fig. 15). Short-wavelength (*S*) cones contain pigment primarily sensitive to short wavelengths with a peak

A. Receptive fields of OFF-center stellate cells in layer 4Cβ of V1

image on the retina is a dark bar at 60û

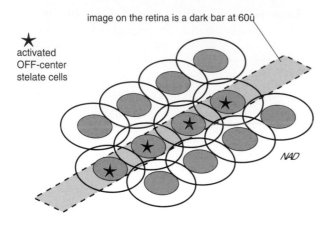

★ activated OFF-center stelate cells

NAD

B. The receptive field of thes OFF simple pyramidal cell is bar shaped and angled at 60û

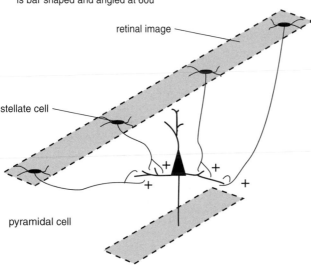

retinal image

stellate cell

pyramidal cell

C. A single orientation column in V1 containing pyramidal cells that respond to dark bars at 60û

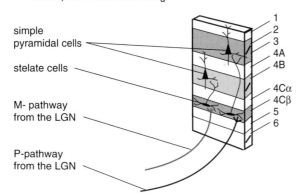

simple pyramidal cells

stelate cells

M- pathway from the LGN

P-pathway from the LGN

1 2 3 4A 4B 4Cα 4Cβ 5 6

FIGURE 12 Receptive field properties of stellate and simple cortical cells. The combined input of several stellate cells with overlapping, circular, center-surround receptive fields (A) establishes a bar-shaped, center-surround receptive field with a preferred orientation in a single

A The Magnocellular or M channel - position and motion detection

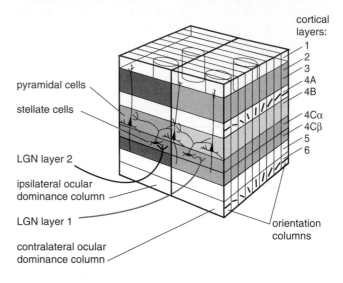

cortical layers: 1 2 3 4A 4B 4Cα 4Cβ 5 6

pyramidal cells

stellate cells

LGN layer 2

ipsilateral ocular dominance column

LGN layer 1

contralateral ocular dominance column

orientation columns

B. The Parvocellular or P channel - form detection

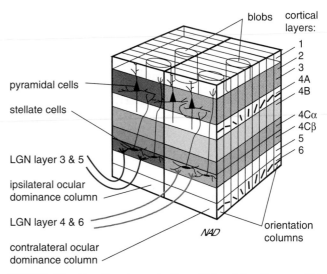

blobs cortical layers: 1 2 3 4A 4B 4Cα 4Cβ 5 6

pyramidal cells

stellate cells

LGN layer 3 & 5

ipsilateral ocular dominance column

LGN layer 4 & 6

contralateral ocular dominance column

NAD

orientation columns

FIGURE 13 Functional organization of M and P pathways within hypercolumns of the primary visual cortex. The major inputs of most LGN neurons are to layer 4C. The M channel (A) synapses in 4Cα, and the P channel (B) synapses primarily in 4Cβ, with a small component (the K channel) projecting directly to the blob regions in layer 2. Thus, color information is processed separately from form.

absorption at approximately 420 nm. Activation of these cones leads to the perception of blue. *M cones* contain middle-wavelength pigment with a peak absorption at 530 nm and contribute to the perception of green; *L cones* have long-wavelength pigment with a peak absorbance at

pyramidal cell (B). Groups of pyramidal cells with the same preferred orientation are clustered in orientation columns within the visual cortex (C).

A. Left cerebral cortex: lateral view and medial view

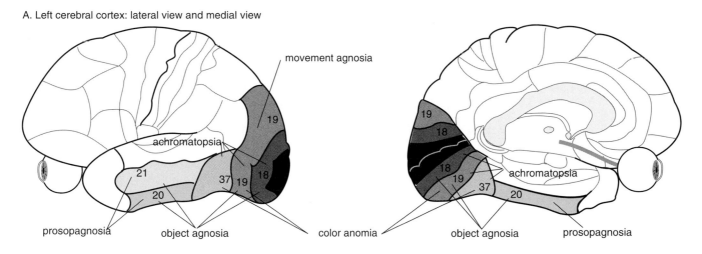

B. Right cerebral cortex: medial view and lateral view

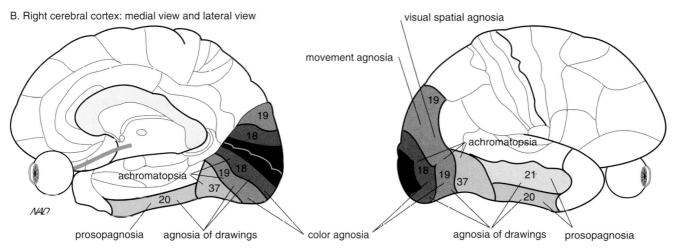

Visual agnosia type	Deficit	Area of lesion
The M channel: Agnosia for depth & movement		
Visual spatial agnosia	Stereoscopic vision	18, 19, 37 on right
Movement agnosia	Discerning object movement	18, 19, 37 bilaterally
The P channel:: Agnosia for form and pattern:		
Object agnosia	Naming, understanding purpose recognition of real objects	18, 19, 20, 21 on left & corpus callosum
Agnosia for drawings	Recognition of drawn objects	18, 19, 20, 21 on right
Prosopagnosia	Recognition of faces	20, 21 bilaterally
The P channel: Agnosia for color:		
Color agnosia	Association of colors with objects	18, 19 on right
Color anomia	Naming colors	Speech zones or connections from 18, 19, 37 on left
Achromatopsia	Distinguishing hues	18, 19, 37

FIGURE 14   The visual agnosias. Visual area 1 (V1) corresponds to Brodmann's area 17 (dark blue). From V1 information passes in separate channels into the visual association areas located in Brodmann's areas 18, 19, and 37 (lighter shades of blue). The M channel projects in a series of steps toward the parietal lobe, carrying spatial and movement information. The P channel passes down toward the temporal lobe as it processes form, pattern and color. In the temporal lobe, the P channel associates vision with higher order functions such as naming objects. The many visual maps in the visual cortex process different aspects of the visual world, and damage to one of these can produce an isolated deficit, leaving the rest of vision intact.

A. Photoreceptor types in the human retina

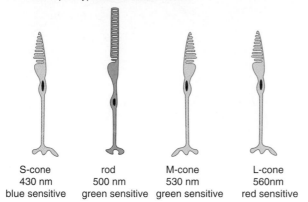

S-cone	rod	M-cone	L-cone
430 nm	500 nm	530 nm	560nm
blue sensitive	green sensitive	green sensitive	red sensitive

B. Absorption spectra of the human rod and three cones.

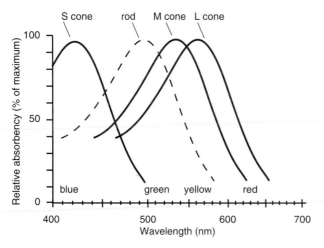

C. Focusing of light in a simplified eye

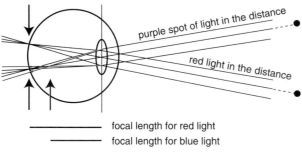

focal length for red light

focal length for blue light

D. objects viewed   appearance   appearance
                  if objects are red   if objects are purple

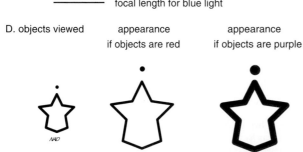

FIGURE 15 Spectral sensitivity of cone photopigments and chromatic aberration in the eye. (A) Four types of photoreceptors in the human retina each contain different photopigments that are maximally

about 560 nm and are the only photopigments sensitive to red light. Short wavelengths are refracted more strongly than long ones; therefore, a full-spectrum image will not be focused in the same plane and will appear blurred. There are two special adaptations to decrease this *chromatic aberration*. First, short-wavelength receptors (blue-sensitive cones and rods containing rhodopsin with peak absorption at 500 nm) are absent from the fovea. Second, an inert pigment, *macular yellow*, which strongly absorbs blue and violet (and hence looks yellow) is deposited in the macula and blocks short-wavelength light from affecting green and red foveal cones.

Color-Opponent Pathways

The three types of cones are coupled in mutually inhibitory, color-opponent fashion to produce color-specific receptive fields in neurons of the visual pathway. Two color-opponent pairs are established: *red versus green* and *blue versus yellow*. Figure 16 illustrates the general circuitry for these pathways and the types of *color-opponent receptive fields* that can result. Ganglion cells in the red/green pathway have single-opponent, ON and OFF, center-surround receptive fields created in part by horizontal cells providing inhibitory "surrounds" to cone/bipolar connections. The four types of ganglion cells in this pathway are classified according to whether they are stimulated or inhibited by red or green light in the center of their receptive fields. Opposite effects are caused by red or green light in the surround portion of the receptive field.

The perception of yellow is generated by combined (rather than opposing) input from red- and green-sensitive cones. Input from yellow (red plus green)-sensitive bipolar cells is compared with input from blue-sensitive cones to form blue versus yellow color-opponent, ON and OFF, ganglion cell types that lack center-surround receptive fields.

Color-coded information is transmitted along P ganglion cells via the lateral geniculate to blob regions of the primary visual cortex (VI). There, information is combined to form *double-opponent cells* that are sensitive to two wavelengths of light in both center and surround

sensitive to different wavelengths. (B) Wavelength absorption for the four types of photopigments in the human retina. Different colors or hues are created by the nervous system depending on the ratio of light absorption by the three photopigments. (C) Rays of light containing a single wavelength of light, such as 600 nm (red), can be focused on the retina through the actions of the lens. If the light is a mixture of long and short wavelengths (for example, purple), the short wavelengths will be more refracted than long wavelengths, and the lens cannot focus both wavelengths simultaneously on the retina; the image will be blurred. (D) A simulation of the appearance of an object (left) when the object is red (middle) or purple (right).

A. Opponent processing in the blue - yellow system

B. Opponent processing in the red - green system

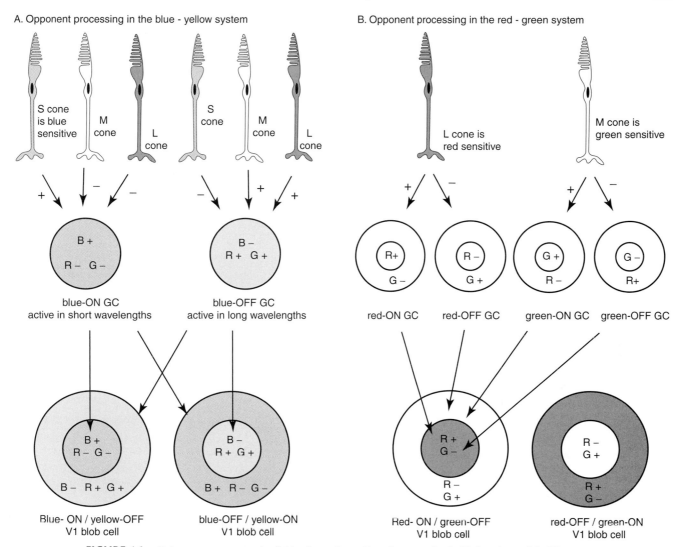

FIGURE 16   Color-opponent receptive fields of ganglion cells and neurons in the blob regions of the V1 cortex. (A) Three types of cones contribute to blue/yellow-opponent receptive fields in ganglion cells (GCs) and cells within blobs of cortical area V1. Excitatory input from blue-sensitive (S) cones and inhibitory input from red-sensitive (L) and green-sensitive (M) cones cause one class of ganglion cells, blue-ON, to be excited in blue light (B+) and to be inhibited in yellow light, as detected by the combined responses of both red- and green-sensitive cones (R-, G-). Another class of ganglion cells, blue-OFF, has the opposite response and is inhibited by blue light and excited by yellow light. Both ganglion cell types send information (via the lateral geniculate nucleus; not shown) to two major types of cortical cells that are clustered in blue/yellow-sensitive V1 blobs and exhibit double-opponent, center-surround receptive fields. (B) Two types of cones contribute to red/green-opponent receptive fields. There are four types of single-opponent ganglion cells with center-surround receptive fields. They send information to two types of red/green-sensitive V1 blob cells that display double-opponent, center-surround receptive fields.

portions of their receptive fields. Understanding this circuitry can be helpful in appreciating certain aspects of color perception. For example, it explains why differently colored backgrounds can affect the apparent color of an object (*e.g.*, why a red vase looks redder when placed on a green background or why it is impossible to perceive the color of an object as red-green [in humans, there is no such color as reddish green]). The latter occurs because red and green input either cancel each other out in the red

versus green pathway or become additive to produce the perception of yellow in the yellow versus blue pathway.

Most nonhuman mammals have only two cone pigments. One is sensitive to shorter wavelengths (S, blue sensitive); the other is sensitive to longer wavelengths of light (L, covering the green to red portions of the spectrum). This more primitive S versus L system has been supplemented in humans through duplication and mutation of the gene for the L pigment to produce two slightly different photopigments, allowing discrimination of multiple hues in the middle (green) to long (red) wavelengths. Genes for red- and green-sensitive pigments are very similar and located on the X chromosome near the 5′ end. Thus, recombination can result in the placement of three genes on one chromosome, leaving only a single L or M gene on the other. A male who inherits a maternal X chromosome containing only the L or M pigment gene will be red or green color blind, respectively.

THE OCULOMOTOR SYSTEM

The *oculomotor system* is comprised of the intrinsic and extrinsic muscles of the eye along with motor nuclei and higher cortical centers that serve to control the position of the eye as well as the shape of the lens and the size of the pupil. The main purpose of the system is to aid in vision by keeping the visual target focused on the fovea, the area of central retina that has the highest visual acuity. Diplopia (double vision), blurred vision, or loss of depth perception can occur as a result of lesions at various points within the neuronal pathways that subserve the oculomotor system. The pathways are generally well defined and well known, so that it is possible to determine the location of a lesion in the oculomotor system by careful analysis of the types of visual deficits expressed and by direct observation of eye position and reflex eye movements. Tests of oculomotor function are an important part of most physical examinations of patients.

Cranial Nerve Innervations of Eye Muscles

Three pairs of extraocular muscles move the globe along three axes. The *medial* and *lateral rectus muscles* rotate the eye in the horizontal plane. The medial rectus muscle is innervated by cranial nerve III (the oculomotor nerve), and the lateral rectus muscle is innervated by cranial nerve VI, the abducens nerve. Both muscles attach to a common tendinous ring around the apex of the orbit and to the eyeball at points parallel to the floor of the orbit. Thus, the force they generate results in horizontal eye movement or horizontal gaze. The medial rectus adducts the eye (*i.e.*, rotates the eye toward the nose along a vertical axis). The lateral rectus abducts the eye (rotates the eye away from the nose) (Fig. 17).

The *superior* and *inferior rectus muscles* are responsible for vertical and torsional movement of the eyeball. Both muscles are innervated by cranial nerve III. The muscles attach to the tendinous ring and to the globe at points 23° off the midsagittal plane. The superior rectus elevates but

A. The muscles of the right eye and their crainal nerves

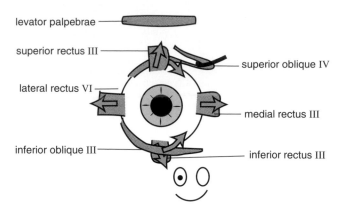

B. Horizontal gaze

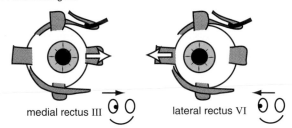

C. Vertical gaze

FIGURE 17 Extraocular muscles and cranial nerves responsible for movements of right eye. (A) Extraocular muscles are controlled by three cranial nerves (III, IV, and VI). The levator palpebrae opens the eyelid and is innervated by cranial nerve III. (B) Horizontal gaze is controlled by paired contraction-relaxation of the medial rectus and lateral rectus muscles. (C) Vertical gaze is controlled by four muscles. Downward gaze is produced by inferior rectus and superior oblique muscles. Because of their placement on the eye, these muscles will also cause lateral and rotational movements, as well (indicated by orientation of the radial lines on the iris). Upward gaze is similarly controlled by the superior rectus and inferior oblique muscles.

also adducts and intorts (*i.e.*, rotates the eye around an anteroposterior axis so that the top of the cornea moves in a circular arc in the nasal direction). The inferior rectus depresses but also adducts and extorts (*i.e.*, the top of the cornea rotates in the temporal direction).

The *superior* and *inferior oblique muscles* also contribute to vertical and rotational movement of the eyeball. The superior oblique muscle is innervated by cranial nerve IV, and the inferior oblique is innervated by cranial nerve III. The superior oblique muscle runs from the tendinous ring through the trochlea and attaches to the eyeball 51° off center. It depresses the eye but also abducts and intorts the eye. The inferior oblique muscle originates from a tendinous area near the lacrimal gland and attaches to the eyeball 51° off center. It elevates but also abducts and extorts.

Figure 17 illustrates the primary movements generated by the extrinsic muscles of the eye. Only the right eye is illustrated. Note that cranial nerves III and VI participate in generating horizontal gaze; cranial nerves III and IV participate in generating vertical gaze. The eyelids are elevated by two groups of muscles: the *levator palpebrae muscle*, which is innervated by cranial nerve III, and the *tarsalis muscle*, which is innervated by sympathetic fibers. The eyelids are closed by the *orbicularis oculi muscle*, which is innervated by cranial nerve VII. Figure 18 diagrams the location of cranial nerves III, IV, and VI and their nuclei within the brain stem. It also illustrates certain topographical relationships of these cranial nerves that may play a role in their susceptibility to certain types lesions. For example, pituitary tumors may compress the oculomotor nerve as it exits the brain stem. Also note that the oculomotor nerve passes near the branch points of two major arteries (posterior cerebral and superior cerebellar), sites that are susceptible to aneurysms. The trochlear nerve is somewhat vulnerable to head trauma because of its long projection course around the lateral aspect of the brain stem. The abducens nerve ascends alongside the basilar artery and is vulnerable to vascular insults involving this artery.

Intrinsic Muscles of the Eye

In addition to its somatic motor component, the oculomotor nucleus contains a parasympathetic component called the *Edinger-Westphal nucleus*. It is located near the midline of the midbrain and sends preganglionic fibers to the ipsilateral *ciliary ganglion*. Postganglionic fibers innervate the ciliary muscle that controls the shape of the lens. Postganglionic fibers of the ciliary ganglion also innervate the pupillary sphincter muscles that constrict the pupil. The *pupillodilator muscle* is innervated by sympathetic neurons.

A. Ventral view of brain stem

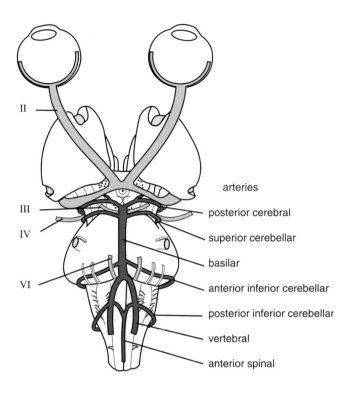

arteries
posterior cerebral
superior cerebellar
basilar
anterior inferior cerebellar
posterior inferior cerebellar
vertebral
anterior spinal

B. Medial view of brain stem

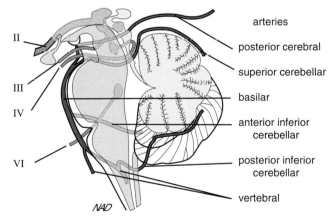

arteries
posterior cerebral
superior cerebellar
basilar
anterior inferior cerebellar
posterior inferior cerebellar
vertebral

FIGURE 18 Cranial nerves and brain stem arteries responsible for eye movement; nerves are shown in blue and blood vessels in gray. The oculomotor nucleus is located under the superior colliculus and gives rise to cranial nerve III that exits ventrally just above the pons. The trochlear nucleus is located under the inferior colliculus. Its axons cross contralaterally and leave the dorsal surface of the brain stem as cranial nerve IV. It then wraps around the lateral brain stem and runs close to nerve III. Note that these two nerves exit between two major arteries and can be damaged by cardiovascular problems. The abducens nucleus is located in the dorsal pons. Its axons exit the ventral surface of the brain stem medially and just below the pons. Cerebral vascular accidents in the nearby anterior inferior cerebellar artery compromise its function.

The actions of the Edinger-Westphal nucleus are important in focusing the object of interest on the fovea. Recall from Fig. 1 that the lens separates the eyeball into two major chambers: The *anterior chamber* lies between the cornea and the lens and is filled with *aqueous humor*; the *vitreal chamber* is located between the lens and the retina and is filled with a highly viscous fluid, the *vitreous humor*, which provides pressure to keep the globe spherical. Light rays are refracted as they pass through these transparent components in the anterior of the eye, particularly the cornea. The curvature of the normal cornea will cause parallel light rays from distant objects to converge at a point roughly 2.4 cm behind it, which is about the distance from the cornea to the retina. A relatively small amount of additional focusing power (approximately 25%) is provided by the lens for bringing distant objects into proper focus. This assumes that the jelly-like lens is being held in a relatively flattened shape by tension on the ligaments that keep it suspended within the circular ciliary muscle of the eye. If the object of interest is closer than about 9 m, additional focusing power must be provided by the lens. This is accomplished through the *accommodation reflex*.

Accommodation Reflex (Near Reflex)

Adaptation of the visual apparatus of the eye for near vision occurs as a reflex action involving cortical, brain stem, and eye structures. The reflex is initiated when visual attention is directed to a nearby object. The initial image of the near object (unfocused) is sent to the occipital cortex via the primary visual pathway. Information is then sent from the visual association areas back to the upper brain stem regions and cranial nerve III. Stimulation of the Edinger-Westphal nucleus causes contraction of ciliary muscle, releasing tension on the suspensory ligaments of the lens and allowing the lens to assume a more spherical shape. The increased curvature of the lens brings the object into focus. Activation of the Edinger-Westphal nucleus also contracts the pupillary constrictor muscle and decreases pupil size, thus blocking certain types of optical aberrations. Stimulation of specific somatic nuclei of cranial III causes both medial rectus muscles to converge the eyes, so that the image falls on the foveal region of both eyes.

Pupillary Light Reflex

When light is directed into the pupil of one eye, both pupils constrict. The response of the illuminated pupil is called the *direct pupillary light reflex*; that of the other pupil is called the *consensual pupillary light reflex*. Light activation of the retina is transmitted, via the optic nerve and tract, to nuclei in the midbrain both on the ipsilateral side, to elicit a direct reflex, and via crossing fibers in a structure called the *posterior commissure*, on the contralateral side, to elicit a consensual reflex. These midbrain areas project to the Edinger-Westphal nuclei, which in turn cause constriction of the pupil. A test of the pupillary light reflex is usually included in standard neurological exams. An altered consensual reflex usually denotes damage to structures localized within the upper

---

Clinical Note

Astigmatism, Hyperopia, Myopia, and Presbyopia
As described earlier, a cornea with normal curvature provides much of the refractive power needed to focus distant objects on the retina without the need for accommodation. This normal condition, called *emmetropia*, will be altered if the curvature of the cornea (or, more rarely, the lens or retinal shape) is irregular, causing portions of the visual field to appear blurred (*astigmatism*). Visual problems are also encountered if the eyeball itself is too short from front to back. In this case, the cornea focuses the image in a plane behind the retina (*hyperopia*, or farsightedness). If the eyeball is too long, the image will be focused in front of the retina (*myopia*, or nearsightedness). These last two conditions can be corrected by convex or concave lenses, respectively. Contact lenses are commonly used to compensate for most forms of astigmatism, as the spherical surface of the contact lens optically replaces the distorted corneal surface.

Another condition that affects virtually all individuals sometime after the age of 40 is *presbyopia*, a condition caused by hardening of the lens that accompanies the aging process. Because the lens generates new lens cells throughout life and none is lost, the lens becomes compacted and loses its elasticity. As a result, the accommodation reflex is greatly attenuated. Bifocal lens are used to compensate for presbyopia by incorporating a concave lens on the top to assist in far vision and a convex lens on the bottom to assist in near vision.

brain stem, whereas loss of the direct reflex may involve only more distal components of the visual or oculomotor pathways.

In addition to simple autonomic reflexes, a number of *complex visual motor reflexes* can be automatically triggered when seeking and tracking a visual target. All eye movements occur in a coordinated fashion in both eyes, in the form of *vergence movements* (when the two eyes move toward each other as part of the accommodation reflex) or in the form of *conjugate movements* (in which the two eyes move in the same direction). These coordinated movements are achieved by joint action of yoked pairs of agonist and antagonist motor nuclei. The coordination is achieved by groups of interneurons located in areas of the reticular formation that surround the motor nuclei of cranial nerves IV and VI. Two specific areas, called *gaze centers*, have been identified. Fibers connect the gaze centers and the appropriate motor nuclei through a midline fiber bundle, the *medial longitudinal funiculus* (MLF).

Horizontal eye movements are coordinated through the *horizontal gaze center* (the paramedian pontine reticular formation, or PPRF) located near cranial nerve IV. Vertical eye movements are coordinated in the *vertical gaze center*, which is thought to be located near cranial nerve VI.

Eye movements are very stereotyped and consist of either *saccades* or *smooth pursuit movements*. Saccades are quick, jerky movements useful in searching for visual targets. This type of movement appears to be preprogrammed by the brain based on the distance the eye has to move to position the desired target on the fovea. Saccades are ballistic in that no corrections in the speed or direction of the movement can be made after the movement is initiated. Smooth pursuit movements are maintained movements that are capable of moving the eyes to continuously follow a moving visual target or moving the eyes across a series of stationary visual targets, as in reading. These movements are not ballistic in that

their speed and direction can be modified continually. Saccades and smooth pursuit movements often occur in combination. A saccade may be used to find a novel target; smooth pursuit movements would then keep it focused on the fovea, adjusting for movement of the object or movement of the observer.

Rhythmic, repeating combinations of a long smooth pursuit followed by a saccade are called *nystagmus*, which is named after the direction of the fast (saccade) component (*e.g.*, left beating nystagmus). Nystagmus would normally be observed in an ice skater who is spinning on the ice. The eyes of the skater would move in smooth pursuit to keep some targeted object focused on the fovea as long as possible and then a saccade would be used to reset the eye position. This reflex movement is called *vestibular nystagmus*, and it requires input from vestibular pathways.

Another condition in which nystagmus normally occurs is when the observer is stationary but the visual field is continuously moving in one direction, as happens when one looks out the window of a moving train. The eyes follow the visual field with a smooth pursuit movement until they reach the limit of the movement range, then a saccade is used to reset. This is called *opticokinetic nystagmus*.

An additional cortical area has been identified that plays an important, although not completely understood, role in initiating and/or influencing smooth pursuits, saccades, and opticokinetic nystagmus. This area is called the *frontal eye field* (Brodmann's area 8) and is located immediately frontal to the premotor area. There is strong evidence that the frontal eye field is necessary for ipsilateral voluntary saccades, but it may also play additional roles in other eye movements, as well. The *cerebellum* provides additional input for the fine-tuning of saccades and smooth pursuit movements.

Pathologic nystagmus occurs when inappropriate movements are observed that do not match the velocity or direction of the intended visual target. Vestibular

Clinical Note

Diplopia and Ptosis

Patients with *ophthalmoplegia* (lesions of cranial nerve III) usually present with an initial complaint of seeing double (*diplopia*). This visual abnormality results from the loss of conjugate eye movements and the inability to maintain the image of the visual target on foveae of both eyes simultaneously. Remember that the visual cortex always

receives a double image, one from each eye, but these images are normally fused by cortical processing into a single three-dimensional image. With loss of conjugate eye movements, the two images are no longer in normal register on the retinotopic map of the visual cortex, and these higher centers cannot fuse the two images. With time, the cortex will try to make adjustments by

Clinical Note (continued)

suppressing the image from the abnormal eye. The patient may then no longer have double vision but will become psychically blind in the affected eye, and stereopsis (depth perception) will be compromised. The complaint of diplopia usually denotes a recent lesion.

Other symptoms of ophthalmoplegia demonstrated by a physical exam may include *strabismus* (the lack of parallelism of the ocular axes). Specifically, the affected pupil will be displaced laterally (*lateral strabismus*) and downward because of the unopposed action of the muscles innervated by cranial nerves IV and VI. Because there is loss of innervation to the levator palpebrae, the eyelid will be partially closed, a condition called *ptosis*. The pupil will be dilated (*mydriasis*) because of the loss of parasympathetic innervation to the sphincter muscle, leaving the unopposed action of the sympathetic innervation of the dilator muscles. The loss of parasympathetic innervation to the eye can also be demonstrated by testing for the pupillary light reflex and accommodation reflex, both of which will be absent in ophthalmoplegia. Because the parasympathetic fibers of the ocular motor nerve tend to lie near the outer surface of the nerve, they are the most vulnerable to compression lesions. Thus, a less severe compression would be reflected primarily in a fixed (unresponsive) and dilated pupil whereas a more severe lesion would cause the loss of all oculomotor nerve function.

Lesions of the oculomotor tract in the midbrain may involve surrounding structures. In the basal midbrain, a single lesion may compromise both the exiting oculomotor fibers and the surrounding motor pathways that transmit information from the motor cortex to muscles on the contralateral side of the body. This would result in *ipsilateral ophthalmoplegia* and *contralateral hemiplegia* (partial paralysis), a condition known as *Weber's syndrome*.

Ptosis can also be caused by loss of sympathetic innervation to the tarsalis muscle of the eyelid, a condition called *Homer's syndrome*. This condition can be distinguished from ophthalmoplegia because sympathetic neurons innervate pupillodilator muscles as well as the tarsalis muscle. A patient with Horner's syndrome would have ptosis (usually not as severe as with ophthalmoplegia) and the pupil would be constricted rather than dilated, as it would be with ophthalmoplegia.

Loss of input from cranial nerve IV can also cause diplopia, particularly when the patient tries to walk down stairs or read, activities that require downward gaze. A patient may also tilt his or her head slightly toward the side of the lesion in an attempt to adjust for the elevated and intorted position of the affected eye. Some loss of superior oblique muscle action may be compensated by actions of unaffected superior and inferior rectus muscles.

A marked *medial strabismus* can result from lesions of the abducens nerve caused by the loss of innervation to the lateral rectus muscles. A patient with medial strabismus tends to keep his head turned toward the side of the lesioned nerve to compensate for the position of the affected eye and thus avoid diplopia.

---

lesions are the most common cause of the pathologic appearance of spontaneous nystagmus.

Suggested Readings

Dowling JE. *The retina: an approachable part of the brain*. Cambridge, MA: Belknap Press of Harvard University Press, 1987.

Fuchs AF, Kaneko CRS, Scudder CA. Brainstem control of saccadic eye movements. *Annu Rev Neurosci* 1985; 8:307–337.

Fukushima K, Kaneko CRS, Fuchs AF. The neuronal substrate of integration in the oculomotor system. *Prog Neurobiol* 1992; 39:609.

Hubel DH. *Eye, brain and vision*, Scientific American library series No. 22, New York: Freeman, 1998.

Land EH. The retinex theory of color vision. *Set Am* 1977; 237(6): 108–128.

Lisberger SG, Morris EJ, Tychsen L. Visual motion processing and sensory-motor integration for smooth pursuit eye movements. *Annu Rev Neurosci* 1987; 10:97–129.

Livingstone M, Hubel D. Segregation of form, color, movement, and depth: anatomy, physiology and perception. *Science* 1988; 240:740–749.

Nathans J. Rhodopsin: structure, function and genetics. *Biochemistry* 1992; 31:4923.

Palczewski K, Benovic J. G-protein-coupled receptor kinases. *Trends Biochem Sci* 1991; 16:387.

Rodieck RW. *The vertebrate retina: principles of structure and function*. San Francisco: Freeman, 1973.

Stone J, Dreher B, Leventhal A. Hierarchal and parallel mechanisms in the organization of visual cortex. *Brain Res Rev* 1979; 1: 345–394.

Stryer L. The molecules of visual excitation. *Sci Am* 1987; 235:42.

The Vestibular System

DIANNA A. JOHNSON

KEY POINTS

- The auditory and vestibular systems both occupy the inner ear cavity and both rely on analogous cell types known as *hair cells* to convey information about sound or position and movement of the head.
- Transduction of mechanical energy into bioelectrical impulses in hair cells is achieved by bending *stereocilia*, opening potassium channels, and causing changes in membrane polarity in the form of *generator potentials* that trigger transmitter release.
- The coordinated response of the *semicircular canals* provides information about head position

and movement, encoded in the pattern of increases and decreases in the tonic firing rates of vestibular neurons.
- Hair cells in the semicircular canals provide information about angular acceleration in three directions; cells in the *utricle* and *saccule* provide information about static head position and about linear acceleration in all directions.
- Information from the *vestibular organs* is sent to the vestibular nuclei and then to motor nuclei of the cerebellum and brain stem in order to coordinate eye movements and maintain posture.

A sense of balance is derived primarily from specialized receptors located in an area near the entryway or vestibule of the inner ear, hence the name *vestibular organ* (Fig. 1). The principal function of the vestibular system is to provide direct information not to the cortex, but to motor centers in the spinal cord, brain stem, thalamus, and cerebellum so that involuntary reflex movements can be initiated and the body can remain balanced.

Vestibular input is also important in coordinating head and body movements and in moving the eyes so that they remain fixed to a point in space when the head is moving. Only in unusual circumstances does information from the vestibular system reach conscious levels (*e.g.*, when the sense of balance is disrupted). In this instance, relay pathways in the brain stem evoke sensations of dizziness and nausea.

The inner ear consists of a series of cavities that form a bony labyrinth in the petrous portion of the temporal bone (see Fig. 1). Within the bony labyrinth lies a series of fluid-filled sacks or tubes formed primarily from simple epithelial cells. Regions of the epithelium are specialized and contain receptors for hearing in the *auditory portion* of the inner ear and for balance in the *vestibular portion*. The vestibular organs consist of three *semicircular canals* and two otolith organs, the *utricle* and the *saccule*.

Auditory and vestibular systems share many functional properties. Both utilize *hair cells* to detect movement. As discussed in Chapter 53, hair cells in the auditory system respond to movement in the fluid of the auditory canal caused by sound waves. Hair cells in the vestibular system respond to movement in the fluid of the vestibular organs caused by head movement or gravity.

STRUCTURE AND FUNCTION OF HAIR CELLS

The term *hair cell* is a misnomer because the "hairs" projecting from the free end of the cell are actually a group of microvilli called *stereocilia*. Each hair cell of the vestibular sensory epithelium also has a single modified

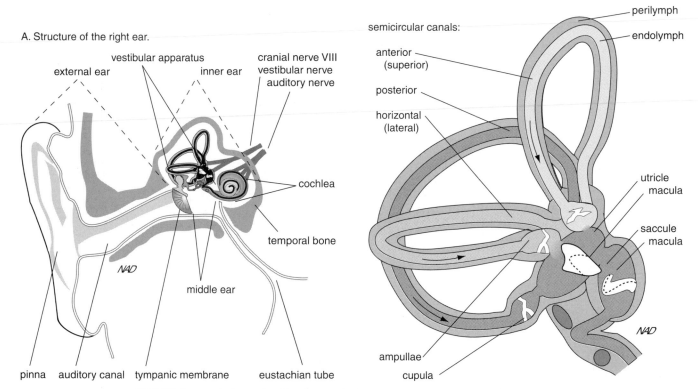

FIGURE 1   Vestibular and auditory organs in the inner ear. (A) Anterior view of the right ear: The external and middle ear are separated by the tympanic membrane or eardrum. The air-filled middle ear contains three tiny ear ossicles that transduce sound vibrations from the tympanic membrane to the oval window in the inner ear and to the cochlea. The vestibular apparatus occupies the vestibule of the inner ear. The vestibular and auditory nerves join to form cranial nerve VIII as it projects to the medulla. (B) Vestibular organs of the right ear: Three semicircular canals and two otolith organs (the utricle and the saccule) are formed by membranous sacs filled with endolymph (blue) and surrounded by perilymph (gray). The cupula within the semicircular canals, and the macula in the otolith organs represent areas of specialized epithelium containing sensory hair cells.

cilium called the *kinocilium* (Fig. 2). Stereocilia vary in length; the longest is next to the kinocilium, and each successive one is shorter. Rows of stereocilia are connected tip to tip by extracellular filaments that act as minute elastic springs.

When stereocilia are deflected toward the kinocilium (or the tallest microvilli, in the case of auditory receptors), tension on the connecting filaments is increased, opening potassium channels within the cell membrane and depolarizing the hair cell. Ordinarily, an increase in potassium conductance would hyperpolarize a cell, but the tips of hair cells are immersed in a special fluid, *endolymph*, containing an unusually high potassium concentration (150 mmol/L) and low sodium concentration (1 mmol/L) compared to that found in the cytoplasm (140 mmol/L and 9 mmol/L, respectively). Thus, opening of potassium channels causes an inward flow of potassium. The resulting depolarization affects the entire hair cell and causes release of the neurotransmitter, *glutamate*, from synapses at the base of the cell. Glutamate depolarizes the adjacent afferent neuron, which generates a series of action potentials.

When stereocilia are deflected away from the kinocilium, the hair cell is hyperpolarized. This happens because, in the normal resting state with stereocilia in an upright position, there is some minimal tension on the connecting fibers, causing some (perhaps 10%) of the potassium channels to remain open. With displacement of stereocilia away from the kinocilium, resting tension is relaxed, potassium channels close, and the cell hyperpolarizes. As a result, less glutamate is released and fewer action potentials are generated in the afferent nerve. Movement in an orthogonal direction from the kinocilium results in no change in tension on the connecting fibers and thus no change in frequency of action potentials in the afferent nerve.

THE SEMICIRCULAR CANALS

The relative position of the kinocilium establishes the *functional polarity* of the hair cell—namely, which direction of movement excites the cell and which inhibits the cell. Directional sensitivity of each of the individual vestibular organs is determined by the specific orientation of hair cells within the organ and the overall shape of the organ. In the semicircular canals, hair cells are clustered in a thickened zone of epithelium, the *ampullary crest*. A gelatinous, diaphragm-like mass, the cupula covers the kinocilia and stereocilia of the hair cells and stretches to the roof of the ampulla. When the head is rotated, the inertia of the fluid in the semicircular canals generates a force against the stereocilia, causing them to bend.

The three pairs of semicircular canals are orthogonally oriented with respect to one another, and matched pairs on either side of the head respond to similar angles of movements. However, the response of each member of the pair is equal and opposite. The functional polarity of the hair cells in the pair of lateral or *horizontal canals* is shown in Fig. 3. In this example, the head turns to the left but the endolymph lags behind because of inertia, causing the fluid to be shifted or displaced to the right. This moves the stereocilia in the left canal in an excitatory direction and stereocilia in the right canal in an inhibitory direction. The opposite occurs when the head is turned to the right.

The *anterior semicircular canal* on one side lies approximately in the same plane as the *posterior canal* on the opposite side, and they act in concert as a functional pair similar to the pair of horizontal canals. The sum effect is that when the head tilts or turns in any of the three ordinal directions, the brain receives two reports: an increase in the firing rate of vestibular nerve fibers from one side and a decrease from the opposite side.

By comparing the activity of all three pairs of canals, the brain can perform a vector analysis and compute the precise head position during any turning movement. The semicircular canals signal changes in *angular* or *rotational acceleration* and do not transmit information about steady-state head position.

OTOLITH ORGANS: THE UTRICLE AND THE SACCULE

The utricle and saccule each contains a region of specialized epithelium, called the *macula*, which is analogous to the ampullary crests of the semicircular canal. Cilia of hair cells within this region project into the *otolithic membrane*, an overlying gelatinous matrix studded with small accretions of calcium called *otoliths*. The macula of the utricle lies roughly in the horizontal plane when the head is held erect, so that otoliths rest directly upon it. If the head is tilted or undergoes linear acceleration in the horizontal plane (*e.g.*, when accelerating in a car), the otoliths deform the gelatinous mass, which in turn bends the cilia of the hair cells. Similar mechanisms operate in the saccule; however, the macula of the saccule is oriented vertically when the head is in its normal position and thus selectively responds to vertically directed linear force (*e.g.*, when riding in an elevator). Because of the constant gravitational pull on the otoliths, hair cells from both of these organs transmit tonic information about the position of the head in space as well as

A. Vestibular hair cell

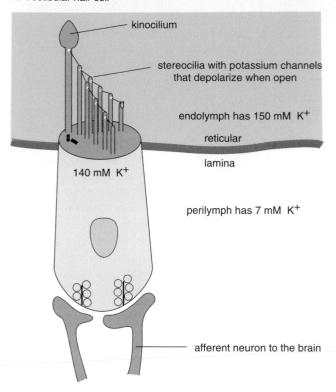

- kinocilium
- stereocilia with potassium channels that depolarize when open
- endolymph has 150 mM K⁺
- reticular
- lamina
- 140 mM K⁺
- perilymph has 7 mM K⁺
- afferent neuron to the brain

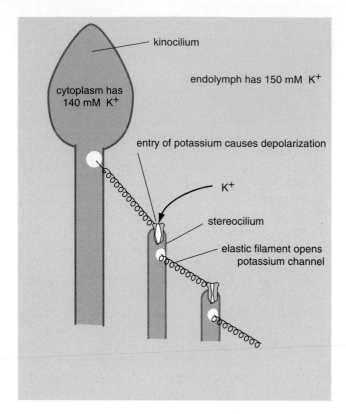

- kinocilium
- endolymph has 150 mM K⁺
- cytoplasm has 140 mM K⁺
- entry of potassium causes depolarization
- K⁺
- stereocilium
- elastic filament opens potassium channel

B. no movement   C. movement toward kinocilium   D. movement away from kinocilium

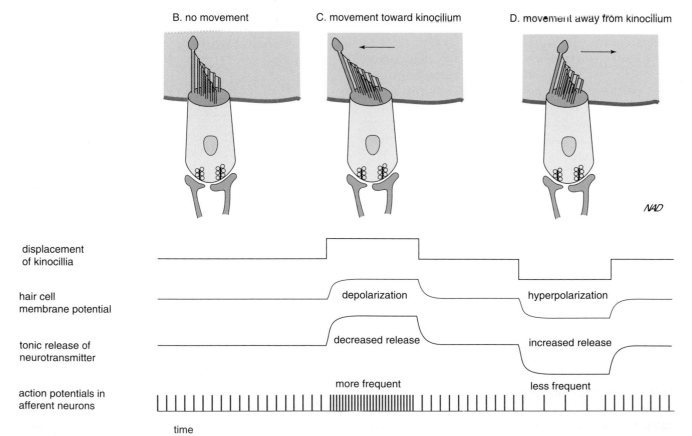

NAD

displacement of kinocillia

hair cell membrane potential   depolarization   hyperpolarization

tonic release of neurotransmitter   decreased release   increased release

action potentials in afferent neurons   more frequent   less frequent

time

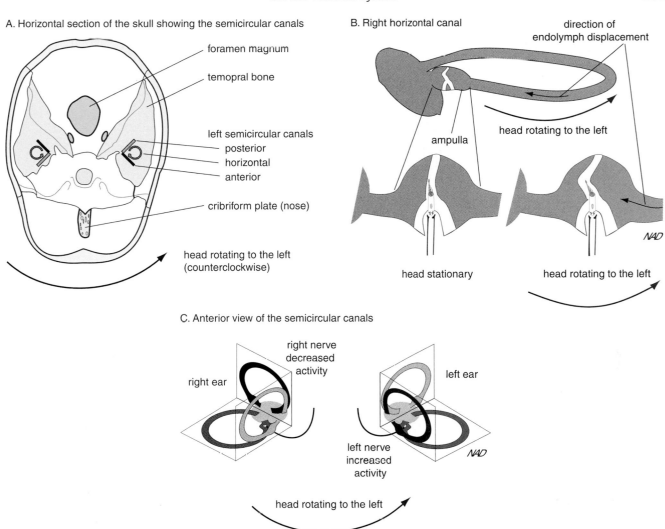

A. Horizontal section of the skull showing the semicircular canals

- foramen magnum
- temporal bone
- left semicircular canals
  - posterior
  - horizontal
  - anterior
- cribriform plate (nose)

head rotating to the left
(counterclockwise)

B. Right horizontal canal

direction of
endolymph displacement

head rotating to the left

ampulla

head stationary

head rotating to the left

NAD

C. Anterior view of the semicircular canals

right nerve
decreased
activity

right ear

left ear

left nerve
increased
activity

NAD

head rotating to the left

FIGURE 3 Paired responses of semicircular canals. (A) Three pairs of semicircular canals as they are positioned in the head; the two horizontal canals (dark blue bars) lie in the same plane and form a functional pair. The same is true for right anterior and left posterior canals (light blue bars). (B) As the head rotates to the left, the endolymph lags behind, dragging the cupula and hair cell tips to the right. (C) Rotation of the head to the left excites hair cells in the left horizontal canal and inhibits hair cells in the right horizontal canal.

linear acceleration due to changes in gravity, acceleration, or deceleration of movement or tilting of the head.

Unlike the simple parallel arrangement of hair cells in the semicircular canals, the organization of hair cell clusters in both the utricle and the saccule is more complex (Fig. 4). A curved border called the *striola* crosses the surface of the macule in both the utricle and the saccule. The functional axes of all hair cells are oriented at right angles to the striola, producing an overall fan-shaped arrangement. Thus, a tilt of the head in any direction will activate the subpopulation of hair cells that has a corresponding axis of polarity, inhibit those with opposite polarity, and have no effect on those orthogonally aligned. These sets of signals from the utricle and saccule allow the brain to continuously monitor the position of the head in space.

FIGURE 2 Structure and function of vestibular hair cells. (A) Hair cells are modified epithelial cells with the tips of the "hairs," or stereocilia, immersed in the unique fluid endolymph, which has a potassium concentration higher than intracellular concentrations. When the cilia are deflected toward the large kinocilium, tiny elastic fibers open potassium channels in the tips of the stereocilia, allowing potassium to move in and depolarizing the cell. (B) to (D). Movement changes the membrane potential, regulates the release of the neurotransmitter (glutamate), and changes the frequency of action potentials in the afferent neurons.

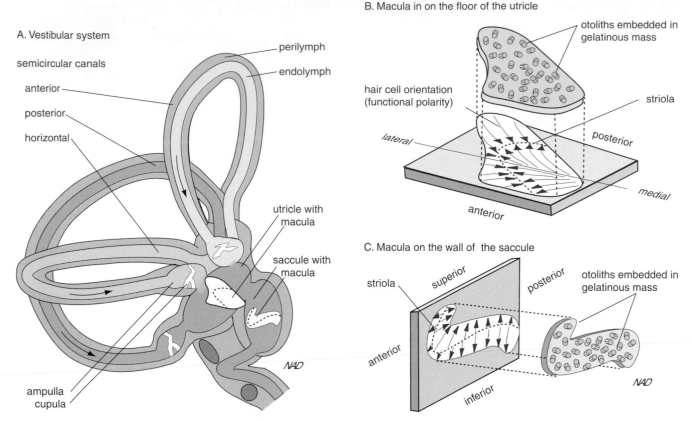

A. Vestibular system

semicircular canals

anterior

posterior

horizontal

perilymph

endolymph

utricle with macula

saccule with macula

ampulla
cupula

B. Macula in on the floor of the utricle

otoliths embedded in gelatinous mass

hair cell orientation (functional polarity)

striola

lateral

posterior

anterior

medial

C. Macula on the wall of the saccule

striola

superior

posterior

otoliths embedded in gelatinous mass

anterior

inferior

FIGURE 4   Structure of the utricle, saccule, and semicircular canals. (A) Structure of the vestibular system showing the utricle and the saccule. (B) The macula, on the floor of the utricle, detects linear acceleration in the horizontal plane. The hair cells lie in rows with their kinocilia pointed in the direction of the arrows. Their tips lie in a gelatinous mass (shown lifted), which is studded with otoliths, tiny calcium-carbonate crystals. Acceleration shifts these little weights, bending the cilia of the hair cells and changing the frequency of action potentials in the afferent neurons. (C) The macula, on the vertical wall of the saccule, detects vertical linear acceleration; it tells which way is up. Similar to the utricle, the macula relies on the movement of otoliths.

VESTIBULAR NUCLEI AND CENTRAL VESTIBULAR PATHWAYS

Hair cells have synaptic endings and release the neurotransmitter glutamate in response to appropriate stimulation. Even so, historically they have not been classified as neurons. By convention, first-order neurons of the vestibular pathway are designated as the afferent fibers originating from cell bodies in the *vestibular ganglion* located near the base of the utricle and saccule (see Fig. 1). The peripheral processes receive postsynaptic stimulation from hair cells, and resulting action potentials are propagated along the peripheral dendrite and on to central axonal processes, both of which are myelinated. The central vestibular fibers join central auditory fibers to form cranial nerve VIII as it projects to the brain stem at the junction between the pons and the medulla.

Cell bodies of the *vestibular nuclei* occupy a substantial area on the lateral aspect of the medulla (Fig. 5). Four divisions of the nucleus are recognized. The lateral vestibular nucleus receives input from all vestibular organs and sends out descending fibers in the *lateral vestibulospinal tract*. These second-order vestibular processes terminate on motor neurons in the spinal cord and provide the excitatory drive to maintain body posture. The lateral vestibular nucleus also receives inhibitory innervation from the cerebellum. This serves to coordinate and control the normal excitatory output of the lateral vestibular nucleus. Patients who have head injuries that damage the incoming cerebellar fibers to the vestibular nuclei suffer a pronounced motor imbalance in the extremities called *decerebrate rigidity*. *Decerebration* (loss of input from higher centers) removes the inhibitory input from the cerebellum and other cortical areas, leaving the motor neurons exposed

A.

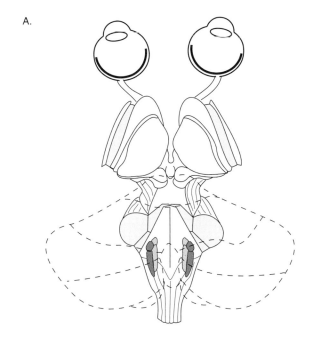

B.

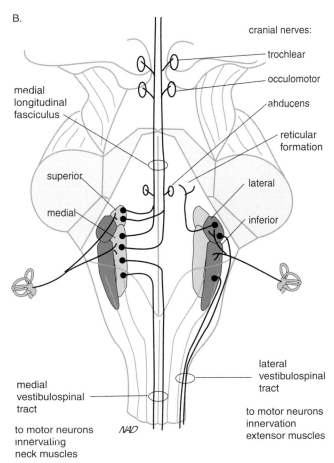

cranial nerves:

trochlear

occulomotor

abducens

reticular
formation

medial
longitudinal
fasciculus

superior

medial

lateral

inferior

lateral
vestibulospinal
tract

medial
vestibulospinal
tract

to motor neurons
innervation
extensor muscles

to motor neurons
innervating
neck muscles

NAD

FIGURE 5 Central vestibular pathways. (A) Position of the vestibular nuclei in a dorsal view of the brain with the cerebral cortex removed and the cerebellum rendered transparent. (B) Major connections of the vestibular nuclei. (Left) Superior and medial vestibular nuclei (light blue). The main input is from the semicircular canals.

to the unopposed excitatory drive of the vestibular nucleus. Under the constant stimulation from motor neurons, muscles of the legs and arms remain rigidly contracted, unable to relax or participate in normal movements.

Medial and superior vestibular nuclei receive information primarily from the semicircular canals. Fibers from the medial nucleus send their second-order fibers in the *medial vestibulospinal tract* to motor neurons located in upper levels of the spinal cord. This pathway elicits reflex movements in neck muscles in order to stabilize head position while walking and, along with other inputs, to coordinate head movements with eye movements. Additional output fibers from the medial nucleus join processes from the superior nucleus in forming a major pathway that travels within the medial longitudinal fasciculus (MLF), one of the most prominent fiber bundles within the brain stem.

As described in Chapter 51, fibers within the MLF project to the three pairs of motor nuclei that control eye movement, enabling the eyes to move appropriately as the head moves. When the head is turned to one side, eyes reflexly rotate in the opposite direction. Without vestibulo-ocular reflexes being mediated through the MLF, visual images are not stabilized on the retina during body movement and visual performance is greatly impaired. Fibers in the MLF are highly myelinated, which is one reason why they serve as a prominent landmark within the less myelinated regions of the brain stem. Myelination increases conduction velocity of the fibers, a much-desired attribute for a reflex pathway such as the MLF, which requires a rapid response time. Unfortunately, the heavy myelination makes it a target for demyelinating diseases such as *multiple sclerosis* and accounts for some of the visual effects (*e.g.*, diplopia and blurred vision) that are often associated with this disease.

The *inferior vestibular nucleus* receives information from all vestibular organs as well as from the cerebellum. It provides second-order fibers for: (1) the *vestibulospinal pathway*, which helps maintain posture, and (2) the *vestibuloreticular pathway*, which terminates in various areas of the reticular formation of the brain stem and helps activate protective reflexes. The latter

---

Both superior and medial nuclei interact to coordinate eye movements with rotational head movements by way of the ascending medial longitudinal fasciculus. The medial nuclei also coordinate neck movements by way of the descending medial vestibulospinal tract. (Right) The lateral and inferior vestibular nuclei (dark blue) receive information from the utricle and saccule as well as the semicircular canals. They send information down the lateral vestibulospinal tract to axial motor neurons controlling posture and balance. The lateral tract also sends commands to the reticular formation for protective reflexes such as the gag reflex. The nuclei also communicate with the cerebellum (not shown.)

Clinical Note

Vertigo and Lesions of Cranial Nerve VIII

Because the sense of balance relies on comparative input from vestibular organs from both ears, damage to either left or right components due to head trauma or disease will lead to an abnormal imbalance of vestibular information. This imbalance can cause a sensation of dizziness (general spatial disorientation) and/or vertigo (the specific perception of a spinning or turning motion of the body in the absence of any actual movement). For example, a tumor of cranial nerve VIII (acoustic neuroma) not only will cause hearing deficits but may also compress the vestibular component of the nerve and reduce the frequency of impulses from the ipsilateral vestibular fibers or block their activity entirely. The vestibular nuclei will then consistently receive a higher impulse frequency from the intact side, which will be interpreted as continual turning of the head away from the side of the lesion. A patient with an acoustic neuroma will often complain of dizziness, nausea, and vertigo. The patient may also exhibit abnormal nystagmus as a result of activation of the reflex eye movements that customarily produce opticokinetic or vestibulokinetic nystagmus.

pathway probably serves as the afferent arm of the gag response to overstimulation of the vestibular organs, which occurs during air or sea sickness. The debilitating dizziness and nausea experienced by astronauts during their first hours in flight result from heightened activity in the vestibuloreticular pathways in response to extremes in linear acceleration coupled with zero gravity.

Suggested Readings

Anniko M. Functional morphology of the vestibular system, in: Jahn AF, Santos-Sacchi, Eds., *Physiology of the ear*. New York: Raven Press, 1988, pp. 457–504.

Gresty MA, Bronstein AM, Brandt T, Dietrich M. Neurology of otolith function. *Brain* 1992; 115:647–673.

Wilson VJ, Jones MG. *Mammalian vestibular physiology*. New York: Plenum Press, 1979

The Auditory System

DIANNA A. JOHNSON

KEY POINTS

- Perceptions of *pitch* and *intensity* are reflections of the frequency and amplitude of audible sounds.
- Sound waves in the auditory canal are transmitted through the eardrum and continue through the three ossicles of the middle ear.
- The *cochlea* of the inner ear consists of fluid-filled ducts that conduct sound waves and cause displacement of stereocilia of auditory hair cells.
- Structural characteristics of the *basilar membrane* in the cochlea result in *tonotopic mapping* of sound pitch so that higher pitches are encoded at the base of the basilar membrane and lower pitches are encoded at its apex.

- *Inner* and *outer auditory hair cells* in the *organ of Corti* transmit sound information to the primary auditory neurons in the *spiral ganglia*.
- Cochlear amplifications of sound are achieved by movement adaptations of outer hair cells.
- Central fibers of the auditory pathway ascend bilaterally to the *superior olivary nucleus* in the brain stem, providing binaural activation of olivary neurons.
- Perception of pitch relies on tonotopy and phase locking of cortical cells arranged in isofrequency bands.
- Interaural comparisons in sound intensity, sound delay, and phase delay permit computation of sound localization in the horizontal plane.

Like the visual system, the auditory system permits analysis of energy (in this case, sound energy) emitted or reflected from objects even when they are located at a considerable distance from the body. At the most primitive level, sound perception can provide a warning system of potentially dangerous objects or predators in the vicinity. The auditory system also provides a mechanism for object recognition. Although a number of animal species have much more acute hearing, humans exhibit a more highly developed sense of auditory recognition involving extensive cortical circuits to interpret the complex vocalizations required for spoken communication. Unfortunately, much more is understood about the mechanical transduction of sound waves and auditory acuity than is currently known about higher cortical functions responsible for auditory recognition.

Sound is defined as audible variations in air pressure, created by alternating areas of compressed air and rarefied air. A plot of the ambient air pressure as sound waves pass a given point demonstrates that these changes occur as smooth, wavelike transitions and that a particular sound produces a characteristic wave form or *frequency* (Fig. 1). Frequency is measured as cycles per second in units called *hertz* (Hz). The human auditory system can respond to sounds between 20 and 20,000 Hz. The sense of *pitch* relates directly to sound frequency; the higher the frequency, the higher the perceived tone. The perceived *loudness* of a sound is a property of the *amplitude* of the sound wave, reflecting the maximum change in air pressure in either direction. It is measured in units called *decibels* (dB), which are calculated as the logarithmic ratio of the test pressure and a reference pressure (minimum audible pressure change generated by a 2000-Hz tone). A test pressure 10 times greater than the reference would have a loudness of 20 dB; a pressure 100 times greater would equal 40 dB. Conversational speech is equivalent to 65 dB. Greater than 100 dB can damage the structures of the inner ear.

Sounds with frequencies above the range of human hearing are called *ultrasound*, which, when mechanically produced, can be used for ultrasonic cleaning or for ultrasonic imaging of internal organs for medical purposes. Some animals (*e.g.*, dogs) have modifications in outer and inner ear structures that allow them to hear these frequencies.

Frequencies below human hearing are called *infrasound*. They can be produced and perceived by some animals, such as whales, who use them for communication over great distances underwater. Infrasound is also inadvertently produced by various mechanical devices, such as air conditioners, aircraft, and automobiles. Intense infrasound does not cause hearing loss as high levels of audible sound do; nonetheless, it can produce resonances in body cavities, including the chest and stomach, and cause dizziness, nausea, headache, and even permanent damage to internal organs.

SOUND TRANSMISSION THROUGH THE MIDDLE AND INNER EARS

Sound waves enter the *auditory canal* after being funneled and reflected by components of the external ear, or *pinna* (Fig. 2). The canal extends 2.5 cm into the skull and ends at the *tympanic membrane* (eardrum), which forms the boundary between the outer and middle ears. The middle ear is an air-filled cavity containing three small bones, the *ossicles*, which are connected in series to the tympanic membrane. Sound waves in the auditory canal cause the tympanic membrane to vibrate and thus transmit the vibrations to the ossicles. The first ossicle, the *malleus* ("hammer"), forms a connection between the tympanic membrane and the *incus* ("anvil"). The incus forms a flexible connection with the *stapes* ("stirrup"), which is attached to the membrane of the *oval window* separating the middle ear from the fluid-filled inner ear. The air in the middle ear is normally self contained, held there by a valve in the eustachian tube connecting the middle ear to the mouth cavity. When the air pressure outside the middle ear is suddenly decreased, as during takeoff in an airplane, the higher air pressure in the middle ear presses (sometimes painfully) against the tympanic and oval window membranes. Temporarily opening the valve by yawning or stretching the jaw allows the pressure to equalize and the pain to subside.

THE COCHLEA

The inner ear houses the *cochlea*, which consists of a bony tube coiled in a tight spiral (see Figs. 2 to 4). The inside of the tube is partitioned longitudinally by two continuous membrane sheets, *Reissner's membrane* and the *basilar membrane*. This divides the main cochlear tube into two large ducts filled with *perilymph*; the *scala vestibuli*, bounded by Reissner's membrane; and the *scala tympani*, bounded by the basilar membrane. The separation between the two membranes forms a third, smaller cavity filled with *endolymph* called the *scala media*, which contains the auditory receptor cells. At the base of the cochlea, each of the two larger scalae is separated from the middle ear by small membranes: the scala vestibuli by the *oval window*, and the scala tympani by the *round window*. At the apex of the cochlea, the two scalae are connected by a hole in the basilar membrane, called the *helicotrema*, which allows the perilymph fluid contents of these two scalae to mix with each other but not with the endolymph fluid of the scala media.

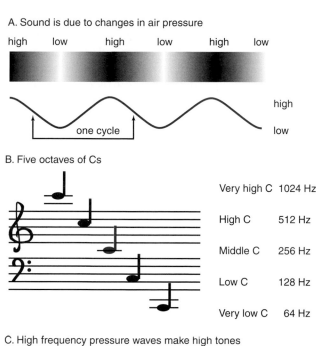

A. Sound is due to changes in air pressure

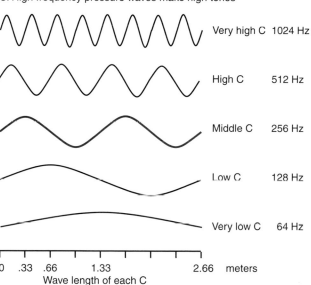

B. Five octaves of Cs

Very high C	1024 Hz
High C	512 Hz
Middle C	256 Hz
Low C	128 Hz
Very low C	64 Hz

C. High frequency pressure waves make high tones

Very high C	1024 Hz
High C	512 Hz
Middle C	256 Hz
Low C	128 Hz
Very low C	64 Hz

0   .33   .66   1.33   2.66   meters
Wave length of each C

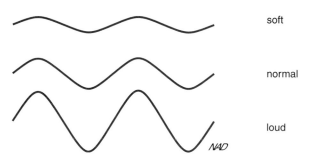

D. High amplitude pressure waves make loud sounds

soft

normal

loud

FIGURE 1   Characteristics of audible sound waves. (A) Sound waves are produced when a vibrating object, such as a bell, compresses and expands air. (B) Musical tones can be represented as notes on a staff or

The oval window transmits sound vibration from the stapes ossicle to the perilymph within the scala vestibuli. From there it spreads through the helicotrema, and the fluid in the scala tympani finally impinges on the round window. Movements of the round window are generally compensatory to those in the oval window, thus providing the necessary give within the otherwise closed fluid system of the cochlea. The Reissner's and basilar membranes form the flexible membrane boundaries of the scala media and allow it to be displaced by movements in the surrounding fluid. A continuous tone transmitted in this fashion causes a standing wave to form in the cochlear membrane system, in particular in the basilar membrane.

A standing wave produced in basilar membrane is not uniform along its full length (see Fig. 4). Structural features of the basilar membrane influence the wave form. The membrane is relatively narrow and stiff at its base, wide and floppy at its apex. Thus, low-frequency waves, associated with low-pitch sound, produce larger displacements in the apex region of the membrane, whereas high-frequency waves, associated with high-pitch sound, produce larger displacements in the basal portion. This means that hair cells located at different locations along the length of the basilar membrane will be more activated by certain pitches than by others. Different locations are maximally deformed by different frequencies. This relationship between position and pitch results in a *tonotopic organization* of the hair cells, with high pitch encoded by hair cells near the base and low pitch encoded by cells near the apex. Positional relationships are maintained among all neurons within the auditory pathway, much as the visual pathway is organized in a retinotopic fashion.

Primary auditory nerve cell bodies located in the *spiral ganglia* near the base of the cochlea send peripheral processes to innervate two types of auditory receptors: inner and outer hair cells. The arrangement of auditory nerve cell bodies within the ganglion and both peripheral and central processes maintain a tonotopic organization.

THE ORGAN OF CORTI

Auditory hair cells are arranged within a structure called the *organ of Corti* that rests on the basilar

as frequency of vibration in hertz (cycles per second.) Doubling the frequency raises the pitch one octave. (C) Pressure oscillations are plotted for these notes over a 7.8-ms interval. Because sound travels 340 m/s in air, one can calculate the wavelength as: wavelength (meters/cycle) = 340 m/s/frequency (Hz). Very low C is produced by a 64-Hz vibration, and only half a cycle occurs in 7.8 ms. Human hearing extends between 20 and 20,000 Hz, and it is the high pitches that are necessary to understand the consonants in speech. (D) Increasing the amplitude of the pressure changes will increase the loudness of the sound.

A. Structure of the right ear

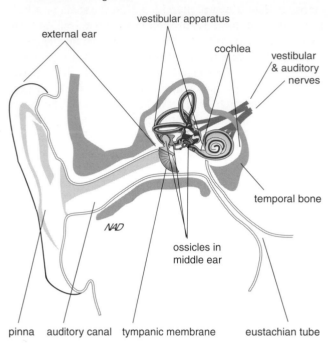

B. Structures of the middle ear

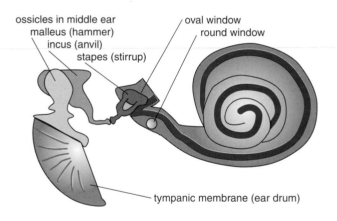

C. Movement of the middle ear structures

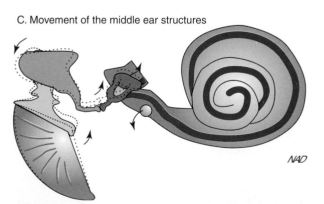

FIGURE 2   Structures of the middle ear. (A) Structure of the right ear showing the placement of the middle ear structures. (B) The three tiniest,

membrane within the scala media (Fig. 5). Multiple rows of *outer hair cells* are separated from a single row of *inner hair cells* by a supporting structure called the *columns of Corti*. Rows of cells are arranged linearly along the curved length of the cochlear structure. Both inner and outer hair cells respond to movement of the fluid in the cochlear duct systems, with movement in one direction causing depolarization and movement in the other direction causing hyperpolarization. As described in Chapter 52, extracellular filaments connect the tips of stereocilia protruding from the surface of hair cells. The stereocilia are of different lengths and are arranged in descending order on each cell. Increased tension on these connecting filaments is generated by movement of endolymph fluid, pushing the stereocilia toward the direction of the longest stereocilium. The increased tension opens potassium channels in the cell membrane. As in the vestibular system, the potassium concentration of endolymph surrounding the stereocilia is high compared to intracellular levels, so that increasing potassium conductance leads to an influx of potassium ions and depolarization. Movement of the stereocilia in the opposite direction releases normal resting tension on the connecting filaments, closes potassium channels, and causes hyperpolarization.

SOUND AMPLIFICATION AND ATTENUATION

The elaborate system of fluid-filled membrane ducts within the cochlea serves to amplify sound vibrations and to provide maximum acuity while at the same time protecting receptor cells from potential damage from high-intensity sounds. The tips of the stereocilia are embedded in a gelatinous meshwork called the *tectoral membrane*, which is connected to the *reticular lamina* at one end and to a bony protuberance called the *modiolus* at the other end (see Fig. 5). This arrangement produces a shearing action on the stereocilia during displacements of the basilar membrane, maximizing the transfer of small to medium movements of fluid while offering some protection against damage from exaggerated movements.

A second amplification mechanism is provided by special properties of the outer hair cells. Depolarization

hardest bones in the body are the ear ossicles. The malleus ("hammer") attaches to the back of the tympanic membrane (eardrum) and to the incus ("anvil"), which attaches to the stapes ("stirrup"). The foot of the stapes fits onto the membrane covering the oval window, which is hidden behind the stapes. (C) Air movements vibrate the large tympanic membrane, which vibrates the ear ossicles, and their lever action vibrates the fluid under the small oval window. The incompressible fluid of the inner ear bulges the membrane of the round window with each vibration. The air pressure on the tympanic membrane is amplified 20 times because of the lever action and the difference in area of the tympanic membrane and the oval window.

A. Cochlea with stapes displaced

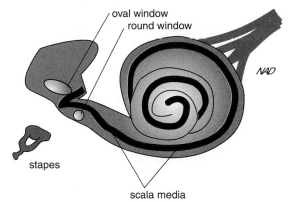

B. Structures of the middle ear

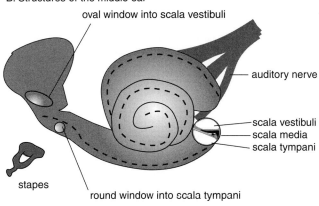

C. Cochlear ducts

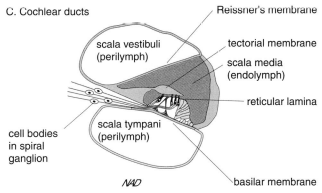

FIGURE 3 Structure of the cochlea. (A) Cochlea with stapes displaced; scala media filled with endolymph is shown in blue. (B) Cochlea with a small wedge removed showing the three canals or scala; the dotted line indicates the relative positions of the basilar membrane within the cochlea. (C) Enlargement of the cochlear ducts shows the hair cells extending into the endolymph.

of the cell by potassium in response to a sound wave stimulates intracellular *motor proteins*, causing the cell to lengthen. Because the hair cell is sandwiched between

the basilar membrane and the tectoral membrane, a lengthening of the cell increases the distance between the two membranes. Subsequently, the basilar membrane bends more and shearing forces generated by the tectoral membrane are increased. The net result is an amplification of the response (*cochlear amplification*) as the subsequent series of sound waves pass over the activated area.

Inner hair cells lack these motor proteins and do not change their length in response to stimulation. They also differ from outer hair cells in the extent to which they are innervated. Though outnumbered 5 to 1 by outer hair cells, the inner hair cells are much more richly innervated. Peripheral fibers from approximately 10 separate spiral ganglia neurons are committed to the innervation of each inner hair cell. The opposite is true for outer hair cells, which are innervated in groups of 10 or more by a single spiral ganglia neuron. Of the spiral ganglia cells, 95% innervate inner hair cells, and the remaining 5% innervate outer hair cells. Thus, the major source of auditory information is the inner hair cell, with perhaps the major function of the outer hair cells being limited to cochlear amplification.

One of the major protective mechanisms for the auditory system, called the *attenuation reflex*, involves not the cochlea per se, but other elements of the middle ear. The *tensor tympani muscle* is attached to the bone of the middle ear and to the malleus. Similarly, the stapedium muscle attaches to the stapes. Both muscles are innervated by brain stem structures that receive input from the auditory pathway as it projects to higher centers. In response to a loud sound, particularly at lower frequencies, motor neurons from the brain stem cause these muscles to contract, thus stabilizing the ossicles and reducing sound conductance to the inner ear. Sound attenuation serves to adapt the ear to continuous sound at high frequency, preventing saturation of the receptors and increasing the dynamic response range of the auditory system. It also serves to protect the structures of the inner ear against potential damage by overstimulation. Because the relay pathway is polysynaptic and involves central neurons, there is a response delay of 50 to 500 ms. Thus, there is no protection against very sudden increases in sound intensity, such as a shotgun blast.

CENTRAL AUDITORY PATHWAYS

The central pathway of the auditory system contains a large number of relay nuclei within the brain stem (Fig. 6). Central fibers from primary sensory neurons in the spiral ganglia project along with vestibular fibers in cranial nerve VIII and synapse first within the *dorsal* and *ventral cochlear nuclei* located near the pontomedullary junction. Ascending fibers from these nuclei project to both the

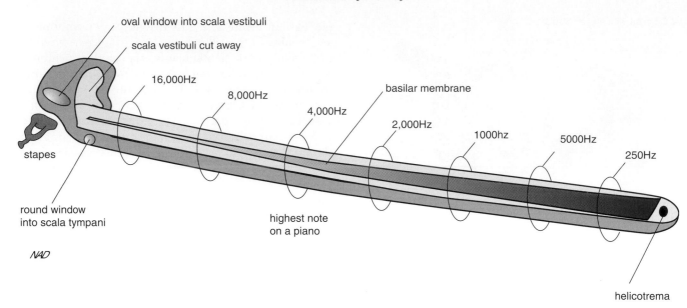

oval window into scala vestibuli

scala vestibuli cut away

16,000Hz

8,000Hz

4,000Hz

basilar membrane

2,000Hz

1000hz

5000Hz

250Hz

stapes

round window
into scala tympani

highest note
on a piano

NAD

helicotrema

FIGURE 4 Structure and functional relationships of the basilar membrane. The cochlea is shown rolled out with the scala vestibuli dissected away, leaving only the basal portion with the oval window. The scala media and the organ of Corti have also been removed, exposing the basilar membrane, with the underlying scala tympani. The basilar membrane (shown in blue) is narrow and stiff at the base and is wide and floppy near the apex, near the helicotrema. Low pitches vibrate the wide part of the membrane; high pitches, the narrow part. Note that the perilymph in the scala vestibuli and tympani are continuous at the opening provided by the helicotrema, allowing perilymph from the two scalae to mix. The endolymph in the scala media is completely isolated from these.

ipsilateral and contralateral *superior olivary nuclei.* Axons from cells in the superior olive travel along with other sensory fibers within the *lateral lemniscus* to reach the *inferior colliculus.* Collicular fibers synapse in the *medial geniculate nucleus* of the thalamus. Geniculate fibers represent the final projection to the *primary auditory cortex* located along the superior lip of the temporal lobe. Auditory fibers are represented bilaterally within the central pathway because of partial decussation of projections at several of the relay points within the brain stem. Thus, central auditory lesions rarely cause deafness in only one ear.

Like the visual cortex, the primary auditory cortex (Brodmann's areas 41 and 42) is composed of six layers: layer IV contains input fibers from the medial geniculate body; cells in layer IV project to small pyramidal cells in layers II and III before projecting to other cortical association areas and to layers V and VI before projecting back to lower brain stem structures. In comparison with the visual system, much less is known about how central pathways process auditory information. In general, we have a basic understanding of only three fundamental attributes of sound: *intensity*, *pitch*, and *location* in space (Table 1).

Table 1   Coding of Sound Intensity, Pitch, and Location

Intensity		
Loud sound	Large amplitude deflection in basilar membrane	More action potentials per neuron and more neurons firing
Soft sound	Small amplitude deflection in basilar membrane	Fewer action potentials per neuron and more neurons firing
Pitch		
High pitch	Vibrates base of basilar membrane	Activates appropriate neurons in tonotopic map
	Hair cells fire at phase-locked, high frequency	Cortical cells in isofrequency bands fire at phase-locked high frequency
Low pitch	Vibrates apex of basilar membrane	Activates appropriate neurons in tonotopic map
	Hair cells fire at phase-locked, low frequency	Cortical cells in isofrequency bands fire at phase-locked low frequency
Location		
Vertical plane	Determined by echo patterns created by pinna	
Horizontal plane	at sound onset	Interaural onset delay
	Change in location of a low-frequency continuous tone	Interaural phase delay
	Change in location of a high-frequency continuous tone	Interaural intensity difference

A. Cochlear ducts

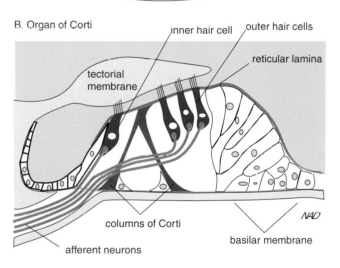

B. Organ of Corti

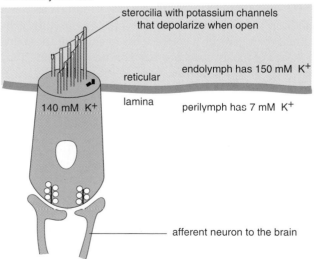

C. Auditory hair cell

SOUND INTENSITY

Information about sound intensity is reflected in both the firing rate of individual neurons and the total number of active neurons. Loud sound waves cause large deflections of appropriate portions of the basilar membrane. This produces correspondingly large numbers of action potentials in individual hair cells within that region, theoretically causing each hair cell to fire once in response to each sound wave that passes. Large deflections also impinge on surrounding regions of the basilar membrane and can depolarize adjacent hair cells to a lesser extent.

PITCH PERCEPTION

As described earlier, encoding information leading to the perception of pitch relies initially on the *tonotopic organization* of hair cells, an organizational plan that is maintained by all subsequent components of the auditory pathway. The tonotopic map of the auditory cortex is composed of groups of adjacent cells that respond maximally to a specific frequency (*characteristic frequency*) that is determined by the specific area of the basilar membrane providing the input. Cells with similar characteristic frequencies are arranged in a series of cortical bands called *isofrequency bands*, with lower frequencies represented most anteriorly within the primary cortical region and higher frequencies located more caudally.

In addition to receiving information from similar regions of the basilar membrane, cells within isofrequency bands share other similarities. Auditory neurons tend to fire at the same phase of any given sound wave. Thus, their response is *phase-locked* to some location on the wave and their firing rate reflects the frequency of the original sound. The firing rate then becomes an additional code for the perceived pitch of the sound to augment the code represented by the position of the cell within the tonotopic map. Neurons within isofrequency bands tend to fire in synchrony at a rate indicative of their characteristic frequency range.

FIGURE 5 Structure of the organ of Corti. (A) Cross section of the cochlear ducts showing the placement of the organ of Corti. (B) Organ of Corti enlarged. The sensory hair cells and endolymph are shown in blue. Reissner's membrane is very flexible and does not contribute to the mechanical properties of the system. Its major function is to separate the scala media from the scala vestibuli. The scala media is separated from the scala tympani by the stiff basilar membrane that establishes the properties of the system. The reticular lamina overlies the cells of the organ of Corti and provides a barrier between the endolymph of the scala media and the perilymph of the scala tympani. (C) The auditory hair cell is similar to the vestibular cell except it is responsive to much higher frequencies. Bending the stereocilia in the direction of the longest stereocilia opens potassium channels and depolarizes the hair cell, causing release of neurotransmitter.

A. Lateral view of cerebral cortex

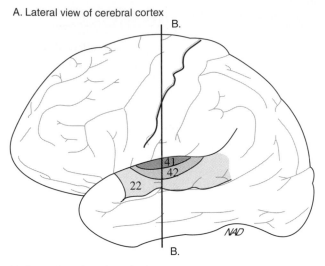

B. Coronal section of cerebral cortex

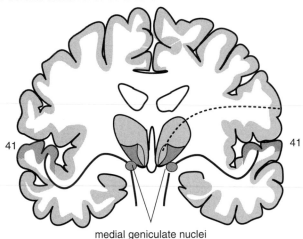

medial geniculate nuclei

C. Dorsal view of brain stem and mid brain

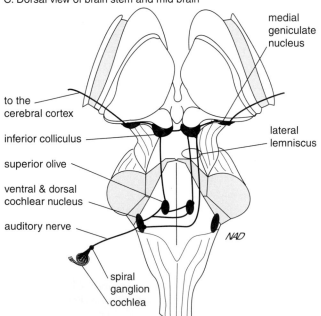

Both phase-locked and tonotopic codes are important because they compensate for the different weaknesses inherent in each type of coding. The tonotopic map works best when comparing different pitches of sounds having equal intensity. For example, it does not work well in differentiating between a soft, low pitch and a much louder, high pitch because both can displace the apex region of the basilar membrane to the same degree.

The phase-lock coding of pitch works best for low frequencies but fails at high frequencies. This is because the firing rate of a neuron is fast enough to match the wave frequency of lower pitches but not that of high pitches. At frequencies above 400 Hz, the intrinsic variability in the timing of the action potential is comparable to the time interval between successive waves. In this case, neurons tend to fire at random phases of the sound wave and provide no code of sound frequency. Thus, the frequency coding of the full audible range of sound relies on phase-locking at low frequencies, on both phase-locking and tonotopy at intermediate frequencies, and on tonotopy alone at high frequencies.

LOCALIZATION OF SOUND

Location of the source of sound plays a large role in central processing in the auditory system. *Localization of sound along the vertical axis* can be achieved using input from only one ear. The folds and bulges in the structure of the outer ear apparently produce reflections of sound waves as they enter the auditory canal. The lack of symmetry of the pinna allows for different echoing effects of sounds based on their angle of entry along a vertical plane. Sound entering the ear from below will have a relatively long time delay between the response to the direct sound and the reflected sound; sound from above will have a relatively short delay time. In this way, echo patterns can provide a code for the location of the sound in space. Similar mechanisms are used by bats (and

---

FIGURE 6 Central auditory pathways. (A) Lateral view of cerebral cortex shows the location of the coronal section below. (B) Coronal section shows the auditory projection from the medial geniculate nucleus to the cerebral cortex. (C) Dorsal view of brain stem and midbrain shows auditory pathways. The auditory-vestibular nerve is cranial nerve VIII, which enters the brain stem at the base of the pons. Auditory information is first processed in the ventral and dorsal cochlear nuclei, and from there the information ascends both ipsilaterally and contralaterally to the superior olive, inferior colliculus, and medial geniculate nucleus. Finally, auditory radiations carry the information to the superior lip of the temporal lobe, designated as Brodmann's areas 41, 42, and 22.

sonar devices on ships) when they emit sounds and analyze echoes in order to locate objects.

Localization of sound in the horizontal plane is more complex and relies on comparisons of subtle differences in input from the two ears. Auditory fibers, as they ascend the brain stem, split in a partial decussation, sending tonotopically matched information from both ears to both left and right superior olivary nuclei. As the first binaural neurons in the auditory pathway, cells in the olivary nuclei represent the first stage at which differences in input from the two ears can be compared. The ears are spaced roughly 20 cm apart. Given the speed of sound at 342 m/s, this creates a time delay (*interaural onset delay*) of 0.06 ms between the time a sound wave reaches one ear and the time it reaches the ear on the opposite side. This calculation pertains to sounds coming from the left or right side; sounds coming from other angles have correspondingly shorter time delays. Sounds coming from straight ahead have no interaural delay.

This system works at sound onset, but it does not work for a continuous tone. Another method, *interaural phase delay*, is required when the source of a continuous tone moves from one location to another. In this case, interaural delay can still be used by comparing differences in the time a given phase of the sound wave reaches each ear. For example, the peak in the sound wave coming from the right will reach the left ear 0.6 ms after it reaches the right ear; however, there is a problem at higher frequencies, particularly those with wave cycles that are shorter than the 20-cm distance between the ears. For example, one cycle of a 2000-Hz sound wave covers only 1.7 cm; thus, the peaks of these cycles are not separated enough in time to allow separate detection by the two ears.

An additional method is used for sound localization of high-frequency tones based on *interaural intensity differences*. The head tends to block the propagation of high-frequency sound waves, creating a sound shadow opposite the sound source. For example, sound coming from one side will be loudest in the ear on the same side but will be partially muffled in the ear on the opposite side because of the head's sound shadow. As with other interaural comparisons described earlier, the interaural intensity difference is greatest for sounds originating from the side, less for sounds at other angles, and absent for sounds from straight ahead or behind. Binaural neurons sensitive to differences in intensity use this information to locate the sound. Because low-frequency sound waves tend to diffract around the head, there is no distinct sound shadow for low frequencies. Localization of the full range of sound frequencies in a horizontal plane is achieved by combined calculations of interaural onset and phase delay for tones of 20 to 2000 Hz and of interaural intensity difference for tones of 2000 to 20,000 Hz.

Suggested Readings

Altschuler R, Hoffman D, Bobbin D, Clopton B. *Neurobiology of hearing*, Vol. 3, *The central auditory system.* New York: Raven Press, 1991.

Corey DP, Hudspeth AJ. Ionic basis of the receptor potential in a vertebrate hair cell. *Nature* 1979; 281:675–677.

Edelman GM, Gall WE, Cowan WM, Eds. *Auditory function: neurobiological bases of hearing.* New York: John Wiley & Sons, 1988.

Hudspeth A. Transduction and tuning by vertebrae hair cells. *Trends Neurosci* 1983; 6:366–389.

Khanna SM, Leonard DGB. Basilar membrane tuning in the cat cochlea. *Science* 1982; 215:305–306.

The Chemical Senses: Smell and Taste

DIANNA A. JOHNSON

KEY POINTS

- The chemical senses rely on specialized receptors that are renewed every few weeks.
- *Olfactory receptor neurons* are the only central nervous system neurons that are directly exposed to and interact with chemical substances in the environment.
- *Odorant binding proteins* serve to enhance sensitivity to odorants at very low concentrations.
- *Odorant receptors* are members of a large receptor superfamily that are coupled to G-protein second-messenger systems.
- *Olfactory receptor neurons* project to the olfactory bulb containing two well-defined inhibitory feedback loops.
- In addition to the primary olfactory pathway, two other pathways send olfactory information

to the cortex to (1) contribute to the perception of flavor and (2) transmit information about pheromones, which influence reproductive behaviors.
- The function of the *gustatory pathway* is to elicit pleasant or aversive reactions to food.
- Recognition of five basic tastes is achieved by the preferential distribution of sensory receptors coupled to different second-messenger systems.
- *Afferent taste fibers* project through three cranial nerves to nuclei closely associated with general visceral sensation.
- The large number of distinguishable tastes is represented by *population responses* of various central taste neurons rather than by activation of individual taste-specific neurons.

The chemical senses, taste and smell, involve a direct interaction of chemical substances in the environment with receptor sites on sensory cells. Thus, compounds must be taken into the body through either aspiration or

ingestion in order to elicit a smell or taste. Perhaps it is because of this direct interaction with potentially harmful substances in the environment that taste and olfactory receptor cells turn over every 4 to 8 weeks. This is not unusual in the case of taste receptor cells because they are modified epithelial cells, not true neurons. Almost all types of epithelial cells normally turn over; however, olfactory cells are true neurons with cell bodies in the nasal mucosa and axons that project directly to brain structures. These olfactory neurons continually reproduce throughout life, more so than any other type of central nervous system (CNS) neurons. New replacement olfactory neurons are generated by stem cells (basal cells) within the olfactory mucosa, and their axons must grow to proper connection sites within the brain. There is hope that studies of the renewal capability of olfactory neurons will provide important leads to understanding how other neurons in the brain could be replaced after cell loss from disease, stroke, or aging. Even with receptor renewal, however, there is a marked decline in the ability to taste and smell during normal aging. Loss of appetite associated with a decline in the chemical senses can be a serious problem leading to an inability to maintain body weight and a decline in general health.

OLFACTORY RECEPTORS

Small, paired patches of olfactory epithelium, each approximately $2\,cm^2$ in size, are located within the uppermost levels of the nasal cavities (Fig. 1). They are arranged in a horizontal line just below the level of the eyes. Three major cell types are present: *olfactory receptor neurons*, *supporting* or *sustentacular cells*, and *basal cells*, which are the stem cells for production of new receptor cells. Interspersed among the cells of the olfactory epithelium are *Bowman's glands*, which produce a layer of

mucus covering the nasal lining. The total mucus content is replaced every 10 minutes and is composed primarily of a mucopolysaccharide solution containing enzymes, antibodies, salts, and special proteins that bind odorous substances. Antibodies in the mucus are particularly important because of the presence of viruses and bacteria in aspirated air. If incorporated into olfactory neurons, these disease vectors can gain direct access to brain tissue by being transported within centrally projecting processes of the olfactory neurons. Hanta virus, present in dust from droppings of infected deer mice, is transmitted to humans in this manner.

Odorant binding proteins present within the mucus bind to odorous molecules, acting like molecular sieves that trap and concentrate the substances and facilitate their interaction with olfactory neurons. Through the actions of binding proteins, the operating sensitivity of the olfactory system is significantly enhanced. Humans can sense certain molecules at a concentration of a few parts per trillion, yet individual olfactory receptor neurons respond only to concentrations that are 1000 times greater. Odorant binding proteins are necessary to concentrate these molecules to reach levels that are within the detectable range of individual receptors.

Odorous molecules vary widely in chemical composition and three-dimensional shape. Humans generally recognize 10,000 separate odors; however, with training, individuals such as whiskey blenders or perfumers can increase that number tenfold. Only 20% of all recognizable odors are pleasant. The rest are unpleasant and represent potentially dangerous substances, thus supporting the assumption that the major role of the olfactory

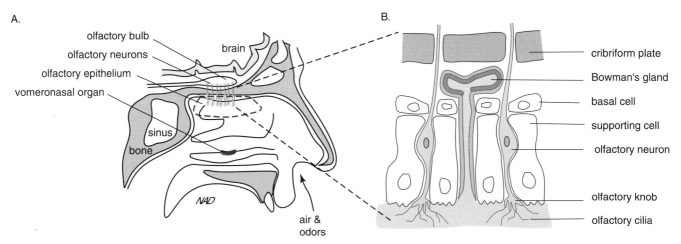

FIGURE 1 Organization of the olfactory epithelium. (A) Olfactory sensory neurons extend from the paired olfactory epithelial patches, through the cribriform plate into the overlying paired olfactory bulbs. The vomeronasal organ also contains olfactory neurons for pheromone detection and sexual function. (B) Olfactory neurons are embedded in the olfactory epithelium with their odor-sensitive cilia extending into the overlying mucus. They are flanked by supporting cells and basal cells, the latter giving rise to new olfactory neurons. Bowman's glands secrete the mucus that traps and transports airborne chemicals.

system is to provide a trigger for protective avoidance behavior.

It does not appear that separate binding proteins within the mucus exist for each recognizable odor; rather, a much smaller set of binding proteins is produced, each of which is capable of binding a wide range of odorants. In contrast, the *odorant receptor proteins* located on membranes of individual receptor cells show a relatively high degree of binding specificity, and it is estimated that more than 1000 distinct types of odorant receptor proteins are produced.

Each neuron produces only one or perhaps very few different types of receptor protein. Sets of odorant-specific neurons are distributed within restricted regions of the olfactory epithelium. Three or more regions have been defined within experimental animals, such as rat, and each region responds to a specified ensemble of odors.

Odorant receptors are members of a receptor superfamily, all with seven transmembrane-spanning alpha-helix regions (Fig. 2). They are produced by members of the largest gene family yet described. The visual pigment,

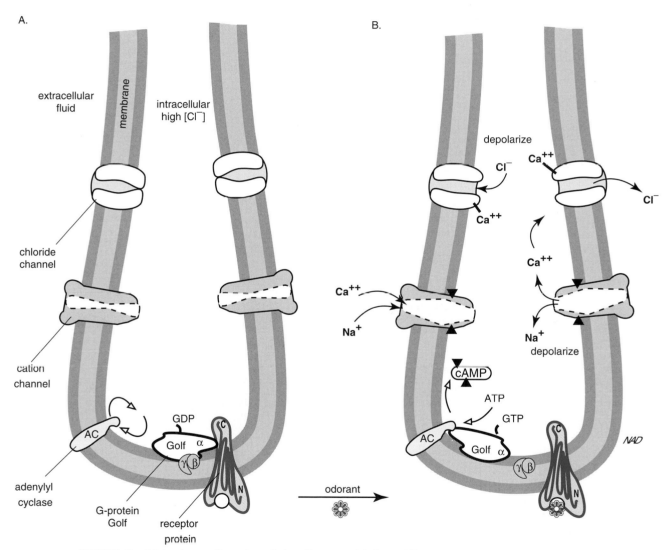

FIGURE 2   Molecular configuration of the olfactory epithelium. Olfactory receptors (shown in blue) are membrane proteins located in the ciliary membrane of olfactory neurons. They are composed of seven membrane-spanning regions and a binding site (white circle) for specific odorants (blue flower). Humans produce about 1000 different receptor proteins, with only one to three types expressed by a given cell. (A) Inactive cilia show a receptor-bound G-protein, called Golf (G-olfactory protein), inactive adenylyl cyclase, and closed cation and chloride channels. (B) When an appropriate odorant binds to the receptor, it activates Golf, which then binds guanosine triphosphate (GTP) and activates adenylyl cyclase to produce cyclic adenosine monophosphate (cAMP). cAMP directly opens cation channels, causing depolarization of the cilia. The entering calcium also opens calcium-sensitive chloride channels. Because of the unusually high levels of intracellular chloride present, chloride leaves the cell, thus causing further depolarization. Depolarization of the cilia spreads to the cell body and leads to action potentials that originate in the initial part of the axon.

rhodopsin, is a member of this same superfamily of proteins, and it shares many common attributes with odorant receptors, including its linkage to a G-protein as well as similarities in the general attributes of the associated second-messenger cascade. The specific G-protein in photoreceptors is transducin; in olfactory receptors, it is called *Golf* (G-protein in olfactory receptors). Binding to the receptor activates Golf, which then stimulates adenylate cyclase to produce cyclic adenosine monophosphate (cAMP). Calcium channels in the neuronal membrane bind cAMP and thereby increase their conductance. This leads to calcium influx, calcium activation of chloride channels, depolarization, and generation of a receptor potential that at threshold levels will fire an action potential. In most neurons, increased chloride conductance leads to hyperpolarization, but in olfactory neurons the opposite occurs. This is because olfactory neurons have an unusually high concentration of intracellular chloride, and increased chloride conductance will thus cause efflux of the negatively charged chloride and concomitant cell depolarization.

THE OLFACTORY BULB

Odorant receptors and associated second-messenger systems are primarily localized within the *ciliary tufts* of olfactory neurons. The tufts extend from the surface of the epithelial plate and are immersed in mucous. The cell body, located in deeper epithelial layers, has a single nonmyelinated axon that must pass through small holes in the *cribriform plate* before synapsing on the *olfactory bulb* located just above the level of the eyes (Fig. 3). These are among the smallest (therefore, slowest conducting) neurons in the body, and they are highly vulnerable to injury from head trauma. Some return of function often occurs because of the regenerative capacity of the neurons, as mentioned earlier.

One of the major structural features of the olfactory bulb is the arrangement of cell processes in bundles called *glomeruli*, which contain the axonal endings of olfactory receptors and the apical dendrites of roughly 100 second-order olfactory neurons. Each glomerulus receives input from 25,000 primary olfactory neurons, and each primary olfactory neuron synapses with several of the secondary neurons within the glomerulus. The end result is that each second-order neuron receives several thousand synaptic inputs from olfactory receptors.

There are two types of second-order olfactory neurons: *mitral* and *tufted cells*. Both receive information from olfactory receptor synapses within the glomerulus. In addition, they make reciprocal synapses with neuronal processes from two types of interneurons: (1) *periglomerular cell* processes within the glomerulus, and (2) *granule*

cell processes in the external plexiform layer located proximal to the glomerular layer. Periglomerular cells provide short feedback loops among the glomeruli, whereas granule cells are part of a long inhibitory feedback loop involving the *olfactory cortex*. In this latter pathway, olfactory signals are sent through olfactory receptors to secondary neurons in the olfactory bulb and on to tertiary cells in the olfactory cortex. These tertiary cells then project back to the bulb and activate granule cells, which in turn release gamma aminobutyric acid (GABA) to inhibit mitral and tufted cells.

The regional patterns of odorant-specific neurons seen in the olfactory epithelium are also seen in the bulb; however, there does not seem to be a one-to-one transfer of odor-specific information. The recognition of a specific odor results from the spatial pattern of olfactory receptors that are activated and that in turn activate a circumscribed group of secondary neurons in the olfactory bulb. It appears that in the olfactory system, as in most other sensory systems, the location of the neurons activated by a specific input provides a code for some aspect of the stimulus. In the visual system, a *retinotopic map* indicates the location of light within the visual field; in the auditory system, a *tonotopic map* indicates the frequency of sound (pitch); and, in the olfactory system, a *odorotopic map* indicates not the location of the odor but the chemical properties of the odorant. Many recognizable odors are actually mixtures, and the olfactory code for these smells seems to be generated in terms of the specific sets of neurons that are activated and the specific regions of the nasal epithelium or olfactory bulb in which they reside.

THE OLFACTORY CORTEX

Mitral cells and tufted cells send central processes within the lateral olfactory tract to the *primary olfactory cortex* located on the inferior surface of the temporal lobe. The neurotransmitters utilized in this pathway appear to be excitatory neuropeptides and perhaps amino acids such as glutamate and aspartate. Mitral cells release cholecystokinin; tufted cells release corticotropin-releasing hormone. This relatively simple sensory pathway is unusual in that it is the only sensory system without a major synaptic relay in the thalamus before projecting to cortical regions.

The primary olfactory cortex occupies the superficial cortical layers of the inferior aspect of the temporal lobe (Fig. 4). It overlies the *hippocampal formation*, with which it is functionally associated, particularly in generating olfactory memory. More anteriorly, the *periamygdaloid region* of the olfactory cortex is connected to the *amygdala* and adjacent structures of the limbic system that provide emotional context to odor recognition. The association

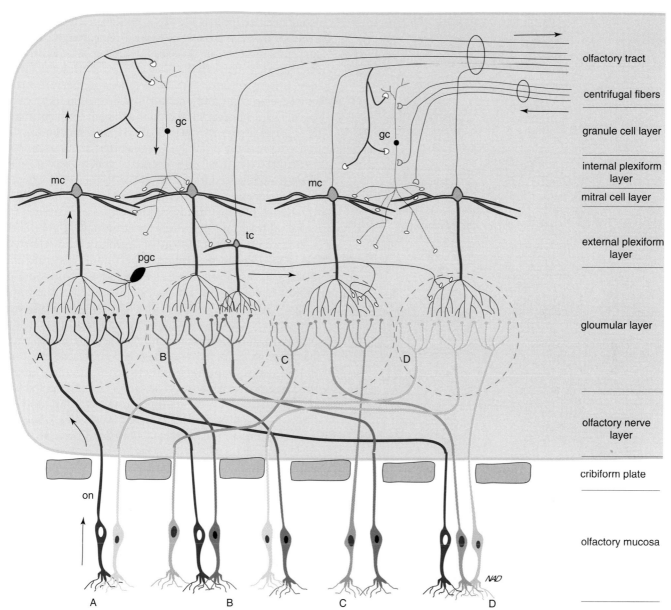

olfactory tract

centrifugal fibers

granule cell layer

internal plexiform layer

mitral cell layer

external plexiform layer

gloumular layer

olfactory nerve layer

cribiform plate

olfactory mucosa

FIGURE 3   Neuronal circuits of the olfactory bulb. Odorant-specific subtypes of olfactory neurons (on) are scattered randomly throughout patches of the olfactory epithelium. Four examples are illustrated in (A) through (D). As their axons project into the olfactory bulb, they become rearranged when they contact the appropriate odorant-specific glomeruli. Four glomeruli are shown, each receiving approximately 25,000 primary axons (three are shown). Two types of second-order neurons, mitral cell (mc) and tufted cell (tc), have their apical dendrites in each glomerulus. Basal dendrites of these cells are located in deeper layers of the olfactory bulb (external plexiform layer), and their axons project to the brain in the olfactory tract. As axons exit the bulb, they send collaterals back to the bulb. The two classes of interneurons, both inhibitory, are shown in black. The periglomerular cell (pgc) of the external plexiform layer forms reciprocal synapses within a given glomerulus and spreads lateral inhibition to surrounding glomeruli. The granule cells (gc) form reciprocal synapses with the basal dendrites of the secondary neurons. They also receive information from mitral and tufted cell collaterals and from the brain via centrifugal fibers.

of specific odors with pleasant or unpleasant sensations is an important feature of the olfaction system and allows it to function in alerting the organism of potentially harmful substances.

There is evidence for a secondary olfactory pathway that, unlike the primary pathway, does have a relay in the thalamus (*medial dorsal nucleus*) before projecting to a cortical region in the *orbitofrontal cortex*. Interestingly,

A. Distribution of olfactory information

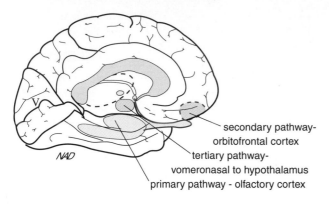

secondary pathway-
orbitofrontal cortex
tertiary pathway-
vomeronasal to hypothalamus
primary pathway - olfactory cortex

B. Primary and bilateral olfactory pathway, medial view enlarged

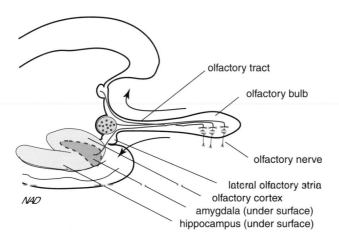

olfactory tract

olfactory bulb

olfactory nerve

lateral olfactory stria
olfactory cortex
amygdala (under surface)
hippocampus (under surface)

C. Secondary olfactory pathway, lateral view

medial dorsal nucleus
of the thalamus
orbitofrontal cortex

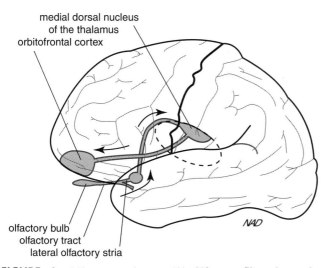

olfactory bulb
olfactory tract
lateral olfactory stria

FIGURE 4 Olfactory pathways. (A) Olfactory fibers leave the olfactory bulb and distribute to a number of CNS areas for further processing. (B) The primary path projects to the temporal lobe adjacent to the hippocampus and amygdala. (C) The secondary pathway projects to the thalamus and then to the frontal cortex.

it has been suggested that this pathway, although relatively small, may actually be more directly responsible for conscious perception of odors, whereas the historically designated primary pathway through the lateral olfactory tract projection to the temporal lobe represents mainly emotion- and memory-related olfactory pathways. The orbitofrontal region is adjacent to the primary taste cortex. Joint projections of taste and smell information are probably combined at the interface of these two cortical areas in order to provide the general sensation of *flavor* from ingested food. Pathways originating from this flavor area project back to the *nucleus of the solitary tract*, which controls visceral autonomic responses in the gastrointestinal tract and gustatory function.

A third olfactory pathway has recently been reported. It has long been known that lower mammals (rats and rabbits) have a *vomeronasal organ* (VNO) consisting of an additional area of olfactory epithelium in the nasal cavity. Until recently, it was thought that the VNO was present only during infancy in humans and that it was lost during later development. New evidence indicates that, though small in area, the VNO is present in adults and provides important olfactory information about odorants, particularly *pheromones*. The VNO is located in a small pit in the midline nasal septum, and its olfactory receptors project to the *accessory olfactory bulb*. Neurons in the accessory bulb send projections to the *hypothalamus*, which controls important reproductive behaviors and gonadosteroid function.

OVERVIEW OF THE GUSTATORY SYSTEM

The elemental role of the gustatory system is to distinguish between food and potential toxins. Two components are required to accomplish this task: (1) a detection system of receptor cells capable of responding to the great diversity of substances in the environment that might be ingested, and (2) neuronal pathways that refer taste information to appropriate cortical structures in order to elicit pleasant or unpleasant sensations. Pleasing sensations associated with food are necessary to maintain the appetite and to initiate appropriate digestive and respiratory responses; unpleasant sensations associated with potential toxins elicit protective reflexes, such as coughing, sneezing, gagging, or vomiting. Some taste responses are inborn (*e.g.*, a preference for sweetness and an aversion to bitter tastes); however, the gustatory system is highly modifiable by experience. Illness that occurs soon after ingestion of a particular food can greatly diminish subsequent preferences for that food. Likewise, tastes can be acquired so that some bitter tastes, such as quinine, are tolerated or even enjoyed. Whether or not a given food is appetizing can also depend on bodily needs. For example, a nutritional deficiency in salt can enhance the appetite and cause a craving for salty food.

Although the sense of taste and smell both involve direct interaction of environmental substances with receptor cells, the two systems utilize somewhat different transduction mechanisms. *Taste receptors* are classified as modified epithelial cells rather than true neurons. They occur in clustered structures called *taste buds*, each of which contains 50 to 100 receptor cells arranged like slices of an orange with a central opening or pore open to the surface of the tongue (Fig. 5). The taste bud also contains basal cells that are the stem cells for the production of new taste receptors. The typical life span of taste receptors is 1 to 2 weeks. Taste receptors synapse on dendrites of *afferent gustatory fibers* projecting into the taste bud.

THE FIVE BASIC TASTES

Humans perceive five basic tastes: *salt, sour, sweet, bitter,* and *umami.* Further distinctions are made based on combinations of these basic sensations. Additional subtleties in flavor can be discerned by the combination of smell and taste and, to a lesser extent, temperature and texture of food.

The five basic tastes are derived from different transduction mechanisms located on taste receptor cells and their specific interactions with different types of molecules. In general, acids elicit a sour taste; salt elicits a salty taste; sugars, some proteins, and amino acid artificial sweeteners such as saccharin and aspartame invoke sweetness; ions such as potassium and magnesium and organic compounds such as quinine and coffee taste bitter. Umami, recently recognized as an additional basic taste, is associated with some amino acids such as glutamate, the common culinary form of which is monosodium glutamate (MSG). These compounds interact with taste receptors and cause an increased release of transmitter, which in turn stimulates primary afferent taste fibers.

There are four mechanisms by which chemicals cause increased transmitter release from taste receptors: (1) direct passage of ions through ion channels, (2) blockage of ionic channels, (3) opening of ionic channels, and (4) activation of second-messenger systems through ligand interactions with membrane receptors (Fig. 6).

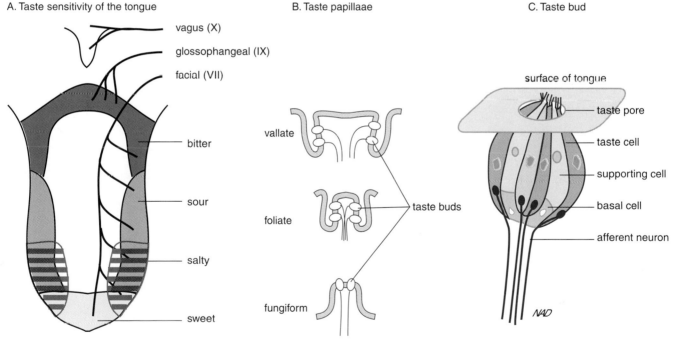

FIGURE 5 Cellular constituents of taste buds. (A) Taste buds are distributed around the edge of the tongue and down the throat. Areas of salt-sensitive taste buds overlap with sweet and sour areas at the front. Fibers from the anterior two-thirds of the tongue send taste information to the brain in the facial nerve (cranial nerve VII). The back of the tongue sends information through the glossopharyngeal nerve (IX), and the palate, pharynx, and epiglottis use the vagus nerve (X). (B) Taste buds are located within specialized papillae. Fungiform papillae cover the areas of the tongue innervated by the facial nerve and contain one to five taste buds each. Foliate papillae, located along the back edge of the tongue, and vallate papillae, located along the posterior midline, contain thousands of taste buds each. (C) Taste buds contain 50 to 150 receptor cells, each with microvilli on their apical surface and synaptic connections with afferent neurons near their base. Taste buds also contain basal cells that replace the short-lived receptor cells every few days. There are also supporting cells that appear to be similar to the glial cells of the central nervous system.

A. Salty taste: Na$^+$ and H$^+$
move through open channels

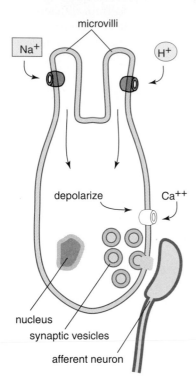

microvilli

Na$^+$

H$^+$

depolarize

Ca^{++}

nucleus

synaptic vesicles

afferent neuron

B. Sour taste:
H$^+$ closes K$^+$ channel

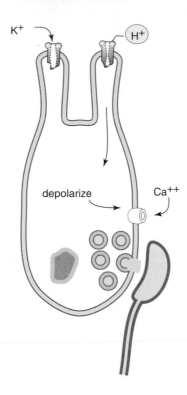

K$^+$

H$^+$

depolarize

Ca^{++}

C. Umami taste
glutamate receptor

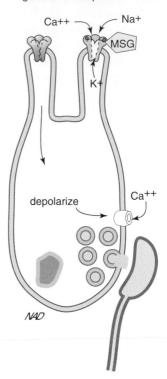

Ca^{++}

Na$^+$

MSG

K$^+$

depolarize

Ca^{++}

NAD

D. Sweet taste: receptor -
G-protein - cAMP - PKA

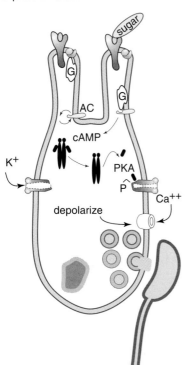

sugar

G

G

AC

cAMP

K$^+$

PKA

P

Ca^{++}

depolarize

E. Bitter - 1 taste::
close K$^+$ channels

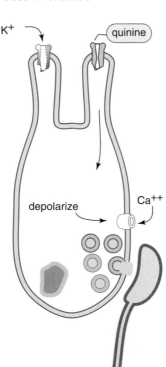

K$^+$

quinine

depolarize

Ca^{++}

F. Bitter - 2 taste:
release internal Ca^{++}

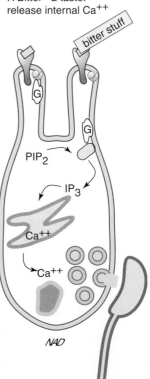

bitter stuff

G

G

PIP$_2$

IP$_3$

Ca^{++}

Ca^{++}

NAD

Salt

Sodium ions are largely responsible for establishing a salty taste. Sodium ions diffuse through the pore of the taste bud and enter the taste receptor cell through sodium-selective channels present in the cell membrane. These channels are characterized by their sensitivity to the drug *amiloride* and their insensitivity to changes in voltage. Thus, they are different from the sodium channels involved in propagation of action potentials. The entry of positively charged ions depolarizes the taste receptor cell, opens voltage-dependent calcium channels, and increases the influx of calcium, thereby allowing calcium to enter the cell and cause release of neurotransmitter.

Sour

A sour taste is associated with acidic substances that affect sour-sensitive taste receptors in two ways: (1) Acids in solution generate hydrogen ions that can permeate the amiloride-sensitive sodium channel described earlier and cause depolarization-stimulated release of neurotransmitter in the same manner as sodium ions. (2) Hydrogen ions also block a potassium-selective channel within the membrane which also causes depolarization because the normal movement of potassium out of the cell is blocked, and more positively charged potassium ions are trapped intracellularly. Foods that cause depolarization and increased transmitter release through both of these mechanisms are perceived as sour; those that cause depolarization only through diffusion of cations through the sodium channel are perceived as salty.

Sweet

Specific *membrane receptor proteins* on some taste cells bind sugars and other sweet-tasting substances. Binding to these receptor sites activates a second-messenger system similar to the one associated with noradrenergic receptors. The receptor is coupled to a G-protein that activates protein kinase A and causes it to phosphorylate and block a potassium-selective channel. As a result, sweet-sensitive taste cells are depolarized.

Bitter

Because many toxic compounds have an unpleasant bitter taste, receptor cells that respond to bitter substances can function as poison detectors; however, some bitter foods are not necessarily unpleasant or toxic (*e.g.*, quinine and caffeine, although one might argue the latter point for caffeine). Several different transduction mechanisms are involved in detection of bitterness in food. Some bitter compounds (*e.g.*, calcium ions and quinine) decrease conductance of potassium-selective channels similar to the mechanisms used for detection of sweetness. Other bitter substances bind to specific membrane receptors that activate second-messenger systems and cause membrane depolarization. One type of bitterness receptor triggers an increased production of the intracellular messenger *inositol triphosphate* (IP_3). In all other transduction mechanisms described earlier, depolarization of the receptor cell causes an increase in calcium influx through voltage-sensitive calcium channels, and calcium ions act as the trigger for release of neurotransmitter. This is not so in the case of the IP_3 transduction mechanism. Here, the membrane potential is not altered; rather, IP_3 causes the release of calcium from internal storage sites which in turn directly stimulates neurotransmitter release.

Umami

The umami taste is not as familiar as the preceding four basic tastes. Nonetheless, it is discernible as a distinctive and delicious taste associated with certain amino acids such as glutamate and perhaps arginine. These amino acids bind to and activate a cation-permeable channel, causing depolarization in a manner similar to glutamate activation of cation channels in the brain.

FIGURE 6 Transduction mechanisms for salt, sour, umami, sweet, and bitter. All taste receptors release neurotransmitters in response to an increase in free intracellular calcium, usually due to depolarization that opens reactive calcium channels. The cause of this depolarization varies with the specific receptor cell. (A) Salty taste is transduced by amiloride-sensitive sodium channels that are always open. When the sodium concentration increases on the surface of the microvilli, sodium ions can enter the channels and depolarize the cell. Hydrogen ions can also enter these channels, thus acid food has a salty taste. (B) Sour taste is elicited when hydrogen ions act directly on open potassium channels, closing them and depolarizing the cell. (C) Umami taste utilizes a cation channel similar to CNS glutamate receptors. Glutamate (e.g., monosodium glutamate, or MSG) binds to the receptor and allows all small cations to pass, including sodium, potassium, and calcium. The net effect of these ionic movements is to depolarize the cell. (D) Sweet tastes, such as sugars and certain proteins, produce depolarization by a chain of events that resemble many CNS neurotransduction systems. Binding of the receptor protein activates a G_s-protein that stimulates adenylyl cyclase to produce cyclic adenosine monophosphate (cAMP), which releases protein kinase A (PKA) from its regulator proteins. PKA then phosphorylates potassium channels, closing them and depolarizing the cell. (E) There are at least two types of bitter receptors; quinine stimulates the bitter 1 receptor by simply closing potassium channels. (F) Other bitter substances bind with a bitter 2 receptor and stimulate the production of inositol triphosphate (IP_3). IP_3 triggers the release of calcium from internal stores, stimulating neurotransmitter release without an intervening depolarization.

A. Lateral view of cerebral cortex

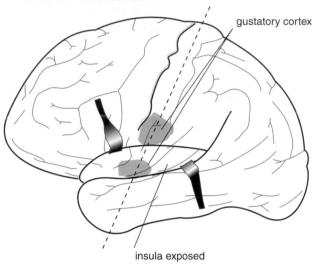

gustatory cortex

insula exposed

B. Coronal section of cerebral cortex

ventral posterior
medial nucleus

gustatory cortex

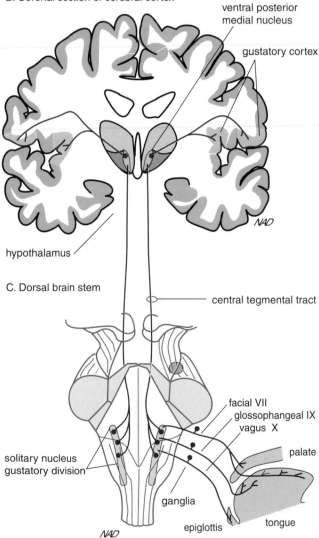

hypothalamus

C. Dorsal brain stem

central tegmental tract

facial VII
glossophangeal IX
vagus X

palate

solitary nucleus
gustatory division

ganglia

epiglottis   tongue

NAD

CENTRAL GUSTATORY PATHWAYS

Taste receptors synapse with dendritic elements of the *primary afferent neurons* of the gustatory pathway, causing the neurons to fire in response to appropriate stimuli above threshold levels. These afferent fibers represent the peripheral projection of the primary sensory neurons of the gustatory pathway, and they arise from cell bodies within three cranial nerve ganglia (Fig. 7). Afferent fibers within the *facial nerve* (cranial nerve VII) arise from the *geniculate ganglia* and innervate the anterior two-thirds of the tongue. Fibers from the *glossopharyngeal nerve* (cranial nerve IX) arise from the *inferior glossopharyngeal ganglia* and innervate the posterior one-third of the tongue. Fibers from the *vagus nerve* (cranial nerve X) arise from the *inferior vagal ganglia* and innervate the smattering of taste buds found in throat regions, including the glottis, epiglottis, and pharynx. Central projections from these three cranial nerve ganglia enter the brain stem along the lateral aspect of the medulla and converge to make synaptic contact with cells in the *rostral* or *gustatory division of the solitary nucleus* of the medulla. Ascending fibers from the solitary nucleus make a synaptic connection in the *ventral posterior medial nucleus of the thalamus*. Thalamic neurons then project to the *primary gustatory cortex* in the insular and orbitofrontal regions.

Compared to other special sensory systems, relatively little is known about central processing of taste information. It has been suggested that taste is not encoded along "labeled lines," with specific neurons designated for each basic taste. Apparently, there is only a crude spatial map of basic tastes, and the discrimination of the many thousands of recognizable tastes is encoded as a population response, with each individual taste neuron having rather broadly tuned response characteristics. Each taste bud is composed of taste receptors, with similar response characteristics reflecting a preference for one or perhaps two basic tastes. Furthermore, taste buds in different regions of the tongue show different preferences (see Fig. 5). Along the lateral edge of the tongue there are three definable areas: (1) the most

FIGURE 7 Central projections of taste buds. Fibers from taste buds enter the ventral surface of the medulla via three pairs of cranial neurons and synapse in the rostral or gustatory division of the solitary nucleus. Secondary neurons carrying gustatory information project to the ventral posterior medial nucleus of the thalamus, and the information is then relayed to the cerebral cortex for conscious appreciation. (A) Lateral view of the cerebral cortex with the lateral sulcus opened to expose the insula; the gustatory cortex is shown in blue. (B) Coronal section taken in the plane indicated by the dashed line in (A). (C) Oral cavity and dorsal view of the brain stem showing the three cranial nerves that carry information from the taste buds. Note that the information is processed ipsilaterally.

anterior, which is responsive to saltiness; (2) the middle area, which overlaps area 1 and is most responsive to sourness; and (3) the posterior area, which responds to bitterness. Taste buds near the tip of the tongue are most responsive to sweet tastes; however, these preferences are best expressed at threshold concentrations of the stimulant. At higher concentrations multiple stimulants are effective in eliciting a response from a given taste bud. Primary and secondary neurons in the gustatory pathway receive input from many individual taste buds, and they do not show a strong spatial map of taste-specific regions of the tongue.

Suggested Readings

Buck LB, Axel R. A novel multigene family may encode odorant receptors: a molecular basis for odor recognition. *Cell* 1991; 5:175–187.

Kauer JS. Coding in the olfactory system, in: Fingerand TE, Silver WL, Eds., *Neurobiology of taste and smell*. New York: Wiley, 1987, pp. 205–231.

McLaughlin S, Margolskee RF. The sense of taste. *Am Sci* 1994; 82:538–545.

Reed RR. Signaling pathways in odor detection. *Neuron* 1992; 8:205–209.

Reed RR. How does the nose know? *Cell* 1990; 60:1–2.

Roper SD. The cell biology of the vertebrate taste receptors. *Annu Rev Neurosci* 1989; 12:329–353.

55

Lower Motor Neurons of the Spinal Cord and Brain Stem

DIANNA A. JOHNSON

KEY POINTS

- *Lower motor neurons* represent the final common pathway that links the nervous system to the appropriate muscle group. Lower motor neurons for muscles of the head and neck are located within cranial nerve nuclei in the brain stem.
- Excitatory and inhibitory interneurons make reciprocal connections among groups of lower motor neurons innervating antagonistic and synergistic muscles.
- *Upper motor neurons* reside at higher levels of the neuronal axis and provide input to interneurons and lower motor neurons, thus creating a hierarchical arrangement of the motor system.
- Lower motor neurons for muscles of the body are arranged *somatotopically* at different levels within the central gray matter of the spinal cord.
- The appropriate pool of motor neurons and the individual muscle fibers they each innervate work synergistically to cause contraction of an entire muscle.
- Intrinsic properties of individual motor neurons and muscle fibers have an important influence on the response characteristics of the motor unit.

There is but one final common pathway between the nervous system and the somatic musculature. The *lower motor neuron* forms the essential neuronal link with the individual muscle cell, and all motor commands must be delivered through it. With loss of the lower motor neuron, muscles cannot respond to any neuronal drive and *flaccid paralysis* results. Lower motor neurons for muscles of the head and neck are located in motor nuclei of the brain stem, with peripheral fibers projecting through cranial nerves. Lower motor neurons for muscles of the body have cell bodies in the ventral horn of the spinal cord and processes that exit in the ventral roots of the spinal nerves. Details of the synaptic interactions at the neuromuscular junction between lower motor neurons and innervated muscle are discussed in Chapter 6.

Lower motor neurons do not act in isolation; rather, they are themselves under continuous control of a number of different inputs generally organized in hierarchical fashion. The lowest control level is represented by direct sensory inputs to the lower motor neuron that can drive simple reflex activity in appropriate muscle groups. The next highest control level is represented by groups of upper motor neurons organized within various motor nuclei of the brain stem or within areas of the gray matter of the spinal cord. Intermediate levels of motor processing can also occur in other motor areas such as the basal ganglia or cerebellum. These motor neurons coordinate more complex muscle reflexes. The highest level of control is represented by upper motor neurons in the motor cortex that elicit conscious movements and learned motor behaviors. Chapters 55 through 59 are devoted to discussions of the different levels within the hierarchical organization of the motor system and how they function in producing the complex repertoire of human movements.

MOTOR NEURONS FOR HEAD AND NECK MUSCLES

Innervation of Extraocular Muscles

Muscles of the head and neck are for the most part highly specialized, and each exhibits unique functional properties. *Extraocular muscles* are innervated by lower motor neurons in cranial nerves III (oculomotor), IV (trochlear), and VI (abducens), which originate from the brain stem. They are among the fastest muscles in the body and are continually active during awake hours as well as during certain segments of the sleep cycle, called *rapid eye movement* (REM) sleep. As described below, the oculomotor system serves as a useful model of many of the generalized properties of motor pathways, including: (1) reciprocal excitatory and inhibitory innervation of synergistic muscle groups, (2) hierarchical organization of cortical centers, and (3) somatotopic organization of cortical motor areas.

As detailed in Chapter 51, activity of lower motor neurons in the *oculomotor, trochlear,* and *abducens nuclei* is under the control of several groups of interneurons. Two groups found in the brain stem, the *horizontal gaze center* and the *vertical gaze center*, yoke the activity of left and right oculomotor nuclei so that they send coordinated signals to left and right eyes, allowing coordinated movement of the two eyes. Experimental stimulation of the left horizontal gaze center will elicit a highly stereotypic eye movement to the left in both eyes. The motor programs from both gaze centers also coordinate neuronal input to the six individual muscles that innervate a single eyeball, so that opposing or antagonistic muscles on either side of the eye do not try to contract at the same time. *Mutual inhibition of antagonistic muscles* is a standard feature of all peripheral motor pathways. It is achieved through *reciprocal inhibitory input* from interneurons to sets of motor neurons supplying antagonistic muscles. The same reciprocal innervation is also necessary for muscles that must act synergistically to carry out complex movements requiring contraction of more than one muscle. In this case, the reciprocal innervation is excitatory.

Two other groups of interneurons, one in the *superior colliculus* of the brain stem and another in an area called the *frontal eye fields* located within the frontal cortex, generate complex patterns of activity that are transmitted to gaze centers and to lower motor neurons of the oculomotor system. These areas initiate complex movements such as smooth pursuit and saccade activity (previously described in Chapter 51). These two types of movement are involved in the visual tracking of moving objects; both are involuntary reflexes. The saccade and smooth pursuit systems offer excellent examples of the *hierarchical arrangement of motor systems*, with successive layers of interneuron circuits each providing an increased level of complexity in the motor pattern generated. The lowest level provides the basic stereotypic range of movement (in this case, movement in the vertical or horizontal plane coordinated by the gaze centers) and higher levels provide more complex patterns (*e.g.*, smooth pursuit movement).

Voluntary control of eye muscles and of all other skeletal muscles of the body originates from the motor cortex, a superficial strip of frontal cortex just anterior to the central sulcus. Just as the sensory cortex is arranged in an orderly fashion, with adjacent parts of the body represented by adjacent cortical regions, so are motor neurons in the motor cortex distributed in an orderly manner. These cells represent the upper motor neurons that provide the excitatory drive either directly to lower motor neurons or indirectly through interneuron circuits in order to elicit voluntary movement. In the case of the oculomotor system, motor commands from the motor cortex must involve the gaze centers because even volitional eye movement must be coordinated. Most individuals cannot move the eyes independently even with conscious effort.

Innervation of Branchiomeric Muscles

Muscles of facial expression are innervated by lower motor neurons within cell bodies located in the facial nucleus and in processes that project through the facial nerve (VII). Muscles of the jaw are innervated by neurons with cell bodies in the *trigeminal motor nucleus*. Muscles of the larynx and pharynx are innervated by motor nuclei of the glossopharyngeal (IX) and vagal (X) nerves. The trapezius and sternomastoid neck muscles are innervated by the accessory nucleus of cranial nerve XI. All of these muscles share a common attribute in that they are embryologically derived from the *branchiomeric arches* and are classified as *branchiomeric muscles*, as opposed to somatic muscles, which are derived from somites. Collections of brain stem interneurons act as motor pattern generators for these motor pathways, providing relatively simple repetitive reflexes, such as chewing, as well as highly complex, integrative movement patterns for protective reflexes, such as the gag, cough, sneeze, and blink reflexes.

Innervation of the Tongue

Somatic muscles of the tongue are innervated by lower motor neurons in the *hypoglossal* (cranial nerve XII) nucleus, and they are involved in protective reflexes of the head region. Cranial nerves IX, X, and XT also participate collectively in the production of speech. Speech is initiated by upper motor neurons located in the inferior gyrus of the frontal lobe, including Brodmann's areas 44 and 45, collectively called *Broca's area*. Studies of neuronal circuits in song birds have provided insights into what may be comparable circuits for speech in humans. Here, again, the hierarchical principle pertains, with pattern generators present at brain stem levels responsible for composing motor command patterns that produce basic sounds, and neurons in higher cortical regions providing the same for more complex words and sentences.

MOTOR NEURONS FOR AXIAL, PROXIMAL, AND DISTAL MUSCLES OF THE BODY

In contrast to the specialized muscle groups of the head and neck region, somatic musculature of the body follows a general organization plan that is applicable throughout. Lower motor neurons for body muscles are located in the ventral or anterior horn of the spinal cord (Fig. 1). These *alpha motor neurons* comprise the best-studied and perhaps best-understood neuronal circuit in the entire nervous system. Their large, easily recognized

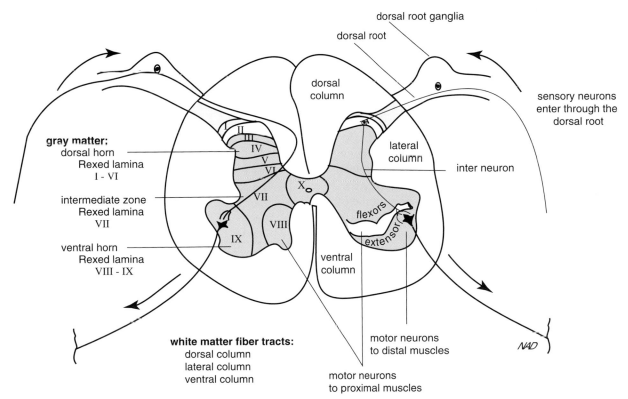

FIGURE 1   Functional distribution of spinal cord motor neurons. Cross-section of a lumbar spinal cord segment. Note that sensory information enters dorsally, whereas motor information exits ventrally. The Rexed laminae, indicated in Roman numerals on the left, are divisions of the spinal gray matter that are based on differences in the histological appearances of the 10 areas. The motor neurons in the ventral horn occur in groups arranged according to their target muscle, as shown on the right. Neurons innervating flexors are more dorsal; those for extensors are ventral; those for distal muscles are lateral (Rexed lamina IV), and those for proximal muscles are medial (Rexed lamina VIII).

cell bodies and relatively simple circuits make them amenable to experimental analysis, and much is known about their functional properties. Axons of alpha motor neurons project through the ventral root of spinal nerves and innervate individual muscle fibers in the trunk (*axial muscles*), arms and legs (*proximal muscles*), and hands, feet, and digits (*distal muscles*). The number of individual muscle fibers innervated by a single neuron varies among muscle groups. Axial muscles, which serve primarily to maintain posture, have relatively sparse innervation, with one alpha motor neuron innervating up to 1000 muscle fibers. Proximal muscles require more innervation to activate the patterns of movement for locomotion. Distal muscles are the most heavily innervated, with a single neuron innervating as few as three or four muscle fibers and affording fine, controlled movements of hands and fingers.

Alpha motor neurons are arranged in functional groups within the ventral horn of the spinal cord. Neurons innervating *flexor muscles* are grouped in the dorsal region of the horn; those innervating *extensor muscles* are found in the ventral region. Neurons innervating axial muscles are located medially within the ventral horn, whereas proximal and distal muscles are innervated by motor neurons in more lateral regions. Thirty paired spinal nerves carry motor fibers to their peripheral targets. Motor fibers exit through the ventral root of the spinal nerve, and after a short projection they merge with the dorsal (sensory) root fibers that carry incoming touch, temperature, pain, and proprioception information from the body. Spinal nerves serve as landmarks to somatotopically separate the spinal cord into 8 cervical, 12 thoracic, 5 lumbar, and 5 sacral levels (see Fig. 1 in Chapter 50). Just as sensory information from the body is divided segmentally into dermatomes as it enters designated spinal levels, motor information exits segmentally, with each spinal nerve serving a designated region of the body called a *myotome*. Each myotome represents the segment of the body that receives motor innervation from an assigned spinal nerve. It should be remembered, however, that segmentation within the spinal cord is not absolute and that motor output as well as sensory input overlaps one or two segments above and below the designated nerve.

Because axial muscles extend the full length of the trunk, motor neurons innervating them are present at all spinal levels. As stated earlier, these neurons are located in the medial portion of the ventral horn, whereas neurons innervating proximal and distal muscles are located more laterally. Because a relatively large number of neurons are required to innervate limbs and digits, the lateral region of the ventral horn is expanded, leading to an overall enlargement of the diameter of the spinal cord in corresponding regions. This results in a *cervical enlargement* at levels C3

to T1 for innervation of muscles of the upper extremities and a *lumbar enlargement* at levels LI to L8 for innervation of lower extremities (see Fig. 1 in Chapter 50).

Each alpha motor neuron innervates a group of individual muscle fibers to form a *motor unit* (Fig. 2). The size of the motor unit depends on functional requirements.

A. A motor unit: one motor neuron and all of its muscle fibers.

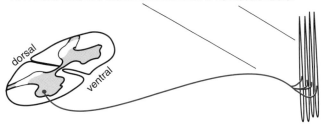

B. A motor neuron pool: All motor neurons to a complete muscle

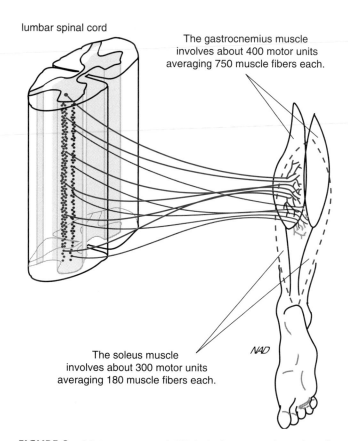

FIGURE 2   Motor neuron pool. (A) A single motor unit consists of a motor neuron and all the muscle fibers it innervates; only four of the hundreds of muscle fibers are illustrated. No muscle fiber receives more than one neuron. (B) The 400 motor neurons of the motor neuron pool innervating the gastrocnemius muscle (shown in blue) are distributed in a column in the lateral part of the ventral horn of the lumbar spinal cord. This pool of neurons supplies approximately 300,000 muscle fibers. The motor neuron pool to the synergistic, soleus muscle is distributed in a column in the adjacent area of the cord (shown in gray).

Antigravity muscles of the leg are composed of large motor units (1000 muscle fibers per neuron); muscles in the fingers which require much finer control are composed of small motor units (two or three fibers per neuron). In most cases, one action potential from an alpha motor neuron will cause a large excitatory postsynaptic potential in the muscle fibers it innervates and a *muscle fiber action potential* will be generated. In response, the muscle fiber will contract and then relax. To evoke a sustained contraction rather than just a twitch, there must be a barrage of action potentials from the nerve. If the action potentials occur at sufficient frequency, there can be a summation effect, because of the buildup of intracellular calcium within the muscle, so that the degree of contraction (*i.e.*, muscle tension), as well as the duration of the fiber contraction, is increased.

To contract an entire muscle, appropriate motor neurons act synergistically as a unit or *motor neuron pool* to activate a critical number of individual fibers within the muscle. Increasing the number of activated motor units will lead to a stronger muscle contraction. In response to increased firing from the neuron pool, the muscle will shorten and cause movement of the limb or digit. Under isometric conditions where the muscle cannot shorten, contraction will lead to increased tension on the muscle.

The responsiveness of a motor neuron is largely influenced by its size. Small neurons (small cell body, small axonal spread, small number of innervated muscle fibers) are the easiest to excite and are therefore the first within the neuron pool to be triggered by appropriate input, and the muscle fibers they innervate will be among the first to be activated. As more excitatory input arrives, larger neurons reach threshold and other fibers are recruited into action. This "size principle" is an important

mechanism for providing the finest control under the lightest load.

Additional variations in muscle function are related to inherent properties of individual muscle cells. *Fast-twitch muscle fibers* rely primarily on glycolysis to provide adenosine triphosphate for metabolic needs. They characteristically fire at high frequency (30 to 60 Hz), and they fatigue relatively quickly. *Slow-twitch muscles* use oxidative phosphorylation and rely on glucose supplied from the circulation to provide adenosine triphosphate. This is reflected in the large number of mitochondria and large amount of myoglobin present, making the tissue appear redder than fast-twitch muscles, which appear relatively white. Slow-twitch fibers are slow to fatigue and fire at lower frequency (10 to 20 Hz). Fast-twitch and slow-twitch fibers are matched with motor neurons that fire at characteristically fast and slow frequencies, respectively. Interestingly, neuronal phenotype (fast or slow) can influence muscle phenotype. Experimentally switching the normal slow-twitch motor neuron innervation of a slow-twitch muscle so that it innervates a fast-twitch muscle will cause the muscle to assume a slow-twitch phenotype. Further evidence for neuronal control over muscle metabolism is seen in the atrophy that occurs in muscles that lose innervation.

Suggested Readings

Donoghue J, Sanes J. Motor areas of the cerebral cortex. *Clin Neurophysiol* 1994; 1:382–396.
Haines DE, Ed. *Fundamental neurosciences.* New York: Churchill Livingstone, 1997.
Henneman E, Somjen G, Carpenter D. Functional significance of cell size in spinal motoneurons. *Neurophysiology* 1965; 28:560–580.

56

Sensory and Motor Pathways Controlling Lower Motor Neurons of the Spinal Cord

DIANNA A. JOHNSON

KEY POINTS

- *Muscle spindles* are nerve endings located within muscle tissue that monitor the length of a muscle as it contracts and elicit a reflex contraction (*stretch* or *myotactic reflex*) when the muscle is stretched beyond a certain limit.
- The *gamma loop system* consists of gamma motor neurons that innervate the small, intrafusal muscles within the muscle spindle organ and serves to adjust the tension on the spindle structure, thereby regulating its sensitivity.
- The *Golgi tendon organ* also monitors muscle length, but it elicits reflexes (the reverse myotactic reflex) that are opposite those of the muscle spindle.
- Basic spinal motor programs are built in part from combinations of myotactic and reverse myotactic reflexes along with reciprocal reflexes resulting from linkages between synergistic and antagonistic motor units.
- Cortical and brain stem regions provide motor instructions to alpha and gamma motor neurons in order to achieve more complex volitional and nonvolitional movements.

THE MUSCLE SPINDLE

The driving force on lower motor neurons comes from three major sources: (1) sensory pathways from the spinal cord and brain stem that trigger reflex actions, (2) interneurons within the spinal cord that interconnect synergistic and antagonistic motor neuron pools, and (3) upper motor neurons in the motor cortex and other motor areas in the brain that provide complex motor commands. One of the major sensory inputs to the lower motor neuron is derived from specialized end organs located within the muscle itself. Two types of specialized sensory endings are present within muscle tissue, and they provide feedback control over firing rates of the alpha motor neuron. The first type, the *muscle spindle*, consists of a group of fine muscle fibers 4 to 10 mm long encapsulated by a fusiform or spindle-shaped connective tissue sheath (Fig. 1). The ends of the sheath are attached to adjacent muscle fibers. The central portion of the individual *intrafusal muscle fiber* is encircled by a specialized

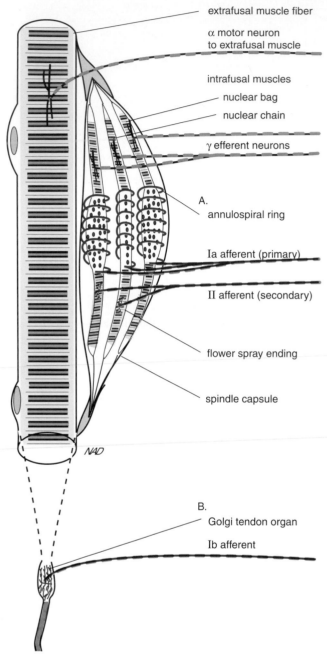

extrafusal muscle fiber

α motor neuron
to extrafusal muscle

intrafusal muscles

nuclear bag

nuclear chain

γ efferent neurons

A.

annulospiral ring

Ia afferent (primary)

II afferent (secondary)

flower spray ending

spindle capsule

NAD

B.

Golgi tendon organ

Ib afferent

FIGURE 1 Organization of the muscle spindle and Golgi tendon organ. Muscle spindles are small, fusiform structures embedded in skeletal muscles (extrafusal muscle fibers) that monitor the length of the muscle. Spindles contain small intrafusal muscles, each with their centrally located nuclei arranged either in a cluster called a *nuclear bag* or in a *nuclear chain*. A spindle contains about four nuclear bags and two nuclear chain fibers enclosed in a capsule. (A) Two types of sensory neurons (shown in blue) innervate the spindle: (1) the annulospiral ring, or Ia afferents, sensitive to both the absolute muscle length and rate of change in length during contractions; and (2) the flower spray, or II afferents, which monitor static length of the muscle. Muscle spindles are also innervated by small myelinated neurons called *gamma afferents* that directly stimulate intrafusal fibers (but not extrafusal muscles) to contract and thereby regulate the tension on the annulospiral ring and flower spray afferents. (B) A second type of sensory structure, the Golgi

mechanoreceptor nerve ending that forms a springlike configuration called the *annulospiral ring*. Changes in the length of fibers within the main muscle mass during contraction or relaxation passively change the tension on the intrafusal fibers through the connective tissue attachment site. This change causes compensatory stretching or relaxation of the sensory endings of the annulospiral ring. Stretching the annulospiral ring activates *mechanosensitive channels*, depolarizes the sensory neuron, and increases its firing rate. The opposite effect is observed when tension on the annulospiral ring is decreased by contraction of the main muscle mass. Thus, the firing rate of the sensory nerve encodes information about the degree of muscle stretch, which is then sent back to the appropriate alpha motor neuron. In this way, the motor neuron is constantly informed about the changes in muscle length that occur in response to any previous stimulation. Because of their small size, intrafusal fibers of the muscle spindle do not make a direct contribution to the force of muscle contraction; nevertheless, their constant monitoring of the efficacy of neuronal activation of the muscle is a central component of many muscle reflexes.

Several different types of intrafusal muscle fibers can be distinguished on the basis of cellular morphology and innervation patterns. *Nuclear chain intrafusal fibers* have nuclei that are arranged in a chainlike configuration, whereas *nuclear bag fibers* have nuclei that are clumped in a baglike fashion near the center of the fiber. Both nuclear chain and nuclear bag fibers are innervated by annulospiral rings from large, myelinated fibers, classified as *group Ia* or *primary afferents*. These sensory neurons adapt rapidly to changes in mechanoreceptor activation and thus act as *rate detectors*. They signal alpha motor neurons about the rate of change in muscle length and, to a lesser degree, the absolute length resulting from contraction or relaxation.

All nuclear chain fibers and one subtype of nuclear bag fibers receive a second type of innervation, classified as *group II afferents*. They are medium-diameter myelinated nerves with endings in a "flower spray" arrangement. They tend to be tonically active and adapt slowly, continuing to fire at a characteristic frequency that encodes the degree of stretch affecting the mechanoreceptors in the annulospiral ring. In this way, the group II afferents provide the alpha motor neuron with information about *static muscle length*, in contrast to the Ia fibers, which primarily provide information about the rate of change in muscle length.

tendon, is inserted between the ends of extrafusal muscle fibers and their attachments to bone. The connective tissue of the Golgi tendon organ is innervated by Ib afferents that monitor the force developed by the muscle during both contraction and passive stretch.

Group Ia fibers from the muscle spindle provide the afferent arm of an important spinal motor reflex, the *stretch* or *myotactic reflex*. Essentially, they provide the pathway by which a muscle automatically contracts immediately after it has been stretched. This reflex can be overridden by input from higher cortical centers in order to carry out purposeful movements; otherwise, it operates as a homeostatic device to resist an imposed stretch. One of its important functions is to maintain a given muscle length in order to balance the body against the pull of gravity.

The myotactic reflex pathway is monosynaptic, consisting of direct excitatory connections between Ia afferents from the muscle spindle and all the alpha motor neurons of the same or homonymous muscle. In response to Ia excitation, the motor neuron pool will fire and cause contraction of the muscle. For example, if an individual is balanced at the top of a stairway and inadvertently leans forward, the resulting stretch on the extensor muscles in the back of the leg will activate Ia fibers within the muscle and initiate a reflex contraction of the same muscle, thus returning the body to its upright position and avoiding a fall. This protective reflex relies only on the simple spinal cord stretch reflex pathway and does not require involvement of brain stem or cortical structures. In addition to its role in the stretch reflex, information from primary Ia fibers also assists higher motor centers in more complex motor computations.

THE GAMMA LOOP SYSTEM

The muscle spindle not only serves as a feedback system to the alpha motor neuron, but has within it a separate feedback system of its own called the *gamma loop system*. This system operates to maintain tension on the intrafusal muscle fibers. To provide information about static or changing muscle length, the annulospiral ring must be under some minimal amount of tension; otherwise, the stretch-sensitive channels remain closed, and no action potentials are generated. A correction for this potential problem is achieved through motor innervation of the small intrafusal muscle fibers themselves, specifically near the ends of the fiber, so that a minimum amount of tension can be maintained on the fibers within the capsule at all times (Fig. 2). The motor neurons innervating the muscle spindle fibers are not the large

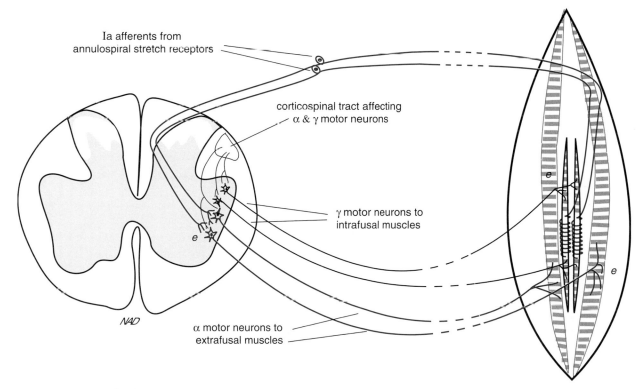

FIGURE 2 Innervation of the muscle spindle by gamma motor neurons. As gamma motor neurons become active, they cause greater contraction of intrafusal fibers and greater stretching of the annulospiral ring. This increased tension on the ring makes it more sensitive to any further change in length caused by contraction or stretching of the extrafusal muscle fibers. Less activity in the gamma motor neurons decreases the sensitivity of the annulospiral ring. Upper motor neurons, most of which have fibers in the corticospinal tract, can control activity in both alpha and gamma motor neurons.

alpha motor neurons that innervate the main muscle mass but a different set of smaller cells, the *gamma motor neurons*. The role of the gamma system is to allow the spindle to maintain high sensitivity over a wide range of muscle lengths during contraction and relaxation. Like alpha motor neurons, the gamma motor neurons are located within the ventral horn of the spinal cord; they also receive substantial input from higher motor centers. A number of descending pathways indirectly influence motor activity through control of the drive on the gamma loop system, thus making muscles more or less susceptible to myotactic reflex responses.

THE GOLGI TENDON ORGAN

The second type of specialized sensory receptor found in muscle tissue is the *Golgi tendon organ*. Its function is to signal the amount of tension generated by muscle contraction. The end organ is composed of braided collagen fibers within a capsule approximately 1 mm in length and 0.1 mm in diameter (Fig. 3). It is innervated by a free nerve ending classified as Ib, slightly smaller than Ia fibers of the muscle spindle. Golgi tendon organs are located at junctions between muscles and their tendinous insertion points. As the muscle contracts, it stretches the capsule of the Golgi tendon organ and causes the Ib fiber to discharge. Information about the amount of tension generated is fed back to interneurons in the spinal cord, forming a polysynaptic feedback loop that inhibits the activity of homologous alpha motor neuron pools. This reflex is called the *reverse myotactic reflex*, and it

prevents the muscle from extreme contraction that could be damaging.

In summary, the muscle spindle and the Golgi tendon organ provide counterbalanced systems for setting overall muscle tone. Group Ia and II afferents from muscle spindles carry information about the static length of the muscle and its rate of change during contraction and relaxation. Because the muscle spindle is connected to the main muscle fibers in parallel, passive stretching or relaxation lengthens the spindle, excites the sensory end organ, and increases the firing rate of Ia and II fibers. Through a monosynaptic feedback pathway (the myotactic reflex), Ia fibers excite alpha motor neurons and cause muscles to contract after they are passively stretched.

In contrast, the Golgi tendon organ is connected to the muscle in series, so that contraction of the muscle causes stretching of the nerve ending and an increased firing rate in the Ib fiber. Through a polysynaptic feedback pathway (the reverse myotactic reflex), Ib fibers inhibit alpha motor neurons and cause muscles to relax after they contract. Both pathways offer protection to the body and to the muscle tissue itself, particularly at extremes of contraction or stretch. At lesser loads, both reflex pathways continually modulate activity of the alpha motor neuron, increasing or decreasing its tendency to fire in response to changes in ongoing muscle activity.

RECIPROCAL REFLEXES

Two important principles should be recognized: (1) The myotactic and reverse myotactic reflexes

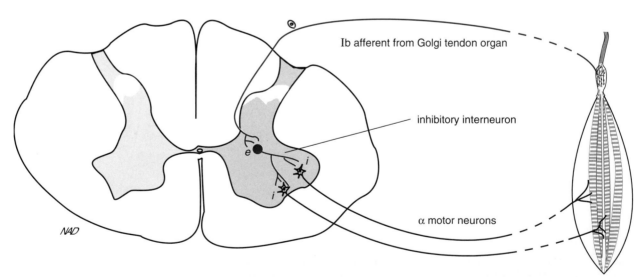

FIGURE 3   Organization and innervation of the Golgi tendon organ. Neurons involved in the Golgi tendon (Ib) reflex measure the force between the muscle and its insertion. When the receptor of the Golgi tendon organ is stretched, either by passive stretching or muscle contraction, it activates an interneuron that inhibits the motor neurons going to the same muscle. Shown in blue are neurons that are excited when the Golgi tendon organ is stretched.

represent key ingredients in most movements of the body, from the most elementary to the most complex. These reflexes might be viewed as representing basic "units" of spinal motor programs. (2) These reflexes do not occur as isolated events; rather, they are ongoing and must be carried out in a coordinated fashion among related muscle groups. This is achieved by a third basic component of spinal motor programs, *reciprocal reflexes* (Fig. 4). For contraction of a selected muscle to succeed in accomplishing the intended movement of the body, synergistic muscles must also be activated to participate in the movement, and antagonistic muscles must be inhibited so they do not interfere. This is accomplished by interneuron networks in the spinal cord that provide excitatory

A. Muscles and neurons involved in a pain withdrawal reflex

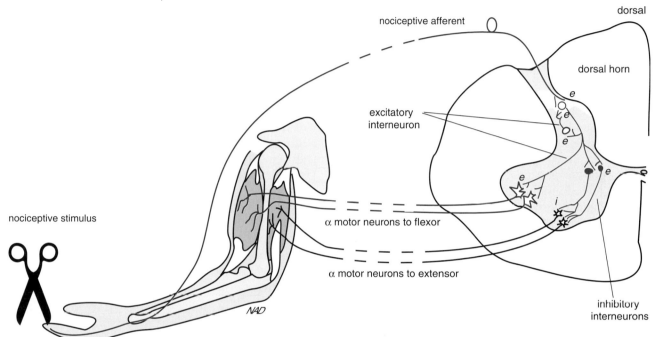

B. Crossed extensor pain reflex and central paths

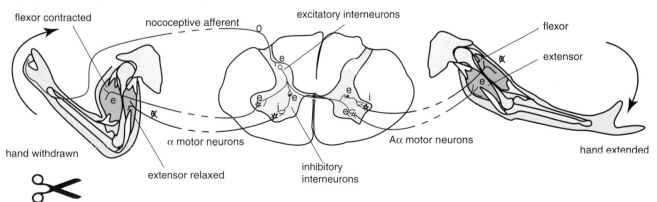

FIGURE 4   Circuitry of the polysynaptic flexor (withdrawal) and crossed extensor reflexes. Neurons that are excited by nociceptive input are shown in blue. (A) Pain withdrawal reflex; stimulated nociceptive fibers send excitatory information to activate interneurons (e) in the spinal cord. Interneurons then stimulate alpha motor neurons that activate flexors, and they inhibit alpha motor neurons that drive extensors. (B) A strong nociceptive stimulus can elicit a crossed extensor response through fibers that cross to the contralateral side to stimulate and inhibit appropriate alpha motor neuron groups. In the legs, this reflex keeps one standing. In the arms, it helps one push away from the irritant and maintain balance.

reciprocal connections among synergistic motor units and inhibitory reciprocal connections between pairs of antagonistic motor units. Cell bodies of these interneurons are dispersed throughout the length of the cord in the ventral horn and intermediate gray matter. Their processes ascend and descend in the *anterolateral funiculi* to reach appropriate levels where they decussate in the anterior white commissure to the contralateral side to link necessary components.

Other types of spinal motor programs provide protective reflexes that quickly remove the body or extremity from the dangerous object (the *withdrawal* or *nociceptive reflex*) or remove the potentially dangerous object from the body (the *wipe* or *scratch reflex*). Nociceptive fibers subserving the withdrawal reflex enter the spinal cord and activate interneurons that form excitatory connections, with flexor motor neurons supplying the entire ipsilateral limb and reciprocal inhibition of extensor muscles to that limb. The result is a rapid flexion of the limb. If the flexion reflex involves the leg, a compensatory extension of the opposite leg must occur to maintain balance. This is achieved automatically through a special type of reciprocal innervation to elicit a *crossed-extensor reflex*.

The scratch reflex is more complex in that it involves repetitive or rhythmic contraction and relaxation of muscles in the arm and fingers in response to a localized noxious stimulus. Even so it is achieved by interneuron circuits residing solely in the spinal cord and is indicative of the sophistication in motor programs that can be achieved without involvement from the brain.

Similar alternating movements are involved in walking. They are reflexive and are also programmed in the spinal cord. They do not require descending commands from upper motor neurons in motor centers of the brain. Though not completely understood, it has been demonstrated in experimental animals that complete transection of the cord at the midthoracic level leaves the hind limbs capable of generating coordinated walking movements. Groups of interneurons in the spinal cord act as central pattern generators, providing a series of alternating commands to control extension and contraction movements that provide the essential building blocks for locomotion. Additional interneuronal connections between the lumbar and cervical spinal segments are necessary to regulate swinging of the arms that accompanies walking.

DESCENDING MOTOR TRACTS IN THE SPINAL CORD

Reflex activity generated from motor programs in the spinal cord represents the foundation of the motor hierarchy. These programs provide the basic plans by which movement can be achieved in a coordinated fashion. In addition, they facilitate the transfer of information about more complex movements and about volitional movement from brain to appropriate groups of lower motor neurons. Motor input to spinal cord levels is received in two major pathways: (1) the *ventromedial pathway*, consisting of four major tracts from various brain stem structures primarily concerned with posture and locomotion, and (2) the *lateral corticospinal tract*, which carries information for volitional movement of proximal and distal muscles under direct cortical control (Fig. 5).

The *vestibulospinal tract* is composed of two subdivisions, each originating from separate vestibular nuclei in the brain stem and descending in slightly different regions of the cord. One of the main functions of this tract is to keep the body, particularly the head region, balanced during movement and to orient the body and head in the direction of new sensory stimuli. Motion in the fluid of the vestibular labyrinth of the middle ear accompanies movement of the head. Activated hair cells then signal specific vestibular nuclei via cranial nerve VIII. Axons from the *lateral vestibular nuclei* form the *lateral vestibulospinal tract*, which projects ipsilaterally in the ventrolateral part of cord throughout its length. This tract terminates in the lateral part of the *intermediate zone* on interneurons that innervate both alpha and gamma motor neurons of extensor muscles in upper and lower extremities.

The *medial vestibulospinal tract* originates in the *medial vestibular nucleus* and descends bilaterally only to midthoracic levels, traveling in a compact myelinated fiber bundle called the *medial longitudinal fasciculus* (MLF). It terminates in the intermediate zone and provides inhibitory input to neurons innervating muscles of the neck and back. The combined activity of the lateral and medial vestibulospinal tracts serves to control balance, posture, and muscle tone; its net effect is to excite antigravity muscles.

The *tectospinal tract* originates from the *superior colliculi*, a structure that forms part of the *tectum*, or roof of the midbrain. The superior colliculus is a coordinating center for visual information, gathering visual input from the retina and the visual cortex, as well as additional sensory information from somatosensory and auditory systems. From these inputs it constructs a spatial map of the immediate environment with respect to the visual field. Cells of the colliculus are retinotopically organized so that a novel sensory stimulus occurring at one point in the visual field leads to stimulation of a corresponding site in the colliculus. In response, collicular neurons become activated and send appropriate signals to motor neurons in the spinal cord, particularly those innervating the head and neck. The head and eyes reflexly move to the appropriate point in space and allow the object of interest to be imaged on the fovea of the retina. The tectospinal

A Four driving forces on lower motor neurons

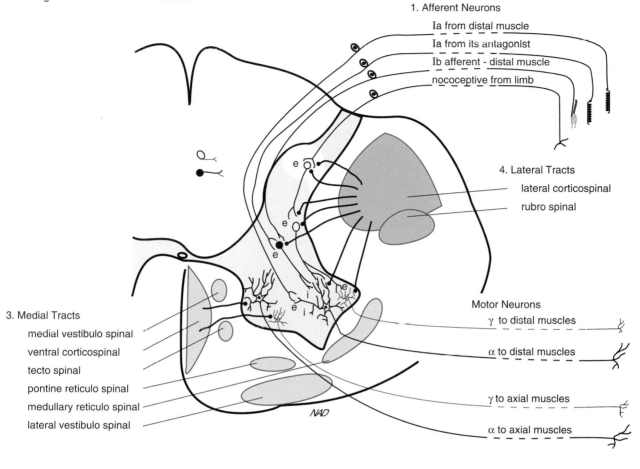

1. Afferent Neurons

Ia from distal muscle

Ia from its antagonist

Ib afferent - distal muscle

nococeptive from limb

4. Lateral Tracts

lateral corticospinal

rubro spinal

Motor Neurons

γ to distal muscles

α to distal muscles

γ to axial muscles

α to axial muscles

3. Medial Tracts

medial vestibulo spinal

ventral corticospinal

tecto spinal

pontine reticulo spinal

medullary reticulo spinal

lateral vestibulo spinal

NAD

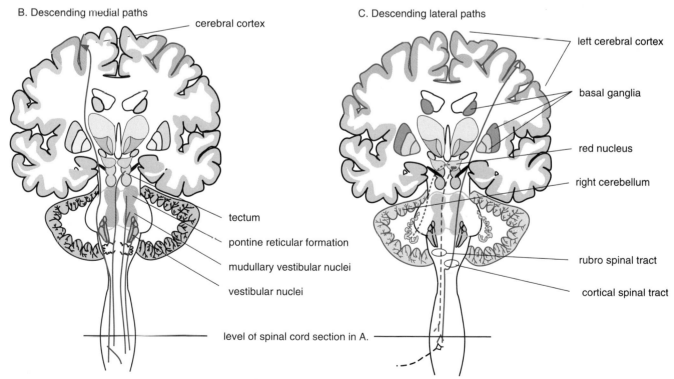

B. Descending medial paths

cerebral cortex

tectum

pontine reticular formation

mudullary vestibular nuclei

vestibular nuclei

level of spinal cord section in A.

C. Descending lateral paths

left cerebral cortex

basal ganglia

red nucleus

right cerebellum

rubro spinal tract

cortical spinal tract

Clinical Note

Lesions and Motor Reflex Abnormalities

Specific lesions within the neuronal components of the somatic motor system can cause characteristic abnormalities in motor reflexes. *Flaccid paralysis* usually denotes the loss of peripheral nerve innervation. Without input from alpha motor neurons of the spinal cord, there is no voluntary movement of the muscle and no resistance to passive movement. No reflexes can be elicited, and atrophy of the muscle will be observed. Unlike components of the central nervous system (CNS), axonal processes of peripheral nerves can regenerate in some cases, particularly if the lesion is localized to peripheral regions of the nerve. Reinnervation can take a year or more, after which time patients may experience a partial or (in rare cases) complete return of function and a reversal in the process of muscle atrophy.

Muscle rigidity is seen in patients with damage to descending spinal tracts that normally inhibit alpha motor neurons or, more commonly, with damage to motor areas in the brain from which these inhibitory tracts originate. Without normal inhibitory input, motor neurons are overly active and cause heightened contraction in somatic muscles. The patient may have difficulty in performing most voluntary and involuntary movements because the balanced (inhibitory versus excitatory) nature of the motor command system

will be disturbed (see Chapter 57). Examination of the patient will demonstrate increased resistance to passive motion in all directions in all muscles because the lesion is central in origin and thus will affect all levels of the spinal cord. A common cause of rigidity is Parkinson's disease, which affects certain motor regions of the basal ganglia (see Chapter 58).

Muscle spasticity is clinically defined as an abnormal increase in muscle tone due to loss of inhibition of gamma motor neurons. Because gamma motor neurons normally regulate the output of muscle stretch receptors, disinhibition of these neurons will lead to an exaggerated (spastic) stretch reflex in the affected muscles. Like muscle rigidity, spasticity reflects loss of descending tracts to the spinal cord. However, in this case the abnormal reflexes characteristically result from a general lesion of the cord, such as spinal transection, which destroys all CNS input to alpha and gamma motor neurons below the level of the lesion. Patients with a severed spinal cord will initially undergo a period of spinal shock, characterized by flaccid paralysis of muscles below the transection. Voluntary control of these muscles is permanently lost, but over time the flaccid paralysis will be followed by muscle spasticity evidenced by a much increased response in tendon reflexes.

---

tract acts in conjunction with the vestibulospinal tract, particularly the medial portion, in order to bring about this orientation response. Because of its relatively small size, the tectospinal tract may be the less important of the two tracts involved in the orientation response.

Two other descending motor tracts originate from the *reticular formation* in the brain stem. Located in the central core of the pons and medulla just anterior to the cerebral aqueduct, the reticular formation is composed of a highly interconnected meshwork of neurons that is not organized in discrete nuclei. This reticular organization serves a number of coordinating functions. As previously described, gaze centers of the oculomotor system are located within the reticular formation. Two motor

centers regulating spinal motor neurons are also recognized. Cells in the *pontine reticular formation* give rise to the *pontine reticulospinal tract*, which projects ipsilaterally along with other fibers in the MLF to terminate on alpha and gamma motor neurons, innervating axial muscles and extensor muscles of the limbs. Through its excitatory input it serves to increase muscle tone.

The *medullary reticulospinal tract* originates in the *medullary reticular formation*, projects bilaterally, and terminates on motor neurons innervating axial and extensor muscles of the limbs. This pathway serves to balance the excitatory drive from the pontine reticulospinal tract by inhibiting motor neurons and decreasing axial and extensor muscle tone. Having separate pathways for excitation

---

FIGURE 5   Driving forces to lower motor neurons. (A) Inputs and outputs from alpha and gamma motor neurons in the spinal cord: (1) Sensory input enters the dorsal aspects of the cord; (2) excitatory and inhibitory interneurons are concentrated in the central gray of the spinal cord and brain stem; (3) medial tracts are comprised of the uncrossed neurons of the corticospinal tract and neurons descending from the brain stem; (4) lateral tracts arise in the cerebellum (relayed through the red nucleus) and the cortex. (B) Coronal section of the brain through the precentral gyrus shows the origin of the descending medial tracts. (C) Coronal section shows the origin of the descending lateral tracts.

and inhibition of the same motor units may seem unnecessary and redundant; nevertheless, there are numerous examples of this type of dual-control organization within the nervous system, leading to the conclusion that the enhanced control and sensitivity it provides must offer significant advantages. The motor cortex provides strong input to the reticular formation and in this way indirectly regulates the ongoing tone of the same motor units that it activates directly through the corticospinal tract.

One of the largest and most important descending tracts in the spinal cord is the *lateral corticospinal tract* or *pyramidal tract* (Fig. 6). It contains 1 million fibers, half of which originate from large pyramidal cells within the primary motor cortex located just anterior to the central

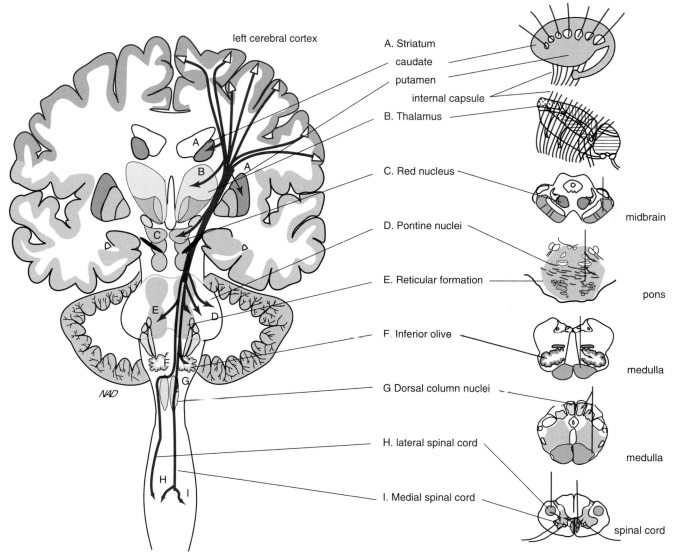

left cerebral cortex

A. Striatum
caudate
putamen
internal capsule
B. Thalamus
C. Red nucleus — midbrain
D. Pontine nuclei
E. Reticular formation — pons
F. Inferior olive — medulla
G Dorsal column nuclei — medulla
H. lateral spinal cord
I. Medial spinal cord — spinal cord

NAD

FIGURE 6 Targets of corticofugal projections. The CNS is shown in coronal section (left), the striatum and thalamus are shown in lateral view, and the brain stem and spinal cord are shown in cross section (right). Descending corticofugal fibers (shown in blue) originate from widespread areas of the motor cortex and coalesce into fiber bundles called the *internal capsule*. They then penetrate nuclear groups in the diencephalon (caudate and putamen) and project through the brain stem. They undergo partial decussation in the medulla and project through the spinal cord. Specific tracts within the main fiber bundles are named according to their targets: The *corticostriate tract* (A, to the caudate and putamen) and the *corticothalamic tract* (B, to the thalamus) are involved in "motor loops" that coordinate movement commands (see Chapter 58). The *corticorubro tract* (C, to the red nucleus) and the *corticopontine tract* (D, to the pontine nuclei) influence cerebellar function (see Chapter 59). The *corticoreticular tract* (E) acts on many of the motor nuclei of the brain stem. The *cortico-olivary tract* (F) and *corticocuneate/gracile tract* (G) modify sensory processing. The *lateral corticospinal tract* (H, descending contralaterally) influences motor control of distal muscles by terminating on neurons in the dorsal, intermediate, and ventral parts of the spinal cord. The *ventral corticospinal tract* (I) descends ipsilaterally and controls proximal muscles spinal cord.

sulcus. The fibers coalesce as they descend in the *internal capsule*, traveling ipsilaterally through the course of the brain stem. They form a pair of thick, pyramidal-shaped bundles on the posterior aspect of the lower medulla. At the junction of the brain stem and spinal cord, most of these fibers decussate and move to the *lateral funiculus* of the cord. Fibers leave the main fiber bundle at all levels of the brain stem and spinal cord, providing conscious control over all cranial nerve motor nuclei and motor segments of the cord. Their role is to initiate voluntary movement.

The cerebellum and the basal ganglia also provide important input to spinal motor neurons, both directly and indirectly through projections involving the motor cortex. These pathways will be described in succeeding chapters after first considering the organization of the motor cortex in more detail.

Suggested Readings

Boyd IA. The isolated mammalian muscle spindle. *Trends Neurosci* 1980; 3:258–265.

Grillner S, Wallen P. Central pattern generators for locomotion, with special reference to vertebrates. *Annu Rev Neurosci* 1985; 8:233–261.

Hullinger M. The mammalian muscle spindle and its central control. *Rev Physiol Biochem Pharmacol* 1984; 101:1–110.

Pearson K. The control of walking. *Sci Am* 1976; 235(6):72–86.

Stein RB. What muscle variable does the nervous system control in limb movements? *Behav Brain Sci* 1992; 5:535.

57

The Motor Cortex

DIANNA A. JOHNSON

KEY POINTS

- Conscious movement is initiated by upper motor neurons in an area of the frontal cortex designated as MI.
- *Pyramidal cells* in the MI issue motor commands based on input from the adjacent motor association area designated as MII, as well as from prefrontal cortex, motor nuclei in the thalamus, and primary (SI) and association (SII) sensory cortex.
- The firing rate of an individual pyramidal cell correlates with the force of the resulting muscle contraction; the average firing rate of a given

cluster of pyramidal cells correlates with the direction of resulting limb movement.
- The MiI motor association area is further divided into a *supplemental motor area*, which coordinates bilaterally coordinated movements, and a *premotor area*, which orients the body for upcoming movement.
- The prefrontal cortex initiates the initial planning stages of the movement, and the sensory association area (SII) focuses attention on parts of the body to be involved in the movement.

The *primary motor cortex* consists of a strip of cortical tissue in the frontal lobe (Fig. 1). On the surface, it is separated from the primary somatosensory cortex by the *central gyrus*. However, these two cortical areas are functionally connected by a bridge of tissue, the *paracentral lobule*, that follows the contour of the central sulcus and links the two areas. Cells in both motor and sensory strips are arranged somatotopically, providing both a motor homunculus and a sensory homunculus that are in register with each other. As discussed in the preceding chapters, the primary sensory cortex is the

primary receiving area for somatosensory information, and its activity correlates with conscious perception of the sensation; however, higher order processing of sensory information occurs in association areas of the parietal cortex. A corollary can be seen in the arrangement of the motor pathway. The primary motor area is the site of initiation of conscious movement; however, events preceding the generation of the initiation signal occur elsewhere in motor association areas in the frontal cortex and in the parietal lobe in combination with sensory processing. Additional components of planned movements occur in

A. Lateral view of right cerebral cortex

B. Medial view of right cerebral cortex

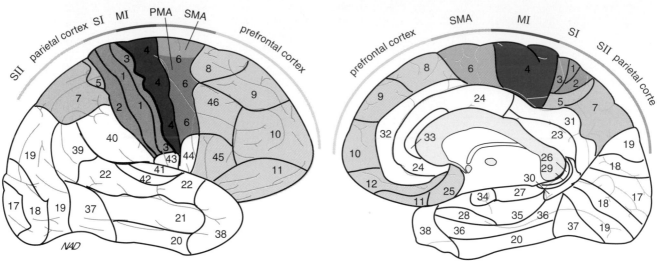

C. Origin of corticofugal paths

D. Somatic destination of motor cortex neurons

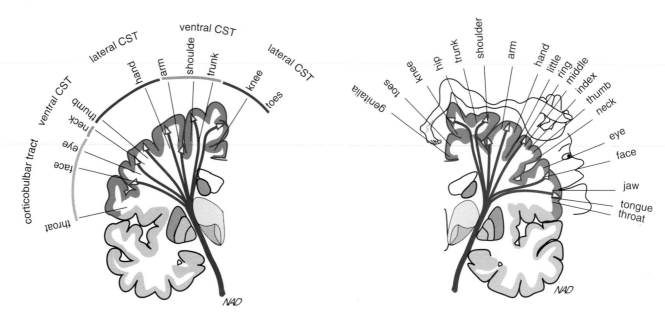

FIGURE 1   Organization of motor and sensory cortex. (A) Lateral and (B) medial views depicting Brodmann's numbering system for cortical areas. Motor I (MI, or area 4) is shown in dark blue. Area 6 contains the premotor area (PMA) laterally and the secondary motor area (SMA) medially. The primary sensory cortex (SI) is shown in gray. Two association areas involved in motor planning are the prefrontal cortex (light blue) and the posterior parietal cortex (SII) (light gray). (C) Origin of the corticofugal paths from the primary motor cortex; the corticobulbar tract supplies the head, the ventral corticospinal tract (CST) supplies the axial and proximal muscles of the body, and the lateral CST supplies the distal muscles of the extremities. (D) Location of MI areas controlling various areas of the body. The homunculus is drawn to show the relative size of cortical representation. Note the huge area devoted to control of the face and hand compared to the area that controls of the rest of the body.

subcortical regions—namely, the basal ganglia and the cerebellum (as discussed in Chapters 58 and 59).

Recognition of the motor strip as being the primary motor area is based on its obligatory role in initiating movement and its low threshold for eliciting a motor response. Several other areas of brain, when experimentally stimulated, can lead to movement, but the motor strip does so at very low thresholds. The output neurons of the motor strip are *pyramidal cells*, the major class of upper motor neurons in the brain. These large cells are located in cortical layer V, and their size and number help distinguish this area from surrounding ones. This type of histologic characterization has been used in establishing the Brodmann classification scheme, which defines 50 different areas of brain. The motor strip is designated as Brodmann's area 4, and the motor association area just anterior to it is designated as Brodmann's area 6. In a more recent classification scheme based on functional differences, area 4 is designated as MI, area 6 as MII, the primary somatosensory area as SI, and the sensory association area just posterior to it as SII.

As pyramidal cell axons exit the motor cortex, they project as corticofugal fibers to lower motor neurons throughout the brain stem and spinal cord. Three major divisions are recognized: (1) the *corticobulbar tract*, which supplies cranial nerve motor nuclei of the head and neck region; (2) the *ventral corticospinal tract*, which innervates lower motor neurons of the axial and proximal muscles of the body; and (3) the *lateral corticospinal tract*, which projects to lower motor neurons of the distal muscles of the extremities. Approximately half of the corticofugal fibers arise from MI. The remainder originate from MII, SI, and SII.

The current theory about pyramidal cell function is that these cells represent primary threshold elements or gatekeepers that make the final decision about when to trigger the motor command. The urge to initiate a purposeful movement does not originate *de novo* from within pyramidal cells in MI; rather, these cells occupy a central position within various input loops, supplying sensory information about current internal and external environmental conditions, as well as information from past experience in the form of sensory and motor memory. Direct inputs carrying various forms of sensory information to MI and MII motor neurons come from nuclei in the thalamus (VL, VA, and CM) and from other cortical areas, including motor association areas in the frontal lobe and areas SI and SII. These are the staging areas for gathering sensory information from widespread regions. Goal-directed movements depend on knowledge of where the body is in space, where it intends to go, and how it intends to get there. Once a plan for movement has been selected, it must be held in memory until the time when instructions are issued to implement the plan. Although MI plays the major role in issuing motor commands, other areas are responsible for the planning stages.

MI PYRAMIDAL CELLS

Electrophysiologic analysis of MI pyramidal cells has demonstrated two ways in which motor commands are encoded. First, the force of muscle contraction is directly related to the firing rate of the appropriate pyramidal cell. High-frequency firing of the upper motor neuron stimulates a high-frequency firing rate of the lower motor neuron, causing summation of intracellular responses in the muscle and increased tension. The direction of the intended movement is encoded in a different way; it involves the combined responses of populations of pyramidal cells. Individual pyramidal cells in MI are directionally selective and fire most vigorously in association with initiation of movement in the preferred direction; however, they are only broadly tuned, meaning that increased firing can occur in association with movements that vary 45° either way from the preferred direction. The firing pattern of any one cell alone is thus a poor predictor of movement direction, but the average output of clusters of broadly tuned, directionally selective cells is highly predictive, providing a combined, or *population movement* vector that matches well with the direction of muscle movement. These clusters of cells are arranged in the cortex so that population vectors are provided for directions covering a full 360°.

HIERARCHICAL ARRANGEMENT OF MOTOR CORTICAL AREAS

Firing of MI upper motor neurons is associated with relatively simple motor commands; more complex movements are linked to upper motor neurons in area MII (Fig. 2). Two separate somatotopic maps are present in MII: The *supplementary motor area* (SMA) is located near the superior medial region of the cortex, and the *premotor area* (PMA) occupies a more lateral position. Pyramidal cells in these areas contribute to corticofugal pathways and are also heavily interconnected to MI. Both areas of MII elicit complex motor responses, but they appear to be involved in somewhat different aspects of generating the motor command, primarily in integrating specific strategies. It has been shown experimentally that the SMA is required for bilaterally coordinated movements. For example, specific cells in the SMA fire in accord with movements in either hand. Individuals with cortical lesions in the SMA area suffer from a condition called *apraxia*. They retain the ability to make simple movements but have a selective inability to perform complex tasks requiring the coordinated actions of two hands, such as buttoning a shirt.

The PMA area is also involved with planning complex movements; however, its control over lower motor neuron activity is indirect. In addition to its reciprocal connections with MI, its major descending projection is to the reticular formation, which in turn projects to spinal neurons controlling axial muscles. This pathway

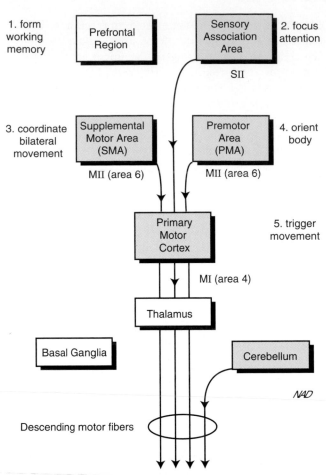

FIGURE 2 Schematic representation of the hierarchical arrangement of the cortical motor pathways. Formulation of a motor plan consists of five basic steps, each of which is associated with a specific cortical region. Three other regions—the thalamus, basal ganglia, and cerebellum—modify the primary motor plan through connecting loops involving these cortical regions. Areas shown in blue provide descending motor fibers that constitute corticobulbar and corticospinal tracts.

acts to orient the body for upcoming movement. Firing of PMA pyramidal cells occurs prior to firing of command neurons in MI. The pattern of firing in PMA is correlated to the direction of the upcoming movement and is maintained until the movement is completed.

Both elements of MII act as higher order motor areas, projecting to MI and providing complex motor commands to lower motor neurons through direct and indirect projections. The highest level in the motor hierarchy is represented by two other regions: *sensory association areas in SII* and the *prefrontal region* which lies anterior to MII. Area SII represents a highly important functional component in the planning of movement, providing focused attention on the parts of the body and the elements in the environment that will likely be involved in the intended motor action. In experiments that visualize neuronal activity in brain areas using positron emission tomography (PET) scanning techniques, the SII region on the

lateral aspect of the parietal lobe appears to be most active when human subjects are asked to think about a specific movement, whether or not the movement is subsequently initiated. MI neurons were not activated under these conditions. Lesions in this area of the brain produce peculiar symptoms, collectively called the *neglect syndrome*, in which the affected individual neglects specific parts of the body corresponding to the areas of the lesioned somatosensory association area. These individuals appear to lose awareness of the affected body regions, and complex motor activity involving this musculature is impaired.

The prefrontal area represents the second major component of the highest motor level. Of the variety of higher cortical functions that have been ascribed to this region, perhaps the best studied involves the role it plays in *delayed-response tasks*. Behavior tests in experimental animals suggest that the prefrontal area plays a pivotal role in temporarily storing information used to guide future action, namely, the formation of a *working memory*. The ability to hold key information in a working memory for a period of time is required in order to weigh the consequences of future actions and to plan accordingly. It is interesting to note that the prefrontal area of primates has a particularly prominent dopaminergic innervation and that depletion of dopamine is thought to contribute to human cognitive disorders, such as *schizophrenia*. Schizophrenics have smaller frontal lobes than normal individuals, and they show altered activation patterns in PET scans of the frontal cortex, particularly when involved in delayed-response motor tasks.

In summary, the primary motor cortex initiates volitional movement not as the first step but as the third step in the hierarchical motor scheme. The prefrontal cortex formulates a working memory that serves in the initial planning of motor events to be completed later; the SII areas of the parietal lobe help focus attention on the event. This information filters to the next step in the scheme, triggering neurons in area MII to plan the coordination of motor commands to accomplish the intended movement. With successful completion of these preliminary steps, MI neurons, through their powerful connections with alpha motor neurons in the cord, set the selected program in motion.

Suggested Readings

Mardsen CD, Rothwell JC, Day BL. The use of peripheral feedback in the control of movement. *Trends Neurosci* 1984; 7:253–257.

Miles FA, Evarts E. Concepts of motor organization. *Annu Rev Psychol* 1979; 30:327–362.

Polit A, Bizzi E. Processes controlling arm movements in monkeys. *Science* 1978; 201:1235–1237.

Stuart DG, Enoka RM. Motoneurons, motor units and the size principle, in: Rosenberg RN, Ed. *The clinical neurosciences*, Vol. 5, *Neurobiology*. New York: Churchill Livingstone, 1983; 471–517.

58

The Basal Ganglia

DIANNA A. JOHNSON

KEY POINTS

- Coordination of cortical motor commands is achieved by motor loop pathways that pass information from the cortex through the basal ganglia and thalamic nuclei and back to the cortex.
- The direct (excitatory) and indirect (inhibitory) pathways of the basal ganglia loop provide opposite, counterbalanced influences on activity of the motor cortex.

- The *dopaminergic nigrostriatal pathway* modulates activity in the direct and indirect pathways, maintaining a critical level of excitatory drive for formulation of coordinated motor commands.
- Specific motor disorders are caused by deficits in the dopaminergic nigrostriatal pathway (*Parkinson's disease*), loss of GABAergic neurons in the corpus striatum (*Huntington's chorea*), or damage to the subthalamus (*hemiballismus*).

The *basal ganglia* are a collection of five pairs of nuclei located in the diencephalon, deep to cortical structures (Fig. 1). One of the nuclei, the *caudate*, has a highly irregular shape. It is long, thin, tapered, and curved into the shape of a ram's horn. It follows the line of the lateral ventricles, with its thin tail adjacent to the temporal horn of the ventricle and its rounded head region adjacent to the lateral aspect of the anterior or frontal horn. The *putamen* and *globus pallidus* collectively are referred to as the *lentiform* or *lenticular* nucleus because of their combined lenslike shape. These two nuclei, along with the head region of the caudate, are interposed between the cortex and its pre- and postsynaptic targets in the brain stem and spinal cord. The bundles of connecting fibers

passing to (*spinocortical* and *bulbar cortical*) and from (*corticofugal*) the cortex are merged in this region, forming a broad band of myelinated processes called the *internal capsule*, which penetrates the group of three basal ganglia, partitioning the caudate from the lentiform (putamen and globus pallidus) nucleus. The contrast between the gray matter of the nuclei and the white myelinated fibers of the internal capsule gives this region of brain a striped appearance. For this reason, the caudate, globus pallidus, and putamen, along with the penetrating fibers of the internal capsule, collectively are referred to as the *corpus striatum*. Because of functional similarities, the caudate and putamen are often considered as a single unit, referred to as the *neostriatum*.

A. Lateral view

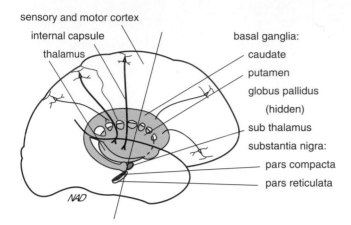

B. Coronal section

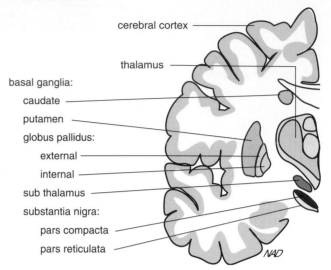

FIGURE 1   Anatomy of basal ganglia and thalamic structures. (A) Lateral view of the brain showing the basal ganglia and thalamus lying under the cerebral cortex. (B) Coronal section taken along the line indicated in (A). The basal ganglia are shown in blue; the cortex and thalamus, in gray. Combined nuclear groups of the basal ganglia form a large, knoblike structure beneath the projecting fibers (internal capsule) from the superficial cortical layers. Divisions of the basal ganglia include the long, C-shaped caudate encircling the flattened putamen. Together, these constitute the input nuclei called the *neostriatum*. The globus pallidus (with internal and external segments) plus the putamen form a lens-shaped mass termed the *lenticular nucleus*. Below these nuclei lie the subthalamus and the substantia nigra (with the subdivisions *pars compacta* and *pars reticulata*). The major targets of the basal ganglia are the ventral anterior, ventral lateral, and centromedian nuclei of the thalamus.

Two additional nuclei of the basal ganglia group, the *subthalamic nucleus* and *substantia nigra*, are located inferior and medial to the corpus striatum, just beneath the thalamus. The basal ganglia represent an important component of the motor system. Previously considered as the *extra-pyramidal system*, the basal ganglia system is now understood to be an integral part of the corticospinal or pyramidal system rather than a stand-alone circuit. Current theory holds that basal ganglia participate in a side loop pathway that fine-tunes motor instructions as they are being programmed in motor area II (MiI) (Fig. 2). This motor loop is responsible for two specific modulatory functions: (1) scaling motor patterns in the context of the task requirements, and (2) controlling the assembly of overall motor plans. For example, motor instructions for writing your name are assembled in cortical brain regions, but the basal ganglia add a scaling factor for the size of the letters so they are appropriate for large writing on a blackboard or small writing in a checkbook. The overall function of the basal ganglia is to enable automatic performance of practiced motor acts. They do not initiate movement; rather, they adjust and update motor

commands in preparation for the next movement in the sequence.

THE MOTOR LOOP: DIRECT PATHWAY

The loop connection between the cortex and the basal ganglia has two branches, a *direct pathway* and an *indirect pathway* (Fig. 3). Although both interconnect the cortex, basal ganglia, and thalamus, the overall effects of these two pathways are opposite and tend to counter-balance each other. Fibers in the direct pathway originate from the entire cerebral cortex and project through the corticostriate pathway to terminate in the neostriatum (caudate and putamen). Information is then sent to the internal portion of the globus pallidus and to the substantia nigra. Fibers exit the basal ganglia and innervate the *ventral anterior* (VA) and *centromedian* (CM) *nuclei* of the thalamus. The loop is completed by thalamocortical fibers that project back to the supplementary motor area (SMA) region of the motor cortex. Of the four synaptic relays in this circuit, the first (cortex to basal ganglia) and last (thalamus to cortex) are

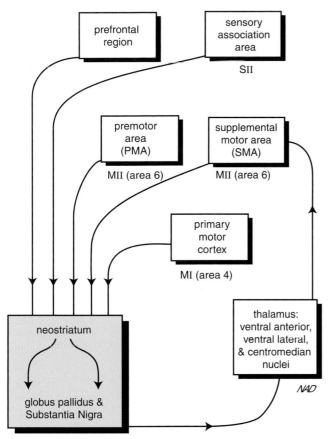

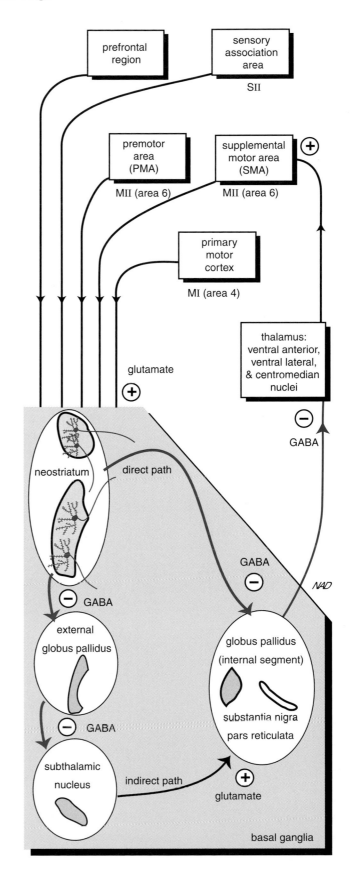

FIGURE 2 The basal ganglia motor loop. Major connections linking motor cortex, basal ganglia (blue box), and thalamus are shown in this simplified scheme. Cortical areas representing all three levels of the motor hierarchy project to the neostriatum (caudate and putamen). Output from the basal ganglia is via the globus pallidus and substantia nigra to three nuclei of the thalamus. The thalamus completes the loop pathway with projections back to the motor cortex, particularly the supplemental motor area.

FIGURE 3 Connections of the direct and indirect pathways in the basal ganglia motor loop. The overall role of the basal ganglia (blue box) is to maintain inhibitory control over the thalamus. Inhibitory influence (−) is shown with blue arrows and excitatory influence (+) is shown with black arrows. Pathways through the basal ganglia are tonically active and coordinate converging excitatory input from the cortex through two separate routes, both of which feed back to the cortex through the thalamus. The direct path includes an inhibitory connection between the neostriatum and the globus pallidus/substantia nigra. Activity in this pathway decreases the level of inhibitory control over the thalamus, which leads to an increase in excitatory feedback to the SMA. The opposite occurs with activity in the indirect pathway. Note that the indirect pathway stimulates the inhibitory fibers to the thalamus, thus decreasing the excitatory feedback to the cortex. The appropriate balance between direct and indirect pathways maintains a suitable excitatory tone within cortical motor pathways.

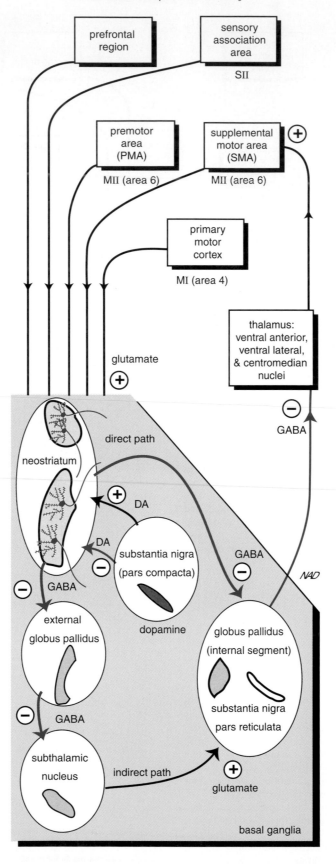

Clinical Note

Basal Ganglia Dysfunction

The principles of excitation–disexcitation/inhibition–disinhibition are particularly important in formulating strategies for specific drug therapies used in treating certain motor disorders such as *Parkinson's disease* (Fig. 5A). For reasons that are not yet understood, there is a pronounced degenerative decline in the number of dopaminergic cells in the substantia nigra in certain individuals. As these dopaminergic cells die, the nigrostriatal pathway degenerates. The resulting motor disorder, Parkinson's disease, affects approximately 1% of the population over the age of 50. Symptoms of the disorder are *rhythmic tremor* at rest, *rigidity*, difficulty in initiating movement (*akinesia*), and slowness in execution of movement (*bradykinesia*). Simple motor programs are less affected than more complex ones. Patients cannot execute simultaneous or sequential motor programs, indicating a defect in the use of motor plans for complex motor acts. They cannot activate motor plans for normal rapid movements. The loss of dopaminergic input to the basal ganglia loop decreases the output from the basal ganglia to the thalamus and removes the excitatory boost necessary for maintaining the critical level of stimulation to the SMA. The treatment for Parkinson's disease is the oral administration of *L-DOPA* (*L-dihydroxyphenylalanine*), the precursor of dopamine. This increases dopamine synthesis in the dopamine-producing cells that survive. The treatment is relatively effective in alleviating the symptoms, but it does not alter the course of the disease or stop further degeneration of substantia nigra cells.

Another neurologic condition that results from basal ganglia dysfunction is *Huntington's chorea* (see Fig. 5B). In this case, decreased muscle tone is combined with *spastic*, or involuntary *choreiform* (dancelike), *movements* of the extremities. Patients also develop dementia, indicating a broader involvement of other brain areas, although motor abnormalities routinely appear first. Death usually occurs within 15 years of onset of symptoms. The condition results from an initial selective loss of GABAergic neurons within the neostriatum. Other striatal neurons degenerate as the disease progresses. Because the direct projection of the basal ganglia to the thalamus is inhibitory, the activity of the thalamus is facilitated (disinhibited) in Huntington's chorea. The abnormal excitation of the motor cortex by the thalamus is thought to be the cause of the uncontrolled movements.

Another motor abnormality associated with basal ganglia structures is *hemiballismus*, a hyperkinetic disorder that results in violent flinging of extremities on one side. The disorder is the result of unilateral damage to the subthalamus, often associated with stroke. With less glutamate being released from the damaged subthalamic nucleus, there is less excitatory drive to the globus pallidus; thus, less GABA is released to inhibit the thalamus. As with Huntington's chorea, less inhibitory control over the thalamus causes more excitation of the SMA.

---

excitatory; the two intervening connections (caudate/putamen to globus pallidus and globus pallidus to thalamus) are inhibitory. Even with two inhibitory synapses within the circuit, the overall system operates as a positive feedback loop. This is achieved by having the two inhibitory synapses arranged in series so that the first inhibitory neuron suppresses activity in the second inhibitory neuron. The decrease in inhibitory input to the next neuron, a process called *disinhibition*, is equivalent to direct excitation.

Circuits employing disinhibition are common elements of central nervous system pathways. As is the case with the basal ganglia motor loop, this type of pathway is usually tonically active and highly sensitive over a broad response range. Tonic activity generated throughout the cortex as a result of incoming sensory information and ongoing motor activity is funneled into the loop, allowing the basal ganglia to sort and summate the information before sending it back to the motor cortex, in particular the SMA. In this way, the

---

FIGURE 4   Connections of the nigrostriatal pathway. Within the basal ganglia motor loop, glutamate is the major excitatory neurotransmitter (+) and γ-aminobutyric acid (GABA) is the major inhibitory neurotransmitter (−). An exception is seen with dopamine, the neurotransmitter for the nigrostriatal pathway. Dopamine from the substantia nigra (pars compacta region) is excitatory to the direct pathway and inhibitory to elements of the indirect pathway. This circuit shifts the balance between the direct and indirect paths, lowering the level of inhibitory control over the thalamus and thus maintaining a high level of excitatory thalamic input to the motor cortex.

A. Hypokenetic disease

prefrontal region

sensory association area
SII

premotor area (PMA)
MII (area 6)

supplemental motor area (SMA)
MII (area 6)   ⊕

primary motor cortex
MI (area 4)

thalamus:
ventral anterior,
ventral lateral,
& centromedian nuclei

glutamate

⊕

⊖
GABA

direct path

neostriatum

⊕ DA

DA

⊖

substantia nigra
(pars compacta)
neuronal death

GABA

⊖

NAD

⊖
GABA

external globus pallidus

⊖   GABA

subthalamic nucleus

indirect path

glutamate   ⊕

globus pallidus
(internal segment)

substantia nigra
pars reticulata

basal ganglia

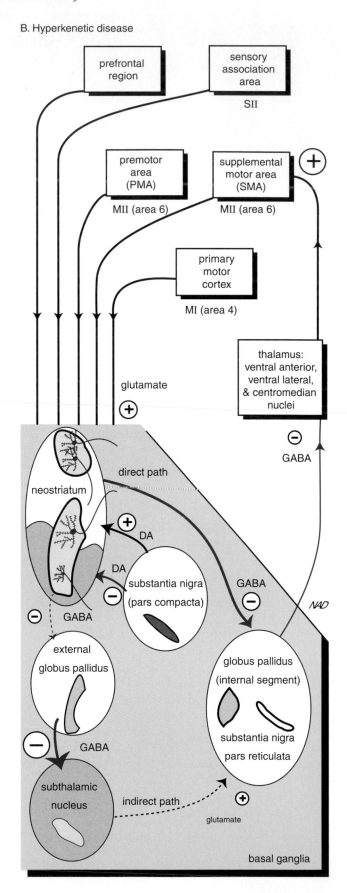

B. Hyperkenetic disease

prefrontal region

sensory association area
SII

premotor area (PMA)
MII (area 6)

supplemental motor area (SMA)
MII (area 6)   ⊕

primary motor cortex
MI (area 4)

thalamus:
ventral anterior,
ventral lateral,
& centromedian nuclei

glutamate

⊕

⊖
GABA

direct path

neostriatum

⊕ DA

DA

⊖

substantia nigra
(pars compacta)

GABA

⊖   GABA

NAD

external globus pallidus

⊖

subthalamic nucleus

indirect path

glutamate   ⊕

globus pallidus
(internal segment)

substantia nigra
pars reticulata

basal ganglia

loop serves to set the appropriate excitatory tone for the SMA.

THE MOTOR LOOP: INDIRECT PATHWAY

Further control of this excitatory feedback loop is achieved by inhibitory actions of the indirect basal ganglia loop. In this pathway, cortical input to the striatum is sent to the external segment of the globus pallidus and then to the subthalamus before converging on the globus pallidus (internal segment)/substantia nigra. Note that the direct and indirect pathways have opposite effects on the globus pallidus/substantia nigra. Thus, the final outcome of the indirect pathway is *disexcitation* (*i.e.*, inhibition) of the cortex, whereas the direct pathway causes cortical stimulation. Although it may be somewhat confusing to sort through the particulars of each synaptic relay in these pathways, it is important to understand the principles of disinhibition and disexcitation and how they affect the overall activity of the motor cortex.

THE NIGROSTRIATAL PATHWAY

A third neuronal component associated with basal ganglia feedback loops is called the *nigrostriatal pathway* (Fig. 4). It projects from the substantia nigra to the caudate and putamen, and its effects are mediated by the neurotransmitter *dopamine*. In essence, dopamine facilitates the motor loop in two ways. First, it provides tonic, excitatory drive to the direct (stimulatory) pathway through the neostriatum, and, second, it inhibits the indirect (inhibitory) pathway. The additive effects of this pathway provide an excitatory bias to the basal ganglia loop and allow the overall level of cortical activity to funnel through the motor loop and provide positive feedback to motor areas.

Suggested Readings

Albin RL, Young AB, Penney JB. The functional anatomy of basal ganglia disorders. *Trends Neurosci* 1989; 12:366–375.

Alexander GE, Crutcher MD. Functional architecture of basal ganglia circuits: neural substrates of parallel processing. *Trends Neurosci* 1990; 13:266–271.

DiFiglia M. Excitotoxic injury of the neostriatum: a model for Huntington's disease. *Trends Neurosci* 1990; 13:286–289.

Gerfen C. The neostriatal mosaic: multiple levels of compartmental organization. TINS 1992; 15:133–139.

Kosik K. Alzheimer's disease: a cell biologic perspective. *Science* 1992; 256:780.

Shepherd GS, Ed. *The synaptic organization of the brain*, 3rd ed. New York: Oxford University Press, 1990.

Yurek DM, Sladek JR, Jr. Dopamine cell replacement: Parkinson's disease. *Annu Rev Neurosci* 1990; 13:415–440.

FIGURE 5   Effects of Parkinson's disease, Huntington's chorea, and hemiballismus on output of the basal ganglia motor loop. (A) In the hypokinetic condition of Parkinson's disease, cells of the substantia nigra degenerate. In the absence of the nigrostriatal pathway, there is increased inhibitory control of the thalamus which leads to a loss of excitatory drive to the motor cortex. With this condition, patients have difficulty in making complex volitional movements. (B) In the hyperkinetic condition of Huntington's chorea, GABAergic cells of the neostriatum degenerate and the inhibitory output to the thalamus is decreased. Disinhibition of the thalamus leads to abnormally high excitatory drive to the motor cortex and causes involuntary, spastic movement. (C) Another hyperkinetic condition, hemiballismus, results from damage to the subthalamic nucleus and a concomitant decrease in inhibitory control over the thalamus. As with Huntington's chorea, the result is uncontrolled movement caused by disinhibition of the thalamus and abnormal excitation of the motor cortex.

The Cerebellum

DIANNA A. JOHNSON

KEY POINTS

- Like the basal ganglia, the cerebellum provides a motor loop pathway for fine-tuning motor commands generated in the cortex, particularly those involving multijoint movement and learning of new motor skills.
- *Purkinje cells* in the *spinocerebellar division* of the cerebellum synapse within deep cerebellar nuclei and send information to the spinal cord to control limb movement.
- *Purkinje cells* in the *vestibulocerebellum* project directly to vestibular nuclei in the brain stem, coordinating their control of postural movements in response to gravity.
- *Purkinje cells* in the *cerebrocerebellum* provide smooth coordination of complex motor programs via synapses in the dentate nucleus that project to the spinal cord via the red nucleus and rubrospinal tract, and to the cortex via the thalamus.

- Projections, called *climbing fibers*, originating from the inferior olivary complex in the medulla, relay sensory information from the spinal cord to Purkinje cells.
- Information from the cortex is sent to the cerebellum through synaptic relays in the pontine nuclei, which project as *mossy fibers*, then to granule cells of the cerebellum, and finally through connections called *parallel fibers* between granule cell axons and Purkinje cells.
- Mossy fiber activity is modulated by climbing fiber input; local inhibitory neurons provide additional control over both granule cells and Purkinje cells.
- Lesions of the cerebellum lead to specific motor deficits associated with loss of coordination, or *ataxia*.

The *cerebellum*, or "little brain," appears as a relatively separate appendage to the cerebrum (telencephalon/diencephalon) and brain stem. There is no anatomic linkage between the cerebellum and cerebrum; the two are separated by a dense sheet of connective tissue, the *tentorium*. Connections are made from the cerebellum to the brain stem through three bridges, or *peduncles* (*superior*, *middle*, and *inferior*). Through these connections, the cerebellum receives both motor and sensory information from all levels of the neural axis and,

Essential Medical Physiology, Third Edition

in turn, sends motor commands back to the spinal cord, brain stem, diencephalon, and motor cortex. Like the basal ganglia, the cerebellum functions as a major relay within motor loop pathways (Fig. 1). Its function is to allow individual components of a given motor plan to be carried out in a smooth fashion, in particular those requiring synergistic, multijoint movement. A general appreciation of cerebellar function can be gained by observing the uncoordinated movements of someone who is intoxicated. The clumsiness, inability to accomplish targeted movement, and loss of equilibrium are largely a direct consequence of depression of cerebellar circuits. The cerebellum also plays a critical role in learning new motor skills (*e.g.*, learning to play a specific piece of music on the piano). With repeated practice of a motor sequence, the cerebellum helps formulate the required programs and then generates appropriate movement sequences on demand without further need for conscious control.

Superficially, the cerebellum looks like a wedge-shaped diminutive cerebrum (Figs. 2 and 3). Gray matter forms the superficial cortical layer, which sends processes to nuclei located within the inner core of the structure. The cerebellar cortex is highly convoluted, more so than the cerebral cortex. The slender compacted folds, called *folia*, are necessary to accommodate the disproportionately large number of cells present. While accounting for only 10% of the volume of the brain, the cerebellum contains 50% of the total number of neurons.

THE SPINOCEREBELLUM

The cerebellum has three general subdivisions. An *intermediate zone* extends down the center of the cerebellum, separating two *lateral hemispheres*. Together with a midline structure called the *vermis*, the intermediate zone comprises the *spinocerebellum*, or *paleocerebellum division*. The spinocerebellum receives sensory information from proprioceptive endings and exteroceptors for touch and pressure from the spinal cord, as well as vestibular, visual, and auditory inputs.

As is the general scheme for all cerebellar regions, output from the spinocerebellum is generated from *Purkinje cells* in the cortex and relayed through deep cerebellar nuclei before exiting through superior and inferior cerebellar peduncles. Two deep cerebellar nuclei, *fastigial* and *interposed nuclei*, are associated with output from the spinocerebellum and send information to the spinal cord through three main routes: (1) indirectly through the reticular formation and the reticulospinal tract, (2) indirectly through the red nucleus and the rubrospinal tract, and (3) directly to spinal motor neurons (see Fig. 2). The spinocerebellum controls descending motor systems

and thus plays a major role in controlling ongoing execution of limb movement.

THE VESTIBULOCEREBELLUM

The *vestibulocerebellum* represents a second functional division of the cerebellum. Also called the *archicerebellum*, it is considered the phylogenetically oldest portion of the cerebellum, and it encompasses a relatively small proportion of the main structure. It resides in the *flocculonodular lobe*, separated from the body of the cerebellum by the *posterolateral fissure*. This division is involved in a relatively simple motor loop formed by reciprocal innervation with vestibular nuclei. The vestibulocerebellum acts as an adjunct to the vestibular nuclei by modifying their output to the spinal cord and their control of postural adjustments to gravity. This is the only cerebellar output that makes no relay within deep cerebellar nuclei. The vestibular nuclei themselves are functionally analogous to the deep cerebellar nuclei, so there is no relay of the Purkinje cell output through nuclei in the cerebellum.

THE CEREBROCEREBELLUM

The final cerebellar division—the *cerebrocerebellum*, or *neocerebellum*—consists of the lateral hemispheres and thus occupies the largest portion of the cerebellum. Input to the cerebrocerebellum is from the *pontine nuclei*, which relay information from the contralateral sensory motor cortex through the middle cerebellar peduncle. Containing 20 million nerve fibers, this is one of the largest tracts in the central nervous system (CNS). Output from cerebrocerebellar cortex is relayed through the *dentate nucleus* with exiting fibers in the superior cerebellar peduncle. Two areas serve as targets for this information: (1) the red nucleus and subsequently spinal motor neurons through the rubrospinal tract, and (2) the thalamus and hence premotor and primary motor cortex. These cerebellar pathways play an important role in the planning and timing of movements, allowing for smooth coordination of the different stages of complex motor programs.

CELLULAR CIRCUITS OF THE CEREBELLUM

The cellular anatomy of the cerebellar cortex is a relatively simple, stereotyped arrangement with the same basic pattern repeating across all divisions (Fig. 4). One output neuron, the *Purkinje cell*, projects to cerebellar nuclei, and there are two inputs. Direct input is provided

A. General motor loop of the cerebral cortex - basal ganglia

B. General motor loop of the cerebral cortex - cerebellum

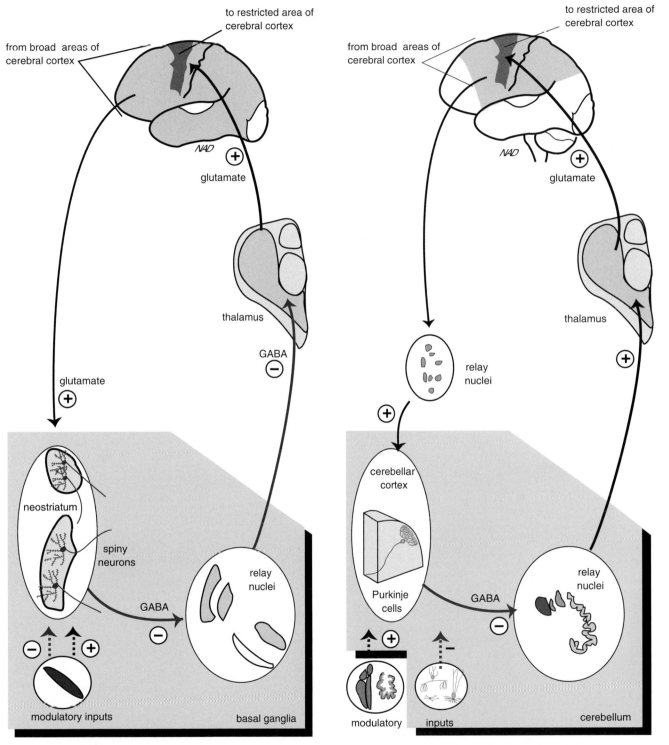

FIGURE 1   Comparison of connections of the basal ganglia (A) and cerebellum (B). The basal ganglia and the cerebellum have many commonalties and influence motor function in similar ways. Excitatory input from broad areas of the cerebral cortex drive spiney neurons in the neostriatal region of the basal ganglia (A) and also relay nuclei in the pons, which then excite Purkinje cells of the cerebellar cortex (B). Both spiney neurons and Purkinje cells release γ-aminobutyric acid (GABA) to inhibit relay neurons in the globus pallidus/substantia nigra (A) or in the deep cerebellar nuclei (B). In turn, these relay nuclei project to the thalamus, where they modulate thalamic inputs to specific motor areas of the cerebral cortex.

A. Medial view of cerebellum as it is attached to the brain stem

B. View of the cerebellar cortex flattened. to show the lobes & areas

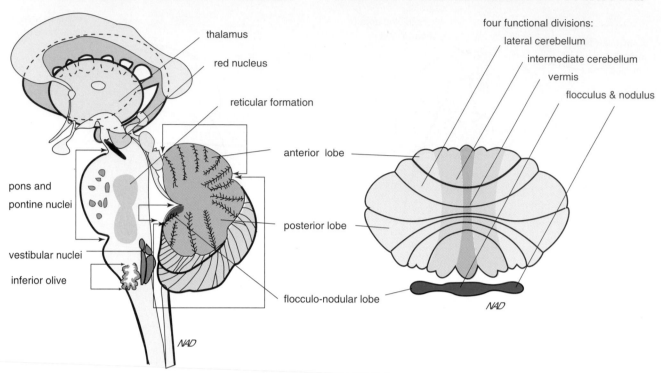

C. Dorsal brain stem and outputs from cerebellum

D. Ventral brain stem and inputs to cerebellum

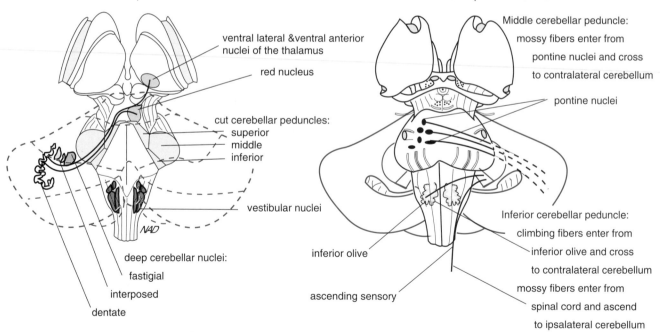

FIGURE 2   Anatomy of the cerebellum. (A) A midsagittal view of the cerebellum and associated structures. (B) Dorsal view of the cerebellar cortex is flattened to reveal the areas that are normally hidden beneath; the individual folia are still folded. If these could be completely smoothed, the cerebellum would be more than 1 m long. The anatomic divisions include the anterior, posterior, and flocculonodular lobes. The functional divisions are shown in shades of blue, and their connections are outlined in Fig. 3. (C) The dorsal brain stem with the cerebellum removed (dashed blue lines) shows the three pairs of cerebellar peduncles; exiting cerebellar fibers project from deep cerebellar nuclei to the thalamus, red nucleus, and brainstem. (D) Ventral brain stem showing the major inputs to the cerebellum. Fibers from the pontine nuclei cross the midline before entering the cerebellum through the massive middle peduncle. The inferior peduncle primarily carries input from the inferior olive and sensory information, particularly from vestibular nuclei.

A. Dorsal view of cerebellum and deep cerebellar nuclei

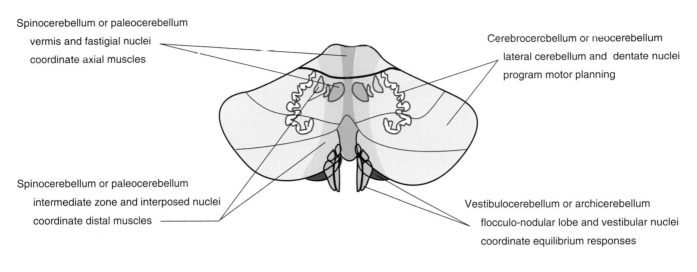

Spinocerebellum or paleocerebellum
  vermis and fastigial nuclei
  coordinate axial muscles

Cerebrocercbellum or neocerebellum
  lateral cerebellum and dentate nuclei
  program motor planning

Spinocerebellum or paleocerebellum
  intermediate zone and interposed nuclei
  coordinate distal muscles

Vestibulocerebellum or archicerebellum
  flocculo-nodular lobe and vestibular nuclei
  coordinate equilibrium responses

B. Motor loop of the cerebrocerebellum

to motor cortex

from broad areas of
cerebral cortex

pontine
nuclei

+

+

+

NAD

glutamate

cerebrocerebellum
lateral hemispheres

GABA
−

dentate
nucleus

VL & VA
nuclei of
thalamus

+

−

somato-
sensory

inhibitory
interneurons

cerebellum

C. Motor loop of the spinocerebellum

sensory information
from spinal cord,
ears and eyes

+

+

spinocerebellum
intermediate zone

vermis

GABA
−

interposed
nucleus

fasitigial
nucleus

red
nucleus

−

inhibitory
interneurons

cerebellum

reticular
formation

FIGURE 3 Connections of the spinocerebellum, vestibulocerebellum, and cerebrocerebellum: (A) relationships among the functional divisions, the deep cerebellar nuclei, and the vestibular nuclei; (B) motor loop of the cerebrocerebellum (see Fig. 1 for details); and (C) motor loop of the spinocerebellum.

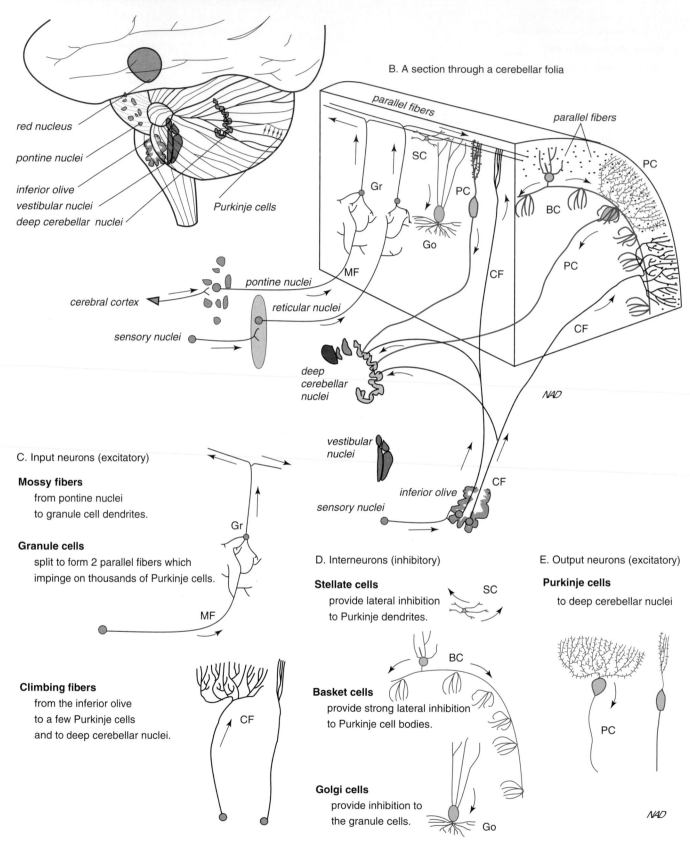

A. Placement of the nuclei associated with the cerebellum

red nucleus
pontine nuclei
inferior olive
vestibular nuclei
deep cerebellar nuclei

Purkinje cells

B. A section through a cerebellar folia

parallel fibers

parallel fibers

SC
Gr
PC
Go
BC
PC
MF
CF
CF
PC

pontine nuclei
cerebral cortex
reticular nuclei
sensory nuclei

deep cerebellar nuclei

NAD

vestibular nuclei

sensory nuclei

inferior olive
CF
CF

C. Input neurons (excitatory)

Mossy fibers
from pontine nuclei
to granule cell dendrites.

Granule cells
split to form 2 parallel fibers which
impinge on thousands of Purkinje cells.

Gr
MF

Climbing fibers
from the inferior olive
to a few Purkinje cells
and to deep cerebellar nuclei.

CF

D. Interneurons (inhibitory)

Stellate cells
provide lateral inhibition
to Purkinje dendrites.

SC

Basket cells
provide strong lateral inhibition
to Purkinje cell bodies.

BC

Golgi cells
provide inhibition to
the granule cells.

Go

E. Output neurons (excitatory)

Purkinje cells
to deep cerebellar nuclei

PC

NAD

by fibers from a large group of cells in the medulla, the *inferior olivary complex*. Fibers enter the cerebellum through the inferior peduncle and wrap around the dendrites of the Purkinje cell. Because of their vinelike appearance, they are called *climbing fibers*. Sensory information from the spinal cord, brain stem, and cerebrum is funneled through the inferior olivary complex and provides strong excitatory drive through synapses of the climbing fibers with Purkinje cells. This synapse is unusual in that it is considered to be one of the strongest, most powerful synaptic connections in the CNS, as it elicits an excitatory postsynaptic potential (EPSP) with every firing of the climbing fiber; however, one climbing fiber innervates one or very few Purkinje cells, and climbing fibers fire at a low frequency of about 1 Hz. This arrangement is not capable of providing the overall excitatory drive of the cerebellum; rather, it functions to modulate the more pervasive influence of a second Purkinje cell input coming from mossy fibers.

Nuclei in the pons send extensive projections, called *mossy fibers*, to the cerebellum, where they synapse not on Purkinje cells directly but on granule cells, which constitute the largest cell population in the brain. Excitatory input to granule cells is then relayed through their processes, called *parallel fibers*, which make excitatory contact with Purkinje cells, each of which receives as many as 200,000 parallel fiber inputs. This input generates small EPSPs that must summate temporally and spatially in order to produce an action potential in the Purkinje cell. Granule cells, because of their extensive input, produce a constant, high-frequency firing of Purkinje cells of about 50 to 100 Hz. The two Purkinje cell inputs are interactive, so that climbing fiber input modulates the effect of mossy fibers on the Purkinje cell. Long-term modulation and interaction between the two inputs are thought to be involved in motor learning.

Three neurons serve as local inhibitory neurons to counterbalance the excitatory pathways within the cerebellum. *Stellate* and *basket cells*, excited by collaterals of parallel fibers from the granule cell, send inhibitory synapses to Purkinje cells in a feedforward loop. *Golgi cells*, also excited by parallel fibers, send inhibitory synapses to granule cells in a feedback loop. Output of the cerebellar cortex is through Purkinje cells that inhibit tonically active cells within the deep cerebellar nuclei.

The unique contribution of the cerebellum to the execution of motor commands is evident in patients with cerebellar lesions. Loss of cerebellar function does not produce paralysis. Instead, affected individuals exhibit abnormal movements with lack of coordination, generally referred to as *cerebellar ataxia*. One aspect of the disorder is characterized by a delay in initiating movements and errors in the rate and regularity of movements. The standard clinical test to demonstrate cerebellar deficits is to have the patient attempt to perform rapid alternating movements, such as tapping with one hand while alternating between the back and the palm of the hand. If cerebellar function is altered, the patient cannot sustain a regular rhythm or steady force. Luckily, the symptoms of cerebellar disease can improve gradually with time, perhaps reflecting the "plasticity" of cerebellar circuits.

Suggested Readings

Flament D, Hore J. Movement and electromyographic disorders associated with cerebellar dysmetria. *Neurophysiology* 1986; 55:1221–1233.

Holmes G. The cerebellum of man. *Brain* 1939; 62:1–30.

Keele SW, Ivry R. Does the cerebellum provide a common computation for diverse tasks? A timing hypothesis. *Ann NY Acad Sci* 1990; 608:179–211.

FIGURE 4   Cellular anatomy of the cerebellar cortex. (A) Lateral view of the cerebellum and brain stem shows the major input nuclei (pontine, inferior olive, and vestibular), as well as the major output target nuclei (deep cerebellar, vestibular, and red nuclei); the orientation of the Purkinje cell dendritic trees is shown in blue along one folium. (B) A wedge of a cerebellar folium shows both cross and longitudinal sections; inhibitory neurons are shown in blue and excitatory neurons are black. Note that the Purkinje cells and all of the interneurons are inhibitory and use GABA as their neurotransmitter. Purkinje cells have a flattened dendritic tree covered with spines. (C) Input neurons include climbing fibers and the mossy fiber/granule cell/parallel fiber relay. The climbing fiber has a configuration similar to the Purkinje cell, except it is not as extensive and lacks spines. Parallel fibers are organized perpendicular to the plane of Purkinje cell dendrites, and they provide excitatory input to all Purkinje cells within a 1-mm distance. (D) Three types of interneurons inhibit Purkinje and granules cells. (E) Purkinje cells are the sole output neurons of the cerebellar cortex. They project to the deep cerebellar nuclei and vestibular nuclei.

60

Neuronal Control of Mood, Emotion, and State of Awareness

DIANNA A. JOHNSON

KEY POINTS

- The autonomic nervous system influences the general emotional status of the individual. The network involved originates from small groups of neurons within the brain stem, with extensive projections to widespread areas throughout the neural axis.
- Pleasurable feelings are associated with activity in the *parasympathetic nervous system*; feelings of anxiety are associated with activity in the *sympathetic nervous system*.
- The *Papez circuit* links the site of emotional awareness, the *cingulate cortex*, with appropriate portions of the autonomic nervous system and thereby triggers the physical expression of the emotion.

- The *hypothalamus* generates aggressive behavior patterns based on input from the Papez circuit and from the *amygdala*.
- Neuronal circuits that result in pleasurable sensations constitute a reward or reinforcement system for certain survival behaviors.
- Neuronal groups within the brain stem that utilize dopamine, norepinephrine, and serotonin provide diffuse modulatory systems that influence pleasure centers and establish the general state of awareness of the brain.
- The *brain stem reticular-activating system* helps establish different functional states of the brain resulting in coordinated patterns of neuronal activity ("brain waves") that can be monitored by means of an electroencephalogram.

In contrast to the rapid, point-to-point transfer of information within sensory and motor pathways, other types of neuronal circuits are present in the nervous system that operate on a much slower time scale and have much broader, less-defined targets. These are described as *diffuse modulatory systems*. One of the best examples is the autonomic nervous system (ANS) (described in

Chapter 9). As implied by its name, the ANS is primarily a reflex, nonvolitional system that is only indirectly influenced by the motor cortex. Preganglionic motor neurons that function in the *sympathetic division* of the ANS are located in the thoracic and upper lumbar portions of the spinal cord gray matter with fibers exiting in the spinal nerves. In the *parasympathetic division*,

---

preganglionic motor neurons are located in brain stem nuclei, with exiting fibers in the cranial nerves, or in the sacral portion of the spinal cord, with fibers exiting accordingly. These preganglionic fibers project to postganglionic motor neurons within sympathetic and parasympathetic ganglia that, in turn, provide dual and, in most cases, antagonistic innervation of glands and smooth muscles. Input to preganglionic neurons comes from autonomic nuclei in the brain stem that act as centers for coordinating autonomic responses of spinal and cranial nerves to regulate breathing rate, heart rate, bladder function, and so on. These autonomic centers receive direct input from the amygdala and the hypothalamus but no direct innervation from the somatic motor cortex. Thus, there is no conscious control of autonomic function (Fig. 1).

The diffuse, modulatory nature of the ANS is best exemplified by the parasympathetic portion of the vagus nerve (cranial nerve X), which originates from a small group of neurons in the vagal motor nucleus of the medulla. These neurons provide preganglionic fibers that synapse in a large number of diverse parasympathetic ganglia controlling muscles of the esophagus, heart, peritoneal cavity, and intestine. Compared to the somatic motor and sensory system, the vagal parasympathetic pathway evokes relatively slow, sustained responses that are coordinated throughout the extensive and broadly distributed group of target tissues. The vagal motor nucleus functions as a master control point, receiving and summating a wide range of sensory inputs and sending signals that in essence elicit a whole-body response. In some cases, all components of the parasympathetic division may react together. For example, vagal stimulation slows the heart and, along with other parasympathetic centers, facilitates food ingestion and digestion.

EMOTIONS RELATED TO AUTONOMIC NERVOUS SYSTEM FUNCTION

Strong emotional feelings are often associated with most autonomic responses. Coordinated parasympathetic responses evoke a "vegetative" state that generates an overall relaxed, pleasurable feeling. They are often linked to eating behavior and can be triggered by food-related sensory stimuli, such as seeing or smelling appetizing food. The opposite is true for coordinated responses from the sympathetic division of the ANS. In this case, sympathetic pathways coordinate protective responses, preparing the individual to flee from a perceived danger or to engage in aggressive behavior. In response to heightened sympathetic activity, the individual will experience feelings of fear, anxiety, and anger.

In the two cases given above, stimuli for parasympathetic and sympathetic responses are external

A. Medial view of areas controlling the autonomic nervous system

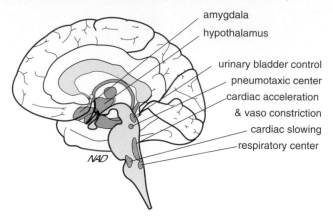

B. Flow of information controlling the autonomic nervous system

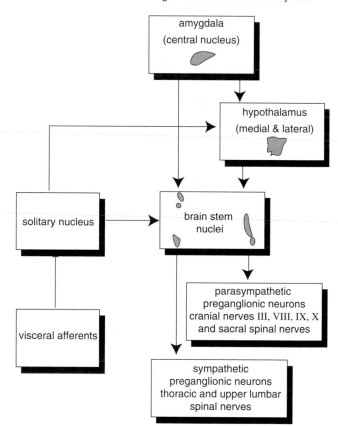

FIGURE 1 Neuronal circuits of mood that link emotional experience with emotional expression. (A) Medial view of the brain; areas controlling the autonomic nervous system are shown in blue. (B) Two major inputs to autonomic nuclei in the brain stem arise from the amygdala and the hypothalamus. A third originates from visceral organs and taste receptors, then projects to the solitary nucleus and then to the brain stem.

(*i.e.*, availability of food or perceived danger). However, it is also possible for internal cues to trigger autonomic reflexes. Recalling past experiences that were particularly pleasurable or anxiety producing can cause strong parasympathetic and sympathetic responses. The autonomic nervous system provides the major mechanism by which elements of mood and emotion are expressed by the body. Other structures in the central nervous system (CNS) participate in generating the emotion, but much of their output is routed through the ANS to appropriate target tissues.

Interestingly, the strong link between the ANS and emotion has led to two opposing theories regarding the emotional experience of fear. The more conventional, the *James-Lange theory*, holds that fear is experienced in response to perception of danger and activation of appropriate cortical areas and that the resulting sympathetic response is triggered after the fright is experienced. The more controversial theory, the *Cannon-Bard theory*, suggests that the presentation of a frightening stimulus leads first to a sympathetic response. The change in the body's sympathetic–parasympathetic tone then is monitored by higher centers in the CNS, and it is this change that is synonymous with the emotional experience of fear. In other words, the sympathetic state itself is the trigger for the emotion. More likely, both theories are correct to some degree, with the perception of an external stimulus and a change in sympathetic tone both contributing to the emotional experience.

THE PAPEZ CIRCUIT EMOTION SYSTEM

One of the cortical areas associated with the general emotional state has been given the name *limbic* (from *limbus*, Latin for "border") *lobe* (Fig. 2). Although not a true lobe, main elements of the limbic lobe—including the *cingulate cortex*, or *cingulum*, and the *hippocampus*—form a C-shaped structure that surrounds the rostral borders of the brain stem. Components of the limbic lobe are linked with two other structures, the *anterior nuclei of the thalamus* and the *hypothalamus*, in a pathway known as the *Papez circuit*. Based on experience with patients having lesions within these areas, it has been proposed that the Papez circuit represents an emotion system responsible for linking the experience and the expression of emotion. The proposed site of emotional experience is the cingulate cortex, which receives sensory information funneled directly through the thalamus and from sensory areas in the cerebral cortex. Output from the cingulate cortex is fed through the *hippocampus* and projects, via a band of fibers called the *fornix*, to the hypothalamus, where it is translated into the expression of emotion through autonomic and other responses. Because the hypothalamus also projects back to the anterior nuclei of the thalamus (mammillothalamic tract), the expression of emotion (through autonomic responses) could add an additional trigger or perhaps a positive-feedback loop for the emotional experience.

The hypothalamus and an associated area, the *amygdala*, play a direct role in mediating aggressive behavior. The amygdala is an almond-shaped nucleus situated at the anterior border of the hippocampus within the temporal lobe (Figs. 2 and 3). It has been demonstrated in several animal species that bilateral ablation of the amygdala reduces fear and decreases certain forms of aggression. Although a rare occurrence, selective damage to the amygdala has been observed in humans. Behavioral testing suggests that those whose amygdala has been damaged have a decreased ability to recognize fear. Consistent with these findings, direct electrical stimulation of the amygdala in humans has been reported to elicit anxiety and fear.

These effects of the amygdala are mediated through connections with the hypothalamus, a major staging area for generation of two distinct types of aggressive behavior in animals. The first, *affective aggression*, is expressed by assuming a threatening or defensive posture, as when a cat hisses and arches its back. The second, *predatory aggression*, involves attacks against a member of a different species for the purpose of obtaining food, as when a cat silently pounces on a mouse. Experimental application of a small stimulating current through electrodes implanted in the medial hypothalamus in cats causes immediate expression of affective aggression, whereas stimulation of lateral regions causes predatory aggression. It has also been demonstrated in primates that the amygdala provides excitatory input to the medial hypothalamus and plays an important role in maintaining an active level of affective aggression normally involved in maintaining a position in the social hierarchy. The amygdala has the opposite effect on the lateral hypothalamus, providing inhibitory input to this region and thus suppressing predatory aggression.

THE EMOTIONAL REWARD PATHWAY

Experimental studies designed to localize specific circuits and identify underlying cellular mechanisms responsible for emotional perception and emotionally driven behavior are particularly challenging. By comparison, sensory and motor pathways are much more amenable to standard techniques of experimental examination. One obvious problem is that the emotional state of experimental animals is difficult to assess. One type of animal behavior experiment has shown that rats will repeatedly self-stimulate electrodes implanted in the brain only when they are placed in certain areas, suggesting the presence of pleasure centers within these regions.

Another source of information that has provided considerable insight into the brain mechanisms of emotion is data gained from patients who require surgical removal of specific brain areas that act as trigger zones for severe

A. Anatomical locations of the Papez circuit components and other limbic structures

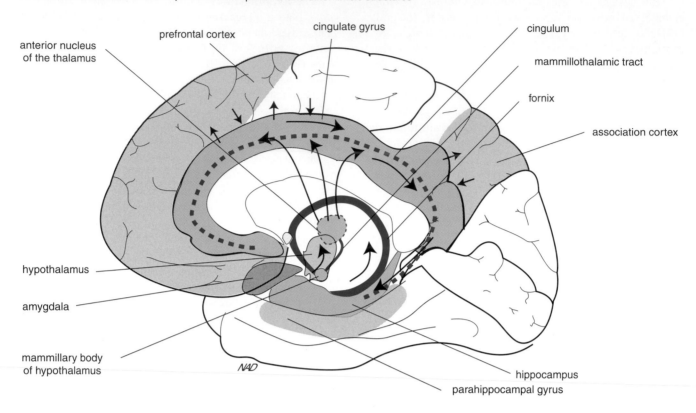

anterior nucleus of the thalamus

prefrontal cortex

cingulate gyrus

cingulum

mammillothalamic tract

fornix

association cortex

hypothalamus

amygdala

mammillary body of hypothalamus

hippocampus

parahippocampal gyrus

NAD

B. Flow diagram of the Papez circuit (blue) and other limbic structures (gray)

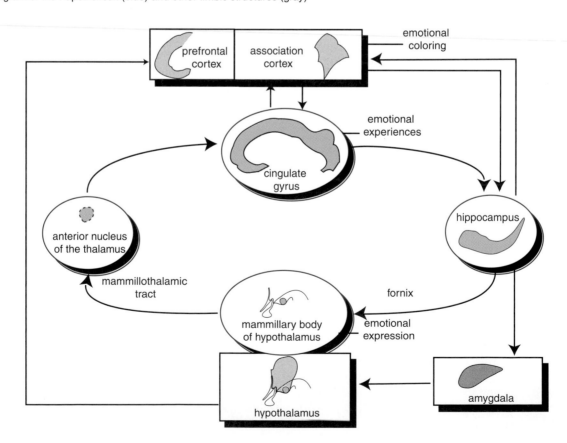

prefrontal cortex

association cortex

emotional coloring

emotional experiences

cingulate gyrus

hippocampus

anterior nucleus of the thalamus

mammillothalamic tract

fornix

mammillary body of hypothalamus

emotional expression

hypothalamus

amygdala

A. Coronal section through the hypothalamus and amygdala

B. Flow of information in aggression

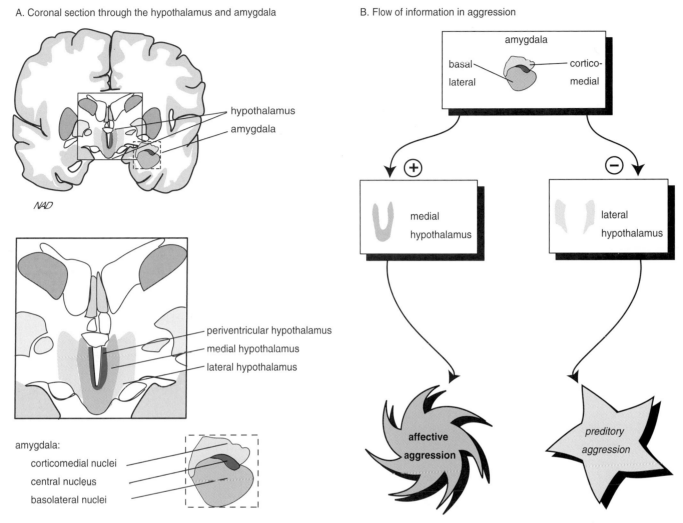

NAD

hypothalamus
amygdala

periventricular hypothalamus
medial hypothalamus
lateral hypothalamus

amygdala:
corticomedial nuclei
central nucleus
basolateral nuclei

amygdala
basal — cortico-
lateral — medial

⊕
medial hypothalamus

⊖
lateral hypothalamus

affective aggression

preditory aggression

FIGURE 3   Neuronal pathways that trigger predatory and affective aggression. (A) Coronal section through the cerebrum (upper panel) with enlargement (lower panel); major divisions of the hypothalamus and amygdala that participate in producing aggressive behavior are shown in blue. (B) Flow of information in the expression of aggression; Affective aggression (left) is displayed during strong emotions such as fear and anger. Predatory aggression (right) is displayed during activities such as hunting.

epileptic seizures. These procedures require that the patient be alert during the surgery while different brain areas are given mild electrical stimulation in order to pinpoint the affected areas and avoid other critical regions. Based on patients' reports, one group of brain areas that consistently evoke intensely pleasurable sensations has been identified. These areas are comparable to those identified in rat and include the *septal area, lateral hypothalamus, ventral tegmental area, dorsal pans,* and a fiber tract that interconnects all these structures, the *medial forebrain bundle.* These pleasure centers may provide the physiologic basis for reinforcement of normal survival behaviors, such as eating, drinking, and sex, as well as for abnormal behaviors, such as use of addictive drugs.

ASCENDING DIFFUSE MODULATORY SYSTEMS

The median forebrain bundle is one of the three *diffuse modulatory systems* that ascend in the brain stem (see Figs. 4 and 5). It contains neuronal fibers from neurons that release *norepinephrine, serotonin,* or *dopamine* as

FIGURE 2   The Papez circuit. The Papez circuit and related structures form a network that links emotional experiences and emotional expression. (A) Anatomic location of the Papez fiber tracts is shown in dark blue, the four relevant nuclei are shown in medium blue, and related structures are shown in gray. (B) Flow diagram of the Papez circuit; the areas of neuronal processing are depicted in ovals, and the connections are indicated by black arrows. Other related structures are shown in rectangles.

Clinical Note

Seizures

Seizures represent the extreme in synchronicity of brain activity and are often associated with some underlying pathological condition. *General seizures* involve the entire cortex; partial seizures involve only limited areas of the cortex. During a seizure, sensory information is blocked and the individual retains no memory of the event. Motor units normally controlled by the areas involved in a seizure will become rigid during clonic activity and will contract rhythmically during clonic activity of the affected motor areas. *"Absence" seizures* occur during childhood and consist of short periods lasting several seconds, during which generalized 3-Hz EEC waves are generated. There is a loss of consciousness, with only subtle motor signs, such as fluttering of the eyelids or facial twitching. Ten percent of the population have one or two isolated seizures during their lifetime without apparent consequence. *Epilepsy* is described as the repeated occurrence of seizure activity and can result from a wide variety of causes, including tumors, trauma, metabolic dysfunction, infection, vascular disease, developmental abnormalities in brain structure, and drug interactions.

neurotransmitters at widely spread innervation sites across the brain. Noradrenergic fibers from the *locus coeruleus* represent a second ascending modulatory system, with perhaps the most diffuse connections known to occur in brain (see Fig. 5). A single locus coeruleus neuron makes as many as 250,000 synapses that spread from the cerebral cortex to the cerebellum. Serotonergic neurons located in the *raphe nuclei* located near the midline of the brain stem provide a third ascending system that innervates most of the brain in a manner similar to the locus coeruleus. Projections from the locus coeruleus and raphe nuclei together form the *reticular-activating system*, which serves to alert the brain to incoming sensory stimuli that are novel. The behavioral response to an abrupt presentation of a new stimulus, such as a loud noise, is an immediate reorientation of the head and perhaps the body toward the novel stimulus and a general increase in awareness and attentiveness. Most of the currently used psychoactive drugs—many of which are analogs of norepinephrine, dopamine, and serotonin—act directly on the neurotransmitter systems that are components of the diffuse modulatory systems subserving the positive reinforcement and the ascending reticular-activating systems.

The reticular-activating system represents but one mechanism that controls the overall functional state of the brain. Its activity helps maintain the awake, interactive state. Different functional states are associated with specific patterns of electrical activity in the cortex that can be recorded using electrodes placed on the scalp. These patterns generate an *electroencephalogram* (EEG) that is the result of large populations of cortical pyramidal cells firing at rates and degrees of synchrony that change in response to different states of arousal. In general, high-frequency firing rates and low-amplitude

A. Self-stimulation sites in the rat brain

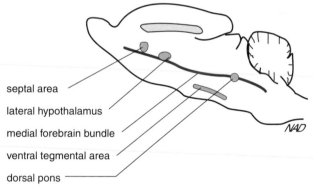

septal area
lateral hypothalamus
medial forebrain bundle
ventral tegmental area
dorsal pons

B. Areas in the human brain comparable to rat self- stimulation sites

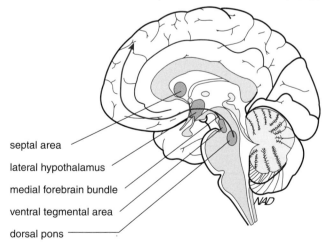

septal area
lateral hypothalamus
medial forebrain bundle
ventral tegmental area
dorsal pons

FIGURE 4   Structures involved in the reward system. (A) Areas of brain identified as pleasure centers are based on the observation that rats self-stimulate electrodes implanted in these regions. (B) Comparable regions in humans have been shown to elicit pleasurable sensations when electrically stimulated during various clinical procedures.

A. Noradrenergic modulatory system

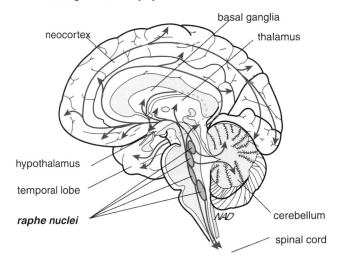

B. Serotonergic modulatory system

C. Dopaminergic modulatory systems

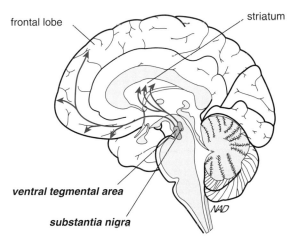

FIGURE 5   Three diffuse modulatory systems of the brain. Brain stem nuclei that give rise to each system are labeled in bold italics on the lower left. (A) A small group of neurons in the locus coeruleus distribute norepinephrine throughout most of the brain, including the neocortex, cerebellum, thalamus, and spinal cord. (B) Neurons producing serotonin arise in the raphe nuclei and have an equally broad distribution. (C) Dopamine neurons arise from two nuclei: (1) the substantia nigra, which projects to the striatum of the basal ganglia, and (2) the ventral tegmental area, which projects to the frontal lobe.

responses are generated by pyramidal cells that are relatively desynchronized in their activity, indicative of responses to continuous sensory input in an awake, alert individual. During slow-wave sleep, pyramidal cells are in a state of reduced responsiveness as sensory input is diminished and they fire more synchronously. EEG patterns show *beta rhythms* (approximately 14 Hz) when the cortex is activated and *theta rhythms* (4 to 7 Hz) or *delta rhythms* (< 4 Hz) during most stages of sleep. *Alpha rhythms* (8 to 13 Hz) are generated during quiet, waking states. Several times during a normal sleep cycle, EEGs change dramatically to patterns more characteristic of the aroused state. This stage of sleep is called *rapid eye movement* (REM) sleep because of the spontaneous eye movements that occur at this time. REM sleep is associated with episodes of dreaming.

Suggested Readings

Erickson RP. On the neural basis of behavior. *Am Scientist* 1984; 72:233–241.

Friedman MI, Stricker EM. The physiological psychology of hunger: a physiological perspective. *Psychol Rev* 1976; 83:409–431.

Stellar JR, Stellar E. *The neurobiology of motivation and reward.* New York: Springer, 1985.

61

Learning and Memory

JOHN H. BYRNE

KEY POINTS

- *Learning* is an experience-dependent generation of enduring internal representations and/or modifications in such representations. *Memory* is the retention of these experience-dependent changes over time.
- There are multiple forms of learning and memory, and multiple but distinct regions of the brain serve as loci for learning and memory.
- *Declarative memory* (explicit memory) encompasses the memory for facts and concepts as well as the memory for events. The medial temporal lobe and diencephalon are critical for declarative memories.
- *Nondeclarative memory* (implicit memory) operates at an unconscious level. It encompasses the memory for skills and habits, priming, examples of associative learning such as

classical conditioning and operant conditioning, and examples of nonassociative learning such as sensitization and habituation. Anatomic loci for nondeclarative memories are diverse.
- At least short-term memory involves changes in existing neural circuits. These changes may involve multiple cellular mechanisms within single neurons.
- Second-messenger systems appear to play a role in mediating cellular changes.
- Changes in the properties of membrane channels are commonly correlated with learning and memory.
- Long-term memory requires new protein synthesis and growth, whereas short-term memory does not.

An important goal of neurobiology is to explain the anatomic, biophysical, and molecular processes occurring in the nervous system that underlie learning and memory. Specifically, what parts of the nervous system are critical for learning? How is information about a learned event acquired and encoded in neuronal terms? How is the information stored, and, once stored, how is it retrieved? Most neuroscientists believe that the answers to these questions lie in understanding how the properties of individual nerve cells and their synaptic connections change when learning occurs. Extensive behavioral, anatomic, physiologic, and molecular analyses over the past decade have revealed what appear to be some general principles of learning and memory. A list of these principles might include the following: (1) there are multiple forms of learning and memory; (2) multiple but distinct regions of the brain serve as loci for learning and memory; (3) at least short-term forms of memory involve changes in existing neural circuits; (4) these changes may involve multiple cellular mechanisms within single neurons; (5) second-messenger systems appear to play a role in mediating cellular changes; (6) changes in the properties of membrane channels are commonly correlated with learning and memory; and (7) long-term memory requires new protein synthesis and growth, whereas short-term memory does not.

WHAT ARE LEARNING AND MEMORY?

Although we all have an intuitive understanding of learning and memory, formulating a rigorous definition can pose some difficulty. One traditional definition of learning is a change in behavior as a result of experience. This definition is less than ideal, however, because some examples of learning do not involve overt behavioral changes. For example, learning may represent a change in an internal state that is behaviorally silent and therefore represents a process that is a *potential* for a change in behavior, rather than an immediate change in behavior. In addition, any definition of learning should distinguish learning from maturational changes and from changes in behavior produced by injury or fatigue. A fairly general definition of learning has been provided by Dudai (1989): *Learning* is an experience-dependent generation of enduring internal representations and/or modification in such representations, whereas *experience* excludes events related to maturation, injury, and fatigue. *Memory* can be defined as the retention of these experience-dependent changes over time. The temporal domains of memory vary considerably, from short-term forms lasting minutes, such as the memory of a telephone number, to long-term forms lasting days, weeks, or lifetimes, such as the memory of a childhood experience. In some cases, a short-term memory can be stabilized into an enduring long-term form. This process is referred to as *consolidation*. Finally, we need to define *retrieval*, the process that allows memory to be accessed. Retrieval is the use of memory in neuronal and behavioral operations (Dudai, 1989).

Amnesia is a disorder of memory which has two broad forms: retrograde and anterograde. *Retrograde amnesia* refers to the loss of memories that were acquired before the amnesia and is usually temporally graded. For example, a patient with head trauma will generally have the greatest loss of memory for events immediately preceding the trauma, whereas more remote memories will be preserved. This temporal gradation is likely due to the interruption of the consolidation process (see earlier). *Anterograde amnesia* refers to an inability to form new memories.

TYPES OF MEMORY AND THEIR ANATOMIC LOCALIZATION

Types of Memory

Psychologists have found that human memory is composed of multiple memory systems (Squire and Knowlton, 2002) (Fig. 1). Although there is some debate as to the precise categorization of these systems, they can generally be divided into two broad categories for humans: *declarative memory* and *nondeclarative memory*. Declarative memory, also known as explicit memory, encompasses the memory for facts and concepts (semantic memory), as well as the memory for events (episodic or autobiographical memory). In humans, declarative memories are associated with conscious recollections of facts and events. Nondeclarative memory, also referred to as implicit memory, operates at an unconscious level and encompasses the memory for skills and habits, priming, examples of associative learning such as classical conditioning, operant conditioning, and examples of nonassociative learning such as sensitization and habituation. Priming is defined as an increased ability to identify or detect a stimulus as a result of its prior exposure. Nonassociative and associative learning are discussed later.

ANATOMIC LOCALIZATION OF MEMORY SYSTEMS

A considerable body of evidence indicates that the various memory systems described earlier are supported by distinct anatomic structures. These conclusions have been derived from a number of strategies. Among these is

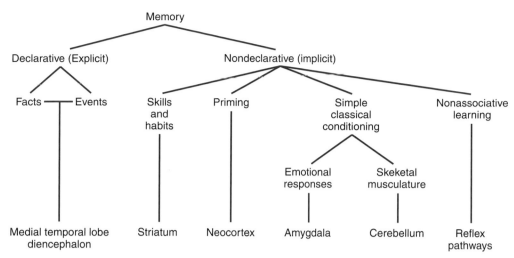

FIGURE 1   Multiple memory systems and their associated brain structures. Memory systems in the brain can be categorized as declarative and nondeclarative. Each system is supported by distinct anatomical regions of the brain. (Modified from Squire, LR, Knowlton, BJ, Memory, hippocampus, and brain systems. In: Gazzanige, MS, ed. *The coginitive neurosciences.* Cambridge: MIT Press, 1995:825–837.)

the use of modern imaging techniques such as positron emission tomography (PET) and functional magnetic resonance imaging (fMRI), which show regions of the brain engaged by specific memory tasks (Fig. 2). Studies of learning and memory deficits in experimental animals with lesions have also helped identify areas of the brain that are involved in learning and memory processes. Finally, brain regions involved in learning and memory have been identified through detailed behavioral assessments of patients with learning and memory deficits produced either by injury (*e.g.*, trauma, tumors, vascular incidents such as stroke) or by surgical removal of regions of the brain which became necessary to treat disorders such as epilepsy.

A particularly revealing early case that has had a major impact on the development of modern views on the distributed representation of memory systems was the study by Brenda Milner and her colleagues of a patient named H.M., who underwent surgery in 1953 to treat severe epileptic seizures. The surgery included removal of the medial temporal region of the brain, which included the amygdala, anterior two-thirds of the hippocampus, and hippocampal gyrus. Following the surgery, H.M. appeared normal in many respects except for a severe anterograde amnesia. Specifically, he could not form any new long-term declarative memories. For example, on the day following an interview, he had no recollection that the interview on the previous day had occurred nor any memory of any events associated with the interview. H.M. also could not add new words to his vocabulary. Thus, he appeared to be unable to form any new memories for events, facts, or concepts. Although he could not form any new long-term episodic or semantic memories, his early childhood memories were intact. Presumably,

these memories are stored in a region of the brain outside the medial temporal region, which was removed during surgery. Of particular interest was the finding that H.M. retained the ability to acquire new skills at a level comparable with normal individuals. For example, H.M. was able to learn a difficult mirror-drawing task. He also learned to play the Tower of Hanoi puzzle. Although he could acquire these new skills (*i.e.*, nondeclarative memory), he had no conscious recollection of ever acquiring them (*i.e.*, no declarative memory).

Studies of the type described earlier indicate that the medial temporal lobe and diencephalon are critical for declarative memories (see Fig. 1). Anatomic loci for nondeclarative memories are more diverse and seem to depend on the particular brain structure or structures engaged by the task. Thus, the learning of certain skills and habits involves the striatum, whereas the learning of certain movements can involve the cerebellum or spinal cord. Conditioning of emotional responses depends on the amygdala (see Fig. 1).

NONDECLARATIVE MEMORY TASKS

Many neurobiologists interested in the cellular analysis of learning and memory have employed paradigms that involve the conditioning of some motor response. These forms of learning are defined by the procedures used to produce them. Among these are associative and nonassociative learning paradigms. Associative learning includes classical conditioning and operant conditioning. In both, two stimuli or events that occur in a temporally paired fashion result in the formation of an association

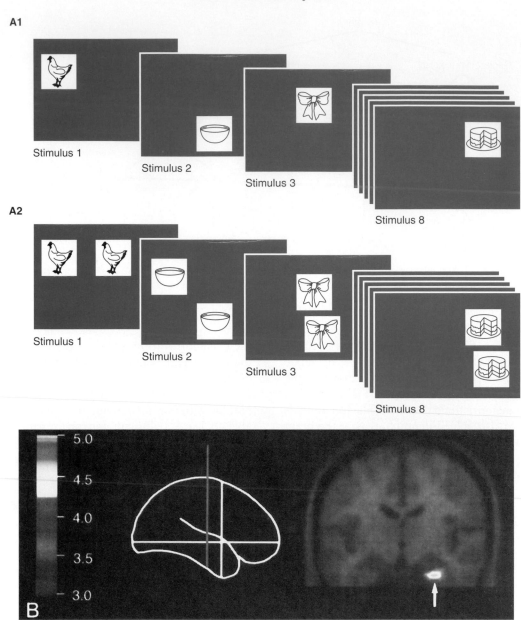

FIGURE 2   Positron emission tomography (PET) analysis of memory localization. (A1) Subjects were instructed to attend to each object in the sequence and remember its location. (A2) Following a 10-minute rest period, the subjects were presented with the objects in two locations and asked to decide which of the two possible locations was correct. (B) During the retrieval phase of the test, the subjects had elevated blood flow (a measure of neural activity) in the right anterior parahippocampal gyrus in the region corresponding to the entorhinal cortex. The extent of blood flow is quantified with the color scale on the left, with red indicating the greatest levels. (Modified from Owen *et al.*, 1996.)

between the two events. In contrast, nonassociative learning is not dependent on pairing. Examples of this form of learning include habituation and sensitization. *Habituation* has been defined as a decrease in a response as a result of repeated stimulation, while *sensitization* has been defined as an increase in response magnitude as a result of a stimulus that increases arousal (Groves and Thompson, 1970; Thompson and Spencer, 1966).

Oftentimes, scientists study nonassociative learning of reflexive behaviors. These learning paradigms are attractive because the stimuli can be precisely controlled and there is generally a well-defined behavioral response. Moreover, the neural pathways of many reflexive behaviors have been described. This knowledge of the circuit enables powerful cell biologic approaches to be applied to the analysis of the underlying mechanisms.

Habituation

Habituation, perhaps the simplest form of nonassociative learning, refers to a decrement of responsiveness caused by repetition of a stimulus. It is generally distinguished from simple fatigue, as responsiveness can be rapidly restored (dishabituated) by the presentation of a novel stimulus to the animal. The parametric features of habituation have been described by Thompson and Spencer (1966).

Sensitization

Sensitization is also a form of nonassociative learning and refers to the enhancement of a behavioral response to a test stimulus as a result of applying a novel stimulus to the animal. The sensitizing stimulus may be of the same modality as the test stimulus and applied at the same site as a test stimulus used to elicit the response, or it may be of a different modality, applied to a different locus. The same stimuli that lead to habituation may also lead to sensitization. Consequently, the strength of a behavioral response elicited by a repeated stimulus may be the net outcome of the two processes of habituation and sensitization (Groves and Thompson, 1970).

Classical Conditioning

Classical conditioning is an example of associative learning in which the presentation of a reinforcing (unconditioned) stimulus is made contingent on that of a preceding (conditioned) stimulus. The change in behavior produced by repeated pairing of the two stimuli can be measured in a number of ways. An example of classical conditioning involves the training procedure originally described by Pavlov (1927) to condition salivation in dogs (Fig. 3). Before training, meat powder (referred to as the *unconditioned stimulus*, or US) reliably elicited salivation (referred to as the *unconditioned response*, or UR). The signal for the presentation of the US is called the *conditioned stimulus* (CS). In the original experiments of Pavlov, the CS was a bell. Traditionally, the CS does not evoke a response similar to the UR and is typically referred to as a *neutral stimulus*. During training, the US was made contingent on the CS by repeatedly pairing the presentations of CS and US. After training, the response to the CS alone had changed such that the bell elicited salivation (the conditioned response, CR). The persistence of a CR after training is called *retention*. When the contingency between CS and US was eliminated by repeatedly presenting the bell in the absence of the meat powder (US), the ability of the CS to elicit the CR (salivation) gradually diminished. This process is called *extinction*.

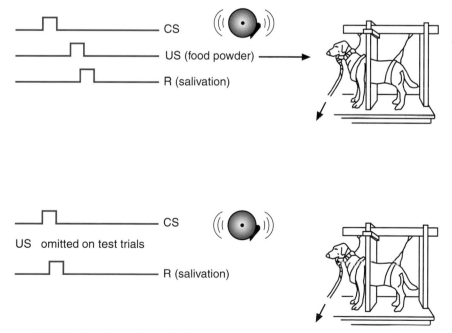

FIGURE 3   Classical conditioning. In the procedure introduced by Pavlov, the production of saliva is monitored continuously. Presentation of meat powder reliably leads to salivation, whereas some neutral stimulus such as a bell initially does not. With repeated pairings of the bell and meat powder, the animal learns that the bell predicts the food and salivates in response to the bell alone. (Modified from Rachlin, 1970.)

In Pavlovian conditioning, contingency is generally established by the close temporal pairing (contiguity) of the CS and US. (For a more detailed discussion see Mackintosh, 1974, 1983; Rescorla, 1967.) Indeed, there is an optimum time interval for conditioning. Shorter or longer intervals between the two stimuli result in less effective conditioning. The interval between CS onset and US onset is called the *CS–US interval* or the *interstimulus interval* (ISI). Most conditioning procedures involve repeated pairings of the CS and US. The interval between these pairings is called the *intertrial interval*.

Specificity of behavioral change to pairing can be most clearly shown by using a differential conditioning procedure. In this procedure, two different conditioned stimuli are used in the same animal; one is specifically paired with the US and is therefore called the CS+, whereas the other, the CS−, is specifically unpaired. Learned changes in behavior can be assessed by comparing the response to the CS+ with that to the CS−.

In Pavlov's experiments, the bell (the CS) did not produce salivation when initially presented alone. In some examples of conditioning procedures, the CS initially produces a small response similar to that evoked by the US. After pairing, the response to the CS is enhanced. This type of conditioning is known as α-conditioning. Both classical conditioning and α-conditioning are similar in that they require a close temporal relationship between the CS and the US; however, in classical conditioning, the CS is initially neutral, whereas in α-conditioning the CS initially produces a weak response that is subsequently enhanced. Some have argued that the distinction between α-conditioning and classical conditioning is somewhat descriptive, because in principle the two could be mediated by identical cellular mechanisms. α-Conditioning and sensitization also resemble one another in that both involve modification of a previously existing response to a stimulus. They differ, however, in their temporal requirements; α-conditioning requires a close temporal association between the CS and US, whereas sensitization does not.

Operant Conditioning

Operant conditioning, or instrumental conditioning, is also a form of associative learning but it differs from classical conditioning in that during this training procedure the reinforcing stimulus is contingent on the performance of a behavior produced by the animal rather than on a CS delivered by the experimenter. The animal therefore learns the consequences of its own behavior and alters that behavior as a result of training. An example of this type of conditioning has been described by Skinner (1938). Before training, a pigeon confined in a small compartment pecks randomly at the walls. During training,

delivery of food reinforcement is made contingent on the animal pecking a single location in the compartment (such as a small disk). Because food and the peck were paired, the pigeon continues to peck at the disk after training, even in the absence of food reinforcement. As in classical conditioning, the learned behavior is extinguished when repeated pecks are no longer followed by food reinforcement. Both operant conditioning and classical conditioning are similar in that both require a close temporal association between two stimuli. Operant conditioning differs from classical conditioning in that the reinforcement is contingent on the animal's response, rather than on the presentation of the CS.

In this brief introduction to conditioning procedures, various simple behavioral modifications have been described. These definitions, however, are operational, and at a mechanistic level (*e.g.*, neural) some of the distinctions between various examples of conditioning may not hold. At the cellular level, basic mechanisms underlying these different examples of learning may be similar.

MECHANISMS OF LEARNING AND MEMORY

Insights from the Study of Simple Animals

Progress in analyzing the mechanisms of learning and memory in the vertebrate nervous system has been hindered by the enormous complexity of the brain. Consequently, many neuroscientists have turned to the study of selected invertebrates in order to exploit their relatively simple nervous systems and large, identifiable neurons that are accessible for detailed anatomical, biophysical, biochemical, and molecular studies.

Behaviors and Neural Circuits

One animal that is well suited for the examination of the molecular, cellular, morphologic, and network mechanisms underlying neuronal plasticity and learning and memory is the marine mollusk *Aplysia*. Neurons and neural circuits that mediate many behaviors in *Aplysia* have been identified. In several cases, these behaviors have been shown to be modified by learning. Moreover, specific loci within neural circuits at which modifications occur during learning have been identified, and aspects of the cellular mechanisms underlying these modifications have been analyzed and modeled (for reviews, see Byrne *et al.*, 1993; Byrne and Kandel, 1996; Hawkins *et al.*, 1993). Defensive reflexes in *Aplysia* (Fig. 4A) exhibit three forms of nonassociative learning: habituation, dishabituation, and sensitization. A single sensitizing stimulus can produce an enhancement that lasts minutes (short-term sensitization), whereas more prolonged training (e.g., multiple stimuli) produces an enhancement that lasts days to weeks (long-term sensitization). *Aplysia* also exhibits classical conditioning and operant conditioning.

A prerequisite for the analysis of the neural and molecular basis of these different forms of learning is an understanding of the neural circuit that controls the behavior. The afferent limb of the withdrawal reflex in Fig. 4A consists of sensory neurons with somata in the central nervous system. The sensory neurons (SNs) monosynaptically excite motor neurons (MNs) that are also located in the central nervous system (Fig. 4B). Activation of the motor neurons leads to contraction of the peripheral muscle and the subsequent withdrawal response. Excitatory and inhibitory interneurons in the withdrawal circuit have also been identified, although these interneurons are not illustrated in Fig. 4B. The sensory neurons appear to be important plastic elements in the neural circuits. Changes in their membrane properties and the strength of their synaptic connections (synaptic efficacy) are associated with sensitization.

Cellular Mechanisms in Sensory Neurons Contributing to Short- and Long-Term Sensitization in Aplysia

Short-Term Sensitization

Short-term sensitization is induced when a single brief train of shocks to the body wall results in the release of modulatory transmitters, such as serotonin (5-HT), from a separate class of interneurons (INs) referred to as *facilitatory neurons* (Fig. 4B). These facilitatory neurons regulate the properties of the sensory neurons and the strength of their connections with postsynaptic interneurons and motor neurons, a process called *heterosynaptic facilitation* (Fig. 4C). The molecular mechanisms contributing to heterosynaptic facilitation are illustrated in Fig. 5. Serotonin binds to at least two types of receptors on the outer surface of the membrane of the sensory neurons. The binding of 5-HT to one class of receptors leads to the activation of adenylyl cyclase, which in turn, leads to an elevation of the intracellular level of the second-messenger adenosine-3′,5′-monophosphate (cyclic adenosine monophosphate, or cAMP) in sensory

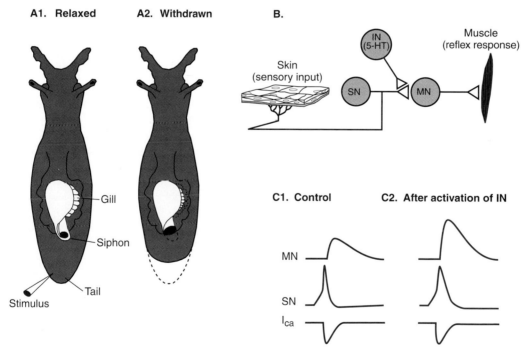

FIGURE 4   Tail-siphon withdrawal reflex of *Aplysia* and model of the neural circuit and its plasticity. (A) Dorsal view of *Aplysia*: (1) relaxed position; (2) stimulus applied to the tail elicits a reflex withdrawal of the tail and siphon. (B) Simplified neural circuit for defensive reflexes in *Aplysia*. A tactile stimulus to the skin activates the afferent terminals of sensory neurons (SNs). These sensory neurons make monosynaptic connections with motor neurons (MNs) as well as interneurons interposed between the SNs and MNs (not shown). Sensitizing stimuli activate facilitatory interneurons (INs) that release modulatory transmitters, one of which is 5-HT. The modulator leads to an alteration of the properties of the SN. (C) Model of heterosynaptic facilitation of the sensorimotor connection that contributes to short- and long-term sensitization in *Aplysia*. An action potential in a SN after the sensitizing stimulus (C2) results in greater transmitter release and hence a larger postsynaptic potential in the MN than an action potential prior to the sensitizing stimulus (C1). For short-term sensitization, the enhancement of transmitter release is due, at least in part, to broadening of the action potential and an enhanced flow of Ca^{2+} into the sensory neuron.

neurons. When cAMP binds to the regulatory subunit of cAMP-dependent protein kinase (protein kinase A, or PKA), the catalytic subunit is freed and can add phosphate groups to specific substrate proteins and hence alter their functional properties. One consequence of this protein phosphorylation is an alteration of the properties of membrane channels. Specifically, the increased levels of cAMP lead to a decrease in the S–K$^+$ current ($I_{K,S}$), a component of the calcium-activated K$^+$ current ($I_{K,Ca}$) and the delayed K$^+$ current ($I_{K,V}$). These changes in membrane currents lead to depolarization of the membrane potential, enhanced excitability, and an increase in the duration of the action potential (Fig. 4C). Cyclic AMP also appears to activate a membrane potential and spike-duration-independent process of facilitation, which is represented in Fig. 5 (large open arrow) as the translocation or mobilization of transmitter vesicles from a storage pool to a releasable pool, making more transmitter-containing vesicles available for release with subsequent action potentials in the sensory neuron. These combined effects contribute to the short-term cAMP-dependent enhancement of transmitter release. Serotonin also appears to act through another class of receptors to increase the level of the second-messenger diacylglycerol (DAG). DAG activates protein kinase C (PKC). PKC, like PKA, activates the spike-duration-independent process of facilitation. In addition, a nifedipine-sensitive Ca^{2+} channel ($I_{ca,Nif}$) and the delayed K$^+$ channel ($I_{K,V}$) are regulated by PKC. Thus, the delayed K$^+$ channel ($I_{K,V}$) is dually regulated by PKC and PKA. The modulation of $I_{K,V}$ makes an important contribution to the increase in duration of the action potential (see Fig. 4C).

The consequences of the activation of these multiple second-messenger systems and modulation of multiple cellular processes occur when test stimuli elicit action potentials in the sensory neuron at various times after the presentation of the sensitizing stimuli (see Fig. 4C). More transmitter is available to be released as a result of the mobilization process, and each action potential is broader, allowing a greater influx of Ca^{2+} to trigger release of the available transmitter. The combined effects of mobilization and spike broadening lead to enhanced release of transmitter from the sensory neuron and consequently a larger postsynaptic potential in the motor neuron. Larger postsynaptic potentials lead to an enhanced activation of interneurons and motor neurons and thus an enhanced behavioral response (*i.e.*, sensitization).

Long-Term Sensitization

Repeating the sensitizing stimuli over a 1.5-hour period leads to the induction of long-term sensitization, the memory of which can persist for days to weeks. Repeated training leads to more prolonged phosphorylation and activation of nuclear regulatory proteins by PKA. Such proteins interact with regulatory regions of DNA and lead to increased transcription of RNA and hence increased synthesis of specific proteins. Some of the resulting proteins may be transcription factors that can activate other genes, some of which may be able to maintain their own activation. Some of the newly synthesized proteins initiate the restructuring of the axon arbor. The sensory neuron can then form additional connections with the same postsynaptic target or make new connections with other cells. As with short-term

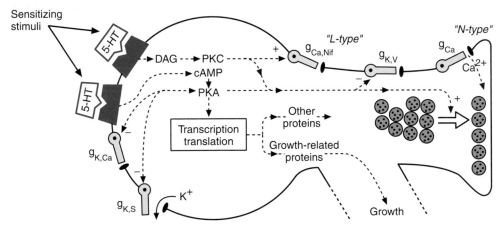

FIGURE 5   Molecular events in the sensory neuron. 5-HT released from the facilitatory neuron (*i.e.*, Fig. 4B) binds to at least two distinct classes of receptors on the outer surface of the membrane of the sensory neuron, which leads to the transient activation of two intracellular second messengers, DAG and cAMP. These second messengers, acting though their respective protein kinases, affect multiple cellular processes, the combined effects of which lead to enhanced transmitter release when a subsequent action potential is fired in the sensory neuron. Long-term alterations are achieved through regulation of protein synthesis and growth. Positive (+) and negative (−) signs indicate enhancement and suppression of cellular processes, respectively (see text for additional details).

sensitization, the enhanced release of transmitter from existing contacts of sensory neurons onto motor neurons and interneurons underlies the long-term enhanced responses of the animal to test stimuli. However, unique to long-term sensitization are the increases in axonal arborization (see Fig. 5) and synaptic contacts that may contribute to the enhanced activation of follower interneurons and motor neurons.

Mechanisms of Classical Conditioning

Mechanisms of classical conditioning have also been examined in *Aplysia*. Classical conditioning procedures, like sensitization procedures, lead to a modulation of neuronal membrane currents and an enhancement of synaptic efficacy (for a review, see Lechner and Byrne, 1997). Interestingly, mechanisms of classical conditioning are at least in part an extension of mechanisms in place that mediate sensitization. Thus, it appears that more complex examples of learning use as building blocks mechanisms for simpler forms of learning.

Operant Conditioning

In addition to being a model system that has revealed a detailed understanding of the mechanisms underlying nonassociative learning (*i.e.*, sensitization) and classical conditioning, *Aplysia* is also a useful model system to investigate the mechanisms underlying operant conditioning (Brembs *et al.*, 2002). In an operant conditioning paradigm, the delivery of a reinforcing stimulus is contingent upon the expression of a designated behavior. In the case of feeding behavior, the operant behavior is ingestive or biting movements and the reinforcement is a stimulus to an afferent nerve which, based on previous work, is a neural pathway enriched in dopamine processes and which signals presence of food in the mouth. Animals are conditioned by delivering a brief stimulus to the nerve each time they display a spontaneous bite. Learning is assessed by measuring the number of spontaneous bites following the conditioning procedure. Animals that had received the operant conditioning procedure displayed a higher number of bites than control animals.

The development of behavioral protocols for operant conditioning allowed for further investigations into the mechanisms underlying the learning. Neural correlates of operant conditioning were found in a single identified neuron involved in the generation of the feeding motor program. The changes consisted of a decreased threshold to elicit a burst of spikes and a decrease in the resting conductance. Both changes would lead to an enhanced probability of generating feeding behavior after conditioning. Of considerable interest was the observation that the effects of behavioral conditioning could be mimicked by pairing activity in the cell isolated in culture with a brief application of dopamine. These data suggest that intrinsic cell-wide plasticity may be one important mechanism underlying this type of learning and that dopamine is a key transmitter mediating the reinforcement. Continued analysis will provide insights into the molecular mechanisms of operant conditioning as well as insights into the mechanistic relationship between operant and classical conditioning.

Long-Term Potentiation

Long-term potentiation (LTP) is a persistent enhancement of synaptic efficacy generally produced as a result of delivering a brief (several-second) high-frequency train (tetanus) of electrical stimuli to an afferent pathway. The great difference between the duration of the tetanus and the duration of the subsequent enhancement is the defining characteristic of LTP. Such long-term synaptic enhancement, lasting at least several hours in *in vitro* preparations and weeks in intact preparations, has received growing attention because of the possibility that it is related to natural mechanisms of learning and memory. For example, there are "cooperative" and "associative" influences on LTP, and there appear to be a number of similarities between neural changes produced by LTP procedures and neural correlates of associative learning (*e.g.*, Pavlovian conditioning). Long-term potentiation has been observed in many regions of the mammalian central nervous system (CNS), in the peripheral nervous system, and in the CNS and neuromuscular junctions of several invertebrates.

Figure 6A illustrates an experimental arrangement for inducing and analyzing LTP. An intracellular recording is made from a postsynaptic neuron that receives monosynaptic excitatory inputs from presynaptic neurons. Brief electric shocks delivered to the afferent pathway lead to the initiation of action potentials in the individual axons in the pathway, and these action potentials propagate to the synaptic terminals. (The rationale for stimulating multiple afferents will become clear when the mechanisms for LTP are discussed later.) The release of transmitter from the multiple afferent terminals produces a summated excitatory postsynaptic potential (EPSP) in the postsynaptic cell. Test stimuli are repeatedly delivered (Fig. 6B1) at a low rate that produces stable EPSPs in the postsynaptic cell (Fig. 6B2). After a stable baseline period, a brief high-frequency tetanus is delivered. Subsequent test stimuli produce enhanced EPSPs. The enhancement is associated with at least two temporal domains. There is a large but transient enhancement that represents a phenomenon known as *post-tetanic potentiation* (PTP). The PTP is followed by a stable and enduring enhancement that persists for

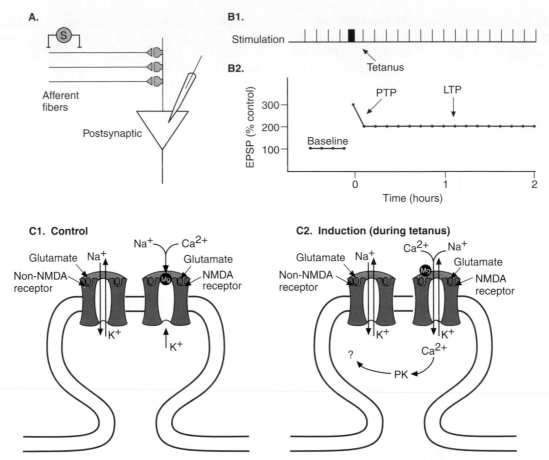

FIGURE 6   Long-term potentiation (LTP). (A) Experimental arrangements for analyzing LTP in the CNS. An intracellular recording is made from a postsynaptic cell (in this case, a pyramidal cell in the CA1 region of the hippocampus), and electric shocks are delivered to an afferent pathway (the Schaeffer collateral pathway) that projects to the postsynaptic neuron. (B) Stimulus protocol and results: (B1) Single weak electric shocks are repeatedly delivered to the afferent pathway. After obtaining several stable baseline responses, a brief high-frequency tetanus is delivered. After the tetanus, the low frequency test stimulation is resumed. (B2) EPSPs are normalized to their control (pretetanus level). The tetanus produces short-term enhancement (post-tetanic potentiation, or PTP) followed by an enduring enhancement (LTP) that persists for at least 2 hours. (C) Mechanisms for the induction of LTP (at the Schaeffer collateral–CA1 pyramidal cell synapse): (C1) The spines of the postsynaptic cell have both NMDA and non-NMDA glutamate receptors. Glutamate released by test stimuli binds to and activates the non-NMDA receptors. Glutamate released by test stimuli also binds to NMDA receptors, but no ions flow through the NMDA channels because they are blocked by Mg^{2+}. (C2) The tetanus produces a large postsynaptic depolarization that displaces the Mg^{2+} from the pore of the NMDA channel. Ca^{2+} can now flow into the spine through the NMDA channel and induce a cascade of biochemical reactions (including activation of Ca^{2+}-dependent protein kinases, or PKs) that lead to a change in synaptic efficacy.

many hours. This enduring enhancement is referred to as LTP.

Although LTP of the form illustrated in Fig. 6B2 has been observed at a number of synapses, the underlying mechanisms for these different examples of LTP can differ. The following discussion will focus on the mechanisms for LTP at a particular synapse (the Schaeffer collateral–CAl pyramidal cell synapse) in the hippocampus. The mechanism at this synapse takes advantage of some of the unique properties of the N-methyl-D-aspartate (NMDA) receptor described in Chapter 6.

The Schaeffer collateral axons make synaptic contacts with the pyramidal cells on specialized dendritic structures called *spines*. The spines have both NMDA and non-NMDA glutamate receptors (Fig. 6C). Test stimuli lead to the release of glutamate from the afferent terminal, which diffuses across the synaptic cleft and binds with both types of receptors. Binding to the non-NMDA receptor leads to an increase in the permeability to Na^+ and K^+ and a subsequent small EPSP. Glutamate also binds to the NMDA receptor but because of the block by Mg^{2+} no permeability changes occur. In contrast, the tetanus produces a large depolarization of the spine because of both temporal summation of the effects of the individual EPSPs produced by the non-NMDA receptors at that spine and spatial summation of the effects of EPSPs produced at spines contacted by the other afferent fibers that are conjointly activated by the nerve shock. The resultant large depolarization of the spine displaces Mg^{2+} from the NMDA receptor, which allows Ca^{2+} influx to occur. As described later, the Ca^{2+} influx through the NMDA receptor is essential for the induction of LTP. The rationale for stimulating multiple afferents (Fig. 6A) should now be clear. In order to remove the Mg^{2+} block of the NMDA channel, depolarization from multiple afferents is necessary.

The Ca^{2+} influx through the NMDA channel activates one or more Ca^{2+}-dependent protein kinases (PKs) (Fig. 6C2), and the phosphorylation of substrate proteins leads to an enduring change in synaptic efficacy. The mechanisms for the induction steps subsequent to the activation of kinases are not fully understood. One possibility is a postsynaptic modification, such as an increase in the number of non-NMDA receptors that are available to bind with transmitter released by the post-tetanus test stimuli. Alternatively, there may be a retrograde messenger released by the postsynaptic cell to directly affect aspects of the release mechanism in the presynaptic neuron. Analyses to distinguish between these possibilities have yielded conflicting results. It is likely that LTP at the Schaeffer collateral–CA1 pyramidal synapse involves modifications of both pre- and postsynaptic processes.

In principle, LTP at this synapse could be induced by activation of a single afferent fiber, but the depolarization produced by a single presynaptic neuron (even with a tetanus) is insufficient to relieve the Mg^{2+} block of the postsynaptic NMDA channel. Thus, other neurons must be activated as well. For example, a tetanus to one afferent pathway in conjunction with a weak test stimulus to a separate pathway would lead to an enhancement of the test pathway even though the test pathway was not tetanized. In this case, the tetanized pathway provides the depolarization to relieve the Mg^{2+} block at the synapse of the test pathway.

Memory Formation of Complex Patterns

The relationship between the synaptic changes and the behavior of conditioned reflexes is generally straightforward, because the locus for the plastic change is part of the mediating circuit. Thus, the change in the strength of the sensorimotor synapse can be related to the memory for sensitization (*e.g.*, see Fig. 4). However, in most other examples of memory, it is considerably less clear how the synaptic changes are induced, and, once induced, how the

information is retrieved. Neurobiologists have turned to artificial neural circuits to gain insights into these issues.

A simple network that can store and "recognize" patterns is illustrated in Fig. 7. The network is artificial, but it is nevertheless inspired by actual circuitry in the CA3 region of the hippocampus. In this example, six different input projections make synaptic connections with the dendrites of each of six postsynaptic neurons (Fig. 7A). The postsynaptic neurons serve as the output of the network. Input projections can carry multiple types of patterned information, and these patterns can be complex. But, to simplify the present discussion, consider that the particular input pathway in Fig. 7A carries information regarding the pattern of neural activity induced by a single brief flash of a spatial pattern of light. For example, activity in the top pathway (line *a*) might represent light falling on the temporal region, whereas activity in the pathway on the bottom (line *f*) might represent light falling on the nasal region of the retina. Thus, the spatial pattern of an image falling upon the retina could be reconstructed from the pattern of neuronal activity over the *n* (in this case, 6) input projections to the network. Three aspects of the circuit endow it with the ability to store and retrieve patterns. First, each of the input lines makes a sufficiently strong connection with its corresponding postsynaptic cell to reliably activate it. Second, each output cell (*z* to *u*) sends an axon collateral that makes an excitatory connection with itself as well as the other five output cells. This pattern of synaptic connectivity leads to a network of 36 synapses (42 when including the 6 input synapses). Third, each of the 36 synaptic connections is modifiable through an LTP-like mechanism (see earlier discussion). Specifically, the strength of a particular synaptic connection is initially weak, but it will increase if the presynaptic and postsynaptic neurons are active at the same time. The circuit configuration with the embedded synaptic "learning rule" leads to an autoassociation or autocorrelation matrix. The autoassociation is derived from the fact that the output is fed back to the input, where it associates with itself.

Now consider the consequences of presenting the patterned input to the network of Fig. 7A. The input pattern will activate the six postsynaptic cells in such a way as to produce an output pattern that will be a replica of the input pattern; in addition, however, the pattern will induce changes in the synaptic strength of the active synapses in the network. For example, synapse 3 will be strengthened, because both the postsynaptic cell, cell *z*, and the presynaptic cell, cell *x*, will be active at the same time. Note also that synapses 1, 5, and 6 will be strengthened. This is so because these input pathways to cell *z* are also active; thus, all synapses that are active at the same time as cell *z* will be strengthened. When the pattern is

A. Before learning   **B. After learning**

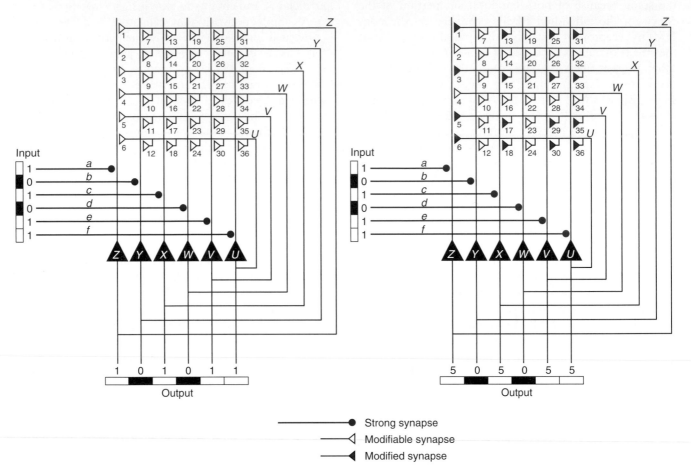

FIGURE 7   Autoassociation network for recognition memory. The artificial circuit consists of six input pathways that make strong connections to each of six output neurons. The output neurons have axon collaterals that make synaptic connections with each of the output cells. (A) A pattern represented by activity in the input lines or axons (*a, b, c, d, e, f*) is presented to the network. The number 1 represents an active axon (*e.g.*, a spike), whereas a 0 represents an inactive axon. The input pathways make strong synapses (filled circles) with the postsynaptic output cells. Thus, the output cells (*u, v, w, x, y, z*) generate a pattern that is a replica of the input pattern. The collateral synapses were initially weak and do not contribute to the output. Nevertheless, the activity in the collaterals that occurred in conjunction (assume minimal delays within the circuit) with the input pattern led to a strengthening of a subset of the 36 synapses. (B) A second presentation of the input produces an output pattern that is an amplified, but otherwise intact, replica of the input. The network can store multiple patterns. Moreover, an incomplete input pattern can be used as a cue to retrieve the complete pattern.

presented again, as in Fig. 7B, the output of the cell will be governed not only by the input but also by the feedback connections, a subset of which were strengthened (Fig. 7B, filled synapses) by the initial presentation of the stimulus. Thus, for output cell *z*, a component of its activity will be derived from input *a* but components will also come from synapses 1, 3, 5, and 6. If each of the initially strong and newly modified synapses is assumed to contribute equally to the firing of output cell *z*, it would be reasonable that the activity would be five times greater than that activity produced by input *a* before the learning. After learning, we see that the output is an

amplified version of the input but that the basic features of the pattern are preserved.

Note that the memory for the pattern does not reside in any one synapse or in any one cell; rather, it is *distributed* throughout the network at multiple sites. The properties of these types of autoassociation networks have been examined by James Anderson, Teuvo Kohonen, David Mar, Edmond Rolls, David Wilshaw, and their colleagues and found to exhibit a number of phenomena that would be desirable for a biologic recognition memory system. For example, the autoassociation networks exhibit pattern completion. If a partial input pattern is

---

Clinical Note

Alzheimer's Disease

Alzheimer's disease was first described by the German physician Alois Alzheimer in 1907. The disease is associated with a progressive decline in cognitive function (*i.e.*, dementia), in general, and a specific impairment in the ability to retrieve old memories (retrograde amnesia) and form new memories (anterograde amnesia). Approximately 4 million individuals in the United States have Alzheimer's disease, and it has been estimated that the cost due to treatment and care of these individuals exceeds $100 billion per year. At the cellular level, the disease is associated with neuronal loss and the presence of neurofibrillary tangles and amyloid plaques. The amyloid plaques are extracellularly localized and consist primarily of β-amyloid, a peptide derived from the transmembrane protein amyloid precursor protein. Neurofibrillary tangles consist of pairs of filaments wound around each other in a helical arrangement, thus giving rise to the term *paired helical filaments* (PHFs). The PHFs contain a hyperphosphorylated form of the microtubule-associated protein, tau. Plaques and tangles are found throughout the brain, but their greatest density is the hippocampus and cerebral cortex. This localization is consistent with the observation that most significant deficits associated with Alzheimer's disease are in episodic and semantic memory. There is no known cure for Alzheimer's disease. Alzheimer's disease is hypothesized to affect cholinergic neurons, thus attempts have been made to treat the disease by enhancing cholinergic neurotransmission. Two approved drugs are Aricept® and Cognex®. Both act by inhibiting acetylcholinesterase (see Chapter 6) and thus potentiating the effects of the acetylcholine released by undamaged cholinergic neurons. Note that a similar strategy is used to treat myasthenia gravis (see Chapter 6).

---

presented, the autoassociation network can complete the pattern; thus, any part of the stored pattern can be used as a cue to retrieve the complete pattern. For the example of Fig. 7, the input pattern was {101011}. This pattern led to an output pattern of {505055}. If the input pattern was degraded to {101000}, the output pattern would be {303022}. There is some change in the strength of firing, but the basic pattern is preserved. Autoassociation networks also exhibit a phenomenon known as *graceful degradation*. The network can still function even if some of the input connections or postsynaptic cells are lost. This property arises from the distributed representation of the memory within the circuit.

Note that the concept of distributed representation of memory crosses multiple levels of organization of memory systems. Multiple brain systems are involved in memory, and memory is distributed in different anatomical regions of the brain (see Fig. 1). Memory is distributed among synapses in a particular memory circuit (see Fig. 7), and memory at any one synapse is represented by multiple cellular changes (see Fig. 5).

References

Brembs B, Lorenzetti FD, Reyes FD, Baxter D, Byrne JH. Operant reward learning in *Aplysia*: neuronal correlates and mechanisms. *Science* 2002; 296: 1706–1709.

Byrne JH, Kandel ER. Presynaptic facilitation revisited: state and time dependence. *J Neurosci* 1996; 16:425–435.

Byrne JH, Zwartjes R, Homayouni R, Critz S, Eskin A. Roles of second messenger pathways in neuronal plasticity and in learning and memory: insights gained from *Aplysia*, in: Nairn AC, Shenolikar S, Eds. *Advances in second messenger and phosphoprotein research*, Vol. 27. New York: Raven Press, 1993, pp. 47–108.

Dudai Y. *The neurobiology of memory: concepts, findings, trends.* New York: Oxford University Press, 1989.

Groves PM, Thompson RF. Habituation: a dual-process theory. *Psychol Rev* 1970; 77:419–450.

Hawkins RD, Kandel ER, Siegelbaum S. Learning to modulate transmitter release: themes and variations in synaptic plasticity. *Ann Rev Neurosci* 1993; 16:625–665.

Lechner HA, Byrne JH. New perspectives on classical conditioning: a synthesis of hebbian and non-hebbian mechanisms. *Neuron*, 1998; 20:355–358.

Lechner HA, Baxter DA, Byrne JH. Classical conditioning of feeding in *Aplysia*: I. Behavioral analysis. *J Neurosci* 2000; 20: 3369–3376.

Mackintosh NJ. *The psychology of animal learning*. New York: Academic, 1983.

Mackintosh NJ. *Conditioning and associative learning*. New York: Oxford University Press, 1974.

Owen AM, Milner B, Petrides M, Evans AC. A specific role for the right parahippocampal gyrus in the retrieval of object-location: a positron emission tomography study. *Cog Neurosci* 1996; 8:588–602.

Pavlov IP. *Conditioned reflexes: an investigation of the physiological activity of the cerebral cortex.* London: Oxford University Press, 1927 (transl. by GV Anrep).

Rachlin H. *Introduction to modern behaviorism*, 2nd ed. San Francisco: Freeman, 1970.

Rescorla RA. Pavlovian conditioning and its proper control procedures. *Psychol Rev* 1967; 74:71–80.

Skinner BF. *The behavior of organisms*. New York: Appleton Century, 1938.

Squire LR, Knowlton BJ. Memory, hippocampus, and brain systems, in: Gazzaniga MS, Ed. *The cognitive neurosciences*. Cambridge: MIT Press, 2002.

Thompson RF, Spencer WA. Habituation: a model phenomenon for the study of neuronal substrates of behavior. *Psychol Rev* 1966; 73:16–43.

Suggested Readings

Baxter DA, Byrne JH. Learning rules from neurobiology, in: Gardner LR, Ed. *The neurobiology of neural networks*. Cambridge: MIT Press, 1983, pp. 71–105.

Bliss TVP, Collinridge G. A synaptic model of memory: long-term potentiation in the hippocampus. *Nature* 1993; 361:31–39.

Byrne JH. Cellular analysis of associative learning. *Physiol Rev* 1987; 67:329–429.

Byrne JH, Crow T. Invertebrate models of learning: *Aplysia* and *Hermissenda*, in: Arbib M, Ed. *Handbook of brain theory and neural network*, 2nd ed. Cambridge: MIT Press/Bradford Books, 2003, pp. 581–585.

Corkin S. Lasting consequences of bilateral medial temporal lobectomy: clinical course and experimental findings in H.M. *Sem Neural* 1984; 4:249–259.

Kandel ER. The molecular biology of memory storage: a dialogue between genes and synapses. *Science* 2001; 294(5544):1030–1038.

Kohonen T. *Self-organization and associative memory*, 3rd ed. Heidelberg: Springer-Verlag, 1989.

Malenka RC, Nicoll RA. Long-term potentiation—a decade of progress? *Science*. 1999; 285:1870–1874.

Milner B, Corkin S, Teuber H. Further analysis of the hippocampal amnesic syndrome: 14-year follow-up study of H.M. *Neuropsychologia* 1968; 6:215–234.

Selkoe DJ, Podlisny MB. Deciphering the genetic basis of Alzheimer's disease. *Ann Rev Genomics Hum Genet* 2002; 67–99.

Wolpaw JR, Schmidt JT, Vaughan TM, Eds. *Activity-driven CNS changes in learning and development*. New York: The New York Academy of Sciences, 1991.

Zucker RS, Regher WG. Short-term synaptic plasticity. *Annu Rev Physiol* 2002; 63:355–405.

INTEGRATIVE PHYSIOLOGY

C H A P T E R

62

Body Temperature Regulation

JAMES A. SCHAFER

KEY POINTS

- The normal body core temperature (about 37°C, 98.6°F) is maintained despite wide variations in environmental temperature and with changes in metabolic heat production. Body temperature has a diurnal variation of about 0.6°C, being lowest in the morning and highest in the late afternoon. With ovulation, the body temperature rises 0.2 to 0.5°C and remains elevated in the second half of the menstrual cycle. The body temperature may rise 2 to 3°C with heavy exercise.

- Heat is transferred to or from the body by the processes of radiation, conduction, and evaporation. Convection enhances the latter two mechanisms. The heat transferred by *conduction* and *radiation* is proportional to the difference between the temperature of the skin and the surroundings, thus either mechanism can result in heat loss or gain. *Evaporation* is solely a heat loss mechanism, resulting in 580 kcal heat loss per 1 L of water evaporated from the skin surface.

- *Heat production* varies from about 80 kcal/hr at rest to more than 1400 kcal/hr with maximal exertion. Heat production may be increased by shivering or thyroid hormone and catecholamines, which accelerate metabolism.

- Cutaneous blood flow is regulated in the range of 0 to 30% of the cardiac output in order to match *body heat loss* to heat production. The resulting change in skin temperature regulates the rate of heat exchange by conduction and

Key Points (*continued*)

radiation. Sympathetic cholinergic nerves regulate sweat production by the cutaneous eccrine glands and regulate heat loss by evaporation.

- Changes in body temperature are sensed by receptors in the *preoptic anterior hypothalamus*, spinal cord, abdomen, skin, and other organs. Information from these receptors is integrated in the hypothalamus, which compares the temperature sensed to a desired set-point. When body temperature deviates from the set-point, the hypothalamus integrates appropriate responses to conserve or dissipate heat.

- Heat pyrexia (*heat stroke*) occurs when body temperature exceeds ~40°C. At this point, hypothalamic regulation ceases and body temperature rises uncontrollably.

- *Heat exhaustion* occurs after excessive loss of salt and water due to sweating, and/or when venous return is compromised by a combination of heavy exertion and maximal cutaneous vasodilation. It can be accompanied by heat cramps and circulatory collapse, although the body temperature is not necessarily significantly elevated.

- *Hypothermia* results in a general slowing of metabolism and of cardiac and neural excitability. The thermoregulatory ability of the hypothalamic integrative center is lost when the body core temperature falls below ~30°C.

- *Fever* is most often seen in the presence of infection. The body's immune cells react to exogenous pyrogens from the infecting organisms by releasing cytokines that act as endogenous pyrogens. These endogenous pyrogens cause the release of arachidonic acid metabolites such as prostaglandin E_2 in the tissues, including the hypothalamus. These autacoids have an effect equivalent to raising the temperature set-point. The response is chills, shivering, cutaneous vasodilation, and increased heat production, often followed in cycles by sweating and cutaneous vasodilation.

The temperature of the human body is maintained within 0.6°C (1°F) of its normal value of 37°C (98.6°F) over a relatively wide range of environmental temperatures and during activity and rest. Body temperature is regulated in this range by means of homeostatic feedback mechanisms that are analogous to those that maintain the balance between the intake and output of solutes and water; however, in this case it is the body temperature rather than a solute concentration or the volume of a fluid compartment that is sensed and regulated. Also, in analogy to the mass-balance relationships that establish the constancy of an electrolyte concentration, the constancy of body temperature depends on a balance of input, output, and production, but, in the case of temperature, the balance is between the input, output, and production of heat, rather than mass. Body temperature is determined by the total quantity of heat in the body and the thermal capacity of the body, which is the rate at which the body temperature rises or falls as its heat content increases or decreases. The thermal capacity varies little from person to person, and it is slightly less than 1°C for every kcal of heat lost or gained per kilogram of body weight. In other words, if a 70-kg man were to retain 70 kcal of heat energy, his body temperature would rise by almost 1°C; if he lost that much heat, it would fall by almost 1°C. This is a comparatively high heat capacity and is due to the fact that 60 to 70% of the body mass is water, which has a very high heat capacity.

Nevertheless, it can be seen that even the metabolic production of heat at rest (70 to 80 kcal/hr) would rapidly raise the body temperature to lethal levels if it were not for the ability of the body to dissipate this heat to the environment.

Because the thermal capacity is fixed, body temperature is directly proportional to the total body heat content, i.e., is the total quantity of heat in the body. As long as this quantity remains constant, the body temperature remains constant; however, this constancy requires a balance of heat input, loss, and production, as shown in Fig. 1. Heat is continually produced by the exothermic biochemical reactions that take place in all

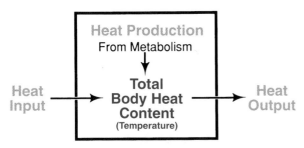

Figure 1   Balance of heat input, output, and production. The total body heat content directly determines the body temperature. Heat is constantly being produced in the body by metabolism. To maintain a constant body temperature, this heat production must be matched by a net loss to the environment. This is accomplished by altering heat input from and output to the environment.

body cells, and heat can be both lost and gained by exchange with the environment. The balance that maintains a constant total body heat content can be expressed by the relation:

$$\text{Heat Loss} = \text{Heat Input} + \text{Heat Production} \quad (1)$$

This equation is completely analogous to the equation for mass balance (Eq. [2] in Chapter 25). The only difference is that the entity that is taken in, produced, or lost by the body is heat energy rather than matter. A constant body temperature is maintained by balancing the rate of heat loss from the body so that it is precisely equal to the sum of the metabolic heat production and any heat input from the environment. Heat exchange with the environment can result in either loss or gain, and both can occur simultaneously. For example, the body may be heated by the visible and infrared radiation from the sun at the same time as it is being cooled by a cold wind. However, because the body is constantly producing heat, even at complete rest (because of the basal metabolic rate), heat loss from the body must always exceed heat input if body temperature is to remain constant. In other words, there must always be a net heat loss to the environment that matches the metabolic heat production.

NORMAL VARIATIONS OF BODY TEMPERATURE

The human is an example of a homeotherm, that is, an animal that maintains a constant, elevated body temperature. The advantage of this regulation is that the body temperature is always near the optimal level for cellular biochemical reactions, thus the overall activity level of the organism can be optimal, regardless of the ambient (environmental) temperature. The drawback is the energy expenditure that is required to maintain a higher body temperature when the ambient temperature is low.

"Normal" body temperature varies slightly among individuals, as well as in the same individual at different times and under different physiologic circumstances. The usual range of normal temperature is 36.2 to 37.8°C (97 to 100°F) when measured rectally and about 0.2 to 0.5°C lower when measured orally. Body temperature shows a diurnal variation of approximately 0.6°C; it is lowest in the early morning just before rising (especially in cold weather) and reaches a maximum in the early evening. Menstruating women have a further monthly variation. The body temperature shows a slight elevation (0.2 to 0.5°C) at the time of ovulation and remains elevated during the second half of the menstrual period. Finally, during hard exercise the body temperature may rise by as much as 2 to 3°C, and even emotional stress may elevate the temperature by up to 2°C.

Even when the body is nude, body temperature can be maintained within the normal range over an environmental temperature range of 10 to 55°C (50 to 130°F) if the air is dry. (Dry air maximizes evaporative heat loss at hot temperatures and minimizes conductive heat loss at low temperatures.) Outside this range, if unprotected, the body loses its regulatory capabilities, and body temperature can approach lethal levels.

EXCHANGE OF HEAT WITH THE ENVIRONMENT

Conduction

As shown in Fig. 2, heat exchange with the environment occurs primarily across the skin by three processes: *conduction* to or from the air molecules or other substances in contact with the skin, *radiation* by infrared rays to or from bodies that are colder or hotter than the skin, and *evaporation* of water vapor in the form of sweat or respiratory secretions. Conductive heat exchange can also occur by ingestion or infusion of hot or cold substances. For example, during hemodialysis or plasma phoresis, the extracorporeal circulation of the blood normally exposes it to a cooler environment, resulting in a decrease in temperature that chills the patient unless the blood is warmed. Radiation and conduction can occur either to or from the skin, leading to either heat gain or heat loss from the body, depending on whether the temperature of the skin surface is less than or greater than that of the environment. On the other hand, evaporation from the skin always results in a heat loss. Conductive and evaporative heat exchange with the air is increased by *convection*, the bulk movement of air around the body. As discussed later, all these heat exchange mechanisms, as well as the rate of metabolic energy production, are regulated by the body's homeostatic mechanisms to ensure the maintenance of a constant total body heat content and, thus, a constant body temperature.

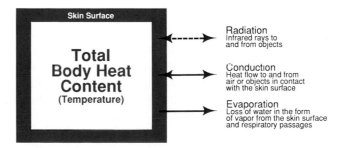

Figure 2 Heat exchange with the environment. The skin serves as an insulating layer between the body core and the environment. Heat exchange with the environment occurs by radiation to and from objects, conduction to and from the air and objects in the environment, and evaporation of water from the skin and respiratory passages.

Heat conduction occurs by the transfer of kinetic energy between the molecules on the surface of the skin and those adjacent to the skin (*e.g.*, air molecules and the molecules in articles of clothing). The rate of heat conduction in kilocalories per hour ($J_{Q,cond}$) between the skin and the surrounding molecules is proportional to the difference in temperature between the skin and these molecules according to the relation:

$$J_{Q,cond} = k_{cond} \cdot \frac{(T_a - T_s)}{L} \qquad (2)$$

where k_{cond} is the *thermal conductivity*, a constant coefficient that describes the heat conduction properties of the layer of air or material immediately adjacent to the skin; L is the thickness of that layer; T_a is the ambient temperature of the environment around the skin; and T_s is the skin temperature. When the skin temperature is greater than that of its surroundings, the heat conduction, $J_{Q,cond}$, will be negative because the difference between these two temperatures is negative. In other words, as common sense would dictate, heat is lost by conduction when the temperature of the skin exceeds that of its surroundings. The equation states that the rate of loss or gain is directly proportional to the temperature difference and inversely proportional to the thickness of the insulating layer.

Radiation

All surfaces at temperatures above absolute zero give off energy in the form of electromagnetic radiation, and the rate of radiation is proportional to the fourth-power temperature of the surface. In the range of biologic and environmental temperatures, this radiation is in the form of infrared rays and visible light. The net heat exchanged between the skin surface and surrounding objects by radiation ($J_{Q,rad}$) is given by the following equation:

$$J_{Q,rad} = k_{rad} \cdot \left(T_a^4 - T_s^4\right) \qquad (3)$$

The constant, k_{rad}, depends on the characteristics of the skin surface and the surfaces of objects in the environment to which heat is lost or from which it is gained. In general, the best radiators or acceptors of radiation are black objects, but this is less marked for infrared wavelengths, which mediate a large proportion of radiative heat exchange in the biologic temperature range. Thus, there is little difference among the races in heat exchange by infrared radiation, although heavily pigmented skin is heated largely by visible radiation, which constitutes a large fraction of the radiation energy from the sun. Although heat exchange by radiation according to Eq. [3] is determined by the 4th power of temperatures (in °K),

it turns out that when the expression $(T_a^4 - T_s^4)$ is expanded mathematically, it is approximately equal to a constant factor times $T_a - T_s$ within the typical range of body and indoor temperatures. Thus, the rate of heat exchange by radiation can be approximated as:

$$J_{Q,rad} = k_{rad}^* \cdot (T_a - T_s) \qquad (4)$$

In this case, k_{rad}^* is not truly a constant (it depends on the temperatures of the two surfaces to a minor degree because of the mathematical approximations made), but it varies little in any practical setting. Because the temperature difference ($T_a - T_s$) is the same regardless of whether it is given in °C or °K, it can be seen that the exchange of heat by radiation, like that by conduction, depends on the temperature difference between the skin surface and that of the surroundings, and it depends on the characteristics of the skin or clothing surface and those of objects in the environment.

The assumptions involved in reducing Eq. [3] to Eq. [4], however, are not accurate when the body exchanges heat with objects of significantly higher or lower temperature. The most pertinent examples are the sun and the night sky. Sunlight has an equivalent temperature of 5800°K, and, despite the fact that the sun occupies a small fraction of the area of the daytime sky, everyone is aware of its warming effect, even in cold weather. At night, when the warming radiation of the sun is absent (except the small amount reflected by the moon) and when there is no cloud cover, the night sky has a temperature of −40 to −50°C. Thus, considerable radiative heat can be lost by the body on a clear night, even in the desert.

Evaporative Heat Loss

Energy is required to convert liquid water to a vapor. This energy is referred to as the *latent heat of vaporization* and is equal to 580 kcal/L of water vaporized. Thus, the rate of heat loss from the body by evaporation ($J_{Q,evap}$) in kilocalories per hour is directly proportional to the rate at which water evaporates from the skin surface and the respiratory passages (J_{H_2O}) in liters per hour:

$$J_{Q,evap} = -(580 \text{ kcal/L}) \cdot J_{H_2O} \qquad (5)$$

Water is constantly being lost from the body by evaporation, even when the ambient temperature is low. Even under sedentary conditions, the basal rate of insensible (loss from the skin plus the expiration of water-saturated air) water loss is about 500 mL/day, resulting in a loss of approximately 300 kcal/day. The loss of water from the skin surface can be dramatically increased by sweating in a hot environment or with exercise. However, this sweat must evaporate for a heat loss to occur, and the water lost in sweat that drips off the skin results in no heat loss from the body. For this reason, the rate of heat loss by evaporation is very much dependent on the relative humidity of the air. When the air surrounding the skin is saturated with water, no evaporation occurs, and no heat can be lost by evaporation. For this reason, heat loss is much less efficient in a hot, humid climate than it is in a hot, dry climate. This effect is expressed by the heat index given in summer weather reports.

Convection

The exchange of heat with the environment by both conduction and evaporation also depends on the bulk movement of the air around the skin surface, which is referred to as *convection*. When the skin is surrounded by motionless air, the air molecules form an insulating layer of sizable thickness (L) around the body, which minimizes the heat exchanged by conduction, as shown in Fig. 3. The effect of the still air layer is to act like a second skin and further insulate the body. Air, because it is a gas, is a relatively good insulator and has a low thermal conductivity, k_{cond}. As shown in Fig. 3, the

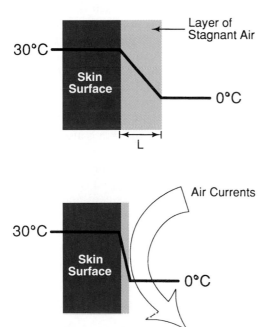

Figure 3 The effect of convection on heat loss to the environment. The air adjacent to the skin forms a stagnant layer that serves as insulation. When the air is still, this insulating layer is relatively thick (large L), and the temperature gradient between the surface of the skin (30°C in this example) and the air farther away from the body (0°C) is shallow. When air currents reduce the thickness of the insulating layer of air, the temperature gradient is steep, and conductive heat transfer increases. Convection also increases evaporative heat loss by mixing drier air into the layer next to the skin.

mixing of air by bulk movement such as a breeze markedly reduces the thickness of the insulating layer of air adjacent to the skin. This reduction in L results in a proportionally larger conductive heat exchange, as indicated by Eq. [2]. This effect of convection increases geometrically with an increase in wind velocity. For example, heat loss with a wind velocity of 8 mph is four times greater than that at 4 mph. On the other hand, convection has no effect on heat exchange by radiation because this exchange depends only on the temperature difference between the actual surface of the skin and the surfaces of objects in the environment.

Convection also increases the rate of evaporation from the skin surface. When the air is still, the layer of air adjacent to the skin becomes nearly saturated with water vapor, resulting in a decrease in evaporation. Convection supplies fresh air with a lesser water content and thus enhances evaporation.

Because of the effects of convection on the rate of heat loss by both conduction and evaporation, much more heat is lost to the environment on a cold, windy day than on an equally cold but still day. This effect gives rise to the wind-chill factor used by meteorologists to indicate the equivalent temperature that would produce the same body cooling on a still day.

Clothing acts to reduce heat exchange with the environment because of its low thermal conductivity (k_{cond}) and its thickness (L). The insulating effect of clothing is due both to the properties of the material of which it is composed and to the air that is trapped in the fabric and serves to increase the layer of still air surrounding the body. The insulating properties are reduced when clothing becomes wet because water, with a higher thermal conductivity, replaces the air. The insulating properties are also reduced by wind, depending on how loosely the material is woven.

Clothing has only a limited ability to reduce heat exchange by radiation. Body heat radiates from the skin surface to the clothing and from the clothing to the environment. Radiation from the skin to the clothing can be reduced in clothing designed for extreme environmental conditions, such as in the arctic or outer space, by layering the inside of the clothing with a highly reflective layer such as gold sputtering.

Total Heat Exchange with the Environment

To calculate the total heat exchange between the body and the environment, the effects of conduction, radiation, and evaporation can be added by combining Eqs. [2], [4], and [5]:

$$J_{Q,total} = \{K \cdot (T_a - T_s)\} - \{(580 \text{ kcal/L}) \cdot J_{H_2O}\} \quad (6)$$

This equation is particularly useful because it demonstrates the primary determinants of the exchange of heat with the environment that are regulated physiologically. The last term in the equation is the heat loss due to evaporation. As noted earlier, the rate of heat loss by evaporation is inversely related to the relative humidity. The first term on the right combines the heat exchange by conduction and radiation because both are proportional to the difference between the temperature of the environment (the ambient temperature) and that of the skin $(T_a - T_s)$. The constant of proportionality K includes several factors that determine heat exchange by the two processes, including the thickness (L) of insulating layers on the surface of the skin, the thermal conductivities of these layers, and their radiative characteristics and those of the surroundings. Convection will increase both terms in the equation by increasing the evaporation of water (J_{H_2O}) and by increasing K because of a decrease in the effective thickness of insulating layers of air and clothing. Convection is normally the greatest contributor to heat loss, especially in the absence of adequate clothing, because it markedly increases both the conductive and evaporative components of heat exchange.

At a given ambient temperature, when other conditions, such as the type of clothing, wind velocity, and relative humidity are constant, Eq. [6] shows that the rate of heat exchange will depend on the temperature of the skin surface (T_s) and the rate of water loss (J_{H_2O}). As discussed in the next sections, the temperature of the skin surface is under physiologic control by regulation of cutaneous blood flow, and the loss of water from the body above the normal insensible rate is determined by the rate of sweat production, which is under autonomic control. Thus, the body regulates heat loss or gain with the environment by controlling two variables: the cutaneous blood flow, which determines the skin temperature, and the rate of sweating, which, at a given relative humidity, is directly proportional to the rate of evaporative heat loss.

PHYSIOLOGIC REGULATION OF HEAT EXCHANGE

Regulation of Skin Temperature

The temperature of the surface of the skin is usually several degrees colder than the core of the body; it also varies from one portion of the body to another. The skin and the subdermal fat layers act as an insulating blanket around the core of the body and protect it from rapid heat loss or heat gain. The insulating properties of skin depend on its thickness, which is primarily determined by the thickness of the layer of fat, a good insulator.

The temperature of the skin surface is also determined by the *cutaneous blood flow* (*i.e.*, the rate of blood flow through the rich capillary network just below the dermis). Cutaneous blood flow is regulated from almost 0 to 30% of the cardiac output. When the cutaneous blood flow is low, the surface temperature of the skin falls, which minimizes heat loss by radiation and conduction. The ability of the skin to reduce these heat losses is limited by the insulating capacity of the skin and the minimal blood flow required to meet the metabolic demands of the tissue. For example, in a very cold climate, the blood flow to the fingers and toes can fall to about 1% of normal. When the cutaneous blood flow is high, the skin temperature can approach the body core temperature, which maximizes the conductive and radiative heat loss.

Cutaneous blood flow is determined by the relative degree of vasoconstriction of precapillary sphincters in the vascular bed just below the dermis. As in other vascular beds, these are under local as well as central regulation (see Chapter 17). Local irritants and heat can cause vasodilation and warming of the immediate skin surface. When a larger area of the skin is warmed, temperature receptors in the skin lead to a reflex increase of blood flow in that area. This response is mediated at the level of the spinal cord. For example, putting your hand in hot water leads to increased blood flow, which is indicated by the redness and relative engorgement of circulation in that hand. Local cooling leads to the opposite reaction.

In response to a generalized change in the temperature of the whole body, the blood flow to the skin is regulated by the hypothalamus. Chilling of the preoptic anterior hypothalamus causes a generalized vasoconstriction in the skin that is mediated by the sympathetic nervous system. Vasoconstriction in the skin is also accompanied by *piloerection*, the raising of body hair that produces the "goose flesh" experienced when one is chilled or frightened. This response is also mediated by the sympathetic nervous system, which produces contraction of smooth muscles associated with the hair follicles. In humans, this response plays only a small role in thermoregulation, but in furry animals, piloerection thickens the insulating layer of the hair and thus reduces heat loss.

Regulation of Sweat Production

The loss of heat by evaporation is regulated by controlling the rate of sweat production by the eccrine sweat glands. These glands are innervated by sympathetic cholinergic nerves, the firing of which can be stimulated by circulating epinephrine and norepinephrine. The latter hormones produce the sweating associated with stress and anxiety. The rate of sweat production can vary from 0 to about 1.5 L/hr in an individual who is not acclimatized and is exercising in a hot environment. When the air is completely dry so that all the sweat evaporates from the skin surface, 1.5 L/hr of sweat production results in a loss of almost 900 kcal/hr.

Regulation of Heat Production

The rate of heat production by a normal 70-kg person can vary from 75 to 80 kcal/hr when sitting still to more than 1400 kcal/hr at maximum rates of exercise. To maintain a constant body temperature, this rate of heat production must be matched by an equal rate of heat loss to the environment. Normally, body temperature regulation is accomplished by physiologic regulation of the rate of heat loss by vasomotor activity in the skin and the rate of sweating, as described earlier. However, the rate of heat production may also be varied to contribute to temperature regulation. Many of these changes in heat production are behavioral. When it is very hot, one naturally reduces one's level of activity and thereby decreases heat production. When it is cold, one can increase heat production by such common behavior as clapping hands and stomping feet.

Physiologically, the most important route of increasing heat production is through *shivering*. Shivering is an asynchronous contraction of the skeletal muscles resulting in increased muscle tone and tremors that occur at a rate of 10 to 20/s. Shivering is produced by descending neural pathways in the lateral columns of the spinal cord that facilitate the spindle stretch reflex arc of major muscle groups and can increase the rate of heat production up to five or six times normal. Shivering is an involuntary response to a fall in the body core temperature, but it can be reduced or stopped by voluntary pathways.

The rate of heat production also depends on the levels of endocrine hormones. Increases in some hormones can result in increased metabolic heat production, so-called *chemical thermogenesis*. Acute exposure to cold has been shown to stimulate catecholamine release from the adrenal medulla. Both circulating epinephrine and norepinephrine and that released by sympathetic nerve endings increase the metabolic rate by uncoupling oxidative phosphorylation and by releasing fat stores in the body. This mechanism can be activated rapidly, but it appears to be most important in animals having large quantities of brown adipose tissue. Humans, with the exception of infants, have little brown adipose tissue; thus, acute chemical thermogenesis can elevate body heat production by at most 10 to 15%.

In individuals who are constantly exposed to a cold climate, there is a chronic increase in thyroxin production that is accompanied by an increase in the size of the thyroid gland. This may account for an increased incidence of goiters in cold climates. Thyroxine also elevates heat production by uncoupling oxidative phosphorylation in many target tissues and can result in a substantial increase in the basal metabolic rate, but it has not been established how much of the total increase in the metabolic rate is due to this mechanism in the normal adaptation to cold.

PHYSIOLOGY AND PATHOPHYSIOLOGY OF RESPONSES TO HEAT AND COLD

Body temperature is regulated by a feedback mechanism that matches net heat loss from the body to the rate of heat production by metabolism. This feedback mechanism is integrated by the hypothalamus, particularly the preoptic anterior area. Although the control of thermoregulatory processes such as shivering and sweating has been attributed to various other regions of the hypothalamus, discrete localization no longer seems valid. In fact, some thermoregulation occurs even in the absence of brain centers above the medulla. Nevertheless, the most sensitive regulation occurs in the hypothalamus and may be ascribed to an "integrative center" that may or may not have an anatomic correlate. The sensors for the feedback regulation of body heat content or temperature, called *thermoreceptors*, are located both in the periphery and in the central nervous system. The preoptic anterior hypothalamus contains temperature-sensitive neurons that increase firing with increasing core temperature. Other neurons respond to cold by increasing their rate of firing. In addition, the spinal cord and the abdominal organs have both heat and cold receptors. The skin also has numerous heat- and cold-responding receptors that convey information about the ambient temperature to the central nervous system.

Because the body core temperature must be maintained within critical limits, it seems that the central thermoreceptors are the primary input to the feedback regulation of body temperature, but the peripheral receptors also appear to play an important role, particularly by allowing the system to anticipate conditions that might compromise maintenance of the body core temperature. For example, just a gust of cold air may cause one to have goose bumps and even shiver before any change in the body core temperature has occurred. Experiments show that heating the preoptic hypothalamic area results in immediate cutaneous vasodilation and sweating; however, the critical hypothalamic temperature at which sweating begins is lower when the skin temperature is higher. Cooling the preoptic hypothalamus results in shivering and cutaneous vasoconstriction, but the temperature threshold for shivering is found to be higher when the skin temperature is lower. In other words, when the body core temperature and the environmental temperature change in the same direction, the physiologic response is enhanced. This dual regulatory system has the advantage of allowing the body to respond more rapidly to changing environmental conditions and with less variation in the body core temperature. For example, when the skin is cold, shivering and cutaneous vasoconstriction begin to occur with very little fall in the core temperature, and, when the skin is warm, vasodilation and sweating occur with little rise in the core temperature. On the other hand, when one is exercising in a cold environment, body core temperature may be elevated by the increased heat production, but sweating is reduced because of the cold skin temperatures.

As shown in Fig. 4, temperature information from the preoptic–anterior hypothalamic area and skin thermoreceptors is conveyed to a hypothalamic integrative center. It appears that the integrated temperature afferent input is compared with a desired temperature, called the *set-point*, in much the same way that room temperature is compared with the temperature setting of a furnace thermostat. When the integrated temperature that is sensed falls below the temperature set-point, heat loss is reduced by vasoconstriction and inhibition of sweating, and heat production is increased by shivering. When the temperature that is sensed rises above the set-point, heat loss is increased by cutaneous vasodilation and sweating, and shivering is inhibited. Heat production may also be reduced by behavioral responses to decrease

the level of motor activity, but obviously cannot be reduced below the basal metabolic rate. Based on the difference between the temperature sensed and the set-point, the integrative center coordinates the efferent responses involving cutaneous vasomotor activity, sweating, shivering, and chemical thermogenesis, as well as behavioral responses coordinated at higher levels.

Adaptation to a Hot Environment

When exposed acutely to a hot environment, the hypothalamic integrative center acts as described earlier to increase heat loss by causing vasodilation and sweating. Both mechanisms can increase heat output from the body as long as the ambient temperature is below the body core temperature; however, when the ambient temperature is higher, the body is constantly *gaining* heat by radiation and conduction, and the only means of effecting a net heat loss to match the rate of heat production is by evaporation. Thus, sweating is a critical determinant of the body's response to higher environmental temperatures. When the relative humidity is high, however, the amount of heat that can be lost by sweating is limited by the rate of evaporation; therefore, body core temperature cannot be kept normal over as high a temperature range in a humid environment. The heat that must be lost is

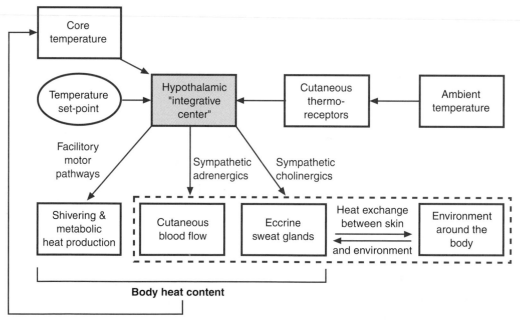

Figure 4   Feedback regulation of body temperature. The integrative center in the hypothalamus compares the body core temperature sensed by the preoptic–anterior hypothalamic thermoreceptors, as modified by input from cutaneous thermoreceptors that sense the environmental (ambient) temperature, with a temperature set-point. It then regulates heat loss and heat production to maintain the body core temperature equal to the set-point. Heat loss from the skin is regulated through control of cutaneous vasomotor activity and sweating. The dashed box surrounds elements involving heat exchange between the skin and the environment. Heat production can be increased by shivering or increased metabolic rate, and decreasing muscle activity can decrease it.

also increased if the body core temperature rises because biochemical reaction rates are accelerated. Metabolic heat production increases about 6% for every °F (10% per °C) of core temperature increase.

When individuals remain for longer periods in a hot climate, there is chronic adaptation in the rate of sweating over a period of 5 to 10 weeks from the normal maximum of 1.5 L/hr to as much as 4 L/hr; over a period of years in the tropics, individuals actually develop increased numbers of sweat ducts. Large fluid losses due to sweating of this magnitude also require adaptation of body fluid and electrolyte regulation to prevent dehydration, which can lead to heat exhaustion, as described later. Individuals adapt to the large losses of salt and water after several weeks in the tropics by increasing their plasma aldosterone levels. Aldosterone not only acts on the renal collecting duct to increase Na^+ reabsorption (see Chapter 29)

but also increases Na^+ reabsorption by the ducts of the eccrine salt glands. Because of the action of aldosterone, the NaCl content of the sweat is reduced. However, the higher aldosterone levels result in greater K^+ secretion into both the sweat and the urine, and this can result in dangerous K^+ depletion when sweating is severe and dietary K^+ intake is low. The K^+ loss can be exacerbated when NaCl is ingested without KCl during periods of heavy sweating.

The body can be subjected to heat stress not only by a hot climate but also by severe exercise in a temperate climate. Body temperature may rise to as high as 39.4 to 40.8°C (103 to 105.5°F) in marathon runners, and this is accompanied by copious sweating and marked cutaneous vasodilation. The loss of salt and water in the sweat as well as the shunting of blood into the skin may result in the heat cramps that often disable athletes. When

---

Clinical Note

Heat Stroke and Heat Exhaustion

In a dry environment, the normal individual can withstand an ambient temperature of up to 150°F for several hours, but, as discussed earlier, heat loss in this circumstance depends on the evaporation of sweat, which is progressively diminished as the relative humidity rises. When the evaporative heat losses cannot match body heat production plus conductive and radiative heat input, the body temperature rises. When it reaches 40 to 41.5°C (104 to 107°F), the temperature integrative center in the hypothalamus ceases to function, and a state of heat *pyrexia* (*heat stroke*) ensues. Heat stroke is a medical emergency because the normal reflexes are lost; in spite of the high body temperature, sweating ceases and death ensues rapidly unless the body is immediately cooled by ice water baths and supportive medical care. Heat stroke is usually observed in the setting of physical exertion in a hot environment, and it may occur very precipitously or be preceded by the symptoms of heat collapse described later. Heat stroke is generally recognized by the presence of dry skin due to the absence of sweating after collapse; however, in an individual who collapses while working or exercising, the skin may still be sweaty. Thus, the most important diagnostic criterion is a body core temperature in excess of 40°C because above this temperature normal physiological temperature regulation is lost and intervention is imperative.

Heat exhaustion (*heat collapse*) also occurs in the setting of a hot environment, but it results from circulatory problems rather than from an increase in body temperature. Severe sweating leads to a decrease in extracellular fluid volume due to NaCl and water losses. The resulting decrease in plasma volume is further complicated by vasodilation, which keeps a larger blood volume in the cutaneous circulation. If venous return is sufficiently compromised, the situation is quite analogous to mild circulatory shock. In this state, the body core temperature is not necessarily elevated, and the collapse might be regarded as protective, although potentially dangerous, because the activity that produces heat ceases. The collapse is preceded by weakness, vertigo, headache, and sometimes by vomiting. The period of collapse is usually brief and the recovery spontaneous if the individual is removed from the hot environment and given fluids. Heat exhaustion and collapse are frequently seen in the elderly, who may be less well hydrated or less aware of their dehydration because of a diminished ability to care for themselves; however, even trained athletes may experience heat collapse, especially when exercising heavily in hot and humid conditions. The chances of developing heat stroke or collapse are also increased by certain drugs that inhibit sweating and/or vasodilation, including atropine, scopolamine, phenothiazides, monoamine oxidase inhibitors, and glutethimide.

extreme, these effects can lead to heat exhaustion (see the clinical note).

Adaptation to a Cold Environment

Acute adaptation to a cold environment involves decreased heat loss by vasoconstriction of the cutaneous vascular bed and inhibition of sweating even during exercise. Shivering and increased muscular activity increase heat production, and generally heat balance can be maintained even by the nude body at temperatures as low as 0 to 4.5°C (32° to 40°F). As discussed earlier, individuals who have adapted to a cold climate over long periods also exhibit a limited ability to increase chemical thermogenesis by elevated catecholamine and thyroxin levels (particularly T3, but also T4; see Chapter 39), but the resulting stimulation of the thyroid gland leads to a greater incidence of goiters in populations in cold

---

Clinical Note

Hypothermia and Cold-Induced Tissue Damage

The body is better able to withstand cooling than heating of the core temperature for limited periods because metabolism slows and irreversible damage is less. When the body temperature falls below 33°C (91°F), mental confusion and sluggishness result, and the thermoregulatory ability of the central nervous system is lost below 30°C (86°F). At this point, shivering stops and consciousness is lost. This is followed by muscular rigidity and collapse. Further cooling leads to slow atrial fibrillation and, eventually, ventricular fibrillation and death.

Hypothermia is frequently seen in derelict alcoholics. Not only do they not have shelter from the cold, but the alcoholic stupor also decreases the generation of heat by muscular activity. Hypothermia is also common following accidental immersion in cold water or wilderness exposure in a cold climate. Immersion can lead to a rapid loss of body heat because the water in contact with the skin is a much better heat conductor than air, and convection is increased by the physical exertion of struggling to swim. In the wilderness, hypothermia can develop rapidly because of changing weather conditions; it usually constitutes the greatest danger to survival. The effects of any temperature drop are compounded by the wind, and evaporative losses can be very high due to rapid respiration and sweating while attempting to escape the situation. The clothing becomes soaked by the perspiration and any precipitation that may be occurring, which decreases the insulation around the body.

Hypothermia is also purposely induced in surgical procedures in which the blood flow to vital organs may be interrupted, as in the case of heart surgery when a heart–lung machine is used. In these situations, hypothermia is induced by packing ice around the patient's body until the desired level of hypothermia is attained. Because the metabolic rate is markedly decreased in hypothermia, tissue damage due to interruptions of the circulation is minimized.

Hypothermia is, of course, treated by rewarming, but this should be done slowly. If heat is applied to the skin rapidly, the resulting cutaneous vasodilation can lead to large volumes of chilled blood entering the general circulation, which further decreases core temperature and can produce fatal cardiac arrhythmias.

There are remarkable accounts of complete recovery from hypothermia in which there has been cardiac arrest or fibrillation for more than an hour. Thus, even if all the usual signs, including apparent rigor mortis, are present, the physician should not rule out the possibility of reversible hypothermia. If there is doubt, the body should be slowly rewarmed to 37°C before the patient is pronounced dead.

In addition to the dangers of systemic hypothermia, local tissue damage due to freezing (frost bite) may occur with prolonged cold exposure. Freezing of the digits or facial tissue is the most common type of frost bite and is facilitated by the vasoconstriction that occurs in response to the cold. Tissue damage can also occur without freezing, as in the condition of immersion foot (trench foot), in which the feet are immersed in cold water for long periods of time. Because of the reduced blood flow in the foot, the nerves and muscles are damaged irreversibly.

Mild hypothermia is also seen in a number of diseases, including myxedema, pituitary insufficiency, Addison's disease, hypoglycemia, and cerebrosvascular disease, and as a consequence of the effects of drugs or alcohol.

climates. Some primitive peoples have adapted to living at or near freezing temperatures with very little clothing. For example, Australian aborigines spend much of the year at low temperatures and yet shiver relatively little; however, when the air is cold, their body temperature falls markedly during the night. Long-distance swimmers are noted for their ability to withstand water temperatures of 15 to 20°C for many hours, as well as for their characteristically heavy layer of cutaneous fat. While they swim in cold water their heat production is increased by the physical activity, and they experience extreme cutaneous vasoconstriction.

Fever

Fever is an elevation of body temperature that can be associated with infection as well as dehydration and thyrotoxicosis. The development of fever appears to involve a resetting of the thermoregulatory set-point to a higher level, as shown in Fig. 5. When this occurs, the hypothalamic regulatory center responds as if the core temperature were too low—cutaneous vasoconstriction and shivering ensue, leading to the feeling of chill that precedes and accompanies fever. When the body core temperature rises to the higher set-point, a new balance of heat loss and heat production is achieved but at a higher body temperature. When a fever "breaks," the set-point falls toward the normal level and the hypothalamic integrative center causes cutaneous vasodilation and sweating, leading to a fall in body temperature to the new set-point. During the course of an infection, the set-point can oscillate up and down, leading to cycles of shivering and sweating.

The fever with infection occurs when the immune system reacts to components of the infecting organism—for example, to lipopolysaccharides of the bacterial cell wall, especially from Gram-negative bacteria. Such substances are called *exogenous pyrogens*. They cause monocytes and macrophages to release cytokines such as interleukins (especially IL-1B and IL-6) and tumor necrosis factor (TNF). These cytokines are referred to as endogenous pyrogens, and they cause the production of

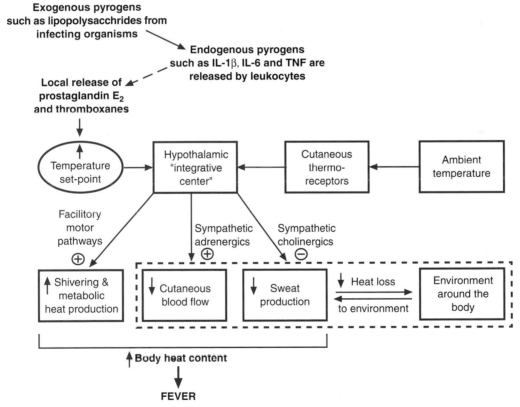

Figure 5   Production of fever. Endogenous pyrogens released from leukocytes in response to exogenous pyrogens released by infectious microorganisms cause the local release of products of arachidonic acid metabolism such as prostaglandin E_2 and thromboxanes. In the hypothalamus, these products cause an increase in the set-point temperature. Because the core temperature is then less than the set-point, the integrative center responds by increasing heat production through shivering, decreasing heat loss by cutaneous vasoconstriction and suppressing sweating. The increased heat production and decreased heat loss lead to an increase in core temperature until the new set-point is reached.

arachidonic acid metabolites, including prostaglandin E_2 and thromboxanes in many tissues, including the hypothalamus. The latter agents act locally in the hypothalamus to produce the change in the temperature set-point; for this reason, aspirin and other drugs that block prostaglandin synthesis reduce fever. However, there are indications that the development of fever is of benefit as a normal body defense in combating some infections.

It is of special importance to note the increased vulnerability of children and the elderly to heat-related illnesses. Children produce more heat from metabolism per kg of body weight than adults, and, because their total body mass is smaller, their core temperature can change very rapidly. Infants and toddlers are also unable to regulate their own fluid intake and they may become rapidly dehydrated in a hot environment. Elderly individuals often have an impaired thirst response to dehydration and/or may be subject to medical conditions and medications that interfere with the normal regulation of salt and water balance. Thus, the elderly are more susceptible to heat exhaustion and heat stroke, and over half of the emergency room visits for these conditions in summer hot spells are typically for patients over 65.

Suggested Readings

Boulant JA. Hypothalamic neurons regulating body temperature, in: Fregly M, Blatteis C, Eds., *Handbook of physiology,* Section 4, *Environmental physiology.* New York: Oxford University Press, 1996, pp. 105–126.

Gage AP, Gonzalez RR. Mechanisms of heat exchange biophysics and physiology, in: Fregly M, Blatteis C, Eds., *Handbook of physiology,* Section 4, *Environmental physiology.* New York: Oxford University Press, 1996, pp. 45–84.

Kluger MJ, Bartfai T, Dinarello CA, Eds., Molecular mechanisms of fever. *Ann NY Acad Sci* 1998; 856:308 pp.

Wexler RK. Evaluation and treatment of heat-related illnesses. *Am Family Phys* 2002; 65:2307–2314.

63

Exercise

FRANK L. POWELL, JR.

KEY POINTS

- Exercise activates the sympathetic nervous system, which prepares the body for "flight or fight."
- ATP generated from anaerobic glycolysis and oxidative phosphorylation provides the energy for muscular work.
- Maximal oxygen consumption ($\dot{V}_{O_2}$max) is the best measure of the capacity to perform dynamic exercise.
- Carbohydrates are the primary metabolic fuel for short bursts or intense heavy exercise. Carbohydrates and fat contribute equally to energy production at the onset of moderate levels exercise, but fat becomes more important during sustained exercise.

- The respiratory system responds to increased oxygen demands during exercise by increasing ventilation to maintain normal arterial P_{O_2} and P_{CO_2} and to minimize acidosis from anaerobic glycolysis.
- Exercise is *not* normally limited by pulmonary gas exchange or mechanics.
- The cardiovascular system supports increased oxygen consumption during exercise by increasing cardiac output, directing blood flow to exercising muscle, and increasing oxygen extraction from blood.
- Exercise training increases $\dot{V}_{O_2}$max by increasing the metabolic capacity of muscles and cardiovascular oxygen delivery to muscle.

OVERVIEW

Exercise can be defined as an increase in skeletal muscle activity; however, the physiological response to exercise involves much more than the muscles themselves and depends on coordinated changes in the other systems covered in previous chapters. Exercise is associated with sports, and everyone can picture a well-trained athlete, with superb physical conditioning that involves the entire body and mind. Of course, exercise is also necessary for the basic activities of daily life, and one of the most serious consequences of disease can be a limitation of exercise capacity. For example, a simple activity such as walking up stairs can be as difficult as running a marathon for someone with chronic lung disease.

The approach to exercise physiology in this chapter is based on the integrative "fight or flight" response of the *autonomic nervous system* (Chapter 9). When we are frightened, for example, the sympathetic nervous system is activated to produce a well-orchestrated suite of physiological changes that prepare us to exercise (*i.e.*, fight our attacker or take flight). The systems that support muscular activity (e.g., respiratory and cardiovascular) are activated, while vegetative and restorative systems (e.g., gastrointestinal and renal) are inhibited. This chapter will not consider the specifics of muscular movement, which is a discipline in its own right (*kinesiology*), or muscle physiology *per se*. For example, the different effects of *isometric* (constant muscle length) versus *isotonic* (constant force) exercise on the cardiovascular system are described here, but the differences in isometric and isotonic muscle function are described in Chapter 7.

METABOLISM

Adenosine triphosphate (ATP) provides the energy for muscle contraction, just as for other cellular functions. ATP can be generated most efficiently by *oxidative phosphorylation*, but it can also be generated by *anaerobic glycolysis*. Anaerobic glycolysis uses glucose or glycogen stores in muscle to generate ATP and is useful at the immediate onset of exercise and for short periods of time. The major muscle store of high-energy phosphate necessary to generate ATP from adenosine diphosphate (ADP) is *phosphocreatine*. Dietary supplementation with creatine phosphate may slightly enhance short bursts of high-intensity exercise, but sustained levels of exercise depend on oxidative phosphorylation. Hence, *maximal oxygen consumption,* or $\dot{V}_{O_2}$ max, is the best measure of absolute exercise capacity.

$\dot{V}_{O_2}$ max is measured by having an individual exercise at increasing workloads while measuring $\dot{V}_{O_2}$. Above a certain workload, $\dot{V}_{O_2}$ will stop increasing, and any additional increases in work are fueled by anaerobic glycolysis. This will increase lactate production, but it is important to note that lactate levels in the blood are controlled by factors other than simply anaerobic production. The term *anaerobic threshold* has been used to describe the point at which arterial lactate levels increase, but this may not correlate perfectly with $\dot{V}_{O_2}$ max, and a better term for this phenomenon is the *lactate threshold*. $\dot{V}_{O_2}$ max is essentially the same for exercise of the legs only or both legs and arms, although it is significantly less when exercise is limited to only the arms; hence, a *bicycle ergometer* is a useful way to measure $\dot{V}_{O_2}$ max in most people. Fitness level and the type of exercise determine how long one can maintain $\dot{V}_{O_2}$ max. For example, a world-class middle-distance runner might sustain $\dot{V}_{O_2}$ max during a 4-minute mile, while an elite long-distance runner might sustain just over 80% $\dot{V}_{O_2}$ max during a marathon. Exercise lasting more than 20 minutes usually occurs at less than 90% of $\dot{V}_{O_2}$ max.

If exercise is started abruptly, $\dot{V}_{O_2}$ will not increase as rapidly as the work level. The difference between the actual work performed and the caloric equivalent of oxygen consumption during the onset of exercise defines the *oxygen debt* or *deficit*; this represents work done during anaerobic glycolysis. Similarly, $\dot{V}_{O_2}$ does not drop immediately to resting levels at the end of exercise but decreases more slowly. However, the extra work being fueled by this oxygen does more than just synthesize glucose from lactate, so many exercise physiologists refer to *excess post-exercise oxygen consumption* (EPOC) instead of oxygen debt. Excess $\dot{V}_{O_2}$ following exercise also results from increased temperature and hormonal stimulation of metabolism, in addition to regenerating phosphocreatine.

Fuels

Fuel selection during exercise is determined by both the duration and intensity of exercise. A short sprint such as a 50-meter dash will be fueled entirely by muscle stores of glycogen and phosphocreatine. At the other extreme of a long trek, oxidation of free fatty acids is the most important fuel. *Protein* is not a major fuel for exercise, comprising less than 2% of the substrate during the first hour of exercise and rising to only 15% after five hours. *Fats* contribute about two-thirds of the energy for low levels of exercise. As the level of $\dot{V}_{O_2}$ increases, *carbohydrate* becomes more important than fat as a substrate, with a crossover at about 35% of $\dot{V}_{O_2}$ max. At the onset of moderate exercise, carbohydrates and fats contribute equally to energy production, but fat metabolism becomes relatively more important as the duration of exercise increases.

The preference for carbohydrate metabolism during short, intense exercise is driven by recruitment of more fast-twitch glycolytic muscle fibers (see Chapter 7) at high levels of exercise and rapid effects of sympathetic stimulation and epinephrine on muscle glycogen breakdown and glycolysis. Lactate from glycolysis also inhibits fat metabolism. Sympathetic stimulation of the pancreas decreases insulin release, but *insulin-independent glucose uptake* in skeletal muscle increases during exercise. Glucose uptake is further enhanced during exercise by recruitment of more insulin-dependent glucose receptors to muscle membranes to cope with the decreased insulin levels. Replenishment of carbohydrates during exercise can improve endurance in trained athletes.

Fat is the primary metabolic fuel during prolonged exercise. *Free fatty acids* are generated from *triglycerides* by *lipases*, which are stimulated by epinephrine, norepinepherine, and glucagon. This response is slower than the sympathetic effects on muscle glycogen. Prolonged exercise generally involves more slow-twitch muscle fibers with more mitochondria and lipolytic enzymes. Also, insulin levels decrease during exercise which reduces the inhibitory effect of insulin on the mobilization of free fatty acids. However, if high levels of carbohydrates are ingested prior to exercise, this may stimulate insulin production and favor carbohydrate over fat metabolism. The hormonal control of metabolic fuels during exercise is also discussed in Chapter 42.

Appetite, or the control of metabolic fuel intake, is related to the levels of energy expended. People increase their caloric intake in direct proportion to increased energy expenditure if they are burning more than 2500 calories per day. Increased glucose uptake (insulin-independent and -dependent) persists for several hours following exercise; this increased uptake lowers blood glucose level and stimulates appetite. Other factors increasing appetite and caloric intake include more rapid emptying of the stomach, increased absorption in the small intestine, and perhaps neural reflexes involving the hypothalamus. Although exercise can increase appetite, the most effective method of weight control is increasing exercise while controlling diet.

RESPIRATION

Figure 1 shows how ventilation increases in proportion to oxygen consumption at different levels of steady-state exercise (e.g., on a bicycle). Above the lactate threshold, ventilation increases more than CO_2 production, so arterial P_{CO_2} decreases and arterial P_{O_2} increases; hence, the respiratory system is able to meet the metabolic demands of the muscles at all levels of exercise, at least in normal humans. However, the mechanical and

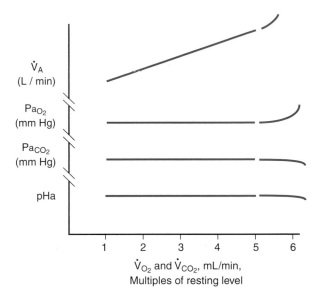

FIGURE 1   Respiratory effects of progressively increasing levels of exercise. Alveolar ventilation increases in proportion to CO_2 production until very high levels of exercise are reached (after the break in the curves), which maintains arterial blood gases at normal levels. However, at very high levels of exercise, lactic acid production decreases pH. This acidosis and possibly other factors provide an excessive stimulus for ventilation so Pa_{CO_2} decreases and Pa_{CO_2} increases.

muscular work of breathing and the efficiency of pulmonary gas exchange can limit gas exchange during disease.

Ventilation and Acid–Base

Figure 2 shows the time course of changes in ventilation during a constant level of submaximal exercise. A rapid initial increase is followed by a slower increase until a steady state is achieved. Current theories for *exercise hyperpnea*, defined as the increase in ventilation during exercise, involve a combination of feedforward mechanisms and feedback from the chemoreceptor and mechanoreceptor reflexes. Feedforward mechanisms (also called *central command*) are neural inputs from higher centers in the brain that directly stimulate the respiratory centers. For example, signals from the motor cortex or hypothalamus to move a muscle can also stimulate ventilation. Feedforward mechanisms have essentially no delay and can explain the rapid increase in ventilation at the onset of exercise. The slower, secondary increase in ventilation is explained by neural mechanisms, such as second-messenger systems activated by neurotransmitters, changes in intracellular calcium that stimulate neurons, and possibly such changes in chemical stimuli to sensory nerves as potassium ions in muscle.

Figure 1 showed how alveolar ventilation increases in perfect proportion to CO_2 production, so Pa_{CO_2} is maintained at the normal resting value during submaximal steady-state exercise. Because arterial blood gases

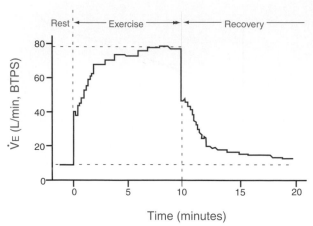

FIGURE 2 Changes in ventilation during a bout of moderate exercise and recovery afterwards. Ventilation increases immediately at the onset of exercise and then continues to increase more slowly to a plateau value during sustained exercise. Similarly, ventilation decreases immediately when exercise stops, but it takes several minutes before ventilation returns to the pre-exercise level. (Adapted from Dejours. *Rev Franc Etude Clin Biol* 1963; 8:439).

do not change significantly, arterial chemoreceptors are not important for hyperpnea at submaximal levels of exercise; in fact, ventilation is normal during steady-state submaximal exercise in patients without carotid bodies. However, most animals except humans actually hyperventilate during exercise, which decreases Pa_{CO_2} and arterial chemoreceptor stimulation. This limits the increase in ventilation and minimizes the work of breathing.

Near maximal levels of exercise, lactic acid starts to appear in the blood which stimulates arterial chemoreceptors. This explains the larger increase in ventilation above the lactate threshold (Fig. 1). Ventilation increases in excess of CO_2 production, decreasing Pa_{CO_2} and increasing Pa_{O_2} near maximum levels of exercise in normal subjects. The respiratory alkalosis helps compensate the metabolic acidosis from lactic acid.

Other afferent signals that may fine tune ventilation during exercise originate from mechanoreceptors in the heart and nonmyelinated sensory nerves in skeletal muscle. Mechanoreceptors from the lung and chest wall can adjust the pattern of breathing during exercise to minimize the work of breathing. Even arterial chemoreceptors can be stimulated by K^+ released from muscle, norepinephrine released from the sympathetic nervous system, and larger oscillations of Pa_{CO_2} during the ventilatory cycle in exercise. However, the ventilatory response to moderate, steady-state exercise is normal without any of these afferent signals, so they are not necessary for exercise hyperpnea.

Pulmonary Mechanics and Gas Exchange

Although exercise increases ventilation more than any other physiological stimulus, mechanical factors do not

limit ventilation or $\dot{V}_{O_2}$ in healthy humans. Peak inspiratory muscle pressure developed during maximal exercise is only about 40% of the maximum capacity because inspiratory muscles operate at decreased sarcomere length (with decreased expiratory volume). However, the forces generated are sufficient to increase ventilation so Pa_{CO_2} does not increase during exercise. Also, despite tenfold increases in ventilatory flow rates with exercise, airway resistance does not increase because of active regulation of both upper and lower airways. The *alae nasae* flare the nostrils to decrease resistance to nose breathing and the *larynx*, which is important for determining resistance to mouth breathing, does not narrow as much during expiration in exercise compared to rest. Exercise also relaxes bronchial smooth muscle by sympathetic stimulation.

The pattern of breathing during exercise involves increased frequency and tidal volume, with expiratory muscles decreasing end expiratory lung volume. As exercise intensity increases, the end expiratory lung volume increases relative to lighter exercise to reduce the possibility for *expiratory flow limitation* (or *dynamic compression*; see Fig. 10 in Chapter 19). Figure 3 illustrates this with flow–volume loops measured during rest and exercise. Figure 3 also shows that there is very little flow limitation at $\dot{V}_{O_2}$ max in healthy, young adults. In contrast, elite athletes may have significant expiratory flow limitation, with over 50% of the tidal volume hitting the maximum expiratory flow rate at the highest levels of

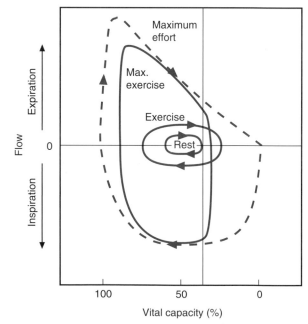

FIGURE 3 Changes in pressure–volume relationships of the lungs during exercise. Lung and tidal volume are adjusted during exercise to reduce the expiratory flow limitation, as described in the text. Maximum curves, shown as dashed lines, are only reached for a portion of the ventilatory cycle during maximum exercise.

exercise and ventilation. This can increase the O_2 cost of ventilation to over 15% of the whole-body $\dot{V}_{O_2}$ in an elite athlete during maximum exercise; however, whole-body $\dot{V}_{O_2}$ increases more than the O_2 cost of breathing, so respiratory mechanics do not limit $\dot{V}_{O_2}$ max in elite athletes, either.

Although mechanical limitations do not cause hypoventilation, pulmonary gas exchange becomes less efficient during heavy exercise for other reasons. During light levels of exercise, gas exchange may actually improve as indicated by a decrease in the alveolar–arterial P_{O_2} (Chapter 21). This is because ventilation–perfusion matching improves when blood flow increases to the tops of the lungs when cardiac output and pulmonary artery pressure increase during exercise (Chapter 18). At higher levels of exercise ($\dot{V}_{O_2} > 3$ L/min), however, all of the pulmonary capillaries are recruited and distended, so further increases in cardiac output and pulmonary artery pressure only lead to interstitial edema. This makes the distribution of blood flow more heterogeneous and worsens ventilation–perfusion matching.

Interstitial edema can also impair diffusion by thickening the blood–gas barrier. The lung diffusing capacity for carbon monoxide (DL_{CO}) decreases in runners after competing in a marathon. At $\dot{V}_{O_2}$ levels > 3 L/min, diffusion limitation explains over half of the alveolar–arterial P_{O_2} difference and becomes more important than ventilation–perfusion matching in determining the efficacy of gas exchange. Decreased transit times for red blood cells in the pulmonary capillaries and the decreased mixed-venous P_{O_2} under these conditions could cause diffusion limitations at high levels of $\dot{V}_{O_2}$ even if the blood–gas barrier was not thickened. Shunt does not change during exercise but an existing shunt can contribute to more arterial hypoxemia because mixed-venous P_{O_2} decreases during exercise.

Cardiovascular System

Cardiac output increases up to 8-fold during exercise in well-trained athletes; however, $\dot{V}_{O_2}$ can increase 10- to 12-fold, so more oxygen must be extracted from the blood during tissue gas exchange. There are also important changes in regional blood flow to focus the circulation on the metabolic demands of the muscles.

Cardiac Output

Cardiac output is increased during exercise by increases in both heart rate and stroke volume. Heart rate increases in response to withdrawal of *vagal* (parasympathetic) tone and increased *sympathetic* stimulation of β_1-adrenergic receptors in pacemaker regions of the heart. Heart rate increases within seconds of the start of exercise or even in anticipation of exercise. Heart rate reaches steady state in 2 to 3 minutes at exercise rates below the anaerobic threshold, in proportion to the increase in $\dot{V}_{O_2}$. Heart rate at $\dot{V}_{O_2}$ max can be estimated for normal subjects as: frequency (beats/min) = 220 − age (years).

Increased heart rate has a *positive inotropic effect* on cardiac contractility due to the *staircase* (or *Treppe*) mechanism (see Fig. 9 in Chapter 11), which increases stroke volume. Sympathetic stimulation of β_1-adrenergic receptors on the ventricles also increases cardiac contractility and stroke volume. Stroke volume reaches the maximum level at about 40% of $\dot{V}_{O_2}$ max, so further increases in cardiac output occur by increasing heart rate.

Myocardial oxygen consumption increases with heart rate and stroke volume, and reaches its maximal level during exercise. The myocardium extracts most of the oxygen from the coronary circulation during exercise so there is no *extraction reserve* (see later discussion). This explains why exercise in patients with coronary artery disease causes *angina pectoris*, a painful response to metabolic changes in the myocardium when flow is limited.

An increase in cardiac output cannot be sustained unless *venous return* also increases. Exercise increases venous return by several mechanisms. Rhythmic contractions during exercise can create a so-called *muscle pump*, which moves blood out of the veins past the one-way valves (see Fig. 8 in Chapter 17). Decreased intrathoracic pressure with large inspiratory efforts also promotes venous return. *Venoconstriction* from sympathetic stimulation of α_1-adrenergic receptors in the veins causes the venous return curve to shift upward, and *arterial vasodilation* in exercising muscle beds (explained below) further increases the slope of the venous return curve (see Fig. 11 in Chapter 14). Finally, the increase in venous return leads to increased stroke volume due to the *Frank-Starling mechanism*, which is an increased contractile response to increased cardiac filling (see Fig. 4 in Chapter 13).

Regional Blood Flow

Blood flow is redistributed during exercise away from vegetative organs and toward actively exercising muscles by both *intrinsic* (local) and *extrinsic* (reflex) mechanisms. In isometric exercise, muscle blood flow may actually decrease with compression of arteries and veins in contracted muscles. In contrast, *active hyperemia*, or an increase in blood flow, occurs in muscles during dynamic exercise. Arterioles dilate in exercising muscles in direct response to local metabolic changes (decreased O_2, increased CO_2 and H^+ ion), adenonsine, and nitric oxide (NO). This local control of muscle blood flow overrides the effect of sympathetic stimulation to

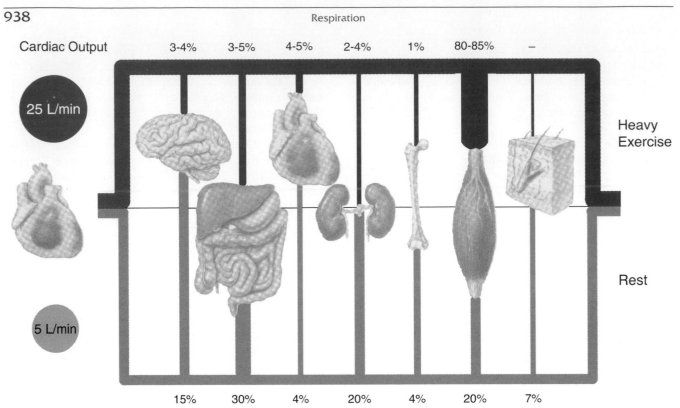

Cardiac Output   3-4%   3-5%   4-5%   2-4%   1%   80-85%   –

25 L/min   Heavy Exercise

5 L/min   Rest

15%   30%   4%   20%   4%   20%   7%

FIGURE 4   Distribution of cardiac output during rest and maximal exercise, when cardiac output increases from 5 to 25 L/min. (Adapted from Powers, SK, Howley, ET., *Exercise physiol.*, 1997.)

vasoconstrict arterioles. Sympathetic activation of vaso-constrictor pathways occurs in response to a *central command* (*i.e.*, feedforward signals) from the motor cortex to medullary cardiovascular centers and from a *muscle chemoreflex*. The muscle chemoreflex is a sympathetic response to stimulation of small-diameter, nonmyelinated sensory nerves in exercising muscles by lactic acid and K^+ ion. Arterial chemoreceptors may be stimulated by H^+ above the anaerobic threshold which would be expected to produce a bradycardia and vaso-constriction, as observed in the *diving reflex*. However, the diving reflex is inhibited by increased ventilation, so arterial chemoreceptors have no significant effects on the cardiovascular system during exercise.

Changes in the distribution of blood flow during dynamic exercise are shown in Fig. 4. Most of the cardiac output is directed toward exercising muscle, including the heart. The brain receives almost the same absolute blood flow at rest and at exercise, so it represents a much smaller percentage of the cardiac output during exercise. However, other organs, such as the kidney and splanchnic bed, actually experience decreased absolute blood flow. Both renal mechanisms for correcting arterial acidosis and gastrointestinal absorption of fuels are too slow to make a significant difference in the acid–base balance or metabolism during exercise. Sympathetic vasoconstriction in these vascular beds explains the decreased flow;

hence, the redistribution of blood flow among organs during exercise depends on an integrated response of local and reflex mechanisms.

Arterial Pressure

Arterial blood pressure increases during exercise, but the increase is much greater in isometric than in dynamic exercise. During isometric exercise, the blood vessels are clamped shut by sustained contractions, and resistance increases. This leads to anaerobic metabolism and strong stimulation of the muscle chemoreflex. Without the benefits of local vasodilation in isometric muscle contraction, the net effect is an increase total peripheral resistance and arterial pressure.

During dynamic exercise, there may be a small (< 20%) increase in mean arterial pressure. Large increases in cardiac output are accompanied by large decreases in peripheral vascular resistance. Vascular resistance decreases because of local metabolites, causing vasodilation and vasodilation of skin vessels to promote cutaneous blood flow and heat loss. The fact that small increases in arterial pressure occur has also been interpreted as a *resetting of the arterial baroreflex*. A similar phenomenon may happen in cardiovascular disease, when the arterial baroreflex becomes less effective at reducing pressure in response to baroreceptor stimulation.

Tissue Gas Exchange

Although O_2 delivery to muscle increases during exercise, it does not increase as much as O_2 demand so O_2 extraction increases. In healthy subjects, hemoglobin oxygen saturation is already nearly 100% in arterial blood, so the only alternative to increase O_2 extraction is to decrease venous oxygen levels. The magnitude of decrease in venous O_2 levels can be calculated from the *Fick principle* (see Chapter 21):

$$\dot{V}_{O_2} = \dot{Q}(Ca_{O_2} - C\bar{v}_{O_2})$$

The limits for decreasing venous oxygen levels are discussed below.

Several factors facilitate tissue gas exchange when O_2 extraction is increased during exercise. Increased blood flow and muscular contraction result in continuous flow in all of the capillaries in a muscle bed. Compared to rest, when some capillaries are perfused only intermittently, this decreases the diffusion distances for O_2. *Hematocrit* and hemoglobin concentration increase during acute exercise, although contraction of the *spleen* to release a reservoir of red blood cells is not as important in humans as it is in dogs and horses. Increased temperature, CO_2, and H^+ ion in exercising muscle will also decrease the affinity of hemoglobin for O_2, facilitating the unloading of O_2 in exercising muscle (*Bohr effect*; see Chapter 20). Finally, *myoglobin* may act as a shuttle to facilitate O_2 transport within myocytes, thereby improving oxygenation near the venous ends of capillaries in muscles. Many of these features are illustrated in Fig. 5 in Chapter 21.

LIMITATIONS OF EXERCISE

$\dot{V}_{O_2}$

What is the physiological explanation for $\dot{V}_{O_2}$ max? This question has fascinated physiologists for over a century. As explained in Chapter 18, O_2 consumption continues at maximum levels in isolated mitochondria until P_{O_2} is much less than that found in venous blood (< 5 Torr; see Fig. 2 in Chapter 18). This suggests that the amount of mitochondria may determine $\dot{V}_{O_2}$ max in an individual, and it is certainly true that well-trained athletes have higher mitochondrial densities in their muscles. However, there is also evidence for gas exchange limitations of maximal exercise. For example, increasing inspired O_2 levels can increase $\dot{V}_{O_2}$ max in well-trained athletes.

Recent evidence suggests that $\dot{V}_{O_2}$ max occurs when venous P_{O_2} reaches a minimum that represents the optimal match between cardiovascular O_2 delivery *and* O_2 diffusion in tissue. When researchers observed that

$P\bar{v}_{O_2}$ during maximal exercise is lower at altitude than it is at sea level, they wondered why the muscles could not reach the same low $P\bar{v}_{O_2}$ values at sea level because this would increase O_2 extraction and increase $\dot{V}_{O_2}$ max. However, further reductions in $P\bar{v}_{O_2}$ would reduce the gradient for O_2 diffusion into the muscle and decrease the O_2 flux. Venous P_{O_2} is the critical value because it defines the lower limit for the gradient driving diffusion of O_2 out of the capillaries during maximal exercise. At $\dot{V}_{O_2}$ max, the mitochondria consume every molecule of O_2 so P_{O_2} outside the capillaries is nearly zero. Increasing $P\bar{v}_{O_2}$ to enhance O_2 diffusion would only decrease O_2 extraction and reduce $\dot{V}_{O_2}$ because of the Fick principle.

Fatigue

Fatigue in muscles or other systems can limit exercise. *Muscle fatigue* is a decrease in force production during intense exercise, when ATP supply does not keep up with ATP demand. Cellular mechanisms decrease ATP utilization (leading to decreased force) to maintain ATP levels and cellular homeostasis. An accumulation of inorganic phosphate and hydrogen ions in muscle fibers interacts with contractile proteins to reduce force production. During prolonged exercise, *neuromuscular fatigue* may result from a failure of *excitation–contraction coupling*, which involves decreased release of calcium from the sarcoplasmic reticulum. Arousal in the central nervous system also affects the state of fatigue by facilitating neural motor outputs, as can increased motivation.

Fatigue will also occur if $\dot{V}_{O_2}$ is limited, as discussed previously. Similarly, metabolic substrates can limit exercise and cause fatigue. Phosphocreatine is the most important substrate for intense exercise that lasts only a few seconds. For exercise lasting less than 3 minutes, most energy comes from anaerobic glycolysis, so lactate buffering to control muscle and blood pH is also important. During aerobic exercise lasting more than an hour, glycogen stores in the muscle and liver are important. Thermoregulation and maintaining body fluids are also important for preventing fatigue during long-term exercise.

TRAINING

Muscle

Endurance training at submaximal levels of $\dot{V}_{O_2}$ increases the oxidative capacity of muscle without causing muscle hypertrophy. Muscle size does not change but there are increases in the number of mitochondria and enzymes for fatty acid oxidation, the citric acid cycle, and the electron transport chain. Capillary density and myoglobin also increase to enhance O_2 transport. Muscles must be activated during training to undergo these changes, so higher levels of $\dot{V}_{O_2}$ are necessary to train fast-twitch glycolytic fibers than slow-twitch oxidative fibers. Activating fast-twitch fibers with heavy training increases mitochondrial enzymes, and the fibers change their biochemistry to a more oxidative pattern.

$\dot{V}_{O_2}$ increases more rapidly after training because there are more mitochondria and the metabolism of glucose and free fatty acids increases faster. ADP does not build up as quickly, resulting in less lactic acid and better preservation of phosphocreatine stores. Increased capillary density and perhaps membrane transporters enhance the transport of free fatty acids from the circulation into muscle. Transporters on mitochondrial membranes improve the uptake of free fatty acids from the myoplasm, and enzymes for fatty acid metabolism in the mitochondria are increased. Glucose uptake is enhanced by insulin-independent and -dependent transporters, which remain elevated for several hours following exercise and are upregulated in trained muscles.

O_2 Transport

The respiratory system does not show significant changes with training. This is consistent with the observation that $\dot{V}_{O_2}$ max is not limited by respiratory factors under most conditions (see above); however, important cardiovascular and tissue changes occur with training to support increased O_2 transport. Cardiac output increases with training and can achieve levels of up to 35 L/min during maximal exercise with normal right atrial pressures in a well-trained individual. This occurs primarily from an increase in stroke volume, with increased end-diastolic volume (preload) and decreased arterial pressure (afterload). Heart rate is lower after exercise training and does not have to increase as much to achieve the same cardiac output. This may involve downregulation of adrenergic receptors on the heart from high levels of catecholamines during exercise. Heart rate recovers more rapidly after exercise in trained athletes because it is not as high during exercise as in untrained people.

Arterial pressure is lower at a given work rate after training, in part because of increases in the cross-sectional area of the vascular bed. Formation of new capillaries is stimulated by *vascular endothelial growth factor* (VEGF) with exercise training. Increased vascularity increases total blood volume and decreases diffusion distances for O_2 in muscle. Myoglobin levels also increase with training which facilitates O_2 diffusion within muscles. Together these changes promote tissue–O_2 exchange so O_2 extraction is increased by muscle, and the arterial–venous O_2 difference increases.

The net result of these changes in O_2 transport, metabolism and muscle function is, on average, a 15% increase in $\dot{V}_{O_2}$ with exercise training. People who have relatively low $\dot{V}_{O_2}$ levels will increase $\dot{V}_{O_2}$ the most with training; however, genetics explains from 40 to over 60% of the difference in $\dot{V}_{O_2}$ max among individuals. The effects of training depend on duration and intensity, and an effective regimen is exercising large muscle masses (e.g., running) at 50 to 75% of $\dot{V}_{O_2}$ max for 20 to 60 minutes, 3 to 5 times per week. Training does not change the amount of muscular work done for a given $\dot{V}_{O_2}$, but it does increase $\dot{V}_{O_2}$ max; hence, any given workload occurs at a lower percentage of $\dot{V}_{O_2}$ max after training.

C H A P T E R

64

Heart Failure and Circulatory Shock

JAMES M. DOWNEY

KEY POINTS

- *Heart failure* is a condition in which the heart is unable to adequately propel blood through the circulation.
- The cardinal sign of heart failure is elevated venous pressure on the side of the heart that is failing. If both sides are failing, venous pressure rises on both sides.
- Increased venous pressure causes distension of the veins and edema (venous congestion). Pulmonary edema can occur quickly, sometimes with fatal consequences.
- Compensations such as increased venous pressure may keep the cardiac output normal in mild failure; however, the ability to increase cardiac output (reserve) is lost.
- An abrupt increase in venous pressure can cause the heart to dilate, which further weakens it due to the law of LaPlace.

- Increased workload causes the heart to hypertrophy; hypertrophy invariably leads to failure.
- In circulatory shock, an inadequate cardiac output causes damage to the organs.
- Causes of shock include low blood volume, loss of autonomic tone, and impaired cardiac function.
- Injury to the periphery occurs in a progressive manner when the cardiac output is reduced. As injury continues, the capillaries become permeable to plasma proteins which causes fluid loss into the tissues. Decreased circulating volume causes decreased venous filling of the heart and a further fall in cardiac output.
- Signs of shock include pale, cool skin; oliguria; thirst; and mental confusion.
- Treatment is directed at restoring the cardiac output as soon as possible, often with fluid resuscitation.

This chapter examines two commonly encountered clinical disorders in an attempt to illustrate how the cardiovascular system is governed by underlying physiological principles. The first example is *heart failure*, where the ability of the heart to pump blood is compromised. The second example is *circulatory shock*, where cardiac output is inadequate to meet the needs of the body. These two examples are often interrelated, as the first can cause the second; hence, both are discussed within a single chapter.

HEART FAILURE

Heart failure is a condition in which the ability of the heart to propel blood through the circulation has been compromised. Heart failure can be acquired in a number of ways, including loss of muscle mass through myocardial infarction, elevated workload, or myopathic processes, including infection. In each of these cases, the end result is the same: decreased stroke volume for any given filling pressure—decreased contractility.

The term *congestive heart failure* actually describes a symptom complex that encompasses all conditions of venous congestion of cardiovascular origin regardless of whether the myocardium is depressed or not. Congestive heart failure can occur with valve dysfunctions and even the high cardiac output states of beriberi and over-transfusion. The current discussion, however, is limited to only those conditions in which the contractility of the heart is depressed.

Elevated Venous Pressure: The Cardinal Sign of Heart Failure

As was explained in Chapter 14, a decrease in contractility moves the cardiac function curve down and to the right, as shown in Fig. 1. The immediate effect is a rise in venous pressure and a decrease in cardiac output. The decreased cardiac output causes these patients to have the characteristic symptoms of a limited capacity for physical activity and chronic fatigue. A rise in venous pressure is a key feature of the failure syndrome and is the most prominent clinical sign of a failing heart. The elevated venous pressure distends the veins and thus raises the capillary pressure, leading to fluid filtration into the tissue. This combination is often referred to as *venous congestion*. In fact, the edema from the venous congestion may be more of a threat to the patient than the reduced cardiac output. There are two sides to the heart, and either may fail independently. Venous pressure rises in the side of the heart that is failing; thus, if the left ventricle fails, then pressure will be elevated in the pulmonary veins. If the right heart is failing, systemic venous pressure will be elevated. If both chambers fail together, as in myopathic diseases, then both systemic and pulmonary venous pressures may rise equally.

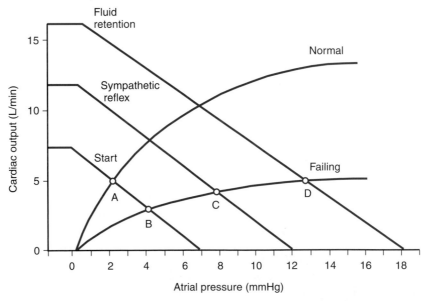

FIGURE 1   The venous function curve and the cardiac output curve are co-plotted for a normal and a failing heart. The starting cardiac output is given by point A. As the heart fails, the cardiac output falls and the venous pressure rises (point B). Sympathetic reflexes constrict the veins and further increase venous pressure (point C). The kidneys retain fluid to increase the blood volume; in this case, they were able to restore the cardiac output to the resting level (point D) but with a further increase in venous pressure to 13 mm Hg. Although this patient's cardiac output would be adequate at rest, he would be unable to increase his cardiac output during physical activity.

When the left ventricle fails, *pulmonary edema* can become a life-threatening complication. The alveoli can quickly fill with edema fluid, causing the patient to literally drown in his own transudate. Pulmonary edema becomes likely only when left atrial pressure exceeds 25 mm Hg. But, the analysis in Fig. 1 would suggest that a decrease in contractility could never increase the venous pressure above the mean circulatory filling pressure of 7 mm Hg. Three factors, however, can cause the pressure to become much higher. The first is the anatomic arrangement that puts the left and right heart in series with each other. Imagine that the heart is suddenly failing because a blood clot has formed in a coronary artery supplying a large portion of the left ventricle. In that case, the left ventricle will be depressed while the right ventricle is relatively unaffected. The right heart will then force blood through the pulmonary circulation and into the left atrium, causing the diastolic pressure (preload) of the left ventricle to continue to rise beyond the mean circulatory filling pressure until the stroke volume again equals that from the right heart. Under those conditions, left atrial pressure can exceed 25 mm Hg within seconds, even though systemic venous pressure may have changed only slightly.

The second process that elevates venous pressure involves the reflex response to lowered arterial pressure. As predicted by Fig. 1, cardiac output will fall with failure and so will arterial pressure. The *baroreceptors* will sense the falling pressure and activate the sympathetic nerves, causing an intense venous constriction that raises the mean circulatory filling pressure and hence venous pressure. The sympathetics also stimulate the heart, causing the right ventricle to beat even more strongly.

The third factor involves the long-term adaptation to a reduced cardiac output. In Chapter 15, we learned that the kidney is the long-term controller of blood pressure and does so by adjusting blood volume. A chronically low cardiac output reduces arterial pressure and will cause *fluid retention* by the kidney. In a healthy individual, fluid retention would stop when the cardiac filling pressure has been sufficiently increased to restore the cardiac output. In the patient with heart failure, however, that mechanism may be ineffective at restoring cardiac output. Rather, the increased blood volume will only shift the venous function curve further to the right (as shown in Fig. 1) and raise venous pressure to dangerous levels.

Estimation of Venous Pressure

Systemic venous pressure is relatively easy to estimate by observing the filling of the jugular veins. The external jugular veins in the neck can be visualized where they run from the clavicle to the angle of the jaw. The clavicle is 10 to 15 cm above the heart, and normally the venous pressure is not enough to support that large a column of blood. As a result, the *jugular veins* in the neck of a healthy person should be collapsed in the upright posture. As venous pressure rises, the jugular veins are seen to be filled over more and more of their length. Systemic venous pressure can be accurately estimated by simply calculating the hydrostatic column from the level of the atrium to the level to which the jugular veins are filled.

Even mild congestion of the pulmonary veins can be detected with a chest x-ray and is seen as opacification of the hilar regions as fluid accumulates in the tissues. Pulmonary venous pressure is not so easy to measure because it is technically very difficult to position a catheter in the pulmonary veins; however, pulmonary venous pressure is an important indicator both of the degree to which the left ventricle has been compromised and of the possibility of a catastrophic pulmonary edema. Therefore, monitoring pulmonary venous pressure in the critically ill patient is often vital. Pulmonary venous pressure can be measured indirectly by insertion of a *Swan–Ganz catheter* and measuring the *pulmonary wedge pressure*. The Swan–Ganz catheter has an inflatable balloon near its tip and is inserted into a peripheral vein and advanced into the right atrium. When partially inflated, the balloon acts as a sail and the flowing blood literally pulls the tip of the catheter through the right ventricle and into the pulmonary arteries. The catheter is then advanced into a small pulmonary arterial branch. When the ballon is inflated it occludes the vessel, stopping the blood flow. If there is no flow through the occluded blood vessel, then there will be no pressure drop across it. Thus, the pressure at the tip of the catheter will suddenly drop to a pressure approximating that at the distal end of the occluded vascular bed, the left atrium.

Compensations in Heart Failure

When the failure is mild, a combination of sympathetic stimulation and fluid retention may be enough to completely restore the cardiac output to normal. When the patient is resting, the only indication that failure is present would be an elevated atrial pressure. This condition is called *compensated failure*. Such a compensation, however, is at the expense of the reserve of the heart, and as a result the ability of the heart to increase cardiac output during exercise would be compromised. In some patients, the decrease in contractility is so severe that no amount of sympathetic stimulation or increased atrial pressure can restore the cardiac output, in which case the cardiac output even at rest will be inadequate.

Fluid retention will continue, even though further elevation of atrial pressure actually decreases the cardiac output and causes a positive-feedback spiral of deterioration. Heart failure in this case is said to be *decompensated*.

The Dilated Heart

The heart is relatively intolerant of a chronically elevated diastolic pressure and can respond by a slippage of the mechanical attachment of the ventricular muscle cells to each other at the intercalated discs. This process can occur in a matter of hours after the sudden appearance of an elevated preload and will cause an increase in diameter and decrease in wall thickness of the ventricular lumen. Such a heart is said to be *dilated*. The actual mass of the dilated heart is unchanged, unlike the case of hypertrophy, for which the mass is increased. Because of the Law of LaPlace (see Chapter 13), a dilated heart is at a severe mechanical disadvantage, and dilation is therefore a serious complication for the failing heart.

Hypertrophy of the Heart Invariably Leads to Failure

We have already explained how coronary artery disease can lead to failure by causing *infarction* (cell death) of part of the ventricular muscle when a coronary artery becomes occluded. Because cardiac muscle does not regenerate, such a lesion will be permanent. Another common source of heart failure is ventricular overload. Like any muscle, the heart tries to adapt to an increased workload by hypertrophy of its individual muscle fibers. Although such an adaptation is usually successful in skeletal muscle, it can lead to disaster in the heart.

Hypertrophy of the heart is undesirable for several reasons. First, the heart is dependent on the venous pressure to stretch the fibers during diastole, and the hypertrophied heart has a thickened wall, which decreases its diastolic compliance; such a condition is illustrated by a shift in the diastolic compliance curve, as shown in Fig. 2. The hypertrophied heart requires an elevated venous pressure to achieve a normal stroke volume. Indeed, diastolic dysfunction may become the primary defect in some heart failure patients.

A second problem is that the remodeling of the muscle cells to contain more myofibrils often results in inferior muscle. Although the exact cause of this lesion is poorly understood, it may involve expression of different isoforms of contractile proteins that replace those normally present in the heart. A third problem is that the heart has a very high oxygen requirement and the coronary circulation does not seem to develop in proportion to the

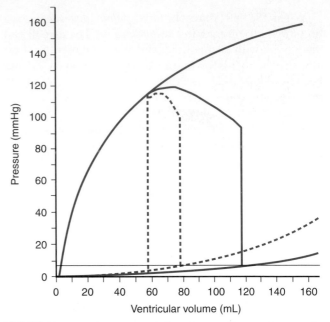

FIGURE 2   Note how a loss of diastolic compliance reduces stroke volume in the concentrically hypertrophied heart. Hypertrophy is shown by the dotted line; the solid line shows normal data. Venous pressure would have to rise to 17 mm Hg to restore stroke volume in this example.

increased muscle mass. As a result, such a heart may always be starved for oxygen and therefore be ischemically depressed.

Hypertrophy Is an Inappropriate Response to an Elevated Workload

When the heart contracts against an increased pressure, as occurs with hypertension or an obstructed outflow tract (*e.g.*, stenosis of the aortic valve), the growth appears to occur inward, tending to obliterate the ventricular lumen. This process normalizes the stress on the individual fibers by thickening the ventricular wall. Such a pattern of hypertrophy is termed *concentric hypertrophy* (Fig. 3). Stretch receptors in the muscle cells detect when the fibers are under an elevated load and stimulate signal transduction pathways involving protein kinases to increase the production of new contractile proteins within minutes of the onset of the stretch. Angiotensin II has also been implicated in contributing to the remodeling process, as its receptors stimulate the protein kinases as well. Drugs that block the action of angiotensin II are particularly effective in patients with hypertension because they not only reduce the blood pressure but also directly interfere with the hypertrophy process within the heart. The thickened ventricular wall of the concentrically hypertrophied heart decreases diastolic compliance

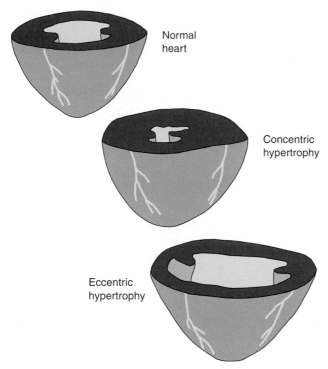

Normal heart

Concentric hypertrophy

Eccentric hypertrophy

FIGURE 3 Transverse section of three ventricles, one normal and two hypertrophied. Chronic pressure overload causes a remodeling of the heart, called *concentric hypertrophy*, in which the heart has a thick wall and small lumen. Chronic volume overload leads to *eccentric hypertrophy*, in which the diameter of the lumen is increased along with the muscle mass.

and the small lumen limits the stroke volume; thus, a total compensation cannot be achieved by hypertrophy. If the underlying cause is not corrected, such a heart will eventually experience a positive feedback spiral to failure, leading to the eventual death of the patient.

When the heart ejects against a normal pressure but has an elevated stroke volume, it responds by keeping the wall thickness relatively constant but increasing the diameter of the lumen. Such a pattern is termed *eccentric hypertrophy* (see Fig. 3). Generally, this pattern of hypertrophy is better tolerated because the diastolic compliance is not lost. In fact, a mild eccentric hypertrophy is often seen in athletes and causes no serious problem. When moderate to severe eccentric hypertrophy occurs in response to a regurgitant aortic or mitral valve, however, the heart will progress to failure. When failure does occur in such a heart, it is difficult to treat. Unlike the case with the heart that hypertrophies in response to a high pressure, correcting the regurgitant valve usually increases rather than decreases the pressure against which the heart must eject, which will exacerbate the failure condition. As a result, the condition must be corrected before the heart becomes too weakened.

Heart failure can occur in the absence of an elevated workload, as occurs in *myopathic* conditions. In these cases, there is a progressive weakening of the heart; in severe cases, the only effective treatment may be a heart transplant. Genetic defects that cause the heart to make one or more defective proteins are now known to underlie some of the cardiomyopathies.

Treatment of the Failing Heart

Treatment of heart failure is directed to correction of the two major problems associated with failure: venous congestion and reduced cardiac output. Systemic venous congestion can cause renal and liver damage as well as discomfort. The threat of pulmonary edema from pulmonary congestion has already been discussed. *Diuretics* and *reduced salt intake* can reduce the accumulated blood volume and venous congestion even though these treatments may actually lower the filling pressure of the heart and thus the stroke volume. One of the oldest cardiac drugs is *digitalis*, which increases the contractility of the heart by causing the muscle cells to accumulate SR calcium (see Chapter 11). Increasing contractility with digitalis increases the cardiac output and lowers the venous pressure. Digitalis in the absence of a diuretic will often cause a brisk diuresis, as the kidney detects the rising cardiac output. Although digitalis and other cardiac stimulants can improve the exercise tolerance in patients with mild failure, they have had little effect on prolonging their lives. Furthermore, in advanced failure these treatments have little beneficial effect.

Recently, *beta-adrenergic receptor blockers* have been tried in these patients with some success. The actual cause of the improvement is not entirely known but is in part related to reduced heart rate, which improves cardiac filling. Some of the benefit may also derive from increased diastolic compliance of the ventricles. Because of the chronically elevated sympathetic tone in these patients and the constant stimulation, the beta-receptors of the heart are downregulated, which blunts the effect of norepinephrine in the ventricle. It is thought that in the presence of low-dose beta blockade the receptors will again proliferate. Beta-receptors will then only be stimulated under conditions of elevated physical activity when enough norepinephrine is released to competitively overcome the beta-blocker. Although contractility at rest may be somewhat depressed, the reserve of the heart is actually increased.

Currently, the most promising treatment for heart failure is the use of *angiotensin converting enzyme (ACE) inhibitors*. ACE inhibitors not only lower the blood pressure and increase cardiac output but also interfere with the detrimental remodeling that occurs in the stressed heart. The important role that angiotensin II plays in the cellular changes that occur in hypertrophy and failure is only beginning to be understood.

CIRCULATORY SHOCK

Shock Is Caused by an Inadequate Cardiac Output

The beginning student often erroneously believes that the heart is the sole determinant of cardiac output. In Chapter 14, we learned that the peripheral circulation also plays an important governing role in the circulation. The peripheral circulation controls the filling pressure of the heart as well as its afterload, two major determinants of the stroke volume. This interaction between the heart and the periphery is vividly illustrated by the shock syndrome. In its simplest form, circulatory shock can occur whenever the cardiac output is inadequate to meet the needs of the periphery. If this continues, the peripheral tissues will incur ischemic injury that will trigger a vicious cycle of events leading to circulatory collapse; this sequence of events is outlined in Fig. 4. Although war has been the scourge of mankind over his entire recorded existence, one benefit of war has been to provide physicians with large numbers of patients in hemorrhagic and traumatic shock to study. As a result, most of the advances in treatment have occurred during such conflicts. To illustrate how far our ability to treat circulatory shock has progressed, at the start of World War I the mortality rate for patients with a fractured femur was 70%, with virtually all of these deaths being due to circulatory shock. A death from such a fracture in a young man would be rare today.

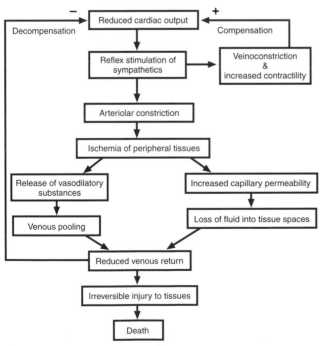

FIGURE 4 A flow diagram showing the chain of events that occur in the failing heart.

The Causes of Circulatory Shock Are Many

The main causes of shock include: (1) inadequate circulating volume, as occurs in hemorrhage, burns, or sepsis; (2) loss of autonomic control of the vasculature, as occurs in central nervous system (CNS) lesions; (3) impaired cardiac function; and (4) elevated peripheral demand, as occurs in sepsis. In the first case, a reduced blood volume lowers cardiac output by reducing ventricular filling pressure. In the second case, loss of vascular tone causes venous pooling and arteriolar dilation, which again reduce the ventricular filling pressure. In the third case, the heart itself is the impediment to the circulation of blood, and in these patients the filling pressure is often elevated. In the fourth case, the cardiac output and filling pressure may be normal but cannot meet the elevated requirements of the periphery.

Circulatory Shock Is Progressive

After World War II, Carl Wiggers simulated hemorrhagic shock in dogs by connecting a large bottle to the femoral artery of an anesthetized dog with a length of flexible tubing. If the bottle was suspended 136 cm above the heart, the dog's blood pressure (100 mm Hg) would be just enough to push blood up the tubing to the reservoir but little would enter. If the reservoir was lowered to 54 cm above the heart, blood quickly flowed out of the dog and into the reservoir. This continued until enough blood had been shed to lower the mean arterial pressure to 40 mm Hg. It was noted that after several hours of this hypotension the blood would begin to flow out of the bottle and back into the dog. When all the blood had returned, the arterial pressure then began to decline and the dog quickly died. This progression of injury in which the animal required more and more volume just to maintain a pressure of 40 mm Hg was referred to as *decompensation*. The decompensation phase is due to injury to the peripheral vascular beds from a prolonged period of inadequate blood flow. During the *compensated phase*, the cardiovascular reflexes attempt to maintain the circulation by vasoconstriction of the systemic arterioles and veins and stimulation of the heart. During decompensation, vascular tone actually falls below normal because of *autoregulatory escape* from the sympathetic constriction and production of vasodilatory cytokines by the ischemically injured peripheral tissues. These substances include *histamine, kinins, nitric oxide, tumor necrosis factor, prostaglandins*, and *interleukins*. As will be seen later, an increase in capillary permeability will also cause fluid to shift into the extravascular space.

Wiggers also noted that after a period of hypotension the bottle could be elevated to restore the cardiac output and blood pressure. If the deficit in cardiac output had been mild, then the dog would recover uneventfully. If the deficit had been severe, however, a normal arterial pressure could not be maintained for long and the dog would again become hypotensive and eventually die. This was termed the *irreversible phase* of shock. Of course, what was irreversible yesterday may be reversible today, for the management of the shock patient has steadily improved over the years.

Blood Pressure Is an Unreliable Indicator of the Shock State

In the preceding example the arterial pressure was forced to fall as the dog was bled. Similarly, the arterial pressure invariably falls in the decompensated phase of circulatory shock. Unfortunately, that observation has erroneously caused many to equate shock with hypotension. It must be stressed that an inadequate cardiac output rather than a low blood pressure is the primary lesion in this syndrome. Because the body has many mechanisms for defending the blood pressure (baroreflex, renin–angiotensin system, carotid bodies, antidiuretic hormone, etc.), the body will meet a sudden drop in cardiac output with an intense peripheral vasoconstriction that may temporarily maintain the blood pressure. The signs of reduced cardiac output can still be seen, however, and will include *pale, cold skin* as well as low urine production (*oliguria*) due to reduced renal blood flow. The circulating catecholamines will cause sweating even though the skin is cold. The patient will often complain of *thirst* as the CNS tries to increase fluid intake in an attempt to restore blood volume. Not all shock states exhibit reduced cardiac output. A good example is *septic shock* (infection), where the oxygen requirements of the body are increased due to the fever and circulating toxins. In the case of sepsis, a cardiac output within normal limits may still be inadequate to meet the high nutritional needs of the periphery.

The Body Tries To Defend the Heart and the Brain

Although most of the peripheral vasculature may reflexively be vasoconstricted, in the early phases of shock the cerebral and coronary circulations are exceptions. These two vital organs retain their normal autoregulation. As blood pressure eventually falls, however, they too will be affected. Confusion and impaired reasoning are often the first signs of inadequate perfusion of the brain. Subendocardial ischemia may be apparent with depressed *ST segments* in the electrocardiogram as perfusion to the heart is compromised. Myocardial

ischemia will occur if blood pressure falls below 50 mm Hg and at even higher pressures if coronary artery disease is present. If the hypotension is severe, myocardial ischemia can depress the heart, which will further lower the blood pressure. Once this positive feedback situation begins, the circulation can collapse suddenly. Injury to other organs can be more insidious. Reduced blood flow to the renal cortex for more than a day can result in death of cortical tissue and renal failure. Furthermore, an increase in pulmonary capillary permeability often results in *adult respiratory syndrome* (ARDS), characterized by poor oxygenation of the blood and reduced pulmonary compliance. ARDS currently accounts for most of the late deaths in otherwise successfully resuscitated patients.

Fluid Resuscitation Has Become the Mainstay of Therapy for Shock

With the exception of cardiogenic shock, the primary reason for inadequate cardiac output in the shock patient is low ventricular filling pressure. As a result, volume expansion, often referred to as *resuscitation*, has been the most successful treatment. Volume expanders fall into three categories: whole blood; cell-free fluids that have a colloid included, such as plasma; or colloid-free fluids, such as lactated Ringer's solution. Surprisingly, administration of colloid-free fluids has yielded the best outcomes. As injury to the peripheral circulation increases, the capillaries become very permeable to the plasma proteins, the colloids that are responsible for osmotically pulling water back into the capillary lumen (see Chapter 16). The failure of the capillaries to restrict the movement of plasma proteins causes massive pooling of fluids in the interstitial spaces. As a result, the colloid becomes ineffective and lactated Ringer's (basically, salt water with some lactic acid added as a nutrient) is just as effective in expanding the patient's blood volume. The body can tolerate dilution of the blood to a hematocrit as low as 20% without serious consequences. As would be expected, fluid resuscitation causes extensive peripheral edema in these patients, but the benefits of increased cardiac output far outweigh any consequence of the swelling.

Current research in shock therapy is focusing on the role that many of the circulating cytokines play in the shock state. It may be that selective blockade of some of them, such as nitric oxide or tumor necrosis factor, may further improve current therapy. Another important field of investigation involves an attempt to understand why the capillaries become so permeable to macromolecules and how this could be reversed. Mechanical attempts at increasing venous return, such as inflatable pants or simple elevation of the legs, have had a minor clinical impact. The amount of blood that can be translocated to the thorax by these devices is small compared to what can be accomplished with simple fluid resuscitation.

Specific Forms of Shock

In its most severe form, heart failure can cause *cardiogenic shock*. When the cardiac output becomes too low, ischemia in the periphery occurs which begins the downward spiral associated with the shock state. What makes cardiogenic shock so difficult to treat is that often the primary lesion itself is often irreparable, as in myocardial infarction or end-stage heart failure. Cardiogenic shock can also result from the abrupt appearance of a valve lesion or filling of the pericardial space with fluid that constricts the heart and opposes diastolic filling (*pericardial tamponade*). Fluid resuscitation will obviously be less effective in cardiogenic shock, although further elevation of preload may still improve cardiac output in selected patients. A *Swan–Ganz catheter* is often used to make sure that only enough fluid is added to achieve an optimal ventricular filling pressure.

As mentioned earlier, hemorrhage reduces blood volume and sepsis causes peripheral pooling of fluids. In *burn* patients, fluid is lost at the site of the burn where raw tissue is exposed. In traumatic shock with extensive blunt tissue injury, internal hemorrhage and the release of toxic substances from the necrotic tissue cause vasodilation and peripheral pooling of blood.

Neurogenic shock may be caused by spinal cord or brain injury. It will be recalled that peripheral vascular control by the CNS is accomplished by varying the tone of the sympathetic nerves that constrict the peripheral vasculature (Chapter 17). Because the sympathetic nerves are unopposed, dilation is accomplished simply by reducing sympathetic tone. The sympathetic nerves course through the spinal cord; hence, transection of the cord will cause a profound vasodilation that will persist for weeks following the event. Hypotension due to a reduction of both filling pressure and peripheral resistance will occur immediately after the lesion, and the shock syndrome will soon follow unless aggressive resuscitation is instituted. *Anaphylactic shock* can occur when an antigen–antibody reaction occurs in the circulation. An example would be transfusion with an incompatible blood type. The reaction stimulates mast cells in the tissues to secrete histamine and related substances that cause vasodilation and increased capillary permeability. In these conditions, the hypotension can be so severe that circulatory collapse occurs within minutes.

Suggested Readings

Braunwald E, Grossman W. Clinical aspects of heart failure, in: Braunwald E, Ed., *Heart disease: a textbook of cardiovascular medicine*. Philadelphia: Saunders, 1992.

Shires GT III, Shires GT, Carrico J. Shock, in: Schwartz SI, Shires GT, Spencer PC, Eds., *Principles of surgery*. New York: McGraw-Hill, 1994, pp. 119–144.

Smith TW, Braunwald E, Kelly RA. The management of heart failure, in: Braunwald E, Ed., *Heart disease: a textbook of cardiovascular medicine*. Philadelphia: Saunders, 1992.

Weil M, von Planta EC, Rackow. Acute circulatory failure, in: Braunwald E, Ed., *Heart disease: a textbook of cardiovascular medicine*. Philadelphia: Saunders, 1992.

65

Diabetic Ketoacidosis

STANLEY G. SCHULTZ

The word *diabetes* derives from the Greek, διαβαινω, meaning "passing through" or "too swift a passage of the matter that is drunk." It is used today to describe the condition of excessive production of urine or polyuria. As discussed earlier (Chapters 28 and 41), there are two forms of diabetes. One is *diabetes insipidus*, which results from impaired secretion or production of the antidiuretic hormone (ADH) due to injury to the hypothalamus or supraoptic nucleii. Diabetes insipidus can also, rarely, be due to impaired function of ADH receptors in the renal distal nephron and is then referred to as *nephrogenic diabetes insipidus*. The word *insipidus* refers to the fact that in this affliction the urine tastes bland or insipid. The other form of diabetes is *diabetes mellitus*, which may arise from a lack of insulin secretion due to autoimmune destruction of pancreatic β-cells (so-called *type I* or *insulin-dependent diabetes mellitus*) or from insulin resistance of target organs (so-called *type II* or *noninsulin-dependent diabetes mellitus*). Some years ago, the former was referred to as juvenile and the latter as adult-onset diabetes mellitus, but, with the increasing incidence of obesity among teenage children, this distinction has become blurred and is no longer in use. In 1679, Dr. Thomas Willis reported that urine from patients afflicted with this illness (which he referred to as the "pissing evil") tasted "honey-like"; hence, the descriptor *mellitus* (from *mellifluous*).

Prior to 1921, severe insulin-dependent diabetes mellitus was a lethal disease, with many of the afflicted dying in a state of severe ketoacidosis. In that year,

Frederick Banting and a medical student, Charles Best, isolated insulin from canine pancreas during a summer research project, the likes of which will probably never be seen again. In 1923, Banting and, ironically, Macleod (the chairman of the department who made available the laboratory space and the use of eight dogs for the historic experiments) shared the Nobel Prize for this life-saving contribution. Today, thanks to this monumental work, severe ketoacidosis can be avoided and managed.

In order to appreciate diabetic ketoacidosis, we will analyze the case of a patient, U.T., who was first hospitalized at the age of 20 years. For several weeks before this hospitalization she had been under considerable academic and financial stress; she was working long hours and often skipped meals. During the week prior to her hospitalization, she lost several pounds of weight and noted a marked increase in thirst and in fluid intake (polydipsia). She would awaken several times per night to void (polyuria). In the days prior to admission, she became increasingly weak, lethargic, and dizzy whenever she tried to stand. She collapsed on several occasions trying to get to the bathroom. She also became nauseated and vomited twice.

On admission to the university infirmary, U.T. was found to have a blood pressure of 90/60 mm Hg when supine, which fell to 65/45 mm Hg when she stood up briefly. Her pulse was 120/min supine and increased to 140/min upon standing. Her respirations were deep and at 30/min, and an alert intern detected the odor of acetone on her breath. Her skin was cold and clammy and demonstrated decreased turgor; her mucous membranes were dry. Urinalysis revealed high levels of glucose and ketone bodies. When she was transferred to the regional medical center hospital, the laboratory data given in Table 1 were obtained. U.T. was suffering from previously undiagnosed insulin-dependent diabetes mellitus, and to understand her condition we must understand the consequences of insulin deficiency on body metabolism.

As discussed in Chapter 41, insulin is an anabolic hormone released from pancreatic β-cells, in part in response to a meal and a subsequent elevation of plasma glucose and amino acid levels. One of the functions of this hormone is to promote the insertion of the carrier for sugars, GLUT4, into the plasma membrane from intracellular stores which permits the energy-independent facilitated uptake of glucose by muscle and adipose tissue (see Chapter 41). Insulin also stimulates glycogenesis and inhibits gluconeogenesis. It follows that, in the absence of insulin, the major body mass of tissue is incapable of consuming glucose, and both glycogenolysis and gluconeogenesis are promoted. This leads to markedly elevated plasma glucose levels (hyperglycemia) ("Water, water everywhere, but not a drop to drink!").

Starved of glucose, the cells turn to lipids as a source of energy. Lypolysis is stimulated, leading to hyperlipemia and flooding of the liver with free fatty acids. AcetylCoA is produced in abundance, which, as discussed in Chapter 41, leads to the formation of acetoacetic acid, β-hydroxybutyric acid, and acetone. Both acetoacetic acid and β-hydroxybutyric acid are weak acids that readily dissociate at physiological pH, leading to a metabolic acidosis; note that U.T.'s plasma pH was at the perilously low level of 7.1. Finally, insulin stimulates protein synthesis and inhibits protein degradation. In the absence of insulin, the reverse takes place, leading to severe protein catabolism, wasting, aminoacidemia, and elevated blood urea nitrogen.

The clinical picture of diabetic ketoacidosis arises from the combination of three pathophysiological conditions resulting from these metabolic derangements: (1) impaired renal ability to elaborate a maximally concentrated urine in spite of a hypertonic plasma and markedly elevated plasma levels of ADH, (2) metabolic acidosis, and (3) markedly increased solute load that must be excreted. The end result is diuresis, volume depletion, electrolyte imbalances, and life-threatening acidosis. We will consider the renal disturbance first.

As discussed in Chapter 26, glucose is freely filtered at the glomerulus and is then reabsorbed by a Na-coupled, carrier-mediated process located in the proximal tubule. As with all carrier-mediated reabsorptive processes, glucose reabsorption is characterized by a tubular maximum (T_m). Normally, the filtered load (glomerular filtration rate [GFR] $\times$ plasma glucose concentration) is well below the T_m so that all filtered glucose is reabsorbed in the first few millimeters of the proximal tubule. If, as in the case of U.T., the plasma glucose concentration is 900 mg/dL ($= 9$ mg/mL) and assuming that the initial GFR is 120 mL/min, then U.T.'s filtered load of glucose is 1080 mg/min (GFR $\times$ plasma glucose concentration). If her T_m for glucose reabsorption initially is the normal 375 mg/min, then 705 mg/min ($1080 - 375$ mg/min) is not reabsorbed and must be excreted in the urine. This amounts to about 1000 g/day, or about 2 lb/day.

Now, as discussed in Chapter 26, the presence of a nonreabsorbable solute in the "leaky" proximal tubule results in the retention of its osmotic equivalent of water and, in turn, a reduction in the fractional reabsorption of Na, urea, and water by the proximal tubule. Thus, more volume enters the descending limb of the loop of Henle and this, in turn (see Chapter 27), results in a washout of the hypertonic renal medulla normally established by the countercurrent mechanism. In short, because of the reduced fractional reabsorption in the proximal tubule, due to the presence of glucose that could not be reabsorbed, the ability of the kidney to elaborate a maximally hypertonic urine-that is, to maximally concentrate filtered solutes-is reduced, and the osmolarity of the final urine may approach isotonicity with the plasma.

It should be noted that U.T.'s plasma level of ADH was very high because the enormous concentration of

Table 1 Laboratory Data for Patient U.T.

		Patient U.T.	Normal
Arterial blood	pH	7.1	7.35–7.45
	PO_2	105	85–95 mm Hg
	P_{CO_2}	21	40 mm Hg
Venous blood (plasma)	Hematocrit	50	38–44%
	Glucose	900	80–120 mg/dL
	Na^+	130	138–145 mEq/L
	K^+	6.0	3.5–5.0 mEq/L
	Cl^-	88	95–105 mEq/L
	HCO_3^-	6.3	24–26 mEq/L
	Albumin	6	3.0–4.0 g/dL
	Blood urea nitrogen	60	5–20 mg/dL
	Creatinine	2.5	0.8–1.4 mg/dL
	Osmolality	360	285–305 mOsm/L
	Renin	3	0.5–1.0 units
	ADH	15	1.0–10 ng/L
Urine	Glucose	600	0 mg/dL
	Osmolality	500	100–1200 mOsm/L
	Creatinine	26.2	50–150 mg/dL
	Flow rate	240	40–100 mL/hr

glucose (50 mM) rendered her plasma hypertonic; thus, the wall of the distal nephron was maximally permeable to water. The submaximal final urine osmolarity is *prima facia* evidence of the washout of the countercurrent multiplier system.

Now, the molecular weight of glucose is 180. Thus, the rate of solute excretion due to glucose *alone* is about 4 mOsm/min (705 mg/min divided by 180 g/mol), or about 5750 mOsm/day. If the normal concentrating ability is 1200 mOsm/L, then this amount of glucose *must* be accompanied by 4.8 L/day of water; however, if, as for U.T. (see Table 1), concentrating ability is reduced to 500 mOsm/L, then the minimum urinary volume is 10 L/day. Thus, an enormous amount of urine is required just to rid the body of nonreabsorbed glucose, and, as we will soon see, this is just the beginning.

As noted previously and discussed in greater detail in Chapter 41, the excessive breakdown of adipose stores and metabolism of lipids result in the production by the liver of large amounts of acetoacetic acid and β-hydroxybutyric acid. Both dissociate at physiological pHs to yield the anions acetoacetate and β-hydroxybutyrate, respectively, as well as H$^+$, which results in a metabolic acidosis. The respiratory response to the decline in plasma pH, sensed largely by peripheral chemoreceptors (Chapters 20 and 31), is an increase in minute volume due to increased respiratory rate and tidal volume ("Kussmaul" breathing), leading to a decline in plasma P$_{CO_2}$ (see Table 1) and partial compensation for the decline in pH. But, the lungs cannot rid the body of excess H$^+$. That task remains for the kidneys and is accomplished in the proximal tubule by acceleration of the apical membrane Na$^+$–H$^+$ exchanger, leading to a complete reabsorption of filtered HCO$_3^-$ and in the distal nephron by acceleration of the apical membrane H$^+$ pump. But, because neither of these carrier-mediated processes can, for energetic reasons, continue to extrude H$^+$ from the cell when the luminal pH falls below 4 (approximately a 10,000-fold H$^+$ gradient across the apical membranes), the urine must be buffered if it is to accommodate much H$^+$. Thus, buffers are elaborated initially in the form of titratable buffers (*i.e.*, H$^+$ bound with other anions to form weak acids). The anions in this case are phosphate and sulfate derived from the breakdown of cell proteins and, most importantly, acetoacetate and β-hydroxybutyrate. If the acidosis persists, NH$_3$ derived from the deamination of glutamine will contribute to the buffering capacity of the urine in the form of a nontitratable buffer (see Chapter 31).

But, while titratable and nontitratable buffers serve to assist, vitally, in ridding the plasma of H$^+$, they add to the renal solute load that must be excreted. Thus, in addition to glucose, we have these buffers and a greater than normal quantity of urea due to enhanced muscle breakdown. All in all, the obligatory solute load that must be excreted is huge while the renal concentrating ability has been compromised. This results in a diuresis that may exceed 12 L/day and, if unmatched by fluid replacement, can lead to hypovolemia, circulatory collapse, and death.

Maternal Adaptations in Pregnancy

H. MAURICE GOODMAN

KEY POINTS

- Homeostatic adaptations in pregnancy provide for the needs of the developing fetus while preserving the ability of the mother to adjust to changing environmental demands.
- Signals from the conceptus adjust the setpoints of feedback mechanisms that govern cardiovascular, renal, and respiratory functions.
- Bidirectional exchange of gases and metabolites between fetal and maternal blood takes place across the large surface area of the placental villi.
- The uteroplacental circulation receives about 20% of the maternal cardiac output.
- Hormonal signals from the developing placenta decrease the sensitivity of vascular smooth muscle to vasoconstrictor agents and lower peripheral resistance and arterial pressure.
- Resting maternal cardiac output increases by 50% as a result of decreased peripheral resistance and increased blood volume.

- Activation of the renin–angiotensin–aldosterone system results in sodium retention and expansion of plasma volume.
- Increased osmoreceptor sensitivity leads to increased thirst and decreased plasma osmolality.
- The increase in plasma volume dilutes erythrocyte and albumin concentrations, thus decreasing the hematocrit and colloid osmotic pressure.
- Decreased glomerular arteriolar resistance increases renal blood flow and glomerular filtration.
- High blood levels of progesterone increase chemoreceptor sensitivity to carbon dioxide and increase inspiratory drive.
- Hyperventilation decreases alveolar and plasma PCO_2 and increases blood pH.
- Decreased PCO_2 in maternal blood facilitates placental gas exchange.

OVERVIEW

Chapter 47 describes the major endocrinological aspects that govern the initiation, maintenance, and termination of pregnancy. Supporting the growth and development of the fetus imposes significant challenges to maternal homeostasis and requires adjustments in the function of virtually every organ system. In this chapter, we consider the responses and adaptations of cardiovascular, renal, and respiratory systems to the demands of pregnancy. Such adaptations must accommodate the metabolic needs of the fetus while preserving the ability of the mother to make homeostatic adjustments to changing environmental demands. At the same time, physiological changes that take place during pregnancy must also prepare the mother to withstand the arduous birthing process and to provide sustenance for her newborn baby. Many of these adjustments are orchestrated by the placenta through the release of hormones and other bioactive substances into the maternal circulation. It is important to note that all of the regulatory and feedback mechanisms that maintain homeostasis in the nonpregnant woman remain operative during pregnancy, but the sensitivity to physiological signals or the set-points of feedback systems are modified. Some changes in maternal physiology in the late stages of pregnancy are adaptations to the physical presence of the expanded mass of the gravid uterus.

CHALLENGES TO MATERNAL HOMEOSTASIS

In the approximately 265 days that elapse between fertilization and delivery of a full-term infant, the mother provides all of the resources required to transform a single pluripotential cell into a complete new individual comprised of more than 600 billion highly specialized cells and weighing about 3.5 kg. Demands on maternal homeostasis to support these remarkable events are not constant but change as pregnancy progresses. Throughout the pregnancy her homeostatic control mechanisms ensure a hospitable environment of constant temperature, oxygen supply, waste disposal, and availability of nutrients, minerals, and vitamins. In addition, she supports growth and development of the placenta and its extraordinary metabolic activity. To house the growing fetus and placenta, her uterus grows from around 50 g to more than 1 kg and from about 7 cm to more than 40 cm in length. It also acquires the contractile capacity required to expel the baby at term. Her breasts grow and develop, and she increases her fat stores in anticipation of providing nourishment for her newborn child. The overall cost to the mother in increased energy expenditure is about 85,000 calories (~300 Cal/day), which must be derived from the diet or fat reserves. Of this, only about one-third is consumed by the fetus to support biosynthetic and metabolic activities. About one-third is consumed by the placenta to support biosynthetic and transfer processes, and much of the remainder fuels the additional workload imposed on the heart, respiratory muscles, and kidneys.

Pregnancy is traditionally dated from the time of the last menstrual period, although ovulation and fertilization occur about 2 weeks later. In describing the progress of pregnancy, it is customary to consider the nine months of gestation in three-month intervals, or *trimesters*. Fetal growth progresses exponentially at a rate of about 1.5% each day until about the middle of the third trimester, when it may slow somewhat (Fig. 1). By the end of the first trimester, the fetus attains only about 2% of its size at birth. Consequently, its metabolic needs are small in the first trimester, and maternal resources are directed toward fueling the rapid growth of the placenta and establishing the infrastructure that will support fetal development in the second trimester and, especially, the third trimester, when the bulk of fetal growth takes place. Metabolic demands of the early weeks are dominated by requirements of the rapidly expanding syncytiotrophoblast and formation of the placenta, which is not complete until about the 12th week. While placental growth slows as pregnancy progresses, its biosynthetic and nutrient transfer activities persist and intensify as term approaches. Even at term, when its mass is less than 15% of that of the fetus, the placenta consumes about 40% of the oxygen extracted from the uterine circulation. Per gram of tissue, the placenta consumes oxygen at about the same rate as the adult brain. About 30% of placental oxygen uptake is consumed in maintaining the ionic gradients that support solute transport, and another 30% is consumed in protein synthesis.

THE PLACENTA

The placenta is a complex organ that (1) anchors the developing fetus to the uterine wall; (2) provides the maternal/fetal interface for the exchange of nutrients, respiratory gases, and fetal wastes; and (3) directs maternal homeostatic adjustments to meet changing fetal needs by secreting hormones and other substances into the maternal circulation. It is a disc-shaped organ that measures about 22 cm in diameter and has an average thickness of about 2.5 cm at the end of pregnancy. The surface facing the developing fetus is called the *chorionic plate*. It is penetrated near its center by the umbilical artery and veins, which branch repeatedly to perfuse the functional units, the tree-like *placental villi*, with fetal blood. The villi are rooted in the chorionic plate and extend toward the *basal plate*, which is comprised of

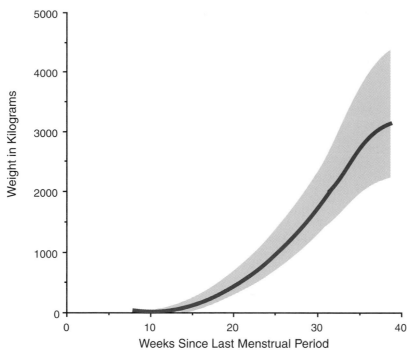

FIGURE 1   Fetal growth during gestation. The shaded area indicates the range of weight of 95% of fetuses.

maternal decidual cells (see Chapter 47) and extravillous syncytiotrophoblast. The mature placenta contains 60 to 70 villous trees, each of which, through repeated branching of the secondary and tertiary villi, gives rise to more than 100,000 intermediate and terminal villi. The villi, whose combined length is estimated to be 90 km, contain increasingly finer branches of arterioles and venules that terminate as clusters of grape-like outgrowths comprised largely of sinusoidal dilated capillaries. The entire villous tree is ensheathed in a continuous layer of syncytiotrophoblast, which overlays a discontinuous layer of the cytotrophoblastic cells. The chorionic plate is fused at its edges with the basal plate to form a hollow cavity, the *intervillous space*, which is perfused with maternal blood that enters through multiple spiral arteries that branch off the radial arteries in the myometrium and exits by way of the uteroplacental veins (Fig. 2).

Maternal and fetal blood do not mix or come in direct contact. Exchange of nutrients and gases takes place principally in the terminal villi across a diffusion barrier comprised of a layer of syncytiotrophoblast, basal lamina, and endothelial cells. By late pregnancy, the diffusion barrier thins to about 5 μm and has a surface area of about 12 m^2. The multifunctional syncytiotrophoblast, which is richly endowed with ion channels, transporters, and exchangers, is the site of both active and passive exchange processes. Bulk movement and mixing of blood in the intervillous space tends to flatten concentration gradients between umbilical venous and maternal blood.

In addition, a relatively high proportion of placental surface is not available for exchange, creating a sizeable dead space. Consequently, high rates of blood flow are needed to compensate for the relative inefficiency of exchange.

The syncytiotrophoblast is also endowed with versatile biosynthetic capacity and is the source of the placental peptide and steroid hormones (see Chapter 47). Large amounts of progesterone, estrogens, human chorionic gonadotropin (hCG), human placental lactogen (hPL), and other secretory products are released by the syncytiotrophoblast directly into the intervillous space and hence the maternal circulation. These placental hormones are largely responsible for orchestrating adjustments in maternal physiology as pregnancy progresses.

THE MATERNAL CIRCULATION

Adequate perfusion of the placenta is an indispensable condition for normal growth and development of the fetus. Any sustained reduction in placental perfusion results in a condition called *intrauterine growth retardation* (IUGR) and may cause premature delivery or delivery of an undersized baby at term. Adjustments in the maternal cardiovascular system enable this critical need to be met without compromising its capacity to respond to the changing demands and challenges of the mother's day-to-day life in an often taxing environment.

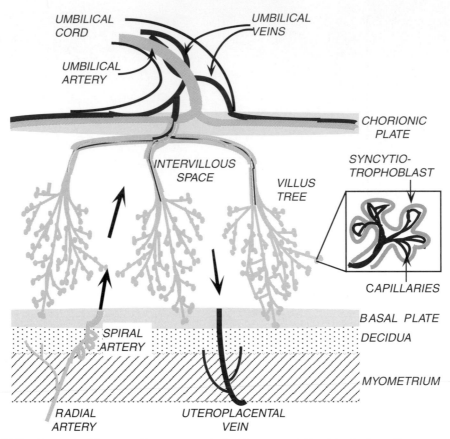

FIGURE 2   Placental villi are treelike structures bathed by maternal blood in the intravillous space that is formed between the basal and chorionic plates. Insert shows twig-like terminal villi consisting of fetal capillaries encased in a sheath of syncytiotrophoblast. Heavy black arrows indicate the direction of maternal blood flow.

To accommodate the required high rates of placental perfusion, maternal cardiac output increases by about 50% at rest. This sustained high output is made possible by a profound decrease in total peripheral resistance and a 50% expansion in blood volume.

Uterine Blood Flow

Blood flow to the uterus increases from about 50 to 100 mL/min in the midluteal phase of the menstrual cycle to more than 1 L/min in the third trimester. At term, uterine blood flow, of which more than 80% is directed to the placental implantation site, accounts for about 15 to 20% of the cardiac output. Increased flow begins shortly after implantation, before formation of the placenta is complete, and rises progressively in parallel with expansion of the uterine mass. Decreased resistance of the small myometrial arteries that lie under and around the implantation site is evident as early as the fifth week and well before similar changes are seen in larger arteries. These early changes are thought to reflect invasion of the spiral arteries by the syncytiotrophoblast and their consequent dilatation. In animal studies, the

absence of autoregulation, as indicated by the lack of reactive hyperemia following a brief arterial occlusion, suggests that arterioles in the gravid uterus may be maximally dilated under basal conditions. The decrease in resistance is so pronounced that some investigators refer to the uteroplacental circulation as an *arterio-venous shunt*.

It may be recalled from the discussion in Chapter 47 that concentrations of estrogens and progesterone in maternal blood plasma increase steadily and dramatically as pregnancy progresses. Evidence from both animal and human studies indicates that estrogens have vasodilatory effects and decrease vascular resistance, especially in the uterus, which is particularly sensitive to this hormonal effect. Estrogens also markedly lower sensitivity of the uterine vasculature to vasoconstrictor agents. Extensive studies of uterine blood flow have been made in pregnant sheep, which have been studied as models for human pregnancy. Infusions of physiologically relevant amounts of angiotensin II or norepinephrine produce little effect on uterine vascular smooth muscle or uteroplacental blood flow, although at the same concentrations these agents cause significant constriction of extrauterine vasculature and increased systemic blood pressure. In fact, infusion of angiotensin II into pregnant ewes actually increases uterine blood flow as a consequence of increased systemic blood pressure. Precisely how the uterine vasculature becomes desensitized to pressor agents has not been established, but some evidence suggests that ovarian and placental hormones stimulate local production of vasodilator agents, including nitric oxide and prostacyclin (PGI_2).

Changes in Cardiac Output

A striking increase in cardiac output is seen within the first 6 weeks of pregnancy. Cardiac output in a typical 55-kg woman increases from about 4.8 L/min before fertilization to more than 7 L/min after the first trimester and remains elevated until delivery. The increase in cardiac output is accompanied by a modest decrease in mean blood pressure resulting primarily from a decline in diastolic pressure. This decrease in blood pressure is seen despite a 50% increase in blood volume and persists until about the middle of the second trimester when blood pressure gradually rises to about or perhaps slightly above the prepregnant level. The increase in cardiac output and the decrease in arterial blood pressure result from a pronounced decrease in total peripheral resistance (Fig. 3).

Vascular Resistance

The decrease in total peripheral resistance is far greater than can be accounted for by the low resistance of the uteroplacental circulation. There is a widespread decrease in tone of arterioles and larger arteries of the mesenteric, limb, cutaneous, and especially the renal (see below) circulations. Overall vasoconstrictor responses to administered angiotensin II and norepinephrine are attenuated, but to a considerably lesser extent than in uterine arterioles. It is important to note, however, that despite the overall decrease in vascular tone, local and systemic vascular regulatory responses remain operative. Reflex adjustments to changes in posture and to the requirements of exercise are only mildly compromised. Despite changing environmental demands, baroreceptor reflexes maintain blood pressure constant, but at a lower set-point.

The basis for the decrease in vascular resistance is a topic of active research. Clearly, the developing embryo must send out some signals to bring about these changes in vascular tone. Infusion of estrogens in nonpregnant human and animal subjects acutely decreases vascular resistance and produces changes that are similar, though less pronounced, to those seen in normal pregnancy. Human vascular endothelial cells express estrogen

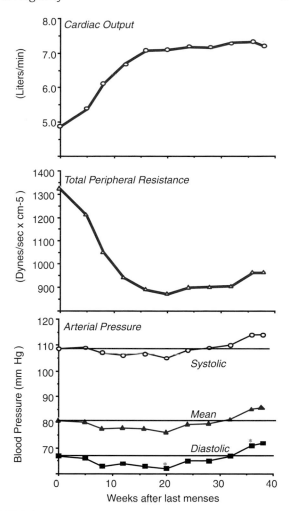

FIGURE 3 Changes in cardiac output, peripheral resistance, and arterial pressures in normal pregnancies. (Drawn from the data of Robson SC, Hunter S, Boys RJ, Dunlop W. *Am J Physiol* 1989; 256:H1060–H1065.)

receptors and, in response to estradiol, rapidly increase their production of vasodilating agents such as nitric oxide and PGI_2 (see Chapter 40). Other studies suggest a similar role for progesterone, alone or in the presence of high levels of estrogens. Most recently, studies in rodents have focused on the vasodepressor effects of another ovarian hormone, relaxin (see Chapter 47). Relaxin is a small peptide that is secreted by the corpus luteum throughout pregnancy. Its concentrations in blood increase in the early days of pregnancy in response to human chorionic gonadotropin (hCG) secreted by the implanting embryo (see Chapter 47). Administration of relaxin to rodents reproduces many of the changes in renal and mesenteric blood flow seen in early pregnancy. In these animals, relaxin stimulates endothelial cell production of nitric oxide, especially in the renal and mesenteric vascular beds, but comparable data are not yet available for humans.

Cardiac Performance

Cardiac output increases as a result of increases in both heart rate and stroke volume (Fig. 4). The acceleration of heart rate is evident a few weeks earlier than the increase in stroke volume and continues to increase gradually until term. The acceleration of heart rate is probably driven by the combined effects of increased venous return, increased blood volume, and decreased arterial pressure. Venous return is increased as a result of decreased arteriolar resistance and increased blood volume (see below). Although increased compliance of the great veins accommodates the increase in volume with little change in central venous pressure, stretch receptors in the walls of the vena cava and right atrium are nevertheless activated and send afferent signals to the vasomotor center, which responds with sympathetic stimulation of the sinoatrial node. This classical Bainbridge reflex (see Chapter 14) accelerates heart rate without increasing arteriolar tone. Decreased activation of baroreceptors in accordance with the decline in arterial pressure probably results in additional impetus to heart rate through decreased vagal inhibitory tone. Studies of baroreceptor reflex responses indicate that parasympathetic input to the heart is diminished in pregnant women. It is also likely that hormones of pregnancy act directly at the level of the vasomotor center in the brain to adjust regulatory set-points.

Stroke volume increases steadily in the first half of pregnancy and thereafter remains constant until late in the third trimester, when it falls somewhat. It may be recalled (see Chapter 13) that stroke volume is determined by cardiac contractility, diastolic volume (preload), and aortic pressure (afterload). Compliance of the heart and aorta increases because of changes in physical properties of the extracellular matrix, a phenomenon called *remodeling*. Such remodeling is thought to occur in response to the high circulating levels of estrogens and progesterone. Increased compliance lowers resistance to ventricular filling and results in increased diastolic volume without a corresponding increase in pressure. Indeed, pregnancy produces a mild increase in heart size without a change in thickness of the ventricular walls. Increased aortic distensibility coupled with decreased peripheral resistance reduce afterload. Because the ejection fraction remains relatively constant throughout pregnancy, end diastolic volume increases and heart size is increased. Myocardial contractility *per se* is little changed, but force and speed of ventricular contraction increase in accordance with the Frank–Starling relationship (see Chapter 13).

BLOOD VOLUME

Blood volume starts to expand before the fourth week of pregnancy and continues to increase until the middle of the third trimester, when it is about 40 to 50% above the prepregnant level (Fig. 5). Although the amount of added volume varies widely from woman to woman, the increase is consistent for each woman in each of her pregnancies and is greater in twin than in single pregnancies. Both plasma and red cell volumes increase, but the plasma volume expands by 50%, while the red cell mass increases by only 20 to 30%. Consequently, the hematocrit declines from about 45% to about 35%, producing the so-called *anemia of pregnancy*. It is not known why expansion of the red cell mass fails to keep pace with the increase in plasma volume. Erythropoietic capacity of bone marrow is not the limiting factor. Further enhancement of red blood cell production is seen after hemorrhagic injury. Similarly, red blood cell production increases further in pregnant women in response to the decrease in oxygen tension (PO_2) encountered with a change in residence from sea level to high altitudes. Pregnancy thus does not interfere with the operation of basic regulatory mechanisms that govern erythropoiesis. Concentrations of erythropoietin (see Chapter 23) in blood plasma increase only modestly during pregnancy.

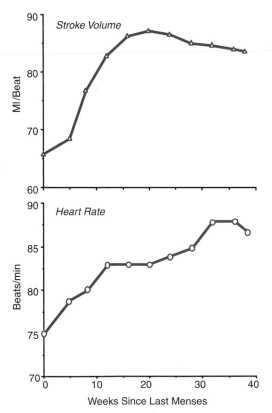

FIGURE 4   Changes in heart rate and stroke volume during normal pregnancy. (Drawn from the data of Robson SC, Hunter S, Boys RJ, Dunlop W. *Am J Physiol* 1989; 256:H1060–H1065.)

Clinical Note

Measurement of Blood Pressure

It is noteworthy that measurements of blood pressure and cardiac output made in the late stages of pregnancy are very much dependent upon maternal position when the measurements are made. The gravid uterus, which occupies a volume of 6 to 7 L, compresses the vena cava and to a lesser extent the abdominal aorta when the subject rests on her back in the supine position. The gravid uterus may thus physically decrease venous return and cardiac output and cause a fall in systemic arterial pressure.

Similarly, when the woman is standing upright, the gravid uterus compresses the femoral veins and interferes with blood flow from the lower extremities to the vena cava. The resulting decline in venous return produces a decline in cardiac output and a fall in blood pressure. Because baroreceptor responses are blunted, the fall in blood pressure may cause some women to faint after prolonged standing or a sudden change in posture. Reliable assessment of blood pressure is usually made with the woman lying on her left side.

It is possible that increased renal blood flow partially compensates for decreased hematocrit in maintaining renal interstitial PO_2 so that erythropoietin-secreting cells are only mildly stimulated. Alternatively, the PO_2 sensitivity of these cells may be decreased by the hormones of pregnancy, so that the set-point for regulation of red cell formation is adjusted downward.

Expansion of the blood volume also dilutes the plasma proteins. The concentration of albumin declines by about 30%. Because albumin is the most abundant plasma protein and the major colloid osmolyte, plasma oncotic pressure also decreases by about 30%. Synthesis of albumin and some other hepatic proteins is thought to be regulated by plasma oncotic pressure; a decrease in oncotic pressure sensed by hepatocytes activates transcription of the albumin gene. Once again it appears that the sensitivity of a regulatory mechanism is reduced by some actions of hormones of pregnancy. Hepatic protein synthetic capacity is not compromised, as evidenced by increased production and secretion of some globulins, fibrinogen, and clotting factors.

The changes in vascular volume and in levels of red blood cells and albumin would be considered pathological in nonpregnant women, but are normal in pregnancy and appear to result from adjustments to the set-points of normally operating feedback regulatory systems. The adaptive advantages of decreased circulating levels of albumin and red blood cells are not known; however, the combined effect of the lower red cell mass and the lower protein content is a decrease in blood viscosity (see Chapter 10) which contributes in at least a minor way to the overall decrease in peripheral resistance shown in Fig. 3. It should be emphasized that, although the hematocrit declines in pregnancy, the total red cell mass increases significantly and may therefore lessen the postpartum impact of the inevitable loss of about 500 mL of blood at delivery.

SODIUM BALANCE

For plasma volume to increase, sodium and water intake and retention must also increase. Hormonal signals to increase sodium conservation are amplified (Fig. 6) early in pregnancy, at about the same time that peripheral resistance begins to decline. It is not known if these events are causally related or are independent responses to signals originating in the fetus. In all, about 900 to 950 mEq of added sodium is retained during gestation. This amount corresponds at term to about

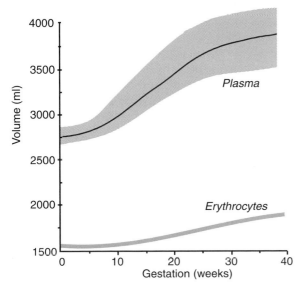

FIGURE 5 Changes in plasma and red cell volume during the course of normal pregnancies. Shaded area indicates range of variation.

Clinical Note

Edema

It should be noted that decreases in arteriolar tone and plasma colloid osmotic pressure change the balance of forces across the capillaries in favor of increased net filtration and expansion of interstitial fluid volume (see Chapter 16).

Consequently, pregnant women are prone to edema, especially in the legs. In late pregnancy, when the gravid uterus compresses the femoral veins and increases venous pressure in the legs, the tendency for net filtration and the accumulation of extravascular fluid is exaggerated.

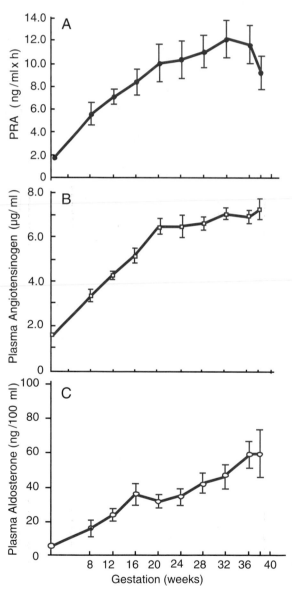

FIGURE 6 Changes in plasma renin activity, angiotensinogen, and aldosterone in the course of pregnancy. Plasma renin activity is given in terms of nanograms of angiotensin produced per hour of incubation of the plasma sample under standardized conditions. (Adapted from August P, Lindheimer MD. In: Laragh JH, Brenner BM, Eds., *Hypertension: pathophysiology, diagnosis, and management.* New York, Raven Press, 1995, pp. 2407–2426.)

6.5 L of extracellular fluid, of which only about half is in the fetus, placenta, and amniotic fluid. The remainder is distributed between maternal vascular and interstitial fluid compartments at a ratio of about 1:3.

The increase in sodium retention over the course of gestation is remarkable when pregnancy-induced changes in renal hemodynamics are also considered. Resistance of the renal vasculature declines to a greater extent than the decrease in total peripheral vascular resistance discussed previously. It may be recalled that the afferent and efferent glomerular arterioles are the major sites of resistance to blood flow in the kidneys (see Chapter 24). Decreased resistance in these vessels results in a greater than 50% increase in effective renal plasma flow (ERPF). Glomerular capillary pressure changes little from the nonpregnant state because resistance is reduced in both the afferent and the efferent glomerular arterioles. Increased flow through the glomerular capillaries at constant hydrostatic pressure and decreased colloid osmotic pressure results in a greater than 50% increase in the glomerular filtration rate (GFR) (see Chapter 24). Consequently, the filtered load of sodium increases from about 20,000 to about 30,000 mEq/day, and reabsorptive mechanisms of the renal tubules are therefore faced with an additional load of 10,000 mEq of sodium just to maintain sodium balance. Their task is made even more difficult by the natriuretic effects of high concentrations of progesterone, which, at high concentrations, may compete for binding to the mineralocorticoid receptor. Nevertheless, a net retention of sodium averaging 3 to 5 mEq of sodium per day between the 8th and 20th weeks must take place to account for the expansion of plasma volume.

The hormone directly responsible for sodium retention is aldosterone (see Chapters 29 and 40), the concentrations of which increase steadily in plasma throughout pregnancy to levels that are far above those seen in nonpregnant women (Fig. 6). Although nonpregnant individuals escape from the sodium-retaining effects of prolonged elevations in aldosterone concentration (see Chapter 40), increased sodium retention is sustained throughout pregnancy. Angiotensin II is the principal stimulus for aldosterone secretion. Its production depends upon secretion of renin by the juxtaglomerular cells of the kidney (Chapters 27 and 40) and the availability of its precursor, angiotensinogen. Hepatic secretion of angiotensinogen is stimulated by high circulating concentrations of estrogens. Increases in angiotensinogen concentration lead to increases in angiotensin formation over the range of physiologically relevant concentrations of renin.

In men and nonpregnant women, the effective blood volume is the regulated variable in the control of renin secretion. Changes in blood volume are sensed by stretch receptors in the walls of great veins in the chest, the atria, and the carotid sinuses, by pressure or stretch receptors in the afferent glomerular arterioles, and by sodium chloride receptors in the maculae densae of the renal tubules (see Chapter 29). It seems paradoxical that renin secretion continues to increase throughout pregnancy despite sustained and increasing expansion of vascular volume and the increased delivery of sodium chloride to the distal tubules that results from the increase in GFR. It is possible that decreased stimulation of pressure receptors due to the decrease in arterial pressure and the increase in compliance of the thoracic vasculature is interpreted by the vasomotor center as a decrease in vascular volume. The physiological response to such an interpretation is increased sympathetic stimulation of the renin-secreting juxtaglomerular cells. Alternatively, hormones of pregnancy may decrease the sensitivity of these stretch receptors or act directly at the hypothalamic level to adjust the set-point for volume regulation. In any event, secretion of renin by the kidneys during pregnancy is not autonomous and is adjusted upward or downward by the same physiological control mechanisms that are operative in nonpregnant women (Fig. 7). For example, the decrease in central venous volume that occurs upon rising from a recumbent to a standing position causes renin secretion to increase in pregnant as well as nonpregnant women. Similarly, renin levels decline after sodium loading and increase upon sodium deprivation whether or not the test subject is pregnant. Thus, the renin/angiotensin system continues to regulate vascular volume throughout pregnancy, but at a higher set-point.

OSMOREGULATION AND THIRST

Retention of sodium is accompanied by retention of water because sodium is the principle osmolyte in extracellular fluid and osmolality of plasma is a tightly controlled variable (see Chapter 28). Increases or decreases in sodium retention are paralleled by changes in water conservation by the kidney in order to maintain constant osmolality of the plasma and hence the extracellular fluid.

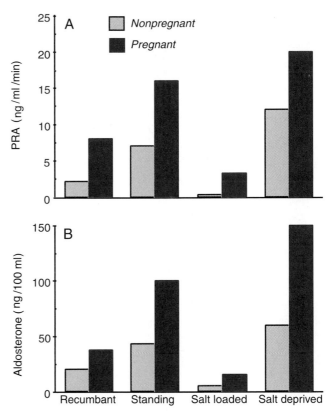

FIGURE 7 Changes in (A) plasma renin activity and (B) aldosterone in response to postural changes and salt loading in 17 pregnant and 11 nonpregnant women. (Drawn from data of Weinberger MH, Kramer NJ, Grim CE, Petersen LP. *J Clin Endocrinol Metab* 1977; 44:69–77.)

Hypothalamic osmoreceptors monitor plasma osmolality and signal secretion of arginine vasopressin (AVP) when osmolality increases. AVP signals the kidneys to conserve water by mechanisms described in Chapter 28, but water conservation can only maintain osmolality constant when water intake is adequate. When plasma osmolality increases, osmoreceptors also signal thirst and drinking behavior and thus coordinate water intake and water conservation.

Under basal conditions the concentration of sodium in the plasma of pregnant women is maintained at 3 to 5 mEq lower than that in plasma of nonpregnant women (Fig. 8). Despite the nearly 10 mOsm difference, plasma concentrations of AVP are quite similar in pregnant and nonpregnant women, indicative of a change in the relationship between osmolality and AVP secretion. The seemingly inappropriately higher concentration of AVP does not imply independence of AVP secretion from plasma osmolality or breakdown of normal feedback relationships. Pregnancy does not interfere with the ability to respond normally to changes in osmolality that follow periods of water deprivation or water loading (Fig. 9). When plasma osmolality was increased in pregnant and nonpregnant women by infusions of

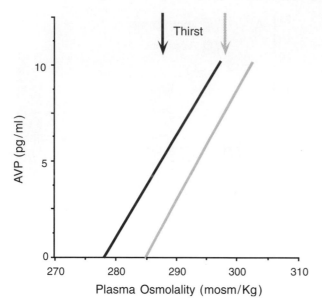

FIGURE 8   Relationship between plasma osmolality and AVP concentrations in eight women before (light blue) and at the end (dark blue) of the first trimester of pregnancy. Arrows indicate plasma osmolality at which a conscious desire to drink (thirst) was experienced. (Drawn from data of Davison JM, Shiells EA, Philips PR, Lindheimer MD. *J Clin Invest* 1988; 81:798–806.)

saline, plasma levels of AVP increased to a similar extent in both groups with each increment in osmolality. While the AVP concentrations changed in parallel in both groups, at each concentration of AVP observed the osmolality was nearly 10 mOsm/L lower in the pregnant women. Pregnancy shifts the relationship between osmolality and thirst to a similar extent. Thus, osmolal regulation remains intact during pregnancy, but operates at a lower set point.

The factors that bring about changes in sensitivity of the osmoreceptor system have not been identified nor is the underlying mechanism understood. A similar but less pronounced shift in the relationship between AVP secretion and plasma osmolality is seen in nonpregnant women during the luteal phase of the menstrual cycle. Progesterone, relaxin, or some other product of the corpus luteum that continues to be secreted throughout pregnancy may thus be responsible for adjusting osmoregulation. In nonpregnant individuals, the relationship between AVP secretion and osmolarity changes with changes in plasma volume. A leftward shift in the AVP/osmolality relationship similar to that seen in pregnant women occurs in hypovolemia (reduced plasma volume), while hypervolemia tends to shift the relationship in the opposite direction. Thus, it seems that the hypervolemia of pregnancy appears as hypovolemia to the volume-regulating mechanisms that govern AVP secretion. Changes in the set-points for volume and osmolal regulation cause regulatory centers to interpret the increase in

plasma volume and the decreased osmolality in pregnancy as normal and to defend these altered states.

When increased plasma volume is sensed by stretch receptors in the right atrium in men and nonpregnant women, secretion of atrial natriuretic factor (ANF) increases (see Chapter 29). ANF combats volume expansion by increasing urinary sodium loss, through direct actions on the kidney and through inhibition of aldosterone production (see Chapter 40). Despite the 50% increase in blood volume, only small, if any, increases in plasma ANF concentrations are seen in normal pregnant women before the third trimester, when a modest rise occurs. Pregnancy therefore appears to decrease the sensitivity of the ANF-secreting cells to increased volume. However, rates of ANF secretion increase or decrease to similar extents in pregnant and nonpregnant women in response to the variations in central venous volume that result from changes in posture or saline

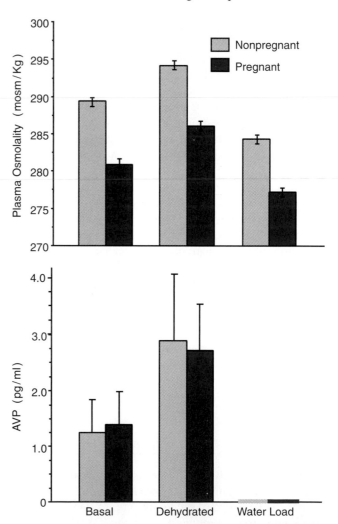

FIGURE 9   Effects of water deprivation and water loading on plasma AVP levels in eight pregnant and eight nonpregnant women. (Drawn from data of Davison JM, Gilmore EA, Dürr J, Robertson GL, Lindheimer MD. *Am J Physiol* 1984; 246:105–109.)

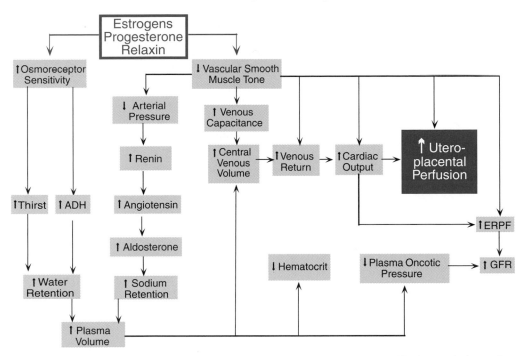

FIGURE 10   Summary of cardiovascular and renal changes in pregnancy. ERPF, effective renal plasma flow; GFR, glomerular filtration rate.

infusions. Thus, as already described for renin and AVP, control mechanisms that govern ANF secretion respond normally to acute changes in volume but defend vascular volume at a higher set point.

Figure 10 summarizes the cardiovascular and renal changes produced in the first trimester and continued throughout gestation. Hormonal signals arising in the trophoblast and/or the ovary and acting directly on vascular smooth muscle and the osmoreceptors initiate the series of changes that ultimately increase placental perfusion. It is likely, but not established, that these peripheral effects are reinforced by complementary effects on regulatory centers in the brain. The adaptive value of most of these changes appears to reside in ensuring adequacy of uteroplacental perfusion while maintaining sufficient

circulatory reserve to service the needs of the mother. The adaptive value of the increase in GFR is unknown, but it may facilitate excretion of fetal wastes and dietary toxins.

RESPIRATORY ADJUSTMENTS

Resting pulmonary ventilation is increased throughout pregnancy. In the menstrual cycle, minute ventilation is greater in the luteal phase than the follicular phase and increases steadily after conception until delivery. By the end of gestation, alveolar ventilation is about 40% greater than the prepregnant rate. This change is brought about by a 20% increase in tidal volume without a change

Clinical Note

Preeclampsia
One of the leading complications of pregnancy is a disorder of unknown etiology called *preeclampsia* (toxemia of pregnancy). Eclampsia is the occurrence of convulsions, coma, and often death following delivery. Preeclampsia is characterized by hypertension, edema, and proteinuria. Inadequacies of uteroplacental blood flow or perhaps immune incompatibilities of mother and fetus have been associated with preeclamptic-like symptoms. All of the signs and symptoms of preeclampsia, which may or may not develop to eclampsia, disappear with delivery. Most of the maternal cardiovascular and renal adjustments discussed above do not occur in preeclampsia.

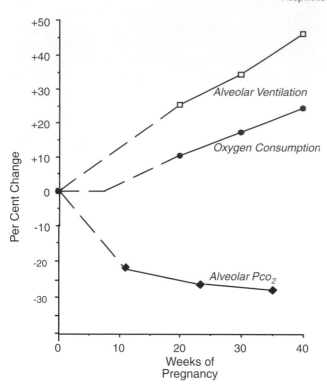

FIGURE 11 Changes in alveolar ventilation, oxygen consumption, and alveolar carbon dioxide content during pregnancy. (Drawn from data of Pernoll ML, Metcalfe J, Kovach PA, Wachtel R, Dunham MJ. *Respir Physiol* 1975; 25:295–310; Pernoll ML, Metcalfe J, Schlenker TL, Welch JE, Matsumoto JA. *Respir Physiol* 1975; 25:285–293.)

in frequency of inspiration. Despite rapid growth and intense metabolic activity of the placenta, little increase in maternal oxygen consumption is evident before the end of the first trimester, because the fetoplacental mass is so small. Thereafter, oxygen consumption and CO_2 production increase steadily in parallel with increasing fetal mass to reach a level that is about 20% above the nonpregnant level (Fig. 11), but hyperventilation continues until the baby is delivered. Because the rate of alveolar ventilation exceeds the metabolic rate, alveolar PCO_2 declines by about 25% and remains at about 30 mm Hg for the remainder of gestation, and PO_2 increases by about 5%. Consequently, arterial PCO_2 declines from about 40 mm Hg to about 32 mm Hg. The plasma bicarbonate concentration falls from about 27 to about 21 mEq/L, while plasma pH increases from 7.40 to about 7.45. Arterial PO_2 increases slightly from about 103 to 107 mm Hg.

As already mentioned, the hyperventilation of pregnancy is achieved by a substantial increase in inspiratory volume. Yet, as pregnancy progresses, growth of the uterine mass might be expected to interfere with the range of motion of the diaphragm and to interfere with expansion of the lungs and breathing movements; however, thoracic volume is maintained at almost the prepregnant level by a change in shape of the chest. Even before the

uterine volume has fully expanded, the angle of the ribs begins to widen, possibly because of relaxation of the costal ligaments, so that the diameter of the chest increases by 5 to 7 cm. In fact, total pulmonary volume is decreased only by about 5%, and the vital capacity is unchanged (Fig. 12). The small decrease in total lung volume is accounted for by a decrease in the functional residual volume that results from the upward pressure exerted by the abdominal contents. This change in functional residual volume increases the efficiency of alveolar ventilation because inspired air is diluted with a smaller volume of residual air in the alveoli. In line with their effects on vascular smooth muscle, hormones of pregnancy also relax airway smooth muscle, which increases the functional dead space but significantly decreases airway resistance.

The hyperventilation of pregnancy is attributable to the high circulating concentrations of progesterone. Ventilatory drive is also increased within a few hours after progesterone is given to nonpregnant subjects of either sex. It may be recalled (Chapter 22) that in the presence of normal oxygen tension, PCO_2 (and the closely linked hydrogen ion concentration) is the variable that is regulated by the respiratory centers. PCO_2 in arterial blood and in cerebrospinal fluid is monitored by arterial chemoreceptors and chemoreceptors in the brain stem. Progesterone increases ventilatory drive by increasing the sensitivity of both central and peripheral chemoreceptors to CO_2. Although pregnant women are hyperresponsive to increases in inhaled CO_2, pregnancy does not reduce their ability to adjust alveolar ventilation upward or downward in response to changes in arterial or inspired PCO_2, indicating that normal feedback regulatory mechanisms remain operative. Because of the heightened sensitivity of the chemoreceptors, PCO_2 at levels that prevail in the nonpregnant woman stimulates ventilation until enough CO_2 is eliminated to establish a new steady

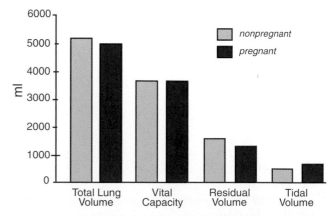

FIGURE 12 Lung volumes in late pregnancy and for nonpregnant women.

TABLE 1   Normal Values of O_2, CO_2, and pH in Maternal and Fetal Blood

	Maternal Artery	Uterine Vein	Fetal Vein	Umbilical Artery
PO_2 (mm Hg)	95	38	30	22
HbO_2 (% saturation)	98	72	75	50
O_2 content (mL/dL)	16.4	11.8	16.2	10.9
O_2 content (nM)	7.3	5.3	7.2	4.5
Hemoglobin (g/dL)	12	12	16	16
O_2 capacity	16.4	16.4	21.9	21.9
PCO_2	32	40	43	48
CO_2 content (nM)	19.6	21.8	25.2	26.3
HCO_3^- (mEq)	18.8	20.7	24	25
pH	7.42	7.35	7.38	7.34

Source: Longo LD. Respiratory gas exchange in the placenta. In: Fahri LE, Tenney SM, Eds., *Handbook of physiology. Section 3: The respiratory system, Vol. IV, Gas exchange,* pp. 351–401. Bethesda, MD: American Physiological Society, Washington, D.C., 1987. With permission.

state at a lower set-point. Progesterone also increases the sensitivity of the chemoreceptors for PO_2, but this change is only evident in intense exercise or in hypoxic conditions such as high altitude. Thus, as already noted for cardiovascular and renal homeostasis, a signal arising in the ovaries and placenta readjusts the set-point for a normally operating negative feedback system.

Resetting the steady-state level of PCO_2 in maternal blood benefits the developing fetus by facilitating the transfer of CO_2 from the fetal to the maternal circulation. PCO_2 equilibrates rapidly across the placental barrier by diffusion of CO_2 in the form of the uncharged, dissolved gas. Its rate of diffusion depends the steepness of the concentration gradient between fetal blood in the umbilical artery and maternal blood in the intervillous space. By lowering the PCO_2 in maternal blood, a steep concentration gradient is created while allowing fetal PCO_2 and blood pH to be maintained at a level that is favorable for rapid cellular growth and development. The small increase in PO_2 is of little consequence for oxygen delivery to the fetus, because maternal hemoglobin is already virtually saturated at the PO_2 that prevails in nonpregnant women. However, the rapid rate of CO_2 transfer across the placental barrier and the resulting transfer of hydrogen ion increase the rate of oxygen delivery to the fetus through operation of the Haldane and Bohr effects (see Chapter 20). The decrease in PCO_2 in fetal capillary blood as it traverses the terminal villi increases the affinity of hemoglobin for oxygen, and hence its degree of saturation at the low partial pressure of oxygen of intervillous blood. At the same time, the increase in PCO_2 in the intervillous space facilitates the unloading of oxygen from maternal hemoglobin. Simultaneously, the diffusion of CO_2 across the placental barrier raises the pH of placental capillary blood and lowers that of intervillous blood to provide a similar effect on hemoglobin loading of maternal and fetal blood.

Gas exchange across the placenta is analogous to gas exchange in the lungs; the arterial supply to the placenta is low in O_2 and high in CO_2; after equilibration with maternal blood in the intervillous space, umbilical venous blood has taken on O_2 and delivered CO_2. Table 1 presents normal values of oxygen, carbon dioxide, and hydrogen ion concentrations in the maternal and fetal circulations in late pregnancy. Although PO_2 and CO_2 equilibrate quickly across the placental barrier, partial pressures of these gases in the umbilical veins differ radically from values in the uterine artery and even the uterine veins. Differences between uterine and umbilical venous blood can be accounted for in part by the relatively high rate of O_2 extraction and CO_2 production by the syncytiotrophoblast and by the relatively large areas of the placenta that are unavailable for exchange. It is important to note that despite the low PO_2 and the metabolic activity of the syncytiotrophoblast, the oxygen content of umbilical venous blood is quite similar to that of maternal arterial blood. This is possible because of the higher content of hemoglobin in fetal than maternal blood and the greater affinity of fetal hemoglobin for oxygen. It is also noteworthy that PCO_2, bicarbonate, and pH in umbilical venous blood are all in the same range as found in arterial blood of nonpregnant subjects (see Chapter 31).

OTHER ADAPTATIONS

Virtually all aspects of maternal physiology are affected by pregnancy, although not all of the changes may be adaptive. Consistent with what has already been described for vascular and airway smooth muscle and for the uterine myometrium (see Chapter 47), contractile tone of smooth muscle throughout the body is decreased by the high levels of ovarian and placental steroid hormones and perhaps relaxin.

Clinical Note

Gall Stones, Gastric Reflux, and Constipation
Decreased tone of the gall bladder results in doubling of its volume in pregnancy and its incomplete emptying in response to cholecystokinin upon eating. Incomplete emptying may contribute to the propensity of pregnant women to develop gall stones. Decreased peristalsis and decreased muscle tone decrease esophageal pressure, which, coupled with increased gastric pressure due to compression produced by the gravid uterus, favors gastric reflux and the heartburn often associated with pregnancy. Decreased motility of intestinal smooth muscle increases the transit time of ingested foods and may increase efficiency of nutrient absorption in the small bowel. Prolongation of transit time in the large bowel along with elevated blood levels of aldosterone result in as much as a 60% increase in water absorption compared to prepregnant levels. While this effect may help to maintain the expanded blood volume, it also accounts for the constipation encountered by many pregnant women.

Relaxation of smooth muscle has been noted throughout the gastrointestinal and urinary systems. Receptors for these hormones are widely distributed, and there is growing evidence for their involvement in nitric oxide production. Production of another potent smooth muscle relaxant, PGI_2, is also increased in many tissues in pregnancy.

Suggested Reading

Chapman AB, Abraham WT, Zamudio S, Coffin C, Merouani A, Young D, Johnson A, Osorio F, Goldberg C, Moore LG, Dahms T, and Schrier RW. Temporal relationships between hormonal and hemodynamic changes in early pregnancy. *Kidney Int* 1998; 54:2056–2063.

Lindheimer MD, Davison JM. Osmoregulation, the secretion of arginine vasopressin and its metabolism during pregnancy. *Eur J Endocrinol* 1995; 132:133–143.

Lindheiner MD, Katz AI. Renal physiology and disease in pregnancy, in: Seldin DW, Giebisch G, Eds., *The Kidney*, 3rd ed., Vol. II, Philadelphia: Lippincott/Williams & Wilkins, 2000, pp. 2597–2644.

Metcalfe J, Bisonnette JM. Gas exchange in pregnancy, in: Fahri LE, Tenney SM, Eds., *Handbook of physiology*, Section 3, *The respiratory system*, Vol. IV, *Gas Exchange*, Bethesda, MD: American Physiological Society, 1987, pp. 341–350.

Rosenfeld CR. Mechanisms regulating angiotensin II responsiveness by the uteroplacental circulation. *Am J Physiol Regul Integr Comp Physiol* 2001; 281:R1025–R1040.

White MM, Zamudio S, Stevens T, Tyler R, Lindenfeld J, Leslie K, Moore LG. Estrogen, progesterone, and vascular reactivity: potential cellular mechanisms. *Endocrine Rev* 1995; 16:739–751.

Index

A

A, *see* Adenine
AA, *see* Afferent arterioles
A bands, cardiac muscle cells, 179
Abdominal muscles, in breathing, 278, 318–319
Abducens nerve, extraocular muscles, 862
Absence seizures, 902
Absolute refractory period
  action potentials, 86–87
  heart, 181
Absorption, *see also* Reabsorption
  balance with secretion, 555, 555f
  calcium, 544–545
  capillary fluid movement, 239
  carbohydrates, 534–535
  chloride by intestines, 551, 551f
  definition, 466
  fat-soluble vitamins, 544
  fluid by leakey epithelia, 553f
  fluid in peritbular capillaries, 348–349
  glucose, galactose, fructose, 534f
  intestinal, calcium, 681, 689
  iron, 545f, 545–546
  isosmotic volume absorption, 378
  lipids, 540, 540f
  pathway, 532–533
  potassium, 547–548
  proteins, 536–537f
  sodium, 550f, 550–551
  sodium chloride, 396f
  solute pathway, 532f
  transcellular and paracellular routes, 549–550
  transport processes, 533
  water, 548, 552f, 552–553
  water-soluble vitamins, 543–544
Absorptive reserve capacity, 549
Accessory olfactory bulb, and vomeronasal organ, 854
Accessory pathway, effect on PR interval, 198–199
Accessory respiratory muscles, 278
Acclimatization, high altitude, 327–329
Accommodation reflex, 828

Accomodation, action potentials, 87
ACE, *see* Angiotensin-converting enzyme
ACE inhibitors, *see* Angiotensin-converting enzyme inhibitors
Acetylcholine
  in acid secretion, 497
  autonomic nervous system, 146
  end-plate potential mechanisms, overview, 104
  gastric secretion, 507f
  iontophoretic application, 103f
  iontophoric application, EPP comparison, 104f
  neuromuscular junction, 102–104
  pacemaker activity, 185
  in pepsin secretion, 511
  SA nodal cell effects, 185f
  synthesis and degradation, 152f
Acetylcholine receptor
  antagonists, 327
  ligand-gated, structure, 105–106
  model, 105f
  myasthenia gravis, 120
Acetylcholinesterase, 103
Acetyl CoA carboxylase
  function, 642
  insulin effects, 651
ACh, *see* Acetylcholine
Achalasia, 485
Acid-base balance, blood
  bicarbonate-pH diagram, 296–297
  disturbance diagnosis, 297–298
  Henderson-Hasselbach equation, 295–296
  metabolic acidosis, 297
  metabolic alkalosis, 297
  respiratory acidosis, 297
  respiratory alkalosis, 297
Acid-base balance, during exercise, 935–936
Acid-base balance, kidney
  anion gap, 457
  bicarbonate generation, 452–454
  bicarbonate reabsorption, 451–452
  buffer role, 448–450
  clinical disturbance overview, 454–455
  common disturbances, 458–461

  metabolic acidosis, 456–457
  metabolic alkalosis, 457–458
  plasma bicarbonate overview, 450–451
  pulmonary regulation of CO_2, 450
  respiratory acidosis, 455–456
  respiratory alkalosis, 456
Acid-base status, steps 1 to 4, 459f, 460f
Acidemia, 454
Acid-labile subunit, 707
Acidosis
  acid-base disturbances, 460
  characteristics, 455–456
  kidney effects, 334
  plasma bicarbonate threshold, 452f
Acinar cells
  pancreas, 514, 516f, 520f
  salivary glands, 499
Acinus
  characteristics, 265
  pancreas, 514
  primary saliva production, 497
  salivary glands, 499
Acrosomal reaction, 759
ACTH, *see* Adrenocorticotropic hormone
Actin
  biochemical interactions with myosin, 125–126, 138–141
  overview, 124, 138
F-Actin, 239
Action potentials
  absolute and relative refractory period, 86–87
  accomodation, 87
  cardiac cells, 180f
  cardiac muscle, 179
  compound, recording, 96f
  contractile cell, 179f
  extracellular recordings, 72
  extracellular voltage creation, 188
  giant squid axon, 77f
  Goldman-Hodgkin-Katz equation, 78
  intracellular recording, 73f, 73–74
  ion channels underlying, specificity, 85
  muscle cells, 135
  muscle fibers, 865